Springer Optimization and Its Applications

Springer Optimization and Its Applications

VOLUME 68

Aims and Scope

Optimization has been expanding in all directions at an astonishing rate during the last few decades. New algorithmic and theoretical techniques have been developed, the diffusion into other disciplines has proceeded at a rapid pace, and our knowledge of all aspects of the field has grown evenmore profound. At the same time, one of the most striking trends in optimization is the constantly increasing emphasis on the interdisciplinary nature of the field. Optimization has been a basic tool in all areas of applied mathematics, engineering, medicine, economics, and other sciences.

The series *Springer Optimization and Its Applications* publishes undergraduate and graduate textbooks, monographs and state-of-the-art expository work that focus on algorithms for solving optimization problems and also study applications involving such problems. Some of the topics covered include nonlinear optimization (convex and nonconvex), network flow problems, stochastic optimization, optimal control, discrete optimization, multi-objective programming, description of software packages, approximation techniques and heuristic approaches.

For further volumes:
www.springer.com/series/7393

Panos M. Pardalos • Pando G. Georgiev •
Hari M. Srivastava
Editors

Nonlinear Analysis

Stability, Approximation, and Inequalities

In honor of Themistocles M. Rassias on the occasion of his 60th birthday

Springer

Editors
Panos M. Pardalos
Center for Applied Optimization
ISE Department
University of Florida
Gainesville, FL
USA

and

Laboratory of Algorithms and Technologies
for Networks Analysis (LATNA)
National Research University
Higher School of Economics
Moscow
Russia

Pando G. Georgiev
Center for Applied Optimization
University of Florida
Gainesville, FL
USA

Hari M. Srivastava
Department of Mathematics & Statistics
University of Victoria
Victoria, British Columbia
Canada

ISSN 1931-6828 Springer Optimization and Its Applications
ISBN 978-1-4614-3497-9 ISBN 978-1-4614-3498-6 (eBook)
DOI 10.1007/978-1-4614-3498-6
Springer New York Heidelberg Dordrecht London

Library of Congress Control Number: 2012940728

Printed on acid-free paper

Springer is part of Springer Science+Business Media (www.springer.com)

With our deepest appreciation,
we dedicate this volume
to the eminent mathematician
Themistocles M. Rassias
on the occasion of his 60th birthday

Themistocles M. Rassias

Preface

This volume is dedicated to Themistocles M. Rassias, on the occasion of his 60th birthday. The articles published here present some recent developments and surveys in Nonlinear Analysis related to the mathematical theories of stability, approximation, and inequalities.

Themistocles M. Rassias was born in Pellana near Sparta in Greece in the year 1951. He is currently a Professor in the Department of Mathematics at the National Technical University of Athens. He received his Ph.D. in Mathematics in the year 1976 from the University of California at Berkeley with Stephen Smale as his thesis advisor.

Rassias' work extends over several fields of Mathematical Analysis. It includes Global Analysis, Calculus of Variations, Nonlinear Functional Analysis, Approximation Theory, Functional Equations, and Mathematical Inequalities and their Applications. Rassias' work has been embraced by several mathematicians internationally and some of his research has been established with the scientific terminology "Hyers–Ulam–Rassias stability", "Cauchy–Rassias stability", "Aleksandrov–Rassias problem for isometric mappings".

The stability theory of functional equations has its roots primarily in the investigations by S.M. Ulam, who posed the fundamental problem for approximate homomorphisms in the year 1940, the stability theorem for the additive mapping due to D.H. Hyers (1941), and the stability theorem for the linear mapping of Th.M. Rassias (1978). Much of the modern stability of functional equations has been influenced by the seminal paper of Th.M. Rassias, entitled "On the stability of the linear mapping in Banach spaces" [*Proceedings of the American Mathematical Society* **72**, 297–300 (1978)], which has provided a theoretical breakthrough. For an extensive discussion of various advances in stability theory of functional equations, the reader is referred to the recently published book of S.-M. Jung, "*Hyers–Ulam–Rassias Stability of Functional Equations in Nonlinear Analysis*", ©Springer, New York, 2011.

In the formulation as well as in the solution of stability problems of functional equations, one frequently encounters the interplay of Mathematical Analysis, Geometry, Algebra, and Topology. Rassias has contributed also to other subjects such as Minimal Surfaces (Plateau problem), Isometric Mappings (Aleksandrov problem), Complex Analysis (Poincaré inequality and Möbius transformations), and Approximation theory (Extremal problems). He has published more than 230 scientific research papers, 6 research books and monographs, and 30 edited volumes on current research topics in Mathematics. He has also published 4 textbooks in Mathematics for Greek university students.

Some of the honors and positions that he has received include "Membership" at the School of Mathematics of the Institute for Advanced Study at Princeton for the academic years 1977–1978 and 1978–1979 (which he did not accept for family reasons); "Research Associate" at the Department of Mathematics of Harvard University (1980) invited by Raoul Bott, "Visiting Research Professor" at the Department of Mathematics of the Massachusetts Institute of Technology (1980) invited by F.P. Peterson; "Accademico Ordinario" of the Accedemia Tiberina Roma (since 1987); "Fellow" of the Royal Astronomical Society of London (since 1991); "Teacher of the year" (1985–1986 and 1986–1987) and "Outstanding Faculty Member" (1989–1990, 1990–1991, and 1991–1992) of the University of La Verne, California (Athens Campus); "Ulam Prize in Mathematics" (2010). In addition to the above, during the last few years, Th.M. Rassias had been bestowed with honorary degrees "Doctor Honoris Causa" from the University of Alba Iulia in Romania (2008) and an "Honorary Doctorate" from the University of Niš in Serbia (2010). In 2003, a volume entitled "*Stability of Functional Equations of Ulam–Hyers–Rassias Type*" was dedicated to the 25 years since the publication of Th. M. Rassias' stability theorem (edited by S. Czerwik, Florida, USA). In 2009, a special issue of the *Journal of Nonlinear Functional Analysis and Applications* (Vol. 14, No. 5) was dedicated to the 30th Anniversary of Th.M. Rassias' stability theorem. In 2007, a special volume of the *Banach Journal of Mathematical Analysis* (Vol. 1, Issues 1 & 2) was dedicated to the 30th Anniversary of Th.M. Rassias' stability theorem. He is an "editor" or "advisory editor" of several international mathematical journals published in the USA, Europe, and Asia. He has delivered lectures at several universities in North America and Europe, including Harvard University, MIT, Yale University, Princeton University, Stanford University, University of Michigan, University of Montréal, Imperial College London, Technion—Israel Institute of Technology (Haifa), Technische Universität Berlin, and the Universität Göttingen.

The contributed papers in the present volume highlight some of the most recent achievements that have been made in Mathematical Analysis.

Rassias' curiosity, enthusiasm as well as his passion for doing research as well as teaching are unlimited. He has served as a mentor in Mathematics to several students at universities where he has taught.

His research work has received up-to-date more than 7,000 citations (see, e.g., the Google Scholar). That is an impressive number of citations for a mathematician. Thus, Rassias has achieved international distinction in the broadest sense.

The reader is referred to the article of Per Enflo and M. Sal Moslehian, An interview with Themistocles M. Rassias, *Banach Journal of Mathematical Analysis* **1**, 252–260 (2007) [see also www.math.ntua.gr/~trassias/].

In what follows, we present a brief outline of the contributed papers in this volume, which are collected in an alphabetical order of the contributors.

In Chap. 1, S. Abramovich deals with Jensen's type inequality, its bounds and refinements, and with eigenvalues of the Sturm–Liouville system.

In Chap. 2, M. Adam and S. Czerwik consider some quadratic difference operators (e.g., Lobaczewski difference operators) and quadratic-linear difference operators (e.g., d'Alembert difference operators and quadratic difference operators) in some special function spaces. They prove a stability result in the sense of Ulam–Hyers–Rassias for the quadratic functional equation in a special class of differentiable functions.

In Chap. 3, C. Affane-Aji and N.K. Govil present a study concerning the location of the zeros of a polynomial starting from the results of Gauss and Cauchy to some of the most recent investigation on the topic.

Chapter 4 by D. Andrica and V. Bulgarean is devoted to isometry groups $Iso_{d_p}(\mathbb{R}^n)$ for $p \geqq 1$, $p \neq 2$ and $p = \infty$, where the metric d_p is appropriately defined.

In Chap. 5, I. Biswas, M. Logares, and V. Muñoz prove that the moduli spaces $\mathscr{M}_\tau(r, \Lambda)$ are, in many cases, rational. Here the moduli spaces are defined by using a concept of τ-stable pairs of rank r and fixed determinant Λ.

In Chap. 6, D. Breaz, Y. Polatoğlu, and N. Breaz investigate a subclass of generalized p-valent Janowski type convex functions and its application to harmonic mappings.

In Chap. 7, J. Brzdęk, D. Popa, and B. Xu present some observations concerning stability of the following linear functional equation:

$$\varphi\big(f^m(x)\big) = \sum_{i=1}^{m} a_i(x)\varphi\big(f^{m-i}(x)\big) + F(x)$$

in the class of functions φ mapping a nonempty set S into a Banach space X over a field $\mathbb{K} \in \{\mathbb{R}, \mathbb{C}\}$, where m is a fixed positive integer and the functions $f : S \to S$, $F : S \to X$ and $a_i : S \to \mathbb{K}$ $(i = 1, \dots, m)$ are given.

In Chap. 8, M.J. Cantero and A. Iserles examine the limiting behavior of solutions to an infinite set of recursions involving q-factorial terms as $q \to 1$.

In Chap. 9, E.A. Chávez and P.K. Sahoo determine the general solutions of the following functional equations:

$$f_1(x+y) + f_2(x+\sigma y) = f_3(x) \quad \text{and}$$
$$f_1(x+y) + f_2(x+\sigma y) = f_3(x) + f_4(y), \quad \big(x, y \in S^n\big),$$

where $f_1, f_2, f_3, f_4 : S^n \to G$ are unknown functions, S is a commutative semigroup, $\sigma : S \to S$ is an endomorphism of order 2, G is a 2-cancellative abelian group, and n is a positive integer.

In Chap. 10, W.-S. Cheung, G. Leng, J. Pečarić, and D. Zhao present recent developments of Bohr-type inequalities.

In Chap. 11, S. Ding and Y. Xing establish some basic norm inequalities, including the Poincaré inequality, weak reverse Hölder inequality, and the Caccioppoli inequality, for conjugate harmonic forms. They also prove the Caccioppoli inequality with the Orlicz norm for conjugate harmonic forms.

In Chap. 12, S.S. Dragomir presents a survey about some recent inequalities related to the celebrated Jensen's result for positive linear or sublinear functionals and convex functions.

In Chap. 13, A. Ebadian and N. Ghobadipour prove the generalized Hyers–Ulam–Rassias stability of bi-quadratic bi-homomorphisms in C^*-ternary algebras and quasi-Banach algebras.

In Chap. 14, E. Elhoucien and M. Youssef apply a fixed point theorem to prove the Hyers–Ulam–Rassias stability of the following quadratic functional equation:

$$f(kx+y)+f\big(kx+\sigma(y)\big)=2k^2f(x)+2f(y).$$

In Chap. 15, M. Fujii, M. Sal Moslehian, and J. Mićić survey several significant results on the Bohr inequality and present its generalizations involving some new approaches.

In Chap. 16, P. Găvruţa and L. Găvruţa provide an introduction to the Hyers–Ulam–Rassias stability of orthogonally additive mappings.

In Chap. 17, M. Eshaghi Gordji, N. Ghobadipour, A. Ebadian, M. Bavand Savadkouhi, and C. Park investigate ternary Jordan homomorphisms on Banach ternary algebras associated with the following functional equation:

$$f\left(\frac{x_1}{2}+x_2+x_3\right)=\frac{1}{2}f(x_1)+f(x_2)+f(x_3).$$

In Chap. 18, F. Habibian, R. Bolghanabadi, and M. Eshaghi Gordji investigate the Hyers–Ulam–Rassias stability of cubic n-derivations from non-archimedean Banach algebras into non-archimedean Banach modules.

In Chap. 19, S.-S. Jin and Y.-H. Lee investigate a fuzzy version of stability for the following functional equation:

$$2f(x+y)+f(x-y)+f(y-x)-f(2x)-f(2y)=0$$

in the sense of M. Mirmostafaee and M.S. Moslehian.

In Chap. 20, K.-W. Jun, H.-M. Kim, and E.-Y. Son prove the generalized Hyers–Ulam stability of the following Cauchy–Jensen functional equation:

$$f(x)+f(y)+nf(z)=nf\left(\frac{x+y}{n}+z\right),$$

in an n-divisible abelian group G for any fixed positive integer $n \geqq 2$.

In Chap. 21, S.-M. Jung applies the fixed point method for proving the Hyers–Ulam–Rassias stability of the gamma functional equation.

In Chap. 22, H.A. Kenary proves the generalized Hyers–Ulam stability in random normed spaces of the following additive-quadratic-cubic-quartic functional equation:

$$f(x+2y)+f(x-2y)=4f(x+y)+4f(x-y)-6f(x)$$
$$+f(2y)+f(-2y)-4f(y)-4f(-y).$$

In Chap. 23, S.V. Konyagin and Yu.V. Malykhin prove the existence of an infinite-dimensional separable Banach space with a basis set such that no arrangement of it forms a Schauder basis.

In Chap. 24, S. Koumandos presents a survey of recent results on positive trigonometric sums.

In Chap. 25, P. Mihăilescu presents a proof of a slightly more general result than the one of Vandiver and Sitaraman, concerning the first case of Fermat's Last Theorem, with consequences for a larger family of Diophantine equations.

In Chap. 26, G.V. Milovanović and M.P. Stanić present a survey of multiple orthogonal polynomials defined by using orthogonality conditions spread out over r different measures. A method for the numerical construction of such polynomials by using the discretized Stieltjes–Gautschi procedure is given.

In Chap. 27, F. Moradlou and G.Z. Eskandani prove the Hyers–Ulam–Rassias stability of C^*-algebra homomorphisms and of generalized derivations on C^*-algebras for the following Cauchy–Jensen functional equation:

$$f\left(\left(\sum_{i=1}^{n} z_i\right)-\left(\sum_{i=1}^{n} x_i\right)\right)+f\left(\left(\sum_{i=1}^{n} z_i\right)-\left(\sum_{i=1}^{n} y_i\right)\right)$$
$$=2f\left(\left(\sum_{i=1}^{n} z_i\right)-\frac{(\sum_{i=1}^{n} x_i)+(\sum_{i=1}^{n} y_i)}{2}\right).$$

In Chap. 28, D. Motreanu and P. Winkert present a survey on the Fučík spectrum of the negative p-Laplacian with different boundary conditions such as the Dirichlet, Neumann, Steklov, and Robin boundary conditions.

In Chap. 29, M. Mursaleen and S.A. Mohiuddine use the notion of almost convergence and statistical convergence in order to prove the Korovkin type approximation theorem by means of the test functions $1, e^{-x}, e^{-2x}$.

In Chap. 30, A. Najati proves the Hyers–Ulam stability of the following functional equation:

$$f(x+y+xy)=f(x+y)+f(xy).$$

In Chap. 31, M.A. Noor, K.I. Noor, and E. Al-Said make use of the projection technique in order to study a new class of quasi-variational inequalities, which they call the extended general nonconvex quasi-variational inequalities, and establish their equivalence with the fixed point problem. They also apply this equivalence to the existence of a solution of the above-named inequalities under some suitable conditions.

In Chap. 32, M.A. Noor, K.I. Noor, and E. Al-Said study a system of general nonconvex variational inequalities involving four different operators. Their results can be viewed as a refinement and improvement of previously known results for variational inequalities.

In Chap. 33, B. Paneah presents a survey on results about a general linear functional operator, which includes Cauchy type functional operators, Jensen type functional operators, and quasiquadratic functional operators.

In Chap. 34, C. Park proves the generalized Hyers–Ulam stability of the following functional equation:

$$2f(x+y)+f(x-y)+f(y-x)=3f(x)+f(-x)+3f(y)+f(-y)$$

in Banach spaces.

In Chap. 35, C. Park, M.E. Gordji, and R. Saadati classify and prove the generalized Hyers–Ulam stability of linear, quadratic, cubic, quartic, and quintic functional equations in complex Banach spaces.

In Chap. 36, A. Prástaro presents results about local and global existence and stability theorems for exotic n-d'Alembert PDEs, previously introduced by the author.

In Chap. 37, V.Yu. Protasov studies the precision of approximation of a function in linear spaces by affine functionals in case their restrictions to every straight line can be approximated by affine functions on that line with a given precision (in the uniform metric).

Chapter 38 is a survey-cum-expository article by H.M. Srivastava who presents a systematic account of some recent developments on univalent and bi-univalent analytic functions, thereby encouraging future researches on these topics in *Geometric Function Theory of Complex Analysis*.

In Chap. 39, Á. Száz presents a detailed survey on the famous Hyers–Ulam stability theorems, Hahn–Banach extension theorems, and their set-valued generalizations. He also reviews the most basic additivity and homogeneity properties of relations and investigates, in greater detail, some elementary operations on relations. These operations and the intersection convolutions of relations allow a new view of relational generalizations of the Hyers–Ulam and the Hahn–Banach theorems.

In Chap. 40, L. Székelyhidi presents a survey on spectral analysis and spectral synthesis over locally compact abelian groups.

In Chap. 41, A. Ungar presents a theory which extracts the Möbius addition in the ball of the Euclidean n-space, from the Möbius transformation of the complex open unit disc, and demonstrates the hyperbolic geometric isomorphism between the resulting Möbius addition and the famous Einstein velocity addition of special relativity theory.

In Chap. 42, B. Yang defines a general Hilbert-type integral operator and studies six particular kinds of this operator with different measurable kernels in several normed spaces.

In Chap. 43, X. Zhao and X. Yang study the stability of the following Pexider type sine functional equation:

$$h(x)k(y)=f^2\left(\frac{x+y}{2}\right)-g^2\left(\frac{x+\sigma y}{2}\right)$$

and extend the results to Banach algebras.

In Chap. 44, Z. Wang and W. Zhang establish some stability results concerning the following additive-quadratic functional equation:

$$f(2x+y)+f(2x-y)=f(x+y)+f(x-y)+2f(2x)-2f(x)$$

in intuitionistic fuzzy normed spaces (IFNS).

We wish to express our deepest appreciation to the above-named mathematicians from the international mathematical community who contributed their papers for publication in this volume on the occasion of the 60th birthday anniversary of Themistocles M. Rassias. In addition, we are also very thankful to Springer for its generous support to this publication.

Gainesville, USA — Panos M. Pardalos
Victoria, Canada — Hari M. Srivastava
Gainesville, USA — Pando G. Georgiev

Themistocles M. Rassias and Stephen Smale at Berkeley, 1990

Themistocles M. Rassias with Henri Cartan in Paris, 1993

Themistocles M. Rassias with Lars V. Ahlfors at Harvard, 1995

Themistocles M. Rassias with Paul Erdős in Zurich, 1994

Themistocles M. Rassias with Vladimir I. Arnold in Paris, 1993

Themistocles M. Rassias with Serge Lang at Berkeley, 1990

Themistocles M. Rassias with Friedrich E.P. Hirzebruch in Bonn, 1987

Themistocles M. Rassias with Israel M. Gelfand in London, 1994

Themistocles M. Rassias with Shizuo Kakutani at Yale, 1990

Themistocles M. Rassias with Jean Dieudonné in Paris, 1989

Contents

Contributors

Shoshana Abramovich Department of Mathematics, University of Haifa, Haifa, Israel

M. Adam Department of Mathematics and Informatics, Higher School of Labour Safety Management in Katowice, Katowice, Poland

Chadia Affane-Aji Department of Mathematics, Tuskegee University, Tuskegee, AL, USA

Eisa Al-Said Mathematics Department, COMSATS Institute of Information Technology, Islamabad, Pakistan

Dorin Andrica Faculty of Mathematics and Computer Science, "Babeş-Bolyai" University, Cluj-Napoca, Romania; Department of Mathematics, College of Science, King Saud University, Riyadh, Saudi Arabia

Indranil Biswas School of Mathematics, Tata Institute of Fundamental Research, Bombay, India

R. Bolghanabadi Research Group of Nonlinear Analysis and Applications (RGNAA), Semnan, Iran

Daniel Breaz Department of Mathematics, "1 Decembrie 1918" University of Alba Iulia, Alba Iulia, Romania

Nicoleta Breaz "1 Decembrie 1918" University of Alba Iulia, Alba Iulia, Romania

Janusz Brzdęk Department of Mathematics, Pedagogical University, Kraków, Poland

Vasile Bulgarean Faculty of Mathematics and Computer Science, "Babeş-Bolyai" University, Cluj-Napoca, Romania

María José Cantero Departamento de Matemática Aplicada, Escuela de Ingeniería y Arquitectura, Universidad de Zaragoza, Zaragoza, Spain

Esteban A. Chávez Department of Mathematics, Duke University, Durham, NC, USA

Wing-Sum Cheung Department of Mathematics, The University of Hong Kong, Pokfulam, Hong Kong

S. Czerwik Institute of Mathematics, Silesian University of Technology, Gliwice, Poland

Shusen Ding Department of Mathematics, Seattle University, Seattle, WA, USA

S.S. Dragomir Mathematics, School of Engineering & Science, Victoria University, Melbourne, Australia; School of Computational and Applied Mathematics, University of the Witwatersrand, Johannesburg, South Africa

Ali Ebadian Department of Mathematics, Urmia University, Urmia, Iran; Department of Mathematics, Payame Noor University (PNU), Tehran, Iran

Elqorachi Elhoucien Department of Mathematics, Faculty of Sciences, University Ibn Zohr, Agadir, Morocco

M. Eshaghi Gordji Center of Excellence in Nonlinear Analysis and Applications (CENAA), Semnan University, Semnan, Iran

G. Zamani Eskandani Faculty of Mathematical Sciences, University of Tabriz, Tabriz, Iran

Masatoshi Fujii Department of Mathematics, Osaka Kyoiku University, Kashiwara, Osaka, Japan

L. Găvruţa Department of Mathematics, "Politehnica" University of Timişoara, Timişoara, Romania

P. Găvruţa Department of Mathematics, "Politehnica" University of Timişoara, Timişoara, Romania

Norouz Ghobadipour Department of Mathematics, Urmia University, Urmia, Iran

Madjid Eshaghi Gordji Department of Mathematics, Semnan University, Semnan, Iran

N.K. Govil Department of Mathematics & Statistics, Auburn University, Auburn, AL, USA

F. Habibian Department of Mathematics, Semnan University, Semnan, Iran

Arieh Iserles Department of Applied Mathematics and Theoretical Physics, Centre for Mathematical Sciences, University of Cambridge, Cambridge, UK

Sun-Sook Jin Department of Mathematics Education, Gongju National University of Education, Gongju, Republic of Korea

Kil-Woung Jun Department of Mathematics, Chungnam National University, Yuseong-gu, Daejeon, Korea

Soon-Mo Jung Mathematics Section, College of Science and Technology, Hongik University, Jochiwon, Republic of Korea

Hassan Azadi Kenary Department of Mathematics, College of Science, Yasouj University, Yasouj, Iran

Hark-Mahn Kim Department of Mathematics, Chungnam National University, Yuseong-gu, Daejeon, Korea

S.V. Konyagin Department of Function Theory, Steklov Mathematical Institute, Russian Academy of Sciences, Moscow, Russia

Stamatis Koumandos Department of Mathematics and Statistics, University of Cyprus, Nicosia, Cyprus

Yang-Hi Lee Department of Mathematics Education, Gongju National University of Education, Gongju, Republic of Korea

Gangsong Leng Department of Mathematics, Shanghai University, Shanghai, P.R. China

Marina Logares Instituto de Ciencias Matemáticas (CSIC-UAM-UC3M-UCM), Madrid, Spain

Y.V. Malykhin Department of Function Theory, Steklov Mathematical Institute, Russian Academy of Sciences, Moscow, Russia

Jadranka Mićić Faculty of Mechanical Engineering and Naval Architecture, University of Zagreb, Zagreb, Croatia

Preda Mihăilescu Mathematisches Institut der Universität Göttingen, Göttingen, Germany

Gradimir V. Milovanović Mathematical Institute of the Serbian Academy of Sciences and Arts, Beograd, Serbia

S.A. Mohiuddine Department of Mathematics, Faculty of Science, King Abdulaziz University, Jeddah, Saudi Arabia

Fridoun Moradlou Department of Mathematics, Sahand University of Technology, Tabriz, Iran

Mohammad Sal Moslehian Department of Pure Mathematics, Center of Excellence in Analysis on Algebraic Structures (CEAAS), Ferdowsi University of Mashhad, Mashhad, Iran

Dumitru Motreanu Département de Mathématiques, Université de Perpignan, Perpignan Cedex, France

Vicente Muñoz Facultad de Matemáticas, Universidad Complutense de Madrid, Madrid, Spain

M. Mursaleen Department of Mathematics, Aligarh Muslim University, Aligarh, India

Abbas Najati Department of Mathematics, University of Mohaghegh Ardabili, Ardabil, Iran

Khalida Inayat Noor Mathematics Department, COMSATS Institute of Information Technology, Islamabad, Pakistan

Muhammad Aslam Noor Mathematics Department, COMSATS Institute of Information Technology, Islamabad, Pakistan

B. Paneah Department of Mathematics, Technion—Israel Institute of Technology, Haifa, Israel

Choonkil Park Department of Mathematics, Hanyang University, Seoul, Republic of Korea

Josip Pečarić Faculty of Textile Technology, University of Zagreb, Zagreb, Croatia

Yasar Polatoğlu Department of Mathematics and Computer Science, Kültür University, Istanbul, Turkey

Dorian Popa Department of Mathematics, Technical University, Cluj-Napoca, Romania

Agostino Prástaro MEMOMAT, University of Roma "La Sapienza", Roma, Italy

V.Y. Protasov Dept. of Mechanics and Mathematics, Moscow State University, Moscow, Russia

Reza Saadati Department of Mathematics and Computer Science, Amirkabir University of Technology, Tehran, Iran

Prasanna K. Sahoo Department of Mathematics, University of Louisville, Louisville, KY, USA

M. Bavand Savadkouhi Department of Mathematics, Semnan University, Semnan, Iran

Eun Young Son Department of Mathematics, Chungnam National University, Yuseong-gu, Daejeon, Korea

H.M. Srivastava Department of Mathematics and Statistics, University of Victoria, Victoria, British Columbia, Canada

Marija P. Stanić Department of Mathematics and Informatics, Faculty of Science, University of Kragujevac, Kragujevac, Serbia

Árpád Száz Institute of Mathematics, University of Debrecen, Debrecen, Pf. 12, Hungary

László Székelyhidi Institute of Mathematics, University of Debrecen, Debrecen, Hungary

Abraham Albert Ungar Department of Mathematics, North Dakota State University, Fargo, ND, USA

Zhihua Wang School of Science, Hubei University of Technology, Wuhan, Hubei, P.R. China; Department of Mathematics, Sichuan University, Chengdu, Sichuan, P.R. China

Patrick Winkert Institut für Mathematik, Technische Universität Berlin, Berlin, Germany

Yuming Xing Department of Mathematics, Harbin Institute of Technology, Harbin, China

Bing Xu Department of Mathematics, Sichuan University, Chengdu, Sichuan, P.R. China

Bicheng Yang Department of Mathematics, Guangdong University of Education, Guangzhou, Guangdong, P.R. China

Xiuzhong Yang College of Mathematics and Information Science, Hebei Normal University, Shijiazhuang, Hebei, P.R. China

Manar Youssef Department of Mathematics, Faculty of Sciences, University Ibn Zohr, Agadir, Morocco

Dandan Zhao Department of Mathematics, The University of Hong Kong, Pokfulam, Hong Kong

Xiaopeng Zhao College of Mathematics and Information Science, Hebei Normal University, Shijiazhuang, Hebei, P.R. China

Chapter 1
Bounds of Jensen's Type Inequality and Eigenvalues of Sturm–Liouville System

Shoshana Abramovich

Abstract In this paper, we deal with Jensen's type inequality, its bounds and refinements, and with eigenvalues of a Sturm–Liouville system. The results are obtained by rearrangements and continuous symmetrization.

Key words Convex functions · Jensen's inequalities · Weighted integrals · Continuous symmetrization · Equimeasurable functions · Sturm–Liouville systems · Eigenvalues

Mathematics Subject Classification 26D15 · 34L15 · 39B62 · 15A42

1.1 Introduction

Jensen's Theorem [7] states that

$$\int \varphi(f(s))\,d\mu(s) \ge \varphi\Big(\int f\,d\mu\Big)$$

holds when $\varphi : \mathbb{R} \to \mathbb{R}$ is convex, μ is a probability measure, and f is a μ-integrable function. The celebrated Jensen's theorems are dealt with in numerous articles; see, for instance, [4, 5, 7, 9] and their references, to quote just a few.

In this paper, we deal with Jensen's type inequality, its bounds and refinements related to the rearrangement of a given function f.

We deal also with a set of equimeasurable functions $f(x,a)$, $-1 \le \alpha \le 3$, $0 \le x \le 1$. For $\alpha = 0, -1, 2, 3$, we get that $f(x,0) = f(x)$ is a given function, $f(x,-1) = f^{-}(x)$ is its decreasing rearrangement, $f(x,1) = f^{**}(x)$ is its symmetrically increasing rearrangement, $f(x,2) = f(1-x)$, and $f(x,3) = f^{+}(x)$ is the increasing rearrangement of $f(x)$, see Fig. 1.1.

Dedicated to Professor Themistocles M. Rassias on the occasion of his 60th birthday.

S. Abramovich (✉)
Department of Mathematics, University of Haifa, Haifa, Israel
e-mail: abramos@math.haifa.ac.il

P.M. Pardalos et al. (eds.), *Nonlinear Analysis*, Springer Optimization and Its Applications 68, In Honor of Themistocles M. Rassias on the Occasion of his 60th Birthday, DOI 10.1007/978-1-4614-3498-6_1, © Springer Science+Business Media, LLC 2012

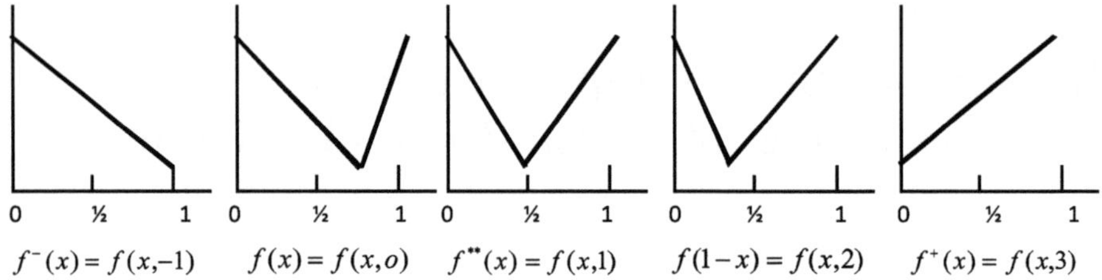

Fig. 1.1 Rearrangements of $f(x)$

For this set of equimeasurable functions, we show that the first eigenvalue of

$$y''(x)+\lambda(\alpha)f(x,\alpha)y(x)=0, \quad y'(0)=y(1)=0, \quad -1\le\alpha\le 3,$$

is increasing in α and so is $\frac{1}{a}\int_0^a \varphi'(x)f(x,\alpha)\,dx-\varphi(\frac{1}{a}\int_0^a f(x,\alpha)\,dx)$ for a convex function $\varphi\in C^2$.

The set of the equimeasurable functions is defined by a procedure named Continuous Symmetrization, as it appears in [8, p. 200] for $0\le\alpha\le 1$ and is extended here to $-1\le\alpha\le 3$. Whereas in [8] the procedure is limited to $0\le\alpha\le 1$, where $f(x,0)$ is the original function f that we start with and $f(x,1)$is the symmetrically increasing rearrangement of $f(x)$, here we extend the set to include the increasing rearrangement of $f(x)$, $f^+(x)=f(x,3)$ and the decreasing rearrangement of $f(x)$, $f^-(x)=f(x,-1)$.

Here are the definitions related to our results on continuous symmetrization.

Definition 1.1 A function $f(x)$ defined on [0, 1] is called left balanced, if for any $x\in[1/2,1]$, $f(1-x)\ge f(x)$ (see [1]).

Definition 1.2 Let $y=f(x)$ be continuous on [0, 1] , not increasing on $[0,l]$ and not decreasing on $[l,1]$. For $x\in[0,l]$ we denote the function inverse to $f(x)$ by $x_1(y)$, and for $x\in[l,1]$ we denote the inverse function by $x_2(y)$. We build a class of functions $f(x,a)$, $-1\le\alpha\le 3$, $0\le x\le 1$, by Continuous Symmetrization procedure. We denote the function inverse to $f(x,a)$ in its decreasing interval by $x_{1\alpha}(y)$, and in the increasing interval we denote the inverse function by $x_{2\alpha}(y)$.

When we deal with a left balanced function, if $f(0)>f(1)$ we add to $x_2(y)$ an interval of definition $f(1)\le y\le f(0)$ for which $x_2(y)=1$.

We agree that if $f(x)$ attains the same constant value k in two intervals $[a,b]$ and $[c,d]$, $a\le b\le c\le d$, and if

$$x_1(k)=(1-m)a+mb, \quad 0\le m\le 1,$$

then for a symmetrization procedure we choose

$$x_2(k)=mc+(1-m)d.$$

With the help of this explanation, the continuous symmetrization procedure goes as follows:

(a) For $-1 \le \alpha \le 0$,

$$\begin{aligned} x_{1\alpha}(y) &= x_1(y) - \alpha\big(1 - x_2(y)\big), && 0 \le x \le (1+\alpha)l - \alpha, \\ x_{2\alpha}(y) &= x_2(y) - \alpha\big(1 - x_2(y)\big), && (1+\alpha)l - \alpha \le x \le 1. \end{aligned} \tag{1.1}$$

(b) For $0 \le \alpha \le 2$,

$$\begin{aligned} x_{1\alpha}(y) &= \left(1 - \frac{\alpha}{2}\right)x_1(y) + \frac{\alpha}{2}\big(1 - x_2(y)\big), && 0 \le x \le (1-\alpha)l + \frac{\alpha}{2}, \\ x_{2\alpha}(y) &= \left(1 - \frac{\alpha}{2}\right)x_2(y) + \frac{\alpha}{2}\big(1 - x_1(y)\big), && (1-\alpha)l + \frac{\alpha}{2} \le x \le 1. \end{aligned} \tag{1.2}$$

(c) For $2 \le \alpha \le 3$,

$$\begin{aligned} x_{1\alpha}(y) &= \big(1 - x_2(y)\big) - (\alpha - 2)\big(1 - x_2(y)\big), && 0 \le x \le (3-\alpha)(1-l), \\ x_{2\alpha}(y) &= \big(1 - x_1(y)\big) - (\alpha - 2)\big(1 - x_2(y)\big), && (3-\alpha)(1-l) \le x \le 1. \end{aligned} \tag{1.3}$$

The continuous symmetrization for $0 \le \alpha \le 1$ and for left balanced functions was discussed in [1].

1.2 Estimation of Jensen's Type Inequalities

We first state a lemma and quote a theorem that we use in the sequel.

Lemma 1.1 *Let $f(x), g(x) \in C^1$, $0 \le x \le a$. Let $z \in C^1$ be an increasing function on $[0, a]$. Let*

$$\int_x^a \big(f(x) - g(x)\big)\,dx \le 0, \qquad \int_0^a \big(f(x) - g(x)\big)\,dx = 0. \tag{1.4}$$

Then

$$\int_0^a z(x)\big(f(x) - g(x)\big)\,dx \le 0. \tag{1.5}$$

Proof Inequality (1.5) simply follows from (1.4):

$$\begin{aligned} &\int_0^a z(x)\big(f(x) - g(x)\big)\,dx \\ &\quad = \left[-z(x)\int_x^a \big(f(x) - g(x)\big)\,dx\right]_0^a + \int_0^a z'(x)\left(\int_x^a \big(f(t) - g(t)\big)\,dt\right)dx \\ &\quad = \int_0^a z'(x)\left(\int_x^a \big(f(t) - g(t)\big)\,dt\right)dx \le 0. \end{aligned}$$

□

Theorem A ([6, Theorem 2]) *Let f and g be integrable functions on $[a,b]$, and let w be a positive integrable function. Suppose that ψ is a strictly increasing function and $\varphi \circ \psi^{-1}$ is concave. Suppose that f is decreasing and that*

$$\int_x^b \psi\big(f(t)\big)w(t)\,dt \geq \int_x^b \psi\big(g(t)\big)w(t)\,dt, \quad \forall x \in [a,b].$$

(a) *If*

$$\int_a^b \psi\big(f(t)\big)w(t)\,dt = \int_a^b \psi\big(g(t)\big)w(t)\,dt,$$

then

$$\int_a^b \varphi\big(f(t)\big)w(t)\,dt \geq \int_a^b \varphi\big(g(t)\big)w(t)\,dt;$$

if g is increasing, the inverse inequality holds.
(b) *If $\varphi \circ \psi^{-1}$ is increasing, then*

$$\int_x^b \varphi\big(f(t)\big)w(t)\,dt \geq \int_x^b \varphi\big(g(t)\big)w(t)\,dt, \quad b \geq x \geq a.$$

Lemma 1.2 *Let $f(x) \in C^1$ be defined on $[0,a]$ and $0 \leq f(x) \leq a$, $a > 0$. If $\varphi(x) \in C^2$ is convex on $[0,a]$ and $\varphi(0) = 0$, then*

$$a\varphi\left(\int_0^a \frac{1}{a} f(x)\,dx\right) \leq \int_0^a \varphi'(x) f(x)\,dx. \tag{1.6}$$

Proof First, we assume that $f : [0,a] \to [0,1]$ and prove that in this case

$$\varphi\left(\int_0^a f(x)\,dx\right) \leq \int_0^a \varphi'(x) f(x)\,dx. \tag{1.7}$$

Denote $F(x) = af(x)$ so that $0 \leq F(x) \leq a$ for $0 \leq x \leq a$. Let $F^-(x) = af^-(x)$ be the decreasing rearrangement of $F(x) = af(x)$, and $V^-(x)$ the inverse function of $F^-(x)$. Using $\varphi(0) = 0$, we obtain:

$$\begin{aligned}
\int_0^a \varphi\big(V^-(x)\big)\,dx &= \int_a^0 \varphi(z)\big(F^-(z)\big)'\,dz = -\int_0^a \varphi(z)\big(F^-(z)\big)'\,dz \\
&= -\big[\varphi(z)\big(F^-(z)\big)\big]_{z=0}^{z=a} + \int_0^a \varphi'(z)\big(F^-(z)\big)\,dz \\
&= a\int_0^a \varphi'(x) f^-(x)\,dx.
\end{aligned} \tag{1.8}$$

In particular, for $\varphi(x) = x$ we get that

$$\int_0^a V^-(x)\,dx = a\int_0^a f^-(x)\,dx = a\int_0^a f(x)\,dx. \tag{1.9}$$

From (1.8) and (1.9), together with Jensen's Inequality, we get that

$$\begin{aligned}\varphi\left(\int_0^a f(x)\,dx\right) &= \varphi\left(\frac{1}{a}\int_0^a V^-(x)\,dx\right)\\ &\le \frac{1}{a}\int_0^a \varphi\bigl(V^-(x)\bigr)\,dx = \frac{1}{a}\int_0^a a\varphi'(x)f^-(x)\,dx\\ &= \int_0^a \varphi'(x)f^-(x)\,dx \qquad (1.10)\end{aligned}$$

holds. We show now that

$$\int_0^a \varphi'(x)f^-(x)\,dx \le \int_0^a \varphi'(x)f(x)\,dx. \qquad (1.11)$$

As $f(x)$ and $f^-(x)$ are equimeasurable when $f^-(x)$ is the decreasing rearrangement of $f(x)$, the inequality

$$\int_x^a f^-(x)\,dx \le \int_x^a f(x)\,dx, \qquad \int_0^a f^-(x)\,dx = \int_0^a f(x)\,dx$$

holds. Therefore, by Lemma 1.1, as $\varphi'(x)$ is nondecreasing, (1.11) holds.

Inequalities (1.11) and (1.10) lead, by Theorem A, to

$$\varphi\left(\int_0^a f(x)\,dx\right) \le \int_0^a \varphi'(x)f^-(x)\,dx \le \int_0^a \varphi'(x)f(x)\,dx.$$

Hence (1.7) is proved for $0 \le f(x) \le 1$, and evidently also (1.6) holds. □

Corollary 1.1 *Under the same conditions on f for a convex function $\phi \in C^2$ on $[0,\infty)$ we get from (1.6) that*

$$a\phi\left(\int_0^a \frac{1}{a}f(x)\,dx\right) \le \int_0^a \phi'(x)f(x)\,dx + \phi(0).$$

If $\phi(0) \le 0$, we get that also in this case (1.6) holds.

Theorem 1.1 *Let $f \in C^1$ be such that $f:[0,1] \to [0,1]$. Let $f^+(x)$ be its increasing rearrangement. Let $\varphi(x) \in C^2$ be a convex function on $[0,1]$, $\varphi(0) \le 0$. Then*

$$\varphi\left(\int_0^1 f(t)\,dt\right) \le \int_0^1 \varphi\bigl(f(t)\bigr)\,dt \le \int_0^1 \varphi'(t)f^+(t)\,dt.$$

Proof The proof follows from the fact that

$$\varphi\left(\int_0^1 f(t)\,dt\right) \le \int_0^1 \varphi\big(f(t)\big)\,dt = \int_0^1 \varphi\big(f^-(t)\big)\,dt$$
$$= \int_0^1 \varphi'(t)u^-(t)\,dt \le \int_0^1 \varphi'(t)f^+(t)\,dt, \tag{1.12}$$

where $u^-(t)$ is the inverse function of $f^-(t)$ the decreasing rearrangement of $f(t)$. The last inequality in (1.12) follows from Lemma 1.1, as $\int_x^1 u^-(t)\,dt \le \int_x^1 f^+(t)\,dt$ and $\int_0^1 u^-(t)\,dt = \int_0^1 f^+(t)\,dt$. □

Theorem 1.2 *Let $f \in C^1, 0 \le x \le 1, 0 \le f(x) \le 1$. Let $u^-(x)$ be the inverse function of $f^-(x)$, the decreasing rearrangement of $f(x)$.*

If

$$\int_x^1 f^-(x)\,dx \le \int_x^1 u^-(x)\,dx, \tag{1.13}$$

then

$$\varphi\left(\int_0^1 f(x)\,dx\right) \le \int_0^1 \varphi'(x)f^-(x)\,dx \le \int_0^1 \varphi\big(f(x)\big)\,dx$$
$$\le \int_0^1 \varphi'(x)f^+(x)\,dx \tag{1.14}$$

when $\varphi : [0,1] \to \mathbb{R}$ is a convex function and $\varphi(0) \le 0$. If $\int_x^1 f^-(x)\,dx \ge \int_x^1 u^-(x)\,dx$, then

$$\varphi\left(\int_0^1 f(x)\,dx\right) \le \int_0^1 \varphi\big(f(x)\big)\,dx \le \int_0^1 \varphi'(x)f^-(x)\,dx$$
$$\le \int_0^1 \varphi(x)f^+(x)\,dx. \tag{1.15}$$

Proof We proved in (1.9) that $\int_0^1 f^-(x)\,dx = \int_0^1 u^-(x)\,dx$. As f and f^- are equimeasurable, we get that from Jensen's Inequality that

$$\varphi\left(\int_0^1 f(x)\,dx\right) = \varphi\left(\int_0^1 f^-(x)\,dx\right) = \varphi\left(\int_0^1 u^-(x)\,dx\right)$$
$$\le \int_0^1 \varphi\big(u^-(x)\big)\,dx = \int_0^1 \varphi'(x)f^-(x)\,dx \le \int_0^1 \varphi'(x)u^-(x)\,dx$$
$$= \int_0^1 \varphi\big(f^-(x)\big)\,dx = \int_0^1 \varphi\big(f(x)\big)\,dx. \tag{1.16}$$

The last inequality in (1.16) follows from Lemma 1.1, (1.13), and also from Theorem A, as $\int_0^1 f^-(x)\,dx = \int_0^1 u^-(x)\,dx$. Together with Theorem 1.2, we get (1.14). □

We can partially extend Theorem 1.2:

Remark 1.1 Let $f \in C^1, 0 \le x \le a, 0 \le f(x) \le M$. Let $u^-(x)$ be the inverse function of $f^-(x)$, the decreasing rearrangement of $f(x)$. Let $\varphi : \mathbb{R} \to \mathbb{R}$ be a convex function and $\varphi(0) \le 0$. If $\int_x^b f_*^-(x)\,dx \le \int_x^b u_*^-(x)\,dx$, where

$$f_*^-(x) = \begin{cases} f^-(x), & x \in [0,a], \\ 0 & x \notin [0,a], \end{cases} \qquad u_*^-(x) = \begin{cases} u^-(x), & x \in [0,M], \\ 0 & x \notin [0,M], \end{cases}$$

and $b = \max\{a, M\}$, then

$$\varphi\left(\int_0^a f(x)\,dx\right) \le \int_0^a \varphi'(x) f^-(x)\,dx \le \int_0^a \varphi\bigl(f(x)\bigr)\,dx.$$

In Example 1.1, we demonstrate Theorem 1.2.

Example 1.1 Let $f(x) = 1 - x^2, 0 \le x \le 1$. The inverse function of $f(x)$ is $u(x) = \sqrt{1-x}$.

It is easy to see that $\int_x^1 \sqrt{1-x}\,dx \ge \int_x^1 (1-x^2)\,dx$ and that $\int_0^1 \sqrt{1-x}\,dx = \int_0^1 (1-x^2)\,dx$.

Therefore, for convex $\varphi(x) \in C^2$ with $\varphi(0) \le 0$ and by the same reasoning as in proving (1.11), we get that $\int_0^1 \varphi'(x)\sqrt{1-x}\,dx \ge \int_0^1 \varphi'(x)(1-x^2)\,dx$, and therefore

$$\varphi\left(\int_0^1 \bigl(1-x^2\bigr)\,dx\right) \le \int_0^1 \varphi'(x)\bigl(1-x^2\bigr)\,dx \le \int_0^1 \varphi\bigl(1-x^2\bigr)\,dx.$$

So in this case, we get a refinement of Jensen's Inequality.

On the other hand,

$$\begin{aligned} \varphi \int_0^1 \sqrt{1-x}\,dx &\le \int_0^1 \varphi(\sqrt{1-x})\,dx = \int_0^1 \varphi'(x)\bigl(1-x^2\bigr)\,dx \\ &\le \int_0^1 \varphi'(x)\sqrt{1-x}\,dx, \end{aligned}$$

which means that in this case Jensen's Inequality is stronger than (1.6).

1.3 Monotonicity of $f(x,\alpha)$ and of Eigenvalues

The use of the results obtained in Sect. 1.2 enables us to show in this section the monotonicity of

$$\int_0^1 \varphi'(x) f(x,\alpha)\,dx - \varphi\left(\int_0^1 f(x,\alpha)\,dx\right) \tag{1.17}$$

and the monotonicity of the first eigenvalue of

$$y''(x) + \lambda(\alpha) f(x,\alpha) y(x) = 0, \quad y'(0) = y(1) = 0, \quad -1 \le \alpha \le 3, \tag{1.18}$$

as a function of α.

The functions $f(x,\alpha)$, $-1 \le \alpha \le 3$, are defined by the continuous symmetrization where $f(x) = f(x,0)$ is left balanced; see Definitions 1.1 and 1.2.

Theorem 1.3 *Let $f(x)$ be continuous on $[0,1]$, not increasing on $[0,l]$, and not decreasing on $[l,1]$, and assume that $f(x)$ is left balanced. Then, for $-1 \le \alpha \le 3$:*

(a) *$f(x,\alpha)$ is continuous on $[0,1]$, not increasing in x on $[0,l(\alpha)]$, and not decreasing in x on $[l(\alpha),1]$, where*

$$l(\alpha) = \begin{cases} (1+\alpha)l - \alpha, & -1 \le \alpha \le 0, \\ (1-\alpha)l + \frac{\alpha}{2}, & 0 \le \alpha \le 2, \\ (3-\alpha)(1-l), & 2 \le \alpha \le 3. \end{cases}$$

(b) *For $-1 \le x \le l(\alpha)$, $f(x,\alpha)$ is not increasing in α, and for $l(\alpha) \le x \le 1$, $f(x,\alpha)$ is not decreasing in α.*
(c) *$\int_x^1 f(x,\alpha)\,dx$, is not decreasing in α.*
(d) *$f(x,\alpha)$ is continuous in α.*

Proof We show that the theorem holds for $-1 \le \alpha \le 0$. The case $0 \le \alpha \le 1$ is fully dealt with in [1, Theorem 1]. The other cases follow similarly.

It is obvious that for the functions inverse to our left balanced $f(x)$, $x_1(y) + x_2(y) \ge 0$ (see also the proof in [1, Theorem 1]).

Part (a): The continuity of $f(x,\alpha)$, $-1 \le \alpha \le 0$ and its monotonicity in $[0,l(\alpha)]$ and in $[l(\alpha),1]$ immediately follow from (1.1).

Part (b): Equations (1.1) imply

$$x_{1\alpha}(y) - x_{1\beta}(y) = (\beta - \alpha)\big(1 - x_2(y)\big), \quad -1 \le \alpha \le \beta \le 0. \tag{1.19}$$

Let $\overline{y}$ be defined by

$$x_{1\alpha}(y) = x_{1\beta}(\overline{y}) = x. \tag{1.20}$$

Then (1.19) and (1.20) give

$$x = x_{1\beta}(\overline{y}) > x_{1\beta}(y). \tag{1.21}$$

As $x_{1\beta}(y)$ is a decreasing function of y, it follows from (1.20) and (1.21) that

$$f(x,\beta) \le f(x,\alpha), \quad \beta \ge \alpha, \quad 0 \le x \le l(\beta). \tag{1.22}$$

By an analogous consideration, we obtain that

$$f(x,\beta) \ge f(x,\alpha), \quad \beta \ge \alpha, \quad l(\beta) \le x \le 1. \tag{1.23}$$

Consequently, by (1.22) and (1.23), $f(x,\alpha)$ is not increasing in α when $0 \le x \le l(\alpha)$ and not decreasing when $l(\alpha) \le x \le 1$. Thus (b) is established.

Part (c): As $f(x,\alpha)$ is not increasing in α when $0 \le x \le l(\alpha)$ and not decreasing when $l(\alpha) \le x \le 1$, from (1.22) and (1.23) it follows that there is an x_0, $l(\beta) \le x_0 \le l(\alpha)$, $\alpha < \beta$ such that

$$\begin{aligned} f(x,\beta) &\le f(x,\alpha), \quad 0 \le x \le x_0, \\ f(x,\beta) &\ge f(x,\alpha), \quad x_0 \le x \le 1. \end{aligned} \tag{1.24}$$

$f(x,\beta)$ and $f(x,\alpha)$ are equimeasurable, therefore $\int_0^1 f(x,\alpha)\,dx = \int_0^1 f(x,\beta)\,dx$, and together with (1.24) we get that $\int_x^1 f(t,\beta)\,dt \ge \int_x^1 f(t,\alpha)\,dt$. Thus (c) is established.

Part (d): For $0 \le \alpha \le 1$ the proof appears in [1]. We prove here the case $-1 \le \alpha \le 0$ which follows step by step the proof for $0 \le \alpha \le 1$ there. The other cases follow similarly.

As $f(x,\alpha)$ is not increasing in α when $0 \le x \le l(\alpha)$, it has a limit from the left and a limit from the right at every point. Let us look at the equality

$$s = x_1\big(f(s,\alpha)\big) - \alpha\big(1 - x_2\big(f(s,\alpha)\big)\big), \quad s < l(\alpha). \tag{1.25}$$

Consider first the case $f(0) = f(1)$, then $x_2(y)$ is increasing and not stationary; therefore,

$$y^+ = f(s,\alpha+0) < f(s,\alpha-0) = y^-$$

cannot occur because it contradicts (1.25). It follows that $f(s,\alpha+0) = f(s,\alpha-0)$, which means continuity of $f(x,\alpha)$ in α, $-1 \le \alpha \le 0$.

If $f(0) > f(1)$ then $f(1,\alpha) = f(0)$ for $-1 \le \alpha \le 0$ by the definition of the symmetrization procedure, but $f(1,0) < f(1,\alpha)$, $-1 \le \alpha \le 0$.

For $0 \le x < 1$, the continuity of α follows as before. Thus the continuity of $f(x,\alpha)$ in α, $-1 \le \alpha \le 0$ is established.

All other cases of α are proved similarly. □

Using Theorem 1.3 we get:

Theorem 1.4 *Let $f : [0,1] \to [0,1]$ be continuous, not increasing on $[0,l]$, and not decreasing on $[l,1]$, and assume that $f(x)$ is left balanced. Let $\varphi \in C^2$ be a convex function on $\mathbb{R}$. Then*

(a) *For* $-1 \le \alpha \le 3$, $\int_0^1 \varphi'(x) f(x,\alpha)\,dx$ *is increasing in* α *when* $f(x,\alpha)$ *is the stage* α *in the rearrangement of* $f(x)$ *by the continuous symmetrization procedure.*
(b) *If* (1.13) *holds, then there is an* α_0, $-1 \le \alpha_0 \le 3$ *such that* $\varphi(\int_0^1 f(x,\alpha_0))\,dx = \int_0^1 \varphi'(x) f(x,\alpha_0)\,dx$.

Proof Part (a). The monotonicity of $\int_0^1 \varphi'(x) f(x,\alpha)\,dx$ is derived directly from Theorem 1.3(c) together with Lemma 1.1.

Part (b). The existence of α_0, $-1 \le \alpha_0 \le 3$ follows from parts (c) and (d) of Theorem 1.3, and Theorem 1.2. □

The last theorem deals with the monotonicity of the first eigenvalue of a Sturm–Liouville system.

Theorem 1.5 *Let* $f : [0,1] \to \mathbb{R}^+$ *be bounded, continuous, not increasing on* $[0,l]$, *and not decreasing on* $[l,1]$, *and assume that* $f(x)$ *is left balanced. Let* $\lambda_1(\alpha)$ *be the first eigenvalue of the system*

$$y''(x) + \lambda(\alpha) f(x,\alpha) y(x) = 0, \quad y'(0) = y(1) = 0. \tag{1.26}$$

Then

$$\lambda_1(\alpha) \le \lambda_1(\beta), \quad -1 \le \alpha < \beta \le 3. \tag{1.27}$$

Proof Let $y_{1,\alpha}(x) \ge 0$, $0 \le x \le 1$, be the first eigenvalue of (1.26). To prove the theorem, we have to establish

$$\int_0^1 \big(f(x,\beta) - f(x,\alpha)\big) y_{1,\beta}^2(x)\,dx \le 0, \quad -1 \le \alpha < \beta \le 3. \tag{1.28}$$

It is known that under the condition of the theorem on $f(x)$, $y_{1,\alpha}(x) \ge 0$ is nonincreasing. Using Lemma 1.1 and Theorem 1.3, (1.28) is obtained (see also [5, Theorem 399]).

Now we use the Rayleigh Ratio for the minimum characterization of the first eigenvalue of (1.26) (see [3] and [2]). We then have

$$\begin{aligned} \lambda_1(\beta) &= \frac{\int_0^1 y_{1,\beta}'^2(x)\,dx}{\int_0^1 f(x,\beta) y_{1,\beta}^2(x)\,dx} \ge \frac{\int_0^1 y_{1,\beta}'^2(x)\,dx}{\int_0^1 f(x,\alpha) y_{1,\beta}^2(x)\,dx} \\ &\ge \min \frac{\int_0^1 v'^2(x)\,dx}{\int_0^1 f(x,\alpha) v^2(x)\,dx} = \lambda_1(\alpha). \end{aligned} \tag{1.29}$$

The first inequality sign in (1.29) follows from (1.28). The minimum is taken over all functions $v(x) \in C^1$, $v'(0) = v(1)$, and $y_{1,\beta}(x)$ clearly belongs to this class. This proves (1.27). □

Remark 1.2 If we replace the boundary conditions with $y(0) = y'(1) = 0$, then $\lambda_1(\alpha)$ is decreasing in α, $-1 \le \alpha \le 3$. Also, if f is right balanced we get analogous results.

References

1. Abramovich, S.: Monotonicity of eigenvalues under symmetrization. SIAM J. Appl. Math. **28**(2), 350–361 (1975)
2. Beesack, P.R., Schwarz, B.: On the zeroes of solutions of second order differential equations. Can. J. Math. **8**, 504–515 (1956)
3. Courant, R., Hilbert, D.: Methods of Mathematical Physics, vol. 1. Interscience, New York (1966)
4. Dragomir, S.S.: Bounds for normalized Jensen functional. Bull. Aust. Math. Soc. **74**(3), 471–478 (2006)
5. Hardy, G.H., Littlewood, J.E., Pólya, G.: Inequalities. Cambridge University Press, London (1952)
6. Pečarić, J., Abramovich, S.: On new majorization theorems. Rocky Mt. J. Math. **27**(3), 903–911 (1997)
7. Pečarić, J., Proschan, F., Tong, Y.L.: Convex Functions Partial Orderings, and Statistical Applications. Academic Press, San Diego (1992)
8. Pólya, G., Szegö, G.: Isoperimetric Inequalities in Mathematical Physics. Princeton University Press, Princeton (1951)
9. Elqorachi, E., Youssef, M., Rassias, T.M.: Hyers–Ulam stability of the quadratic and Jensen functional equations on unbounded domains. J. Math. Sci. Adv. Appl. **4**(2), 287–301 (2010)

Chapter 2
Quadratic Operators and Quadratic Functional Equation

M. Adam and S. Czerwik

Abstract In the first part of this paper, we consider some quadratic difference operators (e.g., Lobaczewski difference operators) and quadratic-linear difference operators (d'Alembert difference operators and quadratic difference operators) in some special function spaces X_λ. We present results about boundedness and find the norms of such operators. We also present new results about the quadratic functional equation. The second part is devoted to the so-called double quadratic difference property in the class of differentiable functions. As an application we prove the stability result in the sense of Ulam–Hyers–Rassias for the quadratic functional equation in a special class of differentiable functions.

Key words Quadratic, d'Alembert, and Lobaczewski difference operators · X_λ spaces · Quadratic functional equation · Stability

Mathematics Subject Classification 39B52 · 39B82 · 47H30

2.1 The X_λ and X_λ^2 Spaces

We shall introduce the spaces X_λ and X_λ^2 (see [7]). A. Bielecki also studied similar spaces in [4] and applied them in the theory of differential equations.

Definition 2.1 Let X and Y be two normed vector spaces and $\lambda \geq 0$. Define

$$X_\lambda := \{f : X \to Y : \|f(x)\| \leq M_f e^{\lambda\|x\|},\ x \in X\},$$

Dedicated to Professor Themistocles M. Rassias on the occasion of his 60th birthday.

M. Adam
Department of Mathematics and Informatics, Higher School of Labour Safety Management in Katowice, Bankowa 8, 40-007 Katowice, Poland
e-mail: madam@wszop.edu.pl

S. Czerwik (✉)
Institute of Mathematics, Silesian University of Technology, Kaszubska 23, 44-100 Gliwice, Poland
e-mail: Stefan.Czerwik@polsl.pl

P.M. Pardalos et al. (eds.), *Nonlinear Analysis*, Springer Optimization and Its Applications 68, In Honor of Themistocles M. Rassias on the Occasion of his 60th Birthday, DOI 10.1007/978-1-4614-3498-6_2,

where M_f is a real constant depending on f. Moreover, for $f \in X_\lambda$

$$\|f\| := \sup_{x \in X} \{ e^{-\lambda \|x\|} \| f(x) \| \}. \tag{2.1}$$

Let us note that the space X_λ with the norm (2.1) was considered by S. Czerwik and K. Dłutek in [10]. It is easy to prove the following

Lemma 2.1 *The space* $(X_\lambda, \|\cdot\|)$, *where* $\|\cdot\|$ *is defined by* (2.1), *is a linear normed space.*

Definition 2.2 Let X and Y be two normed vector spaces and $\lambda \geq 0$. Define

$$X_\lambda^2 := \{ g : X \times X \to Y : \| g(x,y) \| \leq M_g e^{\lambda(\|x\| + \|y\|)}, \ x, y \in X \},$$

where M_g is a real constant depending on g. Moreover, for $g \in X_\lambda^2$

$$\|g\| := \sup_{x,y \in X} \{ e^{-\lambda(\|x\| + \|y\|)} \| g(x,y) \| \}. \tag{2.2}$$

We have

Lemma 2.2 *The space* $(X_\lambda^2, \|\cdot\|)$, *where* $\|\cdot\|$ *is defined by* (2.2), *is a linear normed space.*

2.2 Quadratic Difference Operator in X_λ Spaces

We define the quadratic difference operator $Q(f)$ by

$$Q(f)(x,y) := f(x+y) + f(x-y) - 2f(x) - 2f(y) \tag{2.3}$$

for $x, y \in X$, $f \in X_\lambda$.

Then we have

Theorem 2.1 *The quadratic difference operator* $Q : X_\lambda \to X_\lambda^2$, *given by the formula* (2.3), *is a linear bounded operator satisfying the inequality*

$$\| Q(f) \| \leq 6 \|f\|, \quad f \in X_\lambda. \tag{2.4}$$

Proof First, we shall verify that if $f \in X_\lambda$, then $Q(f) \in X_\lambda^2$. We have

$$\begin{aligned} \| Q(f)(x,y) \| &= \| f(x+y) + f(x-y) - 2f(x) - 2f(y) \| \\ &\leq M_f e^{\lambda(\|x+y\|)} + M_f e^{\lambda(\|x-y\|)} + 2M_f e^{\lambda(\|x\|)} + 2M_f e^{\lambda(\|y\|)} \\ &\leq 6M_f e^{\lambda(\|x\| + \|y\|)}, \end{aligned}$$

thus $Q(f) \in X_\lambda^2$ as claimed.

Clearly, Q is linear. For $f \in X_\lambda$, we obtain

$$\begin{aligned}
\|Q(f)\| &= \sup_{x,y\in X} e^{-\lambda(\|x\|+\|y\|)} \|f(x+y)+f(x-y)-2f(x)-2f(y)\| \\
&\le \sup_{x,y\in X} e^{-\lambda(\|x\|+\|y\|)} \|f(x+y)\| + \sup_{x,y\in X} e^{-\lambda(\|x\|+\|y\|)} \|f(x-y)\| \\
&\quad + 2\sup_{x\in X} e^{-\lambda(\|x\|+\|y\|)} \|f(x)\| + 2\sup_{y\in X} e^{-\lambda(\|x\|+\|y\|)} \|f(y)\| \\
&= \|f\| + \|f\| + 2\|f\| + 2\|f\| = 6\|f\|.
\end{aligned}$$

Therefore,

$$\|Q(f)\| \le 6\|f\|, \quad f \in X_\lambda,$$

which conclude the proof. □

Under some additional assumptions, we can prove some further results. In fact, we have

Theorem 2.2 *Let* $\mathbb{R} \subset X$, $\mathbb{R} \subset Y$ *and* $\|x\| = |x|$ *for* $x \in \mathbb{R}$. *Then*

$$\|Q\| = 6. \tag{2.5}$$

Proof Let $\{x_n\}$ be a strictly decreasing sequence of positive numbers such that

$$\lim_{n\to\infty} x_n = 0.$$

Let us define for $n \in \mathbb{N}$ a function f_n by

$$f_n(x) := \begin{cases} -e^{2\lambda x_n}, & x = x_n, \\ e^{2\lambda x_n}, & x = 2x_n, \\ e^{2\lambda x_n}, & x = 0, \\ 0, & \text{otherwise.} \end{cases}$$

Clearly, we have

$$\|f_n(x)\| \le e^{2\lambda x_n} e^{\lambda\|x\|}, \quad x \in X,$$

so $f_n \in X_\lambda$ for all $n \in \mathbb{N}$. Moreover,

$$e^{-\lambda\|x\|}\|f_n(x)\| = \begin{cases} e^{2\lambda x_n}, & x = 0, \\ e^{2\lambda x_n}, & x = x_n, \\ 1, & x = 2x_n, \\ 0, & \text{otherwise.} \end{cases}$$

Because the sequence $\{x_n\}$ is a decreasing sequence of positive numbers convergent to zero, we obtain that $\|f_n\| = e^{2\lambda x_n}$ for all $n \in \mathbb{N}$. We also have

$$\begin{aligned}\|Q(f_n)\| &= \sup_{x,y\in X} \{e^{-\lambda(\|x\|+\|y\|)} \|f(x+y) + f(x-y) - 2f(x) - 2f(y)\|\} \\ &\geq e^{-\lambda x_n} \|f_n(2x_n) + f_n(0) - 4f_n(x_n)\| = e^{-\lambda x_n} \cdot 6e^{\lambda x_n} = 6.\end{aligned}$$

Thus $\|Q(f_n)\| \geq 6$ for $n \in \mathbb{N}$. We also know from (2.4) that $\|Q\| \leq 6$. Suppose on the contrary that $\|Q\| < 6$. Then there exists $\varepsilon > 0$ such that

$$\|Q(f_n)\| \leq (6-\varepsilon)\|f_n\|, \quad f_n \in X_\lambda.$$

On the other hand, we have for $f_n \in X_\lambda$

$$6 \leq \|Q(f_n)\| \leq (6-\varepsilon)e^{2\lambda x_n}.$$

Taking into account that $x_n \to 0$ as $n \to \infty$, we get $6 \leq 6-\varepsilon$, where $\varepsilon > 0$, which is impossible. Thus we obtain eventually that $\|Q\| = 6$, and the proof is complete. □

For further information on new results concerning the quadratic difference operator on other spaces, see also the papers [9, 11, 12].

2.3 D'Alembert and Lobaczewski Difference Operators in X_λ Spaces

In this section, we shall recall the definition of the quadratic bounded operator. The Lobaczewski difference operator is an interesting example of a quadratic operator.

Here we shall present the ideas and main results obtained by S. Czerwik and K. Król in [13].

Let X and Y be linear spaces over a field $\mathbb{K}$.

Definition 2.3 An operator $Q: X \to Y$ is called quadratic if it satisfies the following equations

$$Q(x+y) + Q(x-y) = 2Q(x) + 2Q(y), \quad x, y \in X, \tag{2.6}$$

$$Q(kx) = k^2 Q(x), \quad x \in X,\ k \in \mathbb{K}. \tag{2.7}$$

Definition 2.4 A quadratic operator $Q: X \to Y$, where X, Y are linear normed spaces over $\mathbb{K}$, is called bounded if there exists an $M \geq 0$ such that

$$\|Q(x)\| \leq M\|x\|^2, \quad x \in X. \tag{2.8}$$

A norm of a quadratic bounded operator $Q: X \to Y$ is defined by

$$\|Q\| := \sup_{x\in X}\{\|Q(x)\| : \|x\| \leq 1\}. \tag{2.9}$$

By $B_Q(X,Y)$ we denote the space of all bounded quadratic operators. It is easy to prove that $B_Q(X,Y)$ with the norm given by (2.9) is a linear normed space.

Let $\mathbb{C}$ denote the set of all complex numbers. For a set X, a symbol $\mathbb{C}^X$ denotes the set of all functions $f\colon X \to \mathbb{C}$.

Definition 2.5 For a linear space X, the Lobaczewski difference operator $L\colon \mathbb{C}^X \to \mathbb{C}^{X^2}$ is defined by

$$L(f)(x,y) := f^2\left(\frac{x+y}{2}\right) - f(x)f(y), \quad x,y \in X. \tag{2.10}$$

One can verify that we have

Remark 2.1 The Lobaczewski difference operator $L\colon \mathbb{C}^X \to \mathbb{C}^{X^2}$ defined by (2.10) is a quadratic operator.

We can also prove that the Lobaczewski operator $L\colon X_\lambda \to X_\lambda^2$, where $Y = \mathbb{C}$, is a quadratic bounded operator. We have even more (see [13]).

Theorem 2.3 *Let $Y = \mathbb{C}$. The Lobaczewski difference operator defined by* (2.10) *belongs to $B_Q(X_\lambda, X_\lambda^2)$, and for all $f \in X_\lambda$ we have*

$$\|L(f)\| \le 2\|f\|^2. \tag{2.11}$$

Under some additional assumptions, we can find the norm of L. In fact, the following is true.

Theorem 2.4 *Let $\mathbb{R}_+ \subset X$, $Y = \mathbb{C}$, and $\|x\| = |x|$ for all $x \in \mathbb{R}_+$. Then*

$$\|L\| = 2, \tag{2.12}$$

where L is given by (2.10).

The proof, similar to the proof of Theorem 2.2, can be found in [13].

Now we shall present results about the d'Alembert difference operator.

Definition 2.6 We denote by B_{LQ} the space

$$B_{LQ}(X,Y) := \{T \in Y^X : \exists L \in B(X,Y) \text{ and } \exists Q \in B_Q(X,Y) \\ \text{such that } T = L + Q\}.$$

Here, of course, $B(X,Y)$ stands for the space of linear bounded operators from X to Y.

For $T = L + Q \in B_{LQ}(X,Y)$, we define

$$\|T\| := \|L\| + \|Q\|.$$

We say that such an operator T is a bounded linear-quadratic operator.

Definition 2.7 Let X be a linear space. The d'Alembert difference operator $A : \mathbb{C}^X \to \mathbb{C}^{X^2}$ is defined by

$$A(f)(x, y) := f(x+y) + f(x-y) - 2f(x)f(y), \quad x, y \in X. \tag{2.13}$$

In the sequel, we present the following

Theorem 2.5 ([13]) *Let $Y = \mathbb{C}$ and X be a normed space. The d'Alembert difference operator $A\colon X_\lambda \to X_\lambda^2$ defined by* (2.13) *belongs to $B_{LQ}(X_\lambda, X_\lambda^2)$, and for all $f \in X_\lambda$ we have*

$$\|A(f)\| \le 2\|f\| + 2\|f\|^2.$$

Proof On account of (2.13), we get $A = L_A + Q_A$, where the linear operator $L_A : X_\lambda \to X_\lambda^2$ and the quadratic operator $Q_A : X_\lambda \to X_\lambda^2$ are given by

$$L_A(f)(x, y) := f(x+y) + f(x-y),$$
$$Q_A(f)(x, y) := -2f(x)f(y).$$

Now, for any $f \in X_\lambda$ we obtain successively

$$\begin{aligned}\|L_A(f)\| &= \sup_{x,y\in X} \{e^{-\lambda(\|x\|+\|y\|)}|f(x+y) + f(x-y)|\} \\ &\le \sup_{x,y\in X} \{e^{-\lambda(\|x\|+\|y\|)}|f(x+y)|\} + \sup_{x,y\in X} \{e^{-\lambda(\|x\|+\|y\|)}|f(x-y)|\} \\ &\le \sup_{x,y\in X} \{e^{-\lambda\|x+y\|}|f(x+y)|\} + \sup_{x,y\in X} \{e^{-\lambda\|x+y\|}|f(x-y)|\} = 2\|f\|.\end{aligned}$$

Therefore, $L_A \in B(X_\lambda, X_\lambda^2)$. We shall now prove that Q_A is bounded and $\|Q_A\| = 2$. Indeed, for $f \in X_\lambda$ we get

$$\begin{aligned}\|Q_A(f)\| &= \sup_{x,y\in X} \{e^{-\lambda\|x+y\|}|2f(x)f(y)|\} \\ &= 2\sup_{x\in X}\{e^{-\lambda\|x\|}|f(x)|\} \cdot \sup_{y\in X}\{e^{-\lambda\|y\|}|f(y)|\} = 2\|f\|^2.\end{aligned}$$

Thus $Q_A \in B_Q(X_\lambda, X_\lambda^2)$ and $\|Q_A\| = 2$. Since $A = L_A + Q_A$, we get that $A \in B_{LQ}(X_\lambda, X_\lambda^2)$ and

$$\|A(f)\| = \|L_A(f) + Q_A(f)\| \le \|L_A(f)\| + \|Q_A(f)\| \le 2\|f\| + 2\|f\|^2,$$

as claimed. □

Under additional assumptions, one can compute the norm of A. Namely, we have

Theorem 2.6 *Let X be a linear normed space, $\mathbb{R}_+ \subset X$, $Y = \mathbb{C}$, and $\|x\| = |x|$ for $x \in \mathbb{R}_+$. Then*

$$\|A\| = 4.$$

The proof, similar to the proof of Theorem 2.2, can be found in [13].

2.4 Quadratic Functional Equation and Functional Equations for Quadratic Differences

At first, we shall give the formula for the general solution of the generalized quadratic functional equation on a group. The result is due to K. Dłutek (see [7]).

Theorem 2.7 *Let G_1 and G_2 be groups with division by two. Let $A, B, C, D\colon G_1 \to G_2$ satisfy the equation*

$$A(x) + B(y) = C(x + y) + D(x - y), \quad x, y \in G_1.$$

Then there exist a quadratic function $K : G_1 \to G_2$ (i.e., a function satisfying the equation (2.6)), additive functions $E, F : G_1 \to G_2$ and constants $S_1, S_2, S_3, S_4 \in G_2$ such that

$$\begin{aligned}
A(x) &= 2K(x) + E(x) + F(x) + S_1,\\
B(x) &= 2K(x) + E(x) - F(x) + S_2,\\
C(x) &= K(x) + E(x) + S_3,\\
D(x) &= K(x) + F(x) + S_4
\end{aligned}$$

for all $x \in G_1$ and $S_1 + S_2 = S_3 + S_4$.

Now we shall state the result concerning the properties of the quadratic difference operator Q on L^P-spaces; for more details, see [11].

Theorem 2.8 *Let (G, Σ, μ) be a complete measurable Abelian group, $\mu(G) < \infty$ and let $(E, \|\cdot\|)$ be a Banach space. If $1 \le p \le \infty$, then the quadratic difference operator*

$$Q\colon L^P_\mu(G, E) \to L^P_{\mu\times\mu}(G \times G, E)$$

given by (2.3) is linear, continuous, and invertible. Moreover, the inverse operator Q^{-1} defined for $h \in Q[L^P_\mu(G, E)]$ is continuous and has the form

$$Q^{-1}h(\cdot) = \big(2\mu(G)\big)^{-1} \int_G h(x, \cdot)\, d\mu(x).$$

For some problems, particularly for the problem of Ulam–Hyers–Rassias stability of functional equations, functional equations for quadratic differences are very useful (see [3, 7, 8]). Let us present a few such equations, which we will need in the proof of Theorem 2.11.

Theorem 2.9 *Let X, Y be Abelian groups and $f\colon X \to Y$ be a function. Then $Q(f)$ given by formula* (2.3) *satisfies the following functional equations*

$$\begin{aligned} &Q(f)(x+y,s+t)+Q(f)(x-y,s-t)+2Q(f)(x,y)+2Q(f)(s,t)\\ &\quad = Q(f)(x+s,y+t)+Q(f)(x-s,y-t)+2Q(f)(x,s)\\ &\qquad +2Q(f)(y,t), \end{aligned} \tag{2.14}$$

$$\begin{aligned} &Q(f)(x+y,s)+Q(f)(x-y,s)+2Q(f)(x,y)\\ &\quad = Q(f)(x+s,y)+Q(f)(x-s,y)+2Q(f)(x,s), \end{aligned} \tag{2.15}$$

$$\begin{aligned} &Q(f)(x+y,t)+Q(f)(x-y,t)+2Q(f)(x,y)\\ &\quad = Q(f)(x,y+t)+Q(f)(x,y-t)+2Q(f)(y,t) \end{aligned} \tag{2.16}$$

for all $x, y, s, t \in X$.

There are also interesting partial differential equations for quadratic differences (see [3, 7, 8]). Let X and Y be normed spaces. The space of all functions $f\colon X \to Y$ that are n-times differentiable will be denoted by $D^n(X, Y)$. By $\partial_k^n f$, $k = 1, 2$, we denote, as usual, the nth partial derivative of $f\colon X \times X \to Y$ with respect to the kth variable.

Theorem 2.10 *Let $f\colon X \to Y$ be a function such that $Q(f) \in D^2(X \times X, Y)$. Then we have*

$$\begin{aligned} &\partial_2^2\big(Q(f)\big)(x+y,0)+\partial_2^2\big(Q(f)\big)(x-y,0)\\ &\quad = 2\partial_1^2\big(Q(f)\big)(x,y)+2\partial_2^2\big(Q(f)\big)(x,0), \end{aligned} \tag{2.17}$$

$$\begin{aligned} &\partial_2^2\big(Q(f)\big)(x+y,0)+\partial_2^2\big(Q(f)\big)(x-y,0)\\ &\quad = 2\partial_2^2\big(Q(f)\big)(x,y)+2\partial_2^2\big(Q(f)\big)(y,0), \end{aligned} \tag{2.18}$$

$$2\partial_{12}^2\big(Q(f)\big)(x,y) = \partial_2^2\big(Q(f)\big)(x+y,0)-\partial_2^2\big(Q(f)\big)(x-y,0) \tag{2.19}$$

for all $x, y \in X$.

From Theorems 2.9 and 2.10, we easily obtain the following corollary.

Corollary 2.1 *Let $f\colon X \to Y$ be a function such that $Q(f) \in D^2(X \times X, Y)$. Then we have*

$$\partial_1\big(Q(f)\big)(0,0) = 0, \tag{2.20}$$

$$\begin{aligned} \partial_{12}^2\big(Q(f)\big)(0,0) &= 0, \\ \partial_2^2\big(Q(f)\big)(0,0) &= 0. \end{aligned} \tag{2.21}$$

Moreover, for all $x \in X$ we have

$$\partial_2\big(Q(f)\big)(x,0) = \partial_2\big(Q(f)\big)(0,0). \tag{2.22}$$

2.5 Double Quadratic Difference Property

In 1940, S.M. Ulam posed the following problem (cf. [29]):

We are given a group $(X,+)$ and a metric group $(Y,+,d)$. Given $\varepsilon > 0$, does there exist a $\delta > 0$ such that if $f: X \to Y$ satisfies the inequality

$$d\big[f(x+y), f(x)+f(y)\big] < \delta \quad \text{for all } x, y \in X,$$

then a homomorphism $A: X \to Y$ exists with

$$d\big[f(x), A(x)\big] < \varepsilon \quad \text{for all } x \in X?$$

One can ask a similar question for other important functional equations. The first partial solution of this problem was given by D.H. Hyers [16] under the assumption that X and Y are Banach spaces. In 1978, Themistocles M. Rassias extended the theorem of Hyers by considering an unbounded Cauchy difference (see [23]). During the last decades, the stability problems of various functional equations have been extensively investigated by many authors (see, e.g., [1, 2, 7, 8, 14, 15, 17, 18, 24–27]).

Assume that X and Y are normed spaces. For a function $f: X \to Y$, we put

$$\|f\|_{\sup} := \sup_{x \in X} \big\|f(x)\big\|.$$

For the quadratic difference, the stability problem can be reformulated as follows. Let $\varepsilon > 0$ be given. Does there exist a $\delta > 0$ such that if $f: X \to Y$ satisfies

$$\big\|Q(f)\big\|_{\sup} < \delta,$$

then there exists a quadratic function $K: X \to Y$ with

$$\|f - K\|_{\sup} < \varepsilon?$$

We can consider Ulam's problem for different norms. In this paper, we are going to prove the stability of the quadratic functional equation in the class of differentiable functions. The same problem for the Cauchy type functional equations was solved by J. Tabor and J. Tabor in [28].

Let X and Y be a real normed space and a real Banach space, respectively. By $\mathbb{N}_0$, $\mathbb{N}$, $\mathbb{R}$ we denote the sets of all nonnegative integers, positive integers, and real numbers, respectively. Let $f: X \to Y$ be an n-times Fréchet differentiable function. By $D^n f$, $n \in \mathbb{N}$, we denote the nth derivative of f, and $D^0 f$ stands for f. By $C^n(X, Y)$ we denote the space of n-times continuously differentiable functions and by $BC^n(X, Y)$ the subspace of $C^n(X, Y)$ consisting of bounded functions. Moreover, $C^0(X, Y)$ and $C^\infty(X, Y)$ stand for the space of continuous functions and the space of infinitely many times continuously differentiable functions, respectively.

Following an idea of J. Tabor and J. Tabor [28], we assume that we are given a norm in $X \times X$ such that $\|(x_1, x_2)\|$ is a function of $\|x_1\|$ and $\|x_2\|$, and the following condition is satisfied

$$\big\|(x, 0)\big\| = \big\|(0, x)\big\| = \|x\|, \quad x \in X.$$

Let $i_1: X \to X \times X$, $i_2: X \to X \times X$ be injections defined by

$$i_1(x) := (x, 0), \quad x \in X,$$
$$i_2(y) := (0, y), \quad y \in X.$$

Let $L: X \times X \to X$ be a bounded linear mapping. It follows directly from the assumed conditions on the norm in $X \times X$ that

$$\|L \circ i_1\| \le \|L\|\|i_1\| = \|L\|,$$
$$\|L \circ i_2\| \le \|L\|\|i_2\| = \|L\|.$$

Therefore, if $F: X \times X \to Y$ is n-times differentiable for $n \in \mathbb{N}$, then

$$\left.\begin{aligned} \|\partial_1 F(x, y)\| &= \|DF(x, y) \circ i_1\| \le \|DF(x, y)\|, \\ \|\partial_2 F(x, y)\| &= \|DF(x, y) \circ i_2\| \le \|DF(x, y)\| \end{aligned}\right\} \tag{2.23}$$

and

$$\left.\begin{aligned} \|\partial_1^{i-2}\partial_2^2 F(x, y)\| &\le \|D^i F(x, y)\|, \\ \|\partial_1^2\partial_2^{i-2} F(x, y)\| &\le \|D^i F(x, y)\| \end{aligned}\right\} \tag{2.24}$$

for all $x, y \in X$ and $i = 2, 3, \ldots, n$.

Let $n \in \mathbb{N}$ and let $f: X \to Y$ be n-times differentiable. Then $Q(f)$ is also n-times differentiable, and by (2.24) we have

$$\big\|Df(x + y) - Df(x - y) - 2Df(y)\big\| \le \big\|D\big(Q(f)\big)(x, y)\big\|, \tag{2.25}$$

$$\big\|D^2 f(x + y) + D^2 f(x - y) - 2D^2 f(y)\big\| \le \big\|D^2\big(Q(f)\big)(x, y)\big\| \tag{2.26}$$

for all $x, y \in X$. Moreover, for $n \ge 3$, we obtain from (2.24)

$$\big\|D^i f(x + y) + D^i f(x - y)\big\| \le \big\|D^i\big(Q(f)\big)(x, y)\big\| \tag{2.27}$$

for all $x, y \in X$ and $i = 3, 4, \ldots, n$.

We will prove that the class $C^n(\mathbb{R}, Y)$ has the so-called double quadratic difference property, i.e., if $f\colon \mathbb{R} \to Y$ is such a function that $Q(f) \in C^n(\mathbb{R} \times \mathbb{R}, Y)$, then there exists exactly one quadratic function $K_0\colon \mathbb{R} \to Y$ such that $f - K_0 \in C^n(\mathbb{R}, Y)$ (see also [3]). The problem of the double difference property for the Cauchy difference $C(f)(x, y) := f(x + y) - f(x) - f(y) \in C^n(X \times X, Y)$ has been investigated in [28]. For more details about the double difference property, the reader is referred to [19].

Lemma 2.3 (See also [3]) *Let $f\colon X \to Y$ be a function such that $Q(f) \in C^2(X \times X, Y)$. Then $K_0\colon X \to Y$ given by the formula*

$$\begin{aligned} K_0(x) = {} & f(x) - f(0) + \frac{1}{2}\partial_2\bigl(Q(f)\bigr)(0,0)(x) \\ & - \frac{1}{2}\int_0^1\int_0^t \partial_2^2\bigl(Q(f)\bigr)(ux,0)\bigl(x^2\bigr)\,du\,dt, \quad x \in X \end{aligned} \tag{2.28}$$

is a quadratic function.

Proof Let $f_1(x) := f(x) - f(0)$ for all $x \in X$. Then $Q(f_1) = Q(f) + 2f(0) \in C^n(X \times X, Y)$ and $Q(f_1)(0,0) = 0$. Moreover, $\partial_2(Q(f_1)) = \partial_2(Q(f))$ and $\partial_2^2(Q(f_1)) = \partial_2^2(Q(f))$. Let us fix arbitrary $x, y \in X$ and consider a function

$$\varphi(t) := Q(f_1)(tx, ty), \quad t \in \mathbb{R}.$$

Obviously, $\varphi \in C^2(\mathbb{R}, Y)$. Then we have

$$D\varphi(t) = \partial_1\bigl(Q(f_1)\bigr)(tx,ty)(x) + \partial_2\bigl(Q(f_1)\bigr)(tx,ty)(y), \quad t \in \mathbb{R}.$$

Hence and from (2.20), we get

$$D\varphi(0) = \partial_2\bigl(Q(f_1)\bigr)(0,0)(y).$$

Therefore, we obtain

$$\begin{aligned} Q(f_1)(x,y) &= \varphi(1) - \varphi(0) = \int_0^1 D\varphi(t)\,dt = \int_0^1\int_0^t D^2\varphi(u)\,du\,dt + D\varphi(0) \\ &= \int_0^1\int_0^t D^2\bigl(Q(f_1)\bigr)(ux,uy)(x,y)\,du\,dt + \partial_2\bigl(Q(f_1)\bigr)(0,0)(y) \\ &= \int_0^1\int_0^t \partial_1^2\bigl(Q(f_1)\bigr)(ux,uy)\bigl(x^2\bigr)\,du\,dt \\ &\quad + 2\int_0^1\int_0^t \partial_{12}^2\bigl(Q(f_1)\bigr)(ux,uy)(xy)\,du\,dt \\ &\quad + \int_0^1\int_0^t \partial_2^2\bigl(Q(f_1)\bigr)(ux,uy)\bigl(y^2\bigr)\,du\,dt + \partial_2\bigl(Q(f_1)\bigr)(0,0)(y). \end{aligned}$$

Thus

$$
\begin{aligned}
Q(f_1)(x,y) &= \int_0^1 \int_0^t \partial_1^2\big(Q(f_1)\big)(ux,uy)\big(x^2\big)\,du\,dt \\
&\quad + 2\int_0^1 \int_0^t \partial_{12}^2\big(Q(f_1)\big)(ux,uy)(xy)\,du\,dt \\
&\quad + \int_0^1 \int_0^t \partial_2^2\big(Q(f_1)\big)(ux,uy)\big(y^2\big)\,du\,dt \\
&\quad + \partial_2\big(Q(f_1)\big)(0,0)(y), \qquad x,y \in X.
\end{aligned} \tag{2.29}
$$

We define the function $K_0\colon X \to Y$ by the formula

$$
\begin{aligned}
K_0(x) &:= f_1(x) + \frac{1}{2}\partial_2\big(Q(f_1)\big)(0,0)(x) \\
&\quad - \frac{1}{2}\int_0^1 \int_0^t \partial_2^2\big(Q(f_1)\big)(ux,0)\big(x^2\big)\,du\,dt, \qquad x \in X.
\end{aligned}
$$

We show that K_0 is a quadratic function. By making use of (2.17), (2.18), (2.19), and (2.29), we obtain for all $x, y \in X$

$$
\begin{aligned}
&K_0(x+y) + K_0(x-y) - 2K_0(x) - 2K_0(y) \\
&\quad = Q(f_1)(x,y) - \partial_2\big(Q(f_1)\big)(0,0)(y) \\
&\qquad - \frac{1}{2}\int_0^1 \int_0^t \partial_2^2\big(Q(f_1)\big)(ux+uy,0)(x+y)^2\,du\,dt \\
&\qquad - \frac{1}{2}\int_0^1 \int_0^t \partial_2^2\big(Q(f_1)\big)(ux-uy,0)(x-y)^2\,du\,dt \\
&\qquad + \int_0^1 \int_0^t \partial_2^2\big(Q(f_1)\big)(ux,0)\big(x^2\big)\,du\,dt \\
&\qquad + \int_0^1 \int_0^t \partial_2^2\big(Q(f_1)\big)(uy,0)\big(y^2\big)\,du\,dt \\
&\quad = \int_0^1 \int_0^t \partial_1^2\big(Q(f_1)\big)(ux,uy)\big(x^2\big)\,du\,dt \\
&\qquad + 2\int_0^1 \int_0^t \partial_{12}^2\big(Q(f_1)\big)(ux,uy)(xy)\,du\,dt \\
&\qquad + \int_0^1 \int_0^t \partial_2^2\big(Q(f_1)\big)(ux,uy)\big(y^2\big)\,du\,dt \\
&\qquad - \frac{1}{2}\int_0^1 \int_0^t \partial_2^2\big(Q(f_1)\big)(ux+uy,0)\big(x^2\big)\,du\,dt
\end{aligned}
$$

$$
\begin{aligned}
&-\int_0^1\int_0^t \partial_2^2\big(Q(f_1)\big)(ux+uy,0)(xy)\,du\,dt \\
&-\frac{1}{2}\int_0^1\int_0^t \partial_2^2\big(Q(f_1)\big)(ux+uy,0)\big(y^2\big)\,du\,dt \\
&-\frac{1}{2}\int_0^1\int_0^t \partial_2^2\big(Q(f_1)\big)(ux-uy,0)\big(x^2\big)\,du\,dt \\
&+\int_0^1\int_0^t \partial_2^2\big(Q(f_1)\big)(ux-uy,0)(xy)\,du\,dt \\
&-\frac{1}{2}\int_0^1\int_0^t \partial_2^2\big(Q(f_1)\big)(ux-uy,0)\big(y^2\big)\,du\,dt \\
&+\int_0^1\int_0^t \partial_2^2\big(Q(f_1)\big)(ux,0)\big(x^2\big)\,du\,dt \\
&+\int_0^1\int_0^t \partial_2^2\big(Q(f_1)\big)(uy,0)\big(y^2\big)\,du\,dt \\
=&\int_0^1\int_0^t \bigg[\partial_1^2\big(Q(f_1)\big)(ux,uy)-\frac{1}{2}\partial_2^2\big(Q(f_1)\big)(ux+uy,0) \\
&-\frac{1}{2}\partial_2^2\big(Q(f_1)\big)(ux-uy,0)+\partial_2^2\big(Q(f_1)\big)(ux,0)\bigg]\big(x^2\big)\,du\,dt \\
&+\int_0^1\int_0^t \big[2\partial_{12}^2\big(Q(f_1)\big)(ux,uy)-\partial_2^2\big(Q(f_1)\big)(ux+uy,0) \\
&+\partial_2^2\big(Q(f_1)\big)(ux-uy,0)\big](xy)\,du\,dt \\
&+\int_0^1\int_0^t \bigg[\partial_2^2\big(Q(f_1)\big)(ux,uy)-\frac{1}{2}\partial_2^2\big(Q(f_1)\big)(ux+uy,0) \\
&-\frac{1}{2}\partial_2^2\big(Q(f_1)\big)(ux-uy,0)+\partial_2^2\big(Q(f_1)\big)(uy,0)\bigg]\big(y^2\big)\,du\,dt=0.
\end{aligned}
$$

Therefore, K_0 is a quadratic function, which completes the proof. □

Theorem 2.11 *Let $n \geq 2$ be a fixed positive integer and let $f\colon \mathbb{R} \to Y$ be a function such that $Q(f) \in C^n(\mathbb{R}\times\mathbb{R}, Y)$. Then there exists a unique quadratic function $K_0\colon \mathbb{R}\to Y$ such that $f - K_0 \in C^n(\mathbb{R}, Y)$ and $D^2(f-K_0)(0)=0$. Moreover, we have for all $x \in \mathbb{R}$*

$$
\begin{aligned}
D(f-K_0)(x) &= \frac{1}{2}\int_0^x \partial_2^2\big(Q(f)\big)(s,0)\,ds - \frac{1}{2}\partial_2\big(Q(f)\big)(0,0), \\
D^2(f-K_0)(x) &= \frac{1}{2}\partial_2^2\big(Q(f)\big)(x,0),
\end{aligned}
$$

$$\|D^k(f-K_0)(x)\| \le \frac{1}{2}\|D^k\big(Q(f)\big)(x,0)\|, \quad k\in\mathbb{N}\backslash\{1\},\ k\le n.$$

Proof Let $f_1(x) := f(x) - f(0)$ for all $x \in \mathbb{R}$. On account of Lemma 2.3, there exists a quadratic function K_0 given by (2.28). Now we prove that $f - K_0$ is a differentiable function. Fix arbitrary $x, h \in \mathbb{R}$, $h \neq 0$. Then we get

$$\begin{aligned}
&\frac{1}{h}\big[f_1(x+h) - K_0(x+h) - \big(f_1(x) - K_0(x)\big)\big]\\
&\quad = \frac{1}{h}\bigg[\frac{1}{2}\int_0^1\int_0^t \partial_2^2\big(Q(f_1)\big)\big(u(x+h),0\big)(x+h)^2\,du\,dt\\
&\qquad - \frac{1}{2}\partial_2\big(Q(f_1)\big)(0,0)(x+h)\\
&\qquad - \frac{1}{2}\int_0^1\int_0^t \partial_2^2\big(Q(f_1)\big)(ux,0)\big(x^2\big)\,du\,dt + \frac{1}{2}\partial_2\big(Q(f_1)\big)(0,0)(x)\bigg]\\
&\quad = \frac{1}{2h}\bigg[\int_0^{x+h}\int_0^v \partial_2^2\big(Q(f_1)\big)(s,0)\,ds\,dv\\
&\qquad - \int_0^x\int_0^v \partial_2^2\big(Q(f_1)\big)(s,0)\,ds\,dv - \partial_2\big(Q(f_1)\big)(0,0)(h)\bigg]\\
&\quad = \frac{1}{2h}\bigg[\int_x^{x+h}\int_0^v \partial_2^2\big(Q(f_1)\big)(s,0)\,ds\,dv - \partial_2\big(Q(f_1)\big)(0,0)(h)\bigg]\\
&\quad = \frac{1}{2h}\bigg[\int_0^1\int_0^{x+th} \partial_2^2\big(Q(f_1)\big)(s,0)(h)\,ds\,dt - \partial_2\big(Q(f_1)\big)(0,0)(h)\bigg]\\
&\quad = \frac{1}{2}\bigg[\int_0^1\int_0^{x+th} \partial_2^2\big(Q(f_1)\big)(s,0)\,ds\,dt - \partial_2\big(Q(f_1)\big)(0,0)\bigg]\\
&\quad \longrightarrow \quad \frac{1}{2}\int_0^1\int_0^x \partial_2^2\big(Q(f_1)\big)(s,0)\,ds\,dt\\
&\qquad - \frac{1}{2}\partial_2\big(Q(f_1)\big)(0,0) = \frac{1}{2}\int_0^x \partial_2^2\big(Q(f_1)\big)(s,0)\,ds - \frac{1}{2}\partial_2\big(Q(f_1)\big)(0,0)
\end{aligned}$$

for $h \to 0$. Hence the function $f - K_0 = f_1 - K_0 + f(0)$ is differentiable at every $x \in \mathbb{R}$, $\partial_2(Q(f_1)) = \partial_2(Q(f))$, $\partial_2^2(Q(f_1)) = \partial_2^2(Q(f))$ and

$$D(f-K_0)(x) = \frac{1}{2}\int_0^x \partial_2^2\big(Q(f)\big)(s,0)\,ds - \frac{1}{2}\partial_2\big(Q(f)\big)(0,0), \quad x\in\mathbb{R}. \tag{2.30}$$

Moreover, since the function $f - K_0$ is differentiable at every $x \in \mathbb{R}$, then there exists also the second difference derivative which is equal to the second derivative (see [21]). We show that

$$D^2(f - K_0)(x) = \frac{1}{2}\partial_2^2\big(Q(f)\big)(x,0), \quad x \in \mathbb{R}.$$

Fix arbitrary $x, h \in \mathbb{R}$ and let a function $\psi: \mathbb{R} \to Y$ be given by

$$\psi(t) := Q(f_1)(x, th), \quad t \in \mathbb{R}.$$

Then $\psi \in C^n(\mathbb{R}, Y)$ and we have

$$\begin{aligned} Q(f_1)(x,h) &= \psi(1) - \psi(0) \\ &= \int_0^1 \int_0^t \partial_2^2\big(Q(f_1)\big)(x, uh)\big(h^2\big)\,du\,dt \\ &\quad + \partial_2\big(Q(f_1)\big)(x,0)(h), \quad x, h \in \mathbb{R}. \end{aligned}$$

From (2.22) we get

$$\partial_2\big(Q(f_1)\big)(x,0) = \partial_2\big(Q(f_1)\big)(0,0), \quad x \in \mathbb{R},$$

hence

$$\begin{aligned} Q(f_1)(x,h) = &\int_0^1 \int_0^t \partial_2^2\big(Q(f_1)\big)(x, uh)\big(h^2\big)\,du\,dt \\ &+ \partial_2\big(Q(f_1)\big)(0,0)(h), \quad x, h \in \mathbb{R}. \end{aligned} \tag{2.31}$$

Next, from (2.18) we obtain

$$\begin{aligned} &\int_0^1 \int_0^t \partial_2^2\big(Q(f_1)\big)(x+uh, 0)\big(h^2\big)\,du\,dt \\ &\quad + \int_0^1 \int_0^t \partial_2^2\big(Q(f_1)\big)(x-uh, 0)\big(h^2\big)\,du\,dt \\ &= 2\int_0^1 \int_0^t \partial_2^2\big(Q(f_1)\big)(x, uh)\big(h^2\big)\,du\,dt \\ &\quad + 2\int_0^1 \int_0^t \partial_2^2\big(Q(f_1)\big)(uh, 0)\big(h^2\big)\,du\,dt, \quad x, h \in \mathbb{R}. \end{aligned} \tag{2.32}$$

Therefore, using (2.31) and (2.32), for each fixed $x, h \in \mathbb{R}$ we have

$$\begin{aligned} \Big\| &f_1(x+h) - K_0(x+h) - 2f_1(x) + 2K_0(x) + f_1(x-h) - K_0(x-h) \\ &- \frac{1}{2}\partial_2^2\big(Q(f_1)\big)(x,0)\big(h^2\big)\Big\| \end{aligned}$$

$$
\begin{aligned}
&= \left\| Q(f_1)(x,h) + 2f_1(h) - 2K_0(h) - \frac{1}{2}\partial_2^2\big(Q(f_1)\big)(x,0)\big(h^2\big) \right\| \\
&= \bigg\| \int_0^1 \int_0^t \partial_2^2\big(Q(f_1)\big)(x,uh)\big(h^2\big)\,du\,dt + \partial_2\big(Q(f_1)\big)(0,0)(h) \\
&\quad + \int_0^1 \int_0^t \partial_2^2\big(Q(f_1)\big)(uh,0)\big(h^2\big)\,du\,dt \\
&\quad - \partial_2\big(Q(f_1)\big)(0,0)(h) - \frac{1}{2}\partial_2^2\big(Q(f_1)\big)(x,0)\big(h^2\big) \bigg\| \\
&= \bigg\| \frac{1}{2}\int_0^1 \int_0^t \partial_2^2\big(Q(f_1)\big)(x+uh,0)\big(h^2\big)\,du\,dt \\
&\quad + \frac{1}{2}\int_0^1 \int_0^t \partial_2^2\big(Q(f_1)\big)(x-uh,0)\big(h^2\big)\,du\,dt \\
&\quad - \int_0^1 \int_0^t \partial_2^2\big(Q(f_1)\big)(x,0)\big(h^2\big)\,du\,dt \bigg\| \\
&= \bigg\| \int_0^1 \int_0^t \Big[\frac{1}{2}\partial_2^2\big(Q(f_1)\big)(x+uh,0) + \frac{1}{2}\partial_2^2\big(Q(f_1)\big)(x-uh,0) \\
&\quad - \partial_2^2\big(Q(f_1)\big)(x,0) \Big]\big(h^2\big)\,du\,dt \bigg\| \\
&\le \|h\|^2 \sup_{u\in[0,1]} \bigg\| \frac{1}{2}\partial_2^2\big(Q(f_1)\big)(x+uh,0) \\
&\quad + \frac{1}{2}\partial_2^2\big(Q(f_1)\big)(x-uh,0) - \partial_2^2\big(Q(f_1)\big)(x,0) \bigg\|.
\end{aligned}
$$

Since $\partial_2^2(Q(f_1)) = \partial_2^2(Q(f))$ is a continuous function,

$$
\frac{1}{2}\partial_2^2\big(Q(f_1)\big)(x+uh,0) + \frac{1}{2}\partial_2^2\big(Q(f_1)\big)(x-uh,0) - \partial_2^2\big(Q(f_1)\big)(x,0) \to 0
$$

for $h \to 0$. Hence we get

$$
D^2(f-K_0)(x) = \frac{1}{2}\partial_2^2\big(Q(f)\big)(x,0), \quad x \in \mathbb{R}. \tag{2.33}
$$

Since $\partial_2^2(Q(f))(x,0) \in C^{n-2}(\mathbb{R}\times\mathbb{R}, Y)$, one has $\partial_2^2(Q(f))(x,0) \in C^{n-2}(\mathbb{R}, Y)$. Finally, $D^2(f-K_0) \in C^{n-2}(\mathbb{R}, Y)$, i.e., $f-K_0 \in C^n(\mathbb{R}, Y)$. Moreover, from (2.21) we also have

$$
D^2(f-K_0)(0) = \frac{1}{2}\partial_2^2\big(Q(f)\big)(0,0) = 0.
$$

To prove the uniqueness of K_0, consider quadratic functions $K_1, K_2 \colon \mathbb{R} \to Y$ such that $f-K_1, f-K_2 \in C^n(\mathbb{R}, Y)$, $D^2(f-K_1)(0) = D^2(f-K_2)(0) = 0$ and conditions (2.30), (2.33) hold. Then $K_1 - K_2$ is a quadratic function and $K_1 - K_2 \in C^n(\mathbb{R}, Y)$. Therefore, for every $x \in \mathbb{R}$, we have

$$\begin{aligned} D^2(K_1 - K_2)(x) &= D^2\big((f - K_2) - (f - K_1)\big)(x) \\ &= D^2(f - K_2)(x) - D^2(f - K_1)(x) = 0. \end{aligned}$$

Since $D^2(K_1 - K_2) = 0$, $D(K_1 - K_2)(x)$ is a constant function for every $x \in \mathbb{R}$. But

$$D(K_1 - K_2)(0) = D(f - K_2)(0) - D(f - K_1)(0) = 0,$$

so $D(K_1 - K_2) = 0$. Analogously, we have that $(K_1 - K_2)(x)$ is a constant function for every $x \in \mathbb{R}$; and since $(K_1 - K_2)(0) = 0$, it yields that $K_1 = K_2$.

It remains to prove that

$$\big\|D^k(f - K_0)(x)\big\| \le \frac{1}{2}\big\|D^k\big(Q(f)\big)(x, 0)\big\|, \quad k \in \mathbb{N}\backslash\{1\},\ k \le n \tag{2.34}$$

for every $x \in \mathbb{R}$. Let $g := f - K_0$. Then $g \in C^n(\mathbb{R}, Y)$, and consequently we have $Q(g) = Q(f) \in C^n(\mathbb{R} \times \mathbb{R}, Y)$. Making use of (2.26) and the fact that $D^2 g(0) = 0$, we obtain

$$\big\|D^2 g(x)\big\| = \big\|D^2 g(x) - D^2 g(0)\big\| \le \frac{1}{2}\big\|D^2\big(Q(g)\big)(x, 0)\big\|, \quad x \in \mathbb{R},$$

which proves (2.34) for $k = 2$. For $3 \le k \le n$, $k \in \mathbb{N}$, condition (2.34) follows directly from (2.27), which completes the proof. □

Corollary 2.2 ([3]) *Under the assumptions of Theorem* 2.11, *we have*

$$\big\|D^k(f - K_0)(0)\big\| \le \frac{1}{2}\big\|D^k\big(Q(f)\big)(0, 0)\big\|, \quad k \in \mathbb{N}_0\backslash\{1\},\ k \le n, \tag{2.35}$$

$$\big\|D^k(f - K_0)\big\|_{\sup} \le \frac{1}{2}\big\|D^k\big(Q(f)\big)\big\|_{\sup}, \quad k \in \mathbb{N}\backslash\{1\},\ k \le n. \tag{2.36}$$

Proof The case $k = 0$ in (2.35) is trivial because obviously $f(0) = -\frac{1}{2}Q(f)(0, 0)$. From (2.34) we obtain (2.35) for $k \ge 2$ and (2.36). The proof is completed. □

Remark 2.2 Let the assumptions of Theorem 2.11 be satisfied and let $\partial_2(Q(f))(0, 0) = 0$. Then the inequality (2.34) (and consequently (2.35) and (2.36)) also holds for $k = 1$.

Proof If $\partial_2(Q(f))(0, 0) = 0$, then from (2.30) we obtain $D(f - K_0)(0) = 0$. Let $g := f - K_0$. Hence $g \in C^n(\mathbb{R}, Y)$, $Q(g) = Q(f) \in C^n(\mathbb{R} \times \mathbb{R}, Y)$, and $C(g) \in C^n(\mathbb{R} \times \mathbb{R}, Y)$. Therefore, on account of (2.23), we get

$$\big\|Dg(x + y) - Dg(y)\big\| \le \big\|D\big(C(g)\big)(x, y)\big\|, \quad x, y \in \mathbb{R}. \tag{2.37}$$

One can easily check that for any function $h\colon \mathbb{R} \to Y$ the following equality holds

$$2C(h)(x,y) + 2C(h)(x,-y) = Q(h)(x,y) + Q(h)(x,-y), \quad x,y \in \mathbb{R},$$

where $C(f)$ denotes the Cauchy difference. Then, in particular, for a function g we obtain

$$D\big(C(g)\big)(x,0) = \frac{1}{2} D\big(Q(g)\big)(x,0) = \frac{1}{2} D\big(Q(f)\big)(x,0), \quad x \in \mathbb{R}.$$

Therefore, by virtue of (2.37) with $y = 0$, from the above equality and the fact that $Dg(0) = 0$, we have

$$\big\|Dg(x)\big\| = \big\|Dg(x) - Dg(0)\big\| \le \big\|D\big(C(g)\big)(x,0)\big\| = \frac{1}{2}\big\|D\big(Q(g)\big)(x,0)\big\|,$$
$$x \in \mathbb{R},$$

which proves the inequality (2.34) for $k = 1$. □

It is still an open problem to prove that the function $f - K_0$ which occurs in Theorem 2.11 is differentiable for every $x \in X$, where X denotes a real normed space.

Corollary 2.3 ([3]) *The quadratic function $K_0\colon \mathbb{R} \to Y$ occurring in Lemma* 2.3 *and Theorem* 2.11 *can be defined by the formula*

$$K_0(x) = \frac{1}{2} \lim_{n\to\infty} n^2 \left[f\left(\frac{x}{n}\right) + f\left(-\frac{x}{n}\right) - 2f(0) \right], \quad x \in \mathbb{R}. \tag{2.38}$$

Theorem 2.11 states, in particular, that the class of infinitely many times differentiable functions has the double quadratic difference property. We may show that the class of analytic functions also has this property.

Corollary 2.4 ([3]) *Let $f\colon \mathbb{R} \to Y$ be a function such that $Q(f)$ is analytic. Then there exists exactly one quadratic function $K\colon \mathbb{R} \to Y$ such that $f - K$ is analytic and $D^2(f - K)(0) = 0$.*

Now we give some auxiliary results which will be used in the sequel.

Lemma 2.4 ([5]) *Let $(G,+)$ be an Abelian group. If a function $f\colon G \to Y$ satisfies the inequality*

$$\big\|f(x+y) + f(x-y) - 2f(x) - 2f(y)\big\| \le \varepsilon, \quad x,y \in G$$

for some $\varepsilon > 0$, then there exists a unique quadratic function $K\colon G \to Y$ such that

$$\big\|f(x) - K(x)\big\| \le \frac{1}{2}\varepsilon, \quad x \in G.$$

Moreover, the function K is given by the formula

$$K(x) = \lim_{n\to\infty} \frac{f(2^n x)}{2^{2n}}, \qquad x \in G.$$

In [6], S. Czerwik provided a generalization of the above result and also proved that if a function $\mathbb{R} \ni t \to f(tx)$ is continuous for each fixed $x \in E$, where E denotes a real normed space, then $K(tx) = t^2 K(x)$ for all $t \in \mathbb{R}$ and $x \in E$.

The following lemma is some kind of an analogue to the Mean Value Theorem for real valued functions.

Lemma 2.5 ([20]) *Let a mapping $T\colon B \to Y$, $B \subset X$, where B is an open set, be two times Fréchet differentiable. Let $x, h \in B$, and let for every $0 \le \alpha \le 1$, $(x + \alpha h) \in B$. Then*

$$\|T(x+h) - T(x) - DT(x)h\| \le \frac{1}{2}\|h\|^2 \sup_{0<\alpha<1} \|D^2T(x+\alpha h)\|.$$

Similarly as for the case of Euler's Theorem for positive homogeneous functions (see [22]), one can prove the following lemma.

Lemma 2.6 *Let $T\colon \mathbb{R} \to Y$ be a homogeneous function of degree 2 such that $T \in C^2(\mathbb{R}, Y)$. Then*

$$T(x) = \frac{1}{2} x^2 D^2 T(x), \qquad x \in \mathbb{R}. \tag{2.39}$$

Proof Since T is a homogeneous function of degree 2, then

$$T(\alpha x) = \alpha^2 T(x), \qquad x, \alpha \in \mathbb{R}.$$

Let us fix an arbitrary $x \in \mathbb{R}$. Differentiating both sides of the above equality with respect to α, we obtain

$$D^2 T(\alpha x) x^2 = 2T(x), \qquad \alpha \in \mathbb{R}.$$

Since $x \in \mathbb{R}$ was chosen arbitrarily, for $\alpha = 1$ we get the equality (2.39), which completes the proof. □

Lemma 2.7 *Let $K\colon \mathbb{R} \to Y$ be a quadratic function and let $f\colon \mathbb{R} \to Y$ be a mapping such that $f \in C^2(\mathbb{R}, Y)$. Assume also that $f - K$ is bounded and*

$$\sup_{(x,y)\in\mathbb{R}\times\mathbb{R}} \|D^2 f(x) - D^2 f(y)\| \le \varepsilon \tag{2.40}$$

for some $\varepsilon > 0$. Then K is differentiable and

$$\sup_{x\in\mathbb{R}} \|D^2 K(x) - D^2 f(x)\| \le \varepsilon. \tag{2.41}$$

Proof Since $f - K$ is bounded, then $Q(f)$ is also bounded and, on account of Lemma 2.4, we have

$$K(x) = \lim_{n\to\infty} \frac{f(2^n x)}{2^{2n}}, \quad x \in \mathbb{R}.$$

Since f is a continuous at each $x \in \mathbb{R}$, K is also continuous and it is of the form (see also [7])

$$K(x) = x^2 K(1), \quad x \in \mathbb{R}.$$

Thus K is differentiable. Let us fix an arbitrary $y \in \mathbb{R}$. Applying Lemma 2.5 to the function

$$T(x) := f(x) - \frac{1}{2}x^2 D^2 f(y), \quad x \in \mathbb{R}$$

and the inequality (2.40), we get for all $x \in \mathbb{R}$

$$\begin{aligned}\left\| f(x) - \frac{1}{2}x^2 D^2 f(y) - f(0) - xDf(0) \right\| &\le \frac{1}{2}|x|^2 \sup_{x\in\mathbb{R}} \left\| D^2 f(x) - D^2 f(y) \right\| \\ &\le \frac{1}{2}\varepsilon |x|^2.\end{aligned}$$

Replacing x by $2^n x$ and dividing both sides of the above inequality by 2^n, we have

$$\left\| \frac{f(2^n x)}{2^{2n}} - \frac{1}{2}x^2 D^2 f(y) - \frac{f(0)}{2^{2n}} - \frac{xDf(0)}{2^n} \right\| \le \frac{1}{2}\varepsilon |x|^2, \quad x \in \mathbb{R}.$$

Letting $n \to \infty$, we conclude that

$$\left\| K(x) - \frac{1}{2}x^2 D^2 f(y) \right\| \le \frac{1}{2}\varepsilon |x|^2, \quad x \in \mathbb{R}.$$

Therefore, by virtue of Lemma 2.6, we obtain

$$\left\| \frac{1}{2}x^2 D^2 K(x) - \frac{1}{2}x^2 D^2 f(y) \right\| \le \frac{1}{2}\varepsilon |x|^2, \quad x \in \mathbb{R},$$

and hence

$$\left\| D^2 K(x) - D^2 f(y) \right\| \le \varepsilon, \quad x \in \mathbb{R}.$$

Since y was arbitrary,

$$\sup_{x\in\mathbb{R}} \left\| D^2 K(x) - D^2 f(x) \right\| \le \varepsilon,$$

which completes the proof. □

Theorem 2.12 *Let $n \geq 2$ be a fixed positive integer and let $f: \mathbb{R} \to Y$ be such a function that $Q(f) \in BC^n(\mathbb{R} \times \mathbb{R}, Y)$. Then there exists a unique quadratic function $K_\infty: \mathbb{R} \to Y$ such that $f - K_\infty \in BC^n(\mathbb{R}, Y)$. Moreover,*

$$\|D^k(f - K_\infty)(0)\| \leq \frac{1}{2}\|D^k(Q(f))(0,0)\|, \quad k \in \mathbb{N}_0\backslash\{2\},\ k \leq n, \tag{2.42}$$

$$\|D^k(f - K_\infty)\|_{\sup} \leq \frac{1}{2}\|D^k(Q(f))\|_{\sup}, \quad k \in \mathbb{N}_0\backslash\{1\},\ k \leq n. \tag{2.43}$$

Proof By virtue of Theorem 2.11, there exists a unique quadratic function $K_0: \mathbb{R} \to Y$ such that $f_1 := f - K_0 \in C^n(\mathbb{R}, Y)$. Then $Q(f_1) = Q(f) \in BC^n(\mathbb{R} \times \mathbb{R}, Y)$. It means, in particular, that $Q(f_1)$ is bounded. By Lemma 2.4, there exists a unique quadratic function $K_1: \mathbb{R} \to Y$ such that

$$\sup_{x \in \mathbb{R}}\|f_1(x) - K_1(x)\| \leq \frac{1}{2} \sup_{(x,y) \in \mathbb{R} \times \mathbb{R}}\|Q(f_1)(x,y)\|. \tag{2.44}$$

We put

$$K_\infty(x) := K_0(x) + K_1(x), \quad x \in \mathbb{R}.$$

Clearly, K_∞ is also a quadratic function and $f - K_\infty = f_1 - K_1 \in C^n(\mathbb{R}, Y)$. From (2.26) we obtain

$$\sup_{(x,y) \in \mathbb{R} \times \mathbb{R}}\|D^2 f_1(x) - D^2 f_1(y)\| \leq \frac{1}{2} \sup_{(x,y) \in \mathbb{R} \times \mathbb{R}}\|D^2(Q(f_1))(x,y)\|. \tag{2.45}$$

Conditions (2.44) and (2.45) mean that the functions f_1 and K_1 satisfy the assumptions of Lemma 2.7 with

$$\varepsilon = \frac{1}{2} \sup_{(x,y) \in \mathbb{R} \times \mathbb{R}}\|D^2(Q(f_1))(x,y)\|.$$

Thus K_1 is differentiable and, by making use of (2.41) and (2.45), we get

$$\begin{aligned}\sup_{x \in \mathbb{R}}\|D^2(f - K_\infty)(x)\| &= \sup_{x \in \mathbb{R}}\|D^2(f_1 - K_1)(x)\| \\ &\leq \frac{1}{2} \sup_{(x,y) \in \mathbb{R} \times \mathbb{R}}\|D^2(Q(f_1))(x,y)\|. \end{aligned} \tag{2.46}$$

For $k = 0$, the inequality (2.42) is obvious; for $k = 1$, it follows from (2.25); and for $k \geq 3$, it is a trivial consequence of (2.27). Making use of (2.44), (2.46), and (2.27), we obtain (2.43), which completes the proof. □

Corollary 2.5 *The quadratic function K_∞ occurring in Theorem 2.12 can be defined by the formula*

$$K_\infty(x) = \lim_{n \to \infty} \frac{f(2^n x)}{2^{2n}}, \quad x \in \mathbb{R}.$$

Proof The formula for K_∞ is a trivial consequence of the fact that the function $f - K_\infty$ is bounded. □

Comparing the formulae for K_0 and K_∞, one can easily notice that these functions are usually different. We will see it in the following example.

Example Let $f: \mathbb{R} \to Y$ be a bounded function such that $f \in C^2(\mathbb{R}, Y)$. Since f is bounded, $K_\infty = 0$. One can easily check that the following equalities hold:

$$\partial_2\big(Q(f)\big)(0,0) = -2Df(0),$$
$$\partial_2^2\big(Q(f)\big)(x,0) = 2D^2f(x) - 2D^2f(0), \quad x \in \mathbb{R}.$$

Therefore, by applying the formula of K_0 given by (2.28), we obtain that

$$K_0(x) = \frac{1}{2}D^2f(0)x^2, \quad x \in \mathbb{R}, \tag{2.47}$$

hence

$$D^2K_0(x) = D^2f(0), \quad x \in \mathbb{R}.$$

Thus $K_0 = K_\infty$ if and only if $D^2f(0) = 0$.

Clearly, one can also obtain the formula (2.47) from (2.38).

2.6 Stability

Let $f: X \to Y$ be a function such that $f \in C^n(X, Y)$ for $n \in \mathbb{N}_0$. In subspaces of $C^n(X, Y)$, one can consider different norms defined in terms of $\|D^i f(0)\|$, $\|D^i f\|_{\sup}$ for $i \le n$. For example, the following norms

$$\|f\| := \sum_{i=0}^{n-1}\big\|D^i f(0)\big\| + \big\|D^n f\big\|_{\sup},$$
$$\|f\| := \sum_{i=0}^{n}\big\|D^i f\big\|_{\sup}, \tag{2.48}$$
$$\|f\| := \max_{i=0,\dots,n}\big\|D^i f\big\|_{\sup}$$

are used very often. Obviously, several other norms can be introduced. We will prove the stability result for the quadratic difference $Q(f)$ in a possibly general setting. We will use the following convention: if $m, n \in \mathbb{N}_0$ and $m > n$, then $\sum_{i=m}^{n} a_i = 0$.

In the sequel, we will use the following assumptions introduced in [28]. Let $n \in \mathbb{N}_0 \cup \{\infty\}$ be fixed. In the set $[0,\infty]^{2n+2}$, we introduce the following order

$$(x_1,\ x_2,\ \dots) \le (y_1,\ y_2,\ \dots)$$

iff $x_i \le y_i$ for $i \in \mathbb{N}$, $i \le 2n+2$.

Let $p\colon [0,\infty]^{2n+2} \to [0,\infty]$ be any function satisfying the following conditions:

(i) $p(x+y) \le p(x) + p(y)$, $x, y \in [0,\infty]^{2n+2}$,
(ii) $p(\alpha x) = \alpha p(x)$, $x, y \in [0,\infty]^{2n+2}$, $\alpha \in [0,\infty]$,
(iii) $x \le y \Longrightarrow p(x) \le p(y)$, $x, y \in [0,\infty]^{2n+2}$.

We additionally assume that $0 \cdot \infty = 0$. From (ii) we obtain that $p(0) = 0$.

We define the mapping $\Phi\colon C^n(X,Y) \to [0,\infty]^{2n+2}$ by the formula

$$\Phi(f) := \bigl(\bigl\| f(0)\bigr\|,\ \|f\|_{\sup},\ \bigl\| Df(0)\bigr\|,\ \|Df\|_{\sup},\ \ldots\bigr)$$

and put

$$S_p(X,Y) := \bigl\{f \in C^n(X,Y) : p\bigl(\Phi(f)\bigr) < \infty\bigr\}.$$

Since $p(0) = 0$, S_p contains at least the zero function. It is easy to notice that S_p is a linear space and that $p \circ \Phi | S_p$ is a seminorm. We will denote this seminorm by $\|\cdot\|_p$. The same notations we will apply for the space $C^n(X \times X, Y)$.

Now we are able to prove the main theorem of this section.

Theorem 2.13 *Let $f\colon \mathbb{R} \to Y$ be a function such that $Q(f) \in S_p(\mathbb{R}\times\mathbb{R}, Y)$ and $\partial_2(Q(f))(0,0) = 0$. We additionally assume that the function p does not depend on the second or fourth and fifth variables. Then there exists a quadratic function $K\colon \mathbb{R} \to Y$ such that $f - K \in S_p(\mathbb{R}, Y)$ and*

$$\|f - K\|_p \le \frac{1}{2}\bigl\|Q(f)\bigr\|_p.$$

Proof Assume that $Q(f) \in C^n(\mathbb{R}\times\mathbb{R}, Y)$. Suppose that p does not depend on the second variable. Then

$$p(0,\ \infty,\ 0,\ \ldots) = p(0,\ 0,\ 0,\ \ldots) = 0.$$

By Theorem 2.11, there exists exactly one quadratic function $K_0\colon \mathbb{R} \to Y$ satisfying conditions (2.35) and (2.36). Then

$$\Phi(f - K_0) \le \frac{1}{2}\bigl[\Phi\bigl(Q(f)\bigr) + (0,\infty,0,\ldots)\bigr],$$

and hence from (i), (ii), and (iii) we have

$$p\bigl(\Phi(f - K_0)\bigr) \le \frac{1}{2}p\bigl(\Phi\bigl(Q(f)\bigr) + (0,\infty,0,\ldots)\bigr) \le \frac{1}{2}p\bigl(\Phi\bigl(Q(f)\bigr)\bigr),$$

i.e.,

$$\|f - K_0\|_p \le \frac{1}{2}\bigl\|Q(f)\bigr\|_p.$$

Suppose now that p does not depend on the fourth and fifth variables. If $Q(f) \in BC^n(\mathbb{R} \times \mathbb{R}, Y)$, then by Theorem 2.12 there exists exactly one quadratic function $K_\infty : \mathbb{R} \to Y$ satisfying conditions (2.42) and (2.43). Hence

$$\Phi(f - K_\infty) \leq \frac{1}{2}\left[\Phi\left(Q(f)\right) + (0, 0, 0, \infty, \infty, 0, \ldots)\right],$$

and consequently from (i), (ii), and (iii) we get

$$p\left(\Phi(f - K_\infty)\right) \leq \frac{1}{2} p\left(\Phi\left(Q(f)\right) + (0, 0, 0, \infty, \infty, 0, \ldots)\right) \leq \frac{1}{2} p\left(\Phi\left(Q(f)\right)\right),$$

i.e.,

$$\|f - K_\infty\|_p \leq \frac{1}{2}\left\|Q(f)\right\|_p.$$

If $Q(f)$ is unbounded, then $\|Q(f)\|_{\sup} = \infty$. By Theorem 2.11, we can find a quadratic function such that the conditions (2.35) and (2.36) hold. Then $\Phi(f - K_0) \leq \frac{1}{2}\Phi(Q(f))$, and hence

$$\|f - K_0\|_p \leq \frac{1}{2}\left\|Q(f)\right\|_p.$$

The proof is completed. □

One can easily notice that if we defined for $n \in \mathbb{N}_0$

$$p(x_1, x_2, \ldots, x_{2n+2}) := \sum_{i=1}^{n} x_{2i-1} + x_{2n+2}, \tag{2.49}$$

then we would obtain stability of the quadratic functional equation in the norm defined by the formula (2.48).

References

1. Aczel, J.: Lectures on Functional Equations and Their Applications. Academic Press, New York (1966)
2. Aczel, J., Dhombres, J.: Functional Equations in Several Variables. Cambridge University Press, Cambridge (1989)
3. Adam, M., Czerwik, S.: On the double quadratic difference property. Int. J. Appl. Math. Stat. **7**, 18–26 (2007)
4. Bielecki, A.: Une remarque sur la méthode de Banach–Cacciopoli–Tikhonov dans la théorie des équations différentielles ordinaires. Bull. Acad. Pol. Sci., Sér. Sci. Math. Astron. Phys. **4**, 261–264 (1956)
5. Cholewa, P.W.: Remarks on the stability of functional equations. Aequ. Math. **27**, 76–86 (1984)
6. Czerwik, S.: On the stability of the quadratic mapping in normed spaces. Abh. Math. Semin. Univ. Hamb. **62**, 59–64 (1992)

7. Czerwik, S.: Functional Equations and Inequalities in Several Variables. World Scientific, New Jersey (2002)
8. Czerwik, S. (ed.): Stability of Functional Equations of Ulam–Hyers–Rassias Type. Hadronic Press, Palm Harbor (2003)
9. Czerwik, S., Dłutek, K.: Superstability of the equation of quadratic functionals in L^P-spaces. Aequ. Math. **63**, 210–219 (2002)
10. Czerwik, S., Dłutek, K.: Cauchy and Pexider operators in some function spaces. In: Rassias, Th.M. (ed.) Functional Equations, Inequalities and Applications, pp. 11–19. Kluwer Academic, Dordrecht (2003)
11. Czerwik, S., Dłutek, K.: Quadratic difference operators in L^P-spaces. Aequ. Math. **67**, 1–11 (2004)
12. Czerwik, S., Dłutek, K.: Stability of the quadratic functional equation in Lipschitz spaces. J. Math. Anal. Appl. **293**, 79–88 (2004)
13. Czerwik, S., Król, K.: The D'Alembert and Lobaczewski difference operators in X_λ spaces. Nonlinear Funct. Anal. Appl. **13**, 395–407 (2008)
14. Forti, G.L.: Hyers–Ulam stability of functional equations in several variables. Aequ. Math. **50**, 143–190 (1995)
15. Ger, R.: A survey of recent results on stability of functional equations. In: Proc. of the 4th ICFEI, Pedagogical University of Cracow, pp. 5–36 (1994)
16. Hyers, D.H.: On the stability of the linear functional equation. Proc. Natl. Acad. Sci. USA **27**, 222–224 (1941)
17. Hyers, D.H., Isac, G., Rassias, Th.M.: Stability of Functional Equations in Several Variables. Birkhäuser, Basel (1998)
18. Jung, S.-M.: Hyers–Ulam–Rassias Stability of Functional Equations in Mathematical Analysis. Hadronic Press, Florida (2000)
19. Laczkovich, M.: Functions with measurable differences. Acta Math. Acad. Sci. Hung. **35**, 217–235 (1980)
20. Luenberger, D.G.: Optimization by Vector Space Methods. PWN, Warsaw (1974) (in Polish)
21. Lusternik, L.A., Sobolew, W.I.: Elements of Functional Analysis. PWN, Warsaw (1959) (in Polish)
22. Maurin, K.: Analysis, Part One: Elements. PWN, Warsaw (1973) (in Polish)
23. Rassias, Th.M.: On the stability of the linear mapping in Banach spaces. Proc. Am. Math. Soc. **72**(2), 297–300 (1978)
24. Rassias, Th.M.: On the stability of functional equations in Banach spaces. J. Math. Anal. Appl. **251**, 264–284 (2000)
25. Rassias, Th.M.: On the stability of functional equations and a problem of Ulam. Acta Appl. Math. **62**, 23–130 (2000)
26. Rassias, Th.M. (ed.): Functional Equations and Inequalities. Kluwer Academic, Norwell (2000)
27. Rassias, Th.M., Tabor, J. (eds.): Stability of Mapping of Hyers–Ulam Type. Hadronic Press, Florida (1994)
28. Tabor, J., Tabor, J.: Stability of the Cauchy type equations in the class of differentiable functions. J. Approx. Theory **98**(1), 167–182 (1999)
29. Ulam, S.M.: Problems in Modern Mathematics. Science Editions. Wiley, New York (1960)

Chapter 3
On the Regions Containing All the Zeros of a Polynomial

Chadia Affane-Aji and N.K. Govil

Abstract Let $p(z) = a_0 + a_1 z + a_2 z^2 + a_3 z^3 + \cdots + a_n z^n$ be a polynomial of degree n, where the coefficients a_k may be complex. Then it is obviously of interest to study problems concerning the location of the zeros of the polynomial $p(z)$. These problems, besides being of theoretical interest, have important applications in many areas, such as signal processing, communication theory, and control theory, and for this reason there is always a need for better and better results. In this paper we make a systematic study of these problems by presenting some results starting from the results of Gauss and Cauchy, who we believe were the earliest contributors in this subject, to some of the most recent ones. Our paper is expository.

Key words Complex polynomials · Location of zeros of polynomials · Complex zeros · Real zeros

Mathematics Subject Classification 30A10 · 30A15 · 26C10

3.1 Introduction

Given a polynomial

$$p(z) = a_0 + a_1 z + a_2 z^2 + a_3 z^3 + \cdots + a_n z^n$$

of degree n, it is well known by the Fundamental Theorem of Algebra that a polynomial of degree n has exactly n zeros; however, the zeros may be coincident.

The Fundamental Theorem of Algebra does not tell us anything about the location of zeros of the polynomial, which plays a vital role in many research areas.

Dedicated to Professor Themistocles M. Rassias on the occasion of his 60th birthday.

C. Affane-Aji
Department of Mathematics, Tuskegee University, Tuskegee, AL 36088, USA
e-mail: affane@mytu.tuskegee.edu

N.K. Govil (✉)
Department of Mathematics & Statistics, Auburn University, Auburn, AL 36849-5310, USA
e-mail: govilnk@auburn.edu

P.M. Pardalos et al. (eds.), *Nonlinear Analysis*, Springer Optimization and Its Applications 68, In Honor of Themistocles M. Rassias on the Occasion of his 60th Birthday,
DOI 10.1007/978-1-4614-3498-6_3,

Therefore, it is obviously of interest to obtain the region that contains all of the zeros or a required number of zeros of a polynomial. These types of problems can mainly be divided into two categories:

- Given an integer p, $1 \le p \le n$, find a region $R = R(a_0, a_1, \ldots, a_n)$ containing at least or exactly p zeros of $p(z)$. For instance, one would like to find the smallest circle $|z| = r$ which will enclose the p zeros of the polynomial.
- Given a region R, find the number $p = p(a_0, a_1, \ldots, a_n)$ such that p zeros lie in the region R; for example, find p zeros whose moduli do not exceed some prescribed value, say r.

The results dealing with the location of zeros of a polynomial, besides being of theoretical interest, have important applications in many areas, such as signal processing, communication theory, and control theory, and for this reason there is always a need for better and better results.

The subject of the location of the zeros of a polynomial is very vast dating back to the time of Gauss and Cauchy, and in this article we discuss some of the results in this subject, starting from the results of Gauss and Cauchy to some of the more recent ones. Due to the limited space, it is not possible to include all the results in this subject, and therefore many important results in this area which we would have liked to include have to be excluded (for a more detailed study of the subject, we refer to the monograph and books written by Dieudonné [12], Marden [22], and Milovanović, Mitrinović, and Rassias [25]).

3.2 Results due to Gauss and Cauchy and Some Related Results

The earliest result concerning the location of the zeros of a polynomial is probably due to Gauss who incidental to his proofs of the Fundamental Theorem of Algebra showed in 1816 that a polynomial

$$P(z) = a_n z^n + a_{n-1} z^{n-1} + \cdots + a_2 z^2 + a_1 z + a_0,$$

with all a_k real, has no zeros outside certain circles $|z| = R$, where

$$R = \max_{1 \le k \le n} \left(n 2^{1/2} |a_k|\right)^{1/k}.$$

However, in the case of arbitrary real or complex a_k, he [14] in 1849 showed that R may be taken as the positive root of the equation

$$z^n - 2^{1/2}\left(|a_1| z^{n-1} + \cdots + |a_n|\right) = 0.$$

As a further indication of Gauss' interest in the location of the zeros of a polynomial, we have his letter (see collected works of Gauss) to Schumacher dated April 2, 1833, in which he tells of having written enough on this topic to fill several volumes, but the only results he published are those in Gauss [14]. Even, his important result stated below

Theorem 3.1 *The zeros of the function* $F(z)=\sum_{j=1}^{p} m_j/(z-z_j)$, *where all* m_j *are real, are the points of the equilibrium in the field of force due to the system of* p *masses* m_j *at the fixed points* z_j *repelling a unit movable mass at* z *according to the inverse distance law.*

On the mechanical interpretation of the zeros of the derivative of a polynomial comes to us only by a brief entry he made presumably in about 1836 in a notebook otherwise devoted to astronomy.

Around 1829, Cauchy [6] (also, see the book of Marden [22, Theorem (27,1), p. 122]) derived more exact bounds for the moduli of the zeros of a polynomial than those given by Gausss, by proving the following

Theorem 3.2 *Let* $p(z)=z^n+\sum_{j=0}^{n-1} a_j z^j$, *be a complex polynomial, then all the zeros of p(z) lie in the disc*

$$\{z:|z|\le\eta\}\subset\{z:|z|<1+A\}, \tag{3.1}$$

where

$$A=\max_{0\le j\le n-1}|a_j|,$$

and η *is the unique positive root of the real-coefficient equation*

$$z^n-|a_{n-1}|z^{n-1}-|a_{n-2}|z^{n-2}-\cdots-|a_1|z-|a_0|=0. \tag{3.2}$$

The result is best possible and the bound is attained when $p(z)$ *is the polynomial on the left hand side of* (3.2).

The proof follows easily from the inequality

$$|p(z)|\ge|z|^n-\left(|a_{n-1}||z|^{n-1}+|a_{n-2}||z|^{n-2}+\cdots+|a_1||z|+|a_0|\right)=0, \tag{3.3}$$

which can be derived easily on applying Triangular Inequality to $p(z)=z^n+\sum_{j=0}^{n-1} a_j z^j$.

From the inequality (3.3) as well follows the following result which is also due to Cauchy [6].

Theorem 3.3 *Let* $p(z)=\sum_{j=0}^{n} a_j z^j$, *be a complex polynomial with* $a_n\ne 0$, *then all the zeros of* $p(z)$ *lie in the disc*

$$\left\{z:|z|<1+\max_{0\le k\le n-1}|a_k/a_n|\right\}.$$

The inequality (3.3) also yields the following result due to Birkhoff [4], which was later proved independently by Cohn [7] and by Berwald [3].

Theorem 3.4 *The zero z_1 of largest modulus of $p(z) = a_0 + a_1 z + \cdots + a_n z^n$, $a_n \neq 0$, satisfies the inequalities*

$$(2^{1/n} - 1)r \leq \alpha \leq |z_1| \leq r \leq \alpha/(2^{1/n} - 1), \tag{3.4}$$

where r is the positive root of (3.2) *and α is defined as*:

$$\alpha = \max_{1\leq k\leq n} \left|a_{n-k}/\{a_n C_k^n\}\right|^{1/k} \leq |z_1|.$$

Here, as usual, C_k^n are the binomial coefficients defined by

$$C_k^n = \frac{n!}{k!(n-k)!}, \qquad 0! = 1. \tag{3.5}$$

The following result is due to Kuniyeda [20], Montel [26], and Tôya [32].

Theorem 3.5 *For any p and q such that*

$$p > 1,\ q > 1, \quad (1/p) + (1/q) = 1, \tag{3.6}$$

the polynomial $p(z) = a_0 + a_1 z + \cdots + a_n z^n$, $a_n \neq 0$, has all its zeros in the circle

$$|z| < \left\{1 + \left[\sum_{j=0}^{n-1} |a_j|^p/|a_n|^p\right]^{q/p}\right\}^{1/q} \leq \left(1 + n^{q/p} M^q\right)^{1/q}, \tag{3.7}$$

where $M = \max|a_j/a_n|$, $j = 0, 1, \ldots, n-1$.

In particular, if we take $p = q = 2$ in inequality (3.7), this reduces to

$$|z| < \left\{1 + \sum_{j=0}^{n-1} |a_j|^2/|a_n|^2\right\}^{1/2}. \tag{3.8}$$

The above inequality (3.8) has been derived in Carmichael-Mason [5], Kelleher [19], and Fujiwara [13].

Note that as $p \to \infty$, the right side of (3.7) approaches the limit $1 + \max_{0\leq j\leq n-1} |a_j|/|a_n|$ and thus Theorem 3.3 can be obtained as a special case of Theorem 3.5.

If we apply inequality (3.8) to the polynomial $(1-z)(a_0 + a_1 z + \cdots + a_n z^n)$, $a_n \neq 0$, we easily get the following result of Williams [34].

Theorem 3.6 *All the zeros of the polynomial $p(z) = a_0 + a_1 z + \cdots + a_n z^n$, $a_n \neq 0$, lie in the disk*

$$|z| \leq \left[1 + \left|\frac{a_0}{a_n}\right|^2 + \left|\frac{a_1 - a_0}{a_n}\right|^2 + \cdots + \left|\frac{a_n - a_{n-1}}{a_n}\right|^2\right]^{1/2}. \tag{3.9}$$

Next, we mention the following result of Walsh [33], which can sometimes be very useful.

Theorem 3.7 *All the zeros of the polynomial* $p(z) = a_0 + a_1 z + \cdots + a_n z^n$, $a_n \neq 0$, *lie in the disk*

$$|z| \leq \sum_{j=1}^{n} |a_{n-j}/a_n|^{1/j}. \tag{3.10}$$

We close this section by stating the following result due to Markovitch [24].

Theorem 3.8 *All the zeros of the polynomial* $h(z) = \sum_{k=0}^{n} a_k b_k z^k$ *lie in the disk* $|z| \leq Mr$, *where* r *is the positive root of the equation*

$$|a_0| + |a_1|z + \cdots + |a_{n-1}|z^{n-1} - |a_n|z^n = 0, \tag{3.11}$$

and $M = \max_{0 \leq k \leq n-1} |b_k/b_{k-1}|^{1/(n-k)}$.

3.3 Grace's Apolarity Theorem and Some Results of Peretz and Rassias

3.3.1 Grace's Apolarity Theorem and Some Related Results

In the beginning of the last century, Grace [16] introduced the following concept of apolar polynomials.

Definition 3.1 Two polynomials $p(z) = \sum_{k=0}^{n} a_k C_k^n z^k$ and $q(z) = \sum_{k=0}^{n} b_k C_k^n z^k$ are said to be apolar if their coefficients satisfy the apolarity condition

$$\sum_{k=0}^{n} (-1)^k C_k^n a_k b_{n-k} = 0, \tag{3.12}$$

where C_k^n are the binomial coefficients defined by $C_k^n = \frac{n!}{k!(n-k)!}$.

In the same paper, Grace [16] proved the following result, known as Grace's Apolarity Theorem, or simply Grace's Theorem, which has been found to be of great use.

Theorem 3.9 *Let the polynomials* $p(z) = \sum_{k=0}^{n} a_k C_k^n z^k$ *and* $q(z) = \sum_{k=0}^{n} b_k C_k^n z^k$ *be apolar. Then any circular domain that contains all the zeros of the polynomial* $p(z)$ *must contain at least one zero of the polynomial* $q(z)$.

Szegö [31] gave an alternative proof of the above theorem of Grace [16], and also gave several applications. Another proof of this theorem was given by Goodman and Schoenberg [15] (also, see Milovanović, Mitrinović, and Rassias [25, p. 188]) for which they use induction on n.

The following applications of Grace's Theorem can be found in Szegö [31] (also in the book of Marden [22], Milovanović, Mitrinović, and Rassias [25], and in paper of Schur [29]).

Theorem 3.10 *If all the zeros of the polynomial* $p(z)=\sum_{k=0}^{n}a_kC_k^nz^k$ *lie in* $|z|<r$ *and all the zeros of the polynomial* $q(z)=\sum_{k=0}^{n}b_kC_k^nz^k$ *lie in* $|z|\le\rho$, *then all the zeros of the polynomial* $\sum_{k=0}^{n}C_k^na_kb_kz^k$ *are in* $|z|<r\rho$.

Theorem 3.11 (Schur–Szegö composite theorem) *If all the zeros of the polynomial* $p(z)=\sum_{k=0}^{n}a_kC_k^nz^k$ *lie in a closed and bounded convex domain* D *and all the zeros of the polynomial* $q(z)=\sum_{k=0}^{n}b_kC_k^nz^k$ *lie in* $[-1,0]$, *then all the zeros of the polynomial* $\sum_{k=0}^{n}C_k^na_kb_kz^k$ *are in* D.

By using Theorem 3.9 of Grace, in his paper Szegö [31] also obtained

Theorem 3.12 *Let the polynomial* $p(z)=z^n+\sum_{j=0}^{n-1}a_jz^j$ *have no zeros in the disk* $|z|\le R$. *Then the "section"* $q(z)=p(z)-z^n=\sum_{j=0}^{n-1}a_jz^j$ *has no zeros in the circular region* $|z|\le R/2$.

Next, we state the following result which is stated in the book of Milovanović, Mitrinović, and Rassias [25, Theorem 1.4.1, p. 197].

Theorem 3.13 *If all the zeros of a polynomial* $p(z)=\sum_{k=0}^{n}a_kz^k$ *lie in a circle* $|z|\le R$, *then for any* a *all the zeros of the polynomial* $p(z)-a$ *lie in the disk* $|z|\le R+|a/a_n|^{1/n}$.

3.3.2 Some Results of Peretz and Rassias

In his book, Marden [22, pp. 68–70] states two theorems which are supposed to be restatements of his results in Marden [23].

Theorem 3.14 *Let* $P(z)=\sum_{k=0}^{m}a_kz^k$, $Q(z)=\sum_{k=0}^{n}b_kz^k$, *and* $R(z)=\sum_{k=0}^{m}a_k\times Q(k)z^k$. *If all the zeros of the polynomial* $P(z)$ *lie in the ring*

$$R_0=\{z:0\le r_1\le|z|\le r_2\le\infty\},\tag{3.13}$$

and if all the zeros of the polynomial $Q(z)$ *lie in the ring*

$$A=\{z:0\le\rho_1\le|z|/|z-m|\le\rho_2\le\infty\},\tag{3.14}$$

then all the zeros of the polynomial $R(z)$ lie in the ring

$$R_n = \{z : 0 \le r_1 \min(1, \rho_1^n) \le |z| \le r_2 \max(1, \rho_2^n)\}. \tag{3.15}$$

Theorem 3.15 *Let $P(z) = \sum_{k=0}^m a_k z^k$, $Q(z) = \sum_{k=0}^n b_k z^k$, and $R(z) = \sum_{k=0}^m a_k \times Q(k) z^k$. If all the zeros of the polynomial $P(z)$ lie in the ring*

$$R_0 = \{z : 0 \le r_1 \le |z| \le r_2 \le \infty\}, \tag{3.16}$$

then all the zeros of the polynomial $R(z)$ lie in the ring

$$r_1 \min[1, |Q(0)/Q(m)|] \le |z| \le r_2 \max[1, |Q(0)/Q(m)|]. \tag{3.17}$$

Theorem 3.14 is a part of Marden's corollary in [23] whereas Theorem 3.15 is not included there.

In 1992, Peretz and Rassias [27] proved that Theorem 3.15 is, in fact, false. For this, they constructed a counterexample, by taking $P(z) = 1 + 2z + z^2 = (1+z)^2$ and $Q(z) = 1 + 2z - z^2$. For these polynomials $n = m = 2$, $Q(0) = 1$, $Q(1) = 2$, and $Q(2) = 1$, and therefore $R(z) = 1 + 4z + z^2$. Note that $P(z)$ has a double zero at $z = -1$ and so we can take $r_1 = r_2 = 1$. Since $Q(0)/Q(2) = 1$, by Theorem 3.15 all the zeros of the polynomial $R(z)$ should lie on $|z| = 1$ while, as can be easily seen, its zeros are $-2 + \sqrt{3}$ and $-2 - \sqrt{3}$, which obviously do not lie on $|z| = 1$.

After establishing that Theorem 3.15 is false, in the same paper Peretz and Rassias [27] prove a correct version of this Theorem 3.15, for which they introduce the following definition.

Definition 3.2 Let $Q(z) = (\beta_1 - z) \cdots (\beta_n - z)$ and m a positive integer. Then

$$Q^+(z) = \prod_{\substack{1 \le j \le n \\ Re(\beta_j) \ge m/2}} (\beta_j - z), \qquad Q^-(z) = \prod_{\substack{1 \le j \le n \\ Re(\beta_j) < m/2}} (\beta_j - z), \tag{3.18}$$

with the understanding that Q^+ or Q^- takes the value 1, if one of the products is empty.

Note that $Q(z) = Q^+(z) Q^-(z)$, and the zeros of Q^+ are those zeros of Q for which $|\beta/(\beta - m)| \ge 1$.

Now, with the above definition, the following theorem of Peretz and Rassias [27] (also see [25, Theorem 1.4.26 on p. 202]) provides a correct version of Theorem 3.15.

Theorem 3.16 *Let $P(z) = \sum_{k=0}^m a_k z^k$, $Q(z) = \sum_{k=0}^n b_k z^k$, and $R(z) = \sum_{k=0}^m a_k \times Q(k) z^k$. If all the zeros of the polynomial $P(z)$ lie in the ring*

$$R_0 = \{z : 0 \le r_1 \le |z| \le r_2 \le \infty\}, \tag{3.19}$$

then all the zeros of the polynomial $R(z)$ lie in the ring

$$r_1\left|Q^-(0)/Q^-(m)\right| \le |z| \le r_2\left|Q^+(0)/Q^+(m)\right|. \tag{3.20}$$

For some results concerning the location of zeros of linear combination of polynomials we refer to the paper of Rubinstein [28].

3.4 Some Recent Generalizations and Refinements of a Theorem of Cauchy

We begin with the Theorem 3.2 of Cauchy [6], and for the sake of completeness we state below this theorem again, although this has already been stated in Sect. 1 of this paper.

Theorem 3.17 (Cauchy, [6]) *Let $p(z) = z^n + \sum_{j=0}^{n-1} a_j z^j$, be a complex polynomial, then all the zeros of $p(z)$ lie in the disc*

$$\{z : |z| \le \eta\} \subset \{z : |z| < 1 + A\}, \tag{3.21}$$

where

$$A = \max_{0 \le j \le n-1} |a_j|,$$

and η is the unique positive root of the equation

$$z^n - |a_{n-1}|z^{n-1} - |a_{n-2}|z^{n-2} - \cdots - |a_1|z - |a_0| = 0. \tag{3.22}$$

The result is best possible and the bound is attained when $p(z)$ is the polynomial on the left hand side of (3.22).

Over the years, the above theorem of Cauchy [6] has been sharpened by many people but here we state the following sharpening of this result, which is due to Joyal, Labelle, and Rahman [18].

Theorem 3.18 *If $B = \max_{0 \le j < n-1} |a_j|$ then all the zeros of the polynomial $p(z) = z^n + \sum_{j=0}^{n-1} a_j z^j$ are contained in the disk*

$$|z| \le \frac{1}{2}\left\{1 + |a_{n-1}| + \sqrt{\left(1 - |a_{n-1}|\right)^2 + 4B}\right\}. \tag{3.23}$$

As is easy to verify that, except in the case when $|a_{n-1}| = B$, the bound obtained by Theorem 3.18 is always sharper than the bound obtained by Theorem 3.17. In case $|a_{n-1}| = B$, both these theorems give the same bound.

It would obviously be of interest to have a result which in every case, including when $|a_{n-1}| = B$, gives a bound sharper than obtainable by Theorem 3.17 due to Cauchy, and in this connection we have the following result of Datt and Govil [8].

Theorem 3.19 *Let $p(z)=z^n+a_{n-1}z^{n-1}+\cdots+a_1z+a_0$ be a polynomial of degree n and*

$$A=\max_{0\le j\le n-1}|a_j|.$$

Then $p(z)$ has all its zeros in the ring shaped region

$$\frac{|a_0|}{2(1+A)^{n-1}(An+1)}\le|z|\le 1+\lambda_0 A, \tag{3.24}$$

where λ_0 is the unique positive root of the equation $x=1-1/(1+Ax)^n$ in the interval $(0,1)$. The upper bound $1+\lambda_0 A$ in the above given region (3.24) is best possible and is attained for the polynomial $p(z)=z^n-A(z^{n-1}+\cdots+z+1)$.

In case we do not wish to solve the equation $x=1-1/(1+Ax)^n$, in order to apply the above result of Datt and Govil [8], we can apply the following result also due to Datt and Govil [8], which in every case clearly gives an improvement over Theorem 3.17 of Cauchy [6].

Theorem 3.20 *Let*

$$p(z)=\sum_{v=0}^{n-1}a_v z^v+z^n$$

be a polynomial of degree n and let $A=\max_{0\le j\le n-1}|a_j|$. Then $p(z)$ has all its zeros in the ring shaped region given by

$$\frac{|a_0|}{2(1+A)^{n-1}(nA+1)}\le|z|\le 1+\left(1-\frac{1}{(1+A)^n}\right)A. \tag{3.25}$$

Since always $(1-\frac{1}{(1+A)^n})<1$, the above Theorem 3.20 in every situation sharpens Theorem 3.17 due to Cauchy.

Theorem 3.18 due to Joyal, Labelle, and Rahman [18] was sharpened by Dewan [9], who proved

Theorem 3.21 *Let $p(z)=z^n+\sum_{k=0}^{n-1}a_k z^k$ be a polynomial of degree n, and let*

$$B=\max_{0\le j\le n-1}|a_j|.$$

Then all the zeros of $p(z)$ lie in the ring shaped region given by

$$R_2\le|z|\le R_1.$$

Here

$$R_1=\frac{1}{2}\left[1+|a_{n-1}|+\left\{\left(1-|a_{n-1}|\right)^2+4B\right\}^{\frac{1}{2}}\right],$$

$$R_2 = \frac{1}{2M_1^2}\left[-R_1^2|b|\left(M_1 - |a_0|\right) + \left\{R_1^4|b|^2\left(M_1 - |a_0|\right)^2 + 4|a_0|R_1^2M_1^3\right\}^{\frac{1}{2}}\right]$$

where

$$M_1 = R_1^n\left[R_1 + 1 + 2|a_{n-1}| + (2n-3)B\right],$$
$$b = a_1 - a_0.$$

In [9], Dewan also proves the following sharpening of Theorem 3.19 of Datt and Govil [8].

Theorem 3.22 *If* $p(z) = z^n + \sum_{k=0}^{n-1} a_k z^k$ *is a polynomial of degree* n *and*

$$A = \max_{0 \le j \le n-1} |a_j|,$$

then $p(z)$ *has all its zeros in the ring shaped region given by*

$$\frac{1}{2M_2^2}\left[-(1+A)^2|b|\left(M_2 - |a_0|\right) + \left\{(1+A)^4|b|^2\left(M_2 - |a_0|\right)^2 + 4|a_0|(1+A)^2M_2^3\right\}^{\frac{1}{2}}\right] \le |z| \le 1 + \lambda_0 A,$$

where λ_0 *is the unique root of the equation* $x = 1 - \frac{1}{(1+Ax)^n}$ *in the interval* $(0, 1)$ *and* $M_2 = (1+A)^n 2(1+nA)$, $b = a_1 - a_0$. *The upper bound* $1+\lambda_0 A$ *in the above ring is best possible and is attained for the polynomial* $p(z) = z^n - A(z^{n-1} + \cdots + z + 1)$.

Although the outer radii of the annulus in Theorem 3.22 and Theorem 3.19 are same, but as can be verified the inner radius of the annulus in Theorem 3.22 is smaller than the inner radius in Theorem 3.19.

The following result of Zeheb [35] is also a refinement of the Theorem 3.17 of Cauchy [6].

Theorem 3.23 *All the zeros of the real polynomial* $p(z) = z^n + \sum_{k=0}^{n-1} a_k z^k$ *lie in the circle* $|z| < 1 + \max\{A_{ij}\}$, *where*

$$A_{ij} = \frac{|a_i a_{j-1} - a_j a_{i-1}|}{|a_i| + |a_j|}, \quad i, j = 0, \ldots, n; \; j > i,$$
$$a_n = 1, \quad a_{-1} = 0.$$

Another result, providing a disk containing all the zeros of a polynomial, is due to Žilović et al. [36].

Theorem 3.24 *All the zeros of the complex polynomial* $p(z) = z^n + \sum_{k=0}^{n-1} a_k z^k$ *lie in the disk*

$$\left\{z \in \mathbb{C} : |z| < \sqrt{1+A}\right\},$$

where

$$A = \max_{0\le k\le n-1} \left\{ \left| a_k^2 + 2(-1)^k (B - C) \right| \right\},$$

$$B = \sum_{\substack{0\le i<j\le [n/2] \\ i+j=k}} a_{2i} a_{2j},$$

$$C = \sum_{\substack{0\le i<j\le [(n-1)/2] \\ i+j=k-1}} a_{2i+1} a_{2j+1}.$$

Here $a_n = 1$ and as usual $[k]$ denotes the integer part of k.

By means of examples, Žilović et al. [36] show that for some polynomials their result gives better bounds than obtainable from several of the above stated results.

Since the beginning, binomial coefficients have appeared in the derivation or as a part of the closed expressions of the bounds. However, Fibonacci's numbers, that is, $F_0 = 0$, $F_1 = 1$, and for $j \ge 2$, $F_j = F_{j-1} + F_{j-2}$ have not appeared either in implicit bounds or explicit bounds for the moduli of the zeros. Diaz-Barrero [10] proved the following result which gives circular domains containing all the zeros of a polynomial where binomial coefficients and Fibonacci's numbers appear. He also gives an example of a polynomial for which the above theorem gives a better bound than the bound obtainable from Theorem 3.17 of Cauchy [6].

Theorem 3.25 *Let $p(z) = \sum_{j=0}^{n} a_j z^j$ $(a_j \ne 0,\ 0 \le j \le n)$ be a complex monic polynomial. Then all its zeros lie in the disk $C_1 = \{z \in \mathbb{C} : |z| \le r_1\}$ or $C_2 = \{z \in \mathbb{C} : |z| \le r_2\}$, where*

$$r_1 = \max_{1\le k\le n} \left\{ \sqrt[k]{\frac{2^{n-1} C_2^{n+1}}{k^2 C_k^n} |a_{n-k}|} \right\},$$

$$r_2 = \max_{1\le k\le n} \left\{ \sqrt[k]{\frac{F_{3n}}{C_k^n 2^k F_k} |a_{n-k}|} \right\}.$$

The proof of the above theorem depends on the identities

$$\sum_{k=1}^{n} k^2 C_k^n = 2^{n-2} n(n+1) \tag{3.26}$$

and

$$\sum_{k=1}^{n} C_k^n 2^k F_k = F_{3n}, \tag{3.27}$$

where F_j are the Fibonacci's numbers, and C_k^n the binomial coefficients defined by

$$C_k^n = \frac{n!}{k!(n-k)!}, \quad 0! = 1. \tag{3.28}$$

The following result, which provides an annulus region containing all the zeros of a polynomial is also due to Diaz-Barrero [11].

Theorem 3.26 *Let $p(z) = \sum_{j=0}^{n} a_j z^j$ $(a_j \neq 0,\ 0 \leq j \leq n)$ be a nonconstant complex polynomial. Then all its zeros lie in the annulus $C = \{z \in \mathbb{C} : r_1 \leq |z| \leq r_2\}$, where*

$$r_1 = \frac{3}{2} \min_{1 \leq j \leq n} \left\{ \frac{2^n F_j C_j^n}{F_{4n}} \left| \frac{a_0}{a_j} \right| \right\}^{1/j},$$

$$r_2 = \frac{2}{3} \max_{1 \leq j \leq n} \left\{ \frac{F_{4n}}{2^n F_j C_j^n} \left| \frac{a_{n-j}}{a_n} \right| \right\}^{1/j}.$$

Here the Fibonacci's numbers F_j and the binomial coefficients C_j^m are as defined above in the previous theorem.

The following result of Kim [21], whose proof depends on the use of the identity

$$\sum_{k=0}^{n} C_k^n = 2^n - 1, \tag{3.29}$$

also provides an annulus containing all the zeros of a polynomial.

Theorem 3.27 *Let $p(z) = \sum_{k=0}^{n} a_k z^k$ $(a_k \neq 0,\ 0 \leq k \leq n)$ be a nonconstant polynomial with complex coefficients. Then all the zeros of $p(z)$ lie in the annulus $A = \{z : r_1 \leq |z| \leq r_2\}$, where*

$$r_1 = \min_{1 \leq k \leq n} \left\{ \frac{C_k^n}{2^n - 1} \left| \frac{a_0}{a_k} \right| \right\}^{1/k}, \qquad r_2 = \max_{1 \leq k \leq n} \left\{ \frac{2^n - 1}{C_k^n} \left| \frac{a_{n-k}}{a_n} \right| \right\}^{1/k}. \tag{3.30}$$

Here again, as usual, C_k^n denote the binomial coefficients.

Theorem 3.17 of Cauchy has also been refined by Sun and Hsieh [30], who proved

Theorem 3.28 *All the zeros of the complex polynomial*

$$p(z) = z^n + \sum_{j=0}^{n-1} a_j z^j$$

lie in the disk

$$\{z : |z| < \eta\} \subset \{z : |z| < 1 + \delta_3\} \subseteq \{z : |z| < 1 + A\},$$

where δ_3 is the unique positive root of the equation

$$Q_3(x) \equiv x^3 + (2 - |a_{n-1}|)x^2 + (1 - |a_{n-1}| - |a_{n-2}|)x - A = 0, \tag{3.31}$$

and

$$A = \max_{0 \le j \le n-1} |a_j|.$$

Using the method similar to that of Sun and Hsieh [30], Jain [17] refined the above result of Sun and Hsieh [30], and proved

Theorem 3.29 *All the zeros of the complex polynomial*

$$p(z) = z^n + \sum_{j=0}^{n-1} a_j z^j$$

lie in the disk

$$\{z : |z| < \eta\} \subset \{z : |z| < 1 + \delta_4\} \subseteq \{z : |z| < 1 + \delta_3\} \subseteq \{z : |z| < 1 + A\},$$

where δ_4 is the unique positive root of the equation

$$\begin{aligned} Q_4(x) \equiv x^4 &+ (3 - |a_{n-1}|)x^3 + (3 - 2|a_{n-1}| - |a_{n-2}|)x^2 \\ &+ (1 - |a_{n-1}| - |a_{n-2}| - |a_{n-3}|)x - A = 0, \end{aligned} \tag{3.32}$$

and $A = \max_{0 \le j \le n-1} |a_j|$ is same as in Theorem 3.28.

In [1], Affane-Aji, Agarwal, and Govil proved the following result which not only includes the above results of Cauchy [6], Sun and Hsieh [30], and Jain [17] as special cases but also provides a tool for obtaining sharper bounds for the location of the zeros of a polynomial.

Theorem 3.30 *All the zeros of the polynomial*

$$p(z) = z^n + \sum_{j=0}^{n-1} a_j z^j$$

lie in the disks

$$\begin{aligned} &\{z : |z| < 1 + \delta_k\} \subseteq \{z : |z| < 1 + \delta_{k-1}\} \subseteq \cdots \\ &\subseteq \{z : |z| < 1 + \delta_1\} \subseteq \{z : |z| < 1 + A\}, \end{aligned}$$

where δ_k is the unique positive root of the kth degree equation

$$Q_k(x) \equiv x^k + \sum_{\nu=2}^{k} \left[C_{k-\nu}^{k-1} - \sum_{j=1}^{\nu-1} C_{k-\nu}^{k-j-1} |a_{n-j}| \right] x^{k+1-\nu} - A = 0. \tag{3.33}$$

Here,

$$A = \max_{0 \le j \le n-1} |a_j|, \quad a_j = 0 \text{ if } j < 0,$$

and for k, a positive integer, C_k^m are the binomial coefficients

$$C_j^m = \frac{m!}{j!(m-j)!}, \quad 0! = 1, \tag{3.34}$$

as defined in (3.28), *and $C_k^m = 0$, if $k < 0$.*

As is easy to verify, for $k = 1$ the above theorem reduces to Theorem 3.17 due to Cauchy [6], for $k = 3$ to the result of Sun and Hsieh [30], and for $k = 4$ it reduces to the result due to Jain [17]. Further, by choosing k sufficiently large, we can make δ_k in the bound to our desired accuracy.

Note that by combining the above theorem with Theorem 3.20 of Datt and Govil [8], one can easily obtain the following result which is a refinement of the above Theorem 3.30.

Theorem 3.31 *All the zeros of the polynomial*

$$p(z) = z^n + \sum_{j=0}^{n-1} a_j z^j$$

lie in the annulus

$$\frac{|a_0|}{2(1+A)^{n-1}(nA+1)} \le |z| \le \{z : |z| < 1 + \delta_k\} \subseteq \{z : |z| < 1 + \delta_{k-1}\} \subseteq \cdots$$
$$\subseteq \{z : |z| < 1 + \delta_1\} \subseteq \{z : |z| < 1 + A\},$$

where δ_k is as defined in Theorem 3.30, *and* $A = \max_{0 \le j \le n-1} |a_j|$.

Similarly, one can obtain a refinement of Theorem 3.30 by combining Theorem 3.30 with Theorem 3.26 of Diaz-Barrero [11].

Recently, Affane-Aji, Biaz, and Govil [2] have proved the following refinement of Theorem 3.30, and constructed examples to show that for some polynomials their theorem, stated below, gives much better bounds than obtainable from Theorem 3.31. More precisely, their result is

Theorem 3.32 *All the zeros of the polynomial*

$$p(z) = z^n + \sum_{j=0}^{n-1} a_j z^j$$

lie in the disks

$$\begin{aligned} R_1 \le |z| \le \{z : |z| < 1 + \delta_k\} &\subseteq \{z : |z| < 1 + \delta_{k-1}\} \subseteq \cdots \\ &\subseteq \{z : |z| < 1 + \delta_1\} \subseteq \{z : |z| < 1 + A\}, \end{aligned}$$

where δ_k *is as defined in Theorem* 3.30, *and*

$$R_1 = \frac{-R^2|a_1|(M - |a_0|) + \sqrt{4R^2M^3|a_0| + \{R^2|a_1|(M - |a_0|)\}^2}}{2M^2}. \tag{3.35}$$

Here $M = \frac{R^{n+1}+(A-1)R^n-AR}{(R-1)}$ *with* $R = 1 + \delta_k$ *and* $A = \max_{0\le j\le n-1} |a_j|$.

Note that $R = 1 + \delta_k > 1$, so for every positive integer k, we have $M > 0$ and $R > 0$. It is obvious that, in general, Theorem 3.32 sharpens Theorem 3.30.

In the same paper, Affane-Aji, Biaz, and Govil [2] also prove the following refinement of Theorem 3.30, which in some cases gives bounds that are sharper than obtainable from Theorems 3.20, 3.26, and 3.31. This they have shown by constructing some examples of polynomials.

Theorem 3.33 *Let* $p(z) = \sum_{j=0}^{n} a_j z^j$ $(a_j \ne 0)$ *be a nonconstant complex polynomial. Then all its zeros lie in the annulus* $C = \{z \in \mathbb{C} : r_1 \le |z| \le r_2\}$, *where*

$$r_1 = \min_{1\le j\le n} \left\{ \frac{jC_j^n}{n2^{n-1}} \left| \frac{a_0}{a_j} \right| \right\}^{1/j}, \tag{3.36}$$

$$r_2 = 1 + \delta_k. \tag{3.37}$$

Here δ_k, *for some positive integer* k, *is as defined in Theorem* 3.30, *and* C_j^n *are the binomial coefficients defined by*

$$C_j^n = \frac{n!}{j!(n-j)!}, \quad 0! = 1. \tag{3.38}$$

References

1. Affane-Aji, C., Agarwal, N., Govil, N.K.: Location of zeros of polynomials. Math. Comput. Model. **50**, 306–313 (2009)
2. Affane-Aji, C., Biaz, S., Govil, N.K.: On annuli containing all the zeros of a polynomial. Math. Comput. Model. **52**, 1532–1537 (2010)

3. Berwald, L.: Elementare Sätze uber die Abgrenzung der Wurzeln einer algebraischen Gleichung. Acta Sci. Math. Litt. Sci. Szeged **6**, 209–221 (1934)
4. Birkhoff, G.D.: An elementary double inequality for the roots of an algebraic equation having greatest value. Bull. Am. Math. Soc. **21**, 494–495 (1914)
5. Carmichael, R.D., Mason, T.E.: Note on the roots of algebraic equations. Bull. Am. Math. Soc. **21**, 14–22 (1914)
6. Cauchy, A.L.: Excercises de Mathematiques. IV Année de Bure Frères, Paris (1829)
7. Cohn, A.: Uber die Anzahl der Wurzeln einer algebraischen Gleichung in einem Kreise. Math. Z. **14**, 110–148 (1922)
8. Datt, B., Govil, N.K.: On the location of zeros of polynomial. J. Approx. Theory **24**, 78–82 (1978)
9. Dewan, K.K.: On the location of zeros of polynomials. Math. Stud. **50**, 170–175 (1982)
10. Diaz-Barrero, J.L.: Note on bounds of the zeros. Mo. J. Math. Sci. **14**, 88–91 (2002)
11. Diaz-Barrero, J.L.: An annulus for the zeros of polynomials. J. Math. Anal. Appl. **273**, 349–352 (2002)
12. Dieudonné, J.: La théorie analytique des polynômes d'une variable. Mémor. Sci. Math. **93** (1938)
13. Fujiwara, M.: A Ueber die Wurzeln der algebraischen Gleichungen. Tôhoku Math. J. **8**, 78–85 (1915)
14. Gauss, K.F.: Beiträge zur Theorie der algebraischen Gleichungen. Abh. Ges. Wiss. Göttingen, vol. 4, Ges. Werke, vol. 3, pp. 73–102 (1850)
15. Goodman, A.W., Schoenberg, I.J.: A proof of Grace's theorem by induction. Honam Math. J. **9**, 1–6 (1987)
16. Grace, J.H.: The zeros of a polynomial. Proc. Camb. Philos. Soc. **11**, 352–357 (1901)
17. Jain, V.K.: On Cauchy's bound for zeros of a polynomial. Turk. J. Math. **30**, 95–100 (2006)
18. Joyal, A., Labelle, G., Rahman, Q.I.: On the location of zeros of polynomials. Can. Math. Bull. **10**, 53–63 (1967)
19. Kelleher, S.B.: Des limites des Zéros d'une polynome. J. Math. Pures Appl. **2**, 169–171 (1916)
20. Kuniyeda, M.: Notes on the roots of algebraic equation. Tôhoku Math. J. **9**, 167–173 (1916)
21. Kim, S.-H.: On the moduli of the zerros of a polynomial. Am. Math. Mon. **112**, 924–925 (2005)
22. Marden, M.: Geometry of Polynomials. Am. Math. Soc. Math. Surveys, vol. 3. Am. Math. Soc., Providence (1966)
23. Marden, M.: The zeros of certain composite polynomials. Bull. Am. Math. Soc. **49**, 93–100 (1943)
24. Markovitch, D.: On the composite polynomials. Bull. Soc. Math. Phys. Serbie **3**(3–4), 11–14 (1951)
25. Milovanovic, G.V., Mitrinovic, D.S., Rassias, Th.M.: Topics in Polynomials: Extremal Problems, Inequalities, Zeros. World Scientific, Singapore (1994)
26. Montel, P.: Sur la limite supérieure des modules des zéros des polynômes. C. R. Acad. Sci. Paris **193**, 974–976 (1931)
27. Peretz, R., Rassias, Th.M.: Some remarks on theorems of M. Marden concerning the zeros of certain composite polynomials. Complex Var. **18**, 85–89 (1992)
28. Rubinstein, Z.: Some results in the location of the zeros of linear combinations of polynomials. Trans. Am. Math. Soc. **116**, 1–8 (1965)
29. Schur, J.: Zwei sätze über algebraische Gleichungen mit lauter rellen Wurzeln. J. Reine Angew. Math. **144**, 75–88 (1914)
30. Sun, Y.J., Hsieh, J.G.: A note on circular bound of polynomial zeros. IEEE Trans. Circuits Syst. I **43**, 476–478 (1996)
31. Szegö, G.: Bemerkungen zu einem Satz von J.H. Grace über die Wurzeln algebraischer Gleichungen. Math. Z. **13**, 28–55 (1922)
32. Tôya, T.: Some remarks on Montel's paper concerning upper limits of absolute values of roots of algebraic equations. Sci. Rep. Tokyo Bunrika Daigaku **A1**, 275–282 (1933)

33. Walsh, J.L.: An inequality for the roots of an algebraic equation. Ann. Math. **25**, 285–286 (1924)
34. Williams, K.P.: Note concerning the roots of an equation. Bull. Am. Math. Soc. **28**, 394–396 (1922)
35. Zeheb, F.: On the largest modulus of polynomial zeros. IEEE Trans. Circuits Syst. I **38**, 333–337 (1991)
36. Žilović, M.S., Roytman, L.M., Combettes, P.L., Swamy, M.N.S.: A bound for the zeros of polynomials. IEEE Trans. Circuits Syst. I **39**, 476–478 (1992)

Chapter 4
Some Remarks on the Group of Isometries of a Metric Space

Dorin Andrica and Vasile Bulgarean

Abstract The main purpose of this paper is to describe the isometry groups $Iso_{d_p}(\mathbb{R}^n)$ for $p \geq 1$, $p \neq 2$, and $p = \infty$, where the metric d_p is given by (4.2). A corollary of the main result contained in Theorem 4.1 and Theorem 4.2 is that in case $p \neq 2$ all these groups are isomorphic and, consequently, they are independent of p. In the last section, the isometry dimension of a finite group with respect to a given metric on the space $\mathbb{R}^n$ is introduced.

Key words Isometry with respect to a metric · Group of isometries · Translations group of the Euclidean n-space · Taxicab metric · Mazur–Ulam theorem · Semi-direct product of groups

Mathematics Subject Classification 51B20 · 51F99 · 51K05 · 51K99 · 51N25

4.1 The Group of Isometries of a Metric Space

Let (X, d) be a metric space. The map $f : X \to X$ is called an *isometry* with respect to the metric d (or a d-*isometry*), if f is surjective and it preserves the distances. That is, for any points $x, y \in X$ the relation $d(f(x), f(y)) = d(x, y)$ holds. From this relation, it follows that f is injective, hence it is bijective. Denote by $Iso_d(X)$ the set of all isometries of the metric space (X, d). It is clear that $(Iso_d(X), \circ)$ is a subgroup of $(S(X), \circ)$, where $S(X)$ denotes the group of all bijective transformations $f : X \to X$. We will call $(Iso_d(X), \circ)$ the *group of isometries* of the metric

Dedicated to Professor Themistocles M. Rassias on the occasion of his 60th birthday.

D. Andrica (✉) · V. Bulgarean
Faculty of Mathematics and Computer Science, "Babeş-Bolyai" University, Cluj-Napoca, Romania
e-mail: dandrica@math.ubbcluj.ro

V. Bulgarean
e-mail: vasilebulgarean@yahoo.com

D. Andrica
Department of Mathematics, College of Science, King Saud University, Riyadh, Saudi Arabia

P.M. Pardalos et al. (eds.), *Nonlinear Analysis*, Springer Optimization and Its Applications 68, In Honor of Themistocles M. Rassias on the Occasion of his 60th Birthday,
DOI 10.1007/978-1-4614-3498-6_4,

space (X, d). A general, important, and complicated problem is to described the group $(Iso_d(X), \circ)$. This problem was formulated in [3] for metric spaces with a metric that is not given by a norm.

Some results towards a solution to this problem are the following. In [27], D.J. Schattschneider found an elementary proof for the property that the group $Iso_{d_1}(\mathbb{R}^2)$ is the semi-direct product of D_4 and $T(2)$, where d_1 is the "Taxicab metric" defined by (4.2) (for $n = 2$ and $p = 1$) and D_4 and $T(2)$ are the symmetry group of the square and the group of translations of $\mathbb{R}^2$, respectively. A similar result holds for the group $Iso_{d_1}(\mathbb{R}^3)$, i.e., this group is isomorphic to the semi-direct product of the groups D_h and $T(3)$, where D_h is the symmetry group of the Euclidean octahedron and $T(3)$ is the group of translations of $\mathbb{R}^3$. This was recently proved by O. Gelisgen, R. Kaya [7]. In fact, the "Taxicab metric" generates many interesting non-Euclidean geometric properties (see the book of E.F. Krause [11] and the papers of G. Chen [4], R. Kaya [9], M. Ozcan and R. Kaya [12]). Another result concerning the isometry group of the plane $\mathbb{R}^2$ with respect to the "Chinese Checker Metric" d_C, where

$$d_C(x, y) = \max\{|x^1 - x^2|, |y^1 - y^2|\} + (\sqrt{2} - 1)\min\{|x^1 - x^2|, |y^1 - y^2|\}, \quad (4.1)$$

was recently obtained by R. Kaya, O. Gelisgen, S. Ekmekci, and A. Bayar [10]. They have showed that this group is isomorphic to the semi-direct product of the dihedral group D_8, the Euclidean symmetry group of a regular octagon, and $T(2)$.

Another interesting problem involving the isometry group of a metric space (X, d) is the following: If $f : X \to X$ is a continuous mapping satisfying the so-called distance 1 preserving property, i.e., $d(x, y) = 1$ implies $d(f(x), f(y)) = 1$, is it then necessarily true that $f \in Iso_d(X)$? This problem concerns the minimal conditions for a mapping $f : X \to X$ to be an isometry of X. In the case of normed spaces, it is connected to the famous Aleksandrov–Rassias problem, and for more details we refer to the papers of S.-M. Jung and Th.M. Rassias [8], B. Mielnik and Th.M. Rassias [14], Th.M. Rassias [18–22], C.-G. Park and Th.M. Rassias [16, 17], Th.M. Rassias and P. Semrl [23], and Th.M. Rassias and S. Xiang [24, 25].

In this paper, we consider $X = \mathbb{R}^n$, and for any real number $p \geq 1$ we define the metric d_p by

$$d_p(x, y) = \left(\sum_{i=1}^{n} |x^i - y^i|^p\right)^{1/p}, \quad (4.2)$$

where $x = (x^1, \ldots, x^n)$, $y = (y^1, \ldots, y^n) \in \mathbb{R}^n$. If $p = \infty$, then the metric d_∞ is defined by

$$d_\infty(x, y) = \max\{|x^1 - y^1|, \ldots, |x^n - y^n|\}. \quad (4.3)$$

In the case $p = 2$, we get the well-known Euclidean metric on $\mathbb{R}^n$. In this case, we have the Ulam's Theorem which states that $Iso_{d_2}(\mathbb{R}^n)$ is isomorphic to the semi-direct product of the orthogonal group $O(n)$ and $T(n)$, where $T(n)$ is the group of translations of $\mathbb{R}^n$ (see, for instance, [5, 26]). The situation $p \neq 2$ is very interesting. The main purpose of this paper is to find the groups $Iso_{d_p}(\mathbb{R}^n)$ for $p \geq 1$

and $p=\infty$. We will prove that in the case $p \neq 2$ all these groups are isomorphic and, consequently, they are independent of p. In the last section, we introduce the isometry dimension of a finite group with respect to a given metric on the space $\mathbb{R}^n$. M. Albertson and D. Boutin [1] have introduced this notion considering the Euclidean n-space, i.e., in this case $d=d_2$, the Euclidean metric. An interesting approach to this case was given by M.M. Patnaik [15].

4.2 The Group $Iso_{d_p}(\mathbb{R}^n)$, $p \neq 2$

In this section, we denote by $S_{d_p}^{n-1}$ the unit sphere of the metric space $(\mathbb{R}^n, d_p)$. The sphere $S_{d_p}^{n-1}$ is defined by

$$S_{d_p}^{n-1} = \{x \in \mathbb{R}^n : |x^1|^p + \cdots + |x^n|^p = 1\}. \tag{4.4}$$

For $p=\infty$, the unit sphere is

$$S_{d_\infty}^{n-1} = \{x \in \mathbb{R}^n : \max\{|x^1|, \ldots, |x^n|\} = 1\}. \tag{4.5}$$

Our main results are the following.

Theorem 4.1 *Let p be a real number, $p \geq 1$ or $p=\infty$. If $f \in Iso_{d_p}(\mathbb{R}^n)$, then f is an affine map.*

Proof The property directly follows from the well-known result of S. Mazur and S. Ulam (see the original reference [13]): Every isometry $f: E \to F$ between real normed spaces is affine. In this case, an isometry is a surjective map satisfying for any $x, y \in E$ the relation $\|f(x)-f(y)\|_F = \|x-y\|_E$. This result was proved by S. Mazur and S. Ulam in 1932. A simple proof was given by J. Väisälä [28], it is based on the ideas of A. Vogt [29] and makes use of reflections at points. We have just to apply this result for the normed spaces $E=F=\mathbb{R}^n$ with the norm $\|\cdot\|_p$ defined by $\|x\|_p = (\sum_{i=1}^n |x^i|^p)^{1/p}$. □

Theorem 4.2 *Let $p \neq 2$ be a real number, $p \geq 1$, and let $f_A : \mathbb{R}^n \to \mathbb{R}^n$ be the linear map defined by the matrix $A \in M_n(\mathbb{R})$. Then $f_A \in Iso_{d_p}(\mathbb{R}^n)$ if and only if A is a permutation matrix, i.e., each row and each column of A has exactly one non-zero entry and this entry is equal to ± 1.*

Proof Let $\|\cdot\|_p$ be the p-norm defined on $\mathbb{R}^n$ by the metric d_p. It is clear that if the matrix A has exactly one non-zero entry which is equal to ± 1, then the linear map f_A satisfies the relation $\|f_A(x)-f_A(y)\|_p = \|x-y\|_p$, that is, f_A is an isometry with respect to the metric d_p.

Conversely, let $A=(a_{ij})$ be the matrix of the linear map f_A and assume that f_A belongs to $Iso_{d_p}(\mathbb{R}^n)$. Because $\|f_A(x)-f_A(0)\|_p = \|x-0\|_p$ and f_A is linear, we get that for any $x \in \mathbb{R}^n$ the relation $\|f_A(x)\|_p = \|x\|_p$ holds. The last relation shows that if $x \in S_{d_p}^{n-1}$, then $f_A(x) \in S_{d_p}^{n-1}$. That is, $Ax^t \in S_{d_p}^{n-1}$, where x^t denotes the transpose of vector $x \in \mathbb{R}^n$. Let $e_1, \ldots, e_n$ be the canonical basis of the space $\mathbb{R}^n$. It is clear that $e_i \in S_{d_p}^{n-1}$ implies $Ae_i^t \in S_{d_p}^{n-1}$, for all $i = 1, \ldots, n$. The last relation is equivalent to

$$\sum_{k=1}^{n} |a_{ki}|^p = 1, \tag{4.6}$$

for all $i = 1, \ldots, n$.

On the other hand, for $i \neq j$, since $e_i, e_j \in S_{d_p}^{n-1}$, we have $2^{\frac{-1}{p}}(\pm e_i \pm e_j) \in S_{d_p}^{n-1}$, hence we get $2^{\frac{-1}{p}} A(\pm e_i \pm e_j)^t \in S_{d_p}^{n-1}$. The last relation shows that for any $i \neq j$ we have

$$\sum_{k=1}^{n} |\pm a_{ki} \pm a_{kj}|^p = 2, \tag{4.7}$$

for any choice of signs $+$ and $-$. It follows that for any $i \neq j$ and for any choice of signs $+$ and $-$, the relation

$$\sum_{k=1}^{n} \left(|\pm a_{ki} \pm a_{kj}|^p - |a_{ki}|^p - |a_{kj}|^p\right) = 0, \tag{4.8}$$

holds. But, if $u, v \geq 0$, then we have the inequality $|u+v|^p \geq |u|^p + |v|^p$, with equality if and only if $uv = 0$. Indeed, if $u + v \neq 0$, then the inequality is equivalent to $1 \geq (\frac{u}{u+v})^p + (\frac{u}{u+v})^p$. The last inequality can be reduced to $1 \geq t^p + (1-t)^p$, where $t = \frac{u}{u+v} \in [0, 1]$, which is clear since $p \geq 1$. If a_{ik} and a_{jk} are both positive, then we choose the signs $+$ and we can apply the previous inequality and get $|a_{ki} + a_{kj}|^p - |a_{ki}|^p - |a_{kj}|^p \geq 0$. If, for instance, $a_{ik} > 0$ and $a_{jk} < 0$, then we choose the signs $+$ and $-$ and we can write the corresponding term of the sum as $|a_{ki} - a_{kj}|^p - |a_{ki}|^p - |-a_{kj}|^p \geq 0$. In any case, for suitable choices of the signs $+$ and $-$, we can obtain all terms of the sum to be positive. Therefore, for any $i \neq j$ and for the corresponding signs $+$ and $-$, we get $|\pm a_{ki} \pm a_{kj}|^p - |a_{ki}|^p - |a_{kj}|^p = 0$. It follows that $a_{ik}a_{jk} = 0$, for $k = 1, \ldots, n$ and for every pair of distinct indices i and j.

It is clear that each row has some non-zero entry to infer that on each row and each column there must be exactly one non-zero entry. This non-zero entry must be ± 1. Consequently, the rows of the matrix A are a permutation of the $\pm e_i$, $i = 1, \ldots, n$, with signs chosen arbitrarily. The total number of such matrices is $2^n n!$ □

4.3 The Group $Iso_{d_\infty}(\mathbb{R}^n)$

At the other extreme, following our paper [2], if we consider the metric d_∞, then the induced norm on $\mathbb{R}^n$ is given by

$$\|x\|_\infty = \max\{|x^1|, \ldots, |x^n|\}, \tag{4.9}$$

where $x = (x^1, \ldots, x^n) \in \mathbb{R}^n$. We have the following duality relation involving the norms $\|\cdot\|_\infty$ and $\|\cdot\|_1$:

$$\|x\|_\infty = \max\{|\langle x, y\rangle| : y \in \mathbb{R}^n, \|y\|_1 = 1\}, \tag{4.10}$$

where $\langle\cdot,\cdot\rangle$ denotes the standard inner product in $\mathbb{R}^n$ inducing the Euclidean norm $\|\cdot\|_2$. Indeed, it is obvious that $|\langle x, y\rangle| = |x^1y^1 + \cdots + x^ny^n| \le \|x\|_\infty\|y\|_1$, and this shows that $\max\{|\langle x, y\rangle| : y \in \mathbb{R}^n, \|y\|_1 = 1\} \le \|x\|_\infty$. For the converse inequality, we note that for any $j = 1, \ldots, n$ the following inequality holds: $\max\{|\langle x, y\rangle| : y \in \mathbb{R}^n, \|y\|_1 = 1\} \ge |\langle x, e_j\rangle| = |x^j|$, hence we get $\max\{|\langle x, y\rangle| : y \in \mathbb{R}^n, \|y\|_1 = 1\} \ge \max\{|x^1|, \ldots, |x^n|\} = \|x\|_\infty$, and we are done.

Now, the relation (4.10) shows that if the linear map f_A preserves the norm $\|\cdot\|_\infty$, then the linear map f_{A^t} preserves the norm $\|\cdot\|_1$, where A^t denotes the transpose of the matrix A. Assume that $A = (a_{ij})$ and apply this property to the vectors $e_1, \ldots, e_n$ of the canonical basis. We get the relations $\max\{|a_{1j}|, \ldots, |a_{nj}|\} = 1$, for $j = 1, \ldots, n$, and $|a_{i1}| + \cdots + |a_{in}| = 1$, for $i = 1, \ldots, n$. Adding the last relations we obtain

$$(|a_{11}| + \cdots + |a_{n1}|) + \cdots + (|a_{1n}| + \cdots + |a_{nn}|) = n. \tag{4.11}$$

But, the relations $\max\{|a_{1j}|, \ldots, |a_{nj}|\} \ge 1$ for $j = 1, \ldots, n$, shows that $|a_{11}| + \cdots + |a_{n1}| \ge 1, \ldots, |a_{1n}| + \cdots + |a_{nn}| \ge 1$. It follows that each term $(|a_{i1}| + \cdots + |a_{in}|)$ in the sum (4.11) contains exactly one term equal to 1 and all other terms are equal to 0. Therefore, the matrix A is also a permutation matrix having the non-zero entries equal to ± 1, and we are done.

The above considerations show that the result in Theorem 4.2 also holds for the metric d_∞.

4.4 Common Conclusions for $Iso_{d_p}(\mathbb{R}^n)$ and $Iso_{d_\infty}(\mathbb{R}^n)$

Considering together the results of Theorem 4.2 and of Sect. 4.3, we obtain:

Theorem 4.3 *Let $p \ne 2$ be a real number, $p \ge 1$ or $p = \infty$. The isometry group $Iso_{d_p}(\mathbb{R}^n)$ is isomorphic to $(S_n \cdot (\mathbb{Z}_2)^n) \cdot T(n)$, where $T(n)$ is the group of translations of the space $\mathbb{R}^n$, S_n denotes the group of permutations of the set $\{1, \ldots, n\}$, and "$\cdot$" denotes the semi-direct product.*

The subgroup of linear isometries of $Iso_{d_p}(\mathbb{R}^2)$ consists of the eight linear maps defined by the following matrices:

$$\begin{bmatrix} 1 & 0 \\ 0 & 1 \end{bmatrix}; \quad \begin{bmatrix} -1 & 0 \\ 0 & 1 \end{bmatrix}; \quad \begin{bmatrix} 1 & 0 \\ 0 & -1 \end{bmatrix}; \quad \begin{bmatrix} -1 & 0 \\ 0 & -1 \end{bmatrix};$$
$$\begin{bmatrix} 0 & 1 \\ 1 & 0 \end{bmatrix}; \quad \begin{bmatrix} 0 & -1 \\ 1 & 0 \end{bmatrix}; \quad \begin{bmatrix} 0 & 1 \\ -1 & 0 \end{bmatrix}; \quad \begin{bmatrix} 0 & -1 \\ -1 & 0 \end{bmatrix}.$$

These linear maps define all the symmetries of the unit sphere $S^1_{d_p}$. For instance, for $p = 1$ the sphere $S^1_{d_1}$ is the boundary of the square with vertices $(1, 0)$, $(-1, 0)$, $(0, 1)$, $(0, -1)$.

The subgroup of linear isometries of $Iso_{d_p}(\mathbb{R}^3)$ consists of the 48 linear maps defined by the corresponding matrices described in Theorem 4.2. Also, these linear maps give all the symmetries of the unit sphere $S^2_{d_p}$. For $p = 1$, the sphere $S^2_{d_1}$ is the boundary of the octahedron with vertices $(\pm 1, 0, 0)$, $(0, \pm 1, 0)$, $(0, 0, \pm 1)$.

The subgroup of linear isometries of $Iso_{d_\infty}(\mathbb{R}^n)$ consists of the $2^n n!$ linear maps defined by the permutation matrices in Theorem 4.2. These maps give all the symmetries of the sphere $S^{n-1}_{d_\infty}$ which is the boundary of the n-cube with vertices at $(\pm 1, \dots, \pm 1)$.

Remark Any linear isometry $f \in Iso_{d_1}(\mathbb{R}^n)$ is a simplicial map on $S^{n-1}_{d_1}$, that is, it maps vertices to vertices, edges to edges, faces to faces, etc. The same property holds for the group $Iso_{d_\infty}(\mathbb{R}^n)$.

4.5 The d-Isometry Dimension of a Finite Group

Let G be a finite group. We call the *d-isometry dimension* of G the last n such that the group can be realized as the group of isometries of a subset of $\mathbb{R}^n$, where d is a given metric on $\mathbb{R}^n$. Let us denote this number by $\delta_d(G)$. M. Albertson and D. Boutin [1] have introduced this notion considering a subset of the Euclidean n-space, i.e., in this case $d = d_2$, the Euclidean metric. In the same paper, M. Albertson and D. Boutin have proved that any group of order n can be realized by a finite subset of the Euclidean n-space containing $n + \binom{n}{2}$ points. In fact, they have proved the inequality $\delta_{d_2}(G) \le |G| - 1$, where $|G|$ denotes the order of group G. M.M. Patnaik [15] has proved the following interesting result (see the book of M. Willard Jr. [30] for basic results concerning the representations of finite groups):

Theorem 4.4 *Let G be a finite group. Then the d_2-isometry dimension $\delta_{d_2}(G)$ is equal to the dimension of a minimal-dimensional faithful real representation of G.*

As a consequence of Theorem 4.4, in [15], the following result is proved:

Corollary 4.1 *Let $G_1, \ldots, G_s$ be finite groups. Then*

$$\delta_{d_2}(G_1 \oplus \cdots \oplus G_s) \leq \delta_{d_2}(G_1) + \cdots + \delta_{d_2}(G_s).$$

Equality cannot always hold, as we can see by taking $s = 2$, $G_1 = \mathbb{Z}_{m_1}$, $G_2 = \mathbb{Z}_{m_2}$, cyclic groups of relatively prime orders m_1 and m_2.

Also in [15], all finite groups with the d_2-isometry dimension equal to 2 or 3 were determined by using the following argument. Let $O_n(\mathbb{R})$ be the group of orthogonal matrices of dimension n. A finite group G has d_2-isometry dimension n if and only if it is isomorphic to a subgroup of $O_n(\mathbb{R})$, and it is not isomorphic to any subgroup of $O_m(\mathbb{R})$ for $m < n$. It follows that the only finite groups with d_2-isometry dimension 2 are cyclic groups $\mathbb{Z}_s$ and the dihedral groups D_s of order $2s$, for $s > 2$. The finite subgroups of $O_3(\mathbb{R})$ are listed in the book of H. Coxeter [6].

Other interesting results involving the computation of the d_2-isometry dimension of some concrete groups are given in [1]. For instance, $\delta_{d_2}(\mathbb{Z}_4 \times \mathbb{Z}_2) = 3$ and $\delta_{d_2}(\mathscr{Q}) \geq 4$, where $\mathscr{Q} = \{\pm 1, \pm i, \pm j, \pm k\}$ denotes the quaternions. In [15], it is mention without proof that $\delta_{d_2}(\mathbb{Z}_2^m) = m$.

As a consequence of our main result about the isometry groups $Iso_{d_p}(\mathbb{R}^n)$ and $Iso_{d_\infty}(\mathbb{R}^n)$, we obtain:

Theorem 4.5 *Let $p \neq 2$ be a real number, $p \geq 1$ or $p = \infty$. The following relations hold*:

$$\delta_{d_p}\big(S_n \cdot (\mathbb{Z}_2)^n\big) \leq n, \tag{4.12}$$

where S_n denotes the group of permutations of the set $\{1, \ldots, n\}$, and "$\cdot$" denotes the semi-direct product.

We can ask a few questions in the same direction as in [1], but for δ_{d_p} instead of δ_{d_2}. We mention here only two problems involving the d-isometry dimension of a finite group.

Problem 4.1 *Let $p \neq 2$ be a real number, $p \geq 1$ or $p = \infty$. Find $\delta_{d_p}(\mathbb{Z}_2^n)$.*

Problem 4.2 *Let $p \neq 2$ be a real number, $p \geq 1$ or $p = \infty$. It is true that*

$$\delta_{d_p}\big(S_n \cdot (\mathbb{Z}_2)^n\big) = n?$$

Acknowledgement The first author is supported by the King Saud University D.S.F.P. Program.

References

1. Albertson, M., Boutin, D.: Realizing finite groups in Euclidean spaces. J. Algebra **225**, 947–955 (2001)
2. Andrica, D., Bulgarean, V.: Note on the group of isometries $Iso_{d_\infty}(\mathbb{R}^n)$. Acta Univ. Apulensis (to appear)

3. Andrica, D., Wiesler, H.: On the isometry groups of a metric space. Semin. Didact. Mat. **5**, 1–4 (1989)
4. Chen, G.: Lines and circles in taxicab geometry. Master thesis, Department of Mathematics and Computer Science, Central Missouri State University (1992)
5. Clayton, W.D.: Euclidean Geometry and Transformations. Addison-Wesley, Reading (1972)
6. Coxeter, H.: Introduction to Geometry. Wiley, New York (1969)
7. Gelisgen, O., Kaya, R.: The taxicab space group. Acta Math. Hung. **122**(1–2), 187–200 (2009)
8. Jung, S.-M., Rassias, Th.M.: On distance-preserving mappings. J. Korean Math. Soc. **41**(4), 667–680 (2004)
9. Kaya, R.: Area formula for taxicab triangles. Pi Mu Epsilon **12**, 213–220 (2006)
10. Kaya, R., Gelisgen, O., Ekmekci, S., Bayar, A.: On the group of the isometries of the plane with generalized absolute metric. Rocky Mt. J. Math. **39**(2) (2009)
11. Krause, E.F.: Taxicab Geometry. Addison-Wesley, Menlo Park (1975)
12. Ozcan, M., Kaya, R.: Area of a triangle in terms of the taxicab distance. Mo. J. Math. Sci. **15**, 178–185 (2003)
13. Mazur, S., Ulam, S.: Sur les transformationes isométriques d'espaces vectoriels normes. C. R. Acad. Sci. Paris **194**, 946–948 (1932)
14. Mielnik, B., Rassias, Th.M.: On the Aleksandrov problem of conservative distances. Proc. Am. Math. Soc. **116**, 1115–1118 (1992)
15. Patnaik, M.M.: Isometry dimension of finite groups. J. Algebra **246**, 641–646 (2001)
16. Park, C.-G., Rassias, Th.M.: The N-isometric isomorphisms in linear n-normed C^*-algebras. Acta Math. Sin. Engl. Ser. **22**(6), 1863–1890 (2006)
17. Park, C.-G., Rassias, Th.M.: Isometries on linear n-normed spaces. J. Inequal. Pure Appl. Math. **7**(5), 168 (2006), 7 pp.
18. Rassias, Th.M.: Is a distance one preserving mapping between metric spaces always an isometry? Am. Math. Mon. **90**, 200 (1983)
19. Rassias, Th.M.: Properties of isometric mappings. J. Math. Anal. Appl. **235**(1), 108–121 (1999)
20. Rassias, Th.M.: Isometries and approximate isometries. Int. J. Math. Math. Sci. **25**(2), 73–91 (2001)
21. Rassias, Th.M.: On the A.D. Aleksandrov problem of conservative distances and the Mazur–Ulam theorem. Nonlinear Anal. **47**(4), 2597–2608 (2001)
22. Rassias, Th.M.: On the Aleksandrov problem for isometric mappings. Appl. Anal. Discrete Math. **1**, 18–28 (2007)
23. Rassias, Th.M., Semrl, P.: On the Mazur–Ulam theorem and the Aleksandrov problem for unit distance preserving mappings. Proc. Am. Math. Soc. **118**, 919–925 (1993)
24. Rassias, Th.M., Xiang, S.: On Mazur–Ulam theorem and mappings which preserve distances. Nonlinear Funct. Anal. Appl. **5**(2), 61–66 (2000)
25. Rassias, Th.M., Xiang, S.: On approximate isometries in Banach spaces. Nonlinear Funct. Anal. Appl. **6**(2), 291–300 (2001)
26. Richard, S.M., George, D.P.: Geometry. A Metric Approach with Models. Springer, New York (1981)
27. Schattschneider, D.J.: The taxicab group. Am. Math. Mon. **91**, 423–428 (1984)
28. Väisälä, J.: A proof of the Mazur–Ulam theorem. Am. Math. Mon. **110**(7), 633–635 (2003)
29. Vogt, A.: Maps which preserve equality of distance. Studia Math. **45**, 43–48 (1973)
30. Willard, M. Jr.: Symmetry Groups and Their Applications. Academic Press, New York (1972)

Chapter 5
Rationality of the Moduli Space of Stable Pairs over a Complex Curve

Indranil Biswas, Marina Logares, and Vicente Muñoz

Abstract Let X be a smooth complex projective curve of genus $g \geq 2$. A pair on X is formed by a vector bundle $E \to X$ and a global non-zero section $\phi \in H^0(E)$. There is a concept of stability for pairs depending on a real parameter τ, giving rise to moduli spaces $\mathscr{M}_\tau(r, \Lambda)$ of τ-stable pairs of rank r and fixed determinant Λ. In this paper, we prove that the moduli spaces $\mathscr{M}_\tau(r, \Lambda)$ are in many cases rational.

Key words Moduli of pairs · Vortex equation · Rationality · Stable rationality

Mathematics Subject Classification 14D20 · 58D27 · 14EM20

5.1 Introduction

Let X be a compact connected Riemann surface of genus $g \geq 2$, which we may interpret as a smooth projective complex curve. Fix a Kähler form ω on X. Consider a C^∞ Hermitian vector bundle $E \to X$ of rank r and degree d (cf. [9]). A unitary connection A on E endows it with a holomorphic structure $\bar{\partial}_A$, given by the $(0, 1)$-part of $A = \partial_A + \bar{\partial}_A$. The connection is said to be *Hermitian–Einstein*, or *Hermitian–*

Dedicated to Professor Themistocles M. Rassias on the occasion of his 60th birthday.

I. Biswas
School of Mathematics, Tata Institute of Fundamental Research, Homi Bhabha Road, Bombay 400005, India
e-mail: indranil@math.tifr.res.in

M. Logares
Instituto de Ciencias Matemáticas (CSIC-UAM-UC3M-UCM), C/Nicolas Cabrera 15, 28049 Madrid, Spain
e-mail: marina.logares@icmat.es

V. Muñoz (✉)
Facultad de Matemáticas, Universidad Complutense de Madrid, Plaza Ciencias 3, 28040 Madrid, Spain
e-mail: vicente.munoz@mat.ucm.es

P.M. Pardalos et al. (eds.), *Nonlinear Analysis*, Springer Optimization and Its Applications 68, In Honor of Themistocles M. Rassias on the Occasion of his 60th Birthday, DOI 10.1007/978-1-4614-3498-6_5,

Yang–Mills, if

$$F_A = c \cdot \mathrm{Id} \cdot \omega.$$

The constant c is constrained by the topology to be $c = \frac{-2\pi\sqrt{-1}}{\mathrm{Vol}_\omega(X)}\frac{d}{r}$. This provides a link between gauge theory and the theory of holomorphic bundles. The fundamental theorem given by the Hitchin–Kobayashi correspondence establishes that a holomorphic structure $\bar{\partial}$ on E arises from a (unique up to unitary gauge automorphism of the bundle) connection A if and only if the holomorphic vector bundle $(E, \bar{\partial})$ is *polystable*; the definition of polystability is recalled below.

For a holomorphic bundle E, we define the slope $\mu(E) := d/r$, where d is its degree and r its rank. We say that E is stable if $\mu(E') < \mu(E)$ for all *holomorphic* proper subbundles $E' \subset E$. A vector bundle E is polystable if it is a direct sum of stable vector bundles of the same slope.

Of much interest is the extension to the case of pairs (E, ϕ) formed by a Hermitian vector bundle E together with a global smooth section $\phi \in \Gamma(E)$. In this case, we look for unitary connections A satisfying a *vortex equation*

$$\begin{cases} \frac{2}{\sqrt{-1}} F_A = (\phi \otimes \phi^* - \tau \cdot \mathrm{Id})\omega, \\ \bar{\partial}_A \phi = 0, \end{cases} \tag{5.1}$$

where ϕ^* is the adjoint of ϕ with respect to the unitary metric, so that $\phi \otimes \phi^* \in \Gamma(\mathrm{End}\, E)$, and τ is a real parameter. In this situation, τ is not constrained. The pair (A, ϕ) induces a holomorphic structure $\bar{\partial}_A$ on E, and ϕ is a *holomorphic section* of the holomorphic vector bundle $(E, \bar{\partial}_A)$.

A holomorphic pair (also called a Bradlow pair) over X is a pair (E, ϕ), where $E \longrightarrow X$ is a holomorphic vector bundle, and $\phi \in H^0(E)$, i.e., a holomorphic section. There is a notion of stability of pairs depending on a parameter $\tau \in \mathbb{R}$. A holomorphic pair is τ-stable whenever the following conditions are satisfied:

- For any subbundle $E' \subset E$, we have $\mu(E') < \tau$;
- For any subbundle $E' \subset E$ such that $\phi \in H^0(E')$, we have $\mu(E/E') > \tau$.

A holomorphic pair is said to be τ-semistable if in the above definition the weak inequalities hold instead of the strict ones. A τ-semistable pair is τ-polystable if it is the direct sum of a τ-stable pair and a polystable vector bundle.

The link between this algebraic–geometric concept and gauge theory comes from a Hitchin–Kobayashi correspondence which establishes that the solutions to (5.1) correspond to polystable holomorphic pairs.

There has been much interest in holomorphic pairs in the last 15 years. A moduli space of τ-stable pairs is the space which parameterizes these objects. There is an algebraic construction of it, [2], which gives it the structure of a quasi-projective complex variety. Let $\mathcal{M}_\tau(r, d)$ be the moduli space of τ-stable pairs of rank r and degree d. For any line bundle Λ over X of degree d, we denote by $\mathcal{M}_\tau(r, \Lambda)$ the moduli space of τ-stable pairs of rank r and fixed determinant $\bigwedge^r E = \Lambda$.

A large number of topological and geometrical properties of $\mathcal{M}_\tau(r, \Lambda)$ have been studied in the literature. In [10], a construction of the moduli space is given using gauge theoretic techniques (more precisely, by reducing the vortex equation to the Hermitian–Einstein equation on a complex surface). This gives the general properties about the possible values of τ for non-emptiness, smoothness, etc. of the moduli space. Thaddeus [18] studied thoroughly the case of rank $r = 2$, computing the Poincaré polynomial of the moduli space and describing its topology quite explicitly. Later this was extended in [17] to compute the Hodge polynomials, and in [14] to rank $r = 3$. The general properties of the mixed Hodge structures of $\mathcal{M}_\tau(r, \Lambda)$ are found in [16]. In [15], a Torelli-type theorem for the moduli spaces $\mathcal{M}_\tau(r, \Lambda)$ is proved; this amounts to the following: the algebraic structure of the moduli space allows one to recover the complex structure of X. We also mention that in [4], the authors compute the Brauer group of these moduli spaces.

The focus of the present paper is another geometrical property of $\mathcal{M}_\tau(r, \Lambda)$, namely the *rationality*. A variety Z is rational if there is a birational rational map $Z \dashrightarrow \mathbb{P}^N$, where $\mathbb{P}^N$ is the complex projective space of dimension N. A birational rational map is an isomorphism between two Zariski open subsets of both spaces. We denote $Z \sim \mathbb{P}^N$.

Let us state now the main results of this paper.

A variety Z is called stably rational if $Z \times \mathbb{P}^n$ is rational for some n, so $Z \times \mathbb{P}^n \sim \mathbb{P}^N$. This notion is weaker than rationality. We have the following result, which we prove in Sect. 5.3.

Theorem 5.1 *Suppose* $(r, g, d) \neq (3, 2, \text{even})$. *Then the variety* $\mathcal{M}_\tau(r, \Lambda)$ *is stably rational for any* τ.

Regarding the rationality of the moduli space of pairs, we have the following, which is proved in Sects. 5.4 and 5.5.

Theorem 5.2 *Let* X *be a smooth complex irreducible and projective curve of genus* $g \geq 2$. *Then, for any* $\tau \in \mathbb{R}$, *rank* r *and line bundle* Λ *of degree* $d > 0$ *over* X, *the moduli space* $\mathcal{M}_\tau(r, \Lambda)$ *is rational in the following cases*:

- $d > rg$,
- $\gcd(r - 1, d) = 1$,
- $\gcd(r, d) = 1, d > r(g - 1)$.

This result is related to the work of Hoffman [11], where by very different techniques, there are some general results which prove the rationality of most moduli spaces $\mathcal{M}_\tau(r, \Lambda)$. The novelty of the proof given here lies in the fact that it uses only elementary techniques.

5.2 Moduli Spaces of Pairs

We collect here some known results about the moduli spaces of pairs; the details can be found in [6, 7, 15, 17], and [18].

Let X be an irreducible smooth projective curve, defined over the field of complex numbers, of genus $g \geq 2$. A *holomorphic pair* (E, ϕ) over X consists of a holomorphic bundle on X and a nonzero holomorphic section $\phi \in H^0(E)$. Let $\mu(E) := \deg(E)/rk(E)$ be the slope of E. Take any $\tau \in \mathbb{R}$. A holomorphic pair (E, ϕ) is called *τ-stable* (respectively, *τ-semistable*) whenever the following conditions are satisfied:

- For any nonzero proper subbundle $E' \subset E$, we have $\mu(E') < \tau$ (respectively, $\mu(E') \leq \tau$);
- For any proper subbundle $E' \subset E$ such that $\phi \in H^0(E')$, we have $\mu(E/E') > \tau$ (respectively, $\mu(E/E') \geq \tau$).

A *critical value* of the parameter $\tau = \tau_c$ is one for which there are strictly τ-semistable pairs. There are only finitely many critical values.

Fix an integer $r \geq 2$, and also fix a holomorphic line bundle Λ over X. Let d be the degree of Λ. We denote by $\mathcal{M}_\tau(r, \Lambda)$ (respectively, $\overline{\mathcal{M}}_\tau(r, \Lambda)$) the moduli space of τ-stable (respectively, τ-polystable) pairs (E, ϕ) of rank $rk(E) = r$ and determinant $\det(E) = \Lambda$. The moduli space $\overline{\mathcal{M}}_\tau(r, \Lambda)$ is a normal projective variety, and $\mathcal{M}_\tau(r, \Lambda)$ is a smooth quasi-projective variety contained in the smooth locus of $\overline{\mathcal{M}}_\tau(r, \Lambda)$.

For non-critical values of the parameter, there are no strictly τ-semistable pairs, so $\mathcal{M}_\tau(r, \Lambda) = \overline{\mathcal{M}}_\tau(r, \Lambda)$, and it is a smooth projective variety. For a critical value τ_c, the variety $\overline{\mathcal{M}}_{\tau_c}(r, \Lambda)$ is in general singular.

Denote $\tau_m := \frac{d}{r}$ and $\tau_M := \frac{d}{r-1}$. The moduli space $\mathcal{M}_\tau(r, \Lambda)$ is empty for $\tau \notin (\tau_m, \tau_M)$. In particular, this forces $d > 0$ for τ-stable pairs. Denote by $\tau_1 < \tau_2 < \cdots < \tau_L$ the collection of all critical values in (τ_m, τ_M). Then the moduli spaces $\mathcal{M}_\tau(r, \Lambda)$ are isomorphic for all values τ in any interval (τ_i, τ_{i+1}), $i = 0, \ldots, L$; here $\tau_0 = \tau_m$ and $\tau_{L+1} = \tau_M$.

However, the moduli space changes when we cross a critical value. Let τ_c be a critical value. Denote $\tau_c^+ := \tau_c + \varepsilon$ and $\tau_c^- := \tau_c - \varepsilon$ for $\varepsilon > 0$ small enough such that (τ_c^-, τ_c^+) does not contain any critical value other than τ_c. We define the *flip loci* $\mathcal{S}_{\tau_c^\pm}$ as the subschemes:

- $\mathcal{S}_{\tau_c^+} = \{(E, \phi) \in \mathcal{M}_{\tau_c^+}(r, \Lambda) \mid (E, \phi) \text{ is } \tau_c^-\text{-unstable}\}$,
- $\mathcal{S}_{\tau_c^-} = \{(E, \phi) \in \mathcal{M}_{\tau_c^-}(r, \Lambda) \mid (E, \phi) \text{ is } \tau_c^+\text{-unstable}\}$.

When crossing τ_c, the variety $\mathcal{M}_\tau(r, \Lambda)$ undergoes a birational transformation:

$$\mathcal{M}_{\tau_c^-}(r, \Lambda) - \mathcal{S}_{\tau_c^-} = \mathcal{M}_{\tau_c}(r, \Lambda) = \mathcal{M}_{\tau_c^+}(r, \Lambda) - \mathcal{S}_{\tau_c^+}.$$

Proposition 5.1 ([14, Proposition 5.1]) *Suppose $r \geq 2$, and let τ_c be a critical value with $\tau_m < \tau_c < \tau_M$. Then*

- $\operatorname{codim} \mathcal{S}_{\tau_c^+} \geq 3$ *except in the case $r = 2$, $g = 2$, d odd and $\tau_c = \tau_m + \frac{1}{2}$ (in which case* $\operatorname{codim} \mathcal{S}_{\tau_c^+} = 2$),
- $\operatorname{codim} \mathcal{S}_{\tau_c^-} \geq 2$ *except in the case $r = 2$ and $\tau_c = \tau_M - 1$ (in which case* $\operatorname{codim} \mathcal{S}_{\tau_c^-} = 1$). *Moreover, we have that* $\operatorname{codim} \mathcal{S}_{\tau_c^-} = 2$ *only for $\tau_c = \tau_M - 2$.*

The codimension of the flip loci is then always positive, hence we have the following corollary:

Corollary 5.1 *The moduli spaces* $\mathscr{M}_\tau(r,\Lambda)$, $\tau \in (\tau_m, \tau_M)$, *are birational.*

For a complex vector space V, by $\mathbb{P}(V)$ we will denote the projective space parameterizing lines in V.

The moduli spaces for the extreme values τ_m^+ and τ_M^- of the parameter are known explicitly. Let $M(r,\Lambda)$ be the moduli space of *stable* vector bundles or rank r and fixed determinant Λ. Define

$$\mathscr{U}_m(r,\Lambda) = \{(E,\phi) \in \mathscr{M}_{\tau_m^+}(r,\Lambda) \mid E \text{ is a stable vector bundle}\}, \tag{5.2}$$

and denote

$$\mathscr{S}_{\tau_m^+} := \mathscr{M}_{\tau_m^+}(r,\Lambda) - \mathscr{U}_m(r,\Lambda).$$

Then there is a map

$$\pi_1 : \mathscr{U}_m(r,\Lambda) \longrightarrow M(r,\Lambda), \qquad (E,\phi) \longmapsto E, \tag{5.3}$$

whose fiber over any E is the projective space $\mathbb{P}(H^0(E))$. When $d \geq r(2g-2)$, and E is stable, we have $H^1(E) = 0$, and hence (5.3) is a projective bundle (cf. [17, Proposition 4.10]).

Regarding the right-most moduli space $\mathscr{M}_{\tau_M^-}(r,\Lambda)$: any τ_M^--stable pair (E,ϕ) sits in an exact sequence

$$0 \longrightarrow \mathscr{O} \stackrel{\phi}{\longrightarrow} E \longrightarrow F \longrightarrow 0,$$

where F is a semistable bundle of rank $r-1$ and $\det(F) = \Lambda$. Let

$$\mathscr{U}_M(r,\Lambda) = \{(E,\phi) \in \mathscr{M}_{\tau_M^-}(r,\Lambda) \mid F \text{ is a stable vector bundle}\},$$

and denote

$$\mathscr{S}_{\tau_M^-} := \mathscr{M}_{\tau_M^-}(r,\Lambda) - \mathscr{U}_M(r,\Lambda).$$

Then there is a map

$$\pi_2 : \mathscr{U}_M(r,\Lambda) \longrightarrow M(r-1,\Lambda), \qquad (E,\phi) \longmapsto E/\phi(\mathscr{O}), \tag{5.4}$$

whose fiber over any $F \in M(r-1,\Lambda)$ is the projective spaces $\mathbb{P}(H^1(F^*))$ (cf. [8, Theorem 7.7]). Note that $H^0(F^*) = 0$, because $d > 0$. So the map in (5.4) is always a projective bundle.

In the particular case of rank $r = 2$, the right-most moduli space is

$$\mathscr{M}_{\tau_M^-}(2,\Lambda) = \mathbb{P}\big(H^1(\Lambda^{-1})\big), \tag{5.5}$$

since $M(1, \Lambda) = \{\Lambda\}$. In particular, Corollary 5.1 shows that all $\mathcal{M}_\tau(2, \Lambda)$ are rational quasi-projective varieties.

We have the following:

Lemma 5.1 ([15, Lemma 5.3]) *Let S be a bounded family of isomorphism classes of strictly semistable bundles of rank r and determinant Λ. Then* $\dim M(r, \Lambda) - \dim S \geq (r-1)(g-1)$.

Proposition 5.2 *The following hold*:

- $\operatorname{codim} \mathcal{S}_{\tau_m^+} \geq 2$ *except in the case* $r = 2$, $g = 2$, d *even* (*in which case* $\operatorname{codim} \mathcal{S}_{\tau_m^+} = 1$).
- *Suppose* $r \geq 3$. *Then* $\operatorname{codim} \mathcal{S}_{\tau_M^-} \geq 2$ *except in the case* $r = 3$, $g = 2$, d *even* (*in which case* $\operatorname{codim} \mathcal{S}_{\tau_M^-} = 1$).

Proof For any $(E, \phi) \in \mathcal{M}_{\tau_m^+}(r, \Lambda)$, the vector bundle E is semistable. Therefore, Lemma 5.1 implies that $\operatorname{codim} \mathcal{S}_{\tau_m^+} \geq (r-1)(g-1)$. Now the first statement follows.

As the dimension of $H^1(F^*)$ is constant, the codimension of $\mathcal{S}_{\tau_M^-}$ in $\mathcal{M}_{\tau_M^-}(r, \Lambda)$ is at least the codimension of a locus of semistable bundles. Applying Lemma 5.1 to $M(r-1, \Lambda)$, we conclude that $\operatorname{codim} \mathcal{S}_{\tau_M^-} \geq (r-2)(g-1)$. Now the second statement follows. □

5.3 Stable Rationality

A variety Z is said to be stably rational if $Z \times \mathbb{P}^n$ is rational for some n. We prove here Theorem 5.1.

Let $Br(\mathcal{M}_\tau(r, \Lambda))$ denote the Brauer group of $\mathcal{M}_\tau(r, \Lambda)$. In [4], the authors computed this group.

Theorem 5.3 ([4, Theorem 1.1]) *Assume that* $(r, g, d) \neq (3, 2, 2)$. *Then*

$$Br\big(\mathcal{M}_\tau(r, \Lambda)\big) = 0.$$

Theorem 5.4 *Let Λ be a line bundle over X. Suppose* $(r, g, d) \neq (3, 2, \text{even})$. *Then the moduli space $\mathcal{M}_\tau(r, \Lambda)$ of τ-stable pairs over X of rank $r \geq 2$ and fixed determinant Λ is stably rational.*

Proof We already know that $\mathcal{M}_\tau(2, \Lambda)$ are rational varieties. So for $r = 2$, the result holds. Also, the birational class of the moduli spaces $\mathcal{M}_\tau(r, \Lambda)$ are independent of τ (for fixed r and Λ).

Now let $r \geq 2$, and fix the line bundle Λ. Let μ be a line bundle on X of degree at least $2g - 2$. Consider the variety

$$M := \big\{(E, \phi, \psi) \mid E \in M(r, \Lambda), \phi \in \mathbb{P}\big(H^0(E \otimes \mu)\big), \psi \in \mathbb{P}\big(H^1(E^*)\big)\big\}.$$

Since $\deg(\Lambda \otimes \mu^r) > r(2g-2)$, it follows that

$$M \longrightarrow \mathcal{U}_m\big(r, \Lambda \otimes \mu^r\big), \qquad (E, \phi, \psi) \mapsto (E \otimes \mu, \phi)$$

(see (5.2)) is a projective bundle.

By Theorem 5.3, $Br(\mathcal{M}_\tau(r, \Lambda \otimes \mu^r)) = 0$. If $(r, g, d) \neq (2, 2, \text{even})$, then Proposition 5.2 says that

$$\operatorname{codim}\big(\mathcal{M}_{\tau_m^+}\big(r, \Lambda \otimes \mu^r\big) - \mathcal{U}_m\big(r, \Lambda \otimes \mu^r\big)\big) \geq 2.$$

By the Purity Theorem [13, VI.5 (Purity)], this implies that $Br(\mathcal{U}_m(r, \Lambda \otimes \mu^r)) = 0$.

So

$$M \text{ is birational to } \mathbb{P}^s \times \mathcal{U}_m\big(r, \Lambda \otimes \mu^r\big) \tag{5.6}$$

for some natural number s. On the other hand, the map

$$M \longrightarrow \mathcal{U}_M(r+1, \Lambda)$$

that sends any (E, ϕ, ψ) to the pair $(\tilde{E}, \tilde{\psi})$ defined by the extension

$$0 \longrightarrow \mathcal{O} \longrightarrow \tilde{E} \longrightarrow E \longrightarrow 0,$$

given by $\psi \in H^1(E^*)$, is again a projective fibration.

For $(r, g, d) \neq (2, 2, \text{even})$, we know that $Br(\mathcal{M}_\tau(r+1, \Lambda)) = 0$ (see Theorem 5.3), and

$$\operatorname{codim}\big(\mathcal{M}_{\tau_M^-}(r+1, \Lambda) - \mathcal{U}_M(r+1, \Lambda)\big) \geq 2$$

by Proposition 5.2. Then the Purity Theorem yields that $Br(\mathcal{U}_M(r+1, \Lambda)) = 0$. So

$$M \text{ is birational to } \mathbb{P}^t \times \mathcal{U}_M(r+1, \Lambda) \tag{5.7}$$

for some natural number t.

From (5.6) and (5.7) it follows that

$$\begin{gathered} \mathbb{P}^s \times \mathcal{M}_\tau\big(r, \Lambda \otimes \mu^r\big) \quad \text{and} \quad \mathbb{P}^t \times \mathcal{M}_\tau(r+1, \Lambda) \\ \text{are birational, for } (r, g, d) \neq (2, 2, \text{even}). \end{gathered} \tag{5.8}$$

Hence, if $(g, d) \neq (2, \text{even})$, we see by an easy induction that $\mathcal{M}_\tau(r+1, \Lambda)$ is stably rational, for any $r+1 \geq 3$.

Finally, if $(g, d) = (2, \text{even})$, we proceed as follows. For $r+1 = 4$, we use a line bundle μ of *odd* degree. Then we already know that $\mathcal{M}_\tau(r, \Lambda \otimes \mu^r)$ is stably rational (because $\deg(\Lambda \otimes \mu^r)$ is odd). From (5.8), the variety $\mathcal{M}_\tau(r+1, \Lambda)$ is also stably rational. For $r+1 \geq 5$, we use induction and (5.8). □

5.4 Rationality for d Large

In this section, we shall suppose that $d/r \geq 2g-1$.

Let $M(r,\Lambda)$ be the moduli space of stable vector bundles of rank r and fixed determinant Λ. Fix a point $x \in X$. Let $M(r,\Lambda(-x))$ be the moduli space of stable vector bundles with rank r and fixed determinant $\Lambda(-x) := \Lambda \otimes \mathscr{O}_X(-x)$.

On $M(r,\Lambda)$, there are three projective bundles associated to the three vector spaces of the following short exact sequence

$$0 \longrightarrow H^0\big(E \otimes \mathscr{O}_X(-x)\big) \longrightarrow H^0(E) \longrightarrow E_x \longrightarrow 0 \tag{5.9}$$

for any $E \in M(r,\Lambda)$. Note that this is exact because $H^1(E \otimes \mathscr{O}_X(-x)) = H^0(E^* \otimes \mathscr{O}_X(x) \otimes K_X)^* = 0$, as $-\frac{d}{r}+1+2g-2 \leq 0$.

First, there is a universal projective bundle $\mathscr{P}$ over $X \times M(r,\Lambda)$. Restricting the universal bundle to $\{x\} \times M(r,\Lambda)$ we get a projective bundle

$$f : \mathscr{P}_x \longrightarrow M(r,\Lambda). \tag{5.10}$$

The fiber of $\mathscr{P}_x$ over any $E \in M(r,\Lambda)$ is the projective space $\mathbb{P}(E_x)$ of lines in E_x.

Secondly, as $\frac{d}{r} \geq 2g-2$, we have the projective bundle

$$\mathscr{P}_0 \longrightarrow M(r,\Lambda),$$

whose fiber over any $E \in M(r,\Lambda)$ is the projective space $\mathbb{P}(H^0(E))$ of lines in $H^0(E)$. Note that we have $H^1(E) = 0$ because $d \geq r(2g-2)$.

Finally, consider as a third projective bundle

$$\mathscr{P}_1 \longrightarrow M(r,\Lambda)$$

whose fiber over any $E \in M(r,\Lambda)$ is the projective space $\mathbb{P}(H^0(E \otimes \mathscr{O}_X(-x)))$, as $\frac{d}{r} - 1 \geq 2g-2$.

From (5.9), there is a natural embedding

$$\mathscr{P}_1 \hookrightarrow \mathscr{P}_0,$$

and a projection

$$\pi : \mathscr{P}_0 - \mathscr{P}_1 \longrightarrow \mathscr{P}_x. \tag{5.11}$$

Recall that the projective bundle $\mathscr{P}_0$ coincides [17, Proposition 4.10] with an open subset of the moduli space of pairs for the extreme value of the parameter $\tau_m^+ = \tau_m + \varepsilon$, $\varepsilon > 0$,

$$\mathscr{U}_m(r,\Lambda) = \big\{(E,\phi) \in \mathscr{M}_{\tau_m^+}(r,\Lambda) \mid E \text{ is a stable vector bundle}\big\}.$$

There is a map

$$\mathscr{P}_0 = \mathscr{U}_m(r,\Lambda) \longrightarrow M(r,\Lambda), \qquad (E,\phi) \mapsto E$$

whose fiber is the projective space $\mathbb{P}(H^0(E))$.

If $\gcd(r, d) = 1$, then the rationality of $\mathscr{M}_\tau(r, \Lambda)$ is easy to deduce, as shown by the following proposition.

Proposition 5.3 *Let* $\gcd(r, d) = 1$, *and* $d > r(2g - 2)$. *Then* $\mathscr{M}_\tau(r, \Lambda)$ *is rational for any* τ.

Proof It is know that when $\gcd(r, d) = 1$, the moduli space $M(r, \Lambda)$ is rational [12, Theorem 1.2]. Since $M(r, \Lambda)$ is a smooth projective rational variety, the Brauer group $Br(M(r, \Lambda)) = 0$. Hence the projective bundle

$$\mathscr{P}_0 \longrightarrow M(r, \Lambda)$$

is the projectivization of a vector bundle over $M(r, \Lambda)$. Since any vector bundle is Zariski locally trivial, it follows that $\mathscr{P}_0$ is birational to $\mathbb{P}^N \times M(r, \Lambda)$ for some N. Therefore, $\mathscr{P}_0$ is rational. Hence $\mathscr{M}_{\tau_m^+}(r, \Lambda)$ is rational (recall that $\mathscr{P}_0$ is a Zariski open subset of $\mathscr{M}_{\tau_m^+}(r, \Lambda)$). So $\mathscr{M}_\tau(r, \Lambda)$, being birational to $\mathscr{M}_{\tau_m^+}(r, \Lambda)$ (see Corollary 5.1), is rational. □

Proposition 5.4 *For any* r *and* Λ, *the Brauer group* $\mathrm{Br}(\mathscr{P}_x)$ *of the variety* $\mathscr{P}_x$ *vanishes. Furthermore, the variety* $\mathscr{P}_x$ *is rational.*

Proof The Brauer group $\mathrm{Br}(M(r, \Lambda))$ is generated by the Brauer class $cl(\mathscr{P}_x)$ of the projective bundle $\mathscr{P}_x$ (cf. [1]). On the other hand, we have an exact sequence

$$\mathbb{Z} \cdot cl(\mathscr{P}_x) \longrightarrow \mathrm{Br}\big(M(r, \Lambda)\big) \longrightarrow \mathrm{Br}(\mathscr{P}_x) \longrightarrow 0$$

(see [1]). Hence $\mathrm{Br}(\mathscr{P}_x) = 0$.

Note that $\mathscr{P}_x$ is an open subset of the moduli space of parabolic bundles with one marked point x and small parabolic weight α with quasi-parabolic structure of the type $E_x \supset l$, where l is a line in E_x. Hence the rationality is given by a theorem of Boden and Yokogawa [5, Theorem 6.2]. □

Theorem 5.5 *If* $d > r(2g - 1)$, *then the moduli space* $\mathscr{M}_\tau(r, \Lambda)$ *is rational, for any* τ.

Proof By the argument in Proposition 5.3, it is enough to see that $\mathscr{P}_0$ is a rational space.

We will show that the projection π in (5.11) is an affine bundle for a vector bundle over $\mathscr{P}_x$.

Consider the projective bundle $f^* \mathscr{P}_0 \longrightarrow \mathscr{P}_x$, where f is the projection in (5.10). Since $\mathrm{Br}(\mathscr{P}_x) = 0$ (see Proposition 5.4), there is a vector bundle

$$\mathscr{W}_0 \longrightarrow \mathscr{P}_x$$

such that $f^* \mathscr{P}_0$ is the projective bundle $\mathbb{P}(\mathscr{W}_0)$ parameterizing the lines in $\mathscr{W}_0$. Fix one such vector bundle W_0.

Consider the projective subbundle $\mathscr{P}_1$ of $\mathscr{P}_0$ in (5.11). The pullback $f^*\mathscr{P}_1 \subset f^*\mathscr{P}_0$ is the projectivization of a unique subbundle

$$\mathscr{W}_1 \subset \mathscr{W}_0. \tag{5.12}$$

Let

$$\mathscr{W} := \mathscr{W}_0/\mathscr{W}_1 \longrightarrow \mathscr{P}_x$$

be the quotient bundle. Note that $\mathbb{P}(\mathscr{W}) = f^*\mathscr{P}_x$; the isomorphism is given by π in (5.11). Let

$$\mathscr{O}_{\mathbb{P}(\mathscr{W})}(-1) \longrightarrow \mathbb{P}(\mathscr{W}) = f^*\mathscr{P}_x \tag{5.13}$$

be the tautological line bundle.

We have a tautological section

$$\sigma : \mathscr{P}_x \longrightarrow f^*\mathscr{P}_x$$

of the projective bundle $f^*\mathscr{P}_x \longrightarrow \mathscr{P}_x$; for any point $z \in \mathscr{P}_x$, the image $\sigma(z)$ is the point of $f^*\mathscr{P}_x$ defined by (z, z). Let

$$\mathscr{L} := \sigma^*\mathscr{O}_{\mathbb{P}(\mathscr{W})}(-1) \longrightarrow \mathscr{P}_x \tag{5.14}$$

be the pullback, where $\mathscr{O}_{\mathbb{P}(\mathscr{W})}(-1)$ is the line bundle in (5.13).

It is straight forward to check that the projection π in (5.11) is an affine bundle for the vector bundle

$$\mathscr{W}_1 \otimes \mathscr{L}^* \longrightarrow \mathscr{P}_x,$$

where $\mathscr{W}_1$ and $\mathscr{L}^*$ are constructed in (5.12) and (5.14), respectively.

The isomorphism classes of affine bundles over a variety Z for a vector bundle $V \longrightarrow Z$ are parameterized by $H^1(Z, V)$. If Z is an affine variety, then $H^1(Z, V) = 0$. Hence affine bundles over an affine variety are trivial (the trivial affine bundle for V is V itself).

Fix a nonempty affine open subset $U_0 \subset \mathscr{P}_x$ such that the vector bundle $(\mathscr{W}_1 \otimes \mathscr{L}^*)|_{U_0}$ is trivial. Since π in (5.11) is an affine bundle for the vector bundle $\mathscr{W}_1 \otimes \mathscr{L}^*$, and $\mathscr{W}_1 \otimes \mathscr{L}^*$ is trivial over U_0, we conclude that $\pi^{-1}(U_0)$ is isomorphic to $U_0 \times \mathbb{C}^N$, where N is the relative dimension of the fibration π. From Proposition 5.4 we know that U_0 is rational. Hence we now conclude that $\mathscr{P}_0$ is rational. □

5.5 Rationality for Small *d*

We want to analyze the cases where $d \leq r(2g-1)$.

We start with the following remark.

Fix a point $x \in X$. By [3, Lemma 2.1], we know that if $d + r \leq r(g-1)$, then a general vector bundle $E \in M_X(r, \Lambda)$ satisfies the condition that

$$H^0\big(E \otimes \mathscr{O}_X(x)\big) = 0. \tag{5.15}$$

Moreover, let $U \subset M_X(r, \Lambda)$ be the subset of the bundles E satisfying (5.15). Then the proof of [3, Lemma 2.1] shows that

$$\operatorname{codim}\bigl(M_X(r,\Lambda) - U\bigr) \geq r(g-1) - d - r + 1.$$

Now let $d' = -d + r(2g-2)$, $\Lambda' = \Lambda^{-1} \otimes K_X^r$, and consider

$$U' = \bigl\{E' = E^* \otimes K_X \mid E \in U\bigr\} \subset M_X\bigl(r, \Lambda'\bigr).$$

Then for any $E' \in U'$,

$$H^1\bigl(E' \otimes \mathcal{O}_X(-x)\bigr) = H^0\bigl(E'^* \otimes K_X \otimes \mathcal{O}_X(x)\bigr)^* = H^0\bigl(E \otimes \mathcal{O}_X(x)\bigr)^* = 0.$$

We rewrite the codimension estimate as

$$\operatorname{codim}\bigl(M_X\bigl(r,\Lambda'\bigr) - U'\bigr) \geq d' - r(g-1) - r + 1. \tag{5.16}$$

We are now ready to prove the following extension of Theorem 5.5.

Theorem 5.6 *If $d > rg$, then the moduli space $\mathcal{M}_\tau(r, \Lambda)$ is rational, for any τ.*

Proof By the previous comments, there is an open subset $U \subset M_X(r, \Lambda)$ where

$$H^1\bigl(E \otimes \mathcal{O}_X(-x)\bigr) = 0$$

for all $E \in U$. Moreover, (5.16) says that

$$\operatorname{codim}\bigl(M_X(r,\Lambda) - U\bigr) \geq d - r(g-1) - r + 1 \geq 2.$$

For all $E \in U$, we have an exact sequence (5.9).

There is a projective bundle

$$\mathcal{P}_0|_U \longrightarrow U$$

whose fiber over any $E \in U$ is the projective space $\mathbb{P}(H^0(E))$. The universal projective bundle (5.10) gives a corresponding projective bundle

$$f|_U : \mathcal{P}_x|_U \longrightarrow U.$$

By Proposition 5.4, the variety $\mathcal{P}_x|_U$ is rational. Moreover, as $\operatorname{codim}(M_X(r,\Lambda) - U) \geq 2$, we have that $\operatorname{codim}(\mathcal{P}_x - \mathcal{P}_x|_U) \geq 2$. Therefore, Proposition 5.4 and the Purity Theorem [13, VI.5 (Purity)] show that $\operatorname{Br}(\mathcal{P}_x|_U) = 0$. Now the arguments in the proof of Theorem 5.5 can be carried out verbatim. □

A simple extra case, which follows by the argument above, is the following:

Corollary 5.2 *Assume $d > r(g-1)$ and $\gcd(r,d) = 1$. Then the moduli space $\mathcal{M}_\tau(r, \Lambda)$ is rational, for any τ.*

Proof This is similar to Proposition 5.3, upon noting that, for $d \geq r(g-1)+1$, the open set

$$U = \{E \in M(r, \Lambda) \mid H^1(E) = 0\}$$

is non-empty and $\operatorname{codim}(M(r, \Lambda) - U) \geq 2$, [3, Lemma 2.1] (see the arguments in the proof of Theorem 5.6). □

Another case that can be covered is the following:

Theorem 5.7 *Let* $\gcd(r-1, d) = 1$ *and* $d > 0$. *Then* $\mathcal{M}_\tau(r, \Lambda)$ *is rational for any* τ.

Proof For this we shall consider the moduli space of pairs $\mathcal{M}_{\tau_M^-}(r, \Lambda)$ for the extreme value of the parameter $\tau_M^- = \tau_M - \varepsilon$, $\varepsilon > 0$. By [8, Sect. 7.2], any τ_M^--stable pair (E, ϕ) sits in an exact sequence

$$0 \longrightarrow \mathcal{O} \xrightarrow{\phi} E \longrightarrow F \longrightarrow 0,$$

where F is a semistable vector bundle of rank $r-1$ with $\det(F) = \Lambda$. Let

$$\mathcal{U}_M(r, \Lambda) := \{(E, \phi) \in \mathcal{M}_{\tau_M^-}(r, \Lambda) \mid F \text{ is a stable vector bundle}\}.$$

Then there is a map

$$\pi_2 : \mathcal{U}_M(r, \Lambda) \longrightarrow M(r-1, \Lambda), \qquad (E, \phi) \longmapsto E/\phi(\mathcal{O}), \tag{5.17}$$

whose fiber over $F \in M(r-1, \Lambda)$ is the projective spaces $\mathbb{P}(H^1(F^*))$ (cf. [8, Theorem 7.7]). Note that $H^0(F^*) = 0$ since $d > 0$. So the morphism in (5.17) is always a projective bundle.

When $\gcd(r-1, d) = 1$, it must be $\mathcal{U}_M(r, \Lambda) = \mathcal{M}_{\tau_M^-}(r, \Lambda)$. Moreover, the moduli space $M(r-1, \Lambda)$ is rational [12, Theorem 1.2]. Since $M(r-1, \Lambda)$ is a smooth projective rational variety, the Brauer group $Br(M(r-1, \Lambda)) = 0$. Hence the projective bundle (5.17) must be a product, i.e., $\mathcal{M}_{\tau_M^-}(r, \Lambda)$ is isomorphic to $\mathbb{P}^N \times M(r-1, \Lambda)$ for some N. Thus $\mathcal{M}_\tau(r, \Lambda)$ is rational for any τ. □

Acknowledgements We thank Norbert Hoffman for kindly pointing us to his work [11]. The second author was supported by (Spanish MICINN) research project MTM2007-67623 and i-MATH. The third author was partially supported by (Spanish MICINN) research project MTM2007-63582.

References

1. Balaji, V., Biswas, I., Gabber, O., Nagaraj, D.S.: Brauer obstruction for a universal vector bundle. C. R. Math. Acad. Sci. Paris **345**, 265–268 (2007)
2. Bertram, A.: Stable pairs and stable parabolic pairs. J. Algebr. Geom. **3**, 703–724 (1994)
3. Biswas, I., Gómez, T., Muñoz, V.: Automorphisms of the moduli spaces of vector bundles over a curve. Expo. Math. (to appear). arXiv:1202.2961

4. Biswas, I., Logares, M., Muñoz, V.: Brauer group of moduli spaces of pairs. Commun. Algebra (to appear). arXiv:1009.5204
5. Boden, H.U., Yokogawa, K.: Rationality of moduli spaces of parabolic bundles. J. Lond. Math. Soc. **59**, 461–478 (1999)
6. Bradlow, S.B., Daskalopoulos, G.: Moduli of stable pairs for holomorphic bundles over Riemann surfaces. Int. J. Math. **2**, 477–513 (1991)
7. Bradlow, S.B., García-Prada, O.: Stable triples, equivariant bundles and dimensional reduction. Math. Ann. **304**, 225–252 (1996)
8. Bradlow, S.B., García-Prada, O., Gothen, P.: Moduli spaces of holomorphic triples over compact Riemann surfaces. Math. Ann. **328**, 299–351 (2004)
9. Craioveanu, M., Puta, M., Rassias, Th.M.: Old and New Aspects in Spectral Geometry. Mathematics and Its Applications, vol. 534. Kluwer Academic, Dordrecht (2001)
10. García-Prada, O.: Dimensional reduction of stable bundles, vortices and stable pairs. Int. J. Math. **5**, 1–52 (1994)
11. Hoffmann, N.: Rationality and Poincaré families for vector bundles with extra structure on a curve. Int. Math. Res. Not. (2007), no. 3. Art. ID rnm010, 30 pp.
12. King, A.D., Schofield, A.: Rationality of moduli of vector bundles on curves. Indag. Math. **10**, 519–535 (1999)
13. Milne, J.S.: Ètale Cohomology. Princeton Mathematical Series, vol. 33. Princeton University Press, Princeton (1980)
14. Muñoz, V.: Hodge polynomials of the moduli spaces of rank 3 pairs. Geom. Dedic. **136**, 17–46 (2008)
15. Muñoz, V.: Torelli theorem for the moduli spaces of pairs. Math. Proc. Camb. Philos. Soc. **146**, 675–693 (2009)
16. Muñoz, V.: Hodge structures of the moduli space of pairs. Int. J. Math. **21**, 1505–1529 (2010)
17. Muñoz, V., Ortega, D., Vázquez-Gallo, M.-J.: Hodge polynomials of the moduli spaces of pairs. Int. J. Math. **18**, 695–721 (2007)
18. Thaddeus, M.: Stable pairs, linear systems and the Verlinde formula. Invent. Math. **117**, 317–353 (1994)

Chapter 6
Generalized p-Valent Janowski Close-to-Convex Functions and Their Applications to the Harmonic Mappings

Daniel Breaz, Yasar Polatoğlu, and Nicoleta Breaz

Abstract Let $A(p,n), n \geq 1, p \geq 1$ be the class of all analytic functions in the open unit disc $\mathbb{D} = \{z||z| < 1\}$ of the form $s(z) = z^p + c_{np+1}z^{np+1} + c_{np+2}z^{np+2} + \cdots$ and let $s(z)$ be an element of $A(p,n)$, if $s(z)$ satisfies the condition $(1 + z\frac{s''(z)}{s'(z)}) = \frac{1+A\varphi(z)}{1+B\varphi(z)}$, then $s(z)$ is a called generalized p-valent Janowski convex function, where A, B are arbitrary fixed real numbers such that $-1 \leq B < A \leq 1$, and $\varphi(z) = z^n \psi(z)$ with $\psi(z)$ being analytic in $\mathbb{D}$ and satisfying the condition $|\psi(z)| < 1$ for every $z \in \mathbb{D}$. The class of generalized p-valent Janowski convex functions is denoted by $C(p,n,A,B)$. Let $s(z)$ be an element of $A(p,n)$, then $s(z)$ is a generalized p-valent Janowski close-to-convex function for $z \in \mathbb{D}$, if there exists a function $\phi(z) \in C(p,n,A,B)$ such that $\frac{s'(z)}{\phi'(z)} = \frac{1+A\varphi(z)}{1+B\varphi(z)}$. $(-1 \leq B \leq A \leq 1, \varphi(z) = z^n \psi(z)$, $\psi(z)$ is analytic and $|\psi(z)| < 1$ for every $z \in \mathbb{D})$. The class of such functions is denoted by $K(p,n,A,B)$.

The aim of this paper is to give an investigation of the class $K(p,n,A,B)$ and its application to the harmonic mappings.

Key words Generalized p-valent Janowski convex function · Generalized p-valent Janowski close-to-convex function · Radius of convexity

Mathematics Subject Classification Primary 30C45 · Secondary 30C55

Dedicated to Professor Themistocles M. Rassias on the occasion of his 60th birthday.

D. Breaz (✉)
Department of Mathematics, "1 Decembrie 1918" University of Alba Iulia, str. N. Iorga, No. 11-13, 510009 Alba Iulia, Romania
e-mail: dbreaz@uab.ro

Y. Polatoğlu
Department of Mathematics and Computer Science, Kültür University, E5 Freeway Bakirköy, 34156 Istanbul, Turkey
e-mail: y.polatoglu@iku.edu.tr

N. Breaz
"1 Decembrie 1918" University of Alba Iulia, str. N. Iorga, No. 11-13, 510009 Alba Iulia, Romania
e-mail: nbreaz@uab.ro

P.M. Pardalos et al. (eds.), *Nonlinear Analysis*, Springer Optimization and Its Applications 68, In Honor of Themistocles M. Rassias on the Occasion of his 60th Birthday, DOI 10.1007/978-1-4614-3498-6_6,

6.1 Introduction

Let Ω be the family of functions $\varphi(z)$ which are regular in $\mathbb{D}$ and satisfying the conditions $\varphi(0)=0$, $|\varphi(z)|<1$ for all $z\in\mathbb{D}$.

Definition 6.1 For arbitrary fixed real numbers $A, B, -1\leq B<A\leq 1$, we denote by $P(p,n,A,B)$, $p\geq 1$, $n\geq 1$, the family of functions $p(z)=p+p_nz^n+p_{n+1}z^{n+1}+p_{n+2}z^{n+2}+\cdots$ which are analytic in $\mathbb{D}$, satisfy the conditions $p(0)=p$, $\operatorname{Re} p(z)>0$ and

$$p(z)=p\frac{1+A\varphi(z)}{1+B\varphi(z)} \tag{6.1}$$

for some $\varphi(z)\in\Omega$ for every $z\in\mathbb{D}$, where $\varphi(z)=z^n\psi(z)$, $\psi(z)$ is analytic in $\mathbb{D}$ and $|\psi(z)|<1$ for every $z\in\mathbb{D}$.

The class $P(1,1,A,B)$ was introduced by W. Janowski in [4].

Next, let $A(p,n)$ be the class of all analytic functions of the form $s(z)=z^p+c_{np+1}z^{np+1}+\cdots$ which are regular in $\mathbb{D}$. In particular, $A(p,1)$ is the class of standard p-valent analytic functions and $A(1,n)$ is the class of all analytic functions for which first $n-1$ coefficients are zero, and $A(1,1)$ is the class of all analytic functions of the standard form.

Definition 6.2 Let $C(p,n,A,B)$ denote the family of functions $s(z)\in A(p,n)$ such that $s(z)$ is in $C(p,n,A,B)$ if and only if

$$1+z\frac{s''(z)}{s'(z)}=p(z) \tag{6.2}$$

for some $p(z)\in P(p,n,A,B)$ and all $z\in\mathbb{D}$ (generalized p-valent Janowski convex functions).

Definition 6.3 Let $s(z)$ be an element of $A(p,n)$. If there exists a function $\phi(z)\in C(p,n,A,B)$ such that

$$\frac{s'(z)}{\phi'(z)}=p(z) \tag{6.3}$$

for some function $p(z)\in P(p,n,A,B)$ for every $z\in\mathbb{D}$, the class of these functions $s(z)$ is denoted by $K(p,n,A,B)$ (generalized p-valent Janowski close-to-convex functions).

Moreover, let $F(z)=z+\alpha_2z^2+\alpha_3z^3+\cdots$ and $G(z)=z+\beta_2z^2+\beta_3z^3+\cdots$ be analytic in $\mathbb{D}$, if there exists a function $\varphi(z)\in\Omega$ such that $F(z)=G(\varphi(z))$ for every $z\in\mathbb{D}$, then we say that $F(z)$ is subordinate to $G(z)$, and we write $F(z)\prec G(z)$. We also note that if $F(z)\prec G(z)$, then $F(\mathbb{D})\subset G(\mathbb{D})$.

Finally, let U be a simply connected domain in the complex plane. A harmonic mapping f has the representation $f=h(z)+\overline{g(z)}$, where $h(z)=z+a_2z^2+a_3z^3+$

$\cdots$ and $g(z) = b_1 z + b_2 z^2 + b_3 z^3 + \cdots$ are analytic in $\mathbb{D}$ and are called the analytic part and co-analytic part of f, respectively. If $J_f(z) = (|h'(z)|^2 - |g'(z)|^2) > 0$ (or $J_f(z) < 0$), then f is called a sense-preserving harmonic mapping in U. The quantity $w(z) = (\frac{g'(z)}{h'(z)})$ is called the second analytic dilatation of f, and it satisfies the condition $|w(z)| < 1$ for every $z \in \mathbb{D}$. The class of all sense-preserving harmonic mappings with $|b_1| < 1$ is denoted by S_H, and the class of all sense-preserving harmonic mappings with $b_1 = 0$ is denoted by S_H^0 [2]. Therefore, we can give the following definition:

Definition 6.4 Let $h(z) = z^p + a_{np+1} z^{np+1} + a_{np+2} z^{np+2} + \cdots$ and $g(z) = b_{np} z^p + b_{np+1} z^{np+1} + b_{np+2} z^{np+2} + \cdots$ be analytic functions in $\mathbb{D}$, and let $J_f(z) = (|h'(z)|^2 - |g'(z)|^2) > 0$, $w(z) = (\frac{g'(z)}{h'(z)})$, $|w(z)| < 1$. Then we say that $f = h(z) + \overline{g(z)}$ is a generalized p-valent sense-preserving harmonic mapping. The class of such functions with $|b_{np}| < 1$ is denoted by $S_H(p,n)$ and for $b_{np} = 0$ is denoted by $S_H^0(p,n)$.

6.2 Main Results

Lemma 6.1 ([4]) *If $q(z) \in \mathscr{P}$ then $p(z) = \frac{(1+A)q(z)+(1-A)}{(1+B)q(z)+(1-B)} \in P(A,B)$, where $\mathscr{P}$ is the class of Caratheodory functions $\mathscr{P} = \{q(z) | q(0) = 1, \operatorname{Re} q(z) > 0, q(z) = 1 + \cdots\}$ and $P(A,B)$ is the class $P(A,B) = \{p(z) | p(0) = 1, p(z) \prec \frac{1+Az}{1+Bz}\}$, and $p(z)$ is analytic.*

Remark 6.1 The proof of the above lemma can be found in [4].

Lemma 6.2 *For integers $p \geq 1$ and $n \geq 1$, let $P(p,n)$ denote the class of functions $p(z) = p + p_n z^n + p_{n+1} z^{n+1} + p_{n+2} z^{n+2} + \cdots$ which are regular in $\mathbb{D}$ and satisfy the conditions $p(0) = p$, $\operatorname{Re} p(z) > 0$ in $\mathbb{D}$. Then*

$$p(z) = p \frac{1 + z^n \psi(z)}{1 - z^n \psi(z)} \tag{6.4}$$

where $\psi(z)$ is an analytic function and satisfies the condition $|\psi(z)| < 1$, for every z in $\mathbb{D}$.

Proof Let $p_1(z)$ be an analytic function and satisfying the conditions $p_1(0) = 1$, $\operatorname{Re} p_1(z) > 0$ in $\mathbb{D}$. Then $p_1(z)$ can be written in the form:

$$p_1(z) = \frac{1 + \varphi(z)}{1 - \varphi(z)} \tag{6.5}$$

where $\varphi(z)$ is analytic in $\mathbb{D}$ and also has one zero with multiplicity equal to n at the origin, hence $\varphi(z) = z^n \psi(z)$, where $\psi(z)$ analytic and satisfies the condition of

Schwarz Lemma for all $z \in \mathbb{D}$. Thus $p_1(z)$ can be written in the form:

$$p_1(z) = \frac{1 + z^n \psi(z)}{1 - z^n \psi(z)}. \tag{6.6}$$

On the other hand, now we consider the function $p(z) = p \cdot p_1(z)$. This function is analytic and satisfies the condition $p(0) = p \cdot p_1(0) = p$, $\operatorname{Re} p(z) = \operatorname{Re}(p \cdot p_1(z)) = p \operatorname{Re} p(z) > 0$ for every $z \in \mathbb{D}$. Using (6.6), we obtain

$$p(z) = p \cdot p_1(z) = p\frac{1 + z^n \psi(z)}{1 - z^n \psi(z)}. \tag{6.7}$$

□

Remark 6.2 We also note that using the subordination principle, then we have

$$p(z) \in P(p,n) \quad \Leftrightarrow \quad p(z) \prec p\frac{1+z^n}{1-z^n}. \tag{6.8}$$

At the same time, we consider the image of the disk $|z| = r$ under the transformation

$$w(z) = \frac{1+z^n}{1-z^n}.$$

Therefore, the image of $|z| = r$ is the disk with the center $C(r) = (\frac{1+r^{2n}}{1-r^{2n}}, 0)$ and radius $\rho(r) = \frac{2r^n}{1-r^{2n}}$.

Lemma 6.3 *If $p(z) \in P(p,n,A,B)$, then*

$$p(z) = p\frac{(1+A)q(z) + (1-A)}{(1+B)q(z) + (1-B)} \tag{6.9}$$

for some $q(z) \in P(p,n)$ every $z \in \mathbb{D}$, and conversely.

On the other hand, the image of the disk $|z| = r$ under the $p(z) \in P(p,n,A,B)$ is the disk with the center $C(r) = (p\frac{1-ABr^{2n}}{1-B^2r^{2n}}, 0)$ and radius $\rho(r) = \frac{p(A-B)r^n}{1-B^2r^{2n}}$.

Remark 6.3 If $p(z) \in P(p,n,A,B)$, then $p(z) \prec p\frac{1-ABr^n}{1-B^2r^n}$, and the image of $|z| = r$ under the transformation $(p\frac{1+Az^n}{1+Bz^n})$ is the disc with the center $c(r) = p\frac{1-ABr^{2n}}{1-B^2r^{2n}}$ and radius $\rho(r) = \frac{p(A-B)r^n}{1-B^2r^{2n}}$.

Theorem 6.1 *Let $s(z)$ be an element of $K(p,n,A,B)$. Then*

$$\frac{s'(z)}{\phi'(z)} = \begin{cases} (1+Bz^n)^{-\frac{P(A-B)}{nB}} \frac{s'(z)}{z^{p-1}} = p\frac{1+Az^n\psi(z)}{1+Bz^n\psi(z)}; & B \neq 0, \\ e^{-\frac{pA}{n}z^n} \frac{s'(z)}{z^{p-1}} = p(1+Az^n\psi(z)); & B = 0, \end{cases} \tag{6.10}$$

where $\phi \in C(p,n,A,B)$ and satisfies (6.3) for every $z \in \mathbb{D}$.

Proof We consider the function

$$\phi(z)=\begin{cases}\int_o^z \xi^{p-1}(1+B\xi^n)^{\frac{p(A-B)}{nB}}\,d\xi; & B\neq 0,\\ \int_0^z \xi^{p-1}e^{\frac{pA}{n}\xi^n}; & B=0,\end{cases} \tag{6.11}$$

then we have

$$1+z\frac{\phi''(z)}{\phi'(z)}=\begin{cases}p\frac{1+Az^n}{1+Bz^n}; & B\neq 0,\\ p(1+Az^n); & B=0.\end{cases} \tag{6.12}$$

This shows that $\phi(z)\in C(p,n,A,B)$. Since

$$\begin{cases}\frac{s'(z)}{\phi'(z)}\prec p\frac{1+Az^n}{1+Bz^n}; & B\neq 0,\\ \frac{s'(z)}{\phi'(z)}\prec p(1+Az^n); & B=0,\end{cases} \tag{6.13}$$

this implies that the general characterization of $K(p,n,A,B)$ is

$$\begin{cases}\frac{s'(z)}{\phi'(z)}=(1+Bz^n)^{\frac{-p(A-B)}{nB}}\frac{s'(z)}{z^{p-1}}=p\frac{1+Az^n\psi(z)}{1+Bz^n\psi(z)}; & B\neq 0,\\ \frac{s'(z)}{\phi'(z)}=e^{-\frac{pA}{n}z^n}\frac{s'(z)}{z^{p-1}}=p(1+Az^n\psi(z)); & B=0.\end{cases} \tag{6.14}$$

□

Corollary 6.1 *If we give particular values to p, n, A, and B, then we obtain the general characterization of the subclass of $K(p,n,A,B)$. For example,*

(i) *When $A=1$, $B=-1$, $n=1$, $p=1$,*

$$(1-z)^2s'(z)=\frac{1+z\psi(z)}{1-z\psi(z)}=\frac{1+\varphi(z)}{1-\varphi(z)}\quad\Rightarrow\quad \operatorname{Re}\big[(1-z)^2s'(z)\big]>0. \tag{6.15}$$

This inequality was found by W. Kaplan in [5].

(ii) *When $A=1$, $B=-1$, $n=1$,*

$$(1-z)^{2p}\frac{s'(z)}{z^{p-1}}=p\frac{1+z\psi(z)}{1-z\psi(z)}\quad\Rightarrow\quad \operatorname{Re}\left[(1-z)^{2p}\frac{s'(z)}{z^{p-1}}\right]>0. \tag{6.16}$$

This result was found by T. Umezava in [6].

(iii) *When $A=1$, $B=-1$, $p=1$,*

$$\left(1-z^n\right)^{\frac{2}{n}}s'(z)=\frac{1+z^n\psi(z)}{1-z^n\psi(z)}, \tag{6.17}$$

which gives $\operatorname{Re}[(1-z^n)^{\frac{2}{n}}s'(z)]>0$. *This is a new characterization of the class $K(1,n,1,-1)$.*

(iv) *When $A=1-2\alpha$, $B=-1$, $p=1$, $n=1$ $(0\le\alpha<1)$,*

$$(1-z)^{2\alpha}s'(z)=\frac{1+(1-2\alpha)z\psi(z)}{1-z\psi(z)}\quad\Rightarrow\quad \operatorname{Re}(1-z)^{2\alpha}s'(z)>\alpha. \tag{6.18}$$

(v) *When* $A=1,\ B=-1$,

$$\left(1-z^n\right)^{\frac{2p}{n}} s'(z)=\frac{1+z\psi(z)}{1-z\psi(z)} \quad \Rightarrow \quad \operatorname{Re}\left(1-z^n\right)^{\frac{2p}{n}} s'(z)>0. \tag{6.19}$$

(vi) *When* $A=1-2\alpha,\ B=-1\ (0\le\alpha<1)$,

$$\left(1-z^n\right)^{\frac{2p\alpha}{n}} s'(z)=\frac{1+(1-2\alpha)z\psi(z)}{1-z\psi(z)} \quad \Rightarrow \quad \operatorname{Re}\left(1-z^n\right)^{\frac{2p\alpha}{n}} s'(z)>\alpha. \tag{6.20}$$

Statements (iv), (v), and (vi) present new characterizations for this class. We also note that if we give specific values to A, B, p, and n, then we obtain a new characterization of the subclasses of close-to-convex functions.

Theorem 6.2 *Let $\phi(z)$ be an element of $C(p,n,A,B)$. Then*

$$\begin{cases} B^{\frac{p-1}{nB}} r^{\frac{p-1}{B}} (1-Br^n)^{\frac{p(A-B)}{nB^2}} \le |\phi'(z)| \le B^{\frac{p-1}{nB}} r^{\frac{p-1}{B}} (1+Br^n)^{\frac{p(A-B)}{nB^2}}; & B\neq 0,\\ r^{p-1}e^{-\frac{pA}{n}r^n} \le |\phi'(z)| \le r^{p-1}e^{\frac{p-1}{n}r^n}; & B=0. \end{cases} \tag{6.21}$$

These results are sharp because the extremal function is

$$\phi'(z)=\begin{cases} z^{p-1}(1+Bz^n)^{\frac{p(A-B)}{nB}}; & B\neq 0,\\ z^{p-1}(1+Az^n); & B=0. \end{cases} \tag{6.22}$$

Proof Let $B\neq 0$. Since $1+z\frac{\phi''(z)}{\phi'(z)} \prec p\frac{1+Az^n}{1+Bz^n}$, using the subordination principle, we have

$$\left|\left(1+z\frac{\phi''(z)}{\phi'(z)}\right)-p\frac{1-ABr^{2n}}{1-B^2r^{2n}}\right| \le \frac{p(A-B)r^n}{1-B^nr^{2n}}. \tag{6.23}$$

After the simple calculations we get

$$\frac{(p-1)-p(A-B)r^n+(B^2-pAB)r^{2n}}{1-B^2r^{2n}} \le \operatorname{Re}\left(z\frac{\phi''(z)}{\phi'(z)}\right)$$
$$\le \frac{(p-1)+p(A-B)r^n+(B^2-pAB)r^{2n}}{1-B^2r^{2n}}. \tag{6.24}$$

On the other hand,

$$\operatorname{Re}\left(z\frac{\phi''(z)}{\phi'(z)}\right)=r\frac{\partial}{\partial r}\log\left|\phi'(z)\right|. \tag{6.25}$$

Using (6.20) in the equality (6.19), we obtain

$$\frac{(p-1)-p(A-B)r^n+(B^2-pAB)r^{2n}}{r(1-B^2r^{2n})}\le\frac{\partial}{\partial r}\log\left|\phi'(z)\right|$$
$$\le\frac{(p-1)+p(A-B)r^n+(B^2-pAB)r^{2n}}{r(1-B^2r^{2n})}. \tag{6.26}$$

Integrating both sides of the inequality (6.26), we obtain (6.21).

Let $B=0$. Since $(1+z\frac{\phi''(z)}{\phi'(z)})\prec p(1+Az^n)$, we have

$$\left|\left(1+z\frac{\phi''(z)}{\phi'(z)}\right)-p\right|\le pAr^n \quad\Rightarrow\quad (p-1)-pAr^n\le\operatorname{Re}\left(z\frac{\phi''(z)}{\phi'(z)}\right)$$
$$\le(p-1)+pAr^n, \tag{6.27}$$

which with (6.20) gives

$$\frac{p-1}{r}-pAr^{n-1}\le\frac{\partial}{\partial r}\log\left|\phi'(z)\right|\le\frac{p-1}{r}+pAr^{n-1}. \tag{6.28}$$

Integrating both sides of the inequality (6.28), we obtain (6.21). □

Corollary 6.2 *Let $s(z)$ be an element of $K(p,n,A,B)$. Then*

$$\begin{cases} pB^{\frac{p-1}{nB}}\frac{1-Ar^n}{1-Br^n}r^{\frac{p-1}{B}}(1-Br^n)^{\frac{p(A-B)}{nB^2}})\le|s'(z)| \\ \quad\le pB^{\frac{p-1}{nB}}\frac{1+Ar^n}{1+Br^n}r^{\frac{p-1}{B}}(1+Br^n)^{\frac{p(A-B)}{nB^2}}; \quad B\neq 0, \\ pr^{p-1}(1-Ar^n)e^{-\frac{pA}{n}r^n}\le|s'(z)|\le pr^{n-1}(1+Ar^n)e^{\frac{pA}{n}r^n}; \quad B=0. \end{cases} \tag{6.29}$$

This corollary is a simple consequence of Theorem 6.1, Theorem 6.2, and the definition of the class $K(p,n,A,B)$. These bounds are sharp because the extremal function can be found in the following manner.

Theorem 6.3 *The radius of convexity of the class $K(p,n,A,B)$ is the smallest positive root of the equation*

$$Q(r)=p(1-r^n)(1-Ar^n)(1+Ar^n(1+Br^n))$$
$$-r^n\cdot n\left[(1+A)(1+Br^n)+(1+B)(1+Ar^n)\right]. \tag{6.30}$$

Proof Let $\phi(z)$ be an element of $C(p,n,A,B)$, then we have

$$\left|\left(1+z\frac{\phi''(z)}{\phi'(z)}\right)-p\frac{1-ABr^{2n}}{1-B^2r^{2n}}\right|\le\frac{p(A-B)r^n}{1-B^2r^{2n}}$$
$$\Rightarrow\quad\frac{p(1-Ar^n)}{1-Br^n}\le\operatorname{Re}\left(1+z\frac{\phi''(z)}{\phi'(z)}\right)\le\frac{p(1+Ar^n)}{1+Br^n}. \tag{6.31}$$

On the other hand, using Lemma 6.3,

$$p(z) = p\frac{(1+A)q(z)+(1-A)}{(1+B)q(z)+(1-B)} \quad \Rightarrow \quad z\frac{p'(z)}{p(z)} = \frac{zq'(z)}{q(z)+\frac{1-A}{1+A}} - \frac{zq'(z)}{q(z)+\frac{1-B}{1+B}}. \tag{6.32}$$

At the same time, since $-1 \le A \le B \le 1$, we have

$$\mu = \frac{1-A}{1+A} \quad \Rightarrow \quad \operatorname{Re}\mu = \beta = \frac{1-A}{1+A} > 0 \quad \text{and}$$
$$\mu = \frac{1-B}{1+B} \quad \Rightarrow \quad \operatorname{Re}\mu = \beta = \frac{1-B}{1+B} > 0. \tag{6.33}$$

Using Bernardi's result from [1], we then get

$$\left|\frac{zq'(z)}{q(z)+\frac{1-A}{1+A}}\right| \le \frac{(1+A)nr^n}{(1-r^n)(1+Ar^n)} \tag{6.34}$$

and

$$\left|\frac{zq'(z)}{q(z)+\frac{1-B}{1+B}}\right| \le \frac{(1+B)nr^n}{(1-r^n)(1+Br^n)}. \tag{6.35}$$

Using (6.33), (6.34), and (6.35), we get

$$z\frac{p'(z)}{p(z)} = \frac{zq'(z)}{q(z)+\frac{1+A}{1-A}} - \frac{zq'(z)}{q(z)+\frac{1-B}{1+B}}$$
$$\Rightarrow \quad \left|z\frac{p'(z)}{p(z)}\right| \le \left|\frac{zq'(z)}{q(z)+\frac{1-A}{1+A}}\right| + \left|\frac{zq'(z)}{q(z)+\frac{1-B}{1+B}}\right|$$
$$\le \frac{nr^n[(1+A)(1+Br^n)+(1+B)(1+Ar^n)]}{(1-r^n)(1+Ar^n)(1+Br^n)}. \tag{6.36}$$

Using the definition of the class of $K(p,n,A,B)$, we obtain

$$\frac{s'(z)}{\phi'(z)} = p(z) \quad \Rightarrow \quad \left(1+z\frac{s''(z)}{s'(z)}\right) = \left(1+z\frac{\phi''(z)}{\phi'(z)}\right) + \left(z\frac{p'(z)}{p(z)}\right)$$
$$\Rightarrow \quad \operatorname{Re}\left(1+z\frac{s''(z)}{s'(z)}\right) \ge \min_{|z|=r,\phi(z)\in K(p,n,A,B)}\left\{\operatorname{Re}\left(1+z\frac{\phi''(z)}{\phi'(z)}\right)\right\} - \left|z\frac{p'(z)}{p(z)}\right|. \tag{6.37}$$

Considering (6.32) and (6.36) together, and after the simple calculations, we get

$$\operatorname{Re}\left(1+z\frac{s''(z)}{s'(z)}\right) \geq \frac{p(1-r^n)(1-Ar^n)(1+Ar^n)(1+Br^n)-r^n n[(1+A)(1+Br^n)+(1+B)(1+Ar^n)]}{(1-r^n)(1-Br^n)(1+Br^n)(1+Ar^n)}. \tag{6.38}$$

The denominator of the above expression on the right hand side of the inequality is positive for $0 \leq r \leq 1$ and

$$\begin{aligned} Q(r) =& p(1-r^n)(1-Ar^n)(1+Ar^n)(1+Br^n) - r^n n\big[(1+A)(1+Br^n) \\ &+ (1+B)(1+Ar^n)\big], \\ Q(0) =& p > 0, \quad Q(1) = -2n(1+A)(1+B) < 0. \end{aligned}$$

Thus the smallest positive root r_0 of the equation $Q(r) = 0$ lies between 0 and 1. Therefore, the inequality

$$\operatorname{Re}\left(1+z\frac{s''(z)}{s'(z)}\right) > 0$$

is valid for $|z| < r_0$. Hence the radius of convexity for $K(p,n,A,B)$ is not less than r_0. □

6.3 Application to the Harmonic Mappings

In this section, we will give an application of generalized p-valent Janowski close-to-convex functions to the harmonic mappings.

Under the guarantee of Lemma 6.2, the image of the domain $\mathbb{D}$ by a generalized p-valent Janowski close-to-convex function is the disc as a subdomain of the right half-plane. At the same time, the right half-plane is a convex domain in the direction of the real and imaginary axes. Therefore, we can apply the J. Clunie and T. Sheil-Small theorem.

Theorem 6.4 ([2]) *A harmonic function $f = h(z) + \overline{g(z)}$ locally univalent in $\mathbb{D}$ is a univalent mapping of $\mathbb{D}$ onto a domain, convex in the direction of the real axis if and only if $[h(z) - g(z)]$ is a conformal univalent mapping of $\mathbb{D}$ onto a convex domain in the direction of the real axis.*

Definition 6.5 Let $f = h(z) + \overline{g(z)}$ be an element of $S_H^0(p,n)$ and $s(z) \in K(p,n,A,B)$. If

$$h(z) - g(z) = s(z) \tag{6.39}$$

then f is called sense-preserving generalized p-valent harmonic Janowski close-to-convex functions. The class of such functions is denoted by $S_H^0K(p,n,A,B)$.

Theorem 6.5 *If $f=[h(z)+\overline{g(z)}]\in S_H^0K(p,n,A,B)$ then*

$$\begin{cases} pB^{\frac{p-1}{nB}}\frac{1-Ar^n}{1-Br^n}r^{\frac{p-1}{B}}(1-Br^n)^{\frac{p(A-B)}{nB^2}}\le|f_z|\le pB^{\frac{p-1}{nB}}\frac{1+Ar^n}{1+Br^n}r^{\frac{p-1}{B}}(1+Br^n)^{\frac{p(A-B)}{nB^2}}; \\ B\neq 0, \\ pr^{p-1}(1-Ar^n)e^{-\frac{pA}{n}r^n}\le|f_z|\le pr^{n-1}(1+Ar^n)e^{\frac{pA}{n}r^n}; \quad B=0; \end{cases} \tag{6.40}$$

$$\begin{cases} pB^{\frac{p-1}{nB}}\frac{|w(z)|}{1+r}r^{\frac{p-1}{B}}(1-Br^n)^{\frac{p(A-B)}{nB^2}}\le|f_{\overline{z}}|\le pB^{\frac{p-1}{nB}}r^{\frac{p-1}{nB}+1}(1+Br^n)^{\frac{p(A-B)}{nB^2}}; \\ B\neq 0, \\ \frac{p|w(z)|r^{p-1}(1-Ar^n)e^{-\frac{Ap}{n}r^n}}{(1+r)}\le|f_{\overline{z}}|\le\frac{pr^p(1+Ar^n)e^{\frac{Ap}{n}r^p}}{(1-r)}; \quad B=0. \end{cases} \tag{6.41}$$

These distortions are sharp.

Proof Let $f=h(z)+\overline{g(z)}$ be an element of $S_H^0K(p,n,A,B)$ and let $s(z)\in C(p,n,A,B)$ then we have

$$s(z)=h(z)-g(z);\ w(z)=\frac{g'(z)}{h'(z)}\quad\Rightarrow\quad f_z=h'(z)$$
$$=\frac{s'(z)}{1-w(z)};\ \overline{f_{\overline{z}}}=g'(z)=\frac{w(z)s'(z)}{1-w(z)} \tag{6.42}$$

where $w(z)$ is the second dilatation of f and satisfies the condition Schwarz lemma. Therefore, we have

$$\frac{|s'(z)|}{1+r}\le\frac{|s'(z)|}{1+|w(z)|}\le|h'(z)|\le\frac{|s'(z)|}{1-|w(z)|}\le\frac{|s'(z)|}{1-r}, \tag{6.43}$$
$$\frac{|s'(z)||w(z)|}{1+r}\le\frac{|s'(z)||w(z)|}{1+|w(z)|}\le|g'(z)|\le\frac{|s'(z)||w(z)|}{1-|w(z)|}\le\frac{r|s'(z)|}{1-r}. \tag{6.44}$$

Using Corollary 6.2 in these inequalities, we obtain (6.40) and (6.41). We also note that these distortions are sharp because the extremal function can be found in the following manner

$$(1+Bz^n)^{-\frac{p(A-B)}{nB}}\frac{s'(z)}{z^{p-1}}=p\frac{1+Az^n}{1+Bz^n};\quad B\neq 0, \tag{6.45}$$

$$e^{-\frac{Ap}{n}z^n}\frac{s'(z)}{z^{p-1}}=p(1+Az^n);\quad B=0, \tag{6.46}$$

$$h'(z)=\frac{s'(z)}{1-w(z)}\quad\Rightarrow\quad h(z)=\int_0^z\frac{s'(\xi)}{1-w(\xi)}d\xi\ g'(z)=\frac{s'(z)w(z)}{1-w(z)}$$
$$\Rightarrow\quad g'(z)=\int_0^z\frac{s'(\xi)w(\xi)}{1-w(\xi)}d\xi, \tag{6.47}$$

and the solution of $h(z)$ and $g(z)$ must be found under the conditions $h(0)=g(0)=0$.

Hence

$$
\begin{aligned}
f(z) = h(z) + \overline{g(z)} &= \int_0^z \frac{s'(\xi)}{1-w(\xi)}\,d\xi + \overline{\int_0^z \frac{s'(\xi)w(\xi)}{1-w(\xi)}\,d\xi} \\
&= \int_0^z \frac{s'(\xi)}{1-w(\xi)}\,d\xi + \overline{\int_0^z \frac{s'(\xi)}{1-w(\xi)}\,d\xi - \int_0^z s'(\xi)\,d\xi} \\
&= \mathrm{Re}\left(\int_0^z \frac{zs'(\xi)}{1-w(\xi)}\,d\xi\right) - \overline{s(z)}.
\end{aligned}
$$

We note that the second dilatation $w(z)$ must be chosen in an appropriate way in order to satisfy the conditions of Schwarz lemma.

Other distortion and growth theorems can be found from Theorem 6.5 by using the corresponding formula from [3]. □

References

1. Bernardi, S.D.: New distortion theorems for functions of positive real part and applications to the univalent convex functions. Proc. Am. Math. Soc. **45**, 113–118 (1974)
2. Clunie, J., Sheil-Small, T.: Harmonic univalent functions. Ann. Acad. Sci. Fenn., Ser. A 1 Math. **9**, 3–25 (1984)
3. Duren, P.: Harmonic Mappings in the Plane. Cambridge University press, Cambridge (2004)
4. Janowski, W.: Some extremal problems for certain families of analytic functions I. Ann. Pol. Math. **XXVII**, 297–326 (1973)
5. Kaplan, W.: Close-to-convex functions. Mich. Math. J. **1**(2), 169–184 (1952)
6. Umezawa, T.: Multivalently close-to-convex functions. Proc. Am. Math. Soc. **8**(5), 869–874 (1957)

Chapter 7
Remarks on Stability of the Linear Functional Equation in Single Variable

Janusz Brzdęk, Dorian Popa, and Bing Xu

Abstract We present some observations concerning stability of the following linear functional equation (in single variable)

$$\varphi\big(f^m(x)\big) = \sum_{i=1}^{m} a_i(x)\varphi\big(f^{m-i}(x)\big) + F(x),$$

in the class of functions φ mapping a nonempty set S into a Banach space X over a field $\mathbb{K} \in \{\mathbb{R}, \mathbb{C}\}$, where m is a fixed positive integer and the functions $f : S \to S$, $F : S \to X$ and $a_i : S \to \mathbb{K}$, $i = 1, \dots, m$, are given. Those observations complement the results in our earlier paper (Brzdęk et al. in J. Math. Anal. Appl. 373:680–689, 2011).

Key words Hyers-Ulam stability · Linear functional equation · Single variable · Banach space · Characteristic root

Mathematics Subject Classification Primary 39B82 · Secondary 39B62

Dedicated to Professor Themistocles M. Rassias on the occasion of his 60th birthday.

J. Brzdęk
Department of Mathematics, Pedagogical University, Podchorążych 2, 30-084 Kraków, Poland
e-mail: jbrzdek@ap.krakow.pl

D. Popa
Department of Mathematics, Technical University, Str. Memorandumului 28, Cluj-Napoca, 400114, Romania
e-mail: Popa.Dorian@math.utcluj.ro

B. Xu (✉)
Department of Mathematics, Sichuan University, Chengdu, Sichuan 610064, P.R. China
e-mail: xb0408@yahoo.com.cn

P.M. Pardalos et al. (eds.), *Nonlinear Analysis*, Springer Optimization and Its Applications 68, In Honor of Themistocles M. Rassias on the Occasion of his 60th Birthday, DOI 10.1007/978-1-4614-3498-6_7,

7.1 Introduction

Let $\mathbb{N}$, $\mathbb{Z}$, $\mathbb{R}$, and $\mathbb{C}$ denote the sets of positive integers, integers, reals, and complex numbers, respectively; moreover, $\mathbb{R}_+ := [0, \infty)$, $\mathbb{N}_0 := \mathbb{N} \cup \{0\}$, and $\mathbb{K} \in \{\mathbb{R}, \mathbb{C}\}$.

Throughout this paper, X is a Banach space over $\mathbb{K}$, S is a nonempty set, $f : S \to S$, $F : S \to X$, $m \in \mathbb{N}$, and $a_j : S \to \mathbb{K}$ for $j = 1, \ldots, m$, unless explicitly stated otherwise. As usual, for each $p \in \mathbb{N}_0$, we write f^p for the pth iterate of f, i.e.,

$$f^0(x) = x, \qquad f^{p+1}(x) = f\big(f^p(x)\big), \quad p \in \mathbb{N}_0,\ x \in S$$

and, only if f is bijective, $f^{-p} = (f^{-1})^p$.

Given a function $\varepsilon_0 : S \to \mathbb{R}_+$, we say that $\varphi_s : S \to X$ is an ε_0-approximate solution (abbreviated to ε_0-solution in the sequel) of the linear functional equation

$$\varphi\big(f^m(x)\big) = \sum_{i=1}^{m} a_i(x)\varphi\big(f^{m-i}(x)\big) + F(x) \tag{7.1}$$

if

$$\left\| \varphi_s\big(f^m(x)\big) - \sum_{i=1}^{m} a_i(x)\varphi_s\big(f^{m-i}(x)\big) - F(x) \right\| \le \varepsilon_0(x), \quad x \in S. \tag{7.2}$$

For information and references on functional equation (7.1), we refer to, e.g., [8, 9, 38, 51, 58, 59] (cf. also [8]); some recent examples of applications can be found, e.g., in [60, Chap. 4].

A simply particular case of functional equation (7.1), with $S \in \{\mathbb{N}_0, \mathbb{Z}\}$, is the difference equation

$$y_{n+m} = \sum_{i=1}^{m} a_i(n) y_{n+m-i} + b_n, \quad n \in S, \tag{7.3}$$

for sequences $(y_n)_{n\in S}$ in X, where $(b_n)_{n\in S}$ is a fixed sequence in X; namely equation (7.1) becomes the difference equation (7.3) with

$$f(n) = n + 1, \qquad y_n := \varphi(n) = \varphi\big(f^n(0)\big), \qquad b_n := F(n), \quad n \in S.$$

In this paper, we give some observations on stability of the functional equation (7.1), which complement the results of our earlier paper [20].

Let us recall that the notion of stability for functional equations was motivated by a problem of S.M. Ulam and a paper of D.H. Hyers [40] in which he published a solution to it (see also [41, 46–48]). However, an earlier result of this kind is due to Gy. Pólya and G. Szegö [67, Teil I, Aufgabe 99] (cf. [36, p. 125]).

The first generalizations of the result in [40] were given by T. Aoki [3] and D.G. Bourgin [11] (without a proof). Further extensions and/or modifications of the

Ulam idea of stability have been proposed in [10, 11, 33, 37, 87] (for some further information, see [28, 31, 34, 44, 52, 53, 81]).

That kind of stability (of functional equations, but not only), called quite often *the Hyers–Ulam stability*, is now a very popular subject of investigation and some quite recent results can be found, e.g., in [2, 4, 15, 16, 23–27, 30, 35, 36, 39, 45, 55–57, 61–63, 78–80, 82–84, 88]).

We should mention here about a crucial role that the paper [68] has played, though in a part it actually rediscovered some ideas of the paper by T. Aoki [3] (quite forgotten at that time somehow). It has drawn the attention of many authors to the Ulam problem anew. That paper, together with numerous further books and papers of Th.M. Rassias (see, e.g., [42–45, 49, 50, 54, 64–66, 69–77]) devoted to the stability of functional as well as differential equations in various classes of mappings, has strongly stimulated the research in this field. Due to this fact, a particular generalization of the Hyers–Ulam stability is now called the *Hyers–Ulam–Rassias* stability (see, e.g., [33, 52, 53, 85]).

Stability of equations in single variable is discussed extensively in [1]. The Hyers–Ulam stability of (7.1) has been investigated so far mainly for $m = 1$ and, except for some results in [17, 85] concerning the case where the coefficient functions $a_1, \ldots, a_m$ are constant and the results in [20], hardly any result for $m > 1$ has been published till now (see, e.g., [1, 5, 12, 32, 86]).

As it was observed by G.L. Forti [32], stability of functional equations in single variable plays a very significant role in the general theory of stability of functional equations; suitable examples can be found in [13, 15, 80]. For more information on functional equations in single variable, we refer to [8, 58].

The results presented in this paper correspond to the outcomes in [6, 7, 21, 22, 29].

7.2 Preliminaries

In what follows, we use a hypothesis concerning the roots of the equation

$$z^m - \sum_{j=1}^{m} a_j(x) z^{m-j} = 0, \tag{7.4}$$

which (for $x \in S$) is the characteristic equation of the functional equation (7.1). The hypothesis reads as follows.

($\mathscr{H}$) *Functions $r_1, \ldots, r_m : S \to \mathbb{C}$ satisfy the condition*

$$\prod_{i=1}^{m} \bigl(z - r_i(x)\bigr) = z^m - \sum_{j=1}^{m} a_j(x) z^{m-j}, \quad x \in S, z \in \mathbb{C}. \tag{7.5}$$

Remark 7.1 Clearly, Hypothesis ($\mathscr{H}$) means that $r_1(x), \ldots, r_m(x) \in \mathbb{C}$ are the complex roots of (7.4) for every $x \in S$. Note that the functions $r_1, \ldots, r_m$ are not unique, but for every $x \in S$ the sequence

$$\big(r_1(x), \dots, r_m(x)\big)$$

is uniquely determined up to a permutation.

Also, it is easily seen that

$$0 \notin a_m(S) \quad \text{if and only if} \quad 0 \notin r_j(S) \quad \text{for } j = 1, \dots, m.$$

We need yet the following definition.

Definition 7.1 We say that a function $\varphi : S \to X$ is f-invariant provided

$$\varphi\big(f(x)\big) = \varphi(x), \quad x \in S.$$

Remark 7.2 It is easily seen that a function $\varphi : S \to X$ is f-invariant if and only if φ is constant on the set

$$[x] := \{y \in S : x\iota y\}$$

for every $x \in S$, where the (equivalence) relation $\iota \subset S^2$ is defined by (cf. [58, p. 14]):

$$\iota := \big\{(x, y) \in S^2 : f^m(x) = f^k(y) \text{ with some } k, m \in \mathbb{N}_0\big\}.$$

Remark 7.3 Under the assumption that $(\mathscr{H})$ holds, $a_1, \dots, a_m$ are f-invariant if and only if $r_1, \dots, r_m$ can be chosen f-invariant (see [20, Remark 3]).

To simplify some statements, we write

$$\prod_{p=1}^{0} \lambda\big(h^p(x)\big) := 1$$

for every $h : S \to S$, $\lambda : S \to \mathbb{K}$, $x \in S$. Moreover, we assume that the restriction to the empty set of any function is injective.

7.3 The Case $m = 1$

We start with some results concerning the case $m = 1$. The next proposition has been proved in [20] (see [20, Lemma 1]); it is a very useful generalization of [84, Theorem 2.1].

Proposition 7.1 *Let* $\varepsilon_0 : S \to \mathbb{R}_+$, $a : S \to \mathbb{K}$,

$$S' := \big\{x \in S : a\big(f^p(x)\big) \neq 0 \textit{ for } p \in \mathbb{N}_0\big\},$$

$$\varepsilon'(x) := \sum_{k=0}^{\infty} \frac{\varepsilon_0(f^k(x))}{\prod_{p=0}^{k} |a(f^p(x))|} < \infty, \quad x \in S',$$

and $\varphi_s : S \to X$ be a function satisfying the inequality

$$\left\| \varphi_s\big(f(x)\big) - a(x)\varphi_s(x) - F(x) \right\| \le \varepsilon_0(x), \quad x \in S. \tag{7.6}$$

Suppose that the function

$$f_0 := f|_{S \setminus S'}$$

is injective,

$$f\big(S \setminus S'\big) \subset S \setminus S' \quad \text{and} \quad a\big(S \setminus S'\big) \subset \{0\}. \tag{7.7}$$

Then the limit

$$\varphi'(x) := \lim_{n\to\infty} \left[\frac{\varphi_s(f^n(x))}{\prod_{j=0}^{n-1} a(f^j(x))} - \sum_{k=0}^{n-1} \frac{F(f^k(x))}{\prod_{j=0}^{k} a(f^j(x))} \right] \tag{7.8}$$

exists for every $x \in S'$ and the function $\varphi : S \to X$, given by:

$$\varphi(x) := \begin{cases} \varphi'(x), & \text{if } x \in S'; \\ F(f_0^{-1}(x)), & \text{if } x \in f(S) \setminus S'; \\ \varphi_s(x) + u(x), & \text{if } x \in S \setminus [S' \cup f(S)], \end{cases} \tag{7.9}$$

with any $u : S \to X$ such that

$$\left\| u(x) \right\| \le \varepsilon_0(x), \quad x \in S,$$

is a solution of the functional equation

$$\varphi\big(f(x)\big) = a(x)\varphi(x) + F(x) \tag{7.10}$$

with

$$\left\| \varphi_s(x) - \varphi(x) \right\| \le \varepsilon'(x), \quad x \in S, \tag{7.11}$$

where

$$\varepsilon'(x) := \begin{cases} \varepsilon_0(f_0^{-1}(x)), & \text{if } x \in f(S) \setminus S'; \\ \varepsilon_0(x), & \text{if } x \in S \setminus [S' \cup f(S)]. \end{cases}$$

Moreover, φ is the unique solution of (7.10) *that satisfies* (7.11) *if and only if*

$$S = S' \cup f(S).$$

Now we give some remarks and examples complementing it.

Remark 7.4 Observe that, in the case

$$\varepsilon_0(x) = \|\varphi_s(f(x)) - a(x)\varphi_s(x) - F(x)\| = \|\varphi_s(f(x)) - F(x)\|, \quad x \in f(S) \setminus S',$$

the formula of Proposition 7.1 defining ε' on the set $f(S) \setminus S'$ is the best possible, because for every solution $\varphi : S \to X$ of (7.10) we have

$$\|\varphi_s(x) - \varphi(x)\| = \|\varphi_s(f(f_0^{-1}(x)) - F(f_0^{-1}(x))\| = \varepsilon_0(f_0^{-1}(x)), \quad x \in f(S) \setminus S'.$$

Moreover, the assumptions that f_0 is injective and

$$f(S \setminus S') \subset S \setminus S'$$

(or some other similar conditions) are necessary in Proposition 7.1. The following two examples show it.

Example 7.1 Let $S = \{-1, 0, 1\}$, $X = \mathbb{K} = \mathbb{R}$,

$$f(x) = x^2, \quad F(x) = \frac{x}{2}, \quad \varphi_s(x) = \frac{x}{2}, \quad \varepsilon_0(x) = |x|, \quad x \in S,$$

and

$$a(1) = a(-1) = 0, \quad a(0) = 1.$$

Then $S' = \{0\}$, (7.6) holds, $\varepsilon'(0) = 0$,

$$f(S \setminus S') \subset S \setminus S',$$

and f_0 is not injective. Next, for every $\varphi : S \to X$ we have:

$$\varphi(f(1)) = \varphi(f(-1))$$

and

$$a(1)\varphi(1) + F(1) = \frac{1}{2} \neq \frac{-1}{2} = a(-1)\varphi(-1) + F(-1),$$

whence (7.10) has no solutions in the class of functions $\varphi : S \to X$.

Example 7.2 Let $S = \{-1, 1\}$, $X = \mathbb{K} = \mathbb{R}$,

$$f(x) = x^2, \quad F(x) = x^2, \quad \varphi_s(x) = x, \quad \varepsilon_0(x) = 2, \quad x \in S,$$

and

$$a(-1) = 0, \quad a(1) = 2.$$

Then $S' = \{1\}$,

$$f(S \setminus S') \cap S' = \{1\},$$

f_0 is injective, (7.6) holds, and $\varepsilon'(1) = 2$. Suppose $\varphi : S \to X$ is a solution of (7.10). Clearly,

$$2\varphi(1) + 1 = a(1)\varphi(1) + F(1) = \varphi\big(f(1)\big)$$
$$= \varphi\big(f(-1)\big) = a(-1)\varphi(-1) + F(-1) = 1,$$

so $\varphi(1) = 0$, and consequently,

$$1 = F(1) = \varphi\big(f(1)\big) - a(1)\varphi(1) = 0.$$

This contradiction means that (7.10) has no solutions in the class of functions $\varphi : S \to X$.

Remark 7.5 Proposition 7.1 yields some information concerning solutions of (7.10). For instance, for every solution $\varphi : S \to X$ of (7.10), inequality (7.6) is satisfied with

$$\varphi_s = \varphi \quad \text{and} \quad \varepsilon_0(x) := 0, \quad x \in S,$$

whence, by Proposition 7.1, (7.9) holds (with $\varphi_s = \varphi$). Since, in the case where $|a(x)| > 1$ and the set

$$\{f^p(x) : p \in \mathbb{N}\}$$

is finite for each $x \in S$, we have $S = S'$ and

$$\lim_{n\to\infty} \frac{\varphi(f^n(x))}{\prod_{j=0}^{n-1} a(f^j(x))} = 0, \quad x \in S,$$

it follows that in this case (7.10) has exactly one solution, given by

$$\varphi(x) := -\lim_{n\to\infty} \sum_{k=0}^{n} \frac{F(f^k(x))}{\prod_{j=0}^{k} a(f^j(x))}, \quad x \in S.$$

An analogous fact can also be derived from the next corollary that has been proved in [20] (see [20, Corollary 1]).

Corollary 7.1 *Let* $a : S \to \mathbb{K}$, $\varepsilon_0 : S \to \mathbb{R}_+$, $\varphi_s : S \to X$ *satisfy* (7.6), f *be bijective,*

$$S'' := \big\{x \in S : a\big(f^{-p}(x)\big) \neq 0 \text{ for } p \in \mathbb{N}\big\},$$

$f(S'') \subset S''$, $a(S \setminus S'') \subset \{0\}$, *and*

$$\varepsilon''(x) := \sum_{k=1}^{\infty} \varepsilon_0\big(f^{-k}(x)\big) \prod_{p=1}^{k-1} \big|a\big(f^{-p}(x)\big)\big| < \infty, \quad x \in S''.$$

Then, for every $x \in S''$, the limit

$$\varphi''(x) := \lim_{n\to\infty}\left[\varphi_s\big(f^{-n}(x)\big)\prod_{j=1}^{n} a\big(f^{-j}(x)\big) + \sum_{k=1}^{n} F\big(f^{-k}(x)\big)\prod_{j=1}^{k-1} a\big(f^{-j}(x)\big)\right] \tag{7.12}$$

exists and the function $\varphi : S \to X$, given by

$$\varphi(x) := \begin{cases} \varphi''(x), & \text{if } x \in S''; \\ F(f^{-1}(x)), & \text{if } x \in S \setminus S'', \end{cases} \tag{7.13}$$

is the unique solution of (7.10) *such that*

$$\|\varphi_s(x) - \varphi(x)\| \le \varepsilon''(x), \quad x \in S, \tag{7.14}$$

where

$$\varepsilon''(x) = \varepsilon_0\big(f^{-1}(x)\big), \quad x \in S \setminus S''.$$

Remark 7.6 Actually, in Corollary 7.1 we have (see [20, Remark 4])

$$S'' = \big\{x \in S : a\big(f^{-p}(x)\big) \neq 0 \textit{ for } p \in \mathbb{N}_0\big\}.$$

The next proposition (see [20, Lemma 2]) presents a somewhat simplified result.

Proposition 7.2 *Assume that f is bijective, $\varepsilon_0 : S \to \mathbb{R}_+$, and $a : S \to \mathbb{K}$ are f-invariant,*

$$\overline{S} := \big\{x \in S : |a(x)| \neq 1\big\},$$

and $\varphi_s : S \to X$ satisfies (7.6). *Then there exists a unique solution $\varphi : \overline{S} \to X$ of* (7.10) *such that*

$$\|\varphi_s(x) - \varphi(x)\| \le \frac{\varepsilon_0(x)}{|1 - |a(x)||}, \quad x \in \overline{S}. \tag{7.15}$$

Remark 7.7 Assume that the assumptions of Proposition 7.2 are valid, $S \neq \overline{S}$, and

$$f(x) = x, \quad F(x) \neq 0, \quad a(x) = 1, \quad x \in S \setminus \overline{S}.$$

Then (7.10) has no solutions in the class of functions X^S. This shows that, in the general situation, the solution $\varphi : \overline{S} \to X$ to (7.10), in the statement of Proposition 7.2, cannot be extended to a solution of (7.10) that maps S into X.

In view of this and the next example, it seems that there is no 'reasonable' general extension of the estimate (7.15) for $x \in S \setminus \overline{S}$, even in the case where (7.10) has solutions that map S into X.

Example 7.3 Let $S = X = \mathbb{R}$ and $\varphi_s(x) = x$ for $x \in \mathbb{R}$. Then

$$\left|\varphi_s(x+1) - \varphi_s(x)\right| = 1, \quad x \in \mathbb{R},$$

which means that (7.6) holds with

$$f(x) = x + 1, \quad a(x) = 1, \quad F(x) = 0, \quad \varepsilon_0(x) = 1, \quad x \in \mathbb{R}.$$

However, for each solution $\varphi : \mathbb{R} \to \mathbb{R}$ of the functional equation

$$\varphi(x+1) = \varphi(x),$$

we have

$$\varphi_s(n) - \varphi(n) = n - \varphi(0), \quad n \in \mathbb{N},$$

and consequently,

$$\sup_{x \in \mathbb{R}} \bigl(\varphi_s(x) - \varphi(x)\bigr) = \infty.$$

Remark 7.8 The form of φ in Proposition 7.2 can be described (see [20, Remark 5]). Namely, the limits

$$\varphi'(x) := \lim_{n\to\infty} \left[\frac{\varphi_s(f^n(x))}{a(x)^n} - \sum_{k=0}^{n-1} \frac{F(f^k(x))}{a(x)^{k+1}}\right],$$

$$\varphi''(y) := \lim_{n\to\infty} \left[\varphi_s\bigl(f^{-n}(y)\bigr)a(y)^n + \sum_{k=1}^{n} F\bigl(f^{-k}(y)\bigr)a(y)^{k-1}\right]$$

exist for every $x \in S_1$, $y \in S_2$ and

$$\varphi(z) := \begin{cases} \varphi'(z), & \text{if } |a(z)| > 1; \\ F(f^{-1}(z)), & \text{if } a(z) = 0; \\ \varphi''(z), & \text{if } 0 < |a(z)| < 1, \end{cases}$$

for each $z \in S$, where

$$S_1 := \bigl\{x \in S : \left|a(x)\right| > 1\bigr\}$$

and

$$S_2 := \bigl\{x \in S : 0 < \left|a(x)\right| < 1\bigr\}.$$

7.4 The General Case

Now we consider the general case, without any restriction on $m \in \mathbb{N}$. The main result in [20] is the following (see [20, Theorem 1]).

Theorem 7.1 *Let* $\varepsilon_0 : S \to \mathbb{R}_+$, $(\mathscr{H})$ *be valid*, $\varphi_s : S \to X$ *be an* ε_0*-solution of* (7.1) (*i.e.*, (7.2) *holds*), r_j *be* f*-invariant for* $j > 1$, $(i_1, \dots, i_m) \in \{-1, 1\}^m$. *Write*

$$s_j := \frac{1}{2}(1 - i_j), \quad j = 1, \dots, m,$$

and

$$S_1 := \{x \in S : r_1(f^{i_1 p}(x)) \neq 0 \text{ for } p \in \mathbb{N}_0\}.$$

Assume that, for each $j \in \{1, \dots, m\}$, *one of the following three conditions holds*:

1° $i_j = 1$ *for* $j = 1, \dots, m$ *and* $0 \notin a_m(S)$;
2° $i_j = 1$ *for* $j = 1, \dots, m$, f *is injective*, $f(S \setminus S_1) \subset S \setminus S_1$, $r_1(S \setminus S_1) \subset \{0\}$;
3° f *is bijective*, $f(S_1) \subset S_1$, *and* $r_1(S \setminus S_1) \subset \{0\}$.

Further, suppose that, for every $j \in \{1, \dots, m\}$,

$$\varepsilon_j(x) := \sum_{k=s_j}^{\infty} \varepsilon_{j-1}(f^{i_j k}(x)) \prod_{p=s_j}^{k-s_j} |r_j(f^{i_j p}(x))|^{-i_j} < \infty, \quad x \in S_j, \tag{7.16}$$

where

$$S_j := \{x \in S : r_j(x) \neq 0\}, \quad j > 1,$$

and, in the case $S \setminus S_j \neq \emptyset$,

$$\varepsilon_j(x) := \begin{cases} \varepsilon_{j-1}(f^{-1}(x)), & \text{if } x \in f(S) \setminus S_j; \\ \varepsilon_{j-1}(x), & \text{if } x \in S \setminus [S_j \cup f(S)], \end{cases}$$

for $x \in S \setminus S_j$, $j \in \{1, \dots, m\}$. *Then* (7.1) *has a solution* $\varphi : S \to X$ *with*

$$\|\varphi_s(x) - \varphi(x)\| \le \varepsilon_m(x), \quad x \in S. \tag{7.17}$$

Moreover, if r_1 *is* f*-invariant and*

$$S \setminus S_j \subset f(S \setminus S_j), \quad j = 1, \dots, m,$$

then (7.1) *has exactly one solution* $\varphi : S \to X$ *such that*

$$\|\varphi_s(x) - \varphi(x)\| \le h(x)\varepsilon_m(x), \quad x \in S,$$

with some f*-invariant function* $h : S \to \mathbb{R}$.

Remark 7.9 Condition (7.16) seems to be quite complicated. However (see [20, Remark 7]), since r_j is f-invariant for every $j > 1$, we have the following simpler expression for $\varepsilon_j(x)$ in Theorem 7.1:

$$\varepsilon_j(x) := \sum_{k=s_j}^{\infty} \varepsilon_{j-1}(f^{i_j k}(x)) |r_j(x)|^{-i_j(k+i_j)} < \infty, \quad x \in S, j > 1,$$

where

$$s_j := \frac{1}{2}(1 - i_j).$$

Conditions (i), (ii) of Proposition 7.3 and the formulas defining φ' and φ'' in Proposition 7.1 and Corollary 7.1 can be simplified analogously.

Remark 7.10 As it is observed in [20, Remark 8], the form of φ in Theorem 7.1 can be determined. For instance, if $\mathbb{K} = \mathbb{C}$, then $\varphi = \varphi_m$, where φ_m can be described by the following procedure.

Let, for $j = 1, \dots, m$, $u_j : S \to X$ be such that

$$\|u_j(x)\| \le \varepsilon_j(x), \quad x \in S \setminus [S_j \cup f(S)],$$

and

$$\psi_m = \varphi_s, \qquad \psi_{j-1}(z) = \psi_j(f(z)) - r_j(z)\psi_j(z), \quad z \in S.$$

For $k = 1, \dots, m$, $x \in S$ write

$$\varphi_k(x) := \begin{cases} \overline{\varphi}_k(x), & \text{if } x \in S_k; \\ \varphi_{k-1}(f^{-1}(x)), & \text{if } x \in f(S) \setminus S_k; \\ \psi_k(x) + u_k(x), & \text{if } x \in S \setminus [S_k \cup f(S)], \end{cases}$$

where $\varphi_0 := F$ and, for $x \in S_k$,

$$\overline{\varphi}_k(x) = \lim_{n\to\infty} \left[\frac{\psi_k(f^{i_k n}(x))}{\prod_{p=0}^{n-1} r_k(f^{i_k(p+s_k)}(x))^{i_k}} - i_k \sum_{j=0}^{n-1} \frac{\varphi_{k-1}(f^{i_k(j+s_k)}(x))}{\prod_{p=s_k}^{j} r_k(f^{i_k p}(x))^{i_k}} \right].$$

Since, for $k > 1$, r_k is f-invariant, the formula for $\overline{\varphi}_k$ can be written in the following simpler form:

$$\overline{\varphi}_k(x) = \lim_{n\to\infty} \left[\frac{\psi_k(f^{i_k n}(x))}{r_k(x)^{i_k n}} - i_k \sum_{j=0}^{n-1} \frac{\varphi_{k-1}(f^{i_k(j+s_k)}(x))}{r_k(x)^{i_k(j+1-s_k)}} \right].$$

The following example (see [20, Example 1]) provides a simple application of Theorem 7.1.

Example 7.4 Let $A \in \mathbb{K} \setminus \{0\}$ and $a : S \to \mathbb{K} \setminus \{0\}$. Then, according to Theorem 7.1, with $m = 2$, $i_1 = i_2 = 1$, and

$$r_1(x) = a(x), \quad r_2(x) = A, \quad a_1(x) = a(x) + A, \quad a_2(x) = -Aa(x), \quad x \in S,$$

for every $\varphi_s : S \to X$ and $\varepsilon_0 : S \to \mathbb{R}$, satisfying the inequality

$$\|\varphi_s(f^2(x)) - (a(x) + A)\varphi_s(f(x)) + Aa(x)\varphi_s(x) - F(x)\| \le \varepsilon_0(x), \quad x \in S,$$

and condition (7.16) for $j = 1, 2$, there exists a solution $\varphi : S \to X$ of the equation

$$\varphi\big(f^2(x)\big) = \big(a(x) + A\big)\varphi\big(f(x)\big) - Aa(x)\varphi(x) + F(x)$$

such that

$$\big\|\varphi_s(x) - \varphi(x)\big\| \le \varepsilon_2(x), \quad x \in S.$$

Moreover, if a is f-invariant, then such a solution is unique.

The uniqueness of φ in Theorem 7.1 is obtained only under the assumption that r_1 is f-invariant. The next proposition (see [20, Lemma 3]) complements, to some degree, that result.

Proposition 7.3 *Assume that $\varrho : S \to \mathbb{R}$, hypothesis ($\mathscr{H}$) holds, r_j is f-invariant for $j > 1$, $\varphi_1, \varphi_2 : S \to X$ are solutions of* (7.1),

$$\begin{aligned} &\big\|\varphi_1(x) - \varphi_2(x)\big\| \le \varrho(x), \quad x \in S, \\ &r_1(x), \dots, r_m(x) \in \mathbb{K}, \quad x \in S, \end{aligned} \tag{7.18}$$

and, for each $j \in \{1, \dots, m\}$, one of the following two conditions is valid.

(i) $r_1(S \setminus S'_1) \subset \{0\}$, *the sequence $\{r_1(f^n(x))\}_{n\in\mathbb{N}}$ is bounded for $x \in S'_1$,*

$$\lim_{n\to\infty} \varrho\big(f^n(x)\big) \prod_{i=0}^{n-1} \big|r_j\big(f^i(x)\big)\big|^{-1} = 0, \quad x \in S'_j,$$

and $S \setminus S'_j \subset f(S \setminus S'_j)$, where

$$S'_j := \big\{x \in S : r_j\big(f^p(x)\big) \ne 0 \text{ for } p \in \mathbb{N}_0\big\}.$$

(ii) *f is bijective, the sequence $\{r_1(f^{-n}(x))\}_{n\in\mathbb{N}}$ is bounded for every $x \in S''$, and*

$$\lim_{n\to\infty} \varrho\big(f^{-n}(x)\big) \prod_{i=1}^{n} \big|r_j\big(f^{-i}(x)\big)\big| = 0, \quad x \in S,$$

where

$$S'' := \big\{x \in S : r_1\big(f^{-p}(x)\big) \ne 0 \text{ for } p \in \mathbb{N}\big\}.$$

Then

$$\varphi_1 = \varphi_2.$$

Remark 7.11 As it is noticed in [20, Remark 6], some kind of conditions, analogous to (i), (ii), cannot be avoided in Proposition 7.3.

Remark 7.12 In some situations, the assumptions of Theorem 7.1 can be satisfied for several different sequences $(i_1, \ldots, i_m)$. But in general, in view of the statement concerning uniqueness of φ, we cannot use this observation to improve the estimate (7.17), because different $(i_1, \ldots, i_m)$ may yield different solutions φ. The following example illustrates this.

Let $\delta > 0$. In the class of functions $\varphi : \mathbb{R} \to \mathbb{R}$, consider the equation

$$\varphi(x+2) = 3\varphi(x+1) - 2\varphi(x) + \frac{\delta}{1+x^2}, \quad x \in \mathbb{R}.$$

Clearly, $r_1(x) \equiv 1$ and $r_2(x) \equiv 2$ are the roots of its characteristic equation

$$z^2 - 3z + 2 = 0.$$

Next $f : \mathbb{R} \to \mathbb{R}$, $f(x) := x + 1$, is bijective,

$$f^{-1}(x) := x - 1, \quad x \in \mathbb{R},$$

both r_1 and r_2 are f-invariant, and the function $\varphi_s(x) \equiv 0$ satisfies

$$\left|\varphi_s(x+2) - 3\varphi_s(x+1) + 2\varphi_s(x) - \frac{\delta}{1+x^2}\right| \le \varepsilon_0(x), \quad x \in \mathbb{R}, \tag{7.19}$$

with

$$\varepsilon_0(x) := \frac{\delta}{1+x^2}, \quad x \in \mathbb{R}.$$

It is easily seen that, for all $x \in \mathbb{R}$, we have (with $i_1 = 1$)

$$\begin{aligned} \varepsilon_1^{(1)}(x) &:= \sum_{t_1=0}^{\infty} \varepsilon_0\big(f^{t_1}(x)\big) \prod_{p=0}^{t_1} \big|r_1\big(f^p(x)\big)\big|^{-1} \\ &= \sum_{t_1=0}^{\infty} \frac{\delta}{1+(x+t_1)^2} \le \delta\left[1 + \int_x^{\infty} \frac{dt}{1+t^2}\right] \\ &\le \delta\left[1 + \frac{\pi}{2} - \arctan x\right] \le (1+\pi)\delta \end{aligned}$$

and (for $i_1 = -1$)

$$\begin{aligned} \varepsilon_1^{(-1)}(x) &:= \sum_{t_1=1}^{\infty} \varepsilon_0\big(f^{-t_1}(x)\big) \prod_{p=1}^{t_1-1} \big|r_1\big(f^{-p}(x)\big)\big| \\ &= \sum_{t_1=1}^{\infty} \frac{\delta}{1+(t_1-x)^2} \le \delta\left[1 + \int_{1-x}^{\infty} \frac{dt}{1+t^2}\right] \end{aligned}$$

$$\leq \delta\left[1+\frac{\pi}{2}-\arctan(1-x)\right] \leq (1+\pi)\delta,$$

whence (for $i_2 = 1$)

$$\varepsilon_2^{(1,1)}(x) := \sum_{t_2=0}^{\infty} \varepsilon_1^{(1)}\big(f^{t_2}(x)\big) \prod_{p=0}^{t_2} \big|r_2\big(f^p(x)\big)\big|^{-1}$$

$$\leq \sum_{t_2=0}^{\infty} 2^{-(t_2+1)}(1+\pi)\delta = (1+\pi)\delta,$$

$$\varepsilon_2^{(-1,1)}(x) := \sum_{t_2=0}^{\infty} \varepsilon_1^{(-1)}\big(f^{t_2}(x)\big) \prod_{p=0}^{t_2} \big|r_2\big(f^p(x)\big)\big|^{-1}$$

$$\leq \sum_{t_2=0}^{\infty} 2^{-(t_2+1)}(1+\pi)\delta = (1+\pi)\delta.$$

Therefore, in view of Theorem 7.1 and Remark 7.10, functions

$$\varphi^{(-1,1)} := -\varepsilon_2^{(-1,1)}, \qquad \varphi^{(1,1)} := \varepsilon_2^{(1,1)}$$

are the unique solutions of equation (7.19) satisfying the estimate (7.17) with

$$\varepsilon_m := \varepsilon_2^{(-1,1)}, \qquad \varepsilon_m := \varepsilon_2^{(1,1)},$$

respectively. Since

$$\varepsilon_2^{(-1,1)}(x) > 0, \qquad \varepsilon_2^{(1,1)}(x) > 0, \quad x \in \mathbb{R},$$

we have $\varphi^{(-1,1)} \neq \varphi^{(1,1)}$.

The next corollary shows that, under some assumptions, solutions of (7.20) generate solutions of (7.21). Clearly, one could ask if different solutions of (7.20) generate different solutions of (7.21). Statement (γ) of the corollary shows that this is the case.

Corollary 7.2 *Assume that* $(\mathscr{H})$ *is valid,* $(i_1, \ldots, i_m) \in \{-1, 1\}^m$, r_j *is* f*-invariant for* $j > 1$, *one of conditions* 1°–3° *of Theorem* 7.1 *holds with*

$$S_1 := \big\{x \in S : r_1\big(f^{i_1 p}(x)\big) \neq 0 \text{ for } p \in \mathbb{N}_0\big\},$$

$\overline{F} : S \to X$, *and, for* $j = 1, \ldots, m$,

$$F_j(x) := \sum_{k=s_j}^{\infty} F_{j-1}\big(f^{i_j k}(x)\big) \prod_{p=s_j}^{k-s_j} \big|r_j\big(f^{i_j p}(x)\big)\big|^{-i_j} < \infty, \quad x \in S_j,$$

where

$$s_j := \frac{1}{2}(1 - i_j), \quad j = 1, \ldots, m,$$

$$S_j := \{x \in S : r_j(x) \neq 0\}, \quad j > 1,$$

$F_0 := \|F\|$, *and, in the case* $S \setminus S_j \neq \emptyset$,

$$F_j(x) = \begin{cases} F_{j-1}(f^{-1}(x)), & \text{if } x \in f(S) \setminus S_j; \\ F_{j-1}(x), & \text{if } x \in S \setminus [S_j \cup f(S)] \end{cases}$$

for $x \in S \setminus S_j$, $j = 1, \ldots, m$. *Then the following three statements are valid.*

(α) *Let* $\psi : S \to X$ *be a solution of the functional equation*

$$\psi\big(f^m(x)\big) = \sum_{i=1}^{m} a_i(x)\psi\big(f^{m-i}(x)\big) + \overline{F}(x) - F(x). \tag{7.20}$$

Then there is a solution $\varphi : S \to X$ *of the equation*

$$\varphi\big(f^m(x)\big) = \sum_{i=1}^{m} a_i(x)\varphi\big(f^{m-i}(x)\big) + \overline{F}(x) \tag{7.21}$$

with

$$\big\|\psi(x) - \varphi(x)\big\| \leq F_m(x), \quad x \in S.$$

Moreover, if r_1 *is* f*-invariant and*

$$S \setminus S_j \subset f(S \setminus S_j), \quad j = 1, \ldots, m,$$

then there exists exactly one solution $\varphi : S \to X$ *of* (7.21) *such that*

$$\big\|\psi(x) - \varphi(x)\big\| \leq h(x)F_m(x), \quad x \in S,$$

with some f*-invariant function* $h : S \to \mathbb{R}$.

(β) *There is a solutions* $\varphi : S \to X$ *of* (7.1) *with*

$$\big\|\varphi(x)\big\| \leq F_m(x), \quad x \in S.$$

Moreover, if r_1 *is* f*-invariant and*

$$S \setminus S_j \subset f(S \setminus S_j), \quad j = 1, \ldots, m,$$

then there exists exactly one solution $\varphi : S \to X$ *of* (7.1) *such that*

$$\big\|\varphi(x)\big\| \leq h(x)F_m(x), \quad x \in S,$$

with some f*-invariant function* $h : S \to \mathbb{R}$.

(γ) *Let $h_1, h_2 : S \to \mathbb{R}$ and r_1 be f-invariant functions,*

$$S \setminus S_j \subset f(S \setminus S_j), \quad j = 1, \dots, m,$$

$\psi_1, \psi_2 : S \to X$ be solutions of (7.20), *$\psi_1 \neq \psi_2$, and $\varphi_1, \varphi_2 : S \to X$ be solutions of* (7.21) *such that*

$$\| \psi_i(x) - \varphi_i(x) \| \leq h_i(x) F_m(x), \quad x \in S, i = 1, 2.$$

Then

$$\varphi_1 \neq \varphi_2.$$

Proof (α) Since

$$\| \psi(f^m(x)) - [a_1(x)\psi(f^{m-1}(x)) + \cdots + a_m(x)\psi(x)] - \overline{F}(x) \| \leq \| F(x) \|, \quad x \in S,$$

by Theorem 7.1 (with $\varphi_s = \psi$ and $\varepsilon_0 = \|F\|$), there is a solution $\varphi : S \to X$ of (7.1) with

$$\| \psi(x) - \varphi(x) \| \leq F_m(x), \quad x \in S;$$

moreover, if r_1 is f-invariant and $S \setminus S_j \subset f(S \setminus S_j)$ for $j = 1, \dots, m$, then we obtain the statement concerning uniqueness of φ.

(β) It follows from the statement (α) with $\psi(x) \equiv 0$ and $\overline{F} = F$.
(γ) Let

$$h_0(x) := \max\{h_1(x), h_2(x)\}, \quad x \in S.$$

In view of Remark 7.2, $h_0 : S \to \mathbb{R}$ is f-invariant. For the proof by contradiction, suppose that $\varphi_1 = \varphi_2$. Note that

$$\varphi_0 := \varphi_2 + \psi_2 - \psi_1$$

is a solution of (7.21) and

$$\| \psi_2(x) - \varphi_0(x) \| = \| \psi_1(x) - \varphi_1(x) \| \leq h_1(x) F_m(x) \leq h_0(x) F_m(x), \quad x \in S.$$

Since, according to statement (α), φ_2 is the unique solution of (7.21) such that

$$\| \psi_2(x) - \varphi_2(x) \| \leq h_2(x) F_m(x) \leq h_0(x) F_m(x), \quad x \in S,$$

we have $\varphi_0 = \varphi_2$, which implies $\psi_1 = \psi_2$. This is a contradiction. □

If we assume that $\varepsilon_0, a_1, \dots, a_m$ are f-invariant and f is bijective, then we obtain the following result, which is much simpler than Theorem 7.1 (see [20, Theorem 2]).

Theorem 7.2 *Suppose that hypothesis* ($\mathscr{H}$) *holds, f is bijective, $\varepsilon_0 : S \to \mathbb{R}$ and $a_1, \ldots, a_m$ are f-invariant,*

$$\widetilde{S} := \{x \in S : |r_j(x)| \neq 1 \text{ for } j = 1, \ldots, m\},$$

and a function $\varphi_s : S \to X$ is an ε_0-solution of (7.1). *Then there is a unique solution $\varphi : \widetilde{S} \to X$ of* (7.1) *such that*

$$\|\varphi_s(x) - \varphi(x)\| \le \frac{\varepsilon_0(x)}{|(1 - |r_1(x)|) \cdots (1 - |r_m(x)|)|}, \quad x \in \widetilde{S}. \tag{7.22}$$

Moreover, φ is the unique solution of (7.1) *such that*

$$\|\varphi_s(x) - \varphi(x)\| \le \varepsilon(x), \quad x \in \widetilde{S},$$

with some f-invariant function $\varepsilon : \widetilde{S} \to \mathbb{R}$.

Remark 7.13 In the case $\mathbb{K} = \mathbb{R}$ and $r_j(S) \subset [0, \infty)$ for $j = 1, \ldots, m$, the estimate (7.22) in Theorem 7.2 is the best possible in the general situation. Namely, let $\varphi_s(x) \equiv 0$ and F be f-invariant (e.g., $F(x) \equiv$ const). Then (7.2) holds with $\varepsilon_0 = \|F\|$. Following the steps described in Remark 7.10, we obtain that

$$\varphi(x) = \frac{F(x)}{(1 - r_1(x)) \cdots (1 - r_m(x))}, \quad x \in \widetilde{S}.$$

Since

$$|r_j(x)| = r_j(x), \quad j = 1, \ldots, m, x \in S,$$

we have

$$\|\varphi_s(x) - \varphi(x)\| = \frac{\varepsilon_0(x)}{|1 - |r_1(x)|| \cdots |1 - |r_m(x)||}, \quad x \in \widetilde{S}.$$

But in some special situations, we can get sometimes much better estimates than (7.22) (see [14, p. 3]).

Clearly, Theorem 7.2 yields the following corollary (see [20, Corollary 2]).

Corollary 7.3 *Suppose that f is bijective, $a_1, \ldots, a_m$ and $\varepsilon_0 : S \to \mathbb{R}$ are f-invariant, $\varphi_s : S \to X$ is an ε_0-solution of* (7.1), ($\mathscr{H}$) *holds, and*

$$l_j := \inf_{x \in S} \big|1 - |r_j(x)|\big| > 0, \quad j \in \{1, \ldots, m\}. \tag{7.23}$$

Then there exists a unique solution $\varphi : S \to X$ of (7.1) *such that*

$$\|\varphi_s(x) - \varphi(x)\| \le \frac{\varepsilon_0(x)}{l_1 \cdots l_m}, \quad x \in S. \tag{7.24}$$

7.5 The Hyers–Ulam Stability

Now we recall some observations from [20] concerning the Hyers–Ulam stability of equation (7.1), i.e., the case where the function ε_0 is constant. To this end, we need the following two definitions (cf. [20, Definitions 2 and 3]).

Definition 7.2 Equation (7.1) is said to be weakly Hyers–Ulam stable (in the class of functions $\psi : S \to X$) provided, for every unbounded function $\psi : S \to X$ with

$$\sup_{x\in S}\left\| \psi\bigl(f^m(x)\bigr) - \sum_{i=1}^{m} a_i(x)\psi\bigl(f^{m-i}(x)\bigr) - F(x)\right\| < \infty,$$

there exists a solution $\varphi : S \to X$ of (7.1) such that

$$\sup_{x\in S}\bigl\|\varphi(x) - \psi(x)\bigr\| < \infty.$$

Definition 7.3 Equation (7.1) is said to be strongly Hyers–Ulam stable (in the class of functions $\psi : S \to X$) provided there exists $\alpha \in \mathbb{R}$ such that, for every $\delta > 0$ and for every $\psi : S \to X$ satisfying

$$\sup_{x\in S}\left\| \psi\bigl(f^m(x)\bigr) - \sum_{i=1}^{m} a_i(x)\psi\bigl(f^{m-i}(x)\bigr) - F(x)\right\| \le \delta,$$

there exists a solution $\varphi : S \to X$ of (7.1) with

$$\sup_{x\in S}\bigl\|\varphi(x) - \psi(x)\bigr\| \le \alpha\delta.$$

Note that strong stability implies the weak one, and an equation that is not weakly Hyers–Ulam stable is not strongly Hyers–Ulam stable either. Corollary 7.3 yields at once the following.

Corollary 7.4 *Suppose that f is bijective and condition* (7.23) *holds. Then, in the case where $a_1, \dots, a_m$ are f-invariant,* (7.1) *is strongly Hyers–Ulam stable.*

Remark 7.14 Assumption (7.23) is necessary in the corollary above because otherwise the equation can be, even, not weakly Hyers–Ulam stable, which the subsequent example shows (see [20, Example 2]).

Example 7.5 Let $\varepsilon > 0$, $f : S \to S$ be a bijection, $r : S \to \mathbb{R} \setminus \{-1, 1\}$ be f-invariant, $\varphi_0 : S \to \mathbb{R}$ be a solution of the equation

$$\varphi\bigl(f(x)\bigr) = \bigl|r(x)\bigr|\varphi(x), \tag{7.25}$$

and $\varphi_s : S \to \mathbb{R}$ be given by the formula:

$$\varphi_s(x) = \varphi_0(x) + \frac{\varepsilon}{1 - |r(x)|}.$$

Then

$$\varphi_s\big(f(x)\big) - \big|r(x)\big|\varphi_s(x) = \varepsilon, \quad x \in S.$$

Suppose that there exists a solution $\varphi : S \to \mathbb{R}$ of (7.25) such that

$$s_0 := \sup_{x\in S}\big|\varphi_s(x) - \varphi(x)\big| < \infty.$$

Write

$$\varrho(x) := 2\max\left\{s_0, \frac{\varepsilon}{|1 - |r(x)||}\right\}, \quad x \in S.$$

Then ϱ is f-invariant and

$$\big|\varphi(x) - \varphi_0(x)\big| \le \big|\varphi(x) - \varphi_s(x)\big| + \big|\varphi_s(x) - \varphi_0(x)\big| \le \varrho(x), \quad x \in S,$$

whence, according to Proposition 7.3, $\varphi_0 = \varphi$. But, in the case where

$$\inf_{x\in S}\big|1 - \big|r(x)\big|\big| = 0,$$

we have

$$s_0 = \sup_{x\in S}\big|\varphi_s(x) - \varphi(x)\big| = \sup_{x\in S}\big|\varphi_s(x) - \varphi_0(x)\big| = \infty,$$

which is a contradiction.

In this way, we have proved that the equation is not weakly Hyers–Ulam stable if

$$\inf_{x\in S}\big|1 - \big|r(x)\big|\big| = 0.$$

One can find many examples of functions f, r, φ_0 satisfying the conditions given above. For instance, it is enough to take $S = \mathbb{R} \setminus \mathbb{Z}$,

$$f(x) = x + 1, \quad r(x) = h\big(x - \lfloor x \rfloor\big), \quad \varphi_0(x) = \big|r(x)\big|^x, \quad x \in S,$$

where $\lfloor x \rfloor$ denotes the integer part (i.e., floor) of x and h is any function mapping (0, 1) onto (0, 1).

7.6 Constant Coefficients

In the special case when the functions $a_1, \ldots, a_m$ are constant, (7.1) becomes the following functional equation

$$\varphi\big(f^m(x)\big) = \sum_{i=1}^{m} a_i \varphi\big(f^{m-i}(x)\big) + F(x) \tag{7.26}$$

with given fixed $a_1, \ldots, a_m \in \mathbb{K}$. Then Theorems 7.1 and 7.2 obtain much simpler forms; namely, we have the subsequent two corollaries (see [20, Corollaries 3 and 4]).

Corollary 7.5 *Let $r_1, \ldots, r_m \in \mathbb{C}$ be the roots of the characteristic equation of* (7.26), *i.e., of the equation*

$$r^m - \sum_{i=1}^{m} a_i r^{m-i} = 0, \tag{7.27}$$

$\varepsilon_0 : S \to \mathbb{R}$, *and $\varphi_s : S \to X$ be an ε_0-solution of* (7.26). *Suppose that one of the following three conditions is valid*:

(i) $a_m \neq 0$ *and, for every $j \in \{1, \ldots, m\}$,*

$$\varepsilon_j(x) = \sum_{k=0}^{\infty} \varepsilon_{j-1}\big(f^k(x)\big)|r_j|^{-k-1} < \infty, \quad x \in S; \tag{7.28}$$

(ii) *f is injective and, for every $j \in \{1, \ldots, m\}$ with $r_j \neq 0$, condition* (7.28) *holds*;

(iii) *f is bijective and, for every $j \in \{1, \ldots, m\}$ with $r_j \neq 0$, there exists $i_j \in \{-1, 1\}$ with*

$$\varepsilon_j(x) = \sum_{k=s_j}^{\infty} \varepsilon_{j-1}\big(f^{i_j k}(x)\big)|r_j|^{-i_j(k+i_j)} < \infty, \quad x \in S, \tag{7.29}$$

where

$$s_j := \frac{1}{2}(1 - i_j).$$

Then there exists a solution $\varphi : S \to X$ of functional equation (7.26) *with*

$$\big\|\varphi_s(x) - \varphi(x)\big\| \le \varepsilon_m(x), \quad x \in S. \tag{7.30}$$

Further, if f is surjective, then there exists exactly one solution $\varphi : S \to X$ of (7.26) *such that*

$$\big\|\varphi_s(x) - \varphi(x)\big\| \le h(x)\varepsilon_m(x), \quad x \in S,$$

with some f-invariant function $h : S \to \mathbb{R}$.

In particular, if ε_0 *is* f*-invariant and*

$$|r_j| \neq 1, \quad j = 1, \dots, m,$$

then

$$\varepsilon_m(x) = \frac{\varepsilon_0(x)}{||r_1| - 1| \cdots ||r_m| - 1|}, \quad x \in S.$$

Corollary 7.6 *Suppose that* $r_1, \dots, r_m \in \mathbb{C}$ *are the roots of the characteristic* (7.27), $\delta > 0$, *and one of the following three conditions hold*:

(i) $|r_j| > 1$ *for every* $j \in \{1, \dots, m\}$;
(ii) f *is injective and* $|r_j| \in \{0\} \cup (1, \infty)$ *for every* $j \in \{1, \dots, m\}$;
(iii) f *is bijective and* $|r_j| \neq 1$ *for every* $j \in \{1, \dots, m\}$.

If a function $\varphi_s : S \to X$ *satisfies*

$$\left\| \varphi_s\big(f^m(x)\big) - \sum_{i=1}^{m} a_i \varphi_s\big(f^{m-i}(x)\big) - F(x) \right\| \leq \delta, \quad x \in S,$$

then (7.26) *has a solution* $\varphi : S \to X$ *such that*

$$\big\|\varphi_s(x) - \varphi(x)\big\| \leq \frac{\delta}{||r_1| - 1| \cdots ||r_m| - 1|}, \quad x \in S. \tag{7.31}$$

Moreover, if f *is surjective, then there exists exactly one solution* $\varphi : S \to X$ *of* (7.26) *such that*

$$\big\|\varphi_s(x) - \varphi(x)\big\| \leq h(x), \quad x \in S,$$

with some f*-invariant function* $h : S \to \mathbb{R}$.

Corollary 7.6 proves that (7.26) is strongly Hyers–Ulam stable under the assumption that characteristic equation (7.27) has no roots of modulus one. The assumption is necessary, which the following two simple examples show.

Example 7.6 Let $\delta > 0$ and

$$f(x) = x + 1, \qquad \varphi_s(x) := \frac{\delta}{2} x^2, \quad x \in \mathbb{K}.$$

Then

$$\big|\varphi_s\big(f^2(x)\big) - 2\varphi_s\big(f(x)\big) + \varphi_s(x)\big| \leq \delta$$

for $x \in \mathbb{K}$. Let $\varphi : \mathbb{K} \to \mathbb{K}$ be a solution of the equation

$$\varphi\big(f^2(x)\big) = 2\varphi\big(f(x)\big) - \varphi(x). \tag{7.32}$$

Write

$$\psi(x) := \varphi(x+1) - \varphi(x), \quad x \in \mathbb{K}.$$

Clearly,

$$\psi(x+1) = \psi(x), \quad x \in \mathbb{K},$$

which yields

$$\varphi(n) = \varphi(0) + \psi(0) + \cdots + \psi(n-1) = \varphi(0) + n\psi(0), \quad n \in \mathbb{N}.$$

Hence it follows that

$$\lim_{n\to\infty} \left|\varphi_s(n) - \varphi(n)\right| = \lim_{n\to\infty} \left|\frac{\delta}{2}n^2 - \varphi(0) - n\psi(0)\right| = \infty,$$

whence

$$\sup_{x\in K} \left|\varphi_s(x) - \varphi(x)\right| = \infty.$$

This means that equation (7.32) is not weakly Hyers–Ulam stable.

Example 7.7 Let $z_0 \in X \setminus \{0\}$, $\delta := \|z_0\|$ and

$$f(x) := x + z_0, \quad \varphi_s(x) := x, \quad x \in S := X.$$

Then

$$\left\|\varphi_s\big(f^2(x)\big) - 3\varphi_s\big(f(x)\big) + 2\varphi_s(x)\right\| = \delta$$

for $x \in X$. Let $\varphi : X \to X$ be a solution of the equation

$$\varphi\big(f^2(x)\big) = 3\varphi\big(f(x)\big) - 2\varphi(x). \tag{7.33}$$

Write

$$\psi(x) := \varphi(x + z_0) - 2\varphi(x), \quad x \in X.$$

Since

$$\psi(x + z_0) = \psi(x), \quad x \in X,$$

it is easy to show that

$$\varphi(nz_0) = 2^n \cdot \big(\varphi(0) + \psi(0)\big) - \psi(0)$$

for $n \in \mathbb{N}$. So

$$\lim_{n\to\infty} \left\|\varphi_s(nz_0) - \varphi(nz_0)\right\| = \lim_{n\to\infty} \left\|nz_0 - 2^n \cdot \big(\varphi(0) + \psi(0)\big) + \psi(0)\right\| = \infty,$$

and consequently

$$\sup_{x\in X}\left\|\varphi_s(x)-\varphi(x)\right\|=\infty.$$

This proves that (7.33) is not weakly Hyers–Ulam stable either.

We end the paper with an example which shows that, in the case where $r_j=0$ for some $j\in\{1,\dots,m\}$, the assumption of injectivity of f is important in Corollary 7.6 (and in Corollary 7.5 and Theorem 7.1, as well).

Example 7.8 Let $m>1$, $a_1,\dots,a_{m-1}\in\mathbb{K}$, $r_1,\dots,r_m\in\mathbb{C}$ be the roots of the equation

$$r^m-\sum_{i=1}^{m}a_i r^{m-i}=0,$$

$r_m=0$, and $|r_j|>1$ for $j\in\{1,\dots,m-1\}$. Suppose there exist $x_1,x_2\in S$ with $F(x_1)\neq F(x_2)$ and $f(x_1)=f(x_2)$. Then, for every $\varphi:S\to X$,

$$\varphi\bigl(f^m(x_1)\bigr)-\sum_{i=1}^{m}a_i\varphi\bigl(f^{m-i}(x_1)\bigr)=\varphi\bigl(f^m(x_2)\bigr)-\sum_{i=1}^{m}a_i\varphi\bigl(f^{m-i}(x_2)\bigr),$$

which means that the equation

$$\varphi\bigl(f^m(x)\bigr)=\sum_{i=1}^{m}a_i\varphi\bigl(f^{m-i}(x)\bigr)+F(x) \tag{7.34}$$

has no solutions $\varphi:S\to X$. But if S is finite, then clearly each function $\varphi_s:S\to X$ is a δ-approximate solution to (7.34) with

$$\delta(y):=\sup_{x\in S}\left\|\varphi_s\bigl(f^m(x)\bigr)-\sum_{i=1}^{m-1}a_i\varphi_s\bigl(f^{m-i}(x)\bigr)-F(x)\right\|,\quad y\in S.$$

For some results on nonstability, we refer to [18, 19]. Let us recall here [18, Theorem 4].

Theorem 7.3 *Let $T\in\{\mathbb{N}_0,\mathbb{Z}\}$ and $r_1,\dots,r_m\in\mathbb{C}$ denote all the roots of the equation*

$$r^m-\sum_{i=1}^{m}a_i r^{m-i}=0. \tag{7.35}$$

Assume that $|r_j|=1$ for some $j\in\{1,\dots,m\}$. Then, for any $\delta>0$, there exists a sequence $(y_n)_{n\in T}$ in X, satisfying the inequality

$$\left\|y_{n+m}-\sum_{i=1}^{m}a_i y_{n+m-i}-b_n\right\|\le\delta,\quad n\in T, \tag{7.36}$$

such that

$$\sup_{n\in T} \|y_n - x_n\| = \infty \tag{7.37}$$

for every sequence $(x_n)_{n\in T}$ *in* X, *fulfilling the recurrence*

$$x_{n+m} = \sum_{i=1}^{m} a_i x_{n+m-i} + b_n, \quad n \in T. \tag{7.38}$$

Moreover, if $r_1, \dots, r_m \in \mathbb{K}$ *or there is a bounded sequence* $(x_n)_{n\in T}$ *in* X *fulfilling* (7.38), *then* $(y_n)_{n\in T}$ *can be chosen unbounded.*

We end this paper with one more nonstability result, which is a simple consequence of Theorem 7.3.

Theorem 7.4 *Suppose that* (7.1) *has a solution in the class of functions mapping* S *into* X, *characteristic equation* (7.35) *has a complex root of modulus* 1, *and there exists* $x_0 \in S$ *such that*

$$f^k(x_0) \neq f^n(x_0), \quad k, n \in \mathbb{N}_0,\ k \neq n,$$

and

$$f(S \setminus S_0) \subset S \setminus S_0,$$

where $S_0 := \{f^n(x_0) : n \in \mathbb{N}_0\}$. *Then, for each* $\delta > 0$, *there is a function* $\psi : S \to X$, *satisfying the inequality*

$$\sup_{x\in S} \left\| \psi\bigl(f^m(x)\bigr) - \sum_{i=1}^{m} a_i \psi\bigl(f^{m-i}(x)\bigr) - F(x) \right\| \le \delta, \tag{7.39}$$

such that

$$\sup_{x\in S} \|\psi(x) - \varphi(x)\| = \infty$$

for arbitrary solution $\varphi : S \to X$ *of* (7.1).

Moreover, if all the roots of characteristic equation (7.35) *are in* $\mathbb{K}$, *then* ψ *can be chosen unbounded.*

Proof Let $(b_n)_{n\in\mathbb{N}_0}$ be a sequence in X given by

$$b_n = F\bigl(f^n(x_0)\bigr), \quad n \in \mathbb{N}_0.$$

Since the characteristic equation (7.35) has a root of modulus 1, according to Theorem 7.3, there exists a sequence $(y_n)_{n\in\mathbb{N}_0}$ in X satisfying

$$\left\| y_{n+m} - \sum_{i=1}^{m} a_i y_{n+m-i} - b_n \right\| \le \delta, \quad n \in \mathbb{N}_0,$$

such that

$$\sup_{n\in\mathbb{N}_0} \|y_n - x_n\| = \infty$$

for every sequence $(x_n)_{n\in\mathbb{N}_0}$ in X satisfying the recurrence

$$x_{n+m} = \sum_{i=1}^{m} a_i x_{n+m-i} + b_n, \quad n \in \mathbb{N}_0. \tag{7.40}$$

Let $\eta : S \to X$ be a solution of (7.1) and define $\psi : A \to X$ by

$$\psi(x) = \begin{cases} y_i, & \text{if } x = f^j(x_0) \text{ for some } j \in \mathbb{N}_0, \\ \eta(x), & \text{otherwise.} \end{cases}$$

One can easily check that ψ satisfies inequality (7.39).

Now let $\varphi : S \to X$ be an arbitrary solution of (7.1). Then

$$\varphi\bigl(f^{n+m}(x_0)\bigr) = \sum_{i=1}^{m} a_i \varphi\bigl(f^{n+m-i}(x_0)\bigr) + F\bigl(f^n(x_0)\bigr), \qquad n \in \mathbb{N}_0.$$

Let $(x_n)_{n\mathbb{N}_0}$ be given by

$$x_n = \varphi\bigl(f^n(x_0)\bigr), \quad n \in \mathbb{N}_0.$$

Then $(x_n)_{n\in\mathbb{N}_0}$ satisfies (7.40). Since

$$\sup_{n\in\mathbb{N}_0} \|y_n - x_n\| = \infty$$

and

$$\|y_n - x_n\| = \bigl\|\psi\bigl(f^n(x_0)\bigr) - \varphi\bigl(f^n(x_0)\bigr)\bigr\|, \quad n \in \mathbb{N}_0,$$

it follows that

$$\sup_{x\in S}\bigl\|\psi(x) - \varphi(x)\bigr\| = \infty.$$

If the roots of the characteristic equation are in $\mathbb{K}$, then, in view of Theorem 7.3, the sequence $(y_n)_{n\in\mathbb{N}_0}$ can be chosen unbounded, and therefore then ψ can be unbounded. □

References

1. Agarwal, R.P., Xu, B., Zhang, W.: Stability of functional equations in single variable. J. Math. Anal. Appl. **288**, 852–869 (2003)
2. Amyari, M., Moslehian, M.S.: Approximate homomorphisms of ternary semigroups. Lett. Math. Phys. **77**, 1–9 (2006)

3. Aoki, T.: On the stability of the linear transformation in Banach spaces. J. Math. Soc. Jpn. **2**, 64–66 (1950)
4. Badora, R., Chmieliński, J.: Decomposition of mappings approximately inner product preserving. Nonlinear Anal. **62**, 1015–1023 (2005)
5. Baker, J.A.: The stability of certain functional equations. Proc. Am. Math. Soc. **112**, 729–732 (1991)
6. Baron, K.: On approximate solutions of a system of functional equations. Ann. Pol. Math. **43**, 305–316 (1983)
7. Baron, K., Jarczyk, W.: On approximate solutions of functional equations of countable order. Aequ. Math. **28**, 22–34 (1985)
8. Baron, K., Jarczyk, W.: Recent results on functional equations in a single variable, perspectives and open problems. Aequ. Math. **61**, 1–48 (2001)
9. Birkhoff, G.D.: Déformations analytiques et fonctions auto-équivalentes. Ann. Inst. Henri Poincaré **9**, 51–122 (1939)
10. Bourgin, D.G.: Approximately isometric and multiplicative transformations on continuous function rings. Duke Math. J. **16**, 385–397 (1949)
11. Bourgin, D.G.: Classes of transformations and bordering transformations. Bull. Am. Math. Soc. **57**, 223–237 (1951)
12. Brydak, D.: On the stability of the functional equation $\varphi[f(x)] = g(x)\varphi(x) + F(x)$. Proc. Am. Math. Soc. **26**, 455–460 (1970)
13. Brzdęk, J.: On a method of proving the Hyers–Ulam stability of functional equations on restricted domains. Aust. J. Math. Anal. Appl. **6**(1), 4 (2009), 10 pp.
14. Brzdęk, J., Jung, S.-M.: A note on stability of a linear functional equation of second order connected with the Fibonacci numbers and Lucas sequences. J. Ineq. Appl. **2010**, 793947 (2010), 10 pp.
15. Brzdęk, J., Pietrzyk, A.: A note on stability of the general linear equation. Aequ. Math. **75**, 267–270 (2008)
16. Brzdęk, J., Popa, D., Xu, B.: The Hyers–Ulam stability of nonlinear recurrences. J. Math. Anal. Appl. **335**, 443–449 (2007)
17. Brzdęk, J., Popa, D., Xu, B.: The Hyers–Ulam stability of linear equations of higher orders. Acta Math. Hung. **120**, 1–8 (2008)
18. Brzdęk, J., Popa, D., Xu, B.: Remarks on stability of the linear recurrence of higher order. Appl. Math. Lett. **23**, 1459–1463 (2010)
19. Brzdęk, J., Popa, D., Xu, B.: On nonstability of the linear recurrence of order one. J. Math. Anal. Appl. **367**, 146–153 (2010)
20. Brzdęk, J., Popa, D., Xu, B.: On approximate solutions of the linear functional equation of higher order. J. Math. Anal. Appl. **373**, 680–689 (2011)
21. Buck, R.C.: On approximation theory and functional equations. J. Approx. Theory **5**, 228–237 (1972)
22. Buck, R.C.: Approximation theory and functional equations II. J. Approx. Theory **9**, 121–125 (1973)
23. Chmieliński, J.: Stability of the orthogonality preserving property in finite-dimensional inner product spaces. J. Math. Anal. Appl. **318**, 433–443 (2006)
24. Chudziak, J.: Approximate solutions of the Gołąb–Schinzel equation. J. Approx. Theory **136**, 21–25 (2005)
25. Chudziak, J.: Stability problem for the Gołąb–Schinzel type functional equations. J. Math. Anal. Appl. **339**, 454–460 (2008)
26. Chudziak, J., Tabor, J.: On the stability of the Gołąb–Schinzel functional equation. J. Math. Anal. Appl. **302**, 196–200 (2005)
27. Chung, J.: Approximately additive Schwartz distributions. J. Math. Anal. Appl. **324**, 1449–1457 (2006)
28. Czerwik, S.: Functional Equations and Inequalities in Several Variables. World Scientific, London (2002)

29. Dankiewicz, K.: On approximate solutions of a functional equation in the class of differentiable functions. Ann. Pol. Math. **49**, 247–252 (1989)
30. Faĭziev, V.A., Rassias, Th.M., Sahoo, P.K.: The space of (ψ, γ)-additive mappings on semigroups. Trans. Am. Math. Soc. **354**, 4455–4472 (2002)
31. Forti, G.L.: Hyers–Ulam stability of functional equations in several variables. Aequ. Math. **50**, 143–190 (1995)
32. Forti, G.L.: Comments on the core of the direct method for proving Hyers–Ulam stability of functional equations. J. Math. Anal. Appl. **295**, 127–133 (2004)
33. Găvruţa, P.: A generalization of the Hyers–Ulam–Rassias stability of approximately additive mappings. J. Math. Anal. Appl. **184**, 431–436 (1994)
34. Ger, R.: A survey of recent results on stability of functional equations. In: Proc. of the 4th International Conference on Functional Equations and Inequalities (Krakow), pp. 5–36. Pedagogical University of Krakow, Krakow (1994)
35. Gilányi, A.: Hyers–Ulam stability of monomial functional equations on a general domain. Proc. Natl. Acad. Sci. USA **96**, 10588–10590 (1999)
36. Gilányi, A., Kaiser, Z., Palés, Z.: Estimates to the stability of functional equations. Aequ. Math. **73**, 125–143 (2007)
37. Gruber, P.M.: Stability of isometries. Trans. Am. Math. Soc. **245**, 263–277 (1978)
38. Hamilton, H.J.: On monotone and convex solutions of certain difference equations. Am. J. Math. **63**, 427–434 (1941)
39. Házy, A.: On the stability of t-convex functions. Aequ. Math. **74**, 210–218 (2007)
40. Hyers, D.H.: On the stability of the linear functional equation. Proc. Natl. Acad. Sci. USA **27**, 222–224 (1941)
41. Hyers, D.H.: Transformations with bounded mth differences. Pac. J. Math. **11**, 591–602 (1961)
42. Hyers, D.H., Isac, G., Rassias, Th.M.: Approximate homomorphisms. Aequ. Math. **44**, 125–153 (1992)
43. Hyers, D.H., Isac, G., Rassias, Th.M.: Topics in Nonlinear Analysis and Applications. World Scientific, Singapore (1997)
44. Hyers, D.H., Isac, G., Rassias, Th.M.: Stability of Functional Equations in Several Variables. Birkhäuser, Boston (1998)
45. Hyers, D.H., Isac, G., Rassias, Th.M.: On the asymptoticity aspect of Hyers–Ulam stability of mappings. Proc. Am. Math. Soc. **126**, 425–430 (1998)
46. Hyers, D.H., Ulam, S.M.: On approximate isometries. Bull. Am. Math. Soc. **51**, 288–292 (1945)
47. Hyers, D.H., Ulam, S.M.: Approximate isometries of the space of continuous functions. Ann. Math. (2) **48**, 285–289 (1947)
48. Hyers, D.H., Ulam, S.M.: Approximately convex functions. Proc. Am. Math. Soc. **3**, 821–828 (1952)
49. Isac, G., Rassias, Th.M.: On the Hyers–Ulam stability of ψ-additive mappings. J. Approx. Theory **72**, 131–137 (1993)
50. Isac, G., Rassias, Th.M.: Stability of ψ-additive mappings: applications to nonlinear analysis. Int. J. Math. Math. Sci. **19**(2), 219–228 (1996)
51. John, F.: Discontinuous convex solutions of difference equations. Bull. Am. Math. Soc. **47**, 275–281 (1941)
52. Jung, S.-M.: Hyers–Ulam–Rassias Stability of Functional Equations in Mathematical Analysis. Hadronic Press, Palm Harbor (2001)
53. Jung, S.-M.: Hyers–Ulam–Rassias Stability of Functional Equations in Nonlinear Analysis. Springer, New York (2011)
54. Jung, S.-M., Rassias, Th.-M.: Ulam's problem for approximate homomorphisms in connection with Bernoulli's differential equation. Appl. Math. Comput. **187**(1), 223–227 (2007)
55. Kim, G.H.: A stability of the generalized sine functional equations. J. Math. Anal. Appl. **331**, 886–894 (2007)

56. Kochanek, T., Lewicki, M.: Stability problem for number-theoretically multiplicative functions. Proc. Am. Math. Soc. **135**, 2591–2597 (2007)
57. Kocsis, I., Maksa, Gy.: The stability of a sum form functional equation arising in information theory. Acta Math. Hung. **79**, 39–48 (1998)
58. Kuczma, M.: Functional Equations in a Single Variable. PWN – Polish Scientific Publishers, Warszawa (1968)
59. Kuczma, M., Choczewski, B., Ger, R.: Iterative Functional Equations. Encyclopedia of Mathematics and Its Applications. Cambridge University Press, Cambridge (1990)
60. Mityushev, V.V., Rogosin, S.V.: Constructive Methods for Linear and Nonlinear Boundary Value Problems for Analytic Functions. Theory and Applications. Monographs and Surveys in Pure and Applied Mathematics, vol. 108. Chapman & Hall/CRC, Boca Raton (2000)
61. Moszner, Z.: Sur les définitions différentes de la stabilité des équations fonctionnelles. Aequ. Math. **68**, 260–274 (2004)
62. Moszner, Z.: On the stability of functional equations. Aequ. Math. **77**, 33–88 (2009)
63. Páles, Zs.: Hyers–Ulam stability of the Cauchy functional equation on square-symmetric groupoids. Publ. Math. (Debr.) **58**, 651–666 (2001)
64. Park, C.-G., Rassias, Th.M.: Hyers–Ulam stability of a generalized Apollonius type quadratic mapping. J. Math. Anal. Appl. **322**, 371–381 (2006)
65. Park, C.-G., Rassias, Th.M.: On a generalized Trif's mapping in Banach modules over a C^*-algebra. J. Korean Math. Soc. **43**(2), 323–356 (2006)
66. Park, C.-G., Rassias, Th.M.: Fixed points and generalized Hyers–Ulam stability of quadratic functional equations. J. Math. Inequal. **1**(4), 515–528 (2007)
67. Pólya, Gy., Szegö, G.: Aufgaben und Lehrsätze aus der Analysis, vol. 108. Springer, Berlin (1925)
68. Rassias, Th.M.: On the stability of the linear mapping in Banach spaces. Proc. Am. Math. Soc. **72**, 297–300 (1978)
69. Rassias, Th.M.: On a modified Hyers–Ulam sequence. J. Math. Anal. Appl. **158**, 106–113 (1991)
70. Rassias, Th.M.: On the stability of the quadratic functional equation and its applications. Stud. Univ. Babeş-Bolyai, Math. **43**(3), 89–124 (1998)
71. Rassias, Th.M.: The problem of S.M. Ulam for approximately multiplicative mappings. J. Math. Anal. Appl. **246**, 352–378 (2000)
72. Rassias, Th.M.: On the stability of functional equations and a problem of Ulam. Acta Appl. Math. **62**, 23–130 (2000)
73. Rassias, Th.M.: On the stability of functional equations in Banach spaces. J. Math. Anal. Appl. **251**, 264–284 (2000)
74. Rassias, Th.M., Semrl, P.: On the behavior of mappings which do not satisfy Hyers–Ulam stability. Proc. Am. Math. Soc. **114**, 989–993 (1992)
75. Rassias, Th.M., Semrl, P.: On the Hyers–Ulam stability of linear mappings. J. Math. Anal. Appl. **173**, 325–338 (1993)
76. Rassias, Th.M., Tabor, J.: What is left of Hyers–Ulam stability? J. Nat. Geom. **1**, 65–69 (1992)
77. Rassias, Th.M., Tabor, J. (eds.): Stability of Mappings of Hyers–Ulam Type. Hadronic Press, Florida (1994)
78. Sikorska, J.: Generalized stability of the Cauchy and Jensen functional equations on spheres. J. Math. Anal. Appl. **345**, 650–660 (2008)
79. Sikorska, J.: On a pexiderized conditional exponential functional equation. Acta Math. Hung. **125**, 287–299 (2009)
80. Sikorska, J.: On a direct method for proving the Hyers–Ulam stability of functional equations. J. Math. Anal. Appl. **372**, 99–109 (2010)
81. Székelyhidi, L.: The stability of the sine and cosine functional equations. Proc. Am. Math. Soc. **110**, 109–115 (1990)
82. Tabor, J., Tabor, J.: General stability of functional equations of linear type. J. Math. Anal. Appl. **328**, 192–200 (2007)

83. Takahasi, S.-E., Miura, T., Takagi, H.: Exponential type functional equation and its Hyers–Ulam stability. J. Math. Anal. Appl. **329**, 1191–1203 (2007)
84. Trif, T.: On the stability of a general gamma-type functional equation. Publ. Math. (Debr.) **60**, 47–61 (2002)
85. Trif, T.: Hyers-Ulam-Rassias stability of a linear functional equation with constant coefficients. Nonlinear Funct. Anal. Appl. **11**(5), 881–889 (2006)
86. Turdza, E.: On the stability of the functional equation $\phi[f(x)] = g(x)\phi(x) + F(x)$. Proc. Am. Math. Soc. **30**, 484–486 (1971)
87. Ulam, S.M.: A Collection of Mathematical Problems. Interscience, New York (1960). Reprinted as: Problems in Modern Mathematics. Wiley, New York (1964)
88. Xu, B., Zhang, W.: Construction of continuous solutions and stability for the polynomial-like iterative equation. J. Math. Anal. Appl. **325**, 1160–1170 (2007)

Chapter 8
On a Curious q-Hypergeometric Identity

María José Cantero and Arieh Iserles

Abstract In this paper, we examine the limiting behavior of solutions to an infinite set of recursions involving q-factorial terms as $q \to 1$. The underlying problem is sensitive to small perturbations and the very existence of a limit, to say nothing of its precise form, is surprising. We determine it by showing that the task at hand is equivalent to the convergence of one set of orthogonal polynomials on the unit circle to another such set, Geronimus polynomials, as $q \to 1$.

Key words Hypergeometric identity · Orthogonal polynomials · OPUC · Geronimus polynomials

Mathematics Subject Classification 42C05 · 16A60 · 30D05

8.1 Statement of the Problem

The subject matter of this paper is a curious fact pertaining to the solution of an infinite triangular set of linear algebraic equations with q-factorial coefficients. Specifically, we concern ourselves with the equations

$$\begin{aligned} &a_0 = 1, \\ &\sum_{\ell=0}^{m} \frac{a_{m-\ell}}{(q,q)_\ell (z,q)_\ell} = \frac{q^m}{(q,q)_m (z,q)_m}, \qquad m = 1, 2, \ldots, \end{aligned} \tag{8.1}$$

Dedicated to Professor Themistocles M. Rassias on the occasion of his 60th birthday.

M.J. Cantero
Departamento de Matemática Aplicada, Escuela de Ingeniería y Arquitectura, Universidad de Zaragoza, Zaragoza, Spain
e-mail: mjcante@unizar.es

A. Iserles (✉)
Department of Applied Mathematics and Theoretical Physics, Centre for Mathematical Sciences, University of Cambridge, Cambridge, UK
e-mail: A.Iserles@damtp.cam.ac.uk

P.M. Pardalos et al. (eds.), *Nonlinear Analysis*, Springer Optimization and Its Applications 68, In Honor of Themistocles M. Rassias on the Occasion of his 60th Birthday, DOI 10.1007/978-1-4614-3498-6_8,

where $z, q \in \mathbb{C}$, $|q| < 1$, and the *q-factorial symbol* $(b, q)_m$ is defined as (see [2])

$$(b,q)_m = \prod_{k=0}^{m-1} \left(1 - q^k b\right), \quad b \in \mathbb{C},\ m \in \mathbb{Z}_+ \cup \{\infty\}.$$

Since the coefficient of a_m in (8.1) is one, the system always has a solution, which can be obtained recursively. Thus,

$$\begin{aligned}
a_1 &= -\frac{1}{1-z},\\
a_2 &= \frac{z}{(1-z)^2(1-qz)},\\
a_3 &= -\frac{(1+q)z^2}{(1-z)^3(1-qz)(1-q^2z)},\\
a_4 &= \frac{[(1+2q+q^2+q^3) - q(1+q+2q^2+q^3)z]z^3}{(1-z)^4(1-qz)^2(1-q^2z)(1-q^3z)},\\
a_5 &= -\frac{(1+q)[(1+2q+q^2+2q^3+q^5) - q(1+2q^2+q^3+2q^4+q^5)z]z^4}{(1-z)^5(1-qz)^2(1-q^2z)(1-q^3z)(1-q^4z)},
\end{aligned}$$

and so on. On the face of it, the expressions are getting increasingly more complex, without any general rule. However, it is our contention in this paper that

$$\lim_{q\to 1} a_m = (-1)^m \frac{(2m-2)!}{(m-1)!m!} \frac{z^{m-1}}{(1-z)^{2m-1}}, \quad m \in \mathbb{N}. \tag{8.2}$$

This identity is surprising, not least because just about everything in (8.1), except for the $\ell = 0$ term, blows up as $q \to 1$. Thus, the terms need to blow up in a perfect balance!

The volatility of (8.1) means that what appear to be very minor and harmless amendments completely change the solution, typically leading to a blow-up as $q \to 1$. The most striking is also the most obvious along the route of seeking to prove (8.2): It is well known that for $q \to 1$ we have $(q,q)_s \approx s!(1-q)^s$ and $(z,q)_s \approx (z)_s$, where $(z)_s = z(z+1)\cdots(z+s-1)$, $s \in \mathbb{Z}_+$, is the *Pochhammer symbol* [4]. Consequently, (8.1) is 'approximated' by

$$\sum_{\ell=0}^{m} \frac{\tilde{a}_{m-\ell}}{\ell!(z)_\ell}(1-q)^{2(m-\ell)} = \frac{q^m}{m!(z)_m}, \quad m \in \mathbb{N}, \tag{8.3}$$

with $\tilde{a} = 1$. However, the solution of (8.3) blows up as $q \to 1$—we do not need to iterate much since already $\tilde{a}_1 = -(1-q)^{-1}z^{-1}$.

Even less drastic changes to (8.1) result either in a blow-up or in a very radical change to its character. Thus, the solution of both

$$\sum_{\ell=0}^{m} \frac{\tilde{a}_{m-\ell}}{(q,q)_\ell (z,q)_\ell} = \frac{1}{(q,q)_m (z,q)_m}, \qquad m \in \mathbb{C}$$

and of

$$\sum_{\ell=0}^{m} \frac{q^\ell \tilde{a}_{m-\ell}}{(q,q)_\ell (z,q)_\ell} = \frac{q^m}{(q,q)_m (z,q)_m}, \qquad m \in \mathbb{N}$$

is, trivially, $\tilde{a}_m \equiv 0$, $\ell \in \mathbb{N}$, while the solution of

$$\sum_{\ell=0}^{m} \frac{\tilde{a}_{m-\ell}}{(q,q)_\ell (z,q)_\ell} = \frac{q^{\frac{1}{2}(m-1)m}}{(q,q)_m (z,q)_m}, \qquad m \in \mathbb{N},$$

blows up as $q \to 1$, since

$$\tilde{a}_2 = -\frac{1}{(1-q^2)(1-z)(1-qz)}.$$

The very fact that the solution of (8.1) stays bounded as $q \to 1$ and that it approaches the fairly complicated expression limit (8.2) is part of the magic of q-hypergeometric functions. The delicate filigree of this set of equations and their orderly progression to an unusual limit is worthy of a Ramanujan. So should be the proof of (8.2): in an ideal world, it would be beautiful, direct, short, and crisp. Unfortunately, such a proof is beyond the wit of the authors. Instead, we present a roundabout proof of (8.2), which is anchored on our work in the theory of *orthogonal polynomials on the unit circle* (OPUC) [1].

8.2 From OPUC to the q-Hypergeometric Identities

A set of monic polynomials $\{\phi_n\}_{n\in\mathbb{Z}_+}$, orthogonal on the unit circle with respect to some measure, can be formally characterized by the set of is *Schur parameters* $a_n = \phi_n(0)$, $n \in \mathbb{Z}_+$ [5]. Specifically, the OPUC $\{\phi_n\}_{n\in\mathbb{Z}_+}$ obeys the recurrence relation

$$a_n\phi_{n+1}(z) = (a_{n+1} + a_n z)\phi_n(z) - \left(1 - |a_n|^2\right)a_{n+1} z \phi_{n-1}(z), \qquad n \in \mathbb{N},$$

with the initial conditions $\phi_0(z) \equiv 1$, $\phi_1(z) = z + a_1$. In [1], we addressed the OPUC with the Schur parameters

$$a_n = \begin{cases} 1, & n = 0, \\ c\alpha^n, & n \in \mathbb{N}, \end{cases} \tag{8.4}$$

where $c, \alpha \in \mathbb{C}$, $0 < |c|, |\alpha| < 1$. Such OPUC fills the space spanned by the arguably the three most important sets of OPUC: *Lebesgue polynomials* $\phi_n(z) = z^n$ $(c = 0)$, *Geronimus polynomials* $(\alpha = 1)$ and *Rogers–Szegő polynomials* $(c = 1)$. The generating function of the OPUC with the parameters (8.4),

$$\Phi_z(t) = \sum_{n=0}^{\infty} \frac{\phi_n(z)}{n!} t^n,$$

obeys the *pantograph-type* functional-differential equation

$$\Phi_z''(t) = (\alpha + z)\Phi_z'(t) - \alpha z \Phi_z(t) + \alpha \tau z \Phi_z(qt), \quad t \geq 0, \tag{8.5}$$

with the initial conditions $\Phi_z(0) = 1$, $\Phi_z'(0) = z + c\alpha$, where $q = |\alpha|^2 \in (0, 1)$ and $\tau = q|c|^2 \in (0, 1)$. Solutions of pantograph-type equations can be expanded into Dirichlet series [3] and this has led in [1] to the explicit expansion

$$\Phi_z(t) = \beta_1(z) \sum_{m=0}^{\infty} \frac{\tau^m e^{\alpha q^m t}}{(q, q)_m (\alpha/z, q)_m} + \beta_2(z) \sum_{m=0}^{\infty} \frac{\tau^m e^{z q^m t}}{(q, q)_m (z/\alpha, q)_m}, \tag{8.6}$$

where β_1 and β_2 are determined by the initial conditions.

Let

$$F(\zeta, \tau, q) = \sum_{m=0}^{\infty} \frac{\tau^m}{(q, q)_m (\zeta, q)_m}, \qquad H(\zeta, \tau, q) = \frac{F(\zeta, q\tau, q)}{F(\zeta, \tau, q)}$$

—both functions clearly converge since $|\tau| \leq q < 1$. Repeatedly differentiating (8.6), it has been proved in [1] that

$$\phi_m(z) = \alpha^m \eta_1(z) \prod_{\ell=1}^{m} H\left(\alpha z^{-1}, q^\ell \tau, q\right) + z^m \eta_2(z) \prod_{\ell=1}^{m} H\left(\alpha^{-1} z, q^\ell \tau, q\right), \quad m \in \mathbb{Z}_+, \tag{8.7}$$

where $\eta_1(z) = \beta_1(z) F(\alpha z^{-1}, \tau, q)$, $\eta_2(z) = \beta_2(z) F(\alpha^{-1} z, \tau, q)$ can also be expressed explicitly in terms of the function H.

Let us consider the case $\alpha \to 1$, hence also $q \to 1$ and $\tau \to |c|^2$. This corresponds to the *Geronimus polynomials* $\{\psi_m\}_{m \in \mathbb{Z}_+}$, with the explicit representation

$$\begin{aligned} \psi_m(z) = {} & \left[\frac{1}{2} - \frac{(1-z) - 2c}{\sqrt{(1-z)^2 + 4|c|^2 z}}\right] \left[\frac{1 + z + \sqrt{(1-z)^2 + 4|c|^2 z}}{2}\right]^m \\ & + \left[\frac{1}{2} + \frac{(1-z) - 2c}{\sqrt{(1-z)^2 + 4|c|^2 z}}\right] \left[\frac{1 + z - \sqrt{(1-z)^2 + 4|c|^2 z}}{2}\right]^m, \\ & m \in \mathbb{Z}_+; \end{aligned} \tag{8.8}$$

see [5, p. 87]. Is it true that, as $\alpha \to 1$, (8.7) tends to (8.8)? For that purpose, it is sufficient to prove that

$$H^\circ(z,c) = \lim_{\alpha\to 1} H\left(\alpha^{-1}z, \tau, q\right) = \frac{1+z-\sqrt{(1-z)^2+4|c|^2 z}}{2z}; \qquad (8.9)$$

see [1]. To this end, let us consider the power series in τ of the function τ. Since

$$H(\zeta,\tau,q) = \sum_{m=0}^{\infty} a_m \tau^m \quad \Rightarrow \quad F(\zeta, q\tau, q) = F(\zeta,\tau,q) \sum_{m=0}^{\infty} a_m \tau^m,$$

a substitution of the power-series definition of F and a straightforward multiplication of infinite series and a comparison of equal powers of τ result in the infinite set (8.1) of recurrence relations for the a_ms. Moreover, expanding the square root in (8.9) in powers of $|c|^2$ yields

$$\begin{aligned}\frac{1+z-\sqrt{(1-z)^2+4|c|^2 z}}{2z} &= 1 - \frac{1}{2}\sum_{m=1}^{\infty} (-1)^m \frac{(-\frac{1}{2})_m}{m!} \frac{4^m z^{m-1} |c|^{2m}}{(1-z)^{2m-1}} \\ &= 1 + \sum_{m=1}^{\infty} (-1)^m \frac{(2m-2)!}{(m-1)!m!} \frac{z^{m-1}}{(1-z)^{2m-1}} |c|^{2m}.\end{aligned}$$

Since $\lim_{\alpha\to 1} \tau = |c|^2$, term by term comparison results in (8.2). In other words, our contention that (8.2) is true is equivalent to the statement that $\lim_{q\to 1} \phi_m(z) = \psi_m(z)$, $m \in \mathbb{N}$.

8.3 From the OUPC to Geronimus Polynomials

Our aim is to demonstrate that (8.9) is true since, by the analysis of the last section, this proves (8.2). Revisiting the work of [1], let

$$R(\zeta,\tau,q) = \frac{F(q\zeta,\tau,q)}{F(\zeta,\tau,q)}, \qquad R^\circ(z,c) = \lim_{\alpha\to 1} R\left(\alpha^{-1}z, \tau, q\right).$$

However,

$$\begin{aligned} F(\zeta,\tau,q) - F(\zeta,q\tau,q) &= \sum_{m=1}^{\infty} \frac{(1-q^m)\tau^m}{(q,q)_m (\zeta,q)_m} = \frac{\tau}{1-\zeta} \sum_{m=0}^{\infty} \frac{\tau^m}{(q,q)_m (q\zeta,q)_m} \\ &= \frac{\tau}{1-\zeta} F(q\zeta,\tau,q) \end{aligned}$$

and, dividing by $F(\zeta,\tau,q)$, we obtain, after elementary algebra,

$$H(\zeta,\tau,q) = 1 - \frac{\tau}{1-\zeta} R(\zeta,\tau,q). \qquad (8.10)$$

Moreover,

$$\begin{aligned}F(\zeta,\tau,q)-F(q\zeta,\tau,q)&=\sum_{m=1}^{\infty}\frac{\tau^m}{(q,q)_m(\zeta,q)_{m+1}}\bigl[(1-q^m\zeta)-(1-\zeta)\bigr]\\&=\zeta\sum_{m=1}^{\infty}\frac{\tau^m}{(q,q)_{m-1}(\zeta,q)_{m+1}}\\&=\frac{\zeta\tau}{(1-\zeta)(1-q\zeta)}F\bigl(q^2\zeta,\tau,q\bigr).\end{aligned}$$

Dividing by $F(\zeta,\tau,q)$, we thus have

$$\begin{aligned}1-R(\zeta,\tau,q)&=\frac{\zeta\tau}{(1-\zeta)(1-q\zeta)}\frac{F(q\zeta,\tau,q)}{F(\zeta,\tau,q)}\times\frac{F(q^2\zeta,\tau,q)}{F(q\zeta,\tau,q)}\\&=\frac{\zeta\tau}{(1-\zeta)(1-q\zeta)}R(\zeta,\tau,q)R(q\zeta,\tau,q).\end{aligned}$$

It is perfectly safe to let $\alpha\to 1$ (hence also $q\to 1$ and $\tau\to |c|^2$) in the last expression, the outcome being the quadratic equation

$$z|c|^2R^{\circ 2}(z,c)+(1-z)^2R^{\circ}(z,c)-(1-z)^2=0.$$

Since $R^{\circ}(0,c)=1$, its solution is

$$R^{\circ}(z,c)=\frac{1-z}{2z|c|^2}\Bigl[-(1-z)+\sqrt{(1-z)^2+4z|c|^2}\Bigr],$$

and substitution in (8.10) results in (8.9). Therefore, the proof of the limit (8.2) follows in a roundabout manner and our work is done.

Acknowledgements The work of the first author was partially supported by the research projects MTM2008-06689-C02-01 and MTM2011-28952-C02-01 from the Ministry of Science and Innovation of Spain and the European Regional Development Fund (ERDF), and by Project E-64 of Diputación General de Aragón (Spain). MJC also wishes to acknowledge financial help of the Spanish Ministry of Education, Programa Nacional de Movilidad de Recursos Humanos del Plan Nacional de I+D+i 2008–2011.

References

1. Cantero, M.J., Iserles, A.: Orthogonal polynomials on the unit circle and functional differential equations. Technical Report 2011/08, DAMTP, University of Cambridge
2. Gasper, G., Rahman, M.: Basic Hypergeometric Series, vol. 96. Cambridge University Press, Cambridge (2004)
3. Iserles, A.: On the generalized pantograph functional-differential equation. Eur. J. Appl. Math. **4**(1), 38 (1993)
4. Rainville, E.D.: Special Functions, vol. 8. Macmillan, New York (1960)
5. Simon, B.: Orthogonal Polynomials on the Unit Circle, vol. 54. Amer. Math. Soc., Providence (2009)

Chapter 9
Jensen and Quadratic Functional Equations on Semigroups

Esteban A. Chávez and Prasanna K. Sahoo

Abstract Let S be a commutative semigroup, $\sigma : S \to S$ an endomorphism of order 2, G a 2-cancellative abelian group, and n a positive integer. One of the goals of this paper is to determine the general solutions of the functional equations $f_1(x+y) + f_2(x+\sigma y) = f_3(x)$ and also $f_1(x+y) + f_2(x+\sigma y) = f_3(x) + f_4(y)$ for all $x, y \in S^n$, where $f_1, f_2, f_3, f_4 : S^n \to G$ are unknown functions. The results of this paper improve and generalize the earlier results due to Ebanks, Kannappan, and Sahoo (Can. Math. Bull. 35:321–327, 1992), and Bae and Park (J. Math. Anal. Appl. 326:1142–1148, 2007), and generalize the works of Sinopoulos (Aequ. Math. 59:255–261, 2000).

Key words Additive function · Biadditive function · Jensen equation · Quadratic function · Quadratic equation · Functional equation on semigroups

Mathematics Subject Classification Primary 39B52

9.1 Introduction

Let $(S, +)$ be a commutative semigroup, $\sigma : S \to S$ an endomorphism of order 2, G a 2-cancellative abelian group, and n a positive integer. In this paper, we determine the general solutions of the functional equations

$$f_1(x+y) + f_2(x+\sigma y) = f_3(x) \tag{9.1}$$

and

$$f_1(x+y) + f_2(x+\sigma y) = f_3(x) + f_4(y) \tag{9.2}$$

Dedicated to Professor Themistocles M. Rassias on the occasion of his 60th birthday.

E.A. Chávez
Department of Mathematics, Duke University, Durham, NC 27708, USA
e-mail: eachav@math.duke.edu

P.K. Sahoo (✉)
Department of Mathematics, University of Louisville, Louisville, KY 40292, USA
e-mail: sahoo@louisville.edu

P.M. Pardalos et al. (eds.), *Nonlinear Analysis*, Springer Optimization and Its Applications 68, In Honor of Themistocles M. Rassias on the Occasion of his 60th Birthday, DOI 10.1007/978-1-4614-3498-6_9,

for all $x, y \in S^n$. The range of the functions in the first equation is a 2-cancellative abelian group, and in the second equation is a uniquely 2-divisible abelian group.

For $n = 1$ and $f_1 = f_2 = f_3 = f_4 = f$, these two equations were studied by Sinopoulos [6]. When $n = 1$, $f_1 = f_2 = f_3 = f$ and $\sigma(y) = -y$, the first equation is the Jensen equation. For an account on Jensen functional equation, the reader is referred to the book by Kuczma [5]. When $f_1 = f_2 = f,\ f_3 = f_4 = 2f$ and $\sigma(y) = -y$, the second equation reduces to the quadratic functional equation $f(x+y) + f(x-y) = 2f(x) + 2f(y)$. The quadratic functional equation is very important as it serves in certain abstract spaces for the definition of norm. It was studied by many authors including Aczél [1]. In 1965, Aczél [1] proved the following result: Let G be an abelian group and let H be an abelian group in which every equation of the form $2x = h \in H$ has one and only one solution $x \in H$. Then, any solution $f : G \to H$ of the quadratic functional equation on G is of the form $f(x) = B(x, x)$, where $B : G \times G \to H$ is a symmetric biadditive form. In 2000, Sinopoulos [6] proved that if $(S, +)$ is a commutative semigroup, G a uniquely 2-divisible abelian group and σ an endomorphism of S such that $\sigma(\sigma x) = x$ for $x \in S$, then the general solution $f : S \to G$ of the quadratic functional equation $f(x+y) + f(x+\sigma y) = 2f(x) + 2f(y)$ is given by $f(x) = B(x, x) + A(x)$ where $B : S \times S \to G$ is an arbitrary symmetric biadditive function with $B(\sigma x, y) = -B(x, y)$ and $A : S \to G$ is an arbitrary additive function with $A(\sigma x) = A(x)$. In 2007, Bae and Park [2] proved that if X and Y are real vector spaces, and a mapping $f : X \times X \to Y$ satisfies the functional equation

$$f(x+y, z+w) + f(x-y, z-w) = 2f(x, z) + 2f(y, w) \quad \forall x, y, z, w \in X \tag{9.3}$$

then there exist two symmetric bi-additive mappings $S_1, S_2 : X \times X \to Y$ and a bi-additive mapping $B : X \times X \to Y$ such that $f(x, y) = S_1(x, x) + B(x, y) + S_2(y, y)$ for all $x, y \in X$. The converse of this result is also true. The functional equation (9.3) is a special case of the functional equation (9.2).

This paper is organized as follows: In Sect. 9.2, we give some terminology and preliminary results which will be used in solving equations (9.1) and (9.2). In Sect. 9.3, we present the general solution of the Jensen functional equation as well as pexiderized Jensen functional equation (9.1) on semigroups. Section 9.4 is devoted to solving the quadratic functional equation on semigroups and its pexiderization (9.2). The results obtained in this section generalizes the works of Bae and Park. Our results in this section generalize the results of Bae and Parks [2] in two different ways: first, we solve a more general functional equation than Bae and Park [2]; second, we solve this more general functional equation on a more general algebraic structure, namely on semigroups. The results of this section also generalize the results of Ebanks, Kannappan, and Sahoo [3] and Sinopoulos [6].

9.2 Notations and Preliminary Results

In the sequel, $(S, +)$ will always denote a commutative semigroup, $(F, +)$ will be an abelian group, $(G, +)$ will be a 2-cancellative abelian group, and $(H, +)$ will be an abelian group uniquely divisible by 2. Furthermore, it will be assumed that $\sigma : S \to S$ is an endomorphism satisfying $\sigma(\sigma(x)) = x$ for all $x \in S$. For notational simplicity, will denote $\sigma(x)$ simply as σx for $x \in S$. Let n be a positive integer. Notice that $(S, +)$ being a commutative semigroup implies that

$$(S^n, +) = \underbrace{(S, +) \times (S, +) \times \cdots \times (S, +)}_{n \text{ times}}$$

is also a commutative semigroup where the sum of two elements in S^n is defined as the individual sum of its n components. As a remark on notation, if $x \in S^n$ we will denote the ith component of x by x_i, so that $x = (x_1, x_2, \ldots, x_n)$. For $x \in S^n$, the notation σx will denote $(\sigma x_1, \sigma x_2, \ldots, \sigma x_n)$. An *additive function* $A : S \to F$ is a function satisfying the functional equation

$$A(x + y) = A(x) + A(y) \tag{9.4}$$

for all $x, y \in S$. A *biadditive function* $B : S^2 \to F$ is a function satisfying the functional equations

$$\begin{cases} B(x + z, y) = B(x, y) + B(z, y), \\ B(x, y + w) = B(x, y) + B(x, w) \end{cases} \tag{9.5}$$

for all $x, y, z, w \in S$. The biadditive function B is *symmetric* if $B(x, y) = B(y, x)$ for all $x, y \in S$.

We will begin by stating, without proof, some basic results proved by Kuczma [4] and Sinopoulos [6]. The following theorem is due to Kuczma [4].

Theorem 9.1 *The general solution $A : S^n \to F$ of the functional equation*

$$A(x + y) = A(x) + A(y) \tag{9.6}$$

for all $x, y \in S^n$ is given by

$$A(x) = \sum_{i=1}^{n} A_i(x_i) \tag{9.7}$$

where $A_i : S \to F$ is an additive function for each $i = 1, 2, \ldots, n$.

The following two theorems are due to Sinopoulos [6].

Theorem 9.2 *The general solution $f : S \to G$ of the functional equation*

$$f(x + y) + f(x + \sigma y) = 2f(x) \tag{9.8}$$

for all $x, y \in S$ is given by

$$f(x) = A(x) + a \tag{9.9}$$

where $A : S \to G$ is an additive function satisfying

$$A(\sigma x) = -A(x) \tag{9.10}$$

for all $x \in S$ and $a \in G$.

Theorem 9.3 *The general solution $f : S \to H$ of the functional equation*

$$f(x+y) + f(x+\sigma y) = 2f(x) + 2f(y) \tag{9.11}$$

for all $x, y \in S$ is given by

$$f(x) = B(x,x) + A(x) \tag{9.12}$$

where $B : S^2 \to H$ is a symmetric biadditive function satisfying

$$B(\sigma x, y) = -B(x,y) \tag{9.13}$$

for all $x, y \in S$ and $A : S \to G$ is an additive function satisfying

$$A(\sigma x) = A(x) \tag{9.14}$$

for all $x \in S$.

Next, we prove some lemmas that will be instrumental for finding the solution of the functional equations (9.1) and (9.2).

Lemma 9.1 *The general solution $A : S^n \to G$ of the functional equations*

$$\begin{cases} A(x+y) = A(x) + A(y), \\ A(\sigma x) = -A(x) \end{cases} \tag{9.15}$$

for all $x, y \in S^n$ is given by (9.7), where $A_p : S \to G$ is an additive function satisfying

$$A_p(\sigma x_p) = -A(x_p) \tag{9.16}$$

for all $x \in S^n$ and each $p = 1, 2, \ldots, n$.

Proof Obviously, since A satisfies (9.6), the solution A is given by (9.7). Moreover, (9.15) and (9.7) imply that

$$A(x+\sigma x) = \sum_{i=1}^{n} A_i(x_i + \sigma x_i) = 0 \tag{9.17}$$

for all $x \in S^n$. Rewrite (9.17) as

$$A_p(x_p + \sigma x_p) = -\sum_{\substack{i=1 \\ i \neq p}}^{n} A_i(x_i + \sigma x_i) \tag{9.18}$$

to notice that the left-hand side of (9.18) is a constant independent of $x_p \in S$. Call this constant a_p. However, for $x, y \in S^n$, notice that

$$a_p = A_p\big((x_p + y_p) + \sigma(x_p + y_p)\big) = A_p(x_p + \sigma x_p) + A_p(y_p + \sigma y_p) = 2a_p,$$

so that $a_p = 0$. Therefore, by (9.18), (9.16) follows. □

Lemma 9.2 *The general solution $A : S^n \to G$ of the functional equations*

$$\begin{cases} A(x + y) = A(x) + A(y), \\ A(\sigma x) = A(x) \end{cases} \tag{9.19}$$

for all $x, y \in S^n$ is given by (9.7), *where $A_p : S \to G$ is an additive function satisfying*

$$A_p(\sigma x_p) = A(x_p) \tag{9.20}$$

for all $x \in S^n$ and each $p = 1, 2, \ldots, n$.

Proof As before, since A satisfies (9.6), the solution A is given by (9.7). Moreover, (9.19) and (9.7) imply that

$$A(\sigma x) - A(x) = \sum_{i=1}^{n} \big[A_i(\sigma x_i) - A_i(x_i)\big] = 0 \tag{9.21}$$

for all $x \in S^n$. Rewrite (9.21) as

$$A_p(\sigma x_p) - A_p(x_p) = -\sum_{\substack{i=1 \\ i \neq p}}^{n} \big[A_i(\sigma x_i) - A_i(x_i)\big] \tag{9.22}$$

to notice that the left-hand side of (9.22) is a constant independent of $x_p \in S$. Call this constant b_p. However, for $x, y \in S^n$, notice that

$$\begin{aligned} b_p = A_p\big(\sigma(x_p + y_p)\big) - A_p(x_p + y_p) &= \big[A_p(\sigma x_p) - A_p(x_p)\big] \\ &\quad + \big[A_p(\sigma y_p) - A_p(y_p)\big] = 2b_p, \end{aligned}$$

so that $b_p = 0$. Therefore, by (9.22), (9.20) follows. □

Definition 9.1 A function μ will be called a sum-fix if it satisfies the functional equation

$$\mu(x + y) = \mu(z + w) \tag{9.23}$$

for all $x, y, z, w \in S$.

In other words, if $s, t \in S$ can be written as $s = x + y$ and $t = z + w$ for $x, y, z, w \in S$, then $\mu(s) = \mu(t)$ but if r cannot be written as the sum of two elements of S, then $\mu(r)$ can take any arbitrary value of F independently of other elements of S.

Lemma 9.3 *The general solution $f_1, f_2, f_3 : S \to F$ of the functional equation*

$$f_1(x+y) = f_2(x) + f_3(y) \tag{9.24}$$

for all $x, y \in S$ is given by

$$\begin{cases} f_1(x) = A(x) + \mu(x), \\ f_2(x) = A(x) + \mu(x+w) - a, \\ f_3(x) = A(x) + a \end{cases} \tag{9.25}$$

where $A : S \to F$ is an additive function satisfying $A(\sigma x) = -A(x)$ for all $x \in S$, $\mu : S \to F$ is a sum-fix function, w is some element in S and $a \in F$.

Proof Set $y = w$ in (9.24) to obtain that

$$f_2(x) = f_1(x+w) - f_3(w). \tag{9.26}$$

Replace (9.26) into (9.24) to obtain

$$f_1(x+y) = f_1(x+w) - f_3(w) + f_3(y). \tag{9.27}$$

Set $x = z$ in (9.27) to obtain that

$$f_3(y) = f_1(z+y) - f_1(z+w) + f_3(w). \tag{9.28}$$

Replace (9.28) into (9.27) to obtain

$$f_1(x+y) = f_1(x+w) + f_1(z+y) - f_1(z+w). \tag{9.29}$$

Define the function $A : S \to F$ by

$$A(x) = f_1(x+z+w) - f_1(z+w). \tag{9.30}$$

Then, putting $x = x + z$ and $y = y + w$ in (9.29), one can write (9.29) in terms of A as

$$A(x+y) = A(x) + A(y), \tag{9.31}$$

so that A is an additive function. Defining $\mu : S \to F$ by

$$\mu(x) = f_1(x) - A(x), \tag{9.32}$$

equation (9.30) may be rewritten as

$$\mu(x+z+w) = \mu(z+w), \tag{9.33}$$

and it is easy to show that (9.33) is a sum-fix, that is, satisfies (9.23). By replacing y by x in (9.28), as well as (9.30) into (9.28) and using (9.23) and rearranging the terms, we can show that

$$f_3(x) - A(x) = f_3(w) - A(w), \tag{9.34}$$

which is a constant independent of $x, w \in S$. Thus,

$$f_3(x) = A(x) + a, \tag{9.35}$$

with $a \in F$. Finally, replacing the solutions of f_1 and f_3 given by (9.32) and (9.35) into (9.26), we get that

$$f_2(x) = \mu(x + w) + A(x) - a \tag{9.36}$$

and the proof of the lemma is now complete. □

9.3 Jensen and Pexiderized Jensen Equations

Now we proceed to determine the general solution of the Jensen functional equation and pexiderized Jensen equation (9.1).

Theorem 9.4 *The general solution $f : S^n \to G$ of the functional equation*

$$f(x + y) + f(x + \sigma y) = 2f(x) \tag{9.37}$$

for all $x, y \in S^n$ is given by

$$f(x) = \sum_{i=1}^{n} A_i(x_i) + a \tag{9.38}$$

where $A_p : S \to G$ is an additive function satisfying $A_p(\sigma x_p) = -A(x_p)$ for all $x \in S^n$ and each $p = 1, 2, \ldots, n$, and $a \in G$.

Proof By (9.9), the general solution of (9.37) is given by $f(x) = A(x) + a$ where A is an additive function satisfying (9.15) and $a \in G$. Therefore, A is given by (9.16), and (9.38) follows. □

Definition 9.2 A function $\ell : S \to F$ will be called σ-conjugate if it satisfies the functional equation

$$\ell(x + y) = \ell(x + \sigma y) \tag{9.39}$$

for all $x, y \in S$.

Theorem 9.5 *Let $(S,+)$ be a commutative semigroup which is uniquely divisible by 2. The general solution $f_1, f_2, f_3 : S \to G$ of the functional equation*

$$f_1(x+y) + f_2(x+\sigma y) = f_3(x) \tag{9.40}$$

for all $x, y \in S$ is given by

$$\begin{cases} f_1(x) = A(x) + \ell(x) + a, \\ f_2(x) = A(x) - \ell(x) + a, \\ f_3(x) = 2A(x) + 2a \end{cases} \tag{9.41}$$

where $A : S \to G$ is an additive function satisfying $A(\sigma x) = -A(x)$ and $\ell : S \to G$ is a σ-conjugate function for all $x \in S$, and $a \in G$.

Proof Replace y by σy in (9.40) to obtain

$$f_1(x+\sigma y) + f_2(x+y) = f_3(x). \tag{9.42}$$

Add (9.40) and (9.42) to obtain that

$$g(x+y) + g(x+\sigma y) = 2 f_3(x), \tag{9.43}$$

where $g : S \to G$ is defined by

$$g(x) = f_1(x) + f_2(x). \tag{9.44}$$

Then, using (9.43) and (9.44), compute

$$\begin{aligned} 2f_3(x+z) + 2f_3(x+\sigma z) &= \big[g(x+z+y) + g(x+z+\sigma y)\big] \\ &\quad + \big[g(x+\sigma z+y) + g(x+\sigma z+\sigma y)\big] \\ &= \big[g\big(x+(z+y)\big) + g\big(x+\sigma(z+y)\big)\big] \\ &\quad + \big[g\big(x+(y+\sigma z)\big) + g\big(x+\sigma(y+\sigma z)\big)\big] \\ &= 2f_3(x+y) + 2f_3(x+\sigma y) \\ &= 4f_3(x). \end{aligned} \tag{9.45}$$

Hence, f_3 satisfies (9.8), and there exists an additive function $A : S \to G$ and $a \in G$ such that

$$f_3(x) = 2A(x) + 2a \tag{9.46}$$

with $A(\sigma x) = -A(x)$ for all $x \in S$. Replace both x and y by $\frac{x}{4} + \sigma \frac{x}{4}$ in (9.43) and use (9.46) to conclude that

$$g\left(\frac{x}{2} + \sigma \frac{x}{2}\right) = 2a. \tag{9.47}$$

Replace both x and y by $\frac{x}{2}$ in (9.43) and use (9.46) and (9.47) to see that

$$g(x) = 2A(x) + 2a. \tag{9.48}$$

Now, subtract (9.40) and (9.42) to obtain (9.39), where $\ell : S \to G$ is a σ-conjugate function defined by

$$2\ell(x) = f_1(x) - f_2(x). \tag{9.49}$$

Finally, the system of (9.44) and (9.49) yield that

$$f_1(x) = A(x) + \ell(x) + a \tag{9.50}$$

and

$$f_2(x) = A(x) - \ell(x) + a. \tag{9.51}$$

This finishes the proof of the theorem. □

The following theorem easily follows from Theorem 9.1 and Theorem 9.5.

Theorem 9.6 *Let $(S, +)$ be a commutative semigroup which is uniquely divisible by 2. The general solution $f_1, f_2, f_3 : S^n \to G$ of the functional equation*

$$f_1(x + y) + f_2(x + \sigma y) = f_3(x) \tag{9.52}$$

for all $x, y \in S^n$ is given by

$$\begin{cases} f_1(x) = \sum_{i=1}^{n} A_i(x_i) + \ell(x) + a, \\ f_2(x) = \sum_{i=1}^{n} A_i(x_i) - \ell(x) + a, \\ f_3(x) = 2\sum_{i=1}^{n} A_i(x_i) + 2a \end{cases} \tag{9.53}$$

where $a \in G$ is a constant, $\ell : S^n \to G$ is a σ-conjugate function for all $x \in S^n$ and $A_i : S \to G$ is an additive function satisfying $A_i(\sigma x_i) = -A_i(x_i)$ for each $i = 1, 2, \ldots, n$.

9.4 Quadratic and Pexiderized Quadratic Equations

In this section, we determine the general solution of the quadratic functional equation and pexiderized quadratic equation (9.2).

Theorem 9.7 *The general solution $B : S^n \times S^n \to H$ of the functional equations*

$$\begin{cases} B(x+z, y) = B(x, y) + B(z, y), \\ B(x, y+w) = B(x, y) + B(x, w) \end{cases} \tag{9.54}$$

for all $x, y, z, w \in S^n$ is given by

$$B(x, y) = \sum_{i=1}^{n} \sum_{j=1}^{n} B_{i,j}(x_i, y_j) \tag{9.55}$$

where $B_{p,q} : S \times S \to H$ is a biadditive function for all $x, y \in S^n$ and each $1 \leq p, q \leq n$.

Proof Since $B(\cdot, y)$ satisfies (9.15) for all $y \in S^n$, by (9.16) it follows that

$$B(x, y) = \sum_{i=1}^{n} B_i(x_i, y) \tag{9.56}$$

where $B_p : S \times S^n \to H$ satisfies that $B_p(\cdot, y)$ is an additive function for all $y \in S^n$, $p = 1, 2, \ldots, n$. Now, use (9.56) to rewrite the second equation of (9.54) as

$$\sum_{i=1}^{n} B_i(x_i, y+w) = \sum_{i=1}^{n} B_i(x_i, y) + \sum_{i=1}^{n} B_i(x_i, w). \tag{9.57}$$

Rearrange the terms of (9.57) to find out that

$$\begin{aligned} &B_p(x_p, y+w) - B_p(x_p, y) - B_p(x_p, w) \\ &\quad = - \sum_{\substack{i=1 \\ i \neq p}}^{n} \big[B_i(x_i, y+w) - B_i(x_i, y) - B_i(x_i, w)\big]. \end{aligned} \tag{9.58}$$

Then, by (9.58), the function $\varphi_p : S^n \times S^n \to H$ defined by

$$\varphi_p(y, w) = B_p(x_p, y+w) - B_p(x_p, y) - B_p(x_p, w) \tag{9.59}$$

is well-defined, that is, does not depend on the particular value of $x_p \in S$. However,

$$\begin{aligned} \varphi_p(y, w) &= B_p(x_p + z_p, y+w) - B_p(x_p + z_p, y) - B_p(x_p + z_p, w) \\ &= \big[B_p(x_p, y+w) - B_p(x_p, y) - B_p(x_p, w)\big] \\ &\quad + \big[B_p(z_p, y+w) - B_p(z_p, y) - B_p(z_p, w)\big] \\ &= \varphi_p(y, w) + \varphi_p(y, w) \end{aligned} \tag{9.60}$$

for all $y, w \in S^n$. Thus, $\varphi_p(y, w) = 0$ for all $y, w \in S^n$ and hence

$$B_p(x_p, y+w) = B_p(x_p, y) + B_p(x_p, w) \tag{9.61}$$

for all $x, y, w \in S^n$. Hence, by (9.7) and (9.61), B_p can be decomposed as

$$B_p(x_p, y) = \sum_{j=1}^{n} B_{p,j}(x_p, y_j), \tag{9.62}$$

where $B_{p,q}(x_p, \cdot)$ is an additive function for all $x_p \in S$; $p, q = 1, 2, \ldots, n$. Next, define the function $\bar{\varphi}_{p,q} : S \times S \to H$ by

$$\bar{\varphi}_{p,q}(x_p, z_p) = B_{p,q}(x_p + z_p, y_q) - B_{p,q}(x_p, y_q) - B_{p,q}(z_p, y_q). \tag{9.63}$$

By (9.62) and the additivity of B_p, it follows that

$$\begin{aligned} &B_{p,q}(x_p + z_p, y_q) - B_{p,q}(x_p, y_q) - B_{p,q}(z_p, y_q) \\ &= -\sum_{\substack{j=1 \\ j \neq q}}^{n} \left[B_{p,j}(x_p + z_p, y_j) - B_{p,j}(x_p, y_j) - B_{p,j}(z_p, y_j)\right], \end{aligned} \tag{9.64}$$

so $\bar{\varphi}_{p,q}$ is well defined, that is, it does not depend on any particular value of $y_q \in S$. Now, use the facts that $B_{p,q}(x_p, \cdot)$ is additive, (9.63) and (9.64) to see that

$$\begin{aligned} \bar{\varphi}_{p,q}(x_p, z_p) &= \bar{\varphi}_{p,q}(x_p, z_p)|_{y_q + w_q} \\ &= \bar{\varphi}_{p,q}(x_p, z_p)|_{y_q} + \bar{\varphi}_{p,q}(x_p, z_p)|_{w_q} \\ &= 2\bar{\varphi}_{p,q}(x_p, z_p) \end{aligned} \tag{9.65}$$

conclude that $\bar{\varphi}_{p,q} \equiv 0$, so that $B_{p,q}$ is also additive on the first component and, therefore, biadditive. □

Corollary 9.1 *The general solution $B : S^n \times S^n \to H$ of the functional equations*

$$\begin{cases} B(x + z, y) = B(x, y) + B(z, y), \\ B(x, y + w) = B(x, y) + B(x, w), \\ B(x, y) = B(y, x), \\ B(\sigma x, y) = -B(x, y) \end{cases} \tag{9.66}$$

for all $x, y, z, w \in S^n$ is given by (9.55) where

1. $B_{p,q} : S \times S \to H$ *is a biadditive function*;
2. $B_{p,q}(x_p, y_q) = B_{q,p}(y_q, x_p)$;
3. $B_{p,p} : S \times S \to H$ *is a symmetric biadditive function*;
4. $B_{p,q}(\sigma x_p, y_q) = -B_{p,q}(x_p, y_q)$, *and*
5. $B_{p,q}(x_p, \sigma y_q) = -B_{p,q}(x_p, y_q)$

for all $x, y \in S^n$ and each $1 \leq p, q \leq n$.

Proof Clearly, the solution of (9.66) will be given by (9.55). By the symmetry of B, it follows that

$$B(x,y)=\sum_{i,j=1}^{n} B_{i,j}(x_i,y_j)=\sum_{i,j=1}^{n} B_{i,j}(y_i,x_j)=B(y,x). \tag{9.67}$$

Now we claim that $B_{p,q}(x_p,y_q)=B_{q,p}(y_q,x_p)$. Rewrite (9.67) as

$$B_{p,q}(x_p,y_q)-B_{q,p}(y_q,x_p)=-\sum_{\substack{i,j=1\\(i,j)\neq(p,q)}}^{n}\left[B_{i,j}(x_i,y_j)-B_{j,i}(y_j,x_i)\right] \tag{9.68}$$

and notice that the right-hand side only depends on $x_p, y_q \in S$. Hence, by replacing each x_i by x_i+z_j and y_j by y_j+w_j with $(i,j)\neq(p,q)$ and using the fact that $B_{i,j}$ is biadditive, it follows that $B_{p,q}(x_p,y_q)-B_{q,p}(y_q,x_p)$ must be a constant. Then, replacing x_p by x_p+z_p and y_q by y_q+w_q and using the fact that $B_{i,j}$ is biadditive, it follows that this constant must be zero, and the claim follows. Consequently, $B_{i,i}$ is symmetric. The proof of $B_{p,q}(\sigma x_p,x_q)=-B_{p,q}(x_p,x_q)$ is very similar to the proof of (9.16). Finally, observe that

$$B_{p,q}(x_p,\sigma y_q)=B_{q,p}(\sigma y_q,x_p)=-B_{q,p}(y_q,x_p)=-B_{p,q}(x_p,y_q) \tag{9.69}$$

to complete the proof. □

Theorem 9.8 *The general solution $f:S^n\to H$ of the functional equation*

$$f(x+y)+f(x+\sigma y)=2f(x)+2f(y) \tag{9.70}$$

for all $x,y\in S^n$ is given by

$$f(x)=\sum_{i=1}^{n}\sum_{j=1}^{i} B_{i,j}(x_i,x_j)+\sum_{i=1}^{n} A_i(x_i) \tag{9.71}$$

where

1. $B_{p,q}: S\times S\to H$ *is a biadditive function;*
2. $B_{p,p}: S\times S\to H$ *is a symmetric biadditive function;*
3. $A_p: S\to H$ *is an additive function;*
4. $B_{p,q}(\sigma x_p,y_q)=-B_{p,q}(x_p,y_q)$;
5. $B_{p,q}(x_p,\sigma y_q)=-B_{p,q}(x_p,y_q)$, *and*
6. $A_p(\sigma x_p)=A_p(x_p)$

for all $x,y\in S^n$ and each $1\le p\le q\le n$.

Proof By (9.12), the general solution of (9.70) is given by $f(x)=B(x,x)+A(x)$ for all $x\in S^n$ where B satisfies (9.54) and A satisfies (9.19). By (9.19), the function

A can be written as (9.20). Also, by (9.54), the function B can be written as (9.55). Finally, take $x = y$ in (9.55) and (9.71) follows. □

The following corollary is an easy consequence of Theorem 9.8.

Corollary 9.2 *Let $(S, +)$ be a commutative semigroup and H an abelian group uniquely divisible by 2. The general solution $f : S \times S \to H$ of the functional equation*

$$f(x+y, z+w) + f(x+\sigma y, z+\sigma w) = 2f(x, z) + 2f(y, w) \tag{9.72}$$

for all $x, y, z, w \in S$ is given by

$$f(x, y) = B_1(x, x) + B(x, y) + B_2(y, y) + A_1(x) + A_2(y) \tag{9.73}$$

where $A_1, A_2 : S \to H$ are additive functions satisfying

$$A_1(\sigma x) = A_1(x), \qquad A_2(\sigma x) = A_2(x),$$

$B_1, B_2 : S \times S \to H$ are symmetric biadditive functions satisfying

$$B_1(\sigma x, y) = -B_1(x, y), \qquad B_2(\sigma x, y) = -B_2(x, y),$$

and $B : S \times S \to H$ is a biadditive function satisfying

$$B(x, \sigma y) = B(\sigma x, y) = -B(x, y)$$

The following corollary improves the result proved by Bae and Park [2].

Corollary 9.3 *Let $(S, +)$ be an abelian group and H an abelian group uniquely divisible by 2. The general solution $f : S \times S \to H$ of the functional equation*

$$f(x+y, z+w) + f(x-y, z-w) = 2f(x, z) + 2f(y, w) \tag{9.74}$$

for all $x, y, z, w \in S$ is given by

$$f(x, y) = B_1(x, x) + B(x, y) + B_2(y, y) \tag{9.75}$$

where $B_1, B_2 : S \times S \to H$ are symmetric biadditive functions, and $B : S \times S \to H$ is a biadditive function.

Now, we present the general solution of pexiderized quadratic equation (9.2).

Theorem 9.9 *Let $(S, +)$ be a commutative semigroup which is uniquely divisible by 2. The general solution $f : S \to H$ of the functional equation*

$$f_1(x+y) + f_2(x+\sigma y) = f_3(x) + f_4(y) \tag{9.76}$$

for all $x, y \in S$ *is given by*

$$\begin{cases} f_1(x) = B(x,x) + A_1(x) + A_2(x) + A_3(x) + \ell(x) + a, \\ f_2(x) = B(x,x) - A_1(x) + A_2(x) + A_3(x) - \ell(x) + a, \\ f_3(x) = 2B(x,x) + 2A_2(x) + 2A_3(x) + 2a - b, \\ f_4(x) = 2B(x,x) + 2A_1(x) + 2A_2(x) + b \end{cases} \tag{9.77}$$

where

1. $B : S \to H$ *is a symmetric biadditive function satisfying* $B(\sigma x, y) = -B(x,y)$;
2. $A_1 : S \to H$ *is an additive function satisfying* $A_1(\sigma x) = -A_1(x)$;
3. $A_2 : S \to H$ *is an additive function satisfying* $A_2(\sigma x) = A_2(x)$;
4. $A_3 : S \to H$ *is an additive function satisfying* $A_3(\sigma x) = -A_3(x)$;
5. $\ell : S \to H$ *is a* σ*-conjugate function, and*
6. $a, b \in H$

for all $x, y \in S$.

Proof Replace y by σy in (9.76) to obtain

$$f_1(x + \sigma y) + f_2(x + y) = f_3(x) + f_4(\sigma y). \tag{9.78}$$

Subtract (9.76) and (9.78) and obtain

$$h(x + y) - h(x + \sigma y) = k(y), \tag{9.79}$$

where $h : S \to H$ satisfies

$$h(x) = f_1(x) - f_2(x) \tag{9.80}$$

and $k : S \to H$ satisfies

$$k(x) = f_4(x) - f_4(\sigma x). \tag{9.81}$$

Replace x by σx in (9.81) to see that

$$k(\sigma x) = -k(x). \tag{9.82}$$

Using (9.79), compute

$$\begin{aligned} k(y + w) + k(y + \sigma w) &= \big[h(x + y + w) - h(x + \sigma y + \sigma w)\big] \\ &\quad + \big[h(x + y + \sigma w) - h(x + \sigma y + w)\big] \\ &= \big[h\big((x + w) + y\big) - h\big((x + w) + \sigma y\big)\big] \\ &\quad + \big[h\big((x + \sigma w) + y\big) - h\big((x + \sigma w) + \sigma y\big)\big] \\ &= 2k(y). \end{aligned} \tag{9.83}$$

Thus, k satisfies (9.8), so there exists an additive function $A_1 : S \to H$ with $A_1(\sigma x) = -A_1(x)$ and $\xi_1 \in H$ such that

$$k(x) = 4A_1(x) + \xi_1. \tag{9.84}$$

However, by using (9.82) in (9.84), it is easy to see that $\xi_1 = 0$, so k is additive. Therefore, we can write (9.81) as

$$f_4(x) - f_4(\sigma x) = 4A_1(x). \tag{9.85}$$

Define $\ell : S \to H$ by

$$2\ell(x) = h(x) - 2A_1(x) \tag{9.86}$$

and use (9.79) and (9.84) to see that ℓ satisfies (9.39), that is, ℓ is σ-conjugate. Next, add (9.76) and (9.78) and obtain

$$g(x+y) + g(x+\sigma y) = 2f_3(x) + f_4(y) + f_4(\sigma y), \tag{9.87}$$

where $g : S \to H$ satisfies

$$g(x) = f_1(x) + f_2(x). \tag{9.88}$$

Substitute $y = z + \sigma z$ in (9.87) to realize that

$$f_3(x) = g(x + z + \sigma z) - f_4(z + \sigma z). \tag{9.89}$$

Replacing (9.89) into (9.87) one gets that

$$g(x+y) + g(x+\sigma y) = 2g(x+z+\sigma z) - 2f_4(z+\sigma z) + f_4(y) + f_4(\sigma y). \tag{9.90}$$

Put $x = w + \sigma w$ in (9.90) to get that

$$\begin{aligned} &g(w+\sigma w+y) + g(w+\sigma w+\sigma y) \\ &\quad = 2g(w+\sigma w+z+\sigma z) - 2f_4(z+\sigma z) + f_4(y) + f_4(\sigma y). \end{aligned} \tag{9.91}$$

Compare equations (9.90) and (9.91) to get an equation of the unknown function g

$$\begin{aligned} g(x+y) + g(x+\sigma y) = 2g(x+z+\sigma z) &+ g(w+\sigma w+y) \\ &+ g(w+\sigma w+\sigma y) - 2g(w+\sigma w+z+\sigma z). \end{aligned} \tag{9.92}$$

Set $x = z + \sigma z$ and $y = w + \sigma w$ in (9.92) and simplify to obtain that

$$2\alpha(z+w) = \alpha(2z) + \alpha(2w), \tag{9.93}$$

where $\alpha : S \to H$ is defined by

$$\alpha(x) = g(x + \sigma x). \tag{9.94}$$

Clearly, we have that

$$\alpha(\sigma x) = \alpha(x). \tag{9.95}$$

Since (9.93) resembles (9.24), by (9.25) it follows that there is an additive function $A_2 : S \to H$ and a constant $a \in H$ satisfying that

$$\alpha(x) = g(x + \sigma x) = 4A_2(x) + 2a. \tag{9.96}$$

Moreover, by using relation (9.95) in (9.96), it is easy to see that

$$A_2(\sigma x) = A_2(x). \tag{9.97}$$

Next, replace y by $y + \sigma y$ in (9.92), use (9.96) rearrange the terms and simplify to find out that

$$g(x + y + \sigma y) - 4A_2(y) = g(x + z + \sigma z) - 4A_2(z). \tag{9.98}$$

Thus, the function $\beta : S \to H$ defined by

$$\beta(x) = g(x + y + \sigma y) - 4A_2(y) - 2a \tag{9.99}$$

is well-defined, that is, it does not depend on the particular value of $y \in S$. Now, some useful identities involving g, A_2 and β will be obtained. Replace x by $x + \sigma x$ in (9.99) and use (9.96) to get that

$$\beta(x + \sigma x) = 4A_2(x). \tag{9.100}$$

Use (9.96) and (9.95) to rewrite the right-hand side of (9.92) to get

$$g(x + y) + g(x + \sigma y) = 2\beta(x) + \beta(y) + \beta(\sigma y) + 4a. \tag{9.101}$$

Replace x by $x + y + \sigma y$ in (9.99) and simplify to yield

$$\beta(x + y + \sigma y) = \beta(x) + 4A_2(y). \tag{9.102}$$

Finally, use (9.99), (9.101) and (9.102) to obtain the functional equation for β as follows:

$$\begin{aligned}\beta(x + y) + \beta(x + \sigma y) &= \big[g\big((x + w + \sigma w) + y\big) \\ &\quad + g\big((x + w + \sigma w) + \sigma y\big)\big] - 8A_2(w) - 4a \\ &= 2\beta(x + w + \sigma w) + \beta(y) + \beta(\sigma y) - 8A_2(w) \\ &= 2\beta(x) + \beta(y) + \beta(\sigma y).\end{aligned} \tag{9.103}$$

Define $B : S \times S \to H$ as

$$4B(x, y) = \beta(x + y) - \beta(x) - \beta(y). \tag{9.104}$$

Using (9.104), (9.100) and (9.102), it is straightforward to see that

$$B(x + \sigma x, y) = 0. \tag{9.105}$$

Furthermore, using (9.104) and (9.103), we can obtain the functional equation for B as follows:

$$\begin{aligned} 4B(x+z, y) + 4B(x+\sigma z, y) &= \big[\beta\big((x+y)+z\big) + \beta\big((x+y)+\sigma z\big)\big] \\ &\quad - \big[\beta(x+z) + \beta(x+\sigma z)\big] - 2\beta(y) \\ &= 2\big[\beta(x+y) - \beta(x) - \beta(y)\big] \\ &= 8B(x, y). \end{aligned} \tag{9.106}$$

Thus, B satisfies (9.8) so, by (9.9), there exist functions $\bar{A} : S \times S \to H$ and $\xi_2 : S \to H$ such that $B(x, y) = \bar{A}(x, y) + \xi(y)$, where $\bar{A}(\cdot, y)$ is additive and $\bar{A}(\sigma x, y) = -\bar{A}(x, y)$ for all $y \in S$. However, by using (9.105) and the definition of $\bar{A}$, it follows that

$$0 = B(x + \sigma x, y) = \bar{A}(x + \sigma x, y) + \xi_2(y) = \bar{A}(x, y) + \bar{A}(\sigma x, y) + \xi_2(y) = \xi_2(y).$$

Thus, since B is clearly symmetric, it follows that B is a biadditive function satisfying that $B(\sigma x, y) = -B(x, y)$. Setting $y = x$ in (9.104), we obtain that

$$4B(x, x) = \beta(2x) - 2\beta(x). \tag{9.107}$$

Similarly, setting $y = x$ in (9.103), we obtain after rearranging terms that

$$\beta(2x) = 3\beta(x) + \beta(\sigma x) - 4A_2(x). \tag{9.108}$$

Replacing (9.108) in (9.107), we get that

$$\beta(x) + \beta(\sigma x) = 4B(x, x) + 4A_2(x). \tag{9.109}$$

Next, define a function $A_3 : S \to H$ by

$$4A_3(x) = \beta(x) - \beta(\sigma x). \tag{9.110}$$

One obvious consequence of (9.107) is the fact that

$$A_3(\sigma x) = -A_3(x). \tag{9.111}$$

Now, we compute using (9.110) and (9.103) to obtain

$$\begin{aligned} 4A_3(x+y) - 4A_3(x+\sigma y) &= \big[\beta(x+y) + \beta(x+\sigma y)\big] \\ &\quad - \big[\beta(\sigma x + y) + \beta(\sigma x + \sigma y)\big] \\ &= 2\big[\beta(x) - \beta(\sigma x)\big] = 8A_3(x). \end{aligned} \tag{9.112}$$

Hence, A_3 satisfies (9.8), so by (9.9), there exist an additive function $\tilde{A} : S \to H$ and a constant $\xi_3 \in H$ satisfying $A_3(x) = \tilde{A}(x) + \xi_3$, where $\tilde{A}(\sigma x) = -\tilde{A}(x)$. However, by using (9.111) and the definition of $\tilde{A}$, it follows that

$$0 = A_3(x + \sigma x) = \tilde{A}(x + \sigma x) + \xi_3 = \tilde{A}(x) + \tilde{A}(\sigma x) + \xi_3 = \xi_3.$$

Thus, A_3 is an additive function. Next, solve the system of (9.109) and (9.110) to find out that

$$\beta(x) = 2\big[A_2(x) + A_3(x) + B(x,x)\big]. \tag{9.113}$$

Express the g terms from (9.91) in terms of B and A_2 using (9.102), (9.96) and (9.109) to get that

$$f_4(y) + f_4(\sigma y) - 4B(y,y) - 4A_2(y) = 2f_4(z + \sigma z) - 8A_2(z). \tag{9.114}$$

Hence, the right and left-hand sides of (9.114) do not depend on $y, z \in S$, so there exists $b \in H$ so that

$$f_4(x) + f_4(\sigma x) = 4B(x,x) + 4A_2(x) + 2b \tag{9.115}$$

and

$$f_4(x + \sigma x) = 4A_2(x) + b. \tag{9.116}$$

In (9.89), using (9.99), (9.116), and (9.113) yields

$$f_3(x) = 2\big[B(x,x) + A_2(x) + A_3(x) + a\big] - b. \tag{9.117}$$

Solve the system of (9.85) and (9.115) to obtain that

$$f_4(x) = 2\big[B(x,x) + A_1(x) + A_2(x)\big] + b. \tag{9.118}$$

Use results (9.113) and (9.109) in (9.101) and simplify to obtain that

$$g(x+y) + g(x+\sigma y) = 4\big[B(x,x) + B(y,y) + A_2(x+y) + A_3(x) + a\big]. \tag{9.119}$$

Replace both x and y by $\frac{x}{2}$ in (9.119) and use (9.96) to obtain

$$g(x) = 2\big[B(x,x) + A_2(x) + A_3(x) + a\big]. \tag{9.120}$$

Finally, use (9.86) and (9.120) to solve the system of (9.80) and (9.88) to obtain that

$$f_1(x) = B(x,x) + A_1(x) + A_2(x) + A_3(x) + \ell(x) + a \tag{9.121}$$

and

$$f_2(x) = B(x,x) - A_1(x) + A_2(x) + A_3(x) - \ell(x) + a. \tag{9.122}$$

Therefore, the general solution of (9.76) is given by (9.121), (9.122), (9.117), and (9.118). This completes the proof of the theorem. □

The next theorem is generalization of the pexiderized quadratic equation in higher dimensions. Since the proof is straightforward applications of the Theorem 9.1 and Theorem 9.9, it will be omitted.

Theorem 9.10 *Let* $(S,+)$ *be a commutative semigroup which is uniquely divisible by* 2. *The general solution* $f : S^n \to H$ *of the functional equation*

$$f_1(x+y)+f_2(x+\sigma y)=f_3(x)+f_4(y) \tag{9.123}$$

for all $x, y \in S^n$ *is given by*

$$\begin{cases} f_1(x)=B(x,x)+A_1(x)+A_2(x)+A_3(x)+\ell(x)+a,\\ f_2(x)=B(x,x)-A_1(x)+A_2(x)+A_3(x)-\ell(x)+a,\\ f_3(x)=2B(x,x)+2A_2(x)+2A_3(x)+2a-b,\\ f_4(x)=2B(x,x)+2A_1(x)+2A_2(x)+b \end{cases} \tag{9.124}$$

where

1. $B : S^n \to H$ *is a symmetric biadditive function satisfying* (9.55);
2. $A_1 : S^n \to H$ *is an additive function satisfying* (9.16);
3. $A_2 : S^n \to H$ *is an additive function satisfying* (9.20);
4. $A_3 : S^n \to H$ *is an additive function satisfying* (9.16);
5. $\ell : S^n \to H$ *is a* σ*-conjugate function, and*
6. $a, b \in H$

for all $x, y \in S$.

References

1. Aczél, J.: The general solution of two functional equations by reduction to functions additive in two variables and with aid of Hamel-bases. Glasnik Mat.-Fiz. Astron. Drustvo Mat. Fiz. Hrvatske **20**, 65–73 (1965)
2. Bae, J.-H., Park, W.-G.: A functional equation originating from quadratic forms. J. Math. Anal. Appl. **326**, 1142–1148 (2007)
3. Ebanks, B.R., Kannappan, P.L., Sahoo, P.K.: A common generalization of functional equations characterizing normed and quasi-inner-product spaces. Can. Math. Bull. **35**, 321–327 (1992)
4. Kuczma, M.: Note on additive functions of several variables. Pr. Nauk. Uniw. Śl. Katow. Nr 30, Pr. Mat. **3**, 75–78 (1973)
5. Kuczma, M.: An Introduction to the Theory of Functional Equations and Inequalities: Cauchy's Equation and Jensen's Inequality. Uniwersytet Slask-P.W.N., Katowice (1985)
6. Sinopoulos, P.: Functional equations on semigroups. Aequ. Math. **59**, 255–261 (2000)

Chapter 10
On Bohr's Inequalities

Wing-Sum Cheung, Gangsong Leng, Josip Pečarić, and Dandan Zhao

Abstract This is an exposition of the recent development of Bohr-type inequalities.

Key words Bohr's inequality · Complex separable Hilbert space

Mathematics Subject Classification 47A30 · 26D15 · 30A10 · 46B20

10.1 Background

The classical Bohr's inequality [3, 12] states that

$$|z_1 + z_2|^2 \leq (1+c)|z_1|^2 + \left(1 + \frac{1}{c}\right)|z_2|^2, \tag{10.1}$$

where $c > 0$, $z_1, z_2 \in \mathbb{C}$ and the equality holds if and only if $z_2 = cz_1$. Over the years, various generalizations of Bohr's inequality have been obtained.

Dedicated to Professor Themistocles M. Rassias on the occasion of his 60th birthday.

W.-S. Cheung (✉) · D. Zhao
Department of Mathematics, The University of Hong Kong, Pokfulam, Hong Kong
e-mail: wscheung@hku.hk

D. Zhao
e-mail: zhdd_80@hotmail.com

G. Leng
Department of Mathematics, Shanghai University, Shanghai 200436, P.R. China
e-mail: gleng@staff.shu.edu.cn

J. Pečarić
Faculty of Textile Technology, University of Zagreb, Pierottijeva 6, Zagreb 10000, Croatia
e-mail: pecaric@element.hr

P.M. Pardalos et al. (eds.), *Nonlinear Analysis*, Springer Optimization and Its Applications 68, In Honor of Themistocles M. Rassias on the Occasion of his 60th Birthday, DOI 10.1007/978-1-4614-3498-6_10,

In the book of J.W. Archbold [2], the following generalization of Bohr's inequality was given:

$$\left|\sum_{i=1}^{n} z_i\right|^2 \le \sum_{i=1}^{n} a_i |z_i|^2, \tag{10.2}$$

where $z_1, \dots, z_n \in \mathbb{C}$ and $a_1, \dots, a_n > 0$ such that $\sum_{i=1}^{n}(1/a_i) = 1$.

Equivalently, in [11] the following inequality was established:

$$\left|\sum_{i=1}^{n} a_i z_i\right|^2 \le \sum_{i=1}^{n} a_i |z_i|^2,$$

for $z_1, \dots, z_n \in \mathbb{C}$ and $a_1, \dots, a_n > 0$ satisfying $\sum_{i=1}^{n} a_i = 1$.

P.M. Vasić and J.D. Kečkić [18] further generalized (10.2) to the following: For $z_1, z_2, \dots, z_n \in \mathbb{C}$, $p_1, p_2, \dots, p_n > 0$ and $r > 1$, we have

$$\left|\sum_{i=1}^{n} z_i\right|^r \le \left(\sum_{i=1}^{n} p_i^{1/(1-r)}\right)^{r-1} \sum_{i=1}^{n} p_i |z_i|^r, \tag{10.3}$$

with the equality holding if and only if $p_1|z_1| = p_2|z_2| = \cdots = p_n|z_n|$ and $z_k \bar{z}_j \ge 0$ $(k, j = 1, 2, \dots, n)$.

In 1961, A. Makowski [10] proved the following inequality:

$$\begin{aligned}(z_1 - z_2)^2 \sin^2 \alpha + (z_1 + z_2)^2 \cos^2 \alpha &\le (1 + c|\cos 2\alpha|) z_1^2 + \left(1 + \frac{1}{c}|\cos 2\alpha| z_2^2\right) \\ &\le (1 + c)|z_1|^2 + \left(1 + \frac{1}{c}\right)|z_2|^2, \end{aligned} \tag{10.4}$$

where $c > 0$ and $z_1, z_2, \alpha \in \mathbb{R}$. This inequality relates to Bohr's inequality (10.1) with $z_1, z_2 \in \mathbb{R}$.

Th.M. Rassias [16, 17] generalized Bohr's inequality (10.1) to the following form

$$\begin{aligned}&(1 + na)|z_1|^2 + \left(1 + (n-1)a + \frac{1}{a}\right)|z_2|^2 + \left(1 + (n-2)a + \frac{2}{a}\right)|z_3|^2 \\ &\quad + \cdots + \left(1 + a + \frac{n-1}{a}\right)|z_n|^2 + \left(1 + \frac{n}{a}\right)|z_{n+1}|^2 \\ &\ge |z_1 + z_2 + \cdots + z_{n+1}|^2, \end{aligned} \tag{10.5}$$

where $a > 0$ and $z_1, \dots, z_{n+1} \in \mathbb{C}$.

In [14] and [9], the following result was obtained: Let $f : [0, +\infty) \to [0, +\infty)$ be a strictly concave and increasing function satisfying $f(0) = 0$, $f(st) \ge f(s)f(t)$, $\lim_{t \to 0^+} \frac{f(t)}{t} = +\infty$ and $\lim_{t \to \infty} \frac{f(t)}{t} = 0$. If $(X, \|\cdot\|)$ is a normed vector space and $x_i \in X$, then for $q_i \in \mathbb{R}$ satisfying $g^{-1}(1/q_i) \ge 1$, where $g(t) = f(t)/t$, $i = 1, \dots, n$, we have

$$f\left(\left\|\sum_{i=1}^{n} x_i\right\|\right) \le \sum_{i=1}^{n} q_i f\left(\|x_i\|\right). \tag{10.6}$$

From (10.6) we can see that if $(X, \|\cdot\|)$ is a normed vector space, $x_i \in X$ and $0 \le r < 1$, $q_i \ge 1$ $(i = 1, \dots, n)$, then [13]

$$\left\|\sum_{i=1}^{n} x_i\right\|^r \le \sum_{i=1}^{n} q_i \|x_i\|^r. \tag{10.7}$$

In 1989, J.E. Pečarić and S.S. Dragomir [13] generalized Bohr's inequality to normed spaces. If $(X, \|\cdot\|)$ is a normed vector space, $f: \mathbb{R}^+ \to \mathbb{R}^+$ is a nondecreasing convex function, $x_i \in X$ and $q_i \ge 0$ $(i = 1, \dots, n)$ such that $Q_n = \sum_{i=1}^{n} q_i > 0$, then

$$f\left(\frac{1}{Q_n}\left\|\sum_{i=1}^{n} q_i x_i\right\|\right) \le \frac{1}{Q_n}\sum_{i=1}^{n} q_i f\left(\|x_i\|\right). \tag{10.8}$$

From (10.7) we can obtain a generalization of (10.3) in normed spaces [13]: For $x_i \in X$, $p_i > 0$, $i = 1, 2, \dots, n$ and $r > 1$, we have

$$\left\|\sum_{i=1}^{n} x_i\right\|^r \le \left(\sum_{i=1}^{n} p_i^{1/(1-r)}\right)^{r-1} \sum_{i=1}^{n} p_i \|x_i\|^r.$$

In the special case where $n = r = 2$, this reduces to

$$\|x_1 + x_2\|^2 \le p\|x_1\|^2 + q\|x_2\|^2, \tag{10.9}$$

where $p, q > 1$ such that $\frac{1}{p} + \frac{1}{q} = 1$. This is a generalization of Bohr's inequality (10.1) in normed spaces.

In [15], J.E. Pečarić and Th.M. Rassias proved that if x_i $(i = 1, \dots, n)$ are elements in an unitary normed vector space X and $a_{ij} > 0$, $1 \le i < j \le n$, then

$$\left\|\sum_{i=1}^{n} x_i\right\|^2 \le \sum_{k=1}^{n}\left(1 + \sum_{j=k+1}^{n} a_{kj} + \sum_{j=1}^{k-1}\frac{1}{a_{jk}}\right)\|x_k\|^2. \tag{10.10}$$

This is a generalization of (10.5) in normed spaces.

There are also other well-known inequalities related to Bohr's inequality:

In [11], D.S. Mitrinović presented that if $z_1, z_2 \in \mathbb{R}$ or $\mathbb{C}$ and $r \ge 0$, then

$$|z_1 + z_2|^r \le C_r\left(|z_1|^r + |z_2|^r\right), \tag{10.11}$$

where

$$C_r = \begin{cases} 1, & 0 \le r \le 1, \\ 2^{r-1}, & r > 1. \end{cases}$$

In [7], D. Delbosco gave the following generalization of (10.11) in normed vector spaces.

Let $(X, \|\cdot\|)$ be a normed vector space and let $r \geq 0$, then for $x_1, x_2, \dots, x_n \in X$, we have

$$\|x_1 + x_2 + \cdots + x_n\|^r \leq C_{r,n}\big(\|x_1\|^r + \|x_2\|^r + \cdots + \|x_n\|^r\big), \tag{10.12}$$

where

$$C_{r,n} = \begin{cases} 1, & 0 \leq r \leq 1, \\ n^{r-1}, & r > 1. \end{cases}$$

J.E. Pečarić and R.R. Janić [14] generalized (10.12) to the following: For any normed vector space $(X, \|\cdot\|)$ and $x_i \in X$, $i = 1, \dots, n$,

(i) If $f : [0, +\infty) \to [0, +\infty)$ is a nondecreasing and convex function, then

$$f\left(\frac{1}{n}\left\|\sum_{i=1}^{n} x_i\right\|\right) \leq \frac{1}{n}\sum_{i=1}^{n} f\big(\|x_i\|\big); \tag{10.13}$$

(ii) If $f : [0, +\infty) \to [0, +\infty)$ is a nondecreasing and concave function with $f(0) = 0$, then

$$f\left(\left\|\sum_{i=1}^{n} x_i\right\|\right) \leq \sum_{i=1}^{n} f\big(\|x_i\|\big). \tag{10.14}$$

Obviously, (10.12) is a special case of (i) in which $f(x) = x^r$, $r > 1$, and (ii) in which $f(x) = x^r$, $0 \leq r < 1$.

Also, we can see that (10.12) can be proved by choosing $f(x) = x^r$, $r > 1$ and $q_i = 1$, $i = 1, 2, \dots, n$ in (10.8) together with (1.7) for $q_i = 1$, $i = 1, 2, \dots, n$.

10.2 Bohr's Inequalities for Hilbert Space Operators

10.2.1 Introduction

In [8], Hirzallah further generalized (10.9) to the context of operator algebras. It was shown that if $\mathbb{H}$ is a complex separable Hilbert space and $B(\mathbb{H})$ is the algebra of all bounded linear operators on $\mathbb{H}$, then for any $A, B \in B(\mathbb{H})$ and conjugate exponents p, q with $q \geq p > 1$,

$$|A - B|^2 + \big|(1 - p)A - B\big|^2 \leq p|A|^2 + q|B|^2,$$

where $|X| := (X^*X)^{1/2}$. It is worthwhile noting that, in [8], only the situation where $q \geq p > 1$, or equivalently, only the situation where $q \geq 2$ and $1 < p \leq 2$ was considered, while the other situations were left unconsidered.

In [4], Cheung and Pečarić continued working in the setting as that in [8], but with the restriction on the conjugate exponents p, q lifted. Meanwhile, the situation of equality was investigated in detail and a connection with the parallelogram law for the Banach algebra $B(\mathbb{H})$ was established. A very interesting inequality was also given as an application of this generalized Bohr's inequalities for Hilbert space operators. In this section, we shall give a brief account on the work of Cheung and Pečarić. For the details of the computations, one is referred to [4].

10.2.2 Bohr's Inequality and the Parallelogram Law in $B(\mathbb{H})$

Let $\mathbb{H}$ be a complex separable Hilbert space and $B(\mathbb{H})$ the algebra of all bounded linear operators on $\mathbb{H}$. For any $X \in B(\mathbb{H})$, write $|X| = (X^*X)^{\frac{1}{2}}$.

Theorem 10.1 *For any $A, B \in B(\mathbb{H})$ and any $p, q \in \mathbb{R}$ with $\frac{1}{p} + \frac{1}{q} = 1$, if $1 < p \leq 2$, then*

(i) $|A - B|^2 + |(1-p)A - B|^2 \leq p|A|^2 + q|B|^2$, *and*
(ii) $|A - B|^2 + |A - (1-q)B|^2 \geq p|A|^2 + q|B|^2$.

Furthermore, in both (i) *and* (ii), *the equality holds if and only if $p = q = 2$ or $(1-p)A = B$.*

Proof (i) We have as in [8]

$$|A - B|^2 = |A|^2 + |B|^2 - \bigl(A^*B + B^*A\bigr)$$

and

$$\bigl|(1-p)A - B\bigr|^2 = (1-p)^2|A|^2 + |B|^2 - (1-p)\bigl(A^*B + B^*A\bigr).$$

By elementary analysis, it is not hard to show that

$$|A - B|^2 + \bigl|(1-p)A - B\bigr|^2 \leq p|A|^2 + q|B|^2,$$

with equality if and only if $p = 2$ (hence $q = 2$) or $\sqrt{p-1}A + \frac{1}{\sqrt{p-1}}B = 0$, that is, $p = q = 2$ or $(1-p)A = B$.

(ii) Similar to (i), since $1 < p \leq 2$, we have $q \geq 2$ and so

$$\begin{aligned} &|A - B|^2 + \bigl|A - (1-q)B\bigr|^2 - p|A|^2 - q|B|^2 \\ &\quad = (q-2)\left[(q-1)|B|^2 + \frac{1}{q-1}|A|^2 + \bigl(A^*B + B^*A\bigr)\right] \\ &\quad = (q-2)\left|\sqrt{q-1}\,B + \frac{1}{\sqrt{q-1}}A\right|^2 \geq 0. \end{aligned} \tag{10.15}$$

Hence

$$|A - B|^2 + \bigl|A - (1-q)B\bigr|^2 \geq p|A|^2 + q|B|^2,$$

with equality if and only if $q = 2$ (hence $p = 2$) or $\sqrt{q-1}\, B + \frac{1}{\sqrt{q-1}}\, A = 0$, that is, $p = q = 2$ or $(1-q)B = A$; or equivalently, $p = q = 2$ or $(1-p)A = B$. □

Remark 10.1 By combining (i) and (ii) in Theorem 10.1, we have, for any $1 < p \leq 2$,

$$|A-B|^2 + \big|(1-p)A - B\big|^2 \leq p|A|^2 + q|B|^2$$
$$\leq |A-B|^2 + \big|A - (1-q)B\big|^2.$$

In particular, if we take $p = q = 2$, then we have

$$|A-B|^2 + |A+B|^2 \leq 2|A|^2 + 2|B|^2 \leq |A-B|^2 + |A+B|^2,$$

that is, the parallelogram law

$$|A-B|^2 + |A+B|^2 = 2|A|^2 + 2|B|^2. \tag{10.16}$$

Equivalently, this is also obtained by directly writing out the equality in (i) or (ii) for the case $p = 2$.

The following are simple consequences of Theorem 10.1.

Corollary 10.1 *For any $A, B \in B(\mathbb{H})$ and any $p, q \in \mathbb{R}$ with $\frac{1}{p} + \frac{1}{q} = 1$, if $p > 2$, then*

(i) $|A-B|^2 + |(1-p)A - B|^2 \geq p|A|^2 + q|B|^2$, *and*
(ii) $|A-B|^2 + |A-(1-q)B|^2 \leq p|A|^2 + q|B|^2$.

Furthermore, in both (i) *and* (ii), *the equality holds if and only if* $(1-p)A = B$.

Corollary 10.2 *For any $A, B \in B(\mathbb{H})$ and any $p, q \in \mathbb{R}$ with $p > 1$ and $\frac{1}{p} + \frac{1}{q} = 1$,*

$$|A+B|^2 \leq p|A|^2 + q|B|^2,$$

with equality if and only if $(p-1)A = B$.

Corollary 10.3 *For any $A, B \in B(\mathbb{H})$ and any $p, q \in \mathbb{R}$ with $\frac{1}{p} + \frac{1}{q} = 1$, if $p < 1$, then*

(i) $|A-B|^2 + |(1-p)A - B|^2 \geq p|A|^2 + q|B|^2$, *and*
(ii) $|A-B|^2 + |A-(1-q)B|^2 \geq p|A|^2 + q|B|^2$.

Furthermore, in both (i) *and* (ii), *the equality holds if and only if* $(1-p)A = B$.

Theorem 10.2 *Let $A, B \in B(\mathbb{H})$ and $\alpha, \beta \in \mathbb{R}$ be nonzero constants.*

(a) *If $\alpha\beta > 0$ with, say, $|\alpha| \geq |\beta| > 0$, then*

$$|A - B|^2 + \frac{1}{\alpha^2}|\beta A + \alpha B|^2 \leq \frac{\alpha + \beta}{\alpha}|A|^2 + \frac{\alpha + \beta}{\beta}|B|^2,$$

with equality if and only if $\alpha = \beta$ or $\beta A + \alpha B = 0$.

(b) *If $\alpha\beta < 0$ with, say, $\alpha > 0 > \beta$,*

(i) *If $\alpha > 0 > \beta \geq -\alpha$, then*

$$|A - B|^2 + \frac{1}{\alpha^2}|\beta A - \alpha B|^2 \leq \frac{\alpha - \beta}{\alpha}|A|^2 - \frac{\alpha - \beta}{\beta}|B|^2,$$

with equality if and only if $\alpha + \beta = 0$ or $\beta A - \alpha B = 0$;

(ii) *If $\alpha > 0 > -\alpha \geq \beta$, then*

$$|A - B|^2 + \frac{1}{\beta^2}|\alpha A - \beta B|^2 \leq -\frac{\alpha - \beta}{\beta}|A|^2 + \frac{\alpha - \beta}{\alpha}|B|^2,$$

with equality if and only if $\alpha + \beta = 0$ or $\alpha A - \beta B = 0$.

Proof (a) If $\alpha \geq \beta > 0$, write

$$p = \frac{\alpha + \beta}{\alpha}, \quad q = \frac{\alpha + \beta}{\beta}.$$

Then Theorem 10.1 applies, and we have

$$|A - B|^2 + \frac{1}{\alpha^2}|\beta A + \alpha B|^2 \leq \frac{\alpha + \beta}{\alpha}|A|^2 + \frac{\alpha + \beta}{\beta}|B|^2,$$

with equality if and only if

$$\alpha = \beta \quad \text{or} \quad \beta A + \alpha B = 0.$$

If $0 > \beta \geq \alpha$, then $-\alpha \geq -\beta \geq 0$ and so from above,

$$|A - B|^2 + \frac{1}{\alpha^2}|\beta A + \alpha B|^2 \leq \frac{\alpha + \beta}{\alpha}|A|^2 + \frac{\alpha + \beta}{\beta}|B|^2,$$

with equality if and only if

$$\alpha = \beta \quad \text{or} \quad \beta A + \alpha B = 0.$$

(b) Follows immediately from (a). □

Interesting inequalities on operators in $B(\mathbb{H})$ can easily be derived from the Bohr-type inequalities obtained above. For this we first observe the following generalization of Adamović's result [1] to $B(\mathbb{H})$.

Lemma 10.1 *For any* $A_i \in B(\mathbb{H})$, $i = 1, \dots, n$,

$$\left|\sum_{i=1}^{n} A_i\right|^2 - \left(\sum_{i=1}^{n} |A_i|\right)^2 = \sum_{1\le i<j\le n} \left[|A_i + A_j|^2 - \left(|A_i| + |A_j|\right)^2\right].$$

Theorem 10.3 *For any* $A_i \in B(\mathbb{H})$, $i = 1, \dots, n$, *and* $p_{ij} > 1$, $q_{ij} \in \mathbb{R}$ *with* $\frac{1}{p_{ij}} + \frac{1}{q_{ij}} = 1$, $1 \le i < j \le n$, *we have*

$$\left|\sum_{i=1}^{n} A_i\right|^2 \le \sum_{k=1}^{n} \left[1 + \sum_{j=k+1}^{n} (p_{kj} - 1) + \sum_{j=1}^{k-1} (q_{jk} - 1)\right] |A_k|^2;$$

Furthermore, the equality holds if and only if $(p_{ij} - 1)A_i = A_j$ *for all* $1 \le i < j \le n$.

Proof By Lemma 10.1, we have

$$\left|\sum_{i=1}^{n} A_i\right|^2 - \sum_{i=1}^{n} |A_i|^2 = \sum_{1\le i<j\le n} \left[|A_i + A_j|^2 - \left(|A_i|^2 + |A_j|^2\right)\right].$$

Applying Corollary 10.2 to $|A_i + A_j|$, the assertion follows. □

Remark 10.2 We easily see that Theorem 10.3 is a generalization of (10.10) to the context of operator algebras.

10.3 Bohr's Inequalities in *n*-Inner Product Spaces

10.3.1 Introduction, Basic Terminologies, and Fundamental Results

As described in Sect. 10.2, Cheung and Pecǎrić [4] have generalized Bohr's inequality to the context of operator algebras with $1/p + 1/q = 1$, $p, q \in \mathbb{R}$. In [5], the results of [4] are further generalized to n-inner product spaces. In this section, we will give a brief account on the results obtained by Cheung et al. in [5]. For the details, one is referred to [5].

First, we recall some basics of n-inner product spaces [6]. Let $n \ge 2$ and X be a linear space of dimension greater than or equal to n over $\mathbb{C}$. A complex-valued function $(\cdot, \cdot|\cdot, \dots, \cdot) : X^{n+1} \to \mathbb{C}$ satisfying the following properties:

(I_1) $(x, x|a_2, \dots, a_n) \ge 0$ and $(x, x|a_2, \dots, a_n) = 0$ if and only if the vectors $x, a_2, \dots, a_n$ are linearly dependent;

(I_2) $(x, x|a_2, \dots, a_n) = (a_2, a_2|x, \dots, a_n)$;

(I_3) $(x, y|a_{i_2}, \ldots, a_{i_n}) = (x, y|a_2, \ldots, a_n)$ for any permutation $(i_2, \ldots, i_n)$ of $(2, \ldots, n)$;
(I_4) $(y, x|a_2, \ldots, a_n) = \overline{(x, y|a_2, \ldots, a_n)}$;
(I_5) $(\alpha x, y|a_2, \ldots, a_n) = \alpha(x, y|a_2, \ldots, a_n)$ for any scalar $\alpha \in \mathbb{C}$;
(I_6) $(x_1 + x_2, y|a_2, \ldots, a_n) = (x_1, y|a_2, \ldots, a_n) + (x_2, y|a_2, \ldots, a_n)$

is called an *n-inner product* on X and $(X, (\cdot, \cdot|\cdot, \ldots, \cdot))$ is called an *n-inner product space*.

In an n-inner product space $(X, (\cdot, \cdot|\cdot, \ldots, \cdot))$, the following extension of Cauchy–Schwarz–Buniakowsky inequality

$$\left|(x, y|a_2, \ldots, a_n)\right| \leq \sqrt{(x, x|a_2, \ldots, a_n)} \cdot \sqrt{(y, y|a_2, \ldots, a_n)}$$

for any $x, y, a_2, \ldots, a_n \in X$ is valid, and it is easy to verify that the real valued function $\|\cdot, \cdot, \ldots, \cdot\| : X^n \to \mathbb{R}$ defined by

$$\|a_1, a_2, \ldots, a_n\| = \sqrt{(a_1, a_1|a_2, \ldots, a_n)}$$

satisfies the following conditions:

(N_1) $\|a_1, a_2, \ldots, a_n\| \geq 0$ and $\|a_1, a_2, \ldots, a_n\| = 0$ if and only if $a_1, a_2, \ldots, a_n$ are linearly dependent;
(N_2) $\|a_{i_1}, a_{i_2}, \ldots, a_{i_n}\| = \|a_1, a_2, \ldots, a_n\|$ for any permutation $(i_1, i_2, \ldots, i_n)$ of $(1, 2, \ldots, n)$;
(N_3) $\|\alpha a_1, a_2, \ldots, a_n\| = |\alpha|\,\|a_1, a_2, \ldots, a_n\|$ for any scalar $\alpha \in \mathbb{C}$;
(N_4) $\|x + y, a_2, \ldots, a_n\| \leq \|x, a_2, \ldots, a_n\| + \|y, a_2, \ldots, a_n\|$.

Any real valued function $\|\cdot, \ldots, \cdot\|$ defined on X^n satisfying conditions $(N_1) \sim (N_4)$ is called an *n-norm* and $(X, \|\cdot, \ldots, \cdot\|)$ is called an *n-normed linear space*.

Throughout this section, X will denote an n-inner product space equipped with the n-norm

$$\|a_1, a_2, \ldots, a_n\| := \sqrt{(a_1, a_1|a_2, \ldots, a_n)}.$$

10.3.2 Bohr's Inequality and the Parallelogram Law

Theorem 10.4 *For any $x, y, a_2, \ldots, a_n \in X$ and $p, q \in \mathbb{R}$ with $\frac{1}{p} + \frac{1}{q} = 1$, if $1 < p \leq 2$, then we have the following*:

1. $\|x - y, a_2, \ldots, a_n\|^2 + \|(1 - p)x - y, a_2, \ldots, a_n\|^2 \leq p\|x, a_2, \ldots, a_n\|^2 + q\|y, a_2, \ldots, a_n\|^2$,
2. $\|x - y, a_2, \ldots, a_n\|^2 + \|x - (1 - q)y, a_2, \ldots, a_n\|^2 \geq p\|x, a_2, \ldots, a_n\|^2 + q\|y, a_2, \ldots, a_n\|^2$.

Furthermore, in both parts 1 *and* 2, *the equality holds if and only if $p = q = 2$ or $px + qy, a_2, \ldots, a_n$ are linearly dependent.*

Proof Part 1. By the identities

$$\begin{aligned}&\|x-y,a_2,\dots,a_n\|^2\\&\quad=\|x,a_2,\dots,a_n\|^2+\|y,a_2,\dots,a_n\|^2-2\operatorname{Re}(x,y|a_2,\dots,a_n)\end{aligned}$$

and

$$\begin{aligned}&\big\|(1-p)x-y,a_2,\dots,a_n\big\|^2\\&\quad=(1-p)^2\|x,a_2,\dots,a_n\|^2+\|y,a_2,\dots,a_n\|^2-2(1-p)\operatorname{Re}(x,y|a_2,\dots,a_n),\end{aligned}$$

it is easy to check that

$$\begin{aligned}&\|x-y,a_2,\dots,a_n\|^2+\big\|(1-p)x-y,a_2,\dots,a_n\big\|^2\\&\quad-p\|x,a_2,\dots,a_n\|^2-q\|y,a_2,\dots,a_n\|^2\le 0,\end{aligned}\tag{10.17}$$

with equality if and only if $p=2$ or $(\sqrt{p-1}\,x+\frac{1}{\sqrt{p-1}}y),a_2,\dots,a_n$ are linearly dependent, that is, $p=q=2$ or $px+qy,a_2,\dots,a_n$ are linearly dependent.

Part 2. The proof is similar to part 1. □

Remark 10.3 By combining parts 1 and 2 in Theorem 10.4, for any $1<p\le 2$, we have

$$\begin{aligned}&\|x-y,a_2,\dots,a_n\|^2+\big\|(1-p)x-y,a_2,\dots,a_n\big\|^2\\&\quad\le\|x-y,a_2,\dots,a_n\|^2+\big\|x-(1-q)y,a_2,\dots,a_n\big\|^2.\end{aligned}$$

In particularly, if $p=q=2$, we have the Parallelogram Law:

$$\begin{aligned}&\|x-y,a_2,\dots,a_n\|^2+\|x+y,a_2,\dots,a_n\|^2\\&\quad=2\|x,a_2,\dots,a_n\|^2+2\|y,a_2,\dots,a_n\|^2.\end{aligned}\tag{10.18}$$

Equivalently, this is also obtained by directly writing out the equality in part 1 or 2 for the case $p=2$.

The following are simple consequences of Theorem 10.4.

Corollary 10.4 *For any $x,y,a_2,\dots,a_n\in X$ and $p,q\in\mathbb{R}$ with $\frac{1}{p}+\frac{1}{q}=1$, if $p>2$, we have*

1. $\|x-y,a_2,\dots,a_n\|^2+\|(1-p)x-y,a_2,\dots,a_n\|^2\ge p\|x,a_2,\dots,a_n\|^2+q\|y,a_2,\dots,a_n\|^2$,
2. $\|x-y,a_2,\dots,a_n\|^2+\|x-(1-q)y,a_2,\dots,a_n\|^2\le p\|x,a_2,\dots,a_n\|^2+q\|y,a_2,\dots,a_n\|^2$.

Furthermore, in both parts 1 and 2, the equality holds if and only if $px+qy,a_2,\dots,a_n$ are linearly dependent.

Corollary 10.5 *For any $x, y, a_2, \dots, a_n \in X$ and $p, q \in \mathbb{R}$ with $p > 1$ and $\frac{1}{p} + \frac{1}{q} = 1$, we have*

$$\|x + y, a_2, \dots, a_n\|^2 \le p\|x, a_2, \dots, a_n\|^2 + q\|y, a_2, \dots, a_n\|^2,$$

with the equality holding if and only if $px - qy, a_2, \dots, a_n$ are linearly dependent.

Corollary 10.6 *For any $x, y, a_2, \dots, a_n \in X$ and $p, q \in \mathbb{R}$ with $\frac{1}{p} + \frac{1}{q} = 1$, if $p < 1$, then we have the following*:

1. $\|x - y, a_2, \dots, a_n\|^2 + \|(1 - p)x - y, a_2, \dots, a_n\|^2 \ge p\|x, a_2, \dots, a_n\|^2 + q\|y, a_2, \dots, a_n\|^2$,
2. $\|x - y, a_2, \dots, a_n\|^2 + \|x - (1 - q)y, a_2, \dots, a_n\|^2 \ge p\|x, a_2, \dots, a_n\|^2 + q\|y, a_2, \dots, a_n\|^2$.

Furthermore, in both parts 1 *and* 2, *the equality holds if and only if $px + qy, a_2, \dots, a_n$ are linearly dependent.*

Theorem 10.5 *For any $x, y, a_2, \dots, a_n \in X$ and $\alpha, \beta \in \mathbb{R}$ be nonzero constants.*

(a) *If $\alpha\beta > 0$ with $|\alpha| \ge |\beta| > 0$, then*

$$\begin{aligned}\|x - y, a_2, \dots, a_n\|^2 &+ \frac{1}{\alpha^2}\|\beta x + \alpha y, a_2, \dots, a_n\|^2 \\ &\le \frac{\alpha + \beta}{\alpha}\|x, a_2, \dots, a_n\|^2 + \frac{\alpha + \beta}{\beta}\|y, a_2, \dots, a_n\|^2,\end{aligned}$$

with the equality if and only if $\alpha = \beta$ or $\beta x + \alpha y, a_2, \dots, a_n$ are linearly dependent.

(b) *Let $\alpha\beta < 0$ with $|\alpha| > 0 > |\beta|$.*

1. *If $\alpha > 0 > \beta \ge -\alpha$, then*

$$\begin{aligned}\|x - y, a_2, \dots, a_n\|^2 &+ \frac{1}{\alpha^2}\|\beta x - \alpha y, a_2, \dots, a_n\|^2 \\ &\le \frac{\alpha - \beta}{\alpha}\|x, a_2, \dots, a_n\|^2 - \frac{\alpha - \beta}{\beta}\|y, a_2, \dots, a_n\|^2,\end{aligned}$$

with the equality if and only if $\alpha = -\beta$ or $\beta x - \alpha y, a_2, \dots, a_n$ are linearly dependent.

2. *If $\alpha > 0 > -\alpha \ge \beta$, then*

$$\begin{aligned}\|x - y, a_2, \dots, a_n\|^2 &+ \frac{1}{\beta^2}\|\alpha x - \beta y, a_2, \dots, a_n\|^2 \\ &\le -\frac{\alpha - \beta}{\beta}\|x, a_2, \dots, a_n\|^2 + \frac{\alpha - \beta}{\alpha}\|y, a_2, \dots, a_n\|^2,\end{aligned}$$

with the equality if and only if $\alpha = -\beta$ or $\alpha x - \beta y, a_2, \dots, a_n$ are linearly dependent.

Proof (a) If $\alpha \geq \beta > 0$, we write

$$p = \frac{\alpha+\beta}{\alpha}, \quad q = \frac{\alpha+\beta}{\beta}.$$

Then Theorem 10.4 applies and we have

$$\begin{aligned} &\|x-y, a_2, \ldots, a_n\|^2 + \frac{1}{\alpha^2}\|\beta x + \alpha y, a_2, \ldots, a_n\|^2 \\ &\leq \frac{\alpha+\beta}{\alpha}\|x, a_2, \ldots, a_n\|^2 + \frac{\alpha+\beta}{\beta}\|y, a_2, \ldots, a_n\|^2, \end{aligned}$$

with the equality if and only if $\alpha = \beta$ or $\beta x + \alpha y, a_2, \ldots, a_n$ are linearly dependent.

If $0 > \beta \geq \alpha$, then $-\alpha \geq -\beta > 0$ and so from above,

$$\begin{aligned} &\|x-y, a_2, \ldots, a_n\|^2 + \frac{1}{\alpha^2}\|\beta x + \alpha y, a_2, \ldots, a_n\|^2 \\ &\leq \frac{\alpha+\beta}{\alpha}\|x, a_2, \ldots, a_n\|^2 + \frac{\alpha+\beta}{\beta}\|y, a_2, \ldots, a_n\|^2, \end{aligned}$$

with the equality if and only if $\alpha = \beta$ or $\beta x + \alpha y, a_2, \ldots, a_n$ are linearly dependent.

(b) Follows immediately from (a). □

Interesting inequalities on operators in the n-inner product space X can easily be derived from the Bohr-type inequalities obtained above. For this, we first observe the following generalization of Adamović's result [1] in an n-inner product space X:

Lemma 10.2 *For any $x_i \in X$, $i = 1, 2, \ldots, n$,*

$$\begin{aligned} &\left\|\sum_{i=1}^n x_i, a_2, \ldots, a_n\right\|^2 - \left(\sum_{i=1}^n \|x_i, a_2, \ldots, a_n\|\right)^2 \\ &= \sum_{1\leq i<j\leq n} \left[\|x_i + x_j, a_2, \ldots, a_n\|^2 - \left(\|x_i, a_2, \ldots, a_n\| + \|x_j, a_2, \ldots, a_n\|\right)^2\right]. \end{aligned}$$

Theorem 10.6 *For any $x_i \in X$, $i = 1, 2, \ldots, n$, and $p_{ij} > 1, q_{ij} \in \mathbb{R}$ with $\frac{1}{p_{ij}} + \frac{1}{q_{ij}} = 1$, $1 \leq i < j \leq n$, we have*

$$\begin{aligned} &\left\|\sum_{i=1}^n x_i, a_2, \ldots, a_n\right\|^2 \\ &\leq \sum_{k=1}^n \left(1 + \sum_{j=k+1}^n (p_{ij}-1) + \sum_{j=1}^{k-1}(q_{ij}-1)\right)\|x_k, a_2, \ldots, a_n\|^2, \end{aligned}$$

the equality holds if and only if $(p_{ij}x_i - q_{ij}x_j), a_2, \dots, a_n$ *are linearly dependent for all* i, j *with* $1 \le i < j \le n$.

Proof By Lemma 10.2, we have

$$\left\| \sum_{i=1}^{n} x_i, a_2, \dots, a_n \right\|^2 - \sum_{i=1}^{n} \|x_i, a_2, \dots, a_n\|^2$$
$$= \sum_{1\le i<j\le n} \left[\|x_i + x_j, a_2, \dots, a_n\|^2 - \left(\|x_i, a_2, \dots, a_n\|^2 + \|x_j, a_2, \dots, a_n\|^2 \right) \right].$$

Applying Corollary 10.5 to $\|x_i + x_j, a_2, \dots, a_n\|$, we have

$$\left\| \sum_{i=1}^{n} x_i, a_2, \dots, a_n \right\|^2 - \sum_{i=1}^{n} \|x_i, a_2, \dots, a_n\|^2$$
$$\le \sum_{1\le i<j\le n} \left[(p_{ij} - 1)\|x_i, a_2, \dots, a_n\|^2 + (q_{ij} - 1)\|x_j, a_2, \dots, a_n\|^2 \right],$$

with the equality if and only if $(p_{ij}x_i - q_{ij}x_j), a_2, \dots, a_n$ are linearly dependent for all i, j with $1 \le i < j \le n$, that is,

$$\left\| \sum_{i=1}^{n} x_i, a_2, \dots, a_n \right\|^2$$
$$\le \sum_{k=1}^{n} \left(1 + \sum_{j=k+1}^{n} (p_{ij} - 1) + \sum_{j=1}^{k-1} (q_{ij} - 1) \right) \|x_k, a_2, \dots, a_n\|^2,$$

with the equality if and only if $(p_{ij}x_i - q_{ij}x_j), a_2, \dots, a_n$ are linearly dependent for all i, j with $1 \le i < j \le n$. □

Remark 10.4 Note that Theorem 10.6 is a generalization of (10.10) to n-inner product spaces.

Acknowledgements The first author's research was supported in part by the Research Grants Council of the Hong Kong SAR, China (Project No. HKU7016/07P). The second author's research was supported in part by the National Natural Science Foundation of China (Grant No. 10971128).

References

1. Adamović, D.D.: Quelques remarques relatives aux Généralisations des Inégalités de Hlawka et de Hornich. Mat. Vesn. **1**(16), 241–242 (1964)
2. Archbold, J.W.: In: Algebra, London (1958)
3. Bohr, H.: Zur Theorie der Fastperiodischen Funktionen I. Acta Math. **45**, 29–127 (1924)

4. Cheung, W.S., Pečarić, J.: Bohr's inequalities for Hilbert space operators. J. Math. Anal. Appl. **323**, 403–412 (2006)
5. Cheung, W.S., Cho, Y.J., Pečarić, J., Zhao, D.D.: Bohr's inequalities in n-inner product spaces. J. Korean Math. Soc., Ser. B **14**, 127–137 (2007)
6. Cho, Y.J., Lin, P.C.S., Kim, S.S., Misiak, A.: Theory of 2-Inner Product Spaces. Nova Science, New York (2001)
7. Delbosco, D.: Sur une Inégalité de la Norme. Publ. Elektroteh. Fak. Univ. Beogr., Mat. Fiz. **678–715**, 206–208 (1980)
8. Hirzallah, O.: Non-commutative operator Bohr inequality. J. Math. Anal. Appl. **282**, 578–583 (2003)
9. Kocić, V.L., Maksimović, D.M.: Variations and generalizations of an inequality due to Bohr. Publ. Elektroteh. Fak. Univ. Beogr., Mat. Fiz. **412–460**, 183–188 (1973)
10. Mąkowski, A.: Bol. Mat. **34**, 1–11 (1961)
11. Mitrinović, D.S.: Analytic Inequalities. Springer, New York (1970)
12. Mitrinović, D.S., Pečarić, J.E., Fink, A.M.: Classical and New Inequalities in Analysis. Kluwer, Dordrecht (1993)
13. Pečarić, J.E., Dragomir, S.S.: A refinement of Jensen inequality and applications. Stud. Univ. Babeş-Bolyai, Math. **34**, 15–19 (1989)
14. Pečarić, J.E., Janić, R.R.: Some remarks on the paper "Sur une inégalité de la norme" of D. Delbosco. Facta Univ. (Niš). Ser. Math. Inform **3**, 39–42 (1988)
15. Pečarić, J.E., Rassias, Th.M.: Variations and generalizations of Bohr's inequality. J. Math. Anal. Appl. **174**, 138–146 (1993)
16. Rassias, Th.M.: A generalization of a triangle-like inequality due to H. Bohr. Abstr. Am. Math. Soc. **5**, 276 (1985)
17. Rassias, Th.M.: On characterizations of inner product spaces and generalizations of the H. Bohr inequality. In: Rassias, Th.M. (ed.) Topics in Mathematical Analysis, Singapore (1989)
18. Vasić, P.M., Kečkić, J.D.: Some inequalities for complex numbers. Math. Balk. **1**, 282–286 (1971)

Chapter 11
Orlicz Norm Inequalities for Conjugate Harmonic Forms

Shusen Ding and Yuming Xing

Abstract We establish some basic norm inequalities, including the Poincaré inequality, weak reverse Hölder inequality, and Caccioppoli inequality, for conjugate harmonic forms. We also prove the Caccioppoli inequality with Orlicz norm for conjugate harmonic forms.

Key words Caccioppoli inequality · The A-harmonic equation and differential forms

Mathematics Subject Classification Primary 35J60 · Secondary 31B05 · 58A10 · 46E35

11.1 Introduction

The L^p theory about differential forms u or du, where u and its conjugate v satisfy the conjugate A-harmonic equation $A(x, du) = d^{\star}v$ or some other versions of the A-harmonic equation, has been very well developed during the recent years, see [1–4]. However, to the best of our knowledge, only little progress has been made in the study of the conjugate form v or $d^{\star}v$ in the last several decades. Differential forms have found many applications in many fields of science, notably the fields of electromagnetism, astronomy, harmonic wavelet analysis, and fluid dynamics, because they describe the behavior of electric, gravitational, and fluid potentials, see [5–7]. Different versions of the inequalities for differential forms have been developed and used in PDEs and analysis during recent years, see [8–18]. In this paper, we will establish some basic inequalities, including the Poincaré inequality, weak

Dedicated to Professor Themistocles M. Rassias on the occasion of his 60th birthday.

S. Ding (✉)
Department of Mathematics, Seattle University, Seattle, WA 98122, USA
e-mail: sding@seattleu.edu

Y. Xing
Department of Mathematics, Harbin Institute of Technology, Harbin, China
e-mail: xyuming@hit.edu.cn

P.M. Pardalos et al. (eds.), *Nonlinear Analysis*, Springer Optimization and Its Applications 68, In Honor of Themistocles M. Rassias on the Occasion of his 60th Birthday,
DOI 10.1007/978-1-4614-3498-6_11,

reverse Hölder inequality, and Caccioppoli inequality, for the conjugate harmonic form. We also prove Caccioppoli inequalities with Orlicz norms for the conjugate harmonic form. The results developed in this paper will provide a powerful tool for scientists and engineers to estimate the conjugate harmonic forms. We will also explore some applications in harmonic analysis and partial differential equations.

Let Ω be a bounded domain in $\mathbf{R}^n$, $n \geq 2$, B and σB be the balls with the same center and $\operatorname{diam}(\sigma B) = \sigma \operatorname{diam}(B)$. We do not distinguish the balls from cubes, throughout this paper. We use $|E|$ to denote the n-dimensional Lebesgue measure of a set $E \subseteq \mathbf{R}^n$. For a function u, the average of u over B is expressed by $u_B = \frac{1}{|B|}\int_B u\,dx$. All integrals involved in this paper are the Lebesgue integrals. We say w is a weight if $w \in L^1_{\mathrm{loc}}(\mathbf{R}^n)$ and $w > 0$ a.e. Differential forms are extensions of differentiable functions in $\mathbf{R}^n$. For example, the function $u(x_1, x_2, \dots, x_n)$ is called a 0-form. A differential 1-form $u(x)$ in $\mathbf{R}^n$ can be expressed as $u(x) = \sum_{i=1}^n u_i(x_1, x_2, \dots, x_n)\,dx_i$, where the coefficient functions $u_i(x_1, x_2, \dots, x_n)$, $i = 1, 2, \dots, n$, are differentiable. Similarly, a differential k-form $u(x)$ can be expressed as

$$u(x) = \sum_I u_I(x)\,dx_I = \sum u_{i_1 i_2 \cdots i_k}(x)\,dx_{i_1} \wedge dx_{i_2} \wedge \cdots \wedge dx_{i_k}, \tag{11.1}$$

where $I = (i_1, i_2, \dots, i_k)$, $1 \leq i_1 < i_2 < \cdots < i_k \leq n$, see [1] and [6] for more properties and applications of differential forms. Let $\wedge^l = \wedge^l(\mathbf{R}^n)$ be the set of all l-forms in $\mathbf{R}^n$, $D'(\Omega, \wedge^l)$ be the space of all differential l-forms in Ω, and $L^p(\Omega, \wedge^l)$ be the l-forms $u(x) = \sum_I u_I(x)\,dx_I$ in Ω satisfying $\int_\Omega |u_I|^p < \infty$ for all ordered l-tuples I, $l = 1, 2, \dots, n$. We express the exterior derivative by d and the Hodge star operator by $\star$. The Hodge codifferential operator $d^\star$ is given by $d^\star = (-1)^{nl+1} \star d \star$, $l = 1, 2, \dots, n$. If $u = \alpha_{i_1 i_2 \cdots i_k}(x_1, x_2, \dots, x_n)\,dx_{i_1} \wedge dx_{i_2} \wedge \cdots \wedge dx_{i_k} = \alpha_I\,dx_I$, $i_1 < i_2 < \cdots < i_k$, is a differential k-form, then

$$\star u = \star \alpha_{i_1 i_2 \cdots i_k}\,dx_{i_1} \wedge dx_{i_2} \wedge \cdots \wedge dx_{i_k} = (-1)^{\sum(I)} \alpha_I\,dx_J, \tag{11.2}$$

where $I = (i_1, i_2, \dots, i_k)$, $J = \{1, 2, \dots, n\} - I$, and $\sum(I) = \frac{k(k+1)}{2} + \sum_{j=1}^k i_j$. For example, in $\wedge^1(\mathbf{R}^3)$, we have

$$\star dx_1 = (-1)^2\,dx_2 \wedge dx_3 = dx_2 \wedge dx_3.$$

We consider here the conjugate A-harmonic equation

$$A(x, du) = d^\star v \tag{11.3}$$

for differential forms, where $A : \Omega \times \Lambda^l(\mathbf{R}^n) \to \Lambda^l(\mathbf{R}^n)$ satisfies the following conditions:

$$\left|A(x, \xi)\right| \leq a|\xi|^{p-1} \quad \text{and} \quad \left\langle A(x, \xi), \xi \right\rangle \geq |\xi|^p \tag{11.4}$$

for almost every $x \in \Omega$ and all $\xi \in \Lambda^l(\mathbf{R}^n)$. Here $a > 0$ is a constant and $1 < p < \infty$ is a fixed exponent associated with (11.3). Applying $d^\star$ to the conjugate A-harmonic equation (11.3) both sides, we have the following A-harmonic equation

$$d^{\star}A(x,du)=0. \tag{11.5}$$

A differential form u is called a closed form if $du=0$ and a differential form v is called a coclosed form if $d^{\star}v=0$. We call u and v a pair of conjugate A-harmonic tensor in Ω if u and v satisfy the conjugate A-harmonic equation (11.3). For example, if we choose operator $A(x,\xi)=\xi$ and $p=2$, then $A(x,\xi)$ satisfies (11.4) and the equation (11.3) reduces to $du=d^{*}v$, which is an analogue of a Cauchy–Riemann system in $\mathbf{R}^n$. Clearly, the A-harmonic equation is not affected by adding a closed form to u and coclosed form to v. Therefore, any type of estimates between u and v must be modulo such forms. Throughout this paper, we always assume that p is the fixed exponent associated with (11.3), $1<p<\infty$ and $p^{-1}+q^{-1}=1$. In the recent years, much progress has been made in the studies of differential forms satisfying different versions of the A-harmonic equation, see [1–4, 8, 9, 13–15, 18] for recent results about L^p norm estimates for solutions of the A-harmonic equation.

11.2 Basic L^p Inequalities

The following weight class was introduced in [16], which is an extension of the several existing classes of weights, such as $A_r^{\lambda}(E)$-weights, $A_r(\lambda, E)$-weights, and $A_r(E)$-weights; see [1] for more results about these weights.

We say that a measurable function $w(x)$ defined on a subset $E\subset\mathbf{R}^n$ satisfies the $A(\alpha,\beta,\gamma;E)$-condition for some positive constants α,β,γ, write $w(x)\in A(\alpha,\beta,\gamma;E)$ if $w(x)>0$ a.e., and

$$\sup_B\left(\frac{1}{|B|}\int_B w^{\alpha}\,dx\right)\left(\frac{1}{|B|}\int_B w^{-\beta}\,dx\right)^{\gamma/\beta}<\infty, \tag{11.6}$$

where the supremum is over all balls $B\subset E$.

We should notice that there are three parameters in the definition of the $A(\alpha,\beta,\gamma;E)$-class. If we choose some special values for these parameters, the $A(\alpha,\beta,\gamma;E)$-class reduces to the existing weight classes. For instance, if $\alpha=\lambda,\beta=1/(r-1)$ and $\gamma=1$ in above definition, the $A(\alpha,\beta,\gamma;E)$-class becomes $A_r(\lambda,E)$-weight, that is, $A_r(\lambda,E)=A(\lambda,1/(r-1),1;E)$. Similarly, $A_r^{\lambda}(E)=A(1,1/(r-1),\lambda;E)$. Also, it is easy to see that the $A(\alpha,\beta,\gamma;E)$-class reduces to the usual $A_r(E)$-weight if $\alpha=\gamma=1$ and $\beta=1/(r-1)$. Moreover, we have proved in [16] that the class of the $A_r(E)$-weights is a proper subset of $A(\alpha,\beta,\gamma;E)$-class.

As usual, a function $\varphi:[0,\infty)\to[0,\infty)$ is called an Orlicz function if φ is continuously increasing with $\varphi(0)=0$. The Orlicz space $L^{\varphi}(\Omega,\mu)$ consists of all measurable functions f on Ω such that $\int_{\Omega}\varphi(\frac{|f|}{\lambda})\,d\mu<\infty$ for some $\lambda=\lambda(f)>0$. $L^{\varphi}(\Omega,\mu)$ is equipped with the nonlinear Luxemburg functional

$$\|f\|_{\varphi(\Omega,\mu)}=\inf\left\{\lambda>0:\int_{\Omega}\varphi\left(\frac{|f|}{\lambda}\right)d\mu\le 1\right\}, \tag{11.7}$$

where the Radon measure μ is defined by $d\mu = g(x)\,dx$ and $g(x) \in A(\alpha, \beta, \gamma; \Omega)$. A convex Orlicz function φ is often called a Young function. If φ is a Young function, then $\|\cdot\|_{\varphi(\Omega,\mu)}$ defines a norm in $L^{\varphi}(\Omega, \mu)$, which is called the Orlicz norm or Luxemburg norm.

Definition 11.1 ([17]) We say a Young function φ lies in the class $G(p, q, C)$, $1 \le p < q < \infty$, $C \ge 1$, if (i) $1/C \le \varphi(t^{1/p})/g(t) \le C$ and (ii) $1/C \le \varphi(t^{1/q})/h(t) \le C$ for all $t > 0$, where g is a convex increasing function and h is a concave increasing function on $[0, \infty)$.

From [17], each of φ, g, and h in the above definition is doubling in the sense that its values at t and $2t$ are uniformly comparable for all $t > 0$, and the consequent fact that

$$C_1 t^q \le h^{-1}\big(\varphi(t)\big) \le C_2 t^q, \qquad C_1 t^p \le g^{-1}\big(\varphi(t)\big) \le C_2 t^p, \tag{11.8}$$

where C_1 and C_2 are constants. Also, for all $1 \le p_1 < p < p_2$ and $\alpha \in \mathbf{R}$, the function $\varphi(t) = t^p \log_+^{\alpha} t$ belongs to $G(p_1, p_2, C)$ for some constant $C = C(p, \alpha, p_1, p_2)$. Here $\log_+(t)$ is defined by $\log_+(t) = 1$ for $t \le e$; and $\log_+(t) = \log(t)$ for $t > e$. Particularly, if $\alpha = 0$, we see that $\varphi(t) = t^p$ lies in $G(p_1, p_2, C)$, $1 \le p_1 < p < p_2$.

We will need the following inequality for solutions to the conjugate A-harmonic equation which appears in [13].

Lemma 11.1 *Let u and v be conjugate A-harmonic tensors in a domain $\Omega \subset \mathbf{R}^n$, $0 < s, t < \infty$, $\sigma > 1$ and $p, q > 1$ with $1/p + 1/q = 1$. Then, there exists a constant C, independent of u and v, such that*

$$\|u - u_Q\|_{s,Q} \le C|Q|^{\beta} \|v - c_1\|_{t,\sigma Q}^{q/p}, \tag{11.9}$$

$$\|v - v_Q\|_{t,Q} \le C|Q|^{-\beta p/q} \|u - c_2\|_{s,\sigma Q}^{p/q} \tag{11.10}$$

for all cubes Q with $\sigma Q \subset \Omega$. Here c_1 is any coclosed form, c_2 is any closed form and $\beta = 1/s + 1/n - (1/t + 1/n)q/p$.

We will need the following Caccioppoli-type inequality for solutions to the A-harmonic equation which appears in [9].

Lemma 11.2 *Let $u \in D'(\Omega, \Lambda^l)$, $l = 0, 1, \ldots, n-1$, be a solution to the A-harmonic equation* (11.3) *on a domain Ω. Then there exists a constant C, independent of u, such that*

$$\|du\|_{p,B} \le C|B|^{-1/n} \|u - c\|_{p,B}$$

for all balls or cubes B with $B \subset \Omega$ and all closed forms c. Here $1 < p < \infty$.

We prove the following weak reverse Hölder inequality for conjugate form v first.

Theorem 11.1 *Let $u \in D'(\Omega, \wedge^{l-1})$ and $v \in D'(\Omega, \wedge^{l+1})$ be a pair of solutions to (11.3) in a bounded domain Ω and $\sigma > 1$ be a constant. For any constants $k_1, k_2 > 0$, Then, there exists a constant C, independent of v, such that*

$$\|v - v_B\|_{k_1,B} \le C|B|^{(k_2-k_1)/k_1k_2}\|v - c_1\|_{k_2,\sigma B} \tag{11.11}$$

for all balls or cubes B with $\sigma B \subset \Omega$ and all coclosed forms c_1.

Proof Using (11.10) with $t = k_1$, we have

$$\|v - v_B\|_{k_1,B} \le C_1|B|^{-\beta p/q}\|u - c_2\|^{p/q}_{s,\sigma_1 B}, \tag{11.12}$$

where c_2 is any closed form, $\sigma_1 > 1$, $\beta = 1/s + 1/n - (1/k_1 + 1/n)q/p$, and $s > 0$ is a constant to be chosen later. Applying the weak reverse Hölder inequality to $u - c_2$, it follows that

$$\|u - c_2\|_{s,\sigma_1 B} \le C_2|B|^{(\tau-s)/s\tau}\|u - c_2\|_{\tau,\sigma_2 B}. \tag{11.13}$$

Choosing $c_2 = u_B$, then substituting (11.13) into (11.12) and using (11.9), we find that

$$\begin{aligned}
\|v - v_B\|_{k_1,B} &\le C_1|B|^{-\beta p/q}\left(C_2|B|^{(\tau-s)/s\tau}\|u - u_B\|_{\tau,\sigma_2 B}\right)^{p/q} \\
&\le C_3|B|^{-\beta p/q}|B|^{\frac{\tau-s}{s\tau}\cdot\frac{p}{q}}\|u - u_B\|^{p/q}_{t,\sigma_2 B} \\
&\le C_3|B|^{-\beta p/q+\frac{\tau-s}{s\tau}\cdot\frac{p}{q}}\left(C_4|B|^{\beta}\|v - c_1\|^{q/p}_{k_2,\sigma_3 B}\right)^{p/q} \\
&\le C_5|B|^{\frac{\tau-s}{s\tau}\cdot\frac{p}{q}}\|v - c_1\|_{k_2,\sigma_3 B}.
\end{aligned} \tag{11.14}$$

Now, select $s > 0$ and $\tau > 0$ such that

$$\frac{1}{s} - \frac{1}{\tau} = \frac{k_2 - k_1}{k_1 k_2}\cdot\frac{q}{p},$$

that is,

$$\frac{\tau - s}{s\tau}\cdot\frac{p}{q} = \frac{k_2 - k_1}{k_1 k_2}. \tag{11.15}$$

Substituting (11.15) into (11.14), we obtain

$$\|v - v_B\|_{k_1,B} \le C_5|B|^{\frac{k_2-k_1}{k_1k_2}}\|v - c_1\|_{k_2,\sigma_3 B}.$$

The proof of Theorem 11.1 is completed. □

Next, we prove the following Poincaré inequality for conjugate form v.

Theorem 11.2 *Let $u \in D'(\Omega, \wedge^{l-1})$ and $v \in D'(\Omega, \wedge^{l+1})$ be a pair of solutions to* (11.3) *in a bounded domain Ω and $\sigma > 1$ be a constant. Then, there exists a constant C, independent of v, such that*

$$\|v - v_B\|_{q,B} \le C|B|^{1/n}\|d^\star v\|_{q,\sigma B} \tag{11.16}$$

for all balls or cubes B with $\sigma B \subset \Omega$.

Proof Choosing $s = p$ and $t = q$ in (11.10) gives

$$\|v - v_B\|_{q,B}^q \le C_1|B|^{(q-p)/n}\|u - c_2\|_{p,\sigma_1 B}^p, \tag{11.17}$$

where c_2 is any closed form and $\sigma_1 > 1$. From [1], it follows that

$$\|du\|_{p,\sigma_1 B}^p \le C_2\|d^\star v\|_{q,\sigma_1 B}^q. \tag{11.18}$$

Applying the Poincaré inequality for u and (11.18), we obtain

$$\begin{aligned}\|u - u_B\|_{p,\sigma_1 B}^p &\le C_3|B|^{p/n}\|du\|_{p,\sigma_1 B}^p \\ &\le C_4|B|^{p/n}\|d^\star v\|_{q,\sigma_1 B}^q.\end{aligned} \tag{11.19}$$

Combining (11.17) and (11.19) yields

$$\|v - v_B\|_{q,B}^q \le C_5|B|^{q/n}\left\|d^\star v\right\|_{q,\sigma_1 B}^q, \tag{11.20}$$

which is equivalent to

$$\|v - v_B\|_{q,B} \le C_6|B|^{1/n}\left\|d^\star v\right\|_{q,\sigma_1 B}.$$

The proof of Theorem 11.2 is completed. □

Now, we prove the following Caccioppoli inequality for the codifferential operator $d^\star$ and v.

Theorem 11.3 *Let $u \in D'(\Omega, \wedge^{l-1})$ and $v \in D'(\Omega, \wedge^{l+1})$ be a pair of solutions to* (11.3) *in a bounded domain Ω and $\sigma > 1$ be a constant. Then, there exists a constant C, independent of v, such that*

$$\left\|d^\star v\right\|_{q,B} \le C|B|^{-1/n}\|v - c_1\|_{q,\sigma B} \tag{11.21}$$

for all balls or cubes B with $\sigma B \subset \Omega$ and all coclosed forms c_1. Here $1 < q < \infty$.

Proof Choosing $s = p, t = q$ and $Q = B$ in Lemma 11.1, we have

$$\|u - u_B\|_{p,B}^p \le C_1|B|^{\frac{p-q}{n}}\|v - c_1\|_{q,\sigma_1 B}^q. \tag{11.22}$$

From (11.3) and (11.4), we obtain

$$\begin{aligned}\int_B |d^\star v|^q\,dx &= \int_B |A(x,du)|^q\,dx \\ &\le C_2 \int_B |du|^{q(p-1)}\,dx \\ &= C_2 \int_B |du|^p\,dx, \end{aligned} \tag{11.23}$$

and hence

$$\|d^\star v\|_{q,B}^q \le C_2 \|du\|_{p,B}^p.$$

From Lemma 11.2, it follows that

$$\|d^\star v\|_{q,B}^q \le C_3\,\mathrm{diam}(B)^{-p}\|u-c\|_{p,\sigma_2 B}^p$$

for any closed form c. Since u_B is a closed form for any ball, we may choose $c = u_B$. Now since $\mathrm{diam}(B) = C_4|B|^{1/n}$, we have

$$\|d^\star v\|_{q,B}^q \le C_5|B|^{-p/n}\|u-u_B\|_{p,\sigma_2 B}^p. \tag{11.24}$$

Finally, a combination of (11.22) and (11.24) yields

$$\|d^\star v\|_{q,B}^q \le C_6|B|^{-q/n}\|v-c_1\|_{q,\sigma_3 B}^q$$

which is the same as

$$\|d^\star v\|_{q,B} \le C|B|^{-1/n}\|v-c_1\|_{q,\sigma_3 B},$$

where $\sigma_3 > \sigma_2 > \sigma_1 > 1$ are constants. We have completed the proof of Theorem 11.3. □

We end this section with the following version of the weak reverse Hölder inequality.

Theorem 11.4 *Let $u \in D'(\Omega, \wedge^{l-1})$ and $v \in D'(\Omega, \wedge^{l+1})$ be a pair of solutions to* (11.3) *in a bounded domain Ω and $\sigma > 1$ be a constant. For any constants $k_1, k_2 > 0$, then, there exists a constant C, independent of v, such that*

$$\|d^\star v\|_{k_1,B} \le C|B|^{(k_2-k_1)/k_1k_2}\|d^\star v\|_{k_2,\sigma B}. \tag{11.25}$$

Proof Note that (11.11) holds for any coclosed form c_1. Hence, we may choose $c_1 = v_B$ in (11.11) and obtain

$$\|v-v_B\|_{k_1,B} \le C|B|^{(k_2-k_1)/k_1k_2}\|v-v_B\|_{k_2,\sigma_1 B} \tag{11.26}$$

for all balls or cubes B with $\sigma_1 B \subset \Omega$. Then, using the Caccioppoli inequality (11.21) with $c_1 = v_B$, inequality (11.26), and the Poincaré inequality (11.16), we find that

$$\begin{aligned}
\|d^\star v\|_{k_1,B} &\leq C_1|B|^{-1/n}\|v - c_1\|_{k_1,\sigma_1 B} \\
&\leq C_1|B|^{-1/n}\|v - v_B\|_{k_1,\sigma_1 B} \\
&\leq C_1|B|^{-1/n}C_2|B|^{(k_2-k_1)/k_1k_2}\|v - v_B\|_{k_2,\sigma_2 B} \\
&\leq C_3|B|^{-1/n}|B|^{(k_2-k_1)/k_1k_2}\big(C_4|B|^{1/n}|d^\star v\|_{k_2,\sigma_3 B}\big) \\
&\leq C_5|B|^{(k_2-k_1)/k_1k_2}|d^\star v\|_{k_2,\sigma_3 B}
\end{aligned}$$

where $\sigma_3 > \sigma_2 > \sigma_1 > 1$ are constants. We have completed the proof of Theorem 11.4. □

11.3 Orlicz Norm Inequalities

The purpose of this section is to develop some estimates which provide upper bounds for the Orlicz norm of $d^\star v$ in terms of the corresponding norm v or $v - c_1$, where v is a differential form satisfying the conjugate A-harmonic equation (11.3) and c_1 is any coclosed form. These kinds of estimates are called the Caccioppoli-type estimates or the Caccioppoli inequalities which have been playing a crucial role in harmonic analysis and the related fields during the last several decades. In many situations, we need to estimate the integral of $d^\star v$.

Using Theorem 11.3 and the method developed in the proof of Theorem 6.2.3 in [1], we can prove the following weak reverse Hölder inequality.

Lemma 11.3 *Let $u \in D'(\Omega, \wedge^{l-1})$ and $v \in D'(\Omega, \wedge^{l+1})$ be a pair of solutions to (11.3) in a bounded domain Ω and $\sigma > 1$ be a constant. For any constants $s, t > 0$, then, there exists a constant C, independent of v, such that*

$$\left(\int_B |d^\star v|^t \, d\mu\right)^{1/t} \leq C\big(\mu(B)\big)^{(s-t)/st}\left(\int_{\sigma B} |d^\star v|^s \, d\mu\right)^{1/s} \tag{11.27}$$

for all balls B with $\sigma B \subset \Omega$, where the Radon measure μ is defined by $d\mu = w(x)\,dx$ and $w \in A(\alpha, \beta, \alpha; \Omega)$, $\alpha > 1$, $\beta > 0$.

We first prove the following Caccioppoli inequality with the Orlicz norms for the conjugate harmonic form.

Theorem 11.5 *Let φ be a Young function in the class $G(s, t, C)$, $1 \leq s < t < \infty$, $C \geq 1$, and Ω be a bounded domain. Assume that $\varphi(|v - c_1|) \in L^1_{\mathrm{loc}}(\Omega, \mu)$, $u \in$*

$D'(\Omega, \wedge^{l-1})$ and $v \in D'(\Omega, \wedge^{l+1})$ are a pair of solutions to (11.3) *in Ω, $\sigma > 1$ is a constant. Then, there exists a constant C, independent of u and v, such that*

$$\int_B \varphi(|d^\star v|)\,d\mu \le C|B|^{-1/n} \int_{\sigma B} \varphi(|v - c_1|)\,d\mu \tag{11.28}$$

for all balls B with $\sigma B \subset \Omega$ and $|B| \ge d_0 > 0$, where c_1 is any coclosed form and the Radon measure μ is defined by $d\mu = w(x)\,dx$ and $w \in A(\alpha, \beta, \alpha; \Omega)$, $\alpha > 1$, $\beta > 0$.

Proof We may assume $w(x) \ge 1$ a.e. in Ω. Otherwise, let $\Omega_1 = \Omega \cap \{x \in \Omega : 0 < w(x) < 1\}$ and $\Omega_2 = \Omega \cap \{x \in \Omega : w(x) \ge 1\}$. Then, $\Omega = \Omega_1 \cup \Omega_2$. We define $W(x)$ by

$$W(x) = \begin{cases} 1, & x \in \Omega_1, \\ w(x), & x \in \Omega_2. \end{cases}$$

Then, $W(x) \ge w(x)$ and it is easy to check that $w(x) \in A(\alpha, \beta, \alpha; \Omega)$ if and only if $W(x) \in A(\alpha, \beta, \alpha; \Omega)$. Thus,

$$\begin{aligned} \left(\int_\Omega |d^\star v|^s\,d\mu\right)^{1/s} &= \left(\int_\Omega |d^\star v|^s w(x)\,dx\right)^{1/s} \\ &\le \left(\int_\Omega |d^\star v|^s W(x)\,dx\right)^{1/s} \end{aligned} \tag{11.29}$$

with $W(x) \ge 1$. Hence, we may suppose that $w(x) \ge 1$ a.e. in Ω. Thus, for any ball $B \subset \Omega$, we have

$$\mu(B) = \int_B d\mu = \int_B w(x)\,dx \ge \int_B dx = |B|. \tag{11.30}$$

Using Jensen's inequality for h^{-1}, (11.8), (ii) in Definition 11.1, and noticing that φ and h are doubling, we obtain

$$\begin{aligned} \int_B \varphi(|d^\star v|)\,d\mu &= h\left(h^{-1}\left(\int_B \varphi(|d^\star v|)\,d\mu\right)\right) \\ &\le h\left(\int_B h^{-1}(\varphi(|d^\star v|))\,d\mu\right) \\ &\le h\left(C_1 \int_B |d^\star v|^t\,d\mu\right) \\ &\le C_2\varphi\left(\left(C_1 \int_B |d^\star v|^t\,d\mu\right)^{1/t}\right) \\ &\le C_3\varphi\left(\left(\int_B |d^\star v|^t\,d\mu\right)^{1/t}\right). \end{aligned} \tag{11.31}$$

From Lemma 11.3, it follows that

$$\left(\int_B |d^\star v|^t d\mu\right)^{1/t} \le C_4(\mu(B))^{(s-t)/st}\left(\int_{\sigma B} |d^\star v|^s d\mu\right)^{1/s} \tag{11.32}$$

for any constants $s, t > 0$. Note that

$$(\mu(B))^{(s-t)/t} \le |B|^{(s-t)/t} \le d_0^{(s-t)/t} \tag{11.33}$$

since $\mu(B) \ge d_0$ and $s < t$. Using (11.8), (i) in Definition 11.1, and using the fact that φ is an increasing function, Jensen's inequality, and noticing that φ and g are doubling, we have

$$\begin{aligned}
\varphi\left(\left(\int_B |d^\star v|^t d\mu\right)^{1/t}\right) &\le \varphi\left(C_5(\mu(B))^{(s-t)/st}\left(\int_{\sigma B} |d^\star v|^s d\mu\right)^{1/s}\right)\\
&\le \varphi\left(C_6|B|^{-1/n}(\mu(B))^{(s-t)/st}\left(\int_{\sigma B} |v-c_1|^s d\mu\right)^{1/s}\right)\\
&\le \varphi\left(\left(C_6^s|B|^{-s/n}(\mu(B))^{(s-t)/t}\int_{\sigma B} |v-c_1|^s d\mu\right)^{1/s}\right)\\
&\le C_7 g\left(C_6^s|B|^{-s/n}(\mu(B))^{(s-t)/t}\int_{\sigma B} |v-c_1|^s d\mu\right)\\
&= C_7 g\left(\int_{\sigma B} C_6^s|B|^{-s/n} d_0^{(s-t)/t}|v-c_1|^s d\mu\right)\\
&\le C_7 \int_{\sigma B} g(C_8|B|^{-s/n}|v-c_1|^s)\, d\mu\\
&\le C_9 \int_{\sigma B} g(|B|^{-s/n}|v-c_1|^s)\, d\mu.
\end{aligned} \tag{11.34}$$

From (i) in Definition 11.1, we find that $g(x) \le C_{10}\varphi(x^{1/s})$. Thus,

$$\int_{\sigma B} g(|B|^{-s/n}|v-c_1|^s)\, d\mu \le C_{10}\int_{\sigma B} \varphi(|B|^{-1/n}|v-c_1|)\, d\mu. \tag{11.35}$$

Combining (11.31), (11.34), and (11.35) yields

$$\int_B \varphi(|d^\star v|)\, d\mu \le C_{11}\int_{\sigma B} \varphi(|B|^{-1/n}|v-c_1|)\, d\mu. \tag{11.36}$$

We have completed the proof of Theorem 11.5.

Note that in the proof of Theorem 11.5, if we replace (11.33) by

$$|B|^{-s/n+(s-t)/t} \le d_0^{-s/n+(s-t)/t}, \tag{3.7$'$}$$

we obtain the following version of the Caccioppoli inequality. □

Corollary 11.1 *Let φ be a Young function in the class $G(s,t,C)$, $1 \le s < t < \infty$, $C \ge 1$, and Ω be a bounded domain. Assume that $\varphi(|v - c_1|) \in L^1_{loc}(\Omega, \mu)$, $u \in D'(\Omega, \wedge^{l-1})$ and $v \in D'(\Omega, \wedge^{l+1})$ are a pair of solutions to* (11.3) *in Ω, $\sigma > 1$ is a constant. Then, there exists a constant C, independent of u and v, such that*

$$\int_B \varphi\big(|d^\star v|\big)\, d\mu \le C \int_{\sigma B} \varphi\big(|v - c_1|\big)\, d\mu \tag{11.37}$$

for all balls B with $\sigma B \subset \Omega$ and $|B| \ge d_0 > 0$, where c_1 is any coclosed form and the Radon measure μ is defined by $d\mu = w(x)\,dx$ and $w \in A(\alpha, \beta, \alpha; \Omega)$, $\alpha > 1$, $\beta > 0$.

If the condition $|B| \ge d_0 > 0$ is dropped in Theorem 11.5, from the proof of Theorem 11.5, we have the following inequality with the Orlicz norms for conjugate harmonic forms.

Corollary 11.2 *Let φ be a Young function in the class $G(s,t,C)$, $1 \le s < t < \infty$, $C \ge 1$, and Ω be a bounded domain. Assume that $\varphi(|v - c_1|) \in L^1_{loc}(\Omega, \mu)$, $u \in D'(\Omega, \wedge^{l-1})$ and $v \in D'(\Omega, \wedge^{l+1})$ are a pair of solutions to* (11.3) *in Ω, $\sigma > 1$ is a constant. Then, there exists a constant C, independent of u and v, such that*

$$\int_B \varphi\big(|d^\star v|\big)\, d\mu \le C \int_{\sigma B} \varphi\big(|B|^{-1/n+(s-t)/st} |v - c_1|\big)\, d\mu \tag{11.38}$$

for all balls B with $\sigma B \subset \Omega$, where c_1 is any coclosed form and the Radon measure μ is defined by $d\mu = w(x)\,dx$ and $w \in A(\alpha, \beta, \alpha; \Omega)$, $\alpha > 1$, $\beta > 0$.

Since each of φ, g, and h in Definition 11.1 is doubling, from the proof of Theorem 11.5 or directly using (11.28) with $w(x) = 1$, we have

$$\int_B \varphi\left(\frac{|d^\star v|}{\lambda}\right) dx \le C \int_{\sigma B} \varphi\left(\frac{|B|^{-1/n} |v - c_1|}{\lambda}\right) dx \tag{11.39}$$

for all balls B with $\sigma B \subset \Omega$ and any constant $\lambda > 0$. From (11.7) and (11.39), the following Caccioppoli inequality with the Orlicz norm

$$\big\| d^\star v \big\|_{\varphi(B)} \le C \big\| |B|^{-1/n} (v - c_1) \big\|_{\varphi(\sigma B)} \tag{11.40}$$

holds under the conditions described in Theorem 11.5.

Choosing $\varphi(x) = x^p \log_+^\alpha x$ in Corollary 11.1, we obtain the following Caccioppoli inequalities with the $L^p(\log_+^\alpha L)$-norms.

Theorem 11.6 *Let $\varphi(x) = x^s \log_+^\alpha x$, $s \ge 1$ and $\alpha \in \mathbf{R}$. Assume that $\varphi(|v - c_1|) \in L^1_{loc}(\Omega, \mu)$, $u \in D'(\Omega, \wedge^{l-1})$ and $v \in D'(\Omega, \wedge^{l+1})$ are a pair of solutions to* (11.3) *in Ω, $\sigma > 1$ is a constant. Then, there exists a constant C, independent of u and v,* *such that*

$$\int_B |d^\star v|^s \log_+^\alpha(|d^\star v|)\,d\mu \le C\int_{\sigma B} |v-c_1|^s \log_+^\alpha(|v-c_1|)\,d\mu \tag{11.41}$$

for all balls B with $\sigma B \subset \Omega$ and $|B| \ge d_0 > 0$, where c_1 is any coclosed form and the Radon measure μ is defined by $d\mu = w(x)\,dx$ and $w \in A(\alpha,\beta,\alpha;\Omega)$, $\alpha > 1$, $\beta > 0$.

We should notice that (11.41) can be written as the following version with the Orlicz norm

$$\|d^\star v\|_{L^s(\log_+^\alpha L)(B,\mu)} \le C\|v-c_1\|_{L^s(\log_+^\alpha L)(\sigma B,\mu)}$$

provided the conditions in Theorem 11.6 are satisfied.

Note that c_1 is any coclosed form in all theorems and corollaries proved above. Hence, we may select $c_1 = 0$ in above theorems and corollaries. For example, choosing $c_1 = 0$ in Theorems 11.5 and Corollary 11.1, respectively.

Corollary 11.3 *Let φ be a Young function in the class $G(s,t,C)$, $1 \le s < t < \infty$, $C \ge 1$, and Ω be a bounded domain. Assume that $\varphi(|v|) \in L^1_{loc}(\Omega,\mu)$, $u \in D'(\Omega,\wedge^{l-1})$ and $v \in D'(\Omega,\wedge^{l+1})$ are a pair of solutions to* (11.3) *in Ω, $\sigma > 1$ is a constant. Then, there exists a constant C, independent of u and v, such that*

$$\int_B \varphi(|d^\star v|)\,d\mu \le C|B|^{-1/n}\int_{\sigma B} \varphi(|v|)\,d\mu \tag{11.42}$$

and

$$\int_B \varphi(|d^\star v|)\,d\mu \le C\int_{\sigma B} \varphi(|v|)\,d\mu \tag{11.43}$$

for all balls B with $\sigma B \subset \Omega$ and $|B| \ge d_0 > 0$, where c_1 is any coclosed form and the Radon measure μ is defined by $d\mu = w(x)\,dx$ and $w \in A(\alpha,\beta,\alpha;\Omega)$, $\alpha > 1$, $\beta > 0$.

Selecting $c_1 = 0$ in Theorem 11.6, we have the following version of the Caccioppoli inequality.

Corollary 11.4 *Let $\varphi(x) = x^s \log_+^\alpha x$, $s \ge 1$ and $\alpha \in \mathbf{R}$. Assume that $\varphi(|v|) \in L^1_{loc}(\Omega,\mu)$, $u \in D'(\Omega,\wedge^{l-1})$ and $v \in D'(\Omega,\wedge^{l+1})$ are a pair of solutions to* (11.3) *in Ω, $\sigma > 1$ be a constant. Then, there exists a constant C, independent of u and v, such that*

$$\int_B |d^\star v|^s \log_+^\alpha(|d^\star v|)\,d\mu \le C\int_{\sigma B} |v|^s \log_+^\alpha(|v|)\,d\mu \tag{11.44}$$

for all balls B with $\sigma B \subset \Omega$ and $|B| \ge d_0 > 0$, where the Radon measure μ is defined by $d\mu = w(x)\,dx$ and $w \in A(\alpha,\beta,\alpha;\Omega)$, $\alpha > 1$, $\beta > 0$.

11.4 Global Inequalities

We will need the following Covering Lemma to extend our local inequalities into the global cases.

Lemma 11.4 ([13]) *Each Ω has a modified Whitney cover of cubes $\mathcal{V} = \{Q_i\}$ such that $\cup_i Q_i = \Omega$, $\sum_{Q_i \in \mathcal{V}} \chi_{\sqrt{\frac{5}{4}Q_i}} \le N\chi_\Omega$ and some $N > 1$, and if $Q_i \cap Q_j \neq \emptyset$, then there exists a cube R (this cube need not be a member of $\mathcal{V}$) in $Q_i \cap Q_j$ such that $Q_i \cup Q_j \subset NR$. Moreover, if Ω is a δ-John domain, then there is a distinguished cube $Q_0 \in \mathcal{V}$ which can be connected with every cube $Q_m \in \mathcal{V}$ by a chain of cubes $Q_0, Q_1, \ldots, Q_k = Q_m$ from $\mathcal{V}$ and such that $Q_m \subset \rho Q_i$, $i = 0, 1, 2, \ldots, k$, for some $\rho = \rho(n, \delta)$.*

Using the above Covering Lemma and Corollary 11.1, we can prove the following global Caccioppoli inequality with Orlicz norm for conjugate harmonic forms.

Theorem 11.7 *Let φ be a Young function in the class $G(s,t,C)$, $1 \le s < t < \infty$, $C \ge 1$, and Ω be a bounded domain. Assume that $\varphi(|v - c_1|) \in L^1(\Omega, \mu)$, $u \in D'(\Omega, \wedge^{l-1})$, and $v \in D'(\Omega, \wedge^{l+1})$ are a pair of solutions to (11.3) in Ω. Then, there exists a constant C, independent of u and v, such that*

$$\int_\Omega \varphi\big(|d^\star v|\big)\, d\mu \le C \int_\Omega \varphi\big(|v - c_1|\big)\, d\mu, \tag{11.45}$$

where $\sigma > 1$ is a constant, c_1 is any coclosed form and the Radon measure μ is defined by $d\mu = w(x)\,dx$ and $w \in A(\alpha, \beta, \alpha; \Omega)$, $\alpha > 1$, $\beta > 0$.

Proof By the Covering Lemma and Corollary 11.1, we obtain

$$\begin{aligned}
\int_\Omega \varphi\big(|d^\star v|\big)\, d\mu &= \int_{\cup_i Q_i} \varphi\big(|d^\star v|\big)\, d\mu \\
&\le \sum_{Q_i \in \mathcal{V}} \int_{Q_i} \varphi\big(|d^\star v|\big)\, d\mu \\
&\le C_1 \sum_{Q_i \in \mathcal{V}} \int_{\sigma Q_i} \varphi\big(|v - c_1|\big)\, d\mu \\
&\le C_1 N \cdot \int_\Omega \varphi\big(|v - c_1|\big)\, d\mu \\
&\le C_2 \cdot \int_\Omega \varphi\big(|v - c_1|\big)\, d\mu.
\end{aligned}$$

The proof of Theorem 11.7 has been completed. □

Similarly, using Theorem 11.6 and the Covering Lemma, we can prove the following norm inequality.

Theorem 11.8 *Let $\varphi(x) = x^s \log_+^\alpha x$, $s \geq 1$ and $\alpha \in \mathbf{R}$. Assume that $\varphi(|v - c_1|) \in L^1(\Omega, \mu)$, $u \in D'(\Omega, \wedge^{l-1})$, and $v \in D'(\Omega, \wedge^{l+1})$ are a pair of solutions to* (11.3) *in Ω, $\sigma > 1$ is a constant. Then, there exists a constant C, independent of u and v, such that*

$$\int_\Omega |d^\star v|^s \log_+^\alpha(|d^\star v|)\, d\mu \leq C \int_\Omega |v - c_1|^s \log_+^\alpha(|v - c_1|)\, d\mu, \tag{11.46}$$

where c_1 is any coclosed form and the Radon measure μ is defined by $d\mu = w(x)\,dx$ and $w \in A(\alpha, \beta, \alpha; \Omega)$, $\alpha > 1$, $\beta > 0$.

Using (11.7), inequalities (11.45) and (11.46) can be written as

$$\|d^\star v\|_{\varphi(\Omega,\mu)} \leq C\|v - c_1\|_{\varphi(\Omega,\mu)} \tag{11.47}$$

and

$$\|d^\star v\|_{L^s(\log_+^\alpha L)(\Omega,\mu)} \leq C\|v - c_1\|_{L^s(\log_+^\alpha L)(\Omega,\mu)}$$

provided the conditions in Theorems 11.7 and 11.8 are satisfied, respectively.

11.5 Applications

As applications of our results proved in this paper, we consider the following examples.

Example 11.1 Let $f(x) = (f_1, f_2, \ldots, f_n)$ be K-quasiregular in $\mathbf{R}^n$, then

$$u = f_l\, df_1 \wedge df_2 \wedge \cdots \wedge df_{l-1}, \tag{11.48}$$

$$v = \star f_{l+1}\, df_{l+2} \wedge \cdots \wedge df_n, \tag{11.49}$$

$l = 1, 2, \ldots, n-1$, are conjugate A-harmonic tensors, that is, u and v satisfy the conjugate A-harmonic equation (11.3), see [1]. Hence, all versions of Caccioppoli-type inequalities proved in this paper hold if v is defined by (11.49). For example, applying (11.43) to v defined above,

$$\int_B \varphi\big(|d^\star(\star f_{l+1} df_{l+2} \wedge \cdots \wedge df_n)|\big)\, d\mu \leq C \int_{\sigma B} \varphi\big(|\star f_{l+1}\, df_{l+2} \wedge \cdots \wedge df_n|\big)\, d\mu \tag{11.50}$$

for all balls B with $\sigma B \subset \Omega$ and $|B| \geq d_0 > 0$, where $\sigma > 1$ is a constant, where $l = 1, 2, \ldots, n-1$ and the Radon measure μ is defined by $d\mu = w(x)\,dx$ and $w \in A(\alpha, \beta, \alpha; \Omega)$, $\alpha > 1$, $\beta > 0$. Using the properties of the operators $d^\star$ and $\star$, (11.50) can be written as

$$\int_B \varphi\big(|d(f_{l+1}\, df_{l+2} \wedge \cdots \wedge df_n)|\big)\, d\mu \leq C \int_{\sigma B} \varphi\big(|f_{l+1}\, df_{l+2} \wedge \cdots \wedge df_n|\big)\, d\mu. \tag{11.51}$$

Particularly, choosing $l = n-2$ in (11.51) yields

$$\int_B \varphi\big(|d(f_{n-1}\,df_n)|\big)\,d\mu \le C\int_{\sigma B}\varphi\big(|f_{n-1}\,df_n|\big)\,d\mu. \tag{11.52}$$

It is easy to see that the integrals on the right hand sides of (11.51) and (11.52) are much easier to evaluate than those on the left hand sides of (11.51) and (11.52). See the following Example 11.2.

Example 11.2 From Sect. 1.8 in [1], we know that

$$f(x) = (f_1, f_2, f_3) = x|x|^\beta = \big(x_1|x|^\beta, x_2|x|^\beta, x_3|x|^\beta\big)$$

is a K-quasiregular mapping in $B = \{(x_1,x_2,x_3)\ 0\le x_1\le 1,\ 0\le x_2\le 1,\ 0\le x_3\le 1\}$. Here $\beta\ne -1$ is a real number. Using Example 11.1 with $l=2$ and $n=3$, we know that $u = f_2\,df_1 = x_2|x|^\beta\,d(x_1|x|^\beta)$ and $v = \star f_3 = \star(x_3|x|^\beta)$ form a pair of conjugate A-harmonic tensors with $p = n/l = 3/2$ and $q = n/(n-l) = 3$. Also, $|v| = |\star(x_3|x|^\beta)| = |x_3|x|^\beta|$. Choosing $\beta = 1$ and applying inequality (11.50) with $d\mu = dx$ yields

$$\begin{aligned}
\int_B \varphi\big(|d^\star(\star(x_3|x|^1))|\big)\,d\mu &\le C_1\int_{\sigma B}\varphi\big(|\star(x_3|x|^1)|\big)\,d\mu\\
&\le C_1\int_{\sigma B}\varphi\big(x_3|x|\big)\,dx\\
&\le C_1\int_{\sigma B}\varphi\big(|x||x|\big)\,dx\\
&\le C_1\int_{\sigma B}\varphi\big(|x|^2\big)\,dx\\
&\le C_1\int_{\sigma B}\varphi(1)\,dx\\
&\le C_1\varphi(1)\int_{\sigma B}dx\\
&\le C_1\varphi(1)|\sigma B|\\
&\le C_2\varphi(1)\sigma^3,
\end{aligned} \tag{11.53}$$

where $\sigma > 1$ is a constant. If φ is given, we can evaluate $\varphi(1)$ on the right side of (11.53).

Remarks (i) We only generalized Caccioppoli inequality into the versions with Orlicz norms. Using the method developed in the proof of Theorem 11.5, we can extend other basic L^p inequalities, such as the Poincaré inequality and weak reverse Hölder inequality established in Sect. 11.2, into the cases of Orlicz norms. (ii) Considering the length of the paper, we only extended two local inequalities into the global cases. Using the same method, we can generalize other local results into the global versions.

References

1. Agarwal, R.P., Ding, S., Nolder, C.A.: Inequalities for Differential Forms. Springer Berlin (2009)
2. Ding, S.: Weighted Hardy–Littlewood inequality for A-harmonic tensors. Proc. Am. Math. Soc. **3**, 1727–1735 (1997)
3. Xing, Y.: Integral inequality for conjugate A-harmonic tensors in john domains. Math. Nachr. **284**, 2003–2010 (2011)
4. Liu, B.: $A_r^{\lambda}(\Omega)$-weighted imbedding inequalities for A-harmonic tensors. J. Math. Anal. Appl. **273**(2), 667–676 (2002)
5. Sachs, S.K., Wu, H.: General Relativity for Mathematicians. Springer, New York (1977)
6. Westenholz, C.: Differential Forms in Mathematical Physics. North Holland, Amsterdam (1978)
7. Serëgin, G.A.: A local estimate of the Caccioppoli inequality type for extremal variational problems of Hencky plasticity. Akad. Nauk SSSR Sibirsk. Otdel., Inst. Mat., Novosibirsk **145**, 127–138 (1988) (Russian)
8. Wang, Y.: Two-weight Poincaré type inequalities for differential forms in $L^s(\mu)$-averaging domains. Appl. Math. Lett. **20**, 1161–1166 (2007)
9. Ding, S.: Two-weight Caccioppoli inequalities for solutions of nonhomogeneous A-harmonic equations on Riemannian manifolds. Proc. Am. Math. Soc. **132**, 2367–2375 (2004)
10. Giaquinta, M., Souček, J.: Caccioppoli's inequality and Legendre–Hadamard condition. Math. Ann. **270**, 105–107 (1985)
11. Perić, I., Žubrinić, D.: Caccioppoli's inequality for quasilinear elliptic operators. Math. Inequal. Appl. **2**, 251–261 (1999)
12. Troianiello, G.M.: Estimates of the Caccioppoli–Schauder type in weighted function spaces. Trans. Am. Math. Soc. **334**, 551–573 (1992)
13. Nolder, C.A.: Hardy–Littlewood theorems for A-harmonic tensors. Ill. J. Math. **43**, 613–631 (1999)
14. Xing, Y., Wang, Y.: BMO and Lipschitz norm estimates for composite operators. Potential Anal. **31**, 335–344 (2009)
15. Ding, S., Nolder, C.A.: Weighted Poincaré-type inequalities for solutions to the A-harmonic equation. Ill. J. Math. **46**, 199–205 (2002)
16. Xing, Y.: A new weight class and Poincaré inequalities with the radon measures. preprint
17. Buckley, S.M., Koskela, P.: Orlicz–Hardy inequalities. Ill. J. Math. **48**, 787–802 (2004)
18. Xing, Y.: Poincaré inequalities with Luxemburg norms in $L^{\varphi}(m)$-averaging domains. J. Inequal. Appl. Article ID 241759 (2010)

Chapter 12
A Survey on Jessen's Type Inequalities for Positive Functionals

S.S. Dragomir

Abstract Some recent inequalities related to the celebrated Jessen's result for positive linear or sublinear functionals and convex functions are surveyed.

Key words Jessen's inequality · Convex functions · Positive linear functionals

Mathematics Subject Classification 26D15 · 26D10

12.1 Introduction

Let L be a linear class of real-valued functions $g : E \to \mathbb{R}$ having the properties

(L1) $f, g \in L$ imply $(\alpha f + \beta g) \in L$ for all $\alpha, \beta \in \mathbb{R}$;
(L2) $\mathbf{1} \in L$, i.e., if $f_0(t) = 1$, $t \in E$ then $f_0 \in L$.

An *isotonic linear functional* $A : L \to \mathbb{R}$ is a functional satisfying

(A1) $A(\alpha f + \beta g) = \alpha A(f) + \beta A(g)$ for all $f, g \in L$ and $\alpha, \beta \in \mathbb{R}$.
(A2) If $f \in L$ and $f \geq 0$, then $A(f) \geq 0$.
The mapping A is said to be *normalized if*
(A3) $A(\mathbf{1}) = 1$.

Isotonic, that is, order-preserving, linear functionals are natural objects in analysis which enjoy a number of convenient properties. Thus, they provide, for example, Jessen's inequality, which is a functional form of Jensen's inequality (see [2] and [14]).

Dedicated to Professor Themistocles M. Rassias on the occasion of his 60th birthday.

S.S. Dragomir (✉)
Mathematics, School of Engineering & Science, Victoria University, Melbourne, Australia
e-mail: Sever.Dragomir@vu.edu.au

S.S. Dragomir
School of Computational and Applied Mathematics, University of the Witwatersrand, Private Bag 3, Wits 2050, Johannesburg, South Africa

P.M. Pardalos et al. (eds.), *Nonlinear Analysis*, Springer Optimization and Its Applications 68, In Honor of Themistocles M. Rassias on the Occasion of his 60th Birthday,
DOI 10.1007/978-1-4614-3498-6_12,

We note that common examples of such isotonic linear functionals A are given by

$$A(g)=\int_E g\,d\mu \text{ or } A(g)=\sum_{k\in E} p_k g_k,$$

where μ is a positive measure on E in the first case and E is a subset of the natural numbers $\mathbb{N}$, in the second ($p_k \geq 0, k \in E$).

We recall Jessen's inequality (see also [12]).

Theorem 12.1 (Jessen's Inequality) *Let $\phi : I \subseteq \mathbb{R} \to \mathbb{R}$ (I is an interval), be a convex function and $f : E \to I$ such that $\phi \circ f$, $f \in L$. If $A : L \to \mathbb{R}$ is an isotonic linear and normalized functional, then*

$$\phi\bigl(A(f)\bigr) \leq A(\phi \circ f). \tag{12.1}$$

A counterpart of this result was proved by Beesack and Pečarić in [2] for compact intervals $I=[\alpha,\beta]$.

Theorem 12.2 (Beesack & Pečarić, 1985, [2]) *Let $\phi : [\alpha,\beta] \subset \mathbb{R} \to \mathbb{R}$ be a convex function and $f : E \to [\alpha,\beta]$ such that $\phi \circ f$, $f \in L$. If $A : L \to \mathbb{R}$ is an isotonic linear and normalized functional, then*

$$A(\phi \circ f) \leq \frac{\beta - A(f)}{\beta-\alpha}\phi(\alpha) + \frac{A(f)-\alpha}{\beta-\alpha}\phi(\beta). \tag{12.2}$$

Remark 12.1 Note that (12.2) is a generalization of the inequality

$$A(\phi) \leq \frac{b-A(e_1)}{b-a}\phi(a) + \frac{A(e_1)-a}{b-a}\phi(b) \tag{12.3}$$

due to Lupaş [13] (see, for example, [2, Theorem A]), which assumed $E=[a,b]$, L satisfies (L1), (L2), $A : L \to \mathbb{R}$ satisfies (A1), (A2), $A(\mathbf{1})=1$, ϕ is convex on E and $\phi \in L$, $e_1 \in L$, where $e_1(x)=x$, $x \in [a,b]$.

The following inequality is well known in the literature as the Hermite–Hadamard inequality

$$\varphi\left(\frac{a+b}{2}\right) \leq \frac{1}{b-a}\int_a^b \varphi(t)\,dt \leq \frac{\varphi(a)+\varphi(b)}{2}, \tag{12.4}$$

provided that $\varphi : [a,b] \to \mathbb{R}$ is a convex function.

Using Theorem 12.1 and Theorem 12.2, we may state the following generalization of the Hermite–Hadamard inequality for isotonic linear functionals ([15] and [16]).

Theorem 12.3 (Pečarić & Beesack, 1991, [15]) *Let $\phi : [a,b] \subset \mathbb{R} \to \mathbb{R}$ be a convex function and $e : E \to [a,b]$ with e, $\phi \circ e \in L$. If $A \to \mathbb{R}$ is an isotonic linear and* *normalized functional, with* $A(e) = \frac{a+b}{2}$, *then*

$$\varphi\left(\frac{a+b}{2}\right) \leq A(\phi \circ e) \leq \frac{\varphi(a)+\varphi(b)}{2}. \tag{12.5}$$

For other results concerning convex functions and isotonic linear functionals, see [5, 12, 15–18] and the recent monograph [11].

12.2 Generalizations of Hermite–Hadamard's Inequalities for Isotonic Linear Functionals

12.2.1 Some Generalizations

The following lemma holds [16]:

Lemma 12.1 *Let* X *be a real linear space and* C *its convex subset. Then the following statements are equivalent for a mapping* $F : X \to \mathbb{R}$:

(i) f *is convex on* C;
(ii) *For all* $x, y \in C$ *the mapping* $g_{x,y} : [0, 1] \to \mathbb{R}$, $g_{x,y}(t) := f(tx + (1-t)y)$ *is convex on* $[0, 1]$.

Proof "(i) $\Rightarrow$ (ii)". Suppose $x, y \in C$ and let $t_1, t_2 \in [0, 1]$, $\lambda_1, \lambda_2 \geq 0$ with $\lambda_1 + \lambda_2 = 1$. Then

$$\begin{aligned} g_{x,y}(\lambda_1 t_1 + \lambda_2 t_2) &= f\big[(\lambda_1 t_1 + \lambda_2 t_2)x + (1 - \lambda_1 t_1 - \lambda_2 t_2)y\big] \\ &= f\big[(\lambda_1 t_1 + \lambda_2 t_2)x + \big[\lambda_1(1-t_1) + \lambda_2(1-t_2)\big]y\big] \\ &\leq \lambda_1 f\big(t_1 x + (1-t_1)y\big) + \lambda_2 f\big(t_2 x + (1-t_2)y\big). \end{aligned}$$

That is, $g_{x,y}$ is convex on $[0, 1]$.

"(ii) $\Rightarrow$ (i)". Now, let $x, y \in C$ and $\lambda_1, \lambda_2 \geq 0$ with $\lambda_1 + \lambda_2 = 1$. Then we have

$$\begin{aligned} f(\lambda_1 x + \lambda_2 y) &= f\big(\lambda_1 x + (1-\lambda_1)y\big) = g_{x,y}(\lambda_1 \cdot 1 + \lambda_2 \cdot 0) \\ &\leq \lambda_1 g_{x,y}(1) + \lambda_2 g_{x,y}(0) = \lambda_1 f(x) + \lambda_2 f(y). \end{aligned}$$

That is, f is convex on C and the statement is proved. □

The following generalization of Hermite–Hadamard's inequality for isotonic linear functionals holds [16]:

Theorem 12.4 (Pečarić & Dragomir, 1991, [16]) *Let* $f : C \subseteq X \to \mathbb{R}$ *be a convex function on* C, L *and* A *satisfy conditions* L1, L2 *and* A1, A2, *and* $h : E \to \mathbb{R}$, $0 \leq h(t) \leq 1$, $h \in L$ *is such that* $g_{x,y} \circ h \in L$ *for* x, y *given in* C. *If* $A(\mathbb{I}) = 1$, *then we have the inequality*

$$f\big(A(h)x+\big(1-A(h)\big)y\big)\le A\big[f\big(hx+(\mathbb{I}-h)y\big)\big]$$
$$\le A(h)f(x)+\big(1-A(h)\big)f(y). \qquad (12.6)$$

Proof Consider the mapping $g_{x,y}:[0,1]\to\mathbb{R}$, $g_{x,y}(s):=f(sx+(1-s)y)$. Then, by the above lemma, we have that $g_{x,y}$ is convex on $[0,1]$. For all $t\in E$, we have

$$g_{x,y}\big(h(t)\cdot 1+\big(1-h(t)\big)\cdot 0\big)\le h(t)g_{x,y}(1)+\big(1-h(t)\big)g_{x,y}(0),$$

which implies that

$$A\big(g_{x,y}(h)\big)\le A(h)g_{x,y}(1)+\big(1-A(h)\big)g_{x,y}(0).$$

That is,

$$A\big[f\big(hx+(\mathbb{I}-h)y\big)\big]\le A(h)f(x)+\big(1-A(h)\big)f(y).$$

On the other hand, by Jessen's inequality applied to $g_{x,y}$, we have

$$g_{x,y}\big(A(h)\big)\le A\big(g_{x,y}(h)\big),$$

which gives

$$f\big(A(h)x+\big(1-A(h)\big)y\big)\le A\big[f\big(hx+(\mathbb{I}-h)y\big)\big],$$

and the proof is completed. □

Remark 12.2 If $h:E\to[0,1]$ is such that $A(h)=\frac{1}{2}$, we get from the inequality (12.6) that

$$f\left(\frac{x+y}{2}\right)\le A\big[f\big(hx+(\mathbb{I}-h)y\big)\big]\le\frac{f(x)+f(y)}{2}, \qquad (12.7)$$

for all x, y in C.

Consequences

(a) If $A=\int_0^1$, $E=[0,1]$, $h(t)=t$, $C=[x,y]\subset\mathbb{R}$, then we recapture from (12.6) the classical inequality of Hermite and Hadamard because

$$\int_0^1 f\big(tx+(1-t)y\big)\,dt=\frac{1}{y-x}\int_x^y f(t)\,dt.$$

(b) If $A=\frac{2}{\pi}\int_0^{\frac{\pi}{2}}$, $E=[0,\frac{\pi}{2}]$, $h(t)=\sin^2 t$, $C\subseteq\mathbb{R}$, then, from (12.7) we get

$$f\left(\frac{x+y}{2}\right)\le\frac{2}{\pi}\int_0^{\frac{\pi}{2}} f\big(x\sin^2 t+y\cos^2 t\big)\,dt\le\frac{f(x)+f(y)}{2},$$

$x, y\in C$, which is a new inequality of Hadamard's type. This is because $\frac{2}{\pi}\int_0^{\frac{\pi}{2}}\sin^2 t\,dt=\frac{1}{2}$.

(c) If $A = \int_0^1$, $E = [0,1]$, $h(t) = t$ and X is a normed linear space, then (12.7) implies that for $f(x) = \|x\|^p$, $x \in X$, $p \geq 1$:

$$\left\| \frac{x+y}{2} \right\|^p \leq \int_0^1 \|tx + (1-t)y\|^p \, dt \leq \frac{\|x\|^p + \|y\|^p}{2}$$

for all $x, y \in X$.

(d) If $A = \frac{1}{n}\sum_{i=1}^n$, $E = \{1, \ldots, n\}$, $\sum_{i=1}^n t_i = \frac{n}{2}$, $C \subseteq \mathbb{R}$, $n \geq 1$, then from (12.7) we also have

$$f\left(\frac{x+y}{2}\right) \leq \frac{1}{n}\sum_{i=1}^n f\big(t_i x + (1-t_i)y\big) \leq \frac{f(x)+f(y)}{2}$$

for all $x, y \in C$, which is a discrete variant of the Hermite–Hadamard inequality.

To give a symmetric generalization of the Hermite–Hadamard inequality, we present the following lemma which is interesting in itself [5].

Lemma 12.2 *Let X be a real linear space and C be its convex subset. If $f : C \to \mathbb{R}$ is convex on C, then for all x, y in C the mapping $g_{x,y} : [0,1] \to \mathbb{R}$ given by*

$$g_{x,y}(t) := \frac{1}{2}\big[f\big(tx + (1-t)y\big) + f\big((1-t)x + ty\big)\big]$$

is also convex on $[0,1]$. In addition, we have the inequality

$$f\left(\frac{x+y}{2}\right) \leq g_{x,y}(t) \leq \frac{f(x)+f(y)}{2} \tag{12.8}$$

for all $x, y \in C$ and $t \in [0,1]$.

Proof Suppose $x, y \in C$ and let $t_1, t_2 \in [0,1]$, $\alpha, \beta \geq 0$ and $\alpha + \beta = 1$. Then

$$\begin{aligned}
g_{x,y}(\alpha t_1 + \beta t_2) &= \frac{1}{2}\big[f\big((\alpha t_1 + \beta t_2)x + (1 - \alpha t_1 - \beta t_2)y\big) \\
&\quad + f\big((1 - \alpha t_1 - \beta t_2)x + (\alpha t_1 + \beta t_2)y\big)\big] \\
&= \frac{1}{2}\big(f\big[\alpha\big(t_1 x + (1-t_1)y\big) + \beta\big(t_2 x + (1-t_2)y\big)\big] \\
&\quad + f\big[\alpha\big((1-t_1) + t_1 xy\big) + \beta\big((1-t_2)x + t_2 y\big)\big]\big) \\
&\leq \frac{1}{2}\big(\alpha f\big[t_1 x + (1-t_1)y\big] + \beta f\big[t_2 x + (1-t_2)y\big] \\
&\quad + \alpha f\big[(1-t_1) + t_1 xy\big] + \beta f\big[(1-t_2)x + t_2 y\big]\big) \\
&= \alpha g_{x,y}(t_1) + \beta g_{x,y}(t_2),
\end{aligned}$$

which shows that $g_{x,y}$ is convex on $[0,1]$.

By the convexity of f, we can state that

$$g_{x,y}(t) \geq f\left[\frac{1}{2}\big(tx + (1-t)y + (1-t)x + ty\big)\right] = f\left(\frac{x+y}{2}\right).$$

In addition,

$$g_{x,y}(t) \leq \frac{1}{2}\big[tf(x) + (1-t)f(y) + (1-t)f(x) + tf(y)\big] \leq \frac{f(x)+f(y)}{2}$$

for all t in [0, 1], which completes the proof. □

Remark 12.3 By the inequality (12.8), we deduce the bounds

$$\sup_{t\in[0,1]} g_{x,y}(t) = \frac{f(x)+f(y)}{2} \quad \text{and} \quad \inf_{t\in[0,1]} g_{x,y}(t) = f\left(\frac{x+y}{2}\right)$$

for all x, y in C.

The following symmetric generalization of the Hermite–Hadamard inequality holds [5]:

Theorem 12.5 (Dragomir, 1992, [5]) *Let $f : C \subseteq X \to \mathbb{R}$ be a convex function on the convex set C, where L and A satisfy the conditions* L1, L2 *and* A1, A2. *Also, $h : E \to \mathbb{R}$, $0 \leq h(t) \leq 1$ $(t \in E)$, and $h \in L$ is such that $f(hx + (1-h)y)$, $f((1-h)x + hy)$ belong to L for x, y fixed in C. If $A(\mathbb{I}) = 1$, then we have the inequality*:

$$\begin{aligned} & f\left(\frac{x+y}{2}\right) \\ & \leq \frac{1}{2}\big[f\big(A(h)x + \big(1-A(h)\big)y\big) + f\big(\big(1-A(h)\big)x + A(h)y\big)\big] \\ & \leq \frac{1}{2}\big(A\big[f\big(hx + (\mathbb{I}-h)y\big)\big] + A\big[f\big((\mathbb{I}-h)x + hy\big)\big]\big) \\ & \leq \frac{f(x)+f(y)}{2}. \end{aligned} \tag{12.9}$$

Proof Let us consider the mapping $g_{x,y} : [0,1] \to \mathbb{R}$ given above. Then, by the above lemma we know that $g_{x,y}$ is convex on [0, 1].

Applying Jensen's inequality to the mapping $g_{x,y}$, we get

$$g_{x,y}\big(A(h)\big) \leq A\big(g_{x,y}(h)\big).$$

However,

$$g_{x,y}\big(A(h)\big) = \frac{1}{2}\big[f\big(A(h)x + \big(1-A(h)\big)y\big) + f\big(\big(1-A(h)\big)x + A(h)y\big)\big]$$

and

$$A(g_{x,y}(h)) = \frac{1}{2}\big(A[f(hx + (\mathbb{I} - h)y)] + A[f((\mathbb{I} - h)x + hy)]\big),$$

and the second inequality in (12.9) is proved.

To prove the first inequality in (12.9), we observe, by (12.8), that

$$f\left(\frac{x+y}{2}\right) \le g_{x,y}(A(h)) \quad \text{as } 0 \le A(h) \le 1,$$

which is exactly the desired outcome.

Finally, by the convexity of f, we observe that

$$\frac{1}{2}\big[f(hx + (\mathbb{I} - h)y) + f((\mathbb{I} - h)x + hy)\big] \le \frac{f(x) + f(y)}{2}$$

on E.

By applying the functional A, since $A(\mathbb{I}) = 1$, we obtain the last part of (12.9). □

Remark 12.4 The above theorem can also be proved by the use of Theorem 12.4 and by Lemma 12.2. We shall omit the details.

Note that, if we choose $A = \int_0^1$, $E = [0, 1]$, $h(t) = t$, $C = [x, y] \subset \mathbb{R}$, we recapture, by (12.9), the Hermite–Hadamard inequality for integrals. This is because

$$\int_0^1 f(tx + (1-t)y)\,dt = \int_0^1 f((1-t)x + ty)\,dt = \frac{1}{y-x}\int_x^y f(t)\,dt.$$

Consequences

(a) Let $h : [0, 1] \to [0, 1]$ be a Riemann integrable function on $[0, 1]$ and $p \ge 1$. Then, for all x, y vectors in the normed space $(X; \|\cdot\|)$ we have the inequality

$$\begin{aligned}
&\left\|\frac{x+y}{2}\right\|^p \\
&\quad \le \frac{1}{2}\left[\left\|\left(1 - \int_0^1 h(t)\,dt\right)x + \left(\int_0^1 h(t)\,dt\right)y\right\|^p\right. \\
&\qquad \left. + \left\|\left(\int_0^1 h(t)\,dt\right)x + \left(1 - \int_0^1 h(t)\,dt\right)y\right\|^p\right] \\
&\quad \le \frac{1}{2}\left[\int_0^1 \|(h(t))x + (1-h(t))y\|^p\,dt + \int_0^1 \|(1-h(t))x + (h(t))y\|^p\,dt\right] \\
&\quad \le \frac{\|x\|^p + \|y\|^p}{2}.
\end{aligned}$$

If we choose $h(t) = t$, we get the inequality obtained above

$$\left\|\frac{x+y}{2}\right\|^p \leq \int_0^1 \|tx + (1-t)y\|^p \, dt \leq \frac{\|x\|^p + \|y\|^p}{2}$$

for all $x, y \in X$.

(b) Let $f : C \subseteq X \to \mathbb{R}$ be a convex function on the convex set C of a linear space X, $t_i \in [0, 1]$ $(i = \overline{1, n})$. Then we have the inequality

$$\begin{aligned} & f\left(\frac{x+y}{2}\right) \\ & \leq \frac{1}{2}\left[f\left(\frac{1}{n}\sum_{i=1}^n t_i x + \frac{1}{n}\sum_{i=1}^n (1-t_i)y\right) + f\left(\frac{1}{n}\sum_{i=1}^n (1-t_i)x + \frac{1}{n}\sum_{i=1}^n t_i y\right)\right] \\ & \leq \frac{1}{2n}\left[\sum_{i=1}^n f\left(t_i x + (1-t_i)y\right) + \sum_{i=1}^n f\left((1-t_i)x + t_i y\right)\right] \\ & \leq \frac{f(x)+f(y)}{2}. \end{aligned}$$

If we put in the above inequality $t_i = \sin^2 \alpha_i$, $\alpha_i \in \mathbb{R}$ $(i = \overline{1, n})$, then we have

$$\begin{aligned} & f\left(\frac{x+y}{2}\right) \\ & \leq \frac{1}{2}\left(f\left[\left(\frac{1}{n}\sum_{i=1}^n \sin^2 \alpha_i\right)x + \left(\frac{1}{n}\sum_{i=1}^n \cos^2 \alpha_i\right)y\right]\right. \\ & \quad \left. + f\left[\left(\frac{1}{n}\sum_{i=1}^n \cos^2 \alpha_i\right)x + \left(\frac{1}{n}\sum_{i=1}^n \sin^2 \alpha_i\right)y\right]\right) \\ & \leq \frac{1}{2n}\sum_{i=1}^n \left(f\left[\left(\sin^2 \alpha_i\right)x + \left(\cos^2 \alpha_i\right)y\right]\right. \\ & \quad \left. + f\left[\left(\cos^2 \alpha_i\right)x + \left(\sin^2 \alpha_i\right)y\right]\right) \\ & \leq \frac{f(x)+f(y)}{2}. \end{aligned}$$

12.2.2 Applications for Special Means

1. For $x, y \geq 0$, let us consider the weighted means

$$A_\alpha(x, y) := \alpha x + (1-\alpha)y$$

and

$$G_\alpha(x, y) := x^\alpha y^{1-\alpha}$$

where $\alpha \in [0, 1]$.

If $h : [0, 1] \to [0, 1]$ is an integrable mapping on $[0, 1]$, then, by Theorem 12.4, we have the inequality

$$A_{\int_0^1 h(t)\,dt}(x, y) \geq \exp\left[\int_0^1 \ln\left[A_{h(t)}(x, y)\right] dt\right] \geq G_{\int_0^1 h(t)\,dt}(x, y). \tag{12.10}$$

If $\int_0^1 h(t)\,dt = \frac{1}{2}$, we get

$$A(x, y) \geq \exp\left[\int_0^1 \ln\left[A_{h(t)}(x, y)\right] dt\right] \geq G(x, y), \tag{12.11}$$

which is a refinement of the classic $A. - G.$ inequality.

In particular, if in this inequality we choose $h(t) = t,\ t \in [0, 1]$, we recapture the well-known result for the identric mean:

$$A(x, y) \geq I(x, y) \geq G(x, y).$$

Now, if we use Theorem 12.5, we can state the following weighted refinement of the classical $A. - G.$ inequality:

$$\begin{aligned} A(x, y) &\geq G\left(A_{\int_0^1 h(t)\,dt}(x, y), A_{\int_0^1 h(t)\,dt}(x, y)\right) \\ &\geq \exp\left[\int_0^1 \ln\left[G\left(A_{h(t)}(x, y), A_{h(t)}(y, x)\right)\right] dt\right] \geq G(x, y). \end{aligned} \tag{12.12}$$

If $\int_0^1 h(t)\,dt = \frac{1}{2}$, then, by (12.12), we get the following refinement of the $A. - G.$ inequality:

$$A(x, y) \geq \exp\left[\int_0^1 \ln\left[G\left(A_{h(t)}(x, y), A_{h(t)}(y, x)\right)\right] dt\right] \geq G(x, y). \tag{12.13}$$

If, in the above inequality we choose $h(t) = t, t \in [0, 1]$, then we get the inequality

$$A(x, y) \geq \exp\left[\int_0^1 \ln\left[G\left(A_t(x, y), A_t(y, x)\right)\right] dt\right] \geq G(x, y). \tag{12.14}$$

2. Some discrete refinements of $A. - G.$ means inequality can also be done.

   If $\bar{x} = (x_1, \ldots, x_n) \in \mathbb{R}_+^n$, we can denote by $G_n(\bar{x})$ the geometric mean of $\bar{x}$, i.e., $G_n(\bar{x}) := (\prod_{i=1}^n x_i)^{\frac{1}{n}}$.

   If $\bar{t} = (t_1, \ldots, t_n) \in [0, 1]^n$, we can define the vector in $\mathbb{R}_+^n$ given by

$$\bar{A}_{\bar{t}}(x, y) := \left(A_{t_1}(x, y), \ldots, A_{t_n}(x, y)\right)$$

where $x, y \geq 0$.

Applying now Theorem 12.4 to the convex mapping $f(x) = -\ln x$ and the linear functional $A := \frac{1}{n}\sum_{i=1}^{n} t_i$, we get the inequality

$$A_{\tilde{t}}(x,y) \geq G_n\big(\bar{A}_{\tilde{t}}(x,y)\big) \geq G_{\tilde{t}}(x,y) \tag{12.15}$$

where $\tilde{t} := \frac{1}{n}\sum_{i=1}^{n} t_i \in [0,1]$ and $x, y \geq 0$.

If we choose t_i so that $\tilde{t} = \frac{1}{2}$, we get

$$A(x,y) \geq G_n\big(\bar{A}_{\tilde{t}}(x,y)\big) \geq G(x,y), \tag{12.16}$$

which is a discrete refinement of the classical $A.-G.$ inequality.

In addition, if we use Theorem 12.5, we can state that

$$\begin{aligned} A(x,y) &\geq G_n\big(A_{\tilde{t}}(x,y), A_{\tilde{t}}(y,x)\big) \\ &\geq G\big(G_n\big(\bar{A}_{\tilde{t}}(x,y)\big), G_n\big(\bar{A}_{\tilde{t}}(y,x)\big)\big) \geq G(x,y), \end{aligned} \tag{12.17}$$

which is another refinement of the $A.-G.$ inequality.

12.3 The Concepts of $m-\Psi$-Convex and $M-\Psi$-Convex Functions

12.3.1 Some Preliminary Results

Assume that the mapping $\Psi : I \subseteq \mathbb{R} \to \mathbb{R}$ (I is an interval) is convex on I and $m \in \mathbb{R}$. We shall say that the mapping $\phi : I \to \mathbb{R}$ is *$m-\Psi$-lower convex* if $\phi - m\Psi$ is a convex mapping on I. We may introduce the class of functions [6]

$$\mathcal{L}(I,m,\Psi) := \big\{\phi : I \to \mathbb{R} | \phi - m\Psi \text{ is convex on } I\big\}. \tag{12.18}$$

Similarly, for $M \in \mathbb{R}$ and Ψ as above, we can introduce the class of *$M-\Psi$-upper convex* functions by [6]

$$\mathcal{U}(I,M,\Psi) := \big\{\phi : I \to \mathbb{R} | M\Psi - \phi \text{ is convex on } I\big\}. \tag{12.19}$$

The intersection of these two classes will be called the class of *$(m,M)-\Psi$-convex functions* and will be denoted by [6]

$$\mathcal{B}(I,m,M,\Psi) := \mathcal{L}(I,m,\Psi) \cap \mathcal{U}(I,M,\Psi). \tag{12.20}$$

Remark 12.5 If $\Psi \in \mathcal{B}(I,m,M,\Psi)$, then $\phi - m\Psi$ and $M\Psi - \phi$ are convex and then $(\phi - m\Psi) + (M\Psi - \phi)$ is also convex which shows that $(M-m)\Psi$ is convex, implying that $M \geq m$ (as Ψ is assumed not to be the trivial convex function $\Psi(t) = 0$, $t \in I$).

The above concepts may be introduced in the general case of a convex subset in a real linear space, but we do not consider this extension here.

In [10], S.S. Dragomir and N.M. Ionescu introduced the concept of *g-convex dominated mappings*, for a mapping $f : I \to \mathbb{R}$. We recall this, by saying, for a given convex function $g : I \to \mathbb{R}$, the function $f : I \to \mathbb{R}$ is *g-convex dominated* iff $g + f$ and $g - f$ are convex mappings on I. In [10], the authors pointed out a number of inequalities for convex dominated functions related to Jensen's, Fuchs', Pečarić's, Barlow–Proschan and Vasić–Mijalković results, etc.

We observe that the concept of g-convex dominated functions can be obtained as a particular case from $(m, M) - \Psi$-convex functions by choosing $m = -1$, $M = 1$, and $\Psi = g$.

The following lemma holds [6].

Lemma 12.3 *Let $\Psi, \phi : I \subseteq \mathbb{R} \to \mathbb{R}$ be differentiable functions on $\mathring{I}$ and Ψ is a convex function on $\mathring{I}$.*

(i) *For $m \in \mathbb{R}$, the function $\phi \in \mathscr{L}(\mathring{I}, m, \Psi)$ iff*

$$m\big[\Psi(x) - \Psi(y) - \Psi'(y)(x - y)\big] \le \phi(x) - \phi(y) - \phi'(y)(x - y) \tag{12.21}$$

for all $x, y \in \mathring{I}$.

(ii) *For $M \in \mathbb{R}$, the function $\phi \in \mathscr{U}(\mathring{I}, M, \Psi)$ iff*

$$\phi(x) - \phi(y) - \phi'(y)(x - y) \le M\big[\Psi(x) - \Psi(y) - \Psi'(y)(x - y)\big] \tag{12.22}$$

for all $x, y \in \mathring{I}$.

(iii) *For $M, m \in \mathbb{R}$ with $M \ge m$, the function $\phi \in \mathscr{B}(\mathring{I}, m, M, \Psi)$ iff both* (12.21) *and* (12.22) *hold.*

Proof Follows by the fact that a differentiable mapping $h : I \to \mathbb{R}$ is convex on $\mathring{I}$ iff $h(x) - h(y) \ge h'(y)(x - y)$ for all $x, y \in \mathring{I}$. □

Another elementary fact for twice differentiable functions also holds [6].

Lemma 12.4 *Let $\Psi, \phi : I \subseteq \mathbb{R} \to \mathbb{R}$ be twice differentiable on $\mathring{I}$ and suppose Ψ is convex on $\mathring{I}$.*

(i) *For $m \in \mathbb{R}$, the function $\phi \in \mathscr{L}(\mathring{I}, m, \Psi)$ iff*

$$m\Psi''(t) \le \phi''(t) \quad \text{for all } t \in \mathring{I}. \tag{12.23}$$

(ii) *For $M \in \mathbb{R}$, the function $\phi \in \mathscr{U}(\mathring{I}, M, \Psi)$ iff*

$$\phi''(t) \le M\Psi''(t) \quad \text{for all } t \in \mathring{I}. \tag{12.24}$$

(iii) *For $M, m \in \mathbb{R}$ with $M \ge m$, the function $\phi \in \mathscr{B}(\mathring{I}, m, M, \Psi)$ iff both* (12.23) *and* (12.24) *hold.*

Proof Follows by the fact that a twice differentiable function $h : I \to \mathbb{R}$ is convex on $\mathring{I}$ iff $h''(t) \geq 0$ for all $t \in \mathring{I}$. □

We consider the p-logarithmic mean of two positive numbers given by

$$L_p(a,b) := \begin{cases} a & \text{if } b = a, \\ [\frac{b^{p+1}-a^{p+1}}{(p+1)(b-a)}]^{\frac{1}{p}} & \text{if } a \neq b, \end{cases}$$

and $p \in \mathbb{R} \setminus \{-1, 0\}$.

The following proposition holds [6].

Proposition 12.1 *Let $\phi : (0,\infty) \to \mathbb{R}$ be a differentiable mapping.*

(i) *For $m \in \mathbb{R}$, the function $\phi \in \mathscr{L}((0,\infty), m, (\cdot)^p)$ with $p \in (-\infty, 0) \cup (1,\infty)$ iff*

$$mp(x-y)\big[L_{p-1}^{p-1}(x,y) - y^{p-1}\big] \leq \phi(x) - \phi(y) - \phi'(y)(x-y) \qquad (12.25)$$

for all $x, y \in (0,\infty)$.

(ii) *For $M \in \mathbb{R}$, the function $\phi \in \mathscr{U}((0,\infty), M, (\cdot)^p)$ with $p \in (-\infty, 0) \cup (1,\infty)$ iff*

$$\phi(x) - \phi(y) - \phi'(y)(x-y) \leq Mp(x-y)\big[L_{p-1}^{p-1}(x,y) - y^{p-1}\big] \qquad (12.26)$$

for all $x, y \in (0,\infty)$.

(iii) *For $M, m \in \mathbb{R}$ with $M \geq m$, the function $\phi \in \mathscr{B}((0,\infty), M, (\cdot)^p)$ with $p \in (-\infty, 0) \cup (1,\infty)$ iff both* (12.25) *and* (12.26) *hold.*

The proof follows by Lemma 12.3 applied to the convex mapping $\Psi(t) = t^p$, $p \in (-\infty, 0) \cup (1,\infty)$ and via some elementary computation. We omit the details.

The following corollary is useful in practice [6].

Corollary 12.1 *Let $\phi : (0,\infty) \to \mathbb{R}$ be a differentiable function.*

(i) *For $m \in \mathbb{R}$, the function ϕ is m-quadratic-lower convex* (*i.e., for $p = 2$*) *iff*

$$m(x-y)^2 \leq \phi(x) - \phi(y) - \phi'(y)(x-y) \qquad (12.27)$$

for all $x, y \in (0,\infty)$.

(ii) *For $M \in \mathbb{R}$, the function ϕ is M-quadratic-upper convex iff*

$$\phi(x) - \phi(y) - \phi'(y)(x-y) \leq M(x-y)^2 \qquad (12.28)$$

for all $x, y \in (0,\infty)$.

(iii) *For $m, M \in \mathbb{R}$ with $M \geq m$, the function ϕ is (m, M)-quadratic-convex if both* (12.27) *and* (12.28) *hold.*

The following proposition holds [6].

Proposition 12.2 *Let $\phi : (0,\infty) \to \mathbb{R}$ be a twice differentiable function.*

(i) *For $m \in \mathbb{R}$, the function $\phi \in \mathscr{L}((0,\infty), m, (\cdot)^p)$ with $p \in (-\infty,0) \cup (1,\infty)$ iff*

$$p(p-1)mt^{p-2} \le \phi''(t) \quad \text{for all } t \in (0,\infty). \tag{12.29}$$

(ii) *For $M \in \mathbb{R}$, the function $\phi \in \mathscr{U}((0,\infty), M, (\cdot)^p)$ with $p \in (-\infty,0) \cup (1,\infty)$ iff*

$$\phi''(t) \le p(p-1)Mt^{p-2} \quad \text{for all } t \in (0,\infty). \tag{12.30}$$

(iii) *For $m, M \in \mathbb{R}$ with $M \ge m$, the function $\phi \in \mathscr{B}((0,\infty), m, M, (\cdot)^p)$ with $p \in (-\infty,0) \cup (1,\infty)$ iff both* (12.29) *and* (12.30) *hold.*

As can be easily seen, the above proposition offers the practical criterion of deciding when a twice differentiable mapping is $(\cdot)^p$-lower- or $(\cdot)^p$-upper-convex with the weights being the constants m and M, respectively.

The following corollary is useful in practice [6].

Corollary 12.2 *Assume that the mapping $\phi : (a,b) \subseteq \mathbb{R} \to \mathbb{R}$ is twice differentiable.*

(i) *If $\inf_{t\in(a,b)} \phi''(t) = k > -\infty$, then ϕ is $\frac{k}{2}$-quadratic-lower-convex on (a,b);*

(ii) *If $\sup_{t\in(a,b)} \phi''(t) = K < \infty$, then ϕ is $\frac{K}{2}$-quadratic-upper-convex on (a,b).*

12.4 Jessen's Inequality for $m - \Psi$-Convex and $M - \Psi$-Convex Functions

12.4.1 A Few Jessen-Type Inequalities

We start with the following result [7].

Theorem 12.6 (Dragomir, 2002, [7]) *Let $\Psi : I \subseteq \mathbb{R} \to \mathbb{R}$ be a convex function and $f : E \to I$ such that $\Psi \circ f$, $f \in L$, and let $A : L \to \mathbb{R}$ be an isotonic linear and normalized functional.*

(i) *If $\phi \in \mathscr{L}(I, m, \Psi)$ and $\phi \circ f \in L$, then we have the inequality*

$$m\big[A(\Psi \circ f) - \Psi\big(A(f)\big)\big] \le A(\phi \circ f) - \phi\big(A(f)\big). \tag{12.31}$$

(ii) *If $\phi \in \mathscr{U}(I, M, \Psi)$ and $\phi \circ f \in L$, then we have the inequality*

$$A(\phi \circ f) - \phi\big(A(f)\big) \le M\big[A(\Psi \circ f) - \Psi\big(A(f)\big)\big]. \tag{12.32}$$

(iii) *If $\phi \in \mathscr{B}(I, m, M, \Psi)$ and $\phi \circ f \in L$, then both* (12.31) *and* (12.32) *hold.*

Proof The proof is as follows.

(i) As $\phi \in \mathscr{L}(I, m, \Psi)$ and $\phi \circ f \in L$, it follows that $\phi - m\Psi$ is convex and $(\phi - m\Psi) \circ f \in L$.

Applying Jessen's inequality for the convex mapping $\phi - m\Psi$, we get

$$(\phi - m\Psi)\big(A(f)\big) \le A\big[(\phi - m\Psi) \circ f\big]. \tag{12.33}$$

However,

$$(\phi - m\Psi)\big(A(f)\big) = \phi\big(A(f)\big) - m\Psi\big(A(f)\big)$$

and

$$A\big[(\phi - m\Psi) \circ f\big] = A(\phi \circ f) - mA(\phi \circ f)$$

and then, by (12.20), we deduce the desired result (12.31).

(ii) Follows in a similar manner by taking into account that $\phi \circ f \in L$ and $\phi \in \mathscr{U}(I, M, \Psi)$ imply $M\Psi - \phi$ is convex and $(M\Psi - \phi) \circ f \in L$.

(iii) Follows by (i) and (ii). □

The following corollary is useful in practice [7].

Corollary 12.3 *Let $\Psi : I \subseteq \mathbb{R} \to \mathbb{R}$ be a twice differentiable convex function on $\mathring{I}$, $f : E \to I$ such that $\Psi \circ f$, $f \in L$, and let $A : L \to \mathbb{R}$ be an isotonic linear and normalized functional.*

(i) *If $\phi : I \to \mathbb{R}$ is twice differentiable and $\phi''(t) \ge m\Psi''(t)$, $t \in \mathring{I}$ (where m is a given real number), then* (12.31) *holds, provided that $\phi \circ f \in L$.*

(ii) *If $\phi : I \to \mathbb{R}$ is twice differentiable and $\phi''(t) \le M\Psi''(t)$, $t \in \mathring{I}$ (where M is a given real number), then* (12.32) *holds, provided that $\phi \circ f \in L$.*

(iii) *If $\phi : I \to \mathbb{R}$ is twice differentiable and $m\Psi''(t) \le \phi''(t) \le M\Psi''(t)$, $t \in \mathring{I}$, then both* (12.31) *and* (12.32) *hold, provided $\phi \circ f \in L$.*

Some particular important cases of the above corollary are embodied in the following propositions [7].

Proposition 12.3 *Assume that the mapping $\phi : I \subseteq \mathbb{R} \to \mathbb{R}$ is twice differentiable on $\mathring{I}$.*

(i) *If $\inf_{t \in \mathring{I}} \phi''(t) = k > -\infty$, then we have the inequality*

$$\frac{k}{2}\big[A(f^2) - [A(f)]^2\big] \le A(\phi \circ f) - \phi\big(A(f)\big) \tag{12.34}$$

provided that $\phi \circ f, f^2, f \in L$.

(ii) *If $\sup_{t \in \mathring{I}} \phi''(t) = K < \infty$, then we have the inequality*

$$A(\phi \circ f) - \phi\big(A(f)\big) \le \frac{K}{2}\big[A(f^2) - [A(f)]^2\big] \tag{12.35}$$

provided that $\phi \circ f, f^2, f \in L$.

(iii) *If* $-\infty < k \leq \phi''(t) \leq K < \infty$, $t \in \mathring{I}$, *then both* (12.34) *and* (12.35) *hold, provided that* $\phi \circ f, f^2, f \in L$.

The proof follows by Corollary 12.3 applied to $\Psi(t) = \frac{1}{2}t^2$ and $m = k$, $M = K$. Another result is the following one [7].

Proposition 12.4 *Assume that the mapping* $\phi : I \subseteq (0, \infty) \to \mathbb{R}$ *is twice differentiable on* $\mathring{I}$. *Let* $p \in (-\infty, 0) \cup (1, \infty)$ *and define* $g_p : I \to \mathbb{R}$, $g_p(t) = \phi''(t)t^{2-p}$.

(i) *If* $\inf_{t \in \mathring{I}} g_p(t) = \gamma > -\infty$, *then we have the inequality*

$$\frac{\gamma}{p(p-1)}\left[A\left(f^p\right) - \left[A(f)\right]^p\right] \leq A(\phi \circ f) - \phi\left(A(f)\right) \tag{12.36}$$

provided that $\phi \circ f, f^p, f \in L$.

(ii) *If* $\sup_{t \in \mathring{I}} g_p(t) = \Gamma < \infty$, *then we have the inequality*

$$A(\phi \circ f) - \phi\left(A(f)\right) \leq \frac{\Gamma}{p(p-1)}\left[A\left(f^p\right) - \left[A(f)\right]^p\right] \tag{12.37}$$

provided that $\phi \circ f, f^p, f \in L$.

(iii) *If* $-\infty < \gamma \leq g_p(t) \leq \Gamma < \infty$, $t \in \mathring{I}$, *then both* (12.36) *and* (12.37) *hold, provided that* $\phi \circ f, f^p, f \in L$.

Proof The proof is as follows.

(i) We have for the auxiliary mapping $h_p(t) = \phi(t) - \frac{\gamma}{p(p-1)}t^p$ that

$$\begin{aligned} h_p''(t) &= \phi''(t) - \gamma t^{p-2} = t^{p-2}\left(t^{2-p}\phi''(t) - \gamma\right) \\ &= t^{p-2}\left(g_p(t) - \gamma\right) \geq 0. \end{aligned}$$

That is, h_p is convex or, equivalently, $\phi \in \mathscr{L}(I, \frac{\gamma}{p(p-1)}, (\cdot)^p)$. Applying Corollary 12.3, we deduce (12.36).

(ii) Goes similarly.

(iii) Follows by (i) and (ii). □

The following proposition also holds [7].

Proposition 12.5 *Assume that the mapping* $\phi : I \subseteq (0, \infty) \to \mathbb{R}$ *is twice differentiable on* $\mathring{I}$. *Define* $l(t) = t^2\phi''(t)$, $t \in I$.

(i) *If* $\inf_{t \in \mathring{I}} l(t) = s > -\infty$, *then we have the inequality*

$$s\left[\ln\left[A(f)\right] - A(\ln f)\right] \leq A(\phi \circ f) - \phi\left(A(f)\right), \tag{12.38}$$

provided that $\phi \circ f, \ln f, f \in L$ *and* $A(f) > 0$.

(ii) *If* $\sup_{t\in\mathring{I}} l(t) = S < \infty$, *then we have the inequality*

$$A(\phi \circ f) - \phi\big(A(f)\big) \le S\big[\ln\big[A(f)\big] - A(\ln f)\big], \tag{12.39}$$

provided that $\phi \circ f, \ln f, f \in L$ *and* $A(f) > 0$.

(iii) *If* $-\infty < s \le l(t) \le S < \infty$ *for* $t \in \mathring{I}$, *then both* (12.38) *and* (12.39) *hold, provided that* $\phi \circ f, \ln f, f \in L$ *and* $A(f) > 0$.

Proof The proof is as follows.

(i) Define the auxiliary mapping $h(t) = \phi(t) + s\ln t$. Then

$$h''(t) = \phi''(t) - \frac{s}{t^2} = \frac{1}{t^2}\big(\phi''(t)t^2 - s\big) \ge 0,$$

which shows that h is convex or, equivalently, $\phi \in \mathscr{L}(I, s, -\ln(\cdot))$. Applying Corollary 12.3, we deduce (12.38).

(ii) Goes similarly.

(iii) Follows by (i) and (ii). □

Finally, the following result also holds [7].

Proposition 12.6 *Assume that the mapping* $\phi : I \subseteq (0,\infty) \to \mathbb{R}$ *is twice differentiable on* $\mathring{I}$. *Define* $\tilde{I}(t) = t\phi''(t)$, $t \in I$.

(i) *If* $\inf_{t\in\mathring{I}} \tilde{I}(t) = \delta > -\infty$, *then we have the inequality*

$$\delta\big[A[f\ln f] - A(f)\ln A(f)\big] \le A(\phi \circ f) - \phi\big(A(f)\big), \tag{12.40}$$

provided that $\phi \circ f, f\ln f, f \in L$ *and* $A(f) > 0$.

(ii) *If* $\sup_{t\in\mathring{I}} \tilde{I}(t) = \Delta < \infty$, *then we have the inequality*

$$A(\phi \circ f) - \phi\big(A(f)\big) \le \Delta\big[A[f\ln f] - A(f)\ln A(f)\big], \tag{12.41}$$

provided that $\phi \circ f, f\ln f, f \in L$ *and* $A(f) > 0$.

(iii) *If* $-\infty < \delta \le \tilde{I}(t) \le \Delta < \infty$ *for* $t \in \mathring{I}$, *then both* (12.40) *and* (12.41) *hold, provided that* $\phi \circ f, f\ln f, f \in L$ *and* $A(f) > 0$.

Proof The proof is as follows.

(i) Define the auxiliary mapping $h(t) = \phi(t) + \delta t\ln t$, $t \in I$. Then

$$h''(t) = \phi''(t) - \frac{\delta}{t} = \frac{1}{t^2}\big[\phi''(t)t - \delta\big] = \frac{1}{t}\big[\tilde{I}(t) - \delta\big] \ge 0,$$

which shows that h is convex or, equivalently, $\phi \in \mathscr{L}(I, \delta, (\cdot)\ln(\cdot))$. Applying Corollary 12.3, we deduce (12.40).

(ii) Goes similarly.

(iii) Follows by (i) and (ii). □

12.4.2 Some Particular Inequalities

We know, by Proposition 12.3, that

$$\frac{1}{2}k\big[A(f^2)-[A(f)]^2\big]\le A(\phi\circ f)-\phi(A(f))$$
$$\le\frac{1}{2}K\big[A(f^2)-[A(f)]^2\big],\tag{12.42}$$

provided that $\phi: I\subseteq\mathbb{R}\to\mathbb{R}$ is twice differentiable on $\mathring{I}$, $-\infty<k\le\phi''(t)\le K<\infty$, $t\in\mathring{I}$, $f:E\to I$, $\phi\circ f$, f^2, $f\in L$, and $A:L\to\mathbb{R}$ is an isotonic linear and normalized functional.

The following inequalities have been established in [7].

1. We assume that $0<m\le f\le M<\infty$, where m, M are real numbers. Then, by (12.42) applied to $\phi:[m,M]\to\mathbb{R}$, $\phi(t)=-\ln t$, we have the inequality

$$\frac{1}{2M^2}\big[A(f^2)-[A(f)]^2\big]\le\ln[A(f)]-A[\ln(f)]$$
$$\le\frac{1}{2m^2}\big[A(f^2)-[A(f)]^2\big],\tag{12.43}$$

provided that $\ln f$, f^2, $f\in L$, and $A(f)>0$.

Note that (12.43) is equivalent to

$$\exp\left[\frac{1}{2M^2}\big[A(f^2)-[A(f)]^2\big]\right]\le\frac{[A(f)]}{\exp[A[\ln(f)]]}$$
$$\le\exp\left[\frac{1}{2m^2}\big[A(f^2)-[A(f)]^2\big]\right].\tag{12.44}$$

2. If we apply (12.42) to $\phi:[m,M]\to\mathbb{R}$, $\phi(t)=t^p$, $p\in(-\infty,0)\cup(1,\infty)$, then we have the inequality

$$\frac{p(p-1)}{2}m^{p-2}\big[A(f^2)-[A(f)]^2\big]$$
$$\le A(f^p)-[A(f)]^p$$
$$\le\frac{p(p-1)}{2}M^{p-2}\big[A(f^2)-[A(f)]^2\big]\tag{12.45}$$

if $p>2$, and

$$\frac{p(p-1)}{2}M^{p-2}\big[A(f^2)-[A(f)]^2\big]$$
$$\le A(f^p)-[A(f)]^p$$

$$\leq \frac{p(p-1)}{2} m^{p-2} \left[A(f^2) - [A(f)]^2\right] \tag{12.46}$$

if $p \in (-\infty, 0) \cup (1, 2)$, provided that $f^2, f^p, f \in L$.

3. If we apply (12.43) to $\phi : [m, M] \to \mathbb{R}$, $\phi(t) = t \ln t$, then we have the inequality

$$\frac{1}{2M}\left[A(f^2) - [A(f)]^2\right] \leq A[f \ln f] - A(f) \ln A(f)$$
$$\leq \frac{1}{2m}\left[A(f^2) - [A(f)]^2\right], \tag{12.47}$$

provided that $f \ln f$, f^2, $f \in L$ and $A(f) > 0$.

Note that the inequality (12.47) is equivalent to

$$\exp\left\{\frac{1}{2M}\left[A(f^2) - [A(f)]^2\right]\right\} \leq \frac{\exp[A(f \ln f)]}{[A(f)]^{A(f)}}$$
$$\leq \exp\left\{\frac{1}{2m}\left[A(f^2) - [A(f)]^2\right]\right\}. \tag{12.48}$$

4. If we assume that $-\infty < m \leq f \leq M < \infty$, and apply the inequality (12.42) to $\phi(t) = e^t$, $t \in \mathbb{R}$, we obtain

$$\frac{1}{2}\exp(m)\left[A(f^2) - [A(f)]^2\right] \leq A\big(\exp(f)\big) - \exp\big(A(f)\big)$$
$$\leq \frac{1}{2}\exp(M)\left[A(f^2) - [A(f)]^2\right], \tag{12.49}$$

provided that $\exp(f)$, f^2, $f \in L$.

Using Proposition 12.4, we may state that

$$\frac{\gamma}{p(p-1)}\left[A(f^p) - [A(f)]^p\right] \leq A(\phi \circ f) - \phi\big(A(f)\big)$$
$$\leq \frac{\Gamma}{p(p-1)}\left[A(f^p) - [A(f)]^p\right], \tag{12.50}$$

provided that $\phi : I \subseteq \mathbb{R} \to \mathbb{R}$ is twice differentiable on $\mathring{I}$, $\gamma \leq \phi''(t)t^{2-p} \leq \Gamma$, $t \in \mathring{I}$, $f : E \to I$, $\phi \circ f$, f^p, $f \in L$, and $A : L \to \mathbb{R}$ is an isotonic linear and normalized functional.

5. If we consider $\phi(t) = -\ln t$ and assume that $0 < m \leq f \leq M < \infty$, then

$$\frac{m^{-p}}{p(p-1)}\left[A(f^p) - [A(f)]^p\right] \leq \ln\big(A(f)\big) - A\big(\ln(f)\big)$$
$$\leq \frac{M^{-p}}{p(p-1)}\left[A(f^p) - [A(f)]^p\right] \tag{12.51}$$

if $p \in (-\infty, 0)$ and

$$\frac{M^{-p}}{p(p-1)}\left[A(f^p) - [A(f)]^p\right] \leq \ln(A(f)) - A(\ln(f)) \leq \frac{m^{-p}}{p(p-1)}\left[A(f^p) - [A(f)]^p\right] \tag{12.52}$$

if $p \in (1, \infty)$, provided that $f^p, \ln f, f \in L$ and $A(f) > 0$.

6. If we consider $\phi(t) = t \ln t$ and assume that $0 < m \leq f \leq M < \infty$, then

$$\frac{m^{1-p}}{p(p-1)}\left[A(f^p) - [A(f)]^p\right] \leq A(f \ln f) - A(f) \ln A(f) \leq \frac{M^{1-p}}{p(p-1)}\left[A(f^p) - [A(f)]^p\right] \tag{12.53}$$

if $p \in (-\infty, 0)$ and

$$\frac{M^{1-p}}{p(p-1)}\left[A(f^p) - [A(f)]^p\right] \leq A(f \ln f) - A(f) \ln A(f) \leq \frac{m^{1-p}}{p(p-1)}\left[A(f^p) - [A(f)]^p\right] \tag{12.54}$$

if $p \in (1, \infty)$, $f \ln f$, f^p, $f \in L$, and $A(f) > 0$.

Finally, by Proposition 12.5, we may state the inequality

$$s\left[\ln\left[A(f)\right] - A(\ln f)\right] \leq A(\phi \circ f) - \phi(A(f)) \leq S\left[\ln\left[A(f)\right] - A(\ln f)\right], \tag{12.55}$$

provided that $\phi : I \subseteq (0, \infty)$ is twice differentiable on $\mathring{I}$, $-\infty < s \leq t^2\phi''(t) \leq S < \infty$, $\phi \circ f$, $\ln f$, $f \in L$ and $A(f) > 0$.

7. If we assume that $0 < m \leq f \leq M < \infty$ and apply (12.55) to $\phi(t) = t \ln t$, we have the inequality

$$m\left[\ln\left[A(f)\right] - A(\ln f)\right] \leq A(f \ln f) - A(f) \ln A(f) \leq M\left[\ln\left[A(f)\right] - A(\ln f)\right] \tag{12.56}$$

provided that $\ln f$, $f \ln f$, $f \in L$ and $A(f) > 0$.

Note that (12.56) is equivalent to

$$\left(\frac{A(f)}{\exp[A(\ln f)]}\right)^m \leq \frac{\exp[A(f \ln f)]}{[A(f)]^{A(f)}} \leq \left(\frac{A(f)}{\exp[A(\ln f)]}\right)^M. \tag{12.57}$$

12.4.3 Applications to Hermite–Hadamard Inequalities

The following integral inequalities were obtained in [7].

(a) Suppose that $\phi : [a,b] \subset \mathbb{R} \to \mathbb{R}$ is twice differentiable and $-\infty < k \le \phi''(t) \le K < \infty$ for $t \in (a,b)$. If in (12.42) we choose $f = e$, i.e., $f(x) = x$, $x \in [a,b]$ and $A(f) := \frac{1}{b-a}\int_a^b f(t)\,dt$ and take into account that

$$A(f^2) - \bigl(A(f)\bigr)^2 = \frac{1}{b-a}\int_a^b x^2\,dx - \left(\frac{1}{b-a}\int_a^b x\,dx\right)^2 = \frac{(b-a)^2}{12},$$

then we get the inequality (see also [11, p. 40])

$$\frac{(b-a)^2}{24}\cdot k \le \frac{1}{b-a}\int_a^b \phi(x)\,dx - \phi\left(\frac{a+b}{2}\right) \le \frac{(b-a)^2}{24}\cdot K. \tag{12.58}$$

(b) Now, if we assume that $\phi : [a,b] \subset (0,\infty) \to \mathbb{R}$ is twice differentiable over $\gamma \le t^{2-p}\phi''(t) \le \Gamma$, $t \in (a,b)$, $p \in (-\infty,0) \cup (1,\infty)$, then by (12.50), in which we choose $f = e$, $A(f) := \frac{1}{b-a}\int_a^b f(t)\,dt$ and taking into account that

$$A(f^p) - \bigl(A(f)\bigr)^p = \frac{1}{b-a}\int_a^b x^p\,dx - \left(\frac{1}{b-a}\int_a^b x\,dx\right)^p = L_p^p(a,b) - A^p(a,b),$$

we get

$$\frac{\gamma}{p(p-1)}\bigl[L_p^p(a,b) - A^p(a,b)\bigr] \le \frac{1}{b-a}\int_a^b \phi(x)\,dx - \phi\left(\frac{a+b}{2}\right) \le \frac{\Gamma}{p(p-1)}\bigl[L_p^p(a,b) - A^p(a,b)\bigr]. \tag{12.59}$$

(c) If $\phi : [a,b] \subset (0,\infty) \to \mathbb{R}$ is twice differentiable and satisfies the condition $s \le t^2\phi''(t) \le S$, $t \in (a,b)$, then by Proposition 12.5 applied to $f = e$, $A(f) := \frac{1}{b-a}\int_a^b f(t)\,dt$ and taking into account that

$$\ln\bigl[A(f)\bigr] - A(\ln f) = \ln\left(\frac{1}{b-a}\int_a^b x\,dx\right) - \frac{1}{b-a}\int_a^b \ln x\,dx = \ln\bigl[A(a,b)\bigr] - \ln I(a,b) = \ln\left[\frac{A(a,b)}{I(a,b)}\right],$$

we get the inequality

$$s\ln\left[\frac{A(a,b)}{I(a,b)}\right] \le \frac{1}{b-a}\int_a^b \phi(x)\,dx - \phi\left(\frac{a+b}{2}\right) \le S\ln\left[\frac{A(a,b)}{I(a,b)}\right] \tag{12.60}$$

or, equivalently,

$$\left[\frac{A(a,b)}{I(a,b)}\right]^{s} \le \frac{\exp[\frac{1}{b-a}\int_a^b \phi(x)\,dx]}{\exp[\phi(\frac{a+b}{2})]} \le \left[\frac{A(a,b)}{I(a,b)}\right]^{S}. \tag{12.61}$$

(d) Finally, if we assume that the twice differentiable function $\phi : [a,b] \subset (0,\infty) \to \mathbb{R}$ satisfies the condition $\delta \le t\phi''(t) \le \Delta$, $t \in (a,b)$, then by Proposition 12.6 and with the same selection of f and A, and taking into account that

$$\begin{aligned}
&A(f\ln f) - A(f)\ln A(f)\\
&= \frac{1}{b-a}\int_a^b x\ln x\,dx - \frac{1}{b-a}\int_a^b x\,dx \cdot \frac{1}{b-a}\int_a^b \ln x\,dx\\
&= \frac{1}{4(b-a)}\left[b^2\ln b^2 - a^2\ln a^2 - \left(b^2-a^2\right)\right] - A(a,b)\ln I(a,b)\\
&= \frac{A(a,b)}{2(b^2-a^2)}\left[b^2\ln b^2 - a^2\ln a^2 - \left(b^2-a^2\right)\right] - A(a,b)\ln I(a,b)\\
&= \frac{A(a,b)}{2}\ln I\left(a^2,b^2\right) - A(a,b)\ln I(a,b)\\
&= \ln\left[\left(\frac{\sqrt{I(a^2,b^2)}}{I(a,b)}\right)^{A(a,b)}\right],
\end{aligned}$$

we may state the inequality

$$\begin{aligned}
\delta\ln\left[\left(\frac{\sqrt{I(a^2,b^2)}}{I(a,b)}\right)^{A(a,b)}\right] &\le \frac{1}{b-a}\int_a^b \phi(x)\,dx - \phi\left(\frac{a+b}{2}\right)\\
&\le \Delta\ln\left[\left(\frac{\sqrt{I(a^2,b^2)}}{I(a,b)}\right)^{A(a,b)}\right],
\end{aligned} \tag{12.62}$$

or, equivalently,

$$\left(\frac{\sqrt{I(a^2,b^2)}}{I(a,b)}\right)^{\delta A(a,b)} \le \frac{\exp[\frac{1}{b-a}\int_a^b \phi(x)\,dx]}{\exp[\phi(\frac{a+b}{2})]} \le \left(\frac{\sqrt{I(a^2,b^2)}}{I(a,b)}\right)^{\Delta A(a,b)}. \tag{12.63}$$

12.5 Lupaş–Beesack–Pečarić Inequality for $m - \Psi$-Convex and $M - \Psi$-Convex Functions

12.5.1 Some Lupaş–Beesack–Pečarić Type Inequalities

We now prove the following result [8].

Theorem 12.7 (Dragomir, 2002, [8]) *Let $\Psi : [\alpha, \beta] \subset \mathbb{R} \to \mathbb{R}$ be a convex function and $f : I \to [\alpha, \beta]$ such that $\Psi \circ f$, $f \in L$, and suppose $A : L \to \mathbb{R}$ is an isotonic linear and normalized functional.*

(i) *If $\phi \in \mathscr{L}(I, m, \Psi)$ and $\phi \circ f \in L$, then we have the inequality*

$$m\left[\frac{\beta - A(f)}{\beta - \alpha}\Psi(\alpha) + \frac{A(f) - \alpha}{\beta - \alpha}\Psi(\beta) - A(\Psi \circ f)\right]$$
$$\leq \frac{\beta - A(f)}{\beta - \alpha}\phi(\alpha) + \frac{A(f) - \alpha}{\beta - \alpha}\phi(\beta) - A(\phi \circ f). \tag{12.64}$$

(ii) *If $\phi \in \mathscr{U}(I, M, \Psi)$ and $\phi \circ f \in L$, then*

$$\frac{\beta - A(f)}{\beta - \alpha}\phi(\alpha) + \frac{A(f) - \alpha}{\beta - \alpha}\phi(\beta) - A(\phi \circ f)$$
$$\leq M\left[\frac{\beta - A(f)}{\beta - \alpha}\Psi(\alpha) + \frac{A(f) - \alpha}{\beta - \alpha}\Psi(\beta) - A(\Psi \circ f)\right]. \tag{12.65}$$

(iii) *If $\phi \in \mathscr{B}(I, m, M, \Psi)$ and $\phi \circ f \in L$, then both* (12.64) *and* (12.65) *hold.*

Proof The proof is as follows.

(i) As $\phi \in \mathscr{L}(I, m, \Psi)$ and $\phi \circ f \in L$, it follows that $\phi - m\Psi$ is convex and $(\phi - m\Psi) \circ f \in L$.

Applying Lupaş–Beesack–Pečarić's inequality for the convex function $\phi - m\Psi$, we get

$$A\big((\phi - m\Psi) \circ f\big) \leq \frac{\beta - A(f)}{\beta - \alpha}(\phi - m\Psi)(\alpha) + \frac{A(f) - \alpha}{\beta - \alpha}(\phi - m\Psi)(\beta). \tag{12.66}$$

However,

$$A\big((\phi - m\Psi) \circ f\big) = A(\phi \circ f) - mA(\Psi \circ f)$$

and then, after some simple computation, (12.66) is equivalent to (12.64).

(ii) Goes likewise, and we omit the details.

(iii) Follows by (i) and (ii). □

The following corollary is useful in practice [8].

Corollary 12.4 *Let $\Psi : I \subseteq \mathbb{R} \to \mathbb{R}$ be a twice differentiable convex function on $\mathring{I}$, $f : E \to I$ such that $\Psi \circ f$, $f \in L$, and suppose $A : L \to \mathbb{R}$ is an isotonic linear and normalized functional.*

(i) *If $\phi : I \to \mathbb{R}$ is twice differentiable, $\phi \circ f \in L$, and $\phi''(t) \geq m\Psi''(t)$, $t \in \mathring{I}$ (where m is a given real number), then* (12.64) *holds.*

(ii) *If $\phi : I \to \mathbb{R}$ is twice differentiable, $\phi \circ f \in L$, and $\phi''(t) \leq M\Psi''(t)$, $t \in \mathring{I}$ (where m is a given real number), then* (12.65) *holds.*

(iii) *If $m\Psi''(t) \leq \phi''(t) \leq M\Psi''(t)$, $t \in \mathring{I}$, then both* (12.64) *and* (12.65) *hold.*

Some particular important cases of the above corollary are embodied in the following propositions [8].

Proposition 12.7 *Assume that the function $\phi : I \subseteq \mathbb{R} \to \mathbb{R}$ is twice differentiable on $\mathring{I}$.*

(i) *If $\inf_{t\in\mathring{I}} \phi''(t) = k > -\infty$, then we have the inequality*

$$\begin{aligned}&\frac{k}{2}\left[(\alpha+\beta)A(f)-\alpha\beta-A\left(f^2\right)\right]\\&\quad\le\frac{\beta-A(f)}{\beta-\alpha}\phi(\alpha)+\frac{A(f)-\alpha}{\beta-\alpha}\phi(\beta)-A(\phi\circ f),\end{aligned}\tag{12.67}$$

provided that $\phi \circ f, f^2, f \in L$.

(ii) *If $\sup_{t\in\mathring{I}} \phi''(t) = K < \infty$, then we have the inequality*

$$\begin{aligned}&\frac{\beta-A(f)}{\beta-\alpha}\phi(\alpha)+\frac{A(f)-\alpha}{\beta-\alpha}\phi(\beta)-A(\phi\circ f)\\&\quad\le\frac{K}{2}\left[(\alpha+\beta)A(f)-\alpha\beta-A\left(f^2\right)\right].\end{aligned}\tag{12.68}$$

provided that $\phi \circ f, f^2, f \in L$.

(iii) *If $-\infty < k \le \phi''(t) \le K < \infty$, $t \in \mathring{I}$, then both* (12.67) *and* (12.68) *hold, provided that $\phi \circ f, f^2, f \in L$.*

Proof The proof is as follows.

(i) Consider the auxiliary mapping $h(t) := \phi(t) - \frac{1}{2}kt^2$. Then $h''(t) = \phi''(t) - k \ge 0$ i.e., h is convex, or, equivalently, $\phi \in \mathcal{L}(I, \frac{1}{2}k, (\cdot)^2)$. Applying Corollary 12.4, we may state

$$\begin{aligned}&\frac{k}{2}\left[\frac{\beta-A(f)}{\beta-\alpha}\alpha^2+\frac{A(f)-\alpha}{\beta-\alpha}\beta^2-A\left(f^2\right)\right]\\&\quad\le\frac{\beta-A(f)}{\beta-\alpha}\phi(\alpha)+\frac{A(f)-\alpha}{\beta-\alpha}\phi(\beta)-A(\phi\circ f),\end{aligned}$$

which is clearly equivalent to (12.67)

(ii) Goes likewise, and we omit the details.

(iii) Follows by (i) and (ii). □

Another result is the following one [8].

Proposition 12.8 *Assume that the mapping $\phi : [\alpha, \beta] \subset (0,\infty) \to \mathbb{R}$ is twice differentiable on (α, β), let $p \in (-\infty, 0) \cup (1,\infty)$, and define $g_p : [\alpha, \beta] \to \mathbb{R}$, $g_p(t) = \phi''(t)t^{2-p}$.*

(i) *If $\inf_{t\in\mathring{I}} g_p(t) = \gamma > -\infty$, then we have the inequality*

$$\frac{\gamma}{p(p-1)}\left[pL_{p-1}^{p-1}(\alpha,\beta)A(f)-\alpha\beta(p-1)L_{p-2}^{p-2}(\alpha,\beta)-A\left(f^p\right)\right]$$
$$\leq\frac{\beta-A(f)}{\beta-\alpha}\phi(\alpha)+\frac{A(f)-\alpha}{\beta-\alpha}\phi(\beta)-A(\phi\circ f), \qquad (12.69)$$

provided that $\phi\circ f, f^p, f\in L$.

(ii) *If* $\sup_{t\in\mathring{I}} g_p(t)=\Gamma<\infty$, *then we have the inequality*

$$\frac{\beta-A(f)}{\beta-\alpha}\phi(\alpha)+\frac{A(f)-\alpha}{\beta-\alpha}\phi(\beta)-A(\phi\circ f)$$
$$\leq\frac{\Gamma}{p(p-1)}\left[pL_{p-1}^{p-1}(\alpha,\beta)A(f)-\alpha\beta(p-1)L_{p-2}^{p-2}(\alpha,\beta)-A\left(f^p\right)\right]. \qquad (12.70)$$

(iii) *If* $-\infty<\gamma\leq g_p(t)\leq\Gamma<\infty$, $t\in\mathring{I}$, *then we have both* (12.69) *and* (12.70).

Proof The proof is as follows.

(i) Consider the auxiliary mapping $h_p(t)=\phi(t)-\frac{\gamma}{p(p-1)}t^p$. Then

$$h_p''(t)=\phi''(t)-\gamma t^{p-2}=t^{p-2}\left(t^{2-p}\phi''(t)-\gamma\right)$$
$$=t^{p-2}\left(g_p(t)-\gamma\right)\geq 0.$$

That is, h_p is convex, or, equivalently, $\phi\in\mathscr{L}(I,\frac{\gamma}{p(p-1)},(\cdot)^p)$. Applying Corollary 12.4, we may state

$$\frac{\gamma}{p(p-1)}\left[\frac{\beta-A(f)}{\beta-\alpha}\alpha^p+\frac{A(f)-\alpha}{\beta-\alpha}\beta^p-A\left(f^p\right)\right]$$
$$\leq\frac{\beta-A(f)}{\beta-\alpha}\phi(\alpha)+\frac{A(f)-\alpha}{\beta-\alpha}\phi(\beta)-A(\phi\circ f),$$

which is equivalent to (12.69).

(ii) Goes likewise.

(iii) Follows by (i) and (ii). □

The following proposition also holds [8].

Proposition 12.9 *Assume that the mapping* $\phi:[\alpha,\beta]\subset(0,\infty)\to\mathbb{R}$ *is twice differentiable on* (α,β). *Define* $l(t)=t^2\phi''(t)$, $t\in[\alpha,\beta]$.

(i) *If* $\inf_{t\in(\alpha,\beta)} l(t)=s>-\infty$, *then we have the inequality*

$$s\left[A(\ln f)+\ln\left[I\left(\frac{1}{\alpha},\frac{1}{\beta}\right)\right]+1-\frac{A(f)}{L(\alpha,\beta)}\right]$$
$$\leq\frac{\beta-A(f)}{\beta-\alpha}\phi(\alpha)+\frac{A(f)-\alpha}{\beta-\alpha}\phi(\beta)-A(\phi\circ f), \qquad (12.71)$$

provided that $\phi \circ f, \ln f$ *and* $f \in L$, *and* $I(\cdot, \cdot)$ *denotes the identric mean, i.e., we recall it*

$$I(u, v) := \begin{cases} u & \text{if } v = u, \\ \frac{1}{e}\left(\frac{u^u}{v^v}\right)^{\frac{1}{u-v}} & \text{if } v \neq u. \end{cases}$$

(ii) *If* $\sup_{t\in(\alpha,\beta)} l(t) = S < \infty$, *then we have the inequality*

$$\frac{\beta - A(f)}{\beta - \alpha}\phi(\alpha) + \frac{A(f) - \alpha}{\beta - \alpha}\phi(\beta) - A(\phi \circ f)$$
$$\leq S\left[A(\ln f) + \ln\left[I\left(\frac{1}{\alpha}, \frac{1}{\beta}\right)\right] + 1 - \frac{A(f)}{L(\alpha, \beta)}\right]. \tag{12.72}$$

(iii) *If* $-\infty < s \leq l(t) \leq S < \infty$ *for* $t \in (\alpha, \beta)$, *then both* (12.71) *and* (12.72) *hold.*

Proof The proof is as follows.

(i) Define the auxiliary function $h(t) = \phi(t) + s \ln t$. Then

$$h''(t) = \phi''(t) - \frac{s}{t^2} = \frac{1}{t^2}\left(\phi''(t)t^2 - s\right) \geq 0,$$

showing that h is convex, or, equivalently, $\phi \in \mathscr{L}(I, s, -\ln(\cdot))$. Applying Corollary 12.4, we may state that

$$s\left[\frac{\beta - A(f)}{\beta - \alpha} \cdot \left[-\ln(\alpha)\right] + \frac{A(f) - \alpha}{\beta - \alpha} \cdot \left[-\ln(\beta)\right] + A(\ln f)\right]$$
$$\leq \frac{\beta - A(f)}{\beta - \alpha}\phi(\alpha) + \frac{A(f) - \alpha}{\beta - \alpha}\phi(\beta) - A(\phi \circ f),$$

which is equivalent to (12.71).

(ii) Goes likewise.

(iii) Follows by (i) and (ii). □

Finally, the following result also holds [8].

Proposition 12.10 *Assume that the mapping* $\phi : [\alpha, \beta] \subset (0, \infty) \to \mathbb{R}$ *is twice differentiable on* (α, β). *Define* $\tilde{l}(t) = t\phi''(t)$, $t \in I$.

(i) *If* $\inf_{t\in(\alpha,\beta)} \tilde{l}(t) = \delta > -\infty$, *then we have the inequality*

$$\delta\left[A(f)\ln I(\alpha, \beta) - \frac{G^2(\alpha, \beta)}{L(\alpha, \beta)} + A(f) - A(f \ln f)\right]$$
$$\leq \frac{\beta - A(f)}{\beta - \alpha}\phi(\alpha) + \frac{A(f) - \alpha}{\beta - \alpha}\phi(\beta) - A(\phi \circ f), \tag{12.73}$$

provided that $\phi \circ f, f \ln f, f \in L$, $G(\alpha, \beta) = \sqrt{ab}$ *is the geometric mean, and* $L(\alpha, \beta)$ *is the logarithmic mean, i.e., we recall it*

$$L(\alpha, \beta) := \begin{cases} \alpha & \text{if } \beta = \alpha, \\ \frac{\beta - \alpha}{\ln \beta - \ln \alpha} & \text{if } \beta \neq \alpha. \end{cases}$$

(ii) *If* $\sup_{t \in (\alpha, \beta)} \tilde{I}(t) = \Delta < \infty$, *then we have the inequality*

$$\frac{\beta - A(f)}{\beta - \alpha}\phi(\alpha) + \frac{A(f) - \alpha}{\beta - \alpha}\phi(\beta) - A(\phi \circ f)$$
$$\leq \Delta\left[A(f) \ln I(\alpha, \beta) - \frac{G^2(\alpha, \beta)}{L(\alpha, \beta)} + A(f) - A(f \ln f)\right]. \qquad (12.74)$$

(iii) *If* $-\infty < \delta \leq \tilde{I}(t) \leq \Delta < \infty$ *for* $t \in (\alpha, \beta)$, *then both* (12.73) *and* (12.74) *hold.*

Proof The proof is as follows.

(i) Define the auxiliary mapping $h(t) = \phi(t) - \delta t \ln t$, $t \in (\alpha, \beta)$. Then

$$h''(t) = \phi''(t) - \frac{\delta}{t} = \frac{1}{t^2}\left[\phi''(t)t - \delta\right] = \frac{1}{t}\left[\tilde{I}(t) - \delta\right] \geq 0,$$

which shows that h is convex or, equivalently, $\phi \in \mathscr{L}(I, \delta, (\cdot)\ln(\cdot))$. Applying Corollary 12.4, we can write

$$\delta\left[\frac{\beta - A(f)}{\beta - \alpha} \cdot [\alpha \ln \alpha] + \frac{A(f) - \alpha}{\beta - \alpha} \cdot [\beta \ln \beta] - A(f \ln f)\right]$$
$$\leq \frac{\beta - A(f)}{\beta - \alpha}\phi(\alpha) + \frac{A(f) - \alpha}{\beta - \alpha}\phi(\beta) - A(\phi \circ f),$$

which is clearly equivalent to (12.73).

(ii) Goes similarly.

(iii) Follows by (i) and (ii). □

12.5.2 Applications of Hermite–Hadamard Inequalities

The following integral inequalities were obtained in [8].

(a) Assume that $\phi : [a, b] \subset \mathbb{R} \to \mathbb{R}$ is a twice differentiable function satisfying the condition $-\infty < k \leq \phi''(t) \leq K < \infty$ for $t \in (a, b)$. If in Proposition 12.7 we choose $A(f) := \frac{1}{b-a}\int_a^b f(t)\,dt$, $f = e$, i.e., $e(x) = x$, $x \in [a, b]$ and take into account that

$$A(f^2) = \frac{b^2 + ab + a^2}{3},$$

then we may state the inequality (see also [11, p. 39])

$$\frac{k(b-a)^2}{12} \le \frac{\phi(b)+\phi(a)}{2} - \frac{1}{b-a}\int_a^b \phi(x)\,dx \le \frac{K(b-a)^2}{12}. \tag{12.75}$$

(b) Now, if we assume that $\phi : [a,b] \subset (0,\infty) \to \mathbb{R}$ is twice differentiable on (a,b) and $-\infty < \gamma \le t^{2-p}\phi''(t) \le \Gamma < \infty$, $t \in (a,b)$, $p \in (-\infty,0) \cup (1,\infty)$, then, applying Proposition 12.8 to integrals, we may state the inequality

$$\begin{aligned}
&\frac{\gamma}{p(p-1)}\big[pL_{p-1}^{p-1}(a,b)A(a,b) - (p-1)G^2(a,b)L_{p-2}^{p-2}(a,b) - L_p^p(a,b)\big]\\
&\le \frac{\phi(b)+\phi(a)}{2} - \frac{1}{b-a}\int_a^b \phi(x)\,dx\\
&\le \frac{\Gamma}{p(p-1)}\big[pL_{p-1}^{p-1}(a,b)A(a,b) - (p-1)G^2(a,b)L_{p-2}^{p-2}(a,b) - L_p^p(a,b)\big].
\end{aligned} \tag{12.76}$$

(c) Suppose that the twice differentiable function $\phi : [a,b] \subset (0,\infty) \to \mathbb{R}$ satisfies the condition $-\infty < s \le t^2\phi''(t) \le S < \infty$. Then by Proposition 12.9 applied to the integral functional, we may state the following inequality

$$\begin{aligned}
s\ln\left[\frac{I(a,b)I\left(\frac{1}{a},\frac{1}{b}\right)}{\exp\left(\frac{A(a,b)-L(a,b)}{L(a,b)}\right)}\right] &\le \frac{\phi(b)+\phi(a)}{2} - \frac{1}{b-a}\int_a^b \phi(x)\,dx\\
&\le S\ln\left[\frac{I(a,b)I\left(\frac{1}{a},\frac{1}{b}\right)}{\exp\left(\frac{A(a,b)-L(a,b)}{L(a,b)}\right)}\right]
\end{aligned} \tag{12.77}$$

or, equivalently,

$$\begin{aligned}
\left[\frac{I(a,b)I\left(\frac{1}{a},\frac{1}{b}\right)}{\exp\left(\frac{A(a,b)-L(a,b)}{L(a,b)}\right)}\right]^s &\le \frac{\exp\left[\frac{\phi(b)+\phi(a)}{2}\right]}{\exp\left[\frac{1}{b-a}\int_a^b \phi(x)\,dx\right]}\\
&\le \left[\frac{I(a,b)I\left(\frac{1}{a},\frac{1}{b}\right)}{\exp\left(\frac{A(a,b)-L(a,b)}{L(a,b)}\right)}\right]^S.
\end{aligned} \tag{12.78}$$

(d) Finally, if we assume that a twice differentiable function $\phi : [a,b] \subset (0,\infty) \to \mathbb{R}$ satisfies the condition $-\infty < \delta \le t\phi''(t) \le 1 < \infty$, then by Proposition 12.10 applied to the integral functional, we may state the following inequality:

$$\begin{aligned}
&\delta A(a,b)\ln\left[\left(\frac{I(a,b)}{\sqrt{I(a^2,b^2)}}\right)\cdot\exp\left(\frac{L(a,b)A(a,b)-G^2(a,b)}{L(a,b)A(a,b)}\right)\right]\\
&\le \frac{\phi(b)+\phi(a)}{2} - \frac{1}{b-a}\int_a^b \phi(x)\,dx
\end{aligned}$$

$$\le \Delta A(a,b)\ln\left[\left(\frac{I(a,b)}{\sqrt{I(a^2,b^2)}}\right)\cdot\exp\left(\frac{L(a,b)A(a,b)-G^2(a,b)}{L(a,b)A(a,b)}\right)\right], \quad (12.79)$$

or, equivalently,

$$\left[\left(\frac{I(a,b)}{\sqrt{I(a^2,b^2)}}\right)\cdot\exp\left(\frac{L(a,b)A(a,b)-G^2(a,b)}{L(a,b)A(a,b)}\right)\right]^{\delta A(a,b)}$$
$$\le \frac{\exp\left[\frac{\phi(b)+\phi(a)}{2}\right]}{\exp\left[\frac{1}{b-a}\int_a^b \phi(x)\,dx\right]}$$
$$\le \left[\left(\frac{I(a,b)}{\sqrt{I(a^2,b^2)}}\right)\cdot\exp\left(\frac{L(a,b)A(a,b)-G^2(a,b)}{L(a,b)A(a,b)}\right)\right]^{\Delta A(a,b)}. \quad (12.80)$$

12.6 A Reverse Inequality

We start with the following result [6] which gives another counterpart for $A(\phi\circ f)$, as did the Lupaş–Beesack–Pečarić result.

Theorem 12.8 (Dragomir, 2001, [6]) *Let $\phi:(\alpha,\beta)\subseteq\mathbb{R}\to\mathbb{R}$ be a differentiable convex function on (α,β), $f:E\to(\alpha,\beta)$ such that $\phi\circ f, f, \phi'\circ f, \phi'\circ f\cdot f\in L$. If $A:L\to\mathbb{R}$ is an isotonic linear and normalized functional, then*

$$0\le A(\phi\circ f)-\phi\big(A(f)\big)\le A\big(\phi'\circ f\cdot f\big)-A(f)\cdot A\big(\phi'\circ f\big)$$
$$\le \frac{1}{4}\big[\phi'(\beta)-\phi'(\alpha)\big](\beta-\alpha) \quad (\textit{if } \alpha,\beta \textit{ are finite}). \quad (12.81)$$

Proof As ϕ is differentiable convex on (α,β), we may write that

$$\phi(x)-\phi(y)\ge\phi'(y)(x-y), \quad \textit{for all } x,y\in(\alpha,\beta), \quad (12.82)$$

from where we obtain

$$\phi\big(A(f)\big)-(\phi\circ f)(t)\ge\big(\phi'\circ f\big)(t)\big(A(f)-f(t)\big) \quad (12.83)$$

for all $t\in E$, as, obviously, $A(f)\in(\alpha,\beta)$.

If we apply to (12.83) the functional A, we may write

$$\phi\big(A(f)\big)-A(\phi\circ f)\ge A(f)\cdot A\big(\phi'\circ f\big)-A\big(\phi'\circ f\cdot f\big),$$

which is clearly equivalent to the first inequality in (12.81).

It is well known that the following Grüss inequality for isotonic linear and normalized functionals holds (see [1])

$$\big|A(hk)-A(h)A(k)\big|\le\frac{1}{4}(M-m)(N-n), \quad (12.84)$$

provided that $h, k \in L$, $hk \in L$ and $-\infty < m \le h(t) \le M < \infty$, $-\infty < n \le k(t) \le N < \infty$, for all $t \in E$.

Taking into account that for finite α, β we have $\alpha < f(t) < \beta$ with ϕ' being monotonic on (α, β), we have $\phi'(\alpha) \le \phi' \circ f \le \phi'(\beta)$, and then, by the Grüss inequality, we may state that

$$A(\phi' \circ f \cdot f) - A(f) \cdot A(\phi' \circ f) \le \frac{1}{4}[\phi'(\beta) - \phi'(\alpha)](\beta - \alpha),$$

and the theorem is completely proved. □

The following corollary holds [6].

Corollary 12.5 *Let $\phi : [a, b] \subset \mathring{I} \subseteq \mathbb{R} \to \mathbb{R}$ be a differentiable convex function on $\mathring{I}$. If $\phi, e_1, \phi', \phi' \cdot e_1 \in L$ ($e_1(x) = x$, $x \in [a, b]$) and $A : L \to \mathbb{R}$ is an isotonic linear and normalized functional, then*

$$\begin{aligned} 0 &\le A(\phi) - \phi(A(e_1)) \le A(\phi' \cdot e_1) - A(e_1) \cdot A(\phi') \\ &\le \frac{1}{4}[\phi'(b) - \phi'(a)](b - a). \end{aligned} \tag{12.85}$$

There are some particular cases which can naturally be considered [6].

1. Let $\phi(x) = \ln x$, $x > 0$. If $\ln f, f, \frac{1}{f} \in L$ and $A : L \to \mathbb{R}$ is an isotonic linear and normalized functional, then

$$0 \le \ln[A(f)] - A[\ln(f)] \le A(f)A\left(\frac{1}{f}\right) - 1, \tag{12.86}$$

provided that $f(t) > 0$ for all $t \in E$ and $A(f) > 0$.

If $0 < m \le f(t) \le M < \infty$, $t \in E$, then, by the second part of (12.81), we have

$$A(f)A\left(\frac{1}{f}\right) - 1 \le \frac{(M - m)^2}{4mM} \quad \text{(which is a known result).} \tag{12.87}$$

Note that the inequality (12.86) is equivalent to

$$1 \le \frac{A(f)}{\exp[A[\ln(f)]]} \le \exp\left[A(f)A\left(\frac{1}{f}\right) - 1\right]. \tag{12.88}$$

2. Let $\phi(x) = \exp(x)$, $x \in \mathbb{R}$. If $\exp(f)$, f, $f \cdot \exp(f) \in L$ and $A : L \to \mathbb{R}$ is an isotonic linear and normalized functional, then

$$\begin{aligned} 0 &\le A[\exp(f)] - \exp[A(f)] \le A[f \exp(f)] - A(f)\exp[A(f)] \\ &\le \frac{1}{4}[\exp(M) - \exp(m)](M - m) \quad (\text{if } m \le f \le M \text{ on } E). \end{aligned} \tag{12.89}$$

12.7 A Further Result for $m-\Psi$-Convex and $M-\Psi$-Convex Functions

12.7.1 Other General Results

We now prove the following result [6].

Theorem 12.9 (Dragomir, 2001, [6]) *Let $\Psi : I \subseteq \mathbb{R} \to \mathbb{R}$ be a differentiable convex function and $f : E \to I$ such that $\Psi \circ f, \Psi' \circ f, \Psi' \circ f \cdot f, f \in L$, and let $A : L \to \mathbb{R}$ be an isotonic linear and normalized functional.*

(i) *If ϕ is differentiable, $\phi \in \mathscr{L}(\mathring{I}, m, \Psi)$ and $\phi \circ f, \phi' \circ f, \phi' \circ f \cdot f \in L$, then we have the inequality*

$$m\big[A\big(\Psi' \circ f \cdot f\big) + \Psi\big(A(f)\big) - A(f) \cdot A\big(\Psi' \circ f\big) - A(\Psi \circ f)\big]$$
$$\leq A\big(\phi' \circ f \cdot f\big) + \phi\big(A(f)\big) - A(f) \cdot A\big(\phi' \circ f\big) - A(\phi \circ f). \quad (12.90)$$

(ii) *If ϕ is differentiable, $\phi \in \mathscr{U}(\mathring{I}, M, \Psi)$ and $\phi \circ f, \phi' \circ f, \phi' \circ f \cdot f \in L$, then we have the inequality*

$$A\big(\phi' \circ f \cdot f\big) + \phi\big(A(f)\big) - A(f) \cdot A\big(\phi' \circ f\big) - A(\phi \circ f)$$
$$\leq M\big[A\big(\Psi' \circ f \cdot f\big) + \Psi\big(A(f)\big) - A(f) \cdot A\big(\Psi' \circ f\big) - A(\Psi \circ f)\big]. \quad (12.91)$$

(iii) *If ϕ is differentiable, $\phi \in \mathscr{B}(\mathring{I}, m, M, \Psi)$ and $\phi \circ f, \phi' \circ f, \phi' \circ f \cdot f \in L$, then both* (12.90) *and* (12.91) *hold.*

Proof The proof is as follows.

(i) As $\phi \in \mathscr{L}(I, m, \Psi)$, then $\phi - m\Psi$ is convex, and we can apply the first part of the inequality (12.81) to $\phi - m\Psi$, getting

$$A\big[(\phi - m\Psi) \circ f\big] - (\phi - m\Psi)\big(A(f)\big)$$
$$\leq A\big[(\phi - m\Psi)' \circ f \cdot f\big] - A(f)A\big((\phi - m\Psi)' \circ f\big). \quad (12.92)$$

However,

$$A\big[(\phi - m\Psi) \circ f\big] = A(\phi \circ f) - mA(\Psi \circ f),$$
$$(\phi - m\Psi)\big(A(f)\big) = \phi\big(A(f)\big) - m\Psi\big(A(f)\big),$$
$$A\big[(\phi - m\Psi)' \circ f \cdot f\big] = A\big(\phi' \circ f \cdot f\big) - mA\big(\Psi' \circ f \cdot f\big)$$

and

$$A\big((\phi - m\Psi)' \circ f\big) = A\big(\phi' \circ f\big) - mA\big(\Psi' \circ f\big),$$

and then, by (12.92), we deduce the desired inequality (12.90).

(ii) Goes likewise, and we omit the details.
(iii) Follows by (i) and (ii). □

The following corollary is useful in practice [6],

Corollary 12.6 *Let $\Psi : I \subseteq \mathbb{R} \to \mathbb{R}$ be a twice differentiable convex function on $\mathring{I}$, $f : E \to I$ such that $\Psi \circ f, \Psi' \circ f, \Psi' \circ f \cdot f, f \in L$, and let $A : L \to \mathbb{R}$ be an isotonic linear and normalized functional.*

(i) *If $\phi : I \to \mathbb{R}$ is twice differentiable, $\phi \circ f, \phi' \circ f, \phi' \circ f \cdot f \in L$, and $\phi''(t) \geq m\Psi''(t)$, $t \in \mathring{I}$, (where m is a given real number), then the inequality* (12.90) *holds.*
(ii) *With the same assumptions, but if $\phi''(t) \leq M\Psi''(t)$, $t \in \mathring{I}$, (where M is a given real number), then the inequality* (12.91) *holds.*
(iii) *If $m\Psi''(t) \leq \phi''(t) \leq M\Psi''(t)$, $t \in \mathring{I}$, then both* (12.90) *and* (12.91) *hold.*

Some particular important cases of the above corollary are embodied in the following proposition [6].

Proposition 12.11 *Assume that the mapping $\phi : I \subseteq \mathbb{R} \to \mathbb{R}$ is twice differentiable on $\mathring{I}$.*

(i) *If $\inf_{t\in\mathring{I}} \phi''(t) = k > -\infty$, then we have the inequality*

$$\frac{1}{2}k\big[A(f^2) - [A(f)]^2\big] \leq A(\phi' \circ f \cdot f) + \phi(A(f)) - A(f) \cdot A(\phi' \circ f) - A(\phi \circ f), \qquad (12.93)$$

provided that $\phi \circ f, \phi' \circ f, \phi' \circ f \cdot f, f^2 \in L$.
(ii) *If $\sup_{t\in\mathring{I}} \phi''(t) = K < \infty$, then we have the inequality*

$$A(\phi' \circ f \cdot f) + \phi(A(f)) - A(f) \cdot A(\phi' \circ f) - A(\phi \circ f) \leq \frac{1}{2}K\big[A(f^2) - [A(f)]^2\big]. \qquad (12.94)$$

(iii) *If $-\infty < k \leq \phi''(t) \leq K < \infty$, $t \in \mathring{I}$, then both* (12.93) *and* (12.94) *hold.*

The proof follows by Corollary 12.6 applied to $\Psi(t) = \frac{1}{2}t^2$ and $m = k$, $M = K$. Another result is the following one [6].

Proposition 12.12 *Assume that the mapping $\phi : I \subseteq (0,\infty) \to \mathbb{R}$ is twice differentiable on $\mathring{I}$. Let $p \in (-\infty, 0) \cup (1, \infty)$ and define $g_p : I \to \mathbb{R}$, $g_p(t) = \phi''(t)t^{2-p}$.*

(i) *If $\inf_{t\in\mathring{I}} g_p(t) = \gamma > -\infty$, then we have the inequality*

$$\frac{\gamma}{p(p-1)}\left[(p-1)\left[A(f^p)-[A(f)]^p\right]-pA(f)\left[A(f^{p-1})-[A(f)]^{p-1}\right]\right]$$
$$\leq A(\phi'\circ f\cdot f)+\phi(A(f))-A(f)\cdot A(\phi'\circ f)-A(\phi\circ f), \qquad (12.95)$$

provided that $\phi\circ f, \phi'\circ f, \phi'\circ f\cdot f, f^p, f^{p-1}\in L$.

(ii) *If* $\sup_{t\in\mathring{I}} g_p(t)=\Gamma<\infty$, *then we have the inequality*

$$A(\phi'\circ f\cdot f)+\phi(A(f))-A(f)\cdot A(\phi'\circ f)-A(\phi\circ f)$$
$$\leq\frac{\Gamma}{p(p-1)}\left[(p-1)\left[A(f^p)-[A(f)]^p\right]\right.$$
$$\left.-pA(f)\left[A(f^{p-1})-[A(f)]^{p-1}\right]\right]. \qquad (12.96)$$

(iii) *If* $-\infty<\gamma\leq g_p(t)\leq\Gamma<\infty$, $t\in\mathring{I}$, *then both* (12.95) *and* (12.96) *hold.*

Proof The proof is as follows.

(i) We have for the auxiliary mapping $h_p(t)=\phi(t)-\frac{\gamma}{p(p-1)}t^p$ that

$$h_p''(t)=\phi''(t)-\gamma t^{p-2}=t^{p-2}\left(t^{2-p}\phi''(t)-\gamma\right)$$
$$=t^{p-2}\left(g_p(t)-\gamma\right)\geq 0.$$

That is, h_p is convex or, equivalently, $\phi\in\mathscr{L}(I,\frac{\gamma}{p(p-1)},(\cdot)^p)$. Applying Corollary 12.6, we get

$$\frac{\gamma}{p(p-1)}\left[pA(f^p)+[A(f)]^p-pA(f)A(f^{p-1})-A(f^p)\right]$$
$$\leq A(\phi'\circ f\cdot f)+\phi(A(f))-A(f)\cdot A(\phi'\circ f)-A(\phi\circ f),$$

which is clearly equivalent to (12.95).

(ii) Goes similarly.

(iii) Follows by (i) and (ii). □

The following proposition also holds [6].

Proposition 12.13 *Assume that the mapping* $\phi: I\subseteq(0,\infty)\to\mathbb{R}$ *is twice differentiable on* $\mathring{I}$. *Define* $l(t)=t^2\phi''(t)$, $t\in I$.

(i) *If* $\inf_{t\in\mathring{I}} l(t)=s>-\infty$, *then we have the inequality*

$$s\left[A(f)A\left(\frac{1}{f}\right)-1-\left(\ln[A(f)]-A[\ln(f)]\right)\right]$$
$$\leq A(\phi'\circ f\cdot f)+\phi(A(f))-A(f)\cdot A(\phi'\circ f)-A(\phi\circ f), \qquad (12.97)$$

provided that $\phi\circ f, \phi^{-1}\circ f, \phi^{-1}\circ f\cdot f, \frac{1}{f}, \ln f\in L$ *and* $A(f)>0$.

(ii) *If* $\sup_{t\in\mathring{I}} l(t) = S < \infty$, *then we have the inequality*

$$A(\phi' \circ f \cdot f) + \phi(A(f)) - A(f) \cdot A(\phi' \circ f) - A(\phi \circ f)$$
$$\leq S\left[A(f)A\left(\frac{1}{f}\right) - 1 - \left(\ln[A(f)] - A[\ln(f)]\right)\right]. \tag{12.98}$$

(iii) *If* $-\infty < s \leq l(t) \leq S < \infty$ *for* $t \in \mathring{I}$, *then both* (12.97) *and* (12.98) *hold.*

Proof The proof is as follows.

(i) Define the auxiliary function $h(t) = \phi(t) + s\ln t$. Then

$$h''(t) = \phi''(t) - \frac{s}{t^2} = \frac{1}{t^2}\left(\phi''(t)t^2 - s\right) \geq 0,$$

which shows that h is convex, or, equivalently, $\phi \in \mathscr{L}(I, s, -\ln(\cdot))$. Applying Corollary 12.6, we may write

$$s\left[-A(\mathbf{1}) - \ln A(f) + A(f)A\left(\frac{1}{f}\right) + A(\ln(f))\right]$$
$$\leq A(\phi' \circ f \cdot f) + \phi(A(f)) - A(f) \cdot A(\phi' \circ f) - A(\phi \circ f),$$

which is clearly equivalent to (12.97).

(ii) Goes similarly.

(iii) Follows by (i) and (ii). □

Finally, the following result also holds [6].

Proposition 12.14 *Assume that the mapping* $\phi : I \subseteq (0,\infty) \to \mathbb{R}$ *is twice differentiable on* $\mathring{I}$. *Define* $\tilde{l}(t) = t\phi''(t)$, $t \in I$.

(i) *If* $\inf_{t\in\mathring{I}} \tilde{l}(t) = \delta > -\infty$, *then we have the inequality*

$$\delta A(f)\left[\ln[A(f)] - A(\ln(f))\right]$$
$$\leq A(\phi' \circ f \cdot f) + \phi(A(f)) - A(f) \cdot A(\phi' \circ f) - A(\phi \circ f), \tag{12.99}$$

provided that $\phi \circ f, \phi\prime \circ f, \phi\prime \circ f \cdot f, \ln f, f \in L$ *and* $A(f) > 0$.

(ii) *If* $\sup_{t\in\mathring{I}} \tilde{l}(t) = \Delta < \infty$, *then we have the inequality*

$$A(\phi' \circ f \cdot f) + \phi(A(f)) - A(f) \cdot A(\phi' \circ f) - A(\phi \circ f)$$
$$\leq \Delta A(f)\left[\ln[A(f)] - A(\ln(f))\right]. \tag{12.100}$$

(iii) *If* $-\infty < \delta \leq \tilde{l}(t) \leq \Delta < \infty$ *for* $t \in \mathring{I}$, *then both* (12.99) *and* (12.100) *hold.*

Proof The proof is as follows.

(i) Define the auxiliary mapping $h(t) = \phi(t) - \delta t \ln t$, $t \in I$. Then

$$h''(t) = \phi''(t) - \frac{\delta}{t} = \frac{1}{t^2}\big[\phi''(t)t - \delta\big] = \frac{1}{t}\big[\tilde{I}(t) - \delta\big] \geq 0,$$

which shows that h is convex or equivalently, $\phi \in \mathscr{L}(I, \delta, (\cdot)\ln(\cdot))$. Applying Corollary 12.6, we get

$$\begin{aligned}&\delta\big[A\big[(\ln f + 1)f\big] + A(f)\ln A(f) - A(f)A(\ln f + 1) - A(f\ln f)\big]\\&\quad \leq A\big(\phi' \circ f \cdot f\big) + \phi\big(A(f)\big) - A(f)\cdot A\big(\phi' \circ f\big) - A(\phi \circ f),\end{aligned}$$

which is equivalent with (12.99).

(ii) Goes similarly.

(iii) Follows by (i) and (ii). □

12.7.2 Some Applications of Bullen's Inequality

The following inequality is well known in the literature as Bullen's inequality (see, for example, [11, p. 10])

$$\frac{1}{b-a}\int_a^b \phi(t)\,dt \leq \frac{1}{2}\left[\frac{\phi(a)+\phi(b)}{2} + \phi\left(\frac{a+b}{2}\right)\right], \tag{12.101}$$

provided that $\phi : [a,b] \to \mathbb{R}$ is a convex function on $[a,b]$. In other words, as (12.138) is equivalent to

$$0 \leq \frac{1}{b-a}\int_a^b \phi(t)\,dt - \phi\left(\frac{a+b}{2}\right) \leq \frac{\phi(a)+\phi(b)}{2} - \frac{1}{b-a}\int_a^b \phi(t)\,dt, \tag{12.102}$$

we can conclude that in the Hermite–Hadamard inequality

$$\frac{\phi(a)+\phi(b)}{2} \geq \frac{1}{b-a}\int_a^b \phi(t)\,dt \geq \phi\left(\frac{a+b}{2}\right) \tag{12.103}$$

the integral mean $\frac{1}{b-a}\int_a^b \phi(t)\,dt$ is closer to $\phi(\frac{a+b}{2})$ than to $\frac{\phi(a)+\phi(b)}{2}$.

Using some of the results pointed out in the previous sections, we may upper and lower bound the *Bullen difference:*

$$B(\phi; a, b) := \frac{1}{2}\left[\frac{\phi(a)+\phi(b)}{2} + \phi\left(\frac{a+b}{2}\right)\right] - \frac{1}{b-a}\int_a^b \phi(t)\,dt$$

(which is positive for convex functions) for different classes of twice differentiable functions ϕ.

Now, if we assume that $A(f) := \frac{1}{b-a}\int_a^b f(t)\,dt$, then for $f = e$, $e(x) = x$, $x \in [a,b]$, we have, for a differentiable function ϕ, that

$$\begin{aligned}
&A(\phi' \circ f \cdot f) + \phi(A(f)) - A(f)\cdot A(\phi' \circ f) - A(\phi \circ f)\\
&= \frac{1}{b-a}\int_a^b x\phi'(x)\,dx + \phi\left(\frac{a+b}{2}\right)\\
&\quad - \frac{a+b}{2}\cdot\frac{1}{b-a}\int_a^b \phi'(x)\,dx - \frac{1}{b-a}\int_a^b \phi(x)\,dx\\
&= \frac{1}{b-a}\left[b\phi(b) - a\phi(a) - \int_a^b \phi(x)\,dx\right] + \phi\left(\frac{a+b}{2}\right)\\
&\quad - \frac{a+b}{2}\cdot\frac{\phi(b)-\phi(a)}{b-a} - \frac{1}{b-a}\int_a^b \phi(x)\,dx\\
&= \frac{\phi(a)+\phi(b)}{2} + \phi\left(\frac{a+b}{2}\right) - \frac{2}{b-a}\int_a^b \phi(x)\,dx\\
&= 2B(\phi; a, b).
\end{aligned}$$

The following integral inequalities were obtained in [6].
(a) Assume that $\phi : [a,b] \subset \mathbb{R} \to \mathbb{R}$ is a twice differentiable function satisfying the property that $-\infty < k \le \phi''(t) \le K < \infty$. Then by Proposition 12.11, we may state the inequality

$$\frac{1}{48}(b-a)^2 k \le B(\phi; a, b) \le \frac{1}{48}(b-a)^2 K. \tag{12.104}$$

This follows by Proposition 12.11 taking into account that

$$\frac{1}{b-a}\int_a^b x^2\,dx - \left(\frac{1}{b-a}\int_a^b x\,dx\right)^2 = \frac{(b-a)^2}{12}.$$

(b) Now assume that a twice differentiable function $\phi : [a,b] \subset (0,\infty) \to \mathbb{R}$ satisfies the property that $-\infty < \gamma \le t^{2-p}\phi''(t) \le \Gamma < \infty$, $t \in (a,b)$, $p \in (-\infty, 0) \cup (1,\infty)$. Then by Proposition 12.12 and taking into account that

$$\begin{aligned}
A(f^p) - (A(f))^p &= \frac{1}{b-a}\int_a^b x^p\,dx - \left(\frac{1}{b-a}\int_a^b x\,dx\right)^p\\
&= L_p^p(a,b) - A^p(a,b),
\end{aligned}$$

and

$$A(f^{p-1}) - (A(f))^{p-1} = L_{p-1}^{p-1}(a,b) - A^{p-1}(a,b),$$

we may state the inequality

$$\frac{\gamma}{p(p-1)}\big[(p-1)\big[L_p^p(a,b)-A^p(a,b)\big]-pA(a,b)\big[L_{p-1}^{p-1}(a,b)-A^{p-1}(a,b)\big]\big]$$
$$\le B(\phi;a,b)$$
$$\le \frac{\Gamma}{p(p-1)}\big[(p-1)\big[L_p^p(a,b)-A^p(a,b)\big]$$
$$-pA(a,b)\big[L_{p-1}^{p-1}(a,b)-A^{p-1}(a,b)\big]\big]. \qquad (12.105)$$

(c) Assume that a twice differentiable function $\phi:[a,b]\subset(0,\infty)\to\mathbb{R}$ satisfies the property that $-\infty<s\le t^2\phi''(t)\le S<\infty$, $t\in(a,b)$, then by Proposition 12.13, and taking into account that

$$A(f)A\big(f^{-1}\big)-1-\ln\big[A(f)\big]+A\ln(f)$$
$$=\frac{A(a,b)}{L(a,b)}-1-\ln A(a,b)+I(a,b)$$
$$=\ln\left[\frac{I(a,b)}{A(a,b)}\cdot\exp\left(\frac{A(a,b)-L(a,b)}{L(a,b)}\right)\right],$$

we get the inequality

$$\frac{s}{2}\ln\left[\frac{I(a,b)}{A(a,b)}\cdot\exp\left(\frac{A(a,b)-L(a,b)}{L(a,b)}\right)\right]$$
$$\le B(\phi;a,b)$$
$$\le\frac{S}{2}\ln\left[\frac{I(a,b)}{A(a,b)}\cdot\exp\left(\frac{A(a,b)-L(a,b)}{L(a,b)}\right)\right]. \qquad (12.106)$$

(d) Finally, if ϕ satisfies the condition $-\infty<\delta\le t\phi''(t)\le\Delta<\infty$, then by Proposition 12.14, we may state the inequality

$$\delta A(a,b)\ln\left[\frac{A(a,b)}{I(a,b)}\right]\le B(\phi;a,b)\le\Delta A(a,b)\ln\left[\frac{A(a,b)}{I(a,b)}\right]. \qquad (12.107)$$

12.8 A Grüss Type Inequality

12.8.1 A Refinement of Grüss Inequality

In 1988, D. Andrica and C. Badea [1] proved the following generalization of the Grüss inequality for isotonic linear functionals.

Theorem 12.10 (Andrica & Badea, 1988, [1]) *If $f,g\in L$ are such that $fg\in L$ and $m\le f\le M$, $n\le g\le N$ where m,M,n,N are given real numbers, then for any* *normalized isotonic linear functional* $A : L \to \mathbb{R}$ *one has the inequality*

$$\left|A(fg) - A(f)A(g)\right| \le \frac{1}{4}(M - m)(N - n). \tag{12.108}$$

The constant $\frac{1}{4}$ *in* (12.108) *is the best possible in the sense that it cannot be replaced by a smaller constant.*

In this paper, we point out a refinement of the Grüss inequality (12.108) for isotonic linear functionals. Applications of the Cauchy–Bunyakowski–Schwartz and Jessen's inequalities are also provided.

The following result due to author holds.

Theorem 12.11 (Dragomir, 2002, [9]) *Let* $f, g \in L$ *be such that* $fg \in L$ *and assume that there exist real numbers* n *and* N *such that*

$$n \le g \le N. \tag{12.109}$$

Then for any normalized isotonic linear functional $A : L \to \mathbb{R}$ *for which* $|f - A(f) \cdot \mathbf{1}| \in L$ *one has the inequality*

$$\left|A(fg) - A(f)A(g)\right| \le \frac{1}{2}(N - n)A\left(\left|f - A(f) \cdot \mathbf{1}\right|\right). \tag{12.110}$$

The constant $\frac{1}{2}$ *in* (12.110) *is the best possible in the sense that it cannot be replaced by a smaller constant.*

Proof Using the linearity property of A, we have

$$\begin{aligned}
A&\left[\left(f - A(f) \cdot \mathbf{1}\right)\left(g - \frac{n+N}{2} \cdot \mathbf{1}\right)\right] \\
&= A\left[\left(f - A(f) \cdot \mathbf{1}\right)g\right] - \frac{n+N}{2}A\left[f - A(f) \cdot \mathbf{1}\right] \\
&= A(fg) - A(f)A(g) - \frac{n+N}{2}\left[A(f) - A(f) \cdot A(\mathbf{1})\right] \\
&= A(fg) - A(f)A(g)
\end{aligned} \tag{12.111}$$

since, by the normality property of A, $A(\mathbf{1}) = 1$.

From (12.109) we may easily deduce that

$$\left|g - \frac{n+N}{2} \cdot \mathbf{1}\right| \le \frac{M-n}{2} \cdot \mathbf{1}. \tag{12.112}$$

It is known that if $h \in L$ so that $|h| \in L$, then, by the monotonicity and linearity of A, one has

$$\left|A(h)\right| \le A\left(|h|\right). \tag{12.113}$$

Using this property, the monotonicity of A, and condition (12.112), we deduce

$$\begin{aligned}
&\left|A\left[(f-A(f)\cdot\mathbf{1})\left(g-\frac{n+N}{2}\cdot\mathbf{1}\right)\right]\right| \\
&\quad\le A\left(\left|(f-A(f)\cdot\mathbf{1})\left(g-\frac{n+N}{2}\cdot\mathbf{1}\right)\right|\right) \\
&\quad\le \frac{N-n}{2}A\left(\left|f-A(f)\cdot\mathbf{1}\right|\right). \qquad (12.114)
\end{aligned}$$

Utilizing (12.111) and (12.114), we deduce the desired result (12.110).

To prove the sharpness of the constant $\frac{1}{2}$, we assume that (12.110) holds with a constant $c>0$ for $A=\frac{1}{b-a}\int_a^b$, $L=L[a,b]$ (the Lebesgue space of integrable functions on $[a,b]$) and g satisfying the condition (12.109) on the interval $[a,b]$, i.e., one has the inequality

$$\begin{aligned}
&\left|\frac{1}{b-a}\int_a^b f(x)g(x)\,dx-\frac{1}{b-a}\int_a^b f(x)\,dx\cdot\frac{1}{b-a}\int_a^b g(x)\,dx\right| \\
&\quad\le c(N-n)\cdot\frac{1}{b-a}\int_a^b\left|f(x)-\frac{1}{b-a}\int_a^b f(y)\,dy\right|dx. \qquad (12.115)
\end{aligned}$$

If we choose $g=f$ and $f:[a,b]\to\mathbb{R}$,

$$f(x)=\begin{cases}-1 & \text{if } x\in[a,\frac{a+b}{2}],\\ 1 & \text{if } x\in(\frac{a+b}{2},b]\end{cases}$$

then

$$\begin{aligned}
\frac{1}{b-a}\int_a^b f^2(x)\,dx-\left(\frac{1}{b-a}\int_a^b f(x)\,dx\right)^2&=1,\\
\frac{1}{b-a}\int_a^b\left|f(x)-\frac{1}{b-a}\int_a^b f(y)\,dy\right|dx&=1,\\
m=-1,\quad M&=1,
\end{aligned}$$

and by (12.115) we deduce $c\ge\frac{1}{2}$. □

The following corollaries are natural consequences of the above result.

Corollary 12.7 *Let $f\in L$ be such that $f^2\in L$ and suppose there exist real numbers m,M such that*

$$m\le f\le M. \qquad (12.116)$$

Then for any $A : L \to \mathbb{R}$, *a normalized isotonic linear functional such that* $|f - A(f) \cdot \mathbf{1}| \in L$, *one has the inequality*

$$0 \leq A(f^2) - [A(f)]^2 \leq \frac{1}{2}(M - m)A(|f - A(f) \cdot \mathbf{1}|). \quad (12.117)$$

The constant $\frac{1}{2}$ *is sharp.*

Corollary 12.8 *Let* $f, g \in L$ *be such that* $fg \in L$ *and* f *satisfy* (12.116) *while* g *satisfies* (12.109). *Then for any normalized isotonic linear functional* $A : L \to \mathbb{R}$ *such that* $|f - A(f) \cdot \mathbf{1}|$, $|g - A(g) \cdot \mathbf{1}| \in L$ *one has the inequality*

$$\begin{aligned} &|A(fg) - A(f)A(g)| \\ &\leq \frac{1}{2}[(M - m)(N - n)]^{\frac{1}{2}}[A(|f - A(f) \cdot \mathbf{1}|)A(|g - A(g) \cdot \mathbf{1}|)]^{\frac{1}{2}}. \end{aligned} \quad (12.118)$$

The constant $\frac{1}{2}$ *is sharp.*

Remark 12.6 Using Hölder's inequality for isotonic linear functionals, we may state the following inequalities as well

$$\begin{aligned} &|A(fg) - A(f)A(g)| \\ &\leq \frac{1}{2}(N - n)A(|f - A(f) \cdot \mathbf{1}|) \text{ if } |f - A(f) \cdot \mathbf{1}| \in L, \\ &\leq \frac{1}{2}(N - n)[A(|f - A(f) \cdot \mathbf{1}|^p)]^{\frac{1}{p}} \text{ if } |f - A(f) \cdot \mathbf{1}|^p \in L,\ p > 1 \\ &\leq \frac{1}{2}(N - n) \sup_{t \in E} |f(t) - A(f)|, \end{aligned} \quad (12.119)$$

provided $f, g \in L$ and $fg \in L$ while g satisfies condition (12.109).

If f and g fulfill the conditions (12.116) and (12.109), then we have the following *refinement of the Grüss inequality* (12.108)

$$\begin{aligned} |A(fg) - A(f)A(g)| &\leq \frac{1}{2}(N - n)A(|f - A(f) \cdot \mathbf{1}|) \\ &\leq \frac{1}{2}(N - n)[A(f^2) - [A(f)]^2]^{\frac{1}{2}} \\ &\leq \frac{1}{4}(M - m)(N - n). \end{aligned} \quad (12.120)$$

The constants $\frac{1}{2}$, $\frac{1}{2}$, and $\frac{1}{4}$ are sharp in (12.120).

The following weighted version of Theorem 12.11 also holds.

Theorem 12.12 (Dragomir, 2002, [9]) *Let $f, g, h \in L$ be such that $h \geq 0$, fh, gh, $fgh \in L$ and there exist real constants n, N such that* (12.109) *holds. Then for any $B : L \to \mathbb{R}$, an isotonic linear functional such that $B(h) > 0$, $h|f - \frac{1}{B(h)} \cdot \mathbf{1}| \in L$, one has the inequality*

$$\left|\frac{B(fgh)}{B(h)} - \frac{B(fh)}{B(h)} \cdot \frac{B(gh)}{B(h)}\right| \leq \frac{1}{2}(N-n)\frac{1}{B(h)}B\left[h\left|f - \frac{1}{B(h)}B(hf)\cdot\mathbf{1}\right|\right]. \tag{12.121}$$

The constant $\frac{1}{2}$ is the best possible.

Proof Apply Theorem 12.10 to the functional $A_h : L \to \mathbb{R}$,

$$A_h(f) := \frac{1}{B(h)}B(hf),$$

which is a normalized isotonic linear functional on L. □

Similar corollaries may be stated from the weighted inequality (12.121), but we omit the details.

12.8.2 Applications to Integral and Discrete Inequalities

Let $(\Omega, \mathscr{A}, \mu)$ be a measurable space consisting of a set Ω, a σ-algebra of subsets of Ω and a countably additive and positive measure μ on $\mathscr{A}$ with values in $\mathbb{R} \cup \{\infty\}$.

For a μ-measurable function $w : \Omega \to \mathbb{R}$ with $w(x) \geq 0$ for μ-a.e. $x \in \Omega$, assume $\int_\Omega w(x)\,d\mu(x) > 0$. Consider the Lebesgue space $L_w(\Omega, \mu) := \{f : \Omega \to \mathbb{R}, f$ is measurable and $\int_\Omega w(x)|f(x)|\,d\mu(x) < \infty\}$.

If $f, g : \Omega \to \mathbb{R}$ are μ-measurable functions and $f, g, fg \in L_w(\Omega, \mu)$, then we may consider the Čebyšev functional

$$\begin{aligned} T_w(f, g) := &\frac{1}{\int_\Omega w(x)\,d\mu(x)} \int_\Omega w(x)f(x)g(x)\,d\mu(x) \\ &- \frac{1}{\int_\Omega w(x)\,d\mu(x)} \int_\Omega w(x)f(x)\,d\mu(x) \\ &\times \frac{1}{\int_\Omega w(x)\,d\mu(x)} \int_\Omega w(x)g(x)\,d\mu(x). \end{aligned} \tag{12.122}$$

We may also consider the functional

$$D_w(f) := \frac{1}{\int_\Omega w(x)\,d\mu(x)} \times \int_\Omega w(x)\left| f(x) - \frac{1}{\int_\Omega w(y)\,d\mu(y)} \int_\Omega w(y) f(y)\,d\mu(y)\right| d\mu(x).$$

Applying Theorem 12.11 to the normalized isotonic linear functional

$$A(f) := \frac{1}{\int_\Omega w(x)\,d\mu(x)} \int_\Omega w(x) f(x)\,d\mu(x),$$

$A : L_w(\Omega,\mu) \to \mathbb{R}$, we may recapture the following result due to Cerone and Dragomir [3]. Note that the proof of this result in [3] is different from the one in Theorem 12.11.

Theorem 12.13 (Cerone & Dragomir, 2002, [3]) *Let $w, f, g : \Omega \to \mathbb{R}$ be μ-measurable functions with $w \geq 0$ μ-a.e. on Ω and $\int_\Omega w(x)\,d\mu(x) > 0$. If $f, g, fg \in L_w(\Omega,\mu)$ and there exist constants n, N such that*

$$-\infty < n \leq g(x) \leq N < \infty \quad \textit{for } \mu\textit{-a.e. } x \in \Omega, \tag{12.123}$$

then we have the inequality

$$\left|T_w(f,g)\right| \leq \frac{1}{2}(N-n) D_w(f). \tag{12.124}$$

The constant $\frac{1}{2}$ is sharp in the sense that it cannot be replaced by a smaller constant.

Remark 12.7 If $\Omega = [a,b]$ and $w(x) = 1$ in Theorem 12.13, then we recapture the result obtained in [4]

$$\left|\frac{1}{b-a}\int_a^b f(x)g(x)\,dx - \frac{1}{b-a}\int_a^b f(x)\,dx \cdot \frac{1}{b-a}\int_a^b g(x)\,dx\right| \leq \frac{1}{2}(N-n)\cdot\frac{1}{b-a}\int_a^b \left| f(x) - \frac{1}{b-a}\int_a^b f(y)\,dy\right| dx \tag{12.125}$$

provided $n \leq g(x) \leq N$ for a.e. $x \in [a,b]$.

Note that the proof in Theorem 12.11 is different from the one in [4], using only the linearity and monotonicity properties of the functional A. We should also remark that in [4] the authors did not show the sharpness of the constant $\frac{1}{2}$.

Now, if we consider the normalized isotonic linear functional

$$A_{\bar{w}}(\bar{x}) := \frac{1}{W_n}\sum_{i=1}^{n} w_i x_i, \tag{12.126}$$

$A_{\bar{w}} : \mathbb{R}^n \to \mathbb{R}$, where $w_i \geq 0$ $(i = \overline{1,n})$ and $W_n := \sum_{i=1}^n w_i > 0$, then, by Theorem 12.11, we may obtain the following discrete inequality obtained by Cerone and Dragomir in [3].

Theorem 12.14 (Cerone & Dragomir, 2002, [3]) *Let* $\bar{a} = (a_1, \ldots, a_n)$, $\bar{b} = (b_1, \ldots, b_n) \in \mathbb{R}$ *be such that there exist constants* $b, B \in \mathbb{R}$ *such that*

$$b \leq b_i \leq B \quad \text{for each } i \in \{1, \ldots, n\}. \tag{12.127}$$

Then one has the inequality

$$\begin{aligned} &\left| \frac{1}{W_n} \sum_{i=1}^n w_i a_i b_i - \frac{1}{W_n} \sum_{i=1}^n w_i a_i \cdot \frac{1}{W_n} \sum_{i=1}^n w_i b_i \right| \\ &\quad \leq \frac{1}{2}(B - b) \frac{1}{W_n} \sum_{i=1}^n w_i \left| a_i - \frac{1}{W_n} \sum_{j=1}^n w_j a_j \right|. \end{aligned} \tag{12.128}$$

The constant $\frac{1}{2}$ *is sharp in* (12.128).

12.8.3 A Counterpart of the (CBS)-Inequality

The following inequality is known in the literature as the Cauchy–Bunyakowski–Schwartz inequality for isotonic linear functionals, or the (CBS)-inequality for short,

$$\left[A(fg)\right]^2 \leq A\left(f^2\right)A\left(g^2\right), \tag{12.129}$$

provided $f, g : E \to \mathbb{R}$ have the property that $fg, f^2, g^2 \in L$ and $A : L \to \mathbb{R}$ is an isotonic linear functional.

Making use of the Grüss inequality (12.121), we may prove the following counterpart of the (CBS)-inequality for isotonic linear functionals.

Theorem 12.15 (Dragomir, 2002, [9]) *Let* $k, l : E \to \mathbb{R}$ *be such that* $k^2, l^2, kl \in L$ *and there exist real constants* $\gamma, \Gamma \in \mathbb{R}$ *such that*

$$\gamma \leq \frac{k}{l} \leq \Gamma. \tag{12.130}$$

Then for any isotonic linear functional $A : L \to \mathbb{R}$ *such that* $|l||A(l^2)k - A(kl)l| \in L$, *one has the inequality*

$$\begin{aligned} 0 &\leq A\left(k^2\right)A\left(l^2\right) - \left[A(kl)\right]^2 \\ &\leq \frac{1}{2}(\Gamma - \gamma)A\left[|l|\left|A\left(l^2\right)k - A(kl)l\right|\right]. \end{aligned} \tag{12.131}$$

The constant $\frac{1}{2}$ *is sharp.*

Proof We choose $f = g = \frac{k}{l}$, $h = l^2$ and $B = A$ in (12.121) to get

$$\begin{aligned} 0 &\le \frac{A(k^2)}{A(l^2)} - \frac{[A(kl)]^2}{[A(l^2)]^2} \\ &\le \frac{1}{2}(\Gamma - \gamma)\frac{1}{A(l^2)}A\left[l^2\left|\frac{k}{l} - \frac{1}{A(l^2)}A(kl)\right|\right], \end{aligned}$$

provided $A(l^2) \neq 0$, which is equivalent to

$$\begin{aligned} 0 &\le A(k^2)A(l^2) - [A(kl)]^2 \\ &\le \frac{1}{2}(\Gamma - \gamma)A(l^2)A\left[\left|kl - \frac{l^2}{A(l^2)}A(kl)\right|\right], \end{aligned}$$

which is clearly equivalent to (12.131). □

The following integral inequality holds.

Corollary 12.9 *Let $w, f, g : \Omega \to \mathbb{R}$ be a μ-measurable function with $w \ge 0$ μ-a.e. on Ω. If $f, g \in L_w^2(\Omega, \mu) := \{f : \Omega \to \mathbb{R}, \int_\Omega w(y) f^2(y)\, d\mu(y) < \infty\}$ and there exist γ, Γ such that*

$$-\infty < \gamma \le \frac{f}{g} \le \Gamma < \infty \quad \text{for } \mu\text{-a.e. } x \in \Omega, \tag{12.132}$$

then one has the inequality

$$\begin{aligned} 0 &\le \int_\Omega w(x) f^2(x)\, d\mu(x) \int_\Omega w(x) g^2(x)\, d\mu(x) \\ &\quad - \left[\int_\Omega w(x) f(x) g(x)\, d\mu(x)\right]^2 \\ &\le \frac{1}{2}(\Gamma - \gamma)\int_\Omega w(x)|g(x)|\left|\left(\int_\Omega w(y) g^2(y)\, d\mu(y)\right) f(x)\right. \\ &\quad \left. - g(x)\int_\Omega w(y) f(y) g(y)\, d\mu(y)\right| d\mu(x) \\ &= \frac{1}{2}(\Gamma - \gamma)\int_\Omega w(x)|g(x)|\left|\int_\Omega w(y) g(y)\begin{vmatrix} f(x) & g(x) \\ f(y) & g(y)\end{vmatrix} d\mu(y)\right| d\mu(x). \end{aligned} \tag{12.133}$$

The constant $\frac{1}{2}$ is sharp.

Remark 12.8 In particular, if $f, g \in L^2(\Omega, \mu)$ and the condition (12.132) holds, then

$$\begin{aligned} 0 &\le \int_\Omega f^2(x)\,d\mu(x) \int_\Omega g^2(x)\,d\mu(x) - \left[\int_\Omega f(x)g(x)\,d\mu(x)\right]^2 \\ &\le \frac{1}{2}(\Gamma - \gamma)\int_\Omega |g(x)| \left| \int_\Omega g(y) \begin{vmatrix} f(x) & g(x) \\ f(y) & g(y) \end{vmatrix} d\mu(y) \right| d\mu(x). \end{aligned} \tag{12.134}$$

The constant $\frac{1}{2}$ is sharp.

The following discrete inequality also holds.

Corollary 12.10 *Let $\bar{a} = (a_1, \dots, a_n)$, $\bar{b} = (b_1, \dots, b_n)$, and $\bar{w} = (w_1, \dots, w_n)$ be the sequences of real numbers such that $w_i \ge 0$, $(i = 1, \dots, n)$, $W_n := \sum_{i=1}^n w_i > 0$ and*

$$\gamma \le \frac{a_i}{b_i} \le \Gamma \quad \text{for each } i \in \{1, \dots, n\}. \tag{12.135}$$

Then one has the inequality

$$\begin{aligned} 0 &\le \sum_{i=1}^n w_i a_i^2 \sum_{i=1}^n w_i b_i^2 - \left(\sum_{i=1}^n w_i a_i b_i\right)^2 \\ &\le \frac{1}{2}(\Gamma - \gamma)\sum_{i=1}^n w_i b_i \left| \sum_{j=1}^n w_j b_j \begin{vmatrix} a_i & b_i \\ a_j & b_j \end{vmatrix} \right|. \end{aligned} \tag{12.136}$$

The constant $\frac{1}{2}$ is sharp.

Remark 12.9 If $\bar{a}, \bar{b}$ satisfy (12.135), then one has the inequality

$$\begin{aligned} 0 &\le \sum_{i=1}^n a_i^2 \sum_{i=1}^n b_i^2 - \left(\sum_{i=1}^n a_i b_i\right)^2 \\ &\le \frac{1}{2}(\Gamma - \gamma)\sum_{i=1}^n b_i \left| \sum_{j=1}^n b_j \begin{vmatrix} a_i & b_i \\ a_j & b_j \end{vmatrix} \right|. \end{aligned} \tag{12.137}$$

The constant $\frac{1}{2}$ is sharp.

12.8.4 A Converse for Jessen's Inequality

In [6], the author has proved the following converse of Jessen's inequality for normalized isotonic linear functionals.

Theorem 12.16 *Let $\Phi : (\alpha, \beta) \subseteq \mathbb{R} \to \mathbb{R}$ be a differentiable convex function on (α, β), let $f : E \to (\alpha, \beta)$ be such that $\Phi \circ f$, f, $\Phi' \circ f$, $(\Phi' \circ f) \cdot f \in L$. If $A : L \to \mathbb{R}$ is an isotonic linear and normalized functional, then*

$$\begin{aligned} 0 &\le A(\Phi \circ f) - \Phi\big(A(f)\big) \\ &\le A\big[(\Phi' \circ f) \cdot f\big] - A(f)A\big(\Phi' \circ f\big) \\ &\le \frac{1}{4}\big[\Phi'(\beta) - \Phi'(\alpha)\big](\beta - \alpha) \quad (\textit{if } \alpha, \beta \textit{ are finite}). \end{aligned} \tag{12.138}$$

We can state the following result improving the inequality (12.138).

Theorem 12.17 (Dragomir, 2001, [6]) *Let $\Phi : [\alpha, \beta] \to \mathbb{R}$ with $-\infty < \alpha < \beta < \infty$, and f, A are as in Theorem* 12.16, *then one has the inequality*

$$\begin{aligned} 0 &\le A(\Phi \circ f) - \Phi\big(A(f)\big) \\ &\le A\big[(\Phi' \circ f) \cdot f\big] - A(f)A\big(\Phi' \circ f\big) \\ &\le \frac{1}{2}\big[\Phi'(\beta) - \Phi'(\alpha)\big]A\big(\big|f - A(f) \cdot \mathbf{1}\big|\big), \end{aligned} \tag{12.139}$$

provided $|f - A(f) \cdot \mathbf{1}| \in L$.

Proof Taking into account that $\alpha \le f \le \beta$ and Φ' is monotonic on $[\alpha, \beta]$, we have $\Phi'(\alpha) \le \Phi' \circ f \le \Phi'(\beta)$. Applying Theorem 12.11, we deduce

$$\begin{aligned} &A\big[(\Phi' \circ f) \cdot f\big] - A(f)A\big(\Phi' \circ f\big) \\ &\quad \le \frac{1}{2}\big[\Phi'(\beta) - \Phi'(\alpha)\big]A\big(\big|f - A(f) \cdot \mathbf{1}\big|\big), \end{aligned}$$

and the theorem is proved. □

The following corollary addressing the integral case also holds.

Corollary 12.11 *Let $\Phi : [\alpha, \beta] \subset \mathbb{R} \to \mathbb{R}$ be a differentiable convex function on (α, β) and let $f : \Omega \to [\alpha, \beta]$ be such that $\Phi \circ f$, f, $\Phi' \circ f$, $(\Phi' \circ f) \cdot f \in L_w(\Omega, \mu)$, where $w \ge 0$ μ-a.e. on Ω with $\int_\Omega w(x)\,d\mu(x) > 0$. Then we have the inequality*

$$\begin{aligned} 0 &\le \frac{1}{\int_\Omega w(x)\,d\mu(x)} \int_\Omega w(x)\Phi\big(f(x)\big)\,d\mu(x) \\ &\quad - \Phi\left(\frac{1}{\int_\Omega w(x)\,d\mu(x)} \int_\Omega w(x)f(x)\,d\mu(x)\right) \\ &\le \frac{1}{\int_\Omega w(x)\,d\mu(x)} \int_\Omega w(x)\Phi'\big(f(x)\big)f(x)\,d\mu(x) \end{aligned}$$

$$
\begin{aligned}
&-\frac{1}{\int_\Omega w(x)\,d\mu(x)}\int_\Omega w(x)\Phi'\big(f(x)\big)\,d\mu(x)\\
&\times\frac{1}{\int_\Omega w(x)\,d\mu(x)}\int_\Omega w(x)f(x)\,d\mu(x)\\
&\le\frac{1}{2}\big[\Phi'(\beta)-\Phi'(\alpha)\big]\frac{1}{\int_\Omega w(x)\,d\mu(x)}\\
&\times\int_\Omega w(x)\left|f(x)-\frac{1}{\int_\Omega w(y)\,d\mu(y)}\int_\Omega w(y)f(y)\,d\mu(y)\right|d\mu(x). \qquad (12.140)
\end{aligned}
$$

Remark 12.10 If $\mu(\Omega)<\infty$ and $\Phi\circ f,\ f,\ \Phi'\circ f,\ (\Phi'\circ f)\cdot f\in L(\Omega,\mu)$, then we have the inequality

$$
\begin{aligned}
0&\le\frac{1}{\mu(\Omega)}\int_\Omega \Phi\big(f(x)\big)\,d\mu(x)-\Phi\left(\frac{1}{\mu(\Omega)}\int_\Omega f(x)\,d\mu(x)\right)\\
&\le\frac{1}{\mu(\Omega)}\int_\Omega \Phi'\big(f(x)\big)f(x)\,d\mu(x)\\
&\quad-\frac{1}{\mu(\Omega)}\int_\Omega \Phi'\big(f(x)\big)\,d\mu(x)\cdot\frac{1}{\mu(\Omega)}\int_\Omega f(x)\,d\mu(x)\\
&\le\frac{1}{2}\big[\Phi'(\beta)-\Phi'(\alpha)\big]\frac{1}{\mu(\Omega)}\int_\Omega\left|f(x)-\frac{1}{\mu(\Omega)}\int_\Omega f(y)\,d\mu(y)\right|d\mu(x).
\end{aligned}
\qquad (12.141)
$$

The case of functions of a real variable is embodied in the following inequality that provides a counterpart for the Jensen's integral inequality

$$
\begin{aligned}
0&\le\frac{1}{b-a}\int_a^b \Phi\big(f(x)\big)\,dx-\Phi\left(\frac{1}{b-a}\int_a^b f(x)\,dx\right)\\
&\le\frac{1}{b-a}\int_a^b \Phi'\big(f(x)\big)f(x)\,dx\\
&\quad-\frac{1}{b-a}\int_a^b \Phi'\big(f(x)\big)\,dx\cdot\frac{1}{b-a}\int_a^b f(x)\,dx\\
&\le\frac{1}{2}\big[\Phi'(\beta)-\Phi'(\alpha)\big]\frac{1}{b-a}\int_a^b\left|f(x)-\frac{1}{b-a}\int_a^b f(y)\,dy\right|dx. \qquad (12.142)
\end{aligned}
$$

The following discrete inequality is valid as well.

Corollary 12.12 *Let $\Phi:[\alpha,\beta]\to\mathbb{R}$ be a differentiable convex function on (α,β). If $x_i\in[\alpha,\beta]$ and $w_i\ge 0$ $(i=1,\dots,n)$ with $W_n>0$, then one has the counterpart* *of Jensen's discrete inequality*

$$\begin{aligned} 0 &\le \frac{1}{W_n}\sum_{i=1}^{n} w_i \Phi(x_i) - \Phi\left(\frac{1}{W_n}\sum_{i=1}^{n} w_i x_i\right) \\ &\le \frac{1}{W_n}\sum_{i=1}^{n} w_i \Phi'(x_i)x_i - \frac{1}{W_n}\sum_{i=1}^{n} w_i \Phi'(x_i)\frac{1}{W_n}\sum_{i=1}^{n} w_i x_i \\ &\le \frac{1}{2}\left[\Phi'(\beta) - \Phi'(\alpha)\right]\frac{1}{W_n}\sum_{i=1}^{n} w_i \left| x_i - \frac{1}{W_n}\sum_{j=1}^{n} w_j x_j\right|. \end{aligned} \tag{12.143}$$

Remark 12.11 In particular, we get the discrete inequality

$$\begin{aligned} 0 &\le \frac{1}{n}\sum_{i=1}^{n} \Phi(x_i) - \Phi\left(\frac{1}{n}\sum_{i=1}^{n} x_i\right) \\ &\le \frac{1}{n}\sum_{i=1}^{n} \Phi'(x_i)x_i - \frac{1}{n}\sum_{i=1}^{n} \Phi'(x_i)\frac{1}{n}\sum_{i=1}^{n} x_i \\ &\le \frac{1}{2}\left[\Phi'(\beta) - \Phi'(\alpha)\right]\frac{1}{n}\sum_{i=1}^{n}\left| x_i - \frac{1}{n}\sum_{j=1}^{n} x_j\right|. \end{aligned} \tag{12.144}$$

12.9 Generalizations of the Hermite–Hadamard Inequality for Isotonic Sublinear Functionals and Related Results

12.9.1 Isotonic Sublinear Functionals

Let L be a linear class of real-valued functions $g : E \to \mathbb{R}$ having the properties:

(L1) $f, g \in L$ imply $(\alpha f + \beta g) \in L$ for all $\alpha, \beta \in \mathbb{R}$;
(L2) $\mathbb{I} \in L$, i.e., if $f(t) = 1$ for all $t \in E$, then $f \in L$.

An *isotonic linear functional* $A : L \to \mathbb{R}$ is a functional satisfying the conditions:

(A1) $A(\alpha f + \beta g) = \alpha A(f) + \beta A(g)$ for all $f, g \in L$ and $\alpha, \beta \in \mathbb{R}$;
(A2) If $f \in L$ and $f \ge 0$, then $A(f) \ge 0$.
The mapping A is said to be *normalized* if
(A3) $A(\mathbb{I}) = 1$.

Isotonic, that is, order-preserving, linear functionals are natural objects in analysis which enjoy a number of convenient properties. Thus, they provide, for example, Jensen's inequality, which is a functional form of Jensen's inequality (see [11, p. 84]) and a functional Hermite–Hadamard inequality.

In this section, we show that these ideas carry over to a sublinear setting [12].

Let E be a non-empty set and K a class of real-valued functions $g : E \to \mathbb{R}$ having the properties:

(K1) $\mathbb{I} \in K$;
(K2) $f, g \in K$ imply $f + g \in K$;
(K3) $f \in K$ implies $\alpha \cdot \mathbb{I} + \beta \cdot f \in K$ for all $\alpha, \beta \in \mathbb{R}$.

We define the family of *isotonic sublinear functionals* $S : K \to \mathbb{R}$ by the properties:

(S1) $S(f+g) \leq S(f) + S(g)$ for all $f, g \in K$;
(S2) $S(\alpha f) = \alpha S(f)$ for all $\alpha \geq 0$ and $f \in K$;
(S3) If $f \geq g$, $f, g \in K$, then $S(f) \geq S(g)$.
An isotonic sublinear functional is said to be *normalized* if
(S4) $S(\mathbb{I}) = 1$ and *totally normalized* if, in addition,
(S5) $S(-\mathbb{I}) = -1$.
We note some immediate consequences. From (K2) and (K3), $f - g$ belongs to K whenever $f, g \in K$, so that from (S1)

$$S(f) = S\big((f-g)+g\big) \leq S(f-g) + S(g)$$

and hence
(S6) $S(f-g) \geq S(f) - S(g)$ if $f, g \in K$.
Moreover, if S is a totally normalized isotonic sublinear functional, then we have
(S7) $S(\alpha \cdot \mathbb{I}) = \alpha$ for all $\alpha \in \mathbb{R}$ and
(S8) $S(f + \alpha \cdot \mathbb{I}) = S(f) + \alpha$ for all $\alpha \in \mathbb{R}$.

Equation (S7) is immediate from (S2) when $\alpha \geq 0$. When $\alpha < 0$, we have

$$S(\alpha \cdot \mathbb{I}) = S\big((-\alpha)\cdot(-\mathbb{I})\big) = (-\alpha)S(-\mathbb{I}) = (-\alpha)(-1) = \alpha.$$

Also, by (S6) and (S7), we have for $\alpha \in \mathbb{R}$

$$S(f - \alpha \cdot \mathbb{I}) \geq S(f) + S(-\alpha \cdot \mathbb{I}) = S(f) - \alpha,$$

which by (S1) and (S7)

$$S(f - \alpha \cdot \mathbb{I}) \leq S(f) + S(-\alpha \cdot \mathbb{I}) = S(f) - \alpha,$$

so that

$$S(f - \alpha \cdot \mathbb{I}) = S(f) - \alpha.$$

Since this holds for all $\alpha \in \mathbb{R}$, we have (S8).

It is clear that every normalized isotonic linear functional is a totally normalized isotonic sublinear functional.

In what follows, we shall present some simple examples of sublinear functionals that are not linear.

Example 12.1 Let $A_1, \ldots, A_n : L \to \mathbb{R}$ be normalized isotonic linear functionals and $p_{i,j} \in \mathbb{R}$ $(i, j \in \{1, \ldots, n\})$ such that

$$p_{i,j} \geq 0 \quad \text{for all } i, j \in \{1, \ldots, n\} \quad \text{and} \quad \sum_{i=1}^{n} p_{i,j} = 1 \quad \text{for all } j \in \{1, \ldots, n\}.$$

Define the mapping $S : L \to \mathbb{R}$ by

$$S(f) = \max_{1 \leq j \leq n} \left\{ \sum_{i=1}^{n} p_{i,j} A_i(f) \right\}.$$

Then S is a totally normalized isotonic sublinear functional on L. As particular cases of this functional, we have the mappings

$$S_0(f) := \max_{1 \leq j \leq n} \{A_i(f)\}$$

and

$$S_Q(f) := \max_{1 \leq j \leq n} \left\{ \frac{1}{Q_j} \sum_{i=1}^{n} q_i A_i(f) \right\}$$

where $q_i \geq 0$ for all $i \in \{1, \ldots, n\}$ and $Q_j > 0$ for $j = 1, \ldots, n$. If we choose $q_i = 1$ for all $i \in \{1, \ldots, n\}$, we also have that

$$S_1(f) := \max_{1 \leq j \leq n} \left\{ \frac{1}{j} \sum_{i=1}^{n} A_i(f) \right\}$$

is a totally normalized isotonic sublinear functional on L.

Example 12.2 If $A_1, \ldots, A_n$ are as above and $A : L \to \mathbb{R}$ is also a normalized isotonic linear functional, then the mapping

$$S_A(f) := \frac{1}{P_n} \sum_{i=1}^{n} p_i \max\{A(f), A_i(f)\}$$

where $p_i \geq 0$ $(1 \leq i \leq n)$ with $P_n = \sum_{i=1}^{n} p_i > 0$, is also a totally normalized isotonic sublinear functional.

The following provide concrete examples.

Example 12.3 Suppose $x = (x_1, \ldots, x_n)$ and $y = (y_1, \ldots, y_n)$ are points in $\mathbb{R}^n$. Then the mappings

$$S(x) := \max_{1 \leq j \leq n} \left\{ \sum_{i=1}^{n} p_{i,j} x_i \right\},$$

where $p_i \geq 0$ and $\sum_{i=1}^{n} p_{i,j} = 1$ for $j \in \{1, \ldots, n\}$,

$$S_0(x) := \max_{1 \leq i \leq n} \{x_i\},$$

and

$$S_Q(x) := \max_{1 \leq j \leq n} \left\{ \frac{1}{Q_j} \sum_{i=1}^{j} q_i A_x \right\},$$

where $q_i \geq 0$ and $Q_j > 0$ for all $i, j \in \{1, \ldots, n\}$, are totally normalized isotonic sublinear functionals on $\mathbb{R}^n$.

Suppose $i_0 \in \{1, \ldots, n\}$ is fixed and $p_i \geq 0$ for all $i \in \{1, \ldots, n\}$, with $P_n > 0$. Then the mapping

$$S_{i0}(x) := \frac{1}{P_n} \sum_{i=1}^{n} p_i \max\{x_{i0}, x_i\}$$

is also totally normalized.

Example 12.4 Denote by $R[a,b]$ the linear space of Riemann integrable functions on $[a,b]$. Suppose that $p \in R[a,b]$ with $p(t) > 0$ for all $t \in [a,b]$. Then the mappings

$$S_p(f) := \sup_{x \in (a,b]} \left[\frac{\int_a^x p(t) f(t)\, dt}{\int_a^x p(t)\, dt} \right]$$

and

$$s_1(f) := \sup_{x \in (a,b]} \left[\frac{1}{x-a} \int_a^x f(t)\, dt \right]$$

are totally normalized isotonic sublinear functionals on $R[a,b]$.

If $C \in [a,b]$, then

$$S_{c,p}(f) := \frac{\int_a^b p(t) \max(f(c), f(t))\, dt}{\int_a^b p(t)\, dt}$$

and

$$s_c(f) := \frac{1}{b-a} \int_a^b \max\bigl(f(c), f(t)\bigr)\, dt$$

are also totally normalized on $R[a,b]$.

12.9.2 Jessen-Type Inequalities for Sublinear Functionals

We can give the following generalization of the well-known Jensen's inequality due to S.S. Dragomir, C.E.M. Pearce, and J.E. Pečarić [12]:

Theorem 12.18 (Dragomir, Pearce & Pečarić, 1995, [12]) *Let $\phi : [\alpha, \beta] \subset \mathbb{R} \to \mathbb{R}$ be a continuous convex function and $f : E \to [\alpha, \beta]$ such that $f, \phi \circ f \in K$. Then, if S is a totally normalized isotonic sublinear functional on K, we have $S(f) \in [\alpha, \beta]$ and*

$$S(\phi \circ f) \geq \phi\big(S(f)\big). \tag{12.145}$$

Proof By (S3) and (S7), $\alpha \cdot \mathbb{I} \leq f \leq \beta \cdot \mathbb{I}$ implies

$$\alpha = S(\alpha \cdot \mathbb{I}) \leq S(f) \leq S(\beta \cdot \mathbb{I}) = \beta$$

so that $S(f) \in [\alpha, \beta]$.
Set $l_1(x) = x$ for all $x \in [\alpha, \beta]$. For an arbitrary but fixed $q > 0$, we have by convexity of ϕ that there exist real numbers $u, v \in \mathbb{R}$ such that

(i) $p \leq \phi$ and
(ii) $p(S(f)) \geq \phi(S(f)) - q$
where

$$p(t) = u \cdot \mathbb{I} + v \cdot l_1(t).$$

If $\alpha < S(f) < \beta$ or if ϕ has a finite derivative in $[\alpha, \beta]$, we can replace (ii) by $p(S(f)) = \phi(S(f))$. Now (i) implies $p \circ f \leq \phi \circ f$. Hence, by (S3)

$$S(\phi \circ f) \geq S(p \circ f) = S(u \cdot \mathbb{I} + v \cdot f).$$

If $v \geq 0$, by (S8) and (S2) we have

$$S(u \cdot \mathbb{I} + v \cdot f) = u + vS(f) = p\big(S(f)\big),$$

while if $v < 0$, by (S6), (S7) and (S2) we have

$$\begin{aligned} S(u \cdot \mathbb{I} + v \cdot f) &= S\big(u \cdot \mathbb{I} - |v| f\big) \geq u - S\big(|v| f\big) \\ &= u - |v| S(f) = u + vS(f) = p\big(S(f)\big). \end{aligned}$$

Therefore, we have in either case

$$S(\phi \circ f) \geq \phi\big(S(f)\big) - q.$$

Since q is arbitrary, the proof is complete. □

Remark 12.12 If $S = A$, a normalized isotonic linear functional on L, then (12.145) becomes the well-known Jessen's inequality.

The following generalizations of Jensen's inequality for isotonic linear functionals also hold:

Let $A_1, \ldots, A_n : L \to \mathbb{R}$ be normalized isotonic linear functionals and $p_{i,j} \in \mathbb{R}$ be such that

$$p_{i,j} \geq 0 \quad \text{and} \quad \sum_{i=1}^{n} p_{i,j} = 1 \quad \text{for all } i, j \in \{1, \ldots, n\}.$$

If $\phi : [\alpha, \beta] \to \mathbb{R}$ is convex and $f : E \to [\alpha, \beta]$ is such that $f, \phi \circ f \in L$ then

$$\max_{1\leq j\leq n} \left\{ \sum_{i=1}^{n} p_{i,j} A_i(\phi \circ f) \right\} \geq \phi\left(\max_{1\leq j\leq n} \left\{ \sum_{i=1}^{n} p_{i,j} A_i(f) \right\} \right).$$

The proof follows by Theorem 12.18 applied to the mapping

$$S(f) := \max_{1\leq j\leq n} \left\{ \sum_{i=1}^{n} p_{i,j} A_i(f) \right\},$$

which is a totally normalized isotonic sublinear functional on L.

Remark 12.13 If $A_1, \ldots, A_n, \phi$ and f are as above, then

$$\max_{1\leq j\leq n} \left\{ A_i(\phi \circ f) \right\} \geq \phi\left(\max_{1\leq j\leq n} \left\{ A_i(f) \right\} \right)$$

and

$$\max_{1\leq j\leq n} \left\{ \frac{1}{Q_j} \sum_{i=1}^{j} q_i A_i(\phi \circ f) \right\} \geq \phi\left(\max_{1\leq j\leq n} \left\{ \frac{1}{Q_j} \sum_{i=1}^{j} q_i A_i(f) \right\} \right)$$

where $q_i \geq 0$ with $Q_j > 0$ for all $i, j \in \{1, \ldots, n\}$.

Corollary 12.13 *If $A_1, \ldots, A_n, \phi$ and f are as shown, $p_i \geq 0$, $i \in \{1, \ldots, n\}$, $P_n > 0$ and $A : L \to \mathbb{R}$ is also a normalized isotonic linear functional, then we have the inequality*

$$\frac{1}{P_n} \sum_{i=1}^{n} p_i \max\left\{ A(\phi \circ f), A_i(\phi \circ f) \right\} \geq \phi\left(\frac{1}{P_n} \sum_{i=1}^{n} p_i \max\left\{ A(f), A_i(f) \right\} \right).$$

The following reverse of Jensen's inequality for sublinear functionals was proved in [12]:

Theorem 12.19 (Dragomir, Pearce & Pečarić, 1995, [12]) *Let $\phi : [\alpha, \beta] \subset \mathbb{R} \to \mathbb{R}$ be a convex function ($\alpha < \beta$) and $f : E \to [\alpha, \beta]$ such that $\phi \circ f, f \in K$. Let $\lambda = \operatorname{sgn}(\phi(\beta) - \phi(\alpha))$. Then, if S is a totally normalized isotonic sublinear functional on K we have*

$$S(\phi \circ f) \leq \frac{\beta\phi(\alpha) - \alpha\phi(\beta)}{\beta - \alpha} + \frac{|\phi(\beta) - \phi(\alpha)|}{\beta - \alpha} S(\lambda f). \tag{12.146}$$

Proof Since ϕ is convex on $[\alpha, \beta]$, we have

$$\phi(v) \le \frac{w-v}{w-u}\phi(u) + \frac{v-u}{w-u}\phi(w),$$

where $u \le v \le w$ and $u < w$ (see also [13, p. 2]).

Set $u = \alpha,\ v = f(t),\ w = \beta$. Then

$$\phi\big(f(t)\big) \le \frac{\beta - f(t)}{\beta-\alpha}\phi(\alpha) + \frac{f(t)-\alpha}{\beta-\alpha}\phi(\beta), \quad t \in E,$$

or, alternatively,

$$\phi \circ f \le \frac{\beta\phi(\alpha) - \alpha\phi(\beta)}{\beta-\alpha} \cdot \mathbb{I} + \frac{\phi(\beta)-\phi(\alpha)}{\beta-\alpha} \cdot f.$$

Applying the functional S and using its properties, we have

$$\begin{aligned} S(\phi \circ f) &\le S\left(\frac{\beta\phi(\alpha) - \alpha\phi(\beta)}{\beta-\alpha} \cdot \mathbb{I} + \frac{\phi(\beta)-\phi(\alpha)}{\beta-\alpha} \cdot f\right) \\ &= \frac{\beta\phi(\alpha) - \alpha\phi(\beta)}{\beta-\alpha} + S\left(\frac{\phi(\beta)-\phi(\alpha)}{\beta-\alpha} \cdot f\right) \\ &= \frac{\beta\phi(\alpha) - \alpha\phi(\beta)}{\beta-\alpha} + \frac{|\phi(\beta)-\phi(\alpha)|}{\beta-\alpha} S(\lambda f). \end{aligned}$$

Hence, the theorem is proved. □

Remark 12.14 If $S = A$, and A is a normalized isotonic linear functional, then, by (12.146), we deduce the inequality

$$A\big(\phi(f)\big) \le \frac{\{(\beta - A(f))\phi(\alpha) + (A(f)-\alpha)\phi(\beta)\}}{(\beta-\alpha)}.$$

This is the result of Lemma 1 from [13]. Note that this last inequality is a generalization of the inequality

$$A(\phi) \le \frac{\{(b - A(l_1))\phi(a) + (A(l_1)-a)\phi(b)\}}{(b-a)}$$

due to A. Lupas [13, Theorem A]. Here, $E = [a, b]$ $(-\infty < a < b < \infty)$, L satisfies (L1), (L2), $A : L \to \mathbb{R}$ satisfies (A1), (A2), $A(\mathbb{I}) = 1, \phi$ is convex on E and $\phi \in L, l_1 \in L$, where $l_1(x) = x, x \in [a, b]$.

By the use of Jensen's and Lupas' inequalities for totally normalized sublinear functionals, we can state the following generalization of the classical Hermite–Hadamard's integral inequality due to S.S. Dragomir, C.E.M. Pearce, and J.E. Pečarić [12].

Theorem 12.20 (Dragomir, Pearce & Pečarić, 1995, [12]) *Let $\phi : [\alpha, \beta] \to \mathbb{R}$ be a convex function and $e : E \to [\alpha, \beta]$ a mapping such that $\phi \circ e$ and e belong to K and let $\lambda := \operatorname{sgn}(\phi(\beta) - \phi(\alpha))$. If S is a totally normalized isotonic sublinear functional on K with*

$$S(\lambda e) = \lambda \cdot \frac{\alpha + \beta}{2} \quad \text{and} \quad S(e) = \frac{\alpha + \beta}{2},$$

then we have the inequality

$$\phi\left(\frac{\alpha + \beta}{2}\right) \leq S(\phi \circ e) \leq \frac{\phi(\alpha) + \phi(\beta)}{2}. \tag{12.147}$$

Proof The first inequality in (12.147) follows by Jensen's inequality (12.145) applied to the mapping e.

By inequality (12.146), we have

$$\begin{aligned} S(\phi \circ e) &\leq \frac{\beta\phi(\alpha) - \alpha\phi(\beta)}{\beta - \alpha} + \frac{(\phi(\beta) - \phi(\alpha))(\beta + \alpha)}{2(\beta - \alpha)} \\ &= \frac{\phi(\alpha) + \phi(\beta)}{2}, \end{aligned}$$

and the statement is proved. □

Remark 12.15 If $S = A, \phi$ is as above and $e : E \to [\alpha, \beta]$ is such that $\phi \circ e,\ e \in L$ and $A(e) = \frac{\alpha+\beta}{2}$, then the Hermite–Hadamard inequality

$$\phi\left(\frac{\alpha + \beta}{2}\right) \leq A(\phi \circ e) \leq \frac{\phi(\alpha) + \phi(\beta)}{2},$$

holds for normalized isotonic linear functionals (see also [16] and [5]).

Remark 12.16 If in the above theorem we assume that $\phi(\beta) \geq \phi(\alpha)$, then we can drop the assumption $S(\lambda e) = \lambda \cdot \frac{\alpha+\beta}{2}$.

Theorem 12.21 (Dragomir, Pearce & Pečarić, 1995, [12]) *Let ϕ, f, and S be defined as in Theorem* 12.19 *with $\phi(\beta) \geq \phi(\alpha)$. Then*

$$S\big(\phi(f)\big) \leq \frac{\{(\beta - S(f))\phi(\alpha) + (S(f) - \alpha)\phi(\beta)\}}{\beta - \alpha}. \tag{12.148}$$

The proof is a simple consequence of Theorem 12.19.

Finally, we have the following result [12]:

Theorem 12.22 (Dragomir, Pearce & Pečarić, 1995, [12]) *Let the hypothesis of Theorem* 12.21 *be fulfilled and let T be an interval which is such that $T \supset \phi([\alpha, \beta])$. If $F(u, v)$ is a real-valued function defined on $T \times T$ and increasing in u, then*

$$\begin{aligned}
&F\big[S(\phi(f)), \phi(S(f))\big] \\
&\quad \le \max_{x\in[a,b]} F\left[\frac{\beta - x}{\beta - a}\phi(\alpha) + \frac{x-\alpha}{\beta-\alpha}\phi(\beta), \phi(x)\right] \\
&\quad = \max_{\theta\in[0,1]} F\big[\theta\phi(\alpha) + (1-\theta)\phi(\beta), \phi\big(\theta\alpha + (1-\theta)\beta\big)\big]. \qquad (12.149)
\end{aligned}$$

Proof By (12.148) and the increasing property of $F(\cdot, y)$, we have

$$\begin{aligned}
F\big[S(\phi(f)), \phi(S(f))\big] &\le F\left[\frac{\beta - S(f)}{\beta - a}\phi(\alpha) + \frac{S(f)-\alpha}{\beta-\alpha}\phi(\beta, \phi(S(f)))\right] \\
&\le \max_{x\in[a,b]} F\left[\frac{\beta - x}{\beta - a}\phi(\alpha) + \frac{x-\alpha}{\beta-\alpha}\phi(\beta), \phi(x)\right].
\end{aligned}$$

Of course, the equality in (12.149) follows immediately from the change of variable $\theta = \frac{\beta - x}{\beta - a}$, so that $x = \theta\alpha + (1-\theta)\beta$ with $0 \le \theta \le 1$. □

12.9.3 Applications to Special Means

1. Suppose that $e \in K$, $p \ge 1$, $e^p \in K$ and S is as above. We can define the mean

   $$L_p(s,e) := \big[S(e^p)\big]^{\frac{1}{p}}.$$

   By the use of Theorem 12.20, we have the inequality

   $$A(\alpha,\beta) \le L_p(s,e) \le \big[A(\alpha^p, \beta^p)\big]^{\frac{1}{p}},$$

   provided that

   $$S(e) = \frac{\alpha+\beta}{2}.$$

   A particular case which generates in its turn the classical L_p-mean is where $S = A$, where A is a linear isotonic functional defined on K.
2. Now, if $e \in K$ is such that $e^{-1} \in K$, we can define the mean as

   $$L(s,e) := \big[S(e^{-1})\big]^{-1}.$$

   If we assume that $S(-e) = -\frac{\alpha+\beta}{2}$ and $S(e) = \frac{\alpha+\beta}{2}$, then, by Theorem 12.20, we have the inequality

   $$H(\alpha,\beta) \le L(S,e) \le A(\alpha,\beta).$$

   A particular case which generalizes in its turn the classical logarithmic mean is where $S = A$, where A is as above.

3. Finally, if we suppose that $e \in K$ is such that $\ln e \in K$, we can also define the mean

$$I(S,e) := \exp\bigl[-S(-\ln e)\bigr].$$

Now, if we assume that $S(-e) = -\frac{\alpha+\beta}{2}$ and $S(e) = \frac{\alpha+\beta}{2}$, then, by Theorem 12.20, we get the inequality:

$$G(\alpha,\beta) \le I(S,e) \le A(\alpha,\beta),$$

which generalizes the corresponding inequality for the identric mean.

References

1. Andrica, D., Badea, C.: Grüss' inequality for positive linear functionals. Period. Math. Hung. **19**(2), 155–167 (1988)
2. Beesack, P.R., Pečarić, J.E.: On Jessen's inequality for convex functions. J. Math. Anal. Appl. **110**, 536–552 (1985)
3. Cerone, P., Dragomir, S.S.: A refinement of the Grüss inequality and applications. Tamkang J. Math. **38**(1), 37–49 (2007). Preprint RGMIA, Res. Rep. Coll. **5**(2) (2002), 14
4. Cheng, X.-L., Sun, J.: A note on the perturbed trapezoid inequality. J. Inequal. Pure Appl. Math. **3**(2), 29 (2002). On line: http://jipam.vu.edu.au
5. Dragomir, S.S.: A refinement of Hadamard's inequality for isotonic linear functionals. Tamkang J. Math. (Taiwan) **24**, 101–106 (1992)
6. Dragomir, S.S.: On a reverse of Jessen's inequality for isotonic linear functionals. J. Inequal. Pure Appl. Math. **2**(3), 36 (2001). On line: http://jipam.vu.edu.au/v2n3/047_01.html
7. Dragomir, S.S.: On the Jessen's inequality for isotonic linear functionals. Nonlinear Anal. Forum **7**(2), 139–151 (2002)
8. Dragomir, S.S.: On the Lupaş–Beesack–Pečarić inequality for isotonic linear functionals. Nonlinear Funct. Anal. Appl. **7**(2), 285–298 (2002)
9. Dragomir, S.S.: A Grüss type inequality for isotonic linear functionals and applications. Demonstr. Math. **36**(3), 551–562 (2003). Preprint RGMIA, Res. Rep. Coll. **5**, 12 (2002). Supplement
10. Dragomir, S.S., Ionescu, N.M.: On some inequalities for convex-dominated functions. L'Anal. Num. Théor. L'Approx. **19**(1), 21–27 (1990)
11. Dragomir, S.S., Pearce, C.E.M.: Selected Topics on Hermite–Hadamard Inequalities and Applications. RGMIA Monographs, Victoria University (2000). http://rgmia.vu.edu.au/monographs.html
12. Dragomir, S.S., Pearce, C.E.M., Pečarić, J.E.: On Jessen's and related inequalities for isotonic sublinear functionals. Acta Sci. Math. (Szeged) **61**, 373–382 (1995)
13. Lupaş, A.: A generalisation of Hadamard's inequalities for convex functions. Univ. Beogr. Elek. Fak. **577–579**, 115–121 (1976)
14. Pečarić, J.E.: On Jessen's inequality for convex functions (III). J. Math. Anal. Appl. **156**, 231–239 (1991)
15. Pečarić, J.E., Beesack, P.R.: On Jessen's inequality for convex functions (II). J. Math. Anal. Appl. **156**, 231–239 (1991)
16. Pečarić, J.E., Dragomir, S.S.: A generalisation of Hadamard's inequality for isotonic linear functionals. Radovi Mat. (Sarajevo) **7**, 103–107 (1991)
17. Pečarić, J.E., Raşa, I.: On Jessen's inequality. Acta Sci. Math. (Szeged) **56**, 305–309 (1992)
18. Toader, G., Dragomir, S.S.: Refinement of Jessen's inequality. Demonstr. Math. **28**, 329–334 (1995)

Chapter 13
On Approximate Bi-quadratic Bi-homomorphisms and Bi-quadratic Bi-derivations in C^*-Ternary Algebras and Quasi-Banach Algebras

Ali Ebadian and Norouz Ghobadipour

Abstract In this paper, we prove the generalized Hyers–Ulam–Rassias stability of bi-quadratic bi-homomorphisms in C^*-ternary algebras and quasi-Banach algebras. Moreover, we investigate stability of bi-quadratic bi-derivations on C^*-ternary algebras and quasi-Banach algebras.

Key words Generalized Hyers–Ulam–Rassias stability · Bi-quadratic bi-homomorphism · Bi-quadratic bi-derivation · Quasi-Banach algebra · C^*-ternary algebra

Mathematics Subject Classification 39B82 · 17A40 · 46B03 · 39B52

13.1 Introduction

The stability problem of functional equations originated from a question of Ulam [59] in 1940, concerning the stability of group homomorphisms. Let $(G_1, \cdot)$ be a group and let $(G_2, *)$ be a metric group with the metric $d(\cdot,\cdot)$. Given $\varepsilon > 0$, does there exist a $\delta > 0$ such that if a mapping $h : G_1 \longrightarrow G_2$ satisfies the inequality $d(h(x \cdot y), h(x) * h(y)) < \delta$ for all $x, y \in G_1$, then there exists a homomorphism $H : G_1 \longrightarrow G_2$ with $d(h(x), H(x)) < \varepsilon$ for all $x \in G_1$? In the other words, under what condition does there exist a homomorphism near an approximate homomorphism? The concept of stability for a functional equation arises when we replace the functional equation by an inequality which acts as a perturbation of the equation. In

Dedicated to Professor Themistocles M. Rassias on the occasion of his 60th birthday.

A. Ebadian (✉) · N. Ghobadipour
Department of Mathematics, Urmia University, Urmia, Iran
e-mail: a.ebadian@urmia.ac.ir

N. Ghobadipour
e-mail: ghobadipour.n@gmail.com

A. Ebadian
Department of Mathematics, Payame Noor University (PNU), Tehran, Iran

P.M. Pardalos et al. (eds.), *Nonlinear Analysis*, Springer Optimization and Its Applications 68, In Honor of Themistocles M. Rassias on the Occasion of his 60th Birthday,
DOI 10.1007/978-1-4614-3498-6_13,

1941, D.H. Hyers [22] gave the first affirmative answer to the question of Ulam for Banach spaces. Let $f : E \longrightarrow E'$ be a mapping between Banach spaces such that

$$\big\| f(x+y) - f(x) - f(y) \big\| \le \delta$$

for all $x, y \in E$, and for some $\delta > 0$. Then there exists a unique additive mapping $T : E \longrightarrow E'$ such that

$$\big\| f(x) - T(x) \big\| \le \delta$$

for all $x \in E$. Moreover, if $f(tx)$ is continuous in t for each fixed $x \in E$, then T is linear. In 1950, T. Aoki [5] was the second author to treat this problem for additive mappings. Finally, in 1978, Th.M. Rassias [52] proved the following theorem.

Theorem 13.1 *Let $f : E \longrightarrow E'$ be a mapping from a normed vector space E into a Banach space E' subject to the inequality*

$$\big\| f(x+y) - f(x) - f(y) \big\| \le \varepsilon \big(\|x\|^p + \|y\|^p \big)$$

for all $x, y \in E$, where ε and p are constants with $\varepsilon > 0$ and $p < 1$. Then there exists a unique additive mapping $T : E \longrightarrow E'$ such that

$$\big\| f(x) - T(x) \big\| \le \frac{2\varepsilon}{2 - 2^p} \|x\|^p$$

for all $x \in E$. If $p < 0$ then inequality (13.3) *holds for all $x, y \neq 0$, and* (13.4) *for $x \neq 0$. Also, if the function $t \mapsto f(tx)$ from $\mathbb{R}$ into E' is continuous for each fixed $x \in E$, then T is linear.*

The stability phenomenon of this kind is called the *Hyers–Ulam–Rassias stability*. In 1991, Z. Gajda [18] answered the question for the case $p > 1$, which was rased by Rassias.

In 1994, a generalization of the Rassias' theorem was obtained by Găvruta as follows [19].

Suppose $(G, +)$ is an abelian group, E is a Banach space, and suppose that the so-called admissible control function $\varphi : G \times G \to \mathbb{R}$ satisfies

$$\tilde{\varphi}(x, y) := 2^{-1} \sum_{n=0}^{\infty} 2^{-n} \varphi\big(2^n x, 2^n y\big) < \infty$$

for all $x, y \in G$. If $f : G \to E$ is a mapping with

$$\big\| f(x+y) - f(x) - f(y) \big\| \le \varphi(x, y)$$

for all $x, y \in G$, then there exists a unique mapping $T : G \to E$ such that $T(x+y) = T(x) + T(y)$ and $\|f(x) - T(x)\| \le \tilde{\varphi}(x, x)$ for all $x, y \in G$.

On the other hand, J.M. Rassias, generalized the Hyers stability result by presenting a weaker condition controlled by a product of different powers of norms (see [43–47]). According to J.M. Rassias Theorem, if it is assumed that there exist constants $\varepsilon \geq 0$ and $p_1, p_2 \in \mathbb{R}$ such that $p = p_1 + p_2 \neq 1$, and $f : E \longrightarrow E'$ is a map from a normed space E into a Banach space E' such that the inequality

$$\left\| f(x+y) - f(x) - f(y) \right\| \leq \varepsilon \|x\|^{p_1} \|y\|^{p_2}$$

for all $x, y \in E$, then there exists a unique additive mapping $T : E \longrightarrow E'$ such that

$$\left\| f(x) - T(x) \right\| \leq \frac{\varepsilon}{2 - 2^p} \|x\|^p,$$

for all $x \in E$. If in addition for every $x \in E$, $f(tx)$ is continuous, then T is linear. Following the techniques of the proof of the corollary of D.H. Hyers [22], it is observed that D.H. Hyers introduced (in 1941) the following Hyers continuity condition: about the continuity of the mapping $f(tx)$ in real t for each fixed x, and then he proved homogeneity of degree one, and therefore the famous linearity. This condition has been assumed further till now, through the complete Hyers direct method, in order to prove linearity for other generalized Hyers–Ulam stability problem forms. During the past few years, several mathematicians have published various generalizations and applications of Hyers–Ulam stability and Hyers–Ulam–Rassias stability to a number of functional equations and mappings, for example, quadratic functional equation, derivations and homomorphisms, ternary derivations, double derivations, multiplicative mappings—superstability, bounded nth differences, mixed functional equations. Several mathematicians have contributed works on these subjects; we mention a few: [6, 10, 14–20, 22–25, 30–42] and [48–55].

Quadratic functional equation was used to characterize inner product spaces [2, 4]. Several other functional equations were also used to characterize inner product spaces. A square norm on an inner product space satisfies the important parallelogram equality

$$\|x+y\|^2 + \|x-y\|^2 = 2\left(\|x\|^2 + \|y\|^2\right).$$

The functional equation

$$f(x+y) + f(x-y) = 2f(x) + 2f(y) \tag{13.1}$$

is related to a symmetric bi-additive function. It is natural that this equation is called a quadratic functional equation. In particular, every solution of the quadratic equation (13.1) is said to be a quadratic function. It is well known that a function f between real vector spaces is quadratic if and only if there exits a unique symmetric bi-additive function B such that $f(x) = B(x,x)$ for all x (see [27]). The bi-additive function B is given by

$$B(x,y) = \frac{1}{4}\left(f(x+y) - f(x-y)\right) \tag{13.2}$$

Hyers–Ulam–Rassias stability problem for the quadratic functional equation (13.1) was proved by Skof for functions $f : A \longrightarrow B$, where A is a normed space and B

a Banach space (see [57]). Cholewa [11] noticed that the theorem of Skof is still true if relevant domain A is replaced an abelian group. In [12], Czerwik proved the Hyers–Ulam–Rassias stability of (13.1). Grabiec [21] has generalized these result mentioned above.

Let X and Y be vector spaces. A mapping $f : X \times X \to Y$ is called bi-quadratic if f satisfies the system of equations

$$\begin{aligned} f(x+y,z)+f(x-y,z) &= 2f(x,z)+2f(y,z), \\ f(x,y+z)+f(x,y-z) &= 2f(x,y)+2f(x,z). \end{aligned} \tag{13.3}$$

Won-Gil Park and Jae-Hyeong Bae [42] proved that a mapping $f : X \times X \to Y$ satisfies (13.3) if and only if it satisfies

$$\begin{aligned} & f(x+y,z+w)+f(x+y,z-w)+f(x-y,z+w)+f(x-y,z-w) \\ & \quad = 4\big[f(x,z)+f(x,w)+f(y,z)+f(y,w)\big]. \end{aligned} \tag{13.4}$$

We recall some basic facts concerning C^*-ternary algebra, quasi-Banach spaces, and some preliminary results.

Ternary algebraic operations were considered in the nineteenth century by several mathematicians such as A. Cayley [9] who introduced the notion of a cubic matrix which, in turn, was generalized by Kapranov, Gelfand, and Zelevinskii in 1990 [26]. The comments on physical applications of ternary structures can be found in [1, 28, 29, 35, 36, 56].

A C^*-ternary algebra is a complex Banach space $\mathcal{A}$, equipped with a ternary product $(x,y,z) \rightarrowtail [xyz]$ of $\mathcal{A}^3$ into $\mathcal{A}$, which is $\mathbb{C}$-linear in the outer variables, conjugate $\mathbb{C}$-linear in the middle variable, and associative in the sense that $[xy[zvw]] = [x[wzy]v] = [[xyz]wv]$, and satisfies $\|[xyz]\| \le \|x\|.\|y\|.\|z\|$ and $\|[xxx]\| = \|x\|^3$ (see [36]). If a C^*-ternary algebra $(A, [\ ])$ has an identity, i.e., an element $e \in \mathcal{A}$ such that $x = [xee] = [eex]$ for all $x \in \mathcal{A}$, then it is routine to verify that $\mathcal{A}$, endowed with $xoy := [xey]$ and $x^* := [exe]$, is a unital C^*- algebra. Conversely, if $(\mathcal{A}, o)$ is a unital C^*- algebra, then $[xyz] := xoy^*oz$ makes $\mathcal{A}$ into a C^*-ternary algebra.

Let $\mathcal{A}$ and $\mathcal{B}$ be a C^*-ternary algebra. A $\mathbb{C}$-linear mapping $H : \mathcal{A} \to \mathcal{B}$ is called a C^*-ternary algebra homomorphism if

$$H\big([abc]\big) = \big[H(a)H(b)H(c)\big]$$

for all $a, b, c \in \mathcal{A}$. A $\mathbb{C}$-linear mapping $\delta : \mathcal{A} \to \mathcal{A}$ is called a C^*-ternary algebra derivation if

$$\delta\big([abc]\big) = \big[\delta(a)bc\big] + \big[a\delta(b)c\big] + \big[ab\delta(c)\big]$$

for all $a, b, c \in \mathcal{A}$ (see [9]).

Let $\mathscr{A}$ and $\mathscr{B}$ be C^*-ternary algebras. A $\mathbb{C}$-bilinear $H : \mathscr{A} \times \mathscr{A} \to \mathscr{B}$ is called a C^*-ternary bi-homomorphism if it satisfies

$$H\big([abc], d\big) = \big[H(a,d)H(b,d)H(c,d)\big],$$
$$H\big(a, [bcd]\big) = \big[H(a,b)H(a,c)H(a,d)\big]$$

for all $a, b, c, d \in \mathscr{A}$. A $\mathbb{C}$-bilinear $\delta : \mathscr{A} \times \mathscr{A} \to \mathscr{A}$ is called a C^*-ternary bi-derivation if it satisfies

$$\delta\big([abc], d\big) = \big[\delta(a,d)bc\big] + \big[a\delta(b,d)c\big] + \big[ab\delta(c,d)\big],$$
$$\delta\big(a, [bcd]\big) = \big[\delta(a,b)cd\big] + \big[b\delta(a,c)d\big] + \big[bc\delta(a,d)\big]$$

for all $a, b, c, d \in \mathscr{A}$ (see [7]).

A quasi-norm is a real-valued function on X satisfying the following:

1. $\|x\| \geq 0$ for all $x \in X$ and $\|x\| = 0$ if and only if $x = 0$.
2. $\|\lambda \cdot x\| = |\lambda| \cdot \|x\|$ for all $\lambda \in \mathbb{R}$ and all $x \in X$.
3. There is a constant $K \geq 1$ such that $\|x + y\| \leq K(\|x\| + \|y\|)$ for all $x, y \in X$.

The pair $(X, \|\cdot\|)$ is called a quasi-normed space if $\|\cdot\|$ is a quasi-norm on X. A quasi-Banach space is a complete quasi-normed space. A quasi-norm $\|\cdot\|$ is called a p-norm $(0 < p \leq 1)$ if

$$\|x + y\|^p \leq \|x\|^p + \|y\|^p$$

for all $x, y \in X$. In this case, a quasi-Banach space is called a p-Banach space. Given a p-norm, the formula $d(x, y) := \|x - y\|^p$ gives us a translation invariant metric on X. By the Aoki–Rolewicz Theorem [58] (see also [8]), each quasi-norm is equivalent to some p-norm. Since it is much easier to work with p-norms, henceforth we restrict our attention mainly to p-norms.

Let $(A, \|\cdot\|)$ be a quasi-normed space. The quasi-normed space $(A, \|\cdot\|)$ is called a quasi-normed algebra if A is an algebra and there is a constant $K > 0$ such that $\|xy\|_B \leq K\|x\|\|y\|$ for all $x, y \in A$. A quasi-Banach algebra is a complete quasi-normed algebra. If the quasi-norm $\|\cdot\|$ is a p-norm then the quasi-Banach algebra is called a p-Banach algebra (see [3]).

This paper is organized as follows: In Sects. 13.2 and 13.3, we investigate the generalized Hyers–Ulam–Rassias stability of bi-quadratic bi-homomorphisms and bi-quadratic bi-derivations in C^*-ternary algebras associated with the functional equation (13.4). In Sects. 13.4 and 13.5, we prove the generalized Hyers–Ulam–Rassias stability of bi-quadratic bi-homomorphisms and bi-quadratic bi-derivations in quasi-Banach algebras associated with the following Jensen-type bi-quadratic functional equation:

$$8f\left(\frac{x+y}{2}, \frac{z+w}{2}\right) + 8f\left(\frac{x-y}{2}, \frac{z-w}{2}\right)$$
$$= f(x,z) + f(x,w) + f(y,z) + f(y,w).$$

13.2 Stability of Bi-quadratic Bi-homomorphisms in C^*-Ternary Algebras

Throughout this section, assume that A is a C^*-ternary algebra with norm $\|\cdot\|_A$ and that B is a C^*-ternary algebra with norm $\|\cdot\|_B$.

Definition 13.1 A mapping $H : A \times A \to B$ is called a C^*-ternary bi-quadratic bi-homomorphism if H satisfies the following properties:

1. H is a bi-quadratic mapping,
2. H is a bi-quadratic homogeneous, that is, $H(\lambda a, \mu b) = \lambda^2\mu^2 H(a, b)$ for all $a, b \in A$ and all $\lambda, \mu \in \mathbb{C}$,
3. $$H\big([a,b,c],d\big) = \big[H(a,d), H(b,d), H(c,d)\big],$$ $$H\big(a,[b,c,d]\big) = \big[H(a,b), H(a,c), H(a,d)\big]$$

   for all $a, b, c, d \in A$.

Definition 13.2 A mapping $H : A \times A \to B$ is called a C^*-ternary bi-quadratic Jordan bi-homomorphism if H satisfies the properties 1 and 2 in Definition 13.1 and

$$H\big([a,a,a],a\big) = H\big(a,[a,a,a]\big) = \big[H(a,a), H(a,a), H(a,a)\big]$$

for all $a \in A$.

Now we investigate the generalized Hyers–Ulam–Rassias stability of bi-quadratic bi-homomorphisms in C^*-ternary algebras.

Theorem 13.2 *Let $f : A \times A \to B$ be a mapping with $f(0,0) = f(a,0) = f(0,b) = 0$ for which there exist functions $\varphi : A^8 \to [0,\infty)$ and $\psi : A^4 \to [0,\infty)$ such that*

$$\sum_{i=0}^{\infty} \frac{1}{16^i}\varphi\big(2^i a, 2^i b, 2^i c, 2^i d, 2^i x, 2^i y, 2^i z, 2^i w\big) < \infty, \tag{13.5}$$

$$\lim_{i\to\infty} \frac{1}{16^i}\psi\big(2^i a, 2^i b, 2^i c, 2^i d\big) = 0, \tag{13.6}$$

$$\begin{aligned}
&\big\| f(\lambda a + \lambda b + \lambda x + \lambda y, \mu c + \mu d + \mu z + \mu w) + f(\lambda a + \lambda b, \mu c - \mu d)\\
&\quad + f(\lambda a - \lambda b, \mu c + \mu d) + f(\lambda a - \lambda b, \mu c - \mu d)\\
&\quad - 4\big[f(\lambda a, \mu c) + f(\lambda a, \mu d) + f(\lambda b, \mu c)\\
&\quad + f(\lambda b, \mu d) - \lambda^2\mu^2 f(x+y, z+w)\big]\big\|\\
&\le \varphi(a,b,c,d,x,y,z,w),
\end{aligned} \tag{13.7}$$

$$\begin{aligned}&\|f([a,b,c],d)-[f(a,d),f(b,d),f(c,d)]\|\\&\quad+\|f(a,[b,c,d]-[f(a,b),f(a,c),f(a,d)])\|\\&\leq\psi(a,b,c,d)\end{aligned}\tag{13.8}$$

for all $a, b, c, d, x, y, z, w \in A$ and all $\lambda, \mu \in \mathbb{T}^1 := \{\lambda \in \mathbb{C};\ |\lambda| = 1\}$. If for each fixed $a, b \in A$ the mapping $t \mapsto f(ta, tb)$ from $\mathbb{R}$ to A is continuous in $t \in \mathbb{R}$, then there is a unique C^*-ternary algebra bi-quadratic bi-homomorphism $H : A \times A \to B$ such that

$$\|f(x,y)-H(x,y)\|\leq\sum_{i=0}^{\infty}\frac{1}{16^i}\varphi(2^ix,2^ix,2^iy,2^iy,0,0,0,0)\tag{13.9}$$

for all $x, y \in A$.

Proof Setting $\lambda = \mu = 1$ and setting $x = y = z = w = 0$, $a = b$ and $c = d$ in (13.7), we have

$$\left\|f(a,c)-\frac{1}{16}f(2a,2c)\right\|\leq\frac{1}{16}\varphi(a,a,c,c,0,0,0,0)$$

for all $a, c \in A$. Replacing a, c by x, y in the above inequality, respectively, we have

$$\left\|f(x,y)-\frac{1}{16}f(2x,2y)\right\|\leq\frac{1}{16}\varphi(x,x,y,y,0,0,0,0)\tag{13.10}$$

for all $x, y \in A$. Thus we obtain

$$\left\|\frac{1}{16^i}f(2^ix,2^iy)-\frac{1}{16^{i+1}}f(2^{i+1}x,2^{i+1}y)\right\|\frac{1}{16^{i+1}}\varphi(2^ix,2^ix,2^iy,2^iy,0,0,0,0)$$

for all $x, y \in A$ and all i. For given integers l, m $(0 \leq l < m)$, we get

$$\begin{aligned}&\left\|\frac{1}{16^l}f(2^lx,2^ly)-\frac{1}{16^m}f(2^mx,2^my)\right\|\\&\quad\leq\sum_{i=l}^{m-1}\frac{1}{16^{i+1}}\varphi(2^ix,2^ix,2^iy,2^iy,0,0,0,0)\end{aligned}\tag{13.11}$$

for all $x, y \in A$. By (13.5), the sequence $\{\frac{1}{16^i}f(2^ix, 2^iy)\}$ is a Cauchy sequence for all $x, y \in A$. Since B is complete, the sequence $\{\frac{1}{16^i}f(2^ix, 2^iy)\}$ converges for all $x, y \in A$. Define $H : A \times A \to B$ by

$$H(x,y):=\lim_{i\to\infty}\frac{1}{16^i}f(2^ix,2^iy)$$

for all $x, y \in A$. Setting $l = 0$ and taking $m \to \infty$ in (13.11), one can obtain the inequality (13.9). Putting $x = y = z = w = 0$ and replacing a, b, c, d by

$2^i a, 2^i b, 2^i c, 2^i d$, respectively, in (13.7) and using (13.5), we see that H satisfies (13.4). By Theorem 4 of [42], we obtain that H is bi-quadratic.

Letting $a = b = c = d = y = w = 0$ in (13.5), we get

$$\|f(\lambda x, \mu z) - \lambda^2\mu^2 f(x, z)\| \le \varphi(0, 0, 0, 0, x, 0, z, 0)$$

for all $x, z \in A$. Replacing z by y in the above inequality, we have

$$\|f(\lambda x, \mu y) - \lambda^2\mu^2 f(x, y)\| \le \varphi(0, 0, 0, 0, x, 0, y, 0) \tag{13.12}$$

for all $x, y \in A$. Replacing x, y by $2^i x, 2^i y$ in (13.12), respectively, and using (13.5) we obtain

$$H(\lambda x, \mu y) = \lambda^2\mu^2 H(x, y)$$

for all $x, y \in A$ and all $\lambda, \mu \in \mathbb{T}^1$. Under the assumption that $f(tx, ty)$ is continuous in $t \in \mathbb{R}$ for each fixed $x, y \in A$, by the same reasoning as in the proofs of [12] and Lemma 2.1 of [7],

$$H(\lambda x, \mu y) = \lambda^2\mu^2 H(x, y)$$

for all $x, y \in A$ and all $\lambda, \mu \in \mathbb{R}$. Hence

$$H(\lambda x, \mu y) = H\left(\frac{\lambda}{|\lambda|}|\lambda|x, \frac{\mu}{|\mu|}|\mu|y\right) = \frac{\lambda^2\mu^2}{|\lambda|^2|\mu|^2} H\big(|\lambda|x, |\mu|y\big) = \lambda^2\mu^2 H(x, y)$$

for all $x, y \in A$ and all $\lambda, \mu \in \mathbb{C}(\lambda, \mu \neq 0)$. This means that H is bi-quadratic homogeneous.

It follows from (13.8) that

$$\begin{aligned}
&\big\|H\big([a, b, c], d\big) - \big[H(a, d), H(b, d), H(c, d)\big]\big\| \\
&\quad + \big\|H\big(a, [b, c, d] - \big[H(a, b), H(a, c), H(a, d)\big]\big)\big\| \\
&= \lim_{i\to\infty} \frac{1}{16^i}\Big(\big\|f\big([2^i a, 2^i b, 2^i c], 2^i d\big) \\
&\quad - \big[f\big(2^i a, 2^i d\big), f\big(2^i b, 2^i d\big), f\big(2^i c, 2^i d\big)\big]\big\| \\
&\quad + \big\|f\big(2^i a, \big[2^i b, 2^i c, 2^i d\big] \\
&\quad - \big[f\big(2^i a, 2^i b\big), f\big(2^i a, 2^i c\big), f\big(2^i a, 2^i d\big)\big]\big)\big\|\Big) \\
&\le \lim_{i\to\infty} \frac{1}{16^i}\psi\big(2^i a, 2^i b, 2^i c, 2^i d\big) = 0
\end{aligned}$$

for all $a, b, c, d \in A$. So

$$\begin{aligned}
H\big([a, b, c], d\big) &= \big[H(a, d), H(b, d), H(c, d)\big], \\
H\big(a, [b, c, d]\big) &= \big[H(a, b), H(a, c), H(a, d)\big]
\end{aligned}$$

for all $a, b, c, d \in A$. If $H' : A \times A \to B$ is another bi-quadratic mapping satisfying (13.9), we obtain

$$\begin{aligned}
\|H(x,y) - H'(x,y)\| &= \frac{1}{16^j}\|H(2^j x, 2^j y) - H'(2^j x, 2^j y)\| \\
&\le \frac{1}{16^j}\|f(2^j x, 2^j y) - H(2^j x, 2^j y)\| \\
&\quad + \frac{1}{16^j}\|f(2^j x, 2^j y) - H'(2^j x, 2^j y)\| \\
&\le \frac{2}{16^j}\sum_{i=0}^{\infty}\frac{1}{16^i}\varphi(2^{i+j}x, 2^{i+j}x, 2^{i+j}y, 2^{i+j}y, 0, 0, 0, 0)
\end{aligned}$$

for all $x, y \in A$. According to (13.5), if $j \to \infty$, then the right hand side of the above inequality tends to 0, so we have $H(x, y) = H'(x, y)$ for all $x, y \in A$. This proves the uniqueness of H.

Thus the mapping H is a unique C^*-ternary algebra bi-quadratic bi-homomorphism satisfying (13.9). □

Corollary 13.1 *Let $p < 4$ and θ be positive real numbers, and let $f : A \times A \to B$ be a mapping such that*

$$\begin{aligned}
&\|f(\lambda a + \lambda b + \lambda x + \lambda y, \mu c + \mu d + \mu z + \mu w) + f(\lambda a + \lambda b, \mu c - \mu d) \\
&\quad + f(\lambda a - \lambda b, \mu c + \mu d) + f(\lambda a - \lambda b, \mu c - \mu d) \\
&\quad - 4[f(\lambda a, \mu c) + f(\lambda a, \mu d) + f(\lambda b, \mu c) \\
&\quad + f(\lambda b, \mu d) - \lambda^2\mu^2 f(x + y, z + w)]\| \\
&\le \theta(\|a\|^p + \|b\|^p + \|c\|^p + \|d\|^p + \|x\|^p + \|y\|^p + \|z\|^p + \|w\|^p), \quad (13.13)
\end{aligned}$$

$$\begin{aligned}
&\|f([a,b,c],d) - [f(a,d), f(b,d), f(c,d)]\| \\
&\quad + \|f(a,[b,c,d] - [f(a,b), f(a,c), f(a,d)])\| \\
&\le \theta(\|a\|^p + \|b\|^p + \|c\|^p + \|d\|^p) \quad (13.14)
\end{aligned}$$

for all $a, b, c, d, x, y, z, w \in A$ and all $\lambda, \mu \in \mathbb{T}^1 := \{\lambda \in \mathbb{C};\ |\lambda| = 1\}$. If for each fixed $a, b \in A$ the mapping $t \mapsto f(ta, tb)$ from $\mathbb{R}$ to A is continuous in $t \in \mathbb{R}$, then there is a unique C^*-ternary algebra bi-quadratic bi-homomorphism $H : A \times A \to B$ such that

$$\|f(x,y) - H(x,y)\| \le \frac{2\theta}{1 - 2^{p-4}}(\|x\|^p + \|y\|^p)$$

for all $x, y \in A$.

Proof The proof follows from Theorem 13.2 by taking

$$\varphi(a,b,c,d,x,y,z,w) := \theta\big(\|a\|^p + \|b\|^p + \|c\|^p + \|d\|^p + \|x\|^p + \|y\|^p + \|z\|^p + \|w\|^p\big),$$

$$\psi(a,b,c,d) := \theta\big(\|a\|^p + \|b\|^p + \|c\|^p + \|d\|^p\big)$$

for all $a,b,c,d,x,y,z,w \in A$. □

Theorem 13.3 *Let $f : A \times A \to B$ be a mapping with $f(0,0) = f(a,0) = f(0,b) = 0$ satisfying (13.7) and (13.8). If there exist functions $\varphi : A^8 \to [0,\infty)$ and $\psi : A^4 \to [0,\infty)$ such that*

$$\sum_{i=1}^{\infty} 16^i \varphi\left(\frac{a}{2^i}, \frac{b}{2^i}, \frac{c}{2^i}, \frac{d}{2^i}, \frac{x}{2^i}, \frac{y}{2^i} \frac{z}{2^i}, \frac{w}{2^i}\right) < \infty, \tag{13.15}$$

$$\lim_{i\to\infty} 16^i \psi\left(\frac{a}{2^i}, \frac{b}{2^i}, \frac{c}{2^i}, \frac{d}{2^i}\right) = 0, \tag{13.16}$$

for all $a,b,c,d,x,y,z,w \in A$, and if for each fixed $a,b \in A$ the mapping $t \mapsto f(ta,tb)$ from $\mathbb{R}$ to A is continuous in $t \in \mathbb{R}$, then there is a unique C^*-ternary algebra bi-quadratic bi-homomorphism $H : A \times A \to B$ such that

$$\big\|H(x,y) - f(x,y)\big\| \le 16 \sum_{i=1}^{\infty} 16^i \varphi\left(\frac{x}{2^i}, \frac{x}{2^i}, \frac{y}{2^i}, \frac{y}{2^i}, 0,0,0,0\right) \tag{13.17}$$

for all $x,y \in A$.

Proof It follows from (13.10) that

$$\left\|16 f\left(\frac{x}{2}, \frac{y}{2}\right) - f(x,y)\right\| \le \varphi\left(\frac{x}{2}, \frac{x}{2}, \frac{y}{2}, \frac{y}{2}, 0,0,0,0\right)$$

for all $x,y \in A$. By induction on n, we shall show that

$$\left\|16^n f\left(\frac{x}{2^n}, \frac{y}{2^n}\right) - f(x,y)\right\| \le 16 \sum_{i=1}^{n} 16^i \varphi\left(\frac{x}{2^i}, \frac{x}{2^i}, \frac{y}{2^i}, \frac{y}{2^i}, 0,0,0,0\right) \tag{13.18}$$

for all $x,y \in A$ and all positive integers n, and that

$$\left\|16^{n+m} f\left(\frac{x}{2^{n+m}}, \frac{y}{2^{n+m}}\right) - 16^m f\left(\frac{x}{2^m}, \frac{y}{2^m}\right)\right\|$$
$$\le 16 \sum_{i=1}^{n} 16^{i+m} \varphi\left(\frac{x}{2^{i+m}}, \frac{x}{2^{i+m}}, \frac{y}{2^{i+m}}, \frac{y}{2^{i+m}}, 0,0,0,0\right) \tag{13.19}$$

for all $n > m$ and all $x,y \in A$. It follows from the convergence (13.15) that the sequence $\{16^n f(\frac{x}{2^n}, \frac{y}{2^n})\}$ is Cauchy. Due to the completeness of B, this sequence is convergent. Set

$$H(x,y) := \lim_{n\to\infty} 16^n f\left(\frac{x}{2^n}, \frac{y}{2^n}\right) \tag{13.20}$$

for all $x, y \in A$. If $m = 0$ and $n \to \infty$ in the inequality (13.19), then by (13.15) and (13.20), we have

$$\left\| H(x,y) - f(x,y) \right\| \le 16 \sum_{i=1}^{\infty} 16^i \varphi\left(\frac{x}{2^i}, \frac{x}{2^i}, \frac{y}{2^i}, \frac{y}{2^i}, 0, 0, 0, 0\right)$$

for all $x, y \in A$. The rest of the proof is similar to the proof of Theorem 13.2. □

Corollary 13.2 *Let $p > 4$ and θ be positive real numbers, and let $f : A \times A \to B$ be a mapping satisfying* (13.13) *and* (13.14). *If for each fixed $a, b \in A$ the mapping $t \mapsto f(ta, tb)$ from $\mathbb{R}$ to A is continuous in $t \in \mathbb{R}$, then there is a unique C^*-ternary algebra bi-quadratic bi-homomorphism $H : A \times A \to B$ such that*

$$\left\| f(x,y) - H(x,y) \right\| \le \frac{32\theta}{2^{p-4} - 1}\left(\|x\|^p + \|y\|^p\right)$$

for all $x, y \in A$.

Remark 13.1 We can formulate similar statement to Theorems 13.2 and 13.3 in which we can use the following condition

$$\begin{aligned} &\left\| f\left([a,a,a],a\right) - \left[f(a,a), f(a,a), f(a,a)\right] \right\| \\ &\quad + \left\| f\left(a, [a,a,a] - \left[f(a,a), f(a,a), f(a,a)\right]\right) \right\| \\ &\le \psi(a,a,a,a) \end{aligned}$$

for all $a \in A$ under suitable conditions on the functions φ and ψ and then obtain the generalized Hyers–Ulam–Rassias stability of bi-quadratic Jordan bi-homomorphisms in C^*-ternary algebras.

13.3 Stability of Bi-quadratic Bi-derivations on C^*-Ternary Algebras: An Alternative Fixed Point Approach

Throughout this section, assume that A is a C^*-ternary algebra with norm $\|\cdot\|_A$.

We use a fixed point method and investigate the generalized Hyers–Ulam–Rassias stability of bi-quadratic bi-derivations on C^*-ternary algebras.

Definition 13.3 A mapping $\delta : A \times A \to A$ is called a C^*-ternary bi-quadratic bi-derivation if δ satisfies the following properties:

1. δ is a bi-quadratic mapping,

2. δ is a bi-quadratic homogeneous, that is, $\delta(\lambda a, \mu b) = \lambda^2 \mu^2 \delta(a, b)$ for all $a, b \in A$ and all $\lambda, \mu \in \mathbb{C}$,
3.
$$\delta\big([a,b,c],d\big) = \big[\delta(a,d),b,c\big] + \big[a,\delta(b,c),d\big] + \big[a,b,\delta(c,d)\big],$$
$$\delta\big(a,[b,c,d]\big) = \big[\delta(a,b),c,d\big] + \big[b,\delta(a,c),d\big] + \big[b,c,\delta(a,d)\big]$$
for all $a, b, c, d \in A$.

Definition 13.4 A mapping $\delta : A \times A \to A$ is called a C^*-ternary bi-quadratic Jordan bi-derivation if δ satisfies the properties 1 and 2, and

$$\delta\big([a,a,a],a\big) = \delta\big(a,[a,a,a]\big) = \big[\delta(a,a),a,a\big] + \big[a,\delta(a,a),a\big] + \big[a,a,\delta(a,a)\big]$$

for all $a \in A$.

We recall a fundamental result in fixed point theory.

Theorem 13.4 (see [13]) *Suppose that a complete generalized metric space $(\mathscr{X}, d)$ and a strictly contractive mapping $J : \mathscr{X} \to \mathscr{X}$ with Lipschitz constant $0 < L < 1$ are given. Then, for a given element $x \in \mathscr{X}$, exactly one of the following assertions is true*:

(i) $d(J^n x, J^{n+1} x) = \infty$ *for all* $n \geq 0$.
(ii) *There exists* n_0 *such that* $d(J^n x, J^{n+1} x) < \infty$ *for all* $n \geq n_0$.

Actually, if (ii) *holds, then the sequence $J^n x$ is convergent to a fixed point x^* of J and*

(iii) x^* *is the unique fixed point of J in* $\Lambda := \{y \in \mathscr{X}, d(J^{n_0} x, y) < \infty\}$;
(iv) $d(y, x^*) \leq \frac{d(y, Jy)}{1-L}$ *for all* $y \in \Lambda$.

Theorem 13.5 *Let $f : A \times A \to A$ be a mapping with $f(0,0) = f(a,0) = f(0,b) = 0$ for which there exist functions $\varphi : A^8 \to [0,\infty)$ and $\psi : A^4 \to [0,\infty)$ satisfying* (13.7), (13.15), (13.16), *and*

$$\begin{aligned} &\big\|f\big([a,b,c],d\big) - \big[f(a,d),b,c\big] - \big[a,f(b,c),d\big] - \big[a,b,f(c,d)\big]\big\| \\ &\quad + \big\|f\big(a,[b,c,d]\big) - \big[f(a,b),c,d\big] - \big[b,f(a,c),d\big] - \big[b,c,f(a,d)\big]\big\| \\ &\leq \psi(a,b,c,d) \end{aligned} \tag{13.21}$$

for all $a, b, c, d \in A$. If for each fixed $a, b \in A$ the mapping $t \mapsto f(ta, tb)$ from $\mathbb{R}$ to A is continuous in $t \in \mathbb{R}$, and there exists an $L < 1$ such that $16\varphi(a,b,c,d,x,y,z,w) \leq L\varphi(2a,2b,2c,2d,2x,2y,2z,2w)$ for all $a,b,c,d,x,y,z,w \in A$, then there is a unique C^*-ternary algebra bi-quadratic bi-derivation $\delta : A \times A \to A$ such that

$$\big\|f(x,y) - \delta(x,y)\big\| \leq \frac{L}{16 - 16L} \varphi(x,x,y,y,0,0,0,0) \tag{13.22}$$

for all $x, y \in A$.

Proof It follows from (13.10) that

$$\left\|16f\left(\frac{x}{2},\frac{y}{2}\right)-f(x,y)\right\| \le \varphi\left(\frac{x}{2},\frac{x}{2},\frac{y}{2},\frac{y}{2},0,0,0,0\right) \le \frac{L}{16}\varphi(x,x,y,y,0,0,0,0) \tag{13.23}$$

for all $x, y \in A$. Consider the set $X := \{g|g : A \times A \to A\}$ and introduce the generalized metric on X

$$d(g,h) := \inf\bigl\{t \in \mathbb{R}^+ : \|g(x,y)-h(x,y)\| \le t\varphi(x,x,y,y,0,0,0,0)\ \forall x, y \in A\bigr\}.$$

It is easy to show that (X, d) is complete.

Now we consider a linear mapping $J : X \to X$ such that

$$Jg(x,y) := 16g\left(\frac{x}{2},\frac{y}{2}\right)$$

for all $x, y \in X$.

Let $g, h \in X$ be given such that $d(g,h) = \varepsilon\varphi(x,x,y,y,0,0,0,0)$. Then

$$\|g(x,y)-h(x,y)\| \le \varepsilon$$

for all $x, y \in X$. Hence

$$\|Jg(x,y)-Jh(x,y)\| = 16\left\|g\left(\frac{x}{2},\frac{y}{2}\right)-h\left(\frac{x}{2},\frac{y}{2}\right)\right\| \le 16\varepsilon\varphi\left(\frac{x}{2},\frac{x}{2},\frac{y}{2},\frac{y}{2},0,0,0,0\right) \le L\varepsilon$$

for all $x, y \in X$. So $d(g,h) = \varepsilon$ implies that $d(Jg, Jh) \le L\varepsilon$. This means that

$$d(Jg, Jh) \le Ld(g,h)$$

for all $g, h \in X$. It follows from (13.27) that $d(f, Jf) \le \frac{L}{16}$.

By Theorem 13.4(iii), J has a unique fixed point in the set $X_1 := \{h \in X : d(f,h) < \infty\}$. Let δ be the fixed point of J, that is,

$$\delta(2x, 2y) = 16\delta(x,y)$$

for all $x, y \in A$ satisfying the condition that there exists a $t \in (0,\infty)$ such that

$$\|\delta(x,y)-f(x,y)\| \le t\varphi(x,x,y,y,0,0,0,0)$$

for all $x, y \in A$. On the other hand, we have

$$\lim_{n\to\infty} d\bigl(J^n f, \delta\bigr) = 0.$$

It follows that

$$\lim_{n\to\infty} 16^n f\left(\frac{x}{2^n}, \frac{y}{2^n}\right) = \delta(x, y) \tag{13.24}$$

for all $x, y \in A$. It follows from $d(f, \delta) \le \frac{1}{16-16L} d(f, Jf)$ that

$$d(f, \delta) \le \frac{L}{16 - 16L}.$$

This implies the inequality (13.22). The rest of the proof is similar to the proof of Theorem 13.3. □

Corollary 13.3 *Let $p > 4$ and θ be positive real numbers, and let $f : A \times A \to A$ be a mapping satisfying* (13.13) *and*

$$\begin{aligned} &\big\| f\big([a, b, c], d\big) - \big[f(a, d), b, c\big] - \big[a, f(b, c), d\big] - \big[a, b, f(c, d)\big]\big\| \\ &\quad + \big\| f\big(a, [b, c, d]\big) - \big[f(a, b), c, d\big] - \big[b, f(a, c), d\big] - \big[b, c, f(a, d)\big]\big\| \\ &\le \theta\big(\|a\|^p + \|b\|^p + \|c\|^p + \|d\|^p\big) \end{aligned} \tag{13.25}$$

for all $a, b, c, d \in A$. If for each fixed $a, b \in A$ the mapping $t \mapsto f(ta, tb)$ from $\mathbb{R}$ to A is continuous in $t \in \mathbb{R}$, then there is a unique C^*-ternary algebra bi-quadratic bi-derivation $\delta : A \times A \to A$ such that

$$\big\| f(x, y) - \delta(x, y)\big\| \le \frac{2\theta}{2^p - 16}\big(\|x\|^p + \|y\|^p\big)$$

for all $x, y \in A$.

Proof Setting

$$\begin{aligned} &\varphi(a, b, c, d, x, y, z, w) \\ &\quad := \theta\big(\|a\|^p + \|b\|^p + \|c\|^p + \|d\|^p + \|x\|^p + \|y\|^p + \|z\|^p + \|w\|^p\big) \end{aligned}$$

yields

$$\psi(a, b, c, d) := \theta\big(\|a\|^p + \|b\|^p + \|c\|^p + \|d\|^p\big)$$

for all $a, b, c, d, x, y, z, w \in A$ in Theorem 13.5. Then taking $L = 2^{4-p}$, we get the desired result. □

Theorem 13.6 *Let $f : A \times A \to A$ be a mapping with $f(0, 0) = f(a, 0) = f(0, b) = 0$ for which there exist functions $\varphi : A^8 \to [0, \infty)$ and $\psi : A^4 \to [0, \infty)$ satisfying* (13.5), (13.6), (13.7), *and* (13.21). *If for each fixed $a, b \in A$ the mapping $t \mapsto f(ta, tb)$ from $\mathbb{R}$ to A is continuous in $t \in \mathbb{R}$, and there exists an $L < 1$ such that $\varphi(2a, 2b, 2c, 2d, 2x, 2y, 2z, 2w) \le 16L\varphi(a, b, c, d, x, y, z, w)$ for* *all $a, b, c, d, x, y, z, w \in A$, then there is a unique C^*-ternary algebra bi-quadratic bi-derivation $\delta : A \times A \to A$ such that*

$$\|f(x, y) - \delta(x, y)\| \leq \frac{L}{1-L}\varphi(x, x, y, y, 0, 0, 0, 0) \tag{13.26}$$

for all $x, y \in A$.

Proof It follows from (13.10) that

$$\left\|f(x, y) - \frac{1}{16}f(2x, 2y)\right\| \leq \frac{1}{16}\varphi(x, x, y, y, 0, 0, 0, 0)$$
$$\leq L\varphi\left(\frac{x}{2}, \frac{x}{2}, \frac{y}{2}, \frac{y}{2}, 0, 0, 0, 0\right) \tag{13.27}$$

for all $x, y \in A$. We set $\Phi(x, y) := \varphi(\frac{x}{2}, \frac{x}{2}, \frac{y}{2}, \frac{y}{2}, 0, 0, 0, 0)$ for all $x, y \in A$. Thus

$$\left\|f(x, y) - \frac{1}{16}f(2x, 2y)\right\| \leq L\Phi(x, y) \tag{13.28}$$

for all $x, y \in A$. Consider the set $X := \{g | g : A \times A \to A\}$ and introduce the generalized metric on X

$$d(g, h) := \inf\{t \in \mathbb{R}^+ : \|g(x, y) - h(x, y)\| \leq t\Phi(x, y) \ \forall x, y \in A\}.$$

It is easy to show that (X, d) is complete.

Now we consider the linear mapping $J : X \to X$ such that

$$Jg(x, y) := \frac{1}{16}g(2x, 2y)$$

for all $x, y \in X$.

Let $g, h \in X$ be given such that $d(g, h) = \varepsilon\Phi(x, y)$. Then

$$\|g(x, y) - h(x, y)\| \leq \varepsilon$$

for all $x, y \in X$. Hence

$$\|Jg(x, y) - Jh(x, y)\| = \frac{1}{16}\|g(2x, 2y) - h(2x, 2y)\| \leq \frac{1}{16}\varepsilon\Phi(2x, 2y) \leq L\varepsilon$$

for all $x, y \in X$. So $d(g, h) = \varepsilon$ implies that $d(Jg, Jh) \leq L\varepsilon$. This means that

$$d(Jg, Jh) \leq Ld(g, h)$$

for all $g, h \in X$. It follows from (13.23) that $d(f, Jf) \leq L$.

By Theorem 13.4(iii), J has a unique fixed point in the set $X_1 := \{h \in X : d(f, h) < \infty\}$. Let δ be the fixed point of J, that is,

$$\delta(2x, 2y) = 16\delta(x, y)$$

for all $x, y \in A$ satisfying the condition that there exists a $t \in (0, \infty)$ such that

$$\|\delta(x, y) - f(x, y)\| \leq t\Phi(x, y)$$

for all $x, y \in A$. On the other hand, we have

$$\lim_{n\to\infty} d\big(J^n f, \delta\big) = 0.$$

It follows that

$$\lim_{n\to\infty} \frac{1}{16^n} f\big(2^n x, 2^n y\big) = \delta(x, y)$$

for all $x, y \in A$. It follows from $d(f, \delta) \leq \frac{1}{1-L} d(f, Jf)$ that

$$d(f, \delta) \leq \frac{L}{1 - L}.$$

This implies the inequality (13.26). The rest of the proof is similar to the proof of Theorem 13.2. □

Corollary 13.4 *Let $p < 4$ and θ be positive real numbers, and let $f : A \times A \to A$ be a mapping satisfying* (13.13) *and* (13.25). *If for each fixed $a, b \in A$ the mapping $t \mapsto f(ta, tb)$ from $\mathbb{R}$ to A is continuous in $t \in \mathbb{R}$, then there is a unique C^*-ternary algebra bi-quadratic bi-derivation $\delta : A \times A \to A$ such that*

$$\|f(x, y) - \delta(x, y)\| \leq \frac{2\theta}{2^{4-p} - 1}\big(\|x\|^p + \|y\|^p\big)$$

for all $x, y \in A$.

Proof Setting

$$\begin{aligned}\varphi(a, b, c, d, x, y, z, w) :=& \theta\big(\|a\|^p + \|b\|^p + \|c\|^p + \|d\|^p + \|x\|^p + \|y\|^p \\ &+ \|z\|^p + \|w\|^p\big),\\ \psi(a, b, c, d) :=& \theta\big(\|a\|^p + \|b\|^p + \|c\|^p + \|d\|^p\big)\end{aligned}$$

for all $a, b, c, d, x, y, z, w \in A$ in Theorem 13.6 Then taking $L = 2^{p-4}$, we get the desired result. □

Remark 13.2 We can formulate similar statement to Theorems 13.5 and 13.6 and then obtain the generalized Hyers–Ulam–Rassias stability of bi-quadratic Jordan bi-derivations on C^*-ternary algebras.

13.4 Stability of Bi-quadratic Bi-homomorphisms Between Quasi-Banach Algebras Associated with Jensen-Type Bi-quadratic Mapping

In this section, we introduce the following Jensen-type bi-quadratic functional equation

$$8f\left(\frac{x+y}{2},\frac{z+w}{2}\right)+8f\left(\frac{x-y}{2},\frac{z-w}{2}\right)$$
$$=f(x,z)+f(x,w)+f(y,z)+f(y,w). \tag{$*$}$$

It is easy to see that the function $f(x,y)=cx^2y^2$ is a solution of the functional equation $(*)$. So in this note, we call the equation $(*)$ a Jensen-type bi-quadratic functional equation. Here our purpose is to establish the generalized Hyers–Ulam–Rassias stability of bi-quadratic bi-homomorphisms between quasi-Banach algebras associated with the functional equation $(*)$.

Throughout this section, assume that A is a quasi-Banach algebra and that B is a p-Banach algebra.

Before taking up the main subject, given $f: A \to B$, we define the difference operator $D_{\lambda,\mu}f: A^8 \to B$ by

$$\begin{aligned} D_{\lambda,\mu}f(x,y,z,w,a,b,c,d) :=& 8f\left(\frac{\lambda(x+y+a+b)}{2},\frac{\mu(z+w+c+d)}{2}\right) \\ &+8f\left(\frac{\lambda(x-y+a-b)}{2},\frac{\mu(z-w+c-d)}{2}\right) \\ &-f(\lambda x,\mu z)-f(\lambda x,\mu w)-f(\lambda y,\mu z) \\ &-f(\lambda y,\mu w)-\lambda^2\mu^2 f(a+b,c+d) \end{aligned}$$

for all $x,y,z,a,b,c,d \in A$ and all $\lambda,\mu \in \mathbb{T}^1 := \{\lambda \in \mathbb{C};\ |\lambda|=1\}$.

Definition 13.5 A mapping $H: A\times A \to B$ is called a bi-quadratic bi-homomorphism if H satisfies the following properties:

1. H is a bi-quadratic mapping,
2. H is a bi-quadratic homogeneous, that is, $H(\lambda a,\mu b)=\lambda^2\mu^2 H(a,b)$ for all $a,b\in A$ and all $\lambda,\mu\in\mathbb{C}$,
3.
$$H(ab,c)=H(a,c)H(b,c),$$
$$H(a,bc)=H(a,b)H(a,c)$$

   for all $a,b,c\in A$.

Definition 13.6 A mapping $H : A \times A \to B$ is called a bi-quadratic Jordan bi-homomorphism if H satisfies the properties 1 and 2 in Definition 13.5 and

$$H(a^2, a) = H(a, a^2) = H(a, a)^2$$

for all $a \in A$.

Theorem 13.7 *Let $f : A \times A \to B$ be a mapping with $f(0,0) = f(x,0) = f(0,y) = 0$ for which there exist functions $\varphi : A^8 \to [0,\infty)$ and $\psi : A^3 \to [0,\infty)$ such that*

$$\sum_{i=0}^{\infty} \frac{1}{16^i} \varphi\bigl(2^i x, 2^i y, 2^i z, 2^i w, 2^i a, 2^i b, 2^i c, 2^i d\bigr) < \infty, \tag{13.29}$$

$$\lim_{i\to\infty} \frac{1}{16^i} \psi\bigl(2^i x, 2^i y, 2^i z\bigr) = 0, \tag{13.30}$$

$$\bigl\| D_{\lambda,\mu} f(x,y,z,w,a,b,c,d)\bigr\| \le \varphi(x,y,z,w,a,b,c,d), \tag{13.31}$$

$$\begin{aligned}&\bigl\| f(xy,z) - f(x,z)f(y,z)\bigr\| + \bigl\| f(x,yz) - f(x,y)f(x,z)\bigr\| \\ &\quad \le \psi(x,y,z)\end{aligned} \tag{13.32}$$

for all $x,y,z,w,a,b,c,d \in A$ and all $\lambda, \mu \in \mathbb{T}^1 := \{\lambda \in \mathbb{C};\ |\lambda| = 1\}$. If for each fixed $a,b \in A$ the mapping $t \mapsto f(ta,tb)$ from $\mathbb{R}$ to A is continuous in $t \in \mathbb{R}$, then there is a bi-quadratic bi-homomorphism $H : A \times A \to B$ such that

$$\bigl\| f(x,y) - H(x,y)\bigr\| \le \sum_{i=0}^{\infty} \frac{1}{16^i} \varphi\bigl(2^i x, 0, 2^i y, 0,0,0,0,0\bigr) \tag{13.33}$$

for all $x, y \in A$.

Proof Setting $\lambda = \mu = 1$ and setting $a = b = c = d = y = w = 0$ in (13.31), we have

$$\left\| 16 f\left(\frac{x}{2}, \frac{z}{2}\right) - f(x,z)\right\| \le \varphi(x,0,z,0,0,0,0,0) \tag{13.34}$$

for all $x, z \in A$. Replacing x, z by $2x, 2y$ in (13.34), respectively, we have

$$\left\| f(x,y) - \frac{1}{16} f(2x,2y)\right\| \le \frac{1}{16}\varphi(2x,0,2y,0,0,0,0,0) \tag{13.35}$$

for all $x, y \in A$. Thus we obtain

$$\left\| \frac{1}{16^i} f\bigl(2^i x, 2^i y\bigr) - \frac{1}{16^{i+1}} f\bigl(2^{i+1} x, 2^{i+1} y\bigr)\right\| \frac{1}{16^{i+1}} \varphi\bigl(2^i x, 0, 2^i y, 0,0,0,0,0\bigr)$$

for all $x, y \in A$ and all i. For given integers l, m $(0 \le l < m)$, we get

$$\left\| \frac{1}{16^l} f(2^l x, 2^l y) - \frac{1}{16^m} f(2^m x, 2^m y) \right\| \le \sum_{i=l}^{m-1} \frac{1}{16^{i+1}} \varphi(2^i x, 0, 2^i y, 0, 0, 0, 0, 0) \tag{13.36}$$

for all $x, y \in A$. By (13.29), the sequence $\{\frac{1}{16^i} f(2^i x, 2^i y)\}$ is a Cauchy sequence for all $x, y \in A$. Since B is complete, the sequence $\{\frac{1}{16^i} f(2^i x, 2^i y)\}$ converges for all $x, y \in A$. Define $H : A \times A \to B$ by

$$H(x, y) := \lim_{i\to\infty} \frac{1}{16^i} f(2^i x, 2^i y)$$

for all $x, y \in A$. Setting $l = 0$ and taking $m \to \infty$ in (13.36), one can obtain the inequality (13.33). By Theorem 13.2, H is bi-quadratic and bi-quadratic homogeneous. It follows from (13.30) and (13.32) that

$$\begin{aligned} &\|H(xy, z) - H(x, z)H(y, z)\| + \|H(x, yz) - H(x, y)H(x, z)\| \\ &= \lim_{i\to\infty} \frac{1}{16^i} \big(\|f(2^i x, 2^i y, 2^i z) - f(2^i x, 2^i y) f(2^i y, 2^i z)\| \\ &\quad + \|f(2^i x, 2^i y, 2^i z) - f(2^i x, 2^i y) f(2^i x, 2^i z)\|\big) \\ &\le \lim_{i\to\infty} \frac{1}{16^i} \psi(2^i x, 2^i y, 2^i z) = 0 \end{aligned}$$

for all $x, y, z \in A$. Hence

$$H(xy, z) = H(x, z)H(y, z),$$
$$H(x, yz) = H(x, y)H(x, z)$$

for all $x, y, z \in A$. If $H' : A \times A \to B$ is another bi-quadratic mapping satisfying (13.33), we obtain

$$\begin{aligned} \|H(x, y) - H'(x, y)\| &= \frac{1}{16^j} \|H(2^j x, 2^j y) - H'(2^j x, 2^j y)\| \\ &\le \frac{1}{16^j} \|f(2^j x, 2^j y) - H(2^j x, 2^j y)\| \\ &\quad + \frac{1}{16^j} \|f(2^j x, 2^j y) - H'(2^j x, 2^j y)\| \\ &\le \frac{2}{16^j} \sum_{i=0}^{\infty} \frac{1}{16^i} \varphi(2^{i+j} x, 0, 2^{i+j} y, 0, 0, 0, 0, 0) \end{aligned}$$

for all $x, y \in A$. According to (13.29), if $j \to \infty$, then the right hand side of above inequality tends to 0, so we have $H(x, y) = H'(x, y)$ for all $x, y \in A$. This proves the uniqueness of H.

Thus the mapping H is a unique bi-quadratic bi-homomorphism satisfying (13.33). □

Corollary 13.5 *Let $p < 4$ and θ be positive real numbers, and let $f : A \times A \to B$ be a mapping such that*

$$\begin{aligned} \left\| D_{\lambda,\mu} f(x, y, z, w, a, b, c, d)\right\| \leq & \theta\big(\|a\|^p + \|b\|^p + \|c\|^p + \|d\|^p \\ & + \|x\|^p + \|y\|^p + \|z\|^p + \|w\|^p\big), \end{aligned} \tag{13.37}$$

$$\begin{aligned} & \max\left\{\left\| f(xy, z) - f(x, z)f(y, z)\right\|, \left\| f(x, yz) - f(x, y)f(x, z)\right\|\right\} \\ & \quad \leq \theta\big(\|x\|^p + \|y\|^p + \|z\|^p\big) \end{aligned} \tag{13.38}$$

for all $x, y, z, w, a, b, c, d \in A$ and all $\lambda, \mu \in \mathbb{T}^1 := \{\lambda \in \mathbb{C};\ |\lambda| = 1\}$. If for each fixed $a, b \in A$ the mapping $t \mapsto f(ta, tb)$ from $\mathbb{R}$ to A is continuous in $t \in \mathbb{R}$, then there is a unique bi-quadratic bi-homomorphism $H : A \times A \to B$ such that

$$\left\| f(x, y) - H(x, y)\right\| \leq \frac{\theta}{1 - 2^{p-4}}\big(\|x\|^p + \|y\|^p\big)$$

for all $x, y \in A$.

Proof The proof follows from Theorem 13.7 by taking

$$\begin{aligned} \varphi(x, y, z, w, a, b, c, d) := & \theta\big(\|a\|^p + \|b\|^p + \|c\|^p + \|d\|^p + \|x\|^p + \|y\|^p \\ & + \|z\|^p + \|w\|^p\big), \\ \psi(x, y, z) := & \theta\big(\|x\|^p + \|y\|^p + \|z\|^p\big) \end{aligned}$$

for all $x, y, z, w, a, b, c, d \in A$. □

Theorem 13.8 *Let $f : A \times A \to B$ be a mapping with $f(0, 0) = f(a, 0) = f(0, b) = 0$ satisfying* (13.31) *and* (13.32). *If there exist functions $\varphi : A^8 \to [0, \infty)$ and $\psi : A^3 \to [0, \infty)$ such that*

$$\sum_{i=1}^{\infty} 16^i \varphi\left(\frac{x}{2^i}, \frac{y}{2^i}, \frac{z}{2^i}, \frac{w}{2^i}, \frac{a}{2^i}, \frac{b}{2^i}\frac{c}{2^i}, \frac{d}{2^i}\right) < \infty, \tag{13.39}$$

$$\lim_{i\to\infty} 16^i \psi\left(\frac{x}{2^i}, \frac{y}{2^i}, \frac{z}{2^i}\right) = 0, \tag{13.40}$$

for all $x, y, z, w, a, b, c, d \in A$, also if for each fixed $a, b \in A$ the mapping $t \mapsto f(ta, tb)$ from $\mathbb{R}$ to A is continuous in $t \in \mathbb{R}$, then there is a unique bi-quadratic *bi-homomorphism* $H : A \times A \to B$ *such that*

$$\|H(x,y) - f(x,y)\| \le 16\sum_{i=1}^{\infty} 16^i \varphi\left(\frac{x}{2^i}, 0, \frac{y}{2^i}, 0, 0, 0, 0, 0\right) \tag{13.41}$$

for all $x, y \in A$.

Proof Replacing z by y in (13.34), we get

$$\left\|16f\left(\frac{x}{2}, \frac{y}{2}\right) - f(x,y)\right\| \le \varphi(x, 0, y, 0, 0, 0, 0, 0) \tag{13.42}$$

for all $x, y \in A$. By induction on n, we have

$$\left\|16^n f\left(\frac{x}{2^n}, \frac{y}{2^n}\right) - f(x,y)\right\| \le \sum_{i=0}^{n-1} 16^i \varphi\left(\frac{x}{2^i}, 0, \frac{y}{2^i}, 0, 0, 0, 0, 0\right) \tag{13.43}$$

for all $x, y \in A$. Thus

$$\left\|16^{n+m} f\left(\frac{x}{2^{n+m}}, \frac{y}{2^{n+m}}\right) - 16^m f\left(\frac{x}{2^m}, \frac{y}{2^m}\right)\right\|$$
$$\le \sum_{i=m}^{m+n-1} 16^i \varphi\left(\frac{x}{2^i}, 0, \frac{y}{2^i}, 0, 0, 0, 0, 0\right) \tag{13.44}$$

for all $x, y \in A$ and all $n > m$. By (13.39), the sequence $\{16^i f(\frac{x}{2^i}, \frac{y}{2^i})\}$ is a Cauchy sequence for all $x, y \in A$. Since B is complete, the sequence $\{16^i f(\frac{x}{2^i}, \frac{y}{2^i})\}$ converges for all $x, y \in A$. Define $H : A \times A \to B$ by

$$H(x,y) := \lim_{i\to\infty} 16^i f\left(\frac{x}{2^i}, \frac{y}{2^i}\right)$$

for all $x, y \in A$. Setting $m = 0$ and taking $n \to \infty$ in (13.44), one can obtain the inequality (13.41). The rest of the proof is similar to the proof of Theorem 13.7. □

Corollary 13.6 *Let* $p > 4$ *and* θ *be positive real numbers, and let* $f : A \times A \to B$ *be a mapping satisfying* (13.37) *and* (13.38). *If for each fixed* $a, b \in A$ *the mapping* $t \mapsto f(ta, tb)$ *from* $\mathbb{R}$ *to* A *is continuous in* $t \in \mathbb{R}$, *then there is a unique bi-quadratic bi-homomorphism* $H : A \times A \to B$ *such that*

$$\|f(x,y) - H(x,y)\| \le \frac{16\theta}{2^{p-4} - 1}\left(\|x\|^p + \|y\|^p\right)$$

for all $x, y \in A$.

Remark 13.3 We can formulate similar statement to Theorems 13.7 and 13.8 and then obtain the generalized Hyers–Ulam–Rassias stability of bi-quadratic Jordan bi-homomorphisms in quasi-Banach algebras.

13.5 Stability of Bi-quadratic Bi-derivations on Quasi-Banach Algebras Associated with Jensen-Type Bi-quadratic Mapping

Throughout this section, assume that A is a quasi-Banach algebra.

Definition 13.7 A mapping $\delta : A \times A \to A$ is called a bi-quadratic bi-derivation if δ satisfies the following properties:

1. δ is a bi-quadratic mapping;
2. δ is a bi-quadratic homogeneous, that is, $\delta(\lambda a, \mu b) = \lambda^2 \mu^2 \delta(a, b)$ for all $a, b \in A$ and all $\lambda, \mu \in \mathbb{C}$;
3.
$$\delta(ab, c) = a\delta(b, c) + \delta(a, c)b,$$
$$\delta(a, bc) = b\delta(a, c) + \delta(a, b)c$$
for all $a, b, c \in A$.

Definition 13.8 A mapping $\delta : A \times A \to A$ is called a bi-quadratic Jordan bi-derivation if δ satisfies the properties 1 and 2 of Definition 13.7 and

$$\delta\big(a^2, a\big) = \delta\big(a, a^2\big) = a\delta(a, a) + \delta(a, a)a$$

for all $a \in A$.

Now, we use a fixed point method and investigate the generalized Hyers–Ulam–Rassias stability of bi-quadratic bi-derivations on quasi-Banach algebras.

Theorem 13.9 *Let $f : A \times A \to A$ be a mapping with $f(0, 0) = f(x, 0) = f(0, y) = 0$ for which there exist functions $\varphi : A^8 \to [0, \infty)$ and $\psi : A^3 \to [0, \infty)$ satisfying* (13.29), (13.30), (13.31), *and*

$$\max\big\{\big\|f(xy, z) - xf(y, z) - f(x, z)y\big\|, \big\|f(xy, z) - xf(y, z) - f(x, z)y\big\|\big\} \\ \leq \psi(x, y, z) \tag{13.45}$$

for all $x, y, z \in A$. If for each fixed $a, b \in A$ the mapping $t \mapsto f(ta, tb)$ from $\mathbb{R}$ to A is continuous in $t \in \mathbb{R}$, and there exists an $L < 1$ such that $\varphi(2a, 2b, 2c, 2d, 2x, 2y, 2z, 2w) \leq 16L\varphi(a, b, c, d, x, y, z, w)$ for all $a, b, c, d, x, y, z, w \in A$, then there is a unique bi-quadratic bi-derivation $\delta : A \times A \to A$ such that

$$\big\|f(x, y) - \delta(x, y)\big\| \leq \frac{L}{1 - L}\varphi(x, 0, y, 0, 0, 0, 0, 0) \tag{13.46}$$

for all $x, y \in A$.

Proof By Theorems 13.6 and 13.7, we have $\delta(x,y) := \lim_{n\to\infty} \frac{1}{16^n} f(2^n x, 2^n y)$ for all $x, y \in A$. It follows from (13.45) that

$$\|f(xy,z) - xf(y,z) - f(x,z)y\| \le \psi(x,y,z), \tag{13.47}$$

$$\|f(xy,z) - xf(y,z) - f(x,z)y\| \le \psi(x,y,z) \tag{13.48}$$

for all $x, y \in A$. By the inequality (13.47), we get

$$\begin{aligned}\|\delta(xy,z) - x\delta(y,z) - \delta(x,z)y\| &= \lim_{n\to\infty} \frac{1}{16^n} \|f(2^n x 2^n y, 2^n z) \\ &\quad - 2^n x f(2^n y, 2^n z) - f(2^n x, 2^n z) 2^n y\| \\ &\le \lim_{n\to\infty} \frac{1}{16^n} \psi(2^n x, 2^n y, 2^n z) = 0\end{aligned}$$

for all $x, y, z \in A$. Hence $\delta(xy,z) = x\delta(y,z) + \delta(x,z)y$ for all $x, y, z \in A$. Moreover, by the inequality (13.48), we obtain $\delta(xy,z) = x\delta(y,z) + \delta(x,z)y$ for all $x, y, z \in A$. The rest of the proof is similar to the proof of Theorems 13.6 and 13.7. □

Corollary 13.7 *Let $p < 4$ and θ be positive real numbers, and let $f : A \times A \to A$ be a mapping satisfying* (13.37) *and*

$$\begin{aligned}&\max\{\|f(xy,z) - xf(y,z) - f(x,z)y\|, \|f(xy,z) - xf(y,z) - f(x,z)y\|\} \\ &\quad \le \theta(\|x\|^p + \|y\|^p + \|z\|^p)\end{aligned} \tag{13.49}$$

for all $x, y, z \in A$. If for each fixed $a, b \in A$ the mapping $t \mapsto f(ta, tb)$ from $\mathbb{R}$ to A is continuous in $t \in \mathbb{R}$, then there is a unique bi-quadratic bi-derivation $\delta : A \times A \to A$ such that

$$\|f(x,y) - \delta(x,y)\| \le \frac{\theta}{2^{4-p} - 1} (\|x\|^p + \|y\|^p)$$

for all $x, y \in A$.

Proof Setting

$$\begin{aligned}\varphi(a,b,c,d,x,y,z,w) &:= \theta(\|a\|^p + \|b\|^p + \|c\|^p + \|d\|^p + \|x\|^p + \|y\|^p \\ &\quad + \|z\|^p + \|w\|^p), \\ \psi(x,y,z) &:= \theta(\|x\|^p + \|y\|^p + \|z\|^p)\end{aligned}$$

for all $a, b, c, d, x, y, z, w \in A$ in Theorem 13.7 and then taking $L = 2^{p-4}$, we get the desired result. □

Theorem 13.10 *Let $f : A \times A \to A$ be a mapping with $f(0,0) = f(x,0) = f(0,y) = 0$ for which there exist functions $\varphi : A^8 \to [0,\infty)$ and $\psi : A^3 \to [0,\infty)$ satisfying* (13.31), (13.39), (13.40), *and* (13.45). *If for each fixed $a, b \in A$ the* *mapping* $t \mapsto f(ta, tb)$ *from* $\mathbb{R}$ *to* A *is continuous in* $t \in \mathbb{R}$, *and there exists an* $L < 1$ *such that* $16\varphi(a,b,c,d,x,y,z,w) \le 16L\varphi(2a,2b,2c,2d,2x,2y,2z,2w)$ *for all* $a,b,c,d,x,y,z,w \in A$, *then there is a unique bi-quadratic bi-derivation* $\delta : A \times A \to A$ *such that*

$$\big\| f(x,y) - \delta(x,y) \big\| \le \frac{L}{16 - 16L} \varphi(x,0,y,0,0,0,0,0) \tag{13.50}$$

for all $x, y \in A$.

Proof The proof is similar to the proof of Theorems 13.7 and 13.9. □

Corollary 13.8 *Let* $p > 4$ *and* θ *be positive real numbers, and let* $f : A \times A \to A$ *be a mapping satisfying* (13.37) *and* (13.49). *If for each fixed* $a, b \in A$ *the mapping* $t \mapsto f(ta, tb)$ *from* $\mathbb{R}$ *to* A *is continuous in* $t \in \mathbb{R}$, *then there is a unique bi-quadratic bi-derivation* $\delta : A \times A \to A$ *such that*

$$\big\| f(x,y) - \delta(x,y) \big\| \le \frac{16\theta}{2^{p-4} - 1} \big(\|x\|^p + \|y\|^p \big)$$

for all $x, y \in A$.

Remark 13.4 We can formulate a similar statement to Theorems 13.9 and 13.10 and then obtain the generalized Hyers–Ulam–Rassias stability of bi-quadratic Jordan bi-derivations on quasi-Banach algebras.

References

1. Abramov, V., Kerner, R., Le Roy, B.: Hypersymmetry a Z_3 graded generalization of supersymmetry. J. Math. Phys. **38**, 1650 (1997)
2. Aczel, J., Dhombres, J.: Functional Equations in Several Variables. Cambridge Univ. Press, Cambridge (1989)
3. Almira, J.M., Luther, U.: Inverse closedness of approximation algebras. J. Math. Anal. Appl. **314**, 30–44 (2006)
4. Amir, D.: Characterizations of Inner Product Spaces. Operator Theory, Advances and Applications vol. 20, pp. 3. Birkhäuser, Basel (1986). vi+200 pp., ISBN:3-7643-1774-4
5. Aoki, T.: On the stability of the linear transformation in Banach spaces. J. Math. Soc. Jpn. **2**, 64–66 (1950)
6. Badora, R.: On approximate derivations. Math. Inequal. Appl. **9**(1), 167–173 (2006)
7. Bae, J.H., Park, W.G.: Approximate bi-homomorphisms and bi-derivations in C^*-ternary algebras. Bull. Korean Math. Soc. **47**(1), 195–209 (2010)
8. Benyamini, Y., Lindenstrauss, J.: Geometric Nonlinear Functional Analysis, vol. 1. Amer. Math. Soc. Colloq. Publ., vol. 48. Amer. Math. Soc., Providence (2000)
9. Cayley, A.: On the 34 concomitants of the ternary cubic. Am. J. Math. **4**(1–4), 1–15 (1881)
10. Chou, C.Y., Tzeng, J.H.: On approximate isomorphisms between Banach $*$-algebras or C^*-algebras. Taiwan. J. Math. **10**(1), 219–231 (2006)
11. Cholewa, P.W.: Remarks on the stability of functional equations. Aequ. Math. **27**, 76–86 (1984)

12. Czerwik, S.: On the stability of the quadratic mapping in normed spaces. Abh. Math. Semin. Univ. Hamb. **62**, 59–64 (1992)
13. Diaz, J.B., Margolis, B.: A fixed point theorem of the alternative for the contractions on generalized complete metric space. Bull. Am. Math. Soc. **74**, 305–309 (1968)
14. Ebadian, A., Ghobadipour, N., Eshaghi Gordji, M.: A fixed point method for perturbation of bimultipliers and Jordan bimultipliers in C^*-ternary algebras. J. Math. Phys. **51**, 103508 (2010)
15. Eshaghi Gordji, M., Ghobadipour, N.: Nearly generalized Jordan derivations. Math. Slovaca **61**(1), 1–8 (2011)
16. Eshaghi Gordji, M., Ghobadipour, N.: Stability of (α, β, γ)-derivations on lie C^*-algebras. Int. J. Geom. Methods Mod. Phys. **7**(7), 1093–1102 (2010)
17. Eshaghi Gordji, M., Rassias, J.M., Ghobadipour, N.: Generalized Hyers–Ulam stability of generalized (n, k)-derivations. Abstr. Appl. Anal. 1–8 (2009)
18. Gajda, Z.: On stability of additive mappings. Int. J. Math. Math. Sci. **14**, 431–434 (1991)
19. Găvruta, P.: A generalization of the Hyers–Ulam–Rassias stability of approximately additive mappings. J. Math. Anal. Appl. **184**, 431–436 (1994)
20. Ghobadipour, N., Ebadian, A., Rassias, Th.M., Eshaghi, M.: A perturbation of double derivations on Banach algebras. Commun. Math. Anal. **11**(1), 51–60 (2011)
21. Grabiec, A.: The generalized Hyers–Ulam stability of a class of functional equations. Publ. Math. (Debr.) **48**, 217–235 (1996)
22. Hyers, D.H.: On the stability of the linear functional equation. Proc. Natl. Acad. Sci. **27**, 222–224 (1941)
23. Hyers, D.H., Rassias, Th.M.: Approximate homomorphisms. Aequ. Math. **44**, 125–153 (1992)
24. Isac, G., Rassias, Th.M.: On the Hyers–Ulam stability of ψ-additive mappings. J. Approx. Theory **72**, 131–137 (1993)
25. Isac, G., Rassias, Th.M.: Stability of ψ-additive mappings: applications to nonlinear analysis. Int. J. Math. Math. Sci. **19**, 219–228 (1996)
26. Kapranov, M., Gelfand, I.M., Zelevinskii, A.: Discrimininants, Resultants and Multidimensional Determinants. Birkhäuser, Berlin (1994)
27. Kannappan, Pl.: Quadratic functional equation and inner product spaces. Results Math. **27**, 368–372 (1995)
28. Kerner, R.: Ternary algebraic structures and their applications in physics. Univ. P.M. Curie preprint, Paris (2000). http://arxiv.org/list/math-ph/0011
29. Kerner, R.: The cubic chessboard, geometry and physics. Class. Quantum Gravity **14**, A203 (1997)
30. Kim, H.-M.: Stability for generalized Jensen functional equations and isomorphisms between C^*-algebras. Bull. Belg. Math. Soc. Simon Stevin **14**(1), 1–14 (2007)
31. Miura, T., Oka, H., Hirasawa, G., Takahasi, S.-E.: Superstability of multipliers and ring derivations on Banach algebras. Banach J. Math. Anal. **1**(1), 125–130 (2007)
32. Miura, T., Hirasawa, G., Takahasi, S.-E.: A perturbation of ring derivations on Banach algebras. J. Math. Anal. Appl. **319**(2), 522–530 (2006)
33. Moslehian, M.S.: Hyers–Ulam–Rassias stability of generalized derivations. Int. J. Math. Math. Sci. (2006). Art. ID 93942, 8 pp.
34. Moslehian, M.S.: Ternary derivations, stability and physical aspects. Acta Appl. Math. **100**(2), 187–199 (2008)
35. Moslehian, M.S., Székelyhidi, L.: Stability of ternary homomorphism via generalized Jensen equation. Resultate Math. **49**(3–4), 289–300 (2006)
36. Moslehian, M.S.: Almost derivations on C^*-ternary rings. Bull. Belg. Math. Soc. Simon Stevin **14**, 135–142 (2007)
37. Park, C.-G.: Linear $*$-derivations on C^*-algebras. Tamsui Oxf. J. Math. Sci. **23**(2), 155–171 (2007)
38. Park, C.G.: Lie $*$-homomorphisms between Lie C^*-algebras and Lie $*$-derivations on Lie C^*-algebras. J. Math. Anal. Appl. **293**, 419–434 (2004)

39. Park, C.: Homomorphisms between Lie JC^*-algebras and Cauchy–Rassias stability of Lie JC^*-algebra derivations. J. Lie Theory **15**, 393–414 (2005)
40. Park, C.: Isomorphisms between C^*-ternary algebras. J. Math. Phys. **47**(10), 103512, 12 pp. (2006)
41. Park, C.-G., Najati, A.: Homomorphisms and derivations in C^*-algebras. Abstr. Appl. Anal. (2007). Art. ID 80630, 12 pp.
42. Park, W.G., Bae, J.H.: On a bi-quadratic functional equation and its stability. Nonlinear Anal. **62**, 643–654 (2005)
43. Rassias, J.M.: On a new approximation of approximately linear mappings by linear mappings. Discuss. Math. **7**, 193–196 (1985)
44. Rassias, J.M.: On the stability of the Euler–Lagrange functional equation. Chin. J. Math. **20**(2), 185–190 (1992)
45. Rassias, J.M.: On approximation of approximately linear mappings by linear mappings. Bull. Sci. Math. (2) **108**(4), 445–446 (1984)
46. Rassias, J.M.: On approximation of approximately linear mappings by linear mappings. J. Funct. Anal. **46**(1), 126–130 (1982)
47. Rassias, J.M.: Solution of a problem of Ulam. J. Approx. Theory **57**(3), 268–273 (1989)
48. Rassias, Th.M.: Functional Equations and Inequalities. Kluwer Academic, Dordrecht (2000)
49. Rassias, Th.M.: On the stability of functional equations and a problem of Ulam. Acta Math. Appl. **62**, 23–130 (2000)
50. Rassias, Th.M.: On the stability of functional equations in Banach spaces. J. Math. Anal. Appl. **251**, 264–284 (2000)
51. Rassias, Th.M.: On the stability of minimum points. Mathematica **45**((68)(1)), 93–104 (2003)
52. Rassias, Th.M.: On the stability of the linear mapping in Banach spaces. Proc. Am. Math. Soc. **72**, 297–300 (1978)
53. Rassias, Th.M.: On the stability of the quadratic functional equation and its applications. Stud. Univ. Babeş-Bolyai, Math. **43**(3), 89–124 (1998)
54. Rassias, Th.M.: The problem of S.M. Ulam for approximately multiplicative mappings. J. Math. Anal. Appl. **246**(2), 352–378 (2000)
55. Rassias, Th.M., Tabor, J.: What is left of Hyers–Ulam stability? J. Nat. Geom. **1**, 65–69 (1992)
56. Sewell, G.L.: Quantum Mechanics and Its Emergent Macrophysics. Princeton Univ. Press, Princeton (2002)
57. Skof, F.: Proprietà locali e approssimazione di operatori. Rend. Semin. Mat. Fis. Milano **53**, 113–129 (1983)
58. Rolewicz, S.: Metric Linear Spaces. PWN–Polish Sci. Publ., Warszawa (1984)
59. Ulam, S.M.: Problems in Modern Mathematics. Wiley, New York (1940), Chap. VI, science ed.

Chapter 14
Fixed Point Approach to the Stability of the Quadratic Functional Equation

Elqorachi Elhoucien and Manar Youssef

Abstract In the present paper, we apply a fixed point theorem to prove the Hyers–Ulam–Rassias stability of the quadratic functional equation

$$f(kx+y)+f\big(kx+\sigma(y)\big)=2k^2f(x)+2f(y), \quad x,y\in E_1$$

from a normed space E_1 into a complete β-normed space E_2, where $\sigma : E_1 \longrightarrow E_1$ is an involution and k is a fixed positive integer larger than 2. Furthermore, we investigate the Hyers–Ulam–Rassias stability for the functional equation in question on restricted domains.

The concept of Hyers–Ulam–Rassias stability originated essentially with the Th.M. Rassias' stability theorem that appeared in his paper "On the stability of linear mapping in Banach spaces" (Proc. Am. Math. Soc. 72:297–300, 1978).

Key words Fixed points · Quadratic functional equation · Stability · Involution

Mathematics Subject Classification 65Q20 · 49K40

14.1 Introduction

In 1940, S.M. Ulam [38] gave a wide ranging talk before the mathematics Club of the University of Wisconsin in which he discussed a number of important unsolved problems. Among those was the question concerning the stability of group homomorphisms:

Given a group G_1, a metric group (G_2,d), a number $\varepsilon>0$, and a mapping $f : G_1 \longrightarrow G_2$ which satisfies $d(f(xy),f(x)f(y))\le\delta$ for all $x,y\in G_1$, do there exist

Dedicated to Professor Themistocles M. Rassias on the occasion of his 60th birthday.

E. Elhoucien (✉) · M. Youssef
Department of Mathematics, Faculty of Sciences, University Ibn Zohr, Agadir, Morocco
e-mail: elqorachi@hotmail.com

M. Youssef
e-mail: manaryoussef1984@gmail.com

P.M. Pardalos et al. (eds.), *Nonlinear Analysis*, Springer Optimization and Its Applications 68, In Honor of Themistocles M. Rassias on the Occasion of his 60th Birthday,
DOI 10.1007/978-1-4614-3498-6_14,

a homomorphism $g : G_1 \longrightarrow G_2$ and a constant $k > 0$, depending only on G_1 and G_2, such that $d(f(x), g(x)) \le k\varepsilon$ for all $x \in G_1$?

In 1941, D.H. Hyers [13] considered the case of approximately additive mappings under the assumption that G_1 and G_2 are Banach spaces.

T. Aoki [1] and Th.M. Rassias [27] provided a generalization of the Hyers' Theorem for additive mappings and for linear mappings, respectively, by allowing the Cauchy difference to be unbounded.

Theorem 14.1 (Th.M. Rassias) *Let $f : E_1 \to E_2$ be a mapping from a normed vector space E_1 into a Banach space E_2. Assume that there exist $\theta > 0$ and $p < 1$ such that*

$$\left\| f(x+y) - f(x) - f(y) \right\| \le \theta\left(\|x\|^p + \|y\|^p\right),$$

for all $x, y \in E_1$ (for all $x, y \in E_1 \backslash \{0\}$ if $p < 0$). Then, the limit

$$a(x) = \lim_{n \to +\infty} \frac{f(2^n x)}{2^n}$$

exists for all $x \in E_1$, and $a : F \to H$ is the unique additive mapping which satisfies

$$\left\| f(x) - a(x) \right\| \le \frac{2\theta}{2 - 2^p} \|x\|^p$$

for all $x \in E_1$ (for all $x \in E_1 \backslash \{0\}$ if $p < 0$). Also, if for each $x \in E_1$ the function $f(tx)$ is continuous in $t \in \mathbb{R}$, then a is $\mathbb{R}$-linear.

This result provided a remarkable generalization of the theorem proved by Hyers. What is more important here is that Rassias' Theorem simulated several mathematicians working with functional equations to investigate this kind of stability for many important functional equations. Taking this fact into consideration, the terminology of Hyers–Ulam–Rassias stability originates from this historical background. Beginning around the year 1980, several results for the Hyers–Ulam–Rassias stability of many functional equations have been proved by several researchers. For more details, we can refer to [6, 7, 9–37].

In 1996, G. Isac and Th.M. Rassias [18] were the first to provide applications of the stability theory of functional equations for the proof of new fixed point theorems with applications.

In [4, 5], L. Cădariu and V. Radu applied the fixed point method to the investigation of the Jensen and the Cauchy functional equations.

Let E_1 be a vector space, let E_2 be a complete normed space, and let $k \ge 2$ be a fixed positive integer. In this paper, we consider the quadratic functional equation

$$f(kx+y) + f\big(kx + \sigma(y)\big) = 2k^2 f(x) + 2f(y), \quad x, y \in E_1 \tag{14.1}$$

where $\sigma : E_1 \to E_1$ is an involution, i.e., $\sigma(x+y) = \sigma(x) + \sigma(y)$, for all $x, y \in E_1$, and $\sigma \circ \sigma(x) = x$, $\forall x \in E_1$.

The functional equation

$$f(kx+y)+f(kx-y)=2k^2f(x)+2f(y), \quad x,y\in E_1 \tag{14.2}$$

corresponds to $\sigma=-I$. The stability problem of (14.2) was proved by J. Lee et al. [19]. Furthermore, J. Lee et al. proved that a mapping $f:E_1\to E_2$ satisfies (14.2) if and only if f is a solution of the classical quadratic functional equation

$$f(x+y)+f(x-y)=2f(x)+2f(y), \quad x,y\in E_1. \tag{14.3}$$

In [9], the authors extended the J. Lee et al. results to the more general equation (14.1). Furthermore, the Hyers–Ulam stability on unbounded domain was also studied.

Using the fixed point method, C. Park et al. [25] proved the Hyers–Ulam stability problem for the quadratic functional (14.2) for $k=2$:

$$f(2x+y)+f(2x-y)=8f(x)+2f(y), \quad x,y\in E_1. \tag{14.4}$$

In 1998, Jung [20] investigated the Hyers–Ulam stability for additive and quadratic mappings on restricted domains (see also [21, 22]). Hyers et al. [14] investigated the Hyers–Ulam stability of the additive mappings on restricted domains. Recently A. Rahimi et al. [26] investigated the Hyers–Ulam–Rassias stability of the quadratic equation on restricted domains. In [9, 24], and [10], the authors studied the Hyers–Ulam stability of the quadratic functional equation (14.1) on unbounded domains.

In this paper, our results are organized as follows: In Sect. 14.2, we apply the fixed point method as in [4] to prove the Hyers–Ulam stability of the functional equation (14.1). In this case, the range of relevant functions is extended to any complete β-normed space. In Sect. 14.3, we investigate the Hyers–Ulam–Rassias stability of the quadratic equation (14.1) on restricted domains.

First, we shall recall two fundamental results in fixed point theory. The reader is referred to the book of D.H. Hyers, G. Isac, and Th.M. Rassias [16] for an extensive account of fixed point theory with several applications.

Theorem 14.2 (Banach's contraction principle) *Let (X,d) be a complete metric space, and consider a mapping $J:X\to X$, which is strictly contractive, that is,*

$$d(Jx,Jy)\le Ld(x,y), \quad \forall x,y\in X$$

for some (Lipschitz constant) $L<1$. Then,

1. *The mapping J has one, and only one, fixed point $x^*=J(x^*)$.*
2. *The fixed point x^* is globally attractive, that is,*

$$\lim_{n\to+\infty} J^n x = x^*$$

for any starting point $x\in X$.

3. *One has the following estimates*:

$$d(J^n x, x^*) \leq L^n d(x, x^*),$$

$$d(J^n x, x^*) \leq \frac{1}{1-L} d(J^n x, J^{n+1} x),$$

$$d(x, x^*) \leq \frac{1}{1-L} d(x, Jx)$$

for all nonnegative integers n and all $x \in X$.

Let X be a set. A function $d : X \times X \to [0, +\infty]$ is called a *generalized metric* on X if d satisfies the following:

1. $d(x, y) = 0$ if and only if $x = y$;
2. $d(x, y) = d(y, x)$ for all $x, y \in X$;
3. $d(x, z) \leq d(x, y) + d(y, z)$ for all $x, y, z \in X$.

Theorem 14.3 (The alternative of fixed point) [8] *Suppose we are given a complete generalized metric space (X, d) and a strictly contractive mapping $J : X \to X$, with the Lipschitz constant $L < 1$. Then, for each given element $x \in X$, either*

$$d(J^n x, J^{n+1} x) = +\infty$$

for all nonnegative integers n or there exists a positive integer n_0 such that

1. $d(J^n x, J^{n+1} x) < +\infty, \forall n \geq n_0$;
2. *The sequence $J^n x$ converges to a fixed point y^* of J;*
3. *y^* is the unique fixed point of J in the set $Y = \{y \in X : d(J^{n_0} x, y) < +\infty\}$;*
4. *$d(y, y^*) \leq \frac{1}{1-L} d(y, Jy)$ for all $y \in Y$.*

Throughout this paper, we fix a real number β with $0 < \beta \leq 1$ and let $\mathbb{K}$ denote either $\mathbb{R}$ or $\mathbb{C}$. Suppose E is a vector space over $\mathbb{K}$. A function $\|\cdot\|_\beta : E \longrightarrow [0, \infty)$ is called a β-norm if and only if it satisfies

1. $\|x\|_\beta = 0$ if and only if $x = 0$;
2. $\|\lambda x\|_\beta = |\lambda|^\beta \|x\|_\beta$ for all $\lambda \in \mathbb{K}$ and all $x \in E$;
3. $\|x + y\|_\beta \leq \|x\|_\beta + \|y\|_\beta$ for all $x, y \in E$.

14.2 Stability Using the Alternative Fixed Point

Let $k \geq 2$ be a fixed positive integer and let $\sigma : E_1 \to E_1$ be an involution. Using the fixed point method, we prove the Hyers–Ulam–Rassias stability of the quadratic functional equation

$$f(kx + y) + f(kx + \sigma(y)) = 2k^2 f(x) + 2f(y), \quad x, y \in E_1. \tag{14.5}$$

Theorem 14.4 *Let E_1 be a vector space over $\mathbb{K}$ and let E_2 be a complete β-normed space over $\mathbb{K}$, where β is a fixed real number with $0 < \beta \le 1$. Suppose $\varphi : E_1 \times E_1 \to \mathbb{R}^+$ is a given function and there exists a constant L, $0 < L < 1$, such that*

$$\varphi(kx, 0) \le k^{2\beta} L\varphi(x, 0) \tag{14.6}$$

for all $x \in E_1$. Furthermore, let $f\colon E_1 \longrightarrow E_2$ be a function with $f(0) = 0$ which satisfies

$$\big\| f(kx + y) + f\big(kx + \sigma(y)\big) - 2k^2 f(x) - 2f(y)\big\| \le \varphi(x, y) \tag{14.7}$$

for all $x, y \in E_1$. If φ satisfies

$$\lim_{n\to\infty} \frac{\varphi(k^n x, k^n y)}{k^{2n\beta}} = 0 \tag{14.8}$$

for all $x, y \in E_1$, then there exists a unique mapping $q : E_1 \to E_2$ which solves (14.5) and

$$\big\| f(x) - q(x)\big\|_\beta \le \frac{1}{2^\beta k^{2\beta}} \frac{1}{1-L} \varphi(x, 0) \tag{14.9}$$

for all $x \in E_1$.

Proof First let us define X to be the set $X := \{g : E_1 \longrightarrow E_2\}$ and introduce the generalized metric on X as follows:

$$d(g, h) = \inf\big\{C \in [0, \infty) : \big\|g(x) - h(x)\big\|_\beta \le C\varphi(x, 0);\ \forall x \in E_1\big\}.$$

It easy to show that (X, d) is complete; see, for example, [6, Theorem 2.5].

Now we define an operator $J : X \longrightarrow X$ such that

$$Jh(x) = \frac{1}{k^2} f(kx), \tag{14.10}$$

for all $x \in E_1$, and we assert that J is strictly contractive on X with the Lipschitz constant L. Given $g, h \in X$, let $C \in [0, \infty)$ be an arbitrary constant with $d(g, h) \le C$, that is,

$$\big\|g(x) - h(x)\big\|_\beta \le C\varphi(x, 0) \tag{14.11}$$

for all $x \in E_1$. It follows from (14.11) and (14.6) that

$$\begin{aligned}\big\|(Jg)(x) - (Jh)(x)\big\|_\beta &= \frac{1}{k^{2\beta}} \big\|g(kx) - h(kx)\big\|_\beta \\ &\le \frac{C}{k^{2\beta}} \varphi(kx, 0) \\ &\le LC\varphi(x, 0)\end{aligned}$$

for all $x \in E_1$. So, $d(Jg, Jh) \leq LC$. Hence we conclude that $d(Jg, Jh) \leq Ld(g, h)$, for all $g, h \in X$.

Now, by letting $y = 0$ in (14.7) and dividing both sides by $2k^2$, we get

$$\frac{1}{2^{\beta}k^{2\beta}}\left\|2f(kx) - 2f\left(k^2x\right)\right\|_{\beta} = \left\|\frac{1}{k^2}f(kx) - f(x)\right\|_{\beta} = \left\|(Jf)(x) - f(x)\right\|_{\beta}$$
$$\leq \frac{1}{2^{\beta}k^{2\beta}}\varphi(x, 0)$$

for all $x \in E_1$, and we claim that

$$d(Jf, f) \leq \frac{1}{2^{\beta}k^{2\beta}} < \infty. \tag{14.12}$$

By Theorem 14.3, there exists a mapping $q : E_1 \longrightarrow E_2$ such that

(i) q is a fixed point of J, that is, $q(kx) = k^2 q(x)$ for all $x \in E_1$. The mapping q is a unique fixed point of J in the set $Y = \{g \in X : d(f, g) < \infty\}$.

(ii) $d(J^n f, q) \to 0$ as $n \to \infty$. This implies that there exists a sequence C_n such that $C_n \to 0$ as $n \to \infty$ and

$$\left\|J^n f(x) - q(x)\right\|_{\beta} \leq C_n\varphi(x, 0) \tag{14.13}$$

for all $x \in E_1$ and all $n \in \mathbb{N}$. Consequently, we obtain

$$q(x) = \lim_{n\to\infty} \frac{1}{k^{2n}} f\left(k^n x\right) \tag{14.14}$$

for all $x \in E_1$.

(iii)

$$d(f, q) \leq \frac{1}{1 - L}d(Jf, f) \leq \frac{1}{2^{\beta}k^{2\beta}}\frac{1}{1 - L}, \tag{14.15}$$

which proves the inequality (14.9).

Now, we will prove that q is a solution of the quadratic functional equation (14.5). It follows from (14.7), (14.14), and (14.8) that

$$\left\|q(kx + y) + q\left(kx + \sigma(y)\right) - 2k^2q(x) - 2q(y)\right\|_{\beta}$$
$$= \lim_{n\to\infty} \frac{1}{k^{2n\beta}}\left\|f\left(kk^n x + k^n y\right) + f\left(kk^n x + \sigma\left(k^n y\right)\right) - 2k^2 f\left(k^n x\right) - 2f\left(k^n y\right)\right\|_{\beta}$$
$$\leq \lim_{n\to\infty} \frac{1}{k^{2n\beta}}\varphi\left(k^n x, k^n y\right) = 0$$

for all $x, y \in E_1$ and this implies the desired result.

Assume now that $q_1 : E_1 \longrightarrow E_2$ is another solution of (14.5) satisfying (14.9), in particular q_1 satisfies $q_1(kx) = k^2q(x)$, so q_1 is a fixed point of J. From the definition of d and the inequality (14.9), the assertion (14.15) is also true with q_1 in place of q. By using Theorem 14.3, we get the uniqueness of q. This ends the proof of the theorem. □

Corollary 14.1 ([25], $k=2$ and $\sigma=-I$) *Let E_1 be a vector space over $\mathbb{K}$ and let E_2 be a complete β-normed space over $\mathbb{K}$, where β is a fixed real number with $0<\beta\le 1$. Suppose $\varphi: E_1\times E_1\to\mathbb{R}^+$ is a given function and there exists a constant L, $0<L<1$, such that*

$$\varphi(2x,0)\le 2^{2\beta}L\varphi(x,0) \tag{14.16}$$

for all $x\in E_1$. Furthermore, let $f: E_1\longrightarrow E_2$ be a function with $f(0)=0$ which satisfies

$$\big\|f(2x+y)+f\big(2x+\sigma(y)\big)-8f(x)-2f(y)\big\|\le\varphi(x,y) \tag{14.17}$$

for all $x,y\in E_1$. If φ satisfies

$$\lim_{n\to+}\frac{\varphi(2^nx,2^ny)}{2^{2n\beta}}=0 \tag{14.18}$$

for all $x,y\in E_1$, then there exists a unique mapping $q: E_1\to E_2$ solution of (14.2) *such that*

$$\big\|f(x)-q(x)\big\|_\beta\le\frac{1}{2^\beta 2^{2\beta}}\frac{1}{1-L}\varphi(x,0) \tag{14.19}$$

for all $x\in E_1$.

Corollary 14.2 *Let $0<p<2$ and θ be a positive real numbers and choose a constant β with $0<\frac{p}{2}<\beta\le 1$. Let E_1 be a vector space over $\mathbb{K}$ and let E_2 be a complete β-normed space over $\mathbb{K}$. If $f: E_1\longrightarrow E_2$ is a mapping such that*

$$\big\|f(kx+y)+f\big(kx+\sigma(y)\big)-2k^2f(x)-2f(y)\big\|\le\theta\big(\|x\|^p+\|y\|^p\big) \tag{14.20}$$

for all $x,y\in E_1$. Then, there exists a unique quadratic mapping $q: E_1\longrightarrow E_2$ which solves (14.5) *and such that*

$$\big\|f(x)-q(x)\big\|\le\frac{\theta}{2^\beta(k^{2\beta}-k^p)}\|x\|^p \tag{14.21}$$

for all $x\in E_1$.

Proof If we set $L=\frac{k^p}{k^{2\beta}}$, then we have $0<L<1$ and take

$$\varphi(x,y)=\theta\big(\|x\|^p+\|y\|^p\big)=k^{2\beta-p}L\theta\big(\|x\|^p+\|y\|^p\big),\quad x,y\in E_1.$$

By putting $x=0$ and $y=0$ in (14.20), we get $\|2k^2f(0)\|_\beta\le 0$, so $f(0)=0$. According to Theorem 14.4, there exists a unique solution $q: E_1\longrightarrow E_2$ of (14.5) such that (14.21) holds for every $x\in E_1$. □

In the following, we will remove the hypothesis $f(0)=0$.

Theorem 14.5 *Let E_1 be a vector space over $\mathbb{K}$ and let E_2 be a complete β-normed space over $\mathbb{K}$, where β is a fixed real number with $0<\beta\leq 1$. Suppose $\varphi: E_1\times E_1\to\mathbb{R}^+$ is a given function and there exists a constant L, $\frac{1}{k^{2\beta}}\leq L<1$, such that*

$$\varphi(kx,0)\leq k^{2\beta}L\varphi(x,0) \tag{14.22}$$

for all $x\in E_1$. Furthermore, let $f\colon E_1\longrightarrow E_2$ be a function which satisfies

$$\left\| f(kx+y)+f\big(kx+\sigma(y)\big)-2k^2f(x)-2f(y)\right\|\leq\varphi(x,y) \tag{14.23}$$

for all $x,y\in E_1$. If φ satisfies

$$\lim_{n\to +}\frac{\varphi(k^nx,k^ny)}{k^{2n\beta}}=0 \tag{14.24}$$

for all $x,y\in E_1$, then there exists a unique mapping $q: E_1\to E_2$ which solves (14.5) and such that

$$\left\| f(x)-q(x)-f(0)\right\|_\beta\leq\frac{1}{2^\beta k^{2\beta}}\frac{1}{1-L}\big(\varphi(0,0)+\varphi(x,0)\big) \tag{14.25}$$

for all $x\in E_1$.

Proof If we put $x=0$, $y=0$ in (14.23), then we obtain $\|2k^2f(0)\|_\beta\leq\varphi(0,0)$. By using the new functions $g(x)=f(x)-f(0)$ and $\psi(x,y)=\varphi(x,y)+\varphi(0,0)$, we obtain

$$\left\| g(kx+y)+g\big(kx+\sigma(y)\big)-2k^2g(x)-2g(y)\right\|\leq\psi(x,y),\quad x,y\in E_1,$$

$\psi(kx,0)=\varphi(0,0)+\varphi(kx,0)\leq\varphi(0,0)+k^{2\beta}L\varphi(x,0)\leq k^{2\beta}L(\varphi(0,0)+\varphi(x,0))=k^{2\beta}L\psi(x,0)$, because $1\leq k^{2\beta}L$. Furthermore, $\frac{\psi(k^nx,k^ny)}{k^{2n\beta}}\to 0$ as $n\to\infty$. Due to Theorem 14.4, we get the rest of the proof. □

In a similar manner, by applying the alternative of fixed point, we can prove the following theorem.

Theorem 14.6 *Let E_1 be a vector space over $\mathbb{K}$ and let E_2 be a complete β-normed space over $\mathbb{K}$, where β is a fixed real number with $0<\beta\leq 1$. Suppose $\varphi: E_1\times E_1\to\mathbb{R}^+$ is a given function and there exists a constant $L, 0<L<1$, such that*

$$\varphi(x,0)\leq\frac{L}{k^{2\beta}}\varphi(kx,0) \tag{14.26}$$

for all $x\in E_1$. Furthermore, let $f\colon E_1\longrightarrow E_2$ be a function with $f(0)=0$ which satisfies

$$\left\| f(kx+y)+f\big(kx+\sigma(y)\big)-2k^2f(x)-2f(y)\right\|\leq\varphi(x,y) \tag{14.27}$$

for all $x, y \in E_1$. *If* φ *satisfies*

$$\lim_{n\to+} k^{2n\beta}\varphi\left(\frac{x}{k^n}, \frac{y}{k^n}\right) = 0 \tag{14.28}$$

for all $x, y \in E_1$, *then there exists a unique mapping* $q : E_1 \to E_2$ *which solves* (14.5) *and such that*

$$\|f(x) - q(x)\|_\beta \leq \frac{1}{2^\beta k^{2\beta}} \frac{L}{(1-L)}\varphi(x, 0) \tag{14.29}$$

for all $x \in E_1$.

Proof We use the same definitions for X and d as in the proof of Theorem 14.4. So, (X, d) is complete. Also, we define the operator $J : X \to X$ by

$$Jh(x) = k^2 f\left(\frac{x}{k}\right) \tag{14.30}$$

for all $x \in E_1$. By mathematical induction, we can show that

$$J^n f(x) = k^{2n} f\left(\frac{x}{k^n}\right) \tag{14.31}$$

for each $n \in \mathbb{N}$.

We apply the same argument as in the proof of Theorem 14.4 and prove that J is a strictly contractive operator. Moreover, we prove that

$$d(Jf, f) \leq \frac{L}{2^\beta k^{2\beta}}. \tag{14.32}$$

According to the fixed point alternative, there exists a function $q : E_1 \to E_2$ which is a fixed point of J, such that

$$q(x) = \lim_{n\to\infty} k^{2n} f\left(\frac{x}{k^n}\right) \tag{14.33}$$

for all $x \in E_1$. Analogously to the proof of Theorem 14.4, we can show that q is a solution of (14.5).

Using Theorem 14.3.4 and (14.32), we get

$$d(f, q) \leq \frac{1}{2^\beta k^{2\beta}} \frac{L}{1-L}, \tag{14.34}$$

which implies the validity of (14.29). The uniqueness of q can be derived by using same argument as in the proof of Theorem 14.4. This completes the proof of this theorem. □

Corollary 14.3 ([25], $k=2$) *Let E_1 be a vector space over $\mathbb{K}$ and let E_2 be a complete β-normed space over $\mathbb{K}$, where β is a fixed real number with $0<\beta\le 1$. Suppose $\varphi: E_1\times E_1\to\mathbb{R}^+$ is a given function and there exists a constant L, $0<L<1$, such that*

$$\varphi(x,0)\le\frac{L}{2^{2\beta}}L\varphi(2x,0) \tag{14.35}$$

for all $x\in E_1$. Furthermore, let $f\colon E_1\longrightarrow E_2$ be a function with $f(0)=0$ which satisfies

$$\big\|f(2x+y)+f(2x-y)-8f(x)-2f(y)\big\|\le\varphi(x,y) \tag{14.36}$$

for all $x,y\in E_1$. If φ satisfies

$$\lim_{n\to+} 2^{2n\beta}\varphi\left(\frac{x}{2^n}x,\frac{y}{2^n}\right)=0 \tag{14.37}$$

for all $x,y\in E_1$, then there exists a unique mapping $q: E_1\to E_2$ which solves (14.3) *and such that*

$$\big\|f(x)-q(x)\big\|_\beta\le\frac{L}{8-8L}\varphi(x,0) \tag{14.38}$$

for all $x\in E_1$.

Corollary 14.4 *Let $p>2$ and θ be a positive real number, and choose a constant β with $0<\beta<\frac{p}{2}$. Let E_1 be a vector space over $\mathbb{K}$ and let E_2 be a complete β-normed space over $\mathbb{K}$. If $f\colon E_1\longrightarrow E_2$ is a mapping which satisfies*

$$\big\|f(kx+y)+f\big(kx+\sigma(y)\big)-2k^2f(x)-2f(y)\big\|\le\theta\big(\|x\|^p+\|y\|^p\big) \tag{14.39}$$

for all $x,y\in E_1$. Then, there exists a unique solution $q\colon E_1\longrightarrow E_2$ of (14.5) *such that*

$$\big\|f(x)-q(x)\big\|_\beta\le\frac{\theta}{2^\beta(k^p-k^{2\beta})}\|x\|^p \tag{14.40}$$

for all $x\in E_1$.

14.3 Stability of the Quadratic Functional Equation (14.5) on Restricted Domains

In this section, using ideas from the papers of F. Skof [37], D.H. Hyers et al. [14], and A. Rahimi et al. [26], the Hyers–Ulam–Rassias stability of the quadratic functional equation (14.5) will be investigated on restricted domains.

In the following theorem, we consider the case: $k\ge 2$.

Theorem 14.7 *Given a real normed-space E_1 and a real Banach space E_2, let $M > 0$, $\varepsilon > 0$, and choose p, k with $0 < p < 2$, $k \geq 2$. Let $f\colon E_1 \longrightarrow E_2$ be a mapping with $f(0) = 0$ satisfying*

$$\left\| f(kx+y) + f\big(kx+\sigma(y)\big) - 2k^2 f(x) - 2f(y) \right\| \leq \delta + \varepsilon\big(\|x\|^p + \|y\|^p\big) \tag{14.41}$$

for all $x, y \in E_1$ such that $\|x\|^p + \|y\|^p \geq M^p$. Then there exists a unique quadratic mapping $Q : E_1 \longrightarrow E_2$ which solves (14.5) and such that

$$\left\| f(x) - Q(x) \right\| \leq \frac{\delta}{2(k^2-1)} + \frac{1}{k^2 - k^p}\frac{\varepsilon}{2}\|x\|^p \tag{14.42}$$

for all $x \in E_1$ with $\|x\| \geq M$.

Proof Letting $y = 0$ in (14.41), we get

$$\left\| f(kx) - k^2 f(x) \right\| \leq \frac{\delta}{2} + \frac{\varepsilon}{2}\|x\|^p, \tag{14.43}$$

for all $x \in E_1$ with $\|x\| \geq M$. If we replace x by $k^n x$ in (14.43) and divide both sides of this inequality by $k^{2(n+1)}$, we obtain

$$\left\| \frac{f(k^{n+1}x)}{k^{2(n+1)}} - \frac{f(k^n x)}{k^{2n}} \right\| \leq \frac{1}{k^{2(n+1)}}\frac{\delta}{2} + \frac{\varepsilon}{2k^2}k^{n(p-2)}\|x\|^p \tag{14.44}$$

for all $x \in E_1$ with $\|x\| \geq M$ and all $n \in \mathbb{N}$. Consequently,

$$\begin{aligned} \left\| \frac{f(k^{n+1}x)}{k^{2(n+1)}} - \frac{f(k^m x)}{k^{2m}} \right\| &\leq \sum_{j=m}^{n} \left\| \frac{f(k^{j+1}x)}{k^{2(j+1)}} - \frac{f(k^j x)}{k^{2j}} \right\| \\ &\leq \frac{\delta}{2}\frac{1}{k^2}\sum_{j=m}^{n}\frac{1}{k^{2j}} + \frac{\varepsilon}{2k^2}\|x\|^p\sum_{j=m}^{n}k^{j(p-2)} \end{aligned} \tag{14.45}$$

for all $x \in E_1$ with $\|x\| \geq M$ and all integers $n \geq m \geq 0$. From (14.45), we deduce that the sequence $\{k^{-2n} f(k^n x)\}$ converges for all $x \in E_1$ with $\|x\| \geq M$, and thus the limit $\psi(x) = \lim_{n\to\infty} \frac{f(k^n x)}{k^{2n}}$ exists when $\|x\| \geq M$.

It's easy to verify that

$$\psi(kx) = k^2\psi(x) \quad \text{when } \|x\| \geq M. \tag{14.46}$$

Letting $m = 0$ and $n \to \infty$ in (14.45), we obtain

$$\left\| f(x) - \psi(x) \right\| \leq \frac{\delta}{2(k^2-1)} + \frac{1}{k^2 - k^p}\frac{\varepsilon}{2}\|x\|^p \tag{14.47}$$

for all $x \in E_1$ with $\|x\| \geq M$.

Now we suppose that $\|x\|$, $\|y\|$, $\|kx+y\|$, $\|kx+\sigma(y)\|$ are all greater than M. Then by (14.41) and the definition of ψ, we get

$$\psi(kx+y)+\psi\big(kx+\sigma(y)\big)=2k^2\psi(x)+2\psi(y). \tag{14.48}$$

Using the methods of [17, 37], and [26], we can extend ψ to the whole space E_1. Given any $x\in E_1$ with $0<\|x\|<M$, let $s=s(x)$ denote the largest integer such that $M\leq k^s\|x\|<k^{\frac{2}{p}}M$. Define the mapping $Q\colon E_1\longrightarrow E_2$ as follows:

$$\begin{cases} Q(0)=0, \\ \frac{\psi(k^s x)}{k^{2s}}, & \text{for } 0<\|x\|<M, \text{ where } s=s(x), \\ \psi(x), & \text{for } \|x\|\geq M. \end{cases} \tag{14.49}$$

Letting $x\in E_1$ with $0<\|x\|<M$ and $s=s(x)$, we have two cases:
Case 1: If $k\|x\|\geq M$, we have from (14.46)

$$Q(kx)=\psi(kx)=\frac{\psi(k^2x)}{k^2}=k^2\frac{\psi(k^sx)}{k^{2s}}=k^2Q(x). \tag{14.50}$$

Case 2: If $0<k\|x\|<M$, then $s-1$ is the largest integer satisfying $M\leq k^{s-1}\|x\|<k^{\frac{2}{p}}M$ and

$$Q(kx)=\frac{\psi(k^sx)}{k^{2(s-1)}}=k^2\frac{\psi(k^sx)}{k^{2s}}=k^2Q(x), \tag{14.51}$$

and we have $Q(kx)=k^2Q(x)$ for all $x\in E_1$ with $0<\|x\|<M$. From (14.46) and the definition of Q, it follows that $Q(kx)=k^2Q(x)$, for all $x\in E_1$.

Given $x\in E_1$ with $x\neq 0$, choose a positive integer m such that $\|k^mx\|\geq M$. By the definition of Q, we have

$$Q(x)=\frac{Q(k^mx)}{k^{2m}}=\frac{\psi(k^mx)}{k^{2m}}, \tag{14.52}$$

and by the definition of ψ, we obtain

$$Q(x)=\lim_{n\to\infty}\frac{f(k^{m+n}x)}{k^{2(m+n)}}=\lim_{n\to\infty}\frac{f(k^nx)}{k^{2n}} \tag{14.53}$$

for all $x\in E_1$ with $x\neq 0$. Since $Q(0)=0$, equation (14.53) is true for $x=0$.

Let $x,y\in E_1$ with $x\neq 0$, $y\neq 0$. From (14.41) and (14.53), we get

$$Q(kx+y)+Q\big(kx+\sigma(y)\big)=2k^2Q(x)+2Q(y). \tag{14.54}$$

If we replace y by $y=\sigma(x)$ in (14.54), we get $Q(x)=Q(\sigma(x))$, for all $x\in E_1$ with $x\neq 0$. Since $Q(0)=0$, the same is true for $x=0$. This implies that (14.54) is true for all $x,y\in E_1$.

Let $T\colon E_1 \longrightarrow E_2$ be another mapping which solves (14.54) and satisfies (14.42). Let $x \in E_1$ with $x \neq 0$ and choose a positive integer m such that $\|k^m x\| \geq M$. Then

$$\begin{aligned}\|Q(x) - T(x)\| &= \frac{1}{k^{2n}} \|Q(k^n x) - T(k^n x)\| \\ &\leq \|Q(k^n x) - f(k^n x)\| + \|T(k^n x) - f(k^n x)\| \\ &\leq \frac{1}{k^{2n}} \frac{\delta}{k^2 - 1} + \frac{\varepsilon}{k^2 - k^p} k^{n(p-2)} \|x\|^p\end{aligned}$$

for all $n \geq m$. By letting $n \to \infty$, we get $Q(x) = T(x)$ for all $x \in E_1$ with $x \neq 0$. Since $Q(0) = 0$, the same is true for $x = 0$, and the proof of Theorem 14.7 is complete. □

In the following theorem, we will remove the hypothesis $f(0) = 0$. In this case, we suppose that σ is a continuous involution, so we have $\sigma(rx) = r\sigma(x)$, for all $r \in \mathbb{R}$, $x \in E_1$.

Theorem 14.8 *Given a real normed-space E_1 and a real Banach space E_2, consider numbers $M \geq 0$, $\varepsilon \geq 0$, $\delta, \geq 0$, and $p, k > 0$ with $0 < p < 2$, $k \geq 2$. Let $f\colon E_1 \longrightarrow E_2$ be a mapping which satisfies* (14.41), *for all $x, y \in E_1$ such that $\|x\|^p + \|y\|^p \geq M^p$, then there exists a unique quadratic mapping $Q : E_1 \longrightarrow E_2$ which solves* (14.5) *and such that*

$$\|f(x) - Q(x)\| \leq \frac{1}{2k^2(k^2 - 1)} \left[(k^2 + 1)\delta + \varepsilon M^p\right] + \frac{\varepsilon}{2(k^2 - k^p)} \|x\|^p, \quad (14.55)$$

for all $x \in E_1$ with $\|x\| \geq M$.

Proof Letting $y = 0$ in (14.41), we get

$$\|f(kx) - k^2 f(x) - f(0)\| \leq \frac{\delta}{2} + \frac{\varepsilon}{2} \|x\|^p, \quad (14.56)$$

for all $x \in E_1$ with $\|x\| \geq M$. Letting $x = 0$ in (14.41), we have two cases:
Case 1: $\sigma = -I$. We choose $z \in E_1$ with $\|z\| = M$ and obtain

$$\|f(z) + f(-z) - k^2 f(0) - 2f(z)\| \leq \delta + \varepsilon M^p. \quad (14.57)$$

If we replace z by $-z$ in (14.57), we get

$$\|f(-z) + f(z) - k^2 f(0) - 2f(-z)\| \leq \delta + \varepsilon M^p. \quad (14.58)$$

Adding (14.57) and (14.58), we obtain

$$\|f(0)\| \leq \frac{1}{2k^2} (\delta + \varepsilon M^p). \quad (14.59)$$

Case 2: If $\sigma \neq -I$. Then there exists a $z_0 \in E_1$ with $z_0 \neq 0$ such that $z_0 + \sigma(z_0) \neq 0$ and we take $z = \frac{z_0+\sigma(z_0)}{\|z_0+\sigma(z_0)\|} M$. Now, from (14.56), (14.59) and the triangle inequality, we have

$$\bigl\| f(kx) - k^2 f(x) \bigr\| \leq \frac{\delta}{2} + \frac{\varepsilon}{2} \|x\|^p + \frac{[\delta + \varepsilon M^p]}{2k^2}. \tag{14.60}$$

The rest of the proof for this case goes through in a similar way. □

In the following, the Hyers–Ulam–Rassias stability on restricted domains for the equation

$$f(x+y) + f\bigl(x + \sigma(y)\bigr) = 2f(x) + 2f(y), \quad x, y \in E, \ (k=1)$$

is investigated.

First, we prove the Hyers–Ulam–Rassias stability of the additive mappings on restricted domains.

Theorem 14.9 *Given a real space E and a real Banach space F, consider numbers $M > 0$, $\theta \geq 0$, and p with $0 < p < 1$. Let the mapping $f\colon E \longrightarrow F$ satisfy the inequality*

$$\bigl\| f(x+y) - f(x) - f(y) \bigr\| \leq \theta\bigl(\|x\|^p + \|y\|^p\bigr) \tag{14.61}$$

for all $x, y \in E$ such that $\|x\|^p + \|y\|^p \geq M^p$. Then there exists a unique additive mapping $A\colon E \longrightarrow F$ such that

$$\bigl\| f(x) - A(x) \bigr\| \leq \frac{2\theta}{2 - 2^p} \|x\|^p + \theta\bigl(4 \times 2^p + 2 \times 3^p + 1\bigr) M^p \tag{14.62}$$

for all $x \in E$.

Proof Assume $\|x\|^p + \|y\|^p < M^p$. If $x = y = 0$, we choose a $z \in E$ with $\|z\| = M$, and we get

$$\bigl\| f(0) - f(0) - f(0) \bigr\| = \bigl\| f(z) - f(0) - f(z) \bigr\| \leq \theta M^p. \tag{14.63}$$

Otherwise, let

$$z = \begin{cases} (\|x\| + M)\frac{x}{\|x\|} & \text{if } \|x\| \geq \|y\|; \\ (\|y\| + M)\frac{y}{\|y\|} & \text{if } \|y\| \geq \|x\|. \end{cases}$$

It is then obvious that $\|z\| \geq M$ and

$$\begin{aligned} &\|x + z\|^p + \|y - z\|^p \geq \max\bigl\{\|x + z\|^p, \|y - z\|^p\bigr\} \geq M^p, \\ &\|y - z\|^p + \|z\|^p \geq \|z\|^p \geq M^p, \\ &\min\bigl\{\|x\|^p + \|z\|^p, \|y\|^p + \|z\|^p\bigr\} \geq \|z\|^p \geq M^p. \end{aligned}$$

Also

$$\max\{\|x+z\|, \|y-z\|\} < 3M, \quad \|z\| < 2M.$$

From (14.61), the above inequalities, and the relation

$$\begin{aligned}
&[f(x+y) - f(x) - f(y)] \\
&\quad = [f(x+y) - f(x+z) - f(y-z)] + [f(x+z) - f(x) - f(z)] \\
&\qquad + [f(y-z) - f(-z) - f(y)] - [f(0) - f(-z) - f(z)] + [f(0)],
\end{aligned}$$

we get

$$\begin{aligned}
\|f(x+y) - f(x) - f(y)\| \leq{}& \theta(4 \times 2^p + 2 \times 3^p + 1)M^p \\
&+ \theta(\|x\|^p + \|y\|^p).
\end{aligned} \tag{14.64}$$

Using ideas from the paper of Th.M. Rassias [27], we get the rest of the proof. □

Using the result of Theorem 14.9, we now prove the Hyers–Ulam–Rassias stability of the quadratic functional equation

$$f(x+y) + f(x+\sigma(y)) = 2f(x) + 2f(y), \quad x, y \in E \tag{14.65}$$

Theorem 14.10 *Let E be a real normed space and F a Banach space. Let numbers $M > 0, \theta \geq 0$ and p with $0 < p < 1$ be chosen. Let $f: E \longrightarrow F$ be a mapping which satisfies*

$$\|f(x+y) + f(x+\sigma(y)) - 2f(x) - 2f(y)\| \leq \theta(\|x\|^p + \|y\|^p) \tag{14.66}$$

for all $x, y \in E$ such that $\|x\|^p + \|y\|^p \geq M^p$. Then there exists a unique quadratic mapping $Q: E \longrightarrow F$ which solves (14.65) *and such that*

$$\begin{aligned}
\|f(x) - Q(x)\| \leq{}& \frac{\theta}{2}\|x\|^p + \frac{\theta}{8}(\|x+\sigma(x)\|^p + \|x-\sigma(x)\|^p) \\
&+ \frac{\theta}{2-2^p}\|x+\sigma(x)\|^p + \frac{\theta}{2}(4 \times 2^p + 2 \times 3^p + 1)M^p \\
&+ \frac{\theta}{2(4-2^p)}\|x-\sigma(x)\|^p \\
&+ \frac{\theta}{24}[16^p + 4 \times 9^p + 8 \times 4^p]M^p
\end{aligned} \tag{14.67}$$

for all $x \in E$ with $\|x\| \geq \frac{M}{2}$.

Proof For $y \in E^+ := \{z \in E/\sigma(z) = z\}$ and $x \in E$, we have from (14.66) the inequality

$$\begin{aligned}\|f(x+y)-f(x)-f(y)\| &= \frac{1}{2}\|f(x+y)+f(x+\sigma(y))-2f(y)-2f(x)\| \\ &\le \frac{\theta}{2}(\|x\|^p+\|y\|^p), \qquad (14.68)\end{aligned}$$

for all $x, y \in E^+$ such that $\|x\|^p + \|y\|^p \ge M^p$.

Since E^+ is an abelian subgroup of E, then from Theorem 14.9, there exists a unique additive mapping $a : E \longrightarrow F$ which satisfies the inequality

$$\|f(x)-a(x)\| \le \theta(4\times 2^p + 2\times 3^p + 1)M^p + \frac{2\theta}{2-2^p}\|x\|^p \qquad (14.69)$$

for all $x \in E^+$.

For $y \in E^- := \{z \in E/\sigma(z) = -z\}$ and $x \in E$, we have from (14.66) the inequality

$$\begin{aligned}&\|f(x+y)+f(x-y)-2f(x)-2f(y)\| \\ &= \|f(x+y)+f(x+\sigma(y))-2f(x)-2f(y)\| \\ &\le \theta(\|x\|^p+\|y\|^p) \qquad (14.70)\end{aligned}$$

for all $x, y \in E^-$ such that $\|x\|^p + \|y\|^p \ge M^p$.

Clearly, E^- is an abelian subgroup of E, so by using [26] there exists a unique mapping $q : E \longrightarrow F$ which solves (14.3) and such that

$$\|f(x)-q(x)\| \le \frac{\theta}{6}[16^p + 4\times 9^p + 8\times 4^p]M^p + \frac{2\theta}{4-2^p}\|x\|^p \qquad (14.71)$$

for all $x \in E^-$.

Letting $x = y$ in (14.66) yields

$$\|4f(x)+f(x+\sigma(x))-f(2x)\| \le 2\theta\|x\|^p \qquad (14.72)$$

for all $x \in E$ with $\|x\| \ge \frac{M}{2^{\frac{1}{p}}}$.

By using

$$\|x\|^p + \|y\|^p \ge (\|x\|+\|y\|)^p \ge \|x+y\|^p \qquad (14.73)$$

for all $x, y \in E$, replacing x by $x - \sigma(x)$ and y by $x + \sigma(x)$ in (14.73), we get

$$\|x+\sigma(x)\|^p + \|x-\sigma(x)\|^p \ge 2^p\|x\|^p. \qquad (14.74)$$

Then by (14.66), we obtain

$$\begin{aligned}&\|f(2x)-f(x+\sigma(x))-f(x-\sigma(x))\| \\ &\le \frac{\theta}{2}(\|x+\sigma(x)\|^p + \|x-\sigma(x)\|^p) \qquad (14.75)\end{aligned}$$

for all $x \in E$ with $\|x\| \ge \frac{M}{2}$.

By adding the result (14.72) to the result (14.75), we get

$$\begin{aligned} &\big\|4f(x) - 2f\big(x+\sigma(x)\big) - f\big(x-\sigma(x)\big)\big\| \\ &\quad \le 2\theta\|x\|^p + \frac{\theta}{2}\big(\big\|x+\sigma(x)\big\|^p + \big\|x-\sigma(x)\big\|^p\big), \end{aligned} \tag{14.76}$$

for all $x \in E$ with $\|x\| \ge \frac{M}{2}$.

By using (14.76), (14.69), and (14.71), we obtain

$$\begin{aligned} &\left\| f(x) - \frac{1}{2}a\big(x+\sigma(x)\big) - \frac{1}{4}q\big(x-\sigma(x)\big)\right\| \\ &\quad \le \left\| f(x) - \frac{1}{2}f\big(x+\sigma(x)\big) - \frac{1}{4}f\big(x-\sigma(x)\big)\right\| \\ &\qquad + \frac{1}{2}\big\|f\big(x+\sigma(x)\big) - a\big(x-\sigma(x)\big)\big\| + \frac{1}{4}\big\|f\big(x-\sigma(x)\big) - q\big(x-\sigma(x)\big)\big\| \\ &\quad \le \frac{\theta}{2}\|x\|^p + \frac{\theta}{8}\big(\big\|x+\sigma(x)\big\|^p + \big\|x-\sigma(x)\big\|^p\big) + \frac{\theta}{2-2^p}\big\|x+\sigma(x)\big\|^p \\ &\qquad + \frac{\theta}{2}\big(4\times 2^p + 2\times 3^p + 1\big)M^p + \frac{\theta}{2(4-2^p)}\big\|x-\sigma(x)\big\|^p \\ &\qquad + \frac{\theta}{24}\big[16^p + 4\times 9^p + 8\times 4^p\big]M^p \end{aligned}$$

for all $x \in E$ with $\|x\| \ge \frac{M}{2}$.

Letting $Q(x) = \frac{1}{2}a(x+\sigma(x)) - \frac{1}{4}q(x-\sigma(x))$, a simple computation shows that Q is a solution of (14.65).

For the uniqueness of Q, we apply the same argument as that in the proof used in [2] and [3]. This ends the proof of Theorem 14.10. □

References

1. Aoki, T.: On the stability of the linear transformation in Banach spaces. J. Math. Soc. Jpn. **2**, 64–66 (1950)
2. Bouikhalene, B., Elqorachi, E., Rassias, Th.M.: On the generalized Hyers–Ulam stability of the quadratic functional equation with a general involution. Nonlinear Funct. Anal. Appl. **12**(2), 247–262 (2007)
3. Bouikhalene, B., Elqorachi, E., Rassias, Th.M.: On the Hyers–Ulam stability of approximately Pexider mappings. Math. Inequal. Appl. **11**, 805–818 (2008)
4. Cădariu, L., Radu, V.: Fixed points and the stability of Jensen's functional equation. J. Inequal. Pure Appl. Math. **4**, article 4 (2003)
5. Cădariu, L., Radu, V.: On the stability of the Cauchy functional equation: A fixed point approach. Grazer Math. Ber. **346**, 43–52 (2004)
6. Cholewa, P.W.: Remarks on the stability of functional equations. Aequ. Math. **27**, 76–86 (1984)

7. Czerwik, S.: On the stability of the quadratic mapping in normed spaces. Abh. Math. Semin. Univ. Hamb. **62**, 59–64 (1992)
8. Diaz, J.B., Margolis, B.: A fixed point theorem of the alternative, for contractions on a generalized complete metric space. Bull. Am. Math. Soc. **74**, 305–309 (1968)
9. Elqorachi, E., Manar, Y., Rassias, Th.M.: Hyers–Ulam stability of the quadratic functional equation. Int. J. Nonlinear Anal. Appl. **1**(2), 11–20 (2010)
10. Elqorachi, E., Manar, Y., Rassias, Th.M.: Hyers–Ulam stability of the quadratic and Jensen functional equations on unbounded domains. J. Math. Sci. Adv. Appl. **4**(2), 287–303 (2010)
11. Găvruta, P.: A generalization of the Hyers–Ulam–Rassias stability of approximately additive mappings. J. Math. Anal. Appl. **184**, 431–436 (1994)
12. Gajda, Z.: On stability of additive mappings. Int. J. Math. Math. Sci. **14**, 431–434 (1991)
13. Hyers, D.H.: On the stability of the linear functional equation. Proc. Natl. Acad. Sci. USA **27**, 222–224 (1941)
14. Hyers, D.H., Isac, G., Rassias, Th.M.: On the asymptoticity aspect of Hyers–Ulam stability of mappings. Proc. Am. Math. Soc. **126**, 425–430 (1998)
15. Hyers, D.H., Rassias, Th.M.: Approximate homomorphisms. Aequ. Math. **44**, 125–153 (1992)
16. Hyers, D.H., Isac, G.I., Rassias, Th.M.: Stability of Functional Equations in Several Variables. Birkhäuser, Basel (1998)
17. Hyers, D.H., Isac, G., Rassias, Th.M.: On the asymptoticity aspect of Hyers–Ulam stability of mappings. Proc. Am. Math. Soc. **126**(2), 425–430 (1998)
18. Isac, G., Rassias, Th.M.: Stability of ψ-additive mappings: applications to nonlinear analysis. Int. J. Math. Math. Sci. **19**, 219–228 (1996)
19. Lee, J.-R., An, J.-S., Park, C.: On the stability of quadratic functional equations. Abs. Appl. Anal. (2008). doi:10.1155/2008/628178
20. Jung, S.-M.: On the Hyers–Ulam stability of the functional equation that have the quadratic property. J. Math. Anal. Appl. **222**, 126–137 (1998)
21. Jung, S.-M.: Stability of the quadratic equation of Pexider type. Abhandlugen aus dem mathematishen Seminar der Universität Hamburg **70**, 175–190 (2000)
22. Jung, S.-M., Sahoo, P.K.: Hyers–Ulam stability of the quadratic equation of Pexider type. J. Korean Math. Soc. **38**(3), 645–656 (2001)
23. Manar, Y., Elqorachi, El., Bouikhalene, B.: Fixed points and Hyers–Ulam–Rassias stability of the quadratic and Jensen functional equations. Nonlinear Funct. Anal. Appl. **15**(2), 273–284 (2010)
24. Manar, Y., Elqorachi, E., Rassias, Th.M.: On the Hyers–Ulam stability of the quadratic and Jensen functional equations on a restricted domain. Nonlinear Funct. Anal. Appl. **15**(4), 647–655 (2010)
25. Park, C., Kim, J.-H.: The stability of a quadratic functional equation with the fixed point alternative. Abs. Appl. Anal. (2009). doi:10.1155/2009/907167
26. Rahimi, A., Najati, A., Bae, J.-H.: On the asymptoticity aspect of Heyrs–Ulam stability of quadratic mappings. J. Inequal. Appl. **2010**, 454875 (2010), 14 pages
27. Rassias, Th.M.: On the stability of linear mapping in Banach spaces. Proc. Am. Math. Soc. **72**, 297–300 (1978)
28. Rassias, Th.M.: On a modified Hyers–Ulam sequence. J. Math. Anal. Appl. **158**, 106–113 (1991)
29. Rassias, Th.M.: Functional Equations and Inequalities. Kluwer Academic, Dordrecht (2001)
30. Rassias, Th.M.: Functional Equations, Inequalities and Applications. Kluwer Academic, Dordrecht (2003)
31. Rassias, Th.M.: The problem of S. M. Ulam for approximately multiplicative mappings. J. Math. Anal. Appl. **246**, 352–378 (2000)
32. Rassias, Th.M.: On the stability of minimum points. Mathematica **45(68)**(1), 93–104 (2003)
33. Rassias, Th.M.: On the stability of the functional equations and a problem of Ulam. Acta Appl. Math. **62**, 23–130 (2000)

34. Rassias, Th.M., Šemrl, P.: On the behavior of mappings which do not satisfy Hyers–Ulam stability. Proc. Am. Math. Soc. **114**, 989–993 (1992)
35. Rassias, Th.M., Šemrl, P.: On the Hyers–Ulam stability of linear mappings. J. Math. Anal. Appl. **173**, 325–338 (1993)
36. Rassias, Th.M., Tabor, J.: Stability of Mappings of Hyers–Ulam Type. Hardronic Press, Palm Harbor (1994)
37. Skof, F.: On the stability of functional equations on a restricted domains and related topics. In: Rassias, Th.M., Tabor, J. (eds.) Stability of Mappings of Hyers–Ulam Type, pp. 141–151. Hardronic Press, Palm Harbor (1994)
38. Ulam, S.M.: A Collection of Mathematical Problems. Interscience, New York (1961). Problems in Modern Mathematics. Wiley, New York (1964)

Chapter 15
Bohr's Inequality Revisited

Masatoshi Fujii, Mohammad Sal Moslehian, and Jadranka Mićić

Abstract We survey several significant results on the Bohr inequality and presented its generalizations in some new approaches. These are some Bohr-type inequalities of Hilbert space operators related to the matrix order and the Jensen inequality. An eigenvalue extension of Bohr's inequality is discussed as well.

Key words Bohr's inequality · Matrix approach · Operator Jensen inequality

Mathematics Subject Classification 47A63 · 26D15

15.1 Bohr Inequalities for Operators

The classical Bohr inequality says that

$$|a+b|^2 \leq p|a|^2 + q|b|^2$$

holds for all scalars a, b and $p, q > 0$ with $1/p + 1/q = 1$ and the equality holds if and only if $(p-1)a = b$, cf. [2]. There have been established many interesting extensions of this inequality in various settings by several mathematicians. Some

---

Dedicated to Professor Themistocles M. Rassias on the occasion of his 60th birthday.

M. Fujii
Department of Mathematics, Osaka Kyoiku University, Kashiwara, Osaka 582-8582, Japan
e-mail: mfujii@cc.osaka-kyoiku.ac.jp

M.S. Moslehian (✉)
Department of Pure Mathematics, Center of Excellence in Analysis on Algebraic Structures (CEAAS), Ferdowsi University of Mashhad, P.O. Box 1159, Mashhad 91775, Iran
e-mail: moslehian@ferdowsi.um.ac.ir
url: http://www.um.ac.ir/~moslehian/

J. Mićić
Faculty of Mechanical Engineering and Naval Architecture, University of Zagreb, Ivana Lučića 5, 10000 Zagreb, Croatia
e-mail: jmicic@fsb.hr
url: http://www.fsb.unizg.hr/matematika/jmicic/

P.M. Pardalos et al. (eds.), *Nonlinear Analysis*, Springer Optimization and Its Applications 68, In Honor of Themistocles M. Rassias on the Occasion of his 60th Birthday, DOI 10.1007/978-1-4614-3498-6_15, © Springer Science+Business Media, LLC 2012

interesting extensions of the classical Bohr inequality were given by Th.M. Rassias in [17]. In 1993, Th.M. Rassias and Pečarić [16] generalized the Bohr inequality by showing that if $(X, \|\cdot\|)$ is a normed linear space, $f : \mathbb{R}_+ \to \mathbb{R}_+$ is a nondecreasing convex function, $p_1 > 0$, $p_j \le 0$ $(j = 2, \ldots, n)$ and $\sum_{j=1}^n p_j > 0$, then

$$f\left(\left\|\sum_{j=1}^n p_j x_j\right\| \Big/ \sum_{j=1}^n p_j\right) \ge \sum_{j=1}^n p_j f(\|x_j\|) \Big/ \sum_{j=1}^n p_j$$

holds for every $x_j \in X$, $j = 1, \ldots, n$.

In 2003, Hirzallah [10] showed that if A, B belong to the algebra $\mathbb{B}(\mathscr{H})$ of all bounded linear operators on a complex (separable) Hilbert space $\mathscr{H}$ and $q \ge p > 1$ with $1/p + 1/q = 1$, then

$$|A - B|^2 + |(p-1)A + B|^2 \le p|A|^2 + q|B|^2, \tag{15.1}$$

where $|C| := (C^*C)^{1/2}$ denotes the absolute value of $C \in \mathbb{B}(\mathscr{H})$. He also showed that if $X \in \mathbb{B}(\mathscr{H})$ such that $X \ge \gamma I$ for some positive number γ, then

$$\gamma \left\| \left| |A - B|^2 \right\| \right| \le \left\| \left| p|A|^2 X + qX|B|^2 \right\| \right|$$

holds for every unitarily invariant norm $|||\cdot|||$. Recall that a unitarily invariant norm $|||\cdot|||$ is defined on a norm ideal $\mathscr{C}_{|||\cdot|||}$ of $\mathbb{B}(\mathscr{H})$ associated with it and has the property $|||UAV||| = |||A|||$ for all unitary U and V and $A \in \mathscr{C}_{|||\cdot|||}$.

In 2006, Cheung and Pečarić [4] extended inequality (15.1) for all positive conjugate exponents $p, q \in \mathbb{R}$. Also the authors of [5] generalized the Bohr inequality to the setting of n-inner product spaces.

In 2007, Zhang [19] generalized inequality (15.1) by removing the condition $q \ge p$ and presented the identity

$$|A - B|^2 + |\sqrt{p/q}A + \sqrt{q/p}B|^2 = p|A|^2 + q|B|^2 \quad \text{for } A, B \in \mathbb{B}(\mathscr{H}).$$

In addition, he proved that for any positive integer k and $A_i \in \mathbb{B}(\mathscr{H})$, $i = 1, \ldots, n$,

$$|t_1 A_1 + \cdots + t_k A_k|^2 \le t_1 |A_1|^2 + \cdots + t_k |A_k|^2 \tag{15.2}$$

holds for every $t_i > 0$, $i = 1, \ldots, k$ such that $\sum_{i=1}^k t_i = 1$.

In 2009, Chansangiam, Hemchote, and Pantaragphong [3] proved that if $A_i \in \mathbb{B}(\mathscr{H})$, α_{ik} and p_i are real numbers, $i = 1, \ldots, n$, $k = 1, \ldots, m$, such that the $n \times n$ matrix $X := (x_{ij})$, defined by $x_{ii} = \sum_{k=1}^m \alpha_{ik}^2 - p_i$ and $x_{ij} = \sum_{k=1}^m \alpha_{ik}\alpha_{jk}$ for $i \ne j$, is positive semidefinite, then

$$\sum_{k=1}^m \left| \sum_{i=1}^n \alpha_{ik} A_i \right|^2 \ge \sum_{i=1}^n p_i |A_i|^2.$$

In 2010, the first author and Zuo [6] had an approach to the Bohr inequality via a generalized parallelogram law for absolute value of operators, i.e.,

$$|A-B|^2+\frac{1}{t}|tA+B|^2=(1+t)|A|^2+\left(1+\frac{1}{t}\right)|B|^2$$

holds for every $A, B \in \mathbb{B}(\mathscr{H})$ and a real scalar $t \neq 0$.

In 2010, Abramovich, J. Barić, and J. Pečarić [1] established new generalizations of Bohr's inequality by applying superquadraticity.

In 2010, the second author and Rajić [15] presented a new operator equality in the framework of Hilbert C^*-modules. Recall that the notion of Hilbert C^*-module is a generalization of the concept of Hilbert space in which the field of scalars $\mathbb{C}$ is replaced by a C^*-algebra $\mathscr{A}$. For every $x \in \mathscr{X}$, the absolute value of x is defined as the unique positive square root of $\langle x, x\rangle \in \mathscr{A}$, that is, $|x| = \langle x, x\rangle^{\frac{1}{2}}$. The authors of [15] extended the operator Bohr inequalities of [4] and [10]. One of their results extending (15.2) of Zhang [19] is as follows. Suppose that $n \geq 2$ is a positive integer, $T_1, \ldots, T_n$ are adjointable operators on $\mathscr{X}$, $T_1^* T_2$ is self-adjoint, $t_1, \ldots, t_n$ are positive real numbers such that $\sum_{i=1}^n t_i = 1$ and $\sum_{i=1}^n t_i |T_i|^2 = I_{\mathscr{X}}$. Assume that for $n \geq 3$, T_1 or T_2 is invertible in the algebra of all adjointable operators on $\mathscr{X}$, operators $T_3, \ldots, T_n$ are self-adjoint and $T_i|T_j| = |T_j|T_i$ for all $1 \leq i < j \leq n$. Then

$$|t_1T_1x_1+\cdots+t_nT_nx_n|^2 \leq t_1|x_1|^2+\cdots+t_n|x_n|^2$$

holds for all $x_1, \ldots, x_n \in \mathscr{X}$.

Vasić and Kečkić [18] obtained a multiple version of the Bohr inequality, which follows from the Hölder inequality. In [14], the second author, Perić, and Pečarić established an operator version of the inequality of Vasić–Kečkić. In 2011, Matharu, the second author, and Aujla [12] gave an eigenvalue extension of Bohr's inequality. In the last section, we present a new approach to the main result of [14]. The interested reader is referred to [7] for many interesting results on the operator inequalities.

15.2 Matrix Approach to Bohr Inequalities

In this section, by utilizing the matrix order we present some Bohr type inequalities. For this, we introduce two notations as follows. For $x = (x_1, \ldots, x_n) \in \mathbb{R}^n$, we define $n \times n$ matrices $\Lambda(x) = x^*x = (x_i x_j)$ and $D(x) = \mathrm{diag}(x_1, \ldots, x_n)$.

Theorem 15.1 *If $\Lambda(a) + \Lambda(b) \leq D(c)$ for $a, b, c \in \mathbb{R}^n$, then*

$$\left|\sum_{i=1}^n a_iA_i\right|^2+\left|\sum_{i=1}^n b_iA_i\right|^2 \leq \sum_{i=1}^n c_i|A_i|^2$$

for arbitrary n-tuple (A_i) in $\mathbb{B}(\mathscr{H})$.

Incidentally, the statement is correct even if the order is replaced by the reverse.

Proof We define a positive linear mapping Φ from $\mathbb{B}(\mathscr{H})^n$ to $\mathbb{B}(\mathscr{H})$ by

$$\Phi(X) = \left(A_1^* \cdots A_n^*\right) X^T (A_1 \cdots A_n),$$

where $\cdot^T$ denotes the transpose operation. Since $\Lambda(a) = (a_1, \dots, a_n)^T (a_1, \dots, a_n)$, we have

$$\Phi\big(\Lambda(a)\big) = \left(\sum_{i=1}^n a_i A_i\right)^* \left(\sum_{i=1}^n a_i A_i\right) = \left|\sum_{i=1}^n a_i A_i\right|^2,$$

so that

$$\left|\sum_{i=1}^n a_i A_i\right|^2 + \left|\sum_{i=1}^n b_i A_i\right|^2 = \Phi\big(\Lambda(a) + \Lambda(b)\big) \le \Phi\big(D(c)\big) = \sum_{i=1}^n c_i |A_i|^2.$$

The remaining part is easily shown in the same way. □

The meaning of Theorem 15.1 will be well explained in the following theorem.

Theorem 15.2 *Let $t \in \mathbb{R}$.*

(i) *If $0 < t \le 1$, then*

$$|A \mp B|^2 + |tA \pm B|^2 \le (1+t)|A|^2 + \left(1 + \frac{1}{t}\right)|B|^2.$$

(ii) *If $t \ge 1$ or $t < 0$, then*

$$|A \mp B|^2 + |tA \pm B|^2 \ge (1+t)|A|^2 + \left(1 + \frac{1}{t}\right)|B|^2.$$

Proof We apply Theorem 15.1 to $a = (1, \mp 1)$, $b = (t, \pm 1)$, and $c = (1+t, 1+1/t)$. Then we consider the order between the corresponding matrices:

$$T = \begin{pmatrix} 1+t & 0 \\ 0 & 1+\frac{1}{t} \end{pmatrix} - \begin{pmatrix} 1 & \mp 1 \\ \mp 1 & 1 \end{pmatrix} - \begin{pmatrix} t^2 & \pm t \\ \pm t & 1 \end{pmatrix} = (1-t)\begin{pmatrix} t & \pm 1 \\ \pm 1 & \frac{1}{t} \end{pmatrix}.$$

Since $\det(T) = 0$, T is positive semidefinite (resp., negative semidefinite) if $0 < t < 1$ (resp., $t > 1$ or $t < 0$). □

As another application of Theorem 15.1, we give a new proof of [19, Theorem 7] as follows.

Theorem 15.3 *If $A_i \in \mathbb{B}(\mathscr{H})$ and $r_i \ge 1$, $i = 1, \dots, n$, with $\sum_{i=1}^n \frac{1}{r_i} = 1$, then*

$$\left|\sum_{i=1}^n A_i\right|^2 \le \sum_{i=1}^n r_i |A_i|^2.$$

In other words, it says that $K(z) = |z|^2$ satisfies the (operator) Jensen inequality

$$\left|\sum_{i=1}^{n} t_i A_i\right|^2 \le \sum_{i=1}^{n} t_i |A_i|^2 \quad \text{for } A_i \in \mathbb{B}(\mathscr{H}) \text{ and } t_i > 0, i = 1, \ldots, n, \text{ with } \sum_{i=1}^{n} t_i = 1.$$

Proof We check the order between the corresponding matrices $D = \operatorname{diag}(r_1, \ldots, r_n)$ and $C = (c_{ij})$ where $c_{ij} = 1$. All principal minors of $D - C$ are nonnegative and it follows that $C \le D$. Really, for natural numbers $k \le n$, put $D_k = \operatorname{diag}(r_{i_1}, \ldots, r_{i_k})$, $C_k = (c_{ij})$ with $c_{ij} = 1, i, j = 1, \ldots, k$ and $R_k = \sum_{j=1}^{k} 1/r_{i_j}$ where $1 \le r_{i_1} < \cdots < r_{i_k} \le n$. Then

$$\det(D_k - C_k) = (r_{i_1} \cdots r_{i_k})(1 - R_k) \ge 0 \quad \text{for arbitrary } k \le n.$$

Hence we have the conclusion of our theorem by Theorem 15.1. □

As another application of Theorem 15.1, we give a new proof of [19, Theorem 7] as follows.

Corollary 15.1 *If $a = (a_1, a_2)$, $b = (b_1, b_2)$, and $p = (p_1, p_2)$ satisfy $p_1 \ge a_1^2 + b_1^2$, $p_2 \ge a_2^2 + b_2^2$, and $(p_1 - (a_1^2 + b_1^2))(p_2 - (a_2^2 + b_2^2)) \ge (a_1 a_2 + b_1 b_2)^2$, then*

$$|a_1 A + a_2 B|^2 + |b_1 A + b_2 B|^2 \le p_1 |A|^2 + p_2 |B|^2$$

for all $A, B \in \mathbb{B}(\mathscr{H})$.

Proof Since the assumption of the above is nothing but the matrix inequality $\Lambda(a) + \Lambda(b) \le D(p)$, Theorem 15.1 implies the conclusion. □

Concluding this section, we observe the monotonicity of the operator function $F(a) = |\sum_{i=1}^{n} a_i A_i|^2$.

Corollary 15.2 *For a fixed n-tuple (A_i) in $\mathbb{B}(\mathscr{H})$, the operator function $F(a) = |\sum_{i=1}^{n} a_i A_i|^2$ for $a = (a_1, \ldots, a_n)$ preserves the order operator inequalities, that is,*

$$\text{if} \quad \Lambda(a) \le \Lambda(b), \quad \text{then} \quad F(a) \le F(b).$$

Proof We prove this putting $F(a) = \Phi(a^* a)$, where Φ is a positive linear mapping as in the proof of Theorem 15.1. □

The following corollary is a 3D version of [19, Lemma 2].

Corollary 15.3 *If $a = (a_1, a_2, a_3)$ and $b = (b_1, b_2, b_3)$ satisfy $|a_i| \le |b_i|$ for $i = 1, 2, 3$ and $a_i b_j = a_j b_i$ for $i \ne j$, then $F(a) \le F(b)$.*

Proof It follows from assumptions that if $i \neq l$ and $j \neq k$, then

$$\begin{vmatrix} a_i a_j - b_i b_j & a_i a_k - b_i b_k \\ a_l a_j - b_l b_j & a_l a_k - b_l b_k \end{vmatrix} = a_k b_j (a_i b_l - b_i a_l) + a_j b_k (b_i a_l - a_i b_l) = 0.$$

This means that all 2nd order principal minors of $\Lambda(b) - \Lambda(a)$ are zero. It follows that $\det(\Lambda(b) - \Lambda(a)) = 0$. Since the diagonal elements satisfy $|a_i| \leq |b_i|$ for $i = 1, 2, 3$, we have the matrix inequality $\Lambda(a) \leq \Lambda(b)$. Now it is sufficient to apply Corollary 15.2. □

15.3 A Generalization of the Operator Bohr Inequality via the Operator Jensen Inequality

As an application of the operator Jensen inequality, in this section we consider a generalization of the operator Bohr inequality. Namely, Jensen's inequality implies Bohr's inequality even in the operator case.

For this, we first target the following inequality which is an extension of the Bohr inequality, precisely, it is a generalized Bohr inequality due to Vasić and Kečkić [18] as follows. If $r > 1$ and $a_1, \ldots, a_n > 0$, then

$$\left|\sum z_i\right|^r \leq \left(\sum a_i^{\frac{1}{1-r}}\right)^{r-1} \sum a_i |z_i|^r \tag{15.3}$$

holds for all $z_1, \ldots, z_n \in \mathbb{C}$.

We note that it follows from Hölder inequality. Actually, $p = \frac{r}{r-1}$ and $q = r$ are conjugate, i.e., $\frac{1}{p} + \frac{1}{q} = 1$. We here set

$$u_i = a_i^{-\frac{1}{q}}; \qquad w_i = u_i^{-1} z_i, \quad i = 1, \ldots, n$$

and use them in the Hölder inequality. Then we have

$$\begin{aligned} \left|\sum_{i=1}^n z_i\right|^r &= \left|\sum_{i=1}^n u_i w_i\right|^r \leq \left(\sum_{i=1}^n |u_i|^p\right)^{\frac{r}{p}} \left(\sum_{i=1}^n |w_i|^q\right)^{\frac{r}{q}} \\ &= \left(\sum_{i=1}^n a_i^{\frac{1}{1-r}}\right)^{r-1} \sum_{i=1}^n a_i |z_i|^r. \end{aligned}$$

Now we propose its operator extension, see also [12, 14].

For the sake of convenience, we recall some notations and definitions.

Let $\mathscr{A}$ be a C^*-algebra of Hilbert space operators and let T be a locally compact Hausdorff space. A field $(A_t)_{t\in T}$ of operators in $\mathscr{A}$ is called a continuous field of operators if the function $t \mapsto A_t$ is norm continuous on T. If μ is a Radon measure on T and the function $t \mapsto \|A_t\|$ is integrable, then one can form the Bochner integral $\int_T A_t \,\mathrm{d}\mu(t)$, which is the unique element in $\mathscr{A}$ such that

$$\varphi\left(\int_T A_t \,\mathrm{d}\mu(t)\right) = \int_T \varphi(A_t)\,\mathrm{d}\mu(t)$$

for every linear functional φ in the norm dual $\mathfrak{A}^*$ of $\mathscr{A}$; cf. [8, Sect. 4.1].

Furthermore, a field $(\varphi_t)_{t\in T}$ of positive linear mappings $\varphi_t : \mathscr{A} \to \mathscr{B}$ between C^*-algebras of operators is called continuous if the function $t \mapsto \varphi_t(A)$ is continuous for every $A \in \mathscr{A}$. If the C^*-algebras include the identity operators (i.e., they are unital C^*-algebras), denoted by the same I, and the field $t \mapsto \varphi_t(I)$ is integrable with integral equal to I, then we say that $(\varphi_t)_{t\in T}$ is unital.

Recall that a continuous real function f defined on a real interval J is called operator convex if $f(\lambda A+(1-\lambda)B) \le \lambda f(A)+(1-\lambda)f(B)$ holds for all $\lambda \in [0, 1]$ and all self-adjoint operators A, B acting on a Hilbert space with spectra in J.

Now, we cite the Jensen inequality for our use below.

Theorem 15.4 [9, Theorem 2.1] *Let f be an operator convex function on an interval J, let T be a locally compact Hausdorff space with a bounded Radon measure μ, and let $\mathscr{A}$ and $\mathscr{B}$ be unital C^*-algebras. If $(\psi_t)_{t\in T}$ is a unital field of positive linear mappings $\psi_t : \mathscr{A} \to \mathscr{B}$, then*

$$f\left(\int_T \psi_t(A_t)\,d\mu(t)\right) \le \int_T \psi_t\big(f(A_t)\big)\,d\mu(t)$$

holds for all bounded continuous fields $(A_t)_{t\in T}$ of self-adjoint elements in $\mathscr{A}$ whose spectra are contained in J.

Theorem 15.5 *Let T be a locally compact Hausdorff space with a bounded Radon measure μ, and let $\mathscr{A}$ and $\mathscr{B}$ be unital C^*-algebras. If $1 < r \le 2$, $a : T \to \mathbb{R}$ is a bounded continuous nonnegative function and $(\phi_t)_{t\in T}$ is a field of positive linear mappings $\psi_t : \mathscr{A} \to \mathscr{B}$ satisfying*

$$\int_T a(t)^{\frac{1}{1-r}}\phi_t(I)\,d\mu(t) \le \int_T a(t)^{\frac{1}{1-r}}\,d\mu(t)I,$$

then

$$\left(\int_T \phi_t(A_t)\,d\mu(t)\right)^r \le \left(\int_T a(t)^{\frac{1}{1-r}}\,d\mu(t)\right)^{r-1}\int_T a(t)\phi_t\big(A_t^r\big)\,d\mu(t)$$

holds for all continuous fields $(A_t)_{t\in T}$ of positive elements in $\mathscr{A}$.

Proof We set $\psi_t = \frac{1}{M}a(t)^{\frac{1}{1-r}}\phi_t$, where $M = \int_T a(t)^{\frac{1}{1-r}}\,d\mu(t) > 0$. Then we have $\int_T \psi_t(I)\,d\mu(t) \le I$. By a routine way, we may assume that $\int_T \psi_t(I)\,d\mu(t) = I$. Since $f(t) = t^r$ is operator convex for $1 < r \le 2$, when we apply Theorem 15.4, we obtain

$$\left(\int_T \frac{1}{M} a(t)^{\frac{1}{1-r}} \phi_t(\tilde{A}_t)\, d\mu(t)\right)^r \le \int_T \frac{1}{M} a(t)^{\frac{1}{1-r}} \phi_t\left(A_t^r\right) d\mu(t)$$

for every bounded continuous field $(\tilde{A}_t)_{t\in T}$ of positive elements in $\mathscr{A}$. Replacing $\tilde{A}_t$ by $a(t)^{-1/(1-r)} A_t$, the above inequality can be written as

$$\left(\int_T \phi_t(A_t)\, d\mu(t)\right)^r \le M^{r-1} \int_T a(t)\phi_t\left(A_t^r\right) d\mu(t),$$

which is the desired inequality. □

Remark 15.1 We note that with the notation as above and

$$\int_T a(t)^{\frac{1}{1-r}} \phi_t(I)\, d\mu(t) \le k \int_T a(t)^{\frac{1}{1-r}}\, d\mu(t) I, \quad \text{for some } k > 0,$$

one has

$$\left(\int_T \phi_t(A_t)\, d\mu(t)\right)^r \le k^{r-1} \left(\int_T a(t)^{\frac{1}{1-r}}\, d\mu(t)\right)^{r-1} \int_T a(t)\phi_t\left(A_t^r\right) d\mu(t).$$

For a typical positive linear mapping $\phi(A) = X^* A X$ for some X, Theorem 15.5 can be written as follows.

Corollary 15.4 *Let T be a locally compact Hausdorff space with a bounded Radon measure μ, and let $\mathscr{A}$ be unital C^*-algebra. If $1 < r \le 2$, $a : T \to \mathbb{R}$ is a bounded continuous nonnegative function and $(X_t)_{t\in T}$ is a bounded continuous field of elements in $\mathscr{A}$ satisfying*

$$\int_T a(t)^{\frac{1}{1-r}} X_t^* X_t\, d\mu(t) \le \int_T a(t)^{\frac{1}{1-r}}\, d\mu(t) I,$$

then

$$\left(\int_T X_t^* A_t X_t\, d\mu(t)\right)^r \le \left(\int_T a(t)^{\frac{1}{1-r}}\, d\mu(t)\right)^{r-1} \int_T a(t) X_t^* A_t^r X_t\, d\mu(t)$$

holds for all continuous fields $(A_t)_{t\in T}$ of positive elements in $\mathscr{A}$.

Similarly, putting a positive linear mapping $\phi(A) = \langle Ax, x\rangle$ for some vector x in a Hilbert space in Theorem 15.5, we obtain the following result.

Corollary 15.5 *Let $(A_t)_{t\in T}$ be a continuous field of positive operator on a Hilbert space $\mathscr{H}$ defined on a locally compact Hausdorff space T equipped with a bounded Radon measure μ.*

If $1 < r \leq 2$, $a : T \to \mathbb{R}$ *is a bounded continuous nonnegative function, and* $(x_t)_{t\in T}$ *is a continuous field of vectors in* $\mathscr{H}$ *satisfying*

$$\int_T a(t)^{\frac{1}{1-r}} \|x_t\|^2 \, d\mu(t) \leq \int_T a(t)^{\frac{1}{1-r}} \, d\mu(t),$$

then

$$\left(\int_T \langle A_t x_t, x_t\rangle \, d\mu(t)\right)^r \leq \left(\int_T a(t)^{\frac{1}{1-r}} \, d\mu(t)\right)^{r-1} \int_T a(t)\langle A_t^r x_t, x_t\rangle \, d\mu(t).$$

The following corollary is a discrete version of Theorem 15.5.

Corollary 15.6 *If* $1 < r \leq 2, a_1, \ldots, a_n > 0$ *and* $\phi_1, \ldots, \phi_n$ *are positive linear mappings* $\phi_i : \mathbb{B}(\mathscr{H}) \to \mathbb{B}(\mathscr{K})$ *satisfying*

$$\sum_{i=1}^n a_i^{\frac{1}{1-r}} \phi_i(I) \leq \sum_{i=1}^n a_i^{\frac{1}{1-r}} I,$$

then

$$\left(\sum_{i=1}^n \phi_i(A_i)\right)^r \leq \left(\sum_{i=1}^n a_i^{\frac{1}{1-r}}\right)^{r-1} \sum_{i=1}^n a_i \phi_i(A_i^r)$$

holds for all bounded continuous fields $(A_t)_{t\in T}$ *of positive elements* $A_1, \ldots, A_n \geq 0$ *in* $\mathbb{B}(\mathscr{H})$.

We can obtain the above inequality in a broader region for r under conditions on the spectra. For this result, we cite a version of Jensen's operator inequality without operator convexity.

Theorem 15.6 [13, Theorem 1] *Let* $A_1, \ldots, A_n$ *be self-adjoint operators* $A_i \in \mathbb{B}(\mathscr{H})$ *with bounds* m_i *and* M_i, $m_i \leq M_i$, $i = 1, \ldots, n$. *Let* $\psi_1, \ldots, \psi_n$ *be positive linear mappings* $\psi_i : \mathbb{B}(\mathscr{H}) \to \mathbb{B}(\mathscr{K})$, $i = 1, \ldots, n$, *such that* $\sum_{i=1}^n \psi_i(1_H) = 1_K$. *If*

$$(m_A, M_A) \cap [m_i, M_i] = \emptyset \quad \text{for } i = 1, \ldots, n, \tag{15.4}$$

where m_A *and* M_A, $m_A \leq M_A$, *are bounds of the self-adjoint operator* $A = \sum_{i=1}^n \psi_i(A_i)$, *then*

$$f\left(\sum_{i=1}^n \psi_i(A_i)\right) \leq \sum_{i=1}^n \psi_i\big(f(A_i)\big)$$

holds for every continuous convex function $f : I \to \mathbf{R}$ *provided that the interval* I *contains all* m_i, M_i.

Theorem 15.7 *Let $A_1, \dots, A_n$ be strictly positive operators $A_i \in \mathbb{B}(\mathscr{H})$ with bounds m_i and M_i, $0 < m_i \le M_i$, $i = 1, \dots, n$. Let $\phi_1, \dots, \phi_n$ be positive linear mappings $\phi_i : \mathbb{B}(\mathscr{H}) \to \mathbb{B}(\mathscr{K})$, $i = 1, \dots, n$.*

If $r \in (-\infty, 0) \cup (1, \infty)$ and $a_1, \dots, a_n > 0$ such that

$$\sum_{i=1}^{n} a_i^{\frac{1}{1-r}} \phi_i(I) \le \sum_{i=1}^{n} a_i^{\frac{1}{1-r}} I,$$

and

$$(m_A, M_A) \cap \left[a(t)^{-1/(1-r)} m_i, a(t)^{-1/(1-r)} M_i\right] = \emptyset \quad \text{for } i = 1, \dots, n,$$

where m_A and M_A, $0 < m_A \le M_A$, are bounds of the strictly positive operator $A = \sum_{i=1}^{n} \phi_i(A_i)$, then

$$\left(\sum_{i=1}^{n} \phi_i(A_i)\right)^r \le \left(\sum_{i=1}^{n} a_i^{\frac{1}{1-r}}\right)^{r-1} \sum_{i=1}^{n} a_i \phi_i\left(A_i^r\right).$$

Proof The proof is quite similar to the one of Theorem 15.5. We omit the details. □

In the rest, we shall prove a matrix analogue of the inequality (15.3). For this, we introduce some usual notations. Let $\mathscr{M}_n$ denote the C^*-algebra of $n \times n$ complex matrices and let $\mathscr{H}_n$ be the set of all Hermitian matrices in $\mathscr{M}_n$. We denote by $\mathscr{H}_n(J)$ the set of all Hermitian matrices in $\mathscr{M}_n$ whose spectra are contained in an interval $J \subseteq \mathbf{R}$. Moreover, we denote by $\lambda_1(A) \ge \lambda_2(A) \ge \cdots \ge \lambda_n(A)$ the eigenvalues of A arranged in the decreasing order with their multiplicities counted.

Matharu, the second author, and Aujla [12] gave a weak majorization inequality and applied it to prove eigenvalue and unitarily invariant norm extensions of (15.3). Their main result reads as follows.

Theorem 15.8 [12, Theorem 2.7] *Let f be a convex function on J, $0 \in J$, $f(0) \le 0$, and let $A \in \mathscr{H}_n(J)$. Then*

$$\sum_{j=1}^{k} \lambda_j\left(f\left(\sum_{i=1}^{\ell} \alpha_i \Phi_i(A)\right)\right) \le \sum_{j=1}^{k} \lambda_j\left(\sum_{i=1}^{\ell} \alpha_i \Phi_i\big(f(A)\big)\right) \quad (1 \le k \le m)$$

holds for positive linear mappings Φ_i, $i = 1, 2, \dots, \ell$ from $\mathscr{M}_n$ to $\mathscr{M}_m$ such that $0 < \sum_{i=1}^{\ell} \alpha_i \Phi_i(I_n) \le I_m$ and $\alpha_i \ge 0$.

The following result is a generalization of [11, Theorem 1].

Corollary 15.7 [12, Corollary 2.8] *Let $A_1, \dots, A_\ell \in \mathscr{H}_n$ and $X_1, \dots, X_\ell \in \mathscr{M}_n$ such that*

$$0 < \sum_{i=1}^{\ell} \alpha_i X_i^* X_i \le I_n,$$

where $\alpha_i > 0$, *and let* f *be a convex function on* $\mathbb{R}$, $f(0) \leq 0$, *and* $f(uv) \leq f(u)f(v)$ *for all* $u, v \in \mathbb{R}$. *Then*

$$\sum_{j=1}^{k} \lambda_j \left(f\left(\sum_{i=1}^{\ell} X_i^* A_i X_i \right) \right) \leq \sum_{j=1}^{k} \lambda_j \left(\sum_{i=1}^{\ell} \alpha_i f\left(\alpha_i^{-1}\right) X_i^* f(A_i) X_i \right) \quad (15.5)$$

holds for $1 \leq k \leq n$.

Proof Let $A \in \mathcal{M}_{\ell n}$ be partitioned as

$$\begin{pmatrix} A_{11} & \cdots & A_{1\ell} \\ \vdots & & \vdots \\ A_{\ell 1} & \cdots & A_{\ell\ell} \end{pmatrix}, \quad A_{ij} \in \mathcal{M}_n, \quad 1 \leq i, \quad j \leq \ell,$$

as an $\ell \times \ell$ block matrix. Consider the linear mappings $\Phi_i : \mathcal{M}_{\ell n} \longrightarrow \mathcal{M}_n, i = 1, \ldots, \ell$, defined by $\Phi_i(A) = X_i^* A_{ii} X_i,\ i = 1, \ldots, \ell$. Then Φ_i's are positive linear mappings from $\mathcal{M}_{\ell n}$ to $\mathcal{M}_n$ such that

$$0 < \sum_{i=1}^{\ell} \alpha_i \Phi_i(I_{\ell n}) = \sum_{i=1}^{\ell} \alpha_i X_i^* X_i \leq I_n.$$

Using Theorem 15.8 for the diagonal matrix $A = \operatorname{diag}(A_{11}, \ldots, A_{\ell\ell})$, we have

$$\sum_{j=1}^{k} \lambda_j \left(f\left(\sum_{i=1}^{\ell} \alpha_i X_i^* A_{ii} X_i \right) \right) \leq \sum_{j=1}^{k} \lambda_j \left(\sum_{i=1}^{\ell} \alpha_i X_i^* f(A_{ii}) X_i \right) \quad (1 \leq k \leq n).$$

Replacing A_{ii} by $\alpha_i^{-1} A_i$ in the above inequality, we get (15.5). □

Now we obtain the following eigenvalue generalization of inequality (15.3).

Theorem 15.9 [12, Theorem 2.9] *Let* $A_1, \ldots, A_\ell \in \mathcal{H}_n$ *and* $X_1, \ldots, X_\ell \in \mathcal{M}_n$ *be such that*

$$0 < \sum_{i=1}^{\ell} p_i^{1/1-r} X_i^* X_i \leq \sum_{i=1}^{\ell} p_i^{1/(1-r)} I_n,$$

where $p_1, \ldots, p_\ell > 0, r > 1$. *Then*

$$\sum_{j=1}^{k} \lambda_j \left(\left| \sum_{i=1}^{\ell} X_i^* A_i X_i \right|^r \right) \leq \left(\sum_{i=1}^{\ell} p_i^{\frac{1}{1-r}} \right)^{r-1} \sum_{j=1}^{k} \lambda_j \left(\sum_{i=1}^{\ell} p_i X_i^* |A_i|^r X_i \right)$$

for $1 \leq k \leq n$.

Proof Apply Corollary 15.7 to the function $f(t) = |t|^r$ and $\alpha_i = \frac{p_i^{1/1-r}}{\sum_{i=1}^{\ell} p_i^{1/(1-r)}}$. □

Acknowledgement The second author would like to thank Tusi Math. Research Group (TMRG), Mashhad, Iran.

References

1. Abramovich, S., Barić, J., Pečarić, J.: Superquadracity, Bohr's inequality and deviation from a mean value. Aust. J. Math. Anal. Appl. **7**, 1 (2010), 9 pp.
2. Bohr, H.: Zur Theorie der Fastperiodischen Funktionen I. Acta Math. **45**, 29–127 (1924)
3. Chansangiam, P., Hemchote, P., Pantaragphong, P.: Generalizations of Bohr inequality for Hilbert space operators. J. Math. Anal. Appl. **356**, 525–536 (2009)
4. Cheung, W.-S., Pečarić, J.: Bohr's inequalities for Hilbert space operators. J. Math. Anal. Appl. **323**(1), 403–412 (2006)
5. Cheung, W.S., Cho, Y.J., Pečarić, J., Zhao, D.D.: Bohr's inequalities in n-inner product spaces. J. Korea Soc. Math. Educ. Ser. B Pure Appl. Math. **14**(2), 127–137 (2007)
6. Fujii, M., Zuo, H.: Matrix order in Bohr inequality for operators. Banach J. Math. Anal. **4**(1), 21–27 (2010)
7. Furuta, T., Mićić Hot, J., Pečarić, J., Seo, Y.: Mond–Pecaric Method in Operator Inequalities. Inequalities for Bounded Selfadjoint Operators on a Hilbert Space. Monographs in Inequalities, vol. 1. Element, Zagreb (2005)
8. Hansen, F., Pedersen, G.K.: Jensen's operator inequality. Bull. Lond. Math. Soc. **35**, 553–564 (2003)
9. Hansen, F., Pečarić, J.E., Perić, I.: Jensen's operator inequality and its converses. Math. Scand. **100**(1), 61–73 (2007)
10. Hirzallah, O.: Non-commutative operator Bohr inequality. J. Math. Anal. Appl. **282**, 578–583 (2003)
11. Kocić, V.Lj., Maksimović, D.M.: Variations and generalizations of an inequality due to Bohr. Univ. Beograd. Publ. Elektrotehn. Fak. Ser. Mat. Fiz. No. 412-460 (1973), pp. 183–188
12. Matharu, J.S., Moslehian, M.S., Aujla, J.S.: Eigenvalue extensions of Bohr's inequality. Linear Algebra Appl. **435**(2), 270–276 (2011)
13. Mićić, J., Pavić, Z., Pečarić, J.: Jensen's inequality for operators without operator convexity. Linear Algebra Appl. **434**, 1228–1237 (2011)
14. Moslehian, M.S., Pečarić, J.E., Perić, I.: An operator extension of Bohr's inequality. Bull. Iran. Math. Soc. **35**(2), 77–84 (2009)
15. Moslehian, M.S., Rajić, R.: Generalizations of Bohr's inequality in Hilbert C^*-modules. Linear Multilinear Algebra **58**(3), 323–331 (2010)
16. Pečarić, J.E., Rassias, Th.M.: Variations and generalizations of Bohr's inequality. J. Math. Anal. Appl. **174**(1), 138–146 (1993)
17. Rassias, Th.M.: On characterizations of inner-product spaces and generalizations of the H. Bohr inequality. In: Rassias, Th.M. (ed.) Topics in Mathematical Analysis. World Scientific, Singapore (1989)
18. Vasić, M.P., Kečkić, D.J.: Some inequalities for complex numbers. Math. Balk. **1**, 282–286 (1971)
19. Zhang, F.: On the Bohr inequality of operators. J. Math. Anal. Appl. **333**, 1264–1271 (2007)

Chapter 16
Hyers–Ulam–Rassias Stability of Orthogonal Additive Mappings

P. Găvruţa and L. Găvruţa

Abstract In this paper, we give an introduction to the Hyers–Ulam–Rassias stability of orthogonally additive mappings. The concept of Hyers–Ulam–Rassias stability originated from Th.M. Rassias' stability theorem that appeared in his paper: On the stability of the linear mapping in Banach spaces, Proc. Am. Math. Soc. 72:297–300, 1978. Our results generalize and simplify the result of R. Ger and J. Sikorska (Bull. Pol. Acad. Sci., Math. 43(2):143–151, 1995). See also Chap. 9 of the book (Hyers et al. in Stability of Functional Equations in Several Variables, Birkhäuser, Basel, 1998).

Key words Stability · Orthogonal additive mappings · ψ-Additive function

Mathematics Subject Classification 65Q20 · 39B55

16.1 Introduction

As an answer to a question posed in 1940 by S.M. Ulam (see [27, 28]), in 1941 D.H. Hyers [14] proved that if $\delta > 0$ and $f : E_1 \to E_2$ is a mapping, with E_1, E_2 being Banach spaces, such that

$$\left\| f(x+y) - f(x) - f(y) \right\| \leq \delta \quad \text{for all } x, y \in E_1,$$

then there exists a unique additive mapping $T : E_1 \to E_2$ such that

$$\left\| f(x) - T(x) \right\| \leq \delta$$

Dedicated to Professor Themistocles M. Rassias on the occasion of his 60th birthday.

P. Găvruţa (✉) · L. Găvruţa
Department of Mathematics, "Politehnica" University of Timişoara, Piaţa Victoriei no. 2, 300006 Timişoara, Romania
e-mail: pgavruta@yahoo.com

L. Găvruţa
e-mail: gavruta_laura@yahoo.com

P.M. Pardalos et al. (eds.), *Nonlinear Analysis*, Springer Optimization and Its Applications 68, In Honor of Themistocles M. Rassias on the Occasion of his 60th Birthday,
DOI 10.1007/978-1-4614-3498-6_16,

for all $x \in E_1$. If $f(tx)$ is continuous in t for each fixed x, then T is a linear mapping. In 1978, Th.M. Rassias [24] gave an important generalization of Hyers' result in the following way:

Consider a real normed space E_1, a Banach space E_2, and a mapping $f : E_1 \to E_2$ such that $f(tx)$ is continuous in t for each fixed x. Assume that there exist $\theta \geq 0$ and $p \in [0, 1)$ such that

$$\left\| f(x+y) - f(x) - f(y) \right\| \leq \theta\left(\|x\|^p + \|y\|^p\right)$$

for any $x, y \in E_1$. Then there exists a unique linear mapping $T : E_1 \to E_2$ such that

$$\left\| f(x) - T(x) \right\| \leq \frac{2\theta}{2 - 2^p}\|x\|^p \quad \text{for any } x \in E_1.$$

G. Isac and Th.M. Rassias [16] obtained a generalization of Rassias' Theorem for ψ-additive mappings:

$$\left\| f(x+y) - f(x) - f(y) \right\| \leq \psi\left(\|x\|\right) + \psi\left(\|y\|\right).$$

In 1994, P. Găvruţa [7] provided a generalization of Th.M. Rassias' Theorem for the unbounded Cauchy difference and introduced the concept of generalized Hyers–Ulam–Rassias stability in the spirit of Th.M. Rassias approach:

Theorem 16.1 *Let G and E be an abelian group and a Banach space, respectively, and let $\varphi : G^2 \to [0, \infty)$ be a function satisfying*

$$\Phi(x, y) = \sum_{k=0}^{\infty} 2^{-k-1}\varphi\left(2^k x, 2^k y\right) < \infty$$

for all $x, y \in G$. If a function $f : G \to E$ satisfies the inequality

$$\left\| f(x+y) - f(x) - f(y) \right\| \leq \varphi(x, y)$$

for any $x, y \in G$, then there exists a unique additive function $A : G \to E$ with

$$\left\| f(x) - A(x) \right\| \leq \Phi(x, x)$$

for all $x \in G$. If, moreover, G is a real normed space and $f(tx)$ is continuous in t for each fixed x in G, then A is a linear function.

For a number of generalizations and extensions of Th.M. Rassias' Theorem, see the books [4, 5, 15, 18] and their references. See also [1, 6, 12, 17, 23].

Some open problems in the field posed by Th.M. Rassias were solved in [2, 8–11].

16.2 Preliminary Results

A number of definitions of orthogonality in a vector space were given in the literature. We consider Hyers–Ulam–Rassias stability of orthogonal additive mappings, where our orthogonality relation is presented in a very general way.

In the following, we denote by X a real vector space and by Y a Banach space.

Definition 16.1 A binary relation $\perp$ on X is called an orthogonality relation if the following properties hold:

1. If $x, y \in X$ and $x \perp y$, then $\alpha x \perp \beta y$ for all $\alpha, \beta \in \mathbb{R}$.
2. For every $x \in X$, there exists $y \in X$ so that $x \perp y$ and

$$x + y \perp x - y.$$

The above definition is more general then the one presented in [25].

Definition 16.2 We say that the mapping $T : X \to Y$ is orthogonally additive if

$$T(x + y) = T(x) + T(y), \tag{16.1}$$

for all $x, y \in X$ with $x \perp y$.

Proposition 16.1 *Let $T : X \to Y$ be an orthogonally additive mapping.*

1. *If T is odd, then $T(2x) = 2T(x)$ for all $x \in X$.*
2. *If T is even, then $T(2x) = 4T(x)$ for all $x \in X$.*

Proof If $x \in X$, by Condition 2 of Definition 16.1, it follows that there exists $y \in X$ such that $x \perp y$ and $x + y \perp x - y$.

From (16.1) it follows that

$$T(2x) = T(x + y) + T(x - y)$$

and also

$$T(x + y) = T(x) + T(y),$$
$$T(x - y) = T(x) + T(-y).$$

Hence

$$T(2x) = 2T(x) + T(y) + T(-y). \tag{16.2}$$

If T is odd, then $T(-y) = -T(y)$, hence from (16.2) it follows that $T(2x) = 2T(x)$.

If T is even, then from Condition 1 of Definition 16.1, it follows that

$$\frac{x + y}{2} \perp \pm\frac{x - y}{2}$$

hence

$$\begin{aligned} T(x) &= T\left(\frac{x+y}{2} + \frac{x-y}{2}\right) \\ &= T\left(\frac{x+y}{2}\right) + T\left(\frac{x-y}{2}\right) \\ &= T\left(\frac{x+y}{2}\right) + T\left(\frac{y-x}{2}\right) = T(y). \end{aligned}$$

From (16.2) it follows that $T(2x) = 4T(x)$. □

Definition 16.3 We say that the function $\psi : X \to [0, \infty)$ is an admissible function if the following conditions hold:

1. $\psi(-x) = \psi(x)$ for all $x \in X$.
2. There exists $\theta \in (0, 2)$ such that

$$\psi(2x) \le \theta\psi(x) \quad \text{for all } x \in X.$$

3. There exists $\eta > 0$ such that if $x, y \in X$, $x \perp y$, then

$$\psi(x) + \psi(y) \le \eta\psi(x+y).$$

4. If $x \perp y$ and $x + y \perp x - y$, then $\psi(x) = \psi(y)$.

We denote by $\mathscr{A}(X) = \mathscr{A}(X, \theta, \eta)$ the set of all admissible functions.

Remark 16.1 If $\psi \in \mathscr{A}(X)$ and $\alpha \ge 0$, then $\alpha\psi \in \mathscr{A}(X)$.

Remark 16.2 If $\psi \in \mathscr{A}(X)$ and $a \ge 2$, then

$$\sum_{n=0}^{\infty} a^{-n}\psi(2^n x) < \infty. \tag{16.3}$$

Indeed, from Property 2 of Definition 16.3 it follows that

$$\psi(2^n x) \le \theta^n \psi(x),$$

hence

$$\frac{\psi(2^n x)}{a^n} \le \psi(x)\left(\frac{\theta}{a}\right)^n$$

and

$$\sum_{n=0}^{\infty}\left(\frac{\theta}{a}\right)^n < \infty \quad \text{since } \theta < 2 \le a.$$

Example 16.1 Let X be a Hilbert space so that $\dim X \geq 2$ and $p \in [0, 1)$. Then the function $\psi : X \to [0, \infty)$ given by

$$\psi(x) = \|x\|^p, \quad x \in X,$$

is in $\mathscr{A}(X)$.

Condition 1 in Definition 16.3 is clear. Relation 2 holds with $\theta = 2^p < 2$. We now prove Condition 3 in Definition 16.3. Consider $x, y \in X$ so that $x \perp y$. Then

$$\begin{aligned}\psi(x+y) &= \left(\|x+y\|^2\right)^{\frac{p}{2}} = \left(\|x\|^2 + \|y\|^2\right)^{\frac{p}{2}} \\ &\geq 2^{\frac{p}{2}-1}\left(\|x\|^p + \|y\|^p\right)\end{aligned}$$

where we use the inequality

$$(a+b)^\alpha \geq 2^{\alpha-1}\left(a^\alpha + b^\alpha\right),$$

for $a, b \geq 0$ and $0 \leq \alpha \leq 1$. Hence Condition 3 holds with $\eta = 2^{1-\frac{p}{2}}$.

We prove Condition 4 in Definition 16.3. If $x \perp y$ and $x + y \perp x - y$, it follows that $\|x\|^2 = \|y\|^2$, hence $\psi(x) = \psi(y)$.

Definition 16.4 We say that a function $f : X \to Y$ is ψ-additive on orthogonal vectors if

$$\left\|f(x+y) - f(x) - f(y)\right\| \leq \psi(x) + \psi(y), \tag{16.4}$$

for all $x, y \in X$, $x \perp y$.

We give our first result on stability.

Proposition 16.2 *Let $a \geq 2$, $\psi \in \mathscr{A}$, and $f : X \to Y$ be such that*

$$\left\|f(2x) - af(x)\right\| \leq \psi(x) \tag{16.5}$$

for all $x \in X$. Then there exists

$$\hat{f}(x) := \lim_{n\to\infty} \frac{f(2^n x)}{a^n}, \tag{16.6}$$

and $\hat{f}$ is the unique mapping that verifies the following two relations

$$\hat{f}(2x) = a\hat{f}(x) \quad \textit{for all } x \in X; \tag{16.7}$$

$$\left\|f(x) - \hat{f}(x)\right\| \leq \frac{1}{a-\theta}\Psi(x) \quad \textit{for all } x \in X. \tag{16.8}$$

Moreover, if f is ψ-additive on orthogonal vectors, then $\hat{f}$ is orthogonally additive.

Proof From (16.5) we have, for a positive integer k,

$$\|f(2^{k+1}x) - af(2^k x)\| \le \psi(2^k x)$$

hence

$$\|a^{-k-1}f(2^{k+1}x) - a^{-k}f(2^k x)\| \le a^{-k-1}\psi(2^k x). \tag{16.9}$$

If $n > m$, by the triangle inequality, it follows that

$$\|a^{-n}f(2^n x) - a^{-m}f(2^m x)\| \le \sum_{k=m}^{n-1} a^{-k-1}\psi(2^k x). \tag{16.10}$$

Since

$$\lim_{m\to\infty} \sum_{k=m}^{n-1} a^{-k-1}\psi(2^k x) = 0$$

(see Remark 16.2), it follows that the sequence $\{a^{-n}f(2^n x)\}$ is a Cauchy sequence. Because Y is a Banach space, we deduce the existence of the limit

$$\hat{f}(x) := \lim_{n\to\infty} a^{-n}f(2^n x).$$

If in equation (16.10) we take $m = 0$ and $n \to \infty$, we obtain

$$\begin{aligned}\|\hat{f}(x) - f(x)\| &\le \sum_{k=0}^{\infty} a^{-k-1}\psi(2^k x)\\ &\le \sum_{k=0}^{\infty} a^{-k-1}\theta^k\psi(x)\\ &= \frac{1}{a-\theta}\psi(x).\end{aligned}$$

If f is ψ-additive on orthogonal vectors, then

$$\|f(x+y) - f(x) - f(y)\| \le \psi(x) + \psi(y)$$

for $x, y \in X, x \perp y$.

From Condition 1 of Definition 16.1 it follows that $2^n x \perp 2^n y$, hence

$$\|f(2^n x + 2^n y) - f(2^n x) - f(2^n y)\| \le \psi(2^n x) + \psi(2^n y),$$

or

$$\|a^{-n}f(2^n x + 2^n y) - a^{-n}f(2^n x) - a^{-n}f(2^n y)\| \le a^{-n}\psi(2^n x) + a^{-n}\psi(2^n y).$$

We let $n \to \infty$ and obtain

$$\|\hat{f}(x+y) - \hat{f}(x) - \hat{f}(y)\| = 0.$$

Uniqueness follows from the following lemma. □

Lemma 16.1 *Let $a \geq 2$, let $\psi_1, \psi_2 \in \mathscr{A}(X)$ and $T_1, T_2 : X \to Y$ be such that*

$$T_i(2x) = aT_i, \quad x \in X, \ i = 1, 2$$

and

$$\|f(x) - T_i(x)\| \leq \psi_i(x), \quad x \in X, \ i = 1, 2.$$

Then $T_1 = T_2$.

Proof We have for $x \in X$ and an integer $n \geq 0$

$$T_i(2^n x) = a^n T_i(x), \quad i = 1, 2.$$

It follows

$$\begin{aligned} \|T_1(x) - T_2(x)\| &= \|a^{-n} T_1(2^n x) - a^{-n} T_2(2^n x)\| \\ &\leq a^{-n} \|T_1(2^n x) - f(2^n x)\| + a^{-n} \|f(2^n x) - T_2(2^n x)\| \\ &\leq a^{-n} \psi_1(2^n x) + a^{-n} \psi_2(2^n x) \underset{n \to \infty}{\longrightarrow} 0, \end{aligned}$$

hence $\|T_1(x) - T_2(x)\| = 0$. □

16.3 Main Results

Theorem 16.2 *Let G be an odd function and ψ-additive on orthogonal vectors. Then there exists*

$$\hat{G}(x) := \lim_{n \to \infty} 2^{-n} G(2^n x),$$

and $\hat{G}$ is the unique odd, orthogonally additive function which verifies

$$\|\hat{G}(x) - G(x)\| \leq \frac{\theta\eta + 4}{2 - \theta} \psi(x) \quad \text{for all } x \in X. \tag{16.11}$$

Proof Consider $x \in X$. From Condition 2 of Definition 16.1, it follows that there exists $y \in X$ such that

$$x \perp y \quad \text{and} \quad x + y \perp x - y.$$

Using Conditions 2 and 3 of Definition 16.3, we obtain

$$\begin{aligned}\|G(2x)-G(x+y)-G(x-y)\| &\le \psi(x+y)+\psi(x-y)\\ &\le \eta\psi(2x)\\ &\le \eta\theta\psi(x),\end{aligned}$$

hence

$$\begin{aligned}\|G(2x)-2G(x)\| \le{}& \|G(2x)-G(x+y)-G(x-y)\|\\ &+\|G(x+y)-G(x)-G(y)\|\\ &+\|G(x-y)-G(x)-G(-y)\|\\ \le{}& \eta\theta\psi(x)+\psi(x)+\psi(y)+\psi(x)+\psi(-y).\end{aligned}$$

Using Condition 4 of Definition 16.3, it follows

$$\|G(2x)-2G(x)\| \le (\eta\theta+4)\psi(x).$$

The conclusion follows from Proposition 16.2 with $a=2$. □

Corollary 16.1 *Let G be an odd function such that*

$$\|G(x+y)-G(x)-G(y)\| \le \varepsilon \quad \text{if } x \perp y.$$

Then there exists a unique odd, orthogonally additive function $\hat{G}$ such that

$$\|\hat{G}(x)-G(x)\| \le 3\varepsilon, \quad x \in X.$$

Corollary 16.2 *We consider $0 \le p < 0$ and X a Hilbert space with* $\dim X \ge 2$. *Let $G: X \to Y$ be an odd function such that*

$$\|G(x+y)-G(x)-G(y)\| \le \varepsilon\big(\|x\|^p+\|y\|^p\big)$$

for $x, y \in X$, $x \perp y$. Then there exists a unique odd orthogonally additive function $\hat{G}$ such that

$$\|\hat{G}(x)-G(x)\| \le \frac{2^{\frac{p}{2}+1}+4}{2-2^p}\varepsilon\|x\|^p, \quad x \in X.$$

Theorem 16.3 *Let $H: X \to Y$ be an even function that is ψ-additive on orthogonal vectors. Then there exists*

$$\hat{H}(x) := \lim_{n\to\infty} 4^{-n} G(x),$$

and $\hat{H}$ is the unique even, orthogonally additive function which verifies

$$\|\hat{H}(x)-H(x)\| \le \frac{\theta\eta+4+2\eta}{4-\theta}\psi(x), \quad \textit{for all } x \in X. \tag{16.12}$$

Proof Consider $x \in X$. From Condition 2 of Definition 16.1, it follows that there exists $y \in X$ such that $x \perp y$ and $x + y \perp x - y$. As in the proof of Theorem 16.2, it follows

$$\|H(2x) - H(x+y) - H(x-y)\| \le \eta\theta\psi(x). \tag{16.13}$$

From Condition 1 of Definition 16.1, it follows that

$$\frac{x+y}{2} \perp \pm\frac{x-y}{2},$$

hence, by Condition 3 of Definition 16.3,

$$\begin{aligned}\left\|H(x) - H\left(\frac{x+y}{2}\right) - H\left(\frac{x-y}{2}\right)\right\| &\le \psi\left(\frac{x+y}{2}\right) + \psi\left(\frac{x-y}{2}\right)\\ &\le \eta\psi(x)\end{aligned}$$

and, by Conditions 3 and 4 of Definition 16.3,

$$\left\|H(y) - H\left(\frac{x+y}{2}\right) - H\left(\frac{y-x}{2}\right)\right\| \le \eta\psi(y) = \eta\psi(x).$$

Using the triangle inequality, we have

$$\|H(x) - H(y)\| \le 2\eta\psi(x), \tag{16.14}$$

since H is even.

From (16.13) and (16.14), we obtain

$$\begin{aligned}\|H(2x) - 4H(x)\| &\le \|H(2x) - H(x+y) - H(x-y)\|\\ &\quad + \|H(x+y) + H(x-y) - 4H(x)\|\\ &\le \eta\theta\psi(x) + \|H(x+y) - H(x) - H(y)\|\\ &\quad + \|H(x-y) - H(x) - H(y)\| + 2\|H(x) - H(y)\|\\ &\le \eta\theta\psi(x) + \psi(x) + \psi(y) + \psi(x) + \psi(-y) + 4\eta\psi(x)\\ &= (\eta\theta + 4 + 4\eta)\psi(x).\end{aligned}$$

The conclusion follows from Proposition 16.2 with $a = 4$. □

Corollary 16.3 *Let $H : X \to Y$ be an even function such that*

$$\|H(x+y) - H(x) - H(y)\| \le \varepsilon \quad \textit{if}\ x \perp y.$$

Then there exists a unique even, orthogonally additive function $\hat{H}$ such that

$$\|\hat{H}(x) - H(x)\| \le \frac{7\varepsilon}{3}, \quad x \in X.$$

Corollary 16.4 *We consider* $0 \le p < 0$ *and* X *a Hilbert space with* $\dim X \ge 2$. *Let* $H : X \to Y$ *be an even function such that*

$$\left\| H(x+y) - H(x) - H(y) \right\| \le \varepsilon\left(\|x\|^p + \|y\|^p\right)$$

for $x, y \in X$, $x \perp y$. *Then there exists a unique even orthogonally additive function* $\hat{H}$ *such that*

$$\left\| \hat{H}(x) - H(x) \right\| \le \frac{2^{\frac{p}{2}+1}(1+2^{1-p})+4}{4-2^p}\varepsilon\|x\|^p, \quad x \in X.$$

Theorem 16.4 *Let* $f : X \to Y$ *be a function* ψ*-additive on orthogonal vectors. Then there exists a unique orthogonally additive mapping* T *such that*

$$\left\| f(x) - T(x) \right\| \le \left(\frac{\theta\eta+4}{2-\theta} + \frac{\theta\eta+4+4\eta}{4-\theta}\right)\psi(x), \quad x \in X.$$

Proof We denote by

$$f_o = \frac{f(x)-f(-x)}{2}, \qquad f_e(x) = \frac{f(x)+f(-x)}{2},$$

the odd and even part of f, respectively. We prove that f_o verifies (16.4):

$$\begin{aligned}
\left\| f_o(x+y) - f_o(x) - f_o(y) \right\| &\le \frac{1}{2}\left\| f(x+y) - f(x) - f(y) \right\| \\
&\quad + \frac{1}{2}\left\| f(-x-y) - f(-x) - f(-y) \right\| \\
&\le \frac{1}{2}\left[\psi(x)+\psi(y)\right] + \frac{1}{2}\left[\psi(-x)+\psi(-y)\right] \\
&= \psi(x)+\psi(y).
\end{aligned}$$

By Theorem 16.2, the function

$$\hat{G}(x) := \lim_{n\to\infty} 2^{-n} f_o\left(2^n x\right)$$

is an odd, orthogonally additive function which verifies

$$\left\| f_o(x) - \hat{G}(x) \right\| \le \frac{\theta\eta+4}{2-\theta}\psi(x), \quad x \in X. \tag{16.15}$$

On the other hand, f_e also verifies (16.4), and hence, by Theorem 16.3, the function

$$\hat{H}(x) := \lim_{n\to\infty} 4^{-n} f_e(x)$$

is an even orthogonally additive function which verifies

$$\left\| f_e(x) - \hat{H}(x) \right\| \le \frac{\theta\eta+4+2\eta}{4-\theta}\psi(x), \quad x \in X. \tag{16.16}$$

We denote $T = \hat{G} + \hat{H}$. We have by (16.15) and (16.16),

$$\|f(x) - T(x)\| \le \left(\frac{\theta\eta + 4}{2-\theta} + \frac{\theta\eta + 4 + 2\eta}{4-\theta}\right)\psi(x), \quad x \in X.$$

We prove that T is unique. Let T' be another orthogonally additive function which verifies the inequality in Theorem 16.4. We denote

$$\widetilde{\psi}(x) = \left(\frac{\theta\eta + 4}{2-\theta} + \frac{\theta\eta + 4 + 2\eta}{4-\theta}\right)\psi(x), \quad x \in X.$$

Clearly, $\widetilde{\psi}(x) \in \mathscr{A}(X)$. We have

$$\begin{aligned}
\|f_o(x) - T_o(x)\| &\le \frac{1}{2}\|f(x) - T(x)\| + \frac{1}{2}\|f(-x) - T(-x)\| \\
&\le \widetilde{\psi}(x), \\
\|f_o(x) - T_o'(x)\| &\le \widetilde{\psi}(x), \\
T_o(2x) &= 2T_o(x), \\
T_o'(2x) &= 2T_o'(x).
\end{aligned}$$

From Lemma 16.1, it follows that $T_o = T_o'$.

Analogously, $T_e = T_e'$. Hence $T = T'$. □

Corollary 16.5 *Let $f : X \to Y$ be a function such that*

$$\|f(x+y) - f(x) - f(y)\| \le \varepsilon, \quad x, y \in X, x \perp y.$$

Then there exists a unique orthogonally additive function $T : X \to Y$ such that

$$\|f(x) - T(x)\| \le \frac{16\varepsilon}{3}, \quad x \in X.$$

Remark 16.3 Corollary 16.5 contains the main result of R. Ger and J. Sikorska [13].

Corollary 16.6 *We consider $0 \le p < 1$ and X a Hilbert space with* $\dim X \ge 2$. *Let $f : X \to Y$ be a function with the following property:*

$$\|f(x+y) - f(x) - f(y)\| \le \left(\|x\|^p + \|y\|^p\right)$$

for $x, y \in X$, $x \perp y$. Then there exists a unique orthogonally additive function T such that

$$\|f(x) - T(x)\| \le \left(\frac{2^{\frac{p}{2}+1} + 4}{2 - 2^p} + \frac{2^{\frac{p}{2}+1}(1 + 2^{1-p}) + 4}{4 - 2^p}\right)\varepsilon\|x\|^p,$$

for all $x \in X$.

For other results on the Hyers–Ulam–Rassias stability for orthogonal mappings, see [3, 19–22, 26].

Acknowledgements The work of the second author is a result of the project "Creşterea calităţii şi a competitivităţii cercetării doctorale prin acordarea de burse" (contract de finantare POSDRU/88/1.5/S/49516). This project is co-funded by the European Social Fund through The Sectorial Operational Programme for Human Resources Development 2007–2013, coordinated by the West University of Timisoara in partnership with the University of Craiova and Fraunhofer Institute for Integrated Systems and Device Technology—Fraunhofer IISB.

References

1. Aoki, T.: On the stability of the linear transformation in Banach spaces. J. Math. Soc. Jpn. **2**, 64–66 (1950)
2. Badea, C.: The general linear equation is stable. Nonlinear Funct. Anal. Appl. **10**(1), 155–164 (2005)
3. Brzdek, J.: On orthogonally exponential and orthogonally additive mappings. Proc. Am. Math. Soc. **125**, 2127–2132 (1997)
4. Czerwik, S.: Functional Equations and Inequalities in Several Variables. World Scientific, New Jersey (2002)
5. Czerwik, S.: Stability of Functional Equations of Ulam–Hyers–Rassias Type. Hadronic Press, Palm Harbor (2003)
6. Gavruta, L., Gavruta, P., Eskandani, G.Z.: Hyers–Ulam stability of frames in Hilbert spaces. Bul. St. Univ. "Politehnica" Timisoara, Ser. Mat.-Fiz. **55**(69), 60–67 (2010)
7. Găvruţa, P.: A generalization of the Hyers–Ulam–Rassias stability of approximately additive mappings. J. Math. Anal. Appl. **184**, 431–436 (1994)
8. Găvruţa, P.: An answer to a question of Th.M. Rassias and J. Tabor on mixed stability of mappings. Bul. Stiint. Univ. Politeh. Timis. Ser. Mat. Fiz. **42**(56), 1–6 (1997)
9. Găvruţa, P.: On the Hyers–Ulam–Rassias Stability of Mappings. In: Recent Progress in Inequalities, vol. 430, pp. 465–469. Kluwer Dordrecht (1998)
10. Găvruţa, P.: On a problem of G. Isac and Th.M. Rassias concerning the stability of mappings. J. Math. Anal. Appl. **261**, 543–553 (2001)
11. Găvruţa, P., Hossu, M., Popescu, D., Caprău, C.: On the stability of mappings and an answer to a problem of Th.M. Rassias. Ann. Math. Blaise Pascal **2**, 55–60 (1995)
12. Găvruţa, P., Găvruţa, L.: A new method for the generalized Hyers–Ulam–Rassias stability. Int. J. Nonlinear Anal. Appl. **1**(2), 11–18 (2010)
13. Ger, R., Sikorska, J.: Stability of the orthogonal additivity. Bull. Pol. Acad. Sci., Math. **43**(2), 143–151 (1995)
14. Hyers, D.H.: On the stability of the linear functional equation. Proc. Natl. Acad. Sci. USA **27**, 222–224 (1941)
15. Hyers, D.H., Isac, G., Rassias, Th.M.: Stability of Functional Equations in Several Variables. Birkhäuser, Basel (1998)
16. Isac, G., Rassias, Th.M.: On the Hyers–Ulam stability of ψ-additive mappings. J. Approx. Theory **72**, 131–137 (1993)
17. Isac, G., Rassias, Th.M.: Stability of ψ-additive mappings: applications to nonlinear analysis. Int. J. Math. Math. Sci. **19**, 219–228 (1996)
18. Jung, S.-M.: Hyers–Ulam–Rassias Stability of Functional Equations in Mathematical Analysis. Hadronic Press, Palm Harbor (2001)
19. Moslehian, M.S.: On the orthogonal stability of the Pexiderized quadratic equation. J. Differ. Equ. Appl. **11**(11), 999–1004 (2005)

20. Moslehian, M.S.: On the stability of the orthogonal Pexiderized Cauchy equation. J. Math. Anal. Appl. **318**, 211–223 (2006)
21. Moslehian, M.S., Rassias, Th.M.: Orthogonal stability of additive type equations. Aequ. Math. **73**, 249–259 (2007)
22. Park, C.-G.: On the stability of the orthogonally quartic functional equation. Bull. Iran. Math. Soc. **31**(1), 63–70 (2005)
23. Rassias, J.M.: Solution of a problem of Ulam. J. Approx. Theory **57**(3), 268–273 (1989)
24. Rassias, Th.M.: On the stability of the linear mapping in Banach spaces. Proc. Am. Math. Soc. **72**, 297–300 (1978)
25. Rätz, J.: On orthogonally additive mappings. Aequ. Math. **28**, 35–49 (1985)
26. Sikorska, J.: Generalized orthogonal stability of some functional equations. J. Inequal. Appl. **2006**, 12404 (2006). 23 pages
27. Ulam, S.M.: A Collection of Mathematical Problems. Interscience, New York (1960)
28. Ulam, S.M.: Problems in Modern Mathematics. Wiley, New York (1964)

Chapter 17
Approximate Ternary Jordan Homomorphisms on Banach Ternary Algebras

Madjid Eshaghi Gordji, N. Ghobadipour, A. Ebadian, M. Bavand Savadkouhi, and Choonkil Park

Abstract Let A and B be two Banach ternary algebras over $\mathbb{R}$ or $\mathbb{C}$. A linear mapping $H : (A, [\ ]_A) \to (B, [\ ]_B)$ is called a ternary Jordan homomorphism if $H([xxx]_A) = [H(x)H(x)H(x)]_B$ for all $x \in A$. In this paper, we investigate ternary Jordan homomorphisms on Banach ternary algebras, associated with the following functional equation

$$f\left(\frac{x_1}{2} + x_2 + x_3\right) = \frac{1}{2}f(x_1) + f(x_2) + f(x_3).$$

Key words Generalized Hyers–Ulam stability · Banach ternary algebra · Ternary Jordan homomorphism · Functional equation

Mathematics Subject Classification Primary 39B52 · 17A40 · 46B03 · 47Jxx

---

Dedicated to Professor Themistocles M. Rassias on the occasion of his 60th birthday.

M.E. Gordji · M.B. Savadkouhi
Department of Mathematics, Semnan University, P.O. Box 35195-363, Semnan, Iran

M.E. Gordji
e-mail: madjid.eshaghi@gmail.com

M.B. Savadkouhi
e-mail: bavand.m@gmail.com

N. Ghobadipour · A. Ebadian
Department of Mathematics, Urmia University, Urmia, Iran

N. Ghobadipour
e-mail: ghobadipour.n@gmail.com

A. Ebadian
e-mail: a.ebadian@gmail.com

C. Park (✉)
Department of Mathematics, Hanyang University, Seoul 133-791, Republic of Korea
e-mail: baak@hanyang.ac.kr

P.M. Pardalos et al. (eds.), *Nonlinear Analysis*, Springer Optimization and Its Applications 68, In Honor of Themistocles M. Rassias on the Occasion of his 60th Birthday, DOI 10.1007/978-1-4614-3498-6_17,

17.1 Introduction and Preliminaries

We say that a functional equation (ξ) is stable if any function g satisfying the equation (ξ) *approximately* is near to true solution of (ξ). Also, we say that a functional equation is *superstable* if every approximately solution is an exact solution of it.

In this paper, we prove the stability and superstability of ternary Jordan homomorphisms on Banach ternary algebras.

Ternary algebraic operations were considered in the nineteenth century by several mathematicians and physicists such as Cayley [8] who introduced the notion of the cubic matrix, which in turn was generalized by Kapranov at el. [21]. The simplest example of such nontrivial ternary operation is given by the following composition rule:

$$\{a,b,c\}_{ijk} = \sum_{l,m,n} a_{nil} b_{ljm} c_{mkn} \quad (i,j,k,\ldots = 1,2,\ldots,\mathbb{N}).$$

Ternary structures and their generalization the so-called n-ary structures raise certain hopes in view of their applications in physics (see [5, 9, 12, 22, 23, 30, 31, 33]).

As it is extensively discussed in [30], the full description of a physical system S implies the knowledge of three basis ingredients: the set of the observables, the set of the states, and the dynamics that describes the time evolution of the system by means of the time dependence of the expectation value of a given observable on a given statue.

A ternary (associative) algebra $(A,[\ ])$ is a linear space A over a scalar field $\mathbb{F}=\mathbb{R}$ or $\mathbb{C}$ equipped with a linear mapping, the so-called ternary product, $[\ ] : A \times A \times A \to A$ such that $[[abc]de] = [a[bcd]e] = [ab[cde]]$ for all $a,b,c,d,e \in A$. This notion is a natural generalization of the binary case. Indeed, if $(A,\odot)$ is a usual (binary) algebra then $[abc] := (a \odot b) \odot c$ induces a ternary product, making A into a ternary algebra which will be called trivial. It is known that unital ternary algebras are trivial, and finitely generated ternary algebras are ternary subalgebras of trivial ternary algebras [6]. There are other types of ternary algebras in which one may consider other versions of associativity. Some examples of ternary algebras are (i) "cubic matrices" introduced by Cayley [8] which were in turn generalized by Kapranov, Gelfand, and Zelevinskii [21]; (ii) the ternary algebra of the polynomials of odd degrees in one variable equipped with the ternary operation $[p_1p_2p_3] = p_1 \odot p_2 \odot p_3$, where $\odot$ denotes the usual multiplication of polynomials. By a Banach ternary algebra we mean a ternary algebra equipped with a complete norm $\|\cdot\|$ such that $\|[abc]\| \le \|a\|\|b\|\|c\|$. If a ternary algebra $(A,[\ ])$ has an identity, i.e., an element e such that $a = [aee] = [eae] = [eea]$ for all $a \in A$, then $a \odot b := [aeb]$ is a binary product for which we have

$$(a \odot b) \odot c = \big[[aeb]ec\big] = \big[ae[bec]\big] = a \odot (b \odot c)$$

and

$$a \odot e = [aee] = a = [eea] = e \odot a,$$

for all $a, b, c \in A$, and so $(A, [\ ])$ may be considered as a (binary) algebra. Conversely, if $(A, \odot)$ is any (binary) algebra, then $[abc] := a \odot b \odot c$ makes A into a ternary algebra with the unit e such that $a \odot b = [aeb]$.

Let A, B be two Banach ternary algebras. A linear mapping $H : (A, [\ ]_A) \to (B, [\ ]_B)$ is called a ternary homomorphism if

$$H\big([xyz]_A\big) = \big[H(x)H(y)H(z)\big]_B$$

for all $x, y, z \in A$.

A linear mapping $H : (A, [\ ]_A) \to (B, [\ ]_B)$ is called a ternary Jordan homomorphism if

$$H\big([xxx]_A\big) = \big[H(x)H(x)H(x)\big]_B$$

for all $x \in A$.

Let A be a Banach (binary) algebra (Then it is well known that A is Banach ternary algebra with the product $[xyz] := x \odot y \odot z$). Let $H : A \to A$ be a (binary) Jordan homomorphism on A. Then $H : A \to A$ is a ternary Jordan homomorphism on A. Hence, there are ternary Jordan homomorphisms which are not ternary homomorphism.

Let A be a Banach (binary) algebra and let $S = \{a \in A : a^3 = 0\}$, suppose A' is the closure of $Lin(S)$. Then A' is a Banach ternary algebra with the trivial product. Every bounded linear map from A' into any Banach ternary algebra is a ternary Jordan homomorphism. For instance, if

$$A = \begin{pmatrix} 0 & \mathbb{C} & \mathbb{C} \\ 0 & 0 & \mathbb{C} \\ 0 & 0 & 0 \end{pmatrix},$$

then $A' = A$, and every linear map from A into any Banach ternary algebra B is a ternary Jordan homomorphism.

The study of stability problems originated from a famous talk given by Ulam [32] in 1940: "Under what condition does there exist a homomorphism near an approximate homomorphism?" In 1941, Hyers [14] answered affirmatively the question of Ulam, and the result can be formulated as follows: If $\varepsilon > 0$ and if $f : E_1 \to E_2$ is a map, with E_1 a normed space, E_2 a Banach spaces such that

$$\big\| f(x+y) - f(x) - f(y) \big\| \le \varepsilon$$

for all $x, y \in E_1$, then there exists a unique additive map $T : E_1 \to E_2$ such that

$$\big\| f(x) - T(x) \big\| \le \varepsilon$$

for all $x \in E_1$. Moreover, if $f(tx)$ is continuous in $t \in \mathbb{R}$ for each fixed $x \in E_1$, then T is linear. This stability phenomenon is called the Hyers–Ulam stability of the additive functional equation $g(x+y) = g(x) + g(y)$.

In 1978, a generalized version of the theorem of Hyers for approximate linear mappings was formulated and proved for the first time by Th.M. Rassias [28] who introduced the unbounded Cauchy difference. Th.M. Rassias' Theorem is stated as follows:

Theorem 17.1 *Let $f : E \to E'$ be a mapping from a normed vector space E into a Banach space E' subject to the inequality*

$$\|f(x+y) - f(x) - f(y)\| \le \varepsilon\big(\|x\|^p + \|y\|^p\big)$$

for all $x, y \in E$, where ε and p are constants with $\varepsilon > 0$ and $p < 1$. Then there exists a unique additive mapping $T : E \to E'$ such that

$$\|f(x) - T(x)\| \le \frac{2\varepsilon}{2-2^p}\|x\|^p$$

for all $x \in E$.

The stability phenomenon that was introduced and proved by Th.M. Rassias is called the *generalized Hyers–Ulam stability*. The stability problems of several functional equations have been extensively investigated by a number of authors and there are many interesting results concerning this problem (see [1–4, 10, 11, 13, 15–17, 25–27, 29]).

Stability of algebraic and topological homomorphisms has been investigated by many mathematicians; for an extensive account on the subject, see [29]. Park [24] studied the stability of Poisson C^*-homomorphisms and $JB*$-homomorphisms associated to the Jensen equation $2f(\frac{x+y}{2}) = f(x) + f(y)$ where f is a mapping between linear spaces. The generalized stability of this equation was studied by Jun and Lee [18] (see also [19]).

A generalization of the Jensen equation is the equation

$$rf\left(\frac{sx+ty}{r}\right) = sf(x) + tf(y),$$

where f is a mapping between linear spaces and r, s, t are given constant values (see [20]). It is easy to see that a mapping $f : X \to Y$ between linear spaces with $f(0) = 0$ satisfies the generalized Jensen equation if and only if it is additive; cf. [7].

The main purpose of the present paper is to offer the generalized Hyers–Ulam stability of ternary Jordan homomorphisms on Banach ternary algebras associated with the following functional equation

$$f\left(\frac{x_1}{2} + x_2 + x_3\right) = \frac{1}{2}f(x_1) + f(x_2) + f(x_3).$$

17.2 Ternary Jordan Homomorphisms

Throughout this section, assume that $(A, [\ ]_A), (B, [\ ]_B)$ are two Banach ternary algebras.

In this section, we investigate ternary Jordan homomorphisms on Banach ternary algebras.

Lemma 17.1 ([24]) *Let X and Y be linear spaces and let $f: X \to Y$ be an additive mapping such that $f(\mu x) = \mu f(x)$ for all $x \in X$ and all $\mu \in \mathbb{T}^1 := \{\lambda \in \mathbb{C}; |\lambda| = 1\}$. Then the mapping f is $\mathbb{C}$-linear.*

Lemma 17.2 *Let $f: A \to B$ be a mapping such that*

$$\left\| f\left(\frac{x_1}{2} + \mu x_2 + x_3\right) - \mu f(x_2) - f(x_3) \right\| \le \frac{1}{2} \| f(x_1) \|, \tag{17.1}$$

for all $x_1, x_2, x_3 \in A$. Then f is $\mathbb{C}$-linear.

Proof Letting $x_1 = x_2 = x_3 = 0$ and $\mu = 1$ in (17.1), we get

$$\| f(0) \| \le \frac{1}{2} \| f(0) \|.$$

So $f(0) = 0$. Letting $x_1 = 0$ and $\mu = 1$ in (17.1), we get

$$\| f(x_2 + x_3) - f(x_2) - f(x_3) \| \le \| f(0) \| = 0$$

for all $x_2, x_3 \in A$. So f is additive.

Letting $x_1 = x_3 = 0$ in (17.1), we get

$$\| f(\mu x_2) - \mu f(x_2) \| \le \| f(0) \|.$$

Hence

$$f(\mu x_2) = \mu f(x_2)$$

for all $x_2 \in A$ and all $\mu \in \mathbb{T}^1$. So by Lemma 17.1, the mapping f is $\mathbb{C}$-linear. □

Now we solve the superstability problem for ternary Jordan homomorphisms as follows.

Theorem 17.2 *Let $p \neq 1$ and θ be nonnegative real numbers, and let $f: A \to B$ be a mapping such that*

$$\left\| f\left(\frac{x_1}{2} + \mu x_2 + x_3\right) - \mu f(x_2) - f(x_3) \right\| \le \frac{1}{2} \| f(x_1) \|, \tag{17.2}$$

for all $\mu \in \mathbb{T}^1$ and all $x_1, x_2, x_3 \in A$,

$$\| f([x_2 x_2 x_2]_A) - [f(x_2) f(x_2) f(x_2)]_B \| \le \theta \|x_2\|^{3p} \tag{17.3}$$

for all $x_2 \in A$. Then the mapping $f: A \to B$ is a ternary Jordan homomorphism.

Proof Assume $p<1$. By Lemma 17.2, the mapping $f:A\to B$ is $\mathbb{C}$-linear. It follows from (17.3) that

$$\begin{aligned}
&\|f([x_2x_2x_2]_A)-[f(x_2)f(x_2)f(x_2)]_B\| \\
&=\frac{1}{n^3}\|f([(nx_2)(nx_2)(nx_2)]_A)-[f(nx_2)f(nx_2)f(nx_2)]_B\| \\
&\le\frac{\theta}{n^3}n^{3p}\|x_2\|^{3p}.
\end{aligned}$$

for all $x_2\in A$. Thus, since $p<1$, by letting n tend to ∞ in the last inequality, we obtain

$$f([x_2x_2x_2]_A)=[f(x_2)f(x_2)f(x_2)]_B$$

for all $x_2\in A$. Hence the mapping $f:A\to B$ is a ternary Jordan homomorphism.

Similarly, one obtains the result for the case $p>1$. □

Theorem 17.3 *Let $p<1$ and θ be nonnegative real numbers, and let $f:A\to B$ be a mapping such that*

$$\begin{aligned}
&\left\|f\left(\frac{x_1}{2}+\mu x_2+x_3\right)-\mu f(x_2)-f(x_3)\right\| \\
&\le\frac{1}{2}\|f(x_1)\|+\theta(\|x_1\|^p+\|x_2\|^p+\|x_3\|^p),
\end{aligned}\tag{17.4}$$

for all $\mu\in\mathbb{T}^1$ and all $x_1,x_2,x_3\in A$,

$$\|f([x_2x_2x_2]_A)-[f(x_2)f(x_2)f(x_2)]_B\|\le\theta\|x_2\|^{3p}\tag{17.5}$$

for all $x_2\in A$. Then there exists a unique ternary Jordan homomorphism $H:A\to B$ satisfying

$$\|H(x_2)-f(x_2)\|\le\frac{2\theta}{2-2^p}\|x_2\|^p\tag{17.6}$$

for all $x_2\in A$.

Proof Setting $\mu=1$ and $x_1=x_2=x_3=0$ in (17.4) yields $f(0)=0$. Let us take $\mu=1$, $x_1=0$ and $x_3=x_2$ in (17.4). Then we obtain

$$\left\|\frac{1}{2}f(2x_2)-f(x_2)\right\|\le\theta\|x_2\|^p,\tag{17.7}$$

for all $x_2\in A$. Now, by induction we get

$$\left\|\frac{1}{2^n}f(2^nx_2)-f(x_2)\right\|\le\theta\|x_2\|^p\sum_{i=0}^{n-1}2^{i(p-1)}.\tag{17.8}$$

In order to show that the functions $H_n(x_2) = \frac{1}{2^n} f(2^n x_2)$ form a convergent sequence, we use the Cauchy convergence criterion. Indeed, replace x_2 by $2^m x_2$ and divide by 2^m in (17.8), where m is an arbitrary positive integer. We find that

$$\left\| \frac{1}{2^{m+n}} f\left(2^{m+n} x_2\right) - \frac{1}{2^m} f\left(2^m x_2\right) \right\| \le \theta \|x_2\|^p \sum_{i=m}^{m+n-1} 2^{i(p-1)}$$

for all positive integers. Hence by the Cauchy criterion, the limit $H(x_2) = \lim_{n\to\infty} H_n(x_2)$ exists for each $x_2 \in A$. By taking the limit as $n \to \infty$ in (17.7), we see that

$$\|H(x_2) - f(x_2)\| \le \theta \|x_2\|^p \sum_{i=0}^{\infty} 2^{i(p-1)}$$

and (17.6) holds for all $x_2 \in A$. Now, we have

$$\begin{aligned}
&\left\| H\left(\frac{x_1}{2} + \mu x_2 + x_3\right) - \mu H(x_2) - H(x_3) \right\| \\
&= \lim_{n\to\infty} \frac{1}{2^n} \left\| f\left(\frac{2^n x_1}{2} + \mu 2^n x_2 + 2^n x_3\right) - \mu f\left(2^n x_2\right) - f\left(2^n x_3\right) \right\| \\
&\le \lim_{n\to\infty} \frac{1}{2^n} \frac{1}{2} \left\| f\left(2^n x_1\right) \right\| + \lim_{n\to\infty} \frac{1}{2^n} \theta \left(\left\|2^n x_1\right\|^p + \left\|2^n x_2\right\|^p + \left\|2^n x_3\right\|^p \right) \\
&= \frac{1}{2} \left\| H(x_1) \right\| + 0
\end{aligned}$$

for all $\mu \in \mathbb{T}^1$ and all $x_1, x_2, x_3 \in A$. So by Lemma 17.2, H is $\mathbb{C}$-linear.

On the other hand,

$$\begin{aligned}
&\left\| H\left([x_2 x_2 x_2]_A\right) - \left[H(x_2) H(x_2) H(x_2)\right]_B \right\| \\
&= \lim_{n\to\infty} \frac{1}{8^n} \left\| f\left(\left[(2^n x_2)(2^n x_2)(2^n x_2)\right]_A\right) - \left[f(2^n x_2) f(2^n x_2) f(2^n x_2)\right]_B \right\| \\
&\le \lim_{n\to\infty} \frac{\theta}{8^n} \left\| 2^n x_2 \right\|^{3p} \\
&= \lim_{n\to\infty} \theta 8^{n(p-1)} \|x_2\|^{3p} = 0
\end{aligned}$$

for all $x_2 \in A$, which means that $H([x_2 x_2 x_2]_A) = [H(x_2) H(x_2) H(x_2)]_B$. Therefore, we conclude that H is a ternary Jordan homomorphism.

Suppose that there exists another ternary Jordan homomorphism $H' : A \to B$ satisfying (17.6). Since $H'(x_2) = \frac{1}{2^n} H'(2^n x_2)$, we see that

$$\begin{aligned}\|H(x_2)-H'(x_2)\| &= \frac{1}{2^n}\|H(2^n x_2)-H'(2^n x_2)\|\\ &\le \frac{1}{2^n}\left(\|f(2^n x_2)-H(2^n x_2)\|+\|f(2^n x_2)-H'(2^n x_2)\|\right)\\ &\le \frac{4\theta}{2-2^p}2^{n(p-1)}\|x_2\|^p,\end{aligned}$$

which tends to zero as $n\to\infty$ for all $x_2\in A$. So that $H'=H$ as claimed, and the proof of the theorem is complete. □

One can easily get the following theorem.

Theorem 17.4 *Let θ be nonnegative real number, and let $f:A\to B$ be a mapping such that*

$$\left\|f\left(\frac{x_1}{2}+\mu x_2+x_3\right)-\mu f(x_2)-f(x_3)\right\|\le\frac{1}{2}\|f(x_1)\|+\theta$$

for all $\mu\in\mathbb{T}^1$ and all $x_1,x_2,x_3\in A$,

$$\|f([x_2x_2x_2]_A)-[f(x_2)f(x_2)f(x_2)]_B\|\le\theta\|x_2\|$$

for all $x_2\in A$. Then there exists a unique ternary Jordan homomorphism $H:A\to B$ satisfying

$$\|H(x_2)-f(x_2)\|\le\theta$$

for all $x_2\in A$.

Proof The proof is similar to the proof of Theorem 17.3. □

Theorem 17.5 *Let $p>1$ and θ be nonnegative real numbers, and let $f:A\to B$ be a mapping such that*

$$\begin{aligned}&\left\|f\left(\frac{x_1}{2}+\mu x_2+x_3\right)-\mu f(x_2)-f(x_3)\right\|\\ &\quad\le\frac{1}{2}\|f(x_1)\|+\theta\left(\|x_1\|^p+\|x_2\|^p+\|x_3\|^p\right),\end{aligned}\tag{17.9}$$

for all $\mu\in\mathbb{T}^1$ and all $x_1,x_2,x_3\in A$,

$$\|f([x_2x_2x_2]_A)-[f(x_2)f(x_2)f(x_2)]_B\|\le\theta\|x_2\|^{3p}\tag{17.10}$$

for all $x_2\in A$. Then there exists a ternary Jordan homomorphism $H:A\to B$ satisfying

$$\|f(x_2)-H(x_2)\|\le\frac{2\theta}{2^p-2}\|x_2\|^p\tag{17.11}$$

for all $x_2\in A$.

Proof Setting $\mu = 1$ and $x_1 = x_2 = x_3 = 0$ in (17.9) yields $f(0) = 0$. Let us take $\mu = 1$, $x_1 = 0$ and $x_3 = x_2$ in (17.9). We obtain

$$\|f(2x_2) - 2f(x_2)\| \leq 2\theta \|x_2\|^p, \tag{17.12}$$

for all $x_2 \in A$. In (17.12), replacing x_2 by $2^{-1}x_2$, we get

$$\|f(x_2) - 2f(2^{-1}x_2)\| \leq 2^{1-p}\theta \|x_2\|^p, \tag{17.13}$$

for all $x_2 \in A$. In (17.13), replacing x_2 by $2^{-1}x_2$ and then result dividing by 2^{-1}, we get

$$\|2f(2^{-1}x_2) - 2^2 f(2^{-2}x_2)\| \leq 2^{2(1-p)}\theta \|x_2\|^p, \tag{17.14}$$

for all $x_2 \in A$. Now, by induction we get

$$\|f(x_2) - 2^n f(2^{-n}x_2)\| \leq \theta \|x_2\|^p \sum_{i=1}^{n} 2^{i(1-p)}. \tag{17.15}$$

In order to show that the functions $H_n(x_2) = 2^n f(2^{-n}x_2)$ form a convergent sequence, we use the Cauchy convergence criterion. Indeed, replace x_2 by $2^{-m}x_2$ and divide by 2^{-m} in (17.15), where m is an arbitrary positive integer. We find that

$$\|2^m f(2^{-m}x_2) - 2^{m+n} f(2^{-(m+n)}x_2)\| \leq \theta \|x_2\|^p \sum_{i=m+1}^{m+n} 2^{i(1-p)}$$

for all positive integers. Hence by the Cauchy criterion, the limit $H(x_2) = \lim_{n\to\infty} H_n(x_2)$ exists for each $x_2 \in A$. By taking the limit as $n \to \infty$ in (17.14), we see that

$$\|f(x_2) - H(x_2)\| \leq \theta \|x_2\|^p \sum_{i=1}^{\infty} 2^{i(1-p)}$$

and (17.11) holds for all $x_2 \in A$. Now, we have

$$\begin{aligned}
&\left\| H\left(\frac{x_1}{2} + \mu x_2 + x_3\right) - \mu H(x_2) - H(x_3) \right\| \\
&= \lim_{n\to\infty} 2^n \left\| f\left(\frac{2^{-n}x_1}{2} + \mu 2^{-n}x_2 + 2^{-n}x_3\right) - \mu f(2^{-n}x_2) - f(2^{-n}x_3) \right\| \\
&\leq \lim_{n\to\infty} 2^n \frac{1}{2} \|f(2^{-n}x_1)\| + \lim_{n\to\infty} 2^n \theta \left(\|2^{-n}x_1\|^p + \|2^{-n}x_2\|^p + \|2^{-n}x_3\|^p\right) \\
&= \frac{1}{2}\|H(x_1)\| + 0
\end{aligned}$$

for all $\mu \in \mathbb{T}^1$ and all $x_1, x_2, x_3 \in A$. So by Lemma 17.2, H is $\mathbb{C}$-linear.

On the other hand,

$$\begin{aligned}
&\left\| H([x_2x_2x_2]_A) - \left[H(x_2)H(x_2)H(x_2)\right]_B \right\| \\
&\quad = \lim_{n\to\infty} 8^n \left\| f\left(\left[(2^{-n}x_2)(2^{-n}x_2)(2^{-n}x_2)\right]_A\right) \right. \\
&\qquad \left. - \left[f(2^{-n}x_2)f(2^{-n}x_2)f(2^{-n}x_2)\right]_B \right\| \\
&\quad \le \lim_{n\to\infty} 8^n\theta \left\|2^{-n}x_2\right\|^{3p} \\
&\quad = \lim_{n\to\infty} \theta 8^{n(1-p)} \|x_2\|^{3p} = 0
\end{aligned}$$

for all $x_2 \in A$, which means that $H([x_2x_2x_2]_A) = [H(x_2)H(x_2)H(x_2)]_B$. Therefore, we conclude that H is a ternary Jordan homomorphism.

Suppose that there exists another ternary Jordan homomorphism $H' : A \to B$ satisfying (17.11). Since $H'(x_2) = 2^{-n}H'(2^{-n}x_2)$, we see that

$$\begin{aligned}
&\left\| H(x_2) - H'(x_2) \right\| = 2^n \left\| H(2^{-n}x_2) - H'(2^{-n}x_2) \right\| \\
&\quad \le 2^n \left(\left\| f(2^{-n}x_2) - H(2^{-n}x_2) \right\| + \left\| f(2^{-n}x_2) - H'(2^{-n}x_2) \right\| \right) \\
&\quad \le \frac{4\theta}{2^p - 2} 2^{n(1-p)} \|x_2\|^p,
\end{aligned}$$

which tends to zero as $n \to \infty$ for all $x_2 \in A$. So $H' = H$ as claimed, and the proof of the theorem is complete. □

17.3 Conclusions

In this paper, we investigated the generalized Hyers–Ulam stability and superstability of ternary Jordan homomorphisms on Banach ternary algebras, associated with the following functional equation

$$f\left(\frac{x_1}{2} + x_2 + x_3\right) = \frac{1}{2} f(x_1) + f(x_2) + f(x_3).$$

References

1. Aoki, T.: On the stability of the linear transformation in Banach spaces. J. Math. Soc. Jpn. **2**, 64–66 (1950)
2. Aczel, J., Dhombres, J.: Functional Equations in Several Variables. Cambridge Univ. Press, Cambridge (1989)
3. Baak, C., Boo, D., Rassias, Th.M.: Generalized additive mapping in Banach modules and isomorphisms between C^*-algebras. J. Math. Anal. Appl. **314**, 150–161 (2006)
4. Savadkouhi, M.B., Gordji, M.E., Rassias, J.M., Ghobadipour, N.: Approximate ternary Jordan derivations on Banach ternary algebras. J. Math. Phys. **50**, 042303 (2009), 9 pp.
5. Bagarello, F., Morchio, G.: Dynamics of mean-field spin models from basic results in abstract differential equations. J. Stat. Phys. **66**, 849–866 (1992)

6. Bazunova, N., Borowiec, A., Kerner, R.: Universal differential calculus on ternary algebras. Lett. Math. Phys. **67**, 195–206 (2004)
7. Boo, D., Oh, S., Park, C., Park, J.: Generalized Jensen's equations in Banach modules over a C^*-algebra and its unitary group. Taiwan. J. Math. **7**, 641–655 (2003)
8. Cayley, A.: On the 34 concomitants of the ternary cubic. Am. J. Math. **4**, 1–15 (1881)
9. Daletskii, Y.L., Takhtajan, L.A.: Leibniz and Lie algebra structures for Nambu algebra. Lett. Math. Phys. **39**, 127 (1997)
10. Gajda, Z.: On stability of additive mappings. Int. J. Math. Math. Sci. **14**, 431–434 (1991)
11. Găvruta, P.: A generalization of the Hyers–Ulam–Rassias stability of approximately additive mappings. J. Math. Anal. Appl. **184**, 431–436 (1994)
12. Haag, R., Kastler, D.: An algebraic approach to quantum field theory. J. Math. Phys. **5**, 848–861 (1964)
13. Hyers, D.H., Isac, G., Rassias, Th.M.: Stability of Functional Equations in Several Variables. Birkhauser, Basel (1998)
14. Hyers, D.H.: On the stability of the linear functional equation. Proc. Natl. Acad. Sci. USA **27**, 222–224 (1941)
15. Hyers, D.H., Rassias, Th.M.: Approximate homomorphisms. Aequ. Math. **44**, 125–153 (1992)
16. Isac, G., Rassias, Th.M.: On the Hyers–Ulam stability of ψ-additive mappings. J. Approx. Theory **72**, 131–137 (1993)
17. Isac, G., Rassias, Th.M.: Stability of ψ-additive mappings: applications to nonlinear analysis. Int. J. Math. Math. Sci. **19**, 219–228 (1996)
18. Jun, K., Lee, Y.: A generalization of the Hyers–Ulam–Rassias stability of Jensen's equation. J. Math. Anal. Appl. **238**, 305–315 (1999)
19. Jung, S.: Hyers–Ulam–Rassias stability of Jensen's equation and its application. Proc. Am. Math. Soc. **126**, 3137–3143 (1998)
20. Jung, S., Moslehian, M.S., Sahoo, P.K.: Stability of generalized Jensen equation on restricted domains (preprint)
21. Kapranov, M., Gelfand, I.M., Zelevinskii, A.: Discriminants, Resultants and Multidimensional Determinants. Birkhäuser, Berlin (1994)
22. Kerner, R.: The cubic chessboard: geometry and physics. Class. Quantum Gravity **14**, A203–A225 (1997)
23. Kerner, R.: Ternary Algebraic Structures and Their Applications in Physics. Pierre et Marie Curie University, Paris (2000)
24. Park, C.: Homomorphisms between Poisson JC^*-algebras. Bull. Braz. Math. Soc. **36**, 79–97 (2005)
25. Park, C., Gordji, M.E.: Comment on "Approximate ternary Jordan derivations on Banach ternary algebras". J. Math. Phys. **51**, 044102 (2010) [Bavand Savadkouhi et al. J. Math. Phys. **51**, 042303 (2009)], 7 pp.
26. Rassias, Th.M.: On the stability of functional equations and a problem of Ulam. Acta Appl. Math. **62**, 23–130 (2000)
27. Rassias, Th.M.: On the stability of functional equations in Banach spaces. J. Math. Anal. Appl. **251**, 264–284 (2000)
28. Rassias, Th.M.: On the stability of the linear mapping in Banach spaces. Proc. Am. Math. Soc. **72**, 297–300 (1978)
29. Rassias, Th.M.: Approximate homomorphisms. Aequ. Math. **44**, 125–153 (1992)
30. Sewell, G.L.: Quantum Mechanics and Its Emergent Macrophysics. Princeton Univ. Press, Princeton (2002)
31. Takhtajan, L.A.: On foundation of the generalized Nambu mechanics. Commun. Math. Phys. **160**, 295 (1994)
32. Ulam, S.M.: Problems in Modern Mathematics. Wiley, New York (1940), Chapter VI, Science ed.
33. Zettl, H.: A characterization of ternary rings of operators. Adv. Math. **48**, 117–143 (1983)

Chapter 18
Approximately Cubic n-Derivations on Non-archimedean Banach Algebras

F. Habibian, R. Bolghanabadi, and M. Eshaghi Gordji

Abstract Let $n > 1$ be an integer, let $\mathscr{A}$ be an algebra, and let X be an A-module. An additive map $D : \mathscr{A} \longrightarrow X$ is called an n-derivation if

$$D\big(\Pi_{i=1}^{n} a_i\big) = D(a_1)a_2 \cdots a_n + a_1 D(a_2)a_3 \cdots a_n + \cdots + a_1 a_2 \cdots a_{n-1} D(a_n)$$

for all $a_1, \ldots, a_n \in \mathscr{A}$. We investigate the Hyers–Ulam–Rassias stability of cubic n-derivations from non-archimedean Banach algebras into non-archimedean Banach modules.

Key words Non-archimedean Banach algebra · Non-archimedean Banach module · Cubic functional equation · Hyers–Ulam–Rassias stability

Mathematics Subject Classification Primary 39B52 · Secondary 39B82 · 46H25

18.1 Introduction and Statement of Results

A definition of stability in the case of homomorphisms between metric groups was proposed in a problem by S.M. Ulam [47] in 1940. Let $(G_1, \cdot)$ be a group and let $(G_2, *)$ be a metric group with the metric $d(\cdot, \cdot)$. Given $\epsilon > 0$, does there exist a $\delta > 0$ such that if a mapping $h : G_1 \longrightarrow G_2$ satisfies the inequality $d(h(x \cdot y), h(x) * h(y)) < \delta$ for all $x, y \in G_1$, then there exists a homomorphism

---

Dedicated to Professor Themistocles M. Rassias on the occasion of his 60th birthday.

F. Habibian
Department of Mathematics, Semnan University, P.O. Box 35195-363, Semnan, Iran
e-mail: habibianf72@yahoo.com

R. Bolghanabadi
Research Group of Nonlinear Analysis and Applications (RGNAA), Semnan, Iran

M. Eshaghi Gordji (✉)
Center of Excellence in Nonlinear Analysis and Applications (CENAA), Semnan University, Semnan, Iran
e-mail: madjid.eshaghi@gmail.com

P.M. Pardalos et al. (eds.), *Nonlinear Analysis*, Springer Optimization and Its Applications 68, In Honor of Themistocles M. Rassias on the Occasion of his 60th Birthday, DOI 10.1007/978-1-4614-3498-6_18,

$H : G_1 \longrightarrow G_2$ with $d(h(x), H(x)) < \epsilon$ for all $x \in G_1$? In this case, the equation of the homomorphism $h(x \cdot y) = h(x) * h(y)$ is called stable. On the other hand, we are looking for situations when the homomorphisms are stable, i.e., if a mapping is an approximate homomorphism, then there exists an exact homomorphism near it.

In 1941, Hyers [26] gave a first affirmative answer to the question of Ulam for Banach spaces as follows:

If E and E' are Banach spaces and $f : E \longrightarrow E'$ is a mapping for which there is an $\varepsilon > 0$ such that $\|f(x+y) - f(x) - f(y)\| \le \varepsilon$ for all $x, y \in E$, then there is a unique additive mapping $L : E \longrightarrow E'$ such that $\|f(x) - L(x)\| \le \varepsilon$ for all $x \in E$.

Hyers' Theorem was generalized by Th.M. Rassias [45] for linear mappings by considering an unbounded Cauchy difference.

The paper of Rassias [45] has provided a lot of influence in the development of what we now call the *generalized Hyers–Ulam stability* or the *Hyers–Ulam–Rassias stability* of functional equations (see [1–25, 27–44]).

Let $\mathbb{K}$ be a field. A non-archimedean absolute value on $\mathbb{K}$ is a function $|\cdot| : \mathbb{K} \to \mathbb{R}$ such that for any $a, b \in \mathbb{K}$ we have

(i) $|a| \ge 0$ and equality holds if and only if $a = 0$,
(ii) $|ab| = |a||b|$,
(iii) $|a + b| \le \max\{|a|, |b|\}$.

Condition (iii) is called the strict triangle inequality. By (ii), we have $|1| = |-1| = 1$. Thus, by induction, it follows from (iii) that $|n| \le 1$ for each integer n. We always assume in addition that $|\cdot|$ is non-trivial, i.e., that there is an $a_0 \in \mathbb{K}$ such that $|a_0| \notin \{0, 1\}$.

Let X be a linear space over a scalar field $\mathbb{K}$ with a non-archimedean non-trivial valuation $|\cdot|$. A function $\|\cdot\| : X \to \mathbb{R}$ is a non-archimedean norm (valuation) if it satisfies the following conditions:

(NA1) $\|x\| = 0$ if and only if $x = 0$;
(NA2) $\|rx\| = |r|\|x\|$ for all $r \in \mathbb{K}$ and $x \in X$;
(NA3) the strong triangle inequality (ultrametric); namely,

$$\|x + y\| \le \max\big\{\|x\|, \|y\|\big\} \quad (x, y \in X).$$

Then $(X, \|\cdot\|)$ is called a non-archimedean space.

It follows from (NA3) that

$$\|x_m - x_\ell\| \le \max\big\{\|x_{j+1} - x_j\| : \ell \le j \le m-1\big\} \quad (m > \ell).$$

Therefore, a sequence $\{x_m\}$ is Cauchy in X if and only if $\{x_{m+1} - x_m\}$ converges to zero in a non-archimedean space. By a complete non-archimedean space we mean one in which every Cauchy sequence is convergent. A non-archimedean Banach algebra is a complete non-archimedean algebra $\mathscr{A}$ which satisfies $\|ab\| \le \|a\|\|b\|$ for all $a, b \in \mathscr{A}$. A non-archimedean Banach space X is a non-archimedean Banach $\mathscr{A}$-bimodule if X is an A-bimodule which satisfies $\max\{\|xa\|, \|ax\|\} \le \|a\|\|x\|$ for all $a \in A$, $x \in X$. For more detailed definitions of non-archimedean Banach algebras, we can refer to [46]. For the history and various aspects of this theory, we refer the reader to [6, 7, 17, 19].

Jan and Kim [30] introduced the following functional equation

$$f(2x+y)+f(2x-y)=2f(x+y)+2f(x-y)+12f(x) \tag{18.1}$$

and they established the general solution and generalized Hyers–Ulam–Rassias stability problem for this functional equation. It is easy to see that the function $f(x)=cx^3$ is a solution of the functional equation (18.1). Thus, it is natural that (18.1) is called a cubic functional equation, and every solution of the cubic functional equation is said to be a cubic function.

Let A be a normed algebra and let X be a Banach A-module. We say that a mapping $D: A \to X$ is a cubic n-derivation if D is a cubic function satisfying

$$D\big(\Pi_{i=1}^n x_i\big) = D(x_1)x_2^3\cdots x_n^3 + x_1^3 D(x_2)x_3^3\cdots x_n^3 + \cdots + x_1^3\cdots x_{n-1}^3 D(x_n) \tag{18.2}$$

for all $x_1,\ldots,x_n \in A$.

Recently, the stability of derivations has been investigated by a number of papers, including [8–15, 18, 20], and references therein. More recently, the third author of the present paper [5] established the stability of ring derivations on non-archimedean Banach algebras. In this paper, we investigate the approximately cubic n-derivations on non-archimedean Banach algebras.

18.2 Main Results

In the following, we suppose that A is a non-archimedean Banach algebra and X is a non-archimedean Banach A-bimodule.

Theorem 18.1 *Let $f: A \longrightarrow X$ be a given mapping with $f(0)=0$ and let $\varphi_1: A\times\cdots\times A \longrightarrow \mathbb{R}^+$, $\varphi_2: A\times A \longrightarrow \mathbb{R}^+$ be mappings such that*

$$\begin{aligned}&\big\|f\big(\Pi_{i=1}^n x_i\big) - f(x_1)x_2^3\cdots x_n^3 - x_1^3 f(x_2)x_3^3\cdots x_n^3 - \cdots - x_1^3\cdots x_{n-1}^3 f(x_n)\big\|\\&\leq \varphi_1(x_1,\ldots,x_n)\end{aligned} \tag{18.3}$$

for all $x_1,\ldots,x_n \in A$. Let

$$\big\|f(2x+y)+f(2x-y)-2f(x+y)-2f(x-y)-12f(x)\big\| \leq \varphi_2(x,y) \tag{18.4}$$

for all $x,y\in A$. Assume that for each $x \in A$

$$\lim_{n\longrightarrow\infty} \max\left\{\frac{1}{|2|^{3k}}\frac{\varphi_2(2^k x,0)}{|2|^4} : 0\leq k\leq n-1\right\},$$

denoted by $\Psi(x,0)$, exists. Suppose

$$\lim_{n\longrightarrow\infty}\frac{\varphi_1(2^n x_1,\ldots,2^n x_n)}{(|2|^{3n})^n} = \lim_{n\longrightarrow\infty}\frac{\varphi_2(2^n x,2^n y)}{|2|^{3n}} = \lim_{n\longrightarrow\infty}\frac{\varphi_2(2^n x,0)}{|2|^{3n}} = 0$$

for all $x_1, \ldots, x_n, x, y \in A$. *Then there exists a unique cubic n-derivation* $D : A \longrightarrow X$ *such that*

$$\|D(x) - f(x)\| \leq \Psi(x, 0) \tag{18.5}$$

for all $x \in A$.

Proof Setting $y = 0$ in (18.4) yields

$$\|2f(2x) - 16f(x)\| \leq \varphi_2(x, 0) \tag{18.6}$$

for all $x \in A$, and then, dividing by $|2|^4$ in (18.6), we obtain

$$\left\| \frac{f(2x)}{2^3} - f(x) \right\| \leq \frac{\varphi_2(x, 0)}{|2|^4} \tag{18.7}$$

for all $x \in A$. In (18.7), replacing x by $2x$ and then dividing by $|2|^3$, we obtain

$$\left\| \frac{f(2^2x)}{2^6} - \frac{f(2x)}{2^3} \right\| \leq \frac{\varphi_2(2x, 0)}{|2|^7}. \tag{18.8}$$

Combining (18.7) and (18.8), and using the strong triangle inequality (NA3), we get

$$\left\| \frac{f(2^2x)}{2^6} - f(x) \right\| \leq \max\left\{ \frac{\varphi_2(2x, 0)}{|2|^7}, \frac{\varphi_2(x, 0)}{|2|^4} \right\}. \tag{18.9}$$

Following the same argument, one can prove by induction that

$$\left\| \frac{f(2^nx)}{2^{3n}} - f(x) \right\| \leq \max\left\{ \frac{1}{|2|^{3k}} \frac{\varphi_2(2^kx, 0)}{|2|^4} : 0 \leq k \leq n-1 \right\}. \tag{18.10}$$

Replacing x by x^{2n-1} and dividing by $|2|^{3n+1}$ in (18.6), we find that

$$\left\| \frac{f(2^nx)}{2^{3n}} - \frac{f(2^{n-1}x)}{2^{3(n-1)}} \right\| \leq \frac{\varphi_2(2^{n-1}x, 0)}{|2|^{3n+1}}$$

for all positive integers n and $x \in A$. Hence $\{\frac{f(2^nx)}{2^{3n}}\}$ is a Cauchy sequence. Since X is complete it follows that $\{\frac{f(2^nx)}{2^{3n}}\}$ is convergent. Set $D(x) = \lim_{n \longrightarrow \infty} \frac{f(2^nx)}{2^{3n}}$. By taking the limit as $n \longrightarrow \infty$ in (18.10), we see that $\|D(x) - f(x)\| \leq \psi(x, 0)$, and (18.5) holds for all $x \in A$. In order to show that D satisfies (18.2), replace x_i by 2^nx_i and in (18.3) and divide by $(|2|^{3n})^n$ to get

$$\left\| \frac{f(2^nx_1 \cdots 2^nx_n)}{(2^{3n})^n} - \frac{f(2^nx_1)(2^nx_2)^3 \cdots (2^nx_n)^3}{(2^{3n})^n} - \cdots \right.$$
$$\left. - \frac{(2^nx_1)^3 \cdots (2^nx_{n-1})^3 f(2^nx_n)}{(2^{3n})^n} \right\| \leq \frac{\varphi_1(2^nx_1, \ldots, 2^nx_n)}{(|2|^{3^n})^n}.$$

Taking the limit as $n \longrightarrow \infty$, we find that D satisfies (18.2). Now, we replace x by $2^n x$ and y by $2^n y$ in (18.4) and divide by $|2|^{3n}$ to get

$$\left\| \frac{f(2\cdot 2^n x + 2^n y)}{2^{3n}} + \frac{f(2\cdot 2^n x - 2^n y)}{2^{3n}} - 2\frac{f(2^n x + 2^n y)}{2^{3n}} \right.$$
$$\left. - 2\frac{f(2^n x - 2^n y)}{2^{3n}} - 12\frac{f(2^n x)}{2^{3n}} \right\| \le \frac{\varphi_2(2^n x, 2^n y)}{|2|^{3n}}.$$

Taking the limit as $n \longrightarrow \infty$, we find that D satisfies (18.1).

Now, suppose that there is another such function $D' : A \to X$ satisfying $D'(2x + y) + D'(2x - y) = 2D'(x + y) + 2D'(x - y) - 12D'(x)$ and $\|D'(x) - f(x)\| \le \psi(x, 0)$. Then for all $x \in A$, we have

$$\begin{aligned}\|D(x) - D'(x)\| &= \lim_{n\longrightarrow\infty} \frac{1}{|2|^{3n}} \|D(2^n x) - D'(2^n x)\| \\ &\le \lim_{n\longrightarrow\infty} \frac{1}{|2|^{3n}} \max\{(\|D(2^n x) - f(2^n x)\|, \\ &\quad \|D'(2^n x) - f(2^n x)\|)\} \\ &\le \lim_{n\longrightarrow\infty} \lim_{k\longrightarrow\infty} \frac{1}{|2|^{3n}} \max\left\{ \frac{\varphi_2(2^j x, 0)}{|2|^{3j}\cdot 2^4} : n \le j \le k + n \right\} = 0,\end{aligned}$$

and therefore $D(x) = D'(x)$. □

Corollary 18.1 *Let θ_1 and θ_2 be nonnegative real numbers, and let p be a real number such that $0 < p < 3$. Suppose that a mapping $f : A \longrightarrow X$ satisfies*

$$\|f(\Pi_{i=1}^n x_i) - f(x_1)x_2^3 \cdots x_n^3 - x_1^3 f(x_2)x_3^3 \cdots x_n^3 - \cdots - x_1^3 \cdots x_{n-1}^3 f(x_n)\| \le \theta_1$$

for $x_1, \ldots, x_n \in A$. Let

$$\|f(2x + y) + f(2x - y) - 2f(x + y) - 2f(x - y) - 12f(x)\| \le \theta_2(\|x\|^p + \|y\|^p)$$

for all $x, y \in A$. Then there exists a unique cubic n-derivation $D : A \longrightarrow X$ such that

$$\|D(x) - f(x)\| \le \lim_{n\to\infty} \max\left\{ \frac{\theta_2\|x\|^p}{|2|^4|2|^{k(3-p)}} \, 0 \le k \le n - 1 \right\}$$

for all $x \in A$.

Proof Let $\varphi_1 : A \times \cdots \times A \longrightarrow \mathbb{R}^+$, $\varphi_2 : A \times A \longrightarrow \mathbb{R}^+$ be mappings such that $\varphi_1(x_1, \ldots, x_n) = \theta_1$ and $\varphi_2(x, y) = \theta_2(\|x\|^p + \|y\|^p)$ for all $x_1, \ldots, x_n, x, y \in A$. We have

$$\lim_{n\longrightarrow\infty} \frac{\varphi_2(2^n x, 2^n y)}{|2|^{3n}} = \lim_{n\longrightarrow\infty} \left(\frac{|2|^{np}}{|2|^{3n}}\right)\varphi_2(x, y) = 0 \quad (x, y \in A),$$

$$\lim_{n\longrightarrow\infty}\frac{\varphi_2(2^n x,0)}{|2|^{3n}}=\lim_{n\longrightarrow\infty}\left(\frac{|2|^{np}}{|2|^{3n}}\right)\varphi_2(x,0)=0\quad (x,y\in A),$$

$$\lim_{n\longrightarrow\infty}\frac{\varphi_1(2^n x_1,\dots,2^n x_n)}{(|2|^{3n})^n}=\lim_{n\longrightarrow\infty}\frac{\theta_1}{(|2|^{3n})^n}=0\quad (x,y\in A).$$

Applying Theorem 18.1, we conclude the required result. □

Corollary 18.2 *Let θ_1 and θ_2 be nonnegative real numbers. Suppose that a mapping $f:A\longrightarrow X$ satisfies*

$$\left\|f\left(\Pi_{i=1}^n x_i\right)-f(x_1)x_2^3\cdots x_n^3-x_1^3 f(x_2)x_3^3\cdots x_n^3-\cdots-x_1^3\cdots x_{n-1}^3 f(x_n)\right\|\le\theta_1,$$

for $x_1,\dots,x_n\in A$. Let

$$\left\|f(2x+y)+f(2x-y)-2f(x+y)-2f(x-y)-12f(x)\right\|\le\theta_2$$

for all $x,y\in A$. Then there exists a unique cubic n-derivation $D:A\longrightarrow X$ such that

$$\left\|D(x)-f(x)\right\|\le\frac{\theta_2}{|2|^4}$$

for all $x\in A$.

Proof Let $\varphi_1:A\times\cdots\times A\longrightarrow\mathbb{R}^+$, $\varphi_2:A\times A\longrightarrow\mathbb{R}^+$ be mappings such that $\varphi_1(x_1,\dots,x_n)=\theta_1$ and $\varphi_2(x,y)=\theta_2$ for all $x_1,\dots,x_n,x,y\in A$. We have

$$\lim_{n\longrightarrow\infty}\frac{\varphi_2(2^n x,2^n y)}{|2|^{3n}}=\lim_{n\longrightarrow\infty}\left(\frac{|2|^{np}}{|2|^{3n}}\right)\varphi_2(x,y)=0\quad (x,y\in A),$$

$$\lim_{n\longrightarrow\infty}\frac{\varphi_1(2^n x_1,\dots,2^n x_n)}{(|2|^{3n})^n}=\lim_{n\longrightarrow\infty}\frac{\theta_1}{(|2|^{3n})^n}=0\quad (x,y\in A),$$

$$\lim_{n\longrightarrow\infty}\frac{\varphi_2(2^n x,0)}{|2|^{3n}}=\lim_{n\longrightarrow\infty}\left(\frac{|2|^{np}}{|2|^{3n}}\right)\varphi_2(x,0)=0\quad (x,y\in A).$$

Applying Theorem 18.1, we conclude the required result. □

In the following, we investigate the superstability of cubic n-derivations.

Corollary 18.3 *Let $0<p<3$ and θ be a positive real number. Suppose $f:A\longrightarrow X$, $\varphi:A\times\cdots\times A\longrightarrow\mathbb{R}^+$ be mappings such that*

$$\begin{aligned}&\left\|f\left(\Pi_{i=1}^n x_i\right)-f(x_1)x_2^3\cdots x_n^3-x_1^3 f(x_2)x_3^3\cdots x_n^3-\cdots-x_1^3\cdots x_{n-1}^3 f(x_n)\right\|\\&\quad\le\varphi(x_1,\dots,x_n)\end{aligned}$$

for all $x_1,\dots,x_n\in A$. Let

$$\left\|f(2x+y)+f(2x-y)-2f(x+y)-2f(x-y)-12f(x)\right\|\le\theta\|y\|^p\quad(18.11)$$

for all $x,y\in A$. Then f is a cubic n-derivation.

Proof Letting $x = y = 0$ in (18.11), we get that $f(0) = 0$. So with $y = 0$ in (18.11), we get $f(2x) = 2^3 f(x)$ for $x \in A$. By applying induction on n, we obtain

$$f(2^n x) = 2^{3n} f(x) \tag{18.12}$$

for all $x \in A$ and $n \in \mathbb{N}$. On the other hand, by Theorem 18.1, the mapping $D : A \longrightarrow X$ defined by

$$D(x) = \lim_{n \to +\infty} \frac{f(2^n x)}{2^{3n}}$$

is a unique cubic n-derivation. Therefore, it follows from (18.12) that $f = D$. So that the mapping $f : A \longrightarrow X$ is a cubic n-derivation. □

Theorem 18.2 *Let $f : A \longrightarrow X$ be a given mapping and let $\varphi_1 : A \times \cdots \times A \longrightarrow \mathbb{R}^+$, $\varphi_2 : A \times A \longrightarrow \mathbb{R}^+$ be mappings such that*

$$\begin{aligned} &\left\| f\left(\Pi_{i=1}^n x_i\right) - f(x_1)x_2^3 \cdots x_n^3 - x_1^3 f(x_2)x_3^3 \cdots x_n^3 - \cdots - x_1^3 \cdots x_{n-1}^3 f(x_n) \right\| \\ &\quad \le \varphi_1(x_1, \ldots, x_n) \end{aligned} \tag{18.13}$$

for all $x_1, \ldots, x_n \in A$. Let

$$\left\| f(2x+y) + f(2x-y) - 2f(x+y) - 2f(x-y) - 12f(x) \right\| \le \varphi_2(x, y) \tag{18.14}$$

for all $x, y \in A$. Assume that $\lim_{n\to\infty} \max\{|2|^{3k-1}\varphi_2(\frac{x}{2^{k+1}}, 0) : 0 \le k \le n-1\} = \psi(x, 0)$ exists, and

$$\begin{aligned} \lim_{n\longrightarrow\infty} \left(|2|^{3n}\right)^n \varphi_1\left(\frac{x_1}{2^n}, \ldots, \frac{x_n}{2n}\right) &= \lim_{n\longrightarrow\infty} |2|^{3n} \varphi_2\left(\frac{x}{2^n}, \frac{y}{2^n}\right) \\ &= \lim_{n\longrightarrow\infty} |2|^{3n} \varphi_2\left(\frac{x}{2^n}, 0\right) = 0 \end{aligned}$$

for all $x_1, \ldots, x_n, x, y \in A$. Then there exists a cubic n-derivation $D : A \to X$ such that

$$\left\| f(x) - D(x) \right\| \le \psi(x, 0) \tag{18.15}$$

for $x \in A$.

Proof Setting $y = 0$ in (18.14), we obtain

$$\left\| 2f(2x) - 16f(x) \right\| \le \varphi_2(x, 0). \tag{18.16}$$

Replacing x by $\frac{x}{2}$ in (18.16) and dividing by 2, one obtains

$$\left\| f(x) - 8f\left(\frac{x}{2}\right) \right\| \le \frac{1}{|2|} \varphi_2\left(\frac{x}{2}, 0\right). \tag{18.17}$$

Again replacing x by $\frac{x}{2}$ in (18.17) and multiplying by $|2|^3$, we obtain

$$\left\|2^3 f\left(\frac{x}{2}\right) - 2^6 f\left(\frac{x}{2^2}\right)\right\| \leq |2|^2 \varphi_2\left(\frac{x}{2^2}, 0\right). \tag{18.18}$$

By using (18.17), (18.18), and the strong triangle inequality (NA3), we obtain

$$\left\|f(x) - 2^6 f\left(\frac{x}{2^2}\right)\right\| \leq \max\left\{\frac{1}{|2|}\varphi_2\left(\frac{x}{2}, 0\right), |2|^2 \varphi_2\left(\frac{x}{2^2}, 0\right)\right\} \tag{18.19}$$

for $x \in A$. Next we prove by induction that

$$\left\|f(x) - 2^{3n} f\left(\frac{x}{2^n}\right)\right\| \leq \max\left\{|2|^{3k-1} \varphi_2\left(\frac{x}{2^{k+1}}, 0\right) : 0 \leq k \leq n-1\right\}. \tag{18.20}$$

Replacing x by $\frac{x}{2^{n-1}}$ and multiplying by $|2|^{3(n-1)}$ in (18.17), we find that

$$\left\|2^{3(n-1)} f\left(\frac{x}{2^{n-1}}\right) - 2^{3n} f\left(\frac{x}{2^n}\right)\right\| \leq \frac{1}{|2|}|2|^{3(n-1)} \varphi_2\left(\frac{x}{2^n}, 0\right) \tag{18.21}$$

for all $x \in A$. Hence by the Cauchy criterion, the limit $D(x) = \lim_{n\to\infty} D_n(x)$ exists for each $x \in A$. By taking the limit as $n \to \infty$ in (18.20), we see that $\|f(x) - D(x)\| \leq \psi(x, 0)$, and (18.15) holds for $x \in A$. To show that D satisfies (18.2), replace x_i by $\frac{x_i}{2^n}$ in (18.13) and multiply by $(|2|^{3n})^n$ to get

$$\begin{aligned}&\left\|(|2|^{3n})^n . f\left(\frac{x_1}{2^n} \cdots \frac{x_n}{2^n}\right) - (|2|^{3n})^n . f\left(\frac{x_1}{2^n}\right)\left(\frac{x_2}{2^n}\right)^3 \cdots \left(\frac{x_2}{2^n}\right)^3 - \cdots \right.\\ &\qquad \left. - (|2|^{3n})^n . \left(\frac{x_1}{2^n}\right)^3 \cdots \left(\frac{x_{n-1}}{2^n}\right)^3 f\left(\frac{x_n}{2^n}\right)\right\|\\ &\quad \leq (|2|^{3n})^n \varphi_1\left(\frac{x_1}{2^n}, \ldots, \frac{x_n}{2^n}\right).\end{aligned}$$

Taking the limit as $n \to \infty$, we find that D satisfies (18.2). If we replace x by $\frac{x}{2^n}$, y by $\frac{y}{2^n}$ in (18.14) and multiply by $|2|^{3n}$, we will have

$$\begin{aligned}&\left\|2^{3n} \cdot f\left(2\frac{x}{2^n} + \frac{y}{2^n}\right) + 2^{3n} \cdot f\left(2\frac{x}{2^n} - \frac{y}{2^n}\right) - 2^{3n} \cdot 2f\left(\frac{x}{2^n} + \frac{y}{2^n}\right)\right.\\ &\qquad \left. - 2^{3n} \cdot 2f\left(\frac{x}{2^n} - \frac{y}{2^n}\right) - 2^{3n} \cdot 12f\left(\frac{x}{2^n}\right)\right\|\\ &\quad \leq |2|^{3n} \varphi_2\left(\frac{x}{2^n}, \frac{y}{2^n}\right).\end{aligned}$$

Taking the limit as $n \to \infty$, we find that D satisfies (18.1). □

The superstability of cubic n-derivations is stated as follows:

Corollary 18.4 *Let $p > 3$ and let θ be a positive real number. Let $f : A \longrightarrow X$, $\varphi : A^n \longrightarrow \mathbb{R}^+$ be mappings such that*

$$\begin{aligned}&\left\| f\left(\Pi_{i=1}^n x_i\right) - f(x_1)x_2^3 \cdots x_n^3 - x_1^3 f(x_2)x_3^3 \cdots x_n^3 - \cdots - x_1^3 \cdots x_{n-1}^3 f(x_n)\right\| \\ &\quad \le \varphi(x_1, \ldots, x_n)\end{aligned}$$

for $x_1, \ldots, x_n \in A$. Let

$$\left\| f(2x+y) + f(2x-y) - 2f(x+y) - 2f(x-y) - 12f(x)\right\| \le \theta \|y\|^p \quad (18.22)$$

for all $x, y \in A$. Then f is a cubic n-derivation.

Proof Setting $x = y = 0$ in (18.22), we get that $f(0) = 0$. Thus setting $y = 0$ in (18.22), we obtain $f(2x) = 2^3 f(x)$ for all $x \in A$. By applying induction on n, one can prove

$$f(x) = 2^{3n} f\left(\frac{x}{2^n}\right) \quad (18.20)$$

for all $x \in A$ and $n \in \mathbb{N}$. On the other hand, by Theorem 18.2, the mapping $D : A \longrightarrow X$ defined by

$$D(x) = \lim_{n\to\infty} 2^{3n} f\left(\frac{x}{2^n}\right)$$

is the unique cubic n-derivation. Therefore, it follows from (18.20) that $f = D$. So the mapping $f : A \longrightarrow X$ is a cubic n-derivation. □

Corollary 18.5 *Let p, q, θ be positive real numbers such that $p + q > 3$. Let $f : A \longrightarrow X$, $\varphi : A \times \cdots \times A \longrightarrow X$ be mappings such that*

$$\begin{aligned}&\left\| f\left(\Pi_{i=1}^n x_i\right) - f(x_1)x_2^3 \cdots x_n^3 - x_1^3 f(x_2)x_3^3 \cdots x_n^3 - \cdots - x_1^3 \cdots x_{n-1}^3 f(x_n)\right\| \\ &\quad \le \varphi(x_1, \ldots, x_n)\end{aligned}$$

for all $x_1, \ldots, x_n \in A$. Let

$$\left\| f(2x+y) + f(2x-y) - 2f(x+y) - 2f(x-y) - 12f(x)\right\| \le \theta \|x\|^q \|y\|^p$$

for all $x, y \in A$. Then f is a cubic n-derivation.

Proof Let $q = 0$, then by Corollary 18.4, we get the result. □

Corollary 18.6 *Let $p > 3$ and θ be a positive real number. Suppose the mapping $f : A \longrightarrow X$ satisfies*

$$\begin{aligned}&\left\| f\left(\Pi_{i=1}^n x_i\right) - f(x_1)x_2^3 \cdots x_n^3 - x_1^3 f(x_2)x_3^3 \cdots x_n^3 - \cdots - x_1^3 \cdots x_{n-1}^3 f(x_n)\right\| \\ &\quad \le \theta \|y\|^p\end{aligned}$$

for all $x_1, \ldots, x_n \in A$. *Let*

$$\left\| f(2x+y) + f(2x-y) - 2f(x+y) - 2f(x-y) - 12f(x) \right\| \le \theta \|y\|^p$$

for all $x, y \in A$. *Then f is a cubic n-derivation.*

Proof Let $\varphi_1 : A \times \cdots \times A \longrightarrow \mathbb{R}^+$ and $\varphi_2 : A \times A \to \mathbb{R}^+$ be mappings such that $\varphi_1(x_1, \ldots, x_n) = \theta \|y\|^p$, $\varphi_2(x, y) = \theta \|y\|^p$, for all $y \in A$. Then by Theorem 18.2, the result follows. □

Acknowledgements The authors would like to thank the Semnan University for its financial support.

References

1. Borelli, C.: On Hyers–Ulam stability for a class of functional equations. Aequ. Math. **54**, 74–86 (1997)
2. Bourgin, D.G.: Class of transformations and bordering transformations. Bull. Am. Math. Soc. **27**, 223–237 (1951)
3. Bavand Savadkouhi, M., Gordji, M.E., Rassias, J.M., Ghobadipour, N.: Approximate ternary Jordan derivations on Banach ternary algebras. J. Math. Phys. **50**, 042303 (2009). 9 pages
4. Ebadian, A., Ghobadipour, N., Gordji, M.E.: A fixed point method for perturbation of bimultipliers and Jordan bimultipliers in C^*-ternary algebras. J. Math. Phys. **51**, 1 (2010), 10 pages. doi:10.1063/1.3496391
5. Eshaghi Gordji, M.: Nearly ring homomorphisms and nearly ring derivations on non-Archimedean Banach algebras. Abstr. Appl. Anal. **2010**, 393247 (2010), 12 pages. doi:10.1155/2010/393247
6. Eshaghi Gordji, M., Bavand Savadkouhi, M.: Stability of cubic and quartic functional equations in non-Archimedean spaces. Acta Appl. Math. **110**, 1321–1329 (2010)
7. Eshaghi Gordji, M., Bavand Savadkouhi, M.: Stability of a mixed type cubic–quartic functional equation in non-Archimedean spaces. Appl. Math. Lett. **23**(10), 1198–1202 (2010)
8. Eshaghi Gordji, M., Bavand Savadkouhi, M.: On approximate cubic homomorphisms. Adv. Differ. Equ. **2009**, 618463 (2009), 11 pages. doi:10.1155/2009/618463
9. Eshaghi Gordji, M., Ghobadipour, N.: Generalized Ulam–Hyers stabilities of quartic derivations on Banach algebras. Proyecciones **29**(3), 209–224 (2010)
10. Eshaghi Gordji, M., Ghaemi, M.B., Kaboli Gharetapeh, S., Shams, S., Ebadian, A.: On the stability of J^*-derivations. J. Geom. Phys. **60**(3), 454–459 (2010)
11. Eshaghi Gordji, M., Ghobadipour, N.: Stability of (α, β, γ)-derivations on Lie C^*-algebras. Int. J. Geom. Methods Mod. Phys. **7**(7), 1–10 (2010). doi:10.1142/S0219887810004737
12. Eshaghi Gordji, M., Habibian, F.: Hyers–Ulam–Rassias stability of quadratic derivations on Banach algebras. Nonlinear Funct. Anal. Appl. **14**(5), 759–766 (2009)
13. Eshaghi Gordji, M., Kaboli Gharetapeh, S., Karimi, T., Rashidi, E., Aghaei, M.: Ternary Jordan derivations on C^*-ternary algebras. J. Comput. Anal. Appl. **12**(2), 463–470 (2010)
14. Eshaghi Gordji, M., Kaboli Gharetapeh, S., Karimi, T., Rashidi, E., Aghaei, M.: Ternary Jordan derivations on C^*-ternary algebras. J. Comput. Anal. Appl. **12**(2), 463–470 (2010)
15. Eshaghi Gordji, M., Karimi, T., Kaboli Gharetapeh, S.: Approximately n-Jordan homomorphisms on Banach algebras. J. Inequal. Appl. **2009**, 870843 (2009), 8 pages
16. Eshaghi Gordji, M., Khodabakhsh, R., Jung, S.M., Khodaei, H.: AQCQ-functional equation in non-Archimedean normed spaces. Abstr. Appl. Anal. **2010**, 741942 (2010), 22 pages. doi:10.1155/2010/741942

17. Eshaghi Gordji, M., Khodaei, H., Khodabakhsh, R.: General quartic-cubic-quadratic functional equation in non-Archimedean normed spaces. U.P.B. Sci. Bull. (Series A) **72**(3), 69–84 (2010)
18. Eshaghi Gordji, M., Rassias, J.M., Ghobadipour, N.: Generalized Hyers–Ulam stability of the generalized (n, k)-derivations. Abstr. Appl. Anal. **2009**, 437931 (2009), 8 pages
19. Eshaghi Gordji, M., Savadkouhi, M.B., Bidkham, M.: Stability of a mixed type additive and quadratic functional equation in non-Archimedean spaces. J. Comput. Anal. Appl. **12**(2), 454–462 (2010)
20. Farokhzad, R., Hosseinioun, S.A.R.: Perturbations of Jordan higher derivations in Banach ternary algebras: an alternative fixed point approach. Int. J. Nonlinear Anal. Appl. **1**(1), 42–53 (2010)
21. Faizev, V.A., Rassias, Th.M., Sahoo, P.K.: The space of (ψ, γ)-additive mappings on semigroups. Trans. Am. Math. Soc. **354**(11), 4455–4472 (2002)
22. Forti, G.L.: An existence and stability theorem for a class of functional equations. Stochastica **4**, 23–30 (1980)
23. Forti, G.L.: Comments on the core of the direct method for proving Hyers–Ulam stability of functional equations. J. Math. Anal. Appl. **295**, 127–133 (2004)
24. Gavruta, P.: An answer to a question of J.M. Rassias concerning the stability of Cauchy functional equation. In: Advances in Equations and Inequalities. Hadronic Math. Ser., pp. 67–71 (1999)
25. Ghobadipour, N., Ebadian, A., Rassias, Th.M., Eshaghi Gordji, M.: A perturbation of double derivations on Banach algebras. Commun. Math. Anal. **11**(1), 51–60 (2011)
26. Hyers, D.H.: On the stability of the linear functional equation. Proc. Natl. Acad. Sci. **27**, 222–224 (1941)
27. Hyers, D.H., Isac, G., Rassias, Th.M.: Stability of Functional Equations in Several Variables. Birkhauser, Boston (1998)
28. Hyers, D.H., Rassias, Th.M.: Approximate homomorphisms. Aequ. Math. **44**, 125–153 (1992)
29. Isac, G., Rassias, Th.M.: On the Hyers–Ulam stability of ψ-additive mappings. J. Approx. Theory **72**, 131–137 (1993)
30. Jun, K.W., Kim, H.M.: The generalized Hyers–Ulam–Rassias stability of a cubic functional equation. J. Math. Anal. Appl. **274**(2), 267–278 (2002)
31. Jung, Y.S., Park, I.S.: On the stability of the functional equation $f(x + y + xy) = f(x) + f(y) + f(x)y + xf(y)$. J. Math. Anal. Appl. **274**(2), 659–666 (2002)
32. Jung, S.M., Rassias, J.M.: A fixed point approach to the stability of a functional equation of the spiral of Theodorus. Fixed Point Theory Appl. (2008, in press)
33. Kim, H.M., Chang, I.S.: stability of the functional equations related to a multiplicative derivation. J. Appl. Math. Comput., Ser. A **11**, 413–421 (2003)
34. Maksa, G.: 18 Problem, In: The 34th International Symposium of Functional Equations, Aequationes Math., Wisla-Jawornik, Poland, June 10–June 19 (1996)
35. Najati, A., Rassias, Th.M.: Stability of homomorphisms and (θ, ϕ)-derivations. Appl. Anal. Discrete Math. **3**(2), 264–281 (2009)
36. Najati, A., Rassias, Th.M.: Stability of a mixed functional equation in several variables on Banach modules. Nonlinear Anal. **72**(3–4), 1755–1767 (2010)
37. Park, C.G., Rassias, Th.M.: Hyers–Ulam stability of a generalized Apollonius type quadratic mapping. J. Math. Anal. Appl. **322**(1), 371–381 (2006)
38. Park, C.-G., Rassias, Th.M.: Homomorphisms in C^*-ternary algebras and JB^*-triples. J. Math. Anal. Appl. **337**, 13–20 (2008)
39. Park, C.-G., Rassias, Th.M.: Homomorphisms and derivations in proper JCQ^*-triples. J. Math. Anal. Appl. **337**(2), 1404–1414 (2008)
40. Páles, Z.: Remark 27, in report on the 34th ISFE. Aequ. Math. **53**, 200–201 (1997)
41. Rassias, Th.M., Tabor, J.: Stability of Mappings of Hyers–Ulam Type. Hadronic Press, Florida (1994)

42. Rassias, Th.M.: On a modified Hyers–Ulam sequence. J. Math. Anal. Appl. **158**, 106–113 (1991)
43. Rassias, Th.M.: On the stability of functional equations originated by a problem of Ulam. Mathematica **44(67)**(1), 39–75 (2002)
44. Rassias, Th.M.: On the stability of functional equations and a problem of Ulam. Acta Appl. Math. **62**(1), 23–130 (2000)
45. Rassias, Th.M.: On the stability of the linear mapping in Banach spaces. Proc. Am. Math. Soc. **72**, 297–300 (1978)
46. Shilkret, N.: Non-archimedian Banach algebras. Ph.D. Thesis, Polytechnic University (1968). 178 pp. ProQuest LLC, Thesis
47. Ulam, S.M.: A Collection of Mathematical Problems. Interscience, New York (1960)

Chapter 19
Fuzzy Stability of a Quadratic-Additive Type Functional Equation

Sun-Sook Jin and Yang-Hi Lee

Abstract In this paper, we investigate a fuzzy version of stability for the functional equation

$$2f(x+y)+f(x-y)+f(y-x)-f(2x)-f(2y)=0$$

in the sense of M. Mirmostafaee and M.S. Moslehian.

Key words Fuzzy normed space · Fuzzy almost quadratic-additive mapping · Quadratic-additive type functional equation

Mathematics Subject Classification Primary 39B52

19.1 Introduction

A classical question in the theory of functional equations is "when is it true that a mapping, which approximately satisfies a functional equation, must be somehow close to an exact solution of the equation?" Such a problem, called *a stability problem of the functional equation*, was formulated by S.M. Ulam [21] in 1940. In the next year, D.H. Hyers [6] gave a partial solution of Ulam's problem for the case of approximate additive mappings. Subsequently, his result was generalized by T. Aoki [1] for additive mappings, and by Th.M. Rassias [19] for linear mappings, by considering the stability problem with unbounded Cauchy differences. During the last decades, the stability problems of functional equations have been extensively investigated by a number of mathematicians, see [4, 5, 7, 9, 10, 12–16, 20].

Dedicated to Professor Themistocles M. Rassias on the occasion of his 60th birthday.

S.-S. Jin · Y.-H. Lee (✉)
Department of Mathematics Education, Gongju National University of Education, Gongju 314-711, Republic of Korea
e-mail: lyhmzi@gjue.ac.kr

S.-S. Jin
e-mail: ssjin@gjue.ac.kr

P.M. Pardalos et al. (eds.), *Nonlinear Analysis*, Springer Optimization and Its Applications 68, In Honor of Themistocles M. Rassias on the Occasion of his 60th Birthday, DOI 10.1007/978-1-4614-3498-6_19,

In 1984, A.K. Katsaras [8] defined a fuzzy norm on a linear space to construct a fuzzy structure on the space. Since then, a few mathematicians have introduced several types of fuzzy norms from different points of view. In particular, T. Bag and S.K. Samanta [2], following Cheng and Mordeson [3], gave an idea of a fuzzy norm in such a manner that the corresponding fuzzy metric is of Kramosil and Michalek type [11]. In 2008, M. Mirmostafaee and M.S. Moslehian [18] obtained a fuzzy version of stability for *the Cauchy functional equation*

$$f(x+y) - f(x) - f(y) = 0. \tag{19.1}$$

In the same year, they [17] proved a fuzzy version of stability for *the quadratic functional equation*

$$f(x+y) + f(x-y) - 2f(x) - 2f(y) = 0. \tag{19.2}$$

A solution of (19.1) is called *an additive mapping* and a solution of (19.2) is called *a quadratic mapping*. In this paper, we consider the functional equation

$$2f(x+y) + f(x-y) + f(y-x) - f(2x) - f(2y) = 0 \tag{19.3}$$

and get a general stability result of it in the fuzzy normed linear space. We call (19.3) *a quadratic-additive type functional equation*. Precisely, we show that if f is a solution of the functional equation (19.3) then the odd part $\frac{f(x)-f(-x)}{2}$ is an additive mapping and the even part $\frac{f(x)+f(-x)}{2}$ is a quadratic function. It is easy to show that every additive mapping and every quadratic mapping are solutions of the functional equation (19.3). So we call a solution of (19.3) *a quadratic-additive mapping*.

19.2 Main Results

We use the definition of a fuzzy normed space given in [2] to exhibit a reasonable fuzzy version of stability for the quadratic-additive type functional equation in the fuzzy normed linear space.

Definition 19.1 ([2]) Let X be a real linear space. A function $N : X \times \mathbb{R} \to [0, 1]$ (the so-called fuzzy subset) is said to be *a fuzzy norm on* X if for all $x, y \in X$ and all $s, t \in \mathbb{R}$,

(N1) $N(x, c) = 0$ for $c \le 0$;
(N2) $x = 0$ if and only if $N(x, c) = 1$ for all $c > 0$;
(N3) $N(cx, t) = N(x, t/|c|)$ if $c \ne 0$;
(N4) $N(x + y, s + t) \ge \min\{N(x, s), N(y, t)\}$;
(N5) $N(x, \cdot)$ is a non-decreasing function on $\mathbb{R}$ and $\lim_{t\to\infty} N(x, t) = 1$.

The pair (X, N) is called *a fuzzy normed linear space*. Let (X, N) be a fuzzy normed linear space. Let $\{x_n\}$ be a sequence in X. Then $\{x_n\}$ is said to be convergent if there exists $x \in X$ such that $\lim_{n\to\infty} N(x_n - x, t) = 1$ for all $t > 0$. In this case, x is called *the limit of the sequence* $\{x_n\}$, and we denote it by $N - \lim_{n\to\infty} x_n = x$. A sequence $\{x_n\}$ in X is called *Cauchy* if for each $\varepsilon > 0$ and each $t > 0$ there exists n_0 such that for all $n \geq n_0$ and all $p > 0$ we have $N(x_{n+p} - x_n, t) > 1 - \varepsilon$. It is known that every convergent sequence in a fuzzy normed space is Cauchy. If each Cauchy sequence is convergent, then the fuzzy norm is said to be *complete* and the fuzzy normed space is called *a fuzzy Banach space*.

Let (X, N) be a fuzzy normed space and (Y, N') a fuzzy Banach space. For a given mapping $f : X \to Y$, we use the abbreviation

$$Df(x, y) := 2f(x + y) + f(x - y) + f(y - x) - f(2x) - f(2y)$$

for all $x, y \in X$. For given $q > 0$, the mapping f is called *a fuzzy q-almost quadratic-additive mapping*, if

$$N'\big(Df(x, y), t + s\big) \geq \min\big\{N\big(x, s^q\big), N\big(y, t^q\big)\big\} \tag{19.4}$$

for all $x, y \in X$ and all $s, t \in (0, \infty)$.

Now we get the general stability result in the fuzzy normed linear space.

Theorem 19.1 *Let q be a positive real number with $q \neq \frac{1}{2}, 1$ and let f be a fuzzy q-almost quadratic-additive mapping from a fuzzy normed space (X, N) into a fuzzy Banach space (Y, N'). Then there is a unique quadratic-additive mapping $F : X \to Y$ such that*

$$N'\big(F(x) - f(x), t\big) \geq \begin{cases} \sup_{t'<t} N(x, (2 - 2^p)^q t'^q) & \text{if } q > 1, \\ \sup_{t'<t} N(x, (\frac{(4-2^p)(2^p-2)}{2})^q t'^q) & \text{if } \frac{1}{2} < q < 1, \\ \sup_{t'<t} N(x, (2^p - 4)^q t'^q) & \text{if } 0 < q < \frac{1}{2} \end{cases} \tag{19.5}$$

for each $x \in X$ and $t > 0$, where $p = 1/q$.

Proof It follows from (19.4) and (N4) that

$$N'\big(f(0), t\big) = N'\big(Df(0, 0), 2t\big) \geq N\big(0, t^q\big) = 1$$

for all $t \in (0, \infty)$. By (N2), we have $f(0) = 0$. We will prove the theorem in three cases, $q > 1$, $\frac{1}{2} < q < 1$, and $0 < q < \frac{1}{2}$.

Case 1 Let $q > 1$ and let $J_n f : X \to Y$ be a mapping defined by

$$J_n f(x) = \frac{1}{2}\big(4^{-n}\big(f(2^n x) + f(-2^n x)\big) + 2^{-n}\big(f(2^n x) - f(-2^n x)\big)\big)$$

for all $x \in X$. Notice that $J_0 f(x) = f(x)$ and

$$J_j f(x) - J_{j+1} f(x) = \frac{2^{j+1} + 1}{2 \cdot 4^{j+1}} Df\big(2^j x, 0\big) + \frac{-2^{j+1} + 1}{2 \cdot 4^{j+1}} Df\big(-2^j x, 0\big) \tag{19.6}$$

for all $x \in X$ and $j \geq 0$. Together with (N3), (N4), and (19.4), this equation implies that if $n+m > m \geq 0$ then

$$\begin{aligned}
&N'\left(J_m f(x) - J_{n+m} f(x), \sum_{j=m}^{n+m-1} \frac{1}{2}\left(\frac{2^p}{2}\right)^j t^p\right) \\
&\geq N'\left(\sum_{j=m}^{n+m-1} \big(J_j f(x) - J_{j+1} f(x)\big), \sum_{j=m}^{n+m-1} \frac{1}{2}\left(\frac{2^p}{2}\right)^j t^p\right) \\
&\geq \min \bigcup_{j=m}^{n+m-1} \left\{N'\left(J_j f(x) - J_{j+1} f(x), \frac{1}{2}\left(\frac{2^p}{2}\right)^j t^p\right)\right\} \\
&\geq \min \bigcup_{j=m}^{n+m-1} \left\{\min\left\{N'\left(\frac{(2^{j+1}+1)Df(2^j x, 0)}{2 \cdot 4^{j+1}}, \frac{(2^{j+1}+1)2^{jp} t^p}{8 \cdot 4^j}\right),\right.\right. \\
&\qquad \left.\left. N'\left(\frac{(-2^{j+1}+1)Df(-2^j x, 0)}{2 \cdot 4^{j+1}}, \frac{(2^{j+1}-1)2^{jp} t^p}{8 \cdot 4^j}\right)\right\}\right\} \\
&\geq \min \bigcup_{j=m}^{n+m-1} \big\{N\big(2^j x, 2^j t\big)\big\} = N(x, t) \qquad (19.7)
\end{aligned}$$

for all $x \in X$ and $t > 0$. Let $\varepsilon > 0$ be given. Since $\lim_{t \to \infty} N(x, t) = 1$, there is a $t_0 > 0$ such that

$$N(x, t_0) \geq 1 - \varepsilon.$$

Observe that for some $\tilde{t} > t_0$, the series $\sum_{j=0}^{\infty} \frac{1}{2}(\frac{2^p}{2})^j \tilde{t}^p$ converges for $p = \frac{1}{q} < 1$. It guarantees that, for an arbitrary given $c > 0$, there exists an $n_0 \geq 0$ such that

$$\sum_{j=m}^{n+m-1} \frac{1}{2}\left(\frac{2^p}{2}\right)^j \tilde{t}^p < c$$

for each $m \geq n_0$ and $n > 0$. Together with (N5) and (19.7), this implies that

$$\begin{aligned}
&N'\big(J_m f(x) - J_{n+m} f(x), c\big) \\
&\geq N'\left(J_m f(x) - J_{n+m} f(x), \sum_{j=m}^{n+m-1} \frac{1}{2}\left(\frac{2^p}{2}\right)^j \tilde{t}^p\right) \\
&\geq N(x, \tilde{t}) \geq N(x, t_0) \geq 1 - \varepsilon
\end{aligned}$$

for all $x \in X$. Hence $\{J_n f(x)\}$ is a Cauchy sequence in the fuzzy Banach space (Y, N'), and so we can define a mapping $F : X \to Y$ by

$$F(x) := N' - \lim_{n \to \infty} J_n f(x).$$

Moreover, if we put $m = 0$ in (19.7), we have

$$N'\big(f(x) - J_n f(x), t\big) \geq N\left(x, \frac{t^q}{\left(\sum_{j=0}^{n-1} \frac{1}{2}\left(\frac{2^p}{2}\right)^j\right)^q}\right) \tag{19.8}$$

for all $x \in X$. Next we will show that F is the desired quadratic additive function. Using (N4), we have

$$\begin{aligned} &N'\big(DF(x,y),t\big) \\ &\geq \min\left\{N'\left(2F(x+y) - 2J_n f(x+y), \frac{t}{6}\right),\right. \\ &\qquad N'\left(F(x-y) - J_n f(x-y), \frac{t}{12}\right), \\ &\qquad N'\left(F(y-x) - J_n f(y-x), \frac{t}{12}\right), \\ &\qquad N'\left(F(2x) - J_n f(2x), \frac{t}{12}\right), \\ &\qquad \left. N'\left(F(2y) - J_n f(2y), \frac{t}{12}\right), N'\left(DJ_n f(x,y), \frac{t}{2}\right)\right\} \end{aligned} \tag{19.9}$$

for all $x, y \in X$ and $n \in \mathbb{N}$. The first five terms on the right hand side of (19.9) tend to 1 as $n \to \infty$ by the definition of F and (N2), and the last term satisfies

$$\begin{aligned} N'\left(DJ_n f(x,y), \frac{t}{2}\right) \geq \min\left\{N'\left(\frac{Df(2^n x, 2^n y)}{2 \cdot 4^n}, \frac{t}{8}\right), N'\left(\frac{Df(-2^n x, -2^n y)}{2 \cdot 4^n}, \frac{t}{8}\right),\right. \\ \left. N'\left(\frac{Df(2^n x, 2^n y)}{2 \cdot 2^n}, \frac{t}{8}\right), N'\left(\frac{Df(-2^n x, -2^n y)}{2 \cdot 2^n}, \frac{t}{8}\right)\right\} \end{aligned}$$

for all $x, y \in X$. By (N3) and (19.4), we obtain

$$\begin{aligned} N'\left(\frac{Df(\pm 2^n x, \pm 2^n y))}{2 \cdot 4^n}, \frac{t}{8}\right) &= N'\left(Df(\pm 2^n x, \pm 2^n y), \frac{4^n t}{4}\right) \\ &\geq \min\left\{N\left(2^n x, \left(\frac{4^n t}{8}\right)^q\right), N\left(2^n y, \left(\frac{4^n t}{8}\right)^q\right)\right\} \\ &\geq \min\left\{N\left(x, \frac{2^{(2q-1)n}}{2^{3q}} t^q\right), N\left(y, \frac{2^{(2q-1)n}}{2^{3q}} t^q\right)\right\} \end{aligned}$$

and

$$N'\left(\frac{Df(\pm 2^n x, \pm 2^n y))}{2 \cdot 2^n}, \frac{t}{8}\right) \geq \min\left\{N\left(x, \frac{2^{(q-1)n}}{2^{3q}} t^q\right), N\left(y, \frac{2^{(q-1)n}}{2^{3q}} t^q\right)\right\}$$

for all $x, y \in X$ and $n \in \mathbb{N}$. Since $q > 1$, together with (N5), we can deduce that the last term of (19.9) also tends to 1 as $n \to \infty$. It follows from (19.9) that

$$N'\big(DF(x,y),t\big) = 1$$

for each $x, y \in X$ and $t > 0$. By (N2), this means that $DF(x,y) = 0$ for all $x, y \in X$. Next we approximate the difference between f and F in a fuzzy sense. For an arbitrary fixed $x \in X$ and $t > 0$, choose $0 < \varepsilon < 1$ and $0 < t' < t$. Since F is the limit of $\{J_n f(x)\}$, there is $n \in \mathbb{N}$ such that

$$N'\big(F(x) - J_n f(x), t - t'\big) \geq 1 - \varepsilon.$$

By (19.8), we have

$$\begin{aligned} N'\big(F(x) - f(x), t\big) &\geq \min\big\{N'\big(F(x) - J_n f(x), t - t'\big), N'\big(J_n f(x) - f(x), t'\big)\big\} \\ &\geq \min\left\{1 - \varepsilon, N\left(x, \frac{t'^q}{\left(\frac{1}{2}\sum_{j=0}^{n-1}\left(\frac{2^p}{2}\right)^j\right)^q}\right)\right\} \\ &\geq \min\big\{1 - \varepsilon, N\big(x, (2 - 2^p)^q t'^q\big)\big\}. \end{aligned}$$

Because $0 < \varepsilon < 1$ is arbitrary, we get the inequality (19.5) in this case. Finally, to prove the uniqueness of F, let $F' : X \to Y$ be another quadratic-additive mapping satisfying (19.5). Then by (19.6), we get

$$\begin{cases} F(x) - J_n F(x) = \sum_{j=0}^{n-1}(J_j F(x) - J_{j+1}F(x)) = 0, \\ F'(x) - J_n F'(x) = \sum_{j=0}^{n-1}(J_j F'(x) - J_{j+1}F'(x)) = 0 \end{cases} \tag{19.10}$$

for all $x \in X$ and $n \in \mathbb{N}$. Together with (N4) and (19.5), this implies that

$$\begin{aligned} &N'\big(F(x) - F'(x), t\big) \\ &\quad = N'\big(J_n F(x) - J_n F'(x), t\big) \\ &\quad \geq \min\left\{N'\left(J_n F(x) - J_n f(x), \frac{t}{2}\right), N'\left(J_n f(x) - J_n F'(x), \frac{t}{2}\right)\right\} \\ &\quad \geq \min\left\{N'\left(\frac{(F - f)(2^n x)}{2 \cdot 4^n}, \frac{t}{8}\right), N'\left(\frac{(f - F')(2^n x)}{2 \cdot 4^n}, \frac{t}{8}\right),\right. \\ &\qquad\qquad N'\left(\frac{(F - f)(-2^n x)}{2 \cdot 4^n}, \frac{t}{8}\right), N'\left(\frac{(f - F')(-2^n x)}{2 \cdot 4^n}, \frac{t}{8}\right), \\ &\qquad\qquad N'\left(\frac{(F - f)(2^n x)}{2 \cdot 2^n}, \frac{t}{8}\right), N'\left(\frac{(f - F')(2^n x)}{2 \cdot 2^n}, \frac{t}{8}\right), \\ &\qquad\qquad \left. N'\left(\frac{(F - f)(-2^n x)}{2 \cdot 2^n}, \frac{t}{8}\right), N'\left(\frac{(f - F')(-2^n x)}{2 \cdot 2^n}, \frac{t}{8}\right)\right\} \\ &\quad \geq \sup_{t' < t} N\big(x, 2^{(q-1)n - 2q}(2 - 2^p)^q t'^q\big) \end{aligned}$$

for all $x \in X$ and $n \in \mathbb{N}$. Observe that, for $q = \frac{1}{p} > 1$, the last term of the above inequality tends to 1 as $n \to \infty$ by (N5). This implies that $N'(F(x) - F'(x), t) = 1$ and so we get

$$F(x) = F'(x)$$

for all $x \in X$ by (N2).

Case 2 Let $\frac{1}{2} < q < 1$ and let $J_n f : X \to Y$ be a mapping defined by

$$J_n f(x) = \frac{1}{2}\left(4^{-n}(f(2^n x) + f(-2^n x)) + 2^n\left(f\left(\frac{x}{2^n}\right) - f\left(-\frac{x}{2^n}\right)\right)\right)$$

for all $x \in X$. Then we have $J_0 f(x) = f(x)$ and

$$\begin{aligned} J_j f(x) - J_{j+1} f(x) = {} & \frac{1}{2 \cdot 4^{j+1}} Df(2^j x, 0) + \frac{1}{2 \cdot 4^{j+1}} Df(-2^j x, 0) \\ & - 2^{j-1} Df\left(\frac{x}{2^{j+1}}, 0\right) + 2^{j-1} Df\left(\frac{-x}{2^{j+1}}, 0\right) \end{aligned}$$

for all $x \in X$ and $j \geq 0$. If $n + m > m \geq 0$, then we have

$$\begin{aligned} & N'\left(J_m f(x) - J_{n+m} f(x), \sum_{j=m}^{n+m-1} \left(\frac{1}{4}\left(\frac{2^p}{4}\right)^j + \frac{1}{2^p}\left(\frac{2}{2^p}\right)^j\right) t^p\right) \\ & \geq \min \bigcup_{j=m}^{n+m-1} \left\{ \min\left\{ N'\left(\frac{Df(2^j x, 0)}{2 \cdot 4^{j+1}}, \frac{2^{jp} t^p}{2 \cdot 4^{j+1}}\right), \right.\right. \\ & \qquad N'\left(\frac{Df(-2^j x, 0)}{2 \cdot 4^{j+1}}, \frac{2^{jp} t^p}{2 \cdot 4^{j+1}}\right), \\ & \qquad N'\left(-2^{j-1} Df\left(\frac{x}{2^{j+1}}, 0\right), \frac{2^{j-1} t^p}{2^{(j+1)p}}\right), \\ & \qquad \left.\left. N'\left(2^{j-1} Df\left(-\frac{x}{2^{j+1}}, 0\right), \frac{2^{j-1} t^p}{2^{(j+1)p}}\right)\right\}\right\} \\ & \geq \min \bigcup_{j=m}^{n+m-1} \left\{ N(2^j x, 2^j t), N\left(\frac{x}{2^{j+1}}, \frac{t}{2^{j+1}}\right)\right\} \\ & = N(x, t) \end{aligned}$$

for all $x \in X$ and $t > 0$. In a similar argument following (19.7) of the previous case, we can define the limit $F(x) := N' - \lim_{n\to\infty} J_n f(x)$ of the Cauchy sequence $\{J_n f(x)\}$ in the Banach fuzzy space Y. Moreover, putting $m = 0$ in the above inequality, we have

$$N'\big(f(x) - J_n f(x), t\big) \ge N\left(x, \frac{t^q}{\big(\sum_{j=0}^{n-1}\big(\frac{1}{4}\big(\frac{2^p}{4}\big)^j + \frac{1}{2^p}\big(\frac{2}{2^p}\big)^j\big)\big)^q}\right) \tag{19.11}$$

for each $x \in X$ and $t > 0$. To prove that F is a quadratic additive function, it is enough to show that the last term of (19.9) in Case 1 tends to 1 as $n \to \infty$. By (N3) and (19.4), we get

$$\begin{aligned}
&N'\left(DJ_n f(x,y), \frac{t}{2}\right)\\
&\ge \min\left\{N'\left(\frac{Df(2^n x, 2^n y)}{2\cdot 4^n}, \frac{t}{8}\right), N'\left(\frac{Df(-2^n x, -2^n y}{2\cdot 4^n}, \frac{t}{8}\right),\right.\\
&\qquad \left. N'\left(2^{n-1} Df\left(\frac{x}{2^n}, \frac{y}{2^n}\right), \frac{t}{8}\right), N'\left(2^{n-1} Df\left(\frac{-x}{2^n}, \frac{-y}{2^n}\right), \frac{t}{8}\right)\right\}\\
&\ge \min\big\{N\big(x, 2^{(2q-1)n-3q} t^q\big), N\big(y, 2^{(2q-1)n-3q} t^q\big),\\
&\qquad \big(x, 2^{(1-q)n-3q} t^q\big), N\big(y, 2^{(1-q)n-3q} t^q\big)\big\}
\end{aligned}$$

for each $x, y \in X$ and $t > 0$. Observe that all the terms on the right hand side of the above inequality tend to 1 as $n \to \infty$, since $\frac{1}{2} < q < 1$. Hence, together with the similar argument after (19.9), we can say that $DF(x,y) = 0$ for all $x, y \in X$. Recall, in Case 1, the inequality (19.5) follows from (19.8). By the same reasoning, we get (19.5) from (19.11) in this case. Now to prove the uniqueness of F, let F' be another quadratic additive mapping satisfying (19.5). Then, together with (N4), (19.5), and (19.10), we have

$$\begin{aligned}
&N'\big(F(x) - F'(x), t\big)\\
&= N'\big(J_n F(x) - J_n F'(x), t\big)\\
&\ge \min\left\{N'\left(J_n F(x) - J_n f(x), \frac{t}{2}\right), N'\left(J_n f(x) - J_n F'(x), \frac{t}{2}\right)\right\}\\
&\ge \min\left\{N'\left(\frac{(F-f)(2^n x)}{2\cdot 4^n}, \frac{t}{8}\right), \left(\frac{(f-F')(2^n x)}{2\cdot 4^n}, \frac{t}{8}\right),\right.\\
&\qquad N'\left(\frac{(F-f)(-2^n x)}{2\cdot 4^n}, \frac{t}{8}\right), N'\left(\frac{(f-F')(-2^n x)}{2\cdot 4^n}, \frac{t}{8}\right),\\
&\qquad N'\left(2^{n-1}\left((F-f)\left(\frac{x}{2^n}\right)\right), \frac{t}{8}\right), N'\left(2^{n-1}\left((f-F')\left(\frac{x}{2^n}\right)\right), \frac{t}{8}\right),\\
&\qquad \left. N'\left(2^{n-1}\left((F-f)\left(\frac{-x}{2^n}\right)\right), \frac{t}{8}\right), N'\left(2^{n-1}\left((f-F')\left(\frac{-x}{2^n}\right)\right), \frac{t}{8}\right)\right\}
\end{aligned}$$

$$\geq \min\left\{\sup_{t'<t} N\left(x, 2^{(2q-1)n-2q}\left(\left(\frac{(4-2^p)(2^p-2)}{2}\right)^q t'^q\right),\right.\right.$$
$$\left.\left.\sup_{t'<t} N\left(x, 2^{(1-q)n-2q}\left(\left(\frac{(4-2^p)(2^p-2)}{2}\right)^q t'^q\right)\right)\right\}$$

for all $x \in X$ and $n \in \mathbb{N}$. Since $\lim_{n\to\infty} 2^{(2q-1)n-2q} = \lim_{n\to\infty} 2^{(1-q)n-2q} = \infty$ in this case, both terms on the right hand side of the above inequality tend to 1 as $n \to \infty$ by (N5). This implies that $N'(F(x) - F'(x),\ t) = 1$ and so $F(x) = F'(x)$ for all $x \in X$ by (N2).

Case 3 Finally, we take $0 < q < \frac{1}{2}$ and define $J_n f : X \to Y$ by

$$J_n f(x) = \frac{1}{2}\left(4^n\left(f\left(2^{-n}x\right) + f\left(-2^{-n}x\right)\right) + 2^n\left(f\left(\frac{x}{2^n}\right) - f\left(-\frac{x}{2^n}\right)\right)\right)$$

for all $x \in X$. Then we have $J_0 f(x) = f(x)$ and

$$J_j f(x) - J_{j+1} f(x) = -\frac{4^j + 2j}{2} Df\left(\frac{x}{2^{j+1}}, 0\right) - \frac{4^j - 2j}{2} Df\left(\frac{-x}{2^{j+1}}, 0\right)$$

which implies that if $n + m > m \geq 0$ then

$$N'\left(J_m f(x) - J_{n+m} f(x), \sum_{j=m}^{n+m-1}\left(\frac{4}{2^p}\right)^j \frac{t^p}{2^p}\right)$$
$$\geq \min \bigcup_{j=m}^{n+m-1}\left\{\min\left\{N'\left(-\frac{4^j + 2j}{2} Df\left(\frac{x}{2^{j+1}}, 0\right), \frac{(4^j + 2^j)t^p}{2 \cdot 2^{(j+1)p}}\right),\right.\right.$$
$$\left.\left. N'\left(-\frac{4^j - 2j}{2} Df\left(-\frac{x}{2^{j+1}}, 0\right), \frac{(4^j - 2^j)t^p}{2 \cdot 2^{(j+1)p}}\right)\right\}\right\}$$
$$\geq \min \bigcup_{j=m}^{n+m-1}\left\{N\left(\frac{x}{2^{j+1}}, \frac{t}{2^{j+1}}\right)\right\}$$
$$= N(x, t)$$

for all $x \in X$ and $t > 0$. Similar to the previous cases, it suggests us to define the mapping $F : X \to Y$ by $F(x) := N' - \lim_{n\to\infty} J_n f(x)$. Putting $m = 0$ in the above inequality, we have

$$N'\left(f(x) - J_n f(x), t\right) \geq N\left(x, \frac{t^q}{\left(\frac{1}{2^p}\sum_{j=0}^{n-1}\left(\frac{4}{2^p}\right)^j\right)^q}\right) \tag{19.12}$$

for all $x \in X$ and $t > 0$. Notice that

$$\begin{aligned}
&N'\left(DJ_nf(x,y),\frac{t}{2}\right)\\
&\geq \min\left\{N'\left(\frac{4^n}{2}Df\left(\frac{x}{2^n},\frac{y}{2^n}\right),\frac{t}{8}\right),N'\left(\frac{4^n}{2}Df\left(\frac{-x}{2^n},\frac{-y}{2^n}\right),\frac{t}{8}\right),\right.\\
&\qquad \left.N'\left(2^{n-1}Df\left(\frac{x}{2^n},\frac{y}{2^n}\right),\frac{t}{8}\right),N'\left(2^{n-1}Df\left(\frac{-x}{2^n},\frac{-y}{2^n}\right),\frac{t}{8}\right)\right\}\\
&\geq \min\left\{N\left(x,2^{(1-2q)n-3q}t^q\right),N\left(y,2^{(1-2q)n-3q}t^q\right),\right.\\
&\qquad \left.N\left(x,2^{(1-q)n-3q}t^q\right),N\left(y,2^{(1-q)n-3q}t^q\right)\right\}
\end{aligned}$$

for each $x, y \in X$ and $t > 0$. Since $0 < q < \frac{1}{2}$, all terms on the right hand side tend to 1 as $n \to \infty$, which implies that the last term of (19.9) tends to 1 as $n \to \infty$. Therefore, we can say that $DF \equiv 0$. Moreover, using a similar argument as that after (19.9) in Case 1, we get the inequality (19.5) from (19.12) in this case. To prove the uniqueness of F, let $F' : X \to Y$ be another quadratic additive function satisfying (19.5). Then by (19.10), we get

$$\begin{aligned}
&N'\left(F(x)-F'(x),t\right)\\
&\geq \min\left\{N'\left(J_nF(x)-J_nf(x),\frac{t}{2}\right),N'\left(J_nf(x)-J_nF'(x),\frac{t}{2}\right)\right\}\\
&\geq \min\left\{N'\left(\frac{4^n}{2}\left((F-f)\left(\frac{x}{2^n}\right)\right),\frac{t}{8}\right),N'\left(\frac{4^n}{2}\left((f-F')\left(\frac{x}{2^n}\right)\right),\frac{t}{8}\right),\right.\\
&\qquad N'\left(\frac{4^n}{2}\left((F-f)\left(-\frac{x}{2^n}\right)\right),\frac{t}{8}\right),N'\left(\frac{4^n}{2}\left((f-F')\left(-\frac{x}{2^n}\right)\right),\frac{t}{8}\right),\\
&\qquad N'\left(2^{n-1}\left((F-f)\left(\frac{x}{2^n}\right)\right),\frac{t}{8}\right),N'\left(2^{n-1}\left((f-F')\left(\frac{x}{2^n}\right)\right),\frac{t}{8}\right),\\
&\qquad \left.N'\left(2^{n-1}\left((F-f)\left(\frac{-x}{2^n}\right)\right),\frac{t}{8}\right),N'\left(2^{n-1}\left((f-F')\left(\frac{-x}{2^n}\right)\right),\frac{t}{8}\right)\right\}\\
&\geq \sup_{t'<t} N\left(x,2^{(1-2q)n-2q}\left(2^p-4\right)^q t'^q\right)
\end{aligned}$$

for all $x \in X$ and $n \in \mathbb{N}$. Observe that, for $0 < q < \frac{1}{2}$, the last term tends to 1 as $n \to \infty$ by (N5). This implies that $N'(F(x) - F'(x), t) = 1$ and $F(x) = F'(x)$ for all $x \in X$ by (N2). □

Remark 19.1 Consider a mapping $f : X \to Y$ satisfying (19.4) for all $x, y \in X$ and a real number $q < 0$. Take any $t > 0$. If we choose a real number s with $0 < 2s < t$, then we have

$$N'\left(Df(x,y),t\right) \geq N'\left(Df(x,y),2s\right) \geq \min\left\{N\left(x,s^q\right),N\left(y,s^q\right)\right\}$$

for all $x, y \in X$. Since $q < 0$, we have $\lim_{s \to 0^+} s^q = \infty$. This implies that

$$\lim_{s \to 0^+} N\big(x, s^q\big) = \lim_{s \to 0^+} N\big(y, s^q\big) = 1$$

and so

$$N'\big(Df(x, y), t\big) = 1$$

for all $x, y \in X$ and $t > 0$. By (N2), it allows us to get $Df(x, y) = 0$ for all $x, y \in X$. In other words, f is itself a quadratic additive mapping if f is a fuzzy q-almost quadratic-additive mapping for the case $q < 0$.

Corollary 19.1 *Let f be an even mapping satisfying all of the conditions of Theorem 19.1. Then there is a unique quadratic mapping $F : X \to Y$ such that*

$$N'\big(F(x) - f(x), t\big) \geq \sup_{t' < t} N\big(x, \big(|4 - 2^p|t'\big)^q\big) \tag{19.13}$$

for all $x \in X$ and $t > 0$, where $p = 1/q$.

Proof Let $J_n f$ be defined as in Theorem 19.1. Since f is an even mapping, we obtain

$$J_n f(x) = \begin{cases} \frac{f(2^n x) + f(-2^n x)}{2 \cdot 4^n} & \text{if } 0 < q < \frac{1}{2}, \\ \frac{1}{2}(4^n(f(2^{-n}x) + f(-2^{-n}x))) & \text{if } q > \frac{1}{2} \end{cases}$$

for all $x \in X$. Notice that $J_0 f(x) = f(x)$ and

$$J_j f(x) - J_{j+1} f(x) = \begin{cases} \frac{1}{2 \cdot 4^{j+1}}(Df(2^j x, 0) + Df(-2^j x, 0) & \text{if } 0 < q < \frac{1}{2}, \\ -2 \cdot 4^{j-1}(Df(\frac{x}{2^{j+1}}, 0) + Df(\frac{-x}{2^{j+1}}, 0)) & \text{if } q > \frac{1}{2} \end{cases}$$

for all $x \in X$ and $j \in \mathbb{N} \cup \{0\}$. From these, using the similar method in Theorem 19.1, we obtain the quadratic-additive mapping F, which is defined by $F(x) = N' - \lim_{n \to \infty} J_n f(x)$, satisfying (19.13). Notice that F is also even and $DF(x, y) = 0$ for all $x, y \in X$. Hence, we get

$$F(x+y) + F(x-y) - 2F(x) - 2F(y) = \frac{1}{2}\big(DF(x, y) - DF(x, 0) - DF(0, y)\big) = 0$$

for all $x, y \in X$. This means that F is a quadratic mapping. □

Corollary 19.2 *Let f be an odd mapping satisfying all of the conditions of Theorem 19.1. Then there is a unique additive mapping $F : X \to Y$ such that*

$$N'\big(F(x) - f(x), t\big) \geq \sup_{t' < t} N\big(x, \big(|2 - 2^p|t'\big)^q\big) \tag{19.14}$$

for all $x \in X$ and $t > 0$, where $p = 1/q$.

Proof Let $J_n f$ be defined as in Theorem 19.1. Since f is an odd mapping, we obtain

$$J_n f(x) = \begin{cases} \frac{f(2^n x)+f(-2^n x)}{2^{n+1}} & \text{if } 0 < q < 1, \\ 2^{n-1}(f(2^{-n}x) + f(-2^{-n}x)) & \text{if } q > 1 \end{cases}$$

for all $x \in X$. Notice that $J_0 f(x) = f(x)$ and

$$J_j f(x) - J_{j+1} f(x) = \begin{cases} \frac{1}{2^{j+2}}(Df(2^j x, 0) - Df(-2^j x, 0)) & \text{if } 0 < q < 1, \\ 2^{j-1}(Df(\frac{-x}{2^{j+1}}, 0) - Df(\frac{x}{2^{j+1}}, 0)) & \text{if } q > 1 \end{cases}$$

for all $x \in X$ and $j \in \mathbb{N} \cup \{0\}$. From these, using a similar method as in Theorem 19.1, we obtain the quadratic-additive mapping F, which is defined by $F(x) = N' - \lim_{n\to\infty} J_n f(x)$, satisfying (19.14). Notice that F is also odd and $DF(x, y) = 0$ for all $x, y \in X$. Hence, we get

$$F(x+y) - F(x) - F(y) = \frac{1}{2}\big(DF(x, y) - DF(x, 0) - DF(0, y)\big) = 0$$

for all $x, y \in X$. This means that F is an additive mapping. □

In the proof of Corollary 19.1 and 19.2, we have shown that the even quadratic-additive function is a quadratic mapping and the odd quadratic-additive mapping is an additive mapping, respectively. From this we easily get that if f is a quadratic-additive mapping then $\frac{f(x)+f(-x)}{2}$ is a quadratic function and $\frac{f(x)-f(-x)}{2}$ is an additive mapping.

On the other hand, we can use Theorem 19.1 to get a classical result in the framework of normed spaces. Let $(X, \|\cdot\|)$ be a normed linear space. Then we can define a fuzzy norm N_X on X by letting

$$N_X(x, t) = \begin{cases} 0, & t \le \|x\|, \\ 1, & t > \|x\| \end{cases}$$

where $x \in X$ and $t \in \mathbb{R}$, see [17]. Suppose that $f : X \to Y$ is a mapping into a Banach space $(Y, |||\cdot|||)$ such that

$$\big|\big|\big|Df(x, y)\big|\big|\big| \le \|x\|^p + \|y\|^p$$

for all $x, y \in X$, where $p > 0$ and $p \ne 1, 2$. Let N_Y be a fuzzy norm on Y. Then we get

$$N_Y\big(Df(x, y), s+t\big) = \begin{cases} 0, & s+t \le |||Df(x, y)|||, \\ 1, & s+t > |||Df(x, y)||| \end{cases}$$

for all $x, y \in X$ and $s, t \in \mathbb{R}$. Consider the case $N_Y(Df(x, y), s+t) = 0$. This implies that

$$\|x\|^p + \|y\|^p \ge \big|\big|\big|Df(x, y)\big|\big|\big| \ge s+t$$

and so either $\|x\|^p \geq s$ or $\|y\|^p \geq t$ in this case. Hence, for $q = \frac{1}{p}$, we have

$$\min\{N_X(x, s^q), N_X(y, t^q)\} = 0$$

for all $x, y \in X$ and $s, t > 0$. Therefore, in every case, the inequality

$$N_Y(Df(x, y), s + t) \geq \min\{N_X(x, s^q), N_X(y, t^q)\}$$

holds. It means that f is a fuzzy q-almost quadratic additive mapping, and by Theorem 19.1, we get the following stability result.

Corollary 19.3 *Let $(X, \|\cdot\|)$ be a normed linear space and let $(Y, |||\cdot|||)$ be a Banach space. If*

$$|||Df(x, y)||| \leq \|x\|^p + \|y\|^p$$

for all $x, y \in X$, where $p > 0$ and $p \neq 1, 2$, then there is a unique quadratic-additive mapping $F : X \to Y$ such that

$$|||F(x) - f(x)||| \leq \begin{cases} \frac{\|x\|^p}{2-2^p} & \text{if } 0 < p < 1, \\ \frac{2\|x\|^p}{(2-2^p)(2^p-4)} & \text{if } 1 < p < 2, \\ \frac{\|x\|^p}{2^p-4} & \text{if } 2 < p \end{cases}$$

for all $x \in X$.

References

1. Aoki, T.: On the stability of the linear transformation in Banach spaces. J. Math. Soc. Jpn. **2**, 64–66 (1950)
2. Bag, T., Samanta, S.K.: Finite dimensional fuzzy normed linear spaces. J. Fuzzy Math. **11**(3), 687–705 (2003)
3. Cheng, S.C., Mordeson, J.N.: Fuzzy linear operator and fuzzy normed linear spaces. Bull. Calcutta Math. Soc. **86**, 429–436 (1994)
4. Czerwik, S.: On the stability of the quadratic mapping in normed spaces. Abh. Math. Semin. Univ. Hamb. **62**, 59–64 (1992)
5. Găvruta, P.: A generalization of the Hyers–Ulam–Rassias stability of approximately additive mappings. J. Math. Anal. Appl. **184**, 431–436 (1994)
6. Hyers, D.H.: On the stability of the linear functional equation. Proc. Natl. Acad. Sci. USA **27**, 222–224 (1941)
7. Jun, K.-W., Lee, Y.-H.: A generalization of the Hyers–Ulam–Rassias stability of the pexiderized quadratic equations, II. Kyungpook Math. J. **47**, 91–103 (2007)
8. Katsaras, A.K.: Fuzzy topological vector spaces II. Fuzzy Sets Syst. **12**, 143–154 (1984)
9. Kim, G.-H.: On the stability of functional equations with square-symmetric operation. Math. Inequal. Appl. **4**, 257–266 (2001)
10. Kim, H.-M.: On the stability problem for a mixed type of quartic and quadratic functional equation. J. Math. Anal. Appl. **324**, 358–372 (2006)
11. Kramosil, I., Michalek, J.: Fuzzy metric and statistical metric spaces. Kybernetica **11**, 326–334 (1975)

12. Lee, Y.-H.: On the Hyers–Ulam–Rassias stability of the generalized polynomial function of degree 2. J. Chuncheong Math. Soc. **22**, 201–209 (2009)
13. Lee, Y.-H.: On the stability of the monomial functional equation. Bull. Korean Math. Soc. **45**, 397–403 (2008)
14. Lee, Y.H., Jun, K.W.: A generalization of the Hyers–Ulam–Rassias stability of Jensen's equation. J. Math. Anal. Appl. **238**, 305–315 (1999)
15. Lee, Y.H., Jun, K.W.: A generalization of the Hyers–Ulam–Rassias stability of Pexider equation. J. Math. Anal. Appl. **246**, 627–638 (2000)
16. Lee, Y.H., Jun, K.W.: On the stability of approximately additive mappings. Proc. Am. Math. Soc. **128**, 1361–1369 (2000)
17. Mirmostafaee, A.K., Moslehian, M.S.: Fuzzy almost quadratic functions. Results Math. **52**, 161–177 (2008)
18. Mirmostafaee, A.K., Moslehian, M.S.: Fuzzy versions of Hyers–Ulam–Rassias theorem. Fuzzy Sets Syst. **159**, 720–729 (2008)
19. Rassias, Th.M.: On the stability of the linear mapping in Banach spaces. Proc. Am. Math. Soc. **72**, 297–300 (1978)
20. Skof, F.: Local properties and approximations of operators. Rend. Semin. Mat. Fis. Milano **3**, 113–129 (1983)
21. Ulam, S.M.: A Collection of Mathematical Problems, p. 63. Interscience, New York (1968)

Chapter 20
Generalized Hyers–Ulam Stability of Cauchy–Jensen Functional Equations

Kil-Woung Jun, Hark-Mahn Kim, and Eun Young Son

Abstract In this paper, we prove the generalized Hyers–Ulam stability of the following Cauchy–Jensen functional equation

$$f(x) + f(y) + nf(z) = nf\left(\frac{x+y}{n} + z\right),$$

in an n-divisible abelian group G for any fixed positive integer $n \geq 2$.

Key words Cauchy–Jensen equations · Generalized Hyers–Ulam stability

Mathematics Subject Classification 39B82

20.1 Introduction

The stability problem of equations originated from a question of Ulam [9] concerning the stability of group homomorphisms.

We are given a group G_1 and a metric group G_2 with metric $\rho(\cdot,\cdot)$. Given $\varepsilon > 0$, does there exist a $\delta > 0$ such that if $f : G_1 \to G_2$ satisfies $\rho(f(xy), f(x)f(y)) < \delta$ for all $x, y \in G_1$, then a homomorphism $h : G_1 \to G_2$ exists with $\rho(f(x), h(x)) < \varepsilon$ for all $x \in G_1$?

In other words, we are looking for situations when the homomorphisms are stable, i.e., if a mapping is almost a homomorphism, then there exists a true homomorphism near it.

Dedicated to Professor Themistocles M. Rassias on the occasion of his 60th birthday.

K.-W. Jun (✉) · H.-M. Kim · E.Y. Son
Department of Mathematics, Chungnam National University, 79 Daehangno, Yuseong-gu, Daejeon 305-764, Korea
e-mail: kwjun@cnu.ac.kr

H.-M. Kim
e-mail: hmkim@cnu.ac.kr

E.Y. Son
e-mail: sey8405@cnu.ac.kr

P.M. Pardalos et al. (eds.), *Nonlinear Analysis*, Springer Optimization and Its Applications 68, In Honor of Themistocles M. Rassias on the Occasion of his 60th Birthday, DOI 10.1007/978-1-4614-3498-6_20,

In 1941, D.H. Hyers [4] considered the case of approximately additive mappings between Banach spaces and proved the following result.

Suppose that E_1 and E_2 are Banach spaces and $f : E_1 \to E_2$ satisfies the following condition: there is an $\varepsilon \geq 0$ such that

$$\| f(x+y) - f(x) - f(y) \| \leq \varepsilon$$

for all $x, y \in E_1$. Then the limit $h(x) = \lim_{n\to\infty} \frac{f(2^n x)}{2^n}$ exists for all $x \in E_1$ and there exists a unique additive mapping $h : E_1 \to E_2$ such that

$$\| f(x) - h(x) \| \leq \varepsilon.$$

Moreover, if $f(tx)$ is continuous in $t \in \mathbb{R}$ for each $x \in E_1$, then the mapping h is linear.

The method which was provided by Hyers, and which produces the additive mapping h, was called a direct method. This method is the most important and most powerful tool for studying the stability of various functional equations.

In 1978, Th.M. Rassias [5] provided a generalization of Hyers Theorem which allows the Cauchy difference to be unbounded.

In 1990, during the 27th International Symposium on Functional Equations, Th.M. Rassias [6] asked the question whether such a theorem can also be proved for $p \geq 1$. In 1991, Z. Gajda [1] following the same approach as in [5], gave an affirmative solution to this question for $p > 1$. It was shown by Z. Gajda [1], as well as by Th.M. Rassias and P. Šemrl [6], that one cannot prove a Th.M. Rassias type theorem when $p = 1$. The counterexamples of Z. Gajda [1], as well as of Th.M. Rassias and P. Šemrl [6], have stimulated several mathematicians to invent new definitions of approximately additive or approximately linear mappings, cf. P. Găvruta [2] and S. Jung [8], who among others studied the stability of functional equations. In 1994, a generalized result of Rassias' theorem was obtained by P. Găvruta in [2].

Next, we present the definitions of the stability of functional inequalities and functional equations. Let G be an n-divisible abelian group $n \in \mathbb{N}$ (i.e., $a \mapsto na : G \to G$ is a surjection) and X be a normed space $\| \cdot \|_X$. Denote by $M(G, X)$ the set of all mappings from G into X, and let $L^\infty(G, X) = \{f : G \to X \mid \|f\|_\infty := \sup_{x\in G} \|f\|_X < \infty\}$.

Definition 20.1 Given mappings $E : M(G, X) \to M(G^r, X)$, $\varphi : M(G, X) \to M(G^r, X)$ and $\psi : G \to \mathbb{R}^+$, if

$$\| E(f)(x_1, x_2, \ldots, x_r) \| \leq \varphi(x_1, x_2, \ldots, x_r) \quad \text{for all } x_1, x_2, \ldots, x_r \in G$$

implies that there exists $g \in M(G, X)$ such that $E(g) = 0$ and $\|f(x) - g(x)\|_\infty \leq \psi(x)$ for all $x \in G$, then we say that the equation $E(f) = 0$ is (φ, ψ)-stable in $M(G, X)$. In this case, we also say that the solutions of the equation $E(f) = 0$ is (φ, ψ)-stable in $M(G, X)$ [3].

Now, for a mapping $f : G \to X$, we consider the following generalized Cauchy–Jensen equation

$$f(x)+f(y)+nf(z)=nf\left(\frac{x+y}{n}+z\right), \quad n \geqq 2 \tag{20.1}$$

for all $x, y, z \in G$, which has been introduced in [3]. First of all, we recall that a mapping f is additive if and only if f satisfies the Cauchy–Jensen equation (20.1) [3].

The generalized Hyers–Ulam stability of functional equation (20.1) has been presented in [3] for a special case $n = 2$. In this paper, we are going to improve the theorems of [3] without using the oddness of approximate additive functions concerning the functional equation (20.1) for the general case.

20.2 Generalized Hyers–Ulam-Stability of Functional Equation (20.1)

From now on, let G be an n-divisible abelian group for some positive integer $n \geqq 2$, $f : G \to Y$, and let Y be a Banach space. Given $f : G \to Y$, we set

$$Df(x,y,z) := f(x)+f(y)+nf(z)-nf\left(\frac{x+y}{n}+z\right)$$

for all $x, y, z \in G$ and for any fixed positive integer $n \geq 2$.

Theorem 20.1 *Let $\varphi : G^3 \to \mathbb{R}^+$ satisfy*

$$\check{\varphi}(x,z) := \sum_{i=0}^{\infty} \frac{1}{2n^{i+1}}\left(\varphi\left(n^{i+1}x, 0, -n^i z\right) + \varphi\left(-n^{i+1}x, 0, n^i z\right)\right) < \infty,$$

for all $x, z \in G$ and $\lim_{k\to\infty} \frac{1}{n^k}\varphi(n^k x, n^k y, n^k z) = 0$ for all $x, y, z \in G$. If $f : G \to Y$ is a mapping such that $f(0) = 0$ and

$$\left\|Df(x,y,z)\right\| \leq \varphi(x,y,z) \tag{20.2}$$

for all $x, y, z \in G$, then there exists a unique additive mapping $h : G \to Y$, defined as $h(x) = \lim_{k\to\infty} \frac{f(n^k x)-f(-n^k x)}{2n^k}$, such that

$$\left\|f(x)-h(x)\right\| \leq \check{\varphi}(x,x) + \frac{\varphi(x,-x,0)}{2} \tag{20.3}$$

for all $x \in G$.

Proof Letting $y = -x$, $z = 0$ in (20.2) and dividing both sides by 2, we have

$$\left\|\frac{f(x)+f(-x)}{2}\right\| \leq \frac{\varphi(x,-x,0)}{2} \tag{20.4}$$

for all $x \in G$. Replacing x by nx and letting $y = 0$ and $z = -x$ in (20.2), we get

$$\|f(nx) + nf(-x)\| \le \varphi(nx, 0, -x) \tag{20.5}$$

for all $x \in G$. Replacing x by $-x$ in (20.5), one has

$$\|f(-nx) + nf(x)\| \le \varphi(-nx, 0, x) \tag{20.6}$$

for all $x \in G$. Put $g(x) = \frac{f(x)-f(-x)}{2}$. Combining (20.5) with (20.6) yields

$$\|ng(x) - g(nx)\| \le \frac{1}{2}\big(\varphi(nx, 0, -x) + \varphi(-nx, 0, x)\big),$$

that is,

$$\left\|g(x) - \frac{1}{n}g(nx)\right\| \le \frac{1}{2n}\big(\varphi(nx, 0, -x) + \varphi(-nx, 0, x)\big) \tag{20.7}$$

for all $x \in G$. It follows from (20.7) that

$$\begin{aligned}\left\|\frac{g(n^l x)}{n^l} - \frac{g(n^m x)}{n^m}\right\| &\le \sum_{k=0}^{m-1}\left\|\frac{1}{n^k}g(n^k x) - \frac{1}{n^{k+1}}g(n^{k+1}x)\right\| \\ &= \sum_{k=l}^{m-1}\frac{1}{n^k}\left\|g(n^k x) - \frac{1}{n}g(n^{k+1}x)\right\| \\ &\le \sum_{k=l}^{m-1}\frac{1}{n^{k+1}}\big(\varphi(n^{k+1}x, 0, -n^k x) + \varphi(-n^{k+1}x, 0, n^k x)\big)\end{aligned} \tag{20.8}$$

for all nonnegative integers m and l with $m > l \ge 0$ and $x \in G$. Since the right hand side of (20.8) tends to zero as $l \to \infty$, we obtain that the sequence $\{\frac{g(n^k x)}{n^k}\}$ is Cauchy for all $x \in G$. Since Y is a Banach space, it follows that the sequence $\{\frac{g(n^k x)}{n^k}\}$ converges in Y. Therefore, we can define a function $h : G \to Y$ by

$$h(x) = \lim_{k\to\infty}\frac{g(n^k x)}{n^k}, \quad x \in G.$$

Moreover, letting $l = 0$ and $m \to \infty$ in (20.8) yields

$$\left\|\frac{f(x) - f(-x)}{2} - h(x)\right\| \le \check{\varphi}(x, x) \tag{20.9}$$

for all $x \in G$. It follows from (20.4) and (20.9) that

$$\|f(x) - h(x)\| \le \check{\varphi}(x, x) + \frac{\varphi(x, -x, 0)}{2}$$

for all $x \in G$. It follows from (20.2) that

$$\begin{aligned}
&\left\| h(x) + h(y) + nh(z) - nh\left(\frac{x+y}{n} + z\right) \right\| \\
&\quad = \lim_{k\to\infty} \frac{1}{n^k} \left\| g(n^k x) + g(n^k y) + ng(n^k z) - ng(n^k\left(\frac{x+y}{n} + z\right) \right\| \\
&\quad = \lim_{k\to\infty} \frac{1}{2n^k} \| Df(n^k x, n^k y, n^k z) - Df(-n^k x, -n^k y, -n^k z) \| \\
&\quad \le \lim_{k\to\infty} \frac{1}{2n^k} (\| Df(n^k x, n^k y, n^k z) \| + \| Df(-n^k x, -n^k y, -n^k z) \|) \\
&\quad \le \lim_{k\to\infty} \frac{1}{2n^k} (\varphi(n^k x, n^k y, n^k z) + \varphi(-n^k x, -n^k y, -n^k z)) = 0
\end{aligned}$$

for all $x, y, z \in G$. This implies that

$$h(x) + h(y) + nh(z) = nh\left(\frac{x+y}{n} + z\right)$$

for all $x, y, z \in G$. Now, it follows that h is additive [3].

Next, let $h' : G \to Y$ be another additive mapping satisfying

$$\| f(x) - h'(x) \| \le \check{\varphi}(x, x) + \frac{\varphi(x, -x, 0)}{2}$$

for all $x \in G$.

Then we have

$$\begin{aligned}
\| h(x) - h'(x) \| &= \left\| \frac{1}{n^k} h(n^k x) - \frac{1}{n^k} h'(n^k x) \right\| \\
&\le \frac{1}{n^k} (\| h(n^k x) - f(n^k x) \| + \| f(n^k x) - h'(n^k x) \|) \\
&\le 2 \frac{1}{n^k} \left(\check{\varphi}(n^k x, n^k x) + \frac{\varphi(n^k x, -n^k x, 0)}{2} \right) \\
&= 2 \sum_{i=0}^{\infty} \frac{1}{n^{i+k+1}} [\varphi(n^{i+k+1} x, 0, -n^{i+k} x) + \varphi(-n^{i+k+1} x, 0, n^{i+k} x)] \\
&\quad + \frac{\varphi(n^k x, -n^k x, 0)}{n^k} \\
&= 2 \sum_{i=k}^{\infty} \frac{1}{n^{i+1}} [\varphi(n^{i+1} x, 0, -n^i x) + \varphi(-n^{i+1} x, 0, n^i x)] + \frac{\varphi(n^k x, -n^k x, 0)}{n^k}
\end{aligned}$$

for all $k \in \mathbb{N}$ and all $x \in G$. Taking the limit as $k \to \infty$, we conclude that

$$h(x) = h'(x)$$

for all $x \in X$. This completes the proof. □

Suppose that X is a normed space in the following corollaries. If we put $\varphi(x, y, z) := \varepsilon(\|x\|^p + \|y\|^q + \|z\|^t)$ and $\varphi(x, y, z) := \varepsilon(\|x\|^p \|y\|^q \|z\|^t)$ in Theorem 20.1, respectively, then we get the Corollaries 20.1 and 20.2.

Corollary 20.1 *Let $0 < p, q, t < 1$, $\varepsilon > 0$. If a mapping $f : X \to Y$ with $f(0) = 0$ satisfies*

$$\big\|Df(x, y, z)\big\| \le \varepsilon\big(\|x\|^p + \|y\|^q + \|z\|^t\big)$$

for all $x, y, z \in X$, then there exists a unique additive mapping $h : X \to Y$ such that

$$\big\|f(x) - h(x)\big\| \le \varepsilon\left[\left(\frac{n^p}{n - n^p} + \frac{1}{2}\right)\|x\|^p + \frac{1}{2}\|x\|^q + \left(\frac{1}{n - n^t} +\right)\|x\|^t\right]$$

for all $x \in X$.

Corollary 20.2 *Let $p + q + t < 1$, $p, q, t > 0$, $\varepsilon > 0$. If a mapping $f : X \to Y$ with $f(0) = 0$ satisfies*

$$\big\|Df(x, y, z)\big\| \le \varepsilon\big(\|x\|^p \|y\|^q \|z\|^t\big)$$

for all $x, y, z \in X$, then f is additive.

The following corollary is an immediate consequence of Theorem 20.1.

Corollary 20.3 *Suppose that a mapping $f : G \to Y$ with $f(0) = 0$ satisfies the inequality*

$$\big\|Df(x, y, z)\big\| \le \varepsilon$$

for all $x, y, z \in G$, where $\varepsilon \ge 0$. Then there exists a unique additive mapping $h : X \to Y$ satisfying the inequality

$$\big\|f(x) - h(x)\big\| \le \frac{\varepsilon}{n - 1} + \frac{\varepsilon}{2}$$

for all $x \in G$ and for any fixed positive integer $n \ge 2$.

We may obtain more simple and sharp approximation than that of Theorem 20.1 for the stability result of equation (20.1) under the oddness condition.

Remark 20.1 Let $\varphi : G^3 \to \mathbb{R}^+$ satisfy $\lim_{k\to\infty} \frac{1}{n^k}\varphi(n^k x, n^k y, n^k z) = 0$ for all $x, y, z \in G$ and

$$\check{\varphi}(x, z) := \sum_{i=0}^{\infty} \frac{1}{n^{i+1}}\varphi\big(n^{i+1}x, 0, -n^i z\big) < \infty,$$

for all $x, z \in G$. Suppose that $f : G \to Y$ is a mapping such that $f(-x) = -f(x)$ for all $x \in G$ and

$$\|Df(x, y, z)\| \leq \varphi(x, y, z)$$

for all $x, y, z \in G$, then there exists a unique additive mapping $h : G \to Y$ such that

$$\|f(x) - h(x)\| \leq \check{\varphi}(x, x)$$

for all $x \in G$.

We can similarly prove another stability theorem under a somewhat different condition as follows:

Remark 20.2 Let $\varphi : G^3 \to \mathbb{R}^+$ satisfy

$$\check{\varphi}(x, y, z) := \sum_{i=0}^{\infty} \frac{1}{n^i} \left(\frac{1}{n} \varphi\big(n^{i+1} x, 0, -n^i z\big) + \varphi\big(n^i x, -n^i y, 0\big) \right) < \infty,$$

for all $x, y, z \in G$ and $\lim_{k\to\infty} \frac{1}{n^k} \varphi(n^k x, n^k y, n^k z) = 0$ for all $x, y, z \in G$. If $f : G \to Y$ is a mapping such that $f(0) = 0$ and

$$\|Df(x, y, z)\| \leq \varphi(x, y, z)$$

for all $x, y, z \in G$, then there exists a unique additive mapping $h : G \to Y$ such that

$$\|f(x) - h(x)\| \leq \check{\varphi}(x, x, x)$$

for all $x \in G$.

Theorem 20.2 *Let $\varphi : G^3 \to \mathbb{R}^+$ satisfy*

$$\tilde{\varphi}(x, z) := \frac{1}{2} \sum_{i=0}^{\infty} n^i \left(\varphi\left(\frac{x}{n^i}, 0, -\frac{z}{n^{i+1}} \right) + \varphi\left(-\frac{x}{n^i}, 0, \frac{z}{n^{i+1}} \right) \right) < \infty,$$

and $\lim_{k\to\infty} n^k \varphi(\frac{x}{n^k}, \frac{y}{n^k}, \frac{z}{n^k}) = 0$ for all $x, y, z \in G$. If $f : G \to Y$ is a mapping such that

$$\|Df(x, y, z)\| \leq \varphi(x, y, z) \tag{20.10}$$

for all $x, y, z \in G$, then there exists a unique additive mapping $h : G \to Y$, defined as $h(x) = \lim_{k\to\infty} n^k [\frac{f(n^{-k}x) - f(-n^{-k}x)}{2}]$, such that

$$\|f(x) - h(x)\| \leq \tilde{\varphi}(x, x) + \frac{\varphi(x, -x, 0)}{2} \tag{20.11}$$

for all $x \in G$.

Proof Putting $x = y = x = 0$ in (20.10), we have $f(0) = 0$ since $\tilde{\varphi}(x,z) < \infty$ and so $\varphi(0,0,0) = 0$. Letting $y = -x$, $z = 0$ in (20.10) and dividing both sides by 2, we have

$$\left\| \frac{f(x) + f(-x)}{2} \right\| \le \frac{\varphi(x,-x,0)}{2} \tag{20.12}$$

for all $x \in G$.

Replacing x by nx and letting $y = 0$ and $z = -x$ in (20.10), we get

$$\left\| f(nx) + nf(-x) \right\| \le \varphi(nx, 0, -x) \tag{20.13}$$

for all $x \in G$. Replacing x by $-x$ in (20.13), we get

$$\left\| f(-nx) + nf(x) \right\| \le \varphi(-nx, 0, x) \tag{20.14}$$

for all $x \in G$. Put $g(x) = \frac{f(x)-f(-x)}{2}$. Using (20.13) and (20.14) yields

$$\left\| ng(x) - g(nx) \right\| \le \frac{1}{2}\big(\varphi(nx, 0, -x) + \varphi(-nx, 0, x)\big)$$

for all $x \in G$. Replacing x by $\frac{x}{n}$, we get

$$\left\| g(x) - ng\left(\frac{x}{n}\right) \right\| \le \frac{1}{2}\left(\varphi\left(x, 0, -\frac{x}{n}\right) + \varphi\left(-x, 0, \frac{x}{n}\right)\right) \tag{20.15}$$

for all $x \in G$. The remainder is similar to the proof of Theorem 20.1. This completes the proof. □

Suppose that X is a normed space in the following corollaries. If we put $\varphi(x,y,z) := \varepsilon(\|x\|^p + \|y\|^q + \|z\|^t)$ and $\varphi(x,y,z) := \varepsilon(\|x\|^p\|y\|^q\|z\|^t)$ in Theorem 20.2, respectively, then we get the Corollaries 20.4 and 20.5.

Corollary 20.4 *Let $p, q, t > 1$, $\varepsilon > 0$. If a mapping $f : X \to Y$ satisfies*

$$\left\| Df(x,y,z) \right\| \le \varepsilon\big(\|x\|^p + \|y\|^q + \|z\|^t\big)$$

for all $x, y, z \in X$, then there exists a unique additive mapping $h : X \to Y$ such that

$$\left\| f(x) - h(x) \right\| \le \varepsilon\left[\left(\frac{n^p}{n^p - n}\right)\|x\|^p + \frac{1}{2}\|x\|^q + \left(\frac{1}{n^t - n} + \frac{1}{2}\right)\|x\|^t\right]$$

for all $x \in X$.

Corollary 20.5 *Let $p + q + t > 1$, $p, q, t > 0$, $\varepsilon > 0$. If a mapping $f : X \to Y$ satisfies*

$$\left\| Df(x,y,z) \right\| \le \varepsilon\big(\|x\|^p\|y\|^q\|z\|^t\big)$$

for all $x, y, z \in X$, then f is additive.

We may obtain more simple and sharp approximation than that of Theorem 20.2 for the stability result under the oddness condition.

Remark 20.3 Let $\varphi : G^3 \to \mathbb{R}^+$ satisfy

$$\tilde{\varphi}(x, z) = \sum_{i=0}^{\infty} n^i \varphi\left(\frac{x}{n^i}, 0, -\frac{z}{n^{i+1}}\right) < \infty,$$

for all $x, z \in G$ and $\lim_{k\to\infty} n^k \varphi(\frac{x}{n^k}, \frac{y}{n^k}, \frac{z}{n^k}) = 0$ for all $x, y, z \in G$. If $f : G \to Y$ is an odd mapping such that

$$\|Df(x, y, z)\| \le \varphi(x, y, z)$$

for all $x, y, z \in G$, then there exists a unique additive mapping $h : G \to Y$ such that

$$\|f(x) - h(x)\| \le \widetilde{\varphi}(x, x)$$

for all $x \in G$.

We can similarly prove another stability theorem under a somewhat different condition as follows:

Remark 20.4 Let $\varphi : G^3 \to \mathbb{R}^+$ satisfy

$$\tilde{\varphi}(x, y, z) := \frac{1}{2}\sum_{i=0}^{\infty} n^i \left(\varphi\left(\frac{x}{n^i}, 0, -\frac{z}{n^{i+1}}\right) + \varphi\left(-\frac{x}{n^i}, -\frac{y}{n^{i+1}}, 0\right)\right) < \infty,$$

and $\lim_{k\to\infty} n^k \varphi(\frac{x}{n^k}, \frac{y}{n^k}, \frac{z}{n^k}) = 0$ for all $x, y, z \in G$. If $f : G \to Y$ is a mapping such that

$$\|Df(x, y, z)\| \le \varphi(x, y, z)$$

for all $x, y, z \in G$, then there exists a unique additive mapping $h : G \to Y$ such that

$$\|f(x) - h(x)\| \le \tilde{\varphi}(x, x, x)$$

for all $x \in G$.

References

1. Gajda, Z.: On the stability of additive mappings. Int. J. Math. Math. Sci. **14**, 431–434 (1991)
2. Găvruta, P.: A generalization of the Hyers–Ulam–Rassias stability of approximately additive mappings. J. Math. Anal. Appl. **184**, 431–436 (1994)
3. Gao, Z.-X., Cao, H.-X., Zheng, W.-T., Xu, L.: Generalized Hyers–Ulam–Rassias stability of functional inequalities and functional equations. J. Math. Inequal. **3**(1), 63–77 (2009)

4. Hyers, D.H.: On the stability of the linear functional equation. Proc. Natl. Acad. Sci. USA **27**, 222–224 (1941)
5. Rassias, Th.M.: On the stability of the linear mapping in Banach spaces. Proc. Am. Math. Soc. **72**, 297–300 (1978)
6. Rassias, Th.M.: The stability of mappings and related topics. In: Report on the 27th ISFE. Aequ. Math., vol. 39, pp. 292–293 (1990)
7. Rassias, Th.M., Šemrl, P.: On the behaviour of mappings which do not satisfy Hyers–Ulam–Rassias stability. Proc. Am. Math. Soc. **114**, 989–993 (1992)
8. Jung, S.: On Hyers–Ulam–Rassias stability of approximately additive mappings. J. Math. Anal. Appl. **204**, 221–226 (1996)
9. Ulam, S.M.: A Collection of the Mathematical Problems. Interscience, New York (1960)

Chapter 21
Fixed Point Approach to the Stability of the Gamma Functional Equation

Soon-Mo Jung

Abstract The gamma function appears occasionally in the physical problems and applications. Especially, the gamma function is useful to develop other functions which have physical applications. It is well known that the gamma function satisfies the following functional equation $f(x+1) = xf(x)$, and hence it is called the gamma functional equation. We will apply the fixed point method for proving the Hyers–Ulam–Rassias stability of the gamma functional equation.

Key words Fix points · Stability · Gamma functional equation

Mathematics Subject Classification 65Q20 · 49K40

21.1 Introduction

In 1940, S.M. Ulam [29] gave a wide ranging talk before the mathematics club of the University of Wisconsin in which he discussed a number of important unsolved problems. Among those was the question concerning the stability of group homomorphisms:

> *Let G_1 be a group and let G_2 be a metric group with the metric $d(\cdot, \cdot)$. Given $\varepsilon > 0$, does there exist a $\delta > 0$ such that if a function $h : G_1 \to G_2$ satisfies the inequality $d(h(xy), h(x)h(y)) < \delta$ for all $x, y \in G_1$, then there exists a homomorphism $H : G_1 \to G_2$ with $d(h(x), H(x)) < \varepsilon$ for all $x \in G_1$?*

Ulam problem for the case of approximately additive functions was solved by D.H. Hyers [9] under the assumption that G_1 and G_2 are Banach spaces. Indeed, Hyers proved that each solution of the inequality $\|f(x+y) - f(x) - f(y)\| \leq \varepsilon$, for all x and y, can be approximated by an exact solution, say an additive function. In this case, the Cauchy additive functional equation, $f(x+y) = f(x) + f(y)$, is said to satisfy the Hyers–Ulam stability.

Dedicated to Professor Themistocles M. Rassias on the occasion of his 60th birthday.

S.-M. Jung (✉)
Mathematics Section, College of Science and Technology, Hongik University, 339-701 Jochiwon, Republic of Korea
e-mail: smjung@hongik.ac.kr

P.M. Pardalos et al. (eds.), *Nonlinear Analysis*, Springer Optimization and Its Applications 68, In Honor of Themistocles M. Rassias on the Occasion of his 60th Birthday, DOI 10.1007/978-1-4614-3498-6_21,

Th.M. Rassias [26] attempted to weaken the condition for the bound of the norm of the Cauchy difference as follows

$$\|f(x+y)-f(x)-f(y)\| \le \varepsilon\big(\|x\|^p+\|y\|^p\big)$$

and derived Hyers' theorem for the stability of the additive mapping as a special case. Thus in [26], a proof of the generalized Hyers–Ulam stability for the linear mapping between Banach spaces was obtained. A particular case of Th.M. Rassias' theorem regarding the Hyers–Ulam stability of the additive mapping was proved by T. Aoki [2]. The stability concept that was introduced by Th.M. Rassias' theorem provided some influence to a number of mathematicians to develop the notion of what is known today with the term Hyers–Ulam–Rassias stability of the linear mappings. Since then, the stability of several functional equations has been extensively investigated by several mathematicians (see, for example, [2, 5–8, 10–15, 19, 24, 27, 28] and the references therein).

The terms Hyers–Ulam–Rassias stability and Hyers–Ulam stability can also be applied to the case of other functional equations, differential equations, and of various integral equations.

The gamma function

$$\Gamma(x)=\int_0^\infty e^{-t}t^{x-1}\,dt \quad (x>0)$$

appears occasionally in the physical problems and applications. Especially, the gamma function is very useful to develop other functions which have physical applications. It is well known that the gamma function satisfies the following functional equation

$$f(x+1)=xf(x), \tag{21.1}$$

and hence it will be called the gamma functional equation (see [22]).

The author has investigated the stability problems for the gamma functional equation (21.1) (see [16–18]).

In this paper, we will adopt the ideas from [4, 20, 21, 25] and prove the Hyers–Ulam–Rassias stability of the gamma functional equation (21.1).

21.2 Preliminaries

For a nonempty set X, we introduce the definition of the generalized metric on X. A function $d : X \times X \to [0,\infty]$ is called a generalized metric on X if and only if d satisfies

(M_1) $d(x,y)=0$ if and only if $x=y$;
(M_2) $d(x,y)=d(y,x)$ for all $x, y \in X$;
(M_3) $d(x,z) \le d(x,y)+d(y,z)$ for all $x, y, z \in X$.

We remark that the only difference of the generalized metric from the usual metric is that the range of the former is permitted to include the infinity.

We now introduce one of fundamental results of fixed point theory. For the proof, we refer to [23]. This theorem will play an important rôle in proving our main theorem.

Theorem 21.1 *Let (X, d) be a generalized complete metric space. Assume that $\Lambda : X \to X$ is a strictly contractive operator with the Lipschitz constant $L < 1$. If there exists a nonnegative integer k such that $d(\Lambda^{k+1}x, \Lambda^k x) < \infty$ for some $x \in X$, then the following are true*:

(a) *The sequence $\{\Lambda^n x\}$ converges to a fixed point x^* of Λ*;
(b) *x^* is the unique fixed point of Λ in*

$$X^* = \{y \in X \mid d(\Lambda^k x, y) < \infty\};$$

(c) *If $y \in X^*$, then*

$$d(y, x^*) \le \frac{1}{1-L} d(\Lambda y, y).$$

21.3 Hyers–Ulam–Rassias Stability

Recently, Cădariu and Radu [4] applied a fixed point method to the investigation of the Cauchy additive functional equation. Using such a clever idea, they could present a proof for the Hyers–Ulam stability of that equation (see [3, 25]).

By using the idea of Cădariu and Radu, we prove our main theorem concerning the Hyers–Ulam–Rassias stability of the gamma functional equation (21.1).

In what follows, we will use the notation $\mathbf{R}^\circ = \mathbf{R} \setminus \{0, -1, -2, \ldots\}$.

Theorem 21.2 *Let $(E, \|\cdot\|)$ be a real (or complex) Banach space and assume that a function $\varphi : \mathbf{R}^\circ \to (0, \infty)$ is given such that there exists a constant L, $0 < L < 1$, with the property*

$$\frac{1}{|x+1|}\varphi(x+1) \le L\varphi(x) \tag{21.2}$$

for all $x \in \mathbf{R}^\circ$. If a function $f : \mathbf{R}^\circ \to E$ satisfies the functional inequality

$$\|f(x+1) - xf(x)\| \le \varphi(x) \tag{21.3}$$

for any $x \in \mathbf{R}^\circ$, then there exists a unique solution function $F : \mathbf{R}^\circ \to E$ of the gamma functional equation (21.1) *with*

$$\|f(x) - F(x)\| \le \frac{1}{1-L}\frac{1}{|x|}\varphi(x) \tag{21.4}$$

for each $x \in \mathbf{R}^\circ$.

Proof Let us define a set $\mathscr{X}$ by

$$\mathscr{X} = \{h : \mathbf{R}^\circ \to E \mid h \text{ is a function}\}$$

and introduce a generalized metric d on $\mathscr{X}$ as follows:

$$d(g,h)=\inf\left\{C\in[0,\infty]\mid \left\|g(x)-h(x)\right\|\le C\frac{1}{|x|}\varphi(x)\text{ for all }x\in\mathbf{R}^\circ\right\}. \quad (21.5)$$

(We will here give a proof for the triangle inequality only. Assume, on the contrary, that $d(g,h)>d(g,k)+d(k,h)$ holds for some $g,h,k\in\mathscr{X}$. Then, by (21.5), there exists an $x_0\in\mathbf{R}^\circ$ with

$$\begin{aligned}\left\|g(x_0)-h(x_0)\right\| &> \left\{d(g,k)+d(k,h)\right\}\frac{1}{|x_0|}\varphi(x_0)\\ &= d(g,k)\frac{1}{|x_0|}\varphi(x_0)+d(k,h)\frac{1}{|x_0|}\varphi(x_0)\\ &\ge \left\|g(x_0)-k(x_0)\right\|+\left\|k(x_0)-h(x_0)\right\|,\end{aligned}$$

a contradiction.)

We assert that $(\mathscr{X},d)$ is complete. We will follow the idea from [21] to prove the completeness of $(\mathscr{X},d)$. Let $\{h_n\}$ be a Cauchy sequence in $(\mathscr{X},d)$. Then, for any $\varepsilon>0$, there exists an integer $N_\varepsilon>0$ such that $d(h_m,h_n)\le\varepsilon$ for all $m,n\ge N_\varepsilon$. It further follows from (21.5) that

$$\forall\varepsilon>0\;\exists N_\varepsilon\in\mathbf{N}\;\forall m,n\ge N_\varepsilon\;\forall x\in\mathbf{R}^\circ:\left\|h_m(x)-h_n(x)\right\|\le\varepsilon\frac{1}{|x|}\varphi(x). \quad (21.6)$$

If x is fixed, (21.6) implies that $\{h_n(x)\}$ is a Cauchy sequence in $(E,\|\cdot\|)$. Since $(E,\|\cdot\|)$ is complete, $\{h_n(x)\}$ converges for each $x\in\mathbf{R}^\circ$. Thus, we can define a function $h:\mathbf{R}^\circ\to E$ by

$$h(x)=\lim_{n\to\infty}h_n(x),$$

and hence h belongs to $\mathscr{X}$.

If we let m increase to infinity, it then follows from (21.6) that

$$\forall\varepsilon>0\;\exists N_\varepsilon\in\mathbf{N}\;\forall n\ge N_\varepsilon\;\forall x\in\mathbf{R}^\circ:\left\|h(x)-h_n(x)\right\|\le\varepsilon\frac{1}{|x|}\varphi(x).$$

Further if we consider (21.5), then we conclude that

$$\forall\varepsilon>0\;\exists N_\varepsilon\in\mathbf{N}\;\forall n\ge N_\varepsilon:d(h,h_n)\le\varepsilon,$$

that is, the Cauchy sequence $\{h_n\}$ converges to h in $(\mathscr{X},d)$. Hence, $(\mathscr{X},d)$ is complete.

Now, let us define an operator $\Lambda:\mathscr{X}\to\mathscr{X}$ by

$$(\Lambda h)(x)=\frac{1}{x}h(x+1)\quad\left(x\in\mathbf{R}^\circ\right) \quad (21.7)$$

for all $h\in\mathscr{X}$. (It is obvious that $\Lambda h\in\mathscr{X}$.)

We assert that Λ is strictly contractive on $\mathscr{X}$. For any $g,h\in\mathscr{X}$, let us choose a $C_{gh}\in[0,\infty]$ satisfying $d(g,h)\le C_{gh}$. Then, using (21.5), we have

$$\left\|g(x)-h(x)\right\|\le C_{gh}\frac{1}{|x|}\varphi(x)\quad\left(x\in\mathbf{R}^\circ\right). \quad (21.8)$$

Using (21.2), (21.7), and (21.8), we get

$$\begin{aligned}\|(\Lambda g)(x) - (\Lambda h)(x)\| &= \left\| \frac{1}{x} g(x+1) - \frac{1}{x} h(x+1) \right\| \\ &\le C_{gh} \frac{1}{|x|} \frac{1}{|x+1|} \varphi(x+1) \\ &\le L C_{gh} \frac{1}{|x|} \varphi(x)\end{aligned}$$

for all $x \in \mathbf{R}^\circ$, that is, $d(\Lambda g, \Lambda h) \le L C_{gh}$. Hence, we may conclude that

$$d(\Lambda g, \Lambda h) \le L d(g, h) \tag{21.9}$$

and we note that $0 < L < 1$.

Moreover, it follows from (21.3) and (21.7) that

$$\|(\Lambda f)(x) - f(x)\| = \left\| \frac{1}{x} f(x+1) - f(x) \right\| \le \frac{1}{|x|} \varphi(x)$$

for every $x \in \mathbf{R}^\circ$. Thus, (21.5) implies that

$$d(\Lambda f, f) \le 1. \tag{21.10}$$

Therefore, it follows from Theorem 21.1(a) that there exists a function $F : \mathbf{R}^\circ \to E$ such that $\Lambda^n f \to F$ in $(\mathscr{X}, d)$ and $\Lambda F = F$. Hence, F is a solution of the gamma functional equation (21.1).

Moreover, Theorem 21.1(c), together with (21.10), implies that

$$d(f, F) \le \frac{1}{1-L} d(\Lambda f, f) \le \frac{1}{1-L},$$

that is, in view of (21.5), the inequality (21.4) is true for each $x \in \mathbf{R}^\circ$.

Finally, let $G : \mathbf{R}^\circ \to E$ be another solution of the gamma equation (21.1) with

$$\|f(x) - G(x)\| \le K \frac{1}{|x|} \varphi(x)$$

for all $x \in \mathbf{R}^\circ$ and for some constant K with $0 < K < \infty$. (It easily follows from (21.1) that G is a fixed point of Λ, that is, $\Lambda G = G$.) Then, since $d(f, G) < \infty$, we have

$$G \in \mathscr{X}^* = \{ g \in \mathscr{X} \mid d(f, g) < \infty \}.$$

So, since both F and G are fixed points of Λ, Theorem 21.1(b) implies that $F = G$, that is, F is unique. □

Remark 21.1 Even though the function $F : \mathbf{R}^\circ \to \mathbf{R}$ is a solution of the gamma functional equation (21.1), F does not necessarily equal to the gamma function Γ on $(0, \infty)$. However, if F is logarithmically convex on $(0, \infty)$ and is a solution of the gamma functional equation (21.1) for $x > 0$, and if $F(1) = 1$, then F necessarily equals to the gamma function Γ on $(0, \infty)$ (see [22]).

Remark 21.2 The main theorem of [16] has been proved by using the direct method, while in this paper we apply a fixed point method for proving Theorem 21.2. The 'old' condition for φ of [16], expressed as

$$\sum_{j=0}^{\infty} \varphi(x+j) \prod_{i=0}^{j} |x+i|^{-1} < \infty,$$

seems to be weaker than our 'new' condition (21.2) of the present paper. Therefore, one of our aims of this paper is to apply a fixed point method for proving that every approximate solution is not far from the exact solution of (21.1).

For a given real number r, we will define

$$\mathbf{R}_{(r)} = \mathbf{R}^\circ \cap (r, \infty).$$

By replacing each $\mathbf{R}^\circ$ with $\mathbf{R}_{(r)}$ in the proof of Theorem 21.2, we can easily prove the following corollary. Hence, we omit its proof.

Corollary 21.1 *Let* $(E, \|\cdot\|)$ *be a real (or complex) Banach space and assume that a function* $\varphi : \mathbf{R}_{(r)} \to (0, \infty)$ *is given such that there exists a constant* L, $0 < L < 1$, *with the property* (21.2) *for all* $x \in \mathbf{R}_{(r)}$. *If a function* $f : \mathbf{R}_{(r)} \to E$ *satisfies the inequality* (21.3) *for any* $x \in \mathbf{R}_{(r)}$, *then there exists a unique solution* $F : \mathbf{R}_{(r)} \to E$ *of the gamma functional equation* (21.1) *satisfying the inequality* (21.4) *for each* $x \in \mathbf{R}_{(r)}$.

21.4 Examples

Let $\varepsilon > 0$ be a constant and let $L = \frac{1}{2}$. If we set $\varphi(x) = \frac{\varepsilon}{x+1}$ for all $x > 0$, then

$$\frac{1}{x+1}\varphi(x+1) = \frac{\varepsilon}{(x+1)(x+2)} < \frac{1}{2}\frac{\varepsilon}{x+1} = L\varphi(x)$$

for any $x > 0$.

According to Corollary 21.1, we have the following example.

Example 21.1 Let $(E, \|\cdot\|)$ be a real (or complex) Banach space. If a function $f : (0, \infty) \to E$ satisfies the inequality

$$\|f(x+1) - xf(x)\| \le \frac{\varepsilon}{x+1}$$

for every $x > 0$, then there exists a unique solution $F : (0, \infty) \to E$ of the gamma functional equation (21.1) with

$$\|f(x) - F(x)\| \le \frac{2\varepsilon}{x(x+1)}$$

for each $x > 0$.

Given constants ε and L with $\varepsilon > 0$ and $\frac{1}{2} \le L < 1$, let us define a function $\varphi : (0, \infty) \to (0, \infty)$ by

$$\varphi(x) = \begin{cases} \varepsilon & \text{if } 0 < x \le 1, \\ L\varepsilon & \text{if } x > 1. \end{cases}$$

Then, we get

$$\frac{1}{x+1}\varphi(x+1) < \begin{cases} \varphi(x+1) = L\varepsilon = L\varphi(x) & \text{if } 0 < x \le 1, \\ \frac{1}{2}\varphi(x+1) = \frac{1}{2}\varphi(x) \le L\varphi(x) & \text{if } x > 1. \end{cases}$$

Hence, the following example is an immediate consequence of Corollary 21.1.

Example 21.2 Let $(E, \|\cdot\|)$ be a real (or complex) Banach space. If a function $f : (0, \infty) \to E$ satisfies the inequality

$$\|f(x+1) - xf(x)\| \le \begin{cases} \varepsilon & \text{if } 0 < x \le 1, \\ L\varepsilon & \text{if } x > 1, \end{cases}$$

then there exists a unique solution $F : (0, \infty) \to E$ of (21.1) with

$$\|f(x) - F(x)\| \le \begin{cases} \frac{1}{1-L}\frac{\varepsilon}{x} & \text{if } 0 < x \le 1, \\ \frac{L}{1-L}\frac{\varepsilon}{x} & \text{if } x > 1. \end{cases}$$

For a constant $\varepsilon > 0$, if we define a constant function $\varphi : (1, \infty) \to (0, \infty)$ by $\varphi(x) = \varepsilon$, then

$$\frac{1}{x+1}\varphi(x+1) = \frac{\varepsilon}{x+1} < \frac{1}{2}\varepsilon = \frac{1}{2}\varphi(x)$$

for all $x > 1$, that is, $\varphi(x) = \varepsilon$ satisfies the condition (21.2) with $L = \frac{1}{2}$.

In view of Corollary 21.1, we now obtain

Example 21.3 Assume that $(E, \|\cdot\|)$ is a real (or complex) Banach space. If a function $f : (1, \infty) \to E$ satisfies the inequality

$$\|f(x+1) - xf(x)\| \le \varepsilon$$

for any $x > 1$, then there exists a unique solution $F : (1, \infty) \to E$ of (21.1) such that

$$\|f(x) - F(x)\| \le \frac{2\varepsilon}{x}$$

for all $x > 1$.

Remark 21.3 Alzer [1] has improved the result of the first author [18] by replacing the upper bound $\frac{3\varepsilon}{x}$ with the best one $\frac{e\varepsilon}{x}$ when the relevant domain is $(0, \infty)$. However, if we restrict ourselves to the case when the relevant domain is $(1, \infty)$, then the upper bound in Example 21.3 for $\|f(x) - F(x)\|$ is less than that of Alzer.

Acknowledgements This research was supported by Basic Science Research Program through the National Research Foundation of Korea (NRF) funded by the Ministry of Education, Science and Technology (No. 2010-0007143).

References

1. Alzer, H.: Remark on the stability of the gamma functional equations. Results Math. **35**, 199–200 (1999)
2. Aoki, T.: On the stability of the linear transformation in Banach spaces. J. Math. Soc. Jpn. **2**, 64–66 (1950)
3. Cădariu, L., Radu, V.: Fixed points and the stability of Jensen's functional equation. J. Inequal. Pure Appl. Math. **4**(1), 4 (2003). http://jipam.vu.edu.au
4. Cădariu, L., Radu, V.: On the stability of the Cauchy functional equation: a fixed point approach. Grazer Math. Ber. **346**, 43–52 (2004)
5. Czerwik, S.: Functional Equations and Inequalities in Several Variables. World Scientific, Singapore (2002)
6. Faizev, V.A., Rassias, Th.M., Sahoo, P.K.: The space of (ϕ, α)-additive mappings on semigroups. Trans. Am. Math. Soc. **354**(11), 4455–4472 (2002)
7. Forti, G.L.: Hyers–Ulam stability of functional equations in several variables. Aequ. Math. **50**, 143–190 (1995)
8. Găvruţă, P.: A generalization of the Hyers–Ulam–Rassias stability of approximately additive mappings. J. Math. Anal. Appl. **184**, 431–436 (1994)
9. Hyers, D.H.: On the stability of the linear functional equation. Proc. Natl. Acad. Sci. USA **27**, 222–224 (1941)
10. Hyers, D.H., Isac, G., Rassias, Th.M.: Topics in Nonlinear Analysis and Applications. World Scientific, Singapore (1997)
11. Hyers, D.H., Isac, G., Rassias, Th.M.: Stability of Functional Equations of Several Variables. Birkhäuser, Basel (1998)
12. Hyers, D.H., Isac, G., Rassias, Th.M.: On the asymptoticity aspect of Hyers–Ulam stability of mappings. Proc. Am. Math. Soc. **126**(2), 425–430 (1998)
13. Hyers, D.H., Rassias, Th.M.: Approximate homomorphisms. Aequ. Math. **44**, 125–153 (1992)
14. Isac, G., Rassias, Th.M.: On the Hyers–Ulam stability of ϕ-additive mappings. J. Approx. Theory **72**, 131–137 (1993)
15. Isac, G., Rassias, Th.M.: Stability of additive mappings: applications to nonlinear analysis. Int. J. Math. Math. Sci. **19**(2), 219–228 (1996)
16. Jung, S.-M.: On the modified Hyers–Ulam–Rassias stability of the functional equation for gamma function. Mathematica (Cluj) **39**(62), 233–237 (1997)
17. Jung, S.-M.: On a general Hyers–Ulam stability of gamma functional equation. Bull. Korean Math. Soc. **34**, 437–446 (1997)
18. Jung, S.-M.: On the stability of gamma functional equation. Results Math. **33**, 306–309 (1998)
19. Jung, S.-M.: Hyers–Ulam–Rassias Stability of Functional Equations in Mathematical Analysis. Hadronic Press, Palm Harbor (2001)
20. Jung, S.-M.: A fixed point approach to the stability of isometries. J. Math. Anal. Appl. **329**, 879–890 (2007)
21. Jung, S.-M.: A fixed point approach to the stability of a Volterra integral equation. Fixed Point Theory Appl. **2007**, 57064 (2007). doi:10.1155/2007/57064
22. Kairies, H.H.: On the optimality of a characterization theorem for the gamma function using the multiplication formula. Aequ. Math. **51**, 115–128 (1996)
23. Margolis, B., Diaz, J.: A fixed point theorem of the alternative for contractions on a generalized complete metric space. Bull. Am. Math. Soc. **74**, 305–309 (1968)

24. Moslehian, M.S., Rassias, Th.M.: Stability of functional equations in non-Archimedian spaces. Appl. Anal. Discrete Math. **1**, 325–334 (2007)
25. Radu, V.: The fixed point alternative and the stability of functional equations. Fixed Point Theory **4**, 91–96 (2003)
26. Rassias, Th.M.: On the stability of the linear mapping in Banach spaces. Proc. Am. Math. Soc. **72**, 297–300 (1978)
27. Rassias, Th.M.: On a modified Hyers–Ulam sequence. J. Math. Anal. Appl. **158**, 106–113 (1991)
28. Rassias, Th.M.: On the stability of functional equations and a problem of Ulam. Acta Appl. Math. **62**, 23–130 (2000)
29. Ulam, S.M.: A Collection of Mathematical Problems. Interscience, New York (1960)

Chapter 22
Random Stability of an AQCQ Functional Equation: A Fixed Point Approach

Hassan Azadi Kenary

Abstract Recently, Hyers–Ulam stability of the following additive-quadratic–cubic–quartic functional equation

$$f(x+2y)+f(x-2y)=4f(x+y)+4f(x-y)-6f(x)$$
$$+f(2y)+f(-2y)-4f(y)-4f(-y)$$

was proved in a Banach space in an earlier work. In this paper, we prove the generalized Hyers–Ulam stability of the above functional equation in random normed spaces.

Key words Fixed points · Random stability · Random normed spaces

Mathematics Subject Classification 65Q20 · 49K40

22.1 Introduction and Preliminaries

The stability problem of functional equations is originated from a question of Ulam [39] concerning the stability of group homomorphisms. Hyers [12] gave a first affirmative partial answer to the question of Ulam for Banach spaces. Hyers's Theorem was generalized by Aoki [2] for additive mappings and by Th.M. Rassias [30] for linear mappings by considering an unbounded Cauchy difference.

Theorem 22.1 (Th.M. Rassias) *Let f be an approximately additive mapping from a normed vector space E into a Banach space E', i.e., f satisfies the inequality*

$$\left|f(x+y)-f(x)-f(y)\right|\le\varepsilon\left(\|x\|^r+\|y\|^r\right)$$

Dedicated to Professor Themistocles M. Rassias on the occasion of his 60th birthday.

H.A. Kenary (✉)
Department of Mathematics, College of Science, Yasouj University, Yasouj 75914-353, Iran
e-mail: azadi@mail.yu.ac.ir

P.M. Pardalos et al. (eds.), *Nonlinear Analysis*, Springer Optimization and Its Applications 68, In Honor of Themistocles M. Rassias on the Occasion of his 60th Birthday, DOI 10.1007/978-1-4614-3498-6_22,

for all $x, y \in E$, *where* ε *and* r *are constants with* $\varepsilon > 0$ *and* $0 \le r < 1$. *Then the mapping* $L : E \to E'$ *defined by* $L(x) := \lim_{n\to\infty} 2^{-n} f(2^n x)$ *is the unique linear mapping which satisfies*

$$\left| f(x) - L(x) \right| \le \frac{2\varepsilon}{2 - 2^r} |x|^r$$

for all $x \in E$.

The paper of Th.M. Rassias [30] has provided a lot of influence in the development of what we call generalized Hyers–Ulam stability or as Hyers–Ulam–Rassias stability of functional equations. A generalization of the Th.M. Rassias theorem was obtained by Găvruta [11] by replacing the unbounded Cauchy difference by a general control function in the spirit of Th.M. Rassias's approach

The functional equation

$$f(x + y) + f(x - y) = 2f(x) + 2f(y)$$

is called a quadratic functional equation. In particular, every solution of the quadratic functional equation is said to be a quadratic mapping. A generalized Hyers–Ulam stability problem for the quadratic functional equation was proved by Skof [38] for mappings $f : X \to Y$, where X is a normed space and Y is a Banach space. Cholewa [7] noticed that the theorem of Skof is still true if the relevant domain X is replaced by an abelian group. Czerwik [8] proved the generalized Hyers–Ulam stability of the quadratic functional equation. The stability problems of several functional equations have been extensively investigated by a number of authors, and there are many interesting results concerning this problem (see [30–35]).

In [15], Jun and Kim considered the following cubic functional equation

$$f(2x + y) + f(2x - y) = 2f(x + y) + 2f(x - y) + 12f(x) \qquad (22.1)$$

which is called a cubic functional equation, and every solution of the cubic functional equation is said to be a cubic mapping.

In [16], Lee et al. considered the following quartic functional equation

$$f(2x + y) + f(2x - y) = 4f(x + y) + 4f(x - y) + 24f(x) - 6f(y) \qquad (22.2)$$

which is called a quartic functional equation, and every solution of the quartic functional equation is said to be a quartic mapping. Quartic functional equations have been investigated in [16].

In this paper, we prove the generalized Hyers–Ulam stability of the following additive-quadratic–cubic–quartic functional equation

$$\begin{aligned} f(x + 2y) + f(x - 2y) &= 4f(x + y) + 4f(x - y) - 6f(x) \\ &\quad - f(2y) - f(-2y) + 4f(y) + 4f(-y) \qquad (22.3) \end{aligned}$$

in random normed spaces.

One can easily show that an odd mapping $f : X \to Y$ satisfies (22.3) if and only if the odd mapping $f : X \to Y$ is an additive-cubic mapping, that is,

$$f(x+2y)+f(x-2y)=4f(x+y)+4f(x-y)-6f(x). \tag{22.4}$$

It was shown in [9, Lemma 2.2] that $g(x) := f(2x)-2f(x)$ and $h(x) := f(2x)-8f(x)$ are cubic and additive, respectively, and that

$$f(x)=\frac{1}{6}g(x)-\frac{1}{6}h(x).$$

One can easily show that an even mapping $f : X \to Y$ satisfies (22.3) if and only if the even mapping $f : X \to Y$ is a quadratic-quartic mapping, that is,

$$\begin{aligned} f(x+2y)+f(x-2y) &= 4f(x+y)+4f(x-y)-6f(x) \\ &\quad +2f(2y)-8f(y). \end{aligned} \tag{22.5}$$

It was shown in [10, Lemma 2.1] that $g(x) := f(2x)-4f(x)$ and $h(x) := f(2x)-16f(x)$ are quartic and quadratic, respectively, and that

$$f(x)=\frac{1}{12}g(x)-\frac{1}{12}h(x).$$

Functional equations of mixed type have been investigated in [9, 10]. Let

$$f_o(x)=\frac{f(x)-f(-x)}{2} \quad \text{and} \quad f_e(x)=\frac{f(x)+f(-x)}{2}.$$

Then f_o is odd and f_e is even. The functions f_o and f_e satisfy the functional equation (22.3). Let $g_o(x) := f_o(2x)-2f_o(x)$ and $h_o(x) := f_o(2x)-8f_o(x)$. Then

$$f_o(x)=\frac{1}{6}g_o(x)-\frac{1}{6}h_o(x).$$

Let $g_e(x) := f_e(2x)-4f_e(x)$ and $h_e(x) := f_e(2x)-16f_o(x)$. Then

$$f_e(x)=\frac{1}{12}g_e(x)-\frac{1}{12}h_e(x).$$

Thus

$$f(x)=\frac{1}{6}g_o(x)-\frac{1}{6}h_o(x)+\frac{1}{12}g_e(x)-\frac{1}{12}h_e(x).$$

Definition 22.1 Let X be a set. A function $d : X \times X \to [0,\infty]$ is called a generalized metric on X if d satisfies

(1) $d(x,y)=0$ if and only if $x=y$,
(2) $d(x,y)=d(y,x)$ for all $x,y \in X$,
(3) $d(x,z) \le d(x,y)+d(y,z)$ for all $x,y,z \in X$.

We recall a fundamental result in fixed point theory.

Theorem 22.2 ([4]) *Let (X,d) be a complete generalized metric space and let $J: X \to X$ be a strictly contractive mapping with Lipschitz constant $L < 1$. Then for each given element $x \in X$, either*

$$d\big(J^n x, J^{n+1} x\big) = \infty \tag{22.6}$$

for all nonnegative integers n or there exists a positive integer n_0 such that

1. $d(J^n x, J^{n+1} x) < \infty$ *for all* $n_0 \geq n_0$;
2. *The sequence* $\{J^n x\}$ *converges to a fixed point* y^* *of* J;
3. y^* *is the unique fixed point of* J *in the set* $Y = \{y \in X \mid d(J^{n_0} x, y) < \infty\}$;
4. $d(y, y^*) \leq \frac{1}{1-L} d(y, Jy)$ *for all* $y \in Y$.

In 1996, Isac and Th.M. Rassias [14] were the first to provide applications of stability theory of functional equations for the proof of new fixed point theorems with applications. By using fixed point methods, the stability problems of several functional equations have been extensively investigated by a number of authors (see [5–29]).

In the sequel, we shall adopt the usual terminology, notions, and conventions of the theory of random normed spaces. Throughout this paper, the space of all probability distribution functions is denoted by Δ^+. Elements of Δ^+ are functions $F: \mathscr{R} \cup [-\infty, +\infty] \to [0,1]$ such that F is left continuous and nondecreasing on $\mathscr{R}$ and $F(0) = 0$, $F(+\infty) = 1$. It is clear that the subset

$$D^+ = \big\{F \in \Delta^+ : l^- F(-\infty) = 1\big\},$$

where $l^- f(x) = \lim_{t \to x^-} f(t)$, is a subset of Δ^+. The space Δ^+ is partially ordered by the usual pointwise ordering of functions, that is, for all $t \in \mathscr{R}$, $F \leq G$ if and only if $F(t) \leq G(t)$.

Definition 22.2 ([37]) A function $T: [0,1]^2 \to [0,1]$ is a continuous triangular norm (briefly a t-norm) if T satisfies the following conditions:

(i) T is commutative and associative;
(ii) T is continuous;
(iii) $T(x,1) = x$ for all $x \in [0,1]$;
(iv) $T(x,y) \leq T(z,w)$ whenever $x \leq z$ and $y \leq w$ for all $x,y,z,w \in [0,1]$.

Three typical examples of continuous t-norms are $T_P(x,y) = xy$, $T_M(x,y) = \min(a,b)$ and $T_L(x,y) = \max\{a+b-1, 0\}$ (the Lukasiewicz t-norm).

Recall that if T is a t-norm and $\{x_n\}$ is a given sequence of numbers in $[0,1]$, $T_{i=1}^n x_i$ is defined recursively by $T_{i=1}^1 x_1$ and $T_{i=1}^n x_i = T(T_{i=1}^{n-1} x_i, x_n)$ for $n \geq 2$. $T_{i=n}^\infty x_i$ is defined as $T_{i=1}^\infty x_{n+i}$.

It is known that for the Lukasiewicz t-norm the following implication holds:

$$\lim_{n\to\infty} (T_L)_{i=1}^{\infty} x_{n+i} = 1 \Leftrightarrow \sum_{n=1}^{\infty}(1 - x_n) < \infty. \tag{22.7}$$

Example 22.1 For any $a \geq 0$, the element $H_a(t)$ of D^+ is defined by

$$H_a(t) = \begin{cases} 0 & \text{if } t \leq a, \\ 1 & \text{if } t > a. \end{cases}$$

We can easily show that the maximal element in Δ^+ is the distribution function $H_0(t)$.

Definition 22.3 ([37]) A random normed space (briefly *RN*-space) is a triple (X, μ, T), where X is a vector space, T is a continuous t-norm, and $\mu : X \to D^+$ is a mapping such that the following conditions hold:

1. $\mu_x(t) = H_0(t)$ for all $t \geq 0$ if and only if $x = 0$;
2. $\mu_{\alpha x}(t) = \mu_x(\frac{t}{|\alpha|})$ for all $\alpha \in \mathscr{R}$, $\alpha \neq 0$, $x \in X$ and $t \geq 0$;
3. $\mu_{x+y}(t+s) \geq T(\mu_x(t), \mu_y(s))$, for all $x, y \in X$ and $t, s \geq 0$.

Definition 22.4 Let (X, μ, T) be an RN-space.

1. A sequence $\{x_n\}$ in X is said to be convergent to $x \in X$ if for all $t > 0$, $\lim_{n\to\infty} \mu_{x_n - x}(t) = 1$.
2. A sequence $\{x_n\}$ in X is said to be a Cauchy sequence in X if for all $t > 0$, $\lim_{n\to\infty} \mu_{x_n - x_m}(t) = 1$.
3. The *RN*-space (X, μ, T) is said to be complete if every Cauchy sequence in X is convergent.

Theorem 22.3 ([37]) *If (X, μ, T) is an RN-space and $\{x_n\}$ is a sequence such that $x_n \to x$, then* $\lim_{n\to\infty} \mu_{x_n}(t) = \mu_x(t)$.

The theory of random normed spaces (RN-spaces) is important as a generalization of deterministic linear normed spaces and also in the study of random operator equations. RN-spaces may also provide us the appropriate tools to study the geometry of nuclear physics and have important application in quantum particle physics. The generalized Hyers–Ulam stability of different functional equations in random normed spaces, RN-spaces and fuzzy normed spaces, has been recently studied in Alsina [1], Mirmostafaee, Mirzavaziri, and Moslehian [23–26], Mihet and Radu [17–20], Mihet, Saadati, and Vaezpour [21, 22], Baktash et al. [3], and Saadati et al. [36].

22.2 Random Stability of Functional Equation (22.3): The Odd Case

Remark 22.1 For a given mapping $f : X \to Y$, we define

$$\eta_f(x, y) = f(x+2y) + f(x-2y) - 4f(x+y) - 4f(x-y) - 6f(x) \\ - f(2y) - f(-2y) + 4f(y) + 4f(-y)$$

for all $x, y \in X$.

In this section, we prove the generalized Hyers–Ulam stability of the functional equation $\eta(x, y) = 0$ in complete random normed spaces in the odd case.

Theorem 22.4 *Let X be a linear space, (Y, μ, T_M) be a complete RN-space, and let Λ be a mapping from X^2 to $D^+(\Lambda(x, y)$ is denoted by $\Lambda_{x,y})$ such that for some $0 < \alpha < \frac{1}{8}$*

$$\Lambda_{2x,2y}(t) \le \Lambda_{x,y}(\alpha t) \tag{22.8}$$

for all $x, y \in X$ and all $t > 0$. Let $f : X \to Y$ be an odd mapping satisfying

$$\mu_{\eta_f(x,y)}(t) \ge \Lambda_{x,y}(t) \tag{22.9}$$

for all $x, y \in X$ and all $t > 0$. Then

$$T(x) := \lim_{n\to\infty} 8^n \left(f\left(\frac{x}{2^{n-1}}\right) - 2f\left(\frac{x}{2^n}\right)\right) \tag{22.10}$$

exists for each $x \in X$ and defines a unique cubic mapping $T : X \to Y$ such that

$$\mu_{f(2x)-2f(x)-T(x)}(t) \ge T_M\left(\Lambda_{x,x}\left(\frac{(1-8\alpha)t}{5\alpha}\right), \Lambda_{2x,x}\left(\frac{(1-8\alpha)t}{5\alpha}\right)\right) \tag{22.11}$$

for all $x \in X$ and all $t > 0$.

Proof Putting $x = y$ in (22.9), we have

$$\mu_{f(3y)-4f(2y)+5f(y)}(t) \ge \Lambda_{y,y}(t) \tag{22.12}$$

for all $y \in X$. Replacing x by $2y$ in (22.9), we get

$$\mu_{f(4y)-4f(3y)+6f(2y)-4f(y)}(t) \ge \Lambda_{2y,y}(t) \tag{22.13}$$

for all $y \in X$. By (22.12) and (22.13), we obtain

$$\begin{aligned} &\mu_{f(4y)-10f(2y)+16f(y)}(5t) \\ &\quad \ge T_M\left(\mu_{f(3y)-4f(2y)+5f(y)}(4t), \mu_{f(4y)-4f(3y)+6f(2y)-4f(y)}(t)\right) \\ &\quad \ge T_M\left(\Lambda_{y,y}(t), \Lambda_{2y,y}(t)\right) \end{aligned} \tag{22.14}$$

for all $y \in X$. Letting $y := \frac{x}{2}$ and $g(x) := f(2x) - 2f(x)$ for all $x \in X$, we get

$$\mu_{g(x)-8g(\frac{x}{2})}(5t) \geq T_M\big(\Lambda_{\frac{x}{2},\frac{x}{2}}(t), \Lambda_{x,\frac{x}{2}}(t)\big) \tag{22.15}$$

for all $x \in X$ and all $t > 0$.

Consider the set

$$S := \{h : X \to Y\} \tag{22.16}$$

and introduce the generalized metric on S by

$$d(h,k) = \inf\big\{u \in \mathscr{R}^+ : \mu_{h(x)-k(x)}(ut) \geq T_M\big(\Lambda_{x,x}(t), \Lambda_{2x,x}(t)\big), \forall x \in X,\ \forall t > 0\big\} \tag{22.17}$$

where, as usual, $\inf \emptyset = +\infty$. It is easy to show that (S,d) is complete (see the proof of Lemma 2.1 in [20]).

Now we consider a linear mapping $J : S \to S$ such that

$$Jg(x) := 8g\left(\frac{x}{2}\right) \tag{22.18}$$

for all $x \in X$ and prove that J is a strictly contractive mapping with the Lipschitz constant 8α.

Let $g, h \in S$ be given such that $d(g,h) < \varepsilon$. Then

$$\mu_{g(x)-h(x)}(\varepsilon t) \geq T_M\big(\Lambda_{x,x}(t), \Lambda_{2x,x}(t)\big) \tag{22.19}$$

for all $x \in X$ and all $t > 0$. Hence

$$\begin{aligned}
\mu_{Jh(x)-Jk(x)}(8\alpha\varepsilon t) &= \mu_{8h(\frac{x}{2})-8k(\frac{x}{2})}(8\alpha\varepsilon t) \\
&= \mu_{h(\frac{x}{2})-k(\frac{x}{2})}(\alpha\varepsilon t) \\
&\geq T_M\big(\Lambda_{\frac{x}{2},\frac{x}{2}}(\alpha t), \Lambda_{x,\frac{x}{2}}(\alpha t)\big) \\
&\geq T_M\big(\Lambda_{x,x}(t), \Lambda_{2x,x}(t)\big)
\end{aligned} \tag{22.20}$$

for all $x \in X$ and all $t > 0$. So $d(g,h) < \varepsilon$ implies that $d(Jg, Jh) < 8\alpha\varepsilon$. This means that

$$d(Jg, Jh) \leq 8\alpha d(g,h) \tag{22.21}$$

for all $g, h \in S$.

It follows from (22.15) that

$$\mu_{g(x)-8g(\frac{x}{2})}(5\alpha t) \geq T_M\big(\Lambda_{x,x}(t), \Lambda_{2x,x}(t)\big) \tag{22.22}$$

for all $x \in X$ and all $t > 0$. So

$$d(g, Jg) \leq 5\alpha < \frac{5}{8} \leq \infty. \tag{22.23}$$

By Theorem 22.2, there exists a mapping $T : X \to Y$ satisfying the following:

1. T is a fixed point of J, that is,

$$T\left(\frac{x}{2}\right) = \frac{1}{8}T(x) \tag{22.24}$$

for all $x \in X$. The mapping T is a unique fixed point of J in the set

$$\Omega = \{h \in S : d(g,h) < \infty\}. \tag{22.25}$$

This implies that T is a unique mapping satisfying (22.24) such that there exists a $u \in (0,\infty)$ satisfying

$$\mu_{g(x)-T(x)}(ut) \geq T_M\big(\Lambda_{x,x}(t), \Lambda_{2x,x}(t)\big) \tag{22.26}$$

for all $x \in X$ and all $t > 0$.

2. $d(J^n g, T) \to 0$ as $n \to \infty$. This implies the equality

$$\lim_{n\to\infty} 8^n g\left(\frac{x}{2^n}\right) = \lim_{n\to\infty} 8^n\left(f\left(\frac{x}{2^{n-1}}\right) - 2f\left(\frac{x}{2^n}\right)\right) = T(x) \tag{22.27}$$

for all $x \in X$. Since $f : X \to Y$ is odd, $T : X \to Y$ is an odd mapping.

3. $d(g,T) \leq \frac{d(g,Jg)}{1-8\alpha}$, which implies the inequality

$$d(g,T) \leq \frac{5\alpha}{1-8\alpha}, \tag{22.28}$$

from which it follows that

$$\mu_{g(x)-T(x)}\left(\frac{5\alpha t}{1-8\alpha}\right) = \mu_{f(2x)-2f(x)-T(x)}\left(\frac{5\alpha t}{1-8\alpha}\right) \geq T_M\big(\Lambda_{x,x}(t), \Lambda_{2x,x}(t)\big) \tag{22.29}$$

for all $x \in X$ and all $t > 0$. This implies that the inequality (22.11) holds.

On the other hand, by (22.8) and (22.9),

$$\begin{aligned}
\mu_{8^n \eta_g(\frac{x}{2^n},\frac{x}{2^n})}(t) &= \mu_{\eta_g(\frac{x}{2^n},\frac{x}{2^n})}\left(\frac{t}{8^n}\right) \\
&= \mu_{\eta_f(\frac{2x}{2^n},\frac{2y}{2^n})-2\eta_f(\frac{x}{2^n},\frac{y}{2^n})}\left(\frac{t}{8^n}\right) \\
&\geq T_M\left(\mu_{\eta_f(\frac{2x}{2^n},\frac{2y}{2^n})}\left(\frac{t}{2\cdot 8^n}\right), \mu_{\eta_f(\frac{x}{2^n},\frac{y}{2^n})}\left(\frac{t}{4\cdot 8^n}\right)\right) \\
&\geq T_M\left(\Lambda_{\frac{2x}{2^n},\frac{2y}{2^n}}\left(\frac{t}{2\cdot 8^n}\right), \Lambda_{\frac{x}{2^n},\frac{y}{2^n}}\left(\frac{t}{4\cdot 8^n}\right)\right) \\
&\geq T_M\left(\Lambda_{2x,2y}\left(\frac{t}{2\cdot(8\alpha)^n}\right), \Lambda_{x,y}\left(\frac{t}{4\cdot(8\alpha)^n}\right)\right)
\end{aligned} \tag{22.30}$$

for all $x, y \in X$, all $t > 0$, and all $n \in \mathcal{N}$. Since $\lim_{n\to\infty} \Lambda_{2x,2y}(\frac{t}{2\cdot(8\alpha)^n}) = 1$ and $\lim_{n\to\infty} \Lambda_{x,x}(\frac{t}{4\cdot(8\alpha)^n}) = 1$ for all $x, y \in X$ and all $t > 0$, by Theorem 22.3, we deduce that

$$\mu_{\eta T(x,y)}(t) = 1 \tag{22.31}$$

for all $x, y \in X$ and all $t > 0$. Thus the mapping $T : X \to Y$ is cubic, as desired. □

Corollary 22.1 *Let $\theta \geq 0$ and let p be a real number with $p > 1$. Let X be a normed vector space with norm $\|\cdot\|$. Let $f : X \to Y$ be an odd mapping satisfying*

$$\mu_{\eta f(x,y)}(t) \geq \frac{t}{t + \theta(\|x\|^p + \|y\|^p)} \tag{22.32}$$

for all $x, y \in X$ and all $t > 0$. Then

$$T(x) := \lim_{n\to\infty} 8^n \left(f\left(\frac{x}{2^{n-1}}\right) - 2f\left(\frac{x}{2^n}\right)\right) \tag{22.33}$$

exists for each $x \in X$ and defines a unique cubic mapping $T : X \to Y$ such that

$$\mu_{f(2x)-2f(x)-T(x)}(t) \geq \frac{(8^p - 8)t}{(8^p - 8)t + 5(1 + 2^p)\theta\|x\|^p}. \tag{22.34}$$

Proof The proof follows from Theorem 22.4 by taking

$$\Lambda_{x,y}(t) = \frac{t}{t + \theta(\|x\|^p + \|y\|^p)} \tag{22.35}$$

for all $x, y \in X$ and all $t > 0$. Then we can choose $\alpha = 8^{-p}$ and we get the desired result. □

Similarly, we can obtain the following. We will omit the proof.

Theorem 22.5 *Let X be a linear space, (Y, μ, T_M) be a complete RN-space, and let Λ be a mapping from X^2 to D^+ ($\Lambda(x, y)$ is denoted by $\Lambda_{x,y}$) such that for some $0 < \alpha < 8$*

$$\Lambda_{\frac{x}{2},\frac{y}{2}}(t) \leq \Lambda_{x,y}(\alpha t) \tag{22.36}$$

for all $x, y \in X$ and all $t > 0$. Let $f : X \to Y$ be an odd mapping satisfying

$$\mu_{\eta f(x,y)}(t) \geq \Lambda_{x,y}(t) \tag{22.37}$$

for all $x, y \in X$ and all $t > 0$. Then

$$T(x) := \lim_{n\to\infty} \left[\frac{f(2^{n+1}x) - 2f(2^n x)}{8^n}\right] \tag{22.38}$$

exists for each $x \in X$ and defines a unique cubic mapping $T : X \to Y$ such that

$$\mu_{f(2x)-2f(x)-T(x)}(t) \geq T_M\left(\Lambda_{x,x}\left(\frac{(8-\alpha)t}{5}\right), \Lambda_{2x,x}\left(\frac{(8-\alpha)t}{5}\right)\right) \tag{22.39}$$

for all $x \in X$ and all $t > 0$.

Corollary 22.2 *Let $\theta \geq 0$ and let p be a real number with $0 < p < 1$. Let X be a normed vector space with norm $\|\cdot\|$. Let $f : X \to Y$ be an odd mapping satisfying*

$$\mu_{\eta_f(x,y)}(t) \geq \frac{t}{t + \theta(\|x\|^p + \|y\|^p)} \tag{22.40}$$

for all $x, y \in X$ and all $t > 0$. Then

$$T(x) := \lim_{n\to\infty}\left[\frac{f(2^{n+1}x) - 2f(2^n x)}{8^n}\right] \tag{22.41}$$

exists for each $x \in X$ and defines a unique cubic mapping $T : X \to Y$ such that

$$\mu_{f(2x)-2f(x)-T(x)}(t) \geq \frac{(8-8^p)t}{(8-8^p)t + 5(1+2^p)\theta\|x\|^p}. \tag{22.42}$$

Proof The proof follows from Theorem 22.5 by taking

$$\Lambda_{x,y}(t) = \frac{t}{t + \theta(\|x\|^p + \|y\|^p)} \tag{22.43}$$

for all $x, y \in X$ and all $t > 0$. Then we can choose $\alpha = 8^p$ and we get the desired result. □

Theorem 22.6 *Let X be a linear space, (Y, μ, T_M) be a complete RN-space, and let Λ be a mapping from X^2 to D^+($\Lambda(x, y)$ is denoted by $\Lambda_{x,y}$) such that for some $0 < \alpha < \frac{1}{2}$*

$$\Lambda_{2x,2y}(t) \leq \Lambda_{x,y}(\alpha t) \tag{22.44}$$

for all $x, y \in X$ and all $t > 0$. Let $f : X \to Y$ be an odd mapping satisfying

$$\mu_{\eta_f(x,y)}(t) \geq \Lambda_{x,y}(t) \tag{22.45}$$

for all $x, y \in X$ and all $t > 0$. Then

$$T(x) := \lim_{n\to\infty} 2^n\left(f\left(\frac{x}{2^{n-1}}\right) - 8f\left(\frac{x}{2^n}\right)\right) \tag{22.46}$$

exists for each $x \in X$ and defines a unique additive mapping $T : X \to Y$ such that

$$\mu_{f(2x)-8f(x)-T(x)}(t) \geq T_M\left(\Lambda_{x,x}\left(\frac{(1-2\alpha)t}{5\alpha}\right), \Lambda_{2x,x}\left(\frac{(1-2\alpha)t}{5\alpha}\right)\right) \tag{22.47}$$

for all $x \in X$ and all $t > 0$.

Proof Let (S,d) be the generalized metric space defined in the proof of Theorem 22.4. Letting $y=\frac{x}{2}$ and $h(x)=f(2x)-8f(x)$ for all $x\in X$ in (22.14), we get

$$\mu_{h(x)-2h(\frac{x}{2})}(5t)\ge T_M\left(\Lambda_{\frac{x}{2},\frac{x}{2}}(t),\Lambda_{x,\frac{x}{2}}(t)\right) \tag{22.48}$$

for all $x\in X$ and all $t>0$. Now we consider the linear mapping $J:S\to S$ such that

$$Jh(x)=2h\left(\frac{x}{2}\right) \tag{22.49}$$

for all $x\in X$. It is easy to see that J is a strictly contractive self-mapping on S with the Lipschitz constant 2α. It follows from (22.48) that $d(h,Jh)\le 5\alpha<\infty$. By Theorem 22.2, there exists a mapping $T:X\to Y$ satisfying the following:
(i) T is a fixed point of J, that is, for all $x\in X$

$$T\left(\frac{x}{2}\right)=\frac{1}{2}T(x). \tag{22.50}$$

Since $f:X\to Y$ is odd, $T:X\to Y$ is odd. The mapping T is a unique fixed point of J in the set

$$M=\{h\in S;\ d(g,h)<\infty\}. \tag{22.51}$$

This implies that T is a unique mapping satisfying (22.50) such that there exists a $u\in(0,\infty)$ satisfying

$$\mu_{g(x)-T(x)}(ut)\ge T_M\left(\Lambda_{x,x}(t),\Lambda_{2x,x}(t)\right) \tag{22.52}$$

for all $x\in X$ and all $t>0$;
(ii) $d(J^nh,T)\to 0$ as $n\to\infty$. This implies that

$$\lim_{n\to\infty}2^nh\left(\frac{x}{2^n}\right)=\lim_{n\to\infty}2^n\left(f\left(\frac{x}{2^{n-1}}\right)-8f\left(\frac{x}{2^n}\right)\right)=T(x) \tag{22.53}$$

for all $x\in X$.
(iii) $d(h,T)\le\frac{d(h,Jh)}{1-2\alpha}$ for every $h\in M$, which implies the inequality

$$d(h,T)\le\frac{5\alpha}{1-2\alpha}. \tag{22.54}$$

This implies that the inequality (22.47) holds. Proceeding as in the proof of the Theorem 22.4, we obtain that the mapping $T:X\to Y$ satisfies $f(x+2y)+f(x-2y)=4f(x+y)+4f(x-y)-6f(x)+f(2y)+f(-2y)-4f(y)-4f(-y)$.

On the other hand,

$$\begin{aligned} T(2x) - 2T(x) &= \left[\lim_{n\to\infty} 2^n h\left(\frac{x}{2^{n-1}}\right) - 2\lim_{n\to\infty} 2^n h\left(\frac{x}{2^n}\right)\right] \\ &= 2\left[\lim_{n\to\infty} 2^{n-1} h\left(\frac{x}{2^{n-1}}\right) - \lim_{n\to\infty} 2^n h\left(\frac{x}{2^n}\right)\right] = 0 \quad (22.55) \end{aligned}$$

for every $x \in X$. So, we obtain that the mapping $T : X \to Y$ is additive. □

22.2.1 Random Stability of Functional Equation (22.3): The Even Case

Using the fixed point method, we prove the generalized Hyers–Ulam stability of the functional equation $\eta_f(x, y) = 0$ in random Banach spaces in the even case.

Theorem 22.7 *Let X be a linear space, (Y, μ, T_M) be a complete RN-space, and let Λ be a mapping from X^2 to D^+ ($\Lambda(x, y)$ is denoted by $\Lambda_{x,y}$) such that for some $0 < \alpha < \frac{1}{16}$*

$$\Lambda_{x,y}(\alpha t) \geq \Lambda_{2x,2y}(t) \tag{22.56}$$

for all $x, y \in X$ and all $t > 0$. Let $f : X \to Y$ be an even mapping with $f(0) = 0$ satisfying

$$\mu_{\eta_f(x,y)}(t) \geq \Lambda_{x,y}(t) \tag{22.57}$$

for all $x, y \in X$ and all $t > 0$. Then

$$T(x) := \lim_{n\to\infty} 16^n \left(f\left(\frac{x}{2^{n-1}}\right) - 4f\left(\frac{x}{2^n}\right)\right) \tag{22.58}$$

exists for each $x \in X$ and defines a unique quartic mapping $T : X \to Y$ such that

$$\mu_{f(2x)-4f(x)-T(x)}(t) \geq T_M\left(\Lambda_{x,x}\left(\frac{(1-16\alpha)t}{5\alpha}\right), \Lambda_{2x,x}\left(\frac{(1-16\alpha)t}{5\alpha}\right)\right) \tag{22.59}$$

for all $x \in X$ and all $t > 0$.

Proof Letting $x = y$ in (22.57), we get

$$\mu_{f(3y)-6f(2y)+15f(y)}(t) \geq \Lambda_{y,y}(t) \tag{22.60}$$

for all $y \in X$ and all $t > 0$. Replacing x by $2y$ in (22.57), we get

$$\mu_{f(4y)-4f(3y)+4f(2y)+4f(y)}(t) \geq \Lambda_{2y,y}(t) \tag{22.61}$$

for all $y \in X$ and all $t > 0$. By (22.60), (22.61), and replacing y by x, we have

$$\begin{aligned}\mu_{f(4x)-20f(2x)+64f(x)}(5t) &\geq T_M\big(\mu_{4(f(3x)-6f(2x)+15f(x))}(4t),\\ &\qquad \mu_{f(4x)-4f(3x)+4f(2x)+4f(x)}(t)\big)\\ &\geq T_M\big(\Lambda_{x,x}(t), \Lambda_{2x,x}(t)\big)\end{aligned} \tag{22.62}$$

for all $x \in X$ and all $t > 0$. Letting $g(x) := f(2x) - 4f(x)$ for all $x \in X$, we get

$$\mu_{g(x)-16g(\frac{x}{2})}(5t) \geq T_M\big(\Lambda_{\frac{x}{2},\frac{x}{2}}(t), \Lambda_{x,\frac{x}{2}}(t)\big) \tag{22.63}$$

for all $x \in X$ and all $t > 0$. Let (S, d) be the generalized metric space defined in the proof of Theorem 22.4. Now we consider a linear mapping $J : S \to S$ such that

$$Jg(x) := 16g\left(\frac{x}{2}\right) \tag{22.64}$$

for all $x \in X$. It is easy to see that J is a strictly contractive self-mapping on S with the Lipschitz constant 16α.

It follows form (22.63) that for all $x \in X$ and all $t > 0$

$$\mu_{g(x)-16g(\frac{x}{2})}(5\alpha t) \geq T_M\big(\Lambda_{x,x}(t), \Lambda_{2x,x}(t)\big). \tag{22.65}$$

So

$$d(g, dg) \leq 5\alpha \leq \frac{5}{16} < \infty. \tag{22.66}$$

The rest of the proof is similar to the proof of Theorem 22.4. □

Corollary 22.3 *Let $\theta \geq 0$ and let p be a real number with $p > 4$. Let X be a normed vector space with norm $\|\cdot\|$. Let $f : X \to Y$ be an even mapping with $f(0) = 0$ satisfying*

$$\mu_{\eta f(x,y)}(t) \geq \frac{t}{t + \theta(\|x\|^p + \|y\|^p)} \tag{22.67}$$

for all $x, y \in X$ and all $t > 0$. Then

$$T(x) := \lim_{n\to\infty} 16^n\left(f\left(\frac{x}{2^{n-1}}\right) - 4f\left(\frac{x}{2^n}\right)\right) \tag{22.68}$$

exists for each $x \in X$ and defines a unique quartic mapping $T : X \to Y$ such that

$$\mu_{f(2x)-4f(x)-T(x)}(t) \geq \frac{(2^p - 16)t}{(2^p - 16)t + 5(1 + 2^p)\theta\|x\|^p}. \tag{22.69}$$

Proof The proof follows from Theorem 22.7 by taking

$$\Lambda_{x,y}(t) = \frac{t}{t + \theta(\|x\|^p + \|y\|^p)} \tag{22.70}$$

for all $x, y \in X$ and all $t > 0$. Then we can choose $\alpha = 2^{-p}$ and we get the desired result. □

Similarly, we can obtain the following. We will omit the proof.

Theorem 22.8 *Let X be a linear space, (Y, μ, T_M) be a complete RN-space, and let Λ be a mapping from X^2 to D^+ ($\Lambda(x, y)$ is denoted by $\Lambda_{x,y}$) such that for some $0 < \alpha < 16$*

$$\Lambda_{x,y}(\alpha t) \geq \Lambda_{\frac{x}{2},\frac{y}{2}}(t) \tag{22.71}$$

for all $x, y \in X$ and all $t > 0$. Let $f : X \to Y$ be an even mapping with $f(0) = 0$ satisfying

$$\mu_{\eta_f(x,y)}(t) \geq \Lambda_{x,y}(t) \tag{22.72}$$

for all $x, y \in X$ and all $t > 0$. Then

$$T(x) := \lim_{n\to\infty}\left[\frac{f(2^{n+1}x) - 4f(2^n x)}{16^n}\right] \tag{22.73}$$

exists for each $x \in X$ and defines a unique quartic mapping $T : X \to Y$ such that

$$\mu_{f(2x)-4f(x)-T(x)}(t) \geq T_M\left(\Lambda_{x,x}\left(\frac{(16-\alpha)t}{5}\right), \Lambda_{2x,x}\left(\frac{(16-\alpha)t}{5}\right)\right) \tag{22.74}$$

for all $x, y \in X$ and all $t > 0$.

Corollary 22.4 *Let $\theta \geq 0$ and let p be a real number with $0 < p < 4$. Let X be a normed vector space with norm $\|\cdot\|$. Let $f : X \to Y$ be an even mapping satisfying $f(0) = 0$ and*

$$\mu_{\eta_f(x,y)}(t) \geq \frac{t}{t + \theta(\|x\|^p + \|y\|^p)} \tag{22.75}$$

for all $x, y \in X$ and all $t > 0$. Then

$$T(x) := \left[\frac{f(2^{n+1}x) - 4f(2^n x)}{16^n}\right] \tag{22.76}$$

exists for each $x \in X$ and defines a unique quartic mapping $T : X \to Y$ such that

$$\mu_{f(2x)-4f(x)-T(x)}(t) \geq \frac{(16-2^p)t}{(16-2^p)t + 5(1+2^p)\theta\|x\|^p}. \tag{22.77}$$

Proof The proof follows from Theorem 22.8 by taking

$$\Lambda_{x,y}(t) = \frac{t}{t + \theta(\|x\|^p + \|y\|^p)} \tag{22.78}$$

for all $x, y \in X$ and all $t > 0$. Then we can choose $\alpha = 2^p$ and we get the desired result. □

Theorem 22.9 *Let X be a linear space, (Y, μ, T_M) be a complete RN-space, and let Λ be a mapping from X^2 to D^+ ($\Lambda(x, y)$ is denoted by $\Lambda_{x,y}$) such that for some $0 < \alpha < \frac{1}{4}$*

$$\Lambda_{x,y}(\alpha t) \le \Lambda_{2x,2y}(t) \tag{22.79}$$

for all $x, y \in X$ and all $t > 0$. Let $f : X \to Y$ be an even mapping with $f(0) = 0$ satisfying

$$\mu_{\eta_f(x,y)}(t) \ge \Lambda_{x,y}(t) \tag{22.80}$$

for all $x, y \in X$ and all $t > 0$. Then

$$T(x) := \lim_{n\to\infty} 4^n\left(f\left(\frac{x}{2^{n-1}}\right) - 16f\left(\frac{x}{2^n}\right)\right) \tag{22.81}$$

exists for each $x \in X$ and defines a unique quadratic mapping $T : X \to Y$ such that

$$\mu_{f(2x)-16f(x)-T(x)}(t) \ge T_M\left(\Lambda_{x,x}\left(\frac{(1-4\alpha)t}{5\alpha}\right), \Lambda_{2x,x}\left(\frac{(1-4\alpha)t}{5\alpha}\right)\right) \tag{22.82}$$

for all $x \in X$ and all $t > 0$.

Proof Let (S, d) be the generalized metric space defined in the proof of Theorem 22.7, $g(x) := f(2x) - 16f(x)$ for all $x \in X$, and let $J : S \to S$ be defined by $Jh(x) := 4h(\frac{x}{2})$. The rest of the proof is similar to the proof of Theorem 22.7. □

Corollary 22.5 *Let $\theta \ge 0$ and let p be a real number with $p > 2$. Let X be a normed vector space with norm $\|\cdot\|$. Let $f : X \to Y$ be an even mapping with $f(0) = 0$ satisfying*

$$\mu_{\eta_f(x,y)}(t) \ge \frac{t}{t + \theta(\|x\|^p + \|y\|^p)} \tag{22.83}$$

for all $x, y \in X$ and all $t > 0$. Then

$$T(x) := \lim_{n\to\infty} 4^n\left(f\left(\frac{x}{2^{n-1}}\right) - 16f\left(\frac{x}{2^n}\right)\right) \tag{22.84}$$

exists for each $x \in X$ and defines a unique quadratic mapping $T : X \to Y$ such that

$$\mu_{f(2x)-16f(x)-T(x)}(t) \ge \frac{(2^p-4)t}{(2^p-4)t + 5(1+2^p)\theta\|x\|^p}. \tag{22.85}$$

Proof The proof follows from Theorem 22.9 by taking

$$\Lambda_{x,y}(t) = \frac{t}{t + \theta(\|x\|^p + \|y\|^p)} \tag{22.86}$$

for all $x, y \in X$ and all $t > 0$. Then we can choose $\alpha = 2^{-p}$ and we get the desired result. □

Similarly, we can obtain the following. We omit the proof.

Theorem 22.10 *Let X be a linear space, (Y, μ, T_M) be a complete RN-space, and let Λ be a mapping from X^2 to D^+ ($\Lambda(x, y)$ is denoted by $\Lambda_{x,y}$) such that for some $0 < \alpha < 4$*

$$\Lambda_{x,y}(\alpha t) \leq \Lambda_{\frac{x}{2},\frac{y}{2}}(t) \tag{22.87}$$

for all $x, y \in X$ and all $t > 0$. Let $f : X \to Y$ be an even mapping with $f(0) = 0$ satisfying

$$\mu_{\eta_f(x,y)}(t) \geq \Lambda_{x,y}(t) \tag{22.88}$$

for all $x, y \in X$ and all $t > 0$. Then

$$T(x) := \lim_{n\to\infty}\left[\frac{f(2^{n+1}x) - 16f(2^n x)}{4^n}\right] \tag{22.89}$$

exists for each $x \in X$ and defines a unique quadratic mapping $T : X \to Y$ such that

$$\mu_{f(2x)-16f(x)-T(x)}(t) \geq T_M\left(\Lambda_{x,x}\left(\frac{(4-\alpha)t}{5}\right), \Lambda_{2x,x}\left(\frac{(4-\alpha)t}{5}\right)\right) \tag{22.90}$$

for all $x \in X$ and all $t > 0$.

Corollary 22.6 *Let $\theta \geq 0$ and let p be a real number with $0 < p < 2$. Let X be a normed vector space with norm $\|\cdot\|$. Let $f : X \to Y$ be an even mapping satisfying $f(0) = 0$ and*

$$\mu_{\eta_f(x,y)}(t) \geq \frac{t}{t + \theta(\|x\|^p + \|y\|^p)} \tag{22.91}$$

for all $x, y \in X$ and all $t > 0$. Then

$$T(x) := \lim_{n\to\infty}\left[\frac{f(2^{n+1}x) - 16f(2^n x)}{4^n}\right] \tag{22.92}$$

exists for each $x \in X$ and defines a unique quadratic mapping $T : X \to Y$ such that

$$\mu_{f(2x)-16f(x)-T(x)}(t) \geq \frac{(4-2^p)t}{(4-2^p)t + 5(1+2^p)\theta\|x\|^p}. \tag{22.93}$$

Proof The proof follows from Theorem 22.10 by taking

$$\Lambda_{x,y}(t) = \frac{t}{t + \theta(\|x\|^p + \|y\|^p)} \tag{22.94}$$

for all $x, y \in X$ and all $t > 0$. Then we can choose $\alpha = 2^p$ and we get the desired result. □

References

1. Alsina, C.: On the stability of a functional equation arising in probabilistic normed spaces. In: General Inequalities. Internationale Schriftenreihe zur Numerischen Mathematik, vol. 80, pp. 263–271. Birkhäuser, Basel (1987)
2. Aoki, T.: On the stability of the linear transformation in Banach spaces. J. Math. Soc. Jpn. **2**, 64–66 (1950)
3. Baktash, E., Cho, Y.J., Jalili, M., Saadati, R., Vaezpour, S.M.: On the stability of cubic mappings and quadratic mappings in random normed spaces. J. Inequal. Appl. **2008**, 902187 (2008). 11 pages
4. Cădariu, L., Radu, V.: Fixed points and the stability of Jensen functional equation. J. Inequal. Pure Appl. Math. **4**(1), 4 (2003)
5. Cadariu, L., Radu, V.: On the stability of the Cauchy functional equation: a fixed point approach. In: Iteration Theory. Grazer Mathematische Berichte, vol. 346, pp. 43–52 (2004). Karl-Franzens-Universität Graz, Graz
6. Cadariu, L., Radu, V.: Fixed point methods for the generalized stability of functional equations in a single variable. Fixed Point Theory Appl. **2008**, 749392 (2008). 15 pages
7. Cholewa, P.W.: Remarks on the stability of functional equations. Aequ. Math. **27**(1–2), 76–86 (1984)
8. Czerwik, S.: On the stability of the quadratic mapping in normed spaces. Abh. Math. Semin. Univ. Hamb. **62**, 59–64 (1992)
9. Eshaghi-Gordji, M., Kaboli-Gharetapeh, S., Park, C., Zolfaghri, S.: Stability of an additive-cubic–quartic functional equation. Adv. Differ. Equ. **2009**, 395693 (2009). 20 pages
10. Eshaghi-Gordji, M., Abbaszadeh, S., Park, C.: On the stability of a generalized quadratic and quartic type functional equation in quasi-Banach spaces. J. Inequal. Appl. **2009**, 153084 (2009). 26 pages
11. Găvruta, P.: A generalization of the Hyers–Ulam–Rassias stability of approximately additive mappings. J. Math. Anal. Appl. **184**(3), 431–436 (1994)
12. Hyers, D.H.: On the stability of the linear functional equation. Proc. Natl. Acad. Sci. USA **27**, 222–224 (1941)
13. Hyers, D.H., Isac, G., Rassias, Th.M.: Stability of Functional Equations in Several Variables. Progress in Nonlinear Differential Equations and Their Applications, vol. 34. Birkhäuser, Boston (1998)
14. Isac, G., Rassias, Th.M.: Stability of ψ-additive mappings: applications to nonlinear analysis. Int. J. Math. Math. Sci. **19**(2), 219–228 (1996)
15. Jun, K.-W., Kim, H.-M.: The generalized Hyers–Ulam–Rassias stability of a cubic functional equation. J. Math. Anal. Appl. **274**(2), 267–278 (2002)
16. Lee, S.H., Im, S.M., Hwang, I.S.: Quartic functional equations. J. Math. Anal. Appl. **307**(2), 387–394 (2005)
17. Miheţ, D.: Fuzzy stability of additive mappings in non-Archimedean fuzzy normed spaces. Fuzzy Sets Syst. (in press)
18. Miheţ, D.: The probabilistic stability for a functional equation in a single variable. Acta Math. Hung. **123**(3), 249–256 (2009)
19. Miheţ, D.: The fixed point method for fuzzy stability of the Jensen functional equation. Fuzzy Sets Syst. **160**(11), 1663–1667 (2009)
20. Mihet, D., Radu, V.: On the stability of the additive Cauchy functional equation in random normed spaces. J. Math. Anal. Appl. **343**(1), 567–572 (2008)
21. Miheţ, D., Saadati, R., Vaezpour, S.M.: The stability of the quartic functional equation in random normed spaces. Acta Appl. Math. (in press)
22. Miheţ, D., Saadati, R., Vaezpour, S.M.: The stability of an additive functional equation in Menger probabilistic ϕ-normed spaces. Math. Slovaca **61**(5), 817–826 (2011)
23. Mirmostafaee, A.K., Mirzavaziri, M., Moslehian, M.S.: Fuzzy stability of the Jensen functional equation. Fuzzy Sets Syst. **159**(6), 730–738 (2008)

24. Mirmostafaee, A.K., Moslehian, M.S.: Fuzzy versions of Hyers–Ulam–Rassias theorem. Fuzzy Sets Syst. **159**(6), 720–729 (2008)
25. Mirmostafaee, A.K., Moslehian, M.S.: Fuzzy approximately cubic mappings. Inf. Sci. **178**(19), 3791–3798 (2008)
26. Mirzavaziri, M., Moslehian, M.S.: A fixed point approach to stability of a quadratic equation. Bull. Braz. Math. Soc. **37**(3), 361–376 (2006)
27. Park, C.: Fixed points and Hyers–Ulam–Rassias stability of Cauchy–Jensen functional equations in Banach algebras. Fixed Point Theory Appl. **2007**, 50175 (2007). 15 pages
28. Park, C.: Generalized Hyers–Ulam stability of quadratic functional equations: a fixed point approach. Fixed Point Theory Appl. **2008**, 493751 (2008). 9 pages
29. Radu, V.: The fixed point alternative and the stability of functional equations. Fixed Point Theory **4**(1), 91–96 (2003)
30. Rassias, Th.M.: On the stability of the linear mapping in Banach spaces. Proc. Am. Math. Soc. **72**(2), 297–300 (1978)
31. Rassias, Th.M.: On the stability of the quadratic functional equation and its applications. Stud. Univ. Babeş-Bolyai, Math. **43**(3), 89–124 (1998)
32. Rassias, Th.M.: The problem of S. M. Ulam for approximately multiplicative mappings. J. Math. Anal. Appl. **246**(2), 352–378 (2000)
33. Rassias, Th.M.: On the stability of functional equations in Banach spaces. J. Math. Anal. Appl. **251**(1), 264–284 (2000)
34. Rassias, Th.M.: On the stability of functional equations and a problem of Ulam. Acta Appl. Math. **62**(1), 23–130 (2000)
35. Rassias, Th.M., Semrl, P.: On the behavior of mappings which do not satisfy Hyers–Ulam stability. Proc. Am. Math. Soc. **114**(4), 989–993 (1992)
36. Saadati, R., Vaezpour, S.M., Cho, Y.J.: A note to paper on the stability of cubic mappings and quartic mappings in random normed spaces. J. Inequal. Appl. **2009**, 214530 (2009). 6 pages
37. Schweizer, B., Sklar, A.: Probabilistic Metric Spaces. North-Holland Series in Probability and Applied Mathematics. North-Holland, New York (1983)
38. Skof, F.: Proprieta locali e approssimazione di operatori. Rendiconti del Seminario Matematico e Fisico di Milano **53**, 113–129 (1983)
39. Ulam, S.M.: A Collection of Mathematical Problems. Interscience Tracts in Pure and Applied Mathematics, vol. 8. Interscience, New York (1960)

Chapter 23
Basis Sets in Banach Spaces

S.V. Konyagin and Y.V. Malykhin

Abstract A set M in a linear normed space X over a field K ($\mathsf{K} = \mathbb{R}$ or $\mathsf{K} = \mathbb{C}$) is called a basis set if every $x \in X$ can be represented as a sum $x = \sum_k c_k e_k$, where $e_k \in M$, $e_k \neq e_l$ $(k \neq l)$, $c_k \in \mathsf{K} \setminus \{0\}$, $\sum_k$ denotes either $\sum_{k=1}^{\infty}$ or $\sum_{k=1}^{N}$, and this representation is unique up to permutations. We prove the existence of an infinite-dimensional separable Banach space X with a basis set M such that no arrangement of M forms a Schauder basis.

Key words Basis · Schauder basis · Rearrangement

Mathematics Subject Classification 46Bxx · 42A20 · 39B52

23.1 Introduction

Consider a linear normed space X over a field K, where $\mathsf{K} = \mathbb{R}$ or $\mathsf{K} = \mathbb{C}$. We call a set $M \subset X$ a *basis set* if every $x \in X$ can be represented as a sum

$$x = \sum_k c_k e_k, \quad \text{where } e_k \in M,\ e_k \neq e_l (k \neq l),\ c_k \in \mathsf{K},\ c_k \neq 0, \tag{23.1}$$

$\sum_k$ denotes either $\sum_{k=1}^{\infty}$ or $\sum_{k=1}^{N}$ ($N \in \mathbb{N} \cup \{0\}$), and representation (23.1) is unique up to permutations.

The goal of this paper is to compare the definition of a basis set with the classical notion of a Schauder basis. Recall that a sequence $(e_1, \ldots, e_k, \ldots)$ is called a

---

Dedicated to Professor Themistocles M. Rassias on the occasion of his 60th birthday.

S.V. Konyagin · Y.V. Malykhin (✉)
Department of Function Theory, Steklov Mathematical Institute, Russian Academy of Sciences, Gubkin str. 8, 119991, Moscow, Russia
e-mail: malykhin@mi.ras.ru

S.V. Konyagin
e-mail: konyagin@mi.ras.ru

P.M. Pardalos et al. (eds.), *Nonlinear Analysis*, Springer Optimization and Its Applications 68, In Honor of Themistocles M. Rassias on the Occasion of his 60th Birthday, DOI 10.1007/978-1-4614-3498-6_23,

Schauder basis in a Banach space X if every $x \in X$ has a unique representation

$$x = \sum_{k=1}^{\infty} c_k e_k, \quad c_k \in \mathsf{K}.$$

It is known that any Schauder basis $(e_1, \dots, e_k, \dots)$ is a minimal system, that is, there is a system $(e_1^*, \dots, e_k^*, \dots)$ in the conjugate space X^* such that

$$\langle e_k^*, e_l \rangle = \delta_{k,l} = \begin{cases} 1, & k = l, \\ 0, & k \neq l. \end{cases}$$

Since $\{e_k\}$ is a minimal system, an equality

$$x = \sum_k c_k e_k$$

for any order of summation implies

$$c_k = \langle e_k^*, x \rangle.$$

Therefore, the set of all elements of a Schauder basis is a basis set. Does every basis set form a Schauder basis after some arrangement? The answer is negative.

Theorem 23.1 *There exists an infinite-dimensional separable Banach space X with a basis set M such that no arrangement of M forms a Schauder basis.*

Our attention was brought to this problem by the exposition of lectures of S.B. Stechkin on approximation theory [1]. There a definition of a basis in a normed space was given, that was, presumably, not equivalent to the definition of a Schauder basis. So we called the latter a "basis set" and tried to find the relationship between them.

Let us give one illustrative example. Consider the space $C(\mathbb{T})$ of 2π-periodic continuous functions and the trigonometric system. It is known that no rearrangement of the trigonometric system is a basis in $C(\mathbb{T})$ (see, for example, [2, 3]). But the question whether it is a basis set is still open. It is equivalent to the well-known problem, posed by P.L. Ul'yanov [4]: *Is it true that for every 2π-periodic continuous function there is a uniformly convergent rearrangement of its trigonometric Fourier series?*

We do not know whether any basis set is a minimal system. By the uniform boundedness principle, if a basis set $\{e_k\}$ is a minimal system, then it is also a uniformly minimal system, that is,

$$\sup_k \|e_k^*\| \cdot \|e_k\| < \infty.$$

There is a uniformly minimal system that is not a basis set. This is a simple corollary of the following result [5]: There exists a function $f \in L(\mathbb{T})$ such that any rearrangement of its Fourier series diverges in $L(\mathbb{T})$.

23.2 Proofs

The construction of the example is based on the following lemma.

Lemma 23.1 *For every $n \in \mathbb{N}$ there exists an n-dimensional normed space X_n with basis $\{e_1, \dots, e_n\}$, satisfying the conditions*:

- *For all $x \in X_n$, $x = \sum_{k=1}^n c_k e_k$, there exists a permutation $\sigma\colon \{1, \dots, n\} \to \{1, \dots, n\}$, such that*

$$\left\| \sum_{k=1}^m c_{\sigma_k} e_{\sigma_k} \right\|_{X_n} \le C_1 \|x\|_{X_n}, \quad m = 1, \dots, n; \tag{23.2}$$

- *For every permutation σ there exist a vector $x = \sum_{k=1}^n c_k e_k$ and a number m such that*

$$\left\| \sum_{k=1}^m c_{\sigma_k} e_{\sigma_k} \right\|_{X_n} > C_2 \sqrt{\log n} \cdot \|x\|_{X_n}. \tag{23.3}$$

Here and later we denote by $C_1, C_2, \dots$ positive absolute constants. Let us derive the theorem from this lemma. We combine spaces X_n, $n = 1, 2, \dots$, into the sum-space

$$X = \left\{ (x_1, \dots, x_n, \dots)\colon x_k \in X_k,\ \|x\|_X := \sum_{n=1}^\infty \|x_n\|_{X_n} < \infty \right\}.$$

The union M of bases in all X_n is a basis set in X. Indeed, pick $x \in X$. For every $n \in \mathbb{N}$ one can find a "good" permutation $\sigma^{(n)}$. Arrange M with respect to $\sigma^{(n)}$ for every n, while n goes from 1 to infinity. This arrangement provides an expansion (23.1) for x; the convergence follows from (23.2). One can see from (23.3) that for any arrangement of M the norm of the partial sum operators is not bounded, hence M cannot form a Schauder basis.

Now we shall prove the lemma. Consider a discrete uniformly bounded orthonormal system $\{e_1, \dots, e_n\}$ in the space K^n:

$$e_k = (e_{k,1}, \dots, e_{k,n}), \quad e_{k,j} \in \mathsf{K},$$

$$\frac{1}{n} \sum_{j=1}^n |e_{k,j}|^2 = 1, \quad k = 1, \dots, n,$$

$$\sum_{j=1}^n e_{k_1,j} \overline{e_{k_2,j}} = 0, \quad 1 \le k_1 < k_2 \le n,$$

$$|e_{k,j}| \le C_3, \quad j = 1, \dots, n,\ k = 1, \dots, n.$$

In the complex case, one can easily take the discrete Fourier basis $\{e_k\}_{k=1}^n$ with $e_{k,j} = \exp(2\pi i jk/n)$.

Each vector $x \in \mathsf{K}^n$ is represented as $x = \sum_{k=1}^{n} \hat{x}_k e_k$, where $\hat{x}_k = \frac{1}{n}\sum_{j=1}^{n} x_j \overline{e_{k,j}}$. Introduce norms $\|x\|_\infty := \max_{1\le j\le n} |x_j|$ and

$$\|x\|_P := \mathsf{E}_\varepsilon \left\| \sum_{k=1}^{n} \varepsilon_k \hat{x}_k e_k \right\|_\infty,$$

where the expectation E_ε is taken over random choices of signs $\varepsilon_k \in \{-1,1\}$ with equal probability. Note that in the continuous case the norm

$$\mathsf{E}_\varepsilon \left\| \sum_{k\in\mathbb{Z}} \varepsilon_k \hat{f}(k) \exp(2\pi i k t) \right\|_\infty$$

was studied by G. Pisier [6], so we can call $\|\cdot\|_P$ the Pisier norm. Finally, let $X_n = \mathsf{K}^n$ with norm $\|x\|_{X_n} := \|x\|_\infty + \|x\|_P$.

"Good" permutations are constructed via S.A. Chobanyan's theorem [7]. It states that for any sequence of vectors $a_1,\dots,a_n$ in a normed space there exists a permutation σ such that

$$\left\| \sum_{k=1}^{m} a_{\sigma_k} \right\| \le 9 \left(\mathsf{E}_\varepsilon \left\| \sum_{k=1}^{n} \varepsilon_k a_k \right\| + \left\| \sum_{k=1}^{n} a_k \right\| \right), \qquad m = 1,\dots,n.$$

Apply this to the space X_n and vectors $a_k = \hat{x}_k e_k$:

$$\begin{aligned} \left\| \sum_{k=1}^{m} \hat{x}_{\sigma_k} e_{\sigma_k} \right\|_{X_n} &\le 9 \left(\mathsf{E}_\varepsilon \left\| \sum_{k=1}^{n} \varepsilon_k \hat{x}_k e_k \right\|_{X_n} + \|x\|_{X_n} \right) \\ &= 9 \left(\mathsf{E}_\varepsilon \left\| \sum_{k=1}^{n} \varepsilon_k \hat{x}_k e_k \right\|_\infty + \mathsf{E}_\varepsilon \left\| \sum_{k=1}^{n} \varepsilon_k \hat{x}_k e_k \right\|_P + \|x\|_{X_n} \right). \end{aligned}$$

The first term equals $\|x\|_P$. The second term can be calculated:

$$\begin{aligned} \mathsf{E}_\varepsilon \left\| \sum_{k=1}^{n} \varepsilon_k \hat{x}_k e_k \right\|_P &= \mathsf{E}_\varepsilon \mathsf{E}_{\varepsilon'} \left\| \sum_{k=1}^{n} \varepsilon_k \varepsilon'_k \hat{x}_k e_k \right\|_\infty \\ &= \mathsf{E}_{\varepsilon''} \left\| \sum_{k=1}^{n} \varepsilon''_k \hat{x}_k e_k \right\|_\infty = \|x\|_P. \end{aligned}$$

Hence, we obtain (23.2) with $C_1 = 27$.

Proceed to the lower bound. Fix a permutation σ. As usual, the norm of the partial sum operator can be written in terms of the Lebesgue function:

$$\sup_{\|x\|_\infty \le 1} \left\| \sum_{k=1}^{m} \hat{x}_{\sigma_k} e_{\sigma_k} \right\|_\infty = \max_{1\le j\le n} \frac{1}{n} \sum_{l=1}^{n} \left| \sum_{k=1}^{m} e_{\sigma_k,j} \overline{e_{\sigma_k,l}} \right|. \tag{23.4}$$

In order to bound the Lebesgue function, one can apply the following theorem of A.M. Olevskii [2, 3]. For every uniformly bounded orthonormal system $(\varphi_k(x))_{k=1}^n$, $|\varphi_k(x)| \le M$, on the segment [0, 1], and any sequence of numbers c_k, $|c_k| \le M$, the following inequality holds:

$$\max_{1\le m\le n} \int_0^1 \left|\sum_{k=1}^m c_k\varphi_k(x)\right| dx \ge C(M)\frac{\log n}{n}\sum_{k=1}^n |c_k|^2. \tag{23.5}$$

($C(M) > 0$ is a constant which depends only on M.) Although the theorem was stated for the real-valued functions, it is also true in the complex case. Indeed, write $\varphi_k = \psi_k + i\theta_k$, $\psi_k(x) \in \mathbb{R}$, $\theta_k(x) \in \mathbb{R}$. If $c_k \in \mathbb{R}$, we can apply (23.5) to the system

$$\widetilde{\varphi}_k(x) = \begin{cases} \sqrt{2}\psi_k(2x), & 0 \le x \le 1/2, \\ \sqrt{2}\theta_k(2x-1), & 1/2 < x \le 1, \end{cases}$$

and obtain that, for some m,

$$\int_0^1 \left|\sum_{k=1}^m c_k\psi_k(x)\right| dx + \int_0^1 \left|\sum_{k=1}^m c_k\theta_k(x)\right| dx \ge C_1(M)\frac{\log n}{n}\sum_{k=1}^n c_k^2.$$

Hence, (23.5) follows for complex (φ_k) and real c_k. If $c_k = |c_k|e^{i\alpha_k} \in \mathbb{C}$, we apply (23.5) to the system $(e^{i\alpha_k}\varphi_k)$ and coefficients $|c_k|$.

Setting $c_k = \overline{\varphi_k(t)}$ one can easily derive that

$$\max_{1\le m\le n} \int_0^1 \left|\sum_{k=1}^m \varphi_k(t)\overline{\varphi_k(x)}\right| dx \ge C_2(M)\log n, \quad t \in E, \tag{23.6}$$

where

$$E = \left\{t\colon \sum_{k=1}^n \left|\varphi_k(t)\right|^2 \ge n/2\right\}, \quad \text{meas}\, E > C_3(M).$$

Let $\varphi_k(x) = e_{k,j}$, $x \in [(j-1)/n, j/n)$. Then the inequality (23.6) implies that for every permutation σ there is a number m such that the value in (23.4) is at least $C_4 \log n$. Hence, for some $x \ne 0$,

$$\left\|\sum_{k=1}^m \hat{x}_{\sigma_k} e_{\sigma_k}\right\|_\infty \ge C_4 \log n \cdot \|x\|_\infty.$$

Now, to establish (23.3), it remains only to prove that $\|x\|_{X_n} \le C_5\sqrt{\log n}\|x\|_\infty$ (let $n > 1$). One can apply, for example, Hoeffding's inequality [8]: If $\xi_1, \dots, \xi_n$ are independent complex-valued random variables with $\mathsf{E}\xi_k = 0$ and $|\xi_k| \le c_k$, then

$$\mathsf{P}\left(\left|\sum_{k=1}^n \xi_k\right| > t\right) \le 4\exp\left(-\frac{t^2}{4\sum_{k=1}^n c_k^2}\right).$$

(One has multipliers 2 instead of 4 in the real case.) Fix coordinate $j \in \{1, \dots, n\}$. Apply Hoeffding's inequality to random variables $\xi_k = \varepsilon_k \hat{x}_k e_{k,j}$, taking into account that $|e_{k,j}| \le C_3$ and $\sum_{k=1}^n |\hat{x}_k|^2 = \|x\|_2^2 \le \|x\|_\infty^2$. Let $t = C_6\sqrt{\log n} \cdot \|x\|_\infty$, then $\mathsf{P}(|\sum_{k=1}^n \varepsilon_k \hat{x}_k e_{k,j}| > t) < n^{-2}$. Hence, with probability at least $1 - n^{-1}$, all coordinates of the sum $\sum_{k=1}^n \varepsilon_k \hat{x}_k e_k$ are at most $C_6\sqrt{\log n} \cdot \|x\|_\infty$, and

$$\|x\|_P = \mathsf{E}_\varepsilon \left\| \sum_{k=1}^n \varepsilon_k \hat{x}_k e_k \right\|_\infty$$
$$\le (1 - n^{-1}) C_6\sqrt{\log n} \cdot \|x\|_\infty + n^{-1} \cdot C_3 n \|x\|_\infty \le C_7\sqrt{\log n} \cdot \|x\|_\infty.$$

References

1. Exposition of lectures of S.B. Stechkin on approximation theory. URO RAN, Ekaterinburg (2010) (Russian)
2. Olevskii, A.M.: Fourier series and Lebesgue functions. Usp. Mat. Nauk **22**(3), 237–239 (1967) (Russian)
3. Olevskii, A.M.: Fourier Series with Respect to General Orthogonal Systems. Springer, Berlin (1975)
4. Ul'yanov, P.L.: Solved and unsolved problems in the theory of trigonometric and orthogonal series. Russ. Math. Surv. **19**(1), 1–62 (1964)
5. Konyagin, S.V.: On divergence of Fourier series with respect to a rearranged trigonometric system. Mat. Zametki **47**(6), 143–145 (1990) (Russian)
6. Pisier, G.: A remarkable homogeneous algebra. Isr. J. Math. **34**, 38–44 (1979)
7. Chobanyan, S.A.: Structure of the set of sums of a conditionally convergent series in a normed space. Math. USSR Sb. **56**(1), 49–62 (1987)
8. Hoeffding, W.: Probability inequalities for sums of bounded random variables. J. Am. Stat. Assoc. **58**, 13–30 (1963)

Chapter 24
Inequalities for Trigonometric Sums

Stamatis Koumandos

Abstract We give a survey of recent results on positive trigonometric sums. Far-reaching extensions and generalizations of classical results are presented. We provide new proofs as well as additional remarks and comments. We also present several other sharp inequalities for trigonometric sums of various types.

Key words Sharp inequalities for trigonometric sums · Positive trigonometric sums

Mathematics Subject Classification Primary 42A05 · Secondary 26D05 · 26D15 · 33C45

24.1 Introduction

Inequalities involving trigonometric sums arise naturally in various problems of Pure and Applied Mathematics. Inequalities that assure nonnegativity or boundedness of partial sums of trigonometric series are of particular interest. Such inequalities are not only of importance within the context of classical Fourier analysis, see, for example, [25, 26, 29, 52, 70, 86], but they have also remarkable applications in other fields such as geometric function theory [20, 23, 44, 45, 48, 54, 55, 61–64, 71, 73, 77–80], approximation theory [1, 28, 37, 39, 76], number theory [16, 38, 75], special functions [2, 18–20, 59], orthogonal polynomials on the unit circle [83], numerical analysis [18, 59, 74], signal processing [40, 41, 69], combinatorial theory [47, 49], and statistics [42].

The problem of establishing positivity of trigonometric sums has been dealt with by many significant mathematicians of the twentieth century, like L. Fejér, D. Jackson, W.H. Young, E. Landau, P. Turàn, L. Vietoris, G. Szegő, and R. Askey who

Dedicated to Professor Themistocles M. Rassias on the occasion of his 60th birthday.

S. Koumandos (✉)
Department of Mathematics and Statistics, University of Cyprus, P.O. Box 20537, 1678 Nicosia, Cyprus
e-mail: skoumand@ucy.ac.cy

P.M. Pardalos et al. (eds.), *Nonlinear Analysis*, Springer Optimization and Its Applications 68, In Honor of Themistocles M. Rassias on the Occasion of his 60th Birthday,
DOI 10.1007/978-1-4614-3498-6_24, © Springer Science+Business Media, LLC 2012

made important contributions and their work will be highlighted in the present paper. For more details on related results and their chronology, we refer to the excellent surveys on the subject [18–20, 72].

Motivated by particular applications, over the last few years, there has been a revived interest in the positivity of certain special trigonometric sums, and some remarkable new results have been obtained, see the recently published research papers [3–15] and [53–64].

The aim of the present paper is to give a systematic account of recent results on positive trigonometric sums and their significance. These results extend, complement, and generalize some results that have deep roots in classical analysis. We give new and simpler proofs of earlier obtained theorems and provide additional remarks and comments. Some closely related new inequalities are also given.

24.2 Inequalities of Fejér–Jackson, Young, and Turàn

From the beginning of the twentieth century, there has been interest in finding trigonometric series with positive partial sums in different mathematical problems and applications. In 1910, Fejér, in connection with the study of Gibbs' phenomenon of Fourier series, conjectured that

$$\sum_{k=1}^{n} \frac{\sin k\theta}{k} > 0 \quad \text{for all } n \in \mathbb{N} \text{ and } 0 < \theta < \pi. \tag{24.1}$$

This conjecture was proved a little later by D. Jackson [50] and independently by T.H. Gronwall [46]. After the publication of these proofs, inequality (24.1) received attention by several mathematicians who gave new and shorter proofs and generalizations of various kinds. It is worth mentioning that E. Landau [68] led to a ten-line proof of (24.1), see also [87, p. 62] or [72, p. 306]. P. Turàn [82] proved a theorem that shows that (24.1) follows from the non-negativity of the classical Fejér kernel

$$\sum_{k=0}^{n} \sin\left(k + \frac{1}{2}\right)\theta \geq 0 \quad \text{for all } n \in \mathbb{N} \text{ and } 0 \leq \theta \leq 2\pi,$$

see also [22].

W.H. Young obtained in [85] an analogue of (24.1) for cosine sums. He showed that

$$1 + \sum_{k=1}^{n} \frac{\cos k\theta}{k} > 0 \quad \text{for all } n \in \mathbb{N} \text{ and } 0 < \theta < \pi. \tag{24.2}$$

Turàn proved in [81], see also [24, pp. 248–249], the following remarkable inequality

$$1 + \sum_{k=1}^{n} \frac{1 \cdot 3 \cdot 5 \cdots (2k-1)}{2 \cdot 4 \cdot 6 \cdots 2k} \cos k\theta > 0 \quad \text{for all } n \in \mathbb{N} \text{ and } 0 < \theta < \pi, \tag{24.3}$$

in order to show that the positivity of the partial sums of a trigonometric series does not imply that it is a Fourier series of a square integrable function.

The sequence of coefficients $\frac{1\cdot 3\cdot 5\cdots(2k-1)}{2\cdot 4\cdot 6\cdots 2k}$ has order of magnitude $k^{-1/2}$ as opposed to the order of magnitude k^{-1} for the coefficients in (24.1) and (24.2). One can see that inequality (24.3) *does not* imply (24.2), by considering the case $n = 1$. It is, however, possible to prove, see Sect. 24.4, the following refinement of (24.3)

$$\frac{1}{2} + \sum_{k=1}^{n} \frac{1\cdot 3\cdot 5\cdots(2k-1)}{2\cdot 4\cdot 6\cdots 2k}\cos k\theta > 0 \quad \text{for all } n \in \mathbb{N} \text{ and } 0 < \theta < \pi, \tag{24.4}$$

which implies (24.2). We also note that there is no analogue of (24.3) and (24.4) for sine sums. It is the aim of the present work to tackle these questions and give far-reaching extensions of both (24.3) and (24.4).

24.3 Vietoris' Inequalities

In 1958, L. Vietoris [84] proved a surprising and quite deep result which provides a substantial improvement of (24.1), (24.2), (24.3), but *not* (24.4). Vietoris gave sufficient conditions on the coefficients of a general class of sine and cosine sums that ensure their simultaneous positivity in $(0, \pi)$. His result is the following.

Theorem 24.1 *Suppose that* $a_0 \geq a_1 \geq \cdots \geq a_n \geq \cdots > 0$ *and*

$$2ka_{2k} \leq (2k-1)a_{2k-1}, \quad \textit{for all } k \geq 1.$$

Then for all positive integers n, *we have*

$$\sum_{k=0}^{n} a_k \cos k\theta > 0, \quad 0 < \theta < \pi \quad \textit{and}, \tag{24.5}$$

$$\sum_{k=1}^{n} a_k \sin k\theta > 0, \quad 0 < \theta < \pi. \tag{24.6}$$

Vietoris observed that (24.5) and (24.6) are equivalent to the corresponding inequalities for the specific case in which $a_k = \gamma_k$, where the sequence γ_k is defined as follows

$$\begin{aligned} &\gamma_0 = \gamma_1 = 1 \quad \text{and} \\ &\gamma_{2k} = \gamma_{2k+1} = \frac{1\cdot 3\cdot 5\cdots(2k-1)}{2\cdot 4\cdot 6\cdots 2k}, \quad k = 1, 2, \ldots. \end{aligned} \tag{24.7}$$

Taking $a_0 = 1$ and $a_k = 1/k$, $k \geq 1$, in Theorem 24.1, we immediately obtain (24.1) and (24.2). We need, however, to do some additional work in order to obtain (24.3). It is shown in [21, p. 299] that Theorem 24.1 implies the following.

Proposition 24.1 *If* $0 \le \rho \le 1/2$ *then for all positive integers* n, *we have*

$$1+\sum_{k=1}^{n}\frac{1\cdot 3\cdot 5\cdots(2k-1)}{2\cdot 4\cdot 6\cdots 2k}\cos\bigl[(2k+\rho)\varphi\bigr]>0, \quad 0<\varphi<\pi. \tag{24.8}$$

Clearly, (24.3) is (24.8) for $\rho = 0$. R. Askey and J. Steinig [21] have given a simplified proof of Theorem 24.1 and have also performed a valuable service in drawing attention to this result. They also presented several applications of Vietoris' Theorem which demonstrate its importance. This Theorem is used to obtain sharp estimates for the location of zeros of a class of trigonometric polynomials whose coefficients satisfy certain growth conditions. Theorem 24.1 has also some remarkable applications in problems dealing with positive quadrature methods. It is worth mentioning that Vietoris' result suggested some more general inequalities for sums of Jacobi polynomials as well as various new summation and transformation formulas for hypergeometric series. Details of these and some historical comments can be found in [18, 19] and [2, p. 371].

Some remarkable applications of Theorem 24.1 in geometric function theory have been obtained in [78] and [80].

In 1995, A.S. Belov [27] proved the following extension of Theorem 24.1.

Theorem 24.2 *Suppose that* a_k, $k = 0, 1, 2\ldots$, *is a monotone sequence of nonnegative real numbers. Then the condition*

$$\sum_{k=1}^{n}(-1)^{k-1}k\,a_k \ge 0 \quad \textit{for all } n \ge 2,\ a_1 > 0, \tag{24.9}$$

is necessary and sufficient for the validity of the inequality

$$\sum_{k=1}^{n} a_k \sin k\theta > 0 \quad \textit{for all } n \in \mathbb{N} \textit{ and } 0<\theta<\pi.$$

Moreover, condition (24.9) *implies that*

$$\sum_{k=0}^{n} a_k \cos k\theta > 0 \quad \textit{for all } n \in \mathbb{N} \textit{ and } 0\le\theta<\pi.$$

It is clear that if a monotone sequence of nonnegative real numbers a_k satisfies condition (24.9) then it is decreasing. It is also obvious that for a nonnegative sequence a_k, condition (24.9) is equivalent to

$$\sum_{k=1}^{n}\bigl((2k-1)a_{2k-1}-2ka_{2k}\bigr)\ge 0 \quad \text{for all } n\ge 1. \tag{24.10}$$

Accordingly, if a sequence a_k satisfies the conditions of Theorem 24.1 then for this sequence (24.9) is valid. Therefore, Theorem 24.1 clearly follows from Theorem 24.2. In order to demonstrate the power of Theorem 24.2 let us consider the following example: Let $a_0 = a_1 = 1$ and $a_k = \frac{1}{k+\alpha}$, $\alpha > 0$. It is easily checked that a_k *does not* satisfy the condition of Theorem 24.1, when $k \geq 2$. On the contrary, an elementary argument shows that for this sequence we have

$$1 + \sum_{k=2}^{n} (-1)^{k-1} \frac{k}{k+\alpha} > 0,$$

for all $\alpha > 0$ and for all $n \in \mathbb{N}$. In view of Theorem 24.2, we deduce that

Proposition 24.2 *For all positive integers n and for all $\alpha > 0$ we have*

$$\sin\theta + \sum_{k=2}^{n} \frac{\sin k\theta}{k+\alpha} > 0, \quad 0 < \theta < \pi$$

and

$$1 + \cos\theta + \sum_{k=2}^{n} \frac{\cos k\theta}{k+\alpha} > 0, \quad 0 < \theta < \pi.$$

Plainly the above inequalities hold true, and for $\alpha = 0$ and they are inequalities (24.1) and (24.2), respectively.

Belov's result is best possible for the positivity on $(0, \pi)$ of sine sums with non-negative and decreasing sequence of coefficients. There are, however, several positive cosine sums for which condition (24.9) fails to hold such as, for example, the results of (24.3) and (24.4). Additional examples of this type will be given in the next section. Here we present some results closely related to Proposition 24.2. G. Gasper proved in [43] the following.

Proposition 24.3 *If $-1 < \alpha \leq A$, then for all positive integers n we have*

$$\frac{1}{1+\alpha} + \sum_{k=1}^{n} \frac{\cos k\theta}{k+\alpha} \geq 0, \quad 0 < \theta < \pi, \tag{24.11}$$

where A is the positive root of the equation

$$9\alpha^7 + 55\alpha^6 - 14\alpha^5 - 948\alpha^4 - 3247\alpha^3 - 5013\alpha^2 - 3780\alpha - 1134 = 0.$$

The result is best possible in the sense that the number A is the largest number for which (24.11) *holds for all n. The numerical value of A is $A = 4.5678018\ldots$. Equality holds in* (24.11) *for some θ only when $n = 3$ and $\alpha = A$.*

The case $\alpha = 0$ of (24.11) is Young's inequality (24.2), while the case $\alpha = 1$ is a classical result due to Rogosinski and Szegő [77].

There is an analogue of Proposition 24.3 for sine sums. This is obtained in [35].

Proposition 24.4 *Suppose that λ_0 is the unique solution of the equation $(1+\lambda)\pi = \tan(\lambda\pi)$ for $0<\lambda<1/2$ and that $\tilde{\alpha}$ is the unique solution of the equation*

$$\sum_{k=1}^{\infty} \frac{2k}{(2k-1+\alpha)(2k+\alpha)(2k+1+\alpha)} = \frac{\sin\lambda_0\pi}{2(1+\lambda_0)\pi}.$$

If $-1<\alpha\le\tilde{\alpha}=2.11023\ldots$, then

$$\sum_{k=1}^{n} \frac{\sin k\theta}{k+\alpha} > 0, \quad 0<\theta<\pi, \tag{24.12}$$

for any odd *positive integer number n. The result is best possible in the sense that the number $\tilde{\alpha}$ is the largest number for which* (24.12) *holds.*

In the case where n is even, inequality (24.12) fails to hold for all $\theta \in (0,\pi)$. Belov's Theorem 24.2 helps explain why this happens. Application of the same Theorem shows that there is no analogue of (24.3) and (24.4) for sine sums. In general, inequality (24.6) cannot hold for all n and $\theta \in (0,\pi)$ when the sequence of coefficients a_k satisfies a condition weaker than (24.9). On the contrary, Vietoris' Theorem gives

$$\sigma_n(\theta) := \sum_{k=1}^{n} \gamma_k \sin k\theta > 0 \quad \text{for all } n\in\mathbb{N} \text{ and } 0<\theta<\pi,$$

where the sequence of coefficients γ_k is as in (24.7). For the sequence γ_k, the relation (24.10) holds for all n as equality and this demonstrates the naturalness of Theorem 24.1 for sine sums. It is, however, of interest to look for an extension of Vietoris sine inequality of a different type. It is possible to find the positive algebraic polynomial $p(\theta)$ of smallest degree such that

$$\sigma_n(\theta) > p(\theta) > 0 \quad \text{for all } n\in\mathbb{N} \text{ and } 0<\theta<\pi. \tag{24.13}$$

It is shown in [14] that if a polynomial $p(\theta)$ satisfies (24.13) then it has to be of degree at least 4. Moreover, all polynomials $p(\theta)$ of fourth degree such that (24.13) holds can be determined. We have, in fact, the following theorem (cf. [14]).

Theorem 24.3 *Let $p(\theta)=\sum_{k=0}^{4}\alpha_k\theta^k$, $\alpha_k\in\mathbb{R}$, $k=0,1,\ldots,4$. Then,* (24.13) *holds if and only if $p(\theta)$ is of the form*

$$p(\theta) = a\theta(\pi-\theta)^3 \quad \textit{and} \quad a\in\left(0, \frac{1}{\pi^3}\right].$$

Using Theorem 24.3 and summation by parts, we can show the following extension of Vietoris sine inequality.

Theorem 24.4 *Suppose that a sequence* (a_k), $k = 1, 2, \ldots$ *satisfies the conditions of Theorem* 24.1. *Then*

$$\sum_{k=1}^{n} a_k \sin k\theta > a_1\theta\left(1 - \frac{\theta}{\pi}\right)^3 \quad \text{for all } 0 < \theta < \pi. \tag{24.14}$$

It is easily seen that condition (24.9) is necessary for the validity of (24.14). It is very likely that for a nonnegative monotone sequence (a_k), $k = 1, 2, \ldots$, condition (24.9) is also sufficient for the truth of inequality (24.14).

Taking $a_k = 1/k$, $k \geq 1$, in Theorem 24.4, we obtain the following functional lower bound for the sums in (24.1)

Corollary 24.1

$$\sum_{k=1}^{n} \frac{\sin k\theta}{k} > \theta\left(1 - \frac{\theta}{\pi}\right)^3, \quad \text{for all } n \in \mathbb{N} \text{ and } 0 < \theta < \pi.$$

Other functional estimates of this type can be found in [5, 32], and [51].

For sharp upper and lower functional estimates of the sums in (24.1) and (24.2), see [3–15, 31, 56, 57], and the references given therein.

24.4 Positivity of Cosine Sums

The Pochhammer symbol $(a)_k$ is defined by

$$(a)_0 = 1, \quad (a)_k = a(a+1)\cdots(a+k-1) = \frac{\Gamma(k+a)}{\Gamma(a)}, \quad k \geq 1.$$

In the case of cosine sums, we have the following extension of Vietoris result:

Theorem 24.5 *Suppose that* $c_{2k} = c_{2k+1} = \frac{(\mu)_k}{k!}$, $k = 0, 1, 2, \ldots$, *with* $0 < \mu < 1$. *For all positive integers n and* $0 \leq \theta < \pi$, *we have*

$$\sum_{k=0}^{n} c_k \cos k\theta > 0, \tag{24.15}$$

precisely when $0 < \mu \leq \mu_0 = 1 - \alpha_0$, *where* α_0 *is the unique solution in* $(0, 1)$ *of the equation*

$$\int_0^{3\pi/2} \frac{\cos t}{t^\alpha}\, dt = 0.$$

The sums in (24.15) *are unbounded below when* $1 \geq \mu > \mu_0$.

The numerical value of μ_0 is $\mu_0 = 0.691556\ldots$.

Note that Vietoris' cosine result is the case $\mu = 1/2$ of the above Theorem because

$$\frac{1\cdot 3\cdot 5\cdots(2k-1)}{2\cdot 4\cdot 6\cdots 2k} = 2^{-2k}\binom{2k}{k} = \frac{(\frac{1}{2})_k}{k!}, \quad k = 1, 2, \ldots.$$

Theorem 24.5 has been first proved in [59]; see also [61] and [63] for some further generalizations and some interesting applications. It is the aim of the present work to provide some additional consequences and refinements of Theorem 24.5. We first give the following.

Theorem 24.6 *Suppose that (a_k), $k = 0, 1, \ldots$, is a decreasing sequence of non-negative numbers such that $a_0 > 0$ and*

$$a_{2k} \leq \frac{k + \mu_0 - 1}{k} a_{2k-1} \quad \text{for all } k \geq 1, \tag{24.16}$$

where μ_0 as in Theorem 24.5. Then for all positive integers n and $0 \leq \theta < \pi$, we have

$$\sum_{k=0}^{n} a_k \cos k\theta > 0.$$

Theorems 24.5 and 24.6 are essentially equivalent and this has been mentioned in both [59] and [61]. The passage from Theorem 24.5 to Theorem 24.6 is based on the standard summation by parts technique which is reflected in the proposition below.

Proposition 24.5 *Let c_k, $k = 0, 1, \ldots$, be a sequence of positive real numbers. We define $S_0(\theta) = c_0$, $S_n(\theta) = c_0 + \sum_{k=1}^{n} c_k \cos k\theta$, for $n \geq 1$. If $S_n(\theta) > 0$ for all $n \geq 1, 0 < \theta < \pi$ and b_k, $k = 0, 1, \ldots$, is a sequence of positive numbers such that*

$$\frac{b_{k+1}}{b_k} \leq \frac{c_{k+1}}{c_k} \quad \text{for all } k = 0, 1, 2\ldots, \tag{24.17}$$

then

$$b_0 + \sum_{k=1}^{n} b_k \cos k\theta > 0, \quad \text{for all } n \in \mathbb{N} \text{ and } 0 < \theta < \pi.$$

Proof Summation by parts gives

$$b_0 + \sum_{k=1}^{n} b_k \cos k\theta = \sum_{k=0}^{n-1} \left(\frac{b_k}{c_k} - \frac{b_{k+1}}{c_{k+1}}\right) S_k(\theta) + \frac{b_n}{c_n} S_n(\theta) > 0$$

because all the terms of the sum on the right hand side of the above equality are positive according to the assumptions of the proposition. The proof is complete. □

It is immediately obvious that the sequence (c_k) in Theorem 24.5 satisfies $c_{2k+1} = c_{2k} = \frac{k+\mu-1}{k} c_{2k-1}$. Therefore, applying Proposition 24.5, we see that Theorem 24.6 follows from Theorem 24.5.

Using Theorem 24.6 and Proposition 24.5, we are able to prove the following closely related result.

Theorem 24.7 *For $\beta \geq 0$, we define $S_0(\theta, \beta) = 1$ and*

$$S_n(\theta, \beta) = \sum_{k=0}^{n} r_k(\beta) \cos k\theta, \quad n = 1, 2, \ldots,$$

where

$$r_{2k}(\beta) = r_{2k+1}(\beta) = \frac{(\frac{1+\beta}{2})_k}{(\frac{2+\beta}{2})_k}, \quad k = 0, 1, 2, \ldots.$$

For all positive integers n and $0 \leq \theta < \pi$, we have

$$S_n(\theta, \beta) \geq 0, \tag{24.18}$$

precisely when $0 \leq \beta \leq \beta_0$, where β_0 is the unique solution in $(2, 3)$ of the equation

$$\min_{\theta \in (0, \pi)} S_6(\theta, \beta) = 0. \tag{24.19}$$

Numerical methods give $\beta_0 = 2.330885\ldots$. Equality holds in (24.18) for some $\theta \in (0, \pi)$ if and only if $n = 6$ and $\beta = \beta_0$.

It should be noted that the case $\beta = 0$ of this Theorem is Vietoris' cosine result. The case $\beta = 1$ has been obtained in [30], while the case $\beta = 2$ in [36]. A proof of the case $\beta = 2.33$ is given in [59]. The result of Theorem 24.7 is best possible.

In view of Proposition 24.5, one needs to establish Theorem 24.7 only for $\beta = \beta_0$. We observe also that $r_{2k+1}(\beta_0) = r_{2k}(\beta_0) = \frac{2k+\beta_0-1}{2k+\beta_0} r_{2k-1}(\beta_0)$. It is easy to verify that the sequence $r_k(\beta_0)$ satisfies condition (24.16) for any $k \geq 2$. This is to say that Theorem 24.7 *is not* an immediate consequence of Theorem 24.6. It is, however, natural to strive to deduce Theorem 24.7 using Theorem 24.6. For this purpose, it is necessary to do some additional work. We first quote some propositions originally discovered in [84] and also used in [59] for the proof of Theorem 24.5.

Proposition 24.6 *Let a_k, $k = 0, 1, \ldots$, be a decreasing sequence of positive real numbers, and let M be an integer such that $1 \leq M \leq N$. The inequality*

$$\sin\frac{\theta}{2} \sum_{k=0}^{N} a_k \cos k\theta \geq \sin\frac{\theta}{2} \sum_{k=0}^{M-1} a_k \cos k\theta - \frac{a_M}{2}\left[1 + \sin\left(M - \frac{1}{2}\right)\theta\right],$$

holds for all real θ.

We define

$$V_M\left(\sin\frac{\theta}{2}\right) := \sin\frac{\theta}{2}\sum_{k=0}^{M-1} a_k \cos k\theta - \frac{a_M}{2}\left[1+\sin\left(M-\frac{1}{2}\right)\theta\right],$$

and this is clearly a sum of powers of $\sin\frac{\theta}{2}$, so substituting $t = \sin\frac{\theta}{2}$, we may write $V_M(t)$ as a polynomial in t of degree not exceeding $2M-1$. These polynomials have the following remarkable properties.

Proposition 24.7 *Let* $a_{2k} = a_{2k+1} = d_k$, $k = 0, 1, 2, \ldots$. *We have for all* $m = 1, 2, \ldots$

(i)

$$V_{2m}\left(\sin\frac{\theta}{2}\right) = V_{2m+1}\left(\sin\frac{\theta}{2}\right)$$
$$= \sin\theta\sum_{k=0}^{m-1} d_k \cos\left(2k+\frac{1}{2}\right)\theta - \frac{d_m}{2}\left[1+\sin\left(2m-\frac{1}{2}\right)\theta\right],$$

(ii)

$$V_{2m+2}\left(\sin\frac{\theta}{2}\right) \geq V_{2m+1}\left(\sin\frac{\theta}{2}\right).$$

The following proposition has been obtained in [84] and also in [21].

Proposition 24.8 *Let* (a_k), $k = 0, 1, \ldots$, *be any decreasing sequence of positive real numbers such that* $a_0 > a_N$. *Then, we have*

$$\sum_{k=0}^{N} a_k \cos k\theta > 0 \quad \textit{for } 0 \leq \theta \leq \frac{\pi}{N}.$$

Next we give a proof of Theorem 24.7.

Proof Let $\varphi(\beta) := \min_{\theta\in(0,\pi)} S_6(\theta, \beta)$. Numerical evaluation yields $\varphi(2.330886) = -8.5071842373 \times 10^{-9}$ and $\varphi(2.330885) = 1.575754587707 \times 10^{-7}$. Hence there is a β_0 such that $2.330885 < \beta_0 < 2.330886 =: \beta_1$ and $\varphi(\beta_0) = 0$.

A summation by parts gives

$$S_n(\theta, b) = \sum_{k=0}^{n-1}\left(\frac{r_k(b)}{r_k(b_0)} - \frac{r_{k+1}(b)}{r_{k+1}(b_0)}\right)S_k(\theta, b_0) + \frac{r_n(b)}{r_n(b_0)}S_n(\theta, b_0). \tag{24.20}$$

We observe that $\frac{r_k(b)}{r_k(b_0)} \geq \frac{r_{k+1}(b)}{r_{k+1}(b_0)}$, $k = 0, 1, \ldots$, if and only if $b < b_0$. Recall that $S_0(\theta, \beta) = 1$, $S_1(\theta, \beta) = 1 + \cos\theta$. Direct computation gives $S_j(\theta, \beta_1) > 0$, for all

$\theta \in (0, \pi)$ when $j = 2, 3, 4, 5$. Applying (24.20) with $b = \beta \le \beta_0$ and $b_0 = \beta_1 > \beta_0$, we infer that

$$S_j(\theta, \beta) > 0 \quad \text{for all } \theta \in (0, \pi) \text{ when } j = 2, 3, 4, 5. \tag{24.21}$$

It follows from the above that there exists a $\theta_0 \in (0, \pi)$ such that $S_6(\theta_0, \beta_0) = 0$. Let $\beta > \beta_0$. Applying (24.20) for $n = 6$, $b = \beta$, $b_0 = \beta_0$, and $\theta = \theta_0$ and using (24.21) with $\beta = \beta_0$, we deduce that $S_6(\theta_0, \beta) < 0$ for all $\beta > \beta_0$, that is, $\varphi(\beta) < 0$ for all $\beta > \beta_0$. In the case where $\beta < \beta_0$, we apply (24.20) with $n = 6$, $b = \beta$, $b_0 = \beta_0$ use (24.21) with $\beta = \beta_0$ together with the observation that by definition $S_6(\theta, \beta_0) \ge 0$ for all $\theta \in (0, \pi)$ to infer that $S_6(\theta, \beta) > 0$ for all $\theta \in (0, \pi)$ whenever $\beta < \beta_0$. Therefore, $\varphi(\beta) > 0$ for all $\beta < \beta_0$, and this in combination with the above proves that equation (24.19) has a unique solution in (2, 3) which is the number β_0. The proof of Theorem 24.7 for $1 \le n \le 6$ is now complete. We also observe that a direct calculation gives $S_7(\theta, \beta_1) > 0$, for all $\theta \in (0, \pi)$. Applying (24.20) for $n \ge 7$, and $b = \beta < \beta_1$, $b_0 = \beta_1$ and recalling that $r_6(b) = r_7(b)$ for all $b > 0$ we see that it remains to prove that

$$S_n(\theta, \beta_1) > 0 \quad \text{for all } \theta \in (0, \pi) \text{ when } n \ge 8. \tag{24.22}$$

It follows from Proposition 24.6 that

$$\sin\frac{\theta}{2} S_N(\theta, \beta_1) \ge V_M\left(\sin\frac{\theta}{2}\right) \quad \text{for all } N \ge M,\ 0 < \theta < \pi. \tag{24.23}$$

By Proposition 24.7 and a direct computation, we see that each polynomial $V_{2m}(\sin\frac{\theta}{2})$ has exactly one zero $\theta_m \in (0, \pi)$ for $m = 4, 5, 6$ and that

$$V_{2m}\left(\sin\frac{\theta}{2}\right) = V_{2m+1}\left(\sin\frac{\theta}{2}\right) > 0 \quad \text{for } \theta_m < \theta < \pi. \tag{24.24}$$

Computation using Maple 14 gives

$$\frac{\pi}{16} < \theta_6 = 0.205441\ldots < \frac{\pi}{15}.$$

Combining this with Proposition 24.8, (24.23), and (24.24), we infer that

$$\begin{aligned} &S_n(\theta, \beta_1) > 0 \quad \text{for all } \theta \in (0, \pi) \text{ when } 12 \le n \le 15, \\ &S_n(\theta, \beta_1) > 0 \quad \text{for all } \theta \in \left[\frac{\pi}{15}, \pi\right) \text{ when } n \ge 16. \end{aligned} \tag{24.25}$$

The cases $8 \le n \le 11$ of (24.22) can be dealt with likewise. In particular, we find that

$$\frac{\pi}{12} < \theta_5 = 0.26859\ldots < \frac{\pi}{11}.$$

Using this we prove (24.22) for $n = 10, 11$. We also have

$$\frac{\pi}{9} < \theta_4 = 0.379739\ldots < \frac{\pi}{8}.$$

By this we get $S_8(\theta, \beta_1) > 0$ for all $\theta \in (0, \pi)$ and that $S_9(\theta, \beta_1) > 0$ for all $\theta \in (\theta_4, \pi)$. By Proposition 24.8, we have $S_9(\theta, \beta_1) > 0$ for all $\theta \in [0, \pi/9]$. In the case where $\pi/9 < \theta \leq \theta_4$, we write $S_9(\theta, \beta_1) = S_{10}(\theta, \beta_1) - r_{10}(\beta_1)\cos 10\theta$ and use the fact that $S_{10}(\theta, \beta_1) > 0$ for all $\theta \in (0, \pi)$ and that $\cos 10\theta < 0$ for all $\theta \in [\pi/9, \theta_4)$. Hence $S_9(\theta, \beta_1) > 0$ for all $\theta \in (0, \pi)$.

In order to establish the remaining cases of (24.22), let $r_k := r_k(\beta_1),\ k = 0, 1, \ldots,\ \delta_k := r_{2k} = r_{2k+1},\ k = 0, 1, \ldots,$ and $c_{2k} = c_{2k+1} = \frac{(\mu_0)_k}{k!} =: d_k,\ k = 0, 1, \ldots,$ where μ_0 is as in Theorem 24.5. Direct computation shows that

$$\begin{aligned} &\delta_j > d_j, \quad j = 1, 2, \ldots, 7, \quad \text{while} \\ &\delta_j < d_j, \quad \text{for all } j \geq 8. \end{aligned} \tag{24.26}$$

We define the sequence $(a_k),\ k = 0, 1, \ldots,$ as follows: $a_{2k} = a_{2k+1} = \lambda_k,\ k = 0, 1, \ldots,$ where

$$\lambda_k = \begin{cases} d_k, & k = 0, 1, 2, \ldots, 7 \\ \delta_k, & k \geq 8. \end{cases}$$

By (24.26), we easily verify that the sequence $(a_k),\ k = 0, 1, \ldots,$ satisfies the conditions of Theorem 24.6. Therefore,

$$T_n(\theta) := \sum_{k=0}^{n} a_k \cos k\theta > 0 \quad \text{for all } \theta \in (0, \pi) \text{ and for all } n \geq 1. \tag{24.27}$$

Let

$$P(\theta) := \sum_{k=2}^{15} (r_k - c_k)\cos k\theta$$

and recall that $r_{2j} - c_{2j} = r_{2j+1} - c_{2j+1} = \delta_j - d_j,\ j \geq 1,\ \delta_0 = d_0 = 1$. Using the first inequality of (24.26), we see that $P(\theta)$ is a strictly decreasing function on $[0, \pi/15]$ and by a direct computation that

$$P(\theta) \geq P\left(\frac{\pi}{15}\right) > 0 \quad \text{for all } \theta \in \left[0, \frac{\pi}{15}\right]. \tag{24.28}$$

Suppose now that $n \geq 16$ and $\theta \in [0, \frac{\pi}{15}]$. Taking into consideration (24.27) and (24.28), we conclude that

$$S_n(\theta,\beta_1)=\sum_{k=0}^{n} r_k \cos k\theta=\sum_{k=0}^{15} r_k \cos k\theta+\sum_{k=16}^{n} r_k \cos k\theta$$

$$>\sum_{k=0}^{15} c_k \cos k\theta+\sum_{k=16}^{n} r_k \cos k\theta=T_n(\theta)>0.$$

This completes the proof of Theorem 24.7. □

By a summation by parts, we are able to prove the following generalization of Theorem 24.7.

Theorem 24.8 *Suppose that (a_k), $k=0,1,\dots$, is a decreasing sequence of non-negative numbers such that $a_0>0$ and*

$$a_{2k}\le \frac{2k+\beta_0-1}{2k+\beta_0}a_{2k-1} \quad \text{for all } k\ge 1, \tag{24.29}$$

where β_0 is as in Theorem 24.7. Then for all positive integers n and $0\le\theta<\pi$, we have

$$\sum_{k=0}^{n} a_k \cos k\theta \ge 0,$$

with equality being true for some $\theta\in(0,\pi)$ if and only if $n=6$ and $a_k=a_0 r_k(\beta_0)$ for $0\le k\le 6$.

The methods of the proof of Theorem 24.7 enable us to prove the following variant of Theorem 24.5.

Theorem 24.9 *Suppose that $c_0=c_1=1$, $c_2=c_3=\frac{4}{5}$, $c_{2k}=c_{2k+1}=\frac{(\mu)_k}{k!}$, $k\ge 2$, with $0<\mu<1$. For all positive integers n and $0\le\theta<\pi$, we have*

$$\sum_{k=0}^{n} c_k \cos k\theta \ge 0, \tag{24.30}$$

precisely when $0<\mu\le\mu_0$, where μ_0 is as in Theorem 24.5. Equality holds in (24.30) if and only if $n=3$ and $\theta=\arccos(1/4)$. The sums in (24.30) are unbounded below when $1\ge\mu>\mu_0$.

Proof In view of Proposition 24.5, we need to establish the theorem only for $\mu=\mu_0$. We first check the cases $1\le n\le 5$ by direct calculation. Suppose next that $n\ge 6$. Since the sequence (c_k), $k=0,1,\dots$, is decreasing, it follows from Proposition 24.6 that, for $n\ge m\ge 1$,

$$\sin\frac{\theta}{2}\sum_{k=0}^{n} c_k \cos k\theta \ge \sin\frac{\theta}{2}\sum_{k=0}^{m-1} c_k \cos k\theta-\frac{c_m}{2}\left[1+\sin\left(m-\frac{1}{2}\right)\theta\right]. \tag{24.31}$$

Moreover, designating the right hand side of (24.31) as $U_m(\sin\frac{\theta}{2})$ and applying Proposition 24.7, we get

$$U_{2k} = U_{2k+1} \leq U_{2k+2}.$$

As in the proof of Theorem 24.7, we see that the polynomial $U_6(\sin\frac{\theta}{2})$ has exactly one zero $\theta_3 \in (0, \pi)$ and

$$U_6\left(\sin\frac{\theta}{2}\right) = U_7\left(\sin\frac{\theta}{2}\right) > 0 \quad \text{for } \theta_3 < \theta < \pi.$$

Computation using Maple 14 gives

$$\frac{\pi}{10} < \theta_3 = 0.34085\ldots < \frac{\pi}{6}.$$

It follows from the above that for all $n \geq 6$

$$\sum_{k=0}^{n} c_k \cos k\theta > 0 \quad \text{for all } \theta \in \left[\frac{\pi}{6}, \pi\right).$$

On the other hand, for $n \geq 6$ and $\theta \in [0, \frac{\pi}{6})$ we have

$$\begin{aligned}\sum_{k=0}^{n} c_k \cos k\theta &= 1 + \cos\theta + \frac{4}{5}\cos 2\theta + \frac{4}{5}\cos 3\theta + \sum_{k=4}^{n} c_k \cos k\theta \\ &> 1 + \cos\theta + \mu_0 \cos 2\theta + \mu_0 \cos 3\theta + \sum_{k=4}^{n} c_k \cos k\theta > 0,\end{aligned}$$

where the last inequality is obtained from Theorem 24.5. Finally, we note that the sums in (24.30) are unbounded below in $(0, \pi)$ if and only if the sums in (24.15) are unbounded below in $(0, \pi)$, and the latter occurs precisely when $1 \geq \mu > \mu_0$, according to Theorem 24.5.

The proof of Theorem 24.9 is complete. □

Applying the results of Theorem 24.5 and Theorem 24.9, we can establish the positivity of other trigonometric sums. Let

$$d_k := c_{2k} = c_{2k+1} = \frac{(\mu)_k}{k!}, \quad k = 0, 1, 2, \ldots, \text{ with } 0 < \mu < 1.$$

It is easy to see that

$$\sum_{k=0}^{2n+1} c_k \cos k\theta = 2\cos\frac{\theta}{2} \sum_{k=0}^{n} d_k \cos\left(2k + \frac{1}{2}\right)\theta. \tag{24.32}$$

Making the transformation $\theta \to \pi - \theta$ in the above, we get

$$\sum_{k=0}^{2n+1} c_k \cos k(\pi - \theta) = 2\sin\frac{\theta}{2}\sum_{k=0}^{n} d_k \sin\left(2k + \frac{1}{2}\right)\theta. \tag{24.33}$$

Adding (24.32) and (24.33), we arrive at

$$2\sum_{k=0}^{n} d_k \cos 2k\theta = \sum_{k=0}^{2n+1} c_k \cos k\theta + \sum_{k=0}^{2n+1} c_k \cos k(\pi - \theta).$$

In view of Theorem 24.5 and the above, we derive the following.

Corollary 24.2 *For all positive integers n and $0 < \theta < \pi$, we have*

$$\text{(i)} \qquad \sum_{k=0}^{n} \frac{(\mu)_k}{k!} \cos\left(2k + \frac{1}{2}\right)\theta > 0,$$

$$\text{(ii)} \qquad \sum_{k=0}^{n} \frac{(\mu)_k}{k!} \sin\left(2k + \frac{1}{2}\right)\theta > 0,$$

$$\text{(iii)} \qquad \sum_{k=0}^{n} \frac{(\mu)_k}{k!} \cos k\theta > 0,$$

precisely when $0 < \mu \le \mu_0$, where μ_0 is as in Theorem 24.5. All three sums are unbounded below when $1 \ge \mu > \mu_0$.

Note that the case $\mu = 1/2$ of (i) and (ii) is Proposition 24.1, while (iii) for $\mu = 1/2$ becomes inequality (24.3).

Inequality (iii) of Corollary 24.2 was first obtained in [62] and applied in the context of starlike functions; compare also the paper [61] for a simpler direct proof of this result and additional comments. In order to prove the unboundedness of the sums in Corollary 24.2, one uses the following asymptotic formula

$$\lim_{n\to\infty}\left(\frac{\theta}{n}\right)^{\mu} \sum_{k=0}^{n} \frac{(\mu)_k}{k!} \cos\left[(2k+\rho)\frac{\theta}{2n}\right] = \frac{1}{\Gamma(\mu)}\int_0^{\theta} \frac{\cos t}{t^{1-\mu}}\,dt, \tag{24.34}$$

$0 \le \rho \le 1/2$; see [58] and [61] for details.

It is clear that inequalities (ii) and (iii) of Corollary 24.2 are equivalent and an analogous result can be obtained in the same way when Theorem 24.9 is applied. We have, in fact, the following.

Corollary 24.3 *Suppose that* $\sigma_0 = 1$, $\sigma_1 = \frac{4}{5}$, $\sigma_k = \frac{(\mu)_k}{k!}$, $k \geq 2$, *with* $0 < \mu < 1$. *For all positive integers* n *and* $0 < \theta < \pi$, *we have*

$$\text{(i)} \qquad \sum_{k=0}^{n} \sigma_k \sin\left(2k + \frac{1}{2}\right)\theta \geq 0,$$

precisely when $0 < \mu \leq \mu_0$, *where* μ_0 *is as in Theorem 24.5. Equality holds in* (i) *if and only if* $n = 1$ *and* $\theta = 2\arccos(\sqrt{6}/4)$.

$$\text{(ii)} \qquad \sum_{k=0}^{n} \sigma_k \cos k\theta > 0,$$

precisely when $0 < \mu \leq \mu_0$. *The sums in both* (i) *and* (ii) *are unbounded below when* $1 \geq \mu > \mu_0$.

As an application of inequality (i) of the corollary above, we obtain the following result which was first proved in [53] in a direct way.

Corollary 24.4 *For all positive integers* n *and* $0 < \theta < \pi$, *we have*

$$\frac{1}{4}\sin\frac{\theta}{2} + \sum_{k=1}^{n} \frac{1}{4k+1} \sin\left(2k + \frac{1}{2}\right)\theta \geq 0,$$

with equality only when $n = 1$ *and* $\theta = 2\arccos(\sqrt{6}/4)$. *The leading factor* $1/4$ *in the above sum is the best possible.*

Proof As the sharp case $n = 1$ is elementary, we assume that $n \geq 2$. Let $\sigma_0 = 1$, $\sigma_1 = \frac{4}{5}$, $\sigma_k = \frac{(\mu_0)_k}{k!}$, $k \geq 2$, and recall that $\mu_0 = 0.691556\ldots$. Let $b_0 = \frac{1}{4}$, $b_1 = \frac{1}{5}$, $b_k = \frac{1}{4k+1}$, $k \geq 2$. Observe that $\frac{b_1}{b_0} = \frac{\sigma_1}{\sigma_0}$, $\frac{b_2}{b_1} = \frac{5}{9} < \frac{5\mu_0(\mu_0+1)}{8} = \frac{\sigma_2}{\sigma_1}$, $\frac{b_{k+1}}{b_k} = \frac{4k+1}{4k+5} < \frac{k+\mu_0}{k+1} = \frac{\sigma_{k+1}}{\sigma_k}$, $k \geq 2$. Now define $S_0(\theta) = \sin\frac{\theta}{2}$, $S_k(\theta) = \sin\frac{\theta}{2} + \sum_{j=1}^{k} \sigma_j \sin(2j + \frac{1}{2})\theta$, $k \geq 1$. Summing by parts and using inequality (i) of Corollary 24.3, we deduce that for $n \geq 2$ we have

$$\begin{aligned}
&\frac{1}{4}\sin\frac{\theta}{2} + \sum_{k=1}^{n} \frac{1}{4k+1} \sin\left(2k + \frac{1}{2}\right)\theta \\
&= \sum_{k=0}^{n} b_k \sin\left(2k + \frac{1}{2}\right)\theta \\
&= \sum_{k=0}^{n-1} \left(\frac{b_k}{\sigma_k} - \frac{b_{k+1}}{\sigma_{k+1}}\right) S_k(\theta) + \frac{b_n}{\sigma_n} S_n(\theta) > 0.
\end{aligned}$$

The proof is complete. □

Next we present some interesting counterparts of (iii) of Corollary 24.2 and (ii) of Corollary 24.3. We first insert the following classical result which is obtained by partial summation and can be found in [24] or [87].

Proposition 24.9 *Suppose that a_k, $k = 0, 1, \ldots$, is a sequence of real numbers. Write $S_0(\theta) = \frac{a_0}{2}$ and $S_n(\theta) = \frac{a_0}{2} + \sum_{k=1}^n a_k \cos k\,\theta$ for $n \ge 1$. We have*

$$S_n(\theta) = \sum_{k=0}^{n-1} \Delta a_k\, D_k(\theta) + a_n\, D_n(\theta)$$
$$= \sum_{k=0}^{n-2} \Delta^2 a_k\, (k+1) F_k(\theta) + n\, \Delta a_{n-1}\, F_{n-1}(\theta) + a_n\, D_n(\theta), \quad (24.35)$$

where

$$\Delta a_k = a_k - a_{k+1}, \qquad \Delta^2 a_k = \Delta a_k - \Delta a_{k+1} = a_k - 2\,a_{k+1} + a_{k+2},$$
$$k = 0, 1, 2, \ldots,$$
$$D_0(\theta) = \frac{1}{2}, \qquad D_n(\theta) = \frac{1}{2} + \sum_{k=1}^n \cos k\theta = \frac{\sin(n+\frac{1}{2})\theta}{2\sin\frac{\theta}{2}}, \qquad n \ge 1,$$

and

$$F_0(\theta) = D_0(\theta) = \frac{1}{2},$$
$$(n+1)F_n(\theta) = \sum_{k=0}^n D_k(\theta) = \frac{1}{2}\left(\frac{\sin(n+1)\frac{\theta}{2}}{\sin\frac{\theta}{2}}\right)^2 \ge 0, \quad n \ge 1.$$

We can now show the following.

Theorem 24.10 *For all positive integers n and $0 < \theta < \pi$, we have*

$$1 + \cos\theta + \sum_{k=2}^n \frac{(\mu)_k}{k!} \cos k\theta > 0, \qquad (24.36)$$

precisely when $0 < \mu \le \mu_0$, where μ_0 is as in Theorem 24.5. The sums are unbounded below when $1 \ge \mu > \mu_0$.

Proof The result is an immediate consequence of Corollary 24.2(iii) or Corollary 24.3(ii) in the case where $0 < \theta \le \pi/2$. In the case where $\pi/2 < \theta < \pi$, we readily check the cases $1 \le n \le 3$. Suppose that $n \ge 4$. We need to establish the result only for $\mu = \mu_0$. We define the sequence $a_0 = 2$, $a_1 = 1$, $a_k = \frac{(\mu_0)_k}{k!}$, $k \ge 2$.

It is easy to see that

$$\Delta^2 a_0 = \frac{\mu_0(\mu_0+1)}{2} > 0, \qquad \Delta^2 a_1 = \frac{1}{6}(\mu_0-1)\big(\mu_0^2-2\mu_0-6\big) > 0,$$

$$\Delta^2 a_k = a_k\,\frac{(1-\mu_0)(2-\mu_0)}{(k+1)(k+2)} > 0, \quad k \ge 2.$$

We use (24.35) to see that, for $n \ge 4$, the sum in (24.36) exceeds

$$\frac{\mu_0(\mu_0+1)}{4} + \frac{1}{3}(\mu_0-1)\big(\mu_0^2-2\mu_0-6\big)\cos^2\frac{\theta}{2} - \frac{a_4}{2\sin\frac{\theta}{2}} > 0.$$

The last inequality can be easily verified. The unboundedness of the sums when $1 \ge \mu > \mu_0$ is proved by applying (24.34) for $\rho = 0$.

This completes the proof of the theorem. □

Remark 24.1 Let $d_k = \frac{(\mu)_k}{k!}$, $k = 0,1,2,\ldots$, and $b_0 = \frac{1}{\mu}$, $b_k = \frac{1}{k^{1-\mu}}$, $k = 1,2,\ldots$, with $0 < \mu < 1$. Clearly, we have $\frac{b_1}{b_0} = \mu = \frac{d_1}{d_0}$, and by Bernoulli's inequality we get

$$\frac{b_{k+1}}{b_k} = \left(1 - \frac{1}{k+1}\right)^{1-\mu} < \frac{k+\mu}{k+1} = \frac{d_{k+1}}{d_k}, \quad k \ge 1.$$

In view of Proposition 24.5 and (iii) of Corollary 24.2, we deduce that, for all positive integers n and $0 < \theta < \pi$, we have

$$\frac{1}{\mu} + \sum_{k=1}^{n} \frac{\cos k\theta}{k^{1-\mu}} > 0, \tag{24.37}$$

for $0 < \mu \le \mu_0$.

In a similar way, it follows from Proposition 24.5 and Theorem 24.10 that for all positive integers n and $0 < \theta < \pi$, we have

$$\frac{1}{\mu} + \frac{1}{\mu}\cos\theta + \sum_{k=2}^{n} \frac{\cos k\theta}{k^{1-\mu}} > 0 \tag{24.38}$$

for $0 < \mu \le \mu_0$.

It is well-known that the sums $\sum_{k=1}^{n} \frac{\cos k\theta}{k^{1-\mu}}$ are uniformly bounded below precisely when $0 < \mu \le \mu_0$, see [87, V1, p. 191]. Moreover, it is shown in [33] that for all positive integers n and $0 < \theta < \pi$

$$1 + \sum_{k=1}^{n} \frac{\cos k\theta}{k^{1-\mu}} > 0, \tag{24.39}$$

for $0 < \mu \le \mu_0$. Accordingly, none of (24.37), (24.38), and (24.39) holds true for all n when $1 \ge \mu > \mu_0$.

The following closely related result has been obtained in [34].

Theorem 24.11 *Let*

$$U_n(\theta,\,\mu) := \mu + \sum_{k=1}^{n} \frac{(\mu)_k}{k!}\cos k\theta.$$

For all positive integers n and $0 \le \theta \le \pi$, *we have*

$$U_n(\theta,\,\mu) \ge 0, \tag{24.40}$$

if and only if $0 < \mu \le \mu_1 = 0.66458\ldots < \mu_0$, *where* μ_1 *is the unique solution* $\mu \in (0, 1)$ *of the equation*

$$\min_{\theta\in[0,\,\pi]} U_7(\theta,\,\mu) = 0. \tag{24.41}$$

Equality holds in (24.40) *for some* $\theta \in (0,\,\pi)$ *if and only if* $\mu = \mu_1$ *and* $n = 7$.

Clearly, the case $\mu = 1/2$ of this theorem is inequality (24.4).

Taking into consideration Remark 24.1, it is easily seen that the above result implies (iii) of Corollary 24.2, (24.36), and (24.39), but *only for* $0 < \mu \le \mu_1 = 0.66458\ldots < \mu_0$.

Here we give a new proof of Theorem 24.11 based on our methods developed in [59, 61–64], and in this section.

Proof As in the proof of Theorem 24.7, we can show that the equation (24.41) has a unique solution μ_1 in the interval (0, 1). By direct computation, we obtain $U_n(\theta,\,\mu_2) > 0$ for all $\theta \in [0,\,\pi]$ when $1 \le n \le 6$, where $\mu_2 := 0.66458 > \mu_1$. We set

$$a_0 = a_0(\mu) := 2\mu, \qquad a_1 = a_1(\mu) := \mu, \qquad a_k = a_k(\mu) := \frac{(\mu)_k}{k!}, \quad k \ge 2.$$

Summation by parts yields

$$U_n(\theta,\,\mu) = \sum_{k=1}^{n-1}\left(\frac{a_k(\mu)}{a_k(\mu_2)} - \frac{a_{k+1}(\mu)}{a_{k+1}(\mu_2)}\right)U_k(\theta,\,\mu_2) + \frac{a_n(\mu)}{a_n(\mu_2)}\,U_n(\theta,\,\mu_2). \tag{24.42}$$

Applying the above formula for $n = 2, \ldots, 6$, we prove that $U_n(\theta,\,\mu) > 0$ for $\theta \in [0,\,\pi]$ for all $\mu < \mu_2$ when $2 \le n \le 6$. Using (24.42) for $n = 7$ and μ_1 instead of μ_2, we show that $U_7(\theta,\,\mu) > 0$ for all $\theta \in [0,\,\pi]$ whenever $\mu < \mu_1$ while $U_7(\theta,\,\mu_1) \ge 0$ for $\theta \in [0,\,\pi]$.

The essential part of the proof amounts to showing

$$U_n(\theta,\,\mu) > 0 \quad \textit{for all } n \ge 8,\ \ \mu = \mu_2 = 0.66458 > \mu_1,\ \text{and } 0 < \theta < \pi. \tag{24.43}$$

Clearly, we have

$$\Delta^2 a_0 = a_2 = \frac{\mu(\mu+1)}{2}, \qquad \Delta^2 a_k = \frac{(1-\mu)(2-\mu)(\mu)_k}{(k+2)!} > 0, \quad k \geq 1.$$

Using (24.35), we obtain

$$U_n(\theta,\,\mu) > \frac{\mu(\mu+1)}{4} - \frac{a_8}{2\sin\frac{\theta}{2}} > 0$$

for $n \geq 8$, $\pi/2 \leq \theta \leq \pi$, and $\mu = 0.66458$.

By Proposition 24.8, we derive that

$$U_n(\theta,\,\mu) > 0, \quad 0 \leq \theta \leq \frac{\pi}{n}.$$

The proof of the remaining cases relies on a general expression for the sums $U_n(\theta,\,\mu)$, namely

$$\begin{aligned} U_n(\theta,\,\mu) = {} & \mu - 1 + \frac{\cos(\frac{\mu}{2}(\pi-\theta))}{(2\sin\frac{\theta}{2})^{\mu}} - \frac{\theta}{2\sin\frac{\theta}{2}}\frac{\cos(\frac{\mu\pi}{2})}{\theta^{\mu}} \\ & + \frac{\theta^{1-\mu}}{2\Gamma(\mu)}\frac{1}{\sin\frac{\theta}{2}}\int_0^{(n+\frac{1}{2})\theta}\frac{\cos t}{t^{1-\mu}}\,dt \\ & - \frac{1}{\Gamma(\mu)}\frac{\theta}{2\sin\frac{\theta}{2}}\sum_{k=n+1}^{\infty}\big(A_k(\theta)+B_k(\theta)\big) \\ & + \sum_{k=n+1}^{\infty}\Delta_k\cos k\theta, \end{aligned} \tag{24.44}$$

where

$$\begin{aligned} A_k(\theta) &:= \int_0^{\frac{1}{2}}\big(L(k,\,t)-M(k,\,t)\big)\cos\big(\theta(k-t)\big)\,dt, \\ B_k(\theta) &:= -2\int_0^{\frac{1}{2}}\sin(\theta t)\,L(k,t)\sin k\theta\,dt, \\ L(k,\,t) &:= \int_0^t \frac{1-\mu}{(k+s)^{2-\mu}}\,ds, \\ M(k,\,t) &:= \int_0^t \frac{1-\mu}{(k-t+s)^{2-\mu}}\,ds, \\ \Delta_k &:= \frac{1}{\Gamma(\mu)\,k^{1-\mu}} - \frac{(\mu)_k}{k!}, \quad k = 1,2,\ldots, \end{aligned}$$

(cf. [62, p. 201] or [63, (3.8)] or [61, (17)]).

For the first remainder term in (24.44), we have the estimates

$$\left|\sum_{k=n+1}^{\infty} A_k(\theta)\right| < \frac{1-\mu}{8}\frac{1}{n^{2-\mu}}, \qquad \left|\sum_{k=n+1}^{\infty} B_k(\theta)\right| < \frac{\theta}{\sin\frac{\theta}{2}}\frac{1-\mu}{12}\frac{1}{n^{2-\mu}}, \tag{24.45}$$

which are valid *for any* $\mu \in (0,\,1)$ and $\theta \in (0,\,\pi/2)$, see [59, Lemma 1] or [62, Lemma 1], or [63, Proposition 1]. Our method of proving (24.43) is mainly based on a sharp estimate for the second remainder term of (24.44). We have that

$$\left|\sum_{k=n+1}^{\infty} \Delta_k \cos k\theta\right| \le \frac{1}{\Gamma(\mu)}\frac{\mu(1-\mu)}{2}\frac{1}{(n+1)^{1-\mu}}, \tag{24.46}$$

which is valid *for any* $\mu \in [1/3,\,1)$ and $\theta \in [\pi/n,\,\pi]$, see [60, Proposition 1] and [64, Theorem 5]. It is worth mentioning here that (24.46) is derived using the sharp inequality

$$x - \frac{\Gamma(x+\mu)}{\Gamma(x+1)}x^{2-\mu} < \frac{\mu(1-\mu)}{2}, \tag{24.47}$$

which, in turn, holds true for all $x > 0$, if and only if $\frac{1}{3} \le \mu < 1$.

Let us denote by $K_n(\theta)$ the integral in (24.44) and recall that this is positive for $\mu = 0.66458$. Moreover, we are able to find sharp lower bounds for it on appropriate intervals. In particular we have:

$$\text{For } \theta \in I_0 := \left(\frac{\pi}{n+\frac{1}{2}},\, \frac{2\pi}{n+\frac{1}{2}}\right), \quad K_n(\theta) \ge \int_0^{3\pi/2} \frac{\cos t}{t^{1-\mu}}\,dt = 0.097798\ldots := \kappa_0;$$

$$\text{For } \theta \in I_1 := \left(\frac{3\pi}{n+\frac{1}{2}},\, \frac{4\pi}{n+\frac{1}{2}}\right), \quad K_n(\theta) \ge \int_0^{7\pi/2} \frac{\cos t}{t^{1-\mu}}\,dt = 0.236885\ldots := \kappa_1;$$

$$\text{For } \theta \in I_2 := \left(\frac{5\pi}{n+\frac{1}{2}},\, \frac{\pi}{2}\right), \quad K_n(\theta) \ge \int_0^{11\pi/2} \frac{\cos t}{t^{1-\mu}}\,dt = 0.298826\ldots := \kappa_2.$$

The numerical values above have been obtained using Maple 14; see also [17] for fast and elementary methods of computation of integrals of this kind. Notice also that for $\theta \in (\frac{2\pi}{n+\frac{1}{2}}, \frac{3\pi}{n+\frac{1}{2}})$ and $\theta \in (\frac{4\pi}{n+\frac{1}{2}}, \frac{5\pi}{n+\frac{1}{2}})$, the desired inequality (24.43) is obvious because of (24.35) and the positivity of $\Delta^2 a_k$.

Next, we observe that

$$\begin{aligned} F(\theta) &:= \frac{\cos(\frac{\mu}{2}(\pi-\theta))}{(2\sin\frac{\theta}{2})^{\mu}} - \frac{\theta}{2\sin\frac{\theta}{2}}\frac{\cos(\frac{\mu\pi}{2})}{\theta^{\mu}} \\ &= \frac{\theta}{2\sin\frac{\theta}{2}}\frac{1}{\theta^{\mu}}\left\{p(\theta)\left(\cos\left(\frac{\mu}{2}(\pi-\theta)\right) - \cos\left(\frac{\mu\pi}{2}\right)\right)\right. \\ &\quad \left. + \left(p(\theta)-1\right)\cos\left(\frac{\mu\pi}{2}\right)\right\} \end{aligned}$$

$$\geq \frac{\theta}{2\sin\frac{\theta}{2}}\left\{\frac{\mu}{2}\theta^{1-\mu}q(\theta)+\theta^{-\mu}\big(p(\theta)-1\big)\cos\left(\frac{\mu\pi}{2}\right)\right\}, \tag{24.48}$$

where

$$p(\theta) := \left(\frac{2\sin\frac{\theta}{2}}{\theta}\right)^{1-\mu},$$

$$q(\theta) := \left(\frac{2\sin\frac{\theta}{2}}{\theta}\right)^{2-\mu}\sin\big(\mu(2\pi-\theta)/4\big).$$

Suppose that $\theta \in I_2$. It follows from the above that

$$\begin{aligned}
&\mu - 1 + F(\theta) + \frac{\theta^{1-\mu}}{2\Gamma(\mu)}\frac{1}{\sin\frac{\theta}{2}}K_n(\theta)\\
&\quad\geq \mu - 1\\
&\qquad + \frac{\theta}{2\sin\frac{\theta}{2}}\left\{\frac{\mu}{2}\theta^{1-\mu}q(\pi/2)+\theta^{-\mu}\left(\big(p(\pi/2)-1\big)\cos\left(\frac{\mu\pi}{2}\right)+\frac{\kappa_2}{\Gamma(\mu)}\right)\right\}\\
&\quad> 0.085.
\end{aligned} \tag{24.49}$$

Let

$$R_n(\theta) := -\frac{1}{\Gamma(\mu)}\frac{\theta}{2\sin\frac{\theta}{2}}\sum_{k=n+1}^{\infty}\big(A_k(\theta)+B_k(\theta)\big)+\sum_{k=n+1}^{\infty}\Delta_k\cos k\theta.$$

It follows from (24.45) and (24.46) that when $\theta \in (0,\pi/2)$ and $n \geq 8$

$$\begin{aligned}
\big|R_n(\theta)\big| &\leq \frac{1}{\Gamma(\mu)}\left[\frac{\pi\sqrt{2}}{4}\frac{1-\mu}{8}\frac{1}{n^{2-\mu}}+\frac{\pi^2}{8}\frac{1-\mu}{6}\frac{1}{n^{2-\mu}}+\frac{\mu(1-\mu)}{2}\frac{1}{(n+1)^{1-\mu}}\right]\\
&< 0.045.
\end{aligned} \tag{24.50}$$

Combine this with (24.44) and (24.49) to conclude that

$$U_n(\theta,\mu) > 0, \quad \text{when } \theta \in I_2 \text{ and } n \geq 8.$$

In the same way, when $\theta \in I_1$ and $n \geq 8$,

$$\begin{aligned}
&\mu - 1 + F(\theta) + \frac{\theta^{1-\mu}}{2\Gamma(\mu)}\frac{1}{\sin\frac{\theta}{2}}K_n(\theta)\\
&\quad\geq \mu - 1\\
&\qquad + \frac{\theta}{2\sin\frac{\theta}{2}}\left\{\frac{\mu}{2}\theta^{1-\mu}q(8\pi/17)+\theta^{-\mu}\left(\big(p(8\pi/17)-1\big)\cos\left(\frac{\mu\pi}{2}\right)+\frac{\kappa_1}{\Gamma(\mu)}\right)\right\}\\
&\quad> 0.049.
\end{aligned} \tag{24.51}$$

From this, (24.44), and (24.50), we infer that

$$U_n(\theta, \mu) > 0, \quad \text{when } \theta \in I_1 \text{ and } n \geq 8.$$

In order to handle the case $\theta \in I_0$, we need to slightly improve the remainder estimate (24.50). We have for $n \geq 28$

$$\begin{aligned}\left|R_n(\theta)\right| \leq \frac{1}{\Gamma(\mu)}\Bigg[&\frac{2\pi}{57\sin(2\pi/57)}\frac{1-\mu}{8}\frac{1}{n^{2-\mu}} + \frac{4\pi^2}{3249\sin^2(2\pi/57)}\frac{1-\mu}{6}\frac{1}{n^{2-\mu}} \\ &+ \frac{\mu(1-\mu)}{2}\frac{1}{(n+1)^{1-\mu}}\Bigg] < 0.0274. \end{aligned} \tag{24.52}$$

When $\theta \in I_0$ and $n \geq 28$, we have

$$\begin{aligned}&\mu - 1 + F(\theta) + \frac{\theta^{1-\mu}}{2\Gamma(\mu)}\frac{1}{\sin\frac{\theta}{2}}K_n(\theta) \\ &\geq \mu - 1 \\ &\quad + \frac{\theta}{2\sin\frac{\theta}{2}}\left\{\frac{\mu}{2}\theta^{1-\mu}q(4\pi/57) + \theta^{-\mu}\left(\left(p(4\pi/57) - 1\right)\cos\left(\frac{\mu\pi}{2}\right) + \frac{\kappa_0}{\Gamma(\mu)}\right)\right\} \\ &> 0.029. \end{aligned} \tag{24.53}$$

It follows from this, (24.44), and (24.52) that

$$U_n(\theta, \mu) > 0, \quad \text{when } \theta \in I_0 \text{ and } n \geq 28.$$

The above method can be adapted to prove positivity of $U_n(\theta, \mu)$ in the remaining cases $8 \leq n \leq 27$ and $\theta \in I_0 = (\frac{\pi}{n+\frac{1}{2}}, \frac{2\pi}{n+\frac{1}{2}})$ by considering additional partitions of I_0. It is more convenient, however, to directly compute the minima of the polynomials $U_n(\theta, \mu)$, $\theta \in I_0$ and $8 \leq n \leq 27$ ($\mu = \mu_2 = 0.66458$). Since we wish to prove positivity of these trigonometric polynomials on a specific interval, we can convert them into algebraic polynomials by setting $x = \cos\theta$ and prove that these polynomials have no zeros in the interval under consideration. As we have already shown, these polynomials are positive at the end points of the interval I_0; therefore, they are positive everywhere in this interval. Application of Sturm's Theorem confirms the result and completes the proof of (24.43). The corresponding calculations can be facilitated by the use of Maple 14.

Finally, suppose that $\theta \in [0, \pi]$ such that $U_7(\theta, \mu_2) \geq 0$. It follows from (24.43) and (24.42) that $U_n(\theta, \mu) > 0$ for all $n \geq 8$ when $\mu < \mu_2$. In the case where $\theta \in [0, \pi]$ and $U_7(\theta, \mu_2) < 0$, using (24.42), we obtain for $n \geq 9$ and $\mu \leq \mu_1 < \mu_2$

$$U_n(\theta,\,\mu) - U_7(\theta,\,\mu) = \sum_{k=8}^{n-1}\left(\frac{a_k(\mu)}{a_k(\mu_2)} - \frac{a_{k+1}(\mu)}{a_{k+1}(\mu_2)}\right)U_k(\theta,\,\mu_2)$$
$$+\frac{a_n(\mu)}{a_n(\mu_2)}U_n(\theta,\,\mu_2) - \frac{a_8(\mu)}{a_8(\mu_2)}U_7(\theta,\,\mu_2) > 0.$$

Note also that

$$U_8(\theta,\,\mu) - U_7(\theta,\,\mu) = \frac{a_8(\mu)}{a_8(\mu_2)}\,U_8(\theta,\,\mu_2) - \frac{a_8(\mu)}{a_8(\mu_2)}\,U_7(\theta,\,\mu_2) > 0.$$

It follows from these that for $n \geq 8$

$$U_n(\theta,\,\mu) > U_7(\theta,\,\mu) \geq 0, \quad \text{when } \mu \leq \mu_1.$$

This completes the proof of the theorem. □

Remark 24.2 In the case where $0 < \mu < \frac{1}{3}$, we have the following counterpart of (24.47) which is obtained in [66]

$$x - \frac{\Gamma(x+\mu)}{\Gamma(x+1)}x^{2-\mu} < \frac{\mu(1-\mu)}{2} + \frac{\mu\,(1-3\mu)\,(1-\mu)\,(2-\mu)}{24x} \quad \text{for all } x > 0.$$

The above inequality is sharp. This entails the following estimate for the second remainder term of (24.44).

$$\left|\sum_{k=n+1}^{\infty} \Delta_k \cos k\theta\right| \leq \frac{1}{\Gamma(\mu)}\frac{\mu(1-\mu)}{2}\left[1 + \frac{(1-3\mu)(2-\mu)}{12(n+1)}\right]\frac{1}{(n+1)^{1-\mu}},$$

which is valid *for any* $\mu \in (0,\,1/3)$ and $\theta \in [\pi/n,\,\pi]$, $n \geq 2$.

Note also that the inequality

$$x - \frac{\Gamma(x+\mu)}{\Gamma(x+1)}x^{2-\mu} > 0, \quad \text{for all } x > 0,$$

holds *for all* $\mu \in (0,\,1)$ (cf. [66]).

Next we give a generalization of Corollary 24.2 which has been established in [61].

Theorem 24.12 *Let* $\rho \in [0,\,1]$. *For all* $\rho \in [0,\,\frac{1}{2}]$, *we have*

$$\sum_{k=0}^{n} \frac{(\mu)_k}{k!}\cos\big[(2k+\rho)\theta\big] > 0, \quad \textit{for all } n \in \mathbb{N} \textit{ and } 0 < \theta < \pi, \tag{24.54}$$

if and only if $0 < \mu \leq \mu_0$ *and the sums in* (24.54) *are unbounded below when* $1 \geq \mu > \mu_0$. *The number* μ_0 *is as in Theorem* 24.5. *In the case where* $\rho \in (\frac{1}{2},\,1]$, *inequality* (24.54) *fails to hold for appropriate* n *and* θ *and any value of* $\mu \in (0,\,1]$.

The point here is that the best possible range of μ for the validity of (24.54) is *independent* of $\rho \in [0, \frac{1}{2}]$. We shall show that this is not the case for the corresponding sine sums. For $\rho \in (0, 1]$, we seek to determine the best possible range of μ so that inequality

$$\sum_{k=0}^{n} \frac{(\mu)_k}{k!} \sin\left[(2k+\rho)\theta\right] > 0, \tag{24.55}$$

holds for all $n = 1, 2, \ldots$ and $0 < \theta < \pi$. As mentioned earlier, (24.54) and (24.55) are equivalent *only when* $\rho = 1/2$.

When studying (24.55), we first consider the limiting case

$$\lim_{n\to\infty} \left(\frac{\theta}{n}\right)^{\mu} \sum_{k=0}^{n} \frac{(\mu)_k}{k!} \sin\left[(2k+\rho)\left(\pi - \frac{\theta}{2n}\right)\right]$$

$$= -\frac{1}{\Gamma(\mu)} \int_0^{\theta} t^{\mu-1} \sin(t - \rho\pi)\, dt. \tag{24.56}$$

Hence a necessary condition for the validity of (24.55) is the non-positivity of the integral in (24.56) for all $\theta > 0$, and in particular for $\theta = (\rho+1)\pi$:

$$I(\mu) := \int_0^{(\rho+1)\pi} t^{\mu-1} \sin(t - \rho\pi)\, dt \le 0.$$

It can be shown, see [65] and compare with [63], that the equation $I(\mu) = 0$ has a unique solution in $(0, 1)$ denoted by $\mu^*(\rho)$ so that

$$I(\mu) > 0, \quad \text{for } \mu > \mu^*(\rho)$$

and

$$I(\mu) < 0, \quad \text{for } \mu < \mu^*(\rho).$$

Therefore, the sums in (24.55) assume negative values for $\mu > \mu^*(\rho)$, $0 < \rho < 1$, appropriate θ and n sufficiently large. It is conjectured, see [63, 64], and [61], that

Conjecture 24.1 For $\rho \in (0, 1]$, inequality (24.55) holds for all $n = 1, 2, \ldots$ and $0 < \theta < \pi$, precisely when $0 < \mu \le \mu^*(\rho)$.

This conjecture is strongly supported by numerical experimentation, and up to now it has been settled for the following special cases: For $\rho = 1$, since $\mu^*(1) = 1$, the result follows from the nonnegativity of the classical Fejér kernel. For $\rho = 1/2$, see [59]. For $\rho = 3/4$, see [63]. For $\rho = 1/4$, see [64]. For ρ in an open neighborhood of $1/5$, see [67].

Note that $\mu^*(\frac{1}{2}) = \mu_0 = 1 - \alpha_0 = 0.691556\ldots$, where μ_0 is the number appearing throughout this section. There are elementary and efficient methods of calculating *any* value of the function $\mu^*(\rho)$ at any required precision (cf. [61]).

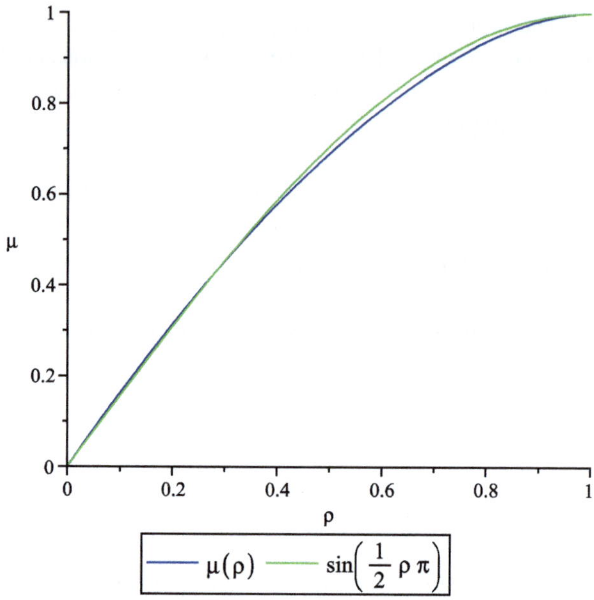

Fig. 24.1 The graphs of $\mu^*(\rho)$ and $\sin(\rho\pi/2)$

It is of interest to note that

$$\left|\mu^*(\rho) - \sin(\rho\pi/2)\right| < 0.02, \quad \rho \in (0, 1).$$

and compare with Fig. 24.1.

It has been recently proved, see [65], that

Theorem 24.13 *The function $\mu^*(\rho)$ is analytic and strictly increasing on* $(0, 1)$.

Inequalities (24.54) and (24.55) have remarkable applications in geometric function theory, see [61–63], and [64]. In particular, let $\mathbb{D} = \{z \in \mathbb{C} : |z| < 1\}$. It can be shown, see [61] (or [66] for a different proof), that for $(\mu, \rho) \in (0, 1]^2$ and $n \in \mathbb{N}$ inequality (24.55) is equivalent to

$$\left|\arg\left[(1-z)^\rho \sum_{k=0}^{n} \frac{(\mu)_k}{k!} z^k\right]\right| < \frac{\rho\pi}{2} \quad \text{for } z \in \mathbb{D}. \tag{24.57}$$

Accordingly, the truth of Conjecture 24.1 would imply that (24.57) is valid precisely when $0 < \mu \le \mu^*(\rho)$. It is easy to see that inequality (24.57) implies that

$$\left|\arg\left[\sum_{k=0}^{n} \frac{(\mu)_k}{k!} z^k\right]\right| < \rho\pi \quad \text{for } z \in \mathbb{D}, \tag{24.58}$$

for the same range of μ.

Inequality (24.58) *does not* hold for $\mu > \mu^*(\rho)$. Indeed, recall that

$$I(\mu) = \int_0^{(\rho+1)\pi} t^{\mu-1}\sin(t-\rho\pi)\,dt > 0 \quad \text{for } \mu > \mu^*(\rho),\ 0<\rho<1. \tag{24.59}$$

Then, define $w := \int_0^{(\rho+1)\pi} t^{\mu-1} e^{it}\,dt$ and observe that $I(\mu) = \operatorname{Im}(e^{-i\rho\pi} w)$. We also have $\operatorname{Im}(w) = \int_0^{(\rho+1)\pi} t^{\mu-1}\sin t\,dt > 0$ for all $\rho \in (0,1)$ and $\mu \in (0,1)$. It follows from this and (24.59) that

$$\rho\pi < \arg w < \pi, \quad \text{for } \mu > \mu^*(\rho),\ 0<\rho<1. \tag{24.60}$$

Suppose that (24.58) holds for $\mu > \mu^*(\rho)$, $0<\rho<1$. Then, for $z = e^{i\frac{\theta}{n}}$, $\theta > 0$ we have

$$-\rho\pi \le \arg\left[\sum_{k=0}^{n} \frac{(\mu)_k}{k!} e^{ik\frac{\theta}{n}}\right] \le \rho\pi.$$

It follows from this and the asymptotic formula

$$\lim_{n\to\infty}\left(\frac{\theta}{n}\right)^{\mu} \sum_{k=0}^{n} \frac{(\mu)_k}{k!} e^{i\frac{k\theta}{n}} = \frac{1}{\Gamma(\mu)} \int_0^{\theta} \frac{e^{it}}{t^{1-\mu}}\,dt$$

that

$$-\rho\pi \le \arg\left[\int_0^{\theta} \frac{e^{it}}{t^{1-\mu}}\,dt\right] \le \rho\pi.$$

Setting $\theta = (\rho+1)\pi$ in the above, we get $-\rho\pi \le \arg w \le \rho\pi$, which contradicts (24.60). Therefore, inequality (24.58) cannot hold for $\mu > \mu^*(\rho)$.

Finally, it should be noted that the polynomial $\sum_{k=0}^{n} \frac{(\mu)_k}{k!} z^k$ is the nth partial sum of the Taylor series expansion at the origin of the function $f_\mu(z) := \frac{1}{(1-z)^\mu}$, $\mu > 0$. The function $f_\mu(z)$ is analytic in $\mathbb{D}$ and satisfies $f_\mu(0) = 1$, and $\operatorname{Re}\frac{zf_\mu'(z)}{f_\mu(z)} > -\frac{\mu}{2}$ for all $z \in \mathbb{D}$. It can be shown that, see [61, 63], and [64], inequalities analogous to (24.57) and (24.58) hold for the partial sums of *any* analytic function $f(z)$ in $\mathbb{D}$ such that $f(0) = 1$ and $\operatorname{Re}\frac{zf'(z)}{f(z)} > -\frac{\mu}{2}$ for all $z \in \mathbb{D}$.

Acknowledgements This research was supported by a grant from the Leventis Foundation (Grant no. 3411-21041).

References

1. Andreani, R., Dimitrov, D.K.: An extremal nonnegative sine polynomial. Rocky Mt. J. Math. **33**, 759–774 (2003)
2. Andrews, G.E., Askey, R., Roy, R.: Special Functions. Cambridge University Press, Cambridge (1999)
3. Alzer, H., Koumandos, S.: Sharp inequalities for trigonometric sums. Math. Proc. Camb. Philos. Soc. **134**, 139–152 (2003)

4. Alzer, H., Koumandos, S.: A sharp bound for a sine polynomial. Colloq. Math. **96**, 83–91 (2003)
5. Alzer, H., Koumandos, S.: Inequalities of Fejér–Jackson type. Monatshefte Math. **139**, 89–103 (2003)
6. Alzer, H., Koumandos, S.: Sharp inequalities for trigonometric sums in two variables. Ill. J. Math. **48**, 887–907 (2004)
7. Alzer, H., Koumandos, S.: Companions of the inequalties of Fejér–Jackson and Young. Anal. Math. **31**, 75–84 (2005)
8. Alzer, H., Koumandos, S.: Sub- and superadditive properties of Fejér's sine polynomial. Bull. Lond. Math. Soc. **38**, 261–268 (2006)
9. Alzer, H., Koumandos, S.: Inequalities for two sine polynomials. Colloq. Math. **105**, 127–134 (2006)
10. Alzer, H., Koumandos, S.: Some monotonic trigonometric sums. Analysis **26**, 429–449 (2006)
11. Alzer, H., Koumandos, S.: A new refinement of Young's inequality. Proc. Edinb. Math. Soc. **50**, 255–262 (2007)
12. Alzer, H., Koumandos, S.: On the partial sums of a Fourier series. Constr. Approx. **27**, 253–268 (2008)
13. Alzer, H., Koumandos, S.: Remarks on a sine polynomial. Arch. Math. **93**, 475–479 (2009)
14. Alzer, H., Koumandos, S., Lamprecht, M.: A refinement of Vietoris' inequality for sine polynomials. Math. Nachr. **283**, 1549–1557 (2010)
15. Alzer, H., Shi, X.: Sharp bounds for trigonometric polynomials in two variables. Anal. Appl. **7**, 341–350 (2009)
16. Arestov, V.V., Kondrat'ev, V.P.: Certain extremal problem for nonnegative trigonometric polynomials. Mat. Zametki **47**, 15–28 (1990)
17. Arias de Reyna, J., Van de Lune, J.: High precision computation of a constant in the theory of trigonometric series. Math. Comput. **78**, 2187–2191 (2009)
18. Askey, R.: Orthogonal Polynomials and Special Functions. Regional Conf. Lect. Appl. Math., vol. 21. SIAM, Philadelphia (1975)
19. Askey, R.: Vietoris's inequalities and hypergeometric series. In: Milovanović, G.V. (ed.) Recent Progress in Inequalities, pp. 63–76. Kluwer, Dordrecht (1998)
20. Askey, R., Gasper, G.: Inequalities for polynomials. In: Baernstein, A. II, Drasin, D., Duren, P., Marden, A. (eds.) The Bieberbach Conjecture. Math. Surveys and Monographs, vol. 21, pp. 7–32. Amer. Math. Soc., Providence (1986)
21. Askey, R., Steinig, J.: Some positive trigonometric sums. Trans. Am. Math. Soc. **187**, 295–307 (1974)
22. Askey, R., Fitch, J., Gasper, G.: On a positive trigonometric sum. Proc. Am. Math. Soc. **19**, 1507 (1968)
23. Barnard, R.W., Pearce, K., Wheeler, W.: On a coefficient conjecture of Brannan. Complex Var. **33**, 51–61 (1997)
24. Bary, N.K.: A Treatise on Trigonometric Series, vol. I. Pergamon, New York (1964)
25. Belov, A.S.: Coefficients of trigonometric series with nonnegative partial sums. Mat. Zametki **41**, 152–158 (1987)
26. Belov, A.S.: Non-Fourier–Lebesgue trigonometric series with nonnegative partial sums. Math. Notes **59**, 18–30 (1996)
27. Belov, A.S.: Examples of trigonometric series with nonnegative partial sums. Math. USSR Sb. **186**, 21–46 (1995) (Russian); **186**, 485–510 (1995) (English translation)
28. Bertoni, V.: The univalent polynomial of suffridge as a summability kernel. Complex Var. Elliptic Equ. **53**, 401–409 (2008)
29. Bisgaard, T.M., Sasvári, Z.: On the positive definiteness of certain functions. Math. Nachr. **186**, 81–99 (1997)
30. Brown, G., Hewitt, E.: A class of positive trigonometric sums. Math. Ann. **268**, 91–122 (1984)

31. Brown, G., Koumandos, S.: On a monotonic trigonometric sum. Monatshefte Math. **123**, 109–119 (1997)
32. Brown, G., Koumandos, S.: A new bound for the Fejér–Jackson sum. Acta Math. Hung. **80**, 21–30 (1998)
33. Brown, G., Wang, K., Wilson, D.C.: Positivity of some basic cosine sums. Math. Proc. Camb. Philos. Soc. **114**, 383–391 (1993)
34. Brown, G., Feng, D., Wang, K.: On positive cosine sums. Math. Proc. Camb. Philos. Soc. **142**, 219–232 (2007)
35. Brown, G., Wang, K.: An extension of the Fejér–Jackson inequality. J. Aust. Math. Soc. A **62**, 1–12 (1997)
36. Brown, G., Yin, Q.: Positivity of a class of cosine sums. Acta Sci. Math. (Szeged) **67**, 221–247 (2001)
37. Brown, J., Goldstein, M., McDonald, J.: A sequence of extremal problems for trigonometric polynomials. J. Math. Anal. Appl. **130**, 545–551 (1988)
38. De la Vallé Poussin, C.J.: Recherches analitiques sur la théorie des nombres premiers. Ann. Soc. Sci. Brux. **20**, 183–256 (1896)
39. Dimitrov, D.K., Merlo, C.A.: Nonnegative trigonometric polynomials. Constr. Approx. **18**, 117–143 (2002)
40. Dumitrescu, B.A.: Positive Trigonometric Polynomials and Signal Processing Applications, 1st edn. Springer, Berlin (2007)
41. Dumitrescu, B.A.: Trigonometric polynomials positive on frequency domains and applications to 2-D FIR filter design. IEEE Trans. Signal Process. **54**, 4282–4292 (2006)
42. Fernández-Durán, J.J.: Circular distributions based on nonnegative trigonometric sums. Biometrics **60**, 499–503 (2004)
43. Gasper, G.: Nonnegative sums of cosine, ultraspherical and Jacobi polynomials. J. Math. Anal. Appl. **26**, 60–68 (1969)
44. Geiger, C., Opfer, G.: Complex Chebyshev polynomials on circular sectors. J. Approx. Theory **24**, 93–118 (1978)
45. Gluchoff, A., Hartmann, F.: Univalent polynomials and non-negative trigonometrics sums. Am. Math. Mon. **195**, 508–522 (1998)
46. Gronwall, T.H.: Über die Gibbssche Erscheinung und die trigonometrischen Summen $\sin x + \frac{1}{2}\sin 2x + \cdots + \frac{1}{n}\sin nx$. Math. Ann. **72**, 228–243 (1912)
47. Guo, V.J.W., Zeng, J.J.: Pairs of lattice paths and positive trigonometric sums. Constr. Approx. **32**, 67–75 (2010)
48. Herzog, F., Piranian, G.: Some properties of the Fejér polynomials. Proc. Am. Math. Soc. **7**, 379–386 (1956)
49. Ismail, M.E.H., Kim, D., Stanton, D.: Lattice paths and positive trigonometric sums. Constr. Approx. **15**, 69–81 (1999)
50. Jackson, D.: Über eine trigonometrische Summe. Rend. Circ. Mat. Palermo **32**, 257–262 (1911)
51. Jagers, A.A.: Problem 73-21, A sine inequality. SIAM Rev. **16**, 550–553 (1974)
52. Katznelson, Y.: Trigonometric series with positive partial sums. Bull. Am. Math. Soc. **71**, 718–719 (1965)
53. Koumandos, S.: On a positive sine sum. Colloq. Math. **71**, 243–251 (1996)
54. Koumandos, S.: Positive trigonometric sums in the theory of univalent functions. In: Papamichael, N., Ruscheweyh, S., Saff, E.B. (eds.) Computational Methods and Function Theory (CMFT'97), pp. 345–357. World. Sci. Publ., River Edge (1999)
55. Koumandos, S.: Monotonic trigonometric sums and coefficients of Bloch functions. Ill. J. Math. **43**, 100–112 (1999)
56. Koumandos, S.: A positive functional bound for certain sine sums. Anal. Math. **26**, 35–52 (2000)
57. Koumandos, S.: Some inequalities for cosine sums. Math. Inequal. Appl. **4**, 267–279 (2001)
58. Koumandos, S.: Positive trigonometric sums and applications. Ann. Math. Inform. **33**, 77–91 (2006)

59. Koumandos, S.: An extension of Vietoris's inequalities. Ramanujan J. **14**, 1–38 (2007)
60. Koumandos, S.: Monotonicity of some functions involving the gamma and psi functions. Math. Comput. **77**, 2261–2275 (2008)
61. Koumandos, S.: Positive trigonometric sums and starlike functions. In: Gautschi, W., Mastroianni, G., Rassias, Th. (eds.) Approximation and Computation. Springer Optimization and its Applications, vol. 42, pp. 157–183 (2011)
62. Koumandos, S., Ruscheweyh, S.: Positive Gegenbauer polynomial sums and applications to starlike functions. Constr. Approx. **23**, 197–210 (2006)
63. Koumandos, S., Ruscheweyh, S.: On a conjecture for trigonometric sums and starlike functions. J. Approx. Theory **149**, 42–58 (2007)
64. Koumandos, S., Lamprecht, M.: On a conjecture for trigonometric sums and starlike functions II. J. Approx. Theory **162**, 1068–1084 (2010)
65. Koumandos, S., Lamprecht, M.: The zeros of certain Lommel functions. Proc. Am. Math. Soc. (to appear)
66. Koumandos, S., Lamprecht, M.: Complete monotonicity and related properties of some special functions. Math. Comput. (to appear)
67. Lamprecht, M.: Topics in geometric function theory and harmonic analysis. Ph.D. Thesis, University of Cyprus (2010)
68. Landau, E.: Über eine trigonometrische Ungleichung. Math. Z. **37**, 36 (1933)
69. Megretski, A.: Positivity of trigonometric polynomials. In: Proceedings of the 42nd IEEE Conference on Decision and Control, Maui, Hawaii, December 2003, pp. 3814–3817 (2003)
70. McDonald, J.N., Siefker, A.: Nonnegative trigonometric sums. J. Math. Anal. Appl. **238**, 580–586 (1999)
71. Milcetich, J.G.: On a coefficient conjecture of Brannan. J. Math. Anal. Appl. **139**, 515–522 (1989)
72. Milovanović, G.V., Mitrinović, D.S., Rassias, Th.M.: Topics in Polynomials: Extremal Problems, Inequalities, Zeros. World Sci., Singapore (1994)
73. Mues, E., Redheffer, R.: Über eine Differentialungleichung m-ter Ordnung im Komplexen. Pac. J. Math. **118**, 487–495 (1985)
74. Peherstorfer, F.: On an extremal problem for nonnegative trigonometric polynomials and the characterization of positive quadrature formulas with Chebyshev weight function. Acta Math. Acad. Sci. Hung. **39**, 107–116 (1982)
75. Pomerance, C.: Remarks on the Pólya–Vinogradov inequality. Integers (Proceedings of the Integers Conference, October 2009) **11A**, 1–11 (2011)
76. Rahman, Q.I.: Inequalities concerning polynomials and trigonometric polynomials. J. Math. Anal. Appl. **6**, 303–324 (1963)
77. Rogosinski, W., Szegő, G.: Über die Abschnitte von Potenzreihen die in einem Kreise beschränkt bleiben. Math. Z. **28**, 73–94 (1928)
78. Ruscheweyh, S.: Coefficient conditions for starlike functions. Glasg. Math. J. **29**, 141–142 (1987)
79. Ruscheweyh, S., Salinas, L.: On starlike functions of order $\lambda \in [\frac{1}{2}, 1)$. Ann. Univ. Mariae Curie-Sklodowska **54**, 117–123 (2000)
80. Ruscheweyh, S., Salinas, L.: Stable functions and Vietoris' Theorem. J. Math. Anal. Appl. **291**, 596–604 (2004)
81. Turàn, P.: Egy Steinhausfele problèmàrol. Mat. Lapok **4**, 263–275 (1953)
82. Turàn, P.: On a trigonometric sum. Ann. Pol. Math. **25**, 155–161 (1953)
83. Van Assche, W.: Orthogonal polynomials in the complex plane and on the real line. Fields Inst. Commun. **14**, 211–245 (1997)
84. Vietoris, L.: Über das Vorzeichen gewisser trigonometrischer Summen. S.-B. Öster. Akad. Wiss. **167**, 125–135 (1958). Teil II: Akad. Wiss. **167**, 192–193 (1959)
85. Young, W.H.: On a certain series of Fourier. Proc. Lond. Math. Soc. (2) **11**, 357–366 (1913)
86. Yudin, V.A.: Trigonometric series with positive partial sums. Math. Notes **53**, 348–350 (1993)
87. Zygmund, A.: Trigonometric Series, 3rd edn. Cambridge University Press, Cambridge (2002)

Chapter 25
On Vandiver's Best Result on FLT1

Preda Mihăilescu

Abstract In a paper from 1934, Vandiver sketched the proof of the claim that the First Case of Fermat's Last Theorem follows from the conjecture presently bearing his name. In 1993, Sitaraman showed that the existing gap in Vandiver's proof could easily be filled by adding a condition on the class group of the pth cyclotomic field. In this paper, we give a proof of a slightly more general result than the one of Vandiver–Sitaraman, with consequences for a larger family of Diophantine equations.

Key words Fermat's last theorem · Diophantine equations · Kummer–Vandiver conjecture · Cyclotomic field

Mathematics Subject Classification 11Dxx · 11A41

25.1 Introduction

Fermarcheology In his paper [13] from 1934, Vandiver announces a new result that he considers to be his most important one concerning the First Case of Fermat's Last Theorem (short: FLT1). The result states that if the odd prime p does not divide the class number h^+ of the maximal real subfield $\mathbb{K}^+$ of the pth cyclotomic field $\mathbb{K} = \mathbb{Q}[\zeta]$, then FLT1 is true; the paper gives only a sketch of the proof. The material is built up with impetus, but the line of the proof becomes sketchy, especially on the last half page, where the prepared argument should lead to the announced claim. The proof is known to be erroneous.

Interestingly, Ribenboim dedicates some space on page 188 of his classical 13 Lectures [8], mentioning that both Iwasawa and Greenberg tried without success to fix Vandiver's error ... but unlike in other of the numerous cases of erroneous statements about FLT reported in his book, there is no information about the error itself.

Dedicated to Professor Themistocles M. Rassias on the occasion of his 60th birthday.

P. Mihăilescu (✉)
Mathematisches Institut der Universität Göttingen, Göttingen, Germany
e-mail: preda@uni-math.gwdg.de

P.M. Pardalos et al. (eds.), *Nonlinear Analysis*, Springer Optimization and Its Applications 68, In Honor of Themistocles M. Rassias on the Occasion of his 60th Birthday, DOI 10.1007/978-1-4614-3498-6_25,

Instead, the chapter ends (*7E*) on a positive tone presenting the following result attributed to Grün: "*If* $p > 3$ *is a prime that does not divide* h_p^+ *and the Bernoulli number* $B_{2kp} \not\equiv 0 \bmod p^3$ *for* $k \in \{2, 3, \ldots, (p-3)/2\}$ *then the first case holds for the exponent* p". This would have been a notable unification of the conditions required for FLTII and FLTI, and adding one condition to $p \nmid h_p^+$ seems no big loss: the happy end of the story is mysteriously saved. Upon opening the 1934 volume of Crelle at the page of Grün's paper, one discovers step by step the arguments for the conditional proof of the ... Second Case of FLT, as they are known from Chap. 9 of Washington's textbook [14]: Grün's work had no tangency with the first case at all.

But optimism tends to move mountains, and Ribenboim was only anticipating by 15 years. In a result announced in 1993 and published in 1996 [9], Sitaraman shows that Vandiver's argument can be saved by adding the following condition: if $A = (\mathscr{C}(\mathbb{Q}[\zeta_p]))_p$ is the pth part of the class group of the pth cyclotomic extension, then $A^p = \{1\}$. This condition together with the conjecture of Kummer–Vandiver do imply the First Case of FLT, and symmetry is recovered: both Cases require the Conjecture plus an additional condition. The Second Case, which is more related to the plus part A^+, requires a condition on the units; the First Case requires a condition on A^-.

There is more to the First Case. Kurihara's proves in [6] by K-theory, that $e_{p-3}A = \{1\}$, and thus the Kummer–Vandiver Conjecture is true for the last component of the class group. Earlier, Banaszak and Gajda had related the even eigenspaces of $e_{2n}A$ to the p-primary torsion group of the group of divisible elements $D(2n)_p \subset K_{4n}(\mathbb{Q})$ and in [1] they observed that $e_{p-2n}A = \{1\}$ for every n and sufficiently large p. Soulé gave in [10] an effective, albeit large lower bound for such p. In view of solely Kurihara's result combined with Vandiver's work, Sitaraman notes that it would suffice to prove that e_3A has exponent p in order to obtain a complete Kummerian proof of the First Case.

Sitaraman has reviewed Vandiver's proof and derived from it the following correct fact: *If* $B_{p-3} \neq 0 \mod p^2$, *then FLT1 holds for* p. In this paper, we give the proof of a more general fact, which extends the result of Vandiver–Sitaraman to the following Diophantine equation:

$$\frac{x^p + y^p}{x + y} = z^p, \quad x, y, z \in \mathbb{Z}, \tag{25.1}$$

$p > 3$ and $p \nmid xy(x^2 - y^2)$. Concretely, we prove the following:

Theorem 25.1 *Let* $p > 3$ *be a prime and suppose that* $B_{p-3} \not\equiv 0 \bmod p^2$. *Then the* (25.1) *has no solutions with* $p \nmid xy(x^2 - y^2)$.

Recently, G. Gras and R. Quême have published on the net a long paper [2] in which they revisit some earlier papers of Vandiver, in which he already used Furtwängler's result, thus connecting properties of solutions to Fermat's equation to properties of some cyclotomic units. In their work, they develop Vandiver's arguments in a geometric direction. It is interesting that they also find that (25.1) is a natural generalization of Fermat's equation, which can be addressed by the same cyclotomic methods as FLT.

25.1.1 Generalities and Notations

The prime p is odd and the pth cyclotomic field is denoted by $\mathbb{K} = \mathbb{Q}[\zeta]$, with ζ a primitive pth root of unity. The Galois group is

$$G = \mathrm{Gal}(\mathbb{K}/\mathbb{Q}) = \left\{\sigma_a : a = 1, 2, \ldots, p-1,\ \zeta \mapsto \zeta^a\right\} \cong (\mathbb{Z}/p\cdot\mathbb{Z})^*.$$

We write simply $N(\cdot)$ for the norm $\mathbf{N}_{\mathbb{K}/\mathbb{Q}}$. If $g \in \mathbb{F}_p$ is a generator of $(\mathbb{Z}/p\cdot\mathbb{Z})^*$, then $\sigma = \sigma_g$ generates G multiplicatively; we write $\jmath = \sigma^{(p-1)/2} = \sigma_{-1} \in G$ for complex multiplication. The p-part of the class group of $\mathbb{K}$ is $A = \mathscr{C}(\mathbb{K})_p$ and $A[p] = \{x \in A : x^p = 1\}$; we write h^+, h^- for the cardinalities of A^+ and A^-, respectively. The groups $E \supseteq C$ are the units (resp., the cyclotomic units) of $\mathbb{K}$, and $U = \mathbb{Z}_p[\zeta_p]$ are the local units. We denote by $\mathbb{H}$ the p-part of the Hilbert class field of $\mathbb{K}$, i.e., the maximal unramified p-abelian extension of $\mathbb{K}$.

For $\sigma \in G$ and $R \in \{\mathbb{F}_p, \mathbb{Z}_p, \mathbb{Z}/(p^m\cdot\mathbb{Z})\}$ we let $\varpi : G \to R$ be the Theichmüller character on G; thus $\varpi(c) \equiv c^{p^{n-1}} \bmod p^n$. The orthogonal idempotents (e.g., [14, §6.3]) $e_k \in R[G]$ are given by

$$e_k = \frac{1}{p-1}\sum_{a=1}^{p-1} \varpi^k(\sigma_a)\cdot\sigma_a^{-1}. \tag{25.2}$$

The group $R[G]$ acts on $A, E/p^N E$, and U, and the orthogonal idempotents induce decompositions of these groups in pairwise disjoint components. If X is a finite abelian p-group on which G acts, then $e_k(\mathbb{Z}_p)$ acts via its approximants to the p^mth order; we shall not introduce additional notations for these approximants. The unramified extension $\mathbb{H}$ decomposes in "components" via

$$\mathbb{H}_k = \mathbb{H}^{(1-e_k)\varphi(A)}, \tag{25.3}$$

which are the subfields with Galois groups $e_k\,\mathrm{Gal}(\mathbb{H}/\mathbb{K}) \cong e_k A$: one considers $e_k \in \mathbb{Z}/(p^n\cdot\mathbb{Z})[G]$ with p^n annihilating A, and φ is the Artin map.

The Stickelberger ideal $I \subset \mathbb{Z}[G]$ annihilates A and it is a $\mathbb{Z}[G]$ submodule of $\mathbb{Z}$-rank $\frac{p+1}{2}$. We refer to [7, §2.2] for more specific properties of I.

The Kummer–Vandiver Conjecture states that $p \nmid h^+$. By mere reflection [14, Theorem 10.10], this implies that $e_{p-2n}A^-$ are $\mathbb{Z}_p$-cyclic. From the theorem of Thaine–Kolyvagin [14, §15], we know additionally that $|e_{2n}A| = |e_{2n}(E/(CE^{p^N}))|$, for sufficiently large N. We shall say that the Kummer–Vandiver conjecture holds for the component $2n$ iff $e_{2n}(E/C)_p = \{1\}$, which is equivalent by Thaine's Theorem, with $e_{2n}A = \{1\}$. Kurihara thus proved in [6], that the Kummer–Vandiver conjecture holds for $e_{p-3}A$. In particular, $\mathbb{H}_3/\mathbb{K}$ is a cyclic extension, possibly trivial.

Let $\sigma \in G$ be a generator and $e'_{2k} \in \mathbb{Z}[G]$ be some lift of $e_{2k} \in \mathbb{Z}/(p^N \cdot \mathbb{Z})[G]$ to $\mathbb{Z}$, and let

$$\eta = \left((1-\zeta)(1-\overline{\zeta})\right)^{1/2} \in C, \quad \eta_{2k} = \eta^{e'_{2k}} \in C. \tag{25.4}$$

The unit η generates C^+ as a $\mathbb{Z}[G]$-module [14, Theorem 8.3] and

$$C = \bigoplus_{k=1}^{\frac{p-1}{3}} \eta_{2k}^{\mathbb{Z}} \cdot C^{p^N}.$$

The *Stickelberger element* $\vartheta = \frac{1}{p}\sum_{c=1}^{p-1} c\sigma_c^{-1} \in \frac{1}{p}\mathbb{Z}[G]$ generates the Stickelberger ideal via

$$I = \vartheta\mathbb{Z}[G] \cap \mathbb{Z}[G].$$

This is a $\frac{p+1}{2}$-dimensional $\mathbb{Z}$-module containing the norm and such that $[\mathbb{Z}[G]^- : I^-]$ is finite, the index being equal to the relative class number $h^- = h(\mathbb{K})/h(\mathbb{K}^+)$, by a Theorem of Iwasawa [14, p. 106]. The ideal I annihilates the class group, and I^- is generated by the Fuchsian elements $\Theta_n = (n - \sigma_n)\vartheta, n = 2, 3, \ldots, (p+1)/2$.

For $\theta = \sum_{c=1}^{p-1} n_c\sigma_c^{-1} \in I$ we write $w(\theta) = \sum_c n_c \in \frac{p-1}{2} \cdot \mathbb{Z}$ for the *weight* of the ideal element. The map

$$\phi : I \to \mathbb{F}_p, \qquad \theta \to \sum_c cn_c \bmod p$$

is the *Fermat quotient map* and $\zeta^\theta = \zeta^{\phi(\theta)}$. For $\theta = \Theta_n$ we have $\phi(\theta) = \frac{n^p - n}{p}$. We refer to [7, §2] for more details about computational aspects related to the Stickelberger ideal.

We shall use in this paper power residue symbols. The one used by Vandiver is the Legendre pth power residue symbol: For $a \in \mathbb{K}$ and $\mathfrak{Q} \subset \mathbb{K}$ a prime ideal, we have

$$\left\{\frac{a}{\mathfrak{Q}}\right\} \equiv a^{(N(\mathfrak{Q})-1)/p} \bmod \mathfrak{Q}.$$

It will be useful to introduce the notation $a \equiv_{\mathfrak{Q}} b$ iff $\{\frac{a}{\mathfrak{Q}}\} = \{\frac{b}{\mathfrak{Q}}\}$ for $a, b \in \mathbb{K}$. Fixing a pth root of unity and for a fixed prime $\mathfrak{Q}$, we have

$$\left\{\frac{a}{\mathfrak{Q}}\right\} = \zeta^{\operatorname{Ind}(a)},$$

an equation defining the *index* $\operatorname{Ind}(a)$ (with respect to $\mathfrak{Q}$). This notation allows using additive notation at a certain point.

The Artin symbol is related to the previous power residue symbol. We define with Hasse [3, II, p. 49] the pth power residue symbol by means of the Artin symbols, thus for $x \in \mathbb{K}$,

$$\left(\frac{x}{\mathfrak{Q}}\right) = \frac{\left(\frac{\mathbb{K}[x^{1/p}]/\mathbb{K}}{\mathfrak{Q}}\right)x^{1/p}}{x^{1/p}}, \tag{25.5}$$

where $(\frac{\mathbb{K}[\alpha^{1/p}]/\mathbb{K}}{\mathfrak{Q}}) \in \mathrm{Gal}(\mathbb{K}[\alpha^{1/p}]/\mathbb{K})$ is the Artin symbol and the fraction on the right hand side determines a unique root of unity, for all possible values of $\alpha^{1/p}$. Whenever both symbols $(\frac{\alpha}{\mathfrak{Q}})$ and $\{\frac{\alpha}{\mathfrak{Q}}\}$ are defined, they are equal.

25.1.2 On the Equation (25.1)

The equation (25.1) is strongly related to the First Case of Fermat's Last Theorem, and it is not hard to see that if it has no solution, it follows also that FLT1 is true. The existence of a non-trivial solution to (25.1) gives raise to some non-trivial ideals of order p, according to a scheme which is classical in the context of FLT1.

Assume that (25.1) has a solution with $p \nmid xy(x^2 - y^2)$ and let $\alpha = x + \zeta y$, $\mathfrak{A} = (\alpha, z)$. One easily verifies that $(\sigma_a(\alpha), \sigma_b(\alpha)) = (1)$ for $a \neq b$, $(ab, p) = 1$, and consequently

$$N(\mathfrak{A}) = z, \qquad \mathfrak{A}^p = (\alpha),$$

see, for instance, [7, §2] or [8] for the related construction in the case of FLT1. In the case of FLT1, it is known that the annihilator ideal of $\mathfrak{A}$ in $\mathbb{F}_p[G]^-$ has large p-rank (e.g., Eichler's Theorem). The same holds in this context too, but we do not need this result. The fact we need is proved in the following lemma

Lemma 25.1 *Assume that* (25.1) *has a solution with* $p \nmid xy(x^2 - y^2)$ *and let* $\mathfrak{A}, \alpha$ *be defined like above. Let* $\beta_0 = \alpha^{(1-\jmath)e_3}$. *Then* $\mathfrak{B} := \mathfrak{A}^{e_3}$ *is not principal.*

Proof Let x, y, z verify (25.1) with $p \nmid xy(x^2 - y^2)$. If Fermat's equation has some non-trivial solution, then one can always find such a pair after eventual permutations in the triple (x, y, z).

Let $u \equiv \frac{y}{x+y} \bmod p$. By substitution, we obtain

$$\beta := \zeta^{-2u}\alpha^{1-\jmath} = (1-\lambda)^{p-2u}\frac{1-\lambda u}{1-\overline{\lambda} u} = 1 + d\lambda^3 + O(\lambda^4),$$

with

$$d \equiv u(u-1)(2u-1)/3 \equiv -\frac{xy(y-x)}{3(x+y)^3} \bmod p.$$

Note that $p \nmid xy(x^2-y^2)$ implies that $(d,p)=1$. Then $\beta^{\sigma-\omega(\sigma)^n} = 1+d(\omega(\sigma)^3-\omega(\sigma)^n)\lambda^3+O(\lambda^4)$. Inserting now the identity $e_3 \equiv \prod_{c=0;c\neq 3}^{p-2}(\sigma-\omega^c(\sigma)) \bmod p$, we see that

$$\beta^{e_3} = 1+dC\lambda^3+O(\lambda^4), \quad \text{with}$$
$$C = \prod_{c=0;c\neq 3}^{p-2}\big(\omega^3(\sigma)-\omega^c(\sigma)\big) = -\omega^{-3}(\sigma) \not\equiv 0 \bmod p. \tag{25.6}$$

Suppose now that $\mathfrak{B}$ is principal, then

$$\beta_0 = \zeta^{2u}(\rho/\overline{\rho})^p,$$

for some $\rho \in \mathbb{K}$ with $\mathfrak{B} = (\rho)$, a condition which is inconsistent with the local development in (25.6) and $d \not\equiv 0 \bmod p$. Therefore, $\mathfrak{B}$ cannot be principal. □

We prove in [7, §2] that in the case when (25.1) has non-trivial solutions, the Fermat quotients

$$\phi(2) = \phi(3) = 0. \tag{25.7}$$

An important result of Furtwängler, which was used by Vandiver, implies in the case of FLT1 that

$$\left\{\frac{\zeta}{\mathfrak{Q}}\right\} = 1 \quad \text{for all prime ideals } \mathfrak{Q} | xy\big(x^2-y^2\big). \tag{25.8}$$

The proof can be found in [8], and we give in [7, §2] an alternative proof using Stickelberger elements. Due to the symmetry of the equation $x^p+y^p+z^p=0$, all the results above stay true upon permuting the unknowns x, y, z.

Using Furtwängler's result, Vandiver proved in 1934:

Theorem 25.2 (Vandiver) *If* $2s$ *is the smallest integer such that* $B_{2s} \equiv 0 \bmod p$, *and* $\mathfrak{Q} \supset \mathfrak{A}$ *is any prime ideal dividing* $\mathfrak{A} = (x+\zeta y, z)$, with x, y, z stemming from a solution of FLT1, *then*

$$\left(\frac{\mathbb{K}[\eta_{2k}^{1/p}]/\mathbb{K}}{\mathfrak{Q}}\right) = 1, \quad 2k = p-3, p-5, \ldots, p+1-2s.$$

In particular, $(\frac{\mathbb{K}[\eta_{2k}^{1/p}]/\mathbb{K}}{\mathfrak{A}}) = 1$ *for all these values.*

The purpose of this paper is to prove the above Theorem under a more general assumption that x, y, z stem from a solution of (25.1), and thus Furtwängler's condition does not necessarily hold. We will thus prove:

Theorem 25.3 (Vandiver2) *If* $2s$ *is the smallest integer such that* $B_{2s} \equiv 0 \bmod p$, *and* $\mathfrak{Q} \supset \mathfrak{A}$ *is any prime ideal dividing* $\mathfrak{A} = (x + \zeta y, z)$, with x, y, z stemming from a solution of (25.1), *then*

$$\left(\frac{\mathbb{K}[\eta_{2k}^{1/p}]/\mathbb{K}}{\mathfrak{Q}}\right) = 1, \quad 2k = p-3, p-5, \ldots, p+1-2s.$$

In particular, $(\frac{\mathbb{K}[\eta_{2k}^{1/p}]/\mathbb{K}}{\mathfrak{A}}) = 1$ *for all these values.*

The proof of this theorem will be provided below. By formulating Vandiver's result in this way, one sees that in the case when $[\mathfrak{A}]$ has trivial image in A/pA, the statement of the theorem cannot be used: this was the error in Vandiver's proof. However, by assuming additionally that $e_{p-2k}A$ is cyclic of order p while $e_{p-2k}[\mathfrak{A}] \neq 1$, Vandiver's Theorem becomes effective. It is indeed known [8] that $[e_3\mathfrak{A}] \neq \{1\}$, see Lemma 25.1; Kurihara's result implies that e_3A must be cyclic. By this assumption, it has exponent p and thus $[\mathbb{H}_3 : \mathbb{K}] = p$, and $\mathrm{Gal}(\mathbb{H}_3/\mathbb{K}) \cong \langle [e_3\mathfrak{A}] \rangle$, since the class $[e_3\mathfrak{A}]$ is non-trivial and the extension has degree p. Therefore, the assumption of a non-trivial solution to (25.1) leads to a contradiction to Theorem 25.3, which requires $(\frac{\mathbb{H}_3/\mathbb{K}}{e_3\mathfrak{A}}) = 1$. This is Theorem 25.1 which thus follows from Theorem 25.3 and Kurihara's result.

25.2 Vandiver's Theorem

Suppose that x, y, z is a non-trivial solution of (25.1), and let $\alpha = (x + \zeta y), \mathfrak{A} = (x + \zeta y, z)$, like in the introduction. We let also $\mathfrak{Q}|\alpha$ be a fixed prime ideal and will write

$$a \equiv_{\mathfrak{Q}} b \quad \Leftrightarrow \quad \left(\frac{a}{\mathfrak{Q}}\right) = \left(\frac{b}{\mathfrak{Q}}\right) \quad \Leftrightarrow \quad \mathrm{Ind}(a) = \mathrm{Ind}(b), \quad a, b \in \mathbb{K}^{\times}.$$

Vandiver's central observation is that the relation (25.8) implies[1] for $1 < c < p$, successively

$$\left\{\frac{\alpha_c}{\mathfrak{Q}}\right\} = \left\{\frac{\alpha + (\zeta^c - \zeta)y}{\alpha}\right\} = \left\{\frac{(\zeta^{c-1} - 1)y}{\mathfrak{Q}}\right\} \quad \text{and}$$

$$1 = \left\{\frac{s^p}{\mathfrak{Q}}\right\} = \left\{\frac{x+y}{\mathfrak{Q}}\right\} = \left\{\frac{\alpha - y\lambda}{\mathfrak{Q}}\right\} = \left\{\frac{y\lambda}{\mathfrak{Q}}\right\}.$$

[1]Note that -1 is a pth power residue, so we may disregard signs in the evaluation of the residue symbols.

We cannot assume that (25.8) holds in relation with (25.1). We let therefore $z = \mathrm{Ind}(\zeta)$, $\pi = \mathrm{Ind}(p)$, $\ell = \mathrm{Ind}(\lambda)$, and $\rho = \mathrm{Ind}(y)$, so the above identities become

$$\left\{\frac{\alpha_c}{\mathfrak{Q}}\right\} = \left\{\frac{\alpha + (\zeta^c - \zeta)y}{\alpha}\right\} = \zeta^{z+\rho} \cdot \left\{\frac{(\zeta^{c-1} - 1)}{\mathfrak{Q}}\right\} \quad \text{and}$$

$$\left\{\frac{x+y}{\mathfrak{Q}}\right\} = \left\{\frac{\alpha - y\lambda}{\mathfrak{Q}}\right\} = \left\{\frac{y\lambda}{\mathfrak{Q}}\right\} = \zeta^{\rho+\ell},$$

thus $\mathrm{Ind}(x+y) = \rho + \ell$ and $\mathrm{Ind}(\alpha_c) = z + \rho + \mathrm{Ind}(\sigma_{c-1}(\lambda))$. The index $\mathrm{Ind}(\sigma_{c-1}(\lambda))$ will lead to the use of Kummer units below.

In our context, the Artin symbol has the advantage of being defined also for $x = \alpha$, since $(\alpha) = \mathfrak{A}^p$ and thus $\mathbb{K}[\alpha^{1/p}]$ is unramified outside p. In our case, $p \nmid xy(x^2 - y^2)$ and thus $(\mathfrak{Q}, p) = 1$, so $\mathfrak{Q}$ is unramified in $\mathbb{K}[\alpha^{1/p}]$. It is, in fact, totally split, as one finds by considering the localization at $\mathfrak{Q}$ and using the fact that $\mathfrak{Q}^p | \alpha$. Therefore, we have

$$\left(\frac{\alpha}{\mathfrak{Q}}\right) = 1. \tag{25.9}$$

Combining (25.9) with the previous relations, one obtains

$$\alpha_c \equiv_{\mathfrak{Q}} (y\zeta) \cdot \left(\zeta^{c-1} - 1\right), \quad 1 < c < p, \ \alpha \equiv_{\mathfrak{Q}} 1. \tag{25.10}$$

Having thus removed the "singularity" at $c = 1$, we may follow Vandiver's strategy, applied to Hasse symbols. The general strategy is to produce linear combinations of the indices $\mathrm{Ind}(\sigma_c(\alpha))$ which vanish. In this way, one may derive conditions which are free of the variables x, y, z. The linear combinations can be, for instance, obtained by applying elements of the Stickelberger ideal. This happens as follows: Let $\alpha = (x + y)\zeta^u \cdot \alpha'$, with $\alpha' \equiv 1 \bmod \lambda^2$. It follows then from general properties of the Stickelberger ideal that

$$\alpha^\theta = (x+y)^{w(\theta)} \zeta^{u\phi(\theta)} \cdot \beta^p \equiv_{\mathfrak{Q}} (y\lambda)^{w(\theta)} \zeta^{u\phi(\theta)}, \quad \beta \in \mathbb{K}. \tag{25.11}$$

One may relate the indices to the ones of cyclotomic units, which was one of Vandiver's favorite themes over more than a decade. The units $\eta_{2k} \in C(\mathbb{K})$ are as defined in (25.4). Let also

$$E_{2k} = \lambda^{(p-1)e_{2k}}, \quad 2k = 2, 4, \ldots, p-3, \tag{25.12}$$

be Kummer's cyclotomic fundamental p-units, in which we removed for simplicity the exponent $(\sigma - 1)(1 + \jmath)$. As a result, E_{2k} are not units, but

$$E_{2k}^{(1+\jmath)(\varpi(\sigma)-1)} = \eta_{2k}^{p-1} \cdot u^p, \quad u \in E',$$

with E' the p-units of $\mathbb{K}$ and $(\varpi(\sigma) - 1, p) = 1$. Let $x_k = \mathrm{Ind}(E_{2k})$.

We deduce from the fact that the orthogonal idempotents yield a decomposition of 1, under application of $\lambda^{(p-1)e_0} = p$ and $\lambda^{(p-1)e_1} = \zeta^{(p-1)/2}$, the following identity

$$\lambda^{-1} \equiv_{\Omega} p \cdot \zeta^{(p-1)/2} \cdot \prod_{k=1}^{(p-3)/2} E_{2k}. \tag{25.13}$$

Note that the action of σ_c on E_{2k} is particularly simple, and it is given by

$$\sigma_c E_{2k} = (E_{2k})^{c^{2k}} \quad \text{for } c \in \mathbb{Z}/(p \cdot \mathbb{Z}).$$

This follows from the property $\sigma_c e_{2k} \equiv c^{2k} e_{2k} \bmod p$ of idempotents.

Next we apply a generic Stickelberger element $\theta = \sum_{c=1}^{p-1} n_c(\theta)\sigma_c^{-1}$ using the previous identities; let $w(\theta)$ denote as usual the weight of θ and $\phi(\theta)$ the Fermat quotient map. Then (25.11) yields

$$\operatorname{Ind}(\alpha^{\theta}) = w(\theta)\operatorname{Ind}(y\lambda) + u\phi(\theta)\operatorname{Ind}(\zeta) = w(\theta)(\rho + \ell) + u\phi(\theta)z;$$

we define

$$\begin{aligned} s_k(\theta) &= \sum_{c=2}^{p-1} n_c(\theta)(1/c-1)^{2k} \in \mathbb{F}_p, \\ s_k(m,\theta) = s_k(\sigma_m\theta) &= \sum_{c=1;c\neq m}^{p-1} n_c(\theta)(m/c-1)^{2k} \in \mathbb{F}_p, \end{aligned} \tag{25.14}$$

and using (25.10), we obtain

$$\begin{aligned} \alpha^{\theta} &\equiv_{\Omega} (\zeta y)^{w(\theta)-n_1(\theta)} \cdot \prod_{c=2}^{p-1} \sigma_{c^{-1}-1}(\lambda)^{n_c(\theta)} \\ &\equiv_{\Omega} (\zeta y)^{w(\theta)-n_1(\theta)} \cdot \frac{\zeta^{(u\phi(\theta)-n_1(\theta))/2}}{p^{w(\theta)-n_1(\theta)}} \cdot \left(\prod_{2k=2}^{p-3} \cdot E_{2k}^{-\sum_c n_c(\theta)\cdot(1/c-1)^{2k}} \right) \\ &\equiv_{\Omega} y^{w(\theta)-n_1(\theta)} \cdot \frac{\zeta^{w(\theta)+(u\phi(\theta)-3n_1(\theta))/2}}{p^{w(\theta)-n_1(\theta)}} \cdot \left(\prod_{2k=2}^{p-3} \cdot E_{2k}^{-\sum_c n_c(\theta)\cdot(1/c-1)^{2k}} \right). \end{aligned}$$

Let $S(\theta) = \operatorname{Ind}(E_{2k}^{-\sum_c n_c(2\theta)\cdot(1/c-1)^{2k}}) = -\sum_{k=1}^{(p-3)/2} s_k(\theta)x_k$. By comparing the last identity with previous expressions for α^{θ}, we deduce

$$\begin{aligned} \operatorname{Ind}(\alpha^{\theta}) &= w(\theta)(\rho+\ell) + u\phi(\theta)z \\ &= \big(w(\theta)-n_1(\theta)\big)(\rho-\pi) + \big(w(\theta) + \big(u\phi(\theta) - 3n_1(\theta)\big)/2\big)z + S(\theta), \end{aligned}$$

thus

$$S(\theta) = w(\theta)(\ell + \pi - z) + \frac{u\phi(\theta)}{2} z + \left(\rho + \frac{3}{2} z - \pi\right) n_1(\theta). \tag{25.15}$$

The map $\Sigma : I \to \mathbb{F}_p : \theta \mapsto S(\theta)$ is a linear functional. We define the kernels

$$\mathscr{S} = \mathrm{Ker}(\Sigma), \quad \mathscr{W} = \mathrm{Ker}(w), \quad \mathscr{F} = \mathrm{Ker}(\phi).$$

We shall prove the following

Lemma 25.2 *The notations being like above, $S(\theta) = 0$ for all $\theta \in I$.*

Proof If the claim is false, then $\mathscr{S} \neq I$ and the spaces $\mathscr{S}, \mathscr{F}$, and $\mathscr{W}$ are three $(p-1)/2 - i(p)$-dimensional subspaces of I/pI, with $\mathscr{F} \neq \mathscr{W}$; here $i(p)$ is the irregularity index.

Suppose first that $\mathscr{S} \not\subset \mathscr{W} \cup \mathscr{F}$. Then (25.15) implies, when setting $w(\theta) = \phi(\theta) = 0$, but $n_1(\theta) \neq 0$, which is always possible by conjugation, that $\pi = \rho + 3z/2$. From this, setting only $w(\theta) = 0$ but $\phi(\theta) \neq 0$—which is possible since $\mathscr{W} \neq \mathscr{F}$—we conclude that $z = 0$, and finally $\ell + \pi = 0$. In particular, (25.15) implies then that $S(\theta) = 0$ for all θ.

Suppose now that $S(\theta) \subset \mathscr{W} \cup \mathscr{F}$. Since all the involved kernels are $(p-1)/2 - i(p)$-dimensional subspaces of I/pI, the inclusion is equivalent to one of $\mathscr{S} \subset \mathscr{W}$ or $\mathscr{S} \subset \mathscr{F}$. The development of S is

$$S(\theta) = \sum_{c=1}^{p-1} n_c(\theta) \sum_{k=1}^{(p-3)/2} (1/c - 1)^{2k} x_k = \sum_{c=1}^{p-1} \varsigma(c) n_c(\theta),$$

$$\varsigma(c) = \sum_{k=1}^{(p-3)/2} (1/c - 1)^{2k} x_k.$$

If $\mathscr{S} \subset \mathscr{W}$, there is a constant $(d, p) = 1$ such that $\varsigma(c) = d$ for all c. Since $\varsigma(1) = 0$ it follows that $d = 0$ and $S(\theta)$ should vanish identically. Assume now that $\mathscr{S} \subset \mathscr{F}$. Then there is also a constant $(d, p) = 1$, such that $\varsigma(c) = cd$ for all c; invoking again the vanishing of $\varsigma(1)$, we deduce in this case too that S must vanish identically. This completes the proof of claim. □

Starting from this fact, we deduce Vandiver's proof of Theorem 25.2. Note that in the case of Fermat's equation, $\mathrm{Ind}(x + y) = \rho + \ell = 0$ and $z = 0$, and it is an exercise left to the reader to show that this implies $S(\theta) = 0$ for all θ. The fundamental phenomenon which arises in Vandiver's computation is the fact that reciprocity leads to some conditions on the power residue symbol of Kummer fundamental units, depending in a *reflected* way upon that Bernoulli numbers that vanish modulo p.

We shall proceed from (25.10) in a more systematic way, following Vandiver's main idea, but not his proof sketch. The contradiction that one expects to achieve in a lucky case can be easily foreseen: suppose that one can prove indeed that

$$\left\{\frac{E_{2n}}{\mathfrak{Q}}\right\} = 1, \tag{25.16}$$

for some $n : B_{2n} \equiv 0 \bmod p$ and for all c and $\mathfrak{Q}|\sigma_c(\alpha)$. Assuming that Vandiver's conjecture holds for this component, the extension $\mathbb{H}_{p-2n}/\mathbb{K}$ is cyclic of degree p and generated, as a Kummer radical, by a cyclotomic unit $E_{2n} \in C$. The relation (25.16) implies by multiplicativity that $\{\frac{E_{2n}}{\mathfrak{A}^{e_{p-2n}}}\} = 1$, and this would imply that the primes of the class $[\mathfrak{A}^{p-2n}] \in A^-$ are all split in $\mathbb{H}_{p-2n}$. However, we assumed additionally that $(e_{p-2n}A)^p$, this leads to the required contradiction.

Fermarcheology Vandiver missed the additional condition and sketched at the end of his paper a quick argument suggesting on base of some previous papers of his own, including [11, 12], that the Bernoulli number $B_{(2n-1)p+1}$ should not vanish assuming the Kummer–Vandiver Conjecture, and thus $p^2 \nmid B_{1,\omega^{-2k}}$, which is equivalent to the missed condition. However, this part of the argument could not be corrected to this day by any of the mathematicians that tried to do so.

Let now $\theta \in I$ be any element. By applying σ_m, we obtain

$$\sum_{k=1}^{(p-3)/2} x_k s_k(m, \theta) = 0, \quad m = 1, 2, \ldots, p-1, \tag{25.17}$$

which is a linear system of equations over $\mathbb{F}_p$.

We consider the system of elements $\theta_a = \sigma_m \sum_{c=1}^{p-1} [\frac{ac}{p}] \sigma_c^{-1}$, the Fuchsian elements, thus $n_c(\theta_a) = [\frac{ac}{p}]$. The following computation ([5, Lemma 1.0], using Propositions 15.2.1 and Proposition 15.2.3 of [4]) will be useful.

Lemma 25.3 *Let $2 \le 2m \le p-1$ be an even integer and $a < p$ a positive integer coprime to p. Then we have the following identities in $\mathbb{F}_p$:*

$$C(a, 2m) := \sum_{c=1}^{p-1} \left[\frac{ac}{p}\right] c^{2m-1} = \frac{a^{2m+1} - a}{2ma^{2m}} B_{2m}, \quad m < p-1, \tag{25.18}$$

$$C(a, p-1) := \sum_{c=1}^{p-1} \left[\frac{ac}{p}\right] c^{p-2} = \frac{a^p - a}{p} = \phi_a. \tag{25.19}$$

Note also that $\sum_c n_c c^{2l} = 0$ for $2 \le 2l \le p-1$ for every $\theta = \sum_c n_c \sigma_c^{-1} \in I$, while $\sum_{c=1}^{p-1} n_c c^0 =: |\theta| = C\frac{p-1}{2}$. The vanishing $I(p\lambda) = 0$ shown in the proof of Lemma 2 implies that $I(\prod_{2k=2}^{p-3} E_{2k}) = 0$. With this, the binomial expansion of

$s_k(m,a)$ yields

$$s_k(m,a) := s_k(\sigma_m\theta_a) = -\sum_{l=1}^{k}\binom{2k}{2l-1}\sum_{c=1}^{p-1} n_c(a)(m/c)^{2l-1}$$
$$= -\sum_{l=1}^{k}\binom{2k}{2l-1} m^{2l-1} c(a, p-2l+1).$$

The substitution $y_l = \sum_{k=l}^{(p-3)/2}\binom{2k}{2l-1}x_k$ allows us to insert the sums $C(a,m)$ defined in (25.18). The system (25.17) becomes

$$\sum_{l=1}^{(p-1)/2} m^{2l-1}\cdot C(a, p-2l+1)y_l = 0, \quad a = 2, 3, \ldots, (p+1)/2. \tag{25.20}$$

Let $X_l = C(a, p-2l+1)y_l$. The above is a regular homogeneous system (with Vandermonde determinant) in the unknowns X_l. We thus have $X_l = 0$ for $l = 1, 2, \ldots, \frac{p-3}{2}$, and consequently $C(a, p-2l+1)y_l = 0$. Letting $(a,p) = 1$, we deduce that

$$y_l = 0 \quad \text{for all } l \text{ such that } B_{p-2l+1} \not\equiv 0 \bmod p. \tag{25.21}$$

For $l = (p-3)/2$ we have $y_{(p-3)/2} = \sum_{2k\geq p-3}\binom{2k}{p-4}x_k = (p-3)x_{p-3} = 0$, so

$$\operatorname{Ind}(E_{p-3}) = 0.$$

In general, if $2s$ is the smallest integer such that $B_{2s} \equiv 0 \bmod p$, one can apply backwards substitution in (25.21), and it follows, using induction and the definition of the y_l, that

$$\operatorname{Ind}(E_{2k}) = 0 \quad \text{for } 2k = p-3, p-5, \ldots, p+1-2s. \tag{25.22}$$

Equation (25.22) holds for every prime $\mathfrak{Q}|\alpha$ and $\alpha = \mathfrak{A}^p$, so by multiplicativity of the Artin symbol,

$$\left(\frac{\eta_{2k}}{\mathfrak{A}}\right) = 1, \quad 2k = p-3, p-5, \ldots, p+1-2s. \tag{25.23}$$

This completes the proof of Theorems 25.1, 25.2 and 25.3.

The Diagonal Nagell Equation is

$$\frac{x^p-1}{x-1} = p^e y^p, \tag{25.24}$$

where $e = 0$ if $x \not\equiv 1 \bmod p$ and $e = 1$ otherwise. We have distinguished in [7] between the First Case in which $x(x^2-1) \not\equiv 0 \bmod p$ and the remaining Second Case, showing that the Second Case is implied by the Kummer–Vandiver conjecture. The Theorem 25.1 then implies in particular:

Corollary 25.1 *If $B_{p-3} \neq 0 \bmod p^2$, then the Diagonal Nagell Equation has no solutions in the First Case.*

25.3 Instead of a Conclusion

Ever since the epochal proof given by Wiles to the conjecture of Taniyama–Shimura, thus confirming also Fermat's Last Theorem, the question is often asked, by friends: *Will there ever be a 'simpler, classical proof'?*—meaning also, a proof accessible to you and me, of course.

Daedalus did fly low and far, yet we are fascinated by Icarus, his son, caught in the temptation of the Light. Why is it so? I do not know. Yet, I see that the last century brought that dream back to us—and first we have seen Jumbos, and only some decades later did men and women with kites or para-gliders fly freely under the sky, without engine, landing hundred kilometers away from their place of departure. Icarus found his way back, but first came the Jumbos. Wiles did fly high and well, and the engines of thought that he prepared will carry more load. But be assured, thinking of Fermat's dream, the spell is broken, the jump is now at hand's reach, not far from where Kummer had suspected it: just prove

1. The Kummer–Vandiver conjecture.
2. That every p^2-primary unit in $\mathbb{K}$ is a global pth power (or, if preferred, $B_{2pn} \not\equiv 0 \bmod p^3, 2n < p$).
3. That $B_{2n} \not\equiv 0 \bmod p^2, 2n < p$.

Possibly, a positive answer to these three problems may help solve also the Fermat equation over $\mathbb{K}^+$, the maximal totally real subfield of the pth cyclotomic extension: Vandiver considered repeatedly this question, too.

The spell being broken, be sure the gliders are just around the corner. *The purpose of this simple paper was to show that once they come, there is a small little that they can bring in Diophantine terms, that Jumbos cannot yet do.* If by that time, flying on your own wings or in an airplane will be the best for you, I do not know: Looking forward to your 65th birthday!

References

1. Banaszak, G., Gajda, W.: On the arithmetic of cyclotomic fields and the K-theory of **Q**. In: Algebraic K-Theory. Contemp. Math., vol. 199. Amer. Math. Soc., Providence (1996)
2. Gras, G., Quême, R.: Some works of Furtwängler and Vandiver revisited and Fermat's last theorem (2011). arXiv:1103.4692
3. Hasse, H.: Algebraische Zahlkörper, 2nd edn. Physica Verlag, Würzburg (1965)
4. Ireland, K., Rosen, M.: A Classical Introduction to Modern Number Theory. Graduate Texts in Mathematics, vol. 84. Springer, Berlin (1990)
5. Jha, V.: The stickelberger ideal in the spirit of Kummer with applications to the first case of Fermat's last theorem. Queens's Papers in Pure and Applied Mathematics 93 (1993)

6. Kurihara, M.: Some remarks on conjectures about cyclotomic fiels and k-groups of $\mathbb{Z}$. Compos. Math. **81**, 223–236 (1992)
7. Mihăilescu, P.: Class number conditions for the Diagonal case of the equation of Nagell and Ljunggren. In: Diophantine Approximation, Festschrift for W. Schmidt's 70th Birthday, pp. 245–273. Springer, Berlin (2008)
8. Ribenboim, P.: 13 Lectures on Fermat's Last Theorem. Springer, Berlin (1979)
9. Sitaraman, S.: Vandiver revisited. J. Number Theory **57**(1), 122–129 (1996)
10. Soulé, C.: Perfect forms and Vandiver's conjecture. J. Reine Angew. Math. **517**, 209–221 (1999)
11. Vandiver, H.S.: Some theorems concerning properly irregular cyclotomic fields. Proc. Natl. Acad. Sci. USA **15**, 202–207 (1929)
12. Vandiver, H.S.: On power characters of singular integers in a properly irregular cyclotomic field. Trans. Am. Math. Soc. **32**, 391–408 (1930)
13. Vandiver, H.S.: Fermat's last theorem and the second factor in the cyclotomic class number. Bull. Am. Math. Soc. **40**, 118–126 (1934)
14. Washington, L.: Introduction to Cyclotomic Fields. Graduate Texts in Mathematics, vol. 83. Springer, Berlin (1996)

Chapter 26
Multiple Orthogonality and Applications in Numerical Integration

Gradimir V. Milovanović and Marija P. Stanić

Abstract In this paper, a brief survey of multiple orthogonal polynomials defined using orthogonality conditions spread out over r different measures are given. We consider multiple orthogonal polynomials on the real line, as well as on the unit semicircle in the complex plane. Such polynomials satisfy a linear recurrence relation of order $r+1$, which is a generalization of the well known three-term recurrence relation for ordinary orthogonal polynomials (the case $r=1$). A method for the numerical construction of multiple orthogonal polynomials by using the discretized Stieltjes–Gautschi procedure are presented. Also, some applications of such orthogonal systems to numerical integration are given. A numerical example is included.

Key words Multiple orthogonal polynomials · Recurrence relations · Numerical integration · Generalized Birkhoff–Young quadrature rules

Mathematics Subject Classification 33D45 · 42C05 · 65D30

26.1 Introduction

Multiple orthogonal polynomials arise naturally in the theory of simultaneous rational approximation, in particular in Hermite–Padé approximation of a system of r (Markov) functions. A good source for information on Hermite–Padé approximation is the book by Nikishin and Sorokin [23, Chap. 4], where the multiple orthogonal

---

Dedicated to Professor Themistocles M. Rassias on the occasion of his 60th birthday.

G.V. Milovanović (✉)
Mathematical Institute of the Serbian Academy of Sciences and Arts, Knez Mihailova 36, p.p. 367, 11001 Beograd, Serbia
e-mail: gvm@mi.sanu.ac.rs

M.P. Stanić
Department of Mathematics and Informatics, Faculty of Science, University of Kragujevac, Radoja Domanovića 12, 34000 Kragujevac, Serbia
e-mail: stanicm@kg.ac.rs

P.M. Pardalos et al. (eds.), *Nonlinear Analysis*, Springer Optimization and Its Applications 68, In Honor of Themistocles M. Rassias on the Occasion of his 60th Birthday, DOI 10.1007/978-1-4614-3498-6_26, © Springer Science+Business Media, LLC 2012

polynomials are called polyorthogonal polynomials. Other good sources of information are the surveys by Aptekarev [1] and de Bruin [5], as well as the papers by Piñeiro [24], Sorokin [26–28], and Van Assche [30].

Historically, Hermite–Padé approximation was introduced by Hermite to prove the transcendence of e. Multiple orthogonal polynomials can be used to give a constructive proof of irrationality and transcendence of certain real numbers (see [30]).

Multiple orthogonal polynomials are a generalization of orthogonal polynomials in the sense that they satisfy $r \in \mathbb{N}$ orthogonality conditions. Let $r \geq 1$ be an integer and let $w_1, w_2, \ldots, w_r$ be r weight functions on the real line such that the support of each w_i is a subset of an interval E_i. Let $\mathbf{n} = (n_1, n_2, \ldots, n_r)$ be a vector of r nonnegative integers, which is called a *multi-index* with length

$$|\mathbf{n}| = n_1 + n_2 + \cdots + n_r.$$

There are two types of multiple orthogonal polynomials (see [32]).

1° *Type I multiple orthogonal polynomials.*

Here we want to find a vector of polynomials $(A_{\mathbf{n},1}, A_{\mathbf{n},2}, \ldots, A_{\mathbf{n},r})$ such that each $A_{\mathbf{n},i}$ is a polynomial of degree $n_i - 1$ and the following orthogonality conditions hold:

$$\sum_{j=1}^{r} \int_{E_j} A_{\mathbf{n},j} x^k w_j(x)\,\mathrm{d}x = 0, \quad k = 0, 1, 2, \ldots, |\mathbf{n}| - 2.$$

2° *Type II multiple orthogonal polynomials.*

A type II multiple orthogonal polynomial is a monic polynomial $P_{\mathbf{n}}$ of degree $|\mathbf{n}|$ which satisfies the following orthogonality conditions:

$$\int_{E_1} P_{\mathbf{n}}(x)\, x^k w_1(x)\,\mathrm{d}x = 0, \quad k = 0, 1, \ldots, n_1 - 1, \tag{26.1}$$

$$\int_{E_2} P_{\mathbf{n}}(x)\, x^k w_2(x)\,\mathrm{d}x = 0, \quad k = 0, 1, \ldots, n_2 - 1, \tag{26.2}$$

$$\vdots$$

$$\int_{E_r} P_{\mathbf{n}}(x)\, x^k w_r(x)\,\mathrm{d}x = 0, \quad k = 0, 1, \ldots, n_r - 1. \tag{26.3}$$

The conditions (26.1)–(26.3) give $|\mathbf{n}|$ linear equations for the $|\mathbf{n}|$ unknown coefficients $a_{k,\mathbf{n}}$ of the polynomial $P_{\mathbf{n}}(x) = \sum_{k=0}^{|\mathbf{n}|} a_{k,\mathbf{n}} x^k$, where $a_{|\mathbf{n}|,\mathbf{n}} = 1$. Since the matrix of coefficients of this system can be singular, we need some additional conditions on the r weight functions to provide the uniqueness of the multiple orthogonal polynomial.

If the polynomial $P_{\mathbf{n}}(x)$ is unique, then $\mathbf{n}$ is a *normal index*. If all indices are normal, then we have a *perfect system*.

For $r = 1$ in both cases, we have the ordinary orthogonal polynomials. In the sequel, we consider only the type II multiple orthogonal polynomials.

There are two distinct cases for which the type II multiple orthogonal polynomials are given (see [32]).

1. *Angelesco systems*—For these systems the intervals E_i on which the weight functions are supported are disjoint, i.e., $E_i \cap E_j = \emptyset$ for $1 \le i \ne j \le r$.
2. *AT systems*—AT systems are such that all the weight functions are supported on the same interval E and the following $|\mathbf{n}|$ functions: $w_1(x),\ xw_1(x), \ldots,$ $x^{n_1-1}w_1(x),\ w_2(x),\ xw_2(x),\ \ldots,\ x^{n_2-1}w_2(x),\ \ldots,\ w_r(x),\ xw_r(x),\ \ldots,$ $x^{n_r-1}w_r(x)$ form a Chebyshev system on E for each multi-index $\mathbf{n}$.

The following two theorems have been proved in [32].

Theorem 26.1 *In an Angelesco system, a type II multiple orthogonal polynomial $P_{\mathbf{n}}(x)$ factors into r polynomials $\prod_{j=1}^{r} q_{n_j}(x)$, where each q_{n_j} has exactly n_j zeros on E_j.*

Theorem 26.2 *In an AT system, a type II multiple orthogonal polynomial $P_{\mathbf{n}}(x)$ has exactly $|\mathbf{n}|$ zeros on E. For the type I vector of multiple orthogonal polynomials, the linear combination $\sum_{j=1}^{r} A_{\mathbf{n},j}(x)w_j(x)$ has exactly $|\mathbf{n}| - 1$ zeros on E.*

For each of the weight functions w_j, $j = 1, 2, \ldots, r$,

$$(f, g)_j = \int_{E_j} f(x)g(x)w_j(x)\,\mathrm{d}x \tag{26.4}$$

denotes the corresponding inner product of f and g.

In the sequel, by $\mathbb{P}_n$ we denote the set of algebraic polynomials of degree at most n, and by $\mathbb{P}$ the set of all algebraic polynomials.

The paper is organized as follows. Section 26.2 is devoted to recurrence relations for some cases of type II multiple orthogonal polynomials. In Sect. 26.3, a numerical procedure for construction of type II multiple orthogonal polynomials based on the discretized Stieltjes–Gautschi procedure [8] is presented. In Sect. 26.4, we transfer the concept of multiple orthogonality to the unit semicircle in the complex plane. Special attention is devoted to the case $r = 2$, for which the coefficients of the recurrence relation for multiple orthogonal polynomials on the semicircle are expressed in terms of the coefficients of recurrence relation for the corresponding type II multiple orthogonal (real) polynomials. Applications of multiple orthogonality to numerical integration are given in Sect. 26.5. Finally, in Sect. 26.6, a numerical example is included.

26.2 Recurrence Relations

It is well known that orthogonal algebraic polynomials satisfy the three-term recurrence relation (see [6, 9, 12]). Such a recurrence relation is one of the most important pieces of information for the constructive and computational use of orthogonal polynomials. Knowledge of the recursion coefficients allows the zeros of orthogonal polynomials to be computed as eigenvalues of a symmetric tridiagonal matrix, and with them the Gaussian quadrature rule, and also allows an efficient evaluation of expansions in orthogonal polynomials.

The type II multiple orthogonal polynomials with nearly diagonal multi-index satisfy recurrence relation of order $r+1$. Let $n \in \mathbb{N}$ and write it as $n = \ell r + j$, with $\ell = [n/r]$ and $0 \le j < r$. The *nearly diagonal multi-index* $\mathbf{s}(n)$ corresponding to n is given by

$$\mathbf{s}(n) = (\underbrace{\ell+1, \ell+1, \ldots, \ell+1}_{j \text{ times}}, \underbrace{\ell, \ell, \ldots, \ell}_{r-j \text{ times}}).$$

Let us denote the corresponding type II multiple (monic) orthogonal polynomials by $P_n(x) = P_{\mathbf{s}(n)}(x)$. Then, the following recurrence relation

$$x P_m(x) = P_{m+1}(x) + \sum_{i=0}^{r} a_{m,r-i} P_{m-i}(x), \quad m \ge 0, \tag{26.5}$$

holds, with initial conditions $P_0(x) = 1$ and $P_i(x) = 0$ for $i = -1, -2, \ldots, -r$ (see [31]).

Setting $m = 0, 1, \ldots, n-1$ in (26.5), we get

$$x \begin{bmatrix} P_0(x) \\ P_1(x) \\ \vdots \\ P_{n-1}(x) \end{bmatrix} = H_n \begin{bmatrix} P_0(x) \\ P_1(x) \\ \vdots \\ P_{n-1}(x) \end{bmatrix} + P_n(x) \begin{bmatrix} 0 \\ \vdots \\ 0 \\ 1 \end{bmatrix},$$

i.e.,

$$H_n \mathbf{P}_n(x) = x \mathbf{P}_n(x) - P_n(x) \mathbf{e}_n, \tag{26.6}$$

where $\mathbf{P}_n(x) = [P_0(x)\ P_1(x)\ \ldots\ P_{n-1}(x)]^T$, $\mathbf{e}_n = [0\ 0\ \ldots\ 0\ 1]^T$, and H_n is the following lower (banded) Hessenberg matrix of order n

$$H_n = \begin{bmatrix} a_{0,r} & 1 & & & & & & \\ a_{1,r-1} & a_{1,r} & 1 & & & & & \\ \vdots & \ddots & \ddots & \ddots & & & & \\ a_{r,0} & \cdots & a_{r,r-1} & a_{r,r} & 1 & & & \\ & a_{r+1,0} & \cdots & a_{r+1,r-1} & a_{r+1,r} & 1 & & \\ & & \ddots & & \ddots & \ddots & \ddots & \\ & & & a_{n-2,0} & \cdots & a_{n-2,r-1} & a_{n-2,r} & 1 \\ & & & & a_{n-1,0} & \cdots & a_{n-1,r-1} & a_{n-1,r} \end{bmatrix}.$$

This kind of matrix has been obtained also in construction of orthogonal polynomials on the radial rays in the complex plane (see [15]).

Let $x_\nu \equiv x_\nu^{(n)}$, $\nu = 1, \ldots, n$, be the zeros of $P_n(x)$. Then (26.6) reduces to the following eigenvalue problem:

$$x_\nu \mathbf{P}_n(x_\nu) = H_n \mathbf{P}_n(x_\nu).$$

Thus, x_ν are the eigenvalues of the matrix H_n and $\mathbf{P}_n(x_\nu)$ are the corresponding eigenvectors. According to (26.6), it is easy to obtain the determinant representation $P_n(x) = \det(xI_n - H_n)$, where I_n is the identity matrix of order n.

For computing zeros of $P_n(x)$ as the eigenvalues of the matrix H_n, we use the EISPACK routine COMQR [25, pp. 277–284]. Notice that this routine needs an upper Hessenberg matrix, i.e., the matrix H_n^T. Also, the MATLAB or MATHEMATICA could be used.

Therefore, the main problem in the construction of the type II multiple orthogonal polynomials in this way is computation of the recurrence coefficients in (26.5), i.e., computation of entries of the Hessenberg matrix H_n. For the simplest case of multiple orthogonality, when $r = 2$, for some classical weight functions (Jacobi, Laguerre, Hermite) one can find explicit formulas for the recurrence coefficients (see [3, 30, 32]). An effective numerical method for constructing the Hessenberg matrix H_n was given in [18].

26.3 Numerical Construction of Multiple Orthogonal Polynomials

In this section, we describe the method for constructing the Hessenberg matrix H_n, presented in [18].

For the calculation of the recurrence coefficient we use some kind of the Stieltjes procedure (cf. [8]), called the *discretized Stieltjes–Gautschi procedure*. At first, we express the elements of H_n in terms of the inner products[1] (26.4), and then we use the corresponding Gaussian rules to discretize these inner products. Of course, we suppose that the type II multiple orthogonal polynomials with respect to the inner products $(\cdot, \cdot)_k$, $k = 1, 2, \ldots, r$, given by (26.4), exist.

Taking $(\cdot, \cdot)_{j+\ell r} = (\cdot, \cdot)_j$, $\ell \in \mathbb{Z}$, for the inner products, the following result holds (see [18, Theorem 4.2]).

Theorem 26.3 *The type II multiple monic orthogonal polynomials* $\{P_n\}$*, with nearly diagonal multi-index, satisfy the recurrence relation*

$$P_{n+1}(x) = (x - a_{n,r})P_n(x) - \sum_{k=0}^{r-1} a_{n,k} P_{n-r+k}(x), \quad n \geq 0, \tag{26.7}$$

[1] Such formulas for coefficients of the three-term recurrence relation for standard orthogonal polynomials on the real line are known as Darboux formulas.

where $P_0(x) = 1$, $P_i(x) = 0$ *for* $i = -1, -2, \ldots, -r$,

$$a_{n,0} = \frac{(xP_n, P_{[(n-r)/r]})_{\nu+1}}{(P_{n-r}, P_{[(n-r)/r]})_{\nu+1}}$$

and

$$a_{n,k} = \frac{(xP_n - \sum_{i=0}^{k-1} a_{n,i} P_{n-r+i}, P_{[(n-r+k)/r]})_{\nu+k+1}}{(P_{n-r+k}, P_{[(n-r+k)/r]})_{\nu+k+1}}, \quad k = 1, 2, \ldots, r.$$

Here, we put $n = \ell r + \nu$, *where* $\ell = [n/r]$ *and* $\nu \in \{0, 1, \ldots, r-1\}$ ($[t]$ *denotes the integer part of* t).

We use alternatively recurrence relation (26.7) and given formulas for coefficients. Knowing P_0 we compute $a_{0,r}$, then knowing $a_{0,r}$ we compute P_1, and then again $a_{1,r}$ and $a_{1,r-1}$, etc. Continuing in this manner, we can generate as many polynomials, and therefore as many of the recurrence coefficients, as are desired.

All of the necessary inner products in the previous formulas can be computed exactly, except for rounding errors, by using the Gauss–Christoffel quadrature rule with respect to the corresponding weight function

$$\int_{E_j} g(t) w_j(t)\, \mathrm{d}t = \sum_{\nu=1}^{N} A_{j,\nu}^{(N)} g\big(\tau_{j,\nu}^{(N)}\big) + R_{j,N}(g), \quad j = 1, 2, \ldots, r. \tag{26.8}$$

Thus, for all calculations we use only the recurrence relation (26.7) for the type II multiple orthogonal polynomials and the Gauss–Christoffel quadrature rules (26.8).

26.4 Multiple Orthogonal Polynomials on the Semicircle

Polynomials orthogonal on the semicircle have been introduced by Gautschi and Milovanović in [11]. Multiple orthogonal polynomials on the semicircle, investigated by Milovanović and Stanić in [19], are a generalization of orthogonal polynomials on the semicircle in the sense that they satisfy $r \in \mathbb{N}$ orthogonality conditions.

We repeat some basic facts about polynomials orthogonal on the semicircle, and then transfer the concept of multiple orthogonality to the semicircle.

Let w be a weight function, which is positive and integrable on the open interval $(-1, 1)$, though possibly singular at the endpoints, and which can be extended to a function $w(z)$ holomorphic in the half disc $D_+ = \{z \in \mathbb{C} : |z| < 1, \operatorname{Im} z > 0\}$. Consider the following two inner products,

$$(f, g) = \int_{-1}^{1} f(x)\overline{g(x)} w(x)\, \mathrm{d}x, \tag{26.9}$$

$$[f, g] = \int_{\Gamma} f(z) g(z) w(z) (\mathrm{i}z)^{-1}\, \mathrm{d}z = \int_0^{\pi} f(\mathrm{e}^{\mathrm{i}\theta}) g(\mathrm{e}^{\mathrm{i}\theta}) w(\mathrm{e}^{\mathrm{i}\theta})\, \mathrm{d}\theta, \tag{26.10}$$

where Γ is the circular part of ∂D_+ and all integrals are assumed to exist, possibly as appropriately defined improper integrals.

The inner product (26.9) is positive definite and therefore generates a unique set of real orthogonal polynomials $\{p_k\}$ (p_k is monic polynomial of degree k). The inner product (26.10) is not Hermitian and the existence of the corresponding orthogonal polynomials, therefore, is not guaranteed.

A system of complex polynomials $\{\pi_k\}$ (π_k is monic of degree k) is called *orthogonal on the semicircle* if $[\pi_k, \pi_\ell] = 0$ for $k \neq \ell$ and $[\pi_k, \pi_\ell] \neq 0$ for $k = \ell$, $k, \ell = 0, 1, 2, \ldots$.

Gautschi, Landau, and Milovanović in [10] have established the existence of orthogonal polynomials $\{\pi_k\}$ assuming only that

$$\mathrm{Re}[1,1] = \mathrm{Re} \int_0^\pi w\big(\mathrm{e}^{\mathrm{i}\theta}\big)\,\mathrm{d}\theta \neq 0.$$

They have represented π_n as a linear (complex) combination of p_n and p_{n-1}, where $\{p_k\}$ is the sequence of the corresponding ordinary orthogonal (real) polynomials with respect to the inner product (26.9):

$$\pi_n(z) = p_n(z) - \mathrm{i}\theta_{n-1} p_{n-1}(z), \quad n \geq 0; \quad p_{-1}(x) = 0,\ p_0(x) = 1.$$

Under certain conditions, the zeros of polynomials orthogonal on the semicircle lie in D_+ (see [10, 11, 13, 14]).

Let C_ε, $\varepsilon > 0$, denote the boundary of D_+ with small circular parts of radius ε and centers at ± 1 spared out. Let $c_{\varepsilon,\pm 1}$ be the circular parts of C_ε with centers at ± 1 and radii ε. We assume that w is such that

$$\lim_{\varepsilon \downarrow 0} \int_{c_{\varepsilon,\pm 1}} g(z) w(z)\,\mathrm{d}z = 0 \quad \text{for all } g \in \mathscr{P},$$

and the following equation holds

$$0 = \int_\Gamma g(z) w(z)\,\mathrm{d}z + \int_{-1}^1 g(x) w(x)\,\mathrm{d}x, \quad g \in \mathscr{P}.$$

It is well known that the real (monic) polynomials $\{p_k(z)\}$, orthogonal with respect to the inner product (26.9), as well as the associated polynomials of the second kind,

$$q_k(z) = \int_{-1}^1 \frac{p_k(z) - p_k(x)}{z - x} w(x)\,\mathrm{d}x, \quad k = 0, 1, 2, \ldots,$$

satisfy a three-term recurrence relation of the form

$$y_{k+1} = (z - a_k) y_k - b_k y_{k-1}, \quad k = 0, 1, 2, \ldots,$$

whit initial conditions $y_{-1} = 0$, $y_0 = 1$ for $\{p_k\}$, and $y_{-1} = -1$, $y_0 = 0$ for $\{q_k\}$.

Definition 26.1 For a positive integer r, a set $W = \{w_1, \dots, w_r\}$ is *an admissible set of weight functions* if for the set W there exists a unique system of the (real) type II multiple orthogonal polynomials, and for each w_j, $j = 1, \dots, r$, there exists a unique system of (monic, complex) orthogonal polynomials relative to the inner product (26.10).

Let $r \geq 1$ be an integer and let $W = \{w_1, w_2, \dots, w_r\}$ be an admissible set of weight functions. Let $\mathbf{n} = (n_1, n_2, \dots, n_r)$ be the multi-index with length $|\mathbf{n}| = n_1 + n_2 + \cdots + n_r$. A multiple orthogonal polynomial on the semicircle is a monic polynomial $\Pi_{\mathbf{n}}(z)$ of degree $|\mathbf{n}|$ that satisfies the following orthogonality conditions:

$$\int_\Gamma \Pi_{\mathbf{n}}(z)\, z^k\, w_j(z)(\mathrm{i}z)^{-1}\, \mathrm{d}z = 0, \quad k = 0, 1, \dots, n_j - 1,\ j = 1, 2, \dots, r. \tag{26.11}$$

For $r = 1$, we have the ordinary orthogonal polynomials on the semicircle.
Let us denote by

$$\begin{aligned} [f, g]_j &= \int_\Gamma f(z) g(z) w_j(z)(\mathrm{i}z)^{-1}\, \mathrm{d}z \\ &= \int_0^\pi f(\mathrm{e}^{\mathrm{i}\theta}) g(\mathrm{e}^{\mathrm{i}\theta}) w_j(\mathrm{e}^{\mathrm{i}\theta})\, \mathrm{d}\theta, \quad j = 1, 2, \dots, r, \end{aligned} \tag{26.12}$$

the corresponding complex inner products.
The equations

$$0 = \int_\Gamma g(z) w_j(z)\, \mathrm{d}z + \int_{-1}^1 g(x) w_j(x)\, \mathrm{d}x \tag{26.13}$$

and

$$\int_\Gamma \frac{g(z) w_j(z)}{\mathrm{i}z}\, \mathrm{d}z = \pi g(0) w_j(0) + \mathrm{i} \,⨍_{-1}^1 \frac{g(x) w_j(x)}{x}\, \mathrm{d}x \tag{26.14}$$

hold for any polynomial g and for all $j = 1, 2, \dots, r$.

We consider only the nearly diagonal multi-indices $\mathbf{s}(n)$ and denote the corresponding multiple orthogonal polynomial on the semicircle by $\Pi_n(z) = \Pi_{\mathbf{s}(n)}(z)$. The corresponding type II multiple orthogonal polynomials (real) $\{P_n\}$ satisfy the recurrence relation (26.7). Also, it is easy to see that for $j = 1, 2, \dots, r$ the associated polynomials of the second kind,

$$Q_n^{(j)}(z) = \int_{-1}^1 \frac{P_n(z) - P_n(x)}{z - x} w_j(x)\, \mathrm{d}x, \quad n = 0, 1, \dots,$$

satisfy the same recurrence relation (but with different initial conditions).

Let us denote by $\mu_k^{(j)}$, $k \in \mathbb{N}_0$, $j = 1, 2, \dots, r$, the moments for the inner products (26.12), i.e.,

$$\mu_k^{(j)} = [z^k, 1]_j = \int_\Gamma z^k w_j(z)(\mathrm{i}z)^{-1}\, \mathrm{d}z, \quad j = 1, 2, \dots, r,\ k \in \mathbb{N}_0.$$

For zero moments we have

$$\mu_0^{(j)} = \int_\Gamma \frac{w_j(z)}{\mathrm{i}z}\,\mathrm{d}z = \pi w_j(0) + \mathrm{i}\,⨍_{-1}^{1} \frac{w_j(x)}{x}\,\mathrm{d}x, \quad j = 1, 2, \ldots, r. \tag{26.15}$$

Let us also denote

$$D_n = \begin{bmatrix} Q_{n-1}^{(1)}(0) - \mathrm{i}\mu_0^{(1)} P_{n-1}(0) & \cdots & Q_{n-r}^{(1)}(0) - \mathrm{i}\mu_0^{(1)} P_{n-r}(0) \\ Q_{n-1}^{(2)}(0) - \mathrm{i}\mu_0^{(2)} P_{n-1}(0) & \cdots & Q_{n-r}^{(2)}(0) - \mathrm{i}\mu_0^{(2)} P_{n-r}(0) \\ \vdots & & \vdots \\ Q_{n-1}^{(r)}(0) - \mathrm{i}\mu_0^{(r)} P_{n-1}(0) & \cdots & Q_{n-r}^{(r)}(0) - \mathrm{i}\mu_0^{(r)} P_{n-r}(0) \end{bmatrix}. \tag{26.16}$$

By using equations (26.13)–(26.14) for appropriately chosen polynomials g and orthogonality conditions (26.11), one can prove existence and uniqueness of multiple orthogonal polynomials on the semicircle with additional conditions that all matrices D_n are regular. The following theorem was proved in [21].

Theorem 26.4 *Let r be a positive integer and $W = \{w_1, \ldots, w_r\}$ be an admissible set of weight functions. Assume in addition that all matrices D_n, given by* (26.16), *are regular. Denoting by $\{P_k\}$ the (real) type II multiple orthogonal polynomials, relative to the set W, we have the following representation*

$$\Pi_k(z) = P_k(z) + \theta_{k,1} P_{k-1}(z) + \theta_{k,2} P_{k-2}(z) + \cdots + \theta_{k,r} P_{k-r}(z).$$

The coefficients $\theta_{k,j}$, $j = 1, 2, \ldots, r$, are the solution of the following system of linear equations

$$\sum_{j=1}^{r} \theta_{k,j}\big(Q_{k-j}^{(m)}(0) - \mathrm{i}\mu_0^{(m)} P_{k-j}(0)\big) = \mathrm{i}\mu_0^{(m)} P_k(0) - Q_k^{(m)}(0), \quad m = 1, 2, \ldots, r.$$

The multiple orthogonal polynomials on the semicircle with nearly diagonal multi-index satisfy the recurrence relation of order $r+1$, too. In a similar way as in the real case, the recurrence coefficients and the multiple orthogonal polynomials on the semicircle could be obtained by using some kind of the discretized Stieltjes–Gautschi procedure. Taking $[f, g]_{j+\ell r} = [f, g]_j$ for each $\ell \in \mathbb{Z}$, the following theorem could be proved (see [19]).

Theorem 26.5 *The multiple orthogonal polynomials on the semicircle $\{\Pi_n\}$, with nearly diagonal multi-index, satisfy the recurrence relation*

$$\Pi_{n+1}(z) = (z - \alpha_{n,r})\Pi_n(z) - \sum_{k=0}^{r-1} \alpha_{n,k} \Pi_{n-r+k}(x), \quad n \ge 0,$$

where $\Pi_0(z) = 1$, $\Pi_{-1}(z) = \Pi_{-2}(z) = \cdots = \Pi_{-r}(z) = 0$,

$$\alpha_{n,0} = \frac{[z\Pi_n, \Pi_{[(n-r)/r]}]_{\nu+1}}{[\Pi_{n-r}, \Pi_{[(n-r)/r]}]_{\nu+1}} \tag{26.17}$$

and

$$\alpha_{n,k} = \frac{[z\Pi_n - \sum_{i=0}^{k-1} \alpha_{n,i}\Pi_{n-r+i}, \Pi_{[(n-r+k)/r]}]_{\nu+k+1}}{[\Pi_{n-r+k}, \Pi_{[(n-r+k)/r]}]_{\nu+k+1}}, \quad k = 1, 2, \ldots, r. \quad (26.18)$$

Here, we put $n = \ell r + \nu$, where $\ell = [n/r]$ and $\nu \in \{0, 1, \ldots, r-1\}$ ($[t]$ denotes the integer part of t).

In order to apply the previous theorem, one has to calculate all of the inner products (26.17)–(26.18), i.e., the integrals of the following type $\int_\Gamma z^j \Pi_l(z) w_k(z) \times (\mathrm{i}z)^{-1}\mathrm{d}z$. For $j \geq 1$, because of (26.13), these integrals could be calculated exactly, except for rounding errors, by using the corresponding Gaussian quadratures. For $j = 0$ one has

$$\int_\Gamma \frac{\Pi_l(z) w_k(z)\, \mathrm{d}z}{\mathrm{i}z} = \mu_0^{(k)} \Pi_l(0) + \mathrm{i} \int_{-1}^{1} \frac{\Pi_l(x) - \Pi_l(0)}{x} w_k(x)\, \mathrm{d}x,$$

and the corresponding Gaussian quadratures and (26.15) could be used.

Knowing the recurrence coefficients, we form a complex lower banded Hessenberg matrix H_n as in the real case. The zeros of the multiple orthogonal polynomials on the semicircle are the eigenvalues of the complex Hessenberg matrix H_n.

26.4.1 Case $r = 2$

Let $W = \{w_1, w_2\}$ be an admissible set of weight functions. The type II (real) multiple orthogonal polynomials satisfy the following recurrence relation

$$P_{k+1}(x) = (x - b_k) P_k(x) - c_k P_{k-1}(x) - d_k P_{k-2}(x), \quad k \geq 0, \qquad (26.19)$$

with initial conditions $P_0(x) = 1$, $P_{-1}(x) = P_{-2} = 0$. The multiple orthogonal polynomials on the semicircle satisfy the following recurrence relation

$$\Pi_{k+1}(z) = (z - \beta_k)\Pi_k(z) - \gamma_k \Pi_{k-1}(z) - \delta_k \Pi_{k-2}(z), \quad k \geq 0, \qquad (26.20)$$

with the initial conditions $\Pi_0(z) = 1$, $\Pi_{-1}(z) = \Pi_{-2}(z) = 0$.

Using Theorem 26.4 for $k \geq 2$, we have the following equation

$$\Pi_k(z) = P_k(z) + \theta_{k,1} P_{k-1}(z) + \theta_{k,2} P_{k-2}(z), \qquad (26.21)$$

where $\theta_{k,1}$ and $\theta_{k,2}$ form the solution of the following system of linear equations

$$\theta_{k,1}\big(Q_{k-1}^{(1)}(0) - \mathrm{i}\mu_0^{(1)} P_{k-1}(0)\big) + \theta_{k,2}\big(Q_{k-2}^{(1)}(0) - \mathrm{i}\mu_0^{(1)} P_{k-2}(0)\big)$$
$$= \mathrm{i}\mu_0^{(1)} P_k(0) - Q_k^{(1)}(0),$$

$$\theta_{k,1}\big(Q_{k-1}^{(2)}(0) - \mathrm{i}\mu_0^{(2)} P_{k-1}(0)\big) + \theta_{k,2}\big(Q_{k-2}^{(2)}(0) - \mathrm{i}\mu_0^{(2)} P_{k-2}(0)\big)$$
$$= \mathrm{i}\mu_0^{(2)} P_k(0) - Q_k^{(2)}(0).$$

Relations between $\theta_{k,1}$, $\theta_{k,2}$ and recurrence coefficients b_k, c_k, d_k were derived in [21]:

$$\theta_{1,1} = b_0 - \frac{\mu_1^{(1)}}{\mu_0^{(1)}}, \qquad \theta_{2,1} = b_0 + b_1 - \frac{\mu_0^{(1)}\mu_2^{(2)} - \mu_2^{(1)}\mu_0^{(2)}}{\mu_0^{(1)}\mu_1^{(2)} - \mu_1^{(1)}\mu_0^{(2)}},$$

$$\theta_{k,1} = b_{k-1} - \frac{d_{k-1}}{\theta_{k-1,2}}, \qquad k \geq 3,$$

$$\theta_{2,2} = c_1 + b_0^2 - b_0 \frac{\mu_0^{(1)}\mu_2^{(2)} - \mu_2^{(1)}\mu_0^{(2)}}{\mu_0^{(1)}\mu_1^{(2)} - \mu_1^{(1)}\mu_0^{(2)}} + \frac{\mu_1^{(1)}\mu_2^{(2)} - \mu_2^{(1)}\mu_1^{(2)}}{\mu_0^{(1)}\mu_1^{(2)} - \mu_1^{(1)}\mu_0^{(2)}},$$

$$\theta_{k,2} = c_{k-1} - d_{k-1}\frac{\theta_{k-1,1}}{\theta_{k-1,2}}, \qquad k \geq 3.$$

Also, in [21], the recurrence coefficients β_k, γ_k, and δ_k were given as functions of b_k, c_k, d_k, $\theta_{k,1}$, and $\theta_{k,2}$:

$$\beta_0 = b_0 - \theta_{1,1},$$

$$\beta_1 = b_1 + \theta_{1,1} - \theta_{2,1}, \qquad \gamma_1 = c_1 + \theta_{1,1} b_0 - \theta_{2,2} - \beta_1\theta_{1,1},$$

$$\gamma_2 = \theta_{2,2} + \theta_{2,1}(b_1 - \theta_{2,1}), \qquad \delta_2 = d_2 - \gamma_2\theta_{1,1} - \beta_2\theta_{2,2} + c_1\theta_{2,1} + b_0\theta_{2,2},$$

$$\delta_3 = \theta_{3,2}(b_1 - \theta_{2,1}),$$

$$\beta_k = \theta_{k,1} + \frac{d_k}{\theta_{k,2}}, \qquad \gamma_k = \theta_{k,2} + d_{k-1}\frac{\theta_{k,1}}{\theta_{k-1,2}}, \qquad \delta_k = d_{k-2}\frac{\theta_{k,2}}{\theta_{k-2,2}}, \qquad k \geq 4.$$

26.5 Applications of Multiple Orthogonality to Numerical Integration

26.5.1 An Optimal Set of Quadrature Rules

Starting with a problem that arises in the evaluation of computer graphics illumination models, Borges [4] has examined the problem of numerically evaluating a set of r definite integrals taken with respect to distinct weight functions, but related to a common integrand and interval of integration. For such a problem, it is not efficient to use a set of r Gauss–Christoffel quadrature rules because valuable information is wasted.

Borges has introduced a performance ratio defined as

$$R = \frac{\text{Overall degree of precision} + 1}{\text{Number of integrand evaluations}}.$$

Taking the set of r Gauss–Christoffel quadrature rules, one has $R = 2/r$ and, hence, $R < 1$ for all $r > 2$.

If we select a set of n distinct nodes, common for all quadrature rules, then the weight coefficients for each of r quadrature rules can be chosen in such a way that $R = 1$. Since the selection of nodes is arbitrary, the quadrature rules may not be the best possible. The aim is to find an optimal set of nodes, by simulating the development of the Gauss–Christoffel quadrature rules.

Let us denote by $W = \{w_1, w_2, \ldots, w_r\}$ an AT system. Following [4, Definition 3], we introduce the following definition.

Definition 26.2 Let W be an AT system (the weight functions w_j, $j = 1, 2, \ldots, r$, are supported on the interval E), $\mathbf{n} = (n_1, n_2, \ldots, n_r)$ be a multi-index, and $n = |\mathbf{n}|$. A set of quadrature rules of the form

$$\int_E f(x) w_j(x)\,\mathrm{d}x \approx \sum_{\nu=1}^{n} A_{j,\nu} f(x_\nu), \qquad j = 1, 2, \ldots, r, \tag{26.22}$$

is *an optimal set with respect to* $(W, \mathbf{n})$ if and only if the weight coefficients, $A_{j,\nu}$, and the nodes, x_ν, satisfy the following equations:

$$\sum_{\nu=1}^{n} A_{j,\nu} x_\nu^{m+n_j-1} = \int_E x^{m+n_j-1} w_j(x)\,\mathrm{d}x, \quad m = 0, 1, \ldots, n;\ j = 1, 2, \ldots, r.$$

The next generalization of the fundamental theorem of Gauss–Christoffel quadrature rules holds (see [18] for proof).

Theorem 26.6 *Let W be an AT system, $\mathbf{n} = (n_1, n_2, \ldots, n_r)$, $n = |\mathbf{n}|$. The quadrature rules* (26.22) *form an optimal set with respect to* $(W, \mathbf{n})$ *if and only if*

1° *They are exact for all polynomials of degree less than or equal to $n - 1$;*
2° *The polynomial $q(x) = \prod_{\nu=1}^{n}(x - x_\nu)$ is the type II multiple orthogonal polynomial $P_{\mathbf{n}}$ with respect to W.*

Remark 26.1 All zeros of the type II multiple orthogonal polynomial $P_{\mathbf{n}}$ are distinct and located in the interval E (Theorem 26.2).

For $r = 1$ in Definition 26.2, we have the Gauss–Christoffel quadrature rule.

According to Theorem 26.6, the nodes of the optimal set of quadrature rules (of Gaussian type) with respect to $(W, \mathbf{n})$ are the zeros of the type II multiple orthogonal polynomial $P_{\mathbf{n}}$ with respect to the given AT system W. When the nodes are known, the weight coefficients $A_{j,\nu}$, $j = 1, 2, \ldots, r$, $\nu = 1, 2, \ldots, n$, can be obtained as the solutions of the following Vandermonde systems of equations

$$V(x_1, x_2, \ldots, x_n) \begin{bmatrix} A_{j,1} \\ A_{j,2} \\ \vdots \\ A_{j,n} \end{bmatrix} = \begin{bmatrix} \mu_0^{(j)} \\ \mu_1^{(j)} \\ \vdots \\ \mu_{n-1}^{(j)} \end{bmatrix}, \quad j = 1, 2, \ldots, r,$$

where

$$\mu_\nu^{(j)} = \int_E x^\nu w_j(x)\,\mathrm{d}x, \quad j = 1, 2, \ldots, r, \; \nu = 0, 1, \ldots, n-1.$$

Each of these Vandermonde systems always has the unique solution because the zeros of the type II multiple orthogonal polynomial $P_{\mathbf{n}}$ are distinct.

For the case of the nearly diagonal multi-indices $\mathbf{s}(n)$, we can compute the nodes x_ν, $\nu = 1, 2, \ldots, n$, of the Gaussian type quadrature rules as eigenvalues of the corresponding banded Hessenberg matrix H_n. Then, from the corresponding recurrence relation, it follows that the eigenvector associated with x_ν is given by $\mathbf{P}_n(x_\nu)$. We can use this fact to compute the weight coefficients $A_{j,\nu}$ by requiring that each rule correctly generate the first n modified moments.

Let us denote by

$$V_n = \left[\mathbf{P}_n(x_1)\; \mathbf{P}_n(x_2)\; \ldots\; \mathbf{P}_n(x_n)\right]$$

the matrix of the eigenvectors of H_n, each normalized so that the first component is equal to 1. Then, the weight coefficients $A_{j,\nu}$ can be obtained by solving systems of linear equations

$$V_n \begin{bmatrix} A_{j,1} \\ A_{j,2} \\ \vdots \\ A_{j,n} \end{bmatrix} = \begin{bmatrix} \mu_0^{*(j)} \\ \mu_1^{*(j)} \\ \vdots \\ \mu_{n-1}^{*(j)} \end{bmatrix}, \quad j = 1, 2, \ldots, r,$$

where

$$\mu_\nu^{*(j)} = \int_E P_\nu(x)\, w_j(x)\,\mathrm{d}x, \quad j = 1, 2, \ldots, r; \; \nu = 0, 1, \ldots, n-1,$$

are modified moments, $P_\nu = P_{\mathbf{s}(\nu)}$. All modified moments can be computed exactly, except for rounding errors, by using the Gauss–Christoffel quadrature rules with respect to the corresponding weight function w_j, $j = 1, 2, \ldots, r$.

In the same way as in the real case, we can generate the optimal set of quadrature rules

$$\int_0^\pi f\left(\mathrm{e}^{\mathrm{i}\theta}\right) w_j\left(\mathrm{e}^{\mathrm{i}\theta}\right) \mathrm{d}\theta \approx \sum_{\nu=1}^{n} \sigma_{j,\nu} f(\zeta_\nu), \quad j = 1, 2, \ldots, r,$$

where for each w_j, $j=1,2,\dots,r$, the corresponding quadrature is exact for all polynomials of degree less than or equal to $n+n_j-1$. The nodes of such an optimal set of quadratures are zeros of the multiple orthogonal polynomial on the semicircle $\Pi_{\mathbf{n}}(z)$, i.e., in the case of the nearly diagonal multi-index, the nodes are the eigenvalues of the Hessenberg matrix H_n. Using the corresponding eigenvectors, we obtain the weight coefficients in a similar way as in the real case.

26.5.2 An Optimal Set of Quadrature Rules with Preassigned Nodes

Let $W=\{w_1,w_2,\dots,w_r\}$ be an AT system. Following Definition 26.2 and ordinary quadrature rules of Gaussian type with preassigned abscissas (see, e.g., [7, Sect. 2.2.1]), we introduce the following definition (see [20]).

Definition 26.3 Let W be an AT system (the weight functions w_j, $j=1,2,\dots,r$, are supported on the interval E), $\mathbf{n}=(n_1,n_2,\dots,n_r)$ be a multi-index, $n=|\mathbf{n}|$. A set of quadrature rules of the form:

$$\int_E f(x)w_j(x)\mathrm{d}x \approx \sum_{i=1}^{k} a_{j,i} f(y_i) + \sum_{\nu=1}^{n} A_{j,\nu} f(x_\nu), \quad j=1,2,\dots,r, \tag{26.23}$$

where the nodes $y_i \in E$, $i=1,2,\dots,k$, are fixed and prescribed in advance, is called *an optimal set of quadrature rules with preassigned nodes* $\{y_i\}_{i=1}^k$ with respect to $(W,\mathbf{n})$ if and only if the weight coefficients, $a_{j,i}$, $A_{j,\nu}$, and the nodes, x_ν, satisfy the following equations:

$$\begin{aligned}\sum_{i=1}^{k} a_{j,i}\, y_i^{m+n_j+k-1} &+ \sum_{\nu=1}^{n} A_{j,\nu}\, x_\nu^{m+n_j+k-1} \\ &= \int_E x^{m+n_j+k-1}\, w_j(x)\,\mathrm{d}x, \quad m=0,1,\dots,n;\end{aligned}$$

for $j=1,2,\dots,r$.

For the set of preassigned nodes $\{y_i\}_{i=1}^k$, we introduce $s(x)$ as a polynomial of degree k, with zeros at y_i, $i=1,2,\dots,k$. Let us denote

$$\widetilde{W}=\{\widetilde{w}_1,\widetilde{w}_2,\dots,\widetilde{w}_r\}, \qquad \widetilde{w}_j(x)=s(x)w_j(x), \quad j=1,2,\dots,r.$$

Theorem 26.7 *Let W be an AT system, $\mathbf{n}=(n_1,n_2,\dots,n_r)$, $n=|\mathbf{n}|$. Suppose that for preassigned nodes, $\{y_i\}_{i=1}^k$, $\widetilde{W}$ is also an AT system. The set of quadrature rules* (26.23) *form an optimal set with preassigned nodes $\{y_i\}_{i=1}^k$ with respect to $(W,\mathbf{n})$ if and only if:*

1° *They are exact for all polynomials of degree less than or equal to* $n+k-1$;
2° *The polynomial* $q(x)=\prod_{\nu=1}^{n}(x-x_\nu)$ *is the type II multiple orthogonal polynomial* $P_{\mathbf{n}}$ *with respect to* $\widetilde{W}$.

Proof Let us suppose first that the quadrature rules (26.23) form the optimal set with preassigned nodes $\{y_i\}_{i=1}^{k}$ with respect to $(W,\mathbf{n})$. In order to prove 1°, we note that for each $j=1,2,\ldots,r$, the corresponding quadrature rule (26.23) is exact for all polynomials from $\mathbb{P}_{n+n_j+k-1}$ and then it is exact for those from $\mathbb{P}_{n+k-1}$. To prove 2°, for $j=1,2,\ldots r$, we assume that $p_j(x)\in\mathbb{P}_{n_j-1}$. Then, $q(x)p_j(x)s(x)\in\mathbb{P}_{n+n_j+k-1}$. Since the corresponding quadrature rule is exact for all such polynomials, it follows that

$$\int_E q(x)p_j(x)s(x)w_j(x)\,\mathrm{d}x=\sum_{i=1}^{k}a_{j,i}q(y_i)p_j(y_i)s(y_i)$$
$$+\sum_{\nu=1}^{n}A_{j,\nu}q(x_\nu)p_j(x_\nu)s(x_\nu).$$

Since $s(y_i)=0$ for $i=1,2,\ldots,k$ and $q(x_\nu)=0$ for $\nu=1,2,\ldots,n$, both sums on the right hand side in the previous equation are identically zero. Thus, we have

$$\int_E q(x)p_j(x)s(x)w_j(x)\,\mathrm{d}x=0,\quad j=0,1,\ldots,r,$$

and 2° follows.

Let us now suppose that for quadrature rules (26.23) 1° and 2° hold.

For $j=1,2,\ldots,r$, let $t_j(x)$ be a polynomial from $\mathbb{P}_{n+n_j+k-1}$. We can write $t_j(x)=u_j(x)\cdot q(x)s(x)+v(x)$, where $u_j(x)\in\mathbb{P}_{n_j-1}$ and $v(x)\in\mathbb{P}_{n+k-1}$. It is easy to see that

$$t_j(y_i)=v(y_i),\quad i=1,2,\ldots,k,\qquad t_j(x_\nu)=v(x_\nu),\quad \nu=1,2,\ldots,n.\quad(26.24)$$

Then, we obtain

$$\int_E t_j(x)w_j(x)\,\mathrm{d}x=\int_E\big[u_j(x)q(x)s(x)+v(x)\big]w_j(x)\,\mathrm{d}x$$
$$=\int_E q(x)u_j(x)s(x)w_j(x)\,\mathrm{d}x+\int_E v(x)\,w_j(x)\,\mathrm{d}x.$$

According to 2°, we have $\int_E q(x)u_j(x)s(x)w_j(x)\,\mathrm{d}x=0$ and, therefore,

$$\int_E t_j(x)\,w_j(x)\,\mathrm{d}x=\int_E v(x)\,w_j(x)\,\mathrm{d}x.$$

Since $v(x) \in \mathbb{P}_{n+k-1}$, it follows from 1° that

$$\int_E v(x)\, w_j(x)\,\mathrm{d}x = \sum_{i=1}^{k} a_{j,i} v(y_i) + \sum_{\nu=1}^{n} A_{j,\nu} v(x_\nu),$$

and hence, using (26.24), we obtain

$$\int_E t_j(x)\, w_j(x)\,\mathrm{d}x = \sum_{i=1}^{k} a_{j,i} v(y_i) + \sum_{\nu=1}^{n} A_{j,\nu} v(x_\nu)$$

$$= \sum_{i=1}^{k} a_{j,i} t_j(y_i) + \sum_{\nu=1}^{n} A_{j,\nu} t_j(x_\nu).$$

This proves that for each $j = 1, 2, \ldots, r$, the corresponding quadrature rule is exact for all polynomials of degree $\le n + n_j + k - 1$. □

According to Theorem 26.7, the nodes x_ν, $\nu = 1, 2, \ldots, n$, of the optimal set of quadrature rules with preassigned nodes (26.23) are the zeros of the type II multiple orthogonal polynomial $P_{\mathbf{n}}$ with respect to the AT system $\widetilde{W}$. In the case of nearly diagonal multi-index, we use the discretized Stieltjes–Gautschi procedure to compute those zeros. When the nodes are known, then for $j = 1, 2, \ldots, r$ we can choose the weight coefficients $a_{j,i}$, $i = 1, 2, \ldots, k$ and $A_{j,\nu}$, $\nu = 1, 2, \ldots, n$, such that they satisfy the following Vandermonde system of equations

$$V(y_1, \ldots, y_k, x_1, \ldots, x_n) \begin{bmatrix} a_{j,1} \\ \vdots \\ a_{j,k} \\ A_{j,1} \\ \vdots \\ A_{j,n} \end{bmatrix} = \begin{bmatrix} \mu_0^{(j)} \\ \mu_1^{(j)} \\ \vdots \\ \mu_{n+k-1}^{(j)} \end{bmatrix}, \quad j = 1, 2, \ldots, r, \tag{26.25}$$

where

$$\mu_i^{(j)} = \int_E x^i\, w_j(x)\,\mathrm{d}x, \quad j = 1, 2, \ldots, r; \quad i = 0, 1, \ldots, n + k - 1,$$

are moments which can be computed exactly, except for rounding errors, by using the Gauss–Christoffel quadrature rules with respect to the corresponding weight function w_j, $j = 1, 2, \ldots, r$.

Each of Vandermonde systems (26.25) has a unique solution if all of the preassigned nodes are distinct from the zeros of type II multiple orthogonal polynomial $P_{\mathbf{n}}$ with respect to $\widetilde{W}$. This is always satisfied in cases when the preassigned nodes are at the end points of the interval E, i.e., in the case of quadrature rules of Gauss–Radau or Gauss–Lobatto type.

26.5.3 *Connections with Generalized Birkhoff–Young Quadrature Rules*

In 1950, Birkhoff and Young [2] proposed a quadrature formula of the form

$$\int_{z_0-h}^{z_0+h} f(z)\,\mathrm{d}z \approx \frac{h}{15}\Big\{24f(z_0)+4\big[f(z_0+h)+f(z_0-h)\big] \\ -\big[f(z_0+\mathrm{i}h)+f(z_0-\mathrm{i}h)\big]\Big\}$$

for numerical integration over a line segment in the complex plane, where $f(z)$ is a complex analytic function in $\{z : |z-z_0| \le r\}$ and $|h| \le r$. This five point quadrature formula is exact for all algebraic polynomials of degree at most five and for its error $R_5^{BY}(f)$ the following estimate [33] can be proved (see also Davis and Rabinowitz [7, p. 136])

$$\big|R_5^{BY}(f)\big| \le \frac{|h|^7}{1890}\max_{z\in S}\big|f^{(6)}(z)\big|,$$

where S denotes the square with vertices $z_0+\mathrm{i}^k h$, $k=0,1,2,3$.

Without loss of generality, the previous quadrature rule can be considered over $[-1,1]$ for analytic functions in the unit disk $\{z : |z| \le 1\}$, so that it becomes

$$\int_{-1}^{1} f(z)\,\mathrm{d}z = \frac{16}{15}f(0)+\frac{4}{15}\big[f(1)+f(-1)\big] \\ -\frac{1}{15}\big[f(\mathrm{i})+f(-\mathrm{i})\big]+R_5(f). \tag{26.26}$$

In 1978, Tošić [29] obtained a significant improvement of (26.26) in the form

$$\int_{-1}^{1} f(z)\,\mathrm{d}z = Af(0)+\frac{1}{6}\left(\frac{7}{5}+\sqrt{\frac{7}{3}}\right)\big[f(r)+f(-r)\big] \\ +\frac{1}{6}\left(\frac{7}{5}-\sqrt{\frac{7}{3}}\right)\big[f(\mathrm{i}r)+f(-\mathrm{i}r)\big]+R_5^T(f),$$

where $r=\sqrt[4]{3/7}$ and

$$R_5^T(f)=\frac{1}{793800}f^{(8)}(0)+\frac{1}{61122600}f^{(10)}(0)+\cdots.$$

This formula was extended by Milovanović and Đorđević [17] to the following quadrature formula of interpolatory type

$$\int_{-1}^{1} f(z)\,dz = Af(0) + C_{11}\big[f(r_1)+f(-r_1)\big] + C_{12}\big[f(ir_1)+f(-ir_1)\big]$$
$$+ C_{21}\big[f(r_2)+f(-r_2)\big] + C_{22}\big[f(ir_2)+f(-ir_2)\big] + R_9(f; r_1, r_2),$$

where $0 < r_1 < r_2 < 1$. They proved that for

$$r_1 = r_1^* = \sqrt[4]{\frac{63-4\sqrt{114}}{143}} \quad \text{and} \quad r_2 = r_2^* = \sqrt[4]{\frac{63+4\sqrt{114}}{143}},$$

this formula has the algebraic degree of precision $p = 13$, with the error-term

$$R_9(f; r_1^*, r_2^*) = \frac{1}{28122661066500} f^{(14)}(0) + \cdots \approx 3.56 \cdot 10^{-14} f^{(14)}(0).$$

In this subsection, we consider a kind of generalized Birkhoff–Young quadrature formulas and give a connection with multiple orthogonal polynomials (cf. [16]). We introduce N-point quadrature formula for weighted integrals of analytic functions in the unit disc $\{z : |z| \le 1\}$,

$$I(f) := \int_{-1}^{1} f(z) w(z)\,dz = Q_N(f) + R_N(f),$$

where $w : (-1, 1) \to \mathbb{R}^+$ is an even positive weight function, for which all moments $\mu_k = \int_{-1}^{1} z^k w(z)\,dz$, $k = 0, 1, \ldots$, exist. For a given fixed integer $m \ge 1$ and for each $N \in \mathbb{N}$, we put $N = 2mn + \nu$ and define the node polynomial

$$\Omega_N(z) = z^\nu \omega_{n,\nu}(z^{2m}) = z^\nu \prod_{k=1}^{n} (z^{2m} - r_k), \quad 0 < r_1 < \cdots < r_n < 1, \tag{26.27}$$

where $n = [N/(2m)]$ and $\nu \in \{0, 1, \ldots, 2m-1\}$.

Now we consider the interpolatory quadrature rule Q_N of the form

$$Q_N(f) = \sum_{j=0}^{\nu-1} C_j f^{(j)}(0) + \sum_{k=1}^{n}\sum_{j=1}^{m} A_{k,j}\big[f(x_k e^{i\theta_j}) + f(-x_k e^{i\theta_j})\big],$$

where

$$x_k = \sqrt[2m]{r_k}, \quad k = 1, \ldots, n; \qquad \theta_j = \frac{(j-1)\pi}{m}, \quad j = 1, \ldots, m.$$

If $\nu = 0$, the first sum in $Q_N(f)$ is empty.

Following [16], we can prove the next result:

Theorem 26.8 *Let m be a fixed positive integer and w be an even positive weight function w on $(-1, 1)$, for which all moments $\mu_k = \int_{-1}^{1} z^k w(z)\,dz$, $k \ge 0$, exist.*

For any $N \in \mathbb{N}$ there exists a unique interpolatory quadrature rule $Q_N(f)$ with a maximal degree of exactness $d_{\max} = 2(m+1)n + s$, where

$$n = \left[\frac{N}{2m}\right], \qquad \nu = N - 2mn, \qquad s = \begin{cases} \nu - 1, & \nu \text{ even}, \\ \nu, & \nu \text{ odd}. \end{cases} \tag{26.28}$$

The node polynomial (26.27) is characterized by the following orthogonality relations

$$\int_{-1}^{1} z^{2k+s+1} \omega_{n,\nu}\big(z^{2m}\big) w(z)\, \mathrm{d}z = 0, \quad k = 0, 1, \ldots, n-1. \tag{26.29}$$

The conditions (26.29) can be expressed in the form

$$\int_{-1}^{1} p_{2k}(z) z^{s+1} \omega_{n,\nu}\big(z^{2m}\big) w(z)\, \mathrm{d}z = 0, \quad k = 0, 1, \ldots, n-1,$$

where $\{p_k\}_{k\in\mathbb{N}_0}$ is a system of polynomials orthogonal with respect to the weight w on $(-1,1)$.

The case with the Chebyshev weight of the first kind $w(z) = 1/\sqrt{1-z^2}$ and $m = 2$ was recently considered by Milovanović, Cvetković, and Stanić [22]. In that case, the previous conditions reduce to

$$\big(T_{2k}, z^{s+1} p_{n,\nu}(z^4)\big) = \int_{-1}^{1} \frac{T_{2k}(z) z^{s+1} p_{n,\nu}(z^4)}{\sqrt{1-z^2}}\, \mathrm{d}z = 0, \quad k = 0, 1, \ldots, n-1,$$

where T_k is the Chebyshev polynomial of the first kind of degree k. The corresponding quadrature rules are

$$Q_{4n+\nu}(f) = \sum_{j=0}^{\nu-1} C_j f^{(j)}(0) + \sum_{k=1}^{n} \big\{A_k\big[f(x_k) + f(-x_k)\big] + B_k\big[f(\mathrm{i}x_k) + f(-\mathrm{i}x_k)\big]\big\},$$

where $\nu = 0, 1, 2, 3$. For $\nu = 0$, the first sum on the right-hand side is empty. Also, in order to have $Q_{4n+\nu}(f) = I(f) = 0$ for $f(z) = z$, it must be $C_1 = 0$, so that $Q_{4n+1}(f) \equiv Q_{4n+2}(f)$.

The parameters of the quadrature formula $Q_{4n+\nu}(f)$ as well as the corresponding maximal degree of exactness $d = 6n + s$, where s is defined by (26.28), are presented in Table 26.1 for $n = 1$ and $\nu = 0, 1, 2, 3$.

By substitution $z^2 = t$, the orthogonality conditions (26.29) can be expressed in the form

$$\int_{0}^{1} t^k \omega_{n,\nu}\big(t^m\big) t^{s/2} w(\sqrt{t})\, \mathrm{d}t = 0, \quad k = 0, 1, \ldots, n-1.$$

This means that the polynomial $t \mapsto \omega_{n,\nu}(t^m)$ of degree mn is orthogonal to $\mathscr{P}_{n-1}$ with respect to the weight function $t^{s/2} w(\sqrt{t})$ on $(0,1)$, and it can be interpreted

Table 26.1 Parameters and the maximal degree of exactness of the generalized Birkhoff–Young–Chebyshev quadrature formula $Q_{4+\nu}(f)$ for $\nu = 0, 1, 2, 3$

ν	x_1	A_1	B_1	C_0	C_2	d
0	$\sqrt[4]{\frac{3}{8}}$	$\frac{\pi}{2}(\frac{1}{2}+\frac{1}{\sqrt{6}})$	$\frac{\pi}{2}(\frac{1}{2}-\frac{1}{\sqrt{6}})$			5
1, 2	$\sqrt[4]{\frac{5}{8}}$	$\frac{3+\sqrt{10}}{20}\pi$	$\frac{3-\sqrt{10}}{20}\pi$	$\frac{2\pi}{5}$		7
3	$\frac{1}{2}\sqrt[4]{\frac{35}{3}}$	$\frac{3(21+2\sqrt{105})}{490}\pi$	$\frac{3(21-2\sqrt{105})}{490}\pi$	$\frac{17\pi}{35}$	$\frac{\pi}{28}$	9

in terms of multiple orthogonal polynomials (see Milovanović [16]). Namely, these conditions are equivalent to

$$\int_0^1 t^{k/m} p_{n,\nu}(t) t^{(s+2)/(2m)-1} w\big(t^{1/(2m)}\big)\,\mathrm{d}t = 0, \quad k = 0, 1, \ldots, n-1.$$

Putting $k = m\ell + j - 1$, $\ell = [k/m]$, we get for each $j = 1, \ldots, m$,

$$\int_0^1 t^{\ell} p_{n,\nu}(t) w_j(t)\,\mathrm{d}t = 0, \quad \ell = 0, 1, \ldots, n_j - 1,$$

where

$$w_j(t) = t^{(s+2j)/(2m)-1} w\big(t^{1/(2m)}\big) \quad \text{and} \quad n_j = 1 + \left[\frac{n-j}{m}\right].$$

Notice that these weight functions, defined on the same interval $E_1 = E_2 = \cdots = E_m = E = (0, 1)$, can be expressed in the form $w_j(t) = t^{(j-1)/m} w_1(t)$, $j = 1, \ldots, m$, where $w_1(t) = t^{(s+2)/(2m)-1} w(t^{1/(2m)})$. Since the Müntz system

$$\{t^{k+(j-1)/m}\}, \quad k = 0, 1, \ldots, n_j - 1; \; j = 1, \ldots, m,$$

is a Chebyshev system on $[0, \infty)$, and also on $E = (0, 1)$, and $w_1(t) > 0$ on E, we conclude that $\{w_j, \; j = 1, \ldots, m\}$ is an AT system on E.

Therefore, according to Theorem 26.2, the unique type II multiple orthogonal polynomial $\omega_{n,\nu}(t) = P_{\mathbf{n}}(t)$ has exactly

$$|\mathbf{n}| := \sum_{j=1}^{m} n_j = \sum_{j=1}^{m} \left(1 + \left[\frac{n-j}{m}\right]\right) = n$$

zeros in $(0, 1)$. Thus, we have the following result [16]:

Theorem 26.9 *Under conditions of Theorem* 26.8, *for any* $N \in \mathbb{N}$ *there exists a unique interpolatory quadrature rule* $Q_N(f)$, *with a maximal degree of exactness*

$$d_{\max} = 2(m+1)n + s,$$

Table 26.2 Recursion coefficients $a_{n,k}$, $k = 0, 1, \ldots, r$, for the type II multiple orthogonal Jacobi polynomials with $r = 3$, $\alpha = 1/2$, $\beta_1 = -1/4$, $\beta_2 = 1/4$, $\beta_3 = 1$; $n \leq 16$

n	$a_{n,3}$	$a_{n,2}$
0	−3.333333333333333(−1)	
1	−1.282051282051282(−1)	2.735042735042735(−1)
2	−8.082010868388577(−2)	2.661439536886072(−1)
3	−1.797818980050774(−1)	2.623762626705582(−1)
4	−1.559462948426531(−1)	2.653111297708491(−1)
5	−1.239638179278716(−1)	2.659979011724685(−1)
6	−1.709146651380284(−1)	2.654960405557197(−1)
7	−1.579012355168128(−1)	2.662346749483896(−1)
8	−1.359869263770880(−1)	2.664756863684053(−1)
9	−1.669363956328833(−1)	2.662496940945655(−1)
10	−1.580814662624477(−1)	2.665641681228860(−1)
11	−1.415386037831715(−1)	2.666771188543775(−1)
12	−1.646602203100053(−1)	2.665436153306013(−1)
13	−1.579776557002493(−1)	2.667136879056348(−1)
14	−1.447181043954951(−1)	2.667770009974078(−1)
15	−1.631843805063865(−1)	2.666880187224645(−1)
16	−1.578284743344368(−1)	2.667934513861474(−1)

n	$a_{n,1}$	$a_{n,0}$
2	2.970182155702518(−2)	
3	1.746702080553980(−2)	−1.086753955083950(−3)
4	4.394216071462117(−2)	7.836954608420134(−4)
5	3.763075042610465(−2)	3.283125040895112(−3)
6	2.909135291014223(−2)	9.936110019727833(−4)
7	4.156697542465302(−2)	1.563768261907128(−3)
8	3.808465719477277(−2)	2.779000545083734(−3)
9	3.222685629651563(−2)	1.386312233788444(−3)
10	4.046385429052276(−2)	1.739912682414390(−3)
11	3.809384444106908(−2)	2.551616250383902(−3)
12	3.367302798492897(−2)	1.555345369944956(−3)
13	3.983305940039197(−2)	1.813811205210830(−3)
14	3.804538508869144(−2)	2.424240420958617(−3)
15	3.450289136608302(−2)	1.649497148723416(−3)
16	3.942565410008006(−2)	1.853693596062819(−3)

Table 26.3 The parameters of the optimal set of quadrature rules in the case of AT system of Jacobi weights for $r = 3$, $\alpha = 1/2$, $\beta_1 = -1/4$, $\beta_2 = 1/4$, $\beta_3 = 1$; $n = 16$

ν	x_ν	$A_{1,\nu}$
1	−9.991207278514688(−1)	2.593845860971087(−2)
2	−9.903618344136677(−1)	7.241932746868121(−2)
3	−9.638312475017886(−1)	1.232021800649184(−1)
4	−9.114418918738332(−1)	1.705746075613651(−1)
5	−8.280210844814640(−1)	2.096867833494312(−1)
6	−7.115498578342734(−1)	2.372180134962362(−1)
7	−5.631228057882331(−1)	2.511227240543505(−1)
8	−3.867448648543452(−1)	2.506499390258389(−1)
9	−1.889812469123731(−1)	2.363793559115719(−1)
10	2.346960814946570(−1)	1.750234363552845(−1)
11	2.153170148057211(−2)	2.101750909061207(−1)
12	4.396510791687633(−1)	1.347552229954631(−1)
13	6.255191622587539(−1)	9.367498172796770(−2)
14	7.821521528902294(−1)	5.613339333990859(−2)
15	9.008275402581413(−1)	2.608777198579168(−2)
16	9.748476093398128(−1)	6.697740217113879(−3)

ν	$A_{2,\nu}$	$A_{3,\nu}$
1	7.595267817320088(−4)	3.896535630665977(−6)
2	7.109430002756949(−3)	2.187023588317833(−4)
3	2.343071830873815(−2)	1.943281045595175(−3)
4	5.076081012916438(−2)	8.240428332930785(−3)
5	8.695782331878246(−2)	2.322282583304115(−2)
6	1.274039941410160(−1)	5.014597864881742(−2)
7	1.659837991885244(−1)	8.919391597730677(−2)
8	1.962854916209727(−1)	1.360251129413198(−1)
9	2.128751628357952(−1)	1.819274220554127(−1)
10	1.944805801244912(−1)	2.277961048571476(−1)
11	2.124257539351905(−1)	2.158470188824615(−1)
12	1.616866751974687(−1)	2.125040277473365(−1)
13	1.194317136600792(−1)	1.719347786751889(−1)
14	7.493654857422641(−2)	1.155852112758036(−1)
15	3.596734227391774(−2)	5.822578930636818(−2)
16	9.412285520984374(−3)	1.567997205810892(−2)

if and only if the polynomial $\omega_{n,\nu}(t)$ *is the type II multiple orthogonal polynomial* $P_{\mathbf{n}}(t)$, *with respect to the weights* $w_j(t) = t^{(s+2j)/(2m)-1}w(t^{1/(2m)})$, *with*

$$n_j = 1 + \left[\frac{n-j}{m}\right], \quad j = 1,\ldots,m.$$

26.6 Numerical Example

As an example we consider the type II multiple orthogonal Jacobi polynomials, i.e., the type II multiple orthogonal polynomials with respect to an AT system consisting of Jacobi weight functions on $[-1, 1]$ with different singularities at -1 and the same singularity at 1. Weight functions are

$$w_j(x) = (1-x)^{\alpha}(1+x)^{\beta_j}, \quad j = 1,2,\ldots,r,$$

where $\alpha, \beta_j > -1$, $j = 1,2,\ldots,r$, and $\beta_i - \beta_l \notin \mathbb{Z}$ whenever $i \neq l$.

In Table 26.2, the coefficients of recurrence relation (26.7) for multiple orthogonal Jacobi polynomials in the case $r = 3$, $\alpha = 1/2$, $\beta_1 = -1/4$, $\beta_2 = 1/4$, $\beta_3 = 1$ for $n \leq 16$ are given (numbers in parentheses denote decimal exponents). The nodes x_ν and the weights $A_{j,\nu}$, $\nu = 1,\ldots,16$, $j = 1,2,3$, of the corresponding optimal set of quadrature rules (26.22) are given in Table 26.3.

Acknowledgements The authors were supported in part by the Serbian Ministry of Education and Science (Project: Approximation of Integral and Differential Operators and Applications, grant number #174015).

References

1. Aptekarev, A.I.: Multiple orthogonal polynomials. J. Comput. Appl. Math. **99**, 423–447 (1998)
2. Birkhoff, G., Young, D.M.: Numerical quadrature of analytic and harmonic functions. J. Math. Phys. **29**, 217–221 (1950)
3. Beckermann, B., Coussement, J., Van Assche, W.: Multiple Wilson and Jacobi-Piñeiro polynomials. J. Approx. Theory **132**(2), 155–181 (2005)
4. Borges, C.F.: On a class of Gauss-like quadrature rules. Numer. Math. **67**, 271–288 (1994)
5. De Bruin, M.G.: Simultaneous Padé approximation and orthogonality. In: Brezinski, C., Draux, A., Magnus, A.P., Maroni, P., Ronveaux, A. (eds.) Proc. Polynômes Orthogoneaux et Applications, Bar-le-Duc, 1984. Lecture Notes in Math., vol. 1171, pp. 74–83. Springer, Berlin (1985)
6. Chihara, T.S.: An Introduction to Orthogonal Polynomials. Gordon and Breach, New York (1978)
7. Davis, P.J., Rabinowitz, P.: Methods of Numerical Integration. Academic Press, New York, San Francisco (1975)
8. Gautschi, W.: Orthogonal polynomials: applications and computation. Acta Numer. **5**, 45–119 (1996)

9. Gautschi, W.: Orthogonal Polynomials: Computation and Approximation. Numerical Mathematics and Scientific Computation. Oxford University Press, Oxford (2004)
10. Gautschi, W., Landau, H.J., Milovanović, G.V.: Polynomials orthogonal on the semicircle, II. Constr. Approx. **3**, 389–404 (1987)
11. Gautschi, W., Milovanović, G.V.: Polynomials orthogonal on the semicircle. J. Approx. Theory **46**, 230–250 (1986)
12. Mastroianni, G., Milovanović, G.V.: Interpolation Processes—Basic Theory and Applications. Springer Monographs in Mathematics. Springer, Berlin (2008)
13. Milovanović, G.V.: Complex orthogonality on the semicircle with respect to Gegenbauer weight: theory and applications. In: Rassias, T.M. (ed.) Topics in Mathematical Analysis. Ser. Pure Math., vol. 11, pp. 695–722. World Sci., Teaneck (1989)
14. Milovanović, G.V.: On polynomials orthogonal on the semicircle and applications. J. Comput. Appl. Math. **49**, 193–199 (1993)
15. Milovanović, G.V.: Orthogonal polynomials on the radial rays in the complex plane and applications. Rend. Circ. Mat. Palermo, Serie II, Suppl. **68**, 65–94 (2002)
16. Milovanović, G.V.: Numerical quadratures and orthogonal polynomials. Stud. Univ. Babeş-Bolyai Math. **56**, 449–464 (2011)
17. Milovanović, G.V., Đorđević, R.Ž.: On a generalization of modified Birkhoff–Young quadrature formula. Univ. Beograd. Publ. Elektrotehn. Fak. Ser. Mat. Fis. **735–762**, 130–134 (1982)
18. Milovanović, G.V., Stanić, M.: Construction of multiple orthogonal polynomials by discretized Stieltjes–Gautschi procedure and corresponding Gaussian quadratures. Facta Univ. Ser. Math. Inform. **18**, 9–29 (2003)
19. Milovanović, G.V., Stanić, M.: Multiple orthogonal polynomials on the semicircle and corresponding quadratures of Gaussian type. Math. Balk. **18**, 373–387 (2004)
20. Milovanović, G.V., Stanić, M.: Multiple orthogonality and quadratures of Gaussian type. Rend. Circ. Mat. Palermo, Serie II, Suppl. **76**, 75–90 (2005)
21. Milovanović, G.V., Cvetković, A.S., Stanić, M.P.: Multiple orthogonal polynomials on the semicircle. Facta Univ. Ser. Math. Inform. **20**, 41–55 (2005)
22. Milovanović, G.V., Cvetković, A.S., Stanić, M.P.: A generalized Birkhoff–Young–Chebyshev quadrature formula for analytic functions. Appl. Math. Comput. **218**, 944–948 (2011)
23. Nikishin, E.M., Sorokin, V.N.: Rational Approximations and Orthogonality, vol. 92. Amer. Math. Soc., Providence (1991)
24. Piñeiro, L.R.: On simultaneous approximations for a collection of Markov functions. Vestn. Mosk. Univ., Ser. I **2**(2), 67–70 (1987). English translation in Moscow Univ. Math. Bull. **42**(2), 52–55 (1987)
25. Smith, B.T., Boyle, J.M., Dongarra, J.J., Garbow, B.S., Ikebe, Y., Klema, V.C., Moler, C.B.: Matrix Eigensystem Routines—EISPACK Guide. Lect. Notes Comp. Science, vol. 6. Springer, Berlin (1976)
26. Sorokin, V.N.: Generalization of classical polynomials and convergence of simultaneous Padé approximants. Tr. Semin. Im. I.G. Petrovskogo **11**, 125–165 (1986). English translation in J. Soviet Math. **45**, 1461–1499 (1986)
27. Sorokin, V.N.: Simultaneous Padé approximation for functions of Stieltjes type. Sib. Mat. Zh. **31**(5), 128–137 (1990). English translation in Sib. Math. J. **31**(5), 809–817 (1990)
28. Sorokin, V.N.: Hermite–Padé approximations for Nikishin systems and the irrationality of $\zeta(3)$. Usp. Mat. Nauk **49**(2), 167–168 (1994). English translation in Russ. Math. Surveys **49**(2), 176–177 (1994)
29. Tošić, Đ.: A modification of the Birkhoff–Young quadrature formula for analytic functions. Univ. Beograd. Publ. Elektrotehn. Fak. Ser. Mat. Fis. **602–633**, 73–77 (1978)
30. Van Assche, W.: Multiple orthogonal polynomials, irrationality and transcendence. In: Berndt, B.C., Gesztesy, F. (eds.) Continued Fractions: From Analytic Number Theory to Constructive Approximation. Contemporary Mathematics, vol. 236, pp. 325–342 (1999)

31. Van Assche, W.: Non-symmetric Linear Difference Equations for Multiple Orthogonal Polynomials. In: CRM Proceedings and Lecture Notes, vol. 25, pp. 391–405. Amer. Math. Soc., Providence (2000)
32. Van Assche, W., Coussement, E.: Some classical multiple orthogonal polynomials. J. Comput. Appl. Math. **127**, 317–347 (2001)
33. Young, D.M.: An error bound for the numerical quadrature of analytic functions. J. Math. Phys. **31**, 42–44 (1952)

Chapter 27
Approximate C^*-Algebra Homomorphisms Associated to an Apollonius–Jensen Type Additive Mapping; A Fixed Point Approach

Fridoun Moradlou and G. Zamani Eskandani

Abstract In this paper, we prove the Hyers–Ulam–Rassias stability of C^*-algebra homomorphisms and of generalized derivations on C^*-algebras for the following Cauchy–Jensen functional equation:

$$f\left(\left(\sum_{i=1}^{n} z_i\right) - \left(\sum_{i=1}^{n} x_i\right)\right) + f\left(\left(\sum_{i=1}^{n} z_i\right) - \left(\sum_{i=1}^{n} y_i\right)\right)$$
$$= 2f\left(\left(\sum_{i=1}^{n} z_i\right) - \frac{(\sum_{i=1}^{n} x_i) + (\sum_{i=1}^{n} y_i)}{2}\right).$$

The concept of Hyers–Ulam–Rassias stability originated from the Th.M. Rassias' stability theorem that appeared in his paper: On the stability of the linear mapping in Banach spaces, Proc. Am. Math. Soc. 72:297–300, 1978.

Key words Cauchy–Jensen functional equation · Fixed point · C^*-algebra homomorphism · Hyers–Ulam–Rassias stability · Generalized derivation

Mathematics Subject Classification Primary 39B72 · 47H10 · 46L05 · 46B03 · 47Jxx

27.1 Introduction and preliminaries

In 1940, Ulam [47] brought up a question in the theory of functional equations which is the following: "When is it true that a function, which approximately satisfies a functional equation $\mathcal{E}$ must be close to an exact solution of $\mathcal{E}$?" If the above

Dedicated to Professor Themistocles M. Rassias on the occasion of his 60th birthday.

F. Moradlou
Department of Mathematics, Sahand University of Technology, Tabriz, Iran
e-mail: moradlou@sut.ac.ir

G.Z. Eskandani (✉)
Faculty of Mathematical Sciences, University of Tabriz, Tabriz, Iran
e-mail: zamani@tabrizu.ac.ir

P.M. Pardalos et al. (eds.), *Nonlinear Analysis*, Springer Optimization and Its Applications 68, In Honor of Themistocles M. Rassias on the Occasion of his 60th Birthday, DOI 10.1007/978-1-4614-3498-6_27,

problem accepts a solution, we say that the equation $\mathcal{E}$ is stable. In 1941, Ulam's problem was solved by Hyers [19] in Banach spaces. This result was generalized by Aoki [2] for additive mappings and by Th.M. Rassias [42] for linear mappings by considering an unbounded Cauchy difference. The paper of Th.M. Rassias [42] has provided a lot of influence in the development of what we now call Hyers–Ulam–Rassias stability of functional equations. P. Găvruta [17] generalized the Th.M. Rassias' result in the spirit of Th.M. Rassias's stability approach. Following the techniques of the proof of the corollary of Hyers [19], we emphasize that Hyers introduced (in 1941) the so-called Hyers continuity condition about the continuity of the mapping, and then he proved homogeneity of degree one and therefore the linearity of the mapping. This condition has been considered further till now, through the complete Hyers direct method, in order to prove linearity for the generalized Hyers–Ulam stability problem for approximate homomorphisms (see [20]). Beginning around the year 1980, the stability problems of a wide class of functional equations and approximate homomorphisms have been extensively investigated by a number of authors, and there are many interesting results concerning this problem (see [4, 6, 9, 11, 13–16, 18, 20, 22–40, 43–45]).

In 2003, Cădariu and Radu applied the fixed point method to the investigation of the Jensen functional equation [7] (see also [8, 9, 41]). They were able to present a short and a simple proof (different of the "*direct method*", initiated by Hyers in 1941) for the generalized Hyers–Ulam stability of Jensen functional equation [7], for Cauchy functional equation [9] and for quadratic functional equation [8].

The following functional equation

$$Q(x+y)+Q(x-y)=2Q(x)+2Q(y) \tag{27.1}$$

is called a *quadratic functional equation*, and every solution of (27.1) is said to be a *quadratic mapping*. F. Skof [46] proved the Hyers–Ulam stability of the quadratic functional equation (27.1) for mappings $f : E_1 \to E_2$, where E_1 is a normed space and E_2 is a Banach space. In [10], S. Czerwik proved the Hyers–Ulam stability of the quadratic functional equation (27.1). C. Borelli and G.L. Forti [5] generalized the stability result of the quadratic functional equation (27.1). Jun and Lee [21] proved the Hyers–Ulam stability of the Pexiderized quadratic equation

$$f(x+y)+g(x-y)=2h(x)+2k(y)$$

for mappings f, g, h, and k. The stability problem of the quadratic equation has been extensively investigated by some mathematicians.

In an inner product space, the equality

$$\|z-x\|^2+\|z-y\|^2=\frac{1}{2}\|x-y\|^2+2\left\|z-\frac{x+y}{2}\right\|^2 \tag{27.2}$$

holds, then it is called the *Apollonius' identity*. If the following functional equation, which was motivated by the above equation, namely

$$Q(z-x)+Q(z-y)=\frac{1}{2}Q(x-y)+2Q\left(z-\frac{x+y}{2}\right), \tag{27.3}$$

holds, then it is called *quadratic* (see [34]). For this reason, the functional equation (27.3) is called a *quadratic functional equation of Apollonius type*, and each solution of the functional equation (27.3) is said to be a *quadratic mapping of Apollonius type*. The quadratic functional equation and several other functional equations are useful to characterize inner product spaces [1].

Recently in [32], C. Park introduced and investigated the following functional equation

$$f\left(\left(\sum_{i=1}^{n} z_i\right) - \left(\sum_{i=1}^{n} x_i\right)\right) + f\left(\left(\sum_{i=1}^{n} z_i\right) - \left(\sum_{i=1}^{n} y_i\right)\right) = 2f\left(\left(\sum_{i=1}^{n} z_i\right) - \frac{(\sum_{i=1}^{n} x_i) + (\sum_{i=1}^{n} y_i)}{2}\right) \tag{27.4}$$

which is called the *generalized Apollonius–Jensen type additive functional equation* and whose solution is said to be a *generalized Apollonius–Jensen type additive mapping*.

We will adopt the idea of Cădariu and Radu [7, 9, 41], to prove the generalized Hyers–Ulam stability results of C^*-algebra homomorphisms as well as to prove the generalized Ulam–Hyers stability of generalized derivations on C^*-algebra, for additive functional equation of n-Apollonius type.

We recall two fundamental results in fixed point theory.

Theorem 27.1 ([7]) *Let (X, d) be a complete metric space and let $J : X \to X$ be strictly contractive, i.e.,*

$$d(Jx, Jy) \le Ld(x, y), \quad \forall x, y \in X$$

for some Lipschitz constant $L < 1$. Then

1. *The mapping J has a unique fixed point $x^* = Jx^*$;*
2. *The fixed point x^* is globally attractive, i.e.,*

$$\lim_{n\to\infty} J^n x = x^*$$

   for any starting point $x \in X$;
3. *One has the following estimates:*

$$d\left(J^n x, x^*\right) \le L^n d\left(x, x^*\right),$$

$$d\left(J^n x, x^*\right) \le \frac{1}{1-L} d\left(J^n x, J^{n+1} x\right),$$

$$d\left(x, x^*\right) \le \frac{1}{1-L} d(x, Jx)$$

   for all nonnegative integers n and all $x \in X$.

Definition 27.1 Let X be a set. A function $d : X \times X \to [0, \infty]$ is called a *generalized metric* on X if d satisfies:

(i) $d(x, y) = 0$ if and only if $x = y$;
(ii) $d(x, y) = d(y, x)$ for all $x, y \in X$;
(iii) $d(x, z) \le d(x, y) + d(y, z)$ for all $x, y, z \in X$.

Theorem 27.2 (See [12]) *Let (X, d) be a complete generalized metric space and let $J : X \to X$ be a strictly contractive mapping with Lipschitz constant $L < 1$. Then for each given element $x \in X$, either*

$$d\big(J^n x, J^{n+1} x\big) = \infty$$

for all nonnegative integers n or there exists a positive integer n_0 such that

1. $d(J^n x, J^{n+1} x) < \infty, \forall n \ge n_0$;
2. *The sequence $\{J^n x\}$ converges to a fixed point y^* of J;*
3. *y^* is the unique fixed point of J in the set $Y = \{y \in X \mid d(J^{n_0} x, y) < \infty\}$;*
4. *$d(y, y^*) \le \frac{1}{1-L} d(y, Jy)$ for all $y \in Y$.*

This paper is organized as follows: In Sect. 27.2, using the fixed point method, we prove the Hyers–Ulam–Rassias stability of C^*-algebra homomorphisms for the generalized Apollonius–Jensen type additive functional equation.

In Sect. 27.3, using the fixed point method, we prove the Hyers–Ulam–Rassias stability of generalized derivations on C^*-algebras for the generalized Apollonius–Jensen type additive functional equation.

Throughout this paper, assume that A is a C^*-algebra with norm $\|\cdot\|_A$ and that B is a C^*-algebra with norm $\|\cdot\|_B$.

27.2 Stability of C^*-Algebra Homomorphisms

For a given mapping $f : A \to B$ and for a fixed positive integer $n \ge 2$, we define

$$\begin{aligned}
&C_\mu f(z_1, \dots, z_n, x_1, \dots, x_n, y_1, \dots, y_n) \\
&:= f\left(\left(\sum_{i=1}^{n} \mu z_i\right) - \left(\sum_{i=1}^{n} \mu x_i\right)\right) + f\left(\left(\sum_{i=1}^{n} \mu z_i\right) - \left(\sum_{i=1}^{n} \mu y_i\right)\right) \\
&\quad - 2\mu f\left(\left(\sum_{i=1}^{n} z_i\right) - \frac{(\sum_{i=1}^{n} x_i) + (\sum_{i=1}^{n} y_i)}{2}\right).
\end{aligned}$$

for all $\mu \in \mathbb{T}^1 := \{\nu \in \mathbb{C} : |\nu| = 1\}$ and all $z_1, \dots, z_n, x_1, \dots, x_n, y_1, \dots, y_n \in A$.

We prove the Hyers–Ulam–Rassias stability of C^*-algebra homomorphisms for the functional equation $C_\mu f(z_1, \dots, z_n, x_1, \dots, x_n, y_1, \dots, y_n) = 0$.

Theorem 27.3 *Let $f : A \to B$ be a mapping satisfying $f(0) = 0$ for which there exists a function $\varphi : A^{3n} \to [0, \infty)$ such that*

$$\sum_{j=0}^{\infty} \frac{1}{2^j}\varphi\left(2^j z_1, \ldots, 2^j z_n, 2^j x_1, \ldots, 2^j x_n, 2^j y_1, \ldots, 2^j y_n\right) < \infty, \tag{27.5}$$

$$\left\| C_\mu f(z_1, \ldots, z_n, x_1, \ldots, x_n, y_1, \ldots, y_n)\right\|_B \le \varphi(z_1, \ldots, z_n, x_1, \ldots, x_n, y_1, \ldots, y_n), \tag{27.6}$$

$$\left\| f(xy) - f(x)f(y)\right\|_B \le \varphi(x, y, \underbrace{0, \ldots, 0}_{3n-2 \text{ times}}), \tag{27.7}$$

$$\left\| f\left(x^*\right) - f(x)^*\right\|_B \le \varphi(\underbrace{x, \ldots, x}_{3n \text{ times}}) \tag{27.8}$$

for all $\mu \in \mathbb{T}^1$ and all $z_1, \ldots, z_n, x_1, \ldots, x_n, y_1, \ldots, y_n \in A$. If there exists an $L < 1$ such that

$$\varphi(\underbrace{x, \ldots, x}_{2n \text{ times}}, \underbrace{-x, \ldots, -x}_{n \text{ times}}) \le 2L\varphi\Bigg(\underbrace{\frac{x}{2}, \ldots, \frac{x}{2}}_{2n \text{ times}}, \underbrace{-\frac{x}{2}, \ldots, -\frac{x}{2}}_{n \text{ times}}\Bigg)$$

for all $x \in A$, then there exists a unique C^*-algebra homomorphism $H : A \to B$ such that

$$\left\| f(x) - H(x)\right\|_B \le \frac{1}{2-2L}\varphi(\underbrace{x, \ldots, x}_{2n \text{ times}}, \underbrace{-x, \ldots, -x}_{n \text{ times}}) \tag{27.9}$$

for all $x \in A$.

Proof Consider the set

$$X := \{g : A \to B, \quad g(0) = 0\}$$

and introduce the *generalized metric* on X:

$$d(g,h) = \inf\Bigg\{C \in \mathbb{R}_+ : \left\|g(x) - h(x)\right\|_B \le C\varphi\Bigg(\underbrace{\frac{x}{n}, \ldots, \frac{x}{n}}_{2n \text{ times}}, \underbrace{-\frac{x}{n}, \ldots, -\frac{x}{n}}_{n \text{ times}}\Bigg), \quad \forall x \in A\Bigg\}.$$

It is easy to show that (X, d) is complete.

Now we consider the linear mapping $J : X \to X$ such that

$$Jg(x) := \frac{1}{2}g(2x)$$

for all $x \in A$.

For any $g, h \in X$, we have

$$
\begin{aligned}
& d(g,h) < C \\
& \Longrightarrow \quad \|g(x)-h(x)\|_B \le C\varphi\Big(\underbrace{\frac{x}{n},\dots,\frac{x}{n}}_{2n \text{ times}},\underbrace{-\frac{x}{n},\dots,-\frac{x}{n}}_{n \text{ times}}\Big), \quad \forall x \in A \\
& \Longrightarrow \quad \left\|\frac{1}{2}g(2x)-\frac{1}{2}h(2x)\right\|_B \le \frac{1}{2}C\varphi\Big(\underbrace{\frac{2x}{n},\dots,\frac{2x}{n}}_{2n \text{ times}},\underbrace{-\frac{2x}{n},\dots,-\frac{2x}{n}}_{n \text{ times}}\Big) \\
& \Longrightarrow \quad \left\|\frac{1}{2}g(2x)-\frac{1}{2}h(2x)\right\|_B \le LC\varphi\Big(\underbrace{\frac{x}{n},\dots,\frac{x}{n}}_{2n \text{ times}},\underbrace{-\frac{x}{n},\dots,-\frac{x}{n}}_{n \text{ times}}\Big) \\
& \Longrightarrow \quad d(Jg,Jh) \le LC.
\end{aligned}
$$

Therefore, we see that

$$d(Jg,Jh) \le Ld(g,h), \quad \forall g,h \in X,$$

which means that J is a strictly contractive self-mapping of X with the Lipschitz constant L.

Letting $\mu = 1$ and $z_1 = \cdots = z_n = x_1 = \cdots = x_n = x$ and $y_1 = \cdots = y_n = -x$ in (27.6), we get

$$\|f(2nx)-2f(nx)\|_B \le \varphi(\underbrace{x,\dots,x}_{2n \text{ times}},\underbrace{-x,\dots,-x}_{n \text{ times}}) \tag{27.10}$$

for all $x \in A$. So

$$\left\|f(x)-\frac{1}{2}f(2x)\right\|_B \le \frac{1}{2}\varphi\Big(\underbrace{\frac{x}{n},\dots,\frac{x}{n}}_{2n \text{ times}},\underbrace{-\frac{x}{n},\dots,-\frac{x}{n}}_{n \text{ times}}\Big)$$

for all $x \in A$. Hence $d(f,Jf) \le \frac{1}{2}$.

By Theorem 27.2, there exists a mapping $H : A \to B$ such that the following hold:

1. H is a fixed point of J, i.e.,

$$H(2x) = 2H(x) \tag{27.11}$$

for all $x \in A$. The mapping H is the unique fixed point of J in the set

$$Y = \{g \in X : d(f,g) < \infty\}.$$

This implies that H is the unique mapping satisfying (27.11) such that there exists $C \in (0, \infty)$ satisfying

$$\|H(x) - f(x)\|_B \le C\varphi\Big(\underbrace{\frac{x}{n}, \dots, \frac{x}{n}}_{2n \text{ times}}, \underbrace{-\frac{x}{n}, \dots, -\frac{x}{n}}_{n \text{ times}}\Big)$$

for all $x \in A$.

2. $d(J^m f, H) \to 0$ as $m \to \infty$. This implies the equality

$$\lim_{m\to\infty} \frac{f(2^m x)}{2^m} = H(x) \tag{27.12}$$

for all $x \in A$.

3. $d(f, H) \le \frac{1}{1-L} d(f, Jf)$, which implies the inequality

$$d(f, H) \le \frac{1}{2-2L}.$$

The latter yields the inequality (27.9).

It follows from (27.5), (27.6), and (27.12) that

$$\begin{aligned}
&\Bigg\| H\Bigg(\Bigg(\sum_{i=1}^n z_i\Bigg) - \Bigg(\sum_{i=1}^n x_i\Bigg)\Bigg) + H\Bigg(\Bigg(\sum_{i=1}^n z_i\Bigg) - \Bigg(\sum_{i=1}^n y_i\Bigg)\Bigg) \\
&\quad - 2H\Bigg(\Bigg(\sum_{i=1}^n z_i\Bigg) - \frac{(\sum_{i=1}^n x_i) + (\sum_{i=1}^n y_i)}{2}\Bigg)\Bigg\|_B \\
&= \lim_{m\to\infty} \frac{1}{2^m} \Bigg\| f\Bigg(\Bigg(\sum_{i=1}^n 2^m z_i\Bigg) - \Bigg(\sum_{i=1}^n 2^m x_i\Bigg)\Bigg) \\
&\quad + f\Bigg(\Bigg(\sum_{i=1}^n 2^m z_i\Bigg) - \Bigg(\sum_{i=1}^n 2^m y_i\Bigg)\Bigg) \\
&\quad - 2f\Bigg(\Bigg(\sum_{i=1}^n 2^m z_i\Bigg) - \Bigg(\sum_{i=1}^n 2^{m-1} x_i\Bigg) + \Bigg(\sum_{i=1}^n 2^{m-1} y_i\Bigg)\Bigg)\Bigg\|_B \\
&\le \lim_{m\to\infty} \frac{1}{2^m} \varphi\big(2^m z_1, \dots, 2^m z_n, 2^m x_1, \dots, 2^m x_n, 2^m y_1, \dots, 2^m y_n\big) = 0
\end{aligned}$$

for all $z_1, \dots, z_n, x_1, \dots, x_n, y_1, \dots, y_n \in A$. So

$$\begin{aligned}
H\Bigg(\Bigg(\sum_{i=1}^n z_i\Bigg) - \Bigg(\sum_{i=1}^n x_i\Bigg)\Bigg) &= -H\Bigg(\Bigg(\sum_{i=1}^n z_i\Bigg) - \Bigg(\sum_{i=1}^n y_i\Bigg)\Bigg) \\
&\quad + 2H\Bigg(\Bigg(\sum_{i=1}^n z_i\Bigg) - \frac{(\sum_{i=1}^n x_i) + (\sum_{i=1}^n y_i)}{2}\Bigg)
\end{aligned}$$

for all $z_1, \dots, z_n, x_1, \dots, x_n, y_1, \dots, y_n \in A$. By Lemma 2.1 of [33], the mapping $H : A \to B$ is Cauchy additive, i.e., $H(x+y) = H(x) + H(y)$ for all $x, y \in A$.

By a similar method to the proof of [32], one can show that the mapping $H : A \to B$ is $\mathbb{C}$-linear.

It follows from (27.7) that

$$\begin{aligned}\|H(xy) - H(x)H(y)\|_B &= \lim_{m\to\infty} \frac{1}{4^m} \|f(4^m xy) - f(2^m x)f(2^m y)\|_B \\ &\le \lim_{m\to\infty} \frac{1}{4^m} \varphi(2^m x, 2^m y, \underbrace{0, \dots, 0}_{3n-2 \text{ times}}) \\ &\le \lim_{m\to\infty} \frac{1}{2^m} \varphi(2^m x, 2^m y, \underbrace{0, \dots, 0}_{3n-2 \text{ times}}) = 0\end{aligned}$$

for all $x, y \in A$. So

$$H(xy) = H(x)H(y)$$

for all $x, y \in A$.

It follows from (27.8) that

$$\begin{aligned}\|H(x^*) - H(x)^*\|_B &= \lim_{m\to\infty} \frac{1}{2^m} \|f(2^m x^*) - f(2^m x)^*\|_B \\ &\le \lim_{m\to\infty} \frac{1}{2^m} \varphi(\underbrace{2^m x, \dots, 2^m x}_{3n \text{ times}}) = 0\end{aligned}$$

for all $x \in A$. So

$$H(x^*) = H(x)^*$$

for all $x \in A$.

Thus $H : A \to B$ is a C^*-algebra homomorphism satisfying (27.9), as desired. □

Corollary 27.1 *Let $r < 1$ and θ be nonnegative real numbers, and let $f : A \to B$ be a mapping satisfying $f(0) = 0$ such that*

$$\|C_\mu f(z_1, \dots, z_n, x_1, \dots, x_n, y_1, \dots, y_n)\|_B \le \theta\left(\sum_{i=1}^{n} \|z_i\|_A^r + \|x_i\|_A^r + \|y_i\|_A^r\right), \tag{27.13}$$

$$\|f(xy) - f(x)f(y)\|_B \le \theta(\|x\|_A^r + \|y\|_A^r), \tag{27.14}$$

$$\|f(x^*) - f(x)^*\|_B \le 3n\theta\|x\|_A^r \tag{27.15}$$

for all $\mu \in \mathbb{T}^1$ *and all* $x, y, z_1, \ldots, z_n, x_1, \ldots, x_n, y_1, \ldots, y_n \in A$. *Then there exists a unique* C^**-algebra homomorphism* $H : A \to B$ *such that*

$$\left\| f(x) - H(x) \right\|_B \le \frac{1}{2 - 2^r} 3n\theta \|x\|_A^r$$

for all $x \in A$.

Proof The proof follows from Theorem 27.3 by taking

$$\varphi(z_1, \ldots, z_n, x_1, \ldots, x_n, y_1, \ldots, y_n) := \theta \left(\sum_{i=1}^{n} \|z_i\|_A^r + \|x_i\|_A^r + \|y_i\|_A^r \right)$$

for all $z_1, \ldots, z_n, x_1, \ldots, x_n, y_1, \ldots, y_n \in A$. Then $L = 2^{r-1}$ and we get the desired result. □

Theorem 27.4 *Let* $f : A \to B$ *be a mapping satisfying* $f(0) = 0$ *for which there exists a function* $\varphi : A^{3n} \to [0, \infty)$ *satisfying* (27.6), (27.7), *and* (27.8) *such that*

$$\sum_{j=0}^{\infty} 2^j \varphi\left(\frac{1}{2^j} z_1, \ldots, \frac{1}{2^j} z_n, \frac{1}{2^j} x_1, \ldots, \frac{1}{2^j} x_n, \frac{1}{2^j} y_1, \ldots, \frac{1}{2^j} y_n \right) < \infty \tag{27.16}$$

for all $z_1, \ldots, z_n, x_1, \ldots, x_n, y_1, \ldots, y_n \in A$. *If there exists an* $L < 1$ *such that*

$$\varphi(\underbrace{x, \ldots, x}_{2n \text{ times}}, \underbrace{-x, \ldots, -x}_{n \text{ times}}) \le \frac{1}{2} L \varphi(\underbrace{2x, \ldots, 2x}_{2n \text{ times}}, \underbrace{-2x, \ldots, -2x}_{n \text{ times}})$$

for all $x \in A$, *then there exists a unique* C^**-algebra homomorphism* $H : A \to B$ *such that*

$$\left\| f(x) - H(x) \right\|_B \le \frac{L}{2 - 2L} \varphi(\underbrace{x, \ldots, x}_{2n \text{ times}}, \underbrace{-x, \ldots, -x}_{n \text{times}}) \tag{27.17}$$

for all $x \in A$.

Proof Similar to the proof of Theorem 27.3, we consider the linear mapping $J : X \to X$ such that

$$Jg(x) := 2g\left(\frac{1}{2} x \right)$$

for all $x \in A$. We can conclude that J is strictly contractive self-mapping of X, with the Lipschitz constant L.

It follows from (27.10) that

$$\begin{aligned}\left\|f(x)-2f\left(\frac{1}{2}x\right)\right\|_B &\le \varphi\Big(\underbrace{\frac{x}{2n},\dots,\frac{x}{2n}}_{2n\text{ times}},\underbrace{-\frac{2x}{n},\dots,-\frac{2x}{n}}_{n\text{ times}}\Big)\\ &\le \frac{L}{2}\varphi\Big(\underbrace{\frac{x}{n},\dots,\frac{x}{n}}_{2n\text{ times}},\underbrace{-\frac{x}{n},\dots,-\frac{x}{n}}_{n\text{ times}}\Big)\end{aligned}$$

for all $x \in A$. Hence $d(f, Jf) \le \frac{L}{2}$.

By Theorem 27.2, there exists a mapping $H : A \to B$ such that

1. H is a fixed point of J, i.e.,

$$H\left(\frac{1}{2}x\right) = \frac{1}{2}H(x) \tag{27.18}$$

for all $x \in A$. The mapping H is the unique fixed point of J in the set

$$Y = \{g \in X : d(f, g) < \infty\}.$$

This implies that H is the unique mapping satisfying (27.18) such that there exists $C \in (0, \infty)$ satisfying

$$\|H(x) - f(x)\|_B \le C\varphi(\underbrace{x,\dots,x}_{2n\text{ times}},\underbrace{-x,\dots,-x}_{n\text{ times}})$$

for all $x \in A$.

2. $d(J^m f, H) \to 0$ as $m \to \infty$. This implies the equality

$$\lim_{n\to\infty} 2^m f\left(\frac{x}{2^m}\right) = H(x)$$

for all $x \in A$.

3. $d(f, H) \le \frac{1}{1-L} d(f, Jf)$, which implies the inequality

$$d(f, H) \le \frac{L}{2-2L},$$

meaning that the inequality (27.17) holds.

The rest of the proof is similar to the proof of Theorem 27.3. □

Corollary 27.2 *Let $r > 1$ and θ be nonnegative real numbers, and let $f : A \to B$ be a mapping satisfying $f(0) = 0$, (27.13), (27.14) and (27.15). Then there exists a unique C^*-algebra homomorphism $H : A \to B$ such that*

$$\|f(x) - H(x)\|_B \le \frac{1}{2^r - 2} 3n\theta \|x\|_A^r$$

for all $x \in A$.

Proof The proof follows from Theorem 27.4 by taking

$$\varphi(z_1, \ldots, z_n, x_1, \ldots, x_n, y_1, \ldots, y_n) := \theta\left(\sum_{i=1}^{n} \|z_i\|_A^r + \|x_i\|_A^r + \|y_i\|_A^r\right)$$

for all $z_1, \ldots, z_n, x_1, \ldots, x_n, y_1, \ldots, y_n \in A$. Then $L = 2^{1-r}$ and we get the desired result. □

27.3 Stability of Generalized Derivations on C^*-Algebras

Definition 27.2 (See [3]) A *generalized derivation* $\delta : A \to A$ is involutive $\mathbb{C}$-linear and fulfills

$$\delta(xyz) = \delta(xy)z - x\delta(y)z + x\delta(yz)$$

for all $x, y, z \in A$.

We prove the Hyers–Ulam–Rassias stability of generalized derivations on C^*-algebras for the functional equation $C_\mu f(z_1, \ldots, z_n, x_1, \ldots, x_n, y_1, \ldots, y_n) = 0$.

Theorem 27.5 *Let $f : A \to A$ be a mapping satisfying $f(0) = 0$ for which there exists a function $\varphi : A^{3n} \to [0, \infty)$ satisfies* (27.5) *such that*

$$\begin{aligned} &\left\|C_\mu f(z_1, \ldots, z_n, x_1, \ldots, x_n, y_1, \ldots, y_n)\right\|_A \\ &\quad \le \varphi(z_1, \ldots, z_n, x_1, \ldots, x_n, y_1, \ldots, y_n), \end{aligned} \tag{27.19}$$

$$\left\|f(xyz) - f(xy)z + xf(y)z - xf(yz)\right\|_A \le \varphi(x, y, z, \underbrace{0, \ldots, 0}_{3n-3 \text{ times}}), \tag{27.20}$$

$$\left\|f(x^*) - f(x)^*\right\|_A \le \varphi(\underbrace{x, \ldots, x}_{3n \text{ times}}) \tag{27.21}$$

for all $\mu \in \mathbb{T}^1$ and all $x, y, z, z_1, \ldots, z_n, x_1, \ldots, x_n, y_1, \ldots, y_n \in A$. If there exists an $L < 1$ such that

$$\varphi(\underbrace{x, \ldots, x}_{2n \text{ times}}, \underbrace{-x, \ldots, -x}_{n \text{ times}}) \le 2L\varphi\Bigg(\underbrace{\frac{x}{2}, \ldots, \frac{x}{2}}_{2n \text{ times}}, \underbrace{-\frac{x}{2}, \ldots, -\frac{x}{2}}_{n \text{ times}}\Bigg)$$

for all $x \in A$. Then there exists a unique generalized derivation $\delta : A \to A$ such that

$$\left\|f(x) - \delta(x)\right\|_A \le \frac{1}{2 - 2L}\varphi(\underbrace{x, \ldots, x}_{2n \text{ times}}, \underbrace{-x, \ldots, -x}_{n \text{ times}}) \tag{27.22}$$

for all $x \in A$.

Proof By the same reasoning as the proof of Theorem 27.3, there exists a unique involutive $\mathbb{C}$-linear mapping $\delta : A \to A$ satisfying (27.22). The mapping $\delta : A \to A$ is given by

$$\delta(x) = \lim_{n\to\infty} \frac{1}{2^n} f(2^n x)$$

for all $x \in A$.

It follows from (27.5) and (27.20) that

$$\begin{aligned}
&\|\delta(xyz) - \delta(xy)z + x\delta(y)z - x\delta(yz)\|_A \\
&= \lim_{n\to\infty} \frac{1}{8^n} \| f(8^n xyz) - f(4^n xy) \cdot 2^n z + 2^n x f(2^n y) \cdot 2^n z - 2^n x f(4^n yz) \|_A \\
&\le \lim_{n\to\infty} \frac{1}{8^n} \varphi(2^n x, 2^n y, 2^n z, \underbrace{0, \ldots, 0}_{3n-3 \text{ times}}) \\
&\le \lim_{n\to\infty} \frac{1}{2^n} \varphi(2^n x, 2^n y, 2^n z, \underbrace{0, \ldots, 0}_{3n-3 \text{ times}}) = 0
\end{aligned}$$

for all $x, y, z \in A$. So

$$\delta(xyz) = \delta(xy)z - x\delta(y)z + x\delta(yz)$$

for all $x, y, z \in A$. Thus $\delta : A \to A$ is a generalized derivation satisfying (27.22). □

Theorem 27.6 *Let $f : A \to A$ be a mapping satisfying $f(0) = 0$ for which there exists a function $\varphi : A^{3n} \to [0, \infty)$ satisfies* (27.16), (27.19), (27.20), *and* (27.21) *for all $x, y, z, z_1, \ldots, z_n, x_1, \ldots, x_n, y_1, \ldots, y_n \in A$. If there exists an $L < 1$ such that*

$$\varphi(\underbrace{x, \ldots, x}_{2n \text{ times}}, \underbrace{-x, \ldots, -x}_{n \text{ times}}) \le \frac{1}{2} L \varphi(\underbrace{2x, \ldots, 2x}_{2n \text{ times}}, \underbrace{-2x, \ldots, -2x}_{n \text{ times}})$$

for all $x \in A$, then there exists a unique generalized derivation $\delta : A \to A$ such that

$$\|f(x) - \delta(x)\|_B \le \frac{L}{2 - 2L} \varphi(\underbrace{x, \ldots, x}_{2n \text{ times}}, \underbrace{-x, \ldots, -x}_{n \text{times}})$$

for all $x \in A$.

Proof The proof is similar to the proofs of Theorems 27.4 and 27.5. □

References

1. Aczél, J., Dhombres, J.: Functional Equations in Several Variables. Cambridge University Press Cambridge (1989)

2. Aoki, T.: On the stability of the linear transformation in Banach spaces. J. Math. Soc. Jpn. **2**, 64–66 (1950)
3. Ara, P., Mathieu, M.: Local Multipliers of C^*-Algebras. Springer, London (2003)
4. Belaid, B., Elhoucien, E., Rassias, Th.M.: On the generalized Hyers–Ulam stability of Swiatak's functional equation. J. Math. Inequal. **1**(2), 291–300 (2007)
5. Borelli, C., Forti, G.L.: On a general Hyers–Ulam stability result. Int. J. Math. Math. Sci. **18**, 229–236 (1995)
6. Bourgin, D.G.: Classes of transformations and bordering transformations. Bull. Am. Math. Soc. **57**, 223–237 (1951)
7. Cădariu, L., Radu, V.: Fixed points and the stability of Jensen's functional equation. J. Inequal. Pure Appl. Math. **4**(1), 4 (2003)
8. Cădariu, L., Radu, V.: Fixed points and the stability of quadratic functional equations. An. Univ. Timişoara, Ser. Mat.-Inform. **41**, 25–48 (2003)
9. Cădariu, L., Radu, V.: On the stability of the Cauchy functional equation: a fixed point approach. Grazer Math. Ber. **346**, 43–52 (2004)
10. Czerwik, S.: The stability of the quadratic functional equation. In: Rassias, Th.M., Tabor, J. (eds.) Stability of Mappings of Hyers–Ulam Type, pp. 81–91. Hadronic Press, Florida (1994)
11. Czerwik, S.: Functional Equations and Inequalities in Several Variables. World Scientific, New Jersey (2002)
12. Diaz, J., Margolis, B.: A fixed point theorem of the alternative for contractions on a generalized complete metric space. Bull. Am. Math. Soc. **74**, 305–309 (1968)
13. Eskandani, G.Z.: On the Hyers–Ulam–Rassias stability of an additive functional equation in quasi-Banach spaces. J. Math. Anal. Appl. **345**, 405–409 (2008)
14. Eskandani, G.Z., Gavruta, P., Rassias, J.M., Zarghami, R.: Generalized Hyers–Ulam stability for a general mixed functional equation in quasi-β-normed spaces. Mediterr. J. Math. **8**, 331–348 (2011)
15. Eskandani, G.Z., Vaezi, H., Dehghan, Y.N.: Stability of mixed additive and quadratic functional equation in non-Archimedean Banach modules. Taiwan. J. Math. **14**, 1309–1324 (2010)
16. Faizev, V.A., Rassias, Th.M., Sahoo, P.K.: The space of (Ψ, γ)-additive mappings on semigroups. Trans. Am. Math. Soc. **354**(11), 4455–4472 (2002)
17. Găvruta, P.: A generalization of the Hyers–Ulam–Rassias stability of approximately additive mappings. J. Math. Anal. Appl. **184**, 431–436 (1994)
18. Găvruta, L., Găvruta, P., Eskandani, G.Z.: Hyers–Ulam stability of frames in Hilbert spaces. Bul. Stiint. Univ. Politeh. Timis. Ser. Mat. Fiz. **55**(69), 2, 60–77 (2010)
19. Hyers, D.H.: On the stability of the linear functional equation. Proc. Natl. Acad. Sci. USA **27**, 222–224 (1941)
20. Hyers, D.H., Isac, G., Rassias, Th.M.: Stability of Functional Equations in Several Variables. Birkhäuser, Basel (1998)
21. Jun, K., Lee, Y.: On the Hyers–Ulam–Rassias stability of a pexiderized quadratic inequality. Math. Inequal. Appl. **4**, 93–118 (2001)
22. Jung, S.-M.: Hyers–Ulam–Rassias Stability of Functional Equations in Mathematical Analysis. Hadronic Press, Florida (2001)
23. Jung, S.-M., Rassias, Th.M.: Generalized Hyers–Ulam stability of Riccati differential equation. Math. Inequal. Appl. **11**(4), 777–782 (2008)
24. Lee, Y.-S., Chung, S.-Y.: Stability of an Euler-Lagrange-Rassias equation in the spaces of generalized functions. Appl. Math. Lett. **21**(7), 694–700 (2008)
25. Miura, T., Takahasi, S.-E., Choda, H.: On the Hyers–Ulam stability of real continuous function valued differentiable map. Tokyo J. Math. **24**, 467–476 (2001)
26. Moradlou, F., Vaezi, H., Eskandani, G.Z.: Hyers–Ulam–Rassias stability of a quadratic and additive functional equation in quasi-Banach spaces. Mediterr. J. Math. **6**(2), 233–248 (2009)
27. Moradlou, F., Vaezi, H., Park, C.: Fixed points and stability of an additive functional equation of n-Apollonius type in C^*-algebras. Abstr. Appl. Anal. (2008). doi:10.1155/2008/672618
28. Moslehian, M.S., Rassias, Th.M.: Generalized Hyers–Ulam stability of mappings on normed Lie triple systems. Math. Inequal. Appl. **11**(2), 371–380 (2008)

29. Najati, A., Eskandani, G.Z.: Stability of a mixed additive and cubic functional equation in quasi–Banach spaces. J. Math. Anal. Appl. **342**, 1318–1331 (2008)
30. Najati, A., Rassias, Th.M.: Stability of a mixed functional equation in several variables on Banach modules. Nonlinear Anal. **72**(3–4), 1755–1767 (2010)
31. Nakmahachalasint, P.: On the generalized Ulam–Găvruta–Rassias stability of mixed-type linear and Euler–Lagrange–Rassias functional equations. Int. J. Math. Math. Sci. 1–10 (2007) (ID 63239)
32. Park, C.: Homomorphisms between Poisson JC^*-algebras. Bull. Braz. Math. Soc. **36**, 79–97 (2005)
33. Park, C.: Hyers–Ulam–Rassias stability of a generalized Apollonius–Jensen type additive mapping and isomorphisms between C^*-algebras. Math. Nachr. **281**(3), 402–411 (2008)
34. Park, C., Rassias, Th.M.: Hyers–Ulam stability of a generalized Apollonius type quadratic mapping. J. Math. Anal. Appl. **322**, 371–381 (2006)
35. Park, C., Rassias, Th.M.: Homomorphisms in C^*-ternary algebras and JB^*-triples. J. Math. Anal. Appl. **337**, 13–20 (2008)
36. Park, C., Rassias, Th.M.: Fixed points and stability of the Cauchy functional equation. Aust. J. Math. Anal. Appl. **6**(1), 14 (2009), 9pp.
37. Park, C.-G., Rassias, Th.M.: Fixed points and generalized Hyers–Ulam stability of quadratic functional equations. J. Math. Inequal. **1**(4), 515–528 (2007)
38. Park, C.-G., Rassias, Th.M.: Homomorphisms and derivations in proper JCQ^*-triples. J. Math. Anal. Appl. **337**(2), 1404–1414 (2008)
39. Popa, D.: Functional inclusions on square-symmetric groupoids and Hyers–Ulam stability. Math. Inequal. Appl. **7**, 419–428 (2004)
40. Prastaro, A., Rassias, Th.M.: Ulam stability in geometry of PDE. Nonlinear Funct. Anal. Appl. **8**, 259–278 (2003)
41. Radu, V.: The fixed point alternative and stability of functional equations. Fixed Point Theory, Cluj-Napoca **IV**(1), 91–96 (2003)
42. Rassias, Th.M.: On the stability of the linear mapping in Banach spaces. Proc. Am. Math. Soc. **72**, 297–300 (1978)
43. Rassias, Th.M.: The problem of S.M. Ulam for approximately multiplicative mappings. J. Math. Anal. Appl. **246**, 352–378 (2000)
44. Rassias, Th.M.: On the stability of minimum points. Mathematica **45(68)**(1), 93–104 (2003)
45. Rassias, Th.M.: On the stability of functional equations in Banach spaces. J. Math. Anal. Appl. **251**, 264–284 (2000)
46. Skof, F.: Local properties and approximations of operators. Rend. Semin. Mat. Fis. Milano **53**, 113–129 (1983)
47. Ulam, S.M.: A Collection of the Mathematical Problems. Interscience, New York (1960)

Chapter 28
The Fučík Spectrum for the Negative p-Laplacian with Different Boundary Conditions

Dumitru Motreanu and Patrick Winkert

Abstract This chapter represents a survey on the Fučík spectrum of the negative p-Laplacian with different boundary conditions (Dirichlet, Neumann, Steklov, and Robin). The close relationship between the Fučík spectrum and the ordinary spectrum is briefly discussed. It is also pointed out that for every boundary condition there exists a first nontrivial curve $\mathscr{C}$ in the Fučík spectrum which has important properties such as Lipschitz continuity, being decreasing and a certain asymptotic behavior depending on the boundary condition. As a consequence, one obtains a variational characterization of the second eigenvalue λ_2 of the negative p-Laplacian with the corresponding boundary condition. The applicability of the abstract results is illustrated to elliptic boundary value problems with jumping nonlinearities.

Key words Fučík spectrum · p-Laplacian · Boundary conditions · Elliptic boundary value problems

Mathematics Subject Classification 47A10 · 35J91 · 35K92 · 35J58

28.1 Introduction

Given a bounded domain $\Omega \subset \mathbb{R}^N$, let T be a selfadjoint linear operator on $L^2(\Omega)$ with compact resolvent and eigenvalues

$$0 < \lambda_0 < \lambda_1 < \cdots < \lambda_k < \cdots .$$

Dedicated to Professor Themistocles M. Rassias on the occasion of his 60th birthday.

D. Motreanu (✉)
Département de Mathématiques, Université de Perpignan, Avenue Paul Alduy 52, 66860 Perpignan Cedex, France
e-mail: motreanu@univ-perp.fr

P. Winkert
Institut für Mathematik, Technische Universität Berlin, Straße des 17. Juni 136, 10623 Berlin, Germany
e-mail: winkert@math.tu-berlin.de

P.M. Pardalos et al. (eds.), *Nonlinear Analysis*, Springer Optimization and Its Applications 68, In Honor of Themistocles M. Rassias on the Occasion of his 60th Birthday, DOI 10.1007/978-1-4614-3498-6_28,

The so-called Fučík spectrum[1] Σ of T is defined as the set of all pairs $(a,b) \in \mathbb{R}^2$ such that the equation

$$Tu = au^+ - bu^- \tag{28.1}$$

has a nontrivial solution. Here we denoted $u^+ = \max(u,0)$ (the positive part of u) and $u^- = \max(-u,0)$ (the negative part of u). Fučík [20] and Dancer [15] were the first authors who recognized that the set Σ plays an important part in the study of semilinear equations of type

$$Tu = g(x,u),$$

where $g : \Omega \times \mathbb{R} \to \mathbb{R}$ is a Carathéodory function with jumping nonlinearities satisfying

$$\frac{g(x,s)}{s} \to a \quad \text{as } s \to +\infty, \qquad \frac{g(x,s)}{s} \to b \quad \text{as } s \to -\infty.$$

Initially, a systematic study of this spectrum was developed by Fučík [21] in the case of the negative Laplacian in one-dimension, i.e., for $N = 1$, with periodic boundary condition. He proved that this spectrum is composed of two families of curves in $\mathbb{R}^2$ emanating from the points (λ_k, λ_k) determined by the eigenvalues λ_k of the negative periodic Laplacian in one-dimension. Afterwards, many authors studied the Fučík spectrum Σ_2 for the negative Laplacian $-\Delta$ with Dirichlet boundary condition on a bounded domain $\Omega \subset \mathbb{R}^N$ (see [2, 5, 14, 24, 25, 28, 29, 34, 35], and the references therein). In this respect, we mention that Dancer [15] proved that the lines $\mathbb{R} \times \{\lambda_1\}$ and $\{\lambda_1\} \times \mathbb{R}$ are isolated in Σ_2, while de Figueiredo and Gossez [16] constructed a first nontrivial curve in Σ_2 passing through (λ_2, λ_2) and characterized it variationally. Here λ_1 and λ_2 respectively denote the first and second eigenvalue of $-\Delta$ with Dirichlet boundary condition. The next step in this direction was to investigate the Fučík spectrum Σ_p of the negative p-Laplacian (or p-Laplace operator) $-\Delta_p$ aiming to extend the results known for $-\Delta$. We recall that $-\Delta_p$ is given by

$$-\Delta_p u = -\operatorname{div}\big(|\nabla u|^{p-2}\nabla u\big), \quad 1 < p < +\infty,$$

which is a nonlinear operator if $p \neq 2$. If $p = 2$, it reduces to the negative Laplacian $-\Delta$. First, Drábek [18] has shown for $p \neq 2$ and in one-dimension that Σ_p has similar properties as in the linear case, i.e., for $p = 2$.

The aim of this chapter is to give an overview about the Fučík spectrum of the negative p-Laplacian $-\Delta_p$ with $1 < p < +\infty$ and different boundary conditions on a bounded domain Ω in $\mathbb{R}^N$.

Let V be a closed subspace of the Sobolev space $W^{1,p}(\Omega)$ such that $W_0^{1,p}(\Omega) \subseteq V \subseteq W^{1,p}(\Omega)$ and let V^* denote the dual space with the duality pairing $\langle \cdot, \cdot \rangle$ between V and V^*. It is well known that the operator $-\Delta_p : V \to V^*$ is bounded,

[1] Svatopluk Fučík (21st October 1944 – 18th May 1979) was a Czech mathematician.

continuous, pseudomonotone, and has the (S_+)-property (i.e., from $u_n \rightharpoonup u$ in V and $\limsup_{n\to+\infty}\langle -\Delta_p u_n, u_n - u\rangle \le 0$ it follows that $u_n \to u$ in V). Note that on $W^{1,p}(\Omega)$ we have

$$\langle -\Delta_p u, v\rangle = \int_\Omega |\nabla u|^{p-2}\nabla u \cdot \nabla v\, dx, \qquad v \in W^{1,p}(\Omega).$$

We refer to [6] for various nonlinear boundary problems involving $-\Delta_p$.

The Fučík spectrum of $-\Delta_p$ depends strongly on the choice of the boundary condition related to Ω. Specifically, the set Σ_p (resp., Θ_p) is called the Fučík spectrum of $-\Delta_p$ with homogeneous Dirichlet (resp., Neumann) boundary condition if for all pairs $(a,b) \in \Sigma_p$ (resp., Θ_p) the equation

$$-\Delta_p u = a\big(u^+\big)^{p-1} - b\big(u^-\big)^{p-1} \quad \text{in } \Omega \tag{28.2}$$

with the boundary condition

$$u = 0 \quad \text{on } \partial\Omega \quad \left(\text{resp., } \frac{\partial u}{\partial \nu} = 0 \quad \text{on } \partial\Omega\right) \tag{28.3}$$

has a nontrivial weak solution. In (28.3), $\partial u/\partial \nu$ stands for the conormal derivative on $\partial\Omega$. If we replace (28.3) by

$$\frac{\partial u}{\partial \nu} = -\beta |u|^{p-2} u \quad \text{on } \partial\Omega$$

with fixed $\beta \ge 0$, we speak of the Fučík spectrum of $-\Delta_p$ with Robin boundary condition denoted by $\widehat{\Sigma}_p$. Finally, we write $\widetilde{\Sigma}_p$ for the Fučík spectrum of $-\Delta_p$ with Steklov boundary condition, which is formed by all $(a,b) \in \mathbb{R}^2$ provided

$$-\Delta_p u = -|u|^{p-2} u \quad \text{in } \Omega, \qquad \frac{\partial u}{\partial \nu} = a\big(u^+\big)^{p-1} - b\big(u^-\big)^{p-1} \quad \text{on } \partial\Omega$$

is solved nontrivially.

These spectra have intensively been studied in the last years. We will present in Sects. 28.2 through 28.5 some of their basic properties. Namely, it will be shown that there exists a close relationship between these spectra and the ordinary spectrum of $-\Delta_p$ subject to different boundary conditions. A fundamental fact is that every Fučík spectrum introduced above contains a first nontrivial curve $\mathscr{C}$ which is Lipschitz continuous and decreasing. However, the asymptotic behavior of these curves is different relative to the imposed boundary condition. Furthermore, we will indicate some applications of these spectra to certain nonlinear elliptic problems with jumping nonlinearities. Subtle phenomena can occur due to the interaction of the involved nonlinearities with these spectra, in particular resonance to spectral elements can appear. We emphasize that these problems and results can be considered beyond the setting of quasilinear elliptic equations. For instance, the field of variational inequalities, as those describing obstacle problems, offers a rich and flexible framework which is highly interesting for its applicability. For different classes of variational inequalities and their applications, we refer to the volume by Pardalos, Rassias, and Khan [33].

28.2 Dirichlet Boundary Condition

The Fučík spectrum of the negative p-Laplacian $-\Delta_p$ with homogeneous Dirichlet boundary condition is defined as the set Σ_p of those $(a,b) \in \mathbb{R}^2$ such that

$$\begin{aligned} -\Delta_p u &= a\big(u^+\big)^{p-1} - b\big(u^-\big)^{p-1} && \text{in } \Omega, \\ u &= 0 && \text{on } \partial\Omega \end{aligned} \tag{28.4}$$

has a nontrivial (weak) solution u, which means that $u \in W_0^{1,p}(\Omega), u \not\equiv 0$, and it satisfies the equation

$$\int_\Omega |\nabla u|^{p-2}\nabla u \cdot \nabla v \, dx = \int_\Omega \big(a\big(u^+\big)^{p-1} - b\big(u^-\big)^{p-1}\big) v \, dx, \quad \forall v \in W_0^{1,p}(\Omega).$$

We note that if $a = b = \lambda$, problem (28.4) reduces to

$$\begin{aligned} -\Delta_p u &= \lambda |u|^{p-2} u && \text{in } \Omega, \\ u &= 0 && \text{on } \partial\Omega, \end{aligned} \tag{28.5}$$

which is called the Dirichlet eigenvalue problem with respect to the negative p-Laplacian $-\Delta_p$. It is known that the first eigenvalue λ_1 of (28.5) is positive, simple, and its corresponding eigenfunctions have constant sign (see Anane [1] and Lindqvist [23]). In fact, the spectrum $\sigma(-\Delta_p)$ of the negative p-Laplacian $-\Delta_p$ associated to (28.5) includes an unbounded sequence of eigenvalues $(\lambda_k), k \in \mathbb{N}$, called the variational eigenvalues, which fulfills

$$0 < \lambda_1 < \lambda_2 \le \cdots \le \lambda_k \le \cdots \to +\infty.$$

The variational eigenvalues satisfy min–max characterizations.

The Fučík spectrum Σ_p of the negative p-Laplacian $-\Delta_p$ with homogeneous Dirichlet boundary condition contains the two lines $\lambda_1 \times \mathbb{R}$ and $\mathbb{R} \times \lambda_1$. Additionally, Σ_p contains the sequence of points $(\lambda_k, \lambda_k), k \in \mathbb{N}$, as can be easily seen from (28.4) and (28.5) by writing $u = u^+ - u^-$. The Fučík spectrum Σ_p has been intensively studied by Cuesta, de Figueiredo, and Gossez [13] in the general case of $1 < p < +\infty$ through a variational approach using the mountain-pass theorem. In order to give a brief overview of their results, let us set for every $s \ge 0$,

$$J_s(u) = \int_\Omega |\nabla u|^p \, dx - s \int_\Omega \big(u^+\big)^p \, dx, \quad u \in W_0^{1,p}(\Omega).$$

The function J_s is of class C^1 on $W_0^{1,p}(\Omega)$. Denote $\tilde{J}_s = J_s|_S$, with S given by

$$S = \left\{ u \in W_0^{1,p}(\Omega) : \int_\Omega |u|^p = 1 \right\}.$$

Since S is a C^1-submanifold of $W_0^{1,p}(\Omega)$, it follows that $\tilde{J}_s$ is of class C^1 on S in the sense of manifolds. Then the curve $s \in \mathbb{R}^+ \mapsto (s + c(s), c(s)) \in \mathbb{R}^2$ described by the min–max values

$$c(s) = \inf_{\gamma \in \Gamma} \max_{u \in \gamma[-1,+1]} \tilde{J}_s(u),$$

where

$$\Gamma = \{\gamma \in C([-1,1], S) : \gamma(-1) = -\varphi_1 \text{ and } \gamma(1) = \varphi_1\}, \tag{28.6}$$

is contained in Σ_p (see [13, Theorem 2.10]). In (28.6), φ_1 denotes the eigenfunction of (28.5) corresponding to λ_1 satisfying $\varphi_1 > 0$ in Ω and $\|\varphi_1\|_p = 1$. Taking into account that Σ_p is symmetric with respect to the diagonal of the plane, it turns out that the curve

$$\mathscr{C} := \{(s + c(s), c(s)), (c(s), s + c(s)) : s \geq 0\} \tag{28.7}$$

is contained in Σ_p. It is shown in [13, Theorem 3.1] that $\mathscr{C}$ given in (28.7) is indeed the first nontrivial curve in Σ_p, which means that the first point in Σ_p belonging to the parallel to the diagonal drawn through a point of $(\mathbb{R}_+ \times \{\lambda_1\}) \times (\{\lambda_1\} \times \mathbb{R})$ must be on $\mathscr{C}$ (see Fig. 28.1). As a consequence, we infer that the curve $\mathscr{C}$ passes through (λ_2, λ_2). In conjunction with the description of $\mathscr{C}$ in (28.7) and the min–max formula for $c(s)$, this yields that λ_2 can be variationally characterized as follows

$$\lambda_2 = \inf_{\gamma \in \Gamma} \max_{u \in \gamma[-1,+1]} \int_\Omega |\nabla u|^p \, dx, \tag{28.8}$$

with Γ introduced in (28.6). Moreover, the curve $\mathscr{C}$ in (28.7) is Lipschitz continuous and decreasing as shown in [13, Proposition 4.1]. Finally, we mention that the limit of $c(s)$ as $s \to +\infty$ is equal to the first eigenvalue λ_1 of (28.5), which is proven in [13, Proposition 4.4].

The work of Cuesta, de Figueiredo, and Gossez [13] was the first paper that gave a complete study of the beginning of the Fučík spectrum of $-\Delta_p$ with homogeneous Dirichlet boundary condition and their variational approach was the starting point for investigating the Fučík spectrum under other boundary conditions (Neumann, Steklov, Robin, see the sections below). The knowledge of the properties of Σ_p, especially the existence of the first nontrivial curve $\mathscr{C}$ and its representation, has demonstrated to be very useful in obtaining multiple solutions results for elliptic equations involving the negative p-Laplacian $-\Delta_p$ and jumping nonlinearities.

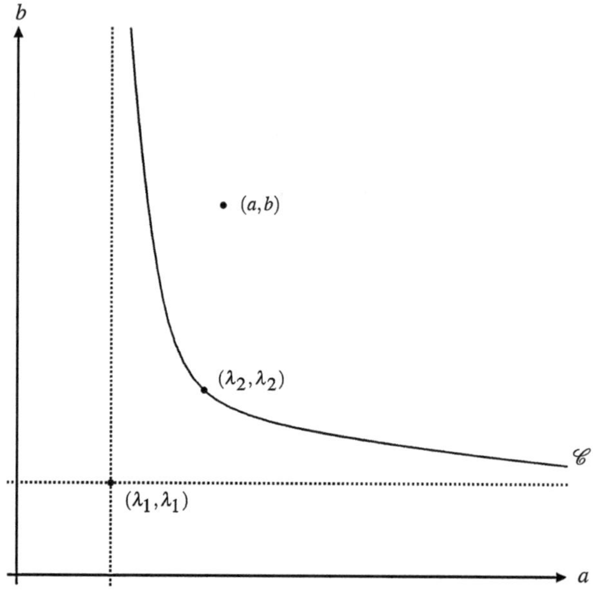

Fig. 28.1 The first nontrivial curve $\mathscr{C}$ of the Fučík spectrum of the negative p-Laplacian with Dirichlet boundary condition. Problem (28.9) has multiple solutions if the pair (a, b) is above the curve $\mathscr{C}$

In order to illustrate the applicability of the Fučík spectrum Σ_p, we consider the following equation with homogeneous Dirichlet boundary condition

$$-\Delta_p u = a\big(u^+\big)^{p-1} - b\big(u^-\big)^{p-1} + g(x,u) \quad \text{in } \Omega, \tag{28.9}$$

where $g : \Omega \times \mathbb{R} \to \mathbb{R}$ is a Carathéodory function satisfying

$$\lim_{t\to 0} \frac{g(x,t)}{|t|^{p-1}} = 0 \quad \text{uniformly for a.a. } x \in \Omega.$$

In Carl and Perera [12], it is proven that problem (28.9) has at least three nontrivial solutions provided the point $(a, b) \in \mathbb{R}^2$ lies above the first nontrivial curve $\mathscr{C}$ in Σ_p constructed in (28.7). Moreover, a complete sign information for the three solutions is available: two solutions have opposite constant sign and the third one is sign-changing (nodal solution). This information is obtained by means of the method of sub-supersolution whose application to problem (28.9) strongly relies on the hypothesis that the point $(a, b) \in \mathbb{R}^2$ is situated above the first nontrivial curve $\mathscr{C}$ in Σ_p. The graphic in Fig. 28.1 marks the position of the point $(a, b) \in \mathbb{R}^2$ entering (28.9) and demonstrates the qualitative behavior of the curve $\mathscr{C}$.

Multiple solutions results concerning problems of type (28.9) and using the representation of the first nontrivial curve $\mathscr{C}$, in particular the characterization of the second eigenvalue λ_2 of $-\Delta_p$ on $W_0^{1,p}(\Omega)$ as stated in (28.8), can be found in numerous publications; see, for example, [7, 10, 30]. We also refer to versions of such results in the case of nonsmooth potential associated to (28.9) (see, e.g., [8, 9, 11]).

28.3 Neumann Boundary Condition

In this section, we give a brief overview of the Fučík spectrum of the negative p-Laplacian $-\Delta_p$ with Neumann boundary condition. In order to avoid misunderstandings, we point out that a Neumann boundary condition stands in this context for a homogeneous Neumann condition. Inhomogeneous Neumann boundary conditions are treated in Sect. 28.4 (Steklov boundary condition) and Sect. 28.5 (Robin boundary condition). Let us first give the relevant definition of this spectrum. The Fučík spectrum of $-\Delta_p$ with Neumann boundary condition, denoted by Θ_p, consists of all pairs $(a,b) \in \mathbb{R}^2$ such that

$$\begin{aligned} -\Delta_p u &= a\big(u^+\big)^{p-1} - b\big(u^-\big)^{p-1} && \text{in } \Omega, \\ \frac{\partial u}{\partial \nu} &= 0 && \text{on } \partial\Omega, \end{aligned} \tag{28.10}$$

is solved nontrivially, meaning that $u \in W^{1,p}(\Omega)$, $u \not\equiv 0$, and verifies the equality

$$\int_\Omega |\nabla u|^{p-2} \nabla u \cdot \nabla v \, dx = \int_\Omega \big(a\big(u^+\big)^{p-1} - b\big(u^-\big)^{p-1}\big) v \, dx, \quad \forall v \in W^{1,p}(\Omega).$$

In (28.10), $\partial u / \partial \nu$ denotes the conormal derivative, that is, $\partial u/\partial \nu = |\nabla u|^{p-2} \nabla u \cdot \nu$, where ν is the unit outward normal to $\partial\Omega$. Problem (28.10) is a special case of the Robin Fučík spectrum that will be introduced in Sect. 28.5. Clearly, in case where $a = b = \lambda$, problem (28.10) becomes the Neumann eigenvalue problem of the negative p-Laplacian given by

$$\begin{aligned} -\Delta_p u &= \lambda |u|^{p-2} u && \text{in } \Omega, \\ \frac{\partial u}{\partial \nu} &= 0 && \text{on } \partial\Omega. \end{aligned} \tag{28.11}$$

As proved in [22], the first eigenvalue $\lambda_1 = 0$ of (28.11) is simple with the corresponding eigenspace $\mathbb{R}$, so all eigenfunctions associated to λ_1 do not change sign in Ω, which does not happen for the higher order eigenvalues. It is easily seen that Θ_p contains in particular $(0,0)$, (λ_2, λ_2) (λ_2 is the second eigenvalue of (28.11)) and the two lines $0 \times \mathbb{R}$ and $\mathbb{R} \times 0$. The nontrivial part of Θ_p is denoted by $\tilde{\Theta}_p$, that is, $\tilde{\Theta}_p = \Theta_p \setminus ((0 \times \mathbb{R}) \cup (\mathbb{R} \times 0))$, which is obviously contained in $\mathbb{R}^+ \times \mathbb{R}^+$.

The basic paper dealing with the Fučík spectrum of the negative Neumann p-Laplacian is due to Arias, Campos, and Gossez [4]. The construction of a first nontrivial curve in $\tilde{\Theta}_p$ can be done similarly to the Dirichlet Fučík spectrum. To this end, for every $s \geq 0$, let $J_s : W^{1,p}(\Omega) \to \mathbb{R}$ be the functional given by

$$J_s(u) = \int_\Omega |\nabla u|^p \, dx - s \int_\Omega \big(u^+\big)^p \, dx$$

and let $\tilde{J}_s$ be its restriction to

$$S=\left\{u\in W^{1,p}(\Omega):\int_\Omega |u|^p=1\right\}.$$

Notice that S is a C^1-submanifold of $W^{1,p}(\Omega)$, so $\tilde{J}_s$ is of class C^1 on S in the sense of manifolds. This enables us to consider the notions of critical points and critical values for the functional $\tilde{J}_s$. Then, the first nontrivial curve $\mathscr{C}$ of Θ_p can be determined as in (28.7), whereas

$$c(s)=\inf_{\gamma\in\Gamma}\max_{u\in\gamma[-1,+1]}\tilde{J}_s(u),$$
$$\Gamma=\left\{\gamma\in C\left([-1,1],S\right):\gamma(-1)=-\varphi_1 \text{ and } \gamma(1)=\varphi_1\right\}.$$

Here we have $\varphi_1=1/|\Omega|^{1/p}$, so $\|\varphi_1\|_p=1$, with $|\Omega|$ denoting the measure of Ω. Arguing as in the case of the Dirichlet Fučík spectrum Σ_p, we see that $\mathscr{C}$ passes through (λ_2,λ_2) (λ_2 denotes the second eigenvalue of (28.11)). Consequently, we get a variational expression of λ_2 as

$$\lambda_2=\inf_{\gamma\in\Gamma}\max_{u\in\gamma[-1,+1]}\int_\Omega |\nabla u|^p\,dx,$$

with Γ introduced above. An important difference between the Dirichlet Fučík spectrum Σ_p and the Neumann Fučík spectrum Θ_p consists in the asymptotic behavior of the first nontrivial curve $\mathscr{C}$. In the Neumann case, to describe the asymptotic properties of the curve $\mathscr{C}$ it is required to consider the situations $p\le N$ and $p>N$ separately. In [4, Theorem 2.3 and Theorem 2.6], it is shown that

$$\lim_{s\to\infty}c(s)=\begin{cases}\lambda_1=0 & \text{if } p\le N,\\ \overline{\lambda} & \text{if } p>N,\end{cases}\tag{28.12}$$

where

$$\overline{\lambda}=\inf\left\{\int_\Omega |\nabla u|^p\,dx : u\in W^{1,p}(\Omega),\ \|u\|_{L^p(\Omega)}=1,\ u \text{ vanishes somewhere in } \overline{\Omega}\right\}.$$

The definition of $\overline{\lambda}$ is meaningful because for $p>N$ the elements $u\in W^{1,p}(\Omega)$ are continuous functions on $\overline{\Omega}$.

An extension of the previous results to the Fučík spectrum of the negative Neumann p-Laplacian with weights has been achieved by Arias, Campos, Cuesta, and Gossez [3]. Therein, for the weights given by the measurable functions $m(x)$ and $n(x)$ on Ω, the authors consider the set Σ of all pairs $(a,b)\in\mathbb{R}^2$ such that

$$\begin{aligned}-\Delta_p u&=am(x)\left(u^+\right)^{p-1}-bn(x)\left(u^-\right)^{p-1} && \text{in } \Omega,\\ \frac{\partial u}{\partial \nu}&=0 && \text{on } \partial\Omega,\end{aligned}\tag{28.13}$$

has a nontrivial solution. Under suitable assumptions on the data it is shown that Σ contains a first nontrivial curve.

Recently, Motreanu and Tanaka [31] used the results presented in the first part of this section to study quasilinear elliptic equations of the form

$$\begin{aligned} -\operatorname{div} A(x, \nabla u) &= f(x, u) \quad \text{in } \Omega, \\ \frac{\partial u}{\partial \nu} &= 0 \qquad\qquad \text{on } \partial\Omega, \end{aligned} \tag{28.14}$$

where, in the principal part of the equation, one has an operator $A \in C^0(\overline{\Omega} \times \mathbb{R}^N, \mathbb{R}^N) \cap C^1(\overline{\Omega} \times (\mathbb{R}^N \setminus \{0\}), \mathbb{R}^N)$ of the form $A(x, y) = a(x, |y|)y$, with $a(x, t) > 0$ for all $(x, t) \in \overline{\Omega} \times (0, +\infty)$, which is strictly monotone with respect to the second variable and fulfills some further regularity assumptions, while $f : \Omega \times \mathbb{R} \to \mathbb{R}$ is a Carathéodory function having a representation similar to (28.9). They prove existence results for multiple solutions to (28.14), the properties of the solution set depending on conditions related to the first nontrivial curve $\mathscr{C}$ in the Neumann Fučík spectrum Θ_p. These results apply in particular to the case of the Neumann p-Laplacian in (28.14), i.e., when $\operatorname{div} A(x, \nabla u) = \Delta_p u$.

28.4 Steklov Boundary Condition

Now we focus on the Steklov Fučík spectrum of $-\Delta_p$ which addresses $-\Delta_p$ with a special nonhomogeneous boundary condition, known as Steklov boundary condition. This spectrum is defined as the set $\widetilde{\Sigma}_p$ of all pairs $(a, b) \in \mathbb{R}^2$ such that

$$\begin{aligned} -\Delta_p u &= -|u|^{p-2}u \qquad\qquad\qquad \text{in } \Omega, \\ \frac{\partial u}{\partial \nu} &= a\left(u^+\right)^{p-1} - b\left(u^-\right)^{p-1} \quad \text{on } \partial\Omega, \end{aligned} \tag{28.15}$$

has a weak solution $u \not\equiv 0$. Let us recall that $u \in W^{1,p}(\Omega)$ is a weak solution of (28.15) if it satisfies the equality

$$\int_\Omega |\nabla u|^{p-2}\nabla u \cdot \nabla v \, dx = -\int_\Omega |u|^{p-2}uv \, dx + \int_{\partial\Omega} \left(a\left(u^+\right)^{p-1} - b\left(u^-\right)^{p-1}\right)v \, d\sigma$$

for all $v \in W^{1,p}(\Omega)$. Here the notation $d\sigma$ stands for the $(N-1)$-dimensional surface measure. The name of this spectrum comes from the fact that if $a = b = \lambda$, (28.15) becomes the so-called Steklov eigenvalue problem, namely

$$\begin{aligned} -\Delta_p u &= -|u|^{p-2}u \quad \text{in } \Omega, \\ \frac{\partial u}{\partial \nu} &= \lambda |u|^{p-2}u \quad\ \text{on } \partial\Omega. \end{aligned} \tag{28.16}$$

The fundamental difference with respect to the Dirichlet and Neumann Fučík spectra is that in the Steklov case a boundary integral is involved, a fact that substantially modifies the analysis regarding the relevant values a and b. The Steklov eigenvalue problem (28.16) was first studied by Martínez and Rossi [26] (see also Lê [22]). They showed that the first eigenvalue is positive, simple, and every eigenfunction corresponding to the first eigenvalue does not change sign in Ω. Actually, we may find an eigenfunction associated to the first eigenvalue λ_1 belonging to $\operatorname{int}(C^1(\overline{\Omega})_+)$, where $\operatorname{int}(C^1(\overline{\Omega})_+)$ denotes the interior of the positive cone $C^1(\overline{\Omega})_+ = \{u \in C^1(\overline{\Omega}) : u(x) \geq 0,\ \forall x \in \Omega\}$ in the Banach space $C^1(\overline{\Omega})$, which is nonempty and given by

$$\operatorname{int}\big(C^1(\overline{\Omega})_+\big) = \big\{u \in C^1(\overline{\Omega}) : u(x) > 0, \forall x \in \overline{\Omega}\big\}.$$

Furthermore, in [19] it is established that there exists a sequence of eigenvalues λ_n of (28.16) such that $\lambda_n \to +\infty$ as $n \to +\infty$. The Steklov Fučík spectrum defined in (28.15) has been studied by Martínez and Rossi [27]. Their approach is mainly based on the ideas of Cuesta, de Figueiredo, and Gossez [13]. Precisely, for each $s \geq 0$, one defines a C^1 functional $J_s : W^{1,p}(\Omega) \to \mathbb{R}$ by

$$J_s(u) = \int_\Omega |\nabla u|^p \, dx + \int_\Omega |u|^p \, dx - s \int_{\partial\Omega} (u^+)^p \, d\sigma.$$

Restricting J_s to

$$S = \left\{u \in W^{1,p}(\Omega) : \int_{\partial\Omega} |u|^p \, d\sigma = 1\right\},$$

one obtains a C^1-functional $\tilde{J}_s$ on the C^1-submanifold S of $W^{1,p}(\Omega)$. Then, the first nontrivial curve in $\tilde{\Sigma}_p$ is expressed as

$$\mathscr{C} = \big\{\big(s + c(s), c(s)\big), \big(c(s), s + c(s)\big) : s \geq 0\big\},$$

where

$$c(s) = \inf_{\gamma \in \Gamma} \max_{u \in \gamma[-1,+1]} \tilde{J}_s(u),$$

$$\Gamma = \big\{\gamma \in C\big([-1,1], S\big) : \gamma(-1) = -\varphi_1 \text{ and } \gamma(1) = \varphi_1\big\}$$

(cf. [27, Theorem 2.1]), where $\varphi_1 \in \operatorname{int}(C^1(\overline{\Omega})_+)$ with $\|\varphi_1\|_p = 1$. In particular, we derive the following variational characterization of the second eigenvalue λ_2 of the Steklov eigenvalue problem (28.16) which results in

$$\lambda_2 = \inf_{\gamma \in \Gamma} \max_{u \in \gamma[-1,+1]} \int_\Omega \big(|\nabla u|^p + |u|^p\big) \, dx. \tag{28.17}$$

As before, the first nontrivial curve $\mathscr{C}$ is Lipschitz continuous and decreasing (cf. [27, Proposition 4.1]). Similar to the Neumann Fučík spectrum, in order to state the asymptotic properties of $\mathscr{C}$, which means, in fact, determining the limit of $c(s)$ as $s \to +\infty$, it is needed to take into account two cases, $p \leq N$ and $p > N$. The following holds (see [27, Theorem 4.1])

$$\lim_{s\to\infty} c(s) = \begin{cases} \lambda_1 & \text{if } p \leq N, \\ \overline{\lambda} > \lambda_1 & \text{if } p > N, \end{cases}$$

where

$$\overline{\lambda} = \inf_{u\in L} \max_{r\in\mathbb{R}} \frac{\|r\varphi_1 + u\|^p_{W^{1,p}(\Omega)}}{\|r\varphi_1 + u\|^p_{L^p(\partial\Omega)}}$$

with

$$L = \left\{u \in W^{1,p}(\Omega) : u \text{ vanishes somewhere on } \partial\Omega\right\}.$$

As an application of the results in [27], consider the following nonlinear elliptic equation subject to Steklov-type boundary condition with perturbation

$$\begin{aligned} -\Delta_p u &= f(x,u) - |u|^{p-2}u && \text{in } \Omega, \\ \frac{\partial u}{\partial \nu} &= a\left(u^+\right)^{p-1} - b\left(u^-\right)^{p-1} + g(x,u) && \text{on } \partial\Omega, \end{aligned} \tag{28.18}$$

for Carathéodory functions $f : \Omega \times \mathbb{R} \to \mathbb{R}$ and $g : \partial\Omega \times \mathbb{R} \to \mathbb{R}$ which are bounded on bounded sets and satisfy

(A) $\lim_{s\to 0} \frac{f(x,s)}{|s|^{p-1}} = 0$ uniformly for a.a. $x \in \Omega$,

(B) $\lim_{s\to 0} \frac{g(x,s)}{|s|^{p-1}} = 0$ uniformly for a.a. $x \in \partial\Omega$,

(C) $\lim_{|s|\to\infty} \frac{f(x,s)}{|s|^{p-2}s} = -\infty$ uniformly for a.a. $x \in \Omega$,

(D) $\lim_{|s|\to\infty} \frac{g(x,s)}{|s|^{p-2}s} = -\infty$ uniformly for a.a. $x \in \partial\Omega$.

(E) There exists $\delta_f > 0$ such that $\frac{f(x,s)}{|s|^{p-2}s} \geq 0$ for all $0 < |s| \leq \delta_f$ and for a.a. $x \in \Omega$.

(F) g satisfies the condition

$$\left|g(x_1,s_1) - g(x_2,s_2)\right| \leq L\left[|x_1 - x_2|^\alpha + |s_1 - s_2|^\alpha\right],$$

for all pairs (x_1,s_1), (x_2,s_2) in $\partial\Omega \times [-M_0, M_0]$, where M_0 is a positive constant and $\alpha_2 \in (0,1]$.

If the point (a,b) is above the first nontrivial curve $\mathscr{C}$ in $\tilde{\Sigma}_p$, problem (28.18) possesses three nontrivial solutions: one solution with positive sign, one solution with negative sign, and the third one being sign-changing (cf. Winkert [38], see also [36] if $a = b = \lambda > \lambda_2$ using the representation in (28.17)). An extension of this result for a nonsmooth problem corresponding to (28.18) can be found in Winkert [37].

28.5 Robin Boundary Condition

Finally, we discuss the Fučík spectrum of $-\Delta_p$ with a Robin boundary condition. To this end, we consider weak solutions $u \in W^{1,p}(\Omega)$ of the problem

$$\begin{aligned} -\Delta_p u &= a(u^+)^{p-1} - b(u^-)^{p-1} && \text{in } \Omega, \\ \frac{\partial u}{\partial \nu} &= -\beta|u|^{p-2}u && \text{on } \partial\Omega, \end{aligned} \tag{28.19}$$

meaning that

$$\int_\Omega |\nabla u|^{p-2}\nabla u \cdot \nabla v\, dx + \beta \int_{\partial\Omega} |u|^{p-2}uv\, d\sigma = \int_\Omega \left(a(u^+)^{p-1} - b(u^-)^{p-1}\right)v\, dx$$

for all $v \in W^{1,p}(\Omega)$. In the formulation of (28.19), the parameter β is supposed to be a fixed, nonnegative constant. The Fučík spectrum of the negative p-Laplacian with Robin boundary condition is defined as the set $\widehat{\Sigma}_p$ of all pairs $(a,b) \in \mathbb{R}^2$ for which a nontrivial solution $u \in W^{1,p}(\Omega)$ of (28.19) exists. Clearly, if $\beta = 0$, it reduces to the Fučík spectrum Θ_p of the negative Neumann p-Laplacian (see Sect. 28.3). As before, the special case $a = b = \lambda$ leads to

$$\begin{aligned} -\Delta_p u &= \lambda|u|^{p-2}u && \text{in } \Omega, \\ \frac{\partial u}{\partial \nu} &= -\beta|u|^{p-2}u && \text{on } \partial\Omega, \end{aligned} \tag{28.20}$$

which is the Robin eigenvalue problem of the negative p-Laplacian.

Problem (28.20) was studied in the important publication of Lê [22] devoted to the eigenvalue problems for the negative p-Laplacian. In the Robin case, he proved similar results as they hold for the other eigenvalue problems. The first eigenvalue in (28.20), denoted as usually by λ_1, is simple, isolated, and can be variationally characterized as follows:

$$\lambda_1 = \inf_{u \in W^{1,p}(\Omega)} \left\{ \int_\Omega |\nabla u|^p\, dx + \beta \int_{\partial\Omega} |u|^p d\sigma : \int_\Omega |u|^p\, dx = 1 \right\}.$$

It is also known that the eigenfunctions corresponding to λ_1 are of constant sign and belong to $C^{1,\alpha}(\overline{\Omega})$ for some $0 < \alpha < 1$.

Recently in [32], the authors of the present text investigated the Fučík spectrum introduced in (28.19) with the aim to complete the picture of the Fučík spectrum involving the negative p-Laplacian by extending to the case of Robin boundary condition the information previously known for Dirichlet problem (see Sect. 28.2), Steklov problem (see Sect. 28.4), and homogeneous Neumann problem (see Sect. 28.3).

The approach in [32] is variational relying on the C^1-functional associated to problem (28.19), which is expressed on $W^{1,p}(\Omega)$ by

$$J(u) = \int_\Omega |\nabla u|^p\, dx + \beta \int_{\partial\Omega} |u|^p\, d\sigma - \int_\Omega \left(a(u^+)^p + b(u^-)^p\right) dx.$$

It is clear that the critical points of J are exactly the (weak) solutions of problem (28.19). In comparison with the corresponding functionals related to the Fučík spectrum for the Dirichlet and Steklov problems, the functional J exhibits an essential difference because its expression does not incorporate the norm of the space $W^{1,p}(\Omega)$, and it is also different from the functional used to treat the Neumann problem because it has the additional boundary term involving β.

The results in [32] can be summarized as follows. Applying various ideas and techniques on the pattern of [4, 13, 27], it is shown that $\widehat{\Sigma}_p$ contains a first nontrivial curve, denoted again by $\mathscr{C}$, and expressed as

$$\mathscr{C} = \big\{\big(s + c(s), c(s)\big), \big(c(s), s + c(s)\big) : s \geq 0\big\},$$

where $c(s)$ is given by

$$c(s) = \inf_{\gamma \in \Gamma} \max_{u \in \gamma[-1,+1]} \tilde{J}_s(u),$$

$$\Gamma = \big\{\gamma \in C\big([-1,1], S\big) : \gamma(-1) = -\varphi_1 \text{ and } \gamma(1) = \varphi_1\big\},$$

(see [32, Theorem 3.3]), with φ_1 standing for the eigenfunction of (28.20) associated to λ_1 which is normalized as $\|\varphi_1\|_{L^p(\Omega)} = 1$ and satisfies $\varphi_1 > 0$ on $\overline{\Omega}$. In the above formula of $c(s)$, $\tilde{J}_s$ is equal to the restriction of the C^1-functional $J_s : W^{1,p}(\Omega) \to \mathbb{R}$ given by

$$J_s(u) = \int_\Omega |\nabla u|^p \, dx + \beta \int_{\partial\Omega} |u|^p \, d\sigma - s \int_\Omega (u^+)^p \, dx$$

to the C^1-submanifold

$$S = \left\{u \in W^{1,p}(\Omega) : \int_\Omega |u|^p \, dx = 1\right\}$$

of $W^{1,p}(\Omega)$. It is shown in [32, Proposition 4.2] that the curve $\mathscr{C}$ is Lipschitz continuous and decreasing. The asymptotic behavior of $\mathscr{C}$ requires, as in the Neumann and Steklov cases, some more considerations. In case $p \leq N$, the following holds:

$$\lim_{s \to +\infty} c(s) = \lambda_1$$

(see [32, Theorem 4.3]). If $p > N$, one can suppose that $\beta > 0$ (the case $\beta = 0$ is included in Sect. 28.3, see (28.12)). In this respect, the key idea is to work with an adequate equivalent norm on the space $W^{1,p}(\Omega)$. So, for $\beta > 0$ one introduces the norm

$$\|u\|_\beta = \|\nabla u\|_{L^p(\Omega)} + \beta \|u\|_{L^p(\partial\Omega)}, \tag{28.21}$$

which is an equivalent norm on $W^{1,p}(\Omega)$ (see also Deng [17, Theorem 2.1]). Then in [32, Theorem 4.4] one obtains that the limit of $c(s)$ as $s \to +\infty$ is

$$\overline{\lambda} = \inf_{u \in L} \max_{r \in \mathbb{R}} \frac{\int_\Omega |\nabla (r\varphi_1 + u)|^p \, dx + \beta \int_{\partial\Omega} |r\varphi_1 + u|^p \, d\sigma}{\int_\Omega |r\varphi_1 + u|^p \, dx},$$

where

$$L = \{u \in W^{1,p}(\Omega) : u \text{ vanishes somewhere in } \overline{\Omega}, u \not\equiv 0\}.$$

Moreover, there holds $\overline{\lambda} > \lambda_1$.

References

1. Anane, A.: Simplicité et isolation de la première valeur propre du p-laplacien avec poids. C. R. Acad. Sci. Paris Sér. I Math. **305**(16), 725–728 (1987)
2. Arias, M., Campos, J.: Radial Fučik spectrum of the Laplace operator. J. Math. Anal. Appl. **190**(3), 654–666 (1995)
3. Arias, M., Campos, J., Cuesta, M., Gossez, J.-P.: An asymmetric Neumann problem with weights. Ann. Inst. Henri Poincaré, Anal. Non Linéaire **25**(2), 267–280 (2008)
4. Arias, M., Campos, J., Gossez, J.-P.: On the antimaximum principle and the Fučik spectrum for the Neumann p-Laplacian. Differ. Integral Equ. **13**(1–3), 217–226 (2000)
5. Các, N.P.: On nontrivial solutions of a Dirichlet problem whose jumping nonlinearity crosses a multiple eigenvalue. J. Differ. Equ. **80**(2), 379–404 (1989)
6. Carl, S., Le, V.K., Motreanu, D.: Nonsmooth Variational Problems and Their Inequalities. Comparison Principles and Applications. Springer Monographs in Mathematics. Springer, New York (2007)
7. Carl, S., Motreanu, D.: Constant-sign and sign-changing solutions for nonlinear eigenvalue problems. Nonlinear Anal. **68**(9), 2668–2676 (2008)
8. Carl, S., Motreanu, D.: Multiple and sign-changing solutions for the multivalued p-Laplacian equation. Math. Nachr. **283**(7), 965–981 (2010)
9. Carl, S., Motreanu, D.: Multiple solutions of nonlinear elliptic hemivariational problems. Pac. J. Appl. Math. **1**(4), 381–402 (2008)
10. Carl, S., Motreanu, D.: Sign-changing and extremal constant-sign solutions of nonlinear elliptic problems with supercritical nonlinearities. Commun. Appl. Nonlinear Anal. **14**(4), 85–100 (2007)
11. Carl, S., Motreanu, D.: Sign-changing solutions for nonlinear elliptic problems depending on parameters. Int. J. Differ. Equ. **2010**, 536236 (2010), pp. 33
12. Carl, S., Perera, K.: Sign-changing and multiple solutions for the p-Laplacian. Abstr. Appl. Anal. **7**(12), 613–625 (2002)
13. Cuesta, M., de Figueiredo, D.G., Gossez, J.-P.: The beginning of the Fučik spectrum for the p-Laplacian. J. Differ. Equ. **159**(1), 212–238 (1999)
14. Dancer, E.N.: Generic domain dependence for nonsmooth equations and the open set problem for jumping nonlinearities. Topol. Methods Nonlinear Anal. **1**(1), 139–150 (1993)
15. Dancer, E.N.: On the Dirichlet problem for weakly non-linear elliptic partial differential equations. Proc. R. Soc. Edinb. A **76**(4), 283–300 (1976/77)
16. de Figueiredo, D.G., Gossez, J.-P.: On the first curve of the Fučik spectrum of an elliptic operator. Differ. Integral Equ. **7**(5–6), 1285–1302 (1994)
17. Deng, S.-G.: Positive solutions for Robin problem involving the $p(x)$-Laplacian. J. Math. Anal. Appl. **360**(2), 548–560 (2009)
18. Drábek, P.: Solvability and Bifurcations of Nonlinear Equations. Longman Scientific & Technical, Harlow (1992)
19. Fernández Bonder, J., Rossi, J.D.: Existence results for the p-Laplacian with nonlinear boundary conditions. J. Math. Anal. Appl. **263**(1), 195–223 (2001)
20. Fučík, S.: Boundary value problems with jumping nonlinearities. Čas. Pěst. Mat. **101**(1), 69–87 (1976)

21. Fučík, S.: Solvability of Nonlinear Equations and Boundary Value Problems. Reidel, Dordrecht (1980)
22. Lê, A.: Eigenvalue problems for the p-Laplacian. Nonlinear Anal. **64**(5), 1057–1099 (2006)
23. Lindqvist, P.: On the equation $\operatorname{div}(|\nabla u|^{p-2}\nabla u)+\lambda|u|^{p-2}u=0$. Proc. Am. Math. Soc. **109**(1), 157–164 (1990)
24. Margulies, C.A., Margulies, W.: An example of the Fučik spectrum. Nonlinear Anal. **29**(12), 1373–1378 (1997)
25. Marino, A., Micheletti, A.M., Pistoia, A.: A nonsymmetric asymptotically linear elliptic problem. Topol. Methods Nonlinear Anal. **4**(2), 289–339 (1994)
26. Martínez, S.R., Rossi, J.D.: Isolation and simplicity for the first eigenvalue of the p-Laplacian with a nonlinear boundary condition. Abstr. Appl. Anal. **7**(5), 287–293 (2002)
27. Martínez, S.R., Rossi, J.D.: On the Fučik spectrum and a resonance problem for the p-Laplacian with a nonlinear boundary condition. Nonlinear Anal. **59**(6), 813–848 (2004)
28. Micheletti, A.M.: A remark on the resonance set for a semilinear elliptic equation. Proc. R. Soc. Edinb. A **124**(4), 803–809 (1994)
29. Micheletti, A.M., Pistoia, A.: A note on the resonance set for a semilinear elliptic equation and an application to jumping nonlinearities. Topol. Methods Nonlinear Anal. **6**(1), 67–80 (1995)
30. Motreanu, D., Tanaka, M.: Sign-changing and constant-sign solutions for p-Laplacian problems with jumping nonlinearities. J. Differ. Equ. **249**(11), 3352–3376 (2010)
31. Motreanu, D., Tanaka, M.: Existence of solutions for quasilinear elliptic equations with jumping nonlinearities under the Neumann boundary condition. Calc. Var. Partial Differ. Equ. **43** (1–2), 231–264 (2012)
32. Motreanu, D., Winkert, P.: On the Fučik spectrum for the p-Laplacian with a robin boundary condition. Nonlinear Anal. **74**(14), 4671–4681 (2011)
33. Pardalos, P.M., Rassias, T.M., Khan, A.A. (Guest eds.): Nonlinear Analysis and Variational Problems. In Honor of George Isac. Springer Optimization and Its Applications, vol. 35. Springer, New York (2010), xxviii+490 pp.
34. Pistoia, A.: A generic property of the resonance set of an elliptic operator with respect to the domain. Proc. R. Soc. Edinb. A **127**(6), 1301–1310 (1997)
35. Schechter, M.: The Fučík spectrum. Indiana Univ. Math. J. **43**(4), 1139–1157 (1994)
36. Winkert, P.: Constant-sign and sign-changing solutions for nonlinear elliptic equations with Neumann boundary values. Adv. Differ. Equ. **15**(5–6), 561–599 (2010)
37. Winkert, P.: Multiple solution results for elliptic Neumann problems involving set-valued nonlinearities. J. Math. Anal. Appl. **377**(1), 121–134 (2011)
38. Winkert, P.: Sign-changing and extremal constant-sign solutions of nonlinear elliptic Neumann boundary value problems. Bound. Value Probl. **2010**, 139126 (2010), pp. 22

Chapter 29
Korovkin Type Approximation Theorem for Almost and Statistical Convergence

M. Mursaleen and S.A. Mohiuddine

Abstract In this paper, we use the notion of almost convergence and statistical convergence to prove the Korovkin type approximation theorem by using the test functions $1, e^{-x}, e^{-2x}$. We also display an interesting example in support of our results.

Key words Almost convergence · Statistical convergence · Positive linear operator · Korovkin type approximation theorem

Mathematics Subject Classification 41A10 · 41A25 · 41A36 · 40H05

29.1 Introduction and Preliminaries

Let c and ℓ_∞ denote the spaces of all convergent and bounded sequences, respectively, and note that $c \subset \ell_\infty$. In the theory of sequence spaces, an application of the well known Hahn–Banach Extension Theorem gave rise to the concept of the Banach limit. That is, the lim functional defined on c can be extended to the whole of ℓ_∞ and this extended functional is known as the Banach limit. In 1948, Lorentz [8] used this notion of a weak limit to define a new type of convergence, known as the almost convergence.

A continuous linear functional φ defined on the space ℓ_∞ is called a *Banach Limit* if (i) $\varphi(x) = \varphi((x_k)) \geq 0$ for $x_k \geq 0$ for each k, (ii) $\varphi(e) = 1$, where $e = (1, 1, 1, \ldots)$, and $\varphi((x_k)) = \varphi((x_{k+1}))$ for all $x = (x_k) \in \ell_\infty$.

Dedicated to Professor Themistocles M. Rassias on the occasion of his 60th birthday.

M. Mursaleen (✉)
Department of Mathematics, Aligarh Muslim University, Aligarh 202002, India
e-mail: mursaleenm@gmail.com

S.A. Mohiuddine
Department of Mathematics, Faculty of Science, King Abdulaziz University, P.O. Box 80203, Jeddah 21589, Saudi Arabia
e-mail: mohiuddine@gmail.com

P.M. Pardalos et al. (eds.), *Nonlinear Analysis*, Springer Optimization and Its Applications 68, In Honor of Themistocles M. Rassias on the Occasion of his 60th Birthday, DOI 10.1007/978-1-4614-3498-6_29,

A sequence $x = (x_k)$ is said to be *almost convergent* to the number L if and only if $\varphi(x) = L$ for all Banach limits φ. A bounded sequence $x = (x_k)$ is almost convergent to the number L if and only if $\lim_{p\to\infty} t_{pn} = L$ uniformly in n, where

$$t_{pn} = \frac{x_n + x_{n+1} + x_{n+2} + \cdots + x_{n+p}}{p+1}.$$

We denote the set of all almost convergent sequences by ac and in this case we write $x_k \to L(ac)$, and L is called the generalized limit (or almost limit) of x; written as $L = ac\text{-}\lim x$. Note that a Banach limit extends the limit functional on c in the sense that $\varphi(x) = \lim x$ for all $x \in c$ and $c \subset ac \subset \ell_\infty$.

If $n = 1$ then almost convergence is reduced to $(C,1)$-convergence, and in this case we write $x_k \to L(C,1)$; where $L = (C,1)\text{-}\lim x$. Note that almost convergence implies $(C,1)$-convergence.

Example 29.1 Define the sequence $z = (z_n)$ by

$$z_n = \begin{cases} 1 & \text{if } n \text{ is odd,} \\ -1 & \text{if } n \text{ is even.} \end{cases}$$

Then x is almost convergent to 0 but not convergent.

Let $C[a,b]$ be the space of all functions f continuous on $[a,b]$. We know that $C[a,b]$ is a Banach space with norm

$$\|f\|_\infty := \sup_{x\in[a,b]} |f(x)|, \quad f \in C[a,b].$$

The classical Korovkin approximation theorem reads as follows [7]:

Let (T_n) be a sequence of positive linear operators from $C[a,b]$ into $C[a,b]$. Then $\lim_n \|T_n(f,x) - f(x)\|_\infty = 0$, *for all* $f \in C[a,b]$ *if and only if* $\lim_n \|T_n(f_i,x) - f_i(x)\|_\infty = 0$, *for* $i = 0,1,2$, *where* $f_0(x) = 1$, $f_1(x) = x$ *and* $f_2(x) = x^2$.

Quite recently, several approximation theorems of such a type were proved in [1] for functions of two variables by using almost convergence of double sequences. For some recent work on this topic, we refer to [9–11] and [12]. Boyanov and Veselinov [3] have proved the Korovkin theorem on $C[0,\infty)$ by using the test functions $1, e^{-x}, e^{-2x}$. In this paper, we generalize the result of Boyanov and Veselinov by using the notion of almost convergence and statistical convergence.

29.2 Main Result

Let $C(I)$ be the Banach space with the uniform norm $\|\cdot\|_\infty$ of all real-valued continuous functions on $I = [0,\infty)$; provided that $\lim_{x\to\infty} f(x)$ is finite. Suppose that $L_n : C(I) \to C(I)$. We write $L_n(f;x)$ for $L_n(f(s);x)$; and we say that L is a positive operator if $L(f;x) \geq 0$ for all $f(x) \geq 0$.

We prove the following generalization of Boyanov and Veselinov [3] for almost convergence.

Theorem 29.1 *Let (T_k) be a sequence of positive linear operators from $C(I)$ into $C(I)$. Then for all $f \in C(I)$*

$$ac\text{-}\lim_{k\to\infty} \left\| T_k(f;x) - f(x) \right\|_\infty = 0 \tag{29.1}$$

if and only if

$$ac\text{-}\lim_{k\to\infty} \left\| T_k(1;x) - 1 \right\|_\infty = 0, \tag{29.2}$$

$$ac\text{-}\lim_{k\to\infty} \left\| T_k\big(e^{-s};x\big) - e^{-x} \right\|_\infty = 0, \tag{29.3}$$

$$ac\text{-}\lim_{k\to\infty} \left\| T_k\big(e^{-2s};x\big) - e^{-2x} \right\|_\infty = 0. \tag{29.4}$$

Proof Since each of $1, e^{-x}, e^{-2x}$ belongs to $C(I)$, conditions (29.2)–(29.4) follow immediately from (29.1). Let $f \in C(I)$. Then there exists a constant $M > 0$ such that $|f(x)| \leq M$ for $x \in I$. Therefore,

$$\big|f(s) - f(x)\big| \leq 2M, \quad -\infty < s, x < \infty. \tag{29.5}$$

It is easy to prove that for a given $\varepsilon > 0$ there is a $\delta > 0$ such that

$$\big|f(s) - f(x)\big| < \varepsilon, \tag{29.6}$$

whenever $|e^{-s} - e^{-x}| < \delta$ for all $x \in I$.

Using (29.5), (29.6), and putting $\psi_1 = \psi_1(s,x) = (e^{-s} - e^{-x})^2$, we get

$$\big|f(s) - f(x)\big| < \varepsilon + \frac{2M}{\delta^2}(\psi_1), \quad \forall |s - x| < \delta.$$

This is,

$$-\varepsilon - \frac{2M}{\delta^2}(\psi_1) < f(s) - f(x) < \varepsilon + \frac{2M}{\delta^2}(\psi_1).$$

Now, applying the operator $T_k(1;x)$ to this inequality, since $T_k(f;x)$ is monotone and linear, we obtain

$$\begin{aligned} T_k(1;x)\left(-\varepsilon - \frac{2M}{\delta^2}(\psi_1)\right) &< T_k(1;x)\big(f(s) - f(x)\big) \\ &< T_k(1;x)\left(\varepsilon + \frac{2M}{\delta^2}(\psi_1)\right). \end{aligned}$$

Note that x is fixed and so $f(x)$ is a constant number. Therefore,

$$-\varepsilon T_k(1;x) - \frac{2M}{\delta^2} T_{j,k}(\psi_1;x) < T_k(f;x) - f(x)T_k(1;x)$$
$$< \varepsilon T_k(1;x) + \frac{2M}{\delta^2} T_k(\psi_1;x). \tag{29.7}$$

But

$$\begin{aligned} T_k(f;x) - f(x) &= T_k(f;x) - f(x)T_k(1;x) + f(x)T_k(1;x) - f(x) \\ &= \big[T_k(f;x) - f(x)T_k(1;x)\big] + f(x)\big[T_k(1;x) - 1\big]. \end{aligned} \tag{29.8}$$

Using (29.7) and (29.8), we have

$$T_k(f;x) - f(x) < \varepsilon T_k(1;x) + \frac{2M}{\delta^2} T_k(\psi_1;x) + f(x)\big(T_k(1;x) - 1\big). \tag{29.9}$$

Now

$$\begin{aligned} T_k(\psi_1;x) &= T_k\big((e^{-s} - e^{-x})^2;x\big) = T_k\big(e^{-2s} - 2e^{-s}e^{-x} + e^{-2x};x\big) \\ &= T_k\big(e^{-2s};x\big) - 2e^{-x} T_k\big(e^{-s};x\big) + \big(e^{-2x}\big)T_k(1;x) \\ &= \big[T_k\big(e^{-2s};x\big) - e^{-2x}\big] - 2e^{-x}\big[T_k\big(e^{-s};x\big) - e^{-x}\big] \\ &\quad + e^{-2x}\big[T_k(1;x) - 1\big]. \end{aligned}$$

Using (29.9), we obtain

$$\begin{aligned} T_k(f;x) - f(x) &< \varepsilon T_k(1;x) + \frac{2M}{\delta^2}\{\big[T_k\big(e^{-2s};x\big) - e^{-2x}\big] \\ &\quad - 2e^{-x}\big[T_k\big(e^{-s};x\big) - e^{-x}\big] + e^{-2x}\big[T_k(1;x) - 1\big]\} \\ &\quad + f(x)\big(T_k(1;x) - 1\big) \\ &= \varepsilon\big[T_k(1;x) - 1\big] + \varepsilon + \frac{2M}{\delta^2}\{\big[T_k\big(e^{-2s};x\big) - e^{-2x}\big] \\ &\quad - 2e^{-x}\big[T_k\big(e^{-s};x\big) - e^{-x}\big] + e^{-2x}\big[T_k(1;x) - 1\big]\} \\ &\quad + f(x)\big(T_k(1;x) - 1\big). \end{aligned}$$

Therefore,

$$\begin{aligned} \big|T_k(f;x) - f(x)\big| &\le \varepsilon + (\varepsilon + M)\big|T_k(1;x) - 1\big| + \frac{2M}{\delta^2}\big|e^{-2x}\big|\big|T_k(1;x) - 1\big| \\ &\quad + \frac{2M}{\delta^2}\big|T_k\big(e^{-2s};x\big) - e^{-2x}\big| + \frac{4M}{\delta^2}\big|e^{-x}\big|\big|T_k\big(e^{-s};x\big) - e^{-x}\big| \\ &\le \varepsilon + \left(\varepsilon + M + \frac{2M}{\delta^2}\right)\big|T_k(1;x) - 1\big| \end{aligned}$$

$$+\frac{2M}{\delta^2}\left|T_k\left(e^{-2s};x\right)-e^{-2x}\right|$$
$$+\frac{4M}{\delta^2}\left|T_k\left(e^{-s};x\right)-e^{-x}\right|, \tag{29.10}$$

since $|e^{-x}| \le 1$ for all $x \in I$. Now, taking $\sup_{x\in I}$, we get

$$\left\|T_k(f;x)-f(x)\right\|_\infty \le \varepsilon + K\left(\left\|T_k(1;x)-1\right\|_\infty + \left\|T_k\left(e^{-s};x\right)-e^{-x}\right\|_\infty + \left\|T_k\left(e^{-2s};x\right)-e^{-2x}\right\|_\infty\right), \tag{29.11}$$

where $K=\max\{\varepsilon+M+\frac{2M}{\delta^2}, \frac{4M}{\delta^2}\}$. Replacing $T_k(f;x)$ by $D_{n,p}(f;x)=\frac{1}{p}\sum_{k=n}^{n+p-1} T_k(f;x)$ in (29.11), letting $p\to\infty$, and using (29.2)–(29.5), we get

$$\lim_{p\to\infty}\left\|D_{n,p}(f;x)-f(x)\right\|_\infty = 0, \text{ uniformly in } n.$$

This completes the proof of the theorem. □

29.3 Statistical and A-Statistical Versions

Let $K \subseteq \mathbb{N}$ and $K_n = \{k \le n : k \in K\}$. Then the *natural density* of K is defined by $\delta(K) = \lim_n n^{-1}|K_n|$ if the limit exists, where $|K_n|$ denotes the cardinality of K_n.

A sequence $x = (x_k)$ of real numbers is said to be *statistically convergent* (cf. [4]) to L provided that for every $\varepsilon > 0$ the set $K_\varepsilon := \{k \in \mathbb{N} : |x_k - L| \ge \varepsilon\}$ has natural density zero, i.e., for each $\varepsilon > 0$,

$$\lim_n \frac{1}{n}\left|\{j \le n : |x_j - L| \ge \varepsilon\}\right| = 0.$$

In this case, we write $L = st\text{-}\lim x$. Note that every convergent sequence is statistically convergent but not conversely. Define the sequence $w = (w_n)$ by

$$w_n = \begin{cases} 1 & \text{if } n = k^2,\ k \in \mathbb{N}, \\ 0 & \text{otherwise.} \end{cases}$$

Then x is statistically convergent to 0 but not convergent.

For a non-negative regular matrix $A = (a_{nk})_{n,k=1}^\infty$, an index set $K = \{k_i\}$ is said to have *A-density* if

$$\delta_A(K) = \lim_n \sum_{k\in K} a_{nk} = \lim_n \sum_i a_{n,k_i}.$$

A sequence x is said to be *A-statistically convergent* to L (cf. [6]) if $\delta_A(K_\varepsilon) = 0$ for every $\varepsilon > 0$. In this case we write $st_A\text{-}\lim x = L$, and $x_k \xrightarrow{st_A} L$.

In the following theorem, we use the notion of statistical convergence analogous to that of [5]. We also display an example to show its importance.

Theorem 29.2 *Let (T_k) be a sequence of positive linear operators from $C(I)$ into $C(I)$. Then for all $f \in C(I)$*

$$st\text{-}\lim_{k\to\infty} \left\| T_k(f;x) - f(x) \right\|_\infty = 0 \tag{29.12}$$

if and only if

$$st\text{-}\lim_{k\to\infty} \left\| T_k(1;x) - 1 \right\|_\infty = 0, \tag{29.13}$$

$$st\text{-}\lim_{k\to\infty} \left\| T_k\left(e^{-s};x\right) - e^{-x} \right\|_\infty = 0, \tag{29.14}$$

$$st\text{-}\lim_{k\to\infty} \left\| T_k\left(e^{-2s};x\right) - e^{-2x} \right\|_\infty = 0. \tag{29.15}$$

Proof For a given $r > 0$ choose $\varepsilon > 0$ such that $\varepsilon < r$. Define the following sets

$$D := \left\{k \le n : \left\| T_k(f) - f \right\|_\infty \ge r\right\},$$

$$D_1 := \left\{k \le n : \left\| T_k(1;x) - 1 \right\|_\infty \ge \frac{r-\varepsilon}{4K}\right\},$$

$$D_2 := \left\{k \le n : \left\| T_k\left(e^{-s};x\right) - e^{-x} \right\|_\infty \ge \frac{r-\varepsilon}{4K}\right\},$$

$$D_3 := \left\{k \le n : \left\| T_k\left(e^{-2s};x\right) - e^{-2x} \right\|_\infty \ge \frac{r-\varepsilon}{4K}\right\}.$$

Then from (29.11), we see that $D \subset D_1 \cup D_2 \cup D_3$, and therefore $\delta(D) \le \delta(D_1) + \delta(D_2) + \delta(D_3)$. Hence conditions (29.13)–(29.15) imply the condition (29.12).

This completes the proof of the theorem. □

In the following, we give an example of a sequence of positive linear operators satisfying the conditions of Theorem 29.2 which does not satisfy the conditions of the Korovkin theorem.

Example 29.2 Consider the sequence of classical Baskakov operators [2]

$$V_n(f;x) := \sum_{k=0}^{\infty} f\left(\frac{k}{n}\right) \binom{n-1+k}{k} x^k (1+x)^{-n-k};$$

where $0 \le x, y < \infty$.

Let $L_n : C(I) \to C(I)$ be defined by

$$L_n(f;x) = (1 + w_n) V_n(f;x),$$

where the sequence (w_n) is defined as above. Since

$$V_n(1;x) = 1,$$

$$V_n(e^{-s};x) = \left(1 + x - xe^{-\frac{1}{n}}\right)^{-n},$$
$$V_n(e^{-2s};x) = \left(1 + x - xe^{\frac{-2}{n}}\right)^{-2n},$$

we have that the sequence (L_n) satisfies the conditions (29.13)–(29.15). Hence by Theorem 29.2, we have

$$st\text{-}\lim_{n\to\infty} \|L_n(f) - f\|_\infty = 0.$$

On the other hand, we get $V_n(f;0) = (1 + w_n)f(0)$, since $V_n(f;0) = f(0)$, and hence

$$\|L_n(f;x) - f(x)\|_\infty \geq |L_n(f;0) - f(0)| = w_n|f(0)|.$$

We see that (L_n) does not satisfy the conditions of the Korovkin theorem, since $\lim_{n\to\infty} w_n$ does not exist.

Similarly, if we define the operator $T_n : C(I) \to C(I)$ by

$$T_n(f;x) = (1 + z_n)V_n(f;x),$$

where the sequence (z_n) is defined as above, then it is easy to see that the sequence (T_n) satisfies the conditions (29.2)–(29.4). Hence by Theorem 29.1, we have

$$ac\text{-}\lim_{n\to\infty} \|T_n(f) - f\|_\infty = 0.$$

But (T_n) does not satisfy the conditions of the Korovkin theorem, since $\lim_{n\to\infty} z_n$ does not exist.

We can further generalize Theorem 29.2 for A-statistical convergence which can be proved similarly.

Theorem 29.3 *Let (T_k) be a sequence of positive linear operators from $C(I)$ into $C(I)$. Then for all $f \in C(I)$*

$$st_A\text{-}\lim_{k\to\infty} \|T_k(f;x) - f(x)\|_\infty = 0 \tag{29.16}$$

if and only if

$$st_A\text{-}\lim_{k\to\infty} \|T_k(1;x) - 1\|_\infty = 0, \tag{29.17}$$
$$st_A\text{-}\lim_{k\to\infty} \|T_k(e^{-s};x) - e^{-x}\|_\infty = 0, \tag{29.18}$$
$$st_A\text{-}\lim_{k\to\infty} \|T_k(e^{-2s};x) - e^{-2x}\|_\infty = 0. \tag{29.19}$$

References

1. Anastassiou, G.A., Mursaleen, M., Mohiuddine, S.A.: Some approximation theorems for functions of two variables through almost convergence of double sequences. J. Comput. Anal. Appl. **13**(1), 37–40 (2011)
2. Becker, M.: Global approximation theorems for Szasz–Mirakjan and Baskakov operators in polynomial weight spaces. Indiana Univ. Math. J. **27**(1), 127–142 (1978)
3. Boyanov, B.D., Veselinov, V.M.: A note on the approximation of functions in an infinite interval by linear positive operators. Bull. Math. Soc. Sci. Math. Roum. **14**(62), 9–13 (1970)
4. Fridy, J.A.: On statistical convergence. Analysis **5**, 301–313 (1985)
5. Gadjiev, A.D., Orhan, C.: Some approximation theorems via statistical convergence. Rocky Mt. J. Math. **32**, 129–138 (2002)
6. Kolk, E.: Matrix summability of statistically convergent sequences. Analysis **13**, 77–83 (1993)
7. Korovkin, P.P.: Linear Operators and Approximation Theory. Hindustan Publ., Delhi (1960)
8. Lorentz, G.G.: A contribution to theory of divergent sequences. Acta Math. **80**, 167–190 (1948)
9. Mohiuddine, S.A.: An application of almost convergence in approximation theorems. Appl. Math. Lett. **24**, 1856–1860 (2011)
10. Mursaleen, M., Alotaibi, A.: Statistical summability and approximation by de la Vallée-Pousin mean. Appl. Math. Lett. **24**, 320–324 (2011)
11. Mursaleen, M., Karakaya, V., Ertürk, M., Gürsoy, F.: Weighted statistical convergence and its application to Korovkin type approximation theorem. Appl. Math. Comput. (2012). doi:10.1016/j.amc.2012.02.068
12. Srivastava, H.M., Mursaleen, M., Khan, A.: Generalized equi-statistical convergence of positive linear operators and associated approximation theorems. Math. Comput. Model. **55**, 2040–2051 (2012)

Chapter 30
On the Stability of an Additive Mapping

Abbas Najati

Abstract In this work, the Hyers–Ulam stability of the functional equation $f(x+y+xy)=f(x+y)+f(xy)$ is proved.

Key words Hyers–Ulam stability · Additive functional equation

Mathematics Subject Classification 39B82 · 39B52

30.1 Introduction

A classical question in the theory of functional equations is the following: "When is it true that a function which approximately satisfies a functional equation ε must be close to an exact solution of ε?" If the problem accepts a solution, we say that the equation ε is stable. The first stability problem concerning group homomorphisms was raised by Ulam [14] in 1940. We are given a group G and a metric group $G^{'}$ with metric $d(\cdot,\cdot)$. Given $\varepsilon>0$, does there exist a $\delta>0$ such that if $f:G\to G'$ satisfies $d(f(xy),f(x)f(y))<\delta$ for all $x,y\in G$, then a homomorphism $h:G\to G^{'}$ exists with $d(f(x),h(x))<\varepsilon$ for all $x\in G$?

Ulam's problem was partially solved by Hyers [5] in 1941. Let E_1 be a normed space, E_2 a Banach space and suppose that a mapping $f:E_1\to E_2$ satisfies the inequality

$$\|f(x+y)-f(x)-f(y)\|\leq\varepsilon$$

for all $x,y\in E_1$, where $\varepsilon>0$ is a constant. Then the limit $T(x)=\lim_{n\to\infty}2^{-n}f(2^nx)$ exists for each $x\in E_1$ and T is the unique additive mapping satisfying

$$\|f(x)-T(x)\|\leq\varepsilon \tag{30.1}$$

Dedicated to Professor Themistocles M. Rassias on the occasion of his 60th birthday.

A. Najati (✉)
Department of Mathematics, University of Mohaghegh Ardabili, Ardabil, Iran
e-mail: a.nejati@yahoo.com

P.M. Pardalos et al. (eds.), *Nonlinear Analysis*, Springer Optimization and Its Applications 68, In Honor of Themistocles M. Rassias on the Occasion of his 60th Birthday, DOI 10.1007/978-1-4614-3498-6_30,

for all $x \in E_1$. Also, if for each x the function $t \longmapsto f(tx)$ from $\mathbb{R}$ to E_2 is continuous on $\mathbb{R}$, then T is linear. If f is continuous at a single point of E_1, then T is continuous everywhere in E_1. Moreover (30.1) is sharp.

In 1978, Th.M. Rassias [10] formulated and proved the following theorem, which implies Hyers' theorem as a special case. Suppose that E and F are real normed spaces with F a complete normed space, $f : E \to F$ is a mapping such that for each fixed $x \in E$ the mapping $t \longmapsto f(tx)$ is continuous on $\mathbb{R}$, and let there exist $\varepsilon > 0$ and $p \in [0, 1)$ such that

$$\big\| f(x+y) - f(x) - f(y) \big\| \le \varepsilon \big(\|x\|^p + \|y\|^p \big) \tag{30.2}$$

for all $x, y \in E$. Then there exists a unique linear mapping $T : E \to F$ such that

$$\big\| f(x) - T(x) \big\| \le \frac{2\varepsilon}{2 - 2^p} \|x\|^p \tag{30.3}$$

for all $x \in E$. The case of the existence of a unique additive mapping had been obtained by T. Aoki [1]. Th.M. Rassias [10] was the first to prove that there exists a unique linear mapping T satisfying (30.3).

In 1990, Th.M. Rassias [11] during the 27th International Symposium on Functional Equations asked the question whether such a theorem can also be proved for $p \ge 1$. In 1991, Gajda [3] gave an affirmative solution to this question for $p > 1$ by following the same approach as in Rassias' paper [10]. It was proved by Gajda [3], as well as by Th.M. Rassias and Šemrl [13] that one cannot prove a Rassias type theorem when $p = 1$. In 1994, P. Găvruta [4] provided a generalization of Rassias' theorem in which he replaced the bound $\varepsilon(\|x\|^p + \|y\|^p)$ in [10] by a general control function $\varphi(x, y)$. The paper of Th.M. Rassias [10] has provided a lot of influence in the development of what we now call the generalized Hyers–Ulam stability of functional equations. During the last decades, several stability problems for various functional equations have been investigated by many mathematicians; we refer the reader to the monographs [2, 6–9, 12].

In this paper, we deal with the functional equation

$$f(x + y + xy) = f(x + y) + f(xy). \tag{30.4}$$

30.2 Solution of Functional Equation (30.4)

Theorem 30.1 *Let X be a vector space and let $f : \mathbb{R} \to X$ be a function. Then f satisfies* (30.4) *if and only if f is additive.*

Proof Let f satisfy (30.4) and $a, b \in \mathbb{R}$ with $a \le 0$. We can find $x, y \in \mathbb{R}$ such that $xy = a, x + y = b$. Since f satisfies (30.4), we get

$$f(a + b) = f(a) + f(b) \tag{30.5}$$

for all $a, b \in \mathbb{R}$ with $a \le 0$. It is clear that $f(0) = 0$. Letting $b = -a$ in (30.5), we get $f(-a) = -f(a)$ for all $a \le 0$. So $f(-a) = -f(a)$ for all $a \in \mathbb{R}$. If $a > 0$, it follows from (30.5) that

$$f(-a+b) = f(-a) + f(b) \tag{30.6}$$

for all $b \in \mathbb{R}$. Replacing b by $b + a$ in (30.6) and using the oddness of f, we get $f(a+b) = f(a) + f(b)$ for all $a, b \in \mathbb{R}$ with $a > 0$. Therefore, f is additive.

Conversely, if f is additive, it is easy to check that f satisfies (30.4). □

30.3 Hyers–Ulam Stability of Functional Equation (30.4) in Banach Spaces

In this section, we investigate the Hyers–Ulam stability problem for the functional equation (30.4). In this section, X is a Banach space.

Theorem 30.2 *Let $\varepsilon \ge 0$ be fixed and let $f : \mathbb{R} \to X$ be a mapping satisfying*

$$\left\| f(x+y+xy) - f(x+y) - f(xy) \right\| \le \varepsilon \tag{30.7}$$

for all $x, y \in \mathbb{R}$. Then there exists a unique additive mapping $A : \mathbb{R} \to X$ satisfying

$$\left\| f(x) - A(x) \right\| \le 3\varepsilon \tag{30.8}$$

for all $x, y \in \mathbb{R}$.

Proof Let $a, b \in \mathbb{R}$ with $a \le 0$. We can find $x, y \in \mathbb{R}$ such that $xy = a, x + y = b$. Since f satisfies (30.7), we get

$$\left\| f(a+b) - f(a) - f(b) \right\| \le \varepsilon \tag{30.9}$$

for all $a, b \in \mathbb{R}$ with $a \le 0$. Putting $b = -a$ in (30.9) yields

$$\left\| f(0) - f(a) - f(-a) \right\| \le \varepsilon \tag{30.10}$$

for all $a \le 0$. So (30.10) holds for all $a \in \mathbb{R}$. It follows from (30.9) that

$$\left\| f(-a+b) - f(-a) - f(b) \right\| \le \varepsilon \tag{30.11}$$

for all $a, b \in \mathbb{R}$ with $a > 0$. Replacing b by $a + b$ in (30.11) and using (30.10), we get

$$\left\| f(a+b) - f(a) - f(b) + f(0) \right\| \le 2\varepsilon \tag{30.12}$$

for all $a, b \in \mathbb{R}$ with $a > 0$. Since $\|f(0)\| \le \varepsilon$, it follows from (30.9) and (30.12) that

$$\left\| f(a+b) - f(a) - f(b) \right\| \le 3\varepsilon \tag{30.13}$$

for all $a, b \in \mathbb{R}$. By the Hyers' theorem, the limit $A(x) = \lim_{n\to\infty} 2^{-n} f(2^n x)$ exists for each $x \in \mathbb{R}$ and A is the unique additive mapping satisfying (30.8). □

Proposition 30.1 *Let $\phi : \mathbb{R} \to \mathbb{R}$ be defined by*

$$\phi(x) := \begin{cases} x & \text{for } |x| < 1, \\ 1 & \text{for } |x| \geq 1. \end{cases}$$

Consider the function $f : \mathbb{R} \to \mathbb{R}$ by the formula

$$f(x) := \sum_{n=0}^{\infty} 2^{-n} \phi(2^n x).$$

Then f satisfies

$$\left| f(x+y+xy) - f(x+y) - f(xy) \right| \leq 12(|x| + |y|) \tag{30.14}$$

for all $x, y \in \mathbb{R}$, and the range of $|f(x) - A(x)|/|x|$ for $x \neq 0$ is unbounded for each additive mapping $A : \mathbb{R} \to \mathbb{R}$.

Proof It is clear that f is bounded by 2 on $\mathbb{R}$. If $|x| + |y| = 0$ or $|x| + |y| \geq \frac{1}{2}$, then

$$\left| f(x+y+xy) - f(x+y) - f(xy) \right| \leq 6 \leq 12\big(|x| + |y|\big).$$

Now suppose that $0 < |x| + |y| < \frac{1}{2}$. Then there exists an integer $k \geq 1$ such that

$$\frac{1}{2^{k+1}} \leq |x| + |y| < \frac{1}{2^k}. \tag{30.15}$$

Therefore,

$$2^m |x+y+xy|, 2^m |x+y|, 2^m |xy| < 1$$

for all $m = 0, 1, \ldots, k-1$. From the definition of f and (30.15), we have

$$\begin{aligned} &\left| f(x+y+xy) - f(x+y) - f(xy) \right| \\ &\quad \leq \sum_{n=k}^{\infty} 2^{-n} \left[\left| \phi\big(2^n(x+y+xy)\big) \right| + \left| \phi\big(2^n(x+y)\big) \right| + \left| \phi\big(2^n(xy)\big) \right| \right] \\ &\quad \leq \frac{6}{2^k} \leq 12\big(|x| + |y|\big). \end{aligned}$$

Therefore, f satisfies (30.14). Let $A : \mathbb{R} \to \mathbb{R}$ be an additive function such that

$$\left| f(x) - A(x) \right| \leq \beta |x|$$

for all $x \in \mathbb{R}$, where $\beta > 0$ is a constant. Then there exists a constant $c \in \mathbb{R}$ such that $A(x) = cx$ for all rational numbers x. So we have

$$\left|f(x)\right| \le \left(\beta + |c|\right)|x| \tag{30.16}$$

for all rational numbers x. Let $m \in \mathbb{N}$ with $m > \beta + |c|$. If x is a rational number in $(0, 2^{1-m})$, then $2^n x \in (0, 1)$ for all $n = 0, 1, \ldots, m - 1$. So

$$f(x) \ge \sum_{n=0}^{m-1} 2^{-n}\phi\left(2^n x\right) = mx > \left(\beta + |c|\right)x$$

which contradicts (30.16). □

30.4 Stability of Functional Equation (30.4) in Topological Vector Spaces

In this section, E is a sequentially complete Hausdorff topological vector space over the field $\mathbb{Q}$ of rational numbers.

Theorem 30.3 *Let V be a nonempty bounded convex subset of E containing the origin. Suppose that $f : \mathbb{R} \to E$ satisfies*

$$f(x + y + xy) - f(x + y) - f(xy) \in V \tag{30.17}$$

for all $x, y \in \mathbb{R}$. Then there exists a unique additive mapping $A : \mathbb{R} \to E$ such that

$$f(x) - A(x) \in \overline{2V - V} \tag{30.18}$$

for all $x \in \mathbb{R}$, where $\overline{2V - V}$ denotes the sequential closure of $2V - V$.

Proof Using the proof of Theorem 30.2, we get

$$f(a + b) - f(a) - f(b) \in V \tag{30.19}$$

for all $a, b \in \mathbb{R}$ with $a \le 0$. Putting $b = -a$ in (30.19), yields

$$f(0) - f(a) - f(-a) \in V \tag{30.20}$$

for all $a \le 0$. So (30.20) holds for all $a \in \mathbb{R}$. It follows from (30.19) that

$$f(-a + b) - f(-a) - f(b) \in V \tag{30.21}$$

for all $a, b \in \mathbb{R}$ with $a > 0$. Replacing b by $a + b$ in (30.21) and using (30.20), we get

$$f(a + b) - f(a) - f(b) + f(0) \in V - V \tag{30.22}$$

for all $a, b \in \mathbb{R}$ with $a > 0$. Since $-f(0) \in V$, V is convex and contains the origin, it follows from (30.19) and (30.22) that

$$f(a+b) - f(a) - f(b) \in 2V - V \tag{30.23}$$

for all $a, b \in \mathbb{R}$. It is easy to prove that

$$\frac{f(2^{n+1}a)}{2^{n+1}} - \frac{f(2^n a)}{2^n} \in \frac{1}{2^{n+1}} W \subseteq W, \tag{30.24}$$

$$\frac{f(2^n a)}{2^n} - f(a) \in \sum_{k=1}^{n} \frac{1}{2^k} W \subseteq W \tag{30.25}$$

for all $a \in \mathbb{R}$ and all integers $n \geq 0$, where $W = 2V - V$. Since V is a nonempty bounded convex subset of E containing the origin, W is a nonempty bounded convex subset of E containing the origin. It follows from (30.24) that

$$\frac{f(2^n a)}{2^n} - \frac{f(2^m a)}{2^m} = \sum_{k=m}^{n-1} \left[\frac{f(2^{k+1}a)}{2^{k+1}} - \frac{f(2^k a)}{2^k} \right] \in \sum_{k=m}^{n-1} \frac{1}{2^{k+1}} W \subseteq \frac{1}{2^m} W \tag{30.26}$$

for all $a \in \mathbb{R}$ and all integers $n > m \geq 0$. Let U be an arbitrary neighborhood of the origin in E. Since W is bounded, there exists a rational number $t > 0$ such that $tW \subseteq U$. Choose $n_0 \in \mathbb{N}$ such that $2^{n_0} t > 1$. Let $a \in \mathbb{R}$ and $m, n \in \mathbb{N}$ with $n \geq m \geq n_0$. Then (30.26) implies that

$$\frac{f(2^n a)}{2^n} - \frac{f(2^m a)}{2^m} \in U. \tag{30.27}$$

Thus, the sequence $\{2^{-n} f(2^n a)\}$ forms a Cauchy sequence in E. By the sequential completeness of E, the limit $A(a) = \lim_{n\to\infty} 2^{-n} f(2^n a)$ exists for each $a \in \mathbb{R}$. So (30.18) follows from (30.25).

To show that $A : \mathbb{R} \to E$ is additive, replace a and b by $2^n a$ and $2^n b$, respectively, in (30.23) and then divide by 2^n to obtain

$$\frac{f(2^n(a+b))}{2^n} - \frac{f(2^n a)}{2^n} - \frac{f(2^n b)}{2^n} \in \frac{1}{2^n} W$$

for all $a \in \mathbb{R}$ and all integers $n \geq 0$. Since $\overline{W}$ is bounded, on taking the limit as $n \to \infty$, we get that A is additive.

To prove the uniqueness of A, assume on the contrary that there is another additive mapping $T : \mathbb{R} \to E$ satisfying (30.18) and there is an $a \in \mathbb{R}$ such that $x = T(a) - A(a) \neq 0$. So there is a neighborhood U of the origin in E such that $x \notin U$, since E is Hausdorff. Since A and T satisfy (30.18), we get $T(b) - A(b) \in \overline{W} - \overline{W}$ for all $b \in \mathbb{R}$. Since W is bounded, $\overline{W} - \overline{W}$ is bounded. Hence there exists a positive integer m such that $\overline{W} - \overline{W} \subseteq mU$. Therefore, $mx = T(ma) - A(ma) \in mU$, which is a contradiction with $x \notin U$. This completes the proof. □

References

1. Aoki, T.: On the stability of the linear transformation in Banach spaces. J. Math. Soc. Jpn. **2**, 64–66 (1950)
2. Czerwik, S.: Functional Equations and Inequalities in Several Variables. World Scientific, New Jersey (2002)
3. Gajda, Z.: On stability of additive mappings. Int. J. Math. Math. Sci. **14**, 431–434 (1991)
4. Găvruta, P.: A generalization of the Hyers–Ulam–Rassias stability of approximately additive mappings. J. Math. Anal. Appl. **184**, 431–436 (1994)
5. Hyers, D.H.: On the stability of the linear functional equation. Proc. Natl. Acad. Sci. USA **27**, 222–224 (1941)
6. Hyers, D.H., Isac, G., Rassias, Th.M.: Stability of Functional Equations in Several Variables. Birkhäuser, Basel (1998)
7. Isac, G., Rassias, Th.M.: On the Hyers–Ulam stability of ψ-additive mappings. J. Approx. Theory **72**, 131–137 (1993)
8. Isac, G., Rassias, Th.M.: Functional inequalities for approximately additive mappings. In: Stability of Mappings of Hyers–Ulam Type, pp. 117–125. Hadronic Press, Palm Harbor (1994)
9. Jung, S.-M.: Hyers–Ulam–Rassias Stability of Functional Equations in Mathematical Analysis. Hadronic Press, Palm Harbor (2001)
10. Rassias, Th.M.: On the stability of the linear mapping in Banach spaces. Proc. Am. Math. Soc. **72**, 297–300 (1978)
11. Rassias, Th.M.: Problem 16; 2, Report of the 27th International Symp. on Functional Equations. Aequ. Math. **39**, 292–293, 309 (1990)
12. Rassias, Th.M. (ed.): Functional Equations, Inequalities and Applications. Kluwer Academic, Dordrecht (2003)
13. Rassias, Th.M., Šemrl, P.: On the behaviour of mappings which do not satisfy Hyers–Ulam stability. Proc. Am. Math. Soc. **114**, 989–993 (1992)
14. Ulam, S.M.: Problems in Modern Mathematics, Chapter VI, Science Editions. Wiley, New York (1964)

Chapter 31
Existence Results for Extended General Nonconvex Quasi-variational Inequalities

Muhammad Aslam Noor, Khalida Inayat Noor, and Eisa Al-Said

Abstract In this paper, we introduce and study a new class of quasi-variational inequalities, which are called the extended general nonconvex quasi-variational inequalities. Using the projection technique, we establish the equivalence between the extended general nonconvex quasi-variational inequalities and the fixed point problem. We use this alternative equivalent formulation to prove the existence of a solution of the extended general quasi-variational inequalities under some suitable conditions. Several special cases are also discussed.

Key words Quasi-variational inequalities · Projection method · Fixed point · Existence

Mathematics Subject Classification 49J40 · 90C33

31.1 Introduction

Quasi-variational inequalities are being used to study a wide class of problems which arise in various branches of pure and applied science, in a unified and general framework. Quasi-variational inequalities combine theoretical and algorithmic advances with new and novel domain of applications. In recent years, considerable interest has been shown in developing various extensions and generalizations of variational inequalities using the novel techniques, both for their own sake and for their applications. There are significant developments of these problems related to

---

Dedicated to Professor Themistocles M. Rassias on the occasion of his 60th birthday.

M.A. Noor (✉) · K.I. Noor · E. Al-Said
Mathematics Department, COMSATS Institute of Information Technology, Park Road, Islamabad, Pakistan
e-mail: noormaslam@hotmail.com

K.I. Noor
e-mail: khalidanoor@hotmail.com

E. Al-Said
e-mail: eisasaid@ksu.edu.sa

P.M. Pardalos et al. (eds.), *Nonlinear Analysis*, Springer Optimization and Its Applications 68, In Honor of Themistocles M. Rassias on the Occasion of his 60th Birthday, DOI 10.1007/978-1-4614-3498-6_31, © Springer Science+Business Media, LLC 2012

nonconvex optimization, iterative methods, and structural analysis. For recent work on the generalized variants of quasi-variational inequalities and their applications, see [1–35].

Motivated and inspired by the research work going on in this field, Noor [26, 27] considered and studied a new class of quasi-variational inequalities, which are called the extended general nonconvex quasi-variational inequalities involving three operators. This class is quite general and a unifying one. It has been shown that the extended general nonconvex quasi-variational inequalities include several known and new classes of variational inequalities as special cases. Using the projection method, we show that the extended general nonconvex quasi-variational inequalities are equivalent to the fixed point problem. This alternative equivalent formulation is used to study the existence of a solution of the extended general quasi-variational inequalities and this is the main of motivation of this paper. For recent applications, formulations, and numerical methods using the neural network technique, see Liu and Cao [7] and Liu and Yang [8]. One can easily show that the different systems of variational inequalities are special cases of the extended general quasi-variational inequalities. Results proved in this paper may stimulate and inspire the readers to discover new and innovative applications of the extended general quasi-variational inequalities in various fields of pure and applied sciences.

31.2 Preliminaries and Basic Results

Let H be a real Hilbert space whose inner product and norm are denoted by $\langle \cdot, \cdot \rangle$ and $\| \cdot \|$, respectively. Let $K(u)$ be a nonempty, closed and convex-valued set in H.

First of all, we recall the following well-known concepts from nonlinear convex analysis and nonsmooth analysis [2, 34].

Definition 31.1 The proximal normal cone of K at $u \in H$ is given by

$$N_K^P(u) := \{\xi \in H : u \in P_K[u + \alpha\xi]\},$$

where $\alpha > 0$ is a constant and

$$P_K[u] = \{u^* \in K : d_K(u) = \|u - u^*\|\}.$$

Here $d_K(\cdot)$ is the usual distance function to the subset K, that is,

$$d_K(u) = \inf_{v \in K} \|v - u\|.$$

The proximal normal cone $N_K^P(u)$ has the following characterization.

Lemma 31.1 *Let K be a nonempty, closed and convex subset in H. Then $\zeta \in N_K^P(u)$, if and only if there exists a constant $\alpha > 0$ such that*

$$\langle \zeta, v - u \rangle \le \alpha \|v - u\|^2, \quad \forall v \in K.$$

Poliquin et al. [34] and Clarke et al. [2] have introduced and studied a new class of nonconvex sets, which are called uniformly prox-regular sets. This class of uniformly prox-regular sets has played an important part in many nonconvex applications such as optimization, dynamic systems, and differential inclusions.

Definition 31.2 For a given $r \in (0, \infty]$, a subset K_r is said to be normalized uniformly r-prox-regular if and only if every nonzero proximal normal to K_r can be realized by an r-ball, that is, $\forall u \in K_r$ and $0 \neq \xi \in N^P_{K_r}(u)$, one has

$$\langle (\xi)/\|\xi\|, v - u\rangle \leq (1/2r)\|v - u\|^2, \quad \forall v \in K_r.$$

It is clear that the class of normalized uniformly prox-regular sets is sufficiently large to include the class of convex sets, p-convex sets, $C^{1,1}$ submanifolds (possibly with boundary) of H, the images under a $C^{1,1}$ diffeomorphism of convex sets, and many other nonconvex sets; see [2, 30, 34]. Obviously, for $r = \infty$, the uniform prox-regularity of K_r is equivalent to the convexity of K. This class of uniformly prox-regular sets has played an important part in many nonconvex applications such as optimization, dynamic systems, and differential inclusions. It is known that if K_r is a uniformly prox-regular set, then the proximal normal cone $N^P_{K_r}(u)$ is closed as a set-valued mapping.

We now recall the well-known proposition which summarizes some important properties of the uniformly prox-regular sets K_r.

Lemma 31.2 *Let K be a nonempty closed subset of H, $r \in]0, \infty]$ and set $K_r = \{u \in H : d_K(u) < r\}$. If K_r is uniformly prox-regular, then*

(i) $\forall u \in K_r$, $P_{K_r}(u) \neq \emptyset$;
(ii) $\forall r' \in]0, r[$, P_{K_r} *is Lipschitz continuous with constant* $\frac{r}{r-r'}$ *on* $K_{r'}$.

It is well known [2, 34] that the union of two disjoint intervals $[a, b]$ and $[c, d]$ is a prox-regular set with $r = \frac{c-b}{2}$. We also consider the following simple examples to give an idea of the importance of the nonconvex sets.

Example 31.1 ([30]) Let $u = (x, y)$ and $v = (t, z)$ belong to the real Euclidean plane. Let $K = \{t^2 + (z-2)^2 \geq 4,\ -2 \leq t \leq 2,\ z \geq -2\}$ be a subset of the Euclidean plane. Then one can easily show that the set K is a prox-regular set K_r.

Example 31.2 ([30]) Let $u = (x, y) \in R^2$, $v = (t, z) \in R^2$ and let $Tu = (-x, 1 - y)$. Let the set K be the union of 2 disjoint squares, say A and B, respectively having the vertices at the points $(0, 1), (2, 1), (2, 3), (0, 3)$ and at the points $(4, 1), (5, 2), (4, 3), (3, 2)$. The fact that K can be written in the form:

$$\{(t, z) \in R^2 : \max\{|t - 1|, |z - 2|\} \leq 1\} \cup \{|t - 4| + |z - 2| \leq 1\}$$

shows that it is a prox-regular set in R^2.

For given three nonlinear operators $T, g, h : H \to H$ and a point-to-set mapping $K_r : u \longrightarrow K_r(u)$ which associates a closed uniformly prox-regular set $K_r(u)$ of H with any element of H, we consider the problem of finding $u \in H : h(u) \in K_r(u)$ such that

$$\langle Tu, g(v) - h(u) \rangle \geq 0, \quad \forall v \in H : g(v) \in K_r(u). \tag{31.1}$$

Inequality of type (31.1) is called the *extended general nonconvex quasi-variational inequality involving three operators.*

We would like to emphasize that extended general quasi-variational inequality (31.1) is equivalent to finding $u \in H : h(u) \in K_r(u)$ such that

$$\langle \rho Tu + h(u) - g(u), g(v) - h(u) \rangle \geq 0, \quad \forall v \in H : g(v) \in K_r(u), \tag{31.2}$$

where $\rho > 0$ is a constant. This equivalent formulation is also useful from the applications point of view. We use this equivalent formulation to study the existence of a solution of the extended general quasi-variational inequalities (31.1).

We now list some special cases of the extended general quasi-variational inequalities (31.1).

I. If $K_r(u) \equiv K(u)$, which is a convex set, then problem (31.1) is equivalent to finding $u \in H, h(u) \in K(u)$ such that

$$\langle Tu, g(v) - h(u) \rangle \geq 0, \quad \forall v \in H : g(v) \in K(u), \tag{31.3}$$

which is known as the extended general quasi-variational inequality, introduced and studied by Noor [19–23]. For the formulation, iterative methods, and their applications in engineering and other discipline, see [7, 8, 19–23, 28].

II. If $g = h$, then problem (31.1) is equivalent to finding $u \in H : g(u) \in K_r(u)$ such that

$$\langle Tu, g(v) - g(u) \rangle \geq 0, \quad \forall v \in H : g(v) \in K_r(u), \tag{31.4}$$

which is known as a general quasi-variational inequality and appears to be a new one. If $g = h$ and $K_r(u) \equiv K(u)$, then problem (31.3) is called the general quasi-variational inequality involving two operators. Furthermore, if $K_r(u) \equiv K$, which is a convex set, then problem (31.3) is known as the general variational inequality, which was introduced and studied by Noor [12] in 1988. It turned out that odd order and nonsymmetric obstacle, free, moving, unilateral, and equilibrium problems arising in various branches of pure and applied sciences can be studied via general variational inequality [7, 8, 11–30].

III. For $g \equiv I$, the identity operator, the extended general quasi-variational inequality (31.1) collapses to: Find $u \in H : h(u) \in K_r(u)$ such that

$$\langle Tu, v - h(u) \rangle \geq 0, \quad \forall v \in K_r(u), \tag{31.5}$$

which is also called the general nonconvex quasi-variational inequality, see Noor et al. [30].

IV. For $h = I$, the identity operator, problem (31.1) is equivalent to finding $u \in K_r(u)$ such that

$$\langle Tu, g(v) - u\rangle \geq 0, \quad \forall v \in H : g(v) \in K_r(u), \tag{31.6}$$

which is also called the general nonconvex quasi-variational inequality.

V. For $g = h = I$, the identity operator, the extended general variational inequality (31.1) is equivalent to finding $u \in K(u)$ such that

$$\langle Tu, v - u\rangle \geq 0, \quad \forall v \in K(u), \tag{31.7}$$

which is known as the classical quasi-variational inequality and was introduced by Benssousan and Lions [1].

From the above discussion, it is clear that the extended general nonconvex quasi-variational inequality (31.1) is the most general and includes several new and previously known classes of variational inequalities as special cases. These variational inequalities have important applications in mathematical programming and engineering sciences. For the recent applications, numerical methods, sensitivity analysis, dynamical systems, and formulation of quasi-variational inequalities and related fields, see [1–32, 34], and the references therein.

If $K_r(u)$ is a nonconvex (uniformly prox-regular) set, then problem (31.1) is equivalent to finding $u \in H : h(u) \in K_r(u)$ such that

$$0 \in \rho Tu + h(u) - g(u) + \rho N^P_{K_r(u)}(u), \tag{31.8}$$

where $N^P_{K_r}(u)$ denotes the normal cone of K_r at u in the sense of nonconvex analysis and $\rho > 0$ is a constant. Problem (31.8) is called the extended general nonconvex quasi-variational inclusion problem associated with extended general nonconvex quasi-variational inequality (31.1). This implies that the extended general nonconvex quasi-variational inequality (31.1) is equivalent to finding a zero of the sum of two monotone operators (31.8). This equivalent formulation plays a crucial and basic part in this paper. We would like to point out that this equivalent formulation allows us to use the projection operator technique for solving the extended general nonconvex quasi-variational inequality (31.1).

Definition 31.3 An operator $T : H \to H$ is said to be

(i) *strongly monotone* if there exists a constant $\alpha > 0$ such that

$$\langle Tu - Tv, u - v\rangle \geq \alpha\|u - v\|^2, \quad \forall u, v \in H.$$

(ii) *Lipschitz continuous* if there exists a constant $\beta > 0$ such that

$$\|Tu - Tv\| \leq \beta\|u - v\|, \quad \forall u, v \in H.$$

If T verifies (i) and (ii), then it follows that $\alpha \leq \beta$.

We would like to point out that the implicit projection operator $P_{K(u)}$ is not nonexpansive. We shall assume that the implicit projection operator $P_{K(u)}$ satisfies the

Lipschitz type continuity condition, which plays an important and fundamental role in the existence theory and in developing numerical methods for solving extended general quasi-variational inequalities.

Assumption 31.1 For all $u, v, w \in H$, the implicit projection operator $P_{K_r(u)}$ satisfies the condition

$$\| P_{K_r(u)}w - P_{K_r(v)}w \| \leq \nu \|u - v\|, \tag{31.9}$$

where $\nu > 0$ is a positive constant.

In many important applications [1–6], the nonconvex convex-valued set $K_r(u)$ can be written as

$$K_r(u) = m(u) + K_r, \tag{31.10}$$

where $m(u)$ is a point–point mapping and K_r is aprox-regular set. In this case, we have

$$P_{K_r(u)}w = P_{m(u)+K_r}(w) = m(u) + P_{K_r}[w - m(u)], \quad \forall u,\ v \in H. \tag{31.11}$$

We note that if $K_r(u)$ is defined by (31.10) and $m(u)$ is a Lipschitz continuous mapping with constant $\gamma > 0$, then, using (31.11), we have

$$\begin{aligned} \|P_{K_r(u)}w - P_{K_r(v)}w\| &= \big\|m(u) - m(v) + P_{K_r}\big[w - m(u)\big] - P_{K_r}[w - m(v)\big\| \\ &\leq \big\|m(u) - m(v)\big\| + \delta\|u - v\| \\ &\leq \{\gamma + \delta\}\|u - v\|, \quad \forall u, v, w \in H, \end{aligned}$$

which shows that Assumption 31.1 holds with $\nu = \{\gamma + \delta\}$.

31.3 Main Results

In this section, we consider the existence of a solution of the extended general quasi-variational inequality (31.1). For this purpose, we recall the following result, which is due to Noor [26, 27].

Lemma 31.3 *The function $u \in H : h(u) \in K_r(u)$ is a solution of the extended general quasi-variational inequality* (31.2) *if and only if $u \in H : h(u) \in K_r(u)$ satisfies the relation*

$$h(u) = P_{K_r(u)}\big[g(u) - \rho Tu\big], \tag{31.12}$$

where $P_{K(u)}$ is the projection operator and $\rho > 0$ is a constant.

Lemma 31.3 implies that the extended general nonconvex quasi-variational inequality (31.1) is equivalent to the implicit fixed point problem (31.12). This alternative equivalent formulation is very useful from the numerical and theoretical points of view. We use this equivalent formulation to study the existence of a solution of (31.1), which is the main motivation of our next result.

We can write (31.12) in the following form:

$$F(u) = u - h(u) + P_{K_r(u)}\big[g(u) - \rho T u\big]. \tag{31.13}$$

In order to prove the existence of a solution of (31.1), it is enough to show that the mapping defined by (31.13) has a fixed point.

Theorem 31.1 *Let the operators* $T, g, h : H \longrightarrow H$ *be strongly monotone with constants* $\alpha > 0$, $\mu_1 > 0$, $\mu_2 > 0$ *and Lipschitz continuous with constants* $\beta > 0, \sigma_1 > 0$, $\sigma_2 > 0$, *respectively. If Assumption* 31.1 *holds and*

$$\left|\rho - \frac{\alpha}{\beta^2}\right| < \frac{\sqrt{\alpha^2 - \beta^2(1-\mu)}}{\beta^2}, \quad \alpha > \beta\sqrt{(1-\mu)} \tag{31.14}$$

$$\mu = \left(\frac{1-(k_1+\nu+\delta k)}{\delta}\right)^2, \quad k_1 + \nu + \delta k < 1$$

$$k = \sqrt{1 - 2\mu_1 + \sigma_1^2} \tag{31.15}$$

$$k_1 = \sqrt{1 - 2\mu_2 + \sigma_2^2}, \tag{31.16}$$

then there exists a solution of the extended general nonconvex quasi-variational inequality (31.1).

Proof Let $u \in$ be a solution of (31.1). Then, from Lemma 31.3, we see the problem of finding the solution of (31.1) is equivalent to finding the fixed point of the mapping $F(u)$, defined by (31.13). For $u_1 \neq u_2 \in H$, and using Assumption 31.1, we have

$$\begin{aligned}
\|F(u_1) - F(u_2)\| &= \|u_1 - u_2 - (h(u_1) - h(u_2))\| \\
&\quad + \|P_{K_r(u_1)}[g(u_1) - \rho T(u_1)] - P_{K_r(u_2)}[g(u_2) - \rho T(u_2)]\| \\
&\le \|u_1 - u_2 - (h(u_1) - h(u_2))\| \\
&\quad + \|P_{K(_r u_1)}[g(u_1) - \rho T(u_1)] - P_{K_r(u_2)}[g(u_1) - \rho T(u_1)]\| \\
&\quad + \|P_{K_r(u_2)}[g(u_1) - \rho T(u_1)] - P_{K_r(u_2)}[g(u_2) - \rho T(u_2)]\| \\
&\le \|u_1 - u_2 - (h(u_1) - h(u_2))\| + \nu\|u_1 - u_2\| \\
&\quad + \delta\|g(u_1) - g(u_2) - \rho(Tu_1 - Tu_2)\| \\
&= \|u_1 - u_2 - (h(u_1) - h(u_2))\| + \nu\|u_1 - u_2\| \\
&\quad + \delta\|u_1 - u_2 - (g(u_1) - g(u_2))\| \\
&\quad + \delta\|u_1 - u_2 - (h(u_1) - h(u_2))\|.
\end{aligned} \tag{31.17}$$

Since the operator T is strongly monotone with constant $\alpha > 0$ and Lipschitz continuous with constant $\beta > 0$, it follows that

$$\begin{aligned}\left\|u_1-u_2-\rho(Tu_1-Tu_2)\right\|^2 &\le \|u_1-u_2\|^2-2\rho\langle Tu_1-Tu_2,u_1-u_2\rangle\\ &\quad+\rho^2\|Tu_1-Tu_2\|^2\\ &\le (1-2\rho\alpha+\rho^2\beta^2)\|u_1-u_2\|^2. \end{aligned}\tag{31.18}$$

In a similar way, we have

$$\left\|u_n-u-\big(g(u_1)-g(u_2)\big)\right\|^2 \le \big(1-2\mu_1+\sigma_1^2\big)\|u_1-u_2\|^2, \tag{31.19}$$

$$\left\|u_n-u-\big(h(u_1)-h(u_2)\big)\right\|^2 \le \big(1-2\mu_2+\sigma_2^2\big)\|u_1-u_2\|^2, \tag{31.20}$$

using the strongly monotonicity constants of $\mu_1 > 0$, $\mu_2 > 0$ and Lipschitz continuity constants $\sigma_1 > 0$, $\sigma_2 > 0$ of the operators g and h, respectively.

From (31.15), (31.17)–(31.20), we have

$$\begin{aligned}\left\|F(u_1)-F(u_2)\right\| &\le \left\{\nu+\delta\sqrt{1-2\mu_1+\sigma_1^2}+\sqrt{1-2\mu_2+\sigma_2^2}\right\}\|u_1-u_2\|\\ &\quad+\sqrt{1-2\alpha\rho+\beta^2\rho^2}\|u_1-u_2\|\\ &=\big(k_1+\nu+\delta k+t(\rho)\big)\|u_1-u\|=\theta\|u_1-u\|,\end{aligned}$$

where

$$t(\rho)=\sqrt{1-2\alpha\rho+\rho^2\beta^2},\qquad \theta=k_1+\nu+\delta k+t(\rho).$$

From (31.14), we see that $\theta < 1$. Thus it follows that the mapping $F(u)$ defined by (31.13) is a contraction mapping, and consequently, it has a fixed point which belongs to $K_r(u)$ satisfying the extended general nonconvex quasi-variational inequality (31.1), the required result. □

31.4 Conclusions

In this paper, we have studied the existence of a solution of the extended general nonconvex quasi-variational inequalities involving three different operators using the fixed point theory in conjunction with projection operators. Several special cases were also discussed. It has been shown that this class of extended general nonconvex quasi-variational inequalities has important and significant applications in various fields of pure and applied sciences. This field offers great opportunities for further research. It is expected that the interplay among all these areas will certainly lead to some innovative, novel, and significant applications in engineering, mathematical and physical sciences.

Acknowledgements The authors would like to thank Dr. S.M. Junaid Zaidi, Rector, CIIT, for providing excellent research facilities. This research is supported by the Visiting Professor Program of King Saud University, Riyadh, Saudi Arabia and Research Grant No. KSU.VPP.108.

References

1. Benssousan, A., Lions, J.L.: Applications des Inequations Variationelles en Control et en Stochastiques. Dunod, Paris (1978)
2. Clarke, F.H., Ledyaev, Y.S., Wolenski, P.R.: Nonsmooth Analysis and Control Theory. Springer, Berlin (1998)
3. Giannessi, F., Maugeri, A.: Variational Inequalities and Network Equilibrium Problems. Plenum, New York (1995)
4. Giannessi, F., Maugeri, A., Pardalos, P.M.: Equilibrium Problems: Nonsmooth Optimization and Variational Inequality Models. Kluwer Academics, Dordrecht (2001)
5. Glowinski, R., Lions, J.L., Trémolières, R.: Numerical Analysis of Variational Inequalities. North-Holland, Amsterdam (1981)
6. Lions, J.L., Stampacchia, G.: Variational inequalities. Commun. Pure Appl. Math. **20**, 493–512 (1967)
7. Liu, Q., Cao, J.: A recurrent neural network based on projection operator for extended general variational inequalities. IEEE Trans. Syst. Man Cybern., Part B, Cybern. **40**, 928–938 (2010)
8. Liu, Q., Yang, Y.: Global exponential system of projection neural networks for system of generalized variational inequalities and related nonlinear minimax problems. Neurocomputing **73**, 2069–2076 (2010)
9. Kravchuk, A.S., Neittaanmaki, P.J.: Variational and Quasi Variational Inequalities in Mechanics. Springer, Dordrecht (2007)
10. Noor, M.A.: On variational inequalities. Ph.D. Thesis, Brunel University, London, UK (1975)
11. Noor, M.A.: An iterative schemes for a class of quasi variational inequalities. J. Math. Anal. Appl. **110**, 463–468 (1985)
12. Noor, M.A.: General variational inequalities. Appl. Math. Lett. **1**, 119–121 (1988)
13. Noor, M.A.: Quasi variational inequalities. Appl. Math. Lett. **1**, 367–370 (1988)
14. Noor, M.A.: Sensitivity analysis for quasi variational inequalities. J. Optim. Theory Appl. **95**, 399–407 (1997)
15. Noor, M.A.: Generalized multivalued quasi variational inequalities (II). Comput. Math. Appl. **35**, 63–78 (1998)
16. Noor, M.A.: New approximation schemes for general variational inequalities. J. Math. Anal. Appl. **251**, 217–229 (2000)
17. Noor, M.A.: Some developments in general variational inequalities. Appl. Math. Comput. **152**, 199–277 (2004)
18. Noor, M.A.: Differentiable nonconvex functions and general variational inequalities. Appl. Math. Comput. **199**, 623–630 (2008)
19. Noor, M.A.: Auxiliary principle technique for extended general variational inequalities. Banach. J. Math. Anal. **2**, 33–39 (2008)
20. Noor, M.A.: Extended general variational inequalities. Appl. Math. Lett. **22**, 182–185 (2009)
21. Noor, M.A.: Sensitivity analysis for general extended variational inequalities. Appl. Math. E-Notes **9**, 17–26 (2009)
22. Noor, M.A.: Some iterative methods for extended general variational inequalities. Albanian J. Math. **2**, 265–275 (2008)
23. Noor, M.A.: Projection iterative methods for extended general variational inequalities. J. Appl. Math. Comput. **32**, 83–95 (2010)
24. Noor, M.A.: On a class of general variational inequalities. J. Adv. Math. Stud. **1**, 31–42 (2008)
25. Noor, M.A.: On merit functions for quasi variational inequalities. J. Math. Inequal. **1**, 259–268 (2007)
26. Noor, M.A.: Some new classes of quasi-variational inequalities. Int. J. Modern Phys. B (2010)
27. Noor, M.A.: Sensitivity analysis of some quasi-variational inequalities. Int. J. Modern Phys. B (2010)
28. Noor, M.A.: Solvability of extended general mixed variational inequalities. Albanian J. Math. **4**, 13–17 (2010)

29. Noor, M.A.: Extended general nonconvex quasi-variational inequalities. Nonlinear Anal. Forum **15**, 33–39 (2010)
30. Noor, M.A.: On an implicit method for nonconvex variational inequalities. J. Optim. Theory Appl. **147** (2010)
31. Noor, M.A., Noor, K.I.: On general quasi-variational inequalities, J. King Saud Univ. Sci. (2010)
32. Noor, M.A., Noor, K.I., Al-Said, E.: Iterative methods for solving general quasi-variational inequalities. Optim. Lett. **4**, 513–530 (2010)
33. Noor, M.A., Noor, K.I., Rassias, Th.M.: Some aspects of variational inequalities. J. Comput. Appl. Math. **47**, 285–312 (1993)
34. Poliquin, R.A., Rockafellar, R.T., Thibault, L.: Local differentiability of distance functions. Trans. Am. Math. Soc. **352**, 5231–5249 (2000)
35. Stampacchia, G.: Formes bilineaires coercitives sur les ensembles convexes. C. R. Acad. Sci., Paris **258**, 4413–4416 (1964)

Chapter 32
Iterative Projection Methods for Solving Systems of General Nonconvex Variational Inequalities

Muhammad Aslam Noor, Khalida Inayat Noor, and Eisa Al-Said

Abstract In this paper, we introduce and consider a new system of general nonconvex variational inequalities involving four different operators. We establish the equivalence between the system of general nonconvex variational inequalities and the fixed points problem using the projection technique. This alternative equivalent formulation is used to suggest and analyze some new explicit iterative methods for this system of nonconvex variational inequalities. We also study the convergence analysis of the new iterative method under certain mild conditions. Several special cases are also considered. Our results can be viewed as a refinement and improvement of the previously known results for variational inequalities.

Key words Iterative algorithms · System of nonconvex variational inequalities with different mappings · Relaxed cocoercive mappings · Lipschitz continuity · Convergence

Mathematics Subject Classification 49J40 · 90C33

32.1 Introduction

In recent years, much attention has been given to a system of variational inequalities, which can be viewed as a general and useful extension of variational inequalities. It is well known that the variational inequalities were introduced by Stampacchia [30]. The techniques and ideas of the system of variational inequalities are being applied

---

Dedicated to Professor Themistocles M. Rassias on the occasion of his 60th birthday.

M.A. Noor (✉) · K.I. Noor · E. Al-Said
Mathematics Department, COMSATS Institute of Information Technology, Park Road, Islamabad, Pakistan
e-mail: noormaslam@hotmail.com

K.I. Noor
e-mail: khalidanoor@hotmail.com

E. Al-Said
e-mail: eisasaid@ksu.edu.sa

P.M. Pardalos et al. (eds.), *Nonlinear Analysis*, Springer Optimization and Its Applications 68, In Honor of Themistocles M. Rassias on the Occasion of his 60th Birthday, DOI 10.1007/978-1-4614-3498-6_32,

in a variety of diverse areas of sciences and proved to be productive and innovative. Using the projection technique, one usually establishes the equivalence between the system of variational inequalities and the fixed point problem. This alternative equivalent formulation has been used to suggest and analyze some iterative methods for solving the system of variational inequalities; see [2, 6, 20–22, 27], and the references therein.

We would like to emphasize that all the results regarding the iterative methods for solving the system of variational inequalities have been considered in the convexity setting. This is because all the techniques are based on the properties of the projection operator over convex sets, which may not hold for nonconvex sets. Noor [16–23] has shown that the concept of a projection technique can be extended to variational inequalities which are considered on the uniformly prox-regular sets. It is well known that the uniformly prox-regular sets are nonconvex sets.

Inspired and motivated by the ongoing research in this area, we introduce and consider a system of general nonconvex variational inequalities involving four different operators. This class of systems includes the system of nonconvex variational inequalities [7, 20] and the classical variational inequalities as special cases. Using essentially the technique of Noor [16–23] in conjunction with the projection operator method, we establish the equivalence between the system of general nonconvex variational inequalities and the fixed-point problem, which is Lemma 32.3. This result can be viewed as the extension of a results of Noor [16–23]. We use this alternative equivalent formulation to suggest and analyze some iterative methods (Algorithm 32.1–Algorithm 32.3) for solving the system of general nonconvex variational inequalities. We also prove the convergence of the proposed iterative methods under suitable conditions, which is the main motivation of Theorem 32.1. Since the new system of general nonconvex variational inequalities includes the system of nonconvex variational inequalities, studied by Moudafi [7] and Noor [20–23] and related optimization problems as special cases, results proved in this paper continue to hold for these problems. For recent generalizations and extensions of these system of general nonconvex variational inclusions/inequalities, see Noor et al. [13–27], and the references therein.

32.2 Formulation and Basic Results

Let H be a real Hilbert space whose inner product and norm are denoted by $\langle\cdot,\cdot\rangle$ and $\|\cdot\|$, respectively. Let K be a nonempty closed and convex set in H. First of all, we recall the following well-known concepts from nonlinear convex analysis and nonsmooth analysis [3, 29].

Definition 32.1 The proximal normal cone of K at $u \in H$ is given by

$$N_K^P(u) := \{\xi \in H : u \in P_K[u + \alpha\xi]\},$$

where $\alpha > 0$ is a constant and

$$P_K[u] = \{u^* \in K : d_K(u) = \|u - u^*\|\}.$$

Here $d_K(\cdot)$ is the usual distance function to the subset K, that is,

$$d_K(u) = \inf_{v \in K} \|v - u\|.$$

The proximal normal cone $N_K^P(u)$ has the following characterization.

Lemma 32.1 *Let K be a nonempty, closed and convex subset in H. Then $\zeta \in N_K^P(u)$, if and only if there exists a constant $\alpha > 0$ such that*

$$\langle \zeta, v - u\rangle \leq \alpha \|v - u\|^2, \quad \forall v \in K.$$

Poliquin et al. [29] and Clarke et al. [3] have introduced and studied a new class of nonconvex sets, which are called uniformly prox-regular sets. This class of uniformly prox-regular sets has played an important part in many nonconvex applications such as optimization, dynamic systems, and differential inclusions.

Definition 32.2 For a given $r \in (0, \infty]$, a subset K_r is said to be normalized uniformly r-prox-regular if and only if every nonzero proximal normal cone to K_r can be realized by an r-ball, that is, $\forall u \in K_r$ and $0 \neq \xi \in N_{K_r}^P$, one has

$$\langle (\xi)/\|\xi\|, v - u\rangle \leq (1/2r)\|v - u\|^2, \quad \forall v \in K_r.$$

It is clear that the class of normalized uniformly prox-regular sets is sufficiently large to include the class of convex sets, p-convex sets, $C^{1,1}$ submanifolds (possibly with boundary) of H, the images under a $C^{1,1}$ diffeomorphism of convex sets, and many other nonconvex sets; see [3, 29]. It is clear that if $r = \infty$, then uniform prox-regularity of K_r is equivalent to the convexity of K. It is known that if K_r is a uniformly prox-regular set, then the proximal normal cone $N_{K_r}^P$ is closed as a set-valued mapping.

We now recall the well known proposition which summarizes some important properties of the uniformly prox-regular sets.

Lemma 32.2 *Let K be a nonempty closed subset of H, $r \in (0, \infty]$ and set $K_r = \{u \in H : d_K(u) < r\}$. If K_r is uniformly prox-regular, then*

(i) $\forall u \in K_r$, $P_{K_r} \neq \emptyset$;
(ii) $\forall r' \in (0, r)$, P_{K_r} *is Lipschitz continuous with constant* $\frac{r}{r-r'}$ *on* $K_{r'}$.
(iii) *The proximal normal cone is closed as a set-valued mapping.*

For given nonlinear operators T_1, T_2, g, h, we consider the problem of finding $x^*, y^* \in K_r$ such that

$$\begin{cases} \langle \rho T_1(y^*) + g(x^*) - g(y^*), g(x) - g(x^*) \rangle \geq 0, \\ \quad \forall x \in H : g(x) \in K_r, \ \rho > 0, \\ \langle \eta T_2(x^*) + h(y^*) - h(x^*), h(x) - h(y^*) \rangle \geq 0, \\ \quad \forall x \in H : h(x) \in K_r, \ \eta > 0, \end{cases} \tag{32.1}$$

which is called the system of general nonconvex variational inequalities.

We now discuss some special cases of the new system of general nonconvex variational inequalities.

I. If $T_1 = T_2 = T$, then the system of general nonconvex variational inequalities (32.1) is equivalent to finding $x^*, y^* \in K_r$ such that

$$\begin{cases} \langle \rho T y^* + g(x^*) - g(y^*), g(x) - g(x^*) \rangle \geq 0, & \forall x \in H : g(x) \in K_r, \\ \langle \eta T x^* + h(y^*) - h(x^*), h(x) - h(x^*) \rangle \geq 0, & \forall x \in H : h(x) \in K_r. \end{cases} \tag{32.2}$$

This system of general nonconvex variational inequalities has been studied by Noor [20].

II. If $\rho = 0, \ x^* = y^*, \ g = h$, then system of general nonconvex variational inequalities (32.1) reduces to finding $x^* \in K_r$ such that

$$\langle Tx^*, g(x) - g(x^*) \rangle \geq 0, \quad \forall x \in H : g(x) \in K_r, \tag{32.3}$$

which is known as the general nonconvex variational inequality, introduced and studied by Noor [16, 17] in recent years.

III. If $K_r \equiv K$, a convex set in H, and $g = I$, the identity operator, then the general noncovex variational inequality is equivalent to finding $x^* \in K$ such that

$$\langle Tx^*, x - x^* \rangle \geq 0, \quad \forall x \in K, \tag{32.4}$$

which is known as the classical variational inequality introduced and studied by Stampacchia [30] in 1964. For appropriate and suitable choice of the operators and the spaces, one can obtain several systems of variational inequalities as special cases of the system of general nonconvex variational inequalities (32.1). This shows that the system of general nonconvex variational inequalities (32.1) is more general and includes several classes of variational inequalities and related optimization problems as special cases. For the recent applications, numerical methods, and formulations of variational inequalities, see [1–30], and the references therein.

32.3 Projection Iterative Methods

In this section, we suggest some explicit iterative algorithms for solving the system of general nonconvex variational inequalities (32.1). First of all, we establish the equivalence between the system of nonconvex variational inequalities and the fixed point problem, which is the main motivation of our next result.

Lemma 32.3 *$x, y \in H : g(x), h(y) \in K_r$ is a solution of (32.1) if and only if $x, y \in H : g(x), h(y) \in K_r$ satisfy the relation*

$$g(x) = P_{K_r}\bigl[g(y) - \rho T_1(y)\bigr], \tag{32.5}$$

$$h(y) = P_{K_r}\bigl[h(x) - \eta T_2(x)\bigr], \tag{32.6}$$

where $\rho > 0$ and $\eta > 0$ are constants.

Proof Let $x, y \in H : g(x), h(y) \in K_r$ be a solution of (32.1) and (32.2). Then, we have

$$0 \in \rho T_1(y) + g(x) - g(y) + N_{K_r}^P\bigl(g(x)\bigr) = \bigl(I + N_{K_r}^P\bigr)\bigl(g(x)\bigr) - \bigl(g(y) - \rho T_1(y)\bigr),$$
$$0 \in \eta T_2(x) + h(y) - h(x) + N_{K_r}^P\bigl(h(y)\bigr) = \bigl(I + N_{K_r}^P\bigr)\bigl(h(y)\bigr) - \bigl(h(x) - \eta T_2(x)\bigr),$$

which implies that

$$g(x) = P_{K_r}\bigl[g(y) - \rho T_1(y)\bigr],$$
$$h(y) = P_{K_r}\bigl[h(x) - \eta T_2(x)\bigr],$$

where we have used the fact that $P_{K_r} = (I + N_{K_r}^P)^{-1}$. □

Lemma 32.3 implies that the system of general nonconvex variational inequalities (32.1) is equivalent to the fixed point problem. This alternative equivalent formulation is used to suggest and analyze a number of iterative methods for solving systems of nonconvex variational inequalities and related optimization problems.

Using Lemma 32.3, we can easily show that finding the solution $x^*, y^* \in H$ of (32.1) is equivalent to finding $(x^*, y^*) \in H$ such that

$$x^* = x^* - g\bigl(x^*\bigr) + P_{K_r}\bigl[g\bigl(y^*\bigr) - \rho T_1\bigl(y^*\bigr)\bigr], \tag{32.7}$$

$$y^* = y^* - h\bigl(y^*\bigr) + P_{K_r}\bigl[h\bigl(x^*\bigr) - \eta T_2\bigl(x^*\bigr)\bigr]. \tag{32.8}$$

We use this alternative equivalent formulation to suggest the following iterative method for solving the system of nonconvex variational inequalities (32.1).

Algorithm 32.1 For an arbitrarily chosen initial point $y_0 \in K_r$, compute the sequences $\{x_n\}$ and $\{y_n\}$ by

$$x_{n+1} = x_{n+1} - g(x_{n+1}) + P_{K_r}\bigl[g(y_n) - \rho T_1(y_n)\bigr], \tag{32.9}$$

$$y_{n+1} = y_{n+1} - h(y_{n+1}) + P_{K_r}\big[h(x_{n+1}) - \eta T_2(x_{n+1})\big]. \tag{32.10}$$

If $T_1 = T_2 = T$, then Algorithm 32.1 reduces to the following.

Algorithm 32.2 For arbitrarily chosen initial points $x_0, y_0 \in K_r$, compute the sequences $\{x_n\}$ and $\{y_n\}$ by

$$x_{n+1} = x_{n+1} - g(x_{n+1}) + P_{K_r}\big[g(y_n) - \rho T(y_n)\big],$$
$$y_{n+1} = y_{n+1} - h(y_{n+1}) + P_{K_r}\big[h(x_{n+1}) - \eta T(x_{n+1})\big],$$

where $a_n \in [0, 1]$ for all $n \geq 0$.

If $T_1 = T_2 = T$ and $g = h = I$, the identity operator, then Algorithm 32.1 reduces to the following.

Algorithm 32.3 For an arbitrarily chosen initial point $y_0 \in K_r$, compute the sequences $\{x_n\}$ and $\{y_n\}$ by

$$x_{n+1} = P_{K_r}\big[y_n - \rho T(y_n)\big],$$
$$y_{n+1} = P_{K_r}\big[x_{n+1} - \eta T(x_{n+1})\big].$$

We would like to emphasize that one can obtain a number of iterative methods for solving a system of (nonconvex) variational inequalities and related optimization problems for appropriate choice of the operators and spaces. This shows that Algorithm 32.1 is quite flexible and general.

Definition 32.3 A mapping $T : H \to H$ is called r-strongly monotone, if there exists a constant $r > 0$ such that

$$\langle Tx - Ty, x - y\rangle \geq r\|x - y\|^2, \quad \forall x, y \in H.$$

Definition 32.4 A mapping $T : H \to H$ is called relaxed γ-cocoercive if there exists a constant $\gamma > 0$ such that

$$\langle Tx - Ty, x - y\rangle \geq -\gamma\|Tx - Ty\|^2, \quad \forall x, y \in H.$$

Definition 32.5 A mapping $T : H \to H$ is called relaxed (γ, r)-cocoercive if there exist constants $\gamma > 0, r > 0$ such that

$$\langle Tx - Ty, x - y\rangle \geq -\gamma\|Tx - Ty\|^2 + r\|x - y\|^2, \quad \forall x, y \in H.$$

The class of relaxed (γ, r)-cocoercive mappings is more general than the class of strongly monotone mappings.

Definition 32.6 A mapping $T : H \to H$ is called μ-Lipschitzian if there exists a constant $\mu > 0$ such that

$$\|Tx - Ty\| \leq \mu\|x - y\|, \quad \forall x, y \in H.$$

32.4 Main Results

In this section, we consider the convergence criteria of Algorithm 32.1 under some suitable mild conditions, and this is the main motivation as well as the main result of this paper. In a similar way, one can study the convergence of other iterative methods for solving problems (32.1)–(32.4).

Theorem 32.1 *Let (x^*, y^*) be a solution of (32.1). Suppose $T_1(\cdot) : H \to H$ is relaxed (γ_1, r_1)-cocoercive and μ_1-Lipschitzian and $T_2(\cdot) : H \to H$ is relaxed (γ_2, r_2)-cocoercive and μ_2-Lipschitzian. Let g be a relaxed (γ_3, r_3)-cocoercive and μ_3-Lipschitz and h be a relaxed (γ_4, r_4)-cocoercive and μ_4-Lipschitzian. If*

$$\left|\rho - \frac{r_1 - \gamma_1\mu_1^2}{\mu_1^2}\right| < \frac{\sqrt{\delta^2(r_1 - \gamma_1\mu_1^2)^2 - \mu_1^2\eta^2}}{\delta\mu_1^2}, \quad \delta r_1 > \delta\gamma_1\mu_1^2 + \mu_1\eta, \tag{32.11}$$

$$\left|\eta - \frac{r_2 - \gamma_2\mu_2^2}{\mu_2^2}\right| < \frac{\sqrt{\delta(r_2 - \gamma_2\mu^2{}_2)^2 - \mu_2^2\xi^2}}{\delta\mu_2^2}, \quad \delta r_2 > \delta\gamma_2\mu_2^2 + \mu_2\xi, \tag{32.12}$$

where

$$\eta^2 = \delta^2 - \big(1 - (1+\delta)k\big)^2,$$

$$\xi^2 = \delta^2 - \big(1 - (1+\delta)k_1\big)^2,$$

$$k = \sqrt{1 - 2\big(r_3 - \gamma_3\mu_3^2\big) + \mu_3^2}, \tag{32.13}$$

$$k_1 = \sqrt{1 - 2\big(r_4 - \gamma_4\mu_4^2\big) + \mu_4^2}, \tag{32.14}$$

then for arbitrarily chosen initial points $x_0, y_0 \in H$, x_n and y_n obtained from Algorithm 32.1 converge strongly to x^* and y^*, respectively.

Proof To prove the result, we first evaluate $\|x_{n+1} - x^*\|$ for all $n \geq 0$. From (32.7), (32.9), and the Lipschitz continuity of the projection operator P_{K_r} with constant $\delta > 0$, we

$$\begin{aligned}
&\|x_{n+1} - x^*\| \\
&\quad = \big\|x_{n+1} - x^* - \big(g(x_{n+1}) - g(x^*)\big) + P_{K_r}\big[g(y_n) - \rho T_1(y_n)\big] \\
&\qquad - P_{K_r}\big[g(y^*) - \rho T_1(y^*)\big]\big\| \\
&\quad \leq \big\|x_{n+1} - x^* - \big(g(x_{n+1}) - g(x^*)\big)\big\| + \big\|P_{K_r}\big[g(y_n) - \rho T_1(y_n)\big] \\
&\qquad - P_{K_r}\big[g(y^*) - \rho T_1(y^*)\big]\big\|
\end{aligned}$$

$$\leq \|x_{n+1} - x^* - (g(x_{n+1}) - g(x^*))\| + \delta\|y_n - y^* - \rho[T_1(y_n) - T_1(y^*)]\| + \delta\|y_n - y^* - (g(y_n) - g(y^*))\|. \tag{32.15}$$

From the relaxed (γ_1, r_1)-cocoercivity and μ_1-Lipschitz continuity of $T_1(\cdot)$, we have

$$\begin{aligned}
&\|y_n - y^* - \rho[T_1(y_n) - T_1(y^*)]\|^2 \\
&= \|y_n - y^*\|^2 - 2\rho\langle T_1(y_n) - T_1(y^*), y_n - y^*\rangle + \rho^2\|T_1(y_n) - T_1(y^*)\|^2 \\
&\leq \|y_n - y^*\|^2 - 2\rho[-\gamma_1\|T_1(y_n) - T_1(y^*)\|^2 + r_1\|y_n - y^*\|^2] \\
&\quad + \rho^2\|T_1(y_n) - T_1(y^*)\|^2 \\
&\leq \|y_n - y^*\|^2 + 2\rho\gamma_1\mu_1^2\|y_n - y^*\|^2 - 2\rho r_1\|y_n - y^*\|^2 + \rho^2\mu_1^2\|y_n - y^*\|^2 \\
&= [1 + 2\rho\gamma_1\mu_1^2 - 2\rho r_1 + \rho^2\mu_1^2]\|y_n - y^*\|^2.
\end{aligned} \tag{32.16}$$

In a similar way, using the (γ_3, r_3)-cocoercivity and μ_3-Lipschitz continuity of the operator g, and (γ_4, r_4)-cocoercivity and μ_4-Lipschitz continuity of the operator h, we have

$$\|y_n - y^* - (g(y_n) - g(y^*))\| \leq k\|y_n - y^*\|, \tag{32.17}$$

$$\|x_n - x^* - (h(y_n) - h(x^*))\| \leq k_1\|x_n - x^*\| \tag{32.18}$$

where k is defined by (32.13) and k_1 is defined by (32.14).

Set

$$\theta_1 = \frac{\delta\{k + [1 + 2\rho\gamma_1\mu_1^2 - 2\rho r_1 + \rho^2\mu_1^2]^{1/2}\}}{1-k}.$$

It is clear from (32.11) that $\theta_1 < 1$.

From (32.15)–(32.17), we have

$$\|x_{n+1} - x^*\| \leq \theta_1\|y_n - y^*\|. \tag{32.19}$$

Similarly, from the relaxed (γ_2, r_2)-cocoercivity and μ_2-Lipschitz continuity of $T_2(\cdot)$, we obtain

$$\begin{aligned}
&\|x_{n+1} - x^* - \eta[T_2(x_{n+1}) - T_2(x^*)]\|^2 \\
&= \|x_{n+1} - x^*\|^2 - 2\eta\langle T_2(x_{n+1}) - T_2(x^*), x_{n+1} - x^*\rangle \\
&\quad + \eta^2\|T_2(x_{n+1}) - T_2(x^*)\|^2 \\
&\leq \|x_{n+1} - x^*\|^2 - 2\eta[-\gamma_2\|T_2(x_{n+1}) - T_2(x^*)\|^2 + r_2\|x_{n+1} - x^*\|^2] \\
&\quad + \eta^2\|T_2(x_{n+1}) - T_2(x^*,)\|^2
\end{aligned}$$

$$
\begin{aligned}
&= \|x_{n+1} - x^*\|^2 + 2\eta\gamma_2 \|T_2(x_{n+1}) - T_2(x^*)\|^2 - 2\eta r_2 \|x_{n+1} - x^*\|^2 \\
&\quad + \eta^2 \|T_2(x_{n+1}) - T_2(x^*)\|^2 \\
&\le \|x_{n+1} - x^*\|^2 + 2\eta\gamma_2\mu_2^2 \|x_{n+1} - x^*\|^2 - 2\eta r_2 \|x_{n+1} - x^*\|^2 \\
&\quad + \eta^2\mu_2^2 \|x_{n+1} - x^*\|^2 \\
&= \left[1 + 2\eta\gamma_2\mu_2^2 - 2\eta r_2 + \eta^2\mu_2^2\right] \|x_{n+1} - x^*\|^2. \qquad (32.20)
\end{aligned}
$$

Hence from (32.8), (32.10), (32.18), (32.20), and the Lipschitz continuity of the projection operator P_{K_r} with constant $\delta > 0$, we have

$$
\begin{aligned}
\|y_{n+1} - y^*\| &= \|y_{n+1} - y^* - (h(y_{n+1}) - h(y^*))\| \\
&\quad + \|P_{K_r}[h(x_{n+1}) - \eta T_2(x_{n+1}) - P_{K_r}[h(x^*) - \eta T_2(x^*)]\| \\
&\le \|y_{n+1} - y^* - (h(y_{n+1}) - h(y^*))\| \\
&\quad + \delta \|h(x_{n+1}) - h(x^*) - \eta(T_2(x_{n+1})] - T_2(x^*))\| \\
&\le \delta \|x_{n+1} - x^* - \eta(T_2(x_{n+1})] - T_2(x^*))\| \\
&\quad + \delta \|x_{n+1} - x^* - (h(x_{n+1}) - h(x^*))\| \\
&\quad + \|y_{n+1} - y^* - (h(y_{n+1}) - h(y^*))\|,
\end{aligned}
$$

from which, we have

$$
\|y_{n+1} - y^*\| \le \theta_2 \|x_{n+1} - x^*\|, \qquad (32.21)
$$

where

$$
\theta_2 = \frac{\delta\{k_1 + [1 + 2\eta\gamma_2\mu_2^2 - 2\eta r_2 + \eta^2\mu_2^2]^{1/2}\}}{1 - k_1}.
$$

From (32.12), it follows that $\theta_2 < 1$.

From (32.19) and (32.21), we obtain that

$$
\begin{aligned}
\|x_{n+1} - x^*\| &\le a_n\theta_1 \|y_n - y^*\| \\
&\le \theta_1 \cdot \theta_2 \|x_n - x^*\| \\
&= \theta_1 \cdot \theta_2 \|x_n - x^*\|.
\end{aligned}
$$

Since the constant $\theta_1\theta_2 < 1$, it follows that $\lim_{n\to\infty} \|x_n - x^*\| = 0$. Hence the result $\lim_{n\to\infty} \|y_n - y^*\| = 0$ follows from (32.21). This completes the proof. □

Remark 32.1 We would like to emphasize that the parameters must satisfy the four conditions in Theorem 32.1 to be compatible and this has been verified in [12, 13, 16, 18, 20–23, 25] for special cases of (nonconvex) variational inequalities and related optimization problems. These conditions have been used in the existence results and also in the studies of the convergence criteria of the iterative methods for solving several classes of (nonconvex) system of variational inequalities. There are several numerical methods for solving the general variational inequalities and related optimization problems in the setting of the classical convexity. To the best of our knowledge, there does not exist numerical methods for solving the nonconvex variational inequalities. We expect that the results proved in this paper will stimulate further research in this fast developing field. The interested researchers may discover some novel and significant applications in the pure and applied sciences. This is another aspect of the future research in this field.

Acknowledgements The authors would like to thank Dr. S.M. Junaid Zaidi, Rector, CIIT, for providing excellent research facilities. This research is supported by the Visiting Professor Program of King Saud University, Riyadh, Saudi Arabia and Research Grant No. KSU.VPP.108.

References

1. Bounkhel, M., Tadj, L., Hamdi, A.: Iterative schemes to solve nonconvex variational problems. J. Inequal. Pure Appl. Math. **4**, 1–14 (2003)
2. Chang, S.S., Lee, H.W.J., Chan, C.K.: Generalized system for relaxed cocoercive variational inequalities in Hilbert spaces. Appl. Math. Lett. **20**, 329–334 (2007)
3. Clarke, F.H., Ledyaev, Y.S., Wolenski, P.R.: Nonsmooth Analysis and Control Theory. Springer, Berlin (1998)
4. Giannessi, F., Maugeri, A., Pardalos, P.M.: Equilibrium Problems, Nonsmooth Optimization and Variational Inequalities Problems. Kluwer Academic, Dordrecht (2001)
5. Glowinski, R., Lions, J.L., Tremolieres, R.: Numerical Analysis of Variational Inequalities. North-Holland, Amsterdam (1981)
6. Huang, Z., Noor, M.A.: An explicit projection method for a system of nonlinear variational inequalities with different (γ, r)-cocoercive mappings. Appl. Math. Comput. **190**, 356–361 (2007)
7. Moudafi, A.: Projection methods for a system of nonconvex variational inequalities. Nonlinear Anal. **71**(1–2), 517–520 (2009)
8. Noor, M.A.: General variational inequalities. Appl. Math. Lett. **1**, 119–121 (1988)
9. Noor, M.A.: Some algorithms for general monotone mixed variational inequalities. Math. Comput. Model. **29**, 1–9 (1999)
10. Noor, M.A.: New approximation schemes for general variational inequalities. J. Math. Anal. Appl. **251**, 217–229 (2000)
11. Noor, M.A.: New extragradient-type methods for general variational inequalities. J. Math. Anal. Appl. **277**, 379–395 (2003)
12. Noor, M.A.: Some developments in general variational inequalities. Appl. Comput. Math. **152**, 199–277 (2004)
13. Noor, M.A.: Iterative schemes for nonconvex variational inequalities. J. Optim. Theory Appl. **121**, 385–395 (2004)
14. Noor, M.A.: Fundamentals of equilibrium problems. Math. Inequal. Appl. **9**, 529–566 (2006)
15. Noor, M.A.: Differentiable nonconvex functions and general variational inequalities. Appl. Math. Comput. **199**, 623–630 (2008)
16. Noor, M.A.: Projection methods for nonconvex variational inequalities. Optim. Lett. **3**, 411–418 (2009)
17. Noor, M.A.: Implicit iterative methods for nonconvex variational inequalities. J. Optim. Theory Appl. **143**, 619–624 (2009)

18. Noor, M.A.: Iterative methods for general nonconvex variational inequalities. Albanian J. Math. **3**, 117–127 (2009)
19. Noor, M.A.: System of nonconvex variational inequalities. J. Adv. Res. Optim. **1**, 1–10 (2009)
20. Noor, M.A.: Some iterative methods for nonconvex variational inequalities. Comput. Math. Model. **21**, 97–109 (2010)
21. Noor, M.A.: General nonconvex variational inequalities and applications. Preprint, Mathematics Department, COMSATS Institute of Information Technology, Islamabad, Pakistan (2009)
22. Noor, M.A.: On a system of general mixed variational inequalities. Optim. Lett. **3**, 437–451 (2009)
23. Noor, M.A.: On an implicit method for nonconvex variational inequalities. J. Optim. Theory Appl. **147**, 97–108 (2010)
24. Noor, M.A.: Principles of Variational Inequalities. Lambert Academic, Saarbrucken (2009)
25. Noor, M.A.: Projection iterative methods for solving some systems of general nonconvex variational inequalities. Appl. Anal. (2011, in press)
26. Noor, M.A., Noor, K.I.: New system of general nonconvex variational inequalities. Appl. Math. E-Notes **10**, 76–85 (2010)
27. Noor, M.A.: Some new systems of general nonconvex variational inequalities involving five different operators. Nonlinear Anal. Forum **15**, 171–179 (2010)
28. Noor, M.A., Noor, K.I., Rassias, Th.M.: Some aspects of variational inequalities. J. Comput. Appl. Math. **47**, 285–312 (1993)
29. Poliquin, R.A., Rockafellar, R.T., Thibault, L.: Local differentiability of distance functions. Trans. Am. Math. Soc. **352**, 5231–5249 (2000)
30. Stampacchia, G.: Formes bilineaires coercivities sur les ensembles convexes. C. R. Acad. Sci. Paris **258**, 4413–4416 (1964)

Chapter 33
On the Asymptotic Behavior of Solutions to General Linear Functional Equations

B. Paneah

Abstract This is a survey of the author's results (Paneah in Aequ. Math. 74(1–2):119–157, 2007; Paneah in Grazer Math. Ber. 351:129–138, 2007; Paneah in Banach J. Math. Anal. 1(1):56–65, 2007; Paneah in Russ. J. Math. Phys. 15(2):291–296, 2008; Paneah in Publ. Math. (Debr.) 75(1–2):251–261, 2009) relating to the asymptotic behavior of approximate solutions to the functional equations $\mathscr{P}F = H_\varepsilon$, $H_\varepsilon = O(\varepsilon)$, depending on a parameter $\varepsilon \to 0$ with

$$\mathscr{P}F(x) = \sum_{j=1}^{N} c_j(x) F \circ a_j(x), \quad x \in D \subset \mathbb{R}^n.$$

This behavior, as it is shown in the above works, is described by the relation

$$F = \Phi + O(\varepsilon).$$

Here the function Φ does not depend on ε and belongs to the kernel of the one-dimensional functional operator $\mathscr{P}_\Gamma$ (restriction of the operator $\mathscr{P}$ to some one-dimensional submanifold $\Gamma \subset D$ subject to determining).

Key words Linear functional equations · Asymptotic behavior · Functional operators · Inverse problems

Mathematics Subject Classification 39Bxx · 62G20 · 93D20

Dedicated to Professor Themistocles M. Rassias on the occasion of his 60th birthday.

B. Paneah (✉)
Department of Mathematics, Technion—Israel Institute of Technology, 32000 Haifa, Israel
e-mail: peter@tx.technion.ac.il

P.M. Pardalos et al. (eds.), *Nonlinear Analysis*, Springer Optimization and Its Applications 68, In Honor of Themistocles M. Rassias on the Occasion of his 60th Birthday,
DOI 10.1007/978-1-4614-3498-6_33,

33.1 Introduction. Formulation of Problem. Old and New

We study the general linear functional operator

$$(\mathscr{P}F)(x) := \sum_{j=1}^{N} c_j(x)(F \circ a_j)(x), \quad x \in D \subset \mathbb{R}.$$

Here $F \in C(I)$ with $I = (-1, 1)$ and $|F|$ norm in C, *coefficients* c_j and *arguments* a_j of $\mathscr{P}$ are continuous functions $D \to \mathbb{R}$ and $D \to I$, respectively; D is a domain with compact closure. The interest to this class of operators is motivated by the fact that many processes and phenomenons in nature or in society are determined by relations connecting the values of a function F at different points (but not values of its derivatives or integrals). It is worth noting that apart from an intrinsic interest this operator recently arose as a necessary technical tool in such diverse fields as Integral geometry, Partial differential equations, Mathematical physics, and even in the combustion theory (see [1–7]). The importance of this class is emphasized by the fact that it contains such popular operators as the Cauchy operator

$$\mathfrak{C}F := F(x_1 + x_2) - F(x_1) - F(x_2) + F(0),$$

the Jensen operator

$$\mathfrak{J}F := F(c_1x_1 + c_2x_2) - c_1F(x_1) - c_2F(x_2),$$

and the quadratic operator

$$\mathfrak{Q}F := F(x_1 + x_2) + F(x_1 - x_2) - 2F(x_1) - 2F(x_2).$$

Practically, all known results related to the general operator $\mathscr{P}$ were concentrated around these three operators and dealt with only two problems in mind:

$\mathscr{H}$-problem: *given an operator* $\mathscr{P}$, *describe the subspace*

$$\ker \mathscr{P} = \{\Phi \mid \mathscr{P}\Phi(x) = 0, \ x \in D\},$$

and

$\mathscr{U}$-problem: *given an operator* $\mathscr{P}$, *prove that for an arbitrary* $\varepsilon > 0$ *if*

$$\left|\mathscr{P}F(x)\right| < \varepsilon \quad \textit{for ALL points } x \in D, \tag{33.1}$$

then

$$\left|(F - \varphi)(t)\right| < c\varepsilon, \quad t \in I, \tag{33.2}$$

with φ *a function from* $\ker \mathscr{P}$ *and* $c > 0$ *a constant independent on* F *and* ε.

We note that inequality (33.2) is equivalent to the relation

$$F = \varphi + O(\varepsilon), \quad \text{as } \varepsilon \to 0,$$

and the latter characterizes an *asymptotic behavior* of a function F as $\varepsilon \to 0$ (but not some mythical "stability" having no relation to the problem in question). The problems of the asymptotic behavior of the solutions to different equations when the parameters guiding these equations tend to zero are very popular as they arise very often on the junction of pure and applied mathematics. Professor S. Ulam was a well-known scientist in the middle of the twentieth century due to his works and interests in applied problems of analysis. It explains easily why it was just Ulam who formulated the $\mathscr{U}$-problem (unsuccessfully mentioning the term *stability*).

I hope that the expression "asymptotic behavior" sounds as solidly as "Ulam stability", but it makes sense, and the Ulam-experts will change the titles of their papers with correct ones.

Concerning solvability of the problems $\mathscr{H}$ and $\mathscr{U}$, it is very simple to describe all achievements of the last 70 years.

In the case $\mathscr{H}$:

If $\mathfrak{C}F(x) = 0$ for all $x \in D$, then $F(t) = \lambda t$,
If $\mathfrak{J}F(x) = 0$ for all $x \in D$, then $F(t) = \lambda t + \mu$,
If $\mathfrak{Q}F(x) = 0$ for all $x \in D$, then $F(t) = \lambda t^2$.

Remark 33.1 Practically all specialists of the $\mathscr{U}$-problem deal with the case $D = \mathbb{R}^n$. Compact domains D do not allow using the Hyers machinery permitting to construct the desired solution. It is assumed also that all functions F are continuous.

In the case $\mathscr{U}$:
There is a finite (very small) set of isolated operators

$$\mathscr{P}F = \sum_{j=0}^{N} c_j F(\overline{a}_j \cdot x), \quad x = (x_1, \dots, x_n) \in \mathbb{R}^n,$$

with constant scalars c_j and vectors $\overline{a}_j \in \mathbb{R}^n$ such that for an arbitrary $\varepsilon > 0$ if

$$\big|\mathscr{P}F(x)\big| < \varepsilon \quad \text{for } all\ x, \tag{33.3}$$

then for some function φ with $\mathscr{P}\varphi = 0$ the relation

$$F = \varphi + O(\varepsilon) \tag{33.4}$$

is valid. (Compare with Theorems 33.1–33.6 and Examples 33.1–33.3 below.)

Remark 33.2 The forms of these operators $\mathscr{P}$ are identical and the proofs are absolutely standard. The common feature for them is applying the above Hyers machinery which remains up to now the main technical tool.

Recently it turned out that all specialists of the $\mathscr{U}$-problem, including Ulam himself, passed by an unexpected fact which is extremely important when dealing with applied problems. It was established (see [6, 8–11]) that the input information in problem $\mathscr{U}$ ($|(\mathscr{P}F)(x)| < \varepsilon$ for *ALL* $x \in D$) in all considered cases is redundant: the same asymptotic relation (33.4) is guaranteed by the significantly weaker condition $|(\mathscr{P}F)(x)| < \varepsilon$ *for the points* $x \in \Gamma \subset D$ *with* Γ *a one-dimensional submanifold subject to determination.*

In the same works, it was clarified that for some operators $\mathscr{P}$ and for appropriate submanifolds $\Gamma \subset D$, $\dim \Gamma = 1$, the asymptotic behavior of a solution F to the equation $\mathscr{P}F = H_\varepsilon$, $|H_\varepsilon|(x) < \varepsilon$, $\varepsilon \to 0$ is described by the relation (33.4) with $\varphi \in \ker \mathscr{P}_\Gamma$, $\mathscr{P}_\Gamma$ being a restriction of the operator $\mathscr{P}$ to Γ. This makes it possible to treat the function φ as an *approximate* solution to the equation $\mathscr{P}F = H_\varepsilon$ localized in a neighborhood of a finite-dimensional subspace $\ker \mathscr{P}_\Gamma$, different from $\ker \mathscr{P}$. Such opportunities are invaluable when dealing with applied problems.

All these observations and particular results related to an extensive class of functional operators $\mathscr{P}$ (see below) lead to a new general problem for the operators $\mathscr{P}$. The solvability of this problem makes it possible to find easily the asymptotic behavior of the solutions to different equations of the form $\mathscr{P}F = H_\varepsilon$, $H_\varepsilon = O(\varepsilon)$.

33.2 Identification Problem for $\mathscr{P}$

Given an operator $\mathscr{P}$, *find a finite-dimensional subspace* $\mathscr{K} \subset C(I)$, *a smooth submanifold* $\Gamma \subset D$ *of a positive codimension and a subspace* $C_{\langle\tau\rangle}(I) \subset C(I)$ *such that the à priori estimate*

$$\inf_{\varphi \in \mathscr{K}} |F - \varphi|_{\langle\tau\rangle} < c|\mathscr{P}_\Gamma F|_{\langle\tau\rangle}, \qquad F \in C_{\langle\tau\rangle}(I), \tag{33.5}$$

holds with a constant c *not depending on* F. *If such a triple* $(\mathscr{K}, \Gamma, C_{\langle\tau\rangle})$ *is found, we say that the identifying problem for the operator* $\mathscr{P}$ *is* $(\mathscr{K}, \Gamma)$*-solvable in the space* $C_{\langle\tau\rangle}$.

In this case, as it follows from (33.5), given an arbitrary $\varepsilon > 0$, if $|\mathscr{P}_\Gamma F|_{\langle\tau\rangle} < \varepsilon/c$ for a function $F \in C_{\langle\tau\rangle}(I)$, then $F = \varphi + h_\varepsilon$ with φ a function from $\mathscr{K}$ and $|h_\varepsilon|_{\langle\tau\rangle} < \varepsilon$. This means in particular that the function φ describes the asymptotic behavior of the function F as $\varepsilon \to 0$ and hence it is an approximate solution to the equation $\mathscr{P}F = H_\varepsilon$, $\varepsilon \to 0$, localizing close to $\mathscr{K}$.

Thus, the essential difference between problem $\mathscr{U}$ and the identification problem is that when searching for an approximate solution F to the identification problem for $\mathscr{P}$ the input information is a smallness of $\mathscr{P}F$ on some submanifold $\Gamma \subset D$, $\dim \Gamma < n$, (to be determined a priori), whereas in the problem $\mathscr{U}$ this smallness is required in the full domain D. It is obvious that when dealing with diverse applied problems this possibility may be of fundamental significance. The other significant difference between both problems in question is that the Ulam problem deals with an approximate solution to the equation $\mathscr{P}F = H_\varepsilon$ lying in the subspace $\mathscr{K} = \ker \mathscr{P}$ only, whereas the identification problem admits such a solution from a wider class of arbitrary fixed finite-dimensional subspaces $\mathscr{K}$.

33.3 The Survey of the Results Related to the Identification Problem

33.3.1 The Cauchy Type Functional Operators

Under this title we join the operators of the type

$$(\mathscr{C}F)(x) = F \circ a(x) - \sum_{j=1}^{N} F \circ a_j(x), \quad x \in D \subset \mathbb{R}^n, \tag{33.6}$$

where

$$a = \sum_{j=1}^{N} a_j \quad \text{everywhere in } D. \tag{33.7}$$

If (33.7) holds only at points x of a curve $\Gamma \subset D$, then the operator $\mathscr{C}$ is called a *weak Cauchy type operator* (*along* Γ). The operator $\mathscr{C}$ has never been studied with respect to any point of view except for the isolated cases involving linear functions $a(x), a_j(x)$.

The operator $\mathfrak{C}$ is the simplest model for a Cauchy and weak Cauchy type operators. To formulate the corresponding result it is necessary to describe a family of curves Γ we deal with when studying the operator $\mathscr{C}$.

Definition 33.1 Given an operator (33.6), a one-dimensional submanifold $\Gamma \subset D$ is called $\mathscr{C}$*-admissible* if it is a non-singular $C_{\langle 1+r \rangle}$-curve, $0 < r < 1$, with the parametric representation $x_j = \zeta_j(t)$, $t \in I$, $1 \le j \le n$, such that the function a maps Γ one-to-one onto I, and the inverse function belongs to the space $C_{\langle 1+r \rangle}(I)$.

Theorem 33.1 (See [6]) *Let $D \subset \mathbb{R}^2$ be a connected bounded domain and $\Gamma \subset D$ a $\mathscr{C}$-admissible curve. Assume that all the functions a and a_j belong to the space $C_{\langle 1+r \rangle}(D)$ and satisfy the conditions*

$$\begin{gathered} \sum_{j,k} a_{j\Gamma} a_{k\Gamma} \neq 0 \quad \text{in } I \setminus \{0\}, \\ a\big(\zeta(0)\big) = 0, \quad \sum_{j,k} \frac{\partial}{\partial \Gamma} a_j\big(\zeta(0)\big) \frac{\partial}{\partial \Gamma} a_k\big(\zeta(0)\big) \neq 0. \end{gathered} \tag{33.8}$$

with $1 \le j, k \le 2$, $\zeta = (\zeta_1, \zeta_2)$.

Then the identification problem for the operator $\mathscr{C}$ is $(\mathscr{K}, \Gamma)$-solvable in the space $C_{\langle r \rangle}$, where $\mathscr{K}$ is the subspace $\{\lambda t\}_{\lambda \in \mathbb{R}}$ of linear functions in the case of Cauchy type operator $\mathscr{C}$, and $\mathscr{K} = \{0\}$ in the case of the weak Cauchy type operator.

Remark 33.3 The latter means the *invertibility of the operator* $\mathscr{C}$ under the above conditions (33.8).

Remark 33.4 We will formulate now a simple corollary of this theorem which is worth comparing with the original Ulam's problem and Hyer's type proofs.

Theorem 33.2 *Let $\mathscr{C}$ be a weak Cauchy operator along Γ, which is a $\mathscr{C}$-admissible $\mathscr{C}_{\langle 1+r\rangle}$-curve in D. Then there is a constant c (depending on $\mathscr{C}$ and Γ) such that any solution F of the equation $\mathscr{C}F = H$ with*

$$|H_\Gamma|_{\langle r\rangle} < \varepsilon$$

satisfies the inequality

$$\left|F(t) - \lambda t\right|_{\langle r\rangle} < c\varepsilon, \quad t \in I,$$

for some λ.

We emphasize again that the main difference between Ulam–Hyers situation and the identification problem is that the smallness of H in Theorem 33.2 is required only on Γ and not in all D as in the Ulam problem. On the other hand, the proofs of both theorems above use general functional analytic methods, rather than manipulations with the specific "near" solutions as in Hyers' approach, which by no means is applicable to the operators $\mathscr{P}$ with nonlinear arguments $a_j(x)$ and variable coefficients $c_j(x)$ (see below).

We give now a pair of examples of the functional equations for which the identification problem is $(\mathscr{K}, \Gamma)$-solvable, by virtue of Theorem 33.1. However, it is not seen how all the previous Hyers-type methods could be adapted to these operators.

Example 33.1 Let $I = [0, 6]$ and $D = \{(x_1, x_2) \mid 0 \le x_1,\, x_2 \le 1\}$. Consider the operator

$$(\mathscr{P}F)(x, y) = F\bigl(x^2 + 2xy + 2y^2 + x^4\bigr) - F\bigl(x^2 + xy + x^4\bigr) - F\bigl(2y^2 + xy\bigr).$$

Take an arbitrary $C_{\langle 1+r\rangle}$-curve Γ in D with the parametric representation

$$\Gamma = \bigl\{x \in D \mid x_1 = \zeta_1(t),\ x_2 = \zeta_2(t),\ 0 \le t \le 1\bigr\},$$

where nondecreasing functions ζ_j satisfy initial conditions

$$\zeta_1(0) = \zeta_2(0) = 0, \qquad \zeta_1(1) = \zeta_2(1) = 1.$$

If Γ satisfies conditions (33.8), then the identification problem is $(\mathscr{K}, \Gamma)$-solvable with $\mathscr{K} = \{\lambda t\}_{\lambda \in \mathbb{R}}$.

In other words, if $|\mathscr{P}_\Gamma F|_{\langle r\rangle} < \varepsilon$, then $|F(t) - \lambda t|_{\langle r\rangle} < c\varepsilon$ with $t \in I$, c constant not depending on F and λ a scalar from $\mathbb{R}$.

Example 33.2 The result of the previous example remains valid for the operator

$$\mathscr{P}F := F\bigl(x_1^2 + x_2 + x_1\bigr) - F\left(\frac{3x_1}{2}\right) - F\left(x_2^2 - 4x_1x_2 - x_1^4 - 4x_1^2 + \frac{3x_1}{2}\right)$$

with

$$\Gamma = \left\{x \mid x_1 = t,\ x_2 = 2t - t^2;\ -1 \le t \le 1\right\}$$

and the above $\mathscr{K}$.

This means that *if* $|\mathscr{P}_\Gamma F| < \varepsilon$, *then* $|F(t) - \lambda t| < c\varepsilon$ *with* $t \in I$ *and some* $\lambda \in \mathbb{R}$. In other words, the asymptotic behavior of the function F for which $|\mathscr{P}_\Gamma F| < \varepsilon$, $\varepsilon \to 0$, is described by the relation $F(t) = \lambda t + O(\varepsilon)$.

33.3.2 The Jensen Type Functional Operators

Under this title in [8] the class of linear functional operators

$$(\mathscr{J} F)(x) := F \circ a(x) - \sum_{j=1}^{N} c_j(x) F \circ a_j(x), \quad x \in D, \tag{33.9}$$

with positive c_j satisfying the conditions

$$\sum_{j=1}^{N} c_j = 1 \tag{33.10}$$

and

$$a(x) = \sum (c_j a_j)(x), \tag{33.11}$$

has been introduced.

If the conditions (33.10) and (33.11) are valid at the points of a curve Γ only, we call $\mathscr{J}$ a *weak* $\mathscr{J}$*-type* operator (along Γ).

The operator $\mathscr{J}$ has never been studied with respect to any point of view except for the particular case $\mathscr{J}$ corresponding to the parameters $N = 2$, $c_1 = c_2 = 1/2$.

To formulate the recent results related to operator (33.9), we need several definitions.

Definition 33.2 Given an operator $\mathscr{P}$, a term $c_j F \circ a_j$ is called the *leading term of* $\mathscr{P}$ if the function a_j maps D *onto* I.

Let Γ be a curve as above and $\zeta : I \to \Gamma$ a one-to-one $C_{\langle 1+r\rangle}$-map. We denote by w_Γ and $\mathscr{P}_\Gamma$ the restriction $w_\Gamma(t) := (w \circ \zeta)(t)$, $t \in I$, of an arbitrary function $w \in C(D)$ and that of the operator

$$\mathscr{P}_\Gamma : F \to \sum c_{j\Gamma}(t)(F \circ a_{j\Gamma})(t), \quad t \in I,$$

respectively.

Definition 33.3 A curve $\Gamma \subset D$ is called $\mathcal{J}$*-admissible* if for the leading term $c_k F \circ a_k$ the coefficient $c_{k\Gamma}$ does not vanish and the $a_{k\Gamma}$ maps Γ one-to-one onto I.

The main result related to the identification problem for operator $\mathcal{J}$ reads as follows.

Theorem 33.3 *Let $\mathcal{J}$ be an operator* (33.6) *with the leading term $c_1 F \circ a_1$, and let $\Gamma \subset D$ be a $\mathcal{J}$-admissible $C_{\langle 1+r \rangle}$-curve, corresponding to this term and satisfying conditions*

$$a_{j\Gamma} = 0 \quad \textit{for all } j = 2, \ldots, n \tag{33.12}$$

and

$$a_1 \circ \zeta(0) = 0, \qquad \left(\frac{\partial}{\partial \Gamma}\right) a_1 \circ \zeta(0) \neq 0. \tag{33.13}$$

If $\mathcal{J}$ is a (weak) Jensen type operator along Γ, then the identification problem for $\mathcal{J}$ is $(\mathcal{K}, \Gamma)$-solvable in the space $C_{\langle 1+r \rangle}$. The subspace $\mathcal{K}$ here coincides with $\ker \mathcal{J} = \{0\}$ *in the case "weak" and with*

$$\ker \mathcal{J}_\Gamma = \bigl\{\varphi \mid \varphi(t) = \alpha t + \beta;\ t \in I,\ \alpha, \beta \in \mathbb{R}\bigr\},$$

otherwise.

The following result may be used as an illustration to this theorem. It will remind the reader that the solvability of the identification problem is equivalent to some specific asymptotic behavior of the solution to the nonhomogeneous equation $\mathcal{J} F = H_\varepsilon$, $H_\varepsilon = O(\varepsilon)$, as $\varepsilon \to 0$.

Example 33.3 Let $I_2 = [0, \mu]$, $D = \{(x, y) \mid 0 \le x, y \le \gamma\}$. Let $\mathcal{P}$ be the operator

$$F \to F\bigl(x\sqrt{1+x^2}e^{x^2 y} + \sqrt{x^2 y^3 + 1}\sin y\bigr) - \frac{1}{3}e^{x^2 y} F\bigl(3x\sqrt{1+x^2}\bigr) - \frac{2}{3}\sqrt{x^2 y^3 + 1} F\left(\frac{3}{2}\sin y\right).$$

At first sight, this operator has no special structure, and hence it is not possible to connect it with one of the already studied operators. But note that with

$$a_1 = 3x\sqrt{1+x^2}, \quad a_2 = \frac{3}{2}\sin y, \quad c_1 = \frac{1}{3}e^{x^2 y}, \quad c_2 = \frac{2}{3}\sqrt{x^2 y^3 + 1},$$

and

$$\gamma = (\ln 2)^{1/3}, \quad \nu = \gamma, \quad \mu = 3\gamma\bigl(1+\gamma^2\bigr)$$

the restriction a_Γ of the function

$$a = x\sqrt{1+x^2}e^{x^2 y} + \sqrt{x^2 y^3 + 1}\sin y$$

to the curve $\Gamma = \{(x, y) \mid x = \nu t, y = 0; 0 \le t \le 1\}$ can be represented in a form

$$a_\Gamma = c_{1\Gamma} a_{1\Gamma} + c_{2\Gamma} a_{2\Gamma},$$

with

$$c_{1\Gamma} + c_{2\Gamma} = 1.$$

It follows that $\mathscr{P}$ is a weak Jensen type operator along Γ. It is easy to show that with the given γ, ν, and μ all the functions a_1, a_2 and a map the domain D into I_2. Furthermore, the Γ is $\mathscr{P}$-admissible, as $c_{1\Gamma} \neq 0$ and the range of the $a_{1\Gamma}$ is $[0, \mu]$. It is clear that relations (33.12) and (33.13) hold for the $a_{1\Gamma}$ and $a_{2\Gamma}$. But in this case by Theorem 33.3, the identification problem for $\mathscr{P}$ is $(\ker \mathscr{P}_\Gamma, \Gamma)$-solvable. *This means, that for all sufficiently small $\varepsilon > 0$ the inequality*

$$|\mathscr{P}_\Gamma F| < \varepsilon$$

implies the following asymptotic behavior of the function F:

$$F(t) = \alpha t + \beta + O(\varepsilon), \quad 0 \le t \le \mu,$$

with α and β some constants. We note that the function $\alpha t + \beta$ solves the equation $\mathscr{P}_\Gamma F = 0$.

33.3.3 Quasiquadratic Functional Operators

The following class of functional operators (see [9–11]) which we demonstrate in this short review consists of the *quasiquadratic* (qq) operators

$$\mathcal{Q}(F) := F(x_1 + x_2) + F(x_1 - x_2) - \alpha_1 F(x_1) - \alpha_2 F(x_2), \quad \{x \mid |x_1 \pm x_2| \le 1\}$$

with α_1, α_2 positive constants. The name *quasiquadratic* has been given to $\mathcal{Q}$ in honor of its very well-known forefather—the quadratic operator

$$\mathfrak{Q}(F) := F(x_1 + x_2) + F(x_1 - x_2) - 2F(x_1) - 2F(x_2).$$

It looks (to me) extremely astonishing but during more than 50 years of the uninterrupted siege of the not too massive functional–operator building, nobody touched coefficients $\alpha_1 = 2$ and $\alpha_2 = 2$. The authors of tenths of books, when referring to $\mathfrak{Q}$, simply rewrote one and the same text:

Theorem 33.4

(i) *If*

$$\mathfrak{Q}F(x) = 0$$

for all points $x \in D$ *and* $F \in C(D)$, *then* $F(t) = \lambda t^2$.

(ii) *If* F *is a continuous function and for an arbitrary real* $\varepsilon > 0$

$$\left|\mathfrak{Q}F(x)\right| < \varepsilon \quad \textit{for all points } x \in D, \tag{33.14}$$

then there is a real λ *such that*

$$\left|F(t) - \lambda t^2\right| \le c\varepsilon, \quad 0 \le t \le 1, \tag{33.15}$$

with a constant $c > 0$ *not depending on* F *and* ε *or, equivalently,*

$$F(t) = \lambda t^2 + O(\varepsilon), \quad \textit{as } \varepsilon \to 0.$$

As a matter of fact, the study of the operator $\mathcal{Q}$ turned out to be a non-trivial, very interesting problem, requiring new methods, new notions, and finally generating a new problem in the theory of linear functional operators. We will give here some results related to $\mathcal{Q}$ and the formulation of the above-mentioned new problem. The corresponding proofs are now in press.

We consider the operator $\mathcal{Q}$ in the domain $D = \{x \mid |x_1 \pm x_2| \le 1\}$ and as Γ we choose the curve

$$\Gamma = \{x \in D \mid x_1 = t + 1, x_2 = t; -1 \le t \le 0\}.$$

Introduce the integer

$$m = \left[\log_2(\alpha_1 + \alpha_2)\right]$$

characterizing the smoothness of functions we work with. The restriction $\mathcal{Q}_\Gamma$ of the operator $\mathcal{Q}$ to Γ, as easily seen, has the form

$$(\mathcal{Q}_\Gamma F)(t) := F(2t+1) + F(1) - \alpha_1 F(t) - \alpha_2 F(t+1), \quad -1 \le t \le 0. \tag{33.16}$$

Let $\Lambda_m = \|\lambda_i^j\|_{i,j=1}^m$ be a matrix of the operator $\mathcal{Q}_\Gamma$ in the space π_m of all polynomials $P_m(t) = \sum_{j=0}^m a_j t^j$ with the basis $\{1, t, \ldots, t^m\}$.

Theorem 33.5 *If* $\alpha_1 + \alpha_2 \ne 2^k$ *for any integer* k, $k \ge 2$, *then the equation* $\mathcal{Q}_\Gamma F = H$ *has a unique solution* $F \in C^m(D)$ *for an arbitrary function* $H \in C^m(D)$, *and the following à priori estimate is valid with a constant* c *independent of* F:

$$|F|_{\langle m\rangle} \le c|\mathcal{Q}_\Gamma F|_{\langle m\rangle}, \quad F \in C^m(D).$$

Theorem 33.6 *Let* $\alpha_1 + \alpha_2 = 2^m$.

(i) *If* $F \in C^m$ *and*

$$\mathcal{Q}_\Gamma F = 0 \tag{33.17}$$

then $F = \sum_{j=0}^{m} a_j t^j$ *with* $\overline{a} = (a_0, a_1, \ldots, a_m)$ *a vector from the subspace* $\ker \Lambda$.

(ii) *The à priori estimate*

$$\left| F - \sum_{j=0}^{m} a_j t^j \right|_{\langle m \rangle} \leq c |\mathcal{Q}_\Gamma F|_{\langle m \rangle}, \quad F \in C^m(D)$$

is valid with a vector $\overline{a} \in \ker \Lambda$. *Equivalently, if*

$$|\mathcal{Q}_\Gamma F|_{\langle m \rangle} < \varepsilon, \tag{33.18}$$

then for some constant $c > 0$ *and for a vector* $\overline{a} \in \ker \Lambda$

$$\left| F - \sum_{j=0}^{m} a_j t^j \right|_{\langle m \rangle} < c\varepsilon, \quad 0 \leq t \leq 1. \tag{33.19}$$

All the results of Theorem 33.5 and Theorem 33.6 were unknown before with the exception of the case $\alpha_1 = \alpha_2 = 2$. But if $\alpha_1 = \alpha_2 = 2$, the result of Theorem 33.4 is significantly weaker than that of Theorem 33.6 because condition (33.14) is supposed to be valid inside a whole domain D whereas analogous condition (33.18) has to be valid only on Γ.

To illustrate the diverse possibilities of our approach, consider in detail the operator $\mathcal{Q}$ in the case $\alpha_1 + \alpha_2 = 4$. This case is studied well when $\alpha_1 = \alpha_2 = 2$, and it has never been discussed for other values of α_j. Consider two situations: $\alpha_1 = 1$, $\alpha_2 = 3$ and $\alpha_1 = 3$, $\alpha_2 = 1$. As above, we choose

$$\Gamma = \{x \mid x_1 = t, x_2 = t + 1; -1 \leq t \leq 0\}$$

and determine a function $w(t) \in C^2$, for which $\mathcal{Q}_\Gamma w = 0$. By (33.16), this function has to satisfy the relation

$$w''(2t+1) - \frac{1}{4}\alpha_1 w''(t) - \frac{1}{4}\alpha_2 w''(t+1) = 0, \quad -1 \leq t \leq 0.$$

As $\alpha_1/4 + \alpha_2/4 = 1$, we can apply the *maximum principle* for functional equations (see [1]) and conclude that

$$w''(t) = \text{const},$$

whence

$$w(t) = a_0 + a_1 t + a_2 t^2.$$

Introduce the vectors

$$\overline{a} = (a_0, a_1, a_2) \quad \text{and} \quad \overline{T} = \left(1, t, t^2\right).$$

Then

$$w(t) = \overline{a} \cdot \overline{T},$$

and it remains to guess those vectors $\overline{a}$ for which

$$\mathcal{Q}_\Gamma(\overline{a} \cdot \overline{T}) = 0.$$

It is easy to check that for all values $t \in I$

$$\mathcal{Q}_\Gamma(\overline{a} \cdot \overline{T}) = \Lambda \overline{a} \cdot \overline{T}$$

with

$$\Lambda = \begin{pmatrix} 2-\alpha_1-\alpha_2 & 2-\alpha_2 & 2-\alpha_2 \\ 0 & 2-\alpha_1-\alpha_2 & 4-2\alpha_2 \\ 0 & 0 & 4-\alpha_1-\alpha_2 \end{pmatrix},$$

and the problem is reduced to searching all vectors $\overline{a}$ from the subspace $\ker \Lambda$. Since $4-\alpha_1-\alpha_2=0$, as a component a_2 of the needed vector $\overline{a}$ an arbitrary constant λ can be chosen, and to determine components a_0, a_1, we have to solve the system of equations

$$\begin{cases} -2a_0 + (2-\alpha_2)a_1 + (2-\alpha_2)\lambda = 0, \\ -2a_1 + (2-\alpha_2)\lambda \cdot 2 = 0. \end{cases}$$

Thus,

$$\begin{aligned} &\text{if } \alpha_1 = 1,\ \alpha_2 = 3,\ \text{then } \overline{a} = (0, -\lambda, \lambda) = \lambda(0, -1, 1) = \lambda \overline{e}_1, \\ &\text{if } \alpha_1 = 2,\ \alpha_2 = 2,\ \text{then } \overline{a} = (0, 0, \lambda) = \lambda(0, 0, 1) = \lambda \overline{e}_2, \\ &\text{if } \alpha_1 = 3,\ \alpha_2 = 1,\ \text{then } \overline{a} = (\lambda, \lambda, \lambda) = \lambda(1, 1, 1) = \lambda \overline{e}_3. \end{aligned}$$

It follows that in the three different situations with parameters α_1, α_2 the set $\ker \mathcal{Q}_\Gamma$ is a one-dimensional subspace spanned by

$$\overline{e}_1 \cdot \overline{T} = -t + t^2, \qquad \overline{e}_2 \cdot \overline{T} = t^2, \quad \text{and} \quad \overline{e}_3 \cdot \overline{T} = 1 + t + t^2,$$

respectively. Therefore, the asymptotic behavior of the solutions to the equations $\mathcal{Q}_\Gamma F = H_\varepsilon$, $\varepsilon \to 0$, is distributed in the following way:

$$F = \lambda(t - t^2) + O(\varepsilon), \qquad F = \lambda t^2 + O(\varepsilon), \qquad F = \lambda(1 + t + t^2) + O(\varepsilon)$$

in the cases $\alpha_1 = 1, \alpha_1 = 2, \alpha_1 = 3$, respectively.

This result makes it possible to formulate a new problem in the theory of the linear functional operators which undoubtedly will be of great interest for those working in applied areas.

Inverse problem Given a family of linear functional operators $\mathcal{P}_\kappa$ parameterized by some index κ, find a "value" $\widetilde{\kappa}$ such that the asymptotic behavior of an approximate solution to the equation $\mathcal{P}_{\widetilde{\kappa}} F = H_\varepsilon$, $H_\varepsilon = O(\varepsilon)$ has a prescribed asymptotic form.

References

1. Paneah, B.: On the solvability of functional equations associated with dynamical systems with two generators. Funct. Anal. Appl. **37**(1), 46–60 (2003)
2. Paneah, B.: Noncommutative dynamical systems with two generators and their applications in analysis. Discrete Contin. Dyn. Syst. **9**(6), 1411–1420 (2003)
3. Paneah, B.: Dynamical approach to some problems in integral geometry. Trans. Am. Math. Soc. **356**(7), 2757–2780 (2004)
4. Paneah, B.: Boundary problems for higher order hyperbolic differential equations in bounded domains. Russ. J. Math. Phys. **11**(4), 456–473 (2004)
5. Paneah, B.: Dynamical systems and functional equations related to boundary problems for hyperbolic differential operators. Dokl. Ross. Akad. Nauk **405**(5), 598–603 (2005) [Dokl., Math. **72**, 949–953 (2005)]
6. Paneah, B.: A new approach to the stability of linear functional operators. Aequ. Math. **78**, 45–61 (2009)
7. Paneah, B.: On the general theory of the Cauchy type functional equations with applications in analysis. Aequ. Math. **74**(1–2), 119–157 (2007)
8. Paneah, B.: On the stability of the linear functional operators structurally associated with the Jensen operator. Grazer Math. Ber. **351**, 129–138 (2007)
9. Paneah, B.: Some remarks on stability and solvability of linear functional equations. Banach J. Math. Anal. **1**(1), 56–65 (2007)
10. Paneah, B.: Identifying functions determined by linear functional operators. Russ. J. Math. Phys. **15**(2), 291–296 (2008)
11. Paneah, B.: The identifying problem related to linear functional operators with linear arguments. Publ. Math. (Debr.) **75**(1–2), 251–261 (2009)
12. Hyers, D., Isac, G., Rassias, Th.: Stability of Functional Equations in Several Variables. Birkhauser, Basel (1999)

Chapter 34
On the Stability of an Additive and Quadratic Functional Equation

Choonkil Park

Abstract In Park et al. (J. Chungcheong Math. Soc. 21:455–466, 2008) considered the following Jensen additive and quadratic type functional equation

$$2f\left(\frac{x+y}{2}\right)+f\left(\frac{x-y}{2}\right)+f\left(\frac{y-x}{2}\right)=f(x)+f(y).$$

In this paper, we investigate the following additive and quadratic functional equation

$$2f(x+y)+f(x-y)+f(y-x)=3f(x)+f(-x)+3f(y)+f(-y). \quad (34.1)$$

Furthermore, we prove the generalized Hyers–Ulam stability of the functional equation (34.1) in Banach spaces.

Key words Additive and quadratic type functional equation · Additive mapping · Quadratic mapping · Generalized Hyers–Ulam stability

Mathematics Subject Classification Primary 39B72 · 46C05

34.1 Introduction and Preliminaries

The stability problem of functional equations was originated from a question of Ulam [16] concerning the stability of group homomorphisms. Hyers [5] gave a first affirmative partial answer to the question of Ulam for Banach spaces. Hyers' Theorem was generalized by Aoki [1] for additive mappings and by Th.M. Rassias [7] for linear mappings by considering an unbounded Cauchy difference. The paper of Th.M. Rassias [7] has provided a lot of influence in the development of what we call *generalized Hyers–Ulam stability* of functional equations. A generalization of the Th.M. Rassias theorem was obtained by Găvruta [4] by replacing the unbounded

Dedicated to Professor Themistocles M. Rassias on the occasion of his 60th birthday.

C. Park (✉)
Department of Mathematics, Hanyang University, Seoul 133-791, Republic of Korea
e-mail: baak@hanyang.ac.kr

P.M. Pardalos et al. (eds.), *Nonlinear Analysis*, Springer Optimization and Its Applications 68, In Honor of Themistocles M. Rassias on the Occasion of his 60th Birthday, DOI 10.1007/978-1-4614-3498-6_34,

Cauchy difference by a general control function in the spirit of Th.M. Rassias' approach.

The square of a norm on an inner product space satisfies the parallelogram equality $\|x+y\|^2+\|x-y\|^2=2\|x\|^2+2\|y\|^2$. The functional equation

$$f(x+y)+f(x-y)=2f(x)+2f(y)$$

is called a *quadratic functional equation.* In particular, every solution of the quadratic functional equation is said to be a *quadratic mapping*. A generalized Hyers–Ulam stability problem for the quadratic functional equation was proved by Skof [15] for mappings $f: X \to Y$, where X is a normed space and Y is a Banach space. Cholewa [2] noticed that the theorem of Skof is still true if the relevant domain X is replaced by an abelian group. In [3], Czerwik proved the generalized Hyers–Ulam stability of the quadratic functional equation. Several functional equations have been investigated in [9–14].

In [8], Th.M. Rassias proved that the norm defined over a real vector space V is induced by an inner product if and only if for a fixed integer $n \geq 2$

$$n\left\|\frac{1}{n}\sum_{i=1}^{n}x_i\right\|^2+\sum_{i=1}^{n}\left\|x_i-\frac{1}{n}\sum_{j=1}^{n}x_j\right\|^2=\sum_{i=1}^{n}\|x_i\|^2$$

holds for all $x_1,\dots,x_n \in V$.

Let V, W be real vector spaces. In [6], it was shown that if a mapping $f: V \to W$ satisfies

$$\sum_{i=1}^{n}f\left(x_i-\frac{1}{n}\sum_{j=1}^{n}x_j\right)=\sum_{i=1}^{n}f(x_i)-nf\left(\frac{1}{n}\sum_{i=1}^{n}x_i\right)$$

for all $x_1,\dots,x_n \in V$, then the mapping $f: V \to W$ satisfies

$$2f\left(\frac{x+y}{2}\right)+f\left(\frac{x-y}{2}\right)+f\left(\frac{y-x}{2}\right)=f(x)+f(y) \qquad (34.2)$$

for all $x, y \in V$. Park et al. [6] proved the generalized Hyers–Ulam stability of the functional equation (34.2) in Banach spaces.

Throughout this paper, let X be a normed vector space with norm $\|\cdot\|$, and Y a Banach space with norm $\|\cdot\|$.

In this paper, we investigate the functional equation (34.1), and prove the generalized Hyers–Ulam stability of the functional equation (34.1) in Banach spaces.

34.2 Quadratic Mapping

It is easily shown that an even mapping $f: V \to W$ satisfies (34.1) if and only if the even mapping $f: V \to W$ is a quadratic mapping, i.e., $f(x+y)+f(x-y)=$

$2f(x)+2f(y)$, and that an odd mapping $f:V\to W$ satisfies (34.1) if and only if the odd mapping $f:V\to W$ is an additive mapping, i.e., $f(x+y)=f(x)+f(y)$.

For a given mapping $f:X\to Y$, we define

$$Df(x,y):=2f(x+y)+f(x-y)+f(y-x)-3f(x)-f(-x)-3f(y)-f(-y)$$

for all $x,y\in X$.

In this section, we prove the generalized Hyers–Ulam stability of the functional equation $Df(x,y)=0$ in Banach spaces for the even case.

Theorem 34.1 *Let $f:X\to Y$ be a mapping for which there exists a function $\varphi: X^2\to[0,\infty)$ such that*

$$\widetilde{\varphi}(x,y):=\sum_{j=1}^{\infty}4^j\varphi\left(\frac{x}{2^j},\frac{y}{2^j}\right)<\infty, \tag{34.3}$$

$$\left\|Df(x,y)\right\|\le\varphi(x,y) \tag{34.4}$$

for all $x,y\in X$. Then there exists a unique quadratic mapping $Q:X\to Y$ such that

$$\left\|f(x)+f(-x)-Q(x)\right\|\le\frac{1}{8}\left(\widetilde{\varphi}(x,x)+\widetilde{\varphi}(-x,-x)\right) \tag{34.5}$$

for all $x\in X$.

Proof Letting $y=x$ in (34.4), we get

$$\left\|2f(2x)-6f(x)-2f(-x)\right\|\le\varphi(x,x) \tag{34.6}$$

for all $x\in X$. Replacing x by $-x$ in (34.6), we get

$$\left\|2f(-2x)-6f(-x)-2f(x)\right\|\le\varphi(-x,-x) \tag{34.7}$$

for all $x\in X$. Let $g(x):=f(x)+f(-x)$ for all $x\in X$. It follows from (34.6) and (34.7) that

$$\left\|g(x)-4g\left(\frac{x}{2}\right)\right\|\le\frac{1}{2}\left(\varphi\left(\frac{x}{2},\frac{x}{2}\right)+\varphi\left(-\frac{x}{2},-\frac{x}{2}\right)\right) \tag{34.8}$$

for all $x\in X$. Hence

$$\begin{aligned}\left\|4^l g\left(\frac{x}{2^l}\right)-4^m g\left(\frac{x}{2^m}\right)\right\|\le{}&\sum_{j=l+1}^{m}\frac{4^j}{8}\varphi\left(\frac{x}{2^j},\frac{x}{2^j}\right)\\&+\sum_{j=l+1}^{m}\frac{4^j}{8}\varphi\left(-\frac{x}{2^j},-\frac{x}{2^j}\right)\end{aligned} \tag{34.9}$$

for all nonnegative integers m and l with $m > l$ and all $x \in X$. It follows from (34.3) and (34.9) that the sequence $\{4^k g(\frac{x}{2^k})\}$ is Cauchy for all $x \in X$. Since Y is complete, the sequence $\{4^k g(\frac{x}{2^k})\}$ converges. So one can define the mapping $Q : X \to Y$ by

$$Q(x) := \lim_{k\to\infty} 4^k g\left(\frac{x}{2^k}\right)$$

for all $x \in X$.

By (34.3) and (34.4),

$$\begin{aligned}\|DQ(x,y)\| &= \lim_{k\to\infty} 4^k \left\| Dg\left(\frac{x}{2^k}, \frac{y}{2^k}\right)\right\| \\ &\le \lim_{k\to\infty} 4^k \left(\varphi\left(\frac{x}{2^k}, \frac{y}{2^k}\right) + \varphi\left(-\frac{x}{2^k}, -\frac{y}{2^k}\right)\right) = 0\end{aligned}$$

for all $x, y \in X$. So $DQ(x,y) = 0$. Since $g : X \to Y$ is even, $Q : X \to Y$ is even. So the mapping $Q : X \to Y$ is quadratic. Moreover, letting $l = 0$ and passing the limit $m \to \infty$ in (34.9), we get (34.5). So there exists a quadratic mapping $Q : X \to Y$ satisfying (34.5).

Now, let $Q' : X \to Y$ be another quadratic mapping satisfying (34.5). Then we have

$$\begin{aligned}\|Q(x) - Q'(x)\| &= 4^q \left\| Q\left(\frac{x}{2^q}\right) - Q'\left(\frac{x}{2^q}\right)\right\| \\ &\le 4^q \left\| Q\left(\frac{x}{2^q}\right) - f\left(\frac{x}{2^q}\right) - f\left(\frac{-x}{2^q}\right)\right\| \\ &\quad + 4^q \left\| Q'\left(\frac{x}{2^q}\right) - f\left(\frac{x}{2^q}\right) - f\left(\frac{-x}{2^q}\right)\right\| \\ &\le 2 \cdot 4^q \widetilde{\varphi}\left(\frac{x}{2^q}, \frac{x}{2^q}\right) + 2 \cdot 4^q \widetilde{\varphi}\left(\frac{-x}{2^q}, \frac{-x}{2^q}\right),\end{aligned}$$

which tends to zero as $q \to \infty$ for all $x \in X$. So we can conclude that $Q(x) = Q'(x)$ for all $x \in X$. This proves the uniqueness of Q. □

Corollary 34.1 *Let $p > 2$ and θ be positive real numbers, and let $f : X \to Y$ be a mapping such that*

$$\|Df(x,y)\| \le \theta(\|x\|^p + \|y\|^p) \tag{34.10}$$

for all $x, y \in X$. Then there exists a unique quadratic mapping $Q : X \to Y$ such that

$$\|f(x) + f(-x) - Q(x)\| \le \frac{2\theta}{2^p - 4}\|x\|^p$$

for all $x \in X$.

Proof Define $\varphi(x, y) = \theta(\|x\|^p + \|y\|^p)$, and apply Theorem 34.1 to get the desired result. □

Theorem 34.2 *Let $f : X \to Y$ be a mapping satisfying $f(0) = 0$ for which there exists a function $\varphi : X^2 \to [0, \infty)$ satisfying (34.4) such that*

$$\widetilde{\varphi}(x, y) := \sum_{j=0}^{\infty} 4^{-j} \varphi\big(2^j x, 2^j y\big) < \infty \tag{34.11}$$

for all $x, y \in X$. Then there exists a unique quadratic mapping $Q : X \to Y$ such that

$$\big\| f(x) + f(-x) - Q(x) \big\| \le \frac{1}{8}\big(\widetilde{\varphi}(x, x) + \widetilde{\varphi}(-x, -x)\big) \tag{34.12}$$

for all $x \in X$.

Proof It follows from (34.8) that

$$\left\| g(x) - \frac{1}{4} g(2x) \right\| \le \frac{1}{8}\big(\varphi(x, x) + \varphi(-x, -x)\big)$$

for all $x \in X$. So

$$\begin{aligned}\left\| \frac{1}{4^l} g\big(2^l x\big) - \frac{1}{4^m} g\big(2^m x\big) \right\| &\le \sum_{j=l}^{m-1} \frac{1}{8 \cdot 4^j} \varphi\big(2^j x, 2^j x\big) \\ &\quad + \sum_{j=l}^{m-1} \frac{1}{8 \cdot 4^j} \varphi\big(-2^j x, -2^j x\big)\end{aligned} \tag{34.13}$$

for all nonnegative integers m and l with $m > l$ and all $x \in X$. It follows from (34.11) and (34.13) that the sequence $\{\frac{1}{4^k} g(2^k x)\}$ is Cauchy for all $x \in X$. Since Y is complete, the sequence $\{\frac{1}{4^k} g(2^k x)\}$ converges. So one can define the mapping $Q : X \to Y$ by

$$Q(x) := \lim_{k\to\infty} \frac{1}{4^k} g\big(2^k x\big)$$

for all $x \in X$.

By (34.4) and (34.11),

$$\begin{aligned}\big\| DQ(x, y) \big\| &= \lim_{k\to\infty} \frac{1}{4^k} \big\| Dg\big(2^k x, 2^k y\big) \big\| \\ &\le \lim_{k\to\infty} \frac{1}{4^k} \big(\varphi\big(2^k x, 2^k y\big) + \varphi\big(-2^k x, -2^k y\big)\big) = 0\end{aligned}$$

for all $x, y \in X$. So $DQ(x, y) = 0$. Since $g : X \to Y$ is even, $Q : X \to Y$ is even. So the mapping $Q : X \to Y$ is quadratic. Moreover, letting $l = 0$ and passing the limit

$m \to \infty$ in (34.13), we get (34.12). So there exists a quadratic mapping $Q : X \to Y$ satisfying (34.12).

The rest of the proof is similar to the proof of Theorem 34.1. □

Corollary 34.2 *Let $p < 2$ and θ be positive real numbers, and let $f : X \to Y$ be a mapping satisfying* (34.10). *Then there exists a unique quadratic mapping $Q : X \to Y$ such that*

$$\|f(x) + f(-x) - Q(x)\| \le \frac{2\theta}{4 - 2^p}\|x\|^p$$

for all $x \in X$.

Proof Define $\varphi(x, y) = \theta(\|x\|^p + \|y\|^p)$, and apply Theorem 34.2 to get the desired result. □

34.3 Additive Mapping

In this section, we prove the generalized Hyers–Ulam stability of the functional equation $Df(x, y) = 0$ in Banach spaces for the odd case.

Theorem 34.3 *Let $f : X \to Y$ be a mapping for which there exists a function $\varphi : X^2 \to [0, \infty)$ such that*

$$\Phi(x, y) := \sum_{j=1}^{\infty} 2^j \varphi\left(\frac{x}{2^j}, \frac{y}{2^j}\right) < \infty, \tag{34.14}$$

$$\|Df(x, y)\| \le \varphi(x, y) \tag{34.15}$$

for all $x, y \in X$. Then there exists a unique additive mapping $A : X \to Y$ such that

$$\|f(x) - f(-x) - A(x)\| \le \frac{1}{4}\big(\Phi(x, x) + \Phi(-x, -x)\big) \tag{34.16}$$

for all $x \in X$.

Proof Letting $y = x$ in (34.15), we get

$$\|2f(2x) - 6f(x) - 2f(-x)\| \le \varphi(x, x) \tag{34.17}$$

for all $x \in X$. Replacing x by $-x$ in (34.17), we get

$$\|2f(-2x) - 6f(-x) - 2f(x)\| \le \varphi(-x, -x) \tag{34.18}$$

for all $x \in X$. Let $h(x) := f(x) - f(-x)$ for all $x \in X$. It follows from (34.17) and (34.18) that

$$\left\|h(x) - 2h\left(\frac{x}{2}\right)\right\| \le \frac{1}{2}\left(\varphi\left(\frac{x}{2}, \frac{x}{2}\right) + \varphi\left(-\frac{x}{2}, -\frac{x}{2}\right)\right) \tag{34.19}$$

for all $x \in X$. Hence

$$\left\|2^l h\left(\frac{x}{2^l}\right) - 2^m h\left(\frac{x}{2^m}\right)\right\| \le \sum_{j=l+1}^{m} \frac{2^j}{4}\varphi\left(\frac{x}{2^j}, \frac{x}{2^j}\right) + \sum_{j=l+1}^{m} \frac{2^j}{4}\varphi\left(-\frac{x}{2^j}, -\frac{x}{2^j}\right) \tag{34.20}$$

for all nonnegative integers m and l with $m > l$ and all $x \in X$. It follows from (34.14) and (34.20) that the sequence $\{2^k h(\frac{x}{2^k})\}$ is Cauchy for all $x \in X$. Since Y is complete, the sequence $\{2^k h(\frac{x}{2^k})\}$ converges. So one can define the mapping $A : X \to Y$ by

$$A(x) := \lim_{k\to\infty} 2^k h\left(\frac{x}{2^k}\right)$$

for all $x \in X$.

By (34.14) and (34.15),

$$\begin{aligned}\|DA(x, y)\| &= \lim_{k\to\infty} 2^k \left\|Dh\left(\frac{x}{2^k}, \frac{y}{2^k}\right)\right\| \\ &\le \lim_{k\to\infty} 2^k \left(\varphi\left(\frac{x}{2^k}, \frac{y}{2^k}\right) + \varphi\left(-\frac{x}{2^k}, -\frac{y}{2^k}\right)\right) = 0\end{aligned}$$

for all $x, y \in X$. So $DA(x, y) = 0$. Since $h : X \to Y$ is odd, $A : X \to Y$ is odd. So the mapping $A : X \to Y$ is additive. Moreover, letting $l = 0$ and passing the limit $m \to \infty$ in (34.20), we get (34.16). So there exists an additive mapping $A : X \to Y$ satisfying (34.16).

The rest of the proof is similar to the proof of Theorem 34.1. □

Corollary 34.3 *Let $p > 1$ and θ be positive real numbers, and let $f : X \to Y$ be a mapping such that*

$$\|Df(x, y)\| \le \theta\left(\|x\|^p + \|y\|^p\right) \tag{34.21}$$

for all $x, y \in X$. Then there exists a unique additive mapping $A : X \to Y$ such that

$$\|f(x) - f(-x) - A(x)\| \le \frac{2\theta}{2^p - 2}\|x\|^p$$

for all $x \in X$.

Proof Define $\varphi(x, y) = \theta(\|x\|^p + \|y\|^p)$, and apply Theorem 34.3 to get the desired result. □

Theorem 34.4 *Let $f : X \to Y$ be a mapping satisfying $f(0) = 0$ for which there exists a function $\varphi : X^2 \to [0, \infty)$ satisfying* (34.15) *such that*

$$\Phi(x, y) := \sum_{j=0}^{\infty} 2^{-j} \varphi(2^j x, 2^j y) < \infty \tag{34.22}$$

for all $x, y \in X$. Then there exists a unique additive mapping $A : X \to Y$ such that

$$\| f(x) - f(-x) - A(x) \| \le \frac{1}{4} (\Phi(x, x) + \Phi(-x, -x)) \tag{34.23}$$

for all $x \in X$.

Proof It follows from (34.19) that

$$\left\| h(x) - \frac{1}{2} h(2x) \right\| \le \frac{1}{4} \varphi(x, x) + \frac{1}{4} \varphi(-x, -x)$$

for all $x \in X$. So

$$\begin{aligned} \left\| \frac{1}{2^l} h(2^l x) - \frac{1}{2^m} h(2^m x) \right\| \le & \sum_{j=l}^{m-1} \frac{1}{2^{j+2}} \varphi(2^j x, 2^j x) \\ & + \sum_{j=l}^{m-1} \frac{1}{2^{j+2}} \varphi(-2^j x, -2^j x) \end{aligned} \tag{34.24}$$

for all nonnegative integers m and l with $m > l$ and all $x \in X$. It follows from (34.22) and (34.24) that the sequence $\{\frac{1}{2^k} h(2^k x)\}$ is Cauchy for all $x \in X$. Since Y is complete, the sequence $\{\frac{1}{2^k} h(2^k x)\}$ converges. So one can define the mapping $A : X \to Y$ by

$$A(x) := \lim_{k \to \infty} \frac{1}{2^k} h(2^k x)$$

for all $x \in X$.

By (34.15) and (34.22),

$$\begin{aligned} \| DA(x, y) \| &= \lim_{k \to \infty} \frac{1}{2^k} \| Dh(2^k x, 2^k y) \| \\ &\le \lim_{k \to \infty} \frac{1}{2^k} (\varphi(2^k x, 2^k y) + \varphi(-2^k x, -2^k y)) = 0 \end{aligned}$$

for all $x, y \in X$. So $DA(x, y) = 0$. Since $h : X \to Y$ is odd, $A : X \to Y$ is odd. So the mapping $A : X \to Y$ is additive. Moreover, letting $l = 0$ and passing the limit $m \to \infty$ in (34.24), we get (34.23). So there exists an additive mapping $A : X \to Y$ satisfying (34.23).

The rest of the proof is similar to the proof of Theorem 34.1. □

Corollary 34.4 *Let $p < 1$ and θ be positive real numbers, and let $f : X \to Y$ be a mapping satisfying* (34.21). *Then there exists a unique additive mapping $A : X \to Y$ such that*

$$\|f(x) - f(-x) - A(x)\| \le \frac{2\theta}{2 - 2^p}\|x\|^p$$

for all $x \in X$.

Proof Define $\varphi(x, y) = \theta(\|x\|^p + \|y\|^p)$, and apply Theorem 34.4 to get the desired result. □

Note that

$$\sum_{j=0}^{\infty} 2^j \varphi\left(\frac{x}{2^j}, \frac{y}{2^j}\right) \le \sum_{j=0}^{\infty} 4^j \varphi\left(\frac{x}{2^j}, \frac{y}{2^j}\right).$$

Combining Theorem 34.1 and Theorem 34.3, we obtain the following result.

Theorem 34.5 *Let $f : X \to Y$ be a mapping satisfying $f(0) = 0$ for which there exists a function $\varphi : X^2 \to [0, \infty)$ satisfying* (34.3) *and* (34.4). *Then there exist an additive mapping $A : X \to Y$ and a quadratic mapping $Q : X \to Y$ such that*

$$\|f(x) - A(x) - Q(x)\| \le \frac{1}{16}\widetilde{\varphi}(x, x) + \frac{1}{16}\widetilde{\varphi}(-x, -x) + \frac{1}{8}\Phi(x, x) + \frac{1}{8}\Phi(-x, -x)$$

for all $x \in X$, where $\widetilde{\varphi}$ and Φ are defined in (34.3) *and* (34.14), *respectively.*

Corollary 34.5 *Let $p > 2$ and θ be positive real numbers, and let $f : X \to Y$ be a mapping satisfying* (34.10). *Then there exist an additive mapping $A : X \to Y$ and a quadratic mapping $Q : X \to Y$ such that*

$$\|f(x) - A(x) - Q(x)\| \le \left(\frac{1}{2^p - 2} + \frac{1}{2^p - 4}\right)\theta\|x\|^p$$

for all $x \in X$.

Proof Define $\varphi(x, y) = \theta(\|x\|^p + \|x\|^p)$, and apply Theorem 34.5 to get the desired result. □

Note that

$$\sum_{j=1}^{\infty} 4^{-j}\varphi(2^j x, 2^j y) \le \sum_{j=1}^{\infty} 2^{-j}\varphi(2^j x, 2^j y).$$

Combining Theorem 34.2 and Theorem 34.4, we obtain the following result.

Theorem 34.6 *Let $f : X \to Y$ be a mapping satisfying $f(0) = 0$ for which there exists a function $\varphi : X^2 \to [0, \infty)$ satisfying* (34.4) *and* (34.22). *Then there exist an additive mapping $A : X \to Y$ and a quadratic mapping $Q : X \to Y$ such that*

$$\|f(x) - A(x) - Q(x)\| \leq \frac{1}{16}\widetilde{\varphi}(x, x) + \frac{1}{16}\widetilde{\varphi}(-x, -x) + \frac{1}{8}\Phi(x, x) + \frac{1}{8}\Phi(-x, -x)$$

for all $x \in X$, where $\widetilde{\varphi}$ and Φ are defined in (34.11) *and* (34.22), *respectively.*

Corollary 34.6 *Let $p < 1$ and θ be positive real numbers, and let $f : X \to Y$ be a mapping satisfying* (34.21). *Then there exist an additive mapping $A : X \to Y$ and a quadratic mapping $Q : X \to Y$ such that*

$$\|f(x) - A(x) - Q(x)\| \leq \left(\frac{1}{2 - 2^p} + \frac{1}{4 - 2^p}\right)\theta\|x\|^p$$

for all $x \in X$.

Proof Define $\varphi(x, y) = \theta(\|x\|^p + \|y\|^p)$, and apply Theorem 34.6 to get the desired result. □

Similarly, we obtain the following.

Corollary 34.7 *Let $1 < p < 2$ and θ be positive real numbers, and let $f : X \to Y$ be a mapping satisfying* (34.10). *Then there exist an additive mapping $A : X \to Y$ and a quadratic mapping $Q : X \to Y$ such that*

$$\|f(x) - A(x) - Q(x)\| \leq \left(\frac{1}{2^p - 2} + \frac{1}{4 - 2^p}\right)\theta\|x\|^p$$

for all $x \in X$.

References

1. Aoki, T.: On the stability of the linear transformation in Banach spaces. J. Math. Soc. Jpn. **2**, 64–66 (1950)
2. Cholewa, P.W.: Remarks on the stability of functional equations. Aequ. Math. **27**, 76–86 (1984)
3. Czerwik, S.: On the stability of the quadratic mapping in normed spaces. Abh. Math. Semin. Univ. Hamb. **62**, 59–64 (1992)
4. Găvruta, P.: A generalization of the Hyers–Ulam–Rassias stability of approximately additive mappings. J. Math. Anal. Appl. **184**, 431–436 (1994)
5. Hyers, D.H.: On the stability of the linear functional equation. Proc. Natl. Acad. Sci. USA **27**, 222–224 (1941)

6. Park, C., Huh, J., Min, W., Nam, D., Roh, S.: Functional equations associated with inner product spaces. J. Chungcheong Math. Soc. **21**, 455–466 (2008)
7. Rassias, Th.M.: On the stability of the linear mapping in Banach spaces. Proc. Am. Math. Soc. **72**, 297–300 (1978)
8. Rassias, Th.M.: New characterizations of inner product spaces. Bull. Sci. Math. **108**, 95–99 (1984)
9. Rassias, Th.M.: On the stability of the quadratic functional equation and its applications. Stud. Univ. Babeş-Bolyai, Math. **XLIII**, 89–124 (1998)
10. Rassias, Th.M.: The problem of S.M. Ulam for approximately multiplicative mappings. J. Math. Anal. Appl. **246**, 352–378 (2000)
11. Rassias, Th.M.: On the stability of functional equations in Banach spaces. J. Math. Anal. Appl. **251**, 264–284 (2000)
12. Rassias, Th.M.: On the stability of functional equations and a problem of Ulam. Acta Appl. Math. **62**, 23–130 (2000)
13. Rassias, Th.M., Šemrl, P.: On the Hyers–Ulam stability of linear mappings. J. Math. Anal. Appl. **173**, 325–338 (1993)
14. Rassias, Th.M., Shibata, K.: Variational problem of some quadratic functionals in complex analysis. J. Math. Anal. Appl. **228**, 234–253 (1998)
15. Skof, F.: Proprietà locali e approssimazione di operatori. Rend. Semin. Mat. Fis. Milano **53**, 113–129 (1983)
16. Ulam, S.M.: Problems in Modern Mathematics. Wiley, New York (1960)

Chapter 35
Classification and Stability of Functional Equations

Choonkil Park, Madjid Eshaghi Gordji, and Reza Saadati

Abstract In this paper, we classify and prove the generalized Hyers–Ulam stability of linear, quadratic, cubic, quartic, and quintic functional equations in complex Banach spaces.

Key words Fixed point · (Linear, quadratic, cubic, quartic, quintic) functional equation · Generalized Hyers–Ulam stability

Mathematics Subject Classification Primary 39B72 · 47H10

35.1 Introduction and Preliminaries

The stability problem of functional equations was originated from a question of Ulam [45] concerning the stability of group homomorphisms. Hyers [17] gave a first affirmative partial answer to the question of Ulam for Banach spaces. Let X and Y be Banach spaces. Assume that $f : X \to Y$ satisfies

$$\left\| f(x+y) - f(x) - f(y) \right\| \le \varepsilon$$

for all $x, y \in X$ and some $\varepsilon \ge 0$. Then there exists a unique additive mapping $T : X \to Y$ such that

---

Dedicated to Professor Themistocles M. Rassias on the occasion of his 60th birthday.

C. Park (✉)
Department of Mathematics, Hanyang University, Seoul 133-791, Republic of Korea
e-mail: baak@hanyang.ac.kr

M.E. Gordji
Department of Mathematics, Semnan University, P.O. Box 35195-363, Semnan, Iran
e-mail: madjid.eshaghi@gmail.com

R. Saadati
Department of Mathematics and Computer Science, Amirkabir University of Technology, 424 Hafez Avenue, Tehran 15914, Iran
e-mail: rsaadati@aut.ac.ir

P.M. Pardalos et al. (eds.), *Nonlinear Analysis*, Springer Optimization and Its Applications 68, In Honor of Themistocles M. Rassias on the Occasion of his 60th Birthday, DOI 10.1007/978-1-4614-3498-6_35,

$$\|f(x) - T(x)\| \le \varepsilon$$

for all $x \in X$.

Th.M. Rassias [36] provided a generalization of Hyers' Theorem which allows the *Cauchy difference to be unbounded.*

Theorem 35.1 (Th.M. Rassias) *Let $f : E \to E'$ be a mapping from a normed vector space E into a Banach space E' subject to the inequality*

$$\|f(x+y) - f(x) - f(y)\| \le \varepsilon\big(\|x\|^p + \|y\|^p\big) \tag{35.1}$$

for all $x, y \in E$, where ε and p are constants with $\varepsilon > 0$ and $p < 1$. Then the limit

$$L(x) = \lim_{n\to\infty} \frac{f(2^n x)}{2^n}$$

exists for all $x \in E$ and $L : E \to E'$ is the unique additive mapping which satisfies

$$\|f(x) - L(x)\| \le \frac{2\varepsilon}{2 - 2^p}\|x\|^p$$

for all $x \in E$. Also, if for each $x \in E$ the mapping $f(tx)$ is continuous in $t \in \mathbb{R}$, then L is $\mathbb{R}$-linear.

The above inequality (35.1) that was introduced for the first time by Th.M. Rassias [36] for the proof of the stability of the linear mapping between Banach spaces has provided a lot of influence in the development of what is now known as the *generalized Hyers–Ulam stability* or the *Hyers–Ulam–Rassias stability* of functional equations. Beginning around the year 1980, the topic of approximate homomorphisms, or the stability of the equation of homomorphism, was studied by a number of mathematicians. Găvruta [11] extended the generalized Hyers–Ulam stability by proving the following theorem in the spirit of Th.M. Rassias' approach.

Theorem 35.2 ([11]) *Let $f : E \to E'$ be a mapping for which there exists a function $\varphi : E \times E' \to [0, \infty)$ such that*

$$\widetilde{\varphi}(x, y) := \sum_{j=0}^{\infty} 2^{-j}\varphi\big(2^j x, 2^j y\big) < \infty,$$

$$\|f(x+y) - f(x) - f(y)\| \le \varphi(x, y)$$

for all $x, y \in E$. Then there exists a unique additive mapping $T : E \to E'$ such that

$$\|f(x) - T(x)\| \le \frac{1}{2}\widetilde{\varphi}(x, x)$$

for all $x \in E$.

Theorem 35.3 ([35]) *Let X be a real normed linear space and Y a real complete normed linear space. Assume that $f : X \to Y$ is an approximately additive mapping for which there exist constants $\theta \geq 0$ and $p \in \mathbb{R} - \{1\}$ such that f satisfies the inequality*

$$\|f(x+y) - f(x) - f(y)\| \leq \theta \cdot \|x\|^{\frac{p}{2}} \cdot \|y\|^{\frac{p}{2}}$$

for all $x, y \in X$. Then there exists a unique additive mapping $L : X \to Y$ satisfying

$$\|f(x) - L(x)\| \leq \frac{\theta}{|2^p - 2|}\|x\|^p$$

for all $x \in X$. If, in addition, $f : X \to Y$ is a mapping such that the transformation $t \to f(tx)$ is continuous in $t \in \mathbb{R}$ for each fixed $x \in X$, then L is an $\mathbb{R}$-linear mapping.

The functional equation

$$f(x+y) + f(x-y) = 2f(x) + 2f(y)$$

is called a *quadratic functional equation*. In particular, every solution of the quadratic functional equation is said to be a *quadratic mapping*. A generalized Hyers–Ulam stability problem for the quadratic functional equation was proved by Skof [44] for mappings $f : X \to Y$, where X is a normed space and Y is a Banach space. Cholewa [5] noticed that the theorem of Skof is still true if the relevant domain X is replaced by an abelian group. Czerwik [6, 7] proved the generalized Hyers–Ulam stability of the quadratic functional equation. The stability problems of several functional equations have been extensively investigated by a number of authors, and there are many interesting results concerning this problem (see [1, 8, 9, 15, 19, 21–33, 37–43]).

We recall two fundamental results in fixed point theory. The reader is referred to the book of D.H. Hyers, G. Isac, and Th.M. Rassias [18] for an extensive account of fixed point theory with several applications.

Theorem 35.4 ([2, 3, 10, 34]) *Let (X, d) be a complete generalized metric space and let $J : X \to X$ be a strictly contractive mapping with Lipschitz constant $L < 1$. Then for each given element $x \in X$, either*

$$d(J^n x, J^{n+1} x) = \infty$$

for all nonnegative integers n or there exists a positive integer n_0 such that

1. $d(J^n x, J^{n+1} x) < \infty, \forall n \geq n_0$;
2. *The sequence $\{J^n x\}$ converges to a fixed point y^* of J;*
3. *y^* is the unique fixed point of J in the set $Y = \{y \in X \mid d(J^{n_0} x, y) < \infty\}$;*
4. *$d(y, y^*) \leq \frac{1}{1-L} d(y, Jy)$ for all $y \in Y$.*

G. Isac and Th.M. Rassias [20] were the first to provide applications of stability theory of functional equations for the proof of new fixed point theorems with applications. By using fixed point methods, the stability problems of several functional equations have been extensively investigated by a number of authors (see [3, 4, 12–16, 25, 26], and [34]).

This paper is organized as follows: In Sect. 35.2, using the fixed point method, we prove the generalized Hyers–Ulam stability of the following functional equations

$$f(x+iy)+f(x-iy)+f(x+y)+f(x-y)=4f(x) \tag{35.2}$$

and $f((1+i)x)=(1+i)^k f(x)$ ($k=1,2,3$, respectively), whose solution is called a *linear mapping, quadratic mapping*, and *cubic mapping*, respectively.

In Sect. 35.3, using the fixed point method, we prove the generalized Hyers–Ulam stability of the following quartic functional equations

$$f(x+iy)+f(x-iy)+f(x+y)+f(x-y)=4f(x)+4f(y) \tag{35.3}$$

and $f((1+i)x)=-4f(x)$, whose solution is called a *quartic mapping*.

In Sect. 35.4, using the fixed point method, we prove the generalized Hyers–Ulam stability of the following quintic functional equations

$$f(x+iy)+f(x-iy)+if(ix+y)+if(ix-y)=0, \tag{35.4}$$

$f((1-i)x)=(1-i)^5 f(x)$ and $f((i-1)x)=(i-1)^5 f(x)$, whose solution is called a *quintic mapping*.

Throughout this paper, assume that X is a complex normed vector space with norm $\|\cdot\|$ and that Y is a complex Banach space with norm $\|\cdot\|$. Let $k=1,2,3$ be fixed.

35.2 Generalized Hyers–Ulam Stability of Linear, Quadratic, and Cubic Functional Equations

For a given mapping $f: X \to Y$, we define

$$Cf(x,y) := f(x+iy)+f(x-iy)+f(x+y)+f(x-y)-4f(x)$$

for all $x, y \in X$.

Using the fixed point method, we prove the generalized Hyers–Ulam stability of the functional equation $Cf(x,y)=0$.

Theorem 35.5 *Let $f: X \to Y$ be a mapping with $f((1+i)x)=(1+i)^k f(x)$ for all $x \in X$ for which there exists a function $\varphi: X^2 \to [0,\infty)$ such that*

$$\sum_{j=0}^{\infty} 2^{-kj}\varphi(2^j x, 2^j y) < \infty, \tag{35.5}$$

$$\|Cf(x,y)\| \le \varphi(x,y) \tag{35.6}$$

for all $x, y \in X$. *If there exists an* $L < 1$ *such that* $\varphi(x,x) \le 2^k L\varphi(\frac{x}{2}, \frac{x}{2})$ *for all* $x \in X$, *then there exists a unique mapping* $Q : X \to Y$ *satisfying* (35.2), $Q((1+i)x) = (1+i)^k Q(x)$ *and*

$$\|f(x) - Q(x)\| \le \frac{1}{(1-L)|4-(1+i)^k|}\varphi(x,x) \tag{35.7}$$

for all $x \in X$.

Proof Consider the set

$$S := \{g : X \to Y\}$$

and introduce the *generalized metric* on S:

$$d(g,h) = \inf\{K \in \mathbb{R}_+ : \|g(x) - h(x)\| \le K\varphi(x,x), \quad \forall x \in X\}.$$

It is easy to show that (S, d) is complete.

Now we consider the linear mapping $J : S \to S$ such that

$$Jg(x) := \frac{1}{2^k} g(2x)$$

for all $x \in X$.

By Theorem 3.1 of [2],

$$d(Jg, Jh) \le Ld(g,h)$$

for all $g, h \in S$.

Letting $y = x$ in (35.6), we get

$$\|f((1+i)x) + f((1-i)x) + f(2x) - 4f(x)\| \le \varphi(x,x)$$

for all $x \in X$. So

$$\left\| f(x) - \frac{1}{2^k} f(2x) \right\| \le \frac{1}{|4-(1+i)^k|}\varphi(x,x) \tag{35.8}$$

for all $x \in X$. Hence $d(f, Jf) \le \frac{1}{|4-(1+i)^k|}$.

By Theorem 35.4, there exists a mapping $Q : X \to Y$ such that

1. Q is a fixed point of J, i.e.,

$$Q(2x) = 2^k Q(x) \tag{35.9}$$

for all $x \in X$. The mapping Q is a unique fixed point of J in the set

$$M = \{g \in S : d(f,g) < \infty\}.$$

This implies that Q is a unique mapping satisfying (35.9) such that there exists $K \in (0, \infty)$ satisfying

$$\|f(x) - Q(x)\| \le K\varphi(x, x)$$

for all $x \in X$.

2. $d(J^n f, Q) \to 0$ as $n \to \infty$. This implies the equality

$$\lim_{n\to\infty} \frac{f(2^n x)}{2^{kn}} = Q(x) \tag{35.10}$$

for all $x \in X$.

3. $d(f, Q) \le \frac{1}{1-L} d(f, Jf)$, which implies the inequality

$$d(f, Q) \le \frac{1}{(1-L)|4-(1+i)^k|}.$$

This implies that the inequality (35.7) holds.

It follows from (35.5), (35.6), and (35.10) that

$$\|CQ(x, y)\| = \lim_{n\to\infty} \frac{1}{2^{kn}} \|Cf(2^n x, 2^n y)\| \le \lim_{n\to\infty} \frac{1}{2^{kn}} \varphi(2^n x, 2^n y) = 0$$

for all $x, y \in X$. So $CQ(x, y) = 0$ for all $x, y \in X$.

It is easy to show that $Q((1+i)x) = (1+i)^k Q(x)$ for all $x \in X$.

Therefore, the mapping $Q : X \to Y$ is a unique mapping satisfying (35.2), (35.7), and $Q((1+i)x) = (1+i)^k Q(x)$ for all $x \in X$, as desired. □

We are going to prove the generalized Hyers–Ulam stability of the functional equation $Cf(x, y) = 0$.

Corollary 35.1 *Let $p < k$ and θ be positive real numbers, and let $f : X \to Y$ be a mapping satisfying $\|Cf(x, y)\| \le \theta(\|x\|^p + \|y\|^p)$ for all $x, y \in X$ and $f((1+i)x) = (1+i)^k f(x)$ for all $x \in X$. Then there exists a unique mapping $Q : X \to Y$ satisfying* (35.2) *and $\|f(x) - Q(x)\| \le \frac{2^{k+1}\theta}{(2^k - 2^p)|4-(1+i)^k|}\|x\|^p$, and $Q((1+i)x) = (1+i)^k Q(x)$ for all $x \in X$.*

Proof The proof follows from Theorem 35.5 by taking

$$\varphi(x, y) := \theta(\|x\|^p + \|y\|^p)$$

for all $x, y \in X$. Then we can choose $L = 2^{p-k}$ and we get the desired result. □

Corollary 35.2 *Let $p < \frac{k}{2}$ and θ be positive real numbers, and let $f : X \to Y$ be a mapping such that*

$$\|Cf(x, y)\| \le \theta \cdot \|x\|^p \cdot \|y\|^p$$

for all $x, y \in X$ *and that* $f((1+i)x) = (1+i)^k f(x)$ *for all* $x \in X$. *Then there exists a unique mapping* $Q : X \to Y$ *satisfying* (35.2) *and* $\|f(x) - Q(x)\| \le \frac{2^k \theta}{(2^k - 4^p)|4-(1+i)^k|}\|x\|^{2p}$ *and* $Q((1+i)x) = (1+i)^k Q(x)$ *for all* $x \in X$.

Proof The proof follows from Theorem 35.5 by taking

$$\varphi(x, y) := \theta \cdot \|x\|^p \cdot \|y\|^p$$

for all $x, y \in X$. Then we can choose $L = 4^{p-\frac{k}{2}}$ and we get the desired result. □

Theorem 35.6 *Let* $f : X \to Y$ *be a mapping with* $f((1+i)x) = (1+i)^k f(x)$ *for all* $x \in X$ *for which there exists a function* $\varphi : X^2 \to [0, \infty)$ *satisfying* (35.6) *such that*

$$\sum_{j=0}^{\infty} 2^{kj} \varphi\left(\frac{x}{2^j}, \frac{y}{2^j}\right) < \infty$$

for all $x, y \in X$. *If there exists an* $L < 1$ *such that* $\varphi(x, x) \le \frac{1}{2^k} L \varphi(2x, 2x)$ *for all* $x \in X$, *then there exists a unique mapping* $Q : X \to Y$ *satisfying* (35.2), $Q((1+i)x) = (1+i)^k Q(x)$, *and*

$$\|f(x) - Q(x)\| \le \frac{L}{(1-L)|4-(1+i)^k|}\varphi(x, x) \tag{35.11}$$

for all $x \in X$.

Proof We consider the linear mapping $J : S \to S$ such that

$$Jg(x) := 2^k g\left(\frac{x}{2}\right)$$

for all $x \in X$.

It follows from (35.8) that

$$\left\| f(x) - 2^k f\left(\frac{x}{2}\right)\right\| \le \frac{2^k}{|4-(1+i)^k|}\varphi\left(\frac{x}{2}, \frac{x}{2}\right) \le \frac{L}{|4-(1+i)^k|}\varphi(x, x)$$

for all $x \in X$. Hence $d(f, Jf) \le \frac{L}{|4-(1+i)^k|}$.

By Theorem 35.4, there exists a mapping $Q : X \to Y$ such that

1. Q is a fixed point of J, i.e.,

$$Q(2x) = 2^k Q(x) \tag{35.12}$$

for all $x \in X$. The mapping Q is a unique fixed point of J in the set

$$M = \{g \in S : d(f, g) < \infty\}.$$

This implies that Q is a unique mapping satisfying (35.12) such that there exists $K \in (0, \infty)$ satisfying

$$\|f(x) - Q(x)\| \le K\varphi(x, x)$$

for all $x \in X$.

2. $d(J^n f, Q) \to 0$ as $n \to \infty$. This implies the equality

$$\lim_{n\to\infty} 2^{kn} f\left(\frac{x}{2^n}\right) = Q(x)$$

for all $x \in X$.

3. $d(f, Q) \le \frac{1}{1-L} d(f, Jf)$, which implies the inequality

$$d(f, Q) \le \frac{L}{(1-L)|4-(1+i)^k|},$$

which implies that the inequality (35.11) holds.

The rest of the proof is similar to the proof of Theorem 35.5. □

Corollary 35.3 *Let $p > k$ and θ be positive real numbers, and let $f : X \to Y$ be a mapping satisfying $\|Cf(x, y)\| \le \theta(\|x\|^p + \|y\|^p)$ for all $x, y \in X$ and $f((1+i)x) = (1+i)^k f(x)$ for all $x \in X$. Then there exists a unique mapping $Q : X \to Y$ satisfying* (35.2), *$\|f(x) - Q(x)\| \le \frac{2^{k+1}\theta}{(2^p-2^k)|4-(1+i)^k|}\|x\|^p$ and $Q((1+i)x) = (1+i)^k Q(x)$ for all $x \in X$.*

Proof The proof follows from Theorem 35.6 by taking

$$\varphi(x, y) := \theta\big(\|x\|^p + \|y\|^p\big)$$

for all $x, y \in X$. Then we can choose $L = 2^{k-p}$ and we get the desired result. □

Corollary 35.4 *Let $p > \frac{k}{2}$ and θ be positive real numbers, and let $f : X \to Y$ be a mapping satisfying $\|Cf(x, y)\| \le \theta \cdot \|x\|^p \cdot \|y\|^p$ for all $x, y \in X$ and $f((1+i)x) = (1+i)^k f(x)$ for all $x \in X$. Then there exists a unique mapping $Q : X \to Y$ satisfying* (35.2) *and $\|f(x) - Q(x)\| \le \frac{2^k\theta}{(2^k-4^p)|4-(1+i)^k|}\|x\|^{2p}$ and $Q((1+i)x) = (1+i)^k Q(x)$ for all $x \in X$.*

Proof The proof follows from Theorem 35.6 by taking

$$\varphi(x, y) := \theta \cdot \|x\|^p \cdot \|y\|^p$$

for all $x, y \in X$. Then we can choose $L = 4^{\frac{k}{2}-p}$ and we get the desired result. □

35.3 Generalized Hyers–Ulam Stability of a Quartic Functional Equation

For a given mapping $f: X \to Y$, we define

$$Df(x,y) := f(x+iy) + f(x-iy) + f(x+y) + f(x-y) - 4f(x) - 4f(y)$$

for all $x, y \in X$. Then $f: \mathbb{C} \to \mathbb{C}$ with $f(x) = x^4$ satisfies (35.3).

Using the fixed point method, we prove the generalized Hyers–Ulam stability of the quartic functional equation $Df(x,y) = 0$.

Theorem 35.7 *Let $f: X \to Y$ be a mapping with $f((1+i)x) = -4f(x)$ for all $x \in X$ for which there exists a function $\varphi: X^2 \to [0,\infty)$ such that*

$$\sum_{j=0}^{\infty} 2^{-4j} \varphi\left(2^j x, 2^j y\right) < \infty, \tag{35.13}$$

$$\left\| Df(x,y) \right\| \le \varphi(x,y) \tag{35.14}$$

for all $x, y \in X$. If there exists an $L < 1$ such that $\varphi(x,x) \le 16L\varphi(\frac{x}{2}, \frac{x}{2})$ for all $x \in X$, then there exists a unique quartic mapping $Q: X \to Y$ such that

$$\left\| f(x) - Q(x) \right\| \le \frac{1}{12 - 12L} \varphi(x,x) \tag{35.15}$$

for all $x \in X$.

Proof Consider the set

$$S := \{g: X \to Y\}$$

and introduce a *generalized metric* on S by

$$d(g,h) = \inf\left\{K \in \mathbb{R}_+ : \|g(x) - h(x)\| \le K\varphi(x,x), \quad \forall x \in X\right\}.$$

It is easy to show that (S,d) is complete.

Now we consider the linear mapping $J: S \to S$ such that

$$Jg(x) := \frac{1}{16} g(2x)$$

for all $x \in X$.

By Theorem 3.1 of [2],

$$d(Jg, Jh) \le Ld(g,h)$$

for all $g, h \in S$.

Letting $y = x$ in (35.14), we get

$$\|f((1+i)x) + f((1-i)x) + f(2x) - 8f(x)\| \le \varphi(x, x)$$

for all $x \in X$. So

$$\left\|f(x) - \frac{1}{16}f(2x)\right\| \le \frac{1}{12}\varphi(x, x) \tag{35.16}$$

for all $x \in X$. Hence $d(f, Jf) \le \frac{1}{12}$.

By Theorem 35.4, there exists a mapping $Q : X \to Y$ such that

1. Q is a fixed point of J, i.e.,

$$Q(2x) = 16Q(x) \tag{35.17}$$

   for all $x \in X$. The mapping Q is a unique fixed point of J in the set

$$M = \{g \in S : d(f, g) < \infty\}.$$

   This implies that Q is a unique mapping satisfying (35.17) such that there exists $K \in (0, \infty)$ satisfying

$$\|f(x) - Q(x)\| \le K\varphi(x, x)$$

   for all $x \in X$.
2. $d(J^n f, Q) \to 0$ as $n \to \infty$. This implies the equality

$$\lim_{n\to\infty} \frac{f(2^n x)}{2^{4n}} = Q(x) \tag{35.18}$$

   for all $x \in X$.
3. $d(f, Q) \le \frac{1}{1-L}d(f, Jf)$, which implies the inequality

$$d(f, Q) \le \frac{1}{12 - 12L}.$$

   This implies that the inequality (35.15) holds.

It follows from (35.13), (35.14), and (35.18) that

$$\|DQ(x, y)\| = \lim_{n\to\infty} \frac{1}{2^{4n}}\|Df(2^n x, 2^n y)\| \le \lim_{n\to\infty} \frac{1}{2^{4n}}\varphi(2^n x, 2^n y) = 0$$

for all $x, y \in X$. So $DQ(x, y) = 0$ for all $x, y \in X$.

It is easy to show that $Q((1+i)x) = -4Q(x)$ for all $x \in X$.

Therefore, the mapping $Q : X \to Y$ is the unique quartic mapping satisfying (35.15), as desired. □

We prove the generalized Hyers–Ulam stability of the quartic functional equation $Df(x, y) = 0$.

Corollary 35.5 *Let $p < 4$ and θ be positive real numbers, and let $f : X \to Y$ be a mapping satisfying $\|Df(x,y)\| \le \theta(\|x\|^p + \|y\|^p)$ for all $x, y \in X$ and $f((1+i)x) = -4f(x)$ for all $x \in X$. Then there exists a unique quartic mapping $Q : X \to Y$ satisfying $\|f(x) - Q(x)\| \le \frac{8\theta}{3(16-2^p)}\|x\|^p$ for all $x \in X$.*

Proof The proof follows from Theorem 35.7 by taking

$$\varphi(x, y) := \theta\big(\|x\|^p + \|y\|^p\big)$$

for all $x, y \in X$. Then we can choose $L = 2^{p-4}$ and we get the desired result. □

Corollary 35.6 *Let $p < 2$ and θ be positive real numbers, and let $f : X \to Y$ be a mapping satisfying $\|Df(x,y)\| \le \theta \cdot \|x\|^p \cdot \|y\|^p$ for all $x, y \in X$ and $f((1+i)x) = -4f(x)$ for all $x \in X$. Then there exists a unique quartic mapping $Q : X \to Y$ satisfying $\|f(x) - Q(x)\| \le \frac{4\theta}{3(16-4^p)}\|x\|^{2p}$ for all $x \in X$.*

Proof The proof follows from Theorem 35.7 by taking

$$\varphi(x, y) := \theta \cdot \|x\|^p \cdot \|y\|^p$$

for all $x, y \in X$. Then we can choose $L = 4^{p-2}$ and we get the desired result. □

Theorem 35.8 *Let $f : X \to Y$ be a mapping with $f((1+i)x) = -4f(x)$ for all $x \in X$ for which there exists a function $\varphi : X^2 \to [0, \infty)$ satisfying* (35.14) *such that*

$$\sum_{j=0}^{\infty} 2^{4j}\varphi\left(\frac{x}{2^j}, \frac{y}{2^j}\right) < \infty$$

for all $x, y \in X$. If there exists an $L < 1$ such that $\varphi(x, x) \le \frac{1}{16}L\varphi(2x, 2x)$ for all $x \in X$, then there exists a unique quartic mapping $Q : X \to Y$ such that

$$\big\|f(x) - Q(x)\big\| \le \frac{L}{12 - 12L}\varphi(x, x) \tag{35.19}$$

for all $x \in X$.

Proof We consider the linear mapping $J : S \to S$ such that

$$Jg(x) := 16g\left(\frac{x}{2}\right)$$

for all $x \in X$.

It follows from (35.16) that

$$\left\|f(x) - 16f\left(\frac{x}{2}\right)\right\| \le \frac{4}{3}\varphi\left(\frac{x}{2}, \frac{x}{2}\right) \le \frac{L}{12}\varphi(x, x)$$

for all $x \in X$. Hence $d(f, Jf) \le \frac{L}{12}$.

By Theorem 35.4, there exists a mapping $Q: X \to Y$ such that

1. Q is a fixed point of J, i.e.,

$$Q(2x) = 16Q(x) \tag{35.20}$$

for all $x \in X$. The mapping Q is a unique fixed point of J in the set

$$M = \{g \in S : d(f, g) < \infty\}.$$

This implies that Q is a unique mapping satisfying (35.20) such that there exists $K \in (0, \infty)$ satisfying

$$\|f(x) - Q(x)\| \le K\varphi(x, x)$$

for all $x \in X$.

2. $d(J^n f, Q) \to 0$ as $n \to \infty$. This implies the equality

$$\lim_{n\to\infty} 2^{4n} f\left(\frac{x}{2^n}\right) = Q(x)$$

for all $x \in X$.

3. $d(f, Q) \le \frac{1}{1-L} d(f, Jf)$, which implies the inequality

$$d(f, Q) \le \frac{L}{12 - 12L},$$

which implies that the inequality (35.19) holds.

The rest of the proof is similar to the proof of Theorem 35.7. □

Corollary 35.7 *Let $p > 4$ and θ be positive real numbers, and let $f : X \to Y$ be a mapping satisfying $\|Df(x, y)\| \le \theta(\|x\|^p + \|y\|^p)$ for all $x, y \in X$ and $f((1+i)x) = -4f(x)$ for all $x \in X$. Then there exists a unique quartic mapping $Q : X \to Y$ satisfying $\|f(x) - Q(x)\| \le \frac{8\theta}{3(2^p-16)}\|x\|^p$ for all $x \in X$.*

Proof The proof follows from Theorem 35.8 by taking

$$\varphi(x, y) := \theta(\|x\|^p + \|y\|^p)$$

for all $x, y \in X$. Then we can choose $L = 2^{4-p}$ and we get the desired result. □

Corollary 35.8 *Let $p > 2$ and θ be positive real numbers, and let $f : X \to Y$ be a mapping such that $\|Df(x, y)\| \le \theta \cdot \|x\|^p \cdot \|y\|^p$ for all $x, y \in X$ and that $f((1+i)x) = -4f(x)$ for all $x \in X$. Then there exists a unique quartic mapping $Q : X \to Y$ satisfying $\|f(x) - Q(x)\| \le \frac{4\theta}{3(16-4^p)}\|x\|^{2p}$ for all $x \in X$.*

Proof The proof follows from Theorem 35.8 by taking

$$\varphi(x, y) := \theta \cdot \|x\|^p \cdot \|y\|^p$$

for all $x, y \in X$. Then we can choose $L = 4^{2-p}$ and we get the desired result. □

35.4 Generalized Hyers–Ulam Stability of a Quintic Functional Equation

For a given mapping $f : X \to Y$, we define

$$Df(x, y) := f(x + iy) + f(x - iy) + if(ix + y) + if(ix - y)$$

for all $x, y \in X$. Then $f : \mathbb{C} \to \mathbb{C}$ with $f(x) = x^5$ satisfies (35.4).

Using the fixed point method, we prove the generalized Hyers–Ulam stability of the quintic functional equation $Df(x, y) = 0$.

Theorem 35.9 *Let $f : X \to Y$ be a mapping with $f((1 - i)x) = (1 - i)^5 f(x)$ and $f((i - 1)x) = (i - 1)^5 f(x)$ for all $x \in X$ for which there exists a function $\varphi : X^2 \to [0, \infty)$ such that*

$$\sum_{j=0}^{\infty} 2^{-5j} \varphi\big(2^j x, 2^j y\big) < \infty, \tag{35.21}$$

$$\big\|Df(x, y)\big\| \le \varphi(x, y) \tag{35.22}$$

for all $x, y \in X$. If there exists an $L < 1$ such that $\varphi(x, x) \le 32L\varphi(\frac{x}{2}, \frac{x}{2})$ for all $x \in X$, then there exists a unique quintic mapping $Q : X \to Y$ such that

$$\big\|f(x) - Q(x)\big\| \le \frac{1}{8 - 8L}\varphi(x, x) \tag{35.23}$$

for all $x \in X$.

Proof Consider the set

$$S := \{g : X \to Y\}$$

and introduce a *generalized metric* on S by

$$d(g, h) = \inf\big\{K \in \mathbb{R}_+ : \big\|g(x) - h(x)\big\| \le K\varphi(x, x), \quad \forall x \in X\big\}.$$

It is easy to show that (S, d) is complete.

Now we consider the linear mapping $J : S \to S$ such that

$$Jg(x) := \frac{1}{32} g(2x)$$

for all $x \in X$.

By Theorem 3.1 of [2],

$$d(Jg, Jh) \le Ld(g, h)$$

for all $g, h \in S$.

Letting $y = x$ in (35.22), we get

$$\left\| f\big((1+i)x\big) + f\big((1-i)x\big) + if\big((1+i)x\big) + if\big((i-1)x\big) \right\| \le \varphi(x, x)$$

for all $x \in X$. So

$$\left\| f(x) - \frac{1}{32} f(2x) \right\| \le \frac{1}{8}\varphi(x, x) \tag{35.24}$$

for all $x \in X$. Hence $d(f, Jf) \le \frac{1}{8}$.

By Theorem 35.4, there exists a mapping $Q : X \to Y$ such that

1. Q is a fixed point of J, i.e.,

$$Q(2x) = 32Q(x) \tag{35.25}$$

for all $x \in X$. The mapping Q is a unique fixed point of J in the set

$$M = \{g \in S : d(f, g) < \infty\}.$$

This implies that Q is a unique mapping satisfying (35.25) such that there exists $K \in (0, \infty)$ satisfying

$$\left\| f(x) - Q(x) \right\| \le K\varphi(x, x)$$

for all $x \in X$.

2. $d(J^n f, Q) \to 0$ as $n \to \infty$. This implies the equality

$$\lim_{n\to\infty} \frac{f(2^n x)}{2^{5n}} = Q(x) \tag{35.26}$$

for all $x \in X$.

3. $d(f, Q) \le \frac{1}{1-L} d(f, Jf)$, which implies the inequality

$$d(f, Q) \le \frac{1}{8 - 8L}.$$

This implies that the inequality 35.23 holds.

It follows from (35.21), (35.22), and (35.26) that

$$\left\| DQ(x, y) \right\| = \lim_{n\to\infty} \frac{1}{2^{5n}} \left\| Df\big(2^n x, 2^n y\big) \right\| \le \lim_{n\to\infty} \frac{1}{2^{5n}} \varphi\big(2^n x, 2^n y\big) = 0$$

for all $x, y \in X$. So $DQ(x, y) = 0$ for all $x, y \in X$.

It is easy to show that $Q((1-i)x) = (1-i)^5 Q(x)$ and $Q((i-1)x) = (i-1)^5 Q(x)$ for all $x \in X$.

Therefore, the mapping $Q : X \to Y$ is the unique quintic mapping satisfying (35.23), as desired. □

We prove the generalized Hyers–Ulam stability of the quintic functional equation $Df(x, y) = 0$.

Corollary 35.9 *Let $p < 5$ and θ be positive real numbers, and let $f : X \to Y$ be a mapping such that $\|Df(x, y)\| \le \theta(\|x\|^p + \|y\|^p)$ for all $x, y \in X$ and that $f((1-i)x) = (1-i)^5 f(x)$ and $f((i-1)^5 x) = (i-1)^5 f(x)$ for all $x \in X$. Then there exists a unique quintic mapping $Q : X \to Y$ satisfying $\|f(x) - Q(x)\| \le \frac{8\theta}{32-2^p}\|x\|^p$ for all $x \in X$.*

Proof The proof follows from Theorem 35.9 by taking

$$\varphi(x, y) := \theta\big(\|x\|^p + \|y\|^p\big)$$

for all $x, y \in X$. Then we can choose $L = 2^{p-5}$ and we get the desired result. □

Corollary 35.10 *Let $p < \frac{5}{2}$ and θ be positive real numbers, and let $f : X \to Y$ be a mapping such that $\|Df(x, y)\| \le \theta \cdot \|x\|^p \cdot \|y\|^p$ for all $x, y \in X$ and that $f((1-i)x) = (1-i)^5 f(x)$ and $f((i-1)^5 x) = (i-1)^5 f(x)$ for all $x \in X$. Then there exists a unique quintic mapping $Q : X \to Y$ satisfying $\|f(x) - Q(x)\| \le \frac{4\theta}{32-4^p}\|x\|^{2p}$ for all $x \in X$.*

Proof The proof follows from Theorem 35.9 by taking

$$\varphi(x, y) := \theta \cdot \|x\|^p \cdot \|y\|^p$$

for all $x, y \in X$. Then we can choose $L = 4^{p-\frac{5}{2}}$ and we get the desired result. □

Theorem 35.10 *Let $f : X \to Y$ be a mapping with $f((1-i)x) = (1-i)^5 f(x)$ and $f((i-1)^5 x) = (i-1)^5 f(x)$ for all $x \in X$ for which there exists a function $\varphi : X^2 \to [0, \infty)$ satisfying (35.22) such that*

$$\sum_{j=0}^{\infty} 2^{5j} \varphi\left(\frac{x}{2^j}, \frac{y}{2^j}\right) < \infty$$

for all $x, y \in X$. If there exists an $L < 1$ such that $\varphi(x, x) \le \frac{1}{32} L\varphi(2x, 2x)$ for all $x \in X$, then there exists a unique quintic mapping $Q : X \to Y$ such that

$$\big\|f(x) - Q(x)\big\| \le \frac{L}{8-8L}\varphi(x, x) \tag{35.27}$$

for all $x \in X$.

Proof We consider the linear mapping $J : S \to S$ such that

$$Jg(x) := 32g\left(\frac{x}{2}\right)$$

for all $x \in X$.

It follows from (35.24) that

$$\left\| f(x) - 32f\left(\frac{x}{2}\right) \right\| \le 4\varphi\left(\frac{x}{2}, \frac{x}{2}\right) \le \frac{L}{8}\varphi(x, x)$$

for all $x \in X$. Hence $d(f, Jf) \le \frac{L}{8}$.

By Theorem 35.4, there exists a mapping $Q : X \to Y$ such that

1. Q is a fixed point of J, i.e.,

$$Q(2x) = 32Q(x) \tag{35.28}$$

for all $x \in X$. The mapping Q is a unique fixed point of J in the set

$$M = \{g \in S : d(f, g) < \infty\}.$$

This implies that Q is a unique mapping satisfying (35.28) such that there exists $K \in (0, \infty)$ satisfying

$$\| f(x) - Q(x)\| \le K\varphi(x, x)$$

for all $x \in X$.

2. $d(J^n f, Q) \to 0$ as $n \to \infty$. This implies the equality

$$\lim_{n\to\infty} 2^{5n} f\left(\frac{x}{2^n}\right) = Q(x)$$

for all $x \in X$.

3. $d(f, Q) \le \frac{1}{1-L} d(f, Jf)$, which implies the inequality

$$d(f, Q) \le \frac{L}{8 - 8L},$$

which implies that the inequality (35.27) holds.

The rest of the proof is similar to the proof of Theorem 35.9. □

Corollary 35.11 *Let $p > 5$ and θ be positive real numbers, and let $f : X \to Y$ be a mapping such that $\|Df(x, y)\| \le \theta(\|x\|^p + \|y\|^p)$ for all $x, y \in X$ and that $f((1-i)x) = (1-i)^5 f(x)$ and $f((i-1)^5 x) = (i-1)^5 f(x)$ for all $x \in X$. Then there exists a unique quintic mapping $Q : X \to Y$ satisfying $\|f(x) - Q(x)\| \le \frac{8\theta}{2^p - 32}\|x\|^p$ for all $x \in X$.*

Proof The proof follows from Theorem 35.10 by taking

$$\varphi(x, y) := \theta\big(\|x\|^p + \|y\|^p\big)$$

for all $x, y \in X$. Then we can choose $L = 2^{5-p}$ and we get the desired result. □

Corollary 35.12 *Let $p > \frac{5}{2}$ and θ be positive real numbers, and let $f : X \to Y$ be a mapping such that $\|Df(x, y)\| \leq \theta \cdot \|x\|^p \cdot \|y\|^p$ for all $x, y \in X$ and that $f((1-i)x) = (1-i)^5 f(x)$ and $f((i-1)^5 x) = (i-1)^5 f(x)$ for all $x \in X$. Then there exists a unique quintic mapping $Q : X \to Y$ satisfying $\|f(x) - Q(x)\| \leq \frac{4\theta}{4^p - 32}\|x\|^{2p}$ for all $x \in X$.*

Proof The proof follows from Theorem 35.10 by taking

$$\varphi(x, y) := \theta \cdot \|x\|^p \cdot \|y\|^p$$

for all $x, y \in X$. Then we can choose $L = 4^{\frac{5}{2}-p}$ and we get the desired result. □

References

1. Bae, J., Park, W.: On the solution of a bi-Jensen functional equation and its stability. Bull. Korean Math. Soc. **43**, 499–507 (2006)
2. Cădariu, L., Radu, V.: Fixed points and the stability of Jensen's functional equation. J. Inequal. Pure Appl. Math. **4**(1), 4 (2003)
3. Cădariu, L., Radu, V.: On the stability of the Cauchy functional equation: a fixed point approach. Grazer Math. Ber. **346**, 43–52 (2004)
4. Cădariu, L., Radu, V.: Fixed point methods for the generalized stability of functional equations in a single variable. Fixed Point Theory Appl. **2008**, 749392 (2008)
5. Cholewa, P.W.: Remarks on the stability of functional equations. Aequ. Math. **27**, 76–86 (1984)
6. Czerwik, S.: On the stability of the quadratic mapping in normed spaces. Abh. Math. Semin. Univ. Hamb. **62**, 59–64 (1992)
7. Czerwik, S.: The stability of the quadratic functional equation. In: Stability of Mappings of Hyers–Ulam Type, pp. 81–91. Hadronic Press, Florida (1994)
8. Czerwik, S.: Functional Equations and Inequalities in Several Variables. World Scientific, New Jersey (2002)
9. Czerwik, S., Krol, K.: Ulam stability of functional equations. Aust. J. Math. Anal. Appl. **6**, 1–15 (2009)
10. Diaz, J., Margolis, B.: A fixed point theorem of the alternative for contractions on a generalized complete metric space. Bull. Am. Math. Soc. **74**, 305–309 (1968)
11. Găvruta, P.: A generalization of the Hyers–Ulam–Rassias stability of approximately additive mappings. J. Math. Anal. Appl. **184**, 431–436 (1994)
12. Găvruta, P., Găvruta, L.: A new method for the generalized Hyers–Ulam–Rassias stability. Int. J. Nonlinear Anal. Appl. **1**, 11–18 (2010)
13. Gordji, M.E., Ghaemi, M.B., Kaboli Gharetapeh, S., Shams, S., Ebadian, A.: On the stability of J^*-derivations. J. Geom. Phys. **60**, 454–459 (2010)
14. Gordji, M.E., Ghanifard, M., Khodaei, H., Park, C.: A fixed point approach to the random stability of a functional equation driving from quartic and quadratic mappings. Discrete Dyn. Nat. Soc. **2010**, 670542 (2010)

15. Gordji, M.E., Khodaei, H.: Solution and stability of generalized mixed type cubic, quadratic and additive functional equation in quasi-Banach spaces. Nonlinear Anal. **71**, 5629–5643 (2009)
16. Gordji, M.E., Najati, A.: Approximately J^*-homomorphisms: A fixed point approach. J. Geom. Phys. **60**, 809–814 (2010)
17. Hyers, D.H.: On the stability of the linear functional equation. Proc. Natl. Acad. Sci. USA **27**, 222–224 (1941)
18. Hyers, D.H., Isac, G., Rassias, Th.M.: Topics in Nonlinear Analysis and Applications. World Scientific, Singapore (1997)
19. Hyers, D.H., Rassias, Th.M.: Approximate homomorphisms. Aequ. Math. **44**, 125–153 (1992)
20. Isac, G., Rassias, Th.M.: Stability of ψ-additive mappings: applications to nonlinear analysis. Int. J. Math. Math. Sci. **19**, 219–228 (1996)
21. Jung, S.: Hyers–Ulam–Rassias Stability of Functional Equations in Mathematical Analysis. Hadronic Press, Palm Harbor (2001)
22. Jung, Y., Chang, I.: The stability of a cubic type functional equation with the fixed point alternative. J. Math. Anal. Appl. **306**, 752–760 (2005)
23. Khodaei, H., Rassias, Th.M.: Approximately generalized additive functions in several variables. Int. J. Nonlinear Anal. Appl. **1**, 22–41 (2010)
24. Lee, S., Im, S., Hwang, I.: Quartic functional equations. J. Math. Anal. Appl. **307**, 387–394 (2005)
25. Park, C.: Fixed points and Hyers–Ulam–Rassias stability of Cauchy–Jensen functional equations in Banach algebras. Fixed Point Theory Appl. **2007**, 50175 (2007)
26. Park, C.: Generalized Hyers–Ulam–Rassias stability of quadratic functional equations: a fixed point approach. Fixed Point Theory Appl. **2008**, 493751 (2008)
27. Park, C., Gordji, M.E.: Comment on approximate ternary Jordan derivations on Banach ternary algebras? [Bavand Savadkouhi et al. J. Math. Phys. **50**, 042303 (2009)]. J. Math. Phys. **51**, 044102 (2010) (7 pages)
28. Park, C., Hong, S., Kim, M.: Jensen type quadratic–quadratic mapping in Banach spaces. Bull. Korean Math. Soc. **43**, 703–709 (2006)
29. Park, C., Hou, J.: Homomorphisms between C^*-algebras associated with the Trif functional equation and linear derivations on C^*-algebras. J. Korean Math. Soc. **41**, 461–477 (2004)
30. Park, C., Park, J.: Generalized Hyers–Ulam stability of an Euler–Lagrange type additive mapping. J. Differ. Equ. Appl. **12**, 1277–1288 (2006)
31. Park, C., Najati, A.: Generalized additive functional inequalities in Banach algebras. Int. J. Nonlinear Anal. Appl. **1**, 54–62 (2010)
32. Park, C., Rassias, Th.M.: Isomorphisms in unital C^*-algebras. Int. J. Nonlinear Anal. Appl. **1**, 1–10 (2010)
33. Park, C., Rassias, Th.M.: On a generalized Trif's mapping in Banach modules over a C^*-algebra. J. Korean Math. Soc. **43**, 323–356 (2006)
34. Radu, V.: The fixed point alternative and the stability of functional equations. Fixed Point Theory Appl. **4**, 91–96 (2003)
35. Rassias, J.M.: On approximation of approximately linear mappings by linear mappings. J. Funct. Anal. **46**, 126–130 (1982)
36. Rassias, Th.M.: On the stability of the linear mapping in Banach spaces. Proc. Am. Math. Soc. **72**, 297–300 (1978)
37. Rassias, Th.M.: On the stability of the quadratic functional equation and its applications. Stud. Univ. Babeş-Bolyai, Math. **XLIII**, 89–124 (1998)
38. Rassias, Th.M.: The problem of S.M. Ulam for approximately multiplicative mappings. J. Math. Anal. Appl. **246**, 352–378 (2000)
39. Rassias, Th.M.: On the stability of functional equations in Banach spaces. J. Math. Anal. Appl. **251**, 264–284 (2000)
40. Rassias, Th.M.: On the stability of functional equations and a problem of Ulam. Acta Appl. Math. **62**, 23–130 (2000)

41. Rassias, Th.M., Šemrl, P.: On the behavior of mappings which do not satisfy Hyers–Ulam stability. Proc. Am. Math. Soc. **114**, 989–993 (1992)
42. Rassias, Th.M., Šemrl, P.: On the Hyers–Ulam stability of linear mappings. J. Math. Anal. Appl. **173**, 325–338 (1993)
43. Rassias, Th.M., Shibata, K.: Variational problem of some quadratic functionals in complex analysis. J. Math. Anal. Appl. **228**, 234–253 (1998)
44. Skof, F.: Proprietà locali e approssimazione di operatori. Rend. Semin. Mat. Fis. Milano **53**, 113–129 (1983)
45. Ulam, S.M.: Problems in Modern Mathematics. Wiley, New York (1960)

Chapter 36
Exotic n-D'Alembert PDEs and Stability

Agostino Prástaro

Abstract In the framework of the PDE's algebraic topology, previously introduced by A. Prástaro, *exotic n-d'Alembert PDEs* are considered. These are n-d'Alembert PDEs, $(d'A)_n$, admitting Cauchy manifolds $N \subset (d'A)_n$ identifiable with exotic spheres, or such that ∂N can be exotic spheres. For such equations, local and global existence theorems and stability theorems are obtained. (See also Prástaro in arXiv:1011.0081, 2010.)

Key words d'Alembert PDEs · Integral bordisms in PDEs · Existence of local and global solutions in PDEs · Conservation laws · Crystallographic groups · Exotic spheres · Singular Cauchy problems · Stability

Mathematics Subject Classification 55N22 · 58J32 · 57R20 · 58C50 · 58J42 · 20H15 · 32Q55 · 32S20

36.1 Introduction

Do exotic PDEs exist where exotic 7-spheres of the same Θ_7-class do not bound smooth solutions?

In some previous works, we studied n-d'Alembert PDEs by using the PDE's algebraic topology, introduced by A. Prástaro. (See [24, 27, 30, 33, 43, 44].) In particular, in [33] the stability properties of such equations are also characterized, showing that the n-d'Alembert equation is an extended crystal PDE, for any $n \geq 2$, and criteria for an extended 0-crystal PDE and a 0-crystal PDE are obtained. Furthermore, we proved that for any $n \geq 2$ one can canonically associate to the n-d'Alembert equation another PDE, namely the *stable extended crystal n-d'Alembert PDE*, having the same regular smooth solutions of the n-d'Alembert equation, but in these solutions, finite-time instabilities do not occur. This allowed avoiding all the problems present

Dedicated to Professor Themistocles M. Rassias on the occasion of his 60th birthday.

A. Prástaro (✉)
MEMOMAT, University of Roma "La Sapienza", Via A. Scarpa, 16, 00161 Roma, Italy
e-mail: prastaro@dmmm.uniroma1.it

P.M. Pardalos et al. (eds.), *Nonlinear Analysis*, Springer Optimization and Its Applications 68, In Honor of Themistocles M. Rassias on the Occasion of his 60th Birthday, DOI 10.1007/978-1-4614-3498-6_36, © Springer Science+Business Media, LLC 2012

in the applications, related to finite instability of solutions. Furthermore, we formulated a workable criterion to recognize asymptotic stability by suitably averaging perturbations. (See [30–34].)

As for higher dimensions, i.e., when $n \geq 7$, the existence of exotic spheres is admitted; it becomes interesting to investigate which implications such phenomena have on the characterization of global solutions of n-d'Alembert PDEs and their stability. In some previous papers, A. Prástaro has studied in some details such phenomena for the Ricci flow equation, which is important to prove the Poincaré conjecture on three dimensional Riemannian manifolds, and its generalizations to higher dimensions. (See [27, 35–40].) Furthermore, in [41] generalizations of such phenomena are considered for any PDE and characterized in the framework of Prástaro's PDE's algebraic topology.

In this paper, we aim to apply this theory to exotic n-d'Alembert PDEs, and to study the interplay between the geometric stability characterization of such equations by using the algebraic topological methods previously introduced in [30–34, 37]. (See also [1–4, 45].)

After this Introduction, the paper splits into two more sections. The first is devoted to the characterization of exotic n-d'Alembert PDEs, and the second to the stability properties of such equations. The main new result is Theorem 36.10 characterizing global solutions of exotic 8-d'Alembert equation. This theorem allows us to answer affirmatively the question put in quotation marks at the beginning of this Introduction. In fact, after Theorem 36.10, we can state that two diffeomorphic exotic 7-spheres, identified with two Cauchy manifolds in $(d'A)_8$ over $\mathbb{R}^8$, bound singular solutions only—they cannot bound smooth solutions. (Compare with the situation for the Ricci flow equation on compact, simply connected 7-dimensional Riemannian manifolds [40].)

36.2 Exotic n-d'Alembert PDEs

In this section, we review some of our recent results about the algebraic topology characterization of PDEs, and that will be useful in the next section.[1] In particular, let us recall the following theorem that relates integral bordism groups of PDEs to subgroups of crystallographic groups. For their proofs, we refer the reader to the original papers.

Remark 36.1 Here and in the following, we shall denote the boundary ∂V of a compact n-dimensional manifold V split as $\partial V = N_0 \bigcup P \bigcup N_1$, where N_0 and N_1 are two disjoint $(n-1)$-dimensional submanifolds of V that are not necessarily

[1]For general information on bordism groups and related problems in differential topology and PDE's geometry, see, e.g., [5, 9–13, 15, 22–29, 31, 47, 48, 50–53]. For crystallographic groups, see the references quoted in [37]. For differential structures and algebraic topology of exotic spheres, see [6–8, 16–21, 35, 39–41, 46].

closed and P is another $(n-1)$-dimensional submanifold of V. For example, if $V = S \times I$, where $I \equiv [0,1] \subset \mathbb{R}$, one has $N_0 = S \times \{0\}$, $N_1 = S \times \{1\}$, $P = \partial S \times I$. In the particular case that $\partial S = \varnothing$, one has also $P = \varnothing$. Let us also recall that by the term *quantum solutions* we mean integral bordisms relating Cauchy hypersurfaces of E_{k+s}, contained in $J^{k+s}_n(W)$, but not necessarily contained in E_{k+s}. (For details, see [22–30].)

Theorem 36.1 [37] *Bordism groups relative to smooth manifolds can be considered as extensions of subgroups of crystallographic groups.*

Definition 36.1 We say that a PDE $E_k \subset J^k_n(W)$ is an *extended* 0*-crystal PDE* if its integral bordism group is zero.[2]

The following theorem relates the integrability properties of a PDE to crystallographic groups.

Theorem 36.2 (Crystal structure of PDEs) [37] *Let $E_k \subset J^k_n(W)$ be a formally integrable and completely integrable PDE such that* $\dim E_k \geq 2n+1$. *Then its integral bordism group $\Omega^{E_k}_{n-1}$ is an extension of a subgroup of some crystallographic group. In this case, we say that E_k is an extended crystal PDE and we define the crystal group of E_k to be the smallest of such crystal groups. The corresponding dimension will be called the crystal dimension of E_k.*

Furthermore, if W is contractible, then E_k is an extended 0*-crystal PDE, when* $\Omega_{n-1} = 0$.

In the following, we relate crystal structure of PDEs to the existence of global smooth solutions, identifying an algebraic-topological obstruction.

Theorem 36.3 [23–25, 37] *Let $E_k \subset J^k_n(W)$ be a formally integrable and completely integrable PDE. Then, in the algebra $\mathbf{H}_{n-1}(E_k) \equiv Map(\Omega^{E_k}_{n-1}; \mathbb{R})$, (Hopf algebra of E_k), there is a subalgebra, (crystal Hopf algebra) of E_k. On such an algebra, we can represent the algebra $\mathbb{R}^{G(d)}$ associated to the crystal group $G(d)$ of E_k. (This justifies the name.) We call crystal conservation laws of E_k the elements of its crystal Hopf algebra.*[3]

Theorem 36.4 [31–34, 37] *Let $E_k \subset J^k_n(W)$ be a formally integrable and completely integrable PDE. Then, the obstruction to finding global smooth solutions of E_k can be identified with the quotient $\mathbf{H}_{n-1}(E_\infty)/\mathbb{R}^{\Omega_{n-1}}$.*

---

[2] Here by the integral bordism group we mean the weak integral bordism group $\Omega^{E_k}_{n-1,w}$.

[3] Recall that $A \equiv Map(\Omega, \mathbb{R})$, Ω a group, has a natural structure of a Hopf algebra if Ω is a finite group. If Ω is not finite, then A has a structure of a Hopf algebra in an extended sense. (See [25].)

We define the crystal obstruction of E_k the above quotient of algebras, and put $cry(E_k) \equiv \mathbf{H}_{n-1}(E_\infty)/\mathbb{R}^{\Omega_{n-1}}$. We call a 0-crystal PDE *an $E_k \subset J^k_n(W)$ such that $cry(E_k) = 0$.*[4]

Corollary 36.1 *Let $E_k \subset J^k_n(W)$ be a 0-crystal PDE. Let $N_0, N_1 \subset E_k$ be two initial and final Cauchy data of E_k such that $X \equiv N_0 \sqcup N_1 \in [0] \in \Omega_{n-1}$. Then there exists a smooth solution $V \subset E_k$ such that $\partial V = X$.*

Definition 36.2 (Exotic PDEs) Let $E_k \subset J^k_n(W)$ be a kth-order PDE on the fiber bundle $\pi : W \to M$, $\dim W = m + n$, $\dim M = n$. We say that E_k is an *exotic PDE* if it admits Cauchy integral manifolds $N \subset E_k$, $\dim N = n - 1$, such that one of the following two conditions is verified.[5]

(i) $\Sigma^{n-2} \equiv \partial N$ is an exotic sphere of dimension $(n-2)$, i.e., Σ^{n-2} is homeomorphic to S^{n-2}, ($\Sigma^{n-2} \approx S^{n-2}$) but not diffeomorphic to S^{n-2}, ($\Sigma^{n-2} \not\cong S^{n-2}$).
(ii) $\varnothing = \partial N$ and $N \approx S^{n-1}$, but $N \not\cong S^{n-1}$.

Example 36.1 The Ricci flow equation can be an exotic PDE for n-dimensional Riemannian manifolds of dimension $n \geq 7$. (See [40].)

Example 36.2 The Navier–Stokes equation can be encoded on the affine fiber bundle $\pi : W \equiv M \times \mathbf{I} \times \mathbb{R}^2 \to M$, $(x^\alpha, \dot{x}^i, p, \theta)_{0\leq\alpha\leq 3, 1\leq i\leq 3} \mapsto (x^\alpha)$. (See [24].) Therefore, Cauchy manifolds are 3-dimensional manifolds. For such a dimension, exotic spheres do not exist. Therefore, the Navier–Stokes equation cannot be an exotic PDE. Similar considerations hold for PDEs of the classical continuum mechanics.

Example 36.3 Let M be an n-dimensional manifold, $n \geq 2$, and let $\pi : E \equiv M \times \mathbb{R} \to M$ be the trivial vector fiber bundle on M. The *n-d'Alembert equation*,[6] $\frac{\partial^n \log f}{\partial x_1 \cdots \partial x_n} = 0$, is an nth-order closed partial differential relation (in the sense of Gromov [11]) on the fiber bundle $\pi : E \equiv M \times \mathbb{R} \to M$, i.e., it defines a subset $Z_n \subset JD^n(E)$ without boundary, $\partial Z_n = \varnothing$. Let $\{x^\alpha, u, u_\alpha, u_{\alpha\beta}, \ldots, u_{\alpha_1\cdots\alpha_n}\}$ be a coordinate system on $JD^n(E)$ adapted to the fiber structure: $\pi_n : JD^n(E) \to M, \overline{\pi}_{n,0} : JD^n(E) \to \mathbb{R}$. Then, $Z_n = F^{-1}(0)$, $F : JD^n(E) \to \mathbb{R}$, where F is a sum of terms of the type:

$$F[s; r|\alpha, \beta_1\beta_2, \ldots, \gamma_1\cdots\gamma_q] \equiv s u^r u_\alpha u_{\beta_1\beta_2} \cdots u_{\gamma_1\cdots\gamma_q},$$

---

[4] An extended 0-crystal PDE $E_k \subset J^k_n(W)$ is not necessarily a 0-crystal PDE. In fact, in order for E_k to be an extended 0-crystal PDE it is enough that $\Omega^{E_k}_{n-1,w} = 0$. This does not necessarily imply that $\Omega^{E_k}_{n-1} = 0$.

[5] In this paper, we will use the same notation adopted in [40]: $\approx$ homeomorphism; $\cong$ diffeomorphism; $\approxeq$ homotopy equivalence; $\simeq$ homotopy.

[6] If $n = 2$ we simply say d'Alembert equation and we will put $(d'A) \equiv (d'A)_2$.

with $\alpha \neq \beta_1 \neq \beta_2 \neq \cdots \neq \gamma_1 \neq \cdots \neq \gamma_q \leq n, s \in \mathbb{Z}, r \in \mathbb{N} \cup \{0\}$. Furthermore, the term in F containing $u_{1\cdots n}$ is just $u_{1\cdots n}u^{n-1}$.[7] Note that F does not have locally constant rank on all Z_n, so Z_n is not a submanifold of $JD^n(E)$. Furthermore, on the open subset $C_n \equiv u^{-1}(\mathbb{R} \setminus 0) \subset JD^n(E)$, one recognizes that F has locally constant rank 1. Hence $Z_n \cap C_n$ is a subbundle of $JD^n(E) \to M$ of dimension $n + \frac{(2n)!}{(n!)^2} - 1$. In the following, by abuse of notation, we shall denote by $(d'A)_n$ either Z_n or $Z_n \cap C_n$.

The n-d'Alembert equation over $M = \mathbb{R}^n$ can be an exotic PDE for n-dimensional manifolds of dimension $n \geq 8$, but one must carefully consider the meaning of smooth Cauchy $(n-1)$-dimensional manifolds there. In fact, it is not possible for any n to embed into the fiber bundle $E = \mathbb{R}^{n+1}$ exotic $(n-1)$-spheres. To be more precise, let us consider the case $n = 8$. Then since $S^7 \subset \mathbb{R}^8 \subset E$, we can embed into E the standard 7-dimensional sphere. On the other hand, it is well known, after some results by E.V. Brieskorn [6, 7], that homotopy 7-spheres Σ^7 can be identified with the intersections of complex hypersurfaces Y_κ, $1 \leq \kappa \leq 28$, in $\mathbb{C}^5$, with a 9-dimensional small sphere X around the singular point at the origin in Y_κ: $\Sigma^7 = Y_\kappa \bigcap X$; see the equations in (36.1).

$$\Sigma^7 = \left\{ (z_1, \ldots, z_5) \in \mathbb{C}^5 \left| \begin{array}{ll} (Y_\kappa) : z_1^2 + z_2^2 + z_3^2 + z_4^3 + z_5^{6\kappa-1} = 0, & 1 \leq \kappa \leq 28 \\ (X) : \sum_j (z_j \bar{z}_j) = 1 & \end{array} \right. \right\}$$
$$\subset \mathbb{C}^5 \cong \mathbb{R}^{10}. \tag{36.1}$$

The intersections $X \bigcap Y_k$, $1 \leq \kappa \leq 28$, have the differential structures identified by $\Theta_7 \cong \mathbb{Z}_{28}$.[8] In other words, exotic 7-spheres are framed manifolds $\Sigma^7 \subset \mathbb{R}^{7+s}$, with $s \geq 3$. Therefore, we cannot embed in the total space $E \equiv \mathbb{R}^9$, of the fiber bundle $\pi : E \to \mathbb{R}^8$, any homotopy 7-sphere. However, this does not exclude that some smooth Cauchy 7-dimensional manifolds in $(d'A)_8$ can be identified with exotic 7-spheres. In fact, since $\dim(d'A)_8 = 12877$, the (Whitney) condition $\dim(d'A)_8 \geq 2 \times 7 + 1 = 15$ is satisfied to embed Σ^7 into $(d'A)_8$. If $N \subset (d'A)_8$ is the image of such an embedding, N cannot in general be diffeomorphic to its image $Y \subset E$ via the canonical projection $\pi_{8,0} : JD^8(E) \to E$. So, in this case we shall talk about singular Cauchy 7-manifolds of $(d'A)_8$. Furthermore, let us emphasize that since the equation defining the open PDE $C_8 \bigcap (d'A)_8$ can be solved with respect to the coordinate $u_{x^1 \ldots x^n}$, we can embed homotopy 7-spheres Σ^7 as smooth integral submanifolds $N \subset (d'A)_8 \subset J_8^8(E)$ such that their Thom–Boardman singular points should not be frozen singularities in the sense introduced in [15]. Therefore, we can state that such 7-dimensional integral manifolds are contained in 8-dimensional integral manifolds $V \subset (d'A)_8$, (singular) solutions of $(d'A)$. Such 7-dimensional integral manifolds are called *admissible Cauchy manifolds* of $(d'A)_8$.

---

[7]For example, for $n = 2$ one has $F = u_{xy}u - u_x u_y$, and for $n = 3$ one has $F = u_{xyz}u^2 - u_{xy}u_z u - u_{xz}u_y u + u_x u_y u_z$.

[8]Θ_n denotes the additive group of diffeomorphism classes of oriented smooth homotopy spheres of dimension n.

36.3 Stability in Exotic n-d'Alembert PDEs

Let us consider, now, the stability of PDEs in the framework of the geometric theory of PDEs. We shall follow the line just drawn in some our previous papers on this subject, where we have unified the integral bordism for PDEs and stability, and related the quantum bordism of PDEs to Ulam stability [49].

Definition 36.3 (Singular solutions of PDEs) Let $\pi : W \to M$ be a fiber bundle with $\dim M = n$ and $\dim W = m + n$. Let $E_k \subset JD^k(W)$ be a PDE and $V \subset E_k$ being a solution of E_k. We say that $p \in V$ is a *singular point* of V of type Σ_i, $i = 0, 1, \ldots, n$, if the canonical map $\pi_k|_V : V \to M$ has a Thom–Boardman singularity of type S_i [5, 31]. Let $\Sigma(V) \subset V$ be the set of singular points of V. Then $V \setminus \Sigma(V) = \bigcup_r V_r$ is the disjoint union of connected components V_r. For each such component, $\pi_k : V_r \to M$ is an immersion and can be represented by means of the image of a kth-order derivative of some section s of π, i.e., $V_r = D^k s(U_r)$, where $U_r \subset M$ is an open subset of M. We also call a solution of E_k any submanifold $V \subset E_k$ that is obtained by projection of $\pi_{k+h,k}$ of some solution $V' \subset (E_k)_{+h} \subset JD^{k+h}(W)$, represented by a smooth integral submanifold of $(E_k)_{+h}$, i.e., $V = \pi_{k+h,k}(V')$. In general, such a solution V is no longer represented by a smooth submanifold of E_k. We say also that instead the smooth manifold $V' \subset (E_k)_{+h}$ *solves singularities* of V, (or *smooths* V). More general solutions are considered taking into account the canonical embedding $JD^k(W) \hookrightarrow J^k_n(W)$, where $J^k_n(W)$ is the k-jet space for n-dimensional submanifolds of W. (For details, see [22–25, 27, 28].)

We define *weak solutions*, solutions $V \subset E_k$, such that the set $\Sigma(V)$ of singular points of V also contains discontinuity points, $q, q' \in V$, with $\pi_{k,0}(q) = \pi_{k,0}(q') = a \in W$, or $\pi_k(q) = \pi_k(q') = p \in M$. We denote such a set by $\Sigma(V)_S \subset \Sigma(V)$, and in such cases we shall talk more precisely of *singular boundary* of V, like $(\partial V)_S = \partial V \setminus \Sigma(V)_S$. However, by abuse of notation, we shall denote $(\partial V)_S$ (resp., $\Sigma(V)_S$) simply by (∂V) (resp., $\Sigma(V)$) if no confusion can arise.

Definition 36.4 (Stable solutions of PDEs) Under the same hypotheses of the above definitions, let $X \to E_k$ be a regular solution, where $X \subset M$ is a smooth n-dimensional compact manifold with boundary ∂X. Then f is *stable* if there is a neighborhood W_f of f in $\underline{Sol}(E_k)$, the manifold of regular solutions of E_k, such that each $f' \in W_f$ is equivalent to f, i.e., f is transformed into f' by vertical symmetries of E_k.

Theorem 36.5 [31] *Let $E_k \subset JD^k(W)$ be a kth-order PDE on the fiber bundle $\pi : W \to M$ in the category of smooth manifolds,* $\dim W = m + n$, $\dim M = n$, $m > 1$. *Let $s : M \to W$ be a section, solution of E_k, and let $\nu : M \to s^*vTW \equiv E[s]$ be a solution of the linearized equation $E_k[s] \subset JD^k(E[s])$. Then a flow $\{\phi_\lambda\}_{\lambda \in J}$ is associated to ν, where $J \subset \mathbb{R}$ is a neighborhood of $0 \in \mathbb{R}$ that transforms V into a new solution $\widetilde{V} \subset E_k$.*

Definition 36.5 Let $E_k \subset J^k_n(W)$, where $\pi : W \to M$ is a fiber bundle, in the category of smooth manifolds. We say that E_k is *functionally stable* if for any compact regular solution $V \subset E_k$ such that $\partial V = N_0 \bigcup P \bigcup N_1$ one has quantum solutions $\widetilde{V} \subset J^{k+s}_n(W)$, $s \geq 0$, such that $\pi_{k+s,0}(\widetilde{N}_0 \sqcup \widetilde{N}_1) = \pi_{k,0}(N_0 \sqcup N_1) \equiv X \subset W$, where $\partial \widetilde{V} = \widetilde{N}_0 \bigcup \widetilde{P} \bigcup \widetilde{N}_1$.

We call the set $\Omega[V]$ of such solutions $\widetilde{V}$ the *full quantum situs* of V. We also call each element $\widetilde{V} \in \Omega[V]$ a *quantum fluctuation* of V.[9]

We call an *infinitesimal bordism* of a regular solution $V \subset E_k \subset JD^k(W)$ an element $\widetilde{V} \in \Omega[V]$, defined in the proof of Theorem 36.5. (See [31].) We denote by $\Omega_0[V] \subset \Omega[V]$ the set of infinitesimal bordisms of V. We call $\Omega_0[V]$ the *infinitesimal situs* of V.

Definition 36.6 Let $E_k \subset J^k_n(W)$, where $\pi : W \to M$ is a fiber bundle, in the category of smooth manifolds. We say that a regular solution $V \subset E_k$, $\partial V = N_0 \bigcup P \bigcup N_1$, is *functionally stable* if the infinitesimal situs $\Omega_0[V] \subset \Omega[V]$ of V does not contain singular infinitesimal bordisms.

Theorem 36.6 [30, 31] *Let $E_k \subset J^k_n(W)$, where $\pi : W \to M$ is a fiber bundle, in the category of smooth manifolds. If E_k is formally integrable and completely integrable, then it is functionally stable as well as Ulam-extended superstable.*

A regular solution $V \subset E_k$ is stable iff it is functionally stable.

Remark 36.2 Let us emphasize that the definition of functionally stable PDE interprets in a pure geometric way the definition of Ulam superstable functional equation just adapted to PDEs. (Compare our geometric approach to the stability of PDE's solutions with the Ljapunov's one in functional analysis [14]).

Definition 36.7 We say that $E_k \subset JD^k(W)$ is a *stable extended crystal PDE* if it is an extended crystal PDE that is functionally stable and all its regular smooth solutions are (functionally) stable.

We say that $E_k \subset JD^k(W)$ is a *stabilizable extended crystal PDE* if it is an extended crystal PDE and a stable extended crystal PDE $^{(S)}E_k \subset JD^{k+s}(W)$ can be canonically associated to E_k. We call $^{(S)}E_k$ just the *stable extended crystal PDE of* E_k.

We have the following criteria for the functional stability of solutions of PDEs and for identifying stable extended crystal PDEs.

Theorem 36.7 (Functional stability criteria) [31] *Let $E_k \subset JD^k(W)$ be a kth-order formally integrable and completely integrable PDE on the fiber bundle $\pi : W \to M$,* $\dim W = m + n$, $\dim M = n$.

---

[9] Let us emphasize that to $\Omega[V]$ belong also (not necessarily regular) solutions $V' \subset E_k$ such that $N'_0 \sqcup N'_1 = N_0 \sqcup N_1$, where $\partial V' = N'_0 \bigcup P' \bigcup N'_1$.

1. *If the symbol* $g_k = 0$, *then all the smooth regular solutions* $V \subset E_k \subset JD^k(W)$ *are functionally stable with respect to any non-weak perturbation. So* E_k *is a stable extended crystal.*
2. *If* E_k *is of finite type, i.e.,* $g_{k+r} = 0$, *for* $r > 0$, *then all the smooth regular solutions* $V \subset E_{k+r} \subset JD^{k+r}(W)$ *are functionally stable with respect to any non-weak perturbation. So* E_k *is a stabilizable extended crystal with stable extended crystal* ${}^{(S)}E_k = E_{k+r}$.
3. *If* $V \subset (E_k)_{+\infty} \subset JD^\infty(W)$ *is a smooth regular solution, then* V *is functionally stable with respect to any non-weak perturbation. So any formally integrable and completely integrable PDE* $E_k \subset JD^k(W)$ *is a stabilizable extended crystal with stable extended crystal* ${}^{(S)}E_k = (E_k)_{+\infty}$.

Remark 36.3 Let us also remark that in evolutionary PDEs, i.e., PDEs built on a fiber bundle $\pi : W \to M$ over a "space-time" M, $\{x^\alpha, y^j\}_{0\leq\alpha\leq n, 1\leq j\leq m} \mapsto \{x^\alpha\}_{0\leq\alpha\leq n}$, where $x^0 = t$ represents the time coordinate, one can consider "asymptotic stability", i.e., the behavior of perturbations of global solutions for $t \to \infty$. In such cases, we can recast our formulation on the corresponding compactified space-times. (For details, see [31, 32].)

From above results one can see that, in general, the functional stability of smooth regular solutions is a very strong requirement. However, the above theorems give us workable criteria to obtain subequations of E_k whose smooth regular solutions have assured functional stability.

A weaker requirement than functional stability is also useful. This is related to a concept of "averaged stability".

In fact, we have the following definition.

Definition 36.8 Let $E_k \subset JD^k(W)$ be a formally integrable and completely integrable PDE on a fiber bundle $\pi : W \to M$, and let $V = D^k s(M) \subset E_k$ be a regular smooth solution of E_k. Let $\xi : M \to E_k[s]$ be the general solution of $E_k[s]$. Let us assume that there is a Euclidean structure on the fiber of $E[s] \to M$. Then, we say that V is *average asymptotic stable* if the function of time $\mathfrak{p}(t)$ defined by the formula

$$\mathfrak{p}(t) = \frac{1}{2vol(B_t)} \int_{B_t} \xi^2 \eta \tag{36.2}$$

has the following behavior: $\mathfrak{p}(t) = \mathfrak{p}(0)e^{-ct}$ for some real number $c > 0$. We call $\tau_0 = 1/c_0$ the *characteristic stability time* of the solution V. If $\tau_0 = \infty$, it means that V is average unstable.[10]

We have the following criterion of average asymptotic stability.

[10]In the following, if there are no reasons for confusion, we shall also call a stable solution a smooth regular solution of a PDE $E_k \subset JD^k(W)$ that is average asymptotic stable.

Theorem 36.8 (Criterion of average asymptotic stability) [31] *A regular global smooth solution s of E_k is average stable if the following conditions are satisfied:*

$$\overset{\bullet}{\mathfrak{p}}(t) \leq c\mathfrak{p}(t), \quad c \in \mathbb{R}^+, \forall t, \tag{36.3}$$

where

$$\overset{\bullet}{\mathfrak{p}}(t) = \frac{1}{2vol(B_t)} \int_{B_t} \left(\frac{\delta \xi^2}{\delta t}\right)\eta = \frac{1}{vol(B_t)} \int_{B_t} \left(\frac{\delta \xi}{\delta t}.\xi\right)\eta. \tag{36.4}$$

Here ξ represents the general solution of the linearized equation $E_k[s]$ of E_k at the solution s. Let us denote by c_0 the infimum of the positive constants c such that inequality (36.3) *is satisfied. Then we call $\tau_0 = 1/c_0$ the characteristic stability time of the solution V. If $\tau_0 = \infty$ means that V is unstable.*[11]

Furthermore, condition (36.3) *is satisfied if the operator $\frac{\delta}{\delta t}$ is self-adjoint on the set of solutions of the linearized equation $E_k[s] \subset J^n_n(E[s])$, where $E[s] \equiv s^* vTW$.*

Theorem 36.9 (The extended crystal structure of the n-d'Alembert equation and stability [33])

1. *For the n-d'Alembert equation one has the following properties*:
   (i) *The n-d'Alembert equation is an extended crystal PDE for any $n \geq 2$. If M is p-connected, $p \in \{0, 1, \dots, n-1\}$, it becomes an extended* 0*-crystal iff $\Omega_{n-1} = 0$. In particular, for $n = 2$ it becomes a* 0*-crystal.*
   (ii) *The n-d'Alembert equation is functionally stable.*
   (iii) *Smooth regular solutions of the n-d'Alembert equation, present, in general, instabilities at finite times. However, the n-d'Alembert equation can be stabilized and its stable extended crystal PDE is its ∞-prolongation $((d'A)_n)_{+\infty}$. There all smooth regular solutions are functionally stable, i.e., they do not present finite time instabilities.*
2. *In the case $n = 2$, with M non-simply connected, $(d'A)$ remains an extended crystal PDE, but no longer an extended* 0*-crystal PDE. For example, this happens if M is a bidimensional torus T^2 which is a connected, orientable, non-simply connected surface. Then, $\Omega_1^{(d'A)} \cong \mathbb{Z}_2 \oplus \mathbb{Z}_2$ (For a proof, see* [33].*) So, the d'Alembert equation on the torus is neither an extended* 0*-crystal PDE nor a* 0*-crystal PDE. The crystal group of such an equation is $G(2) = \mathbb{Z} \rtimes D_4 = p4m$. Its crystal dimension is* 2.

   In the case $n = 2$, with $M = \mathbb{R}^2$, we can build solutions with the methods of characteristics, that are average unstable.
3. *Let us consider the* 3*-d'Alembert equation on the non-simply connected, orientable,* 3*-dimensional manifold $M = \mathbb{R}P^3$. In this case, one has $\Omega_2^{(d'A)} \cong \mathbb{Z}_2 \oplus \mathbb{Z}_2$. Thus this is another example where one has $(d'A)_3$ that is an extended crystal PDE, but it cannot be an extended* 0*-crystal PDE and or a* 0*-crystal PDE.*

[11] τ_0 has just the physical dimension of a time.

Thus this equation has the same crystal group and crystal dimension of equation considered in the above example.

Proof Even if these results are proved in [33], let us review their proofs here, in order to better understand the following ones.

1.(i) The n-d'Alembert equation $(d'A)_n \subset JD^n(E)$ is a nth-order PDE, formally integrable, and completely integrable on the trivial vector fiber bundle $\pi : E \equiv M \times \mathbb{R} \to M$.[12] (See [44].) This means that we can locally reproduce all the results obtained for the n-d'Alembert equation on $\mathbb{R}^n$. (See [24, 44].) A local solution passes through any point $q \in (d'A)_n$. Furthermore, the set of local solutions of the n-d'Alembert equation on n-dimensional manifolds contains the set of the local functions that can be represented as $f(x^1, \dots, x^n) = f_1(x^2, \dots, x^n) \cdots f_n(x^1, \dots, x^{n-1})$. This follows directly from previous considerations and results contained in [24, 43, 44]. Now, the set $\mathfrak{Sol}_{\mathrm{loc}}(d'A)_n$, $n \geq 2$, of all local solutions of the equation $\frac{\partial^n \log f}{\partial x_n \cdots \partial x_1} = 0$, considered on an n-dimensional manifold M, is larger than the set of all local functions f that can be represented as $f(x^1, \dots, x^n) = f_1(x^2, \dots, x^n) \cdots f_n(x^1, \dots, x^{n-1})$. (See [43, 44].) In the following, we shall consider the n-d'Alembert equation given as a submanifold $(d'A)_n$ of the jet space $J^n_n(E)$ by means of the embedding $(d'A)_n \hookrightarrow JD^n(E) \hookrightarrow J^n_n(E)$. The characterization of global solutions of $(d'A)_n$ is made by means of its integral bordism groups. One has $\Omega_p^{(d'A)_n} \cong \Omega_p((d'A)_n)$, for $p \in \{0, \dots, n-1\}$. This follows from the fact that the n-d'Alembert equation is formally integrable and completely integrable. (See [44].) We get $\Omega_p^{(d'A)_n} \cong \underline{\Omega}_p(M) \cong \bigoplus_{r,s,r+s=p} H_r(M; \mathbb{Z}_2) \otimes_{\mathbb{Z}_2} \Omega_s$, $p \in \{0, \dots, n-1\}$. In the particular case when $\dim M = 2$ and M is p-connected, $p \in \{0, 1\}$, the integral bordism group $\Omega_1^{(d'A)} = 0$. Thus $(d'A)$ is an extended 0-crystal PDE. Furthermore, one can also prove that for such a case there are no obstructions coming from the integral characteristic numbers. In fact, all the conservation laws on closed 1-dimensional smooth integral manifolds are zero [24]. Then one has *cry*$(d'A) = 0$, for p-connected M, $p \in \{0, 1\}$. Thus in this case $(d'A)$ becomes a 0-crystal.

(ii) The n-d'Alembert equation is functionally stable since it is formally integrable and completely integrable. (See Theorem 36.6.)

(iii) The functional instabilities come from the fact that the symbol of the n-d'Alembert equation is not zero. In fact, one has

$$\dim(g_n)_q = \frac{(2n-1)!}{n!(n-1)!} - 1, \quad \forall q \in \left(d'A\right)_n. \tag{36.5}$$

Furthermore, in the ∞-prolongation $((d'A)_n)_{+\infty} \subset J^\infty_n(W)$, we get all the smooth solutions of $(d'A)_n$, and there, since the corresponding symbol is zero, $((g_n))_{+\infty} = 0$, admissible singular (non-weak) perturbations do not exist. Thus, $((d'A)_n)_{+\infty}$ is necessarily the stable extended crystal of $(d'A)_n$. Therefore, $(d'A)_n$ is a stabilizable PDE.

[12] $(d'A)_n$ considered in this theorem is a submanifold of $JD^n(E)$, hence it coincides with $Z_n \cap C_n$.

2. We have proved in [43, 44] that $(d'A)$ admits the following characteristic strips:

$$\begin{cases} \zeta_1 \equiv u[\partial y + u_y \partial u + u_{yx}\partial u_x + u_{yy}\partial u_y] + u_{xx}u_y\partial u_{xx} + u_{yy}u_x\partial u_{yx}, \\ \zeta_2 \equiv u[\partial x + u_x\partial u + u_{yx}\partial u_y + u_{xx}\partial u_x] + u_{xx}u_y\partial u_{xy} + u_{yy}u_x\partial u_{yy}. \end{cases} \tag{36.6}$$

These respectively generate characteristic 1-dimensional distributions in the following sub-equations ${}_i(d'A) \subset (d'A)$, $i = 1, 2$:

$$ {}_1(d'A) : \begin{Bmatrix} u_{xx} = 0 \\ uu_{xy} - u_x u_y = 0 \end{Bmatrix}; \qquad {}_2(d'A) : \begin{Bmatrix} u_{yy} = 0 \\ uu_{xy} - u_x u_y = 0 \end{Bmatrix}. \tag{36.7}$$

For such equations, the above mentioned 1-dimensional distributions are respectively characteristic distributions. Therefore, for such equations we can build characteristic solutions that are, of course, also solutions of $(d'A)$. For example, we have proved in [44] that the solution generated by ζ_1 is given by the following formula:

$$u(x, y) = \left(\frac{\beta}{2}y^2 + \alpha y + 1\right)h(x), \tag{36.8}$$

where $\alpha, \beta \in \mathbb{R}$ and $h(x)$ is an arbitrary function on one real variable. Let us now investigate if such a solution is average stable. The parametric equations for the characteristic flow on such a solution, say $V \subset (d'A)$, are given by the following differential system:

$$\begin{cases} \dot{x} = 0, \\ \dot{y} = u, \\ \dot{u} = uu_y. \end{cases} \tag{36.9}$$

The general solution of the linearized equation $(d'A)[V] \subset JD^2(E[s])$ can be obtained from the general symmetry vector field for $(d'A)$, given in [44]. Then we get

$$\begin{cases} \xi = [s(y) + r(x)]u\partial u, \\ u(x, y) = (\frac{\beta}{2}y^2 + \alpha y + 1)h(x), \end{cases} \tag{36.10}$$

where s and r are arbitrary functions. Let us denote by $\xi(x, y)$ the component of the vertical vector field ξ. Then one explicitly has $\xi(x, y) = [s(y) + r(x)](\frac{\beta}{2}y^2 + \alpha y + 1)h(x)$. From the arbitrariness of the functions r, s, and h, one can see that $\xi(x, y)$ can have singular points. So the solution (36.8) is not stable in $(d'A)$. Furthermore, it is not asymptotically stable since $\lim_{y\to\infty} \xi(x, y) = \infty$. In order to investigate whether it is average stable, let us consider the differential operator $\frac{\delta\xi}{\delta t}$ on $(d'A)[V]$. One has $\frac{\delta\xi}{\delta t} = (\partial t.\xi) + (\partial x.\xi)\dot{x} + (\partial y.\xi)\dot{y} = (\partial y.\xi)u(x, y)$. For its adjoint, one has $\frac{\delta^*\phi}{\delta t} = -(\partial y.(u(x, y)\phi)) = -(\partial y.\phi)u(x, y) - (\partial y.u(x, y))\phi$. Thus, the operator $\frac{\delta\xi}{\delta t}$ is not self-adjoint on the solution in (36.8), hence such a solution is not average stable.

3. This follows directly from previous parts. □

Theorem 36.10 (Stability in exotic 8-d'Alembert PDEs over $\mathbb{R}^8$) *Let us consider* $(d'A)_8$ *over* $\mathbb{R}^8$. *The integral singular bordism group* $\Omega_{7,s}^{(d'A)_8}$ *of the 8-d'Alembert PDE over* $\mathbb{R}^8$ *is* $\Omega_{7,s}^{(d'A)_8} = \mathbb{Z}_2$. *If we consider admissible Cauchy manifolds* $N \subset (d'A)_8$, *identified with 7-dimensional homotopy spheres,* (homotopy equivalence full admissibility hypothesis), *then one has* $\Omega_{7,s}^{(d'A)_8} \cong \Omega_7^{(d'A)_8} = 0$, *hence* $(d'A)_8$ *becomes an extended* 0*-crystal PDE, but also a* 0*-crystal PDE. The bordism classes in* $\Omega_7^{(d'A)_8}$ *are identified by Cauchy manifolds represented by diffeomorphic homotopy spheres. In particular, in the homotopy equivalence full admissibility hypothesis, starting from an admissible Cauchy manifold* $N_0 \subset (d'A)_8$, *identified with* S^7, *one can arrive at a singular solution to any other admissible Cauchy manifold* $N_1 \subset (d'A)_8$. *Such a solution is unstable. Moreover, there exists a smooth solution* V *such that* $V = N_0 \sqcup N_1$, *iff* $N_0 \cong N_1$. *Such a solution can be stabilized.*

Proof In fact, $\Omega_{7,s}^{(d'A)_8} \cong \bigoplus_{0\le r,s\le 7} \Omega_r \otimes_{\mathbb{Z}_2} H_s(M;\mathbb{Z}_2)$. Taking into account that for $M \cong \mathbb{R}^8$ one has $H_r(M;\mathbb{Z}_2) = 0$ for $0 < r \le 7$, and $H_0(M;\mathbb{Z}_2) = \mathbb{Z}_2$, and that $\Omega_7 \cong \mathbb{Z}_2$, we get $\Omega_{7,s}^{(d'A)_8} = \mathbb{Z}_2$. If we consider admissible Cauchy 7-dimensional homotopy spheres only, we have that they have necessarily all integral characteristic numbers, i.e., the evaluations on such manifolds of all the conservation laws give the same numbers. (For a proof, one can copy a similar proof given in [1, 2, 40] for the Ricci flow equation.) Therefore, they belong to the same singular integral bordism class, i.e., $\Omega_{7,s}^{(d'A)_8} = 0$. Since one has the short exact sequence

$$0 \longrightarrow K_{7,s}^{(d'A)_8} \longrightarrow \Omega_7^{(d'A)_8} \longrightarrow \Omega_{7,s}^{(d'A)_8} \longrightarrow 0,$$

we get that, under the *homotopy equivalence full admissibility hypothesis*, one has $\Omega_7^{(d'A)_8} \cong K_{7,s}^{(d'A)_8}$. Let us emphasize that even if the number of differentiable structures on 7-dimensional spheres is 28, smooth Cauchy-manifolds-exotic-7-spheres cannot be contained in $((d'A)_8)_{+\infty}$ since they are singular integral manifolds. So smooth Cauchy manifolds contained in $((d'A)_8)_{+\infty}$ can be identified with S^7 only. Furthermore, taking into account that smooth solutions, bording smooth Cauchy manifolds, necessitate identifying diffeomorphisms between the corresponding sectional submanifolds, it follows that $\Omega_7^{(d'A)_8} = 0$ must be true, too. Therefore, we also get $cry((d'A)_8) = 0$.

For the previous arguments it is important to state that the space of conservation laws is not zero. □

Lemma 36.1 *The space of conservation laws of* $(d'A)_n$ *is not zero.*

Proof In fact, conservation laws of $(d'A)_n$ are $(n-1)$-differential forms $\omega = \omega_i\, dx^1 \wedge \cdots \wedge \widehat{dx^i} \wedge \cdots \wedge dx^n$ on $((d'A)_n)_{+\infty}$ such that for any smooth integral n-manifold $V \subset (d'A)_n$, a solution of $(d'A)_n$, one has $d\omega|_V = 0$. Then, one can take the $(n-1)$-differential forms ω given in (36.11):

$$\begin{cases}\omega=\omega_i\,dx^1\wedge\cdots\wedge\widehat{dx^i}\wedge\cdots\wedge dx^n,\\ \omega_i=\omega(x^1,\dots,\widehat{x^i},\dots,x^n,I_{\alpha i})_{\alpha\neq i,|\alpha|\geq 0},\\ I_{\alpha i}\equiv(\partial x_\alpha.(\partial x_1\cdots\widehat{\partial x_i}\cdots\partial x_n.\log f)),\end{cases}\tag{36.11}$$

where $f:M\to\mathbb{R}$ is a smooth function on M and ω_i are arbitrary smooth functions of their arguments. The "widehat" over the symbols means the absence of the underlying symbols. In fact, one has the following:

$$\begin{cases}d\omega=\sum_{1\leq i\leq n}\sum_{p\neq i}(\partial x_p.\omega_i)\,dx^p\wedge dx^1\wedge\cdots\wedge\widehat{dx^i}\cdots\wedge dx^n\\ \qquad+\sum_{1\leq i\leq n}\sum_{\alpha\neq i;|\alpha|\geq 0}(\partial I_{i\alpha}.\omega_i)(\partial x_i.I_{i\alpha})\,dx^1\wedge\cdots\wedge dx^n\\ \quad=\sum_{\alpha\neq i;|\alpha|\geq 0}[\sum_{1\leq i\leq n}(\partial I_{i\alpha}.\omega_i)](\partial x_\alpha.(\partial x_1\cdots\partial x_n.\log f))\,dx^1\wedge\cdots\wedge dx^n.\end{cases}\tag{36.12}$$

Now $d\omega|_V=0$ if $V\subset(d'A)_n$ is a smooth $(n+1)$-dimensional integral manifold, (singular) solution of $(d'A)_n$. In fact, if f satisfies the equation $(\partial x_1\cdots\partial x_n.\log f)=0$, then $(\partial x_\alpha.(\partial x_1\cdots\partial x_n.\log f))=0$, for $|\alpha|\geq 0$. This directly follows from the prolongations of the formally integrable and completely integrable n-d'Alembert equation. For example, for $n=2$ we get for $(d'A)_2$ and its first prolongation $((d'A)_2)_{+1}$ the equations given in (36.13).

$$(d'A)_2:\{f_{xy}f-f_xf_y=0\};\qquad ((d'A)_2)_{+1}:\begin{Bmatrix}f_{xy}f-f_xf_y=0,\\ f_{xxy}f-f_{xx}f_y=0,\\ f_{xyy}f-f_{yy}f_x=0\end{Bmatrix}.\tag{36.13}$$

On the other hand, we have

$$\begin{cases}(\partial x(\partial x\partial y.\log f))=(f_{xxy}f-f_{xx}f_y)f-2(f_{xy}f-f_xf_y)f_x,\\ (\partial y(\partial x\partial y.\log f))=(f_{xyy}f-f_{yy}f_x)f-2(f_{xy}f-f_xf_y)f_y.\end{cases}\tag{36.14}$$

Therefore, on the 2-d'Alembert equation one has

$$\begin{cases}(\partial x(\partial x\partial y.\log f))|_V=0,\\ (\partial y(\partial x\partial y.\log f))|_V=0.\end{cases}\tag{36.15}$$

This process can be iterated on all the prolongation orders. □

To conclude the proof of Theorem 36.10, it is enough to consider that on a finite order, where singular solutions live, the symbol of the 8-d'Alembert equation is not zero. Thus these solutions are unstable. Instead, smooth solutions can be stabilized since these can be identified with smooth integral manifolds of the infinity prolongation $((d'A)_8)_{+\infty}$ where the symbol is zero. (See Theorem 36.7.)

Corollary 36.2 *Under the homotopy equivalence full admissibility hypothesis, $(d'A)_8$ admits a singular global attractor in the sense introduced in* [39], *i.e., all the* *admissible Cauchy manifolds belong to the same integral singular bordism class of* $(d'A)_8$. *Furthermore, in the sphere full admissibility hypothesis, i.e., when we consider admissible all the smooth Cauchy manifolds identifiable via diffeomorphisms with* S^7, $(d'A)_8$ *admits a smooth global attractor in the sense that all the smooth admissible Cauchy manifolds belong to the same integral smooth bordism class of* $(d'A)_8$.

Acknowledgements I would like to thank Editors for their kind invitation to contribute my paper to this book, dedicated to Themistocles M. Rassias on occasion of his 60th birthday.

Work partially supported by MIUR Italian grants "PDE's Geometry and Applications".

References

1. Agarwal, R.P., Prástaro, A.: Geometry of PDEs. III(I): webs on PDEs and integral bordism groups. The general theory. Adv. Math. Sci. Appl. **17**(1), 239–266 (2007)
2. Agarwal, R.P., Prástaro, A.: Geometry of PDEs. III(II): webs on PDEs and integral bordism groups. Applications to Riemannian geometry PDEs. Adv. Math. Sci. Appl. **17**(1), 267–281 (2007)
3. Agarwal, R.P., Prástaro, A.: Singular PDE's geometry and boundary value problems. J. Nonlinear Convex Anal. **9**(3), 417–460 (2008)
4. Agarwal, R.P., Prástaro, A.: On singular PDE's geometry and boundary value problems. Appl. Anal. **88**(8), 1115–1131 (2009)
5. Boardman, J.M.: Singularities of differentiable maps. Publ. Math. IHÉS **33**, 21–57 (1967)
6. Brieskorn, E.V.: Examples of singular normal complex spaces which are topological manifolds. Proc. Natl. Acad. Sci. **55**(6), 1395–1397 (1966)
7. Brieskorn, E.V.: Beispiele zur Differentialtopologie von Singularitäten. Invent. Math **2**, 1–14 (1966)
8. Dubrovin, B.A., Fomenko, A.T., Novikov, S.P.: Modern Geometry-Methods and Applications. Part I; Part II; Part III Springer, New York (1990). Original Russian edition: Sovremennaja Geometria: Metody i Priloženia. Nauka, Moskva (1979)
9. Goldshmidt, H.: Integrability criteria for systems of non-linear partial differential equations. J. Differ. Geom. **1**, 269–307 (1967)
10. Golubitsky, M., Guillemin, V.: Stable Mappings and Their Singularities. Springer, New York (1973)
11. Gromov, M.: Partial Differential Relations. Springer, Berlin (1986)
12. Hirsch, M.: Differential Topology. Springer, New York (1976)
13. Krasilšhchik, I.S., Lychagin, V., Vinogradov, A.M.: Geometry of Jet Spaces and Nonlinear Partial Differential Equations. Gordon and Breach Science, Amsterdam (1986)
14. Ljapunov, A.M.: Stability of Motion; with an Contribution by V.A. Pliss and an Introduction by V.P. Basov. Mathematics in Science and Engineering, vol. 30. Academic Press, New York (1966)
15. Lychagin, V., Prástaro, A.: Singularities of Cauchy data, characteristics, cocharacteristics and integral cobordism. Diff. Geom. Appl. **4**, 283–300 (1994)
16. Milnor, J.: On manifolds homeomorphic to the 7-sphere. Ann. Math. **64**(2), 399–405 (1956)
17. Moise, E.: Affine structures in 3-manifolds. V. The triangulation theorem and Hauptvermuntung. Ann. Math. Second Ser. **56**, 96–114 (1952)
18. Moise, E.: Geometric Topology in Dimension 2 and 3. Springer, Berlin (1977)
19. Nash, J.: Real algebraic manifolds. Ann. Math. **56**(2), 405–421 (1952)
20. Perelman, G.: The entropy formula for the Ricci flow and its geometry applications (2002). arXiv:math/0211159v1
21. Perelman, G.: Ricci flow with surgery on three-manifolds. arXiv:math/0303109v1
22. Prástaro, A.: Quantum geometry of PDEs. Rep. Math. Phys. **30**(3), 273–354 (1991)

23. Prástaro, A.: Quantum and integral (co)bordisms in partial differential equations. Acta Appl. Math. **51**, 243–302 (1998)
24. Prástaro, A.: (Co)bordism groups in PDEs. Acta Appl. Math. **59**(2), 111–202 (1999)
25. Prástaro, A.: (Co)bordism groups in quantum PDEs. Acta Appl. Math. **64**(2–3), 111–217 (2000)
26. Prástaro, A.: Quantized Partial Differential Equations. World Scientific, Singapore (2004)
27. Prástaro, A.: Geometry of PDEs. I: integral bordism groups in PDE's. J. Math. Anal. Appl. **319**, 547–566 (2006)
28. Prástaro, A.: Geometry of PDEs. II: variational PDE's and integral bordism groups. J. Math. Anal. Appl. **321**, 930–948 (2006)
29. Prástaro, A.: Geometry of PDEs. IV: Navier–Stokes equation and integral bordism groups. J. Math. Anal. Appl. **338**(2), 1140–1151 (2008)
30. Prástaro, A.: (Un)stability and bordism groups in PDEs. Banach J. Math. Anal. **1**(1), 139–147 (2007)
31. Prástaro, A.: Extended crystal PDEs stability. I: The general theory. Math. Comput. Model. **49**(9–10), 1759–1780 (2009)
32. Prástaro, A.: Extended crystal PDEs stability. II: The extended crystal MHD-PDEs. Math. Comput. Model. **49**(9–10), 1781–1801 (2009)
33. Prástaro, A.: On the extended crystal PDE's stability. I: The n-d'Alembert extended crystal PDEs. Appl. Math. Comput. **204**(1), 63–69 (2008)
34. Prástaro, A.: On the extended crystal PDEs stability. II: Entropy-regular-solutions in MHD-PDEs. Appl. Math. Comput. **204**(1), 82–89 (2008)
35. Prástaro, A.: Surgery and bordism groups in quantum partial differential equations. I: The quantum Poincaré conjecture. Nonlinear Anal. Theory Methods Appl. **71**(12), 502–525 (2009)
36. Prástaro, A.: Surgery and bordism groups in quantum partial differential equations. II: Variational quantum PDEs. Nonlinear Anal. Theory Methods Appl. **71**(12), 526–549 (2009)
37. Prástaro, A.: Extended crystal PDEs. arXiv:0811.3693
38. Prástaro, A.: Quantum extended crystal super PDEs. Nonlinear Anal. Real World Appl. (in press). arXiv:0906.1363
39. Prástaro, A.: Exotic heat PDEs. Commun. Math. Anal. **10**(1), 64–81 (2011). arXiv:1006.4483
40. Prástaro, A.: Exotic heat PDEs. II, arXiv:1009.1176. To appear in the book: Essays in Mathematics and Its Applications—Dedicated to Stephen Smale. (Eds. P.M. Pardalos and Th.M. Rassias). Springer, New York
41. Prástaro, A.: Exotic PDEs. arXiv:1101.0283
42. Prástaro, A.: Exotic n-d'Alembert PDEs and stability (2010). arXiv:1011.0081
43. Prástaro, A., Rassias, Th.M.: A geometric approach to an equation of J. d'Alembert. Proc. Am. Math. Soc. **123**(5), 1597–1606 (1995)
44. Prástaro, A., Rassias, Th.M.: A geometric approach of the generalized d'Alembert equation. J. Comput. Appl. Math. **113**(1–2), 93–122 (2000)
45. Prástaro, A., Rassias, Th.M.: Ulam stability in geometry of PDEs. Nonlinear Funct. Anal. Appl. **8**(2), 259–278 (2003)
46. Smale, S.: Generalized Poincaré conjecture in dimension greater than four. Ann. Math. **74**(2), 391–406 (1961)
47. Switzer, A.S.: Algebraic Topology-Homotopy and Homology. Springer, Berlin (1976)
48. Tognoli, A.: Su una congettura di nash. Ann. Sc. Norm. Super. Pisa **27**, 167–185 (1973)
49. Ulam, S.M.: A Collection of Mathematical Problems. Interscience, New York (1960)
50. Wall, C.T.C.: Determination of the cobordism ring. Ann. Math. **72**, 292–311 (1960)
51. Wall, C.T.C.: Surgery on Compact Manifolds. London Math. Soc. Monographs, vol. 1. Academic Press, New York (1970)
52. Wall, C.T.C.: In: Ranicki, A.A. (ed.) Surgery on Compact Manifolds, 2nd edn. Amer. Math. Soc. Surveys and Monographs, vol. 69. Amer. Math. Soc., Providence (1999)
53. Warner, F.W.: Foundations of Differentiable Manifolds and Lie Groups. Scott, Foresman, Glenview (1971)

Chapter 37
Stability of Affine Approximations on Bounded Domains

V.Y. Protasov

Abstract Let f be an arbitrary function defined on a convex subset G of a linear space. Suppose the restriction of f on every straight line can be approximated by an affine function on that line with a given precision $\varepsilon > 0$ (in the uniform metric); what is the precision of approximation of f by affine functionals globally on G? This problem can be considered in the framework of stability of linear and affine maps. We show that the precision of the global affine approximation does not exceed $C(\log d)\varepsilon$, where d is the dimension of G, and C is an absolute constant. This upper bound is sharp. For some bounded domains $G \subset \mathbb{R}^d$, it can be improved. In particular, for the Euclidean balls the upper bound does not depend on the dimension, and the same holds for some other domains. As auxiliary results we derive estimates of the multivariate affine approximation on arbitrary domains and characterize the best affine approximations.

Key words Approximation · Affine functionals · Stability · Refinement theorem

Mathematics Subject Classification 41A50 · 41A63 · 39A30 · 52A20

37.1 Introduction

Consider an arbitrary function $f : G \to \mathbb{R}$, where G is a convex subset of a linear space V. Assume that the restriction of f to every straight line $l \subset V$, $l \cap G \neq \emptyset$ can be approximated (in the uniform metric) with a given precision $\varepsilon > 0$ by an affine function on l. Is it true that there is an affine functional $\varphi : G \to \mathbb{R}$ approximating f uniformly on the whole domain G with a precision $C(\varepsilon)$, where $C(\varepsilon) \to 0$ as $\varepsilon \to 0$? Clearly, if a function $f : G \to \mathbb{R}$ is affine on each straight line, then it is affine on the whole domain G. The question is whether this property is stable under

Dedicated to Professor Themistocles M. Rassias on the occasion of his 60th birthday.

V.Y. Protasov (✉)
Dept. of Mechanics and Mathematics, Moscow State University, Vorobyovy Gory, 119992, Moscow, Russia
e-mail: v-protassov@yandex.ru

P.M. Pardalos et al. (eds.), *Nonlinear Analysis*, Springer Optimization and Its Applications 68, In Honor of Themistocles M. Rassias on the Occasion of his 60th Birthday, DOI 10.1007/978-1-4614-3498-6_37,

small perturbations: if a function f is "almost affine" on every straight line, then it is globally "almost affine"?

This problem is related to the concept of stability of affine and linear maps that have been thoroughly studied in many papers. The stability means that any "almost linear" map can be approximated by a linear one. Such problems originated with Ulam and Hyers [7, 21] and became very popular in the literature due to many applications (see [1, 4, 5, 8] and references therein). There are several classes of stability depending on the sense we understand the almost linearity. For example, Ulam–Hyers–Rassias stability for Cauchy equation [11, 15, 16], Lipschitz stability [13, 14, 19], etc., see [3, 10, 17, 18]. Tools elaborated in the literature to prove stability (the convex separation technique, fixed point principle, selections of multi-valued maps, etc.) do not work for our problem, and we use a different approach based on the idea of refinability. Let us first introduce some notation.

A functional $\varphi : V \to \mathbb{R}$ is affine if $\varphi((1-t)\mathbf{x}+t\mathbf{y}) = (1-t)\varphi(\mathbf{x})+t\varphi(\mathbf{y}), t \in \mathbb{R}$. Every affine functional is a sum of a linear functional and a constant.

Definition 37.1 Let $f : G \to \mathbb{R}$ be an arbitrary function on a convex set G and ε be a positive number; we say that f possesses the ε-property if for any straight line l there exists an affine function φ_l on l such that $|f(\mathbf{x}) - \varphi_l(\mathbf{x})| \le \varepsilon$ for every $\mathbf{x} \in l \cap G$.

Affine functions are precisely those possessing the 0-property. Observe that the ε-property does not depend on affine transforms of the domain G, or on addition of affine functionals to f. Suppose A is an affine bijection of V and β is an affine functional on V; then the function $f(A\cdot)+\beta(\cdot)$ possesses the ε-property if and only if $f(\cdot)$ does.

For any set K we denote $\|f\|_K = \sup_{\mathbf{x}\in K} |f(\mathbf{x})|$, and $E(f, K)$ is the infimum of $\|f - \varphi\|_K$ over all affine functionals φ. Our problem can be formulated as follows: Find a function $C(\varepsilon)$ such that for every function $f : G \to \mathbb{R}$ that possesses the ε-property we have $E(f, G) \le C(\varepsilon)$. For the entire space V, i.e., in case $G = V$, this problem is rather simple, and the answer is given by the following

Proposition 37.1 *Let V be an arbitrary linear space. If a function $f : V \to \mathbb{R}$ possesses the ε-property on V, then $E(f, V) \le \varepsilon$.*

The proof is given in the Appendix. For bounded convex sets G, the problem becomes more interesting and more difficult. We first restrict ourselves to finite-dimensional spaces. It can be shown easily that for every $d \ge 2$ there exists a constant C_d such that for any $G \subset \mathbb{R}^d$ we have $E(f, G) \le C_d\varepsilon$, whenever f possesses the ε-property on G. It is not difficult to derive a polynomial upper bound for C_d, for example, $C_d \le Cd^2$. Do there exist bounds for C_d independent of the dimension? In general the answer is negative. In Theorem 37.2, we show that $C_d \le C \log_2 d$, where C is an absolute constant (C is close to 2 for large d). This logarithmic bound is asymptotically sharp and is attained, for example, if G is a simplex or is an L_1-ball in $\mathbb{R}^d$ (Propositions 37.4 and 37.5). That is why for any infinite-dimensional space V

there are no functions $C(\varepsilon)$ with the required property (Corollary 37.5). Therefore, the problem becomes estimating the constants $C_d(G)$ for concrete convex domains $G \subset \mathbb{R}^d$. This is the subject of Sect. 37.5. It is shown in Theorem 37.3 that if G is a Euclidean ball in $\mathbb{R}^d$, then for every function $f : G \to \mathbb{R}$ with the ε-property we have $E(f, G) < 22\varepsilon$ (regardless of the dimension!). The same holds for all ellipsoids (this is obvious). Moreover, if G can be sandwiched between two homothetic ellipsoids with the ratio $\gamma > 1$, then this constant does not exceed $E(f, G) < 24\gamma\varepsilon$. Thus, $C_d(G)$ depends essentially on the geometry of the domain. It does not actually depend on the local properties (such as the smoothness of the boundary, etc.), but rather on global geometrical properties of the set G. Let us stress that we do not make any assumptions on the approximated function f, it may be discontinuous or non-measurable.

The paper is organized as follows. In Sect. 37.2, we prove the main auxiliary result, Theorem 37.1, that establishes an upper bound for $E(f, G)$ and characterizes affine functionals of the best multivariate approximation. Section 37.3 deals with some special properties of a function that follow from the ε-property. Using those results in Sect. 37.4, we prove that $E(f, G) \le C(\log_2 d)\varepsilon$ for any convex set $G \subset \mathbb{R}^d$. This logarithmic upper bound is attained for entropy-type functions f on simplices and on cross-polytopes G (Propositions 37.4 and 37.5). In Sect. 37.5, we prove the upper bounds for Euclidean balls and for ellipsoid-like domains. Finally, in Sect. 37.6, several open problems are formulated. Some long or technical proofs are placed in the Appendix.

Throughout the paper, we write $\mathscr{A}(V)$ and $\mathscr{L}(V)$, respectively, for the spaces of affine and linear functionals on V, $|\cdot|$ for the Euclidean norm in $\mathbb{R}^d$. For any function g, we denote $\|g\|_K = \sup_{\mathbf{x}\in K} |g(\mathbf{x})|$; for affine functions g, we also use the notation $\|g\| = \sup_{\mathbf{x}\in\mathbb{R}^d, |\mathbf{x}|\le 1} |g(\mathbf{x})|$. As usual, $\mathrm{co}(M)$ denotes the convex hull of a set M, $\mathrm{diam}(M)$ is its diameter, and $\mathrm{dist}\{M_1, M_2\} = \inf_{\mathbf{x}_i \in M_i} |\mathbf{x}_1 - \mathbf{x}_2|$. Some of absolute constants will be denoted by C, they may take different values. Elements of vector spaces are written in bold letters.

37.2 A Criterion of the Best Affine Approximation

In this section, we obtain a lower bound for the value $E(f, K)$ and prove a criterion for the best affine approximations. This is done for any function f defined on an arbitrary set $K \subset \mathbb{R}^d$. In the next section, we use those results as the main tool for proving the fundamental theorems. The main idea is not actually new: A problem of uniform approximation on some domain can be reduced to approximation on a finite set of points. For multivariate continuous functions on compact sets, the corresponding results can be easily deduced from well-known results of the approximation theory (alternance, refinement theorem, etc.) We are going to generalize them for arbitrary sets and functions. This aspect is discussed in Remark 37.1. Everywhere in this section, $K \subset \mathbb{R}^d$ is an arbitrary set and f is an arbitrary bounded real-valued function on it.

Proposition 37.2 *Let $f : K \to \mathbb{R}$ be an arbitrary function and $\alpha > 0$ be a given number. Suppose there exists a number $C > 0$ such that for every $\rho > 0$ there are finite nonempty subsets A, B of K and an affine functional φ such that $\|\varphi\| \le C$ $\operatorname{dist}\{\operatorname{co}(A), \operatorname{co}(B)\} < \rho$, and*

$$f(\mathbf{a}_i) - \varphi(\mathbf{a}_i) > \alpha, \quad \mathbf{a}_i \in A; \qquad f(\mathbf{b}_j) - \varphi(\mathbf{b}_j) < -\alpha, \quad \mathbf{b}_j \in B; \quad (37.1)$$

then $E(f, K) \ge \alpha$.

Proof Suppose on the contrary that for some function $g \in \mathscr{A}(\mathbb{R}^d)$ we have $\|f - g\|_K \le q$, where $q < \alpha$. For an arbitrary $\rho > 0$, we have the corresponding sets A, B and a function $\varphi \in \mathscr{A}(\mathbb{R}^d)$. Take points $\mathbf{a} \in \operatorname{co}(A), \mathbf{b} \in \operatorname{co}(B)$ such that $|\mathbf{a} - \mathbf{b}| < \rho$. Since

$$g(\mathbf{a}_i) - \varphi(\mathbf{a}_i) = \big(f(\mathbf{a}_i) - \varphi(\mathbf{a}_i)\big) - \big(f(\mathbf{a}_i) - g(\mathbf{a}_i)\big) > \alpha - q, \quad \mathbf{a}_i \in A,$$

it follows that

$$(g - \varphi)(\mathbf{a}) = (g - \varphi)\left(\sum_{i=1}^{m} t_i \mathbf{a}_i\right) = \sum_{i=1}^{m} t_i \big(g(\mathbf{a}_i) - \varphi(\mathbf{a}_i)\big) > \alpha - q,$$

where $\sum_{i=1}^{m} t_i = 1$, $t_i \ge 0$, $i = 1, \dots, m$. Applying the same argument to the points $\mathbf{b}_j \in B$, we obtain $g(\mathbf{b}) - \varphi(\mathbf{b}) < -(\alpha - q)$. Now take a difference:

$$(g - \varphi)(\mathbf{a}) - (g - \varphi)(\mathbf{b}) > 2(\alpha - q).$$

Therefore, $\|g - \varphi\| |\mathbf{a} - \mathbf{b}| > 2(\alpha - q)$. Since $\|g - \varphi\| \le \|g\| + \|\varphi\| \le \|g\| + C$ and $|\mathbf{a} - \mathbf{b}| < \rho$, we have $(\|g\| + C)\rho > 2(\alpha - q)$, which is impossible for sufficiently small ρ. □

In practice, lower bounds for $E(f, G)$ can be easily derived by the following special case of Proposition 37.2:

Corollary 37.1 *If for a given function $f : K \to \mathbb{R}$ there are $\varphi \in \mathscr{A}(\mathbb{R}^d)$ and finite nonempty sets $A, B \subset K$ with intersecting convex hulls such that* (37.1) *holds, then $E(f, K) > \alpha$.*

Thus, we have sufficient conditions to ensure that f cannot be well approximated by affine functionals, i.e., that the value $E(f, G)$ exceeds a given number $\alpha > 0$. To verify those conditions, it suffices to check inequalities (37.1) at finitely many points $\mathbf{a}_i, \mathbf{b}_j$, for some $\varphi \in \mathscr{A}(\mathbb{R}^d)$. The main difficulty, of course, is to find such points and the functional φ. A question arises if those conditions are invertible. In other words, do such finite systems of points always exist, and how many points do they contain? An affirmative answer is given in Theorem 37.1 below. Such systems exist, and the total number of points is bounded by $d + 1$ (or $d + 2$, for the condition of

Corollary 37.1). Before we formulate the statement, let us make one simple observation.

Every bounded function $f : K \to \mathbb{R}$ admits the best approximating affine functional $\varphi \in \mathscr{A}(\mathbb{R}^d)$, i.e., $\|f - \varphi\|_K = E(f, K)$. To prove this, we assume, without loss of generality, that the affine hull of K is $\mathbb{R}^d$, otherwise the problem is reduced to a smaller dimension. Hence $\mathrm{co}(K)$ contains a Euclidean ball of radius $r > 0$. Therefore, for any affine operator g one has $\|g\|_K \geq r\|g\|$, and consequently $\|f - g\|_K \geq r\|g\| - \|f\|_K$. Since f is bounded, choosing $M > 0$ large enough, we may restrict ourselves to the set $\mathscr{V}_M = \{g \in \mathscr{A}(\mathbb{R}^d) \mid \|g\| \leq M\}$. The functional $\eta(g) = \|f - g\|_K = \sup_{\mathbf{x} \in K} |f(\mathbf{x}) - g(\mathbf{x})|$ is lower semicontinuous on $\mathscr{A}(\mathbb{R}^d)$, hence it attains its minimum on the compact set $\mathscr{V}_M$.

The following theorem characterizes the best approximating functional φ for a function f.

Theorem 37.1 *Let $f : K \to \mathbb{R}$ be an arbitrary bounded function defined on a set $K \subset \mathbb{R}^d$. A functional $\varphi \in \mathscr{A}(\mathbb{R}^d)$ is the best approximating for f if and only if for every $\alpha < \|f - \varphi\|_K$ one of the two following conditions is satisfied:*

(a) *There are numbers $m, m' \geq 1$, $m + m' \leq d + 2$, and two subsets $A, B \subset K$ consisting of m and m' points, respectively, such that* $\mathrm{co}(A) \cap \mathrm{co}(B) \neq \emptyset$ *and* (37.1) *holds.*

(b) *There are numbers $m, m' \geq 1$, $m + m' \leq d + 1$, such that for every $\rho > 0$ there exist two subsets $A, B \subset K$ consisting of m and m' points, respectively, such that* $\mathrm{dist}\{\mathrm{co}(A), \mathrm{co}(B)\} < \rho$ *and* (37.1) *holds.*

The proof is in the Appendix. The main idea is to consider the value $\|f - \varphi\|_K$ as a function of φ defined on the set $\mathscr{A}(\mathbb{R}^d)$. This function is convex, hence the minimum is attained at the point $\varphi = 0$ precisely when the subdifferential of this function contains zero. On the other hand, this function is a pointwise supremum of linear functionals, and hence, by Dubovitskii–Milyutin theorem, its subdifferential is a closure of the convex hull of derivatives of those linear functionals. That closure contains zero if and only if for any $\alpha < \|f\|_K$ there are two finite systems of points possessing property (37.1), whose convex hulls are arbitrarily close to each other. This is a scheme of the proof; all the details with necessary explanations and references are in the Appendix.

Note that assumption (a) is basically weaker than (b), and the only advantage of (b) is the smaller number of points. Thus, either there are sets A, B of total cardinality $\leq d + 2$ satisfying (37.1) such that their convex hulls intersect, or there are such sets A, B with arbitrarily close convex hulls, but in this case their total cardinality can be reduced to $d + 1$.

Let us also remark that for continuous functions f on a compact set K assumption (b) obviously implies (a) (if we do not count the number of points). In this case, Theorem 37.1 yields the following simple criterion for the functionals of best approximation:

Corollary 37.2 *Let f be a continuous function on a compact set $K \subset \mathbb{R}^d$. A functional $\varphi \in \mathscr{A}(\mathbb{R}^d)$ is the best approximation for f if and only if there exist two sets of points $A, B \subset K$ of total cardinality at most $d+2$, whose convex hulls intersect, such that*

$$\begin{aligned} f(\mathbf{a}_i) - \varphi(\mathbf{a}_i) &= \|f - \varphi\|_K, \quad \mathbf{a}_i \in A \quad \textit{and} \\ f(\mathbf{b}_j) - \varphi(\mathbf{b}_j) &= -\|f - \varphi\|_K, \quad \mathbf{b}_j \in B. \end{aligned} \tag{37.2}$$

Remark 37.1 Corollary 37.2 has a simpler and more elementary proof than Theorem 37.1. Actually, the existence of a finite system of points satisfying (37.2) follows easily from the so-called refinement theorem originated with Levin [12] and Ioffe and Tikhomirov [9]. Some prototypes of the refinement theorem appeared much earlier, in the works of Lyusternik in the early 1950s, and can be traced back to Vallée Poussin and Tchebychev (see [9] for general discussion of this aspect). The refinement theorem extends the notion of Tchebychev's alternance from univariate polynomials to multivariate convex functions. However, that theorem is applicable only for lower semicontinuous functions f on compact domains K. That is why, to prove Theorem 37.1 for general functions on arbitrary sets K, we have to apply a different technique.

37.3 Auxiliary Facts on the ε-Property

In this section, we derive several consequences of the ε-property of a function. They will be used in the sequel to prove the fundamental theorems. The following inequalities follow directly from the ε-property.

Lemma 37.1 *Let f possess the ε-property and let $\mathbf{x} = (1-t)\mathbf{a} + t\mathbf{b}$, where $\mathbf{a}, \mathbf{b} \in G$, $t \in [0,1]$. Then*

$$f(\mathbf{b}) = t^{-1}\big(f(\mathbf{x}) - (1-t)f(\mathbf{a})\big) + \omega_1(\mathbf{a},\mathbf{b},t)\varepsilon, \quad \textit{where } |\omega_1| < \frac{2}{t}, \tag{37.3}$$

and

$$f(\mathbf{x}) = (1-t)f(\mathbf{a}) + tf(\mathbf{b}) + \omega_2(\mathbf{a},\mathbf{b},t)\varepsilon, \quad \textit{where } |\omega_2| < 2. \tag{37.4}$$

Corollary 37.3 *Suppose f possesses the ε-property; then for each $\mathbf{x} \in [\mathbf{a},\mathbf{b}]$ we have*

$$f(\mathbf{x}) \le \max\{f(\mathbf{a}), f(\mathbf{b})\} + 2\varepsilon.$$

Thus, we have several inequalities for the values of the function f on a line. The next simple result concerns the values on parallel lines.

Lemma 37.2 *Let f possess the ε-property and let $[\mathbf{a}_1, \mathbf{a}_2]$ and $[\mathbf{b}_1, \mathbf{b}_2]$ be parallel segments such that $(\mathbf{a}_2 - \mathbf{a}_1) = \lambda(\mathbf{b}_2 - \mathbf{b}_1)$; then*

$$f(\mathbf{a}_2) - f(\mathbf{a}_1) = \lambda\big(f(\mathbf{b}_2) - f(\mathbf{b}_1)\big) + \omega_3(\mathbf{a}_1, \mathbf{a}_2, \mathbf{b}_1, \mathbf{b}_2)\varepsilon, \tag{37.5}$$

where $|\omega_3| < 4\ (1 + \lambda)$.

Proof Let $\mathbf{x}$ be the point of intersection of the segments $[\mathbf{a}_1, \mathbf{b}_2]$ and $[\mathbf{a}_2, \mathbf{b}_1]$; then $\mathbf{x}$ divides both these segments in the ratio λ. Let t be such that $t : (1 - t) = \lambda$. Applying (37.3) to those two segments and taking a difference, we arrive at (37.5). □

In the next two results, G is a bounded convex subset of $\mathbb{R}^d$.

Proposition 37.3 *A function possessing the ε-property on a domain G is bounded on it.*

Proof Put the origin O at some interior point of G and take basis vectors $\mathbf{b}_1, \dots, \mathbf{b}_d$ of length h. We assume h is small enough so that the cube $\mathscr{C}$ with vertices $(\pm\mathbf{b}_1, \dots, \pm\mathbf{b}_d)$ is contained in G. For any $\mathbf{x} = \sum_i x_i \mathbf{b}_i$ inside this cube, we consequently apply (37.5) and obtain

$$|\mathbf{x}| \leq \sum_{i=1}^{d} \big(|x_i| \big| f(\mathbf{b}_i) - f(0)\big| + (4 + 4x_i)\varepsilon\big) \leq \sum_{i=1}^{d} \big| f(\mathbf{b}_i) - f(0)\big| + 8d.$$

Thus, f is bounded on the cube $\mathscr{C}$. Choosing now a positive constant μ such that $G \subset \mu\mathscr{C}$ and applying (37.3) for an arbitrary point $\mathbf{b} \in G$ and for $\mathbf{a} = 0, \mathbf{x} = \mu^{-1}\mathbf{b}$, we conclude that $f(\mathbf{b})$ is uniformly bounded for all $\mathbf{b} \in G$. □

Finally, we need the following consequence of the ε-property that will be referred to as ε-continuity.

Lemma 37.3 *Let f possess the ε-property; then there is a constant C depending only on G such that for any $\mathbf{a}, \mathbf{b} \in G$ such that $|\mathbf{a} - \mathbf{b}| \leq \rho$ we have*

$$\big| f(\mathbf{a}) - f(\mathbf{b})\big| \leq C\rho + (4 + 4C\rho)\varepsilon. \tag{37.6}$$

Proof Without loss of generality, it can be assumed that G is of full dimension d. Then it contains some ball of radius $r > 0$. Hence, G contains a segment of length $2r$ parallel to the segment $[\mathbf{a}, \mathbf{b}]$. Applying (37.5) for these two segments and for $\lambda = |\mathbf{a} - \mathbf{b}|/(2r)$, we obtain (37.6) with $C = \max\{\frac{1}{2r}, \frac{\|f\|_G}{r}\}$. □

Note that a function with the ε-property may be discontinuous everywhere, for example, $f(\mathbf{x}) = 2\varepsilon D(\mathbf{x})$, where $D(\mathbf{x})$ is the Dirichlet function on the segment [0, 1]. Nevertheless, by Lemma 37.3, it is always ε-continuous, i.e., possesses property (37.6): its modulus of continuity $\omega_f(\rho)$ does not exceed $4\varepsilon + C_0\rho$, where C_0 is an absolute constant. The factor 4, in general, cannot be reduced.

Thus, a function with the ε-property is uniformly bounded and ε-continuous.

37.4 Approximation on General Domains

We start with the first fundamental theorem which states that for every convex set $G \subset \mathbb{R}^d$ and for every function f that possesses the ε-property on it we have $E(f,G) \le C \log d$, where C is an absolute constant. Then we show that this upper bound cannot be improved.

Theorem 37.2 *For an arbitrary convex set $G \subset \mathbb{R}^d$ and for an arbitrary function $f : G \to \mathbb{R}$ that possess the ε-property, we have $E(f,G) \le 2(1+\log_2 d)\varepsilon$.*

In the proof, we use two auxiliary results. The first one deals with the case when G is a simplex. As usual, $\lceil x \rceil$ denotes the smallest integer that is bigger than or equal to x.

Lemma 37.4 *If a function f possesses the ε-property on a k-dimensional simplex Δ and vanishes at all its vertices, then $\|f\|_\Delta \le 2\lceil \log_2(k+1)\rceil \varepsilon$.*

Proof It suffices to consider the case $\varepsilon = 1$. Denote by $\mathbf{x}_1, \dots, \mathbf{x}_{k+1}$ the vertices of the simplex Δ, and by μ_k the supremum of values $\|f\|_\Delta$ for all functions f satisfying the assumptions of the lemma. For $k = 1$ we have $2\lceil \log_2(k+1)\rceil = 2$, and the assertion obviously follows from the ε-property. Consider now an arbitrary odd $k = 2r+1$ and take any point $\mathbf{a} = \sum_{i=1}^{2r+2} t_i \mathbf{x}_i \in \Delta$. Writing $s = \sum_{i=1}^{r+1} t_i$, $\mathbf{a}_1 = s^{-1}\sum_{i=1}^{r+1} t_i \mathbf{x}_i$, $\mathbf{a}_2 = (1-s)^{-1}\sum_{i=r+2}^{2r+2} t_i \mathbf{x}_i$, we have $\mathbf{a} = s\mathbf{a}_1 + (1-s)\mathbf{a}_2$. Observe that $\mathbf{a}_j \in \Delta_j$, $j = 1,2$, where Δ_1 is the simplex with vertices $\mathbf{x}_1, \dots, \mathbf{x}_{r+1}$ and Δ_2 is the simplex with vertices $\mathbf{x}_{r+2}, \dots, \mathbf{x}_{2r+2}$. These simplices are both r-dimensional, hence $f(\mathbf{a}_j) \le \mu_r$, $j = 1,2$. Since $\mathbf{a} \in [\mathbf{a}_1, \mathbf{a}_2]$, and f possesses the ε-property with $\varepsilon = 1$, Corollary 37.3 yields $f(\mathbf{a}) \le \max\{f(\mathbf{a}_1), f(\mathbf{a}_2)\} + 2 \le \mu_r + 2$. If $k = 2r$, then we consider an r-dimensional simplex $\Delta_1 = \mathrm{co}\{\mathbf{x}_1, \dots, \mathbf{x}_{r+1}\}$ and an $(r-1)$-dimensional one $\Delta_2 = \mathrm{co}\{\mathbf{x}_{r+2}, \dots, \mathbf{x}_{2r+1}\}$. Similarly, we find points $\mathbf{a}_j \in \Delta_j$, $j = 1,2$ such that $\mathbf{a} \in [\mathbf{a}_1, \mathbf{a}_2]$. Applying the ε-property (Corollary 37.3) and the fact that $\mu_{r-1} \le \mu_r$, we obtain

$$f(\mathbf{a}) \le \max\bigl\{f(\mathbf{a}_1), f(\mathbf{a}_2)\bigr\} + 2 \le \max\{\mu_r, \mu_{r-1}\} + 2 = \mu_r + 2.$$

Thus, the sequence $\{\mu_k\}_{k\in\mathbb{N}}$ satisfies the inequalities $\mu_1 \le 2$ and $\mu_{2r} \le \mu_r + 2$, $\mu_{2r+1} \le \mu_r + 2$, $r \in \mathbb{N}$. By induction one easily shows that $\mu_k \le 2\lceil \log_2(k+1)\rceil$. □

The proof of the following technical lemma is straightforward, and we omit it.

Lemma 37.5 *For arbitrary natural numbers m, m', we have*

$$\lceil \log_2 m \rceil + \lceil \log_2 m' \rceil \le 2\log_2\bigl(m + m' - 1\bigr).$$

Proof of Theorem 37.2 It suffices to consider the case $\varepsilon = 1$. Without loss of generality, it can be assumed that the best approximating function φ is identically zero. For an arbitrary positive constant $\alpha < E(f,G)$, there are two sets of points A, B possessing either property (**a**) or (**b**) from Theorem 37.1. Assume first property (**b**). Let $\mathbf{a} \in \mathrm{co}(A)$ and $\mathbf{b} \in \mathrm{co}(B)$ be such that $|\mathbf{a} - \mathbf{b}| < \rho$. If the dimension r of the set $\mathrm{co}(A)$ is smaller than $m - 1$, where m is the cardinality of A, then by the Caratheodory theorem there are $r + 1$ points from A, whose convex hull contains **a**. Hence, one can remove all other points of A, and all the assumptions remain valid. Thus, removing, if necessary, some extra points, it may be assumed that the dimensions of $\mathrm{co}(A)$ and $\mathrm{co}(B)$ are $m - 1$ and $m' - 1$, respectively, i.e., these sets are simplices. Therefore, there is an affine function h such that $h(\mathbf{a}_i) = f(\mathbf{a}_i)$, $i = 1, \ldots, m$. Clearly, $h(\mathbf{a}) \geq \min_{i=1,\ldots,m} f(\mathbf{a}_i) > \alpha$. The function $h - f$ possesses the ε-property on the simplex $\Delta = \mathrm{co}(A)$ and vanishes at each of its vertices. Whence, by Lemma 37.4, $(h - f)(\mathbf{a}) \leq 2\lceil \log_2 m \rceil$, and consequently, $f(\mathbf{a}) > \alpha - 2\lceil \log_2 m \rceil$. Similarly, $f(\mathbf{b}) < -\alpha + 2\lceil \log_2 m' \rceil$. Taking a difference, we obtain

$$f(\mathbf{a}) - f(\mathbf{b}) > 2\alpha - 2\lceil \log_2 m \rceil - 2\lceil \log_2 m' \rceil.$$

On the other hand, Lemma 37.3 for $\varepsilon = 1$ yields that there is a constant C such that $f(\mathbf{a}) - f(\mathbf{b}) < 4 + 5C\rho$. Thus,

$$2 + \frac{5}{2}C\rho > \alpha - \lceil \log_2 m \rceil - \lceil \log_2 m' \rceil.$$

Taking the limit as $\rho \to 0$ and $\alpha \to E(f,G)$, we obtain $E(f,G) \leq \lceil \log_2 m \rceil + \lceil \log_2 m' \rceil + 2$. Applying now Lemma 37.5 and taking into account that $m + m' - 1 \leq d$, we complete the proof.

Assume now property (**a**). Let $\mathbf{x} = \mathrm{co}(A) \cap \mathrm{co}(B)$. Arguing as above, we conclude that $f(\mathbf{x}) > \alpha - 2\lceil \log_2 m \rceil$ and $f(\mathbf{x}) < -\alpha + 2\lceil \log_2 m' \rceil$; therefore, $\alpha < \lceil \log_2 m \rceil + \lceil \log_2 m' \rceil$. Taking the limit as $\alpha \to E(f,G)$, we obtain $E(f,G) \leq \lceil \log_2 m \rceil + \lceil \log_2 m' \rceil$, where $m + m' \leq d + 2$. One of the numbers m, m', say, m, is bigger than 1. Since $\lceil \log_2 m \rceil \leq \lceil \log_2 (m-1) \rceil + 1$, and $(m-1) + m' \leq d + 1$, we see that

$$\lceil \log_2 m \rceil + \lceil \log_2 m' \rceil \leq \lceil \log_2 (m-1) \rceil + \lceil \log_2 m' \rceil + 1 < 2(1 + \log_2 d),$$

and the theorem follows. □

In the proof of Theorem 37.2, we saw that if property (**a**) holds, then $E(f,G) \leq \lceil \log_2 m \rceil + \lceil \log_2 m' \rceil$, where $m + m' \leq d + 2$. In particular, this is always the case for continuous functions f (Corollary 37.2), when we can replace Lemma 37.5 by the following simple inequality:

$$\lceil \log_2 m \rceil + \lceil \log_2 m' \rceil \leq 2\lceil \log_2 (m + m' - 2) \rceil, \\ m, m' \in \mathbb{N},\ m + m' \geq 3. \tag{37.7}$$

Consequently, Theorem 37.2 can be slightly improved for continuous functions:

Corollary 37.4 *For an arbitrary convex compact set $G \subset \mathbb{R}^d$, $d \geq 2$, and for a continuous function $f : G \to \mathbb{R}$ that possesses the ε-property, we have $E(f, G) \leq 2\lceil \log_2 d \rceil \varepsilon$.*

Now let us show that the upper bound $E(f, G) \leq C \log d$ is attained in $\mathbb{R}^d$ for some polyhedra G and for "entropy-type" functions f.

Consider an odd function $p(t)$ on the segment $[-1, 1]$ that for positive t coincides with the entropy function: $p(t) = t \ln t$. Thus, $p(0) = 0$ and $p(t) = -t \ln(-t)$, $t \in [-1, 0)$. We need the following simple property of the entropy function, whose proof is in the Appendix.

Lemma 37.6 *For every segment $[a, b] \subset [-1, 1]$ we have $E(p, [a, b]) \leq \frac{b-a}{2e}$, if $0 \notin (a, b)$, and $E(p, [a, b]) \leq \frac{b-a}{e}$, if $0 \in (a, b)$.*

Proposition 37.4 *If G is a d-dimensional simplex, then there is a function $f : G \to \mathbb{R}$ possessing the ε-property, such that $E(f, G) \geq \frac{e \ln 2}{2}(\log_2(d + 1))\varepsilon$.*

Proof Let $G = \{\mathbf{x} = (x_1, \ldots, x_{d+1}) \in \mathbb{R}^{d+1} \mid \sum_{i=1}^{d+1} x_i = 1, x_i \geq 1, i = 1, \ldots, d + 1\}$ and $f(\mathbf{x}) = \sum_{i=1}^{d+1} x_i \ln x_i$. If a line intersects the simplex G in some segment $[\mathbf{a}, \mathbf{b}]$, then applying Lemma 37.6 for every i and taking the sum over all $i = 1, \ldots, d + 1$, we see that f possesses the ε-property with $\varepsilon \leq \sum_{i=1}^{d+1} \frac{|b_i - a_i|}{2e} \leq \frac{2}{2e} = \frac{1}{e}$. On the other hand, for $\alpha = \frac{1}{2} \ln(d + 1)$ and $\varphi = -\alpha$ (an identical constant), we have that the function $f - \varphi$ takes the value α at each vertex of the simplex, and the value $-\alpha$ at its center. Hence, by Corollary 37.1, we conclude that $E(f, G) \geq \alpha$, and therefore $E(f, G) \geq \frac{e \ln 2}{2}(\log_2(d + 1))\varepsilon$. □

Remark 37.2 The constant in the example of Proposition 37.4 is $\frac{e \ln 2}{2} = 0.942\ldots$, hence $E(f, G) \geq 0.942(\log_2(d + 1))\varepsilon$. This is less than half of the value of the upper bound from Theorem 37.2, which is approximately 2. It seems to be a challenging problem to evaluate the sharp constant in that inequality (we formulate it in Sect. 37.6).

In many problems of convex geometry, a simplex is the "worst" convex body, in the sense that some geometrical inequalities become equalities precisely for simplices. However, for our problem this is not the case. It appears that even for centrally-symmetric convex bodies the ratio $E(f, G)/\varepsilon$ may grow logarithmically with the dimension. Let $\mathscr{S}_d = \{\mathbf{x} \in \mathbb{R}^d \mid \sum_{i=1}^{d} |x_i| \leq 1\}$ be a unit d-dimensional cross-polytope, i.e., the unit ball of $\mathbb{R}^d$ with the L_1-norm.

Proposition 37.5 *There is a function $f : \mathscr{S}_d \to \mathbb{R}$ possessing the ε-property such that*

$$E(f, \mathscr{S}_d) \geq \frac{e \ln 2}{4}(\log_2 d)\varepsilon.$$

Proof Let $f(\mathbf{x}) = \sum_{i=1}^{d} p(x_i)$, where the function $p(t)$ is defined in Lemma 37.6. If a line intersects $\mathscr{S}_d$ in some segment $[\mathbf{a}, \mathbf{b}]$, then applying Lemma 37.6 for every i and taking the sum over all $i = 1, \ldots, d$, we see that f possesses the ε-property with $\varepsilon \le \sum_{i=1}^{d} \frac{|b_i - a_i|}{e} \le \frac{2}{e}$. On the face $\Delta = \{\mathbf{x} \in \mathscr{S}_d \mid \sum_{i=1}^{d} x_i = 1\}$, which is a $(d-1)$-dimensional simplex, we have $E(f, G) \ge \frac{1}{2} \ln d \ge \frac{e \ln 2}{4} (\log_2 d)\varepsilon$. □

Proposition 37.4 implies that for infinite-dimensional spaces the ε-property, in general, does not guarantee the approximability by affine functionals.

Corollary 37.5 *Suppose V is an infinite-dimensional space and $M > 0$ is a given number; then for every $\varepsilon > 0$ there is a convex set $G \subset V$ and a function $f : G \to \mathbb{R}$ with the ε-property, for which $E(f, G) > M$.*

Proof Take $d + 1$ independent elements of V. Their convex hull is a d-dimensional simplex on which, by Proposition 37.4, there is a function f with the ε-property, and $E(f, G) \ge C(\log_2(d+1))\varepsilon$, where C is an absolute constant. For sufficiently large d, we have $E(f, G) > M$. □

37.5 Special Domains. Estimates Independent of the Dimension

In view of the results of Sect. 37.4, functions with the ε-property can be approximated by affine functionals with the precision $C(\log d)\varepsilon$. This estimate holds for all d-dimensional domains G. However, for some concrete domains it can possibly be improved. In this section, we show that for Euclidean balls there is a uniform estimate which does not depend on the dimension. The same is true for all "ellipsoid-like" domains in $\mathbb{R}^d$ and in the Hilbert space. For the d-dimensional cube, this problem exhibited unexpected resistance, and the answer is still unknown (see Problem 3 in Sect. 37.6).

Theorem 37.3 *If G is a d-dimensional Euclidean ball, then for any function $f : G \to \mathbb{R}$ with the ε-property we have $E(f, G) \le 4(4 + \sqrt{2})\varepsilon$.*

To prove the theorem, we need several auxiliary results and notation. For a function $f : G \to \mathbb{R}$, we denote by $E_-(f, G) = \inf_{g \in \mathscr{L}(\mathbb{R}^d)} \sup_{\mathbf{x} \in G} (f(\mathbf{x}) - g(\mathbf{x}))$ the best approximation of f from below by linear functionals. For this value we have basically the same results that were established for $E(f, G)$ in Sect. 37.2. The proofs of the following assertions are very similar to the proofs of Proposition 37.2, Corollary 37.1, and Theorem 37.1 respectively, and we omit them.

Proposition 37.6 *Let $f : K \to \mathbb{R}$ be an arbitrary function and $\alpha > 0$ a given number. Suppose there exists a number $C > 0$ such that for every $\rho > 0$ there is a finite nonempty set $A \subset K$ and a linear functional φ such that $\|\varphi\| \le C$ $\mathrm{dist}\{0, \mathrm{co}(A)\} < \rho$, and*

$$f(\mathbf{a}_i) - \varphi(\mathbf{a}_i) > \alpha, \quad \mathbf{a}_i \in A; \tag{37.8}$$

then $E_-(f,K) \geq \alpha$.

The following special case offers a convenient way to estimate $E_-(f,G)$ from below.

Corollary 37.6 *If for a given function* $f : K \to \mathbb{R}$ *there is a functional* $\varphi \in \mathscr{L}(\mathbb{R}^d)$ *and a finite nonempty set* $A \subset K$ *such that* $0 \in \mathrm{co}(A)$ *and* (37.8) *holds, then* $E_-(f,K) > \alpha$.

Every function f has a linear functional of the best approximation from below, for which $\sup_{\mathbf{x}\in K}(f(\mathbf{x}) - \varphi(\mathbf{x})) = E_-(f,K)$. The existence is proved in the same way as for affine functionals in Sect. 37.2. The following analogue of Theorem 37.1 provides a criterion of the best approximation.

Proposition 37.7 *Suppose* $f : K \to \mathbb{R}$ *is an arbitrary function defined on a set* $K \subset \mathbb{R}^d$; *then a functional* $\varphi \in \mathscr{L}(\mathbb{R}^d)$ *is the best approximation from below for* f *if and only if for every* $\alpha < E_-(f,K)$ *and for every* $\rho > 0$ *there is a set* $A \subset K$ *of* $n \leq d+1$ *points such that* $\mathrm{dist}\{0, \mathrm{co}(A)\} < \rho$, *and* (37.8) *holds.*

Also in the proof of Theorem 37.3 we need the following technical lemma (proved in the Appendix).

Lemma 37.7 *Let points* $\mathbf{x}_1, \ldots, \mathbf{x}_{d+1} \in \mathbb{R}^d$ *be such that* $\frac{1}{2} \leq |\mathbf{x}_i| \leq 1$, $i = 1, \ldots, d+1$, *and* $\sum_{i=1}^{d+1} t_i \mathbf{x}_i = 0$, *where* $\sum_i t_i = 1$, $t_i \geq 0$, $i = 1, \ldots, d+1$. *Then there exist* i, j *such that* $(\mathbf{x}_i, \mathbf{x}_j) < 0$ *and* $t_i, t_j \geq \kappa_d$, *where the constant* κ_d *depends only on the dimension* d.

We denote by $\mathscr{B}_d$ a unit d-dimensional Euclidean ball centered at the origin. The proof of Theorem 37.3 contains many technical details, but its crucial idea can be described easily. Assume $f(0) = 0$ and the linear functional of the best approximation from below for f is identically zero. If the value $E_-(f, \mathscr{B}_d)$ exceeds α, then, by Proposition 37.7, there is a simplex Δ such that the value of f at all its vertices exceeds α, and $\mathrm{dist}\{0, \Delta\}$ is arbitrarily small. Assume for the moment that it vanishes, i.e., $0 \in \Delta$. Then we use the following geometrical fact: if a simplex Δ lies in $\mathscr{B}_d$ and contains the origin, then it contains an edge $\mathbf{x}_1\mathbf{x}_2$ such that the distance from its midpoint $\mathbf{x}$ to the origin is smaller than $\frac{\sqrt{2}}{2}$. Using Lemma 37.1, one shows that $f(\frac{\mathbf{x}_1}{|\mathbf{x}_i|})$ exceeds $\alpha + \delta$, whenever α is large enough ($\delta > 0$ is some constant). Replacing one of the vertices $\mathbf{x}_1$ or $\mathbf{x}_2$ by $\mathbf{x}$ and adding a suitable linear functional to f, we obtain a new simplex for which the value of f at all vertices exceeds $\alpha + C\delta$. Repeating this procedure n times, we get a simplex for which the value of f at all vertices exceeds $\alpha + Cn\delta$. By Corollary 37.6, this means that $E_-(f, \mathscr{B}_d) > \alpha + Cn\delta$, which tends to infinity as $n \to \infty$. The contradiction proves that α cannot be large. Then, using the central symmetry, we show that $E(f, \mathscr{B}_d)$ cannot be large either.

Proof of Theorem 37.3 It can be assumed that $G = \mathscr{B}_d, \varepsilon = 1$, and (with possible addition of a constant) $f(0) = 0$. We also assume that $\varphi \equiv 0$, where $\varphi \in \mathscr{L}(\mathbb{R}^d)$ is the best approximating functional for f from below. Let us first show that $E_-(f, \mathscr{B}_d) \leq 4(3+\sqrt{2})$. If, on the contrary, $\sup_{|\mathbf{x}|\leq 1} f(\mathbf{x}) > 4(3+\sqrt{2})$, then, by Proposition 37.7, for any $\alpha \in (4(3+\sqrt{2}), \sup_{|\mathbf{x}|\leq 1} f(\mathbf{x}))$ and for an arbitrary $\rho > 0$ there is a set $A = \{\mathbf{a}_1, \ldots, \mathbf{a}_n\} \subset \mathscr{B}_d$ such that $n \leq d+1, \operatorname{dist}\{0, \operatorname{co}(A)\} < \rho$ and $f(\mathbf{a}_i) > \alpha,\ i = 1, \ldots, n$. By compactness, it may be assumed that each point $\mathbf{a}_i$ converges to some point $\mathbf{x}_i$ as $\rho \to 0$. By the ε-continuity (Lemma 37.3), we have $f(\mathbf{x}_i) > \alpha - 4$. Writing $X = \{\mathbf{x}_1, \ldots, \mathbf{x}_n\}$, we obviously have $0 \in \operatorname{co}(X)$. Without loss of generality, it can be assumed that $n = d+1$ and that the convex hull of any proper subset of X does not contain zero, otherwise the problem is reduced from $\mathbb{R}^d$ to the linear span of that subset. Thus, the set $\Delta_1 = \operatorname{co}(X)$ is a simplex, and 0 is its interior point. Let us show that $|\mathbf{x}_i| \geq \frac{1}{2}$ for all i. If $|\mathbf{x}_i| < \frac{1}{2}$ for some i, then applying (37.3) to the points $0, \mathbf{x}_i$, and $\frac{1}{|\mathbf{x}_i|}\mathbf{x}_i$, we get $f(\frac{\mathbf{x}_i}{|\mathbf{x}_i|}) \geq 2(\alpha-4)-4 > \alpha - 4$, since $\alpha > 8$. Hence, if $|\mathbf{x}_i| < \frac{1}{2}$, one may replace that point by $\frac{1}{|\mathbf{x}_i|}\mathbf{x}_i$, and all the assumptions remain valid. Applying Lemma 37.7 to the points $\mathbf{x}_1, \ldots, \mathbf{x}_{d+1}$, we select two of them, say, $\mathbf{x}_1$ and $\mathbf{x}_2$, for which $(\mathbf{x}_1, \mathbf{x}_2) < 0$ and $t_1, t_2 \geq \kappa_d$, where t_i are the barycentric coefficients: $0 = \sum_i t_i \mathbf{x}_i$. The point $\frac{\mathbf{x}_1+\mathbf{x}_2}{\sqrt{2}}$ belongs to $\mathscr{B}_d$ because

$$\left|\frac{\mathbf{x}_1+\mathbf{x}_2}{\sqrt{2}}\right|^2 = \frac{|\mathbf{x}_1|^2 + |\mathbf{x}_2|^2 + 2(\mathbf{x}_1, \mathbf{x}_2)}{2} < \frac{|\mathbf{x}_1|^2 + |\mathbf{x}_2|^2}{2} \leq 1.$$

Invoking (37.4), we obtain $f(\frac{\mathbf{x}_1+\mathbf{x}_2}{2}) > \alpha - 6$. Applying now (37.3), we get

$$f\left(\frac{\mathbf{x}_1+\mathbf{x}_2}{\sqrt{2}}\right) \geq \sqrt{2}\left(f\left(\frac{\mathbf{x}_1+\mathbf{x}_2}{2}\right) - 2\right) \geq \sqrt{2}(\alpha - 8).$$

By the assumption $\alpha > 4(3+\sqrt{2}) + \delta$, where $\delta > 0$ is some constant. Therefore, $\sqrt{2}(\alpha - 8) > \alpha - 4 + (\sqrt{2}-1)\delta$. Writing $\mathbf{x}_0 = \frac{\mathbf{x}_1+\mathbf{x}_2}{\sqrt{2}}$, we obtain $f(\mathbf{x}_0) > \alpha - 4 + (\sqrt{2}-1)\delta$. Assume without loss of generality that $t_1 \leq t_2$. We have

$$0 = \sum_{i=1}^{d+1} t_i \mathbf{x}_i = \sqrt{2} t_1 \mathbf{x}_0 + (t_2 - t_1)\mathbf{x}_2 + \sum_{i=3}^{d+1} t_i \mathbf{x}_i.$$

Consequently, $0 \in \Delta_2 = \operatorname{co}\{\mathbf{x}_0, \mathbf{x}_2, \mathbf{x}_3, \ldots, \mathbf{x}_{d+1}\}$, and the corresponding barycentric combination $0 = s_0\mathbf{x}_0 + s_2\mathbf{x}_2 + \cdots + s_{d+1}\mathbf{x}_{d+1}$ has the following coefficients: $s_0 = \frac{\sqrt{2}t_1}{s}, s_2 = \frac{t_2-t_1}{s}, s_i = \frac{t_i}{s}, i = 3, \ldots, d+1$, where $s = (\sqrt{2}-1)t_1 + \sum_{i=2}^{d+1} t_i = (\sqrt{2}-2)t_1 + 1$. Since $t_1 \geq \kappa_d$, we have $s_0 \geq \sqrt{2}\kappa_d$.

Define now a linear functional $\beta(\mathbf{x})$ by equalities $\beta(\mathbf{x}_i) = h, i = 2, \ldots, d+1$, where the constant $h > 0$ will be determined later. Such a functional exists since $0 \notin \operatorname{co}\{\mathbf{x}_2, \ldots, \mathbf{x}_{d+1}\}$ (the assumption on the set X). For the point $\mathbf{x}_0$, we have

$$\beta(\mathbf{x}_0) = -\frac{1}{s_0}\sum_{i=2}^{d+1} s_i \beta(\mathbf{x}_i) = -h\frac{1}{s_0}\sum_{i=2}^{d+1} s_i = -h\frac{1-s_0}{s_0}.$$

Solving the equation

$$(\sqrt{2}-1)\delta - h\frac{1-\sqrt{2}\kappa_d}{\sqrt{2}\kappa_d} = h,$$

we get $h = (2-\sqrt{2})\kappa_d\delta$. Since $s_0 \geq \sqrt{2}\kappa_d$ and $f(\mathbf{x}_0) > \alpha - 4 + (\sqrt{2}-1)\delta$, it follows that for this h we have $f(\mathbf{x}_i) + \beta(\mathbf{x}_i) > \alpha - 4 + h, i = 0, 2, 3, \ldots, d+1$.

Thus, we obtain a linear functional β and a simplex Δ_2 containing the origin for which the value of $f + \beta$ at each vertex exceeds $\alpha - 4 + h$. Repeating the same argument for Δ_2, we get the value $\alpha - 4 + 2h$, etc. After nth iteration we obtain a linear functional β_n and a simplex Δ_n containing the origin such that the value of the function $f + \beta_n$ at each of its vertices exceeds $\alpha - 4 + nh$. Corollary 37.6 implies now that $E_-(f, \mathscr{B}_d) > \alpha - 4 + nh$, which gives a contradiction as $n \to \infty$.

Thus, there is a function $g \in \mathscr{L}(\mathbb{R}^d)$ such that $f(\mathbf{x}) - g(\mathbf{x}) \leq 4(3+\sqrt{2})$ for all $\mathbf{x} \in \mathscr{B}_d$. Now we can estimate $E(f, \mathscr{B}_d)$. From (37.3) it follows that

$$-\big(f(\mathbf{x}) - g(\mathbf{x})\big) \leq f(-\mathbf{x}) - g(-\mathbf{x}) + 4 \leq 4(4+\sqrt{2}).$$

Whence, $|f(\mathbf{x}) - g(\mathbf{x})| \leq 4(4+\sqrt{2}), \mathbf{x} \in \mathscr{B}_d$. □

Remark 37.3 From the proof of Theorem 37.3 it follows that there is an affine functional φ such that $f(0) = \varphi(0)$ and $\|f - \varphi\|_G \leq 4(4+\sqrt{2})\varepsilon$.

The result of Theorem 37.3 apparently holds for all ellipsoids as well. Moreover, if a set G can be sandwiched between two ellipsoids homothetic with the ratio $\gamma > 1$, then the value $E(f, G)$ can be estimated by γ, uniformly for all dimensions. For an arbitrary convex domain $G \subset \mathbb{R}^d$, we define the constant $\gamma(G)$ as the infimum of real numbers $k \geq 1$ for which there are ellipsoids E_1, E_2 homothetic with coefficient k with respect to their common center and such that $E_1 \subset G \subset E_2$.

Proposition 37.8 *For an arbitrary function $f : G \to \mathbb{R}$ with the ε-property, we have*

$$E(f, G) \leq (18+4\sqrt{2})\gamma(G)\varepsilon.$$

Proof For any $k > \gamma(G)$, there are ellipsoids E_1, E_2 homothetic with coefficient k with respect to their center such that $E_1 \subset G \subset E_2$. After a suitable affine transform it can be assumed that $E_1 = \mathscr{B}_1$, so E_2 is a ball of radius k centered at the origin. Theorem 37.3 implies that there is $\varphi \in \mathscr{A}(\mathbb{R}^d)$ such that $\|f - \varphi\|_{\mathscr{B}_d} \leq 4(4+\sqrt{2})\varepsilon$, and, moreover, $f(0) - \varphi(0) = 0$ (Remark 37.3). For an arbitrary point $\mathbf{x} \in G$, the point $\mathbf{y} = \frac{1}{k}\mathbf{x}$ belongs to $\mathscr{B}_d$, whence $|f(\mathbf{y}) - \varphi(\mathbf{y})| \leq 4(4+\sqrt{2})\varepsilon$. Since the functional $f - \varphi$ possesses the ε-property on $\mathscr{B}_d$, one can apply (37.3) for $f - \varphi$ and get

$$\big|f(\mathbf{x}) - \varphi(\mathbf{x})\big| = \big|f(k\mathbf{y}) - \varphi(k\mathbf{y})\big| \leq k\big(\big|f(\mathbf{y}) - \varphi(\mathbf{y})\big| + 2\varepsilon\big) \leq k(18+4\sqrt{2})\varepsilon.$$

Taking now the limit as $k \to \gamma(G)$, we conclude the proof. □

Remark 37.4 It is well-known that $\gamma(G) \le d$ for every convex set $G \subset \mathbb{R}^d$, and $\gamma(G) \le \sqrt{d}$ for a centrally-symmetric set [6]. However, the estimates for $E(f, G)$ obtained by Proposition 37.8 with these values of γ are worse than the estimate from Theorem 37.2. So, it makes sense to apply Proposition 37.8 for "ellipsoid-like" domains, which have small constants γ. Note that such domains may have non-smooth boundary (for example, they may be polyhedra). That is why the smoothness of the boundary, or other local properties of the domain G, does not play a role in the estimation of $E(f, G)$.

The only geometrical property of the Euclidean ball used in the proof of Theorem 37.3 is the following: Every simplex contained in the ball and covering its center has an edge such that the distance from its midpoint to the center of the ball is smaller than $\frac{\sqrt{2}}{2}$. This constant does not depend on the dimension, which leads to the absolute constant in the bound for $E(f, G)$. Basically, for every convex body that possesses this property with some constant $q < 1$, the value $E(f, G)$ can be estimated by q, regardless of the dimension.

It is an interesting question which geometrical properties of convex domains are responsible for the upper bounds of $E(f, G)$.

Remark 37.5 If the function f is continuous, then the estimates from Theorem 37.3 and Proposition 37.8 can be slightly improved, by using continuity instead of the ε-continuity. When, in the beginning of the proof of Theorem 37.3, we spot a converging sequence of sets A and pass to the limiting set X, there is no need to invoke Lemma 37.3. For a continuous function, we could just conclude that $f(\mathbf{a}_i) \to f(\mathbf{x}_i)$ as $\mathbf{a}_i \to \mathbf{x}_i$. Whence, we obtain $f(\mathbf{x}_i) > \alpha$ instead of $f(\mathbf{x}_i) > \alpha - 4$. This yields the following final estimate. In Theorem 37.3 for continuous f, we have $E(f, G) \le 4(3+\sqrt{2})\varepsilon$. In Proposition 37.8 for continuous f, we have $E(f, G) \le (14+4\sqrt{2})\varepsilon$.

Corollary 37.7 *The results of Theorem* 37.3 *and of Proposition* 37.8 *for bounded functions f hold in the separable Hilbert space.*

Proof Let G be a ball in a Hilbert space H and f be a function possessing the ε-property on it. Let $\{L_k\}_{k\in\mathbb{N}}$ be an embedded system of finite-dimensional closed subspaces of H, such that the closure of $\bigcup_{k\in\mathbb{N}} L_k$ coincides with H. For any $\alpha < 4(4+\sqrt{2})\varepsilon$ and for each k there exists an affine functional φ_k on $G_k = G \cap L_k$ such that $\|f - \varphi_k\|_{G_k} \le \alpha$. Since f is bounded on G, it follows that all φ_k are bounded uniformly. Hence, passing to subsequences we may assume that the sequence φ_j converges on every G_k as $j \to \infty$. The limiting functional φ is well-defined, affine, bounded on G, and satisfies $\|f - \varphi\| \le \alpha$.

The proof of Proposition 37.8 for a Hilbert space is literally the same as for finite-dimensional spaces. □

37.6 Open Problems

In this section, we leave several open problems on affine approximation of functions with the ε-property.

Problem 1 *What is the sharp constant in Theorem* 37.2 *for a given dimension* d*?*

Problem 2 *What is the sharp constant in Theorem* 37.3*?*

Problem 3 *Does there exist an absolute constant* C *such that for any function* f *with the* ε*-property on the* d*-dimensional cube* $\mathscr{C}_d$ *one has* $E(f, \mathscr{C}_d) \le C\varepsilon$*?*

Problem 4 *The same question as in Problem* 3, *but for the* L_p*-ball in* $\mathbb{R}^d$, $p \in (1, +\infty)$.

For $p = 1$ the answer is negative (Proposition 37.5), for $p = 2$ it is affirmative (Theorem 37.3).

Appendix

Proof of Proposition 37.1 For every straight line $l \subset V$, there is an affine functional $\varphi_l : l \to \mathbb{R}$ such that $\|f - \varphi_l\| \le \varepsilon$. This functional is unique, up to an addition of a constant. Indeed, if functionals φ_l and $\tilde{\varphi}_l$ possess this property, then $\|\varphi_l - \tilde{\varphi}_l\|_l \le 2\varepsilon$, hence the affine functional $\varphi_l - \tilde{\varphi}_l$ is identically constant.

Consider now the function $\varphi(\mathbf{x}) = \varphi_l(\mathbf{x}) - \varphi_l(0)$, where l is a line passing through the points 0 and $\mathbf{x}$. This function is well-defined (does not depend on the choice of φ_l) and homogeneous. Let us show that φ is linear. It suffices to prove its additivity. Observe first that since $|f(\mathbf{x}) - \varphi_l(\mathbf{x})| \le \varepsilon$ and $|f(0) - \varphi_l(0)| \le \varepsilon$, it follows that

$$\|f - \varphi\|_V \le 2\varepsilon. \tag{37.9}$$

Take arbitrary independent $\mathbf{x}, \mathbf{y} \in V$ and consider the two-dimensional plane L spanned by them. We need to show that $\varphi(\mathbf{x}+\mathbf{y}) = \varphi(\mathbf{x}) + \varphi(\mathbf{y})$. Let ψ be a linear functional on L such that $\psi(\mathbf{x}) = \varphi(\mathbf{x})$ and $\psi(\mathbf{y}) = \varphi(\mathbf{y})$. Replacing the function f on L by $f - \psi$, and the function φ by $\varphi - \psi$, we assume that $\varphi(\mathbf{x}) = \varphi(\mathbf{y}) = 0$. Since the three points $\mathbf{x} + \mathbf{y}, 2\mathbf{x}, 2\mathbf{y}$ are collinear, from the ε-property it follows that $|f(\mathbf{x} + \mathbf{y})| \le \max\{|f(2\mathbf{x})|, |f(2\mathbf{y})|\} + 2\varepsilon$. Furthermore, (37.9) implies that $|f(2\mathbf{x})| \le 2\varepsilon$ and $|f(2\mathbf{y})| \le 2\varepsilon$. Thus, $|f(\mathbf{x}+\mathbf{y})| \le 4\varepsilon$, and, invoking (37.9) again, we conclude that $|\varphi(\mathbf{x} + \mathbf{y})| \le 6\varepsilon$. Similarly, $|\varphi(\lambda\mathbf{x} + \lambda\mathbf{y})| \le 6\varepsilon$ for every $\lambda \in \mathbb{R}$, which, due to the homogeneity, means that $\varphi(\mathbf{x}+\mathbf{y}) = 0$. Thus, φ is additive on V, and hence it is linear.

Replacing now f by $f - \varphi$ on V we assume $\varphi \equiv 0$. By (37.9), the function f is uniformly bounded on V. Let us add a constant to f so that $\sup_{\mathbf{x}\in V} f(\mathbf{x}) =$

$-\inf_{\mathbf{x}\in V} f(\mathbf{x})$. If this supremum is greater than ε, then there are points $\mathbf{z}_1, \mathbf{z}_2 \in V$ such that $f(\mathbf{z}_1) > \varepsilon$, $f(\mathbf{z}_2) < -\varepsilon$. However, in this case the function φ_l, where l is a line connecting z_1 and z_2, cannot be identically constant, otherwise either $(f-\varphi_l)(\mathbf{z}_1)$ or $(f-\varphi_l)(\mathbf{z}_2)$ exceeds ε by modulus. Consequently, φ_l grows to $+\infty$ on l, and hence so does f. This contradiction proves that $\sup_{\mathbf{x}\in\mathbb{R}^d} f(\mathbf{x}) \le \varepsilon$, and therefore $\|f-\varphi\|_V \le \varepsilon$. □

Proof of Theorem 37.1 Sufficiency follows immediately from Proposition 37.2 and Corollary 37.1.

(*Necessity*). We realize the proof for bounded sets K because we only need this case. The proof for general sets is similar. Replacing f by $f-\varphi$, it can be assumed that $\varphi \equiv 0$. The functional $\eta(g) = \|f-g\|_K$ is convex and closed on $\mathscr{A}(\mathbb{R}^d)$, therefore it attains its minimum at the point $\varphi = 0$ iff $0 \in \partial\eta(0)$, where $\partial\eta$ is the subdifferential (see [20]). Since $\eta(g) = \sup_{\mathbf{x}\in K} |f(\mathbf{x}) - g(\mathbf{x})|$, the set $\mathscr{V}_M$ is convex and compact, and the functions $|f(\mathbf{x}) - g(\mathbf{x})|$ are uniformly Lipschitz continuous in g on the set $\mathscr{V}_M$, so by the generalized Dubovitskii–Milyutin theorem [2] we have

$$\partial\eta(0) = \Big[\lim_{\alpha\to\|f\|_K-0} \mathrm{co}\big\{\partial\big(|f(\mathbf{x}) - g(\mathbf{x})|\big)\big|_{g=0} \,\big|\, \mathbf{x}\in K, |f(\mathbf{x})| > \alpha\big\}\Big]$$
$$= \Big[\lim_{\alpha\to\|f\|_K-0} \mathrm{co}\big\{-\delta_{\mathbf{x}}(\cdot) \,\big|\, \mathbf{x}\in K,\ f(\mathbf{x}) > \alpha\big\} \cup \big\{\delta_{\mathbf{x}}(\cdot) \,\big|\, \mathbf{x}\in K,\ f(\mathbf{x}) < -\alpha\big\}\Big],$$

where $[\cdot]$ denotes the closure, and $\delta_{\mathbf{x}} : \mathscr{A}(\mathbb{R}^d) \to \mathbb{R}$ is the delta-function, $\delta_{\mathbf{x}}(g) = g(\mathbf{x})$. Since $0 \in \partial\eta(0)$, we see that for every α such that $0 < \alpha < \|f\|_K$ the convex hull of the set

$$\big\{-\delta_{\mathbf{x}}(\cdot) \,\big|\, \mathbf{x}\in K,\ f(\mathbf{x}) > \alpha\big\} \cup \big\{\delta_{\mathbf{x}}(\cdot) \,\big|\, \mathbf{x}\in K,\ f(\mathbf{x}) < -\alpha\big\}$$

comes arbitrarily close to the origin. Thus, for every $\varepsilon > 0$ there are convex combinations of finitely many points from this set, whose norms are less than ε. Since the space dual to $\mathscr{A}(\mathbb{R}^d)$ is of dimension $d+1$, from the Caratheodory theorem, it follows that there are such sets of cardinality $\le d+2$. Thus, there exist points $\mathbf{a}_1, \ldots, \mathbf{a}_m, \mathbf{b}_1, \ldots, \mathbf{b}_{m'} \in K$, where $m + m' \le d+2$, and nonnegative numbers $\{t_i\}_{i=1}^m$, $\{s_i\}_{j=1}^{m'}$ with $\sum_i t_i + \sum_j s_j = 1$, such that $f(\mathbf{a}_i) > \alpha$, $f(\mathbf{b}_j) < -\alpha$, and

$$\left|\left(-\sum_{i=1}^m t_i\delta_{\mathbf{a}_i} + \sum_{j=1}^{m'} s_j\delta_{\mathbf{b}_j}\right)(g)\right| = \left|-\sum_{i=1}^m t_i g(\mathbf{a}_i) + \sum_{j=1}^{m'} s_j g(\mathbf{b}_j)\right| < \varepsilon$$

for any $g \in \mathscr{A}(\mathbb{R}^d)$ such that $\|g\| \le 1$. Without loss of generality it can be assumed that $\sum_i t_i \ge 1/2$. Writing $T = \sum_i t_i$ and $t_i' = t_i/T$, $s_j' = s_j/T$, we obtain

$$\left|-\sum_i t_i' g(\mathbf{a}_i) + \sum_j s_j' g(\mathbf{b}_j) + \Big(1 - \sum_j s_j'\Big) g(\mathbf{b}_1) - \Big(1 - \sum_j s_j'\Big) g(\mathbf{b}_1)\right|$$
$$< T^{-1}\varepsilon \le 2\varepsilon$$

for $\|g\| \le 1$. Substituting $g \equiv 1$, we see that $|(1-\sum_j s'_j)| < 2\varepsilon$. On the other hand,

$$\mathbf{a} = \sum_i t'_i \mathbf{a}_i \in \mathrm{co}(A) \quad \text{and} \quad \mathbf{b} = \sum_j s'_j \mathbf{b}_j + \Big(1 - \sum_j s'_j\Big)\mathbf{b}_1 \in \mathrm{co}(B).$$

Thus, for every g such that $\|g\| \le 1$ we have

$$\Big|-g(\mathbf{a}) + g(\mathbf{b}) + \Big(1 - \sum_j s'_j\Big) g(\mathbf{b}_1)\Big| < 2\varepsilon.$$

Since $\|g\| \le 1$, it follows that $\|g\|_K \le C = \max\{1, \mathrm{diam}(K)\}$. Therefore,

$$\big|-g(\mathbf{a}) + g(\mathbf{b})\big| < 2\varepsilon + \Big|1 - \sum_j s'_j\Big| C < 2\varepsilon(1+C), \qquad g \in \mathscr{A}\big(\mathbb{R}^d\big), \|g\| \le 1.$$

On the other hand, there is $g \in \mathscr{A}(\mathbb{R}^d)$, $\|g\| \le 1$, such that $|-g(\mathbf{a}) + g(\mathbf{b})| = |\mathbf{a} - \mathbf{b}|$. Thus, $|\mathbf{a} - \mathbf{b}| < 2(1+C)\varepsilon$. In particular, for $\varepsilon = \frac{\rho}{2(1+C)}$ we obtain $\mathrm{dist}\{\mathrm{co}(A), \mathrm{co}(B)\} < \rho$.

It remains to show that either condition (**a**) is satisfied, or $m + m' \le d + 1$ (by now we have proved that $m + m' \le d + 2$). We take points $\mathbf{x} \in \mathrm{co}(A)$ and $\mathbf{y} \in \mathrm{co}(B)$ such that $|\mathbf{x} - \mathbf{y}| = \mathrm{dist}\{\mathrm{co}(A), \mathrm{co}(B)\}$. If $\mathbf{x} = \mathbf{y}$, then we have condition (**a**). If $|\mathbf{x} - \mathbf{y}| > 0$, then either one of the points $\mathbf{x}$ or $\mathbf{y}$ lies on the boundary of the corresponding set, or the vector $\mathbf{x} - \mathbf{y}$ is orthogonal to the affine spans of these sets. In the first case, when, say, $\mathbf{x}$ is on the boundary of the polyhedron $\mathrm{co}(A)$, then we take the smallest face of that polyhedron containing $\mathbf{x}$. Replacing the set A by the set of vertices of this face, we reduce the total number of points of A and B at least to $d + 1$, after which property (**b**) is satisfied. Consider the second case, when the vector $\mathbf{x} - \mathbf{y}$ is orthogonal to the affine spans of A and B. Denote these spans by $\tilde{A}$ and $\tilde{B}$, respectively. It follows that both $\tilde{A}$ and $\tilde{B}$ are parallel to one hyperplane. Hence, either $\dim(\tilde{A}) + \dim(\tilde{B}) \le d - 1$, or there is a straight line parallel to both $\tilde{A}$ and $\tilde{B}$. In the first case, by the Caratheodory theorem, one can choose at most $\dim(\tilde{A}) + 1$ points of the set A and at most $\dim(\tilde{B}) + 1$ points of the set B so that the convex hulls of these sets still contain $\mathbf{x}$ and $\mathbf{y}$, respectively. The total number of points is reduced to $\dim(\tilde{A}) + \dim(\tilde{B}) + 2 \le d + 1$, which proves (**b**). Otherwise, if there is a straight line l parallel to both $\tilde{A}$ and $\tilde{B}$, the two lines passing through $\mathbf{x}$ and $\mathbf{y}$ parallel to l intersect the polyhedra $\mathrm{co}(A)$ and $\mathrm{co}(B)$ by some segments $[\mathbf{x}_1, \mathbf{x}_2]$ and $[\mathbf{y}_1, \mathbf{y}_2]$, respectively. The distance between those two segments equals to $|\mathbf{x} - \mathbf{y}|$, and it is attained at one of the ends of these segments, say, at $\mathbf{x}_1$. Thus, $\mathrm{dist}\{\mathbf{x}_1, [\mathbf{y}_1, \mathbf{y}_2]\} = |\mathbf{x} - \mathbf{y}|$. However, $\mathbf{x}_1$ lies on the boundary of $\mathrm{co}(A)$, and we again come to the first case. □

Proof of Lemma 37.6 Let $a, b \ge 0$. Since the function $p(t)$ is convex on the segment $[0, 1]$, we have $E(p, [a, b]) = \frac{1}{2} \sup_{t \in [a,b]} (\xi_{a,b}(t) - p(t))$, where $\xi_{a,b}$ is the affine function such that $\xi_{a,b}(a) = p(a), \xi_{a,b}(b) = p(b)$. It is shown easily that $\sup_{t \in [a,b]} (\xi_{a,b}(t) - p(t)) \le \sup_{t \in [0,b-a]} (\xi_{a,b}(t) - p(t)) = \frac{b-a}{e}$, hence

$E(p,[a,b]) \leq \frac{b-a}{2e}$. The case $a,b \leq 0$ follows from the symmetry. Finally, in the case $a < 0 < b$, assuming $|b| \geq |a|$, we have

$$E\big(p,[a,b]\big) \leq \sup_{t\in[a,b]} \big|\xi_{0,b}(t) - p(t)\big| = \sup_{t\in[0,b]} \big(\xi_{0,b}(t) - p(t)\big) = \frac{b}{e} \leq \frac{b-a}{e}. \qquad \square$$

Proof of Lemma 37.7 We have

$$0 = \left(\sum_{i=1}^{d+1} t_i \mathbf{x}_i\right)^2 = \sum_{i=1}^{d+1} t_i^2 |\mathbf{x}_i|^2 + 2\sum_{i\neq j} t_i t_j (\mathbf{x}_i, \mathbf{x}_j)$$

$$\geq \frac{1}{4}\sum_{i=1}^{d+1} t_i^2 + (d+1)d \min_{i\neq j}\big\{t_i t_j (\mathbf{x}_i, \mathbf{x}_j)\big\}.$$

On the other hand, $\frac{1}{4}\sum_{i=1}^{d+1} t_i^2 \geq \frac{1}{4(d+1)}$. Therefore,

$$(d+1)d \min_{i\neq j}\big\{t_i t_j (\mathbf{x}_i, \mathbf{x}_j)\big\} \leq -\frac{1}{4(d+1)}.$$

Whence, there are i and j such that $t_i t_j (\mathbf{x}_i, \mathbf{x}_j) \leq -\frac{1}{4d(d+1)^2}$. Taking into account that $t_i \leq 1, t_j \leq 1$ and $|(\mathbf{x}_i, \mathbf{x}_j)| \leq 1$, we see that both t_i and t_j are at least $\kappa_d = \frac{1}{4d(d+1)^2}$. $\square$

References

1. Badora, R., Ger, R., Pales, Z.: Additive selections and the stability of the Cauchy functional equations. ANZIAM J. **44**, 323–337 (2003)
2. Dubovitskii, A.Ya., Milyutin, A.A.: Extremum problems in the presence of restrictions. USSR Comput. Math. Math. Phys. **5**(3), 1–80 (1965)
3. Forti, G.L.: The stability of homeomorphisms and amenability, with applications to functional equations. Abh. Math. Semin. Univ. Hamb. **57**, 215–226 (1987)
4. Găvruta, P.: A generalization of the Hyers–Ulam–Rassias stability of approximately additive mappings. J. Math. Anal. Appl. **184**(3), 431–436 (1994)
5. Gajda, Z.: On stability of additive mappings. Int. J. Math. Sci. **14**, 431–434 (1991)
6. de Guzman, M.: Differentiation of Integrals in R^n. Springer, Berlin (1975)
7. Hyers, D.H.: On the stability of the linear functional equation. Proc. Natl. Acad. Sci. USA **27**, 222–224 (1941)
8. Hyers, D.H., Isac, G., Rassias, Th.M.: Stability of Functional Equations in Several Variables. Birkhäuser, Boston (1998)
9. Ioffe, A.D., Tikhomirov, V.M.: Theory of Extremal Problems. Elsevier/North-Holland, Amsterdam (1979)
10. Johnson, B.E.: Approximately multiplicative maps between Banach algebras. J. Lond. Math. Soc., II. Ser. **37**(2), 294–316 (1988)
11. Jung, S.-M.: On the Hyers–Ulam–Rassias stability of approximately additive mappings. J. Math. Anal. Appl. **204**(1), 221–226 (1996)

12. Levin, V.L.: On the subdifferentials of convex functionals. Usp. Mat. Nauk **25**, 183–184 (1970)
13. Mako, Z., Páles, Zs.: On the Lipschitz perturbation of monotonic functions. Acta Math. Hung. **113**(1–2), 1–18 (2006)
14. Protasov, V.Yu.: On linear selections of convex set-valued maps. Funct. Anal. Appl. **45**, 46–55 (2011)
15. Rassias, Th.M.: On the stability of functional equations and the problem of Ulam. Acta Appl. Math. **62**, 23–130 (2000)
16. Rassias, Th.M., Šemrl, P.: On the behavior of mappings which do not satisfy Hyers–Ulam stability. Proc. Am. Math. Soc. **114**, 989–993 (1992)
17. Šemrl, P.: The stability of approximately additive functions. In: Rassias, Th.M., Tabor, J. (eds.) Stability of Mappings of Hyers-Ulam Type, pp. 135–140. Hadronic Press, Florida (1994)
18. Tabor, J., Tabor, J.: Local stability of the Cauchy and Jensen equations in function spaces. Aequ. Math. **58**, 311–320 (1999)
19. Tabor, J., Yost, D.: Applications of inverse limits to extensions of operators and approximation of Lipschitz functions. J. Approx. Theory **116**, 257–267 (2002)
20. Tikhomirov, V.: Convex analysis. In: Gamkrelidze, R.V. (ed.) Analysis II: Convex Analysis and Approximation Theory. Springer, Berlin (1990)
21. Ulam, S.M.: A Collection of Mathematical Problems. Interscience, New York (1960)

Chapter 38
Some Inequalities and Other Results Associated with Certain Subclasses of Univalent and Bi-Univalent Analytic Functions

H.M. Srivastava

Abstract In recent year, various interesting properties and characteristics (including, for example, coefficient bounds and coefficient inequalities) of many different subclasses of univalent and bi-univalent analytic functions have been systematically investigated. The main object of this essentially survey-cum-expository article is first to present a brief account of some important contributions to the theory of univalent and bi-univalent analytic functions, which have been made in several recent works. References to other more recent investigations involving many closely-related function classes are also provided for motivating and encouraging future researches on these topics in Geometric Function Theory of Complex Analysis.

Key words Analytic functions of real or complex orders · Univalent functions · Bi-univalent functions · Taylor–Maclaurin series · Inverse functions · Koebe function · Starlike functions · Convex functions · Bi-starlike functions · Bi-convex functions · Strongly bi-starlike functions · Coefficient bounds · Close-to-convex functions · Schwarz function · Integral operators · Univalence criteria

Mathematics Subject Classification Primary 30C45 · 30C50 · Secondary 34-99 · 44-99

38.1 Introduction and Definitions

Throughout this article, we let $\mathbb{R} = (-\infty, \infty)$ be the set of *real* numbers, $\mathbb{C}$ be the set of *complex* numbers and $\mathbb{N}$ given by

$$\mathbb{N} := \{1, 2, 3, \dots\} = \mathbb{N}_0 \setminus \{0\} \quad \big(\mathbb{N}_0 := \{0, 1, 2, 3, \dots\}\big)$$

Dedicated to Professor Themistocles M. Rassias on the occasion of his 60th birthday.

H.M. Srivastava (✉)
Department of Mathematics and Statistics, University of Victoria, Victoria, British Columbia V8W 3R4, Canada
e-mail: harimsri@math.uvic.ca

P.M. Pardalos et al. (eds.), *Nonlinear Analysis*, Springer Optimization and Its Applications 68, In Honor of Themistocles M. Rassias on the Occasion of his 60th Birthday, DOI 10.1007/978-1-4614-3498-6_38,

be the set of *positive* integers, it being understood (as usual) that

$$\mathbb{C}^* := \mathbb{C} \setminus \{0\} \quad \text{and} \quad \mathbb{N}^* := \mathbb{N} \setminus \{1\}.$$

Suppose also that $\mathscr{A}$ denotes the class of functions $f(z)$ *normalized* by the following Taylor–Maclaurin series expansion:

$$f(z) = z + \sum_{n=2}^{\infty} a_n z^n \quad (z \in \mathbb{U}), \tag{38.1}$$

which are *analytic* in the *open* unit disk

$$\mathbb{U} := \{z : z \in \mathbb{C} \text{ and } |z| < 1\},$$

$\mathbb{C}$ being, as already mentioned above, the set of *complex* numbers. Thus, equivalently, $\mathscr{A}$ denotes the class of functions $f(z)$ which are analytic in $\mathbb{U}$ and normalized by

$$f(0) = f'(0) - 1 = 0.$$

Suppose also that $\mathscr{S}$ denotes the subclass of functions in $\mathscr{A}$ which are *univalent* in $\mathbb{U}$ (for details, see [17, 45, 46]; see also the recent works [2, 15, 40–42, 46, 47]).

Some of the numerous important and well-investigated subclasses of the univalent function class $\mathscr{S}$ include (for example) the class $\mathscr{S}^*(\kappa)$ of *starlike functions of order* κ in $\mathbb{U}$ and the class $\mathscr{K}(\kappa)$ of *convex functions of order* κ in $\mathbb{U}$. By definition, we have

$$\mathscr{S}^*(\kappa) := \left\{ f : f \in \mathscr{S} \text{ and } \Re\left(\frac{zf'(z)}{f(z)}\right) > \kappa \ (z \in \mathbb{U};\ 0 \leqq \kappa < 1) \right\} \tag{38.2}$$

and

$$\mathscr{K}(\kappa) := \left\{ f : f \in \mathscr{S} \text{ and } \Re\left(1 + \frac{zf''(z)}{f'(z)}\right) > \kappa \ (z \in \mathbb{U};\ 0 \leqq \kappa < 1) \right\}. \tag{38.3}$$

It follows easily from the definitions (38.2) and (38.3) that

$$f(z) \in \mathscr{K}(\kappa) \quad \Longleftrightarrow \quad zf'(z) \in \mathscr{S}^*(\kappa). \tag{38.4}$$

Furthermore, for the relatively more familiar classes $\mathscr{S}^*$ and $\mathscr{K}$ of *starlike functions* in $\mathbb{U}$ and *convex functions* in $\mathbb{U}$, respectively, we have

$$\mathscr{S}^* := \mathscr{S}^*(0) \quad \text{and} \quad \mathscr{K} := \mathscr{K}(0).$$

In particular, the class $\mathscr{K}$ of convex functions in $\mathbb{U}$ consists of functions that map the unit disk $\mathbb{U}$ into a convex region.

It is well known that every function $f \in \mathscr{S}$ has an inverse f^{-1}, defined by

$$f^{-1}\big(f(z)\big) = z \quad (z \in \mathbb{U})$$

and

$$f\big(f^{-1}(w)\big) = w \quad \left(|w| < r_0(f);\ r_0(f) \geqq \frac{1}{4}\right).$$

In fact, the inverse function f^{-1} is given by

$$f^{-1}(w) = w - a_2w^2 + \big(2a_2^2 - a_3\big)w^3 - \big(5a_2^3 - 5a_2a_3 + a_4\big)w^4 + \cdots. \tag{38.5}$$

A function $f \in \mathscr{A}$ is said to be *bi-univalent* in $\mathbb{U}$ if both $f(z)$ and $f^{-1}(z)$ are univalent in $\mathbb{U}$. We denote by Σ the class of bi-univalent functions in $\mathbb{U}$ given by the Taylor–Maclaurin series expansion (38.1). Some of the examples of functions in the class Σ are listed here as follows:

$$\frac{z}{1-z}, \quad -\log(1-z), \quad \frac{1}{2}\log\left(\frac{1+z}{1-z}\right),$$

and so on. However, the familiar Koebe function is not a member of the function class Σ. Other common examples of functions in $\mathscr{S}$ such as

$$z - \frac{z^2}{2} \quad \text{and} \quad \frac{z}{1-z^2}$$

are also not members of the function class Σ.

Lewin [20] first investigated the bi-univalent function class Σ and showed that

$$|a_2| < 1.51.$$

Subsequently, Brannan and Clunie [10] conjectured that

$$|a_2| \leqq \sqrt{2}.$$

Netanyahu [26], on the other hand, showed that

$$\max_{f\in\Sigma} |a_2| = \frac{4}{3}.$$

The coefficient estimate problem for each of the following Taylor–Maclaurin coefficients:

$$|a_n| \quad \big(n \in \mathbb{N} \setminus \{1,2\};\ \mathbb{N} := \{1,2,3,\ldots\}\big)$$

is presumably still an *open* problem.

Brannan and Taha [12] (see also [48]) introduced certain subclasses of the bi-univalent function class Σ similar to the familiar subclasses $\mathscr{S}^*(\alpha)$ and $\mathscr{C}(\alpha)$ (see [11]) of the univalent function class $\mathscr{S}$. Thus, following Brannan and Taha [12] (see also [48]), a function $f \in \mathscr{A}$ is in the class $\mathscr{S}^*_\Sigma[\alpha]$ $(0 < \alpha \leqq 1)$ of *strongly bi-starlike* functions of order α if each of the following conditions is satisfied:

$$f \in \Sigma \quad \text{and} \quad \left|\arg\left(\frac{zf'(z)}{f(z)}\right)\right| < \frac{\alpha\pi}{2} \quad (z \in \mathbb{U};\ 0 < \alpha \leqq 1) \tag{38.6}$$

and

$$\left|\arg\left(\frac{zg'(w)}{g(w)}\right)\right| < \frac{\alpha\pi}{2} \quad (w \in \mathbb{U};\ 0 < \alpha \leqq 1), \tag{38.7}$$

where g is the extension of f^{-1} to $\mathbb{U}$. The classes $\mathscr{S}_{\Sigma}^{*}(\kappa)$ and $\mathscr{K}_{\Sigma}(\kappa)$ of bi-starlike functions of order κ and bi-convex functions of order κ, corresponding (respectively) to the functions classes $\mathscr{S}^{*}(\kappa)$ and $\mathscr{K}(\kappa)$ defined by (38.2) and (38.3), were also introduced analogously. For each of the function classes $\mathscr{S}_{\Sigma}^{*}(\kappa)$ and $\mathscr{K}_{\Sigma}(\kappa)$, they found *non-sharp* estimates on the first two Taylor–Maclaurin coefficients $|a_2|$ and $|a_3|$ (for details, see [12] and [48]).

In the present article, we first introduce the two novel subclasses

$$\mathscr{H}_{\Sigma}^{\alpha} \quad (0 < \alpha \leqq 1) \quad \text{and} \quad \mathscr{H}_{\Sigma}(\beta) \quad (0 \leqq \beta < 1)$$

of the above-defined function class Σ and proceed to find, in Sect. 38.2 of this article, estimates on the coefficients $|a_2|$ and $|a_3|$ for functions in these new subclasses of the function class Σ.

Definition 38.1 A function $f(z)$ given by (38.1) is said to be in the function class $\mathscr{H}_{\Sigma}^{\alpha}$ $(0 < \alpha \leqq 1)$ if the following conditions are satisfied:

$$f \in \Sigma \quad \text{and} \quad \left|\arg\big(f'(z)\big)\right| \leqq \frac{\alpha\pi}{2} \quad (z \in \mathbb{U};\ 0 < \alpha \leqq 1) \tag{38.8}$$

and

$$\left|\arg\big(g'(w)\big)\right| < \frac{\alpha\pi}{2} \quad (w \in \mathbb{U};\ 0 < \alpha \leqq 1), \tag{38.9}$$

where the function g is given by

$$g(w) = w - a_2 w^2 + \big(2a_2^2 - a_3\big)w^3 - \big(5a_2^3 - 5a_2a_3 + a_4\big)w^4 + \cdots. \tag{38.10}$$

Definition 38.2 A function $f(z)$ given by (38.1) is said to be in the function class $\mathscr{H}_{\Sigma}(\beta)$ $(0 \leqq \beta < 1)$ if the following conditions are satisfied:

$$f \in \Sigma \quad \text{and} \quad \Re\big(f'(z)\big) > \beta \quad (z \in \mathbb{U};\ 0 \leqq \beta < 1) \tag{38.11}$$

and

$$\Re\big(g'(w)\big) > \beta \quad (w \in \mathbb{U};\ 0 \leqq \beta < 1), \tag{38.12}$$

where the function g is defined by (38.10).

In Sect. 38.3 of this article, we introduce and investigate two interesting subclasses

$$\mathscr{M}_g(n, \lambda, b) \quad \text{and} \quad \mathscr{M}_g(n, \lambda, b; u)$$

of analytic functions of *complex* order, which are defined by means of the familiar Sălăgean derivative operator

$$D^n \quad \big(n \in \mathbb{N}_0 := \mathbb{N} \cup \{0\};\ \mathbb{N} = \{1,2,3,\ldots\}\big).$$

In Sect. 38.4 of this article, we discuss some recent extensions of univalence criteria for several families of integral operators. Finally, in the concluding section (Sect. 38.5), we briefly indicate some more recent *further* developments on the subjects of this article. The importance of analytic, geometric, and other inequalities associated with (among others) various families of polynomials and functions cannot be overemphasized (see, for example, [22, 34], and [35]).

38.2 Coefficient Inequalities and Coefficient Bounds for the Function Classes $\mathscr{H}_\Sigma^\alpha$ and $\mathscr{H}_\Sigma(\beta)$

We first state and prove the following result (see also Srivastava et al. [43]).

Theorem 38.1 *Let* $f(z)$ *given by* (38.1) *be in the function class* $\mathscr{H}_\Sigma^\alpha$. *Then*

$$|a_2| \leqq \alpha\sqrt{\frac{2}{\alpha+2}} \quad \text{and} \quad |a_3| \leqq \frac{\alpha(3\alpha+2)}{3}. \tag{38.13}$$

Proof Following the work of Srivastava et al. [43], we begin by writing the argument inequalities in (38.8) and (38.9) of Definition 38.1 in their following *equivalent* forms:

$$f'(z) = \big[Q(z)\big]^\alpha \quad \text{and} \quad g'(w) = \big[L(w)\big]^\alpha,$$

respectively, where $Q(z)$ and $L(w)$ satisfy the following inequalities:

$$\Re\big(Q(z)\big) > 0 \quad (z \in \mathbb{U}) \quad \text{and} \quad \Re\big(L(w)\big) > 0 \quad (w \in \mathbb{U}).$$

Furthermore, the functions $Q(z)$ and $L(w)$ have the forms:

$$Q(z) = 1 + c_1 z + c_2 z^2 + \cdots \tag{38.14}$$

and

$$L(w) = 1 + l_1 w + l_2 w^2 + \cdots,$$

respectively. Now, equating the coefficients of $f'(z)$ with $[Q(z)]^\alpha$ and the coefficients of $g'(w)$ with $[L(w)]^\alpha$, we get

$$2a_2 = \alpha c_1, \tag{38.15}$$

$$3a_3 = \alpha c_2 + \frac{\alpha(\alpha-1)}{2} c_1^2, \tag{38.16}$$

$$-2a_2 = \alpha l_1 \tag{38.17}$$

and

$$3\left(2a_2^2 - a_3\right) = \alpha l_2 + \frac{\alpha(\alpha-1)}{2} l_1^2. \tag{38.18}$$

From (38.15) and (38.17), we get

$$c_1 = -l_1 \quad \text{and} \quad 8a_2^2 = \alpha^2\left(c_1^2 + l_1^2\right). \tag{38.19}$$

Moreover, from (38.16) and (38.18), we observe that

$$6a_2^2 - \left(\alpha c_2 + \frac{\alpha(\alpha-1)}{2}\, c_1^2\right) = \alpha l_2 + \frac{\alpha(\alpha-1)}{2}\, l_1^2.$$

By a rearrangement together with the second identity in (38.19), we get

$$6a_2^2 = \alpha(c_2 + l_2) + \frac{\alpha(\alpha-1)}{2}\left(l_1^2 + c_1^2\right) = \alpha(c_2 + l_2) + \alpha(\alpha-1)\frac{4a_2^2}{\alpha^2}.$$

Consequently, we obtain

$$a_2^2 = \frac{\alpha^2}{2(\alpha+2)}(c_2 + l_2),$$

which, in conjunction with the following well-known inequalities (cf. [5, p. 41]):

$$|c_2| \leqq 2 \quad \text{and} \quad |l_2| \leqq 2,$$

gives us the desired estimate on $|a_2|$ as asserted in (38.13).

Next, with a view to estimating the bound on $|a_3|$, we subtract (38.15) from (38.14), and we thus find that

$$6a_3 - 6a_2^2 = \alpha c_2 + \frac{\alpha(\alpha-1)}{2}\, c_1^2 - \left(\alpha l_2 + \frac{\alpha(\alpha-1)}{2} l_1^2\right).$$

Upon substituting the value of a_2^2 from (38.19) and observing that

$$c_1^2 = l_1^2,$$

it follows that

$$a_3 = \frac{1}{4}\alpha^2 c_1^2 + \frac{1}{6}\alpha(c_2 - l_2).$$

The familiar inequalities (cf. [17, p. 41]),

$$|c_2| \leqq 2 \quad \text{and} \quad |l_2| \leqq 2,$$

now yield

$$|a_3| \leqq \frac{1}{4}\alpha^2 \cdot 4 + \frac{1}{6}\alpha \cdot 4 = \frac{\alpha(3\alpha+2)}{3}.$$

This completes the proof of Theorem 38.1. □

Our demonstration of Theorem 38.2 below, solving the corresponding coefficient problems for the bi-univalent function class $H_\Sigma(\beta)$ $(0 \leqq \beta < 1)$, given by Definition 38.2, is much akin to that of Theorem 38.1 above (see, for details, Srivastava et al. [43]).

Theorem 38.2 *Let* $f(z)$ *given by* (38.1) *be in the function class* $\mathscr{H}_\Sigma(\beta)$ $(0 \leqq \beta < 1)$. *Then*

$$|a_2| \leqq \sqrt{\frac{2(1-\beta)}{3}} \quad \textit{and} \quad |a_3| \leqq \frac{(1-\beta)(5-3\beta)}{3}.$$

38.3 The Subclasses $\mathscr{M}_g(n,\lambda,b)$ and $\mathscr{M}_g(n,\lambda,b;u)$ of Analytic Functions of Complex Order

For functions $f(z)$ in the class $\mathscr{S}^*(\alpha)$ given by (38.2), Robertson [36] proved some coefficient bounds which were subsequently and extensively investigated by (among others) Nasr and Aouf [24] and Altintaş et al. [2–9] in the contexts of various interesting subclasses of analytic functions of *complex* order. Here, in this section, we introduce and investigate the following two subclasses:

$$\mathscr{M}_g(n,\lambda,b) \quad \text{and} \quad \mathscr{M}_g(n,\lambda,b;u)$$

of analytic functions of *complex* order, which are defined by means of the familiar *Sălăgean derivative operator* D^n $(n \in \mathbb{N}_0)$ (for details, see Sălăgean [38]), where

$$D^0 f(z) = f(z) \quad \text{and} \quad D^1 f(z) = zf'(z)$$

and, in general,

$$D^n f(z) := D\big(D^{n-1} f(z)\big) \quad (n \in \mathbb{N})$$

or, equivalently,

$$D^n f(z) = z + \sum_{j=2}^{\infty} j^n a_j z^j \quad (n \in \mathbb{N}_0;\ f \in \mathscr{A}).$$

In recent years, several authors obtained many interesting results for various subclasses of analytic functions involving the Sălăgean derivative operator D^n (see, among other recent works, [42]). For example, Deng [9] defined a function class $\mathscr{B}(n,\lambda,\alpha,b)$ by

$$\Re\left(1+\frac{1}{b}\left[\frac{z[(1-\lambda)D^n f(z)+\lambda D^{(n+1)}f(z)]'}{(1-\lambda)D^n f(z)+\lambda D^{n+1}f(z)}-1\right]\right)>\alpha$$
$$\left(0\leqq\alpha<1;\ 0\leqq\lambda\leqq 1;\ n\in\mathbb{N}_0;\ b\in\mathbb{C}\setminus\{0\}\right)$$

and also investigated the subclass $\mathscr{T}(n,\lambda,\alpha,b;u)$ of the analytic function class $\mathscr{A}$, which consists of functions $f(z)\in\mathscr{A}$ satisfying the following *nonhomogeneous Cauchy–Euler differential equation*:

$$z^2\frac{d^2w}{dz^2}+2(1+u)z\frac{dw}{dz}+u(1+u)w=(1+u)(2+u)h(z),$$

where

$$w=f(z)\in\mathscr{A},\qquad h(z)\in\mathscr{B}(n,\lambda,\alpha,b)\quad\text{and}\quad u\in\mathbb{R}\setminus(-\infty,-1].$$

In the same work [9], coefficient inequalities and coefficient bounds for the subclasses $\mathscr{B}(n,\lambda,\alpha,b)$ and $\mathscr{T}(n,\lambda,\alpha,b;u)$ of analytic functions of complex order were obtained.

Now, by making use of the Sălăgean derivative operator D^n, we introduce each of the following subclasses of analytic functions of *complex* order b.

Definition 38.3 Let $g:\mathbb{U}\to\mathbb{C}$ be a convex function such that

$$g(0)=1\quad\text{and}\quad\Re\big(g(z)\big)>0\quad(z\in\mathbb{U}).$$

We denote by $\mathscr{M}_g(n,\lambda,b)$ the class of functions given by

$$\mathscr{M}_g(n,\lambda,b):=\Bigg\{f:f\in\mathscr{A}\text{ and }$$
$$1+\frac{1}{b}\left(\frac{z[(1-\lambda)D^n f(z)+\lambda D^{(n+1)}f(z)]'}{(1-\lambda)D^n f(z)+\lambda D^{n+1}f(z)}-1\right)\in g(\mathbb{U})\ (z\in\mathbb{U})\Bigg\}$$
$$\left(0\leqq\lambda\leqq 1;\ n\in\mathbb{N}_0;\ b\in\mathbb{C}\setminus\{0\}\right).$$

Definition 38.4 A function $f\in\mathscr{A}$ is said to be in the class $\mathscr{M}_g(n,\lambda,b;u)$, if it satisfies the following nonhomogeneous Cauchy–Euler differential equation:

$$z^2\frac{d^2w}{dz^2}+2(1+u)z\frac{dw}{dz}+u(1+u)w=(1+u)(2+u)h(z)$$
$$\left(w=f(z)\in\mathscr{A};\ h(z)\in\mathscr{M}_g(n,\lambda,b);\ u\in\mathbb{R}\setminus(-\infty,-1]\right).$$

Remark 38.1 There are many choices of the function g which provide interesting subclasses of analytic functions of complex order. In particular, if we let

$$g(z) = \frac{1+(1-2\alpha)z}{1-z} \quad (0 \leqq \alpha < 1;\ z \in \mathbb{U}),$$

it is easy to verify that the function $g(z)$ is convex in $\mathbb{U}$ and satisfies the hypotheses of Definition 38.3. If $f \in \mathscr{M}_g(n, \lambda, b)$, then

$$\Re\left(1 + \frac{1}{b}\left[\frac{z[(1-\lambda)D^n f(z) + \lambda D^{(n+1)} f(z)]'}{(1-\lambda)D^n f(z) + \lambda D^{n+1} f(z)} - 1\right]\right) > \alpha \quad (z \in \mathbb{U}),$$

that is,

$$f \in \mathscr{B}(n, \lambda, \alpha, b).$$

Remark 38.2 In view of Remark 38.1, by taking

$$g(z) = \frac{1+(1-2\alpha)z}{1-z} \quad (0 \leqq \alpha < 1;\ z \in \mathbb{U})$$

in Definitions 38.3 and 38.4, we easily observe that the function classes

$$\mathscr{M}_g(n, \lambda, b) \quad \text{and} \quad \mathscr{M}_g(n, \lambda, b; u)$$

become the aforementioned function classes

$$\mathscr{B}(n, \lambda, \alpha, b) \quad \text{and} \quad \mathscr{T}(n, \lambda, \alpha, b; u),$$

respectively.

By using the principle of subordination between analytic functions (see Definition 38.5 below; see also [21, 42], and [47]), we obtain coefficient bounds for functions in the subclasses

$$\mathscr{M}_g(n, \lambda, b) \quad \text{and} \quad \mathscr{M}_g(n, \lambda, b; u)$$

of analytic functions of complex order, which are introduced by Definitions 38.3 and 38.4. The results presented here unify and extend the corresponding results obtained earlier by Nasr and Aouf [24], Altintaş et al. [2–9] and Deng [16].

Definition 38.5 For two functions f and g, analytic in $\mathbb{U}$, we say that the function $f(z)$ is *subordinate* to $g(z)$ in $\mathbb{U}$, and write

$$f(z) \prec g(z) \quad (z \in \mathbb{U}),$$

if there exists a Schwarz function $\mathfrak{w}(z)$, analytic in $\mathbb{U}$ with

$$\mathfrak{w}(0) = 0 \quad \text{and} \quad \left|\mathfrak{w}(z)\right| < 1 \quad (z \in \mathbb{U}),$$

such that

$$f(z) = g\big(\mathfrak{w}(z)\big) \quad (z \in \mathbb{U}).$$

In particular, if the function g is univalent in $\mathbb{U}$, the above subordination is equivalent to

$$f(0) = g(0) \quad \text{and} \quad f(\mathbb{U}) \subset g(\mathbb{U}).$$

The proofs of the main results (Theorems 38.3 and 38.4 below) are based essentially upon the following lemma due to Rogosinski [37] (for details, see Srivastava et al. [46]).

Lemma 38.1 *Let the function g given by*

$$g(z) = \sum_{k=1}^{\infty} g_k z^k$$

be convex in $\mathbb{U}$. Also let the function f given by

$$f(z) = \sum_{k=1}^{\infty} a_k z^k$$

be holomorphic in $\mathbb{U}$. If

$$f(z) \prec g(z) \quad (z \in \mathbb{U}),$$

then

$$|a_k| \leqq |g_1| \quad (k \in \mathbb{N}).$$

Theorem 38.3 *Let the function $f(z)$ be given by* (38.1). *If $f \in \mathcal{M}_g(n, \lambda, b)$ then*

$$|a_j| \leqq \frac{\prod_{k=0}^{j-2}(k + |g'(0)||b|)}{j^n(1 - \lambda + j\lambda)(j-1)!} \quad \big(j \in \mathbb{N}^* := \mathbb{N} \setminus \{1\} = \{2, 3, 4, \ldots\}\big).$$

Theorem 38.4 *Let the function $f(z) \in \mathcal{A}$ be given by* (38.1). *If $f \in \mathcal{M}_g(n, \lambda, b; u)$, then*

$$|a_j| \leqq \frac{(1+u)(2+u)\prod_{k=0}^{j-2}(k + |g'(0)||b|)}{j^n(1 - \lambda + j\lambda)(j-1)!(j+u)(j+1+u)} \quad \big(j \in \mathbb{N}^*;\ u \in \mathbb{R} \setminus (-\infty, -1]\big).$$

In view of Remark 38.1, if we set

$$g(z) = \frac{1 + (1 - 2\alpha)z}{1 - z} \quad (0 \leqq \alpha < 1;\ z \in \mathbb{U})$$

in Theorems 38.3 and 38.4, respectively, we can readily deduce the following two corollaries.

Corollary 38.1 *Let the function $f \in \mathscr{A}$ be given by* (38.1). *If $f \in \mathscr{B}(n, \lambda, \alpha, b)$, then*

$$|a_j| \leqq \frac{\prod_{k=0}^{j-2}[k + 2|b|(1-\alpha)]}{j^n(1-\lambda+j\lambda)(j-1)!} \quad \left(j \in \mathbb{N}^*\right).$$

Corollary 38.2 *Let the function $f \in \mathscr{A}$ be given by* (38.1). *If $f \in \mathscr{T}(n, \lambda, \alpha, b; u)$, then*

$$|a_j| \leqq \frac{(1+u)(2+u)\prod_{k=0}^{j-2}(k + 2|b|(1-\alpha\pi))}{j^n(1-\lambda+j\lambda)(j-1)!(j+u)(j+1+u)} \quad \left(j \in \mathbb{N}^*;\ u \in \mathbb{R} \setminus (-\infty, -1]\right).$$

Remark 38.3 Corollaries 38.1 and 38.2 were obtained by Deng [16]. It should be observed here that, by using Theorems 38.1 and 38.2, we are able to derive these results much more easily.

38.4 Univalence Criteria Involving Certain Families of Integral Operators

In many recent investigations (see, for example, the works by Pascu [29], Ozaki and Nunokawa [28], and Pescar and Breaz [32]), several interesting theorems dealing with univalence criteria were proven. Here, in this section, we consider two general families of integral operators given by Definition 38.6 below.

Definition 38.6 The first family of integral operators, studied by Breaz and Breaz [13], is defined as follows:

$$F_{n,\alpha}(z) := \left(\left[n(\alpha-1)+1\right]\int_0^z \prod_{j=1}^{n}\left(\frac{g_j(u)}{u}\right)^{1/\alpha} u^{n(\alpha-1)}du\right)^{1/[n(\alpha-1)+1]}$$
$$(g_j \in \mathscr{A};\ j = 1, \ldots, n). \tag{38.20}$$

The second family of integral operators was introduced by Breaz and Breaz [14] and it has the following form (see also a recent investigation on this subject by Breaz et al. [15]):

$$G_{n,\alpha}(z) := \left(\left[n(\alpha-1)+1\right]\int_0^z \prod_{j=1}^{n}\left[g_j(u)\right]^{\alpha-1}du\right)^{1/[n(\alpha-1)+1]}$$
$$(g_j \in \mathscr{A};\ j = 1, \ldots, n). \tag{38.21}$$

Remark 38.4 For $n = 1$, the integral operator $F_{n,\alpha}$ defined by (38.20) reduces to the operator $F_{1,\alpha}$ which is related closely to some known integral operators investigated earlier in *Univalent Function Theory* (for details, see [44] and [45]). The operator $F_{1,\alpha}$ was studied by Pescar [30]. Upon setting $n = 1 = \alpha$ in (38.20), we are led to the integral operator $F_{1,1}$ which was studied by Alexander [1].

Remark 38.5 For $n = 1$, the integral operator $G_{1,\alpha}(z)$ defined by (38.21) was studied by Moldoveanu and Pascu [23]. Furthermore, in their special case when $n = 1$ and $\alpha = a + ib$ $(a, b \in \mathbb{R})$, the integral operators in (38.20) and (38.21) obviously reduce to the integral operators in the aforementioned theorems proven by Pescar and Breaz [32].

In order to prove the main results of this section (Theorems 38.5 and 38.6 below), we need to make use of the following lemma (for details, see Srivastava et al. [40]).

Lemma 38.2 (General Schwarz Lemma [25]) *Let the function $f(z)$ be regular in the disk*

$$\mathbb{U}_R = \{z : z \in \mathbb{C} \text{ and } |z| < R\},$$

with

$$|f(z)| < M \quad (z \in \mathbb{U}_R)$$

for a fixed $M > 0$. If $f(z)$ has one zero with multiplicity order bigger than m for $z = 0$, then

$$|f(z)| \leqq \frac{M}{R^m} |z|^m \quad (z \in \mathbb{U}_R). \tag{38.22}$$

The equality in (38.22) *can hold true only if*

$$f(z) = e^{i\theta} \left(\frac{M}{R^m}\right) z^m,$$

where θ is a constant.

Theorem 38.5 *Let $M \geqq 1$ and suppose that each of the functions $g_j \in \mathscr{A}$ $(j = 1, \ldots, n)$ satisfies the following inequality*:

$$\left|\frac{z^2 f'(z)}{[f(z)]^2} - 1\right| \leqq 1 \quad (z \in \mathbb{U}). \tag{38.23}$$

Also let $\alpha = a + ib$ $(a, b \in \mathbb{R})$ be a complex number with the components a and b constrained by

$$a \in \left(0, \sqrt{(2M+1)n}\right]$$

and

$$a^4 + a^2 b^2 - \left[(2M+1)n\right]^2 \geqq 0.$$

If

$$\left|g_j(z)\right| \leqq M \quad \left(z \in \mathbb{U};\ j \in \{1,\dots,n\}\right),$$

then the function $F_{n,\alpha}(z)$ *defined by* (38.20) *is in the univalent function class* $\mathcal{S}$ *in* $\mathbb{U}$.

Theorem 38.6 *Let* $M \geqq 1$ *and suppose that each of the functions* $g_j \in \mathcal{A}$ $(j = 1,\dots,n)$ *satisfies the inequality* (38.23). *Also let* $\alpha = a + ib$ $(a, b \in \mathbb{R})$ *be a complex number with the components* a *and* b *constrained by*

$$a \in \left[\frac{(2M+1)n}{(2M+1)n+1}, \frac{(2M+1)n}{(2M+1)n-1}\right], \qquad b \in \left[0, \frac{1}{\sqrt{[(2M+1)n]^2-1}}\right]$$

and

$$\left[(a-1)^2 + b^2\right]\left[(2M+1)n\right]^2 - a^2 \leqq 0.$$

If

$$\left|g_j(z)\right| \leqq M \quad \left(z \in \mathbb{U};\ j \in \{1,\dots,n\}\right),$$

then the function $G_{n,\alpha}(z)$ *defined by* (38.21) *is in the univalent function class* $\mathcal{S}$ *in* $\mathbb{U}$.

Corollaries 38.3 and 38.4 below follow from Theorem 38.5 upon setting $M = 1$ and $n = 1$, respectively.

Corollary 38.3 *Let each of the functions* $g_j \in \mathcal{A}$ $(j = 1,\dots,n)$ *satisfy the inequality* (38.23). *Also let* $\alpha = a + ib$ $(a, b \in \mathbb{R})$ *be a complex number with the components* a *and* b *constrained by*

$$a \in \left(0, \sqrt{3n}\right] \quad \text{and} \quad a^4 + a^2b^2 - (3n)^2 \geqq 0.$$

If

$$\left|g_j(z)\right| \leqq 1 \qquad \left(z \in \mathbb{U}; \quad j \in \{1,\dots,n\}\right),$$

then the function $F_{n,\alpha}(z)$ *defined by*

$$F_{n,\alpha}(z) := \left(\left[n(\alpha-1)+1\right]\int_0^z \prod_{j=1}^{n}\left(\frac{g_j(u)}{u}\right)^{1/\alpha} u^{n(\alpha-1)}\,du\right)^{1/[n(\alpha-1)+1]}$$

$$(g_j \in \mathcal{A}; \quad j = 1,\dots,n).$$

is in the univalent function class $\mathcal{S}$ *in* $\mathbb{U}$.

Corollary 38.4 *Let* $M \geqq 1$ *and suppose that the function* $g \in \mathcal{A}$ *satisfies the inequality* (38.23). *Also let* $\alpha = a + ib$ $(a, b \in \mathbb{R})$ *be a complex number with the* *components a and b constrained by*

$$a \in (0, \sqrt{2M+1}] \quad \textit{and} \quad a^4 + a^2b^2 - (2M+1)^2 \geqq 0.$$

If

$$|g(z)| \leqq M \quad (z \in \mathbb{U}),$$

then the function $\mathfrak{F}_\alpha(z)$ *defined by*

$$\mathfrak{F}_\alpha(z) = \left[(a+ib)\int_0^z u^{a+ib-1}\left(\frac{g(u)}{u}\right)^{1/(a+ib)} du\right]^{1/(a+ib)} \qquad (\alpha = a+ib)$$

is in the univalent function class $\mathcal{S}$ *in* $\mathbb{U}$.

Remark 38.6 Corollary 38.4 provides an extension of a result due to Pescar and Breaz [12].

Remark 38.7 If we set $M = n = 1$ in Theorem 38.5, we obtain another result due to Pescar and Breaz [32].

The following results (Corollaries 38.5 and 38.6 below) can be deduced from Theorem 38.5 by putting $M = 1$ and $n = 1$, respectively.

Corollary 38.5 *Let each of the functions* $g_j \in \mathcal{A}$ $(j = 1, \ldots, n)$ *satisfy the inequality* (38.23). *Also let* $\alpha = a + ib$ $(a, b \in \mathbb{R})$ *be a complex number with the components a and b constrained by*

$$a \in \left[\frac{3n}{3n+1}, \frac{3n}{3n-1}\right], \qquad b \in \left[0, \frac{1}{\sqrt{9n^2-1}}\right]$$

and

$$9\left[(a-1)^2 + b^2\right]n^2 - a^2 \leqq 0.$$

If

$$|g_j(z)| \leqq 1 \quad \left(z \in \mathbb{U};\ j \in \{1, \ldots, n\}\right),$$

then the function $G_{n,\alpha}(z)$ *defined by*

$$G_{n,\alpha}(z) := \left(\left[n(\alpha-1)+1\right]\int_0^z \prod_{j=1}^n \left[g_j(u)\right]^{\alpha-1} du\right)^{1/[n(\alpha-1)+1]}$$

$$(g_j \in \mathcal{A};\ j = 1, \ldots, n).$$

is in the univalent function class $\mathcal{S}$ *in* $\mathbb{U}$.

Corollary 38.6 *Let $M \geqq 1$ and suppose that the function $g \in \mathscr{A}$ satisfies the inequality* (38.23). *Also let $\alpha = a + ib$ $(a, b \in \mathbb{R})$ be a complex number with the components a and b constrained by*

$$a \in \left[\frac{2M+1}{2M+2}, \frac{2M+1}{2M}\right], \qquad b \in \left[0, \frac{1}{2\sqrt{M(M+1)}}\right]$$

and

$$\left[(a-1)^2 + b^2\right](2M+1)^2 - a^2 \leqq 0.$$

If

$$\left|g(z)\right| \leqq M \quad (z \in \mathbb{U};\ M \geqq 1),$$

then the function $\mathfrak{G}_\alpha(z)$ defined by

$$\mathfrak{G}_\alpha(z) = \left((a+ib)\int_0^z \left[g(u)\right]^{a+ib-1} du\right)^{1/(a+ib)} \qquad (\alpha = a + ib)$$

is in the univalent function class $\mathscr{S}$ in $\mathbb{U}$.

Remark 38.8 Corollary 38.6 provides an extension of one of the aforementioned theorems due to Pescar and Breaz [32].

Remark 38.9 If, in Theorem 38.6, we set $M = n = 1$, we obtain another result due to Pescar and Breaz [32].

38.5 Further Developments and Concluding Remarks and Observations

In this concluding section of our article, we find it to be worthwhile to first observe that several other interesting criteria for univalence are provided in the recent works [31] and [33]. Moreover, in a very recent work [51], one can find some coefficient bounds and other related results for a general family of analytic and close-to-convex functions in the open unit disk $\mathbb{U}$, which is motivated essentially by some of the earlier investigations reported in [18, 19], and [27].

Some interesting extensions of the results associated with certain analytic and bi-univalent function classes, which we have already reported in Sect. 38.2 above, were given recently by Xu et al. [49]. The results of Xu et al. [49] are based upon the following definition.

Definition 38.7 Let the functions $h, p : \mathbb{U} \to \mathbb{C}$ be so constrained that

$$\min\left\{\Re\big(h(z)\big), \Re\big(p(z)\big)\right\} > 0 \quad (z \in \mathbb{U}) \quad \text{and} \quad h(0) = p(0) = 1.$$

Also let the function $f(z)$, defined by (38.1), be in the analytic function class $\mathscr{A}$. We say that $f \in \mathscr{H}_{\Sigma}^{h,p}$ if the following conditions are satisfied:

$$f \in \Sigma \quad \text{and} \quad f'(z) \subset h(\mathbb{U}) \quad (z \in \mathbb{U}) \tag{38.24}$$

and

$$g'(w) \subset p(\mathbb{U}) \quad (w \in \mathbb{U}), \tag{38.25}$$

where the function $g(w)$ is given by (38.10).

Remark 38.10 There are many choices of the functions $h(z)$ and $p(z)$ which provide interesting subclasses of the analytic function class $\mathscr{A}$. For example, if we let

$$h(z) = p(z) = \left(\frac{1+z}{1-z}\right)^{\alpha} \quad (z \in \mathbb{U};\ 0 < \alpha \leqq 1)$$

or

$$h(z) = p(z) = \frac{1+(1-2\beta)z}{1-z} \quad (z \in \mathbb{U};\ 0 \leqq \beta < 1),$$

it is easy to verify that the functions $h(z)$ and $p(z)$ satisfy the hypotheses of Definition 38.1. If $f \in \mathscr{H}_{\Sigma}^{h,p}$, then

$$f \in \Sigma \quad \text{and} \quad \left|\arg\big(f'(z)\big)\right| \leqq \frac{\alpha\pi}{2} \quad (z \in \mathbb{U};\ 0 < \alpha \leqq 1)$$

and

$$\left|\arg\big(g'(w)\big)\right| \leqq \frac{\alpha\pi}{2} \quad (w \in \mathbb{U};\ 0 < \alpha \leqq 1)$$

or

$$f \in \Sigma \quad \text{and} \quad \Re\big(f'(z)\big) > \beta \quad (z \in \mathbb{U};\ 0 \leqq \beta < 1)$$

and

$$\Re\big(g'(w)\big) > \beta \quad (z \in \mathbb{U};\ 0 \leqq \beta < 1),$$

where the function g is given by (38.4). This means that

$$f \in \mathscr{H}_{\Sigma}^{\alpha} \quad (0 < \alpha \leqq 1) \quad \text{or} \quad f \in \mathscr{H}_{\Sigma}^{\beta}\ (0 \leqq \beta < 1).$$

We now state and prove a few general results involving the bi-univalent function class $\mathscr{H}_{\Sigma}^{h,p}$ given by Definition 38.7, which generalize as well as improve the related work of Srivastava et al. [43] (for details, see [49] and Sect. 38.2 above).

Theorem 38.7 *Let the function $f(z)$ given by the Taylor–Maclaurin series expansion* (38.1) *be in the bi-univalent function class $\mathscr{H}_{\Sigma}^{h,p}$. Then*

$$|a_2| \leqq \sqrt{\frac{|h''(0)| + |p''(0)|}{12}} \quad \textit{and} \quad |a_3| \leqq \frac{|h''(0)|}{6}. \tag{38.26}$$

Proof First of all, we write the argument inequalities in (38.24) and (38.25) in their equivalent forms as follows:

$$f'(z) = h(z) \quad (z \in \mathbb{U}) \quad \text{and} \quad g'(w) = p(w) \quad (w \in \mathbb{U}),$$

respectively, where the functions $h(z)$ and $p(z)$ satisfy the conditions of Definition 38.7. Furthermore, the functions $h(z)$ and $p(w)$ have the following Taylor–Maclaurin series expansions:

$$h(z) = 1 + h_1 z + h_2 z^2 + \cdots$$

and

$$p(w) = 1 + p_1 w + p_2 w^2 + \cdots,$$

respectively. Now, upon equating the coefficients of $f'(z)$ with those of $h(z)$ and the coefficients of $g'(w)$ with those of $p(w)$, we get

$$2a_2 = h_1, \tag{38.27}$$

$$3a_3 = h_2, \tag{38.28}$$

$$-2a_2 = p_1, \tag{38.29}$$

and

$$3(2a_2^2 - a_3) = p_2. \tag{38.30}$$

From (38.27) and (38.29), we find that

$$h_1 = -p_1 \quad \text{and} \quad 8a_2^2 = h_1^2 + p_1^2.$$

Also, from (38.28) and (38.30), we obtain

$$6a_2^2 = h_2 + p_2,$$

which gives us the desired estimate on the coefficient $|a_2|$ as asserted in (38.26).

Next, in order to find the bound on the coefficient $|a_3|$, we subtract (38.30) from (38.28). We thus get

$$6a_3 - 6a_2^2 = h_2 - p_2. \tag{38.31}$$

Upon substituting the value of a_2^2 from (38.31) into (38.28), it follows that

$$a_3 = \frac{h_2}{3},$$

as claimed. This obviously completes the proof of Theorem 38.7. □

In light of Remark 38.10, if we set

$$h(z) = p(z) = \left(\frac{1+z}{1-z}\right)^{\alpha} \quad (z \in \mathbb{U};\ 0 < \alpha \leqq 1)$$

and

$$h(z) = p(z) = \frac{1+(1-2\beta)z}{1-z} \quad (z \in \mathbb{U};\ 0 \leqq \beta < 1)$$

in Theorem 38.7, we can readily deduce Corollaries 38.7 and 38.8, respectively, which we merely state here *without* proof.

Corollary 38.7 *Let the function* $f(z)$ *given by the Taylor–Maclaurin series expansion* (38.1) *be in the bi-univalent function class* $\mathscr{H}_{\Sigma}^{\alpha}$ $(0 < \alpha \leqq 1)$. *Then*

$$|a_2| \leqq \frac{\sqrt{6}}{3}\alpha \quad \textit{and} \quad |a_3| \leqq \frac{2}{3}\alpha^2.$$

Remark 38.11 It is easily proved that

$$\frac{\sqrt{6}}{3}\alpha \leqq \alpha\sqrt{\frac{2}{\alpha+2}} \quad \text{and} \quad \frac{2}{3}\alpha^2 \leqq \frac{\alpha(3\alpha+2)}{3} \quad (0 < \alpha \leqq 1),$$

which, in conjunction with Corollary 38.7, obviously yields an improvement of Theorem 38.1.

Corollary 38.8 *Let the function* $f(z)$ *given by the Taylor–Maclaurin series expansion* (38.1) *be in the bi-univalent function class* $\mathscr{H}_{\Sigma}^{\beta}$ $(0 \leqq \beta < 1)$. *Then*

$$|a_2| \leqq \sqrt{\frac{2(1-\beta)}{3}} \quad \textit{and} \quad |a_3| \leqq \frac{2(1-\beta)}{3}.$$

Remark 38.12 It is fairly straightforward to verify that

$$\frac{2(1-\beta)}{3} \leqq \frac{(1-\beta)(5-3\beta)}{3} \quad (0 \leqq \beta < 1),$$

which, in conjunction with Corollary 38.8, leads us to an improvement of Theorem 38.2.

We next observe that, in several recent papers (the first one by Srivastava et al. [39] and the subsequent one by Xu et al. [50]), one can find coefficient bounds associated with various subclasses of analytic functions of *complex* order $\gamma \in \mathbb{C}^* := \mathbb{C} \setminus \{0\}$ (see also Sect. 38.3 above). For the sake of the interested reader, we choose to summarize here the works by Srivastava et al. [39] and Xu et al. [50] as follows.

Definition 38.8 (See [39]) Let $\mathcal{S}(\lambda, \gamma, A, B)$ denote the class of functions given by

$$\mathcal{S}(\lambda, \gamma, A, B) = \left\{ f : f \in \mathcal{A} \quad \text{and} \quad 1 + \frac{1}{\gamma}\left(\frac{zf'(z) + \lambda z^2 f''(z)}{\lambda z f'(z) + (1-\lambda) f(z)} - 1\right) \prec \frac{1+Az}{1+Bz} \quad (z \in \mathbb{U}) \right\}$$
$$\big(0 \leqq \lambda \leqq 1;\ \gamma \in \mathbb{C}^*;\ -1 \leqq B < A \leqq 1\big). \tag{38.32}$$

Definition 38.9 (See [39]) A function $f(z) \in \mathcal{A}$ is said to be in the class $\mathcal{K}(\lambda, \gamma, A, B, m; u)$ if it satisfies the following *nonhomogeneous Cauchy–Euler type differential equation of order* m:

$$z^m \frac{d^m w}{dz^m} + \binom{m}{1}(u+m-1) z^{m-1} \frac{d^{m-1} w}{dz^{m-1}} + \cdots + \binom{m}{m} w \prod_{j=0}^{m-1}(u+j) = g(z) \prod_{j=0}^{m-1}(u+j+1)$$
$$\big(w = f(z) \in \mathcal{A};\ g(z) \in \mathcal{S}(\lambda, \gamma, A, B);\ u \in \mathbb{R} \setminus (-\infty, -1];\ m \in \mathbb{N}^*\big). \tag{38.33}$$

Theorem 38.8 (See [39]) *Let the function* $f(z)$ *be defined by* (38.1). *If* $f \in \mathcal{S}(\lambda, \gamma, A, B)$, *then*

$$|a_n| \leqq \frac{\prod_{k=0}^{n-2}\left(k + \frac{2|\gamma|(A-B)}{1-B}\right)}{(n-1)![1+\lambda(n-1)]} \quad \big(n \in \mathbb{N}^*\big).$$

Theorem 38.9 (See [39]) *Let the function* $f(z)$ *be defined by* (38.1). *If* $f \in \mathcal{K}(\lambda, \gamma, A, B, m; u)$, *then*

$$|a_n| \leqq \frac{\prod_{k=0}^{n-2}\left(k + \frac{2|\gamma|(A-B)}{1-B}\right) \prod_{j=0}^{m-2}(u+j+1)}{(n-1)![1+\lambda(n-1)] \prod_{j=0}^{m-1}(u+j+n)} \quad \big(m, n \in \mathbb{N}^*\big).$$

Definition 38.10 (See [50]) Let $g : \mathbb{U} \to \mathbb{C}$ be a convex function such that

$$g(0) = 1 \quad \text{and} \quad \Re\big[g(z)\big] > 0 \quad (z \in \mathbb{U}).$$

We denote by $\mathscr{S}_g(\lambda,\gamma)$ the class of functions given by

$$\mathscr{S}_g(\lambda,\gamma)=\left\{f: f\in\mathscr{A} \quad\text{and}\quad 1+\frac{1}{\gamma}\left(\frac{zf'(z)+\lambda z^2 f''(z)}{\lambda z f'(z)+(1-\lambda)f(z)}-1\right)\right.$$
$$\left.\in g(\mathbb{U})\ (z\in\mathbb{U})\right\}$$
$$\left(0\leqq\lambda\leqq 1;\ \gamma\in\mathbb{C}^*\right).$$

Definition 38.11 (See [50]) A function $f\in\mathscr{A}$ is said to be in the class $\mathscr{K}_g(\lambda,\gamma,m;u)$ if it satisfies the following *nonhomogeneous Cauchy–Euler differential equation*:

$$z^m\frac{d^m w}{dz^m}+\binom{m}{1}(u+m-1)z^{m-1}\frac{d^{m-1}w}{dz^{m-1}}+\cdots+\binom{m}{m}w\prod_{j=0}^{m-1}(u+j)$$
$$=h(z)\prod_{j=0}^{m-1}(u+j+1)$$
$$\left(w=f(z)\in\mathscr{A};\ h(z)\in\mathscr{S}_g(\lambda,\gamma);\ u\in\mathbb{R}\setminus(-\infty,-1];\ m\in\mathbb{N}^*\right). \tag{38.34}$$

Remark 38.13 There are many choices of the function $g(z)$ which provide interesting subclasses of analytic functions of complex order $\gamma\in\mathbb{C}^*$. In particular, if we let

$$g(z)=\frac{1+Az}{1+Bz}\quad(-1\leqq B<A\leqq 1;\ z\in\mathbb{U}), \tag{38.35}$$

it is fairly easy to verify that $g(z)$ is a convex function in $\mathbb{U}$ and satisfies the hypotheses of Definition 38.10. Clearly, therefore, the function class $\mathscr{S}_g(\lambda,\gamma)$, with the function $g(z)$ given by (38.35), coincides with the function class $\mathscr{S}(\lambda,\gamma,A,B)$ given by Definition 38.8.

Remark 38.14 In view of Remark 38.13, if the function $g(z)$ is given by (38.35), it is easily observed that the function classes

$$\mathscr{S}_g(\lambda,\gamma)\quad\text{and}\quad\mathscr{K}_g(\lambda,\gamma,m;u)$$

reduce to the aforementioned function classes

$$\mathscr{S}(\lambda,\gamma,A,B)\quad\text{and}\quad\mathscr{K}(\lambda,\gamma,A,B,m;u),$$

respectively (see Definitions 38.8 and 38.9).

The coefficient inequalities asserted by Theorems 38.10 and 38.11 below correspond to the Definitions 38.10 and 38.11, respectively. The demonstration of these results is based upon Lemma 38.3 below (see, for details, [50]).

Lemma 38.3 (See [38]) *Let the function* $\mathfrak{g}(z)$ *given by*

$$\mathfrak{g}(z) = \sum_{k=1}^{\infty} \mathfrak{b}_k z^k \quad (z \in \mathbb{U})$$

be convex in $\mathbb{U}$. *Also let the function* $\mathfrak{f}(z)$ *given by*

$$\mathfrak{f}(z) = \sum_{k=1}^{\infty} \mathfrak{a}_k z^k \quad (z \in \mathbb{U})$$

be holomorphic in $\mathbb{U}$. *If*

$$\mathfrak{f}(z) \prec \mathfrak{g}(z) \quad (z \in \mathbb{U}),$$

then

$$|\mathfrak{a}_k| \leqq |\mathfrak{b}_1| \quad (k \in \mathbb{N}).$$

Theorem 38.10 *Let the function* $f(z)$ *be defined by* (38.1). *If* $f \in \mathscr{S}_g(\lambda, \gamma)$, *then*

$$|a_n| \leqq \frac{\prod_{k=0}^{n-2}(k + |g'(0)||\gamma|)}{(n-1)![1 + \lambda(n-1)]} \quad \big(n \in \mathbb{N}^*\big).$$

Theorem 38.11 *Let the function* $f(z) \in \mathscr{A}$ *be defined by* (38.1). *If* $f \in \mathscr{K}_g(\lambda, \gamma, m; u)$, *then*

$$|a_n| \leqq \frac{\prod_{k=0}^{n-2}(k + |g'(0)| \cdot |\lambda|) \prod_{j=0}^{m-2}(u + j + 1)}{(n-1)![1 + \lambda(n-1)] \prod_{j=0}^{m-1}(u + j + n)} \quad \big(m, n \in \mathbb{N}^*\big)$$

$$\big(0 \leqq \lambda \leqq 1;\ \gamma \in \mathbb{C}^*;\ u \in \mathbb{R} \setminus (-\infty, -1]\big).$$

In view of Remarks 38.13 and 38.14, if we let the function $g(z)$ in Theorems 38.10 and 38.11 be given by (38.35), we can readily deduce the following Corollaries 38.9 and 38.10, respectively, which we choose to merely state here *without* proofs.

Corollary 38.9 *Let the function* $f \in \mathscr{A}$ *be defined by* (38.1). *If* $f \in \mathscr{S}(\lambda, \gamma, A, B)$, *then*

$$|a_n| \leqq \frac{\prod_{k=0}^{n-2}(k + |\gamma|(A - B))}{(n-1)![1 + \lambda(n-1)]} \quad \big(n \in \mathbb{N}^*\big).$$

Corollary 38.10 *Let the function* $f \in \mathscr{A}$ *be defined by* (38.1). *If* $f \in \mathscr{K}(\lambda, \gamma, A, B, m; u)$, *then*

$$|a_n| \leqq \frac{\prod_{k=0}^{n-2}(k + |\lambda|(A - B)) \prod_{j=0}^{m-2}(u + j + 1)}{(n-1)![1 + \lambda(n-1)] \prod_{j=0}^{m-1}(u + j + n)} \quad \big(m, n \in \mathbb{N}^*\big)$$

$$\left(0 \leqq \lambda \leqq 1;\ \gamma \in \mathbb{C}^*;\ u \in \mathbb{R} \setminus (-\infty, -1]\right).$$

Remark 38.15 It is easy to see that

$$\left[k + |\gamma|(A - B)\right] \leqq \left(k + \frac{2|\gamma|(A - B)}{1 - B}\right)$$

$$\left(k \in \mathbb{N}_0;\ -1 \leqq B < A \leqq 1;\ \gamma \in \mathbb{C}^*\right),$$

which, in conjunction with Corollaries 38.9 and 38.10, obviously yields significant improvements over Theorems 38.8 and 38.9 (see also the aforementioned earlier work by Srivastava et al. [39] for several *further* corollaries and consequences Theorems 38.8 and 38.9).

Acknowledgements It is a great pleasure for me to dedicate this article to Prof. Dr. Themistocles Michael Rassias on the occasion of his 60th birthday. The present investigation was supported, in part, by the *Natural Sciences and Engineering Research Council of Canada* under Grant OGP0007353.

References

1. Alexander, J.W.: Functions which map the interior of the unit circle upon simple regions. Ann. Math. (Ser. 2) **17**, 12–22 (1915)
2. Altıntaş, O., Irmak, H., Owa, S., Srivastava, H.M.: Coefficient bounds for some families of starlike and convex functions of complex order. Appl. Math. Lett. **20**, 1218–1222 (2007)
3. Altintaş, O., Irmak, H., Srivastava, H.M.: Fractional calculus and certain starlike functions with negative coefficients. Comput. Math. Appl. **30**(2), 9–15 (1995)
4. Altintaş, O., Özkan, Ö.: Starlike, convex and close-to-convex functions of complex order. Hacettepe Bull. Nat. Sci. Eng. Ser. B **28**, 37–46 (1991)
5. Altintaş, O., Özkan, Ö.: On the classes of starlike and convex functions of complex order. Hacettepe Bull. Nat. Sci. Eng. Ser. B **30**, 63–68 (2001)
6. Altintaş, O., Özkan, Ö., Srivastava, H.M.: Neighborhoods of a class of analytic functions with negative coefficients. Appl. Math. Lett. **13**(3), 63–67 (1995)
7. Altintaş, O., Özkan, Ö., Srivastava, H.M.: Majorization by starlike functions of complex order. Complex Var. Theory Appl. **46**, 207–218 (2001)
8. Altintaş, O., Özkan, Ö., Srivastava, H.M.: Neighborhoods of a certain family of multivalent functions with negative coefficient. Comput. Math. Appl. **47**, 1667–1672 (2004)
9. Altintaş, O., Srivastava, H.M.: Some majorization problems associated with p-valently starlike and convex functions of complex order. East Asian Math. J. **17**, 175–183 (2001)
10. Brannan, D.A., Clunie, J.G. (eds.): Aspects of Contemporary Complex Analysis. Proceedings of the NATO Advanced Study Institute (University of Durham, Durham; July 1–20, 1979). Academic Press, New York (1980)
11. Brannan, D.A., Clunie, J., Kirwan, W.E.: Coefficient estimates for a class of star-like functions. Can. J. Math. **22**, 476–485 (1970)
12. Brannan, D.A., Taha, T.S.: On some classes of bi-unvalent functions. In: Mazhar, S.M., Hamoui, A., Faour, N.S. (eds.) Mathematical Analysis and Its Applications, Kuwait, February 18–21, 1985. KFAS Proceedings Series, vol. 3, pp. 53–60. Pergamon, Oxford (1988); see also Studia Univ. Babeş-Bolyai Math. **31**(2), 70–77 (1986)
13. Breaz, D., Breaz, N.: The univalent conditions for an integral operator on the classes $\mathscr{S}(p)$ and $\mathscr{T}_2$. J. Approx. Theory Appl. **1**, 93–98 (2005)

14. Breaz, D., Breaz, N.: Univalence of an integral operator. Mathematica (Cluj) **47**(70), 35–38 (2005)
15. Breaz, D., Breaz, N., Srivastava, H.M.: An extension of the univalent condition for a family of integral operators. Appl. Math. Lett. **22**, 41–44 (2009)
16. Deng, Q.: Certain subclass of analytic functions with complex order. Appl. Math. Comput. **208**, 359–362 (2009)
17. Duren, P.L.: Univalent Functions. Grundlehren der Mathematischen Wissenschaften, vol. 259. Springer, New York (1983)
18. Gao, C., Zhou, S.: On a class of analytic functions related to the starlike functions. Kyungpook Math. J. **45**, 123–130 (2005)
19. Kowalczyk, J., Leś-Bomba, E.: On a subclass of close-to-convex functions. Appl. Math. Lett. **23**(10), 1147–1151 (2010)
20. Lewin, M.: On a coefficient problem for bi-univalent functions. Proc. Am. Math. Soc. **18**, 63–68 (1967)
21. Miller, S.S., Mocanu, P.T.: Differential Subordination: Theory and Applications. Series on Monographs and Textbooks in Pure and Applied Mathematics, vol. 225. Dekker, New York (2000)
22. Milovanović, G.V., Mitrinović, D.S., Rassias, Th.M.: Topics in Polynomials: Extremal Problems, Inequalities, Zeros. World Scientific, Singapore (1994)
23. Moldoveanu, S., Pascu, N.N.: Integral operators which preserve the univalence. Mathematica (Cluj) **32**(55), 159–166 (1990)
24. Nasr, M.A., Aouf, M.K.: Radius of convexity for the class of starlike functions of complex order. Bull. Fac. Sci. Assiut Univ., Sect. a Nat. Sci. **12**, 153–159 (1983)
25. Nehari, Z.: Conformal Mapping. McGraw-Hill, New York (1952). Reprinted by Dover Publications Incorporated, New York (1975)
26. Netanyahu, E.: The minimal distance of the image boundary from the origin and the second coefficient of a univalent function in $|z| < 1$. Arch. Ration. Mech. Anal. **32**, 100–112 (1969)
27. Owa, S., Nunokawa, M., Saitoh, H., Srivastava, H.M.: Close-to-convexity, starlikeness, and convexity of certain analytic functions. Appl. Math. Lett. **15**, 63–69 (2002)
28. Ozaki, S., Nunokawa, M.: The Schwarzian derivative and univalent functions. Proc. Am. Math. Soc. **33**, 392–394 (1972)
29. Pascu, N.N.: On a univalence criterion. II. In: Itinerant Seminar on Functional Equations, Approximation and Convexity (Cluj-Napoca, 1985), pp. 153–154. Preprint 86–6, University "Babeş-Bolyai", Cluj-Napoca, 1985
30. Pescar, V.: New criteria for univalence of certain integral operators. Demonstr. Math. **33**, 51–54 (2000)
31. Pescar, V.: On the univalence of an integral operator. Appl. Math. Lett. **23**, 615–619 (2010)
32. Pescar, V., Breaz, D.: Some integral operators and their univalence. Acta Univ. Apulensis, Mat.-Inform. **15**, 147–152 (2008)
33. Pescar, V., Breaz, D.: On an integral operator. Appl. Math. Lett. **23**, 625–629 (2010)
34. Rassias, Th.M. (ed.): Constantin Carathéodory (1873–1950): An International Tribute, vols. I and II. World Scientific, Singapore (1991)
35. Rassias, Th.M., Srivastava, H.M. (eds.): Analytic and Geometric Inequalities and Applications. Series on Mathematics and Its Applications, vol. 478. Kluwer Academic, Dordrecht (1999)
36. Robertson, M.S.: On the theory of univalent functions. Ann. Math. (Ser. 1) **37**, 374–408 (1936)
37. Rogosinski, W.: On the coefficients of subordinate functions. Proc. Lond. Math. Soc. (Ser. 2) **48**, 48–82 (1943)
38. Sălăgean, G.Ş.: Subclass of univalent functions. In: Complex Analysis: Fifth Romanian–Finnish Seminar, Part 1, Bucharest, 1981. Lecture Notes in Mathematics, vol. 1013, pp. 362–372. Springer, Berlin (1983)
39. Srivastava, H.M., Altıntaş, O., Kırcı Serenbay, S.: Coefficient bounds for certain subclasses of starlike functions of complex order. Appl. Math. Lett. **24**, 1359–1363 (2011)

40. Srivastava, H.M., Deniz, E., Orhan, H.: Some general univalence criteria for a family of integral operators. Appl. Math. Comput. **215**, 3696–3701 (2010)
41. Srivastava, H.M., Eker, S.S.: Some applications of a subordination theorem for a class of analytic functions. Appl. Math. Lett. **21**, 394–399 (2008)
42. Srivastava, H.M., Eker, S.S., Şeker, B.: A certain convolution approach for subclasses of analytic functions with negative coefficients. Integral Transforms Spec. Funct. **20**, 687–699 (2009)
43. Srivastava, H.M., Mishra, A.K., Gochhayat, P.: Certain subclasses of analytic and bi-univalent functions. Appl. Math. Lett. **23**, 1188–1192 (2010). doi:10.1016/j.aml.2010.05.009
44. Srivastava, H.M., Owa, S. (eds.): Univalent Functions, Fractional Calculus, and Their Applications. Halsted, New York (1989)
45. Srivastava, H.M., Owa, S. (eds.): Current Topics in Analytic Function Theory. World Scientific, Singapore (1992)
46. Srivastava, H.M., Xu, Q.-H., Wu, G.-P.: Coefficient estimates for certain subclasses of spiral-like functions of complex order. Appl. Math. Lett. **23**, 763–768 (2010)
47. Srivastava, H.M., Yang, D.-G., Xu, N.-E.: Subordinations for multivalent analytic functions associated with the Dziok-Srivastava operator. Integral Transforms Spec. Funct. **20**, 581–606 (2009)
48. Taha, T.S.: Topics in univalent function theory. Ph.D. Thesis. University of London (1981)
49. Xu, Q.-H., Gui, Y.-C., Srivastava, H.M.: Coefficient estimates for a certain subclass of analytic and bi-univalent functions. Appl. Math. Lett. **25**, 990–994 (2012)
50. Xu, Q.-H., Gui, Y.-C., Srivastava, H.M.: Coefficient estimates for certain subclasses of analytic functions of complex order. Taiwan. J. Math. **15**, 2377–2386 (2011)
51. Xu, Q.-H., Srivastava, H.M., Li, Z.: A certain subclass of analytic and close-to-convex functions. Appl. Math. Lett. **24**, 396–401 (2011)

Chapter 39
The Hyers–Ulam and Hahn–Banach Theorems and Some Elementary Operations on Relations Motivated by Their Set-Valued Generalizations

Árpád Száz

Abstract In the first part of this paper, we provide several historical facts on the famous Hyers–Ulam stability theorems, Hahn–Banach extension theorems, and their set-valued generalizations with numerous references.

These generalizations will clearly show that the essence of the above mentioned theorems is nothing but the statement of the existence of a certain homogeneous, additive, or linear selection function of a particular relation.

In the second part of this paper, motivated by the above generalizations, we briefly review the most basic additivity and homogeneity properties of relations and investigate, in greater detail, some elementary operations on relations.

More concretely, for any relation F on one group X to another Y, we define two relations $-F$ and $\check{F}$ on X to Y such that $\check{F}(x) = F(-x)$ and $(-F)(x) = -F(x)$ for all $x \in X$. Moreover, we also define $\hat{F} = -\check{F}$ and $F^{\Delta} = F \cap \hat{F}$.

Furthermore, if in particular Y is a vector space over $\mathbb{Q}$, then for any $k \in \mathbb{Z}$, with $k \neq 0$, we also define a relation F_k on X to Y such that $F_k(x) = k^{-1} F(kx)$ for all $x \in X$. Moreover, we also define $F^{\star} = \bigcap_{n=1}^{\infty} F_n$ and $F^{*} = F^{\Delta\star}$.

The above operations and the intersection convolutions of relations, which can only be sketched here, will certainly allow of instructive treatments of some hoped-for common relational generalizations of the Hyers–Ulam and Hahn–Banach theorems.

Key words Relations on groups · Partial and global negatives · Hyers transforms · Intersection convolutions

Mathematics Subject Classification 03E20 · 26E25 · 39B82 · 46A22

Dedicated to Professor Themistocles M. Rassias on the occasion of his 60th birthday.

Á. Száz (✉)
Institute of Mathematics, University of Debrecen, 4010 Debrecen, Pf. 12, Hungary
e-mail: szaz@science.unideb.hu

P.M. Pardalos et al. (eds.), *Nonlinear Analysis*, Springer Optimization and Its Applications 68, In Honor of Themistocles M. Rassias on the Occasion of his 60th Birthday,
DOI 10.1007/978-1-4614-3498-6_39,

39.1 Historical Notes and Motivations

A. The Hyers–Ulam Stability Theorems In 1925, Pólya and Szegő [187, Aufgabe 99, pp. 17, 171] proved a strict inequality form of the following statement in two rather difficult ways.

Theorem A.1 *If* $(a_n)_{n=1}^{\infty}$ *is a sequence of real numbers such that*

$$|a_{n+m} - a_n - a_m| \le 1$$

for all $n, m \in \mathbb{N}$, *then there exists a number* b *such that, for all* $n \in \mathbb{N}$, *we have*

$$|a_n - bn| \le 1.$$

Remark A.2 Neither the above authors nor the mathematics community could recognize the significance of this theorem at that time. It was first cited by Kuczma [139, p. 424] in 1985 at the suggestion of R. Ger. However, Maligranda [149] still did not mention it.

According to Ger [78], his attention to Theorem A.1 was first drawn by M. Laczkovich at an undetectable conference, who indicated that the real-valued particular case of Hyers's stability theorem can easily be derived from Theorem A.1.

Hyers [111] in 1941, giving a partial answer to a general problem formulated by S.M. Ulam during a talk before a Mathematical Colloquium at the University of Wisconsin in 1940, proved, in a quite simple way, a slightly weaker particular case the following stability theorem in a Banach space.

Theorem A.3 *If* f *is an* ε*-approximately additive function of a commutative semigroup* X *to a Banach space* Y, *for some* $\varepsilon > 0$, *in the sense that*

$$\|f(x+y) - f(x) - f(y)\| \le \varepsilon$$

for all $x, y \in X$, *then there exists an additive function* g *of* X *to* Y *such that* g *is* ε*-near to* f *in the sense that, for all* $x \in X$, *we have*

$$\|f(x) - g(x)\| \le \varepsilon.$$

Remark A.4 It is easy to see that the number b and the additive function g are uniquely determined in the above theorems.

Namely, if b and g are as in Theorems A.1 and A.3, then because of the necessary homogeneity properties of the corresponding norms and the additive function g we have

$$b = \lim_{n\to\infty} n^{-1} a_n \quad \text{and} \quad g(x) = \lim_{n\to\infty} f_n(x),$$

for all $x \in X$, where $f_n(x) = n^{-1} f(nx)$.

To define the function g, Hyers originally used the subsequence $(f_{2^n})_{n=1}^{\infty}$, since its pointwise convergence can be more easily verified. Moreover, it can also be well used when f is assumed to be only ε-approximately 2-homogeneous in the sense that $\|f(2x)-2f(x)|\le\varepsilon$ for all $x\in X$. In this case, it can also be shown that the sequence $(f_{2^n})_{n=1}^{\infty}$ is rapidly uniformly convergent [253].

Remark A.5 Because of the $\mathbb{N}$-homogeneity of additive functions [252], it is clear that Theorem A.1 is a particular case of Theorem A.3. Moreover, by M. Laczkovich, the $Y=\mathbb{R}$ particular case of Theorem A.3 can be easily derived from Theorem A.1.

To see this, suppose that f is as in Theorem A.3 and $Y=\mathbb{R}$. Moreover, for any $x\in X$ and $n\in\mathbb{N}$, define

$$a_n(x)=\varepsilon^{-1}f(nx).$$

Then, by the ε-approximate additivity of f, we have

$$\left|a_{n+m}(x)-a_n(x)-a_m(x)\right|\le 1$$

for all $x\in X$ and $n,m\in\mathbb{N}$. Thus, by Theorem A.1, for each $x\in X$ there exists a real number $b(x)$ such that

$$\left|a_n(x)-b(x)n\right|\le 1$$

for all $n\in\mathbb{N}$. Now, by taking $g(x)=\varepsilon b(x)$, we can see that

$$\left|f(nx)-g(x)n\right|\le\varepsilon$$

for all $n\in\mathbb{N}$. Hence, we can immediately infer that

$$\left|f(x)-g(x)\right|\le\varepsilon \quad\text{and}\quad g(x)=\lim_{n\to\infty}f_n(x),$$

where $f_n(x)=n^{-1}f(nx)$. Thus, to complete the proof, it remains to note only that, by the ε-approximate additivity of f, for any $x,y\in X$ and $n\in\mathbb{N}$ we also have

$$\left|f_n(x+y)-f_n(x)-f_n(y)\right|\le n^{-1}\varepsilon.$$

Namely, hence by letting $n\to\infty$ we can already infer that $g(x+y)=g(x)+g(y)$.

Remark A.6 Forti [57] in 1987, having in mind some abstract theorems of Székelyhidi [269] and Gajda [65], proved that if X is a semigroup such that the Hyers theorem holds for any real-valued (complex-valued) function of X, then the same theorem holds also for any function of X to an arbitrary real (complex) Banach space. Thus, by Remark A.5, Theorem A.1 is actually equivalent to Theorem A.3.

In this respect, it is also worth mentioning that Schwaiger [225] in 1988 proved that if Y is a normed space such that the Hyers's theorem holds for every function f of $\mathbb{N}$ to Y, then Y is necessary complete. (See also Forti and Schwaiger [60] for some more general results.)

Remark A.7 Hyers's stability theorem has later been generalized by several authors in various directions. For instance, Aoki [5], Th.M. Rassias [197] and Găvruţă [72] replaced ε by the more general quantities $\varepsilon(\|x\|^p + \|y\|^p)$ and $\varphi(x, y)$, respectively.

The possibility of a particular case of latter generalization, when $\varphi(x, y)$ is a suitable function of $\|x\|$ and $\|y\|$, was already remarked, but not accomplished by Bourgin [30] in 1951, who reviewed, but not cited Aoki's paper. Moreover, Forti [55] and Grabiec [92] proved much more general theorems.

Remark A.8 Forti [57] already remarked that for the most part of Hyers's theorem the domain X of f may be an arbitrary semigroup. And only the additivity of the function g requires X to be commutative.

Weaker sufficient conditions were also considered by Rätz [209] and Páles [177]. (See also Tabor [271] and Volkmann [279].) Furthermore, Székelyhidi [268] noticed that the existence of an invariant mean is also sufficient. (See also Kazhdan [130].)

Invariant means were later also used by Forti [57], Gajda [67], Badora [8] and Badora, Ger and Páles [11]. Moreover, Gajda, A. Smajdor, and W. Smajdor [71], Páles [176], Badora [10], and Huang and Li [109] applied Hahn–Banach type theorems to obtain stability results.

Remark A.9 Meantime, some negative results have also been established. Paganoni [172] in 1980 and Forti and Schwaiger [60] in 1989 observed that, by defining $f(k) = [k/2]$ and $g(k) = [k\sqrt{2}]$ for all $k \in \mathbb{Z}$, we can get 1-approximately additive functions of $\mathbb{Z}$ to itself such that for any additive functions of φ and ψ of $\mathbb{Z}$ to $\mathbb{Z}$ and $\mathbb{Q}$, respectively, the differences $f - \varphi$ and $g - \psi$ are unbounded.

Moreover, by using the free group generated by two elements, Forti [56] in 1985 showed that the commutativity of X in Theorem A.3 cannot be omitted even if X is a group. For a more detailed treatment of Forti's function, see Bahyrycz [7].

In this respect, it is also worth mentioning that Špakula and Zlatoš [236] in 2004, by using a result of Kazhdan [130], showed that there are compact, commutative metric groups X and Y, the latter being endowed with an invariant metric, such that for each $\varepsilon > 0$ there exists a continuous ε-approximately additive function f of X to Y such that $d(f, g) \geq 1$ for every additive function g of X to Y.

Remark A.10 Furthermore, we can also note that the Hyers sequence $(f_{2^n})_{n=1}^{\infty}$ itself has also been generalized by several authors.

The interested reader is referred to Th.M. Rassias [198], Găvruta, Hossu, Popescu, and Căprău [75], Lee and Jun [142], and Gilányi, Kaiser, and Páles [80]. Moreover, in set-valued settings, we refer to Gajda and Ger [69], Popa [188], Nikodem and Popa [168], Lu and Park [146], and to the papers of the present author [251, 255].

Remark A.11 Finally, we note that the numerous investigations, motivated by Hyers's theorem and the influential mathematical results and proposed open problems and conjectures of Themistocles M. Rassias, have led to an enormous theory of the stability of functional equations and inequalities.

The interested reader can get a wide overview on the subject by consulting the books by Hyers, Isac, and Rassias [115], Jung [125], and Czerwik [42], and the surveys by Bourgin [30], Hyers [112], Hyers and Rassias [113], Rassias and Tabor [207], Ger [78], Forti [58], Jung [126], Rassias [199]–[204], Székelyhidi [270], Sánchez and Castillo [220], Moszner [153], and Czerwik and Król [44].

Remark A.12 Lately, instead of Hyers's direct method, fixed point theorems have also been widely used to obtain stability results for functional equations. (For the origins and some recent developments, see Baker [13], Radu [193], Mihet [150], Park and Rassias [179, 180], P. Găvruta and L. Găvruta [74], Takahasi, Miura, and Takagi [276].)

In this respect, it is also worth mentioning that recently the Hyers–Ulam stability of recurrences, differential, integral and operational equations and inequalities has also been intensively investigated by several authors. However, the corresponding papers will not be included in the extensive references since in the sequel we shall only be interested in set-valued generalizations of the stability of additivity and homogeneity properties.

B. Set-Valued Generalizations of the Hyers–Ulam Theorems Hyers's theorem was transformed into set-valued settings by W. Smajdor [231] and Gajda and Ger [69] in 1986 and 1987, respectively, by making use the following observations.

Remark B.1 If f and g are as in Theorem A.3 and $B=\{y\in Y:\ \|y\|\le\varepsilon\}$, then

$$g(x)-f(x)\in B \quad \text{and} \quad f(x+y)-f(x)-f(y)\in B,$$

and hence

$$g(x)\in f(x)+B \quad \text{and} \quad f(x+y)\in f(x)+f(y)+B$$

for all $x, y\in X$.

Therefore, by defining $F(x)=f(x)+B$ for all $x\in X$, we can get a set-valued function F of X to Y such that g is a selection of F and F is subadditive. That is, $g(x)\in F(x)$ and $F(x+y)\subset F(x)+F(y)$ for all $x, y\in X$.

Thus, the essence of Hyers's theorem is nothing but the statement of the existence of an additive selection function of a certain subadditive set-valued function.

In particular, Gajda and Ger [69] proved the following generalization of Theorem A.3. (See also Gajda [67, Theorem 4.2].)

Theorem B.2 *If F is a subadditive set-valued function of a commutative semigroup X to a Banach space Y such that the values of F are nonempty, closed and convex, and moreover*

$$\sup\big\{\operatorname{diam}\big(F(x)\big):\ x\in X\big\}<+\infty,$$

then F has an additive selection function f.

Remark B.3 Moreover, they have also proved an extension of this theorem to a separated, sequentially complete topological vector space Y.

For this, it was necessary to introduce first an appropriate notion of the diameter of a subset of Y relative to a balanced neighborhood of the origin in Y.

Remark B.4 The importance of the observations of W. Smajdor, Gajda, and Ger was soon recognized by Hyers and Rassias [113], Rassias [200], Hyers, Isac, and Rassias [115, pp. 204–231], and Czerwik [42, pp. 301–329].

Moreover, by using the direct method of Gajda and Ger, Popa [188, 189], Nikodem and Popa [168], Piao [184], Lu and Park [146], and the present author [251, 255] extended the results of Gajda and Ger.

Remark B.5 In particular, in [251], by using relations and relators instead of set-valued functions and topologies, we have proved the subsequent improvement and generalization of Theorem B.2.

Unfortunately, the publication of [251] has been rejected by the editors of the journal Mathematical Inequalities and Applications after an almost three-year-long consideration. In the meantime, they did not answered most of my letters and sometimes acted as if they had not received my manuscript.

Theorem B.6 *If F is a closed-valued, 2-subhomogeneous, subadditive relation of a commutative semigroup X to a separated, sequentially complete vector relator space $Y(\mathscr{S})$ such that the sequence $(2^{-n}F(2^n x))_{n=1}^{\infty}$ is infinitesimal for all $x \in X$, then F has an additive selection function f.*

Remark B.7 Here, $\mathscr{S}$ is a nonempty family of relations on the vector space Y which is, to some extent, compatible with the linear operations in Y.

And the infinitesimality of a sequence $(A_n)_{n=1}^{\infty}$ of subsets of $Y(\mathscr{S})$ means only that for each $S \in \mathscr{S}$ there exist $y \in Y$ and $n \in \mathbb{N}$ such that $A_n \subset S(y)$.

Remark B.8 Now, we can also state that the additive functions f given in the above theorems are uniquely determined.

Namely, if F and f are as in Theorem B.6, then because of the $\mathbb{N}$-homogeneity of f, the infinitesimality conditions on F, and the separatedness of Y we necessarily have

$$\{f(x)\} = \bigcap_{n=1}^{\infty} F_n(x)$$

for all $x \in X$, where $F_n(x) = n^{-1}F(nx)$.

Remark B.9 In this respect, it is also worth mentioning that if F is a relation on a groupoid X to a vector space Y, and Φ is only an $\mathbb{N}$-superhomogeneous partial selection relation of F, then we already have

$$\Phi(x) = (n^{-1}n)\Phi(x) = n^{-1}(n\Phi(x)) \subset n^{-1}\Phi(nx) \subset n^{-1}F(nx) = F_n(x)$$

for all $n \in \mathbb{N}$ and $x \in X$, and thus

$$\Phi(x) \subset \bigcap_{n=1}^{\infty} F_n(x) = \left(\bigcap_{n=1}^{\infty} F_n\right)(x)$$

for all $x \in X$. Therefore, Φ is also a partial selection relation of the intersection $F^{\star}$ of the relations F_n.

Note that if in particular X is a group and Φ is $\mathbb{Z}^*$-superhomogeneous, with $\mathbb{Z}^* = \mathbb{Z} \setminus \{0\}$, then we can quite similarly see that Φ is also a partial selection relation of the intersection F^* of the relations F_k defined such that $F_k(x) = k^{-1}F(kx)$ for all $x \in X$ and $k \in \mathbb{Z}^*$.

Remark B.10 Moreover, to motivate our forthcoming considerations, it is also worth mentioning that if F is a relation on one groupoid X to another Y and Φ is a restriction of a semi-subadditive selection relation Ψ of F, then for any $x \in X$ and $u \in D_F$ and $v \in D_\Phi$, with $x = u + v$, we have

$$\Psi(x) = \Psi(u+v) \subset \Psi(u) + \Psi(v) \subset F(u) + \Phi(v),$$

and thus

$$\Psi(x) \subset \bigcap \big\{ F(u) + \Phi(v) : \ u \in D_F, \ v \in D_\Phi, \ x = u + v \big\}.$$

Therefore, Ψ is also a partial selection relation of the intersection convolution $F * \Phi$ of F and Φ considered by the present author in [243] and [258]. (See also [29, 46, 47, 53, 86, 254].)

A natural totalization of a set-valued function of Zs. Páles, presented by Gajda and Ger [69, p. 282] and Hyers, Isac, and Rassias [115, p. 210], shows that the diameter and infinitesimality condition on F cannot be left out from Theorems B.2 and B.6 even if $X = \mathbb{R}$.

Example B.11 For any $x \in \mathbb{R}$, define

$$F(x) = \mathbb{R} \quad \text{if } x < 0 \quad \text{and} \quad F(x) = \big[x^2, +\infty\big[\quad \text{if } x \geq 0.$$

Then, F is a closed- and convex-valued subadditive relation on $\mathbb{R}$ such that F does not have an additive selection function.

To check the latter statement, assume on the contrary that f is an additive selection function of F. Then, in particular, f is $\mathbb{N}$-homogeneous. Therefore, by using Remark B.9, we can immediately arrive at the contradiction that

$$f(1) \in F_n(1) = n^{-1}F(n) = n^{-1}\big[n^2, +\infty\big[= [n, +\infty[,$$

and thus $n \leq f(1)$ for all $n \in \mathbb{N}$. If Φ is a superadditive ($\mathbb{N}$-superhomogeneous) partial selection relation of F, then we can quite similarly see that $D_\Phi \subset]-\infty, 0]$.

Remark B.12 Joining my early investigations on additive relations, my younger brother G. Száz, in a work prepared for a student competition in Hungary in 1971, proved that every linear relation F of one vector space X to another Y has a linear selection function f.

Thus, F can be written in the useful form $F(x) = f(x) + F(0)$ for all $x \in X$. That is, F has a linear representing selection function f. Thus, by using quotient spaces, the investigation of linear relations can, in principle, be reduced to that of linear functions. However, the theory of quotient spaces actually rests on that of linear equivalence relations.

Remark B.13 The importance of linear relations lies mainly in the fact that the inverse, closure, completion, and adjoint of a linear relation are linear relations.

Moreover, several theorems on linear functions, such as the open mapping and closed graph theorems, for instance, can be, most naturally, generalized in terms of linear relations. Namely, they are easy particular cases of the convex ones.

Linear set-valued functions and relations were certainly first investigated by Berge [24, p. 133] in 1959 and Arens [6] in 1961. (See also Kelley and Namioka [132, p. 101] and Coddington [39].) However, Lee and Nashed [143] and Cross [41, p. 23] attribute them to the works of J. von Neumann [161, 162] in 1932 and 1950.

In this respect, it is also worth mentioning that by Lee and Nashed [143] linear selections of linear relations in Hilbert and Banach spaces were already given by Arens [6] and Coddington–Dijksma [40]. However, the methods applied by the latter authors, Lee and Nashed [143] and Cross [41, p. 15] greatly differ from those of G. Száz and Á. Száz [243, 249, 261].

Remark B.14 In 1971, G. Száz also established the subsequent example which shows that, in contrast to the linear ones, an additive relation need not have an additive selection function.

However, this observation is exclusively attributed to Godini [90] in the extensive literature on set-valued functions and the stability of functional equations. (See, for instance, Baker [12, p. 321], Baron [18, p. 8], and Sablik [219, p. 182].)

Several algebraic properties of additive relations (or more generally, subobjects of product objects) were already established by Lorenzen [145] in 1954 and Lambek [140] in 1958. (See also MacLane [147, 148, pp. 51–63], and Whitehead [282, pp. 722–727].)

However, the study of additive and convex relations in the extensive theory of functional equations could only become a standard subject with the pioneering books by W. Smajdor [233], Nikodem [164], Hyers, Isac, and Th.M. Rassias [115], and Czerwik [42].

Example B.15 For each $n \in \mathbb{N}$, define $s_n = \sum_{k=0}^{n-1} k/n!$. Then, there exists an additive relation F of $\mathbb{R}$ to itself such that

$$F(m/n!) = ms_n + \mathbb{Z}$$

for all $n \in \mathbb{N}$ and $m \in \mathbb{Z}$. Moreover, F does not have an additive selection function.

To prove the latter statement, assume on the contrary that f is an additive selection function of F. Then, by using a slightly more delicate argument as in Example B.11, we can see that

$$f(1) \in F_{1/n!}(1) = n!F(1/n!), \quad \text{and thus} \quad f(1)/n! \in F(1/n!)$$

for all $n \in \mathbb{N}$. Hence, by defining

$$\alpha_n = \inf\bigl\{|y|: \ y \in F(1/n!)\bigr\},$$

we can see that $\alpha_n \le |f(1)|/n!$, and thus $n!\alpha_n \le |f(1)|$ for all $n \in \mathbb{N}$.

On the other hand, by using that $F(1/n!) = s_n + \mathbb{Z} = \{s_n + k: \ k \in \mathbb{Z}\}$, we can see that

$$\alpha_n = \inf\bigl\{|s_n + k|: \ k \in \mathbb{Z}\bigr\}$$

for all $n \in \mathbb{N}$. Moreover, by induction, we can easily see that $s_n < 1/2$ for all $n \in \mathbb{N}$ with $n > 3$. Therefore,

$$|s_n + k| = |k + s_n| \ge \bigl||k| - |s_n|\bigr| \ge |k| - s_n > 1/2$$

for all $n \in \mathbb{N}$ and $k \in \mathbb{Z}$ with $n > 3$ and $k \neq 0$. Hence, we can already see that $\alpha_n = s_n$, and thus $n!s_n \le |f(1)|$ for all $n \in \mathbb{N}$ with $n > 3$. And this is a contradiction since $n!s_n = \sum_{k=0}^{n-1} k!$ for all $n \in \mathbb{N}$.

Remark B.16 Later, the above results of G. Száz, which had not been appreciated by the referees of the competition, were presented in our joint paper [261], and his Ph.D. Thesis [Additív és lineáris relációk, University of Budapest, 1974]. Fortunately, in those happy old days of peace, we could publish papers in the Publ. Math. Debrecen.

Godini [90], Kuczma [139], Holá, Kupka, and Maličky [104]–[105], W. Smajdor [233], Nikodem and Popa [164]–[167], Castillo and Ruiz-Cobo [38], and Cross [41] cited our paper. However, for instance, A. Smajdor [228], Lee and Nashed [143], Adasch [3], Sablik [219] Páles [173], Abreu and Etcheberry [1], Czerwik [42], Hassi, Sebestyén, De Snoo,and Szafraniec [97], Sandovici, Snoo, and Winkler [223], and Álvarez [4] did not mention our paper.

Remark B.17 By proving a Hahn–Banach type extension theorem, Páles [173] could give a necessary and sufficient condition in order that a certain set-valued function F of one locally convex Hausdorff topological vector space X to another Y could have a continuous linear selection.

Sufficient conditions for the existence of additive selections for additive and super-additive set-valued functions have formerly been given by Rådström [192], Godini [90], Przeslawski [191], Nikodem [163], A. Smajdor [228], and Gajda [67, p. 53].

Moreover, by using the technique of generalized invariant means, Badora, Ger, and Páles [11] have proved a very general additive section theorem which includes the following theorem as a particular case.

Theorem B.18 *Assume that F is a set-valued function of a commutative semigroup X to a locally convex Hausdorff space Y such that the values of F are nonempty, closed, convex, and weakly compact. Moreover, suppose that there exists a function f of X to Y such that*

$$f(x+y) - f(y) \in F(x)$$

for all $x, y \in X$. Then, there exists an additive selection function g of F.

Remark B.19 To see the necessity of this curious condition on F, note that if g is as above, then $g(x+y) - g(y) = g(x) + g(y) - g(y) = g(x) \in F(x)$ for all $x, y \in X$.

C. The Hahn–Banach Extension Theorems By Saccoman [221] and Buskes [36], the origins of the Hahn–Banach theorems go back to the early papers of Riesz [211] in 1907 and Helly [99] in 1912.

Riesz solved moment problems inspired by the works of D. Hilbert and E. Schmidt on integrable functions. Helly simplified and generalized the results of Riesz by using sequence spaces. (See also Fuchssteiner and Horváth [62].)

Hahn [94] in 1927, having in mind integral equations and referring to some later works of Riesz and Helly, proved the following more abstract theorem, with a superfluous completeness assumption, in a surprisingly elegant presentation.

Theorem C.1 *Let V be a linear subspace of a real normed space X and φ be a real continuous linear function of V. Then, there exists a real continuous linear function f of X that extends φ and has the same norm as φ.*

Remark C.2 Note that now the norm $\|\varphi\| = \sup\{|\varphi(x)|: \ \|x\| \leq 1\}$ is finite by the assumed continuity of φ. (Hahn originally assumed boundedness instead of continuity and used "Steigung" D instead of "Norm" $\|\cdot\|$.)

To obtain the results of Riesz and Helly as corollaries, Banach [14] in 1929 rediscovered Hahn's theorem. Moreover, he proved the following more powerful theorem which was later included in his famous book [15], where Hahn's paper was already cited.

Theorem C.3 *Let X be a real vector space and p be a real sublinear function of X in the sense that p is subadditive and positively homogeneous. Assume that V is a linear subspace of X and φ is a real linear function of V that is dominated by p in the sense that $\varphi(v) \leq p(v)$ for all $v \in V$. Then, there exists a linear function f of X that extends φ and is still dominated by p.*

Remark C.4 Now, since f is odd, we actually have $-p(-x) \le f(x) \le p(x)$, and hence $|f(x)| \le \max\{p(x), p(-x)\}$ for all $x \in X$.

Therefore, Theorem C.1 can be immediately derived from Theorem C.3 by taking $p(x) = \|\varphi\| \|x\|$ for all $x \in X$. Note that this p is actually an equivalent norm on X whenever $\varphi \neq 0$.

Moreover, since p is 2-homogeneous, we also have $p(0) = 0$. Therefore, by taking $V = \{0\}$ and $\varphi = \{(0, 0)\}$, the following important corollary can also be immediately derived from Theorem C.3.

Corollary C.5 *If X and p are as in Theorem C.3, then there exists a real linear function f of X such that $f \le p$.*

Remark C.6 The theorems of Hahn and Banach have later been generalized by a great number authors in an enormous variety of directions.

The interested reader can get many interesting insights in the subjects from the excellent surveys by Fuchssteiner and Horváth [62], Buskes [36], and Narici and Beckenstein [159].

For instance, Banach and Mazur [16] in 1933 showed that the theorem of Hahn is no longer true for linear functions with values in Euclidean spaces. (See also Saccoman [222] and the references therein.)

Moreover, Murray [154] in 1936, Soukhomlinov [235] and Bohnenblust and Sobczyk [27] in 1938 showed that Hahn's theorem can be extended to complex-valued linear functions.

However, it is now more important to note that Nachbin [155] in 1950 proved the following far reaching generalization of Hahn's theorem.

Theorem C.7 *Let X and Y be real normed spaces such that the family of all closed balls in Y has the binary intersection property. Moreover, assume that φ is a continuous linear function of a subspace V of X to Y. Then, there exists a continuous linear extension f of φ to X that has the same norm as φ.*

Remark C.8 By Nachbin, a collection of sets is said to have the binary intersection property if every family of its mutually intersecting members has a nonempty intersection. The binary intersection property of the family of all closed balls in $\mathbb{R}$ was already used by Helley in 1912.

Nachbin originally also proved a certain converse to the above theorem. An analogous result in the non-Archimedean case was given by Ingleton [116] in 1952. Moreover, Holbrook [107] in 1975 extended the result of Nachbin to normed spaces over complex and quaternion scalars by using an appropriate generalization of the binary intersection property.

However, it is now more important to note that, by extending the results of Kaufman [129] and Kranz [137], Fuchssteiner [61] proved the following remarkable generalization of Corollary C.5.

Theorem C.9 *Let X be a commutative preordered semigroup, and moreover $\bar{\mathbb{R}} = \mathbb{R} \cup \{-\infty\}$. Assume that p is an increasing subadditive and q is an arbitrary superadditive function of X to $\bar{\mathbb{R}}$ such that $q \le p$. Then, there exists an increasing additive function f of X to $\bar{\mathbb{R}}$ such that $q \le f \le p$.*

Remark C.10 Note that thus the inequality relation in X may, in particular, be symmetric. Or more specifically, it can even be the equality relation in X. Therefore, Corollary C.5 can be immediately derived from Theorem C.9 by taking $q(x) = -\infty$ for all $x \in X$.

Remark C.11 If p and q are arbitrary functions of a commutative semigroup X to $\bar{\mathbb{R}}$, then by an important consequence of an abstract Rodé type separation theorem of Nikodem, Páles, and Wasowicz [170] the following assertions are equivalent:

(i) $q \le f \le p$ for some additive function f of X to $\bar{\mathbb{R}}$;
(ii) $\sum_{i=1}^{n} q(x_i) \le \sum_{j=1}^{m} p(y_j)$ for any finite sequences $(x_i)_{i=1}^{n}$ and $(y_j)_{j=1}^{m}$ in X with $\sum_{i=1}^{n} x_i = \sum_{j=1}^{m} y_j$.

Note that in particular if p is subadditive, q is superadditive and $q \le p$, then condition (ii) automatically holds. Therefore, the corresponding particular case of Theorem C.9 is equivalent to the above result.

In view of this fact, it would be of some interest to prove a preordered generalization of the above result of Nikodem, Páles, and Wasowicz. Invariant generalizations of the results of Fuchssteiner have already been given by Boccuto and Candeloro [26] and Gajda [68].

By using the sandwich Theorem C.9, Fuchssteiner [61] and the present author [259] have proved particular cases of the following generalization of Theorem C.3 which is likely to be also true.

Theorem C.12 *Let X be a commutative preordered semigroup and p be a subadditive function of X to $\bar{\mathbb{R}}$. Moreover, assume that V is a subsemigroup of X and φ is a function of V to $\bar{\mathbb{R}}$. Then, the following assertions are equivalent:*

(i) *φ can be extended to an increasing additive function f of X to $\bar{\mathbb{R}}$ such that $f \le p$;*
(ii) *φ is additive and $\varphi(v) \le p(x) + \varphi(w)$ for all $x \in X$ and $v, w \in V$ with $v \le x + w$.*

Remark C.13 To prove the implication (i) $\Longrightarrow$ (ii), we can at once note that if f is as in (i), then φ is also increasing, additive, and $\varphi \le p$ on V.

Moreover, if x, v and w are as in (ii), then we necessarily have

$$\varphi(v) = f(v) \le f(x + w) = f(x) + f(w) \le p(x) + \varphi(w).$$

Remark C.14 Hence, we can immediately infer that

$$\varphi(v) \le \inf\{p(s) + \varphi(t) : \ s \in X, \ t \in V, \ v \le s + t\}$$

for all $v \in V$. This observation was also used by Fuchssteiner [61] and Gajda [68].

However, they did not observed that, by using a straightforward generalization of the infimal convolution of Moreau [151] and Strömberg [237], the second condition of (ii) can be briefly expressed by writing that $\varphi \leq p * \varphi$ on V.

Unfortunately, by writing our papers [87, 88, 259], we did not observe that a particular case of the infimal convolution was already used by Rodrígues-Salinas [215, Definición 5], despite that we already had a letter and a paper of J. Horváth [108] on the outstanding results of Professor B. Rodrígues-Salinas.

Remark C.15 The function q defined by

$$q(x) = (p * \varphi)(x) = \inf\{p(u) + \varphi(v) : \ u \in X, \ v \in V, \ x \leq u + v\},$$

for all $x \in X$, can easily seen to be increasing even if p is not assumed to be subadditive.

Moreover, if in particular X has a zero element and $p(0) \leq 0$, then by the definition of q we can at once see that

$$q(v) \leq p(0) + \varphi(v) \leq 0 + \varphi(v) = \varphi(v)$$

for all $v \in V$, and thus $q \leq \varphi$ on V. Therefore, if $\varphi \leq p * \varphi$ also holds on V, then q is actually an extension of φ, and thus in particular φ is also increasing.

Remark C.16 On the other hand, if the zero element 0 of X is contained in V and φ is additive, then because of $\varphi(0) = \varphi(0) + \varphi(0)$, we have either $\varphi(0) = 0$ or $\varphi(0) = -\infty$, and hence $\varphi(0) \leq 0$. Therefore, by the definition of q, we also have

$$q(x) \leq p(x) + \varphi(0) \leq p(x) + 0 = p(x)$$

for all $x \in X$, and thus $q \leq p$.

Moreover, by [259, Theorem 2.5], we can also state that q is subadditive. Therefore, q is, in general, a better control function for φ than p.

Finally we note that, having in mind the corresponding particular case of Theorem C.12, Glavosits and Száz [89] have proved the following generalization of Hyers's theorem.

Theorem C.17 *If f is an ε-approximately additive function of a commutative semigroup X to a Banach space Y, for some $\varepsilon \geq 0$, and φ is a 2-homogeneous function of a subsemigroup V of X to Y which is δ-near to f, for some $\delta \geq 0$, then φ can be extended to an additive function ψ of X to Y which is ε-near to f.*

Remark C.18 To see that the above theorem is more general than that of Hyers, note that if in particular X has a zero element 0, then

$$\|f(0)\| = \|f(0+0) - f(0) - f(0)\| \leq \varepsilon.$$

Thus, $\varphi = \{(0,0)\}$ is an additive function of the subgroup $\{0\}$ of X to Y such that φ is ε-near to f. Therefore, by the Theorem C.17 there exists an additive function ψ of X to Y which is ε-near to f.

Note that, in view of Theorem C.12, it would also be of some interest to prove a preordered generalization of Theorem C.17.

D. Set-Valued Generalizations of the Hahn–Banach Theorems The theorems of Banach, Nachbin, and Ingleton were transformed into a common set-valued setting by Rodríguez-Salinas and Bou [217] in 1974 by making use of the following observations.

Remark D.1 If p and φ are as in Theorem C.3 and $q(x) = -p(-x)$ for all $x \in X$, then the set-valued function F, defined by

$$F(x) = \big[q(x), p(x)\big]$$

for all $x \in X$, is $\mathbb{R}$-homogeneous and subadditive. Moreover, φ is a partial selection function of F.

To see the required homogeneity property of F, note that if $x \in X$ and $\lambda \in \mathbb{R}$ is such that $\lambda > 0$, then because of $p(\lambda x) = \lambda p(x)$ we also have

$$q(\lambda x) = -p\big(-(\lambda x)\big) = -p\big(\lambda(-x)\big) = -\big(\lambda p(-x)\big) = \lambda\big(-p(-x)\big) = \lambda q(x).$$

Therefore,

$$F(\lambda x) = \big[q(\lambda x), p(\lambda x)\big] = \big[\lambda q(x), \lambda p(x)\big] = \lambda\big[q(x), p(x)\big] = \lambda F(x).$$

Moreover, we also have

$$F(-x) = \big[q(-x), p(-x)\big] = \big[-p(x), p(-x)\big] = -\big[-p(-x), p(x)\big] = -F(x).$$

Therefore,

$$F\big((-\lambda)x\big) = F\big(\lambda(-x)\big) = \lambda F(-x) = \lambda\big(-F(x)\big) = (-\lambda)F(x).$$

Now, to complete the proof of the $\mathbb{R}$-homogeneity of F, it remains only to note that $F(0) = [q(0), p(0)] = [0,0] = \{0\}$, and thus $F(0x) = F(0) = \{0\} = 0F(x)$ also holds.

Remark D.2 If φ is as in Theorem C.7 and $p(x) = \|\varphi\|\,\|x\|$ for all $x \in X$ and $q(y) = \|y\|$ for all $y \in Y$, then the set-valued function F, defined by

$$F(x) = \big\{y \in Y :\ q(y) \leq p(x)\big\}$$

for all $x \in X$, is $\mathbb{R}$-homogeneous and subadditive. Moreover, φ is a partial selection function of F.

To prove the subadditivity of F, note that if $x_1, x_2 \in X$ and $y \in F(x_1 + x_2)$, then

$$q(y) \leq p(x_1 + x_2) \leq p(x_1) + p(x_2).$$

Therefore, if $p(x_1) + p(x_2) \neq 0$, then by defining

$$y_1 = p(x_1)\big(p(x_1) + p(x_2)\big)^{-1} y \quad \text{and} \quad y_2 = p(x_2)\big(p(x_1) + p(x_2)\big)^{-1} y,$$

we have not only $y = y_1 + y_2$, but also

$$q(y_i) = p(x_i) q(y)\big(p(x_1) + p(x_2)\big)^{-1} \leq p(x_i),$$

and hence $y_i \in F(x_i)$ for $i = 1, 2$. Therefore, $y = y_1 + y_2 \in F(x_1) + F(x_2)$.

While, if $p(x_1) + p(x_2) = 0$, then $q(y) = 0$. Therefore, by defining $y_1 = 0$ and $y_2 = y$, we have not only $y = y_1 + y_2$, but also $q(y_i) = 0 \leq p(x_i)$, and hence $y_i \in F(x_i)$ for $i = 1, 2$. Thus, $y = y_1 + y_2 \in F(x_1) + F(x_2)$ is again true. Consequently, $F(x_1 + x_2) \subset F(x_1) + F(x_2)$, and thus F is subadditive.

Remark D.3 The above remarks show that the essence of the theorems of Banach and Nachbin is nothing but the statement that a certain linear partial selection function φ of a certain set-valued function F can be extended to a total linear selection function f of F.

Having in mind this fact, Rodríguez-Salinas and Bou [217] proved the following common generalization of the most basic Hahn–Banach type linear extension theorems.

Theorem D.4 *Let X and Y be vector spaces over the same field K, and suppose that $\mathscr{A}$ is a nonempty, translation-invariant family of nonempty subsets of Y having the binary intersection property. Moreover, assume that F is a K-homogeneous subadditive function of X to $\mathscr{A}$ and φ is a linear partial selection function of F. Then φ can be extended to a total linear selection function f of F.*

Remark D.5 Normed spaces with Nachbin's extension property have been investigated by Kelley [131], Hasumi [98], Hustad [110], and Holbrook [107]. However, the significance of the above theorem has only been acknowledged by Horváth [108], Fuchssteiner and Horváth [62], and Fuchssteiner and Lusky [63, p. 75].

Buskes [36, p. 27] appreciates only the work of Ioffe [117] as a natural continuation of the investigations of Nachbin [155] and Ingleton [116]. He only mentions the paper [217] of Rodríguez-Salinas and Bou with reference to the works of Ioffe [117] and Fuchssteniner and Lusky [63, p. 75]. While, Narici and Beckenstein do not even include it in the references of [159].

Remark D.6 In 1981, Ioffe [117] proved a certain converse to the theorem of Rodríguez-Salinas and Bou by establishing the equivalence of the binary intersection property of a certain family $\mathscr{A}$ of sets and the linear extension property of certain $\mathscr{A}$-valued functions, called fans. These are subadditive counterparts of the superadditive convex processes of Rockafellar [214].

Moreover, in 1992 Gajda, A. Smajdor, and W. Smajdor [71], being unaware of the result of Rodríguez-Salinas and Bou, proved the following

Theorem D.7 *Let X be a commutative group and Y be a vector space over $\mathbb{Q}$. And assume that $\mathscr{A}$ is a nonempty family of nonempty subsets of Y, having the binary intersection property, which is invariant under translations by the elements of Y and multiplications by multiplicative inverses of the members of $\mathbb{Z}^* = \mathbb{Z} \setminus \{0\}$. Moreover, suppose that F is a $\mathbb{Z}^*$-subhomogeneous, subadditive function of X to $\mathscr{A}$. Then, every additive partial selection function φ of F can be extended to a total additive selection function f of F.*

Remark D.8 By using this Hahn–Banach type extension theorem, the above authors could prove Hahn–Banach type generalizations of several Hyers–Ulam type stability theorems.

In this respect, it is also worth mentioning that Páles [176] in 1998, Badora [10] in 2006, and Huang and Li [109] in 2009 also proved some general Hyers–Ulam type stability theorems with the help of Hahn–Banach type theorems. However, none of these authors mentions the paper [71] of Gajda, A. Smajdor, and W. Smajdor.

In 1998, by introducing the intersection convolution of relations, Theorem D.4 was also generalized by the present author [243] in the following more convenient relational form.

Theorem D.9 *Let X and Y be vector spaces over the same field K. Assume that F is a K^*-subhomogeneous relation of X to Y. Then, the following assertions are equivalent:*

(i) *$F * \varphi$ is a total relation on X to Y for any linear partial selection function φ of F;*
(ii) *Every linear partial selection relation Φ of F can be extended to a total linear selection relation Ψ of $F + \Phi(0)$.*

Remark D.10 Theorem D.4 can be easily derived from this theorem, by noticing that if F is as in Theorem D.4 and φ is as in (i), with domain D_φ, then for any $x \in X$ and $v, t \in D_\varphi$ we have

$$\begin{aligned}
0 \in F(t-v) - \varphi(t-v)&\\
= F\big((x-v)-(x-t)\big) - \varphi(t-v)&\\
\subset F(x-v) + F\big(-(x-t)\big) - \big(\varphi(t)+\varphi(-v)\big)&\\
= F(x-v) - F(x-t) - \big(\varphi(t)-\varphi(v)\big)&\\
= F(x-v) + \varphi(v) - \big(F(x-t)+\varphi(t)\big),&
\end{aligned}$$

and hence $(F(x-v)+\varphi(v)) \cap (F(x-t)+\varphi(t)) \neq \emptyset$. Therefore,

$$(F * \varphi)(x) = \bigcap_{v \in D_\varphi} \big(F(x-v)+\varphi(v)\big) \neq \emptyset,$$

and thus Theorem D.9 can be applied to get the required assertion.

Remark D.11 Moreover, Theorem D.9 can also be used to easily prove that if F is a $K \setminus \{0\}$-superhomogeneous, superadditive relation of one vector space X to another Y over K, then every linear partial selection relation Φ of F can be extended to a total linear selection relation Ψ of $F + \Phi(0)$.

Namely, if φ is as in (i), then for any $x \in X$ and $v \in D_\varphi$ we have

$$F(x-v)+\varphi(v) \subset F(x-v)+F(v) \subset F(x)$$

and

$$\begin{aligned} F(x) &= F(x) - \varphi(v) + \varphi(v) \\ &= F(x) + \varphi(-v) + \varphi(v) \\ &\subset F(x) + F(-v) + \varphi(v) \subset F(x-v) + \varphi(v), \end{aligned}$$

and thus $F(x-v)+\varphi(v) = F(x)$. Therefore,

$$(F * \varphi)(x) = \bigcap_{v \in D_\varphi} \big(F(x-v)+\varphi(v)\big) = \bigcap_{v \in D_\varphi} F(x) = F(x) \neq \emptyset,$$

and thus Theorem D.9 can be applied to get the required assertion.

Remark D.12 The theorem of Gajda, A. Smajdor, and W. Smajdor [71] was carried over to concave set-valued functions by W. Smajdor and Szczawińska [234] in 1995.

Later, it was also used by Badora [8] in 1993 and Czerwik [42, pp. 333–338] in 2002. However, neither of the above authors mentions the paper [217] of Rodríguez-Salinas and Bou. Moreover, Czerwik [42] does not cite the corresponding papers of the present author, too.

Finally, we note that, by using the convolutional method, Glavosits and Száz [86] have proved the following relational generalization of Theorem D.7.

Theorem D.13 *Let X be a commutative group and Y be a vector space over $\mathbb{Q}$. Assume that $\mathscr{A}$ is a family subsets of Y, having the binary intersection property, which is invariant under translations by the elements of Y and multiplications by the multiplicative inverses of the members of $\mathbb{N}$. Moreover, suppose that F is an odd, $\mathbb{N}$-subhomogeneous, subadditive relation of X to Y such that $F(x) \in \mathscr{A}$ for all $x \in X$. Then, each odd, $\mathbb{N}$-semi-subhomogeneous, superadditive partial selection relation Φ of F can be extended to a total, $\mathbb{Z}^*$-homogeneous, additive selection relation Ψ of $F + \Phi(0)$.*

Remark D.14 In view of Theorem C.12, it would be of some interest to prove a generalization of the latter theorem for commutative preordered semigroups.

39.2 A Few Basic Facts on Relations

A subset F of a product set $X \times Y$ is called a relation on X to Y. If in particular $F \subset X^2$, then we may simply say that F is a relation on X. In particular, $\Delta_X = \{(x,x): \ x \in X\}$ is called the identity relation on X.

If F is a relation on X to Y, then for any $x \in X$ and $A \subset X$ the sets $F(x) = \{y \in Y: \ (x,y) \in F\}$ and $F[A] = \bigcup_{a \in A} F(a)$ are called the images of x and A under F, respectively.

Instead of $y \in F(x)$ sometimes we shall also write xFy. Moreover, the sets $D_F = \{x \in X: \ F(x) \neq \emptyset\}$ and $R_F = F[X] = F[D_F]$ will be called the domain and range of F, respectively.

If in particular $D_F = X$, then we say that F is a relation of X to Y, or that F is a total relation on X to Y. While, if $R_F = Y$, then we say that F is a relation on X onto Y.

If F is a relation on X to Y, then $F = \bigcup_{x \in X} \{x\} \times F(x) = \bigcup_{x \in D_F} \{x\} \times F(x)$. Therefore, a relation F on X to Y can be naturally defined by specifying $F(x)$ for all $x \in X$, or by specifying D_F and $F(x)$ for all $x \in D_F$.

For instance, if F is a relation on X to Y, then the inverse relation F^{-1} of F can be naturally defined such that $F^{-1}(y) = \{x \in X: \ y \in F(x)\}$ for all $y \in Y$. Thus, we also have $F^{-1} = \{(y,x): \ (x,y) \in F\}$.

Moreover, if in addition G is a relation on Y to Z, then the composition relation $G \circ F$ of G and F can be naturally defined such that $(G \circ F)(x) = G[F(x)]$ for all $x \in X$. Thus, we also have $(G \circ F)[A] = G[F[A]]$ for all $A \subset X$.

Now, a relation F on X may be called reflexive, symmetric and transitive if $\Delta_{D_F} \subset F$, $F^{-1} = F$ and $F \circ F \subset F$, respectively. Moreover, for instance, F may be called anti-symmetric if $F \cap F^{-1} \subset \Delta_X$.

Note that if either $F^{-1} \subset F$ or $F \subset F^{-1}$, then F is already symmetric. Moreover, if F is reflexive and transitive, then F is idempotent in the sense that $F \circ F = F$. While, if F is reflexive and anti-symmetric, then $F \cap F^{-1} = \Delta_{D_F}$.

As usual, a transitive (symmetric) reflexive relation is called a preorder (tolerance) relation. Moreover, a symmetric (anti-symmetric) preorder relation is called an equivalence (partial order) relation.

In particular, a relation f on X to Y is called a function if for each $x \in D_f$ there exists $y \in Y$ such that $f(x) = \{y\}$. In this case, by identifying singletons with their elements, we may simply write $f(x) = y$ in place of $f(x) = \{y\}$.

If F is a relation on X to Y and $A, B \subset X$, then in general we only have $F[A] \setminus F[B] \subset F[A \setminus B]$. Moreover, if $A_i \subset X$ for all $i \in I$, then in general we only have $F[\bigcap_{i \in I} A_i] \subset \bigcap_{i \in I} F[A_i]$ and $F[\bigcup_{i \in I} A_i] = \bigcup_{i \in I} F[A_i]$.

However, if in particular $F = f^{-1}$ for some function f on Y to X, then differences and intersections are also preserved under F. Moreover, if in addition

$D_f = Y$, then we also have $F[A^c] = F[A]^c$ where c means complementation with respect to X and Y, respectively.

In this respect, it is also worth mentioning that if F and G are relations on X to Y and $A \subset X$, then in general we only have $F[A] \setminus G[A] \subset (F \setminus G)[A]$. Moreover, if F_i is a relation on X to Y for all $i \in I$, then in general we only have $(\bigcap_{i \in I} F_i)[A] \subset \bigcap_{i \in I} F_i[A]$ and $(\bigcup_{i \in I} F_i)[A] = \bigcup_{i \in I} F_i[A]$.

However, if in particular $A = \{x\}$, for some $x \in X$, then the corresponding equalities are also true. Moreover, for any $x \in X$, we also have $F^c(x) = F(x)^c$, where c means complementation with respect to $X \times Y$ and Y, respectively.

Moreover, we note that if F is a relation on X to Y, then a subset Φ of F is called a partial selection relation of F. Thus, we also have $D_\Phi \subset D_F$. Therefore, a partial selection relation Φ of F may be called total if $D_\Phi = D_F$.

In the sequel, the total selection relations of a relation F will be usually be simply called the selection relations of F. Thus, the Axiom of Choice can be briefly expressed by saying that every relation F has a selection function.

If F is a relation on X to Y and $U \subset D_F$, then the relation $F|U = F \cap (U \times Y)$ is called the restriction of F to U. Moreover, if F and G are relations on X to Y such that $D_F \subset D_G$ and $F = G|D_F$, then G is called an extension of F.

39.3 A Few Basic Facts on Groupoids and Vector Spaces

A function $\star$ of a set X to itself is called an unary operation in X. Moreover, a function $*$ of X^2 to X is called a binary operation in X. In these cases, for any $x, y \in X$, we usually write $x^\star$ and $x * y$ in place of $\star(x)$ and $*((x, y))$, respectively.

An ordered pair $X(+) = (X, +)$, consisting of a set X and a binary operation $+$ in X is called a groupoid. Instead of groupoids, it is usually sufficient to consider only semigroups (associative grupoids) or even monoids (semigroups with zero).

However, several definitions on semigroups can be naturally extended to groupoids. For instance, if X is a groupoid, then for any $x \in X$ and $n \in \mathbb{N}$, with $n \neq 1$, we may naturally define $nx = (n-1)x + x$ with the convention that $1x = x$.

Thus, if in particular X is a semigroup, then for any $x, y \in X$ and $n, m \in \mathbb{N}$, we have $(n+m)x = nx + mx$, $(nm)x = n(mx)$ and $n(x+y) = nx + ny$, whenever x and y commute in the sense that $x + y = y + x$.

If in particular X is a groupoid with zero, then for any $x \in X$ we may also naturally define $0x = 0$. Moreover, if more specially X is a group, then for any $x \in X$ and $n \in \mathbb{N}$ we may also naturally define $(-n)x = -(nx)$. Thus, we also have $(-n)x = n(-x)$. Moreover, the counterparts of the above rules remain true. Thus, a commutative group X is already a module over $\mathbb{Z}$.

If X is a groupoid, then for any $n \in \mathbb{N}$ and $A, B \subset X$ we may also naturally define $nA = \{na\colon\ a \in A\}$ and $A + B = \{a + b\colon\ a \in A, b \in B\}$. Thus, for instance, $2A$ can be easily confused with the possibly strictly larger set $A + A$, which may also be naturally denoted by $2A$.

Moreover, if in particular X is a group, then for any $k \in \mathbb{Z}$ and $A \subset X$ we may also define $kA = \{ka\colon\ a \in A\}$. And, for any $A, B \subset X$, we may also write $-A =$

$(-1)A$ and $A - B = A + (-B)$ despite that the family $\mathscr{P}(X)$ is, in general, only a monoid.

If more specially X is a vector space over K, then for any $\lambda \in K$ and $A \subset X$ we may also define $\lambda A = \{\lambda a : a \in A\}$. Note that thus only two axioms of a vector space may fail to hold for $\mathscr{P}(X)$. Namely, only the one point subsets of X can have additive inverses. Moreover, in general we only have $(\lambda + \mu)A \subset \lambda A + \mu A$.

If X is a vector space over K, then for any $A \subset X$ and $\lambda \in K \setminus \{0\}$ we have $\lambda^{-1}A = \{x \in X : \lambda x \in A\}$. Therefore, if, for instance, X is only a group, then for any $A \subset X$ and $k \in \mathbb{Z} \setminus \{0\}$ we may naturally define $k^{-1}A = \{x \in X : ka \in A\}$.

However, in this more general case, several useful rules of computations with sets in vector spaces are no longer true. For instance, in general we only have $k(k^{-1}A) \subset A \subset k^{-1}(kA)$.

A subset A of a groupoid X is called left-translation-invariant if $x + A = A$ for all $x \in X$. Note that if in particular X is a group and either $x + A \subset A$ for all $x \in X$ or $A \subset x + A$ for all $x \in X$, then A is already left-translation-invariant.

Moreover, a subset of a groupoid X is called normal if $x + A = A + x$ for all $x \in X$. Note that if in particular X is a group and either $x + A \subset A + x$ for all $x \in X$ or $A + x \subset x + A$ for all $x \in X$, then A is already normal.

On the other hand, a subset A of groupoid X is called a subgroupoid if $A + A \subset A$. Moreover, if in particular X is a group, then A is called symmetric if $-A = A$. Note that if either $-A \subset A$ or $A \subset -A$, then A is already symmetric.

Furthermore, a subset A of a vector space X over $\mathbb{K} = \mathbb{Q}$, $\mathbb{R}$ or $\mathbb{C}$ is called λ-convex, for some $\lambda \in [0,1] \cap \mathbb{K}$, if $\lambda A + (1 - \lambda)A \subset A$. Moreover, A is called Λ-convex, for some $\Lambda \subset [0,1] \cap \mathbb{K}$, if it is λ-convex for all $\lambda \in \Lambda$.

If F and G are relations on a set X to groupoid Y and $(F+G)(x) = F(x)+G(x)$ for all $x \in X$, then the relation $F + G$ is called the pointwise sum of F and G. If in particular X is also a gruopoid, then this can be easily confused with the global sum $\{(x + z, y + w) : (x, y) \in F, (z, w) \in G\}$ of F and G which may also be naturally denoted by $F + G$.

If F is a relation on a set X to a group Y and $(-F)(x) = -F(x)$ for all $x \in X$, then the relation $-F$ is called the pointwise negative of F. If in particular X is also a group, then this can be easily confused with the global negative $\{(-x, -y) : (x, y) \in F\}$ of F which may also be naturally denoted by $-F$.

Quite similarly, if, for instance, F is a relation on a set X to a groupoid Y and $(nF)(x) = nF(x)$ for all $x \in X$ and some $n \in \mathbb{N}$, then the relation nF is called the pointwise multiple of F by n. If in particular X is also a groupoid, then this can be easily confused with the global multiple $\{(nx, ny) : (x, y) \in F\}$ of F by n which may also be naturally denoted by nF.

The global and pointwise algebraic operations on relations have been mainly studied in [82, 83, 266]. In particular, it is noteworthy that if F and G are relations on one groupoid X to another Y, and $F + G$ is the global sum of sum of F and G, then $(F + G)(x) = \bigcup\{F(u) + G(u) : u, v \in X, x = u + v\}$ for all $x \in X$. Therefore, in contrast to the intersection convolution of relations [243, 258], the union convolution need not be introduced.

39.4 Constant-Like and Translation Relations

Definition 39.1 A relation F on a groupoid X to a set Y is called constant-like if $F(y) \subset F(x+y)$ for all $x, y \in X$.

Remark 39.1 By using the notation uFv instead $v \in F(u)$, the above inclusion can be expressed by writing that yFz implies $(x+y)Fz$ for all $x \in X$. Thus, the inequality $\leq$ is a constant-like relation on $[-\infty, 0]$.

Moreover, it is also worth noticing that if in particular Y is a groupoid with zero, then by using the global sum of relations, the above inclusion can also be expressed by writing that $(x, 0) + F \subset F$ for all $x \in X$.

Constant-like relations were first considered in [83] under the name of pointwise translation relations. Their introduction is mainly motivated by the next obvious theorem and the forthcoming Definition 39.2 and Remark 39.13.

Theorem 39.1 *If F is a relation on a group X to a set Y, then the following assertions are equivalent*:

(i) *F is constant-like*;
(ii) *$F(x) = F(0)$ for all $x \in X$*;
(iii) *$F(x+y) \subset F(y)$ for all $x, y \in X$.*

From the above theorem, it is clear that in particular we also have

Corollary 39.1 *If F is a constant-like relation on a group X to a set Y, then either $D_F = \emptyset$ or $D_F = X$.*

The following example also shows that the implication (i) $\Longrightarrow$ (ii) need not be true if X is only a monoid.

Example 39.1 If in particular $X = [0, +\infty]$ and $F(x) = [0, x]$ for all $x \in X$, then F is a constant-like relation on X despite that it is strictly increasing in the sense that $F(x)$ is a proper subset of $F(y)$ for all $x, y \in X$ with $x < y$.

In this respect, it is also worth mentioning that the inverse of a constant-like relation is not, in general, constant-like.

Theorem 39.2 *If F is a relation on a set X to a groupoid Y, then the following assertions are equivalent*:

(i) *F^{-1} is a constant-like*;
(ii) *$y + F(x) \subset F(x)$ for all $x \in X$ and $y \in Y$.*

From this theorem, it is clear that in particular we also have

Corollary 39.2 *If F is a relation on a set X to a group Y, then F^{-1} is constant-like if and only if the values of F are left-translation-invariant.*

More specially, by Theorems 39.1 and 39.2, we can also state

Corollary 39.3 *If F is a relation on one group X to another Y, then both F and F^{-1} are constant-like if and only if either $F = \emptyset$ or $F = X \times Y$.*

Moreover, as some immediate consequence of the corresponding definitions, we can also easily establish the following theorems.

Theorem 39.3 *If F is a constant-like relation on a groupoid X to a set Y and G is an arbitrary relation on Y to a set Z, then $G \circ F$ is also a constant-like relation.*

Theorem 39.4 *The family of all constant-like relations on a groupoid X to a set Y is closed under unions and intersections.*

Remark 39.2 If in particular F is a constant-like relation on a group X to a set Y, then $F^c = X^2 \setminus F$ is also a constant-like relation on X to Y.

Now, concerning the pointwise and global algebraic operations on constant-like relations, for instance, we can easily establish the following two theorems.

Theorem 39.5 *If F and G are constant-like relations on one groupoid X to another Y, then their pontwise sum $F + G$ is also a constant-like relation.*

Remark 39.3 If F is as in the above theorem, then we can also state that the pointwise multiple nF, with $n \in \mathbb{N}$, is also a constant-like relation.

Theorem 39.6 *If F is a constant-like and G is an arbitrary relation on one semigroup X to a groupoid Y, then their global sum $F + G$ is also a constant-like relation.*

Remark 39.4 Note that if in particular F and G are constant-like relations on a group X to a groupoid Y, then $(F + G)(x) = F(0) + G(0)$ for all $x \in X$. Therefore, the global and the pointwise sums of F and G coincide.

Definition 39.2 A relation F on a groupoid X is called a translation relation if $x + F(y) \subset F(x + y)$ for all $x, y \in X$.

Remark 39.5 Note that the above inclusion can be expressed by writing that yFz implies $(x + y)F(x + z)$ for all $x \in X$. Thus, in particular the inequality $\leq$ on $\mathbb{R}$ is a translation relation.

Moreover, it is also worth noticing that, by using the global sum of relations, the above inclusion can also be expressed by writing that $\Delta_X + F \subset F$. Thus, if in particular X has a zero element, then the corresponding equality is also true.

Translation functions, under the names "centralizers" and "multipliers" too, have been used by several authors to construct extensions of semigroups, rings and modules. (See, for instance, Larsen [141] and Száz [238, 247], and the references therein.)

While, translation relations were first studied by the present author in [245] to consider compatible relators on groupoids and vector spaces [257]. Their introduction can also be motivated by the following simple theorem and the forthcoming Remark 39.13.

Theorem 39.7 *If F is a relation on a group X, then the following assertions are equivalent*:

(i) *F is a translation relation*;
(ii) *$F(x)=x+F(0)$ for all $x \in X$*;
(iii) *$F(x+y) \subset x+F(y)$ for all $x, y \in X$.*

Remark 39.6 Assertion (iii) can be expressed by writing that for any $x, y, z \in X$, with $(x+y)Fz$, there exists $w \in X$ with yFw such that $x+w=z$.

In this respect, it is also worth noticing that, by using the pointwise sum of relations, assertion (ii) can be expressed by writing that $F=\Delta_X+X \times F(0)$.

From the above theorem, it is clear that in particular we also have

Corollary 39.4 *If F is a translation relation on a group X, then either $D_F=\emptyset$ or $D_F=X$.*

By Theorem 39.7, we may naturally introduce the following

Definition 39.3 A translation relation F on groupoid with zero is called normal if $F(0)$ is a normal subset of X (i.e., $x+F(0)=F(0)+x$ for all $x \in X$).

Concerning translation relations, in [245, 248] we have also proved the following theorems.

Theorem 39.8 *If F is a translation relation on a groupoid X, then F^{-1} is also a translation relation.*

Theorem 39.9 *If F is a normal translation relation on a group X, then $F^{-1}(x)=-F(-x)$ for all $x \in X$.*

Remark 39.7 The equality $F^{-1}(0)=-F(0)$ is true even if F is not normal.

Theorem 39.10 *If F and G are translation relations on a groupoid X, then $G \circ F$ is also a translation relation.*

Theorem 39.11 *If F is a normal and G is an arbitrary translation relation on a group X, then $(G \circ F)(x + y) = F(x) + G(y)$ for all $x, y \in X$.*

Remark 39.8 The equality $(G \circ F)(0) = F(0) + G(0)$ is true even if F is an arbitrary relation on X.

Corollary 39.5 *If F and G are as in Theorem* 39.11, *then $F + G = G \circ F = F \circ G$, where $F + G$ means now the global sum of F and G.*

Theorem 39.12 *The family of all translation relations on a groupoid X is closed under unions and intersections.*

Remark 39.9 If in particular F is a translation relation on a group, then $F^c = X^2 \setminus F$ is also a translation relation.

Theorem 39.13 *If F is a translation relation and G is an arbitrary relation on a groupoid X, then the global sum $F + G$ is also a translation relation.*

Remark 39.10 In contrast to the above theorems, the pointwise operations on relations lead out from the family of all translation relations on a group or a vector space X.

39.5 Further Additivity Properties of Relations

A particular case of the following properties was already used by Fechner [51] in 2007 to formulate a general definition for the Hyers–Ulam stability of conditional Cauchy equations.

Definition 39.4 Let F be a relation on one groupoid X to another Y and let Ω be a relation on X. Then, F is called

(i) Ω-subadditive if $F(x + y) \subset F(x) + F(y)$ for all $(x, y) \in \Omega$;
(ii) Ω-superadditive if $F(x) + F(y) \subset F(x + y)$ for all $(x, y) \in \Omega$.

Remark 39.11 Now, in particular the relation F may be naturally called subadditive (superadditive) if it is X^2-subadditive (X^2-superadditive). Moreover, F may be naturally called additive if it both subadditive and superadditive.

Remark 39.12 Note that thus F is superadditive if and only if xFz and yFw imply that $(x + y)F(z + w)$. Thus, the inequality $\leq$ on $\mathbb{R}$ is superadditive.

Moreover, by using the global sum of relations, we can also at once see that F is superadditive if and only if $F + F \subset F$. That is, F is a subgroupoid of $X \times Y$.

Remark 39.13 In this respect, it is also worth noticing that if in particular F is a superadditive relation on a groupoid X to a groupoid Y with zero such that $F^{-1}(0) = X$, then F is already constant-like.

While, if in particular F is of total and reflexive superadditive relation on a groupoid X, then F is already a translation relation. Moreover, from Theorem 39.11 we can see that a normal translation relation F on a group X is superadditive if and only if it is transitive.

Note that if F is a D_F^2-superadditive relation on one groupoid X to another Y, then F is already superadditive. However, the corresponding assertion is not true for subadditive relations. Therefore, we shall also need the following

Definition 39.5 A relation F on one groupoid X to another Y is called

(i) semi-subadditive if it is D_F^2-subadditive;
(ii) left-quasi-subadditive if it is $D_F \times X$-subadditive;
(iii) right-quasi-subadditive if it is $X \times D_F$-subadditive.

Remark 39.14 Now, the relation F may be naturally called quasi-subadditive if it both left-quasi-subadditive and right-quasi-subadditive.

Moreover, F may be naturally called quasi-additive if it is both quasi-subaditive and superadditive. Later, we shall see that quasi-additivity is also a quite important additivity property.

In the sequel, by using some more special ground sets, we shall also need some further reasonable weakenings of global sub- and super-additivity.

Definition 39.6 A relation F on a groupoid X with zero to an arbitrary groupoid Y is called

(i) left-zero-subadditive if it is $\{0\} \times X$-subadditive;
(ii) left-zero-superadditive if it is $\{0\} \times X$-superadditive.

Remark 39.15 The right-zero-subadditive and right-zero-superadditive relations are defined analogously by using the relation $X \times \{0\}$.

Now, the relation F may, for instance, be naturally called zero-subadditive if it is both left-zero-subadditive and right-zero-subadditive.

By the corresponding definitions, we evidently have the following

Theorem 39.14 *A relation F on one groupoid X with zero to another Y, then*

(i) *F is zero-subadditive if $0 \in F(0)$;*
(ii) *F is zero-superadditive if $F(0) \subset \{0\}$.*

Hence, it is clear that in particular we also have

Corollary 39.6 *If F is a relation on one groupoid X with zero to another Y, such that either F is superadditive and $0 \in F(0)$ or F is subadditive and $F(0) \subset \{0\}$, then F is zero-additive.*

Definition 39.7 A relation F on a group X to a groupoid Y is called inversion-subadditive (resp., inversion-superadditive) it is Ω-subadditive (resp., Ω-superadditive) with $\Omega = \{(x, -x) : x \in X\}$.

Remark 39.16 Now, the relation F may also be naturally called inversion-semi-subadditive if it is $\Omega | D_F$-subadditive with the above Ω.

Moreover, F may be naturally called inversion-additive (inversion-semi-additive) if is both inversion-subadditive (inversion-semi-subadditive) and inversion-superadditive.

If F is inversion-semi-subadditive, we have not only $F(0) \subset F(x) + F(-x)$, but also $F(0) \subset F(-x) + F(x)$ for all $x \in D_F$. Namely, if $F(0) \neq \emptyset$, then by the first inclusion we also have $F(-x) \neq \emptyset$, and thus $-x \in D_F$ for all $x \in D_F$.

Quite similarly, we can also easily prove the following

Theorem 39.15 *If F is an inversion-subadditive relation on a group X to a groupoid Y such that $0 \in D_F$, then F is total.*

To establish the basic homogeneity properties of subadditive and superadditive relations, we shall also need the following

Definition 39.8 For some $n \in \mathbb{N}$, a relation F on one groupoid X to another Y is called

(i) n-subhomogeneous if $F(nx) \subset nF(x)$ for all $x \in X$;
(ii) n-superhomogeneous if $nF(x) \subset F(nx)$ for all $x \in X$.

Remark 39.17 Now, the relation F may be naturally called n-semi-subhomogeneous if $F(nx) \subset nF(x)$ for all $x \in D_F$.

Moreover, the relation F may, for instance, be naturally called n-semi-homogeneous if it is both n-semi-subhomogeneous and n-superhomogeneous.

And the relation F may, for instance, be naturally called A-subhomogeneous, for some $A \subset \mathbb{N}$, if it is n-subhomogeneous for all $n \in A$.

By induction, we can easily prove the following

Theorem 39.16 *If F is a superadditive relation on one groupoid X to another Y, then F is $\mathbb{N}$-superhomogeneous.*

From this theorem, it is clear that in particular we also have

Corollary 39.7 *If f is a semi-additive function on one groupoid X to another Y, then f is $\mathbb{N}$-semi-homogeneous.*

Concerning the $\mathbb{N}$-subhomogeneity of subadditive relations, we can only prove

Theorem 39.17 *If F is a subadditive, $\mathbb{N}^{-1}$-convex-valued relation on a groupoid to a vector space over Y over $\mathbb{Q}$, then F is $\mathbb{N}$-subhomogeneous.*

Proof Namely, if $n \in \mathbb{N}$ such that $F(nx) \subset nF(x)$, then we also have

$$\begin{aligned} F\big((n+1)x\big) &= F(nx+x) \subset F(nx) + F(x) \subset nF(x) + F(x) = F(x) + nF(x) \\ &= (n+1)\big((n+1)^{-1}F(x) + \big(1-(n+1)^{-1}\big)F(x)\big) \subset (n+1)F(x). \end{aligned}$$

□

Now as an immediate consequence of the above two theorems, we can also state

Corollary 39.8 *If F is an additive, $\mathbb{N}^{-1}$-convex-valued relation on a groupoid X to a vector space Y over $\mathbb{Q}$, then F is $\mathbb{N}$-homogeneous.*

By using an analogue of Definition 39.8, we can easily establish the following

Theorem 39.18 *If F is a relation on one groupoid X with zero to another Y, then*

(i) *F is zero-superhomogeneous if $0 \in F(0)$;*
(ii) *F is zero-subhomogeneous if either $0 \notin D_F$ or $F(0) \subset \{0\}$ and $D_F = X$.*

Remark 39.18 Note that if $D_F = X$ is not required in (ii), then we can only state that F is zero-semi-subhomogeneous.

Now, as an immediate consequence of Theorems 39.15 and 39.18, we can also state

Corollary 39.9 *If X and F is an inversion-subadditive relation on a group X to a groupoid with zero such that either $0 \notin D_F$ or $F(0) \subset \{0\}$, then F is zero-subhomogeneous.*

39.6 Further Homogeneity Properties of Relations

Definition 39.9 A relation F on a group X to a set Y is called even if $F(-x) = F(x)$ for all $x \in X$.

Moreover, a relation F of one group X to another Y is called odd if $F(-x) = -F(x)$ for all $x \in X$.

Remark 39.19 Now, the relation F may also be naturally called semi-subeven (semi-subodd) if $F(-x) \subset F(x)$ $(F(-x) \subset -F(x))$ for all $x \in D_F$.

However, by the following obvious theorems, some further similar weakenings of Definition 39.9 need not be introduced.

Theorem 39.19 *If F is a relation on one group X to a set Y, then the following assertions are equivalent:*

(i) *F is even;*
(ii) *$F(-x) \subset F(x)$ for all $x \in X$;*
(iii) *$F(x) \subset F(-x)$ for all $x \in D_F$.*

Theorem 39.20 *If F is a relation on one group X to another Y, then the following assertions are equivalent:*

(i) *F is odd;*
(ii) *$F(-x) \subset -F(x)$ for all $x \in X$;*
(iii) *$-F(x) \subset F(-x)$ for all $x \in D_F$.*

The subsequent theorems, whose proofs are again omitted, will already indicate that odd relations are more important than the even ones.

Theorem 39.21 *If f is an inversion-semi-additive function on one group X to another Y, then f is odd.*

Corollary 39.10 *If f is a semi-additive function on one group X to another Y, with a symmetric domain, then f is odd.*

Now, by using an analogue of Definition 39.8, we can also easily prove the following

Theorem 39.22 *If F is an odd, n-subhomogeneous (n-superhomogeneous) relation on one group X to another Y, for some $n \in \mathbb{N}$, then F is $-n$-subhomogeneous ($-n$-superhomogeneous).*

Now, as an immediate consequence of this theorem, we can also state

Corollary 39.11 *If F is an odd, $\mathbb{N}$-subhomogeneous ($\mathbb{N}$-superhomogeneous) relation on one group X to another Y, then F is $\mathbb{Z} \setminus \{0\}$-subhomogeneous ($\mathbb{Z} \setminus \{0\}$-superhomogeneous).*

Moreover, as an immediate consequence of this corollary and Theorem 39.16, we can also state the following

Theorem 39.23 *If F is an odd, superadditive relation on one group X to another Y, then F is $\mathbb{Z} \setminus \{0\}$-superhomogeneous.*

Remark 39.20 Note that if in addition $0 \in F(0)$, then by Theorem 39.18 we can also state that F is $\mathbb{Z}$-superhomogeneous.

By the above results, it is clear that in particular we also have the following

Theorem 39.24 *If f is a semi-additive function on one group X to another Y, with a symmetric domain, then f is $\mathbb{Z}$-semi-homogeneous.*

On the other hand, as an immediate consequence of Corollary 39.11 and Theorem 39.17, we can also state the following

Theorem 39.25 *If F is an odd, subadditive $\mathbb{N}^{-1}$-convex-valued relation on a group X to a vector space Y over $\mathbb{Q}$, then $\mathbb{Z} \setminus \{0\}$-subhomogeneous.*

Remark 39.21 Note that if in addition $F(0) \subset \{0\}$, then by Corollary 39.9 we can also state that F is $\mathbb{Z}$-subhomogeneous.

Analogously to Definition 39.8, we may also naturally introduce the following

Definition 39.10 For some $\lambda \in K$, a relation F on one vector space X over K to another Y is called

(i) λ-subhomogeneous if $F(\lambda x) \subset \lambda F(x)$ for all $x \in X$;
(ii) λ-superhomogeneous if $\lambda F(x) \subset F(\lambda x)$ for all $x \in X$.

Remark 39.22 Now, the relation F may be naturally called λ-semi-subhomogeneous if $F(\lambda x) \subset \lambda F(x)$ for all $x \in D_F$.

Moreover, the relation F may, for instance, be naturally called λ-homogeneous if it is both λ-subhomogeneous and λ-superhomogeneous.

And the relation F may, for instance, be naturally called A-subhomogeneous, for some $A \subset K$, if it is λ-subhomogeneous for all $\lambda \in A$.

Analogously to Theorem 39.22, we can easily establish the following

Theorem 39.26 *If F is an odd λ-subhomogeneous (λ-superhomogeneous) relation on one vector space X over K to another Y, for some $\lambda \in K$, then F is $-\lambda$-subhomogeneous ($-\lambda$-superhomogeneous).*

Now, as an immediate consequence of this theorem, we can also state

Corollary 39.12 *If F is an odd $\mathbb{K}_+ \setminus \{0\}$-subhomogeneous ($\mathbb{K}_+ \setminus \{0\}$-superhomogeneous) relation on one vector space X over $\mathbb{K} = \mathbb{Q}$ or $\mathbb{R}$ to another Y, then F is $\mathbb{K} \setminus \{0\}$-subhomogeneous ($\mathbb{K} \setminus \{0\}$-superhomogeneous).*

Moreover, we can also easily prove the following

Theorem 39.27 *If F is a λ-subhomogeneous (λ-superhomogeneous) relation on one vector space X over K to another Y, for some $\lambda \in K \setminus \{0\}$, then F is λ^{-1}-superhomogeneous (λ^{-1}-subhomogeneous).*

Now, as an immediate consequence of this theorem, we can also state

Corollary 39.13 *If F is an A-subhomogeneous (A-superhomogeneous) relation on one vector space X over K to another Y, for some $A \subset K \setminus \{0\}$ with $A^{-1} \subset A$, then F is A-homogeneous.*

Remark 39.23 Important particular cases are when $K = \mathbb{K}$ and $A = \mathbb{Q}_+ \setminus \{0\}$, $\mathbb{R}_+ \setminus \{0\}$, or $\mathbb{K} \setminus \{0\}$.

In the sequel, F will, for instance, be briefly called subhomogeneous if it is only $K \setminus \{0\}$-subhomogeneous. Namely, the 0-subhomogeneity is a too restrictive property.

Now, because of Corollary 39.13, we may also naturally introduce the following

Definition 39.11 A relation F on one vector space X over K to another Y is called

(i) sublinear if it is both homogeneous and subadditive;
(ii) superlinear if it is both homogeneous and superadditive.

Remark 39.24 Quite similarly, the relation F may be naturally called linear if it is both homogeneous and additive.

Moreover, the relation F may, for instance, be naturally called quasi-linear if it is both homogeneous and quasi-additive.

In the next section, we shall see that a nonempty relation F on one vector space X over K to another Y is quasi-linear if and only if it superlinear, or equivalently, it is a linear subspace of the product space $X \times Y$. Thus, our present terminology slightly differs from the earlier one [249, 261].

39.7 Quasi-odd Relations and Odd-Like Selections

Definition 39.12 A relation F on a group X to a groupoid Y with zero is called quasi-odd if $0 \in F(x) + F(-x)$ for all $x \in D_F$.

Remark 39.25 Thus, an odd relation is, in particular, quasi-odd. Moreover, each reflexive relation on a group, with a symmetric domain, is quasi-odd.

Furthermore, we can note that if F is an inversion-semi-subadditive relation on a group X to a groupoid Y with zero such that $0 \in F(0)$, then F is quasi-odd.

Now, as an improvement of [84, Theorem 5.7], we can also prove

Theorem 39.28 *If F is a nonvoid, quasi-odd and superadditive relation on a group X to a monoid Y, then $0 \in F(0)$ and F is quasi-additive.*

Proof If $x \in D_F$, then $0 \in F(x) + F(-x) \subset F(0)$. Moreover,

$$F(x+y) = \{0\} + F(x+y) \subset F(x) + F(-x) + F(x+y) \subset F(x) + F(y)$$

for all $y \in X$. The case $x \in X$ and $y \in D_F$ can be treated quite similarly. □

From this theorem, it is clear that in particular we also have

Corollary 39.14 *If F is a nonempty superlinear relation on one vector space X over K to another Y, then $0 \in F(0)$ and F is quasi-linear.*

Concerning quasi-odd relations, we can also easily establish the following

Theorem 39.29 *If F is a relation on one group X to another Y, then the following assertions are equivalent*:

(i) *F is quasi-odd*;
(ii) $-F(x) \cap F(-x) \neq \emptyset$ *for all* $x \in D_F$.

Definition 39.13 A partial selection relation Φ of a relation F on one group X to another Y is called odd-like if $-\Phi(x) \subset F(-x)$ for all $x \in D_\Phi$.

Remark 39.26 Note that if Φ is a odd partial selection relation of F, then $-\Phi(x) = \Phi(-x) \subset F(-x)$ for all $x \in D_\Phi$. Therefore, Φ is odd-like.

Moreover, if Φ is a partial selection relation of F and F is odd, then $-\Phi(x) \subset -F(x) = F(-x)$ for all $x \in D_\Phi$. Therefore, Φ is again odd-like.

Now, in addition to Theorem 39.29, we can also easily establish the following

Theorem 39.30 *If F is a relation on one group X to another Y, then the following assertions are equivalent*:

(i) *F is quasi-odd*;
(ii) *F has an odd-like selection function φ.*

Remark 39.27 In [84], by using Zorn's lemma, we proved that if F is a relation on one group X to another Y, then the following assertions are equivalent:

(i) F has an odd selection function φ;
(ii) F is quasi-odd and for any $x \in D_F$, with $2x = 0$, there exists $y \in F(x)$ such that $2y = 0$.

Definition 39.14 A selection relation Φ of a relation F on a groupoid X with zero to an arbitrary one Y is called

(i) left-representing $F(x) = \Phi(x) + F(0)$ for all $x \in X$;
(ii) right-representing if $F(x) = F(0) + \Phi(x)$ for all $x \in X$.

Remark 39.28 Now, a selection relation Φ of F may be naturally called representing if it both left-representing and right-representing. However, this terminology differs from the earlier one [84, 261].

Now, as an improvement of [84, Theorem 4.8], we can also prove the following

Theorem 39.31 *If F is a right-zero-superadditive and inversion-superadditive relation on one group X to another Y and Φ is an odd-like selection relation of F, then Φ is a left-representing selection relation of F.*

Proof For any $x \in D_F$, we have $\Phi(x) + F(0) \subset F(x) + F(0) \subset F(x)$ and

$$F(x) = \{0\} + F(x) \subset \Phi(x) - \Phi(x) + F(x) \subset \Phi(x)$$
$$+ F(-x) + F(x) \subset \Phi(x) + F(0).$$

Therefore, $F(x) = \Phi(x) + F(0)$ for all $x \in D_F$. Hence, since $F(x) = \emptyset$ for all $x \in X \setminus D_F$, it is clear that the required assertion is also true. □

Remark 39.29 If φ is a selection function of a left-zero-superadditive relation F on a groupoid X with zero to a group Y such that $F(x) \subset \varphi(x) + F(0)$ for all $x \in D_F$ and $-\varphi[D_F] \subset \varphi[D_F]$, then it can be shown that φ is actually a representing selection function of F.

However, it is now more important to note that, as an immediate consequence of Theorems 39.30 and 39.31, we can also state

Corollary 39.15 *If F is a quasi-odd, right-zero-superadditive and inversion-superadditive relation on one group X to another Y, then F has a left-representing selection function φ.*

Hence, it is clear that in particular we also have

Corollary 39.16 *If F is a quasi-odd and inversion-superadditive relation on one group X to another Y such that $F(0) \subset \{0\}$, then F is a function.*

Remark 39.30 Some deeper sufficient conditions in order that a relation should be a function have been given by Nikodem and Popa [167].

39.8 Operations on Subadditive and Superadditive Relations

By using the corresponding definitions, we can easily prove the following

Theorem 39.32 *If F_i is an Ω-subadditive relation on one groupoid X to another Y for some $\Omega \subset X \times Y$ and all $i \in I$, then $F = \bigcup_{i \in I} F_i$ is also Ω-subadditive.*

Now, as some immediate consequence of this theorem, we can also state

Corollary 39.17 *If F_i is a subadditive relation on one groupoid X to another Y for all $i \in I$, then $F = \bigcup_{i \in I} F_i$ is also subadditive.*

Corollary 39.18 *If F_i is a zero-subadditive relation on a groupoid X with zero to an arbitrary groupoid Y for all $i \in I$, then $F = \bigcup_{i \in I} F_i$ is also zero-superadditive.*

Corollary 39.19 *If F_i is an inversion-subadditive relation on a group X to a groupoid Y for all $i \in I$, then $F = \bigcup_{i \in I} F_i$ is also inversion-subadditive.*

The following example show that the unions of additive relations need not be additive.

Example 39.2 For any $x \in \mathbb{R}$, define $f_1(x) = \{0\}$ and $f_2(x) = \{x\}$. Then, f_1 and f_2 are additive functions on $\mathbb{R}$ such that the relation $F = f_1 \cup f_2$ is not additive.

Namely, for any $x \in \mathbb{R}$, we have

$$F(x) = (f_1 \cup f_2)(x) = f_1(x) \cup f_2(x) = \{0\} \cup \{x\} = \{0, x\}.$$

Thus, for instance, $F(-1 + 1) = F(0) = \{0\}$, but $F(-1) + F(1) = \{0, -1\} + \{0, 1\} = \{-1, 0, 1\}$.

Analogously to Theorem 39.32, we can also easily prove the following

Theorem 39.33 *If F_i is an Ω-superadditive relation on one groupoid X to another Y for some $\Omega \subset X \times Y$ and all $i \in I$, then $F = \bigcap_{i \in I} F_i$ is also Ω-superadditive.*

Now, as some immediate consequence of this theorem, we can also state

Corollary 39.20 *If F_i is a superadditive relation on one groupoid X to another Y for all $i \in I$, then $F = \bigcap_{i \in I} F_i$ is also superadditive.*

Corollary 39.21 *If F_i is a zero-superadditive relation on a groupoid X with zero to an arbitrary groupoid Y for all $i \in I$, then $F = \bigcap_{i \in I} F_i$ is also zero-superadditive.*

Corollary 39.22 *If F_i is an inversion-superadditive relation on a group X to a groupoid Y for all $i \in I$, then $F = \bigcap_{i \in I} F_i$ is also inversion-superadditive.*

The following example show that the intersections of additive relations need not also be additive.

Example 39.3 For any $x \in \mathbb{R}$, define $F_1(x) = [0, +\infty[$ and $F_2(x) = [x, +\infty[$. Then, F_1 and F_2 are additive relations on $\mathbb{R}$ such that the relation $F = F_1 \cap F_2$ is not additive.

Namely, for any $x \in \mathbb{R}$, we have

$$F(x) = (F_1 \cap F_2)(x) = F_1(x) \cap F_2(x) = [0, +\infty[\cap [x, +\infty[$$
$$= \begin{cases} [0, +\infty[& \text{if} \quad x < 0, \\ [x, +\infty[& \text{if} \quad 0 \leq x. \end{cases}$$

Thus, for instance, $F(-2+1) = F(-1) = [0, +\infty[$, but $F(-2) + F(1) = [0, +\infty[+ [1, +\infty[= [1, +\infty[$.

Remark 39.31 Note that relations F_1, F_2, and F considered in the latter example are just the epigraphs of the functions f_1 and f_1 given in Example 39.2, and the function f defined by $f(x) = (f_1 \vee f_2)(x) = \sup(f_1 \cup f_2)(x) = \sup\{0, x\}$.

In addition to Theorem 39.33, we can also easily prove the following

Theorem 39.34 *If F is a superadditive relation on one groupoid X to another Y, then F^{-1} is also superadditive.*

Unfortunately, the inverse of a subadditive relation need not be subadditive. However, as an immediate consequence of Theorems 39.34 and 39.28, we can also state

Corollary 39.23 *If F is a superadditive relation on a monoid X to a group Y such that F^{-1} is quasi-odd, then F^{-1} is quasi-additive.*

Remark 39.32 A simple computation shows that F^{-1} is quasi-odd if and only if $R_F = \bigcup\{-F(x) \cap F(y) : \; x, y \in X, x + y = 0\}$.

If in particular X is also a group, then the above condition can be briefly expressed by writing that $R_F = \bigcup_{x \in X} -F(x) \cap F(-x)$. Hence, by Theorem 39.29, we can see that the inverse of quasi-odd relation is not, in general, quasi-odd.

Fortunately, the inverse of an odd relation is always odd. Therefore, as an immediate consequence of Theorems 39.34 and 39.28, we can also state

Corollary 39.24 *If F is an odd, superadditive relation on one group X to another Y, then F^{-1} is odd and quasi-additive.*

Concerning subadditive and superadditive relations, we can also easily establish the following theorems.

Theorem 39.35 *If F is a subadditive (superadditive) relation on one groupoid X to another Y and G is a subadditive (superadditive) relation on Y to a groupoid Z, then $G \circ F$ is also subadditive (superadditive).*

Theorem 39.36 *If F and G are subadditive (superadditive) relations on a groupoid X to a commutative semigroup Y, then their pointwise sum $F+G$ is also subadditive (superadditive).*

Remark 39.33 If F is as in the above theorem, we can also state that the pointwise multiple nF of F with every $n \in \mathbb{N}$ is also subadditive (superadditive).

Theorem 39.37 *If F and G are superadditive relations on a groupoid to a commutative semigroup Y, then their global sum $F + G$ is also superadditive.*

Remark 39.34 If F is as in the above theorem, we can also state that the global multiple nF of F with every $n \in \mathbb{N}$ is also superadditive.

39.9 Partial and Total Negatives of Relations

Notation 39.1 *Let X and Y be groups, and assume that F is a relation on X to Y. Moreover, for any $x \in X$, define*

$$\check{F}(x) = F(-x) \quad \textit{and} \quad \hat{F}(x) = -F(-x).$$

Remark 39.35 Thus, $\check{F}$ and $\hat{F}$ are relations on X to Y such that

$$D_{\check{F}} = D_{\hat{F}} = -D_F = \{-x: \ x \in D_F\},$$
$$\check{F} = \big\{(-x, y): \ (x, y) \in F\big\} \quad \text{and} \quad \hat{F} = \big\{(-x, -y): \ (x, y) \in F\big\}.$$

Namely, for instance, for any $x \in X$ and $y \in Y$ we have

$$(x, y) \in \hat{F} \iff y \in \hat{F}(x) \iff y \in -F(-x)$$
$$\iff -y \in F(-x) \iff (-x, -y) \in F \iff -(x, y) \in F.$$

Therefore, $\hat{F}$ is just the global negative of F investigated in [82, 83, 266].

Remark 39.36 The global negative of F has to be carefully distinguished from the more usual pointwise negative $-F$ of F, defined such that $(-F)(x) = -F(x)$, for all $x \in X$, and thus $-F = \{(x, -y): \ (x, y) \in F\}$.

Namely, for instance, if $\Delta = \Delta_X$ is the identity function of X, then we can at once see that $\hat{\Delta} = \Delta$. But, $-\Delta = \Delta$ if and only if $-x = x$, or equivalently $2x = 0$ for all $x \in X$.

Note that the definitions of the pointwise negative $-F$ and the partial negative $\check{F}$ of F do not require X and Y, respectively, to be groups. However, in the sequel, we shall mainly be interested in the total negative $\hat{F}$ of F.

Now, as some simple but important consequences of the above definitions, we can also easily establish the following two theorems.

Theorem 39.38 *We have*

(i) $(-F)^{\vee} = -\check{F} = \hat{F}$;
(ii) $(-F)^{\wedge} = -\hat{F} = \check{F}$.

Hint. From (i), we get $-\hat{F} = -(-\check{F}) = \check{F}$ and $(-F)^{\wedge} = (-(-F))^{\vee} = \check{F}$.

Theorem 39.39 *We have*

(i) $\check{\check{F}} = \hat{\hat{F}} = F$;
(ii) $\hat{\check{F}} = \check{\hat{F}} = -F$.

Hint. By using Theorem 39.38 and $\check{\check{F}} = F$, we can see that $\hat{\check{F}} = -\check{\check{F}} = -F$, $\check{\hat{F}} = (-\check{F})^{\vee} = -\check{\check{F}} = -F$ and $\hat{\hat{F}} = -\check{\hat{F}} = -(-F) = F$.

From (i), it is clear that in particular we also have

Corollary 39.25 *The operations* $\vee$ *and* $\wedge$ *are injective. Moreover,* $\vee^{-1} = \vee$ *and* $\wedge^{-1} = \wedge$.

Now, in addition to Theorems 39.19 and 39.20, we can also easily prove the following two theorems.

Theorem 39.40 *The following assertions are equivalent*:

(i) F *is even*; (ii) $\check{F} = F$; (iii) $\hat{F} = -F$;
(iv) $-F$ *is even*; (v) $\check{F}$ *is even*; (vi) $\hat{F}$ *is even.*

Proof By definitions, (i) and (ii) are equivalent. Moreover, by Theorem 39.38, (ii) and (iii) are also equivalent.

Furthermore, by Theorems 39.39 and 39.38, we have

$$\check{\check{F}} = \check{F} \iff F = \check{F},$$

$$\check{\hat{F}} = \hat{F} \iff -F = \hat{F} \iff \check{F} = F,$$

$$(-F)^{\vee} = -F \iff \hat{F} = -F \iff \check{F} = F.$$

Therefore, by the equivalence of (i) and (ii), assertions (iv), (v), and (iii) are also equivalent to (i). □

From the equivalence of (i) and (iii), by Remark 39.35, we can immediately get

Corollary 39.26 F *is even if and only if its global and pointwise negatives coincide.*

Analogously to Theorem 39.40, we can also easily prove the following

Theorem 39.41 *The following assertions are equivalent*:

(i) F *is odd*; (ii) $\check{F} = -F$; (iii) $\hat{F} = F$;

(iv) $-F$ *is odd*; (v) $\check{F}$ *is odd*; (vi) $\hat{F}$ *is odd.*

From Theorem 39.41, by Theorem 39.21, we can immediately get

Corollary 39.27 *If* ϕ *is an inversion-semiadditive function on* X *to* Y, *then*

(i) $\hat{\phi} = \phi$;
(ii) $\check{\phi} = -\phi$.

Remark 39.37 Thus, if ϕ is as above, and moreover $\phi \subset F$, then we can also state that $\phi = \hat{\phi} \subset \hat{F}$, and thus $\phi \subset F \cap \hat{F}$.

This shows that to determine some extensions of ϕ, in addition to $\hat{F}$, it is also necessary to investigate the relation $F^{\Delta} = F \cap \hat{F}$. However, before doing this, we shall first establish some further basic properties of $\hat{F}$.

39.10 Homogeneity and Additivity Properties of $\hat{F}$

Theorem 39.42 *For any* $k \in \mathbb{Z}$, *the following assertions are equivalent*:

(i) F *is* k*-subhomogeneous* (k*-superhomogeneous*);
(ii) $\hat{F}$ *is* k*-subhomogeneous* (k*-superhomogeneous*).

Proof If F is k-subhomogeneous, then for any $x \in X$ we have

$$\check{F}(kx) = F\big(-(kx)\big) = F\big(k(-x)\big) \subset kF(-x) = k\check{F}(x).$$

Hence, by Theorem 39.38, we can already see that

$$\hat{F}(kx) = -\check{F}(kx) \subset -\big(k\check{F}(x)\big) = k\big(-\check{F}(x)\big) = k\hat{F}(x).$$

Therefore, $\hat{F}$ is also k-subhomogeneous. Now, the converse implication is immediate from the fact that $F = \hat{\hat{F}}$. □

Corollary 39.28 *For any* $k \in \mathbb{Z}$, *the following assertions are equivalent*:

(i) F *is* k*-homogeneous*;
(ii) $\hat{F}$ *is* k*-homogeneous.*

Remark 39.38 Note that if in particular X and Y are vector spaces over $\mathbb{Q}$, then the same assertions holds with $r \in \mathbb{Q}$ in place of $k \in \mathbb{Z}$.

Theorem 39.43 *The following assertions are equivalent*:

(i) F *is subadditive* (*superadditive*);
(ii) $\hat{F}$ *is subadditive* (*superadditive*).

Proof If F is subadditive, then for any $x, y \in X$, we have

$$\check{F}(x+y) = F\big(-(x+y)\big) = F\big(-y+(-x)\big) \subset F(-y) + F(-x) = \check{F}(y) + \check{F}(x).$$

Hence, by Theorem 39.38, we can already see that

$$\hat{F}(x+y) = -\check{F}(x+y) \subset -\big(\check{F}(y)+\check{F}(x)\big) = -\check{F}(x) + \big(-\check{F}(y)\big) = \hat{F}(x) + \hat{F}(y).$$

Therefore, $\hat{F}$ is subadditive. Now, the converse implication is immediate from the fact that $F = \hat{\hat{F}}$. □

Corollary 39.29 *The following assertions are equivalent*:

(i) F *is additive*;
(ii) $\hat{F}$ *is additive*.

In addition to Theorem 39.43, it is also worth proving the following two, more particular theorems.

Theorem 39.44 *The following assertions are equivalent*:

(i) F *is constant-like*;
(ii) $\hat{F}$ *is constant-like*.

Proof If (i) holds, then by Theorem 39.1, we have $F(x) = F(0)$ for all $x \in X$. Hence, we can see that $\check{F}(x) = F(-x) = F(0) = \check{F}(0)$ for all $x \in X$. Now, by Theorem 39.38, we can already see that $\hat{F}(x) = (-\check{F})(x) = -\check{F}(x) = -\check{F}(0) = (-\check{F})(0) = \hat{F}(0)$ for all $x \in X$. Therefore, again by Theorem 39.1, $\hat{F}$ is also constant-like. Now, the converse implication is immediate from the fact that $F = \hat{\hat{F}}$. □

Remark 39.39 Note that if (i) holds, then F is even, and thus by Theorem 39.40, we have $\check{F} = F$ and $\hat{F} = -F$.

Theorem 39.45 *If in particular X is commutative and $Y = X$, then the following assertions are equivalent*:

(i) F *is a translation relation*;
(ii) $\hat{F}$ *is a translation relation*.

Proof If (i) holds, then by Theorem 39.7 we have $F(x) = x + F(0)$ for all $x \in X$. Hence, we can see that $\hat{F}(x) = -F(-x) = -(-x + F(0)) = x + (-F(0)) = x +$

$\hat{F}(0)$ for all $x \in X$. Therefore, again by Theorem 39.7, (ii) also holds. Now, the converse implication is immediate from the fact that $F = \hat{\hat{F}}$. □

Remark 39.40 If (i) holds, then by Theorem 39.9 we can also see that $\hat{F}(x) = -F(-x) = F^{-1}(x)$ for all $x \in X$, and thus $\hat{F} = F^{-1}$. Therefore, F is odd if and only if F is symmetric.

In the sequel, we shall also need the following

Theorem 39.46 *For any $A \subset X$, we have $\hat{F}[A] = -F[-A]$.*

Proof If $y \in \hat{F}[A]$, then there exists $x \in A$ such that $y \in \hat{F}(x)$. Hence, we can see that $y \in -F(-x) \subset -F[-A]$. Therefore, $\hat{F}[A] \subset -F[-A]$. Hence, by writing $\hat{F}$ in place of F and $-A$ in place of A, and using Theorem 39.39, we can infer that $F[-A] = \hat{\hat{F}}[-A] \subset -\hat{F}[-(-A)] = -\hat{F}[A]$. Therefore, $-F[-A] \subset \hat{F}[A]$, and thus the required equality is also true. □

39.11 Compatibility of ∧ with the Basic Operations on Relations

Theorem 39.47 *The operation $\wedge$ preserves inclusions, unions, intersections, differences, and complements.*

Proof It is enough to prove only that the operation $\wedge$ preserves complements and intersections, since the remaining assertions are consequences.

For this, note that for any $x \in X$ we have

$$\left(F^c\right)^{\vee}(x) = F^c(-x) = F(-x)^c = \check{F}(x)^c.$$

Hence, by Theorem 39.38, we can already see that

$$\left(F^c\right)^{\wedge}(x) = -\left(F^c\right)^{\vee}(x) = -\check{F}(x)^c = \left(-\check{F}(x)\right)^c = \hat{F}(x)^c = \hat{F}^c(x).$$

Namely, the map $y \mapsto -y$, where $y \in Y$, is injective and onto Y, and thus it preserves complements with respect to Y. Therefore, we also have $(F^c)^{\wedge} = \hat{F}^c$. □

Now, as an immediate consequence of Theorems 39.41 and 39.47, we can also state

Corollary 39.30 *The family of all odd relations on X to Y is closed under unions, intersections, differences, and complements.*

Remark 39.41 Thus, there exist a largest odd relation contained in F and a smallest odd relation containing F.

Theorem 39.48 *We have*

(i) $(F^{-1})^{\wedge} = \hat{F}^{-1} = -(-F)^{-1}$;
(ii) $(G \circ F)^{\wedge} = \hat{G} \circ \hat{F}$ *for any relation* G *on* Y *to another group* Z.

Proof For any $y \in Y$ and $x \in X$ we have

$$x \in \big(F^{-1}\big)^{\wedge}(y) \iff x \in -F^{-1}(-y) \iff -x \in F^{-1}(-y)$$
$$\iff -y \in F(-x) \iff y \in -F(-x) \iff y \in \hat{F}(x)$$
$$\iff x \in \hat{F}^{-1}(y),$$

and

$$x \in \hat{F}^{-1}(y) \iff y \in \hat{F}(x) \iff y \in -F(-x) \iff y \in (-F)(-x)$$
$$\iff -x \in (-F)^{-1}(y) \iff x \in -(-F)^{-1}(y)$$
$$\iff x \in \big(-(-F)^{-1}\big)(y).$$

Therefore, $(F^{-1})^{\wedge}(y) = \hat{F}^{-1}(y) = (-(-F)^{-1})(y)$ for all $y \in Y$, and thus (i) is true.

Moreover, if G is as in (ii), then by Theorem 39.46 we have

$$(G \circ F)^{\wedge}(x) = -(G \circ F)(-x) = -G\big[F(-x)\big]$$
$$= -G\big[-(-F(-x))\big] = -G\big[-\hat{F}(x)\big] = \hat{G}\big[\hat{F}(x)\big] = (\hat{G} \circ \hat{F})(x)$$

for all $x \in X$. Therefore, (ii) is also true. □

Remark 39.42 By using quite similar arguments, we can also easily see that

(i) $(F^{-1})^{\vee} = (-F)^{-1}$;
(ii) $\check{F}^{-1} = -F^{-1}$.

Hence, by using Theorem 39.38, we can derive the first statement of Theorem 39.48.

Now, as an immediate consequence of Theorems 39.48 and 39.41, we can also state

Corollary 39.31 *The following assertions hold*:

(i) F^{-1} *is odd if and only if* F *is odd.*
(ii) *If* F *is odd, then* $G \circ F$ *is also odd for any odd relation* G *on* Y *to another group* Z.

Remark 39.43 From Remark 39.42, by using Theorem 39.40, we can quite similarly see that F^{-1} is even if and only if $-F = F$. That is, F is symmetric-valued. Now, we can also note if F is even, then F is odd if only if F^{-1} is even.

Concerning the operation $\wedge$, we can also easily prove the following

Theorem 39.49 *We have*

(i) $(kF)^{\wedge} = k\hat{F}$ *for all* $k \in \mathbb{Z}$;
(ii) $(F+G)^{\wedge} = \hat{F} + \hat{G}$ *for any relation* G *on* X *to* Y *whenever* Y *is commutative.*

Now, as an immediate consequence of Theorems 39.49 and 39.41, we can also state

Corollary 39.32 *The following assertions hold*:

(i) *If* F *is odd, then* kF *is also odd for all* $k \in \mathbb{Z}$.
(ii) *If* F *is odd, then* $F+G$ *is also odd for any odd relation* G *on* X *to* Y *whenever* Y *is commutative.*

Remark 39.44 Note that if in particular Y is a vector space over $\mathbb{Q}$, then the same assertions hold with $r \in \mathbb{Q}$ in place of $k \in \mathbb{Z}$.

Finally, we note that by using Theorems 39.47, 39.48, and 39.39, we can also easily establish the following

Theorem 39.50 *If in particular* $Y = X$, *then*

(i) $\hat{F}$ *is reflexive* $\iff$ F *is reflexive*;
(ii) $\hat{F}$ *is transitive* $\iff$ F *is transitive*;
(iii) $\hat{F}$ *is symmetric* $\iff$ F *is symmetric*;
(iv) $\hat{F}$ *is anti-symmetric* $\iff$ F *is anti-symmetric.*

Proof If F is reflexive, then $\Delta_{D_F} \subset F$. Hence, by Theorem 39.47, it follows that $\hat{\Delta}_{D_F} \subset \hat{F}$. Moreover, by the corresponding definitions, we can see that $\hat{\Delta}_{D_F} = \Delta_{-D_F} = \Delta_{D_{\hat{F}}}$. Therefore, $\Delta_{D_{\hat{F}}} \subset \hat{F}$, and thus $\hat{F}$ is also reflexive.

While, if F is anti-symmetric, then $F \cap F^{-1} \subset \Delta_X$. Hence, by Theorems 39.48 and 39.47, we can infer that $\hat{F} \cap \hat{F}^{-1} = \hat{F} \cap (F^{-1})^{\wedge} = (F \cap F^{-1})^{\wedge} \subset \hat{\Delta}_X = \Delta_X$. Therefore, $\hat{F}$ is also anti-symmetric.

Now, if $\hat{F}$ is reflexive (anti-symmetric), then from the equality $F = \hat{\hat{F}}$ we can see that F is also reflexive (anti-symmetric). Therefore, (i) and (iv) are true. □

Remark 39.45 In the $X = Y$ particular case, we can quite similarly see that

(i) $\hat{F}$ is idempotent if and only if F is idempotent;
(ii) if G is another relation on X, then $\hat{F}$ and $\hat{G}$ are commuting if and only if F and G are commuting.

39.12 Intersections of F with Its Partial and Total Negatives

Notation 39.2 *In addition to Notation* 39.1, *we also define*

$$F^{\triangledown} = F \cap \check{F}, \qquad F^{\vartriangle} = F \cap \hat{F}, \qquad F^{\diamond} = \check{F} \cap \hat{F};$$

$$F^{\bullet} = -F \cap F, \qquad F^{\blacktriangledown} = -F \cap \check{F}, \qquad F^{\blacktriangle} = -F \cap \hat{F}.$$

Remark 39.46 Moreover, for instance, we may also naturally define

$$F^{\Diamond} = F^{\triangledown} \cap F^{\vartriangle}, \qquad F^{\blacklozenge} = F^{\blacktriangledown} \cap F^{\blacktriangle}, \qquad F^{\bigstar} = F^{\triangledown} \cap F^{\blacktriangle}.$$

Remark 39.47 By using the corresponding definitions, we can easily see that

(i) $F^{\Diamond} = F \cap F^{\diamond} = \check{F} \cap F^{\vartriangle} = \hat{F} \cap F^{\triangledown} = F \cap \check{F} \cap \hat{F}$;
(ii) $F^{\blacklozenge} = -F \cap F^{\diamond} = \check{F} \cap F^{\blacktriangle} = \hat{F} \cap F^{\blacktriangledown} = -F \cap \check{F} \cap \hat{F}$;
(iii) $F^{\bigstar} = F^{\diamond} \cap F^{\bullet} = F^{\vartriangle} \cap F^{\blacktriangledown} = -F \cap F^{\Diamond} = F \cap F^{\blacklozenge} = -F \cap F \cap \check{F} \cap \hat{F}$.

However, in the sequel, we shall only investigate the interrelationships among the operations considered in Remark 39.36 and Notations 39.1 and 39.2. For this, we shall first prove the following

Theorem 39.51 *We have*

(i) $(-F)^{\bullet} = -F^{\bullet} = F^{\bullet}$;
(ii) $(-F)^{\diamond} = -F^{\diamond} = F^{\diamond}$;
(iii) $(-F)^{\triangledown} = -F^{\triangledown} = F^{\blacktriangle}$;
(iv) $(-F)^{\vartriangle} = -F^{\vartriangle} = F^{\blacktriangledown}$;
(v) $(-F)^{\blacktriangledown} = -F^{\blacktriangledown} = F^{\vartriangle}$;
(vi) $(-F)^{\blacktriangle} = -F^{\blacktriangle} = F^{\triangledown}$.

Proof By the corresponding properties of $-$ and Theorem 39.38, we have

$$-F^{\bullet} = -(-F \cap F) = F \cap (-F) = F^{\bullet},$$

$$(-F)^{\bullet} = -(-F) \cap (-F) = F \cap (-F) = F^{\bullet};$$

$$(-F)^{\diamond} = (-F)^{\vee} \cap (-F)^{\wedge} = \hat{F} \cap \check{F} = F^{\diamond},$$

$$-F^{\diamond} = -(\check{F} \cap \hat{F}) = -\check{F} \cap (-\hat{F}) = \hat{F} \cap \check{F} = F^{\diamond}.$$

Therefore, (i) and (ii) are true.

Moreover, quite similarly, we also have

$$(-F)^{\triangledown} = -F \cap (-F)^{\vee} = -F \cap \hat{F} = F^{\blacktriangle},$$

$$-F^{\triangledown} = -(F \cap \check{F}) = -F \cap (-\check{F}) = -F \cap \hat{F} = F^{\blacktriangle};$$

$$(-F)^{\vartriangle} = -F \cap (-F)^{\wedge} = -F \cap \check{F} = F^{\blacktriangledown},$$

$$-F^{\vartriangle} = -(F \cap \hat{F}) = -F \cap (-\hat{F}) = -F \cap \check{F} = F^{\blacktriangledown}.$$

Therefore, (iii) and (iv) are also true.

Now, from (iv) and (iii), we can see that

$$-F^{\blacktriangledown} = -\left(-F^{\vartriangle}\right) = F^{\vartriangle}, \qquad (-F)^{\blacktriangledown} = \left(-(-F)\right)^{\vartriangle} = F^{\vartriangle};$$
$$-F^{\blacktriangle} = -\left(-F^{\triangledown}\right) = F^{\triangledown}, \qquad (-F)^{\blacktriangle} = \left(-(-F)\right)^{\triangledown} = F^{\triangledown}.$$

Therefore, (v) and (vi) are also true. □

Analogously to the above theorem, we can also easily prove the following theorems.

Theorem 39.52 *We have*

(i) $\check{F}^{\bullet} = F^{\bullet\vee} = F^{\diamond}$;
(ii) $\check{F}^{\diamond} = F^{\diamond\vee} = F^{\bullet}$;
(iii) $\check{F}^{\triangledown} = F^{\triangledown\vee} = F^{\triangledown}$;
(iv) $\check{F}^{\vartriangle} = F^{\vartriangle\vee} = F^{\blacktriangledown}$;
(v) $\check{F}^{\blacktriangledown} = F^{\blacktriangledown\vee} = F^{\vartriangle}$;
(vi) $\check{F}^{\blacktriangle} = F^{\blacktriangle\vee} = F^{\blacktriangle}$.

Theorem 39.53 *We have*

(i) $\hat{F}^{\bullet} = F^{\bullet\wedge} = F^{\diamond}$;
(ii) $\hat{F}^{\diamond} = F^{\diamond\wedge} = F^{\bullet}$;
(iii) $\hat{F}^{\triangledown} = F^{\triangledown\wedge} = F^{\blacktriangle}$;
(iv) $\hat{F}^{\vartriangle} = F^{\vartriangle\wedge} = F^{\vartriangle}$;
(v) $\hat{F}^{\blacktriangledown} = F^{\blacktriangledown\wedge} = F^{\blacktriangledown}$;
(vi) $\hat{F}^{\blacktriangle} = F^{\blacktriangle\wedge} = F^{\triangledown}$.

Theorem 39.54 *We have*

(i) $F^{\triangledown\triangledown} = F^{\triangledown}$;
(ii) $F^{\triangledown\vartriangle} = F^{\vartriangle\triangledown} = F^{\bigstar}$;
(iii) $F^{\triangledown\bullet} = F^{\bullet\triangledown} = F^{\bigstar}$;
(iv) $F^{\triangledown\diamond} = F^{\diamond\triangledown} = F^{\bigstar}$;
(v) $F^{\triangledown\blacktriangledown} = F^{\blacktriangledown\triangledown} = F^{\blacktriangle}$;
(vi) $F^{\triangledown\blacktriangle} = F^{\blacktriangle\triangledown} = F^{\blacktriangle}$.

Proof To prove (iii) and (iv), note that, by Theorems 39.51, 39.52, and 39.53 and Remark 39.47, we have

$$F^{\triangledown\bullet} = -F^{\triangledown} \cap F^{\triangledown} = F^{\blacktriangle} \cap F^{\triangledown} = F^{\bigstar}, \qquad F^{\bullet\triangledown} = F^{\bullet} \cap F^{\bullet\vee} = F^{\bullet} \cap F^{\diamond} = F^{\bigstar};$$
$$F^{\triangledown\diamond} = F^{\triangledown\vee} \cap F^{\triangledown\wedge} = F^{\triangledown} \cap F^{\blacktriangle} = F^{\bigstar}, \qquad F^{\diamond\triangledown} = F^{\diamond} \cap F^{\diamond\vee} = F^{\diamond} \cap F^{\bullet} = F^{\bigstar}.$$

□

Theorem 39.55 *We have*

(i) $F^{\vartriangle\vartriangle} = F^{\vartriangle}$; (ii) $F^{\vartriangle\bullet} = F^{\bullet\vartriangle} = F^{\bigstar}$;

(iii) $F^{\vartriangle\diamond} = F^{\diamond\vartriangle} = F^{\bigstar}$; (iv) $F^{\vartriangle\blacktriangledown} = F^{\blacktriangledown\vartriangle} = F^{\blacktriangledown}$;

(v) $F^{\vartriangle\blacktriangle} = F^{\blacktriangle\vartriangle} = F^{\bigstar}$.

Theorem 39.56 *We have*

(i) $F^{\diamond\diamond} = F^{\bullet}$; (ii) $F^{\diamond\bullet} = F^{\bullet\diamond} = F^{\diamond}$;

(iii) $F^{\diamond\blacktriangledown} = F^{\blacktriangledown\diamond} = F^{\bigstar}$; (iv) $F^{\diamond\blacktriangle} = F^{\blacktriangle\diamond} = F^{\bigstar}$.

Theorem 39.57 *We have*

(i) $F^{\bullet\bullet} = F^{\bullet}$; (ii) $F^{\bullet\blacktriangledown} = F^{\blacktriangledown\bullet} = F^{\bigstar}$; (iii) $F^{\bullet\blacktriangle} = F^{\blacktriangle\bullet} = F^{\bigstar}$.

Theorem 39.58 *We have*

(i) $F^{\blacktriangledown\blacktriangledown} = F^{\vartriangle}$; (ii) $F^{\blacktriangle\blacktriangle} = F^{\triangledown}$; (iii) $F^{\blacktriangledown\blacktriangle} = F^{\blacktriangle\blacktriangledown} = F^{\bigstar}$.

Proof To prove (iii), note that, by Theorems 39.51, 39.53, and 39.52 and Remark 39.47, we have

$$F^{\blacktriangledown\blacktriangle} = -F^{\blacktriangledown} \cap F^{\blacktriangledown\wedge} = F^{\vartriangle} \cap F^{\blacktriangledown} = F^{\bigstar},$$
$$F^{\blacktriangle\blacktriangledown} = -F^{\blacktriangle} \cap F^{\blacktriangle\vee} = F^{\triangledown} \cap F^{\blacktriangle} = F^{\bigstar}.$$
□

Remark 39.48 Note that, by Theorems 39.52 and 39.53, for instance, we have

$$\check{F}^{\blacklozenge} = \check{F}^{\blacktriangledown} \cap \check{F}^{\blacktriangle} = F^{\vartriangle} \cap F^{\blacktriangle} \quad \text{and} \quad \hat{F}^{\blacklozenge} = \hat{F}^{\blacktriangledown} \cap \hat{F}^{\blacktriangle} = F^{\blacktriangledown} \cap F^{\triangledown}.$$

Therefore, the relations $F^{\triangledown} \cap F^{\blacktriangledown}$ and $F^{\vartriangle} \cap F^{\blacktriangle}$ should also be denoted somehow.

Remark 39.49 By using the corresponding definitions, we can easily see that

(i) $F^{\triangledown} \cap F^{\blacktriangledown} = -F \cap F^{\triangledown} = F \cap F^{\blacktriangledown} = F^{\bullet} \cap \check{F} = -F \cap F \cap \check{F}$;
(ii) $F^{\vartriangle} \cap F^{\blacktriangle} = -F \cap F^{\vartriangle} = F \cap F^{\blacktriangle} = F^{\bullet} \cap \hat{F} = -F \cap F \cap \hat{F}$.

39.13 The Even and Odd Cores of Relations

The importance of the relations $F^{\vartriangle}$ and $F^{\blacktriangledown}$ is quite obvious from the next

Theorem 39.59 *The following assertions are equivalent*:

(i) F *is quasi-odd*; (ii) $D_F \subset D_{F^\vartriangle}$;

(iii) $D_{F^\vartriangle} = D_F$; (iv) $D_F \subset D_{F^\blacktriangledown}$;

(v) $D_{F^\blacktriangledown} = D_F$.

Proof By the corresponding definitions, for any $x \in X$, we have

$$F^\blacktriangledown(x) = (-F \cap \check{F})(x) = (-F)(x) \cap \check{F}(x) = -F(x) \cap F(-x).$$

Hence, by Theorem 39.29, we can see that F is quasi-odd if and only if $F^\blacktriangledown(x) \neq \emptyset$, i.e., $x \in D_{F^\blacktriangledown}$ for all $x \in D_F$. Therefore, F is quasi-odd if and only if $D_F \subset D_{F^\blacktriangledown}$. Thus, (i) and (iv) are equivalent.

Moreover, by the corresponding definitions, we can see that $F^\blacktriangledown \subset -F$, and thus $D_{F^\blacktriangledown} \subset D_{-F} = D_F$. Therefore, (iv) and (v) are also equivalent.

Moreover, by Theorem 39.51, we can see that $F^\vartriangle = -F^\blacktriangledown$, and thus $D_{F^\vartriangle} = D_{-F^\blacktriangledown} = D_{F^\blacktriangledown}$. Therefore, (ii) and (iii) are also equivalent to (iv) and (v), respectively. □

From the above theorem, it is clear that in particular we have

Corollary 39.33 *The following assertions are equivalent*:

(i) F *is total and quasi-odd*; (ii) $F^\vartriangle$ *is total*;

(iii) $F^\blacktriangledown$ *is total*.

Now, by using the latter corollary and Theorems 39.51, 39.52, and 39.53, we can also easily prove the following

Theorem 39.60 *If F is total, then following assertions are equivalent*:

(i) F *is quasi-odd*; (ii) $-F$ *is quasi-odd*;

(iii) $\check{F}$ *is quasi-odd*; (iv) $\hat{F}$ *is quasi-odd*.

Proof By Corollary 39.33 and Theorems 39.52, 39.53, and 39.51, we can see that

$$\check{F} \text{ is quasi-odd} \iff \check{F}^\blacktriangledown \text{ is total} \iff F^\vartriangle \text{ is total},$$
$$\hat{F} \text{ is quasi-odd} \iff \hat{F}^\vartriangle \text{ is total} \iff F^\vartriangle \text{ is total},$$
$$-F \text{ is quasi-odd} \iff (-F)^\blacktriangledown \text{ is total} \iff F^\vartriangle \text{ is total}.$$

Therefore, again by Corollary 39.33, the required assertions are also equivalent. □

Remark 39.50 Note that, by Theorem 39.55, $D_{F^{\triangle\triangle}} = D_{F^{\triangle}}$ and $D_{F^{\blacktriangledown\triangle}} = D_{F^{\blacktriangledown}}$. Therefore, by Theorem 39.59, $F^{\triangle}$ and $F^{\blacktriangledown}$ are always quasi-odd.

However, the latter facts are of no particular importance for us since by the following results these relations are actually always odd.

Theorem 39.61 *The following assertions hold*:

(i) $F^{\triangle}$ *is the largest odd relation contained in* $F(\hat{F})$;
(ii) $F^{\triangledown}$ *is the largest even relation contained in* $F(\check{F})$.

Proof By definition, we evidently have $F^{\triangle} = F \cap \hat{F} \subset F$. Moreover, by Theorem 39.53, we have $F^{\triangle\wedge} = F^{\triangle}$. Therefore, by Theorem 39.41, $F^{\triangle}$ is always odd.

On the other hand, if G is an odd relation on X to Y such that $G \subset F$, then by Theorems 39.41 and 39.47 we have $G = \hat{G} \subset \hat{F}$. Therefore, we also have $G \subset F \cap \hat{F} = F^{\triangle}$.

This proves the first part of (i). The proof of the first part of (ii) is quite similar. Moreover, by Theorems 39.53 and 39.52, we have $F^{\triangle} = \hat{F}^{\triangle}$ and $F^{\triangledown} = \check{F}^{\triangledown}$. Therefore, the second parts of (i) and (ii) are also true. □

From the above theorem, by using Theorems 39.51 and 39.53, we can immediately derive

Corollary 39.34 *The following assertions hold*:

(i) $F^{\blacktriangle}$ *is the largest even relation contained in* $-F(\hat{F})$;
(ii) $F^{\blacktriangledown}$ *is the largest odd relation contained in* $-F(\check{F})$.

Proof By Theorems 39.51 and 39.53, we have $F^{\blacktriangle} = (-F)^{\triangledown}$ and $F^{\blacktriangle} = \hat{F}^{\triangledown}$. Therefore, by Theorem 39.61, (i) is true. The proof of (ii) is quite similar. □

Remark 39.51 By using quite similar arguments as in the proof of Theorem 39.61, we can also see that $F \cup F^{\wedge}$ is the smallest odd relation containing $F(\hat{F})$.

This shows that we should better write $F_{\triangledown}$ for $F \cap \hat{F}$ and $F^{\triangle}$ for $F \cup \hat{F}$. However, in this paper we shall only be interested in intersections of relations.

Now, in addition to Theorem 39.41, we can also easily prove the following

Theorem 39.62 *The following assertions are equivalent*:

(i) F *is odd*; (ii) $F^{\triangle} = F$; (iii) $F^{\triangle} = \hat{F}$; (iv) $F^{\blacktriangledown} = \check{F}$.

Proof If (i) holds, then by Theorem 39.41 we have $F^{\triangle} = F \cap \hat{F} = F \cap F = F$, and thus (ii) also holds. While, if (ii) holds, then by Theorem 39.61 we can see that (i) also holds. Therefore, (i) and (ii) are equivalent.

Now, by Theorems 39.41, 39.53, and 39.52, we can also see that

$$F \text{ is odd} \iff \hat{F} \text{ is odd} \iff \hat{F}^{\vartriangle} = \hat{F} \iff F^{\vartriangle} = \hat{F},$$
$$F \text{ is odd} \iff \check{F} \text{ is odd} \iff \check{F}^{\vartriangle} = \check{F} \iff F^{\blacktriangledown} = \check{F}.$$

Therefore, (iii) and (iv) are also equivalent to (i). □

In this respect, it is also worth noticing that we also have the following

Theorem 39.63 *If F is odd, then $F^{\diamond} = F^{\triangledown} = F^{\blacktriangle} = F^{\bullet}$.*

Proof Namely, by Theorem 39.41, we have $F^{\diamond} = \check{F} \cap \hat{F} = -F \cap F = F^{\bullet}$, $F^{\blacktriangle} = -F \cap \hat{F} = -F \cap F = F^{\bullet}$ and $F^{\triangledown} = F \cap \check{F} = F \cap (-F) = F^{\bullet}$. □

Now, as an immediate consequence of the above results and Theorem 39.21, we can also state

Corollary 39.35 *If ϕ is an inversion-semiadditive function on one group X to another Y, then*

(i) $\phi^{\vartriangle} = \phi$;
(ii) $\phi^{\diamond} = \phi^{\triangledown} = \phi^{\blacktriangle} = \phi^{\bullet}$;
(iii) $\phi^{\blacktriangledown} = \check{\phi}$.

Moreover, in addition to Theorem 39.62, we can also easily prove

Theorem 39.64 *If F is odd, then $F^{\bullet}$, $F^{\diamond}$, $F^{\triangledown}$ and $F^{\blacktriangle}$ are also odd.*

Proof Now, by Theorems 39.53 and 39.63, we have $F^{\bullet\wedge} = F^{\diamond} = F^{\bullet}$. Therefore, by Theorem 39.41, $F^{\bullet}$ is odd. Hence, by Theorem 39.63, it is clear that the remaining assertions are also true. □

Quite similarly, in addition to Theorem 39.40, we can also prove the following three theorems.

Theorem 39.65 *The following assertions are equivalent:*

(i) *F is even*;
(ii) $F^{\triangledown} = F$;
(iii) $F^{\triangledown} = \check{F}$;
(iv) $F^{\blacktriangle} = \hat{F}$.

Theorem 39.66 *If F is even, then $F^{\diamond} = F^{\vartriangle} = F^{\blacktriangledown} = F^{\bullet}$.*

Theorem 39.67 *If F is even, then $F^{\bullet}$, $F^{\diamond}$, $F^{\vartriangle}$, and $F^{\blacktriangledown}$ are also even.*

The following example shows that if $F^{\vartriangle}$ is both odd and even, then in contrast to Theorems 39.41 and 39.40 F need not be either odd or even.

Example 39.4 For any $x \in \mathbb{R}$, define

$$F(x) = \{0\} \quad \text{if } x < 0 \quad \text{and} \quad F(x) = [-x, x] \quad \text{if } x \geq 0.$$

Then, then F is a non-odd and non-even, symmetric-valued relation on $\mathbb{R}$ such that $\hat{F}(x) = -F(-x) = F(-x)$, and thus

$$\hat{F}(x) = [x, -x] \quad \text{if } x < 0 \quad \text{and} \quad \hat{F}(x) = \{0\} \quad \text{if } x \geq 0.$$

Hence, we can see that $F^{\vartriangle}(x) = (F \cap \hat{F})(x) = F(x) \cap \hat{F}(x) = \{0\}$ for all $x \in \mathbb{R}$. Therefore, $F^{\vartriangle}$ is, in particular, both odd and even.

Remark 39.52 Note that if F is symmetric-valued, i.e., $-F = F$, then $F^{\bullet} = F$, $\check{F} = \hat{F} = F^{\diamond}$ and $F^{\triangledown} = F^{\vartriangle} = F^{\blacktriangledown} = F^{\blacktriangle}$.

39.14 Homogeneity and Additivity Properties of $F^{\vartriangle}$ and Compatibility Properties of $\vartriangle$

Theorem 39.68 *If F is k-subhomogeneous (k-superhomogeneous), for some $k \in \mathbb{Z}$, then $F^{\vartriangle}$ is also k-subhomogeneous (k-superhomogeneous).*

Proof If F is k-subhomogeneous, then by Theorem 39.42 $\hat{F}$ is also k-subhomogeneous. Hence, we can easily see that $F^{\vartriangle} = F \cap \hat{F}$ is also k-subhomogeneous. □

Corollary 39.36 *If F is k-homogeneous, for some $k \in \mathbb{Z}$, then $F^{\vartriangle}$ is also k-homogeneous.*

Remark 39.53 Note that if in particular X and Y are vector spaces over $\mathbb{Q}$, then the same assertions hold with $r \in \mathbb{Q}$ in place of $k \in \mathbb{Z}$.

Theorem 39.69 *If F is superadditive, then $F^{\vartriangle}$ is also superadditive.*

Proof Now, by Theorem 39.43, $\hat{F}$ is also superadditive. Hence, by Corollary 39.20, we can see that $F^{\vartriangle} = F \cap \hat{F}$ is also superadditive. □

Remark 39.54 In view of Theorems 39.43 and 39.69, it would be of some interest to find an additive (or only a subadditive) relation F on $\mathbb{R}$ such that $F^{\vartriangle}$ be non-subadditive.

By Theorems 39.44 and 39.4 and Remark 39.39, it is clear that in particular we have the following

Theorem 39.70 *If F is constant-like, then $F^{\vartriangle}$ is also constant-like. Moreover, we have $F^{\vartriangle} = F^{\bullet}$.*

Moreover, by Theorems 39.45 and 39.12 and Remark 39.40, it is clear that in particular we also have the following

Theorem 39.71 *If in particular X is commutative and $Y = X$, and moreover F is a translation relation, then F^{Δ} is also a translation relation. Moreover, we have $F^{\Delta} = F \cap F^{-1}$.*

Remark 39.55 From Example 39.4, we can see that if F is a total relation on $\mathbb{R}$ such that F^{Δ} is constant function, then F need not even be a translation relation.

Now, in contrast to Theorems 39.47, 39.48, and 39.49, we can only prove the following three, less convenient, theorems.

Theorem 39.72 *The map Δ preserves inclusions and intersections.*

Proof For this, note that if F_i is a relation on X to Y for all $i \in I$ and $F = \bigcap_{i \in I} F_i$, then by Theorem 39.47 and the corresponding properties of the intersections, we have

$$F^{\Delta} = F \cap \hat{F} = \Big(\bigcap_{i \in I} F_i\Big) \cap \Big(\bigcap_{i \in I} \hat{F}_i\Big) = \bigcap_{i \in I} (F_i \cap \hat{F}_i) = \bigcap_{i \in I} F_i^{\Delta}. \qquad \square$$

Remark 39.56 If $F = \bigcup_{i \in I} \hat{F}_i$, then by using Theorem 39.47 we can only prove that $\bigcup_{i \in I} F_i^{\Delta} \subset F^{\Delta}$.

Remark 39.57 In this respect, it is also worth noticing that, by Theorem 39.53, we have $F^{\Delta} \cup \hat{F}^{\Delta} = F^{\Delta} \cup F^{\Delta} = F^{\Delta} = F \cap \hat{F}$.

While, by Theorems 39.47 and 39.39, we have $(F \cup \hat{F})^{\Delta} = (F \cup \hat{F}) \cap (F \cup \hat{F})^{\wedge} = (F \cup \hat{F}) \cap (\hat{F} \cup \hat{\hat{F}}) = (F \cup \hat{F}) \cap (\hat{F} \cup F) = F \cup \hat{F}$. Thus, in general $F^{\Delta} \cup \hat{F}^{\Delta}$ is a proper subset of $(F \cup \hat{F})^{\Delta}$.

Theorem 39.73 *We have*

(i) $(F^{-1})^{\Delta} = (F^{\Delta})^{-1}$;
(ii) $G^{\Delta} \circ F^{\Delta} \subset (G \circ F)^{\Delta}$ *for any relation G on Y to another group Z.*

Proof By Theorem 39.48 and the corresponding property of inversion, we have

$$\big(F^{-1}\big)^{\Delta} = F^{-1} \cap \big(F^{-1}\big)^{\wedge} = F^{-1} \cap \hat{F}^{-1} = (F \cap \hat{F})^{-1} = \big(F^{\Delta}\big)^{-1}$$

Moreover, if G is as in (ii), then by using Theorem 39.48 and the monotonicity property of composition we can see that

$$\begin{aligned} G^{\Delta} \circ F^{\Delta} &= (G \cap \hat{G}) \circ (F \cap \hat{F}) \\ &\subset (G \circ F) \cap (\hat{G} \circ \hat{F}) = (G \circ F) \cap (G \circ F)^{\wedge} = (G \circ F)^{\Delta}. \end{aligned} \qquad \square$$

Remark 39.58 From Theorems 39.59 and 39.73, we can see that: F^{-1} is quasi-odd $\iff D_{F^{-1}} = D_{(F^{-1})^{\triangle}} \iff D_{F^{-1}} = D_{(F^{\triangle})^{-1}} \iff R_F = R_{F^{\triangle}} \iff R_F = R_{F \cap \hat{F}} \iff F[X] = (F \cap \hat{F})[X]$. That is,

$$F[X] = \bigcup_{x \in X} (F \cap \hat{F})(x) = \bigcup_{x \in X} (F \cap \hat{F})(-x)$$
$$= \bigcup_{x \in X} F(-x) \cap \hat{F}(-x) = \bigcup_{x \in X} F(-x) \cap \big(-F(x)\big).$$

Theorem 39.74 *We have*

(i) $kF^{\triangle} \subset (kF)^{\triangle}$ *for any* $k \in \mathbb{Z}$;
(ii) $F^{\triangle} + G^{\triangle} \subset (F+G)^{\triangle}$ *for any relation* G *on* X *to* Y *whenever* Y *is commutative.*

Remark 39.59 Note that if in particular Y is a vector space over $\mathbb{Q}$, then the corresponding equality holds with $r \in \mathbb{Q} \setminus \{0\}$ in place of $k \in \mathbb{Z}$.

39.15 The Hyers Transforms of Relations

Notation 39.3 *Let* X *be a group and* Y *be a vector space over* $\mathbb{Q}$. *And define* $\mathbb{Z}^* = \mathbb{Z} \setminus \{0\}$. *Moreover, assume that* F *is a relation on* X *to* Y. *And, for any* $k \in \mathbb{Z}^*$ *and* $x \in X$, *define*

$$F_k(x) = k^{-1} F(kx).$$

Remark 39.60 Note that thus F_k is a relation on X to Y such that

$$D_{F_k} = k^{-1} D_F = \{x \in X : \ kx \in D_F\}, \qquad F_k = k^{-1} F = \big\{(x, y) : \ k(x, y) \in F\big\}.$$

Namely, for any $x \in X$ and $y \in Y$, we have

$$(x, y) \in F_k \iff y \in F_k(x) \iff y \in k^{-1} F(kx)$$
$$\iff ky \in F(kx) \iff (kx, ky) \in F \iff k(x, y) \in F.$$

Remark 39.61 Moreover, note that the definition of F_k does not really need Y to be a vector space. And the definition of F_n, for $n \in \mathbb{N}$, does not require X and Y to be a groups.

In this respect, it is also worth noticing that if in particular X is also a vector space over $\mathbb{Q}$, then we may also naturally define $F_r(x) = r^{-1} F(rx)$ for all $x \in X$ and $r \in \mathbb{Q}$ with $r \neq 0$.

Remark 39.62 In the sequel, the relation F_k, or rather the family $(F_n)_{n \in \mathbb{N}}$ or $(F_k)_{k \in \mathbb{Z}^*}$ will be called Hyers transform of F.

Subsequences of the sequence $(F_n)_{n=1}^{\infty}$, in the functional case, have formerly been utilized by Hyers [111], Th.M. Rassias [198], Găvruta, Hossu, Popescu, and Căprău [75] and Lee and Jun [142].

While, the relational case has mainly been considered by Gajda and Ger [69], Popa [188], Nikodem and Popa [168], Lu and Park [146], and the present author [251, 255].

The importance of the Hyers transform is already quite obvious from the following

Theorem 39.75 *For any $k \in \mathbb{Z}^*$,*

(i) *F is k-subhomogeneous if and only if $F_k \subset F$;*
(ii) *F is k-superhomogeneous if and only if $F \subset F_k$.*

Proof For any $x \in X$, we have $F(kx) \subset kF(x) \iff k^{-1}F(kx) \subset F(x) \iff F_k(x) \subset F(x)$. Thus, in particular, (i) is true. The proof of (ii) is quite similar. □

Hence, it is clear that in particular we also have

Corollary 39.37 *For any $k \in \mathbb{Z}^*$, the relation F is k-homogeneous if and only if $F_k = F$.*

Moreover, as an immediate consequence of Theorems 39.16 and 39.75, we can also state

Corollary 39.38 *If ϕ is a semi-additive function on X to Y, then $\phi \subset \phi_k$ for all $k \in \mathbb{Z}^*$.*

Remark 39.63 Note that if $x \in X \setminus D_\phi$, then $\phi(x) = \emptyset$. However, for some $k \in \mathbb{Z}^*$, we may have $kx \in D_\phi$, and thus $F_k(x) = k^{-1}\phi(kx) \neq \emptyset$.

Of course, if in particular X is also a vector space over $\mathbb{Q}$ and D_Φ is a subspace of X, then $kx \in D_\phi$ implies $x \in D_\Phi$. Therefore, the equality $\phi_k = \phi$ is also true.

Analogously to Theorem 39.47, we can also easily prove the following

Theorem 39.76 *For any $k \in \mathbb{Z}^*$, the map $F \mapsto F_k$ preserves inclusions, unions, intersections, differences, and complements.*

Proof For this, note that if F_i is a relation on X to Y for all $i \in I$ and $F = \bigcap_{i \in I} F_i$, then for any $x \in X$ we have $F(x) = \bigcap_{i \in I} F_i(x)$, and thus

$$F_k(x) = k^{-1}F(kx) = k^{-1}\bigcap_{i\in I} F_i(kx)$$

$$= \bigcap_{i\in I} k^{-1}F_i(kx) = \bigcap_{i\in I} (F_i)_k(x) = \Big(\bigcap_{i\in I} (F_i)_k\Big)(x).$$

Namely, the map $y \mapsto k^{-1}y$, where $y \in Y$, is injective, and thus it preserves intersections. Therefore, we also have $F_k = \bigcap_{i\in I}(F_i)_k$. □

From this theorem, by Corollary 39.37, it is clear that in particular we also have

Corollary 39.39 *For any $k \in \mathbb{Z}^*$, the family of all k-homogeneous relations is closed under unions, intersections, differences, and complements.*

Remark 39.64 Thus, for any $k \in \mathbb{Z}^*$, there exist a largest k-homogeneous relation contained in F and a smallest k-homogeneous relation containing F.

However, it is now more important to note that by using the corresponding definitions and some former results, we can also easily prove the following

Theorem 39.77 *For any $k \in \mathbb{Z}^*$, we have*

(i) $(-F)_k = -F_k$;

(ii) $(\hat{F})_k = (F_k)^\wedge = F_{-k}$;

(iii) $(\check{F})_k = (F_k)^\vee = -F_{-k}$;

(iv) $(F^\Delta)_k = (F_k)^\Delta = F_k \cap F_{-k}$.

Proof For any $x \in X$, we have

$$(-F)_k(x) = k^{-1}(-F)(kx) = k^{-1}\big(-F(kx)\big)$$
$$= -\big(k^{-1}F(kx)\big) = -F_k(x) = (-F_k)(x).$$

Therefore, (i) is true.

Moreover, for any $x \in X$, we have

$$(\check{F})_k(x) = k^{-1}\check{F}(kx) = k^{-1}F(-kx)$$
$$= k^{-1}F\big(k(-x)\big) = F_k(-x) = (F_k)^\vee(x),$$
$$(\check{F})_k(x) = k^{-1}F(-kx) = -(-k)^{-1}F\big((-k)x\big) = -F_{-k}(x) = (-F_{-k})(x).$$

Therefore, $(\check{F})_k = (F_k)^\vee$ and $(\check{F})_k = -F_{-k}$, and thus (iii) is also true.

Now, by using Theorem 39.38, we can also see that

$$(\hat{F})_k = (-\check{F})_k = -(\check{F})_k = -(F_k)^\vee = (F_k)^\wedge,$$
$$(\hat{F})_k = (-\check{F})_k = -(-F_{-k}) = F_{-k}.$$

Therefore, (ii) is also true.

Moreover, by using (ii) and Theorem 39.76, we can also see that

$$(F_k)^\Delta = F_k \cap (F_k)^\wedge = F_k \cap F_{-k},$$
$$\left(F^\Delta\right)_k = (F \cap \hat{F})_k = F_k \cap (\hat{F})_k = F_k \cap (F_k)^\wedge = (F_k)^\Delta.$$

Therefore, (iv) is also true. □

Remark 39.65 Because of this theorem, we may simply write $\check{F}_k$, $\hat{F}_k$ and F_k^Δ in place of $(\check{F})_k$, $(\hat{F})_k$ and $(F^\Delta)_k$, respectively.

Now, as a very particular case of Theorem 39.65, we also state

Corollary 39.40 *We have*

(i) $\hat{F} = F_{-1}$; (ii) $F^\Delta = F \cap F_{-1}$.

Moreover, as an immediate consequence of Theorems 39.77, 39.41, and 39.40, we can also state

Corollary 39.41 *If F is odd (even), then F_k is also odd (even) for all $k \in \mathbb{Z}^*$.*

However, it is now more important to note that we also have the following

Theorem 39.78 *For any $k, l \in \mathbb{Z}^*$, we have $(F_k)_l = F_{kl} = (F_l)_k$.*

Proof For any $x \in X$, we have

$$(F_k)_l(x) = l^{-1} F_k(lx) = l^{-1}\left(k^{-1} F\left(k(lx)\right)\right) = (kl)^{-1} F\left((kl)x\right) = F_{kl}(x).$$

Therefore, the required equalities are also true. □

Corollary 39.42 *If $k, l \in \mathbb{Z}^*$, then for any $x \in X$ we have $F_k(lx) = l F_{kl}(x)$.*

Proof By Theorem 39.78, $F_k(lx) = ll^{-1} F_k(lx) = l(F_k)_l(x) = l F_{kl}(x)$. □

39.16 Homogeneity, Additivity and Compatibility Properties of the Hyers Transform

Theorem 39.79 *If F is l-subhomogeneous (superhomogeneous), for some $l \in \mathbb{Z}^*$, then F_k is also l-subhomogeneous (superhomogeneous) for all $k \in \mathbb{Z}^*$.*

Proof If F is l-subhomogeneous, then by Theorem 39.75, we have $F_l \subset F$. Hence, by Theorems 39.78 and 39.76, we can see that $(F_k)_l = (F_l)_k \subset F_k$ for all $k \in \mathbb{Z}^*$. Therefore, again by Theorem 39.75, F_k is l-subhomogeneous for all $k \in \mathbb{Z}^*$. □

Corollary 39.43 *If F is l-homogeneous, for some $l \in \mathbb{Z}^*$, then F_k is also l-homogeneous for all $k \in \mathbb{Z}^*$.*

Remark 39.66 If in particular X is also a vector space over $\mathbb{Q}$ and F is r-homogeneous, for some $r \in \mathbb{Q}$, then F_k is also r-homogeneous for all $k \in \mathbb{Z}^*$.

Theorem 39.80 *If $k, l, p \in \mathbb{Z}^*$ such that $l = pk$, and moreover F is p-subhomogeneous (p-superhomogeneous), then $F_l \subset F_k$ ($F_k \subset F_l$).*

Proof If F is p-subhomogeneous, then by Theorem 39.75 we have $F_p \subset F$. Hence, by Theorems 39.78 and 39.76, we can see that $F_l = F_{pk} = (F_p)_k \subset F_k$. Therefore, the first statement of the theorem is true. The second statement can be proved quite similarly. □

Corollary 39.44 *If F is $\mathbb{N}$-subhomogeneous ($\mathbb{Z}^*$-subhomogeneous) and $(k_n)_{n=1}^{\infty}$ is a sequence in $\mathbb{N}$ ($\mathbb{Z}^*$) such that k_n divides k_{n+1} for all $n \in \mathbb{N}$, then the sequence $(F_{k_n})_{n=1}^{\infty}$ is decreasing.*

Remark 39.67 By induction, we can see that the above condition on the sequence $(k_n)_{n=1}^{\infty}$ means only that there exists a sequence $(l_n)_{n=1}^{\infty}$ in $\mathbb{N}$ ($\mathbb{Z}^*$) such that $k_n = k_1 l_2 l_3 \cdots l_n$ for all $n \in \mathbb{N}$ with $n > 1$.

Thus, in particular, we may naturally take $k_n = 2^n$ for all $n \in \mathbb{N}$, or $k_n = n!$ for all $n \in \mathbb{N}$.

Theorem 39.81 *If in particular X is commutative and F is subadditive (superadditive), then F_k is also subadditive (superadditive) for all $k \in \mathbb{Z}^*$.*

Proof If F is subadditive, then for any $k \in \mathbb{Z}^*$ and $x, y \in X$ we have

$$\begin{aligned} F_k(x+y) &= k^{-1}F\big(k(x+y)\big) = k^{-1}F(kx+ky) \subset k^{-1}\big(F(kx)+F(ky)\big) \\ &= k^{-1}F(kx) + k^{-1}F(ky) = F_k(x) + F_k(y). \end{aligned}$$

Therefore, F_k is also subadditive. □

Corollary 39.45 *If in particular X is commutative and F is additive, then F_k is also additive for all $k \in \mathbb{Z}^*$.*

In addition to Theorem 39.81, it is also worth mentioning that the following two more particular theorems are also true.

Theorem 39.82 *If F is constant-like, then, for each $k \in \mathbb{Z}^*$, F_k is also constant like.*

Theorem 39.83 *If in particular $X = Y$ and F is a translation relation, then, for each $k \in \mathbb{Z}^*$, F_k is also a translation relation.*

Proof By Theorem 39.7, we have $F(x) = x + F(0)$, and hence

$$F_k(x) = k^{-1}F(kx) = k^{-1}\big(kx + F(0)\big) = x + k^{-1}F(k0) = x + F_k(0)$$

for all $x \in X$. Therefore, again by Theorem 39.7, F_k is also a translation relation. □

From the above two theorems, by Corollary 39.37, it is clear that in particular we also have

Corollary 39.46 *If F is as in Theorem 39.83 or 39.82, then F is k-homogeneous, for some $k \in \mathbb{Z}^*$, if and only if $kF(0) = F(0)$.*

Proof Namely, by Corollary 39.37, F_k is k-homogeneous if and only if $F_k = F$. That is, by Theorems 39.83 and 39.7,

$$x + k^{-1}F(0) = x + F_k(0) = F_k(x) = F(x) = x + F(0)$$

for all $x \in X$, or equivalently, $k^{-1}F(0) = F(0)$, i.e., $kF(0) = F(0)$. □

Now, as a counterpart of Theorem 39.46, we can also prove the following

Theorem 39.84 *For any $k \in \mathbb{Z}^*$ and $A \subset X$, we have $F_k[A] = k^{-1}F[kA]$.*

Proof By using the corresponding definitions and the fact that unions are preserved under relations, we can see that

$$F_k[A] = \bigcup_{x\in A} F_k(x) = \bigcup_{x\in A} k^{-1}F(kx) = k^{-1}\bigcup_{x\in A} F(kx) = k^{-1}F[kA].$$ □

Now, analogously to Theorems 39.48 and 39.49, we can also easily prove the following two theorems.

Theorem 39.85 *For any $k \in \mathbb{Z}^*$, we have*

(i) $(F^{-1})_k = (F_k)^{-1}$ *if in particular X is also a vector space over $\mathbb{Q}$;*
(ii) $(G \circ F)_k = G_k \circ F_k$ *for any relation G on Y to another vector space Z over $\mathbb{Q}$.*

Proof If the condition of (i) holds, then for any $x \in X$ and $y \in Y$ we have

$$\begin{aligned} x \in \big(F^{-1}\big)_k(y) &\iff x \in k^{-1}F^{-1}(ky) \iff kx \in F^{-1}(ky) \\ &\iff ky \in F(kx) \iff y \in k^{-1}F(kx) \iff y \in F_k(x) \\ &\iff x \in (F_k)^{-1}(y). \end{aligned}$$

Therefore, $(F^{-1})_k(y) = (F_k)^{-1}(y)$ for all $y \in Y$, and thus (i) is also true.

Moreover, if G is as in (ii), then by Theorem 39.84 we have

$$\begin{aligned}(G \circ F)_k(x) &= k^{-1}(G \circ F)(kx) = k^{-1}G\big[F(kx)\big] = k^{-1}G\big[k\big(k^{-1}F(kx)\big)\big] \\ &= k^{-1}G\big[kF_k(x)\big] = G_k\big[F_k(x)\big] = (G_k \circ F_k)(x)\end{aligned}$$

for all $x \in X$. Therefore, (ii) is also true. □

From this theorem, by Corollary 39.37, it is clear that in particular we also have

Corollary 39.47 *The following assertions hold*:

(i) *If in particular X is also a vector space over $\mathbb{Q}$ and F is k-homogeneous, for some $k \in \mathbb{Z}^*$, then F^{-1} is also k-homogeneous.*
(ii) *If F is k-homogeneous, for some $k \in \mathbb{Z}^*$, then $G \circ F$ is also k-homogeneous for any k-homogeneous relation G on Y to another vector space Z over $\mathbb{Q}$.*

Theorem 39.86 *For any $k \in \mathbb{Z}^*$, we have*

(i) $(rF)_k = rF_k$ *for any* $r \in \mathbb{Q}$,
(ii) $(F + G)_k = F_k + G_k$ *for any relation G on X to Y.*

From this theorem, by Corollary 39.37, it is clear that in particular we also have

Corollary 39.48 *The following assertions hold*:

(i) *If F is k-homogeneous, for some $k \in \mathbb{Z}^*$, then rF is also k-homogeneous for all $r \in \mathbb{Q}$.*
(ii) *If F is k-homogeneous, for some $k \in \mathbb{Z}^*$, then $F + G$ is also k-homogeneous for any k-homogeneous relation G on X to Y.*

Finally, we note that by using Theorems 39.76 and 39.85, we can also easily establish the following

Theorem 39.87 *If in particular $Y = X$, then for any $k \in \mathbb{Z}^*$*

(i) *F_k is reflexive if F is reflexive*;
(ii) *F_k is transitive if F is transitive*;
(iii) *F_k is symmetric if F is symmetric*;
(iv) *F_k is anti-symmetric if F is anti-symmetric.*

Remark 39.68 In the $X = Y$ particular case, for any $k \in \mathbb{Z}^*$, we can also at once state that

(i) F_k is idempotent if F is idempotent.
(ii) if G is another relation on X, then F_k and G_k are commuting if F and G are commuting.

39.17 Two Superhomogenizations of Relations

Notation 39.4 *In addition to Notation* 39.3, *we also define*

$$F^\star = \bigcap_{n\in\mathbb{N}} F_n \quad \textit{and} \quad F^* = \bigcap_{k\in\mathbb{Z}^*} F_k.$$

Remark 39.69 Thus, by the corresponding definitions, we evidently have

$$F^* \subset F^\star \subset F_1 = F \quad \text{and} \quad F^* \subset F_1 \cap F_{-1} = F \cap \hat{F} = F^\Delta.$$

Remark 39.70 Note that, in contrast to the relations F_k, the relations $F^\star$ and F^* may be very partial even if F is total.

Therefore, it will be an important task to give some sufficient conditions on F in order that the above relations could be total.

The appropriateness of Notation 39.4 is already quite obvious from the following two theorems which give only very particular answers to the above problem.

Theorem 39.88 *The following assertions are equivalent*:

(i) F *is* $\mathbb{N}$-*superhomogeneous*; (ii) $F \subset F^\star$;

(iii) $F = F^\star$.

Proof By the corresponding definitions and Theorem 39.75, we can see that: F is $\mathbb{N}$-superhomogeneous $\iff$ F is n-superhomogeneous for all $n\in\mathbb{N}$ $\iff$ $F \subset F_n$ for all $n\in\mathbb{N}$ $\iff$ $F \subset \bigcap_{n\in\mathbb{N}} F_n$ $\iff$ $F \subset F^\star$. Therefore, (i) and (ii) are equivalent. Moreover, by Remark 39.69, it is clear that (ii) and (iii) are also equivalent. □

Now, as an immediate consequence of the above theorem and Corollary 39.7, we can also state

Corollary 39.49 *If* ϕ *is a semi-additive function on* X *to* Y, *then* $\phi^\star = \phi$.

Analogously to Theorem 39.88, we can also easily prove the following

Theorem 39.89 *The following assertions are equivalent*:

(i) F *is* $\mathbb{Z}^*$-*superhomogeneous*; (ii) $F \subset F^*$;

(iii) $F = F^*$.

Hence, by Theorem 39.83, it is clear that in particular we also have

Corollary 39.50 *If ϕ is a semi-additive function on X to Y, with a symmetric domain, then $\phi^* = \phi$.*

In accordance with Theorem 39.72, we can now only prove the following

Theorem 39.90 *The maps $\star$ and $*$ preserve inclusions and intersections.*

Remark 39.71 If $F = \bigcup_{i \in I} F_i$, then we can only prove that $\bigcup_{i \in I} (F_i)^* \subset F^*$.

Now, as an immediate consequence of Theorems 39.90, 39.88, and 39.89, we can also state

Corollary 39.51 *The family of all $\mathbb{N}$-superhomogeneous ($\mathbb{Z}^*$-superhomogeneous) relations on X to Y is closed under intersections.*

Remark 39.72 Thus, there exists a smallest $\mathbb{N}$-superhomogeneous ($\mathbb{Z}^*$-superhomogeneous) relation on X to Y containing F.

However, it is now more important to note that by using our former results we can also easily prove the following

Theorem 39.91 *We have*

(i) $(-F)^\star = -F^\star$;

(ii) $\check{F}^\star = F^{\star\vee} = -\hat{F}^\star$;

(iii) $\hat{F}^\star = F^{\star\wedge} = \bigcap_{n \in \mathbb{N}} F_{-n}$;

(iv) $F^{\Delta\star} = F^{\star\Delta} = F^*$;

(v) $(F_k)^\star = (F^\star)_k = \bigcap_{n \in \mathbb{N}} F_{nk}$;

(vi) $F^{\star\star} = F^\star$.

Proof By Theorem 39.77 and the corresponding property of the operation $-$, we have

$$(-F)^\star = \bigcap_{n \in \mathbb{N}} (-F)_n = \bigcap_{n \in \mathbb{N}} -F_n = -\bigcap_{n \in \mathbb{N}} F_n = -F^\star.$$

Therefore, (i) is true.

Moreover, by Theorems 39.77 and 39.47, we have

$$\hat{F}^\star = \bigcap_{n \in \mathbb{N}} \hat{F}_n = \bigcap_{n \in \mathbb{N}} (F_n)^\wedge = \bigcap_{n \in \mathbb{N}} F_{-n}, \qquad \hat{F}^\star = \bigcap_{n \in \mathbb{N}} (F_n)^\wedge = \Big(\bigcap_{n \in \mathbb{N}} F_n\Big)^\wedge = F^{\star\wedge}.$$

Therefore, (iii) is also true.

Now, by using (i) and Theorem 39.38, we can also easily see that

$$\check{F}^\star = (-\hat{F})^\star = -\hat{F}^\star \quad \text{and} \quad F^{\star\vee} = -F^{\star\wedge} = -\hat{F}^\star.$$

Therefore, (ii) is also true.

On the other hand, by using (iii), we can see that

$$F^{*}=\bigcap_{k\in\mathbb{Z}^{*}}F_{k}=\Big(\bigcap_{n\in\mathbb{N}}F_{n}\Big)\cap\Big(\bigcap_{n\in\mathbb{N}}F_{-n}\Big)=F^{\star}\cap F^{\star\wedge}=F^{\star\Delta}.$$

Moreover, by using (iii) and Theorem 39.90, we can also see that

$$F^{*}=F^{\star}\cap F^{\star\wedge}=F^{\star}\cap\hat{F}^{\star}=(F\cap\hat{F})^{\star}=F^{\Delta\star}.$$

Therefore, (iv) is also true.

Moreover, if $k\in\mathbb{Z}^{*}$, then by using Theorems 39.78 and 39.76 we can see that

$$(F_{k})^{\star}=\bigcap_{n\in\mathbb{N}}(F_{k})_{n}=\bigcap_{n\in\mathbb{N}}F_{nk}=\bigcap_{n\in\mathbb{N}}(F_{n})_{k}=\Big(\bigcap_{n\in\mathbb{N}}F_{n}\Big)_{k}=\big(F^{\star}\big)_{k}.$$

Therefore, (v) is also true.

Now, if $m\in\mathbb{N}$, then using (v) we can see that

$$F^{\star}=\bigcap_{n\in\mathbb{N}}F_{n}\subset\bigcap_{n\in\mathbb{N}}F_{nm}=\big(F^{\star}\big)_{m},\quad\text{and thus}\quad F^{\star}\subset\bigcap_{m\in\mathbb{N}}\big(F^{\star}\big)_{m}=F^{\star\star}.$$

Moreover, by Remark 39.69 and Theorem 39.90, it is clear that $F^{\star\star}\subset F^{\star}$. Therefore, (vi) is also true. □

Remark 39.73 Because of (v), we may write $F_{k}^{\star}$ in place of $(F_{k})^{\star}$ and $(F^{\star})_{k}$.

Moreover, as an immediate consequence of Theorems 39.91, 39.41, and 39.40, we can state

Corollary 39.52 *If F is odd (even), then $F^{\star}$ is also odd (even).*

Now, by using Theorem 39.91 and our former results, we can also easily prove the following

Theorem 39.92 *We have*

(i) $(-F)^{*}=-F^{*}$; (ii) $\check{F}^{*}=F^{*\vee}=-F^{*}$;

(iii) $\hat{F}^{*}=F^{*\wedge}=F^{*}$; (iv) $F^{\Delta *}=F^{*\Delta}=F^{*}$;

(v) $(F_{k})^{*}=(F^{*})_{k}$; (vi) $F^{\star *}=F^{*\star}=F^{**}=F^{*}$.

Proof By using Theorems 39.91 and 39.51, we can see that

$$(-F)^{*}=(-F)^{\Delta\star}=\big(-F^{\Delta}\big)^{\star}=-F^{\Delta\star}=-F^{*}.$$

Therefore, (i) is true.

Moreover, by using Theorems 39.91 and 39.53, we can also see that

$$\hat{F}^{*} = \hat{F}^{\Delta\star} = F^{\Delta\star} = F^{*} \quad \text{and} \quad F^{*\wedge} = F^{\star\Delta\wedge} = F^{\star\Delta} = F^{*}.$$

Therefore, (iii) is also true. Moreover, we can note that proof of (iv) is quite similar.

Now, by using (i) and (iii) and Theorem 39.38, we can also see that

$$\check{F}^{*} = (-\hat{F})^{*} = -\hat{F}^{*} \quad \text{and} \quad F^{*\vee} = -F^{*\wedge} = -\hat{F}^{*}.$$

Therefore, (ii) is also true.

Moreover, if $k \in \mathbb{Z}^{*}$, then by using Theorems 39.91 and 39.77, we can see that

$$(F_k)^{*} = (F_k)^{\Delta\star} = \big((F_k)^{\Delta}\big)^{\star} = \big((F^{\Delta})_k\big)^{\star} = \big(F^{\Delta\star}\big)_k = \big(F^{*}\big)_k.$$

Therefore, (v) is also true.

Finally, by Theorem 39.91, we can also see that

$$F^{**} = F^{\star\star\Delta} = F^{\star\Delta} = F^{*} \quad \text{and} \quad F^{**} = F^{\Delta\star\star} = F^{\Delta\star} = F^{*},$$

and $F^{**} = F^{\Delta**} = F^{\Delta\star} = F^{*}$. Therefore, (vi) is also true. □

Remark 39.74 In addition to (i) and (ii), it is also worth noticing that, by Theorems 39.91 and 39.51, we also have

$$F^{*} = F^{\star\Delta} = -F^{\star\blacktriangledown} \quad \text{and} \quad F^{*} = F^{\Delta\star} = \big(-F^{\blacktriangledown}\big)^{\star} = -F^{\blacktriangledown\star}.$$

Therefore, $F^{\star\blacktriangledown} = F^{\blacktriangledown\star} = -F^{*}$ is also true.

Remark 39.75 Because of (v), we may write F_k^{*} in place of $(F_k)^{*}$ and $(F^{*})_k$.

Moreover, as an immediate consequence of Theorems 39.92, 39.41, and 39.40, we can also state

Corollary 39.53 *F^{*} is always odd. Moreover, F^{*} is even if and only if it is symmetric-valued.*

Remark 39.76 Note that if in particular F is symmetric-valued, i.e., $-F = F$, then by Theorem 39.92 we also have $-F^{*} = (-F)^{*} = F^{*}$. Therefore, F^{*} is also symmetric-valued.

However, it is now more important to note that, by using Theorems 39.91, 39.88, 39.92, and 39.89, we can also easily prove the following

Theorem 39.93 *The following assertions hold*:

(i) *$F^{\star}$ is the largest $\mathbb{N}$-superhomogeneous relation contained in F*;
(ii) *F^{*} is the largest $\mathbb{Z}^{*}$-superhomogeneous relation contained in F.*

Proof By Remark 39.69, we have $F^* \subset F$. Moreover, from Theorem 39.92, we know that $F^{**} = F^*$. Therefore, by Theorem 39.89, F^* is always $\mathbb{Z}^*$-superhomogeneous.

Therefore, to complete the proof of (ii) we need only note that if G is a $\mathbb{Z}^*$-superhomogeneous relation on X to Y such that $G \subset F$, then by Theorems 39.89 and 39.90 we also have $G = G^* \subset F^*$. □

Remark 39.77 By using Theorems 39.75, 39.76, and 39.78, we can easily see that $\bigcup_{n\in\mathbb{N}} F_n$ ($\bigcup_{k\in\mathbb{Z}^\star} F_k$) is the smallest $\mathbb{N}$-subhomogeneous ($\mathbb{Z}^*$-subhomogeneous) relation containing F.

However, in view of Remark 39.72 and Theorem 39.93, it would be more interesting to determine the smallest $\mathbb{N}$-superhomogeneous ($\mathbb{Z}^*$-superhomogeneous) relation containing F.

39.18 Homogeneity and Additivity Properties of $F^\star$ and F^* and Compatibility Properties of $\star$ and $*$

In addition to Theorem 39.93, we can also prove the following

Theorem 39.94 *If F is k-subhomogeneous, for some $k \in \mathbb{Z}^\star$, then $F^\star$ and F^* are also k-subhomogeneous.*

Proof By Theorem 39.75, we have $F_k \subset F$. Hence, by using Theorems 39.91, 39.92, and 39.76, we can infer that $(F^\star)_k = (F_k)^\star \subset F^\star$ and $(F^*)_k = (F_k)^* \subset F^*$. Thus, again by Theorem 39.75, the required assertions are also true. □

Remark 39.78 If in particular X is also a vector space over $\mathbb{Q}$ and F is r-homogeneous, for some $r \in \mathbb{Q} \setminus \{0\}$, then it can be easily seen that $F^\star$ and F^* are also r-homogeneous.

Now, as an immediate consequence of Theorems 39.94 and 39.93, we can also state

Corollary 39.54 *If F is $\mathbb{N}$-subhomogeneous ($\mathbb{Z}^*$-subhomogeneous), then $F^\star$ is $\mathbb{N}$-homogeneous (F^* is $\mathbb{Z}^*$-homogeneous).*

However, it is now more interesting that, by using Theorem 39.80, we can also prove the following

Theorem 39.95 *If F is $\mathbb{N}$-subhomogeneous ($\mathbb{Z}^*$-subhomogeneous) and $A \subset \mathbb{N}$ ($A \subset \mathbb{Z}^*$) such that, for each $k \in \mathbb{N}$ ($k \in \mathbb{Z}^*$), there exists $l \in A$ such that k divides l, then*

$$F^\star = \bigcap_{l\in A} F_l \quad \left(F^* = \bigcap_{l\in A} F_l\right).$$

Proof Because of $A \subset \mathbb{N}$, it is clear that

$$F^\star = \bigcap_{k \in \mathbb{N}} F_k \subset \bigcap_{l \in A} F_l.$$

Moreover, by the hypothesis of the theorem, for each $k \in \mathbb{N}$, there exists $l_k \in A$ such that k divides l_k. Hence, because of $\{l_k\}_{k \in \mathbb{N}} \subset A$ and Theorem 39.80, we can see that

$$\bigcap_{l \in A} F_l \subset \bigcap_{k \in \mathbb{N}} F_{l_k} \subset \bigcap_{k \in \mathbb{N}} F_k = F^\star.$$

Therefore, the first statement of the theorem is true. The second statement can be proved quite similarly. □

Remark 39.79 Note that if $A \subset \mathbb{N}$, then there exists a sequence $(l_n)_{n=1}^{\infty}$ in A such that $A = \{l_n\}_{n=1}^{\infty}$. In this case, the hypothesis of the theorem means that, for each $k \in \mathbb{N}$, there exist $n_k \in \mathbb{N}$ such that k divides l_{n_k}. That is, there exists $p_k \in \mathbb{N}$ such that $l_{n_k} = p_k k$.

Thus, in particular, for any sequence $(p_n)_{n=1}^{\infty}$ in $\mathbb{N}$, we may naturally take $l_n = np_n$ for all $n \in \mathbb{N}$. More specially, we can take $l_n = pn$ for some $p \in \mathbb{N}$ and all $n \in \mathbb{N}$, or $l_n = n!$ for all $n \in \mathbb{N}$. However, in contrast to Hyers's method, we cannot take $l_n = 2^n$ for all $n \in \mathbb{N}$.

Analogously to Theorem 39.69, we can now only prove the following

Theorem 39.96 *If in particular X is commutative and F is superadditive, then $F^\star$ and F^* are also superadditive.*

Proof Now, by Theorem 39.81, F_k is also superadditive for all $k \in \mathbb{N}$. Hence, by Corollary 39.20, we can see that the required assertions are true. □

Quite similarly, by Theorems 39.82, 39.83, 39.4, and 39.12, we can also state the following two more particular theorems.

Theorem 39.97 *If F is constant-like, then $F^\star$ and F^* are also constant-like.*

Theorem 39.98 *If in particular $X = Y$ and F is a translation relation, then $F^\star$ and F^* are also translation relations.*

Now, in addition to Theorem 39.90, we can only prove the following theorems.

Theorem 39.99 *We have*

(1) $(F^{-1})^\star = (F^\star)^{-1}$ *and* $(F^{-1})^* = (F^*)^{-1}$ *if in particular X is also a vector space over $\mathbb{Q}$;*

(2) $G^\star \circ F^\star \subset (G \circ F)^\star$ *and* $G^* \circ F^* \subset (G \circ F)^*$ *for any relation* G *on* Y *to another vector space* Z *over* $\mathbb{Q}$.

Proof If X and G are as above, then by Theorem 39.85 and the corresponding properties of inversion and composition, we have

$$(F^{-1})^\star = \bigcap_{n\in\mathbb{N}} (F^{-1})_n = \bigcap_{n\in\mathbb{N}} F_n^{-1} = \Big(\bigcap_{n\in\mathbb{N}} F_n\Big)^{-1} = (F^\star)^{-1},$$

$$\begin{aligned} G^\star \circ F^\star &= \Big(\bigcap_{n\in\mathbb{N}} G_n\Big) \circ \Big(\bigcap_{n\in\mathbb{N}} G_n\Big) \subset \bigcap_{n\in\mathbb{N}} G_n \circ F_n \\ &= \bigcap_{n\in\mathbb{N}} (G \circ F)_n = (G \circ F)^\star. \end{aligned}$$

Therefore, the required assertions are true for $\star$. □

From this theorem, by Theorems 39.88 and 39.89, it is clear that in particular we also have

Corollary 39.55 *The following assertions hold*:

(i) *If in particular* X *is also a vector space over* $\mathbb{Q}$ *and* F *is* $\mathbb{N}$*-superhomogeneous* ($\mathbb{Z}^*$*-superhomogeneous*), *then* F^{-1} *is also* $\mathbb{N}$*-superhomogeneous* ($\mathbb{Z}^*$*-superhomogeneous*).
(ii) *If* F *is* $\mathbb{N}$*-superhomogeneous* ($\mathbb{Z}^*$*-superhomogeneous*), *then* $G \circ F$ *is also* $\mathbb{N}$*-superhomogeneous* ($\mathbb{Z}^*$*-superhomogeneous*) *for any* $\mathbb{N}$*-superhomogeneous* ($\mathbb{Z}^*$*-superhomogeneous*) *relation* G *on* Y *to another vector space* Z *over* $\mathbb{Q}$.

Theorem 39.100 *We have*

(i) $(rF)^\star = rF^\star$ *and* $(rF)^* = rF^*$ *for any* $r \in \mathbb{Q}$ *with* $r \neq 0$;
(ii) $F^\star + G^\star \subset (F+G)^\star$ *and* $F^* + G^* \subset (F+G)^*$ *for any relation* G *on* X *to* Y.

From this theorem, by Theorems 39.88 and 39.89, it is clear that in particular we also have

Corollary 39.56 *The following assertions hold*:

(i) *If* F *is* $\mathbb{N}$*-superhomogeneous* ($\mathbb{Z}^*$*-superhomogeneous*), *then* rF *is also* $\mathbb{N}$*-superhomogeneous* ($\mathbb{Z}^*$*-superhomogeneous*) *for all* $r \in \mathbb{Q}$ *with* $r \neq 0$.
(ii) *If* F *is* $\mathbb{N}$*-superhomogeneous* ($\mathbb{Z}^*$*-superhomogeneous*), *then* $F + G$ *is also* $\mathbb{N}$*-superhomogeneous* ($\mathbb{Z}^*$*-superhomogeneous*) *for any* $\mathbb{N}$*-superhomogeneous* ($\mathbb{Z}^*$*-superhomogeneous*) *relation* G *on* X *to* Y.

Finally, we note that by using Theorems 18.8 and 39.99, we can also easily establish the following

Theorem 39.101 *If in particular $Y = X$, then*

(i) $F^\star$ *and* F^* *are reflexive if* F *is reflexive;*
(ii) $F^\star$ *and* F^* *are transitive if* F *is transitive;*
(iii) $F^\star$ *and* F^* *are symmetric if* F *is symmetric;*
(iv) $F^\star$ *and* F^* *are anti-symmetric if* F *is anti-symmetric.*

39.19 A Few Basic Facts on the Intersection Convolutions of Relations

Because of the page limit, the intersection convolutions of relations cannot be treated here. We can only note some basic facts on them.

In the light of Remarks B.10 and C.14, in addition to Notations 39.1, and 39.2, we may also naturally introduce the following

Notation 39.5 *Let X be a commutative preordered group and Y be a commutative group. And assume that F and G are relations on X to Y.*

Moreover, for any $x \in X$, define

$$(F \star G)(x) = \bigcap \big\{ F(u) + G(v) : \ u \in D_F, \ v \in D_G, \ u + v \le x \big\},$$

$$(F * G)(x) = \bigcap \big\{ F(u) + G(v) : \ u \in D_F, \ v \in D_G, \ x \le u + v \big\}.$$

Remark 39.80 Thus, if in particular the inequality relation in X is symmetric, or more specially it is just the equality relation in X, then $(F \star G)(x) = (F * G)(x)$ for all $x \in X$, and thus $F \star G = F * G$.

More generally, it can also be shown that $(F * G)^\wedge = \hat{F} \star \hat{G}$, and thus $F * G = (\hat{F} \star \hat{G})^\wedge$. Therefore, the properties of $*$ can be derived from those of $\star$ and $\wedge$. However, it is sometimes more convenient to apply duality.

Remark 39.81 Concerning the above operations, it is also worth noticing that

$$\begin{aligned} &\big((F \star G) \cup (F * G)\big)(x) \\ &\quad = (F \star G)(x) \cup (F * G)(x) \\ &\qquad \subset \bigcap \big\{ F(u) + G(v) : \ x = u + v, \ u \in D_F, \ v \in D_G \big\} \\ &\quad = \bigcap \big\{ F(x - v) + G(v) : \ v \in (x - D_F) \cap D_G \big\}. \end{aligned}$$

Remark 39.82 Moreover, if in particular F is total and decreasing, then we can also easily see that

$$(F \star G)(x) = \bigcap_{v \in D_G} \big(F(x - v) + G(v) \big).$$

Therefore, the relation $F \star G$ is not only a subset, but also a natural generalization of the ordinary intersection convolution of F and G investigated in [29, 46, 47, 53, 86, 243, 254, 258]. (See also Beg [23], Moreau [151], Strömberg [237], and [81, 87, 88, 181, 256, 259].)

Concerning the operation $\star$, for instance, we can also prove the following

Theorem 39.102 *If D_G is a subgroup of X and G is superadditive, then for any $x, y \in X$ we have*

$$(F \star G)(x) + G(y) \subset (F \star G)(x + y).$$

Hence, by using a similar argument as in the proof of Theorem 39.28, we can immediately derive

Corollary 39.57 *If D_G is a subgroup of X and G is quasi-odd and superadditive, then for any $x \in X$ and $y \in D_G$ we have*

$$(F \star G)(x + y) = (F \star G)(x) + G(y).$$

Finally, we note that the following theorem is also true.

Theorem 39.103 *If in particular X is a preordered vector space over $\mathbb{Q}$ and Y is a vector space over $\mathbb{Q}$, then for any $n \in \mathbb{N}$ we have*

(i) $(F \star G)_n = (F_n \star G_n)$; (ii) $(F \star G)_{-n} = (F_{-n} * G_{-n})$.

Hence, by using Corollary 39.37, we can immediately derive

Corollary 39.58 *If X and Y are as in Theorem 39.103 and F and G are n-homogeneous, for some $n \in \mathbb{N}$, then $F \star G$ and $F * G$ are also n-homogeneous.*

Acknowledgements The author is indebted to J. Horváth and Th.M. Rassias for several valuable pieces of advice.

Moreover, the author would also like to thank R. Ger, M. Sablik, Zs. Páles, and G. Horváth for some helpful discussions.

References[1]

1. Abreu, J., Etcheberry, A.: Hahn–Banach and Banach–Steinhaus theorems for convex processes. Period. Math. Hung. **20**, 289–297 (1989)

[1] Unfortunately, the author does not have access to the items [14, 39, 40, 43, 99, 103, 112, 126, 137, 138, 158, 160, 162, 183, 185, 208, 210, 277]. Moreover, he cannot actually read works written in French, Italian, and Spanish.

2. Aczél, J., Dhombres, J.: Functional Equations in Several Variables. Cambridge University Press, Cambridge York (1989)
3. Adasch, N.: Der satz über offene lineare relationen in topologischen vectoräumen. Note Mat. **11**, 1–5 (1991)
4. Álvarez, T.: On the Browder essential sprectum of a linear relation. Publ. Math. (Debr.) **73**, 145–154 (2008)
5. Aoki, T.: On the stability of the linear transformation in Banach spaces. J. Math. Soc. Jpn. **2**, 491–495 (1950)
6. Arens, R.: Operational calculus of linear relations. Pac. J. Math. **11**, 9–23 (1961)
7. Bahyrycz, A.: Forti's example of an unstable homomorphism equation. Aequ. Math. **74**, 310–313 (2007)
8. Badora, R.: On some generalized invariant means and their application to the stability of the Hyers–Ulam type. Ann. Pol. Math. **58**, 111–126 (1993)
9. Badora, R.: On approximately additive functions. Ann. Math. Sil. **8**, 111–126 (1994)
10. Badora, R.: On the Hahn–Banach theorem for groups. Arch. Math. **86**, 517–528 (2006)
11. Badora, R., Ger, R., Páles, Zs.: Additive selections and the stability of the Cauchy functional equation. ANZIAM J. **44**, 323–337 (2003)
12. Baker, J.A.: On some mathematical characters. Glas. Mat. **25**, 319–328 (1990)
13. Baker, J.A.: The stability of certain functional equations. Proc. Am. Math. Soc. **112**, 729–732 (1991)
14. Banach, S.: Sur les fonctionelles linéaires I–II. Stud. Math. **1**, 211–216, 223–239 (1929)
15. Banach, S.: Théorie des Opérations Linéaires. Druk M. Garasiński, Warsawa (1932)
16. Banach, S., Mazur, S.: Zur theorie der linearen dimension. Stud. Math. **4**, 100–112 (1933)
17. Baron, K.: Functions with differences in subspaces. In: Proceedings of the 18th International Symposium on Functional Equations, Faculty of Mathematics, pp. 8–9. University of Waterloo (1980)
18. Baron, K.: A remark on the stability of the Cauchy equation. Rocz. Nauk.-Dydakt. Pr. Mat. **11**, 7–12 (1985)
19. Baron, K., Kannappan, Pl.: On the Pexider difference. Fundam. Math. **134**, 247–254 (1990)
20. Baron, K., Volkmann, P.: On functions close to homomorphisms between square symmetric structures. Seminar LV **14**, 1–12 (2002). http://www.mathematik.uni-karlsruhe.de/~semlv
21. Baron, K., Simon, A., Volkmann, P.: On functions having Cauchy differences in some prescribed sets. Aequ. Math. **52**, 254–259 (1996)
22. Baron, K., Sablik, M., Volkmann, P.: On decent solutions of a functional congruence. Rocz. Nauk.-Dydakt. Pr. Mat. **17**, 27–40 (2000)
23. Beg, I.: Fuzzy multivalued functions. Bull. Allahabad Math. Soc. **21**, 41–104 (2006)
24. Berge, C.: Topological Spaces Including a Treatment of Multi-Valued Functions, Vector Spaces and Convexity. Oliver and Boyd, London (1963)
25. Berz, E.: Sublinear functions on $\mathbb{R}$. Aequ. Math. **12**, 200–206 (1975)
26. Boccuto, A., Candeloro, D.: Sandwich theorems and applications to invariant measures. Atti Sem. Math. Fis. Univ. Modena **38**, 511–524 (1990)
27. Bohnenblust, H.F., Sobczyk, A.: Extensions of functionals on complex linear spaces. Bull. Am. Math. Soc. **44**, 91–93 (1938)
28. Boros, Z.: Stability of the Cauchy equation in ordered fields. Math. Pannon. **11**, 191–197 (2000)
29. Boros, Z., Száz, Á.: Reflexivity, transitivity, symmetry, and antisymmetry of the intersection convolution of relations. Rostock. Math. Kolloq. **63**, 55–62 (2008)
30. Bourgin, D.G.: Clases of transformations and bordering transformations. Bull. Am. Math. Soc. **57**, 223–237 (1951)
31. Brzdek, J.: On the Cauchy difference. Glas. Mat. **27**, 263–269 (1992)
32. Brzdek, J.: On a method of proving the Hyers–Ulam stability of functional equations on restricted domains. Aust. J. Math. Anal. Appl. **6**, 1–10 (2009). Article 4
33. Brzdek, J., Tabor, J.: Stability of the Cauchy congruence on restricted domain. Arch. Math. **82**, 546–550 (2004)

34. Brzdek, J., Popa, D., Xu, B.: Selections of set-valued maps satisfying a linear inclusion in a single variable. Nonlinear Anal. **74**, 324–330 (2011)
35. Burai, P., Száz, Á.: Relationships between homogeneity, subadditivity and convexity properties. Publ. Elektroteh. Fak. Univ. Beogr., Mat. **16**, 77–87 (2005)
36. Buskes, G.: The Hahn–Banach theorem surveyed. Diss. Math. **327**, 1–49 (1993)
37. Chu, H.-Y., Kang, D.S., Rassias, Th.M.: On the stability of a mixed n-dimensional quadratic functional equation. Bull. Belg. Math. Soc. **15**, 9–24 (2008)
38. Castillo, E., Ruiz-Cobo, M.R.: Functional Equations and Modelling in Science and Engineering. Dekker, New York (1992)
39. Coddington, E.A.: Extension theory of formally normal and symmetric subspaces. Mem. Am. Math. Soc. **134**, 5–7 (1973)
40. Coddington, E.A., Dijksma, A.: Adjoint subspaces in Banach spaces with applications to ordinary differential subspaces. Ann. Mat. Pura Appl. **118**, 1–118 (1978)
41. Cross, R.: Multivalued Linear Operators. Dekker, New York (1998)
42. Czerwik, S.: Functional Equations and Inequalities in Several Variables. World Scientific, London (2002)
43. Czerwik, S. (ed.): Stability of Functional Equations of Ulam–Hyers–Rassias Type. Hadronic Press, Palm Harbor (2003)
44. Czerwik, S., Król, K.: Ulam stability of functional equations. Aust. J. Math. Anal. Appl. **6**, 1–15 (2009). Article 6
45. Dacić, R.: On multi-valued functions. Publ. Inst. Math. (Belgr.) **9**, 5–7 (1969)
46. Dascăl, J., Száz, Á.: Inclusion properties of the intersection convolution of relations. Ann. Math. Inform. **36**, 47–60 (2009)
47. Dascăl, J., Száz, Á.: A necessary condition for the extensions of subadditive partial selection relations. Tech. Rep., Inst. Math., Univ. Debrecen **1**, 1–13 (2009)
48. Faziev, V.A., Rassias, Th.M.: The space of (ψ, ν)-pseudocharacters on semigroups. Nonlinear Funct. Anal. Appl. **5**, 107–126 (2000)
49. Faziev, V.A., Rassias, Th.M., Sahoo, P.K.: The space of (ψ, ν)-additive mappings onsemigroups. Trans. Am. Math. Soc. **354**, 4455–4472 (2002)
50. Farkas, T., Száz, Á.: Minkowski functionals of summative sequences of absorbing and balanced sets. Bul. Ştiinţ. Univ. Baia Mare, Ser. B, Fasc. Mat.-Inform. **16**, 323–334 (2000)
51. Fechner, W.: Separation theorems for conditional functional equations. Ann. Math. Sil. **21**, 31–40 (2007)
52. Fechner, W.: On an abstract version of a functional inequality. Math. Inequal. Appl. **11**, 381–392 (2008)
53. Figula, A., Száz, Á.: Graphical relationships between the infimum and the intersection convolutions. Math. Pannon. **21**, 23–35 (2010)
54. Findlay, G.D.: Reflexive holomorphic relations. Can. Math. Bull. **3**, 131–132 (1960)
55. Forti, G.L.: An existence and stability theorem for a class of functional equations. Stohastica **4**, 23–30 (1980)
56. Forti, G.L.: Remark 11. Aequ. Math. **29**, 90–91 (1985)
57. Forti, G.L.: The stability of homomorphisms and amenability, with applications to functional equations. Abh. Math. Sem. Univ. Hamburg **57**, 215–226 (1987)
58. Forti, G.L.: Hyers–Ulam stability of functional equations in several variables. Aequ. Math. **50**, 143–190 (1995)
59. Forti, G.L.: Comments on the core of the direct method for proving Hyers–Ulam stability of functional equations. J. Math. Anal. Appl. Aequ. Math. **295**, 127–133 (2004)
60. Forti, G.L., Schwaiger, J.: Stability of homomorphisms and completeness. C. R. Math. Rep. Acad. Sci. Can. **11**, 215–220 (1989)
61. Fuchssteiner, B.: Sandwich theorems and lattice semigroups. J. Funct. Anal. **16**, 1–14 (1974)
62. Fuchssteiner, B., Horváth, J.: Die Bedeutung der Schnitteigenschaften beim Hahn–Banachschen Satz, Jahrbuch Überblicke Math, pp. 107–121. Bibliograph. Inst., Mannheim (1979). (The publication of an expanded English version of this paper in the Publ. Math. Debrecen was prevented in 1997 by the Editorial Board headed by L. Tamássy)

63. Fuchssteiner, B., Lusky, W.: Convex Cones. North-Holland, New York (1981)
64. Fullerton, R.E.: An intersection property for cones in a linear space. Proc. Am. Math. Soc. **9**, 558–561 (1958)
65. Gajda, Z.: On stability of the Cauchy equation on semigroups. Aequ. Math. **36**, 76–79 (1988)
66. Gajda, Z.: On stability of additive mappings. Int. J. Math. Sci. **14**, 431–434 (1991)
67. Gajda, Z.: Invariant means and representations of semigroups in the theory of functional equations. Pr. Nauk. Uniw. Ślask. Katowic. **1273**, 1–81 (1992)
68. Gajda, Z.: Sandwich theorems and amenable semigroups of transformations. Grazer Math. Ber. **316**, 43–58 (1992)
69. Gajda, Z., Ger, R.: Subadditive multifunctions and Hyers–Ulam stability. In: Walter, W. (ed.) General Inequalities 5. Internat. Ser. Numer. Math., vol. 80, pp. 281–291. Birkhäuser, Basel (1987)
70. Gajda, Z., Kominek, Z.: On separation theorems for subadditive and superadditive functionals. Stud. Math. **100**, 25–38 (1991)
71. Gajda, Z., Smajdor, A., Smajdor, W.: A theorem of the Hahn–Banach type and its applications. Ann. Pol. Math. **57**, 243–252 (1992)
72. Găvruţă, P.: A generalization of the Hyers–Ulam–Rassias stability of approximately additive mappings. J. Math. Anal. Appl. **184**, 431–436 (1994)
73. Găvruţă, P.: On a problem of G. Isac and Th.M. Rassias concerning the stability of mappings. J. Math. Anal. Appl. **261**, 543–553 (2001)
74. Găvruta, P., Găvruta, L.: A new method for the generalized Hyers–Ulam–Rassias stability. Int. J. Nonlinear Anal. Appl. **1**, 11–18 (2010)
75. Găvruta, P., Hossu, M., Popescu, D., Căprău, C.: On the stability of mappings and an answer to a problem of Th.M. Rassias. Ann. Math. Blaise Pascal **2**, 55–60 (1995)
76. Ger, R.: The singular case in the stability behaviour of linear mappings. Grazer Math. Ber. **316**, 59–70 (1992)
77. Ger, R.: On a factorization of mappings with a prescribed behavior of the Cauchy difference. Ann. Math. Sil. **8**, 147–155 (1994)
78. Ger, R.: A survey of recent results on stability of functional equations. In: Proceedings of the 4th International Conference on Functional Equations and Inequalities, pp. 5–36. Pedagogical University of Krakow (1994)
79. Ger, R., Volkmann, P.: On sums of linear and bounded mappings. Abh. Math. Semin. Univ. Hamb. **68**, 103–108 (1998)
80. Gilányi, A., Kaiser, Z., Páles, Zs.: Estimates to the stability of functional equations. Aequ. Math. **73**, 125–143 (2007)
81. Glavosits, T., Kézi, Cs.: On the domain of oddness of an infimal convolution. Miskolc Math. Notes **12**, 31–40 (2011)
82. Glavosits, T., Száz, Á.: Pointwise and global sums and negatives of binary relations. An. St., Univ. Ovidius Constanta **11**, 87–94 (2003)
83. Glavosits, T., Száz, Á.: Pointwise and global sums and negatives of translation relations. An. St., Univ. Ovidius Constanta **12**, 27–44 (2004)
84. Glavosits, T., Száz, Á.: On the existence of odd selections. Adv. Stud. Contemp. Math. (Kyungshang) **8**, 155–164 (2004)
85. Glavosits, T., Száz, Á.: General conditions for the subadditivity and superadditivity of relations. Scientia, Ser. A, Math. Sci. **11**, 31–43 (2005)
86. Glavosits, T., Száz, Á.: Constructions and extensions of free and controlled additive relations. Tech. Rep., Inst. Math., Univ. Debrecen **1**, 1–49 (2010)
87. Glavosits, T., Száz, Á.: The infimal convolution can be used to easily prove the classical Hahn–Banach theorem. Rostock. Math. Kolloq. **65**, 71–83 (2010)
88. Glavosits, T., Száz, Á.: The generalized infimal convolution can be used naturally prove some dominated monotone additive extension theorems. Tech. Rep., Inst. Math., Univ. Debrecen **4**, 1–26 (2010)
89. Glavosits, T., Száz, Á.: A Hahn–Banach type generalization of the Hyers–Ulam theorem. An. St., Univ. Ovidius Constanta **19**, 139–144 (2011)

90. Godini, G.: Set-valued Cauchy functional equation. Rev. Roum. Math. Pures Appl. **20**, 1113–1121 (1975)
91. Godini, G.: An approach to generalizing Banach spaces: normed almost linear spaces. Rend. Circ. Mat. Palermo, Suppl. **5**, 33–50 (1984)
92. Grabiec, A.: The generalized Hyers–Ulam stability of a class of functional equations. Publ. Math. (Debr.) **48**, 217–235 (1996)
93. Gselmann, E., Száz, Á.: An instructive treatment of a generalization of Găvruţă's stability theorem. Sarajevo J. Math. **6**, 3–21 (2010)
94. Hahn, H.: Über lineare Gleichungsysteme in linearen Räumen. J. Reine Angew. Math. **157**, 214–229 (1927)
95. Harrop, R., Weston, J.D.: An intersection property in locally convex spaces. Proc. Am. Math. Soc. **7**, 535–538 (1956)
96. Harte, R.E.: A generalization of the Hahn–Banach theorem. J. Lond. Math. Soc. **40**, 283–287 (1965)
97. Hassi, S., Sebestyén, Z., De Snoo, H.S.V., Szafraniec, F.H.: A canonical decomposition for linear operators and linear relations. Acta Math. Hung. **115**, 281–307 (2007)
98. Hasumi, M.: The extension property of complex Banach spaces. Tohoku Math. J. **10**, 135–142 (1958)
99. Helly, E.: Über lineare Functionaloperationen. Sitzungsber. Acad. Wiss. Wien **121**, 265–297 (1912)
100. Henney, D.: The structure of set-valued additive functions. Port. Math. **26**, 463–471 (1967)
101. Henney, D.: Properties of set-valued additive functions. Am. Math. Mon. **75**, 203–206 (1968)
102. Henney, D.: Representations of set-valued additive functions. Aequ. Math. **3**, 230–235 (1969)
103. Hochstadt, H.: Eduard Helly, father of the Hahn–Banach theorem. Math. Intell. **2**, 123–125 (1980)
104. Holá, L'.: Some properties of almost continuous linear relations. Acta Math. Univ. Comen. **50–51**, 61–69 (1987)
105. Holá, L'., Kupka, I.: Closed graph and open mapping theorems for linear relations. Acta Math. Univ. Comen. **46–47**, 157–162 (1985)
106. Holá, L'., Maličký, P.: Continuous linear selectors of linear relations. Acta Math. Univ. Comen. **48–49**, 153–157 (1986)
107. Holbrook, J.A.R.: Concerning the Hahn–Banach theorem. Proc. Am. Math. Soc. **50**, 322–327 (1975)
108. Horváth, J.: Some selected results of Professor Baltasar Rodríguez-Salinas. Rev. Mat. Univ. Complut. Madr. **9**, 23–72 (1996)
109. Huang, J., Li, Y.: The Hahn–Banach theorem on arbitrary groups. Kyungpook Math. J. **49**, 245–254 (2009)
110. Hustad, O.: A note on complex $\mathscr{P}_1$ spaces. Isr. J. Math. **16**, 117–119 (1973)
111. Hyers, D.H.: On the stability of the linear functional equation. Proc. Natl. Acad. Sci. USA **27**, 222–224 (1941)
112. Hyers, D.H.: The stability of homomorphisms and related topics. In: Rassias, Th.M. (ed.) Global Analysis on manifolds. Teubner-Texte Math., vol. 57, pp. 140–153, Leipzig (1983)
113. Hyers, D.H., Rassias, Th.M.: Approximate homomorphisms. Aequ. Math. **44**, 125–153 (1992)
114. Hyers, D.H., Isac, G., Rassias, Th.M.: Topics in Nonlinear Analysis and Applications. World Scientific, London (1997)
115. Hyers, D.H., Isac, G., Rassias, Th.M.: Stability of Functional Equations in Several Variables. Birkhäuser, Boston (1998)
116. Ingleton, A.W.: The Hahn–Banach theorem for non-Archimedean-valued fields. Proc. Camb. Philol. Soc. **48**, 41–45 (1952)
117. Ioffe, A.D.: A new proof of the equivalence of the Hahn–Banach extension and the least upper bound properties. Proc. Am. Math. Soc. **82**, 385–389 (1981)

118. Ioffe, A.D.: Nonsmooth analysis: differential calculus of nondifferentiable mappings. Trans. Am. Math. Soc. **266**, 1–56 (1981)
119. Isac, G., Rassias, Th.M.: On the Hyers–Ulam stability of ψ-additive mappings. J. Approx. Theory **72**, 131–137 (1993)
120. Isac, G., Rassias, Th.M.: Functional inequalities for approximately additive mappings. In: Rassias, Th.M., Tabor, J. (eds.) Stability of Mappings of Hyers–Ulam Type, pp. 117–125. Hadronic Press, Palm Harbor (1994)
121. Isac, G., Rassias, Th.M.: Stability of ψ-additive mappings: applications to non-linear analysis. Int. J. Math. Math. Sci. **19**, 219–228 (1996)
122. Jameson, G.: Ordered Linear Spaces. Lect. Notes Math., vol. 141. Springer, Berlin (1970)
123. Jun, K.-W., Shin, D.-S., Kim, B.-D.: On Hyers–Ulam–Rassias stability of the Pexider equation. J. Math. Anal. Appl. **239**, 20–29 (1999)
124. Jung, S.-M.: Hyers–Ulam–Rassias stability of functional equations. Dyn. Syst. Appl. **6**, 541–565 (1997)
125. Jung, S.-M.: Hyers–Ulam–Rassias Stability of Functional Equations in Mathematical Analysis. Hadronic Press, Palm Harbor (2001)
126. Jung, S.-M.: Survey on the Hyers–Ulam–Rassias stability of functional equations. In: Czerwik, S. (ed.) Stability of Functional Equations of Ulam–Hyers–Rassias Type, pp. 91–117. Hadronic Press, Palm Harbor (2003)
127. Kaiser, Z., Páles, Zs.: An example of a stable functional equation when the Hyers method does not work. J. Inequal. Pure Appl. Math. **6**, 1–11 (2005)
128. Kaufman, R.: Extension of functionals and inequalities on an Abelian semi-group. Proc. Am. Math. Soc. **17**, 83–85 (1966)
129. Kaufman, R.: Interpolation of additive functionals. Stud. Math. **27**, 269–272 (1966)
130. Kazhdan, D.: On ε-representations. Isr. J. Math. **43**, 315–323 (1982)
131. Kelley, J.L.: Banach spaces with the extension property. Trans. Am. Math. Soc. **72**, 323–326 (1952)
132. Kelley, J.L., Namioka, I.: Linear Topological Spaces. Van Nostrand, New York (1963)
133. Khodaei, H., Rassias, Th.M.: Approximately generalized additive functions in several variables. Int. J. Nonlinear Anal. Appl. **1**, 22–41 (2010)
134. Kim, G.K.: On the stability of functional equations with square-symmetric operations. Math. Inequal. Appl. **4**, 257–266 (2001)
135. Kominek, Z.: On Hyers–Ulam stability of the Pexider equation. Demonstr. Math. **37**, 373–376 (2004)
136. Kotarski, W.: A remark on the Hahn–Banach theorem. Rad. Mat. **3**, 105–109 (1987)
137. Kranz, P.: Additive functionals on abelian semigroups. Ann. Soc. Math. Pol. **16**, 239–246 (1972)
138. Kranz, P.: Extensions of additive functionals and semicharacters on commutative semigroups. Semigroup Forum **18**, 293–305 (1979)
139. Kuczma, M.: An Introduction to the Theory of Functional Equations and Inequalities. Polish Sci. Publ. and Univ. Ślaski, Warszawa (1985)
140. Lambek, J.: Goursat theorem and the Zassenhaus lemma. Can. J. Math. **10**, 45–56 (1958)
141. Larsen, R.: An Introduction to the Theory of Multipliers. Sringer, Berlin (1971)
142. Lee, Y.-H., Jun, K.-W.: On the stability of approximately additive mappings. Proc. Am. Math. Soc. **128**, 1361–1369 (1999)
143. Lee, S.J., Nashed, M.Z.: Algebraic and topological selections of multi-valued linear relations. Ann. Scuola Norm. Sup. Pisa **17**, 111–126 (1990)
144. Lee, S.J., Nashed, M.Z.: Normed linear relations: domain decomposability, adjoint subspaces, and selections. Linear Algebra Appl. **153**, 135–159 (1991)
145. Lorenzen, P.: Über die Korrespondenzen Einer. Struct. Mat. Zeitshr. **60**, 61–65 (1954)
146. Lu, G., Park, Ch.: Hyers–Ulam stability of additive set-valued functional equations. Appl. Math. Lett. (2011). doi:10.1016/j.aml.2011.02.024
147. MacLane, S.: An algebra of additive relations. Proc. Natl. Acad. Sci. USA **47**, 1043–1051 (1961)

148. MacLane, S.: Homology. Springer, Berlin (1963)
149. Maligranda, L.: A result of Tosio Aoki about a generalization of Hyers–Ulam stability of additive functions – a question of priority. Aequ. Math. **75**, 289–296 (2008)
150. Mihet, D.: The Hyers–Ulam stability for two functional equation in a single variable. Banach J. Math. Anal. **2**, 48–52 (2008)
151. Moreau, J.J.: Inf-convolution, sous-additivité, convexité des fonctions numériques. J. Math. Pures Appl. **49**, 109–154 (1970)
152. Moslehian, M.S., Rassias, Th.M.: Stability of functional equations in non-Archimedean spaces. Appl. Anal. Discrete Math. **1**, 325–334 (2007)
153. Moszner, Z.: On the stability of functional equations. Aequ. Math. **77**, 33–88 (2009)
154. Murray, F.J.: Linear transformations in L_p $(p > 1)$. Trans. Am. Math. Soc. **39**, 83–100 (1936)
155. Nachbin, L.: A theorem of the Hahn–Banach type for linear transformations. Trans. Am. Math. Soc. **68**, 28–46 (1950)
156. Najati, A., Rassias, Th.M.: Stability of homomorphisms and (θ, φ)-derivations. Appl. Anal. Discrete Math. **3**, 264–281 (2009)
157. Najati, A., Rassias, Th.M.: Stability of a mixed functional equation in several variables on Banach modules. Nonlinear Anal. **72**, 1755–1767 (2010)
158. Narici, L.: On the Hahn–Banach theorem. In: Advanced Courses of Mathematical Analysis II, pp. 87–122. World Sci., Hackensack (2007)
159. Narici, L., Beckenstein, E.: The Hahn–Banach theorem: the life and times. Topol. Appl. **77**, 193–211 (1997)
160. Narici, L., Beckenstein, E.: The Hahn–Banach theorem and the sad life of E. Helly. In: Advanced Courses of Mathematical Analysis III, pp. 97–110. World Sci., Hackensack (2008)
161. Von Neumann, J.: Über adjungierte Functional-operatoren. Ann. Math. **33**, 294–310 (1932)
162. Von Neumann, J.: Functional Operators: The Geometry of Orthogonal Spaces. Ann. Math. Stud., vol. 22. Princeton University Press, Princeton (1950)
163. Nikodem, K.: Additive selections of additive set-valued functions. Univ. u Novom Sadu, Zb. Rad. Prirod.-Mat. Fak., Ser. Mat. **18**, 143–148 (1988)
164. Nikodem, K.: K-convex and K-concave set-valued functions. Zeszyty Nauk. Politech. Lódz. Mat. **559**, 1–75 (1989)
165. Nikodem, K.: The stability of the Pexider equation. Ann. Math. Sil. **5**, 91–93 (1991)
166. Nikodem, K.: Remarks on additive injective set-valued functions. An. Univ. Oradea, Fasc. Mat. **11**, 175–180 (2004)
167. Nikodem, K., Popa, D.: On single-valuedness of set-valued maps satisfying linear inclusions. Banach J. Math. Anal **3**, 44–51 (2009)
168. Nikodem, K., Popa, D.: On selections of general linear inclusions. Publ. Math. (Debr.) **75**, 239–249 (2009)
169. Nikodem, K., Wasowicz, Sz.: A sandwich theorem and Hyers–Ulam stability of affine functions. Aequ. Math. Sil. **49**, 160–164 (1995)
170. Nikodem, K., Páles, Zs., Wasowicz, Sz.: Abstract separation theorems of Rodé type and their applications. Ann. Pol. Math. **72**, 207–217 (1999)
171. Nikodem, K., Sadowska, E., Wasowicz, Sz.: Note on separation by subadditive and sublinear functions. Ann. Math. Sil. **14**, 7–21 (2000)
172. Paganoni, L.: Soluzione di una equazione funzionale su dominio ristretto. Boll. Unione Mat. Ital. **17-B**, 979–993 (1980)
173. Páles, Zs.: Linear selections for set-valued functions and extension of bilinear forms. Arch. Math. (Basel) **62**, 427–432 (1994)
174. Páles, Zs.: Separation with symmetric bilinear forms and symmetric selections of set-valued functions. Publ. Math. (Debr.) **46**, 321–331 (1995)
175. Páles, Zs.: Separation by semidefinite bilinear forms. Int. Ser. Numer. Math. **123**, 259–267 (1997)
176. Páles, Zs.: Generalized stability of the Cauchy functional equations. Aequ. Math. **56**, 222–232 (1998)

177. Páles, Zs.: Hyers–Ulam stability of the Cauchy functional equation on square-symmetric groupoids. Publ. Math. (Debr.) **58**, 651–666 (2001)
178. Páles, Zs., Székelyhidi, L.: On approximate sandwich and decomposition theorems. Ann. Univ. Sci. Budapest Sect. Comput. **23**, 59–70 (2004)
179. Park, Ch., Rassias, Th.M.: Fixed points and generalized Hyers–Ulam stability of quadratic functional equations. J. Math. Inequal. **1**, 515–528 (2007)
180. Park, Ch., Rassias, Th.M.: Fixed points and stability of the Cauchy functional equation. Aust. J. Math. Anal. Appl. **6**, 14 (2009). pp. 1–9
181. Pataki, G.: On a generalized infimal convolution of set functions. Manuscript
182. Peng, J., Lee, H.W.J., Rong, W., Yang, X.: A generalization of Hahn–Banach extension theorem. J. Math. Anal. Appl. **302**, 441–449 (2005)
183. Pérez-Garcia, C.: The Hahn–Banach extension property in p-adic analysis. In: P-adic Functional Analysis. Lect. Notes in Pure and Appl. Math., vol. 137, pp. 127–140. Dekker, New York (1992)
184. Piao, Y.J.: The existence and uniqueness of additive selections for (α, β)–(β, α) type subadditive set-valued maps. J. Northeast Norm. University **41**, 33–40 (2009)
185. Plappert, P.: Sandwich theorem for monotone additive functions. Semigroup Forum **51**, 347–355 (1995)
186. Plewnia, J.: A generalization of the Hahn–Banach theorem. Ann. Pol. Math. **58**, 47–51 (1993)
187. Pólya, Gy., Szegő, G.: Aufgaben und Lehrsätze aus der Analysis I. Springer, Berlin (1925)
188. Popa, D.: Additive selections of (α, β)-subadditive set valued maps. Glas. Mat. **36**, 11–16 (2001)
189. Popa, D.: Functional inclusions on square-symmetric grupoids and Hyers–Ulam stability. Math. Inequal. Appl. **7**, 419–428 (2004)
190. Popa, D., Vornicescu, N.: Locally compact set-valued solutions for the general linear equation. Aequ. Math. **67**, 205–215 (2004)
191. Przeslawski, K.: Linear and lipschitz continuous selectors for the family of convex sets in Euclidean vector spaces. Bull. Acad. Pol. Sci. Ser. Sci. Math. **33**, 31–33 (1985)
192. Rådström, H.: One-parameter semigroups of subsets of a real linear space. Ark. Mat. **4**, 87–97 (1960)
193. Radu, V.: The fixed point alternative and the stability of functional equations. Fixed Point Theory **4**, 91–96 (2003)
194. Rassias, J.M.: On approximation of approximately linear mappings by linear mappings. J. Funct. Anal. **46**, 126–130 (1982)
195. Rassias, J.M.: On approximation of approximately linear mappings by linear mappings. Bull. Sci. Math. **108**, 445–446 (1984)
196. Rassias, J.M.: Solution of a problem of Ulam. J. Approx. Theory **57**, 268–273 (1989)
197. Rassias, Th.M.: On the stability of the linear mapping in Banach spaces. Proc. Am. Math. Soc. **72**, 279–300 (1978)
198. Rassias, Th.M.: On a modified Hyers–Ulam sequence. J. Math. Anal. Appl. **158**, 106–113 (1991)
199. Rassias, Th.M.: On the stability of the quadratic functional equation and its applications. Stud. Univ. Babeş-Bolyai, Math. **43**, 89–124 (1998)
200. Rassias, Th.M.: Stability and set-valued functions. In: Analysis and Topology, pp. 585–614. World Sci., River Edge (1998)
201. Rassias, Th.M.: On the stability of functional equations and a problem of Ulam. Acta Appl. Math. **62**, 23–130 (2000)
202. Rassias, Th.M.: The problem of S.M. Ulam for approximately multiplicative mappings. J. Math. Anal. Appl. **246**, 352–378 (2000)
203. Rassias, Th.M.: On the stability of functional equations in Banach spaces. J. Math. Anal. Appl. **251**, 264–284 (2000)
204. Rassias, Th.M.: On the stability of functional equations originated by a problem of Ulam. Mathematica (Cluj) **44**, 39–75 (2002)

205. Rassias, Th.M., Šemrl, P.: On the behavior of mappings which do not satisfy Hyers–Ulam stability. Proc. Am. Math. Soc. **114**, 989–993 (1992)
206. Rassias, Th.M., Šemrl, P.: On the Hyers–Ulam stability of linear mappings. J. Math. Anal. Appl. **173**, 325–338 (1993)
207. Rassias, Th.M., Tabor, J.: What is left of Hyers–Ulam stability? J. Nat. Geom. **1**, 65–69 (1992)
208. Rassias, Th.M., Tabor, J. (eds.): Stability of Mappings of Hyers–Ulam Type. Hadronic Press, Palm Harbor (1994)
209. Rätz, J.: On approximately additive mappings. In: Beckenbach, E.F. (ed.) General Inequalities 2, Oberwolfach, 1978. Int. Ser. Num. Math., vol. 47, pp. 233–251. Birkhäuser, Basel (1980)
210. Revenko, A.V.: On the extension of linear functionals. Ukr. Math. Visn. **6**, 113–125 (2009) (Russian)
211. Riesz, F.: Sur les systémes orthogonaux de fonctions. C. R. Acad. Sci. Paris **144**, 615–619 (1907)
212. Riesz, F.: Untersuchungen über Systeme integrierbare Funktionen. Math. Ann. **69**, 449–497 (1910)
213. Robinson, S.M.: Normed convex processes. Trans. Am. Math. Soc. **174**, 127–140 (1992)
214. Rockafellar, R.T.: Monotone processes of convex and concave type. Mem. Amer. Math. Soc. **77** (1967)
215. Rodríguez-Salinas, B.: Generalizacíon sobre módulos del teorema de Hahn–Banach y sus aplicaciones. Collect. Math. **14**, 105–151 (1962)
216. Rodríguez-Salinas, B.: Algunos problemas y teoremas de extension de aplicaciones lineales. Rev. R. Acad. Cienc. Exactas Fís. Nat. Madr. **65**, 677–704 (1971)
217. Rodríguez-Salinas, B., Bou, L.: A Hahn–Banach theorem for arbitrary vector spaces. Boll. Unione Mat. Ital. **10**, 390–393 (1974)
218. Roth, W.: Hahn–Banach type theorems for locally convex cones. J. Aust. Math. Soc. **68**, 104–125 (2000)
219. Sablik, M.: A functional congruence revisited. Grazer Math. Ber. **316**, 181–200 (1992)
220. Sánchez, F.C., Castillo, J.M.F.: Banach space techniques underpinning a theory for nearly additive mappings. Diss. Math. **404**, 1–73 (2002)
221. Saccoman, J.J.: Extension theorems by Helly and Riesz revisited. Riv. Mat. Univ. Parma **16**, 223–230 (1990)
222. Saccoman, J.J.: On the extension of linear operators. Int. J. Math. Math. Sci. **28**, 621–623 (2001). http:/ijmms.hindawi.com
223. Sandovici, A., de Snoo, H., Winkler, H.: Ascent, descent, nullity, defect, and related notions for linear relations in linear spaces. Linear Algebra Appl. **423**, 456–497 (2007)
224. Saveliev, P.: Lomonosov's invariant subspace theorem for multivalued linear operators. Proc. Am. Math. Soc. **131**, 825–834 (2002)
225. Schwaiger, J.: Remark 12. Aequ. Math. **35**, 120–121 (1988)
226. Simons, S.: Extended and sandwich versions of the Hahn–Banach theorem. J. Math. Anal. Appl. **21**, 112–122 (1968)
227. Simons, S.: From Hahn–Banach to Monotonicity. Springer, Berlin (2008)
228. Smajdor, A.: Additive selections of superadditive set-valued functions. Aequ. Math. **39**, 121–128 (1990)
229. Smajdor, A.: The stability of the Pexider equation for set-valued functions. Rocz. Nauk.-Dydakt. Pr. Mat. **13**, 277–286 (1993)
230. Smajdor, A., Smajdor, W.: Affine selections of convex set-valued functions. Aequ. Math. **51**, 12–20 (1996)
231. Smajdor, W.: Subadditive set-valued functions. Glas. Mat. **21**, 343–348 (1986)
232. Smajdor, W.: Superadditive set-valued functions and Banach–Steinhaus theorem. Rad. Mat. **3**, 203–214 (1987)
233. Smajdor, W.: Subadditive and subquadratic set-valued functions. Pr. Nauk. Univ. Śląsk. Katowic. **889**, 1–73 (1987)

234. Smajdor, W., Szczawińska, J.: A theorem of the Hahn–Banach type. Demonstr. Math. **28**, 155–160 (1995)
235. Soukhomlinov, G.A.: On the extension of linear functionals in complex and quaternion linear spaces. Mat. Sb. **3**, 353–358 (1938)
236. Špakula, J., Zlatoš, P.: Almost homomorphisms of compact groups. Ill. J. Math. **48**, 1183–1189 (2004)
237. Strömberg, T.: The operation of infimal convolution. Diss. Math. **352**, 1–58 (1996)
238. Száz, Á.: Convolution multipliers and distributions. Pac. J. Math. **60**, 267–275 (1975)
239. P185R1: Aequ. Math. **22**, 308–309 (1981)
240. Száz, Á.: Generalized preseminormed spaces of linear manifolds and bounded linear relations. Univ. u Novom Sadu Zb. Rad. Prirod. Mat. Fak. Ser. Mat. **14**, 49–78 (1984)
241. Száz, Á.: Projective generation of preseminormed spaces by linear relations. Studia Sci. Math. Hung. **23**, 297–313 (1988)
242. Száz, Á.: Pointwise limits of nets of multilinear maps. Acta Sci. Math. (Szeged) **55**, 103–117 (1991)
243. Száz, Á.: The intersection convolution of relations and the Hahn–Banach type theorems. Ann. Pol. Math. **69**, 235–249 (1998)
244. Száz, Á.: An extension of Banach's closed graph theorem to linear relations. Leaflets Math. Pécs, 144–145 (1998)
245. Száz, Á.: Translation relations, the building blocks of compatible relators. Math. Montisnigri **12**, 135–156 (2000)
246. Száz, Á.: Preseminorm generating relations. Publ. Elektroteh. Fak. Univ. Beogr., Mat. **12**, 16–34 (2001)
247. Száz, Á.: Partial multipliers on partially ordered sets. Novi Sad J. Math. **32**, 25–45 (2002)
248. Száz, Á.: Relationships between translation and additive relations. Acta Acad. Paed. Agriensis, Sect. Math. (N.S.) **30**, 179–190 (2003)
249. Száz, Á.: Linear extensions of relations between vector spaces. Comment. Math. Univ. Carol. **44**, 367–385 (2003)
250. Száz, Á.: Lower semicontinuity properties of relations in relator spaces. Tech. Rep., Inst. Math., Univ. Debrecen **4**, 1–52 (2006)
251. Száz, Á.: An extension of an additive selection theorem of Z. Gajda and R. Ger. Tech. Rep., Inst. Math., Univ. Debrecen **8**, 1–24 (2006)
252. Száz, Á.: Applications of fat and dense sets in the theory of additive functions. Tech. Rep., Inst. Math., Univ. Debrecen **3**, 1–29 (2007)
253. Száz, Á.: An instructive treatment of a generalization of Hyers's stability theorem. In: Rassias, Th.M., Andrica, D. (eds.) Inequalities and Applications, pp. 245–271. Cluj University Press, Romania (2008)
254. Száz, Á.: Relationships between the intersection convolution and other important operations on relations. Math. Pannon. **20**, 99–107 (2009)
255. Száz, Á.: Applications of relations and relators in the extensions of stability theorems for homogeneous and additive functions. Aust. J. Math. Anal. Appl. **6**, 16 (2009), pp. 66
256. Száz, Á.: A reduction theorem for a generalized infimal convolution. Tech. Rep., Inst. Math., Univ. Debrecen **11**, 1–4 (2009)
257. Száz, Á.: Foundations of the theory of vector relators. Adv. Stud. Contemp. Math. (Kyungshang), **20**, 139–195 (2010)
258. Száz, Á.: The intersection convolution of relations. Creative Math. Inf. **19**, 209–217 (2010)
259. Száz, Á.: The infimal convolution can be used to derive extension theorems from the sandwich ones. Acta Sci. Math. (Szeged) **76**, 489–499 (2010)
260. Száz, Á.: Relation theoretic operations on box and totalization relations. Tech. Rep., Inst. Math., Univ. Debrecen **13**, 1–22 (2010)
261. Száz, Á., Száz, G.: Additive relations. Publ. Math. (Debr.) **20**, 172–259 (1973)
262. Száz, Á., Száz, G.: Linear relations. Publ. Math. (Debr.) **27**, 219–227 (1980)
263. Száz, Á., Száz, G.: Absolutely linear relations. Aequ. Math. **21**, 8–15 (1980)

264. Száz, Á., Száz, G.: Quotient spaces defined by linear relations. Czechoslov. Math. J. **32**, 227–232 (1982)
265. Száz, Á., Száz, G.: Multilinear relations. Publ. Math. (Debr.) **31**, 163–164 (1984)
266. Száz, Á., Túri, J.: Pointwise and global sums and negatives of odd and additive relations. Octogon **11**, 114–125 (2003)
267. Szczawińska, J.: Selections of biadditive set-valued functions. Ann. Math. Sil. **8**, 227–240 (1994)
268. Székelyhidi, L.: Remark 17. Aequ. Math. **29**, 95–96 (1985)
269. Székelyhidi, L.: Note on Hyers's theorem. C. R. Math. Rep. Acad. Sci. Can. **8**, 127–129 (1986)
270. Székelyhidi, L.: Ulam's problem, Hyers's solution—and to where they led. In: Rassias, Th.M. (ed.) Functional Equations and Inequalities. Math. Appl., vol. 518, pp. 259–285. Kluwer Acad., Dordrecht (2000)
271. Tabor, J.: Remark 18. Aequ. Math. **29**, 96 (1985)
272. Tabor, J. Jr.: Ideally convex sets and hyers theorem. Funkc. Ekvacioj **43**, 121–125 (2000)
273. Tabor, J. Jr.: Hyers theorem and the cocycle property. In: Daróczy, Z., Páles, Zs. (eds.) Funtional Equations—Results and Advances, pp. 275–291. Kluwer, Boston (2002)
274. Tabor, J., Tabor, J.: Restricted stability and shadowing. Publ. Math. (Debr.) **73**, 49–58 (2008)
275. Tabor, J., Tabor, J.: Stability of the Cauchy functional equation in metric groupoids. Aequ. Math. **76**, 92–104 (2008)
276. Takahasi, S.-E., Miura, T., Takagi, H.: On a Hyers–Ulam–Aoki–Rassias type stability and a fixed point theorem. J. Nonlinear Convex Anal. **11**, 423–439 (2010)
277. Ulam, S.M.: Collection of Mathematical Problems. Interscience, New York (1960)
278. Ursescu, C.: Multifunctions with convex closed graph. Czechoslov. Math. J. **25**, 438–441 (1975)
279. Volkmann, P.: On the stability of the Cauchy equation. Leaflets Math. Pécs, 150–151 (1998)
280. Volkmann, P.: Zur Rolle der ideal konvexen Mengen bei der Stabilität der Cauchyschen Funktionalgleichung. Seminar LV **6**, 1–6 (1999). http://www.mathematik.uni-karlsruhe.de/~semlv
281. Volkmann, P.: Instabilität einer zu $f(x+y) = f(x) + f(y)$ äquivalenten Functionalgleichung. Seminar LV **23**, 1 (2006). http://www.mathematik.uni-karlsruhe.de/~semlv
282. Whitehead, W.: Elements of Homotopy Theory. Springer, Berlin (1978)
283. Zălinescu, C.: Hahn–Banach extension theorems for multifunctions. Math. Methods Oper. Res. **68**, 493–508 (2008)

Chapter 40
Spectral Analysis and Spectral Synthesis

László Székelyhidi

Abstract Spectral analysis and spectral synthesis deal with the description of translation invariant function spaces over locally compact Abelian groups. One considers the space $\mathscr{C}(G)$ of all complex valued continuous functions on a locally compact Abelian group G, which is a locally convex topological linear space with respect to the point-wise linear operations (addition, multiplication with scalars) and to the topology of uniform convergence on compact sets. A variety is a closed translation invariant subspace of this space. Continuous homomorphisms of G into the additive topological group of complex numbers and into the multiplicative topological group of nonzero complex numbers, respectively, are called *additive* and *exponential functions*, respectively. A function is a *polynomial* if it belongs to the algebra generated by the continuous additive functions. An *exponential monomial* is the product of a polynomial and an exponential. It turns out that exponential functions, or more generally, exponential monomials can be considered as basic building bricks of varieties. A given variety may or may not contain any exponential function or exponential monomial. If it contains an exponential function, then we say that *spectral analysis* holds for the variety. An exponential function in a variety can be considered as a kind of *spectral value* and the set of all exponential functions in a variety is called the *spectrum* of the variety. It follows that spectral analysis for a variety means that the spectrum of the variety is nonempty. On the other hand, the set of all exponential monomials contained in a variety is called the *spectral set* of the variety. It turns out that if an exponential monomial belongs to a variety, then the exponential function appearing in the representation of this exponential monomial belongs to the variety, too. Hence, if the spectral set of a variety is nonempty, then also the spectrum of the variety is nonempty and spectral analysis holds. There is, however, an even stronger property of some varieties, namely, if the spectral set of the variety spans a dense subspace of the variety. In this case, we say that *spectral synthesis* holds for the variety. It follows that for nonzero varieties spectral synthesis implies spectral analysis. If spectral analysis (resp., spectral synthesis) holds for ev-

---

Dedicated to Professor Themistocles M. Rassias on the occasion of his 60th birthday.

L. Székelyhidi (✉)
Institute of Mathematics, University of Debrecen, Debrecen, Hungary
e-mail: lszekelyhidi@gmail.com

P.M. Pardalos et al. (eds.), *Nonlinear Analysis*, Springer Optimization and Its Applications 68, In Honor of Themistocles M. Rassias on the Occasion of his 60th Birthday,
DOI 10.1007/978-1-4614-3498-6_40,

ery variety on an Abelian group, then we say that *spectral analysis* (resp., *spectral synthesis*) holds on the Abelian group. A famous and pioneer result of L. Schwartz exhibits the situation by stating that if the group is the reals with the Euclidean topology, then spectral values do exist, that is, any nonzero variety contains an exponential function. In other words, in this case the spectrum is nonempty, spectral analysis holds. Furthermore, spectral synthesis also holds in this situation: there are sufficiently many exponential monomials in the variety in the sense that their linear hull is dense in the subspace. In this survey paper, we present a summary of the relevant results in spectral analysis and spectral synthesis including the most recent developments.

Key words Spectral analysis · Spectral synthesis · Locally compact groups

Mathematics Subject Classification 43A45 · 43A60 · 43A65

40.1 Introduction

Spectral analysis and spectral synthesis deal with the description of translation invariant function spaces over locally compact Abelian groups. We consider the space $\mathscr{C}(G)$ of all complex valued continuous functions on a locally compact Abelian group G, which is a locally convex topological linear space with respect to the point-wise linear operations (addition, multiplication with scalars) and to the topology of uniform convergence on compact sets. Continuous homomorphisms of G into the additive topological group of complex numbers and into the multiplicative topological group of nonzero complex numbers, respectively, are called *additive* and *exponential functions*, respectively. A function is a *polynomial* if it belongs to the algebra generated by the continuous additive functions. An *exponential monomial* is the product of a polynomial and an exponential.

It turns out that exponential functions, or more generally, exponential monomials can be considered as basic building bricks of varieties. A given variety may or may not contain any exponential function or exponential monomial. If it contains an exponential function, then we say that *spectral analysis* holds for the variety. An exponential function in a variety can be considered as a kind of *spectral value* and the set of all exponential functions in a variety is called the *spectrum* of the variety. It follows that spectral analysis for a variety means that the spectrum of the variety is nonempty. On the other hand, the set of all exponential monomials contained in a variety is called the *spectral set* of the variety. It turns out that if an exponential monomial belongs to a variety, then the exponential function appearing in the representation of this exponential monomial belongs to the variety, too. Hence, if the spectral set of a variety is nonempty, then also the spectrum of the variety is nonempty and spectral analysis holds. There is, however, an even stronger property of some varieties, namely, if the spectral set of the variety spans a dense subspace of the variety. In this case we say that *spectral synthesis* holds for the variety. It follows, that for nonzero varieties spectral synthesis implies spectral analysis.

If spectral analysis (resp., spectral synthesis) holds for every variety on an Abelian group, then we say that *spectral analysis* (resp., *spectral synthesis*) holds on the group. A famous and pioneer result of L. Schwartz [13] exhibits the situation by stating that if the group is the reals with the Euclidean topology, then spectral values do exist, that is, any nonzero variety contains an exponential function. In other words, in this case the spectrum is nonempty, spectral analysis holds. Furthermore, spectral synthesis also holds in this situation: there are sufficiently many exponential monomials in the variety in the sense that their linear hull is dense in the subspace.

Theorem 40.1 (L. Schwartz, 1947) *If V is a closed linear translation invariant space of complex valued continuous functions on the real line, then the linear combinations of exponential monomials of the form $x \mapsto x^k e^{\lambda x}$ are dense in V.*

An interesting particular case is presented by discrete Abelian groups. Here the problem seems to be purely algebraic: all complex functions are continuous, and convergence is meant in the point-wise sense. The archetype is the additive group of integers: in this case, the closed translation invariant function spaces can be characterized by systems of homogeneous linear difference equations with constant coefficients. It is known that these function spaces are spanned by exponential monomials corresponding to the characteristic values of the equation, together with their multiplicities. In this sense, the classical theory of homogeneous linear difference equations with constant coefficients can be considered as spectral analysis and spectral synthesis on the additive group of integers.

The next simplest case is the case of systems of homogeneous linear difference equations with constant coefficients in several variables. The corresponding—nontrivial—result by M. Lefranc [9] settles this case.

Theorem 40.2 (M. Lefranc, 1958) *Spectral synthesis holds on $\mathbb{Z}^n$ for any positive integer n.*

Obviously, this theorem implies that spectral synthesis holds on any finitely generated free Abelian group. This result has been extended by the present author for any finitely generated Abelian group in [16].

Theorem 40.3 (2001) *Spectral synthesis holds on any finitely generated Abelian group.*

At this point, the reader may ask the natural question: What about general Abelian groups? In his 1965 paper [3], R.J. Elliot presented a theorem on spectral synthesis for arbitrary Abelian groups. However, in 1987 Z. Gajda in [4] called my attention to the fact that the proof of Elliot's theorem had several gaps. Since then several efforts have been made to solve the problem of discrete spectral analysis and spectral synthesis on Abelian groups. In the subsequent paragraphs, we present the development of this theory until the present status.

40.2 Functional Equations

The study of varieties on a locally compact Abelian group is closely related to the study of systems of functional equations. Namely, it turns out that the solution space of a wide class of systems of functional equations on locally compact Abelian groups forms a variety. These classes of functional equations are the so-called *convolution type functional equations*.

Let $\mathscr{M}_c(G)$ denote the space of all complex valued compactly supported Radon measures on G, equipped with the pointwise operations and the weak topology. Then $\mathscr{M}_c(G)$ is a locally convex topological vector space. If μ is a measure, then the solutions $f : G \to \mathbb{C}$ of the convolution equation

$$f * \mu = 0 \tag{40.1}$$

form a variety. This is the *solution space* of the previous equation, denoted by $V(\mu)$. If G is discrete, then $\mathscr{M}_c(G)$ is the space of all complex valued finitely supported measures on G, and equations of the form (40.1) are exactly what we call *finite difference equations*.

More generally, let Λ be a set of measures in $\mathscr{M}_c(G)$ and let $V(\Lambda)$ denote the set of all functions in $\mathscr{C}(G)$ for which (40.1) holds for each μ in Λ. Then, clearly, $V(\Lambda)$ is a variety. This variety will be denoted by $\Lambda^\perp$. This is the *solution space* of the system of convolution type functional equations

$$f * \mu = 0, \quad \mu \in \Lambda. \tag{40.2}$$

Conversely, let V be a set in $\mathscr{C}(G)$ and let $\Lambda(V)$ denote the set of all measures μ in $\mathscr{M}_c(G)$ for which (40.1) holds for each f in V. Then $\Lambda(V)$ is an ideal in $\mathscr{M}_c(G)$, the so-called *annihilator* of V, which is denoted by $V^\perp$. The next theorem easily follows from the Hahn–Banach theorem.

Theorem 40.4 *If V is a variety in $\mathscr{C}(G)$, then $V^{\perp\perp} = V$. If Λ is a closed ideal in $\mathscr{M}_c(G)$, then $\Lambda^{\perp\perp} = \Lambda$.*

This theorem implies that each variety in $\mathscr{C}(G)$ is actually the solution space of a system of convolution type functional equations. Indeed, as $V = (V^\perp)^\perp$, V is the solution space of the system of convolution type functional equations which correspond to the measures in $V^\perp$. This means that the study of varieties is equivalent to the study of the systems of convolution type functional equations. This idea has been worked out in the monograph [14].

Let G be a locally compact Abelian group and let Λ and Γ be sets of measures in $\mathscr{M}_c(G)$. We say that Λ *implies* Γ, if $V(\Lambda)$ is a subset of $V(\Gamma)$. We say that Λ *is equivalent to* Γ, if $V(\Lambda)$ is a equal to $V(\Gamma)$.

Theorem 40.5 *Let G be a locally compact Abelian group and suppose that spectral synthesis holds on G. Let Λ and Γ be sets of compactly supported complex Radon measures on G. Then Λ implies Γ if and only if the spectral set of Λ is a subset of* *the spectral set of Γ. Moreover, Λ is equivalent to Γ if and only if the spectral set of Λ is equal to the spectral set of Γ.*

For the proof of this theorem, see [15].

40.3 Non-finitely Generated Discrete Abelian Groups

After the 1987 remark of Z. Gajda, one could ask the natural question: Can spectral synthesis hold on non-finitely generated discrete Abelian groups? Of course, the same question can be formulated concerning spectral analysis: Can spectral analysis hold on non-finitely generated discrete Abelian groups? This latter problem has close connection with the classical Wiener Tauberian theorem. A possible formulation of one version of this theorem on locally compact Abelian groups is the following: If G is a locally compact Abelian group, then any nonzero closed translation invariant subspace of $L^\infty(G)$ contains a character. It is easy to see that the essentially bounded nonzero exponential monomials are exactly the characters. Hence the statement of the Wiener Tauberian Theorem can be reformulated.

Theorem 40.6 (Wiener Tauberian Theorem) *On any locally compact Abelian group spectral analysis holds for the nonzero varieties in $L^\infty(G)$.*

Hence, in some sense, this theorem can be considered as a kind of spectral analysis theorem. On discrete Abelian groups, the first general result in this direction for varieties of unbounded functions was the following (see [17]).

Theorem 40.7 *Spectral analysis holds on any discrete Abelian torsion group.*

The proof of this theorem heavily depends on the fact that on commutative torsion groups the nonzero exponential monomials are exactly the characters (see [17, Theorem 3]).

At this point, we can answer our second question above. Namely, as there are, obviously, Abelian torsion groups, which are not finitely generated, hence there are non-finitely generated discrete Abelian groups on which spectral analysis holds. Nevertheless, the problem of finding non-finitely generated discrete Abelian groups on which spectral synthesis holds remains open. Actually, so far we have no example of a discrete Abelian group on which spectral analysis or spectral synthesis fails to hold. A counterexample due to the present author for Elliot's theorem was presented at the 41st International Symposium on Functional Equations, Noszvaj, Hungary, 2003. This counterexample depends on the following observation (see [18]).

Theorem 40.8 *Let G be an Abelian group. If there exists a symmetric bi-additive function $B : G \times G \to \mathbb{C}$ such that the variety V generated by the quadratic function $x \mapsto B(x, x)$ is of infinite dimension, then spectral synthesis fails to hold for V.*

Proof Let $f(x) = B(x,x)$ for all x in G. From the equation

$$f(x+y) = B(x+y, x+y) = B(x,x) + 2B(x,y) + B(y,y) \tag{40.3}$$

we see that the translation invariant subspace generated by f is generated by the functions 1, f, and all the additive functions of the form $x \mapsto B(x,y)$, where y runs through G. Hence our assumption on B is equivalent to the condition that there are infinitely many functions of the form $x \mapsto B(x,y)$ with y in G, which are linearly independent. This also implies that there is no positive integer n such that B can be represented in the form

$$B(x,y) = \sum_{k=1}^{n} a_k(x) b_k(y),$$

where $a_k, b_k : G \to \mathbb{C}$ are additive functions $(k = 1, 2, \ldots, n)$. Indeed, the existence of a representation of this form would mean that the number of linearly independent additive functions of the form $x \mapsto B(x,y)$ is at most n.

It is clear that any translate of f, hence any function g in V, satisfies

$$\Delta_y^3 g(x) = 0 \tag{40.4}$$

for all x, y in G: this can be checked directly for f. Here the operator Δ_y is defined, as usual, by

$$\Delta_y f(x) = f(x+y) - f(x)$$

for each function f, and real numbers x, y. Hence any exponential m in V satisfies the same equation, which implies

$$m(x)\big(m(y) - 1\big)^3 = 0$$

for all x, y in G, and this means that m is identically 1. It follows that any exponential monomial in V is a polynomial. By the results in [2] (see also [14]) and by (40.4), g can be uniquely represented in the following form:

$$g(x) = A(x,x) + c(x) + d$$

for all x in G, where $A : G \times G \to \mathbb{C}$ is a symmetric bi-additive function, $c : G \to \mathbb{C}$ is additive, and d is a complex number. Here "uniqueness" means that the "monomial terms" $x \mapsto A(x,x)$, $x \mapsto c(x)$, and d are uniquely determined (see [14]). In particular, any polynomial p in V has a similar representation, which means that it can be written in the form

$$p(x) = \sum_{k=1}^{n} \sum_{l=1}^{m} c_{kl}\, a_k(x) b_l(x) + c(x) + d = p_2(x) + c(x) + d$$

with some positive integers n, m, additive functions $a_k, b_l, c : G \to \mathbb{C}$, and constants c_{kl}, d. Suppose that p_2 is not identically zero. By assumption, p is the pointwise limit of a net formed by linear combinations of translates of f, which means by functions of the form (40.3). Linear combinations of functions of the form (40.3) can be written as

$$\varphi(x) = cB(x,x) + A(x) + D,$$

with some additive function $A : G \to \mathbb{C}$ and constants c, D. Any net formed by these functions has the form

$$\varphi_\gamma(x) = c_\gamma B(x,x) + A_\gamma(x) + D_\gamma.$$

From pointwise convergence,

$$\lim_\gamma \frac{1}{2} \Delta_y^2 \varphi_\gamma(x) = \frac{1}{2} \Delta_y^2 p(x) = p_2(y)$$

follows for all x, y in G. On the other hand,

$$\lim_\gamma \frac{1}{2} \Delta_y^2 \varphi_\gamma(x) = B(y,y) \lim_\gamma c_\gamma,$$

holds for all x, y in G, hence the limit $\lim_\gamma c_\gamma = c$ exists and is different from zero, which gives $B(x,x) = \frac{1}{c} p_2(x)$ for all x in G, and this is impossible.

We infer that any exponential monomial φ in V is actually a polynomial of degree at most 1, which satisfies

$$\Delta_y^2 \varphi(x) = 0 \tag{40.5}$$

for each x, y in G, hence any function in the closed linear hull of the exponential monomials in V satisfies this equation. However, f does not satisfy (40.5), hence the linear hull of the exponential monomials in V is not dense in V. □

From this theorem, we derive the following result [18].

Theorem 40.9 *If G is the additive group of the reals with the discrete topology, then spectral synthesis does not hold on G.*

This theorem provides a counterexample for Elliot's theorem. At the same time, we obtain a necessary condition for the validity of spectral synthesis on discrete Abelian groups [18].

Theorem 40.10 *If spectral synthesis holds on a discrete Abelian group, then its torsion free rank is finite.*

By this theorem, Lefranc's result is the best possible for free Abelian groups: spectral synthesis holds exactly on the finitely generated ones. In [18], the following reasonable conjecture has been formulated.

Conjecture 40.1 *Spectral synthesis holds on a discrete Abelian group if and only if its torsion free rank is finite.*

40.4 The Torsion Free Rank

In this section, we exhibit the role and importance of the torsion free rank in the spectral problems. Let G be an Abelian group. The *torsion free rank* of G is the cardinality of a maximal linearly independent subset of G. For instance, the torsion free rank of a torsion group is 0, the torsion free rank of $\mathbb{Z}$ is 1, the torsion free rank of $\mathbb{Z}^\kappa$ is κ, for any cardinality κ. The torsion free rank of a finitely generated discrete Abelian group is finite. In the following theorem, we give a simple characterization of the torsion free rank (see [19]).

Theorem 40.11 *The torsion free rank of an Abelian group is equal to the dimension of the linear space consisting of all complex additive functions of the group in the sense that either both are finite and equal, or both are infinite.*

Proof Let G be an Abelian group and let $k = r_0(G) \le +\infty$. Then G has a subgroup isomorphic to $\mathbb{Z}^k$. If k is infinite then this is equal to the non-complete direct product of k copies of $\mathbb{Z}$. We will identify this subgroup with $\mathbb{Z}^k$. Obviously, $\mathbb{Z}^k$ has at least k linearly independent complex additive functions; for instance, we can take the projections onto the different factors of the product group. On the other hand, it is well known that any homomorphism of a subgroup of an Abelian group into a divisible Abelian group can be extended to a homomorphism of the whole group. As the additive group of complex numbers is obviously divisible, the above mentioned linearly independent complex additive functions of $\mathbb{Z}^k$ can be extended to complex homomorphisms of the whole group G, and the extensions are clearly linearly independent, too. Hence the dimension of the linear space of all complex additive functions of G is not less then the torsion free rank of G.

Now we suppose that $k < +\infty$. Let Φ denote the natural homomorphism of G onto the factor group with respect to $\mathbb{Z}^k$. As it is a torsion group, hence for each element g of G there is a positive integer n such that

$$0 = n\Phi(g) = \Phi(ng),$$

thus ng belongs to the kernel of Φ, which is $\mathbb{Z}^k$. This means that there exist integers $m_1, m_2, \ldots, m_k$ such that

$$ng = (m_1, m_2, \ldots, m_k).$$

Suppose now that there are $k+1$ linearly independent complex additive functions $a_1, a_2, \ldots, a_{k+1}$ on G. Then there exist elements $g_1, g_2, \ldots, g_{k+1}$ in G such that the $(k+1) \times (k+1)$ matrix $(a_i(g_j))$ is regular. For $l = 1, 2, \ldots, k$ we let e_l denote the vector in $\mathbb{C}^k$ whose lth coordinate is 1, the others are 0. By our above consideration, there are integers $m_l^{(j)}$, n_j for $l = 1, 2, \ldots, k$ and $j = 1, 2, \ldots, k+1$ such that

$$n_j g_j = \bigl(m_1^{(j)}, m_2^{(j)}, \ldots, m_k^{(j)}\bigr).$$

Hence we have

$$\begin{aligned} a_i(n_j g_j) &= a_i\bigl(m_1^{(j)}, m_2^{(j)}, \ldots, m_k^{(j)}\bigr) \\ &= m_1^{(j)} a_i(e_1) + m_2^{(j)} a_i(e_2) + \cdots + m_k^{(j)} a_i(e_k), \end{aligned}$$

and therefore

$$a_i(g_j) = \sum_{l=1}^{k} \frac{m_l^{(j)}}{n_j} a_i(e_l)$$

holds for $i, j = 1, 2, \ldots, k+1$. This means that the linearly independent columns of the matrix $(a_i(g_j))$ are linear combinations of the columns of the matrix $(a_i(e_l))$ for $i = 1, 2, \ldots, k+1$; $l = 1, 2, \ldots, k$. But this is impossible because the latter matrix has only k columns, hence its rank is at most k.

We have shown that if the torsion free rank of G is the finite number k then the dimension of the linear space consisting of all complex additive functions of G is at most k, hence the theorem is proved. □

The following characterization of Abelian groups with finite torsion free rank may explain the role of this concept in the spectral analysis and synthesis problems—especially, in the light of Theorem 40.8.

Theorem 40.12 *The torsion free rank of the Abelian group G is finite if and only if any bi-additive function $B : G \times G \to \mathbb{C}$ has the form*

$$B(x, y) = a_1(x)b_1(y) + a_2(x)b_2(y) + \cdots + a_n(x)b_n(y)$$

for x, y in G, where $a_i, b_i : G \to \mathbb{C}$ are additive functions $(i = 1, 2, \ldots, n)$.

The proof of this theorem can be found in [19].

Using the concept of torsion free rank, M. Laczkovich and G. Székelyhidi were able to characterize those discrete Abelian groups on which spectral analysis holds (see [7]).

Theorem 40.13 (M. Laczkovich, G. Székelyhidi, 2005) *Spectral analysis holds on a discrete Abelian group if and only if the torsion free rank of the group is less than the continuum.*

This theorem provides another counterexample for Elliot's theorem. Indeed, if c denotes the continuum cardinality, then the torsion free rank of the Abelian group $\mathbb{Z}^c$ is not less than the continuum, hence, by this theorem, on $\mathbb{Z}^c$ spectral analysis fails to hold. It follows that on $\mathbb{Z}^c$ spectral synthesis fails to hold, too.

So far we still have not seen a non-finitely generated discrete Abelian group on which spectral synthesis holds. The following theorem shows that there are groups of this type (see [1]).

Theorem 40.14 *Spectral synthesis holds on any Abelian torsion group.*

Obviously there are non-finitely generated Abelian torsion groups, hence, for instance, the non-complete direct product of infinitely many copies of any nontrivial Abelian torsion group provides an example for a non-finitely generated Abelian group on which spectral synthesis holds. Thus, at this moment we have only one basic open problem: What about Conjecture 40.1? We note that an affirmative answer to the following question would solve Conjecture 40.1 in the positive.

Question 40.1 *Is it true that if spectral synthesis holds on two discrete Abelian groups, then it holds on their direct product, as well?*

Indeed, it is clear that the torsion free rank of the product of two Abelian groups with finite torsion free rank is finite, too. Unfortunately, there is no simple direct way to answer Question 40.1. However, the following theorem gives a decisive solution for the problem of discrete spectral synthesis (see [8]).

Theorem 40.15 *Spectral synthesis holds on a discrete Abelian group if and only if its torsion free rank is finite.*

We note that, obviously, this theorem gives an affirmative answer to Question 40.1, too. Further, the theorem implies that there are discrete Abelian groups on which spectral analysis holds, but spectral synthesis fails to hold.

40.5 Non-discrete Abelian Groups

Suppose now that G is a locally compact Abelian group. If the topology on G is non-discrete, then so far, there are only a limited number of results about spectral analysis and spectral synthesis. On the one hand, the classical result of L. Schwartz in [13] completely solves both problems in the real case. Another important result in this respect has been published by D.I. Gurevič in [5]. Actually, he showed that spectral synthesis does not hold on $\mathbb{R}^2$. Hence, at least the answer to Question 40.1 is negative in the non-discrete case. Nevertheless, the problem of spectral analysis on $\mathbb{R}^2$ is still unsolved. In the subsequent paragraphs, we present some partial results which hold in the non-discrete case, too.

A nonzero variety in $\mathscr{C}(G)$ is called *decomposable* if it is the sum of two subvarieties, both of them different from it. Otherwise it is called *indecomposable*. Clearly, if V is a finite dimensional variety, which is the sum of two subvarieties, both of them different from it, then both summands have smaller dimension than that of V.

The following characterization of exponential monomials is very useful (see [20]).

Theorem 40.16 *Let G be a locally compact Abelian group. A continuous complex valued function on G is an exponential monomial if and only if it generates a finite* *dimensional indecomposable variety. Further, a continuous complex valued function on G is an exponential monomial if and only if it is included in a finite dimensional indecomposable variety.*

Finite dimensional varieties always contain indecomposable subvarieties. This follows from well-known results on classical functional equations as it is proved in [20] (see also [11, 12]). This yields the following result.

Theorem 40.17 *Let G be a locally compact Abelian group. Then spectral synthesis—hence also spectral analysis—holds in every finite dimensional nonzero variety in $\mathscr{C}(G)$.*

Using this theorem the basic problem of spectral analysis and spectral synthesis can be reformulated (see [20]).

Theorem 40.18 *Let G be a locally compact Abelian group and let V be a variety in $\mathscr{C}(G)$. Spectral analysis holds in V if and only if V has a nonzero finite dimensional subvariety. Spectral synthesis holds in V if and only if V is the sum of its finite dimensional subvarieties.*

40.6 Non-Abelian Groups

Using Theorem 40.16, one can define exponential monomials on arbitrary—not necessarily commutative—locally compact groups: a continuous complex valued function is called an *exponential monomial*, if it belongs to a finite dimensional indecomposable variety. Obviously, we say that *spectral analysis holds in a variety*, if there is a nonzero exponential monomial in the variety, and *spectral synthesis holds in a variety*, if the linear hull of the set of all exponential monomials in the variety is dense in the variety. An analogue of Theorem 40.17 is the following theorem (see [20]).

Theorem 40.19 *Let G be a locally compact group. Then spectral synthesis—hence also spectral analysis—holds for each finite dimensional variety in $\mathscr{C}(G)$.*

Also an analogue of Theorem 40.18 can be derived as in [20].

Theorem 40.20 *Let G be a locally compact group and let V be a variety in $\mathscr{C}(G)$. Spectral analysis holds in V if and only if V has a nonzero finite dimensional subvariety. Spectral synthesis holds in V if and only if V is the sum of its finite dimensional subvarieties.*

As a consequence we can formulate the following theorem (see [20]).

Theorem 40.21 *Let G be a locally compact group. Spectral analysis holds over G if and only if each variety in $\mathscr{C}(G)$ has a nonzero finite dimensional subvariety. Spectral synthesis holds over G if and only if each variety in $\mathscr{C}(G)$ is the sum of finite dimensional varieties.*

This theorem makes it possible to deal with the case of compact groups. For this investigation, we shall use the classical results of the theory of almost periodic functions. These considerations enlighten the close connection between the theory of spectral analysis and synthesis and the theory of almost periodic functions.

Following [6], given a group G, a function $f : G \to \mathbb{C}$ is called *almost periodic*, if the set of its translates is relatively compact in the Banach space $\mathscr{B}(G)$ of all bounded complex valued functions, equipped with the sup-norm. If G is a locally compact topological group, then the set of all continuous almost periodic functions $\mathscr{A}(G)$ on G forms a translation invariant closed subspace of $\mathscr{C}(G) \cap \mathscr{B}(G)$, that is, a variety.

In [10, paragraph 13], the author deals with *modules* of almost periodic functions. Actually, by a module he means a linear subspace of $\mathscr{A}(G)$. An *invariant module* is a translation invariant subspace and a closed invariant module is exactly a variety. A module is called *finite* if it is finite dimensional, and it is called *irreducible* if it has no proper submodule. The fundamental theorem of almost periodic functions follows (see [10, Hauptsatz on p. 47]).

Theorem 40.22 *Each closed invariant submodule in $\mathscr{A}(G)$ is the sum of finite irreducible invariant submodules.*

In our terminology, this theorem reads as follows.

Theorem 40.23 *Each variety in $\mathscr{A}(G)$ is the sum of finite dimensional varieties, which have no proper subvarieties.*

Now we can easily derive the following result.

Theorem 40.24 *Spectral synthesis—hence also spectral analysis—holds over compact groups.*

Proof If G is a compact group, then every continuous complex valued function on G is almost periodic (see [10, Satz 1. on p. 154]), that is, $\mathscr{A}(G) = \mathscr{C}(G)$. Hence, by the previous theorem, the proof is complete. □

References

1. Bereczky, Á., Székelyhidi, L.: Spectral synthesis on torsion groups. J. Math. Anal. Appl. **304**(2), 607–613 (2005)

2. Djokovič, D.Z.: A representation theorem for $(X_1 - 1)(X_2 - 1) \cdots (X_n - 1)$ and its applications. Ann. Pol. Math. **22**, 189–198 (1969)
3. Elliot, R.J.: Two notes on spectral synthesis for discrete Abelian groups. Math. Proc. Camb. Philos. Soc. **61**, 617–620 (1965)
4. Gajda, Z.: Private communication. Hamburg–Rissen (1987)
5. Gurevič, D.I.: Counterexamples to a problem of L. Schwartz. Funkc. Anal. Ego Prilož. **9**(2), 29–35 (1975). (English translation: Funct. Anal. Appl. **9**, (2), 116–120 (1975))
6. Hewitt, E., Ross, K.: Abstract Harmonic Analysis I, II. Die Grundlehren der Mathematischen Wissenschaften, vol. 115. Springer, Berlin (1963)
7. Laczkovich, M., Székelyhidi, G.: Harmonic analysis on discrete Abelian groups. Proc. Am. Math. Soc. **133**(6), 1581–1586 (2005)
8. Laczkovich, M., Székelyhidi, L.: Spectral synthesis on discrete Abelian groups. Math. Proc. Camb. Philos. Soc. **143**(01), 103–120 (2007)
9. Lefranc, M.: L'analyse harmonique dans $\mathbb{Z}^n$. C. R. Acad. Sci. Paris **246**, 1951–1953 (1958)
10. Maak, W.: Fastperiodische Funktionen. Die Grundlehren der Mathematischen Wissenschaften, vol. 61. Springer, Berlin (1950)
11. McKiernan, M.A.: The matrix equation $a(x \circ y) = a(x) + a(x)a(y) + a(y)$. Aequ. Math. **15**, 213–223 (1977)
12. McKiernan, M.A.: Equations of the form $H(x \circ y) = \sum_i f_i(x)g_i(y)$. Aequ. Math. **16**, 51–58 (1977)
13. Schwartz, L.: Théorie génerale des fonctions moyenne-périodiques. Ann. Math. **48**(4), 857–929 (1947)
14. Székelyhidi, L.: Convolution Type Functional Equations on Topological Abelian Groups. World Scientific, Singapore (1991)
15. Székelyhidi, L.: On convolution type functional equations. Math. Pannon. **10**(2), 271–275 (1999)
16. Székelyhidi, L.: On discrete spectral synthesis. In: Daróczy, Z. Páles, Zs. (eds.) Functional Equations – Results and Advances, pp. 263–274. Kluwer Academic, Boston (2001)
17. Székelyhidi, L.: A Wiener Tauberian theorem on discrete Abelian torsion groups. Ann. Acad. Paedag. Cracov. Studia Math. I **4**, 147–150 (2001)
18. Székelyhidi, L.: The failure of spectral synthesis on some types of discrete Abelian groups. J. Math. Anal. Appl. **291**, 757–763 (2004)
19. Székelyhidi, L.: Polynomial functions and spectral synthesis. Aequ. Math. **70**(1–2), 122–130 (2005)
20. Székelyhidi, L.: Spectral synthesis problems on locally compact groups. Monatshefte Math. **161**(2), 223–232 (2010)

Chapter 41
Möbius Transformation and Einstein Velocity Addition in the Hyperbolic Geometry of Bolyai and Lobachevsky

Abraham Albert Ungar

Abstract In this chapter, dedicated to the 60th Anniversary of Themistocles M. Rassias, Möbius transformation and Einstein velocity addition meet in the hyperbolic geometry of Bolyai and Lobachevsky. It turns out that Möbius addition that is extracted from Möbius transformation of the complex disc and Einstein addition from his special theory of relativity enable the introduction of Cartesian coordinates and vector algebra as novel tools in the study of hyperbolic geometry.

Key words Möbius transformation · Einstein velocity addition · Hyperbolic geometry

Mathematics Subject Classification 51M10 · 35Q76 · 83A05

41.1 Introduction

Einstein addition law of relativistically admissible velocities is isomorphic to Möbius addition that is extracted from the common Möbius transformation of the complex open unit disc. Accordingly, both Einstein addition and Möbius addition in the open unit ball of the Euclidean n-space possess the structure of a gyrovector space that forms a natural powerful generalization of the common vector space structure. Einstein and Möbius gyrovector spaces continue to attract research interest as novel algebraic settings for hyperbolic geometry, giving rise to the incorporation of Cartesian coordinates and vector algebra into the study of the hyperbolic geometry of Bolyai and Lobachevsky [68, 73]. Outstanding novel results and elegant compatibility with well-known results in hyperbolic geometry make the novel gyrovector space approach to analytic hyperbolic geometry [70] an obvious contender for augmenting the traditional way of studying hyperbolic geometry synthetically.

Dedicated to Professor Themistocles M. Rassias on the occasion of his 60th birthday.

A.A. Ungar (✉)
Department of Mathematics, North Dakota State University, Fargo, ND 58108, USA
e-mail: Abraham.Ungar@ndsu.edu

P.M. Pardalos et al. (eds.), *Nonlinear Analysis*, Springer Optimization and Its Applications 68, In Honor of Themistocles M. Rassias on the Occasion of his 60th Birthday,
DOI 10.1007/978-1-4614-3498-6_41,

Professor Themistocles M. Rassias' special predilection and contribution to the study of Möbius transformations is revealed in his work in the areas of Möbius transformations, including [23–26] and [56, 61, 62], along with essential mathematical developments found, for instance, in [7, 12, 42, 45–48]. The latter contain essential research on geometric transformations including Möbius transformations.

The initial purpose of this article, dedicated to the 60th Anniversary of Themistocles Rassias, is to extract Möbius addition in the ball $\mathbb{R}_c^n$ of the Euclidean n-space $\mathbb{R}^n$, $n \in \mathbb{N}$, from the Möbius transformation of the complex open unit disc, and to demonstrate the hyperbolic geometric isomorphism between the resulting Möbius addition and the famous Einstein velocity addition of special relativity theory. We will then see that

1. Möbius addition in the ball $\mathbb{R}_c^n$ forms the algebraic setting for the Cartesian–Poincaré ball model of hyperbolic geometry, and
2. Einstein addition in the ball $\mathbb{R}_c^n$ forms the algebraic setting for the Cartesian–Beltrami–Klein ball model of hyperbolic geometry, just as the common
3. Vector addition in the space $\mathbb{R}^n$ forms the algebraic setting for the standard Cartesian model of Euclidean geometry.

Remarkably, Items 1–3 enable Möbius addition in $\mathbb{R}_c^n$, Einstein addition in $\mathbb{R}_c^n$, and the standard vector addition in $\mathbb{R}^n$ to be studied comparatively, as in [72].

Counterintuitively, Einstein velocity addition law of relativistically admissible velocities is neither commutative nor associative. The breakdown of commutativity in Einstein addition seemed undesirable to Émile Borel in 1909. According to the historian of relativity physics Scott Walter [76, Sect. 10], the famous mathematician and a former doctoral student of Poincaré, Émile Borel (1871–1956), was renowned for his work on the theory of functions, in which a chair was created for him at the Sorbonne in 1909. In the years following his appointment, he took up the study of relativity theory. Borel "fixed" the seemingly "defective" result that Einstein velocity addition law is noncommutative. According to Walter, Borel's version of commutativized relativistic velocity addition involves a significant modification of Einstein's relativistic velocity composition law.

Contrasting Borel, in this article we commutativize the Einstein velocity addition law by composing Einstein addition with an appropriate Thomas precession in a natural way suggested by analogies with the classical parallelogram addition law and supported experimentally by cosmological observations of stellar aberration.

Historically, the link between Einstein's special theory of relativity and the non-Euclidean style was developed during the period 1908–1912 by Varičak, Robb, Wilson and Lewis, and Borel [76]. The subsequent development that followed 1912 appeared about 80 years later, in 2001, as the renowned historian Scott Walter describes in [77]:

> Over the years, there have been a handful of attempts to promote the non-Euclidean style for use in problem solving in relativity and electrodynamics, the failure of which to attract any substantial following, compounded by the absence of any positive results must give pause to anyone considering a similar undertaking. Until recently, no one was in a position to offer an improvement on the tools available since 1912. In his [2001] book, Ungar

> furnishes the crucial missing element from the panoply of the non-Euclidean style: an elegant nonassociative algebraic formalism that fully exploits the structure of Einstein's law of velocity composition. The formalism relies on what the author calls the "missing link" between Einstein's velocity addition formula and ordinary vector addition: Thomas precession ...
>
> Scott Walter, 2002 [77]

Indeed, the special relativistic effect known as *Thomas precession* is mathematically abstracted into an operator called a *gyrator*, denoted "gyr". The latter, in turn, justifies the prefix "gyro" that we extensively use in *gyrolanguage*, where we prefix a gyro to any term that describes a concept in Euclidean geometry and in associative algebra to mean the analogous concept in hyperbolic geometry and in nonassociative algebra. Thus, for instance, Einstein's velocity addition is neither commutative nor associative, but it turns out to be both *gyrocommutative* and *gyroassociative*, giving rise to the algebraic structures known as gyrogroups and gyrovector spaces. Remarkably, the mere introduction of the gyrator turns Euclidean geometry, the geometry of classical mechanics, into hyperbolic geometry, the geometry of relativistic mechanics.

The breakdown of commutativity in Einstein velocity addition law seemed undesirable to the famous mathematician Émile Borel. Borel's resulting attempt to "repair" the seemingly "defective" Einstein velocity addition in the years following 1912 is described by Walter in [76, p. 117]. Here, however, we see that there is no need to repair Einstein velocity addition law for being noncommutative since it suggestively gives rise to the gyroparallelogram law of gyrovector addition, which turns out to be commutative. The compatibility of the gyroparallelogram addition law of Einsteinian velocities with cosmological observations of stellar aberration is explained in [68, Chap. 13] and mentioned in [73, Sect. 10.2]. The extension of the gyroparallelogram addition law of $k = 2$ summands in $\mathbb{R}_c^n$ to a corresponding k-dimensional gyroparallelepiped (gyroparallelotope) addition law of $k > 2$ summands is presented in this article and, with proof, in [68, Theorem 10.6].

41.2 Möbius Addition

The most general Möbius transformation of the complex open unit disc

$$\mathbb{D} = \{z \in \mathbb{C} : |z| < 1\} \tag{41.1}$$

in the complex plane $\mathbb{C}$ is given by the polar decomposition [1, 34],

$$z \mapsto e^{i\theta} \frac{a+z}{1+\overline{a}z} = e^{i\theta}(a \oplus_{\mathrm{M}} z) \tag{41.2}$$

Möbius addition $\oplus_{\mathrm{M}}$ in the disc is extracted from (41.2), allowing the generic Möbius transformation of the disc to be viewed as a Möbius left *gyrotranslation*

$$z \mapsto a \oplus_{\mathrm{M}} z = \frac{a+z}{1+\overline{a}z} \tag{41.3}$$

followed by a rotation. Here $\theta \in \mathbb{R}$ is a real number, $a, z \in \mathbb{D}$, $\overline{a}$ is the complex conjugate of a, and $\oplus_{\mathrm{M}}$ represents Möbius addition in the disc.

Möbius addition $a \oplus_{\mathrm{M}} z$ and subtraction $a \ominus_{\mathrm{M}} z = a \oplus_{\mathrm{M}} (-z)$ are found useful in the geometric viewpoint of complex analysis; see, for instance, [60, 64], [34, pp. 52–53, 56–57, 60], and the Schwarz–Pick Lemma in [21, Theorem 1.4, p. 64]. However, prior to the appearance of [63] in 2001 these were not considered 'addition' and 'subtraction' since it has gone unnoticed that, being *gyrocommutative* and *gyroassociative*, they share analogies with the common vector addition and subtraction, as we will see in the sequel.

Möbius addition $\oplus_{\mathrm{M}}$ is neither commutative nor associative. The breakdown of commutativity in Möbius addition is "repaired" by the introduction of a *gyrator*

$$\mathrm{gyr} : \mathbb{D} \times \mathbb{D} \to \mathrm{Aut}(\mathbb{D}, \oplus_{\mathrm{M}}) \tag{41.4}$$

that generates gyroautomorphisms according to the equation

$$\mathrm{gyr}[a,b] = \frac{a \oplus_{\mathrm{M}} b}{b \oplus_{\mathrm{M}} a} = \frac{1 + a\overline{b}}{1 + \overline{a}b} \in \mathrm{Aut}(\mathbb{D}, \oplus_{\mathrm{M}}) \tag{41.5}$$

where $\mathrm{Aut}(\mathbb{D}, \oplus_{\mathrm{M}})$ is the automorphism group of the Möbius groupoid $(\mathbb{D}, \oplus_{\mathrm{M}})$. Here a groupoid is a nonempty set with a binary operation, and an automorphism of the groupoid $(\mathbb{D}, \oplus_{\mathrm{M}})$ is a bijective self-map $f : \mathbb{D} \to \mathbb{D}$ of the set $\mathbb{D}$ that respects its binary operation $\oplus_{\mathrm{M}}$, that is, $f(a \oplus_{\mathrm{M}} b) = f(a) \oplus_{\mathrm{M}} f(b)$ for all $a, b \in \mathbb{D}$. Being gyrations, the automorphisms $\mathrm{gyr}[a,b]$ are also called *gyroautomorphisms*.

The inverse of the automorphism $\mathrm{gyr}[a,b]$ is clearly $\mathrm{gyr}[b,a]$,

$$\mathrm{gyr}^{-1}[a,b] = \mathrm{gyr}[b,a]. \tag{41.6}$$

The gyration definition in (41.5) suggests the following *gyrocommutative law* of Möbius addition in the disc,

$$a \oplus_{\mathrm{M}} b = \mathrm{gyr}[a,b](b \oplus_{\mathrm{M}} a). \tag{41.7}$$

The resulting gyrocommutative law (41.7) is not terribly surprising since it is generated by definition, but we are not finished.

Coincidentally, the gyroautomorphism $\mathrm{gyr}[a,b]$ that repairs in (41.7) the breakdown of commutativity, repairs the breakdown of associativity in $\oplus_{\mathrm{M}}$ as well, giving rise to the following *left and right gyroassociative law* of Möbius addition

$$\begin{aligned} a \oplus_{\mathrm{M}} (b \oplus_{\mathrm{M}} z) &= (a \oplus_{\mathrm{M}} b) \oplus_{\mathrm{M}} \mathrm{gyr}[a,b]z, \\ (a \oplus_{\mathrm{M}} b) \oplus_{\mathrm{M}} z &= a \oplus_{\mathrm{M}} \big(b \oplus_{\mathrm{M}} \mathrm{gyr}[b,a]z\big) \end{aligned} \tag{41.8}$$

for all $a, b, z \in \mathbb{D}$. Moreover, Möbius gyroautomorphisms possess their own rich structure obeying, for instance, the two elegant identities

$$\begin{aligned} \mathrm{gyr}[a \oplus_{\mathrm{M}} b, b] &= \mathrm{gyr}[a,b], \\ \mathrm{gyr}[a, b \oplus_{\mathrm{M}} a] &= \mathrm{gyr}[a,b] \end{aligned} \tag{41.9}$$

called the left and the right loop property.

In order to extend Möbius addition from the disc to the ball, we identify complex numbers of the complex plane $\mathbb{C}$ with vectors of the Euclidean plane $\mathbb{R}^2$ in the usual way,

$$\mathbb{C} \ni u = u_1 + iu_2 = (u_1, u_2) = \mathbf{u} \in \mathbb{R}^2. \tag{41.10}$$

Then

$$\begin{aligned} \bar{u}v + u\bar{v} &= 2\mathbf{u}\cdot\mathbf{v}, \\ |u| &= \|\mathbf{u}\| \end{aligned} \tag{41.11}$$

give the inner product and the norm in $\mathbb{R}^2$, so that Möbius addition in the disc $\mathbb{D}$ of the complex plane $\mathbb{C}$ becomes Möbius addition in the disc

$$\mathbb{R}^2_{c=1} = \{\mathbf{v} \in \mathbb{R}^2 : \|\mathbf{v}\| < c = 1\} \tag{41.12}$$

of the Euclidean plane $\mathbb{R}^2$. Indeed,

$$\begin{aligned} \mathbb{D} \ni u \oplus v &:= \frac{u+v}{1+\bar{u}v} \\ &= \frac{(1+u\bar{v})(u+v)}{(1+\bar{u}v)(1+u\bar{v})} \\ &= \frac{(1+\bar{u}v+u\bar{v}+|v|^2)u + (1-|u|^2)v}{1+\bar{u}v+u\bar{v}+|u|^2|v|^2} \\ &= \frac{(1+2\mathbf{u}\cdot\mathbf{v}+\|\mathbf{v}\|^2)\mathbf{u} + (1-\|\mathbf{u}\|^2)\mathbf{v}}{1+2\mathbf{u}\cdot\mathbf{v}+\|\mathbf{u}\|^2\|\mathbf{v}\|^2} \\ &=: \mathbf{u} \oplus \mathbf{v} \in \mathbb{R}^2_{c=1} \end{aligned} \tag{41.13}$$

for all $u, v \in \mathbb{D}$ and all $\mathbf{u}, \mathbf{v} \in \mathbb{R}^2_{c=1}$. The last equation in (41.13) is a vector equation, so that its restriction to the ball of the Euclidean two-dimensional space is a mere artifact. Suggestively, we thus arrive at the following definition of Möbius addition in the ball of any real inner product space.

Definition 41.1 (Möbius Addition in the Ball) Let $\mathbb{V} = (\mathbb{V}, +, \cdot)$ be a real inner product space with a binary operation $+$ and a positive definite inner product $\cdot$ ([37, p. 21]; following [33], also known as Euclidean space) and let $\mathbb{V}_s$ be the s-ball of $\mathbb{V}$,

$$\mathbb{V}_s = \{\mathbf{v} \in \mathbb{V} : \|\mathbf{v}\| < s\} \tag{41.14}$$

for any fixed $s > 0$. Möbius addition $\oplus_M$ is a binary operation in $\mathbb{V}_s$ given by the equation

$$\mathbf{u} \oplus_M \mathbf{v} = \frac{(1+\frac{2}{s^2}\mathbf{u}\cdot\mathbf{v}+\frac{1}{s^2}\|\mathbf{v}\|^2)\mathbf{u} + (1-\frac{1}{s^2}\|\mathbf{u}\|^2)\mathbf{v}}{1+\frac{2}{s^2}\mathbf{u}\cdot\mathbf{v}+\frac{1}{s^4}\|\mathbf{u}\|^2\|\mathbf{v}\|^2} \tag{41.15}$$

where $\cdot$ and $\|\cdot\|$ are the inner product and norm that the ball $\mathbb{V}_s$ inherits from its space $\mathbb{V}$.

In the limit of large s, $s \to \infty$, the ball $\mathbb{V}_s$ in Definition 41.1 expands to the whole of its space $\mathbb{V}$, and Möbius addition in $\mathbb{V}_s$ reduces to the vector addition, +, in $\mathbb{V}$. Accordingly, the right hand side of (41.15) is known as a Möbius translation [49, p. 129]. An earlier study of Möbius translation in several dimensions, using the notation $-\mathbf{u} \oplus_{\mathrm{M}} \mathbf{v} =: T_{\mathbf{u}}\mathbf{v}$, is found in [2] and in [3], where it is attributed to Poincaré. Both Ahlfors [2] and Ratcliffe [49], who studied the Möbius translation in several dimensions, did not call it a Möbius addition since it has gone unnoticed at the time that Möbius translation is regulated by algebraic laws analogous to those that regulate vector addition.

Möbius addition $\oplus_{\mathrm{M}}$ in the open unit ball $\mathbb{V}_s$ of any real inner product space $\mathbb{V}$ is thus a most natural extension of Möbius addition in the open complex unit disc. Like the Möbius disc groupoid $(\mathbb{D}, \oplus_{\mathrm{M}})$, the Möbius ball groupoid $(\mathbb{V}_s, \oplus_{\mathrm{M}})$ turns out to be a gyrocommutative gyrogroup, defined in Definitions 41.2–41.3 in Sect. 41.4, as one can check straightforwardly by computer algebra. Interestingly, the gyrocommutative law of Möbius addition was already known to Ahlfors [2, Eq. 39]. The accompanied gyroassociative law of Möbius addition, however, had gone unnoticed.

Möbius addition satisfies the gamma identity

$$\gamma_{\mathbf{u}\oplus_{\mathrm{M}}\mathbf{v}} = \gamma_{\mathbf{u}}\gamma_{\mathbf{v}}\sqrt{1 + \frac{2}{s^2}\mathbf{u}\cdot\mathbf{v} + \frac{1}{s^4}\|\mathbf{u}\|^2\|\mathbf{v}\|^2} \tag{41.16}$$

for all $\mathbf{u}, \mathbf{v} \in \mathbb{V}_s$, where $\gamma_{\mathbf{u}}$ is the gamma factor

$$\gamma_{\mathbf{u}} = \frac{1}{\sqrt{1 - \frac{\|\mathbf{u}\|^2}{s^2}}} \tag{41.17}$$

in the s-ball $\mathbb{V}_s$.

The gamma factor appears also in Einstein velocity addition of relativistically admissible velocities, and it is known in special relativity theory as the Lorentz gamma factor. The gamma factor $\gamma_{\mathbf{v}}$ is real if and only if $\mathbf{v} \in \mathbb{V}_s$. Hence, the gamma identity (41.16) demonstrates that $\mathbf{u}, \mathbf{v} \in \mathbb{V}_s \Rightarrow \mathbf{u} \oplus_{\mathrm{M}} \mathbf{v} \in \mathbb{V}_s$ so that, indeed, Möbius addition $\oplus_{\mathrm{M}}$ is a binary operation in the ball $\mathbb{V}_s$.

41.3 Einstein Velocity Addition

Let c be any positive constant, let $(\mathbb{R}^n_c, +, \cdot)$ be the Euclidean n-space, and let

$$\mathbb{R}^n_c = \left\{\mathbf{v} \in \mathbb{R}^n_c : \|\mathbf{v}\| < c\right\} \tag{41.18}$$

be the c-ball of all relativistically admissible velocities of material particles. It is the open ball of radius c, centered at the origin of $\mathbb{R}^n$, consisting of all vectors $\mathbf{v}$ in $\mathbb{R}^n$ with magnitude $\|\mathbf{v}\|$ smaller than c.

Einstein velocity addition in the c-ball of all relativistically admissible velocities is given by the equation [18], [40, p. 55], [50, Eq. 2.9.2], [63],

$$\mathbf{u} \oplus \mathbf{v} = \frac{1}{1+\frac{\mathbf{u}\cdot\mathbf{v}}{c^2}} \left\{ \mathbf{u} + \frac{1}{\gamma_{\mathbf{u}}}\mathbf{v} + \frac{1}{c^2}\frac{\gamma_{\mathbf{u}}}{1+\gamma_{\mathbf{u}}}(\mathbf{u}\cdot\mathbf{v})\mathbf{u} \right\} \tag{41.19}$$

satisfying the *gamma identity*

$$\gamma_{\mathbf{u}\oplus\mathbf{v}} = \gamma_{\mathbf{u}}\gamma_{\mathbf{v}}\left(1+\frac{\mathbf{u}\cdot\mathbf{v}}{c^2}\right) \tag{41.20}$$

for all $\mathbf{u},\mathbf{v} \in \mathbb{R}^n_c$, where $\gamma_{\mathbf{u}}$ is the gamma factor (41.17),

$$\gamma_{\mathbf{u}} = \frac{1}{\sqrt{1-\frac{\|\mathbf{u}\|^2}{c^2}}} \tag{41.21}$$

in the c-ball $\mathbb{R}^n_c$.

In physical applications, $\mathbb{R}^n = \mathbb{R}^3$ is the Euclidean 3-space, which is the space of all classical, Newtonian velocities, and $\mathbb{R}^n_c = \mathbb{R}^3_c \subset \mathbb{R}^3$ is the c-ball of $\mathbb{R}^3$ of all relativistically admissible, Einsteinian velocities. Furthermore, the constant c represents in physical applications the vacuum speed of light.

Einstein addition (41.19) of relativistically admissible velocities was introduced by Einstein in his 1905 paper [15], [16, p. 141] that founded the special theory of relativity. We may note here that the Euclidean 3-vector algebra was not so widely known in 1905 and, consequently, was not used by Einstein. Einstein calculated in [15] the behavior of the velocity components parallel and orthogonal to the relative velocity between inertial systems, which is as close as one can get without vectors to the vectorial version (41.19).

In full analogy with vector addition and subtraction, we use the abbreviation $\mathbf{u} \ominus \mathbf{v} = \mathbf{u} \oplus (-\mathbf{v})$ for Einstein subtraction, so that, for instance, $\mathbf{v} \ominus \mathbf{v} = \mathbf{0}$, $\ominus\mathbf{v} = \mathbf{0} \ominus \mathbf{v} = -\mathbf{v}$ and, in particular,

$$\ominus(\mathbf{u} \oplus \mathbf{v}) = \ominus\mathbf{u} \ominus \mathbf{v} \tag{41.22}$$

and

$$\ominus\mathbf{u} \oplus (\mathbf{u} \oplus \mathbf{v}) = \mathbf{v} \tag{41.23}$$

for all $\mathbf{u},\mathbf{v}$ in the ball. Identity (41.22) is called the *automorphic inverse property*, and Identity (41.23) is called the *left cancellation law* of Einstein addition [65, 68, 70]. Einstein addition does not obey the immediate right counterpart of the left cancellation law (41.23) since, in general,

$$(\mathbf{u} \oplus \mathbf{v}) \ominus \mathbf{v} \neq \mathbf{u}. \tag{41.24}$$

However, this seemingly lack of a right cancellation law will be repaired in (41.47), following the emergence of a second gyrogroup binary operation in Definition 41.4 below, which we introduce in order to capture analogies with classical results.

In the Newtonian limit of large c, $c \to \infty$, the ball $\mathbb{R}^n_c$ expands to the whole of its space $\mathbb{R}^n$, as we see from (41.18), and Einstein addition $\oplus$ in $\mathbb{R}^n_c$ reduces to the common vector addition $+$ in $\mathbb{R}^n$, as we see from (41.19) and (41.21).

Einstein addition is noncommutative. Indeed, $\|\mathbf{u}\oplus\mathbf{v}\| = \|\mathbf{v}\oplus\mathbf{u}\|$, but, in general,

$$\mathbf{u}\oplus\mathbf{v} \neq \mathbf{v}\oplus\mathbf{u} \tag{41.25}$$

for $\mathbf{u}, \mathbf{v} \in \mathbb{R}^n_c$. Moreover, Einstein addition is also nonassociative since, in general,

$$(\mathbf{u}\oplus\mathbf{v})\oplus\mathbf{w} \neq \mathbf{u}\oplus(\mathbf{v}\oplus\mathbf{w}) \tag{41.26}$$

for $\mathbf{u}, \mathbf{v}, \mathbf{w} \in \mathbb{R}^n_c$.

It seems that following the breakdown of commutativity and associativity in Einstein addition some mathematical regularity has been lost in the transition from Newton velocity addition in $\mathbb{R}^n$ to Einstein velocity addition (41.19) in $\mathbb{R}^n_c$. This is, however, not the case since, as we will see in Sect. 41.4, the gyrator comes to the rescue [43, 44, 63, 65, 68, 70, 77]. Indeed, we will find in Sect. 41.4 that the mere introduction of gyrations endows the Einstein groupoid $(\mathbb{R}^n_c, \oplus)$ with a grouplike rich structure [57] that we call a *gyrocommutative gyrogroup*. Furthermore, we will find in Sect. 41.5 that Einstein gyrogroups admit scalar multiplication that turns them into Einstein gyrovector spaces. The latter, in turn, form the algebraic setting for the Cartesian–Beltrami–Klein ball model of hyperbolic geometry, just as Euclidean vector spaces $\mathbb{R}^n$ form the algebraic setting for the standard Cartesian model of Euclidean geometry.

When the nonzero vectors $\mathbf{u}, \mathbf{v} \in \mathbb{R}^n_c \subset \mathbb{R}^n$ are parallel in $\mathbb{R}^n$, $\mathbf{u}\|\mathbf{v}$, that is, $\mathbf{u} = \lambda\mathbf{v}$ for some $0 \neq \lambda \in \mathbb{R}$, Einstein addition reduces to the Einstein addition of parallel velocities [78, p. 50],

$$\mathbf{u}\oplus\mathbf{v} = \frac{\mathbf{u}+\mathbf{v}}{1+\frac{1}{c^2}\|\mathbf{u}\|\,\|\mathbf{v}\|}, \qquad \mathbf{u}\|\mathbf{v} \tag{41.27}$$

which was confirmed experimentally by Fizeau's 1851 experiment [39]. Owing to its simplicity, some books on special relativity present Einstein velocity addition in its restricted form (41.27) rather than its general form (41.19).

The restricted Einstein addition (41.27) is both commutative and associative. Accordingly, the restricted Einstein addition is a group operation, as Einstein noted in [15]; see [16, p. 142]. In contrast, Einstein made no remark about group properties of his addition law of velocities that need not be parallel. Indeed, the general Einstein addition (41.19) is not a group operation but, rather, a gyrocommutative gyrogroup operation, a structure that was discovered more than 80 years later, in 1988 [55], and is presented in Definitions 41.2–41.3 in Sect. 41.4.

41.4 Einstein Gyrogroups and Gyrations

A description of the 3-space rotation, which since 1926 [54] is named after Thomas, is found in Silberstein's 1914 book [51]. In 1914, Thomas precession did not have a name, and Silberstein called it in his 1914 book a "certain space-rotation" [51, p. 169]. An early study of Thomas precession, made by the famous mathematician Émile Borel in 1913, is described in his 1914 book [6] and, more recently, in [52]. According to Belloni and Reina [5], Sommerfeld's route to Thomas precession dates back to 1909. However, prior to Thomas' discovery the relativistic peculiar 3-space rotation had a most uncertain physical status [76, p. 119]. The only knowledge Thomas had in 1925 about the peculiar relativistic gyroscopic precession [29] came from De Sitter's formula describing the relativistic corrections for the motion of the moon, found in Eddington's book [14], which was just published at that time [63, Sect. 1, Chap. 1].

The physical significance of the peculiar rotation in special relativity emerged in 1925 when Thomas relativistically re-computed the precessional frequency of the doublet separation in the fine structure of the atom, and thus rectified a missing factor of 1/2. This correction has come to be known as the *Thomas half* [9]. Thomas' discovery of the relativistic precession of the electron spin on Christmas 1925 thus led to the understanding of the significance of the relativistic effect that became known as *Thomas precession*. Llewellyn Hilleth Thomas died in Raleigh, NC, on April 20, 1992. A paper [8] dedicated to the centenary of the birth of Llewellyn H. Thomas (1902–1992) describes the Bloch gyrovector of quantum information and computation.

For any $\mathbf{u}, \mathbf{v} \in \mathbb{R}^n_c$, let $\mathrm{gyr}[\mathbf{u}, \mathbf{v}] : \mathbb{R}^n_c \to \mathbb{R}^n_c$ be the self-map of $\mathbb{R}^n_c$ given in terms of Einstein addition $\oplus$, (41.19), by the equation [55]

$$\mathrm{gyr}[\mathbf{u}, \mathbf{v}]\mathbf{w} = \ominus(\mathbf{u} \oplus \mathbf{v}) \oplus \{\mathbf{u} \oplus (\mathbf{v} \oplus \mathbf{w})\} \tag{41.28}$$

for all $\mathbf{w} \in \mathbb{R}^n_c$. The self-map $\mathrm{gyr}[\mathbf{u}, \mathbf{v}]$ of $\mathbb{R}^n_c$, which takes $\mathbf{w} \in \mathbb{R}^n_c$ into $\ominus(\mathbf{u} \oplus \mathbf{v}) \oplus \{\mathbf{u} \oplus (\mathbf{v} \oplus \mathbf{w})\} \in \mathbb{R}^n_c$, is the gyration generated by $\mathbf{u}$ and $\mathbf{v}$. Being the mathematical abstraction of the relativistic Thomas precession, the gyration has an interpretation in hyperbolic geometry [75] as the negative hyperbolic triangle defect [68, Theorem 8.55].

In the Newtonian limit, $c \to \infty$, Einstein addition $\oplus$ in $\mathbb{R}^n_c$ reduces to the common vector addition + in $\mathbb{R}^n$, which is associative. Accordingly, in this limit the gyration $\mathrm{gyr}[\mathbf{u}, \mathbf{v}]$ in (41.28) reduces to the identity map of $\mathbb{R}^n$, called the trivial map. Hence, as expected, Thomas gyrations $\mathrm{gyr}[\mathbf{u}, \mathbf{v}]$, $\mathbf{u}, \mathbf{v} \in \mathbb{R}^n_c$, vanish (that is, they become *trivial*) in the Newtonian limit.

It follows from the gyration equation (41.28) that gyrations measure the extent to which Einstein addition deviates from associativity, where associativity corresponds to trivial gyrations.

The gyration equation (41.28) can be manipulated (with the help of computer algebra) into the equation

$$\mathrm{gyr}[\mathbf{u}, \mathbf{v}]\mathbf{w} = \mathbf{w} + \frac{A\mathbf{u} + B\mathbf{v}}{D} \tag{41.29}$$

where

$$\begin{aligned}
A &= -\frac{1}{c^2}\frac{\gamma_{\mathbf{u}}^2}{(\gamma_{\mathbf{u}}+1)}(\gamma_{\mathbf{v}}-1)(\mathbf{u}\cdot\mathbf{w}) + \frac{1}{c^2}\gamma_{\mathbf{u}}\gamma_{\mathbf{v}}(\mathbf{v}\cdot\mathbf{w}),\\
&\quad + \frac{2}{c^4}\frac{\gamma_{\mathbf{u}}^2\gamma_{\mathbf{v}}^2}{(\gamma_{\mathbf{u}}+1)(\gamma_{\mathbf{v}}+1)}(\mathbf{u}\cdot\mathbf{v})(\mathbf{v}\cdot\mathbf{w})\\
B &= -\frac{1}{c^2}\frac{\gamma_{\mathbf{v}}}{\gamma_{\mathbf{v}}+1}\{\gamma_{\mathbf{u}}(\gamma_{\mathbf{v}}+1)(\mathbf{u}\cdot\mathbf{w}) + (\gamma_{\mathbf{u}}-1)\gamma_{\mathbf{v}}(\mathbf{v}\cdot\mathbf{w})\},\\
D &= \gamma_{\mathbf{u}}\gamma_{\mathbf{v}}\left(1+\frac{\mathbf{u}\cdot\mathbf{v}}{c^2}\right) + 1 = \gamma_{\mathbf{u}\oplus\mathbf{v}} + 1 > 1
\end{aligned} \tag{41.30}$$

for all $\mathbf{u}, \mathbf{v}, \mathbf{w} \in \mathbb{R}_c^n$.

Allowing $\mathbf{w} \in \mathbb{R}^n \supset \mathbb{R}_c^n$ in (41.29)–(41.30), that is, extending the domain of $\mathbf{w}$ from $\mathbb{R}_c^n$ to $\mathbb{R}^n$, gyrations $\mathrm{gyr}[\mathbf{u},\mathbf{v}]$ are expendable to linear maps of $\mathbb{R}^n$ for all $\mathbf{u}, \mathbf{v} \in \mathbb{R}_c^n$.

In each of the three special cases when (i) $\mathbf{u} = \mathbf{0}$, or (ii) $\mathbf{v} = \mathbf{0}$, or (iii) $\mathbf{u}$ and $\mathbf{v}$ are parallel in $\mathbb{R}_c^n \subset \mathbb{R}^n$, $\mathbf{u}\|\mathbf{v}$, we have $A\mathbf{u} + B\mathbf{v} = \mathbf{0}$ so that $\mathrm{gyr}[\mathbf{u},\mathbf{v}]$ is trivial,

$$\begin{aligned}
\mathrm{gyr}[\mathbf{0},\mathbf{v}]\mathbf{w} &= \mathbf{w},\\
\mathrm{gyr}[\mathbf{u},\mathbf{0}]\mathbf{w} &= \mathbf{w},\\
\mathrm{gyr}[\mathbf{u},\mathbf{v}]\mathbf{w} &= \mathbf{w}, \qquad \mathbf{u}\|\mathbf{v}
\end{aligned} \tag{41.31}$$

for all $\mathbf{u}, \mathbf{v} \in \mathbb{R}_c^n$, and all $\mathbf{w} \in \mathbb{R}^n$.

It follows from (41.29) that

$$\mathrm{gyr}[\mathbf{v},\mathbf{u}]\big(\mathrm{gyr}[\mathbf{u},\mathbf{v}]\mathbf{w}\big) = \mathbf{w} \tag{41.32}$$

for all $\mathbf{u}, \mathbf{v} \in \mathbb{R}_c^n$, $\mathbf{w} \in \mathbb{R}^n$, so that gyrations are invertible linear maps of $\mathbb{R}^n$, the inverse of $\mathrm{gyr}[\mathbf{u},\mathbf{v}]$ being $\mathrm{gyr}[\mathbf{v},\mathbf{u}]$ for all $\mathbf{u}, \mathbf{v} \in \mathbb{R}_c^n$.

Gyrations keep the inner product of elements of the ball $\mathbb{R}_c^n$ invariant, that is,

$$\mathrm{gyr}[\mathbf{u},\mathbf{v}]\mathbf{a} \cdot \mathrm{gyr}[\mathbf{u},\mathbf{v}]\mathbf{b} = \mathbf{a}\cdot\mathbf{b} \tag{41.33}$$

for all $\mathbf{a}, \mathbf{b}, \mathbf{u}, \mathbf{v} \in \mathbb{R}_c^n$. Hence, $\mathrm{gyr}[\mathbf{u},\mathbf{v}]$ is an *isometry* of $\mathbb{R}_c^n$, keeping the norm of elements of the ball $\mathbb{R}_c^n$ invariant,

$$\big\|\mathrm{gyr}[\mathbf{u},\mathbf{v}]\mathbf{w}\big\| = \|\mathbf{w}\|. \tag{41.34}$$

Accordingly, $\mathrm{gyr}[\mathbf{u},\mathbf{v}]$ represents a rotation of the ball $\mathbb{R}_c^n$ about its origin for any $\mathbf{u}, \mathbf{v} \in \mathbb{R}_c^n$.

The invertible self-map $\mathrm{gyr}[\mathbf{u},\mathbf{v}]$ of $\mathbb{R}_c^n$ respects Einstein addition in $\mathbb{R}_c^n$,

$$\mathrm{gyr}[\mathbf{u},\mathbf{v}](\mathbf{a}\oplus\mathbf{b}) = \mathrm{gyr}[\mathbf{u},\mathbf{v}]\mathbf{a} \oplus \mathrm{gyr}[\mathbf{u},\mathbf{v}]\mathbf{b} \tag{41.35}$$

for all $\mathbf{a}, \mathbf{b}, \mathbf{u}, \mathbf{v} \in \mathbb{R}_c^n$, so that $\mathrm{gyr}[\mathbf{u},\mathbf{v}]$ is an automorphism of the Einstein groupoid $(\mathbb{R}_c^n, \oplus)$. We recall that an automorphism of a groupoid $(\mathbb{R}_c^n, \oplus)$ is a bijective self-map of the groupoid $\mathbb{R}_c^n$ that respects its binary operation, that is, it satisfies (41.35).

Under bijection composition the automorphisms of a groupoid $(\mathbb{R}^n_c, \oplus)$ form a group known as the automorphism group, and denoted $\mathrm{Aut}(\mathbb{R}^n_c, \oplus)$. Being special automorphisms, gyrations $\mathrm{gyr}[\mathbf{u}, \mathbf{v}] \in \mathrm{Aut}(\mathbb{R}^n_c, \oplus)$, $\mathbf{u}, \mathbf{v} \in \mathbb{R}^n_c$, are also called *gyroautomorphisms*, gyr being the gyroautomorphism generator called the *gyrator*.

The gyroautomorphisms $\mathrm{gyr}[\mathbf{u}, \mathbf{v}]$ regulate Einstein addition in the ball $\mathbb{R}^n_c$, giving rise to the following nonassociative algebraic laws that "repair" the breakdown of commutativity and associativity in Einstein addition:

$$\begin{aligned} \mathbf{u} \oplus \mathbf{v} &= \mathrm{gyr}[\mathbf{u}, \mathbf{v}](\mathbf{v} \oplus \mathbf{u}) && \text{(Gyrocommutativity)}, \\ \mathbf{u} \oplus (\mathbf{v} \oplus \mathbf{w}) &= (\mathbf{u} \oplus \mathbf{v}) \oplus \mathrm{gyr}[\mathbf{u}, \mathbf{v}]\mathbf{w} && \text{(Left Gyroassociativity)}, \\ (\mathbf{u} \oplus \mathbf{v}) \oplus \mathbf{w} &= \mathbf{u} \oplus (\mathbf{v} \oplus \mathrm{gyr}[\mathbf{v}, \mathbf{u}]\mathbf{w}) && \text{(Right Gyroassociativity)} \end{aligned} \tag{41.36}$$

for all $\mathbf{u}, \mathbf{v}, \mathbf{w} \in \mathbb{R}^n_c$. It is clear from the identities in (41.36) that the gyroautomorphisms $\mathrm{gyr}[\mathbf{u}, \mathbf{v}]$ measure of the failure of commutativity and associativity in Einstein addition.

Owing to the gyrocommutative law in (41.36), the gyrator is recognized as the familiar Thomas precession of special relativity theory. The gyrocommutative law was already known to Silberstein in 1914 [51] in the following sense. The Thomas precession generated by $\mathbf{u}, \mathbf{v} \in \mathbb{R}^3_c$ is the unique rotation that takes $\mathbf{v} \oplus \mathbf{u}$ into $\mathbf{u} \oplus \mathbf{v}$ about an axis perpendicular to the plane of $\mathbf{u}$ and $\mathbf{v}$ through an angle $< \pi$ in $\mathbb{R}^n_c$, thus giving rise to the gyrocommutative law. Obviously, Silberstein did not use the terms "Thomas precession" and "gyrocommutative law" since these terms have been coined later, respectively, following Thomas' 1926 paper [54], and by the author in 1991 [57, 59] following the discovery of the gyrocommutative and the gyroassociative laws of Einstein addition in [55]. Thus, contrasting the discovery before 1914 of what we presently call the gyrocommutative law of Einstein addition, the gyroassociative laws of Einstein addition, left and right, were discovered by the author about 75 years later, in 1988 [55].

Thomas precession has purely kinematical origin, as emphasized in [67], so that the presence of Thomas precession is not connected with the action of any force.

A most important and useful property of gyrations is the so called *reduction property* (left and right),

$$\begin{aligned} \mathrm{gyr}[\mathbf{u} \oplus \mathbf{v}, \mathbf{v}] &= \mathrm{gyr}[\mathbf{u}, \mathbf{v}] && \text{(Left Reduction Property)}, \\ \mathrm{gyr}[\mathbf{u}, \mathbf{v} \oplus \mathbf{u}] &= \mathrm{gyr}[\mathbf{u}, \mathbf{v}] && \text{(Right Reduction Property)} \end{aligned} \tag{41.37}$$

for all $\mathbf{u}, \mathbf{v} \in \mathbb{R}^n_c$. The left loop property will prove useful in (41.46) below in solving a basic gyrogroup equation.

Identities (41.36)–(41.37) are the basic identities of the gyroalgebra of Einstein addition. They can be verified straightforwardly by computer algebra, as explained in [63, Sect. 8].

The grouplike groupoid $(\mathbb{R}^n_c, \oplus)$ that regulates Einstein addition, $\oplus$, in the ball $\mathbb{R}^n_c$ of the Euclidean n-space $\mathbb{R}^n$ is a *gyrocommutative gyrogroup* called an *Einstein gyrogroup*. Einstein gyrogroups and gyrovector spaces are studied in [63, 65, 68, 70]. Gyrogroups are not peculiar to Einstein addition [69]. Rather, they are abound in the theory of groups [17, 19, 20], loops [27], quasigroup [28, 35], and Lie groups [30–32].

Thus, the type of structure arising in the study of Einstein velocity addition (and Möbius addition) is of rather frequent occurrence and hence merits an axiomatic approach. Taking the key features of Einstein velocity addition law as axioms, and guided by analogies with groups, we are led to the following formal definition of gyrogroups.

Definition 41.2 (Gyrogroups) A groupoid is a nonempty set with a binary operation. A groupoid $(G, \oplus)$ is a gyrogroup if its binary operation satisfies the following axioms. In G there is at least one element, 0, called a left identity, satisfying

$$\text{(G1)} \qquad 0 \oplus a = a$$

for all $a \in G$. There is an element $0 \in G$ satisfying axiom (G1) such that for each $a \in G$ there is an element $\ominus a \in G$, called a left inverse of a, satisfying

$$\text{(G2)} \qquad \ominus a \oplus a = 0.$$

Moreover, for any $a, b, c \in G$ there exists a unique element $\mathrm{gyr}[a, b]c \in G$ such that the binary operation obeys the left gyroassociative law

$$\text{(G3)} \qquad a \oplus (b \oplus c) = (a \oplus b) \oplus \mathrm{gyr}[a, b]c.$$

The map $\mathrm{gyr}[a, b] : G \to G$ given by $c \mapsto \mathrm{gyr}[a, b]c$ is an automorphism of the groupoid $(G, \oplus)$, that is,

$$\text{(G4)} \qquad \mathrm{gyr}[a, b] \in \mathrm{Aut}(G, \oplus),$$

and the automorphism $\mathrm{gyr}[a, b]$ of G is called the gyroautomorphism, or the gyration, of G generated by $a, b \in G$. The operator $\mathrm{gyr} : G \times G \to \mathrm{Aut}(G, \oplus)$ is called the gyrator of G. Finally, the gyroautomorphism $\mathrm{gyr}[a, b]$ generated by any $a, b \in G$ possesses the left reduction property

$$\text{(G5)} \qquad \mathrm{gyr}[a, b] = \mathrm{gyr}[a \oplus b, b].$$

The first pair of the gyrogroup axioms are like the group axioms. The last pair present the gyrator axioms and the middle axiom links the two pairs.

As in group theory, we use the notation $a \ominus b = a \oplus (\ominus b)$ in gyrogroup theory as well.

In full analogy with groups, some gyrogroups are gyrocommutative according to the following definition.

Definition 41.3 (Gyrocommutative Gyrogroups) A gyrogroup $(G, \oplus)$ is gyrocommutative if its binary operation obeys the gyrocommutative law

$$\text{(G6)} \qquad a \oplus b = \mathrm{gyr}[a, b](b \oplus a)$$

for all $a, b \in G$.

First gyrogroup properties are studied in [73, Chap. 1], and more gyrogroup theorems are studied in [63, 65, 68]. Thus, for instance, as in group theory, any gyrogroup possesses a unique identity element which is both left and right, and any element of a gyrogroup possesses a unique inverse.

In order to illustrate the power and elegance of the gyrogroup structure, we solve below the two basic gyrogroup equations (41.38) and (41.45).

Let us consider the gyrogroup equation

$$\mathbf{a} \oplus \mathbf{x} = \mathbf{b} \tag{41.38}$$

in a gyrogroup $(G, \oplus)$ for the unknown $\mathbf{x}$. If $\mathbf{x}$ exists, then by the right gyroassociative law (41.36) and by (41.31), we have

$$\begin{aligned} \mathbf{x} &= \mathbf{0} \oplus \mathbf{x} \\ &= (\ominus\mathbf{a} \oplus \mathbf{a}) \oplus \mathbf{x} \\ &= \ominus\mathbf{a} \oplus \big(\mathbf{a} \oplus \mathrm{gyr}[\mathbf{a}, \ominus\mathbf{a}]\mathbf{x}\big) \\ &= \ominus\mathbf{a} \oplus (\mathbf{a} \oplus \mathbf{x}) \\ &= \ominus\mathbf{a} \oplus \mathbf{b} \end{aligned} \tag{41.39}$$

noting that $\mathrm{gyr}[\mathbf{a}, \ominus\mathbf{a}]$ is trivial by (41.31).

Thus, if a solution to (41.38) exists, it must be given uniquely by

$$\mathbf{x} = \ominus\mathbf{a} \oplus \mathbf{b} \tag{41.40}$$

Conversely, if $\mathbf{x} = \ominus\mathbf{a} \oplus \mathbf{b}$, then $\mathbf{x}$ is indeed a solution to (41.38) since by the left gyroassociative law and (41.31) we have

$$\begin{aligned} \mathbf{a} \oplus \mathbf{x} &= \mathbf{a} \oplus (\ominus\mathbf{a} \oplus \mathbf{b}) \\ &= \big(\mathbf{a} \oplus (\ominus\mathbf{a})\big) \oplus \mathrm{gyr}[\mathbf{a}, \ominus\mathbf{a}]\mathbf{b} \\ &= \mathbf{0} \oplus \mathbf{b} \\ &= \mathbf{b} \end{aligned} \tag{41.41}$$

Substituting the solution (41.40) in its equation (41.38) and replacing $\mathbf{a}$ by $\ominus\mathbf{a}$ we recover the left cancellation law (41.23) for Einstein addition

$$\ominus\mathbf{a} \oplus (\mathbf{a} \oplus \mathbf{b}) = \mathbf{b}. \tag{41.42}$$

The gyrogroup operation (or, addition) of any gyrogroup has an associated dual operation, called the *gyrogroup cooperation* (or, *coaddition*), which is defined below.

Definition 41.4 (The Gyrogroup Cooperation (Coaddition)) Let $(G, \oplus)$ be a gyrogroup with gyrogroup operation (or, addition) $\oplus$. The gyrogroup cooperation (or, coaddition) $\boxplus$ is a second binary operation in G given by the equation

$$\mathbf{a} \boxplus \mathbf{b} = \mathbf{a} \oplus \mathrm{gyr}[\mathbf{a}, \ominus\mathbf{b}]\mathbf{b} \tag{41.43}$$

for all $\mathbf{a}, \mathbf{b} \in G$.

Replacing b by $\ominus b$ in (41.43) we have the *cosubtraction* identity

$$\mathbf{a} \boxminus \mathbf{b} := \mathbf{a} \boxplus (\ominus\mathbf{b}) = \mathbf{a} \ominus \mathrm{gyr}[\mathbf{a}, \mathbf{b}]\mathbf{b} \tag{41.44}$$

for all $\mathbf{a}, \mathbf{b} \in G$.

To motivate the introduction of the gyrogroup cooperation and to illustrate the use of the left reduction property (G5), we solve the equation

$$\mathbf{x} \oplus \mathbf{a} = \mathbf{b} \tag{41.45}$$

for the unknown $\mathbf{x}$ in a gyrogroup $(G, \oplus)$.

Assuming that a solution $\mathbf{x}$ to (41.45) exists, we have the following chain of equations

$$\begin{aligned}
\mathbf{x} &= \mathbf{x} \oplus \mathbf{0} \\
&= \mathbf{x} \oplus (\mathbf{a} \ominus \mathbf{a}) \\
&= (\mathbf{x} \oplus \mathbf{a}) \oplus \mathrm{gyr}[\mathbf{x}, \mathbf{a}](\ominus\mathbf{a}) \\
&= (\mathbf{x} \oplus \mathbf{a}) \ominus \mathrm{gyr}[\mathbf{x}, \mathbf{a}]\mathbf{a} \\
&= (\mathbf{x} \oplus \mathbf{a}) \ominus \mathrm{gyr}[\mathbf{x} \oplus \mathbf{a}, \mathbf{a}]\mathbf{a} \\
&= \mathbf{b} \ominus \mathrm{gyr}[\mathbf{b}, \mathbf{a}]\mathbf{a} \\
&= \mathbf{b} \boxminus \mathbf{a}
\end{aligned} \tag{41.46}$$

where the gyrogroup cosubtraction, (41.44), which captures here an obvious analogy, comes into play. Hence, if a solution $\mathbf{x}$ to the gyrogroup equation (41.45) exists, it must be given uniquely by (41.46). One can show that the latter is indeed a solution to (41.45) [68, Sect. 2.4].

The gyrogroup cooperation is introduced into gyrogroups in order to capture useful analogies between gyrogroups and groups, and to uncover duality symmetries with the gyrogroup operation. Thus, for instance, the gyrogroup cooperation uncovers the seemingly missing right counterpart of the left cancellation law (41.23), giving rise to the right cancellation law,

$$(\mathbf{b} \boxminus \mathbf{a}) \oplus \mathbf{a} = \mathbf{b} \tag{41.47}$$

for all $\mathbf{a}, \mathbf{a}$ in G, which is obtained by substituting the result of (41.46) into (41.45).

Remarkably, the right cancellation law (41.47) can be dualized, giving rise to the dual right cancellation law

$$(\mathbf{b} \ominus \mathbf{a}) \boxplus \mathbf{a} = \mathbf{b}. \tag{41.48}$$

As an example, and for later reference, we note that it follows from the right cancellation law (41.47) that

$$\mathbf{d} = (\mathbf{b} \boxplus \mathbf{c}) \ominus \mathbf{a} \quad \Longleftrightarrow \quad \mathbf{b} \boxplus \mathbf{c} = \mathbf{d} \boxplus \mathbf{a} \tag{41.49}$$

for $\mathbf{a}, \mathbf{b}, \mathbf{c}, \mathbf{d}$ in any gyrocommutative gyrogroup $(G, \oplus)$.

An elegant gyrocommutative gyrogroup identity that involves the gyrogroup cooperation, verified in [68, Theorem 3.12], is

$$\mathbf{a} \oplus (\mathbf{b} \oplus \mathbf{a}) = \mathbf{a} \boxplus (\mathbf{a} \oplus \mathbf{b}). \tag{41.50}$$

A gyrogroup cooperation is commutative if and only if the gyrogroup is gyrocommutative [65, Theorem 3.4] [68, Theorem 3.4]. Hence, in particular, Einstein coaddition is commutative. Indeed, Einstein coaddition, $\boxplus$, in an Einstein gyrogroup $(\mathbb{R}^n_c, \oplus)$, defined in (41.43), can be written as [68, Eq. 3.195]

$$\begin{aligned}
\mathbf{u} \boxplus \mathbf{v} &= \frac{\gamma_{\mathbf{u}} + \gamma_{\mathbf{v}}}{\gamma_{\mathbf{u}}^2 + \gamma_{\mathbf{v}}^2 + \gamma_{\mathbf{u}}\gamma_{\mathbf{v}}(1 + \frac{\mathbf{u}\cdot\mathbf{v}}{s^2}) - 1}(\gamma_{\mathbf{u}}\mathbf{u} + \gamma_{\mathbf{v}}\mathbf{v}) \\
&= \frac{\gamma_{\mathbf{u}} + \gamma_{\mathbf{v}}}{(\gamma_{\mathbf{u}} + \gamma_{\mathbf{v}})^2 - (\gamma_{\mathbf{u}\ominus\mathbf{v}} + 1)}(\gamma_{\mathbf{u}}\mathbf{u} + \gamma_{\mathbf{v}}\mathbf{v}) \\
&= 2 \otimes \frac{\gamma_{\mathbf{u}}\mathbf{u} + \gamma_{\mathbf{v}}\mathbf{v}}{\gamma_{\mathbf{u}} + \gamma_{\mathbf{v}}} \\
&= 2 \otimes \frac{\gamma_{\mathbf{u}}\mathbf{u} + \gamma_{\mathbf{v}}\mathbf{v}}{2 + (\gamma_{\mathbf{u}} - 1) + (\gamma_{\mathbf{v}} - 1)}
\end{aligned} \tag{41.51}$$

for $\mathbf{u}, \mathbf{v} \in G$, demonstrating that it is commutative, as expected. The symbol $\otimes$ in (41.51) represents *scalar multiplication* so that, for instance, $2 \otimes \mathbf{v} = \mathbf{v} \oplus \mathbf{v}$, for all $\mathbf{v}$ in a gyrogroup $(G, \oplus)$, as explained in Sect. 41.5 below. It turns out that Einstein coaddition $\boxplus$ is more than just a commutative binary operation in the ball. Remarkably, it forms the (hyperbolic) *gyroparallelogram addition law* in the ball, illustrated in Fig. 41.6.

The extreme sides of (41.51) suggest that the application of Einstein coaddition to three summands is given by the following *gyroparallelepiped addition law*

$$\mathbf{u} \boxplus_3 \mathbf{v} \boxplus_3 \mathbf{w} := 2 \otimes \frac{\gamma_{\mathbf{u}}\mathbf{u} + \gamma_{\mathbf{v}}\mathbf{v} + \gamma_{\mathbf{w}}\mathbf{w}}{2 + (\gamma_{\mathbf{u}} - 1) + (\gamma_{\mathbf{v}} - 1) + (\gamma_{\mathbf{w}} - 1)} \tag{41.52}$$

for $\mathbf{u}, \mathbf{v}, \mathbf{w} \in G$, the ternary operation $\boxplus_3$ being *Einstein coaddition of order three*.

Einstein coaddition (41.52) of three summands is commutative and associative in the generalized sense that it is a symmetric function of the summands. The gyroparallelepiped that results from the gyroparallelepiped law (41.52) is studied in detail in [68, Sects. 10.9–10.12].

We may note that by (41.51)–(41.52) we have $\mathbf{u} \boxplus_3 \mathbf{v} \boxplus_3 \mathbf{0} = \mathbf{u} \boxplus \mathbf{v}$, as expected. However, unexpectedly we have $\mathbf{u} \boxplus_3 \mathbf{v} \boxplus_3 (\ominus\mathbf{v}) \neq \mathbf{u}$, in general.

The extension of (41.52) to the Einstein coaddition of k summands, $k > 3$, is now straightforward, giving rise to the higher dimensional *gyroparallelotope law* in $\mathbb{R}^n_c$,

$$\mathbf{v}_1 \boxplus_k \mathbf{v}_2 \boxplus_k \cdots \boxplus_k \mathbf{v}_k := 2 \otimes \frac{\sum_{i=1}^{k} \gamma_{\mathbf{v}_i} \mathbf{v}_i}{2 + \sum_{i=1}^{k} (\gamma_{\mathbf{v}_i} - 1)} \tag{41.53}$$

for $\mathbf{v}_k \in G$, $k \in \mathbb{N}$, where $\boxplus_k$ is a k-ary operation called *Einstein coaddition of order* k. An interesting study of parallelotopes in Euclidean geometry is found in [10].

In the Euclidean limit $c \to \infty$, (i) gamma factors tend to 1, and (ii) the hyperbolic scalar multiplication, $\otimes$, of a gyrovector (see Sect. 41.6) by 2 tends to the common scalar multiplication of a vector by 2. Hence, in the Euclidean limit, the right-hand side of (41.53) tends to the vector sum $\sum_{i=1}^{k} \mathbf{v}_i$ in $\mathbb{R}^n$, as expected.

41.5 Einstein Gyrovector Spaces

Let $k \otimes \mathbf{v}$ be the Einstein addition of k copies of $\mathbf{v} \in \mathbb{R}^n_c$, that is $k \otimes \mathbf{v} = \mathbf{v} \oplus \mathbf{v} \cdots \oplus \mathbf{v}$ (k terms). Then, it follows from Einstein addition (41.19) and straightforward algebra that [58]

$$k \otimes \mathbf{v} = c \frac{\left(1 + \frac{\|\mathbf{v}\|}{c}\right)^k - \left(1 - \frac{\|\mathbf{v}\|}{c}\right)^k}{\left(1 + \frac{\|\mathbf{v}\|}{c}\right)^k + \left(1 - \frac{\|\mathbf{v}\|}{c}\right)^k} \frac{\mathbf{v}}{\|\mathbf{v}\|}. \tag{41.54}$$

The definition of scalar multiplication in an Einstein gyrovector space requires analytically continuing k off the positive integers, thus obtaining the following definition [59]:

Definition 41.5 An Einstein gyrovector space $(\mathbb{R}^n_c, \oplus, \otimes)$ is an Einstein gyrogroup $(\mathbb{R}^n_c, \oplus)$, $\mathbb{R}^n_c \subset \mathbb{R}^n_c$, with scalar multiplication $\otimes$ given by the equation

$$\begin{aligned} r \otimes \mathbf{v} &= c \frac{\left(1 + \frac{\|\mathbf{v}\|}{c}\right)^r - \left(1 - \frac{\|\mathbf{v}\|}{c}\right)^r}{\left(1 + \frac{\|\mathbf{v}\|}{c}\right)^r + \left(1 - \frac{\|\mathbf{v}\|}{c}\right)^r} \frac{\mathbf{v}}{\|\mathbf{v}\|} \\ &= c \tanh\left(r \tanh^{-1} \frac{\|\mathbf{v}\|}{c} \right) \frac{\mathbf{v}}{\|\mathbf{v}\|} \end{aligned} \tag{41.55}$$

where r is any real number, $r \in \mathbb{R}$, $\mathbf{v} \in \mathbb{R}^n_c$, $\mathbf{v} \neq \mathbf{0}$, and $r \otimes \mathbf{0} = \mathbf{0}$, and with which we use the notation $\mathbf{v} \otimes r = r \otimes \mathbf{v}$.

Einstein gyrovector spaces are studied in [68, Sect. 6.18] and [70]. Einstein scalar multiplication does not distribute over Einstein addition, but it possesses other properties of vector spaces. For any positive integer n, and for all real numbers

$r, r_1, r_2 \in \mathbb{R}$, and $\mathbf{v} \in \mathbb{R}_c^n$, we have

$$\begin{aligned} n \otimes \mathbf{v} &= \mathbf{v} \oplus \cdots \oplus \mathbf{v} && (n \text{ terms}), \\ (r_1 + r_2) \otimes \mathbf{v} &= r_1 \otimes \mathbf{v} \oplus r_2 \otimes \mathbf{v} && \text{(Scalar Distributive Law)}, \\ (r_1 r_2) \otimes \mathbf{v} &= r_1 \otimes (r_2 \otimes \mathbf{v}) && \text{(Scalar Associative Law)}, \\ r \otimes (r_1 \otimes \mathbf{v} \oplus r_2 \otimes \mathbf{v}) &= r \otimes (r_1 \otimes \mathbf{v}) \oplus r \otimes (r_2 \otimes \mathbf{v}) && \text{(Monodistributive Law)} \end{aligned}$$

in any Einstein gyrovector space $(\mathbb{R}_c^n, \oplus, \otimes)$.

Any Einstein gyrovector space $(\mathbb{R}_c^n, \oplus, \otimes)$ inherits the common inner product and the norm from its vector space $\mathbb{R}^n$. These turn out to be invariant under gyrations, that is,

$$\begin{aligned} \mathrm{gyr}[\mathbf{a}, \mathbf{b}]\mathbf{u} \cdot \mathrm{gyr}[\mathbf{a}, \mathbf{b}]\mathbf{v} &= \mathbf{u} \cdot \mathbf{v}, \\ \|\mathrm{gyr}[\mathbf{a}, \mathbf{b}]\mathbf{v}\| &= \|\mathbf{v}\| \end{aligned} \tag{41.56}$$

for all $\mathbf{a}, \mathbf{b}, \mathbf{u}, \mathbf{v} \in \mathbb{R}_c^n$.

Unlike vector spaces, Einstein gyrovector spaces $(\mathbb{R}_c^n, \oplus, \otimes)$ do not possess the distributive law since, in general,

$$r \otimes (\mathbf{u} \oplus \mathbf{v}) \neq r \otimes \mathbf{u} \oplus r \otimes \mathbf{v} \tag{41.57}$$

for $r \in \mathbb{R}$ and $\mathbf{u}, \mathbf{v} \in \mathbb{R}_c^n$. One might suppose that there is a price to pay in mathematical regularity when replacing ordinary vector addition with Einstein addition, but this is not the case as demonstrated in [63, 65, 68], and as noted by S. Walter in [77].

Owing to the break down of the distributive law in gyrovector spaces, the following gyrovector space identity, called the *Two-Sum Identity* [68, Theorem 6.7], proves useful:

$$2 \otimes (\mathbf{u} \oplus \mathbf{v}) = \mathbf{u} \oplus (2 \otimes \mathbf{v} \oplus \mathbf{u}). \tag{41.58}$$

In full analogy with the common Euclidean distance function, Einstein addition admits the *gyrodistance* function

$$d_\oplus(A, B) = \|\ominus A \oplus B\| \tag{41.59}$$

that obeys the gyrotriangle inequality [68, Theorem 6.9]

$$d_\oplus(A, B) \leq d_\oplus(A, P) \oplus d_\oplus(P, B) \tag{41.60}$$

for any points $A, B, P \in \mathbb{R}_c^n$ in an Einstein gyrovector space $(\mathbb{R}_c^n, \oplus, \otimes)$. The gyrodistance function is invariant under the group of motions of its Einstein gyrovector space, that is, under *left gyrotranslations* and rotations of the space [68, Sect. 4]. The gyrotriangle inequality (41.60) reduces to a corresponding gyrotriangle equality,

$$d_\oplus(A, B) = d_\oplus(A, P) \oplus d_\oplus(P, B) \tag{41.61}$$

if and only if point P lies between points A and B, that is, point P lies on the gyrosegment AB, as shown in Fig. 41.2. Accordingly, the gyrodistance function is

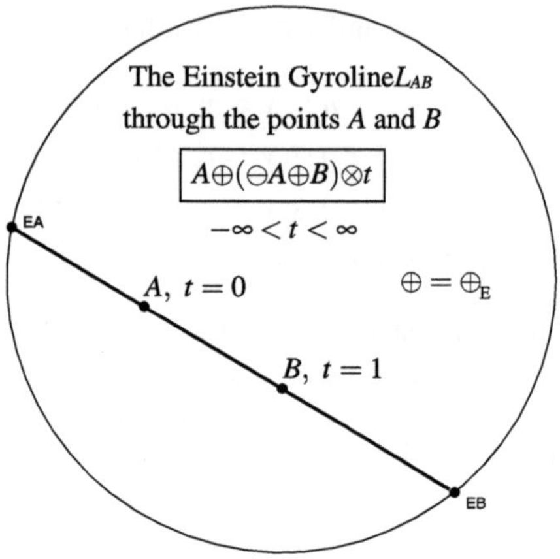

Fig. 41.1 [The Einstein Gyroline] The unique gyroline L_{AB} in an Einstein gyrovector space $(\mathbb{R}^n_c, \oplus, \otimes)$ through two given points A and B. The case of the Einstein gyrovector plane, when $\mathbb{R}^n_c = \mathbb{R}^2_{c=1}$ is the real open unit disc, is shown

gyroadditive on gyrolines, as demonstrated in (41.61) and illustrated graphically in Fig. 41.2.

Furthermore, the Einstein gyrodistance function (41.59) in any n-dimensional Einstein gyrovector space $(\mathbb{R}^n_c, \oplus, \otimes)$ possesses a familiar Riemannian line element. It gives rise to the Riemannian line element ds_e^2 of the Einstein gyrovector space with its *gyrometric* (41.59),

$$\begin{aligned} ds_e^2 &= \left\|(\mathbf{x}+d\mathbf{x})\ominus\mathbf{x}\right\|^2 \\ &= \frac{c^2}{c^2-\mathbf{x}^2}d\mathbf{x}^2 + \frac{c^2}{(c^2-\mathbf{x}^2)^2}(\mathbf{x}\cdot d\mathbf{x})^2 \end{aligned} \tag{41.62}$$

for $\mathbf{x}\in\mathbb{R}^n_c$, where $d\mathbf{x}^2 = d\mathbf{x}\cdot d\mathbf{x}$, as shown in [68, Theorem 7.6].

Remarkably, the Riemannian line element ds_e^2 in (41.62) turns out to be the well-known line element that the Italian mathematician Eugenio Beltrami introduced in 1868 in order to study hyperbolic geometry by a Euclidean disc model, now known as the Beltrami–Klein disc [38, p. 220], [4]. An English translation of his historically significant 1868 essay on the interpretation of non-Euclidean geometry is found in [53]. The significance of Beltrami's 1868 essay rests on the generally known fact that it was the first to offer a concrete interpretation of hyperbolic geometry by interpreting "straight lines" as geodesics on a surface of a constant negative curvature. Beltrami, thus, constructed a Euclidean disc model of the hyperbolic plane [38, 53], which now bears his name along with the name of Klein.

We have thus found that the Beltrami–Klein ball model of hyperbolic geometry is regulated algebraically by Einstein gyrovector spaces with their gyrodistance function (41.59) and Riemannian line element (41.62), just as the standard model of Euclidean geometry is regulated algebraically by vector spaces with their Euclidean distance function and the Riemannian line element $ds^2 = d\mathbf{x}^2$.

In full analogy with Euclidean geometry, the unique Einstein gyroline L_{AB}, Fig. 41.1, that passes through two given points A and B in an Einstein gyrovec-

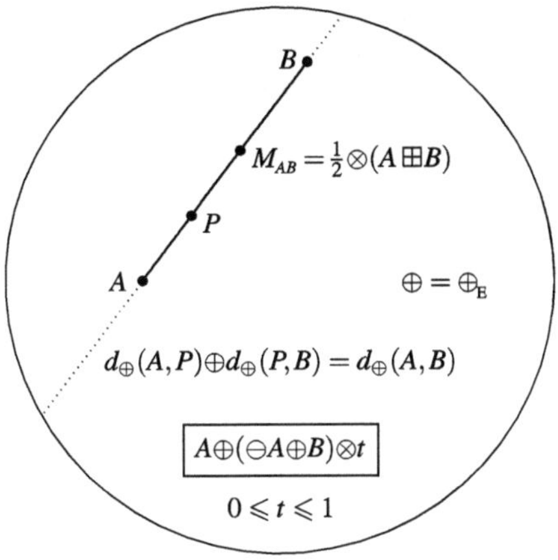

Fig. 41.2 The gyrosegment AB that links the points A and B in $(\mathbb{R}^n_c, \oplus, \otimes)$, with one of its generic points P and its gyromidpoint M_{AB}. The point P lies between A and B and, hence, obeys the gyrotriangle equality, (41.61)

tor space $\mathbb{R}^n_c = (\mathbb{R}^n_c, \oplus, \otimes)$ is given by the parametric equation

$$L_{AB}(t) = A \oplus (\ominus A \oplus B) \otimes t \tag{41.63}$$

with the parameter $t \in \mathbb{R}$. The gyroline L_{AB} passes through the point A when $t = 0$ and, owing to the left cancellation law (41.23), it passes through the point B when $t = 1$.

Einstein gyrolines in the ball $\mathbb{R}^n_c$ are chords of the ball, as shown in Fig. 41.1. These chords of the ball turn out to be the familiar geodesics of the Beltrami–Klein ball model of hyperbolic geometry [38]. Accordingly, Einstein gyrosegments are Euclidean segments, as shown in Fig. 41.2. The result that Einstein gyrosegments are Euclidean segments is well exploited in [72, 73] in the use of hyperbolic barycentric coordinates for the determination of various hyperbolic triangle centers. It enables one to determine points of intersection of gyrolines by common methods of linear algebra.

The gyromidpoint M_{AB} of gyrosegment AB, shown in Fig. 41.2, is the unique point of the gyrosegment that satisfies the equation $d_\oplus(M_{AB}, A) = d_\oplus(M_{AB}, B)$. It is given by each of the following equations [70, Theorem 3.33],

$$M_{AB} = A \oplus (\ominus A \oplus B) \otimes \frac{1}{2} = \frac{\gamma_A A + \gamma_B B}{\gamma_A + \gamma_B} = \frac{1}{2} \otimes (A \boxplus B) \tag{41.64}$$

in full analogy with Euclidean midpoints, shown in Fig. 41.5. One may note that the extreme right equation in (41.64) appears in (41.51) in an equivalent form.

The endpoints of a gyroline in an Einstein gyrovector space $(\mathbb{R}^n_c, \oplus, \otimes)$ are the points where the gyroline approaches the boundary of the ball $\mathbb{R}^n_c$. Following (41.63), the endpoints E_A and E_B of the gyroline $L_{AB}(t)$ in Fig. 41.1 are

$$\begin{aligned} E_A &= \lim_{t\to-\infty} \{A \oplus (\ominus A \oplus B) \otimes t\}, \\ E_B &= \lim_{t\to\infty} \{A \oplus (\ominus A \oplus B) \otimes t\}. \end{aligned} \tag{41.65}$$

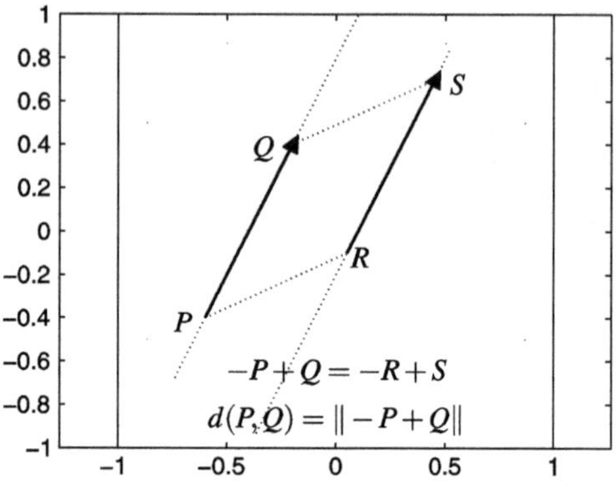

Fig. 41.3 Two equivalent vectors in a Euclidean vector plane $(\mathbb{R}^2, +, \cdot)$. The two vectors are parallel and have equal values and, hence, equal lengths

Explicit expressions for the gyroline endpoints in Einstein gyrovector spaces are presented in (41.162), p. 766.

41.6 Vectors and Gyrovectors

Elements of a real inner product space $\mathbb{V} = (\mathbb{V}, +, \cdot)$, called points and denoted by capital italic letters, A, B, P, Q, etc., give rise to vectors in $\mathbb{V}$, denoted by bold roman lowercase letters $\mathbf{u}, \mathbf{v}$, etc. Any two ordered points $P, Q \in \mathbb{V}$ give rise to a unique rooted vector $\mathbf{v} \in \mathbb{V}$, rooted at the point P. It has a tail at the point P and a head at the point Q, and it has the value $-P + Q$,

$$\mathbf{v} = -P + Q. \tag{41.66}$$

The length of the rooted vector $\mathbf{v} = -P + Q$ is the distance between the points P and Q, given by the equation

$$\|\mathbf{v}\| = \| - P + Q\|. \tag{41.67}$$

Two rooted vectors $-P + Q$ and $-R + S$ are equivalent if they have the same value, that is,

$$-P + Q \sim -R + S \quad \text{if and only if} \quad -P + Q = -R + S. \tag{41.68}$$

The relation $\sim$ in (41.68) between rooted vectors is reflexive, symmetric, and transitive, so that it is an equivalence relations that gives rise to equivalence classes of rooted vectors.

Two equivalent rooted vectors in a Euclidean vector plane are shown in Fig. 41.3. Being equivalent in Euclidean geometry, the two vectors in Fig. 41.3 are parallel and they possess equal lengths.

To liberate rooted vectors from their roots we define a *vector* to be an equivalence class of rooted vectors. The vector $-P + Q$ is thus a representative of all rooted vectors with value $-P + Q$. Accordingly, the two vectors in Fig. 41.3 are equal.

A point $P \in \mathbb{V}$ is identified with the vector $-O + P$, O being the arbitrarily selected origin of the space $\mathbb{V}$. Hence, the algebra of vectors can be applied to points as well. Naturally, geometric and physical properties regulated by a vector space are independent of the choice of the origin.

Let $A, B, C \in \mathbb{V}$ be three non-collinear points, and let

$$\begin{aligned} \mathbf{u} &= -A + B, \\ \mathbf{v} &= -A + C \end{aligned} \tag{41.69}$$

be two vectors in $\mathbb{V}$ that possess the same tail, A. Furthermore, let D be a point of $\mathbb{V}$ given by the parallelogram condition

$$D = B + C - A. \tag{41.70}$$

The quadrangle (also known as a quadrilateral; see [11, p. 52]) $ABDC$ turns out to be a parallelogram in Euclidean geometry, Fig. 41.5, since its two diagonals, AD and BC, intersect at their midpoints, that is,

$$\frac{1}{2}(A + D) = \frac{1}{2}(B + C). \tag{41.71}$$

Clearly, the midpoint equality (41.71) is equivalent to the parallelogram condition (41.70).

The vector addition of the vectors **u** and **v** that generate the parallelogram $ABDC$, according to (41.69), gives the vector **w** by the parallelogram addition law, Fig. 41.5,

$$\mathbf{w} := -A + D = (-A + B) + (-A + C) = \mathbf{u} + \mathbf{v}. \tag{41.72}$$

Here, by definition, **w** is the vector formed by the diagonal AD of the parallelogram $ABDC$, as shown in Fig. 41.5.

Vectors in the space $\mathbb{V}$ are, thus, equivalence classes of ordered pairs of points, Fig. 41.3, which add according to the parallelogram law, Fig. 41.5.

Gyrovectors emerge in an Einstein gyrovector space $(\mathbb{V}_c, \oplus, \otimes)$ in a way fully analogous to the way vectors emerge in the space $\mathbb{V}$, where $\mathbb{V}_c$ is the c-ball of the space $\mathbb{V}$, see (41.14).

Elements of $\mathbb{V}_c$, called points and denoted by capital italic letters, A, B, P, Q, etc., give rise to gyrovectors in $\mathbb{V}_c$, denoted by bold roman lowercase letters **u**, **v**, etc. Any two ordered points $P, Q \in \mathbb{V}_c$ give rise to a unique rooted gyrovector $\mathbf{v} \in \mathbb{V}_c$, rooted at the point P. It has a tail at the point P and a head at the point Q, and it has the value $\ominus P \oplus Q$,

$$\mathbf{v} = \ominus P \oplus Q. \tag{41.73}$$

The gyrolength of the rooted gyrovector $\mathbf{v} = \ominus P \oplus Q$ is the gyrodistance between the points P and Q, given by the equation

$$\|\mathbf{v}\| = \| \ominus P \oplus Q\|. \tag{41.74}$$

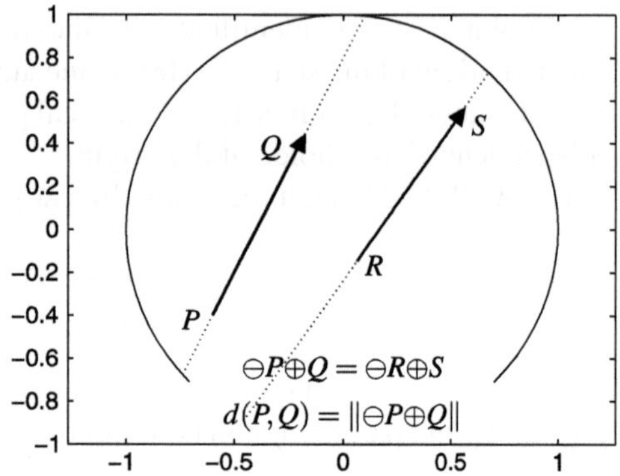

Fig. 41.4 Two equivalent gyrovectors in an Einstein gyrovector plane $(\mathbb{R}^2_c, \oplus, \otimes)$. The two gyrovectors have equal values and, hence, equal gyrolengths

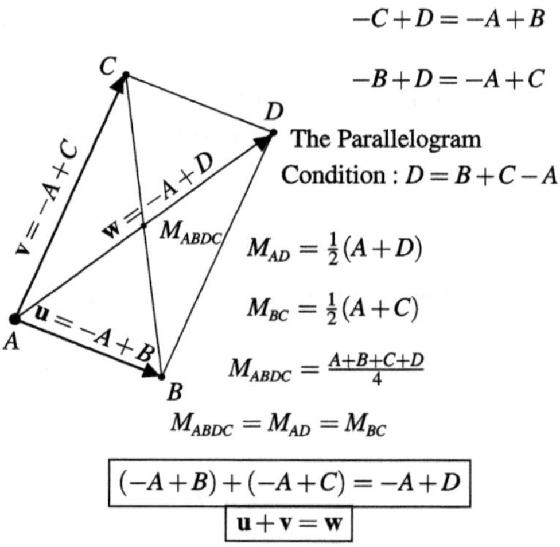

Fig. 41.5 The Euclidean parallelogram and its addition law in a Euclidean vector plane $(\mathbb{R}^2, +, \cdot)$. The diagonals AD and BC of parallelogram $ABDC$ intersect each other at their midpoints. The midpoints of the diagonals AD and BC are, respectively, M_{AD} and M_{BC}, each of which coincides with the parallelogram center M_{ABDC}

Two rooted gyrovectors $\ominus P \oplus Q$ and $\ominus R \oplus S$ are equivalent if they have the same value, that is,

$$\ominus P \oplus Q \sim \ominus R \oplus S \quad \text{if and only if} \quad \ominus P \oplus Q = \ominus R \oplus S. \tag{41.75}$$

The relation $\sim$ in (41.75) between rooted gyrovectors is reflexive, symmetric, and transitive, so that it is an equivalence relation that gives rise to equivalence classes of rooted gyrovectors.

Two equivalent rooted gyrovectors in an Einstein gyrovector plane are shown in Fig. 41.4. Being equivalent in hyperbolic geometry, the two gyrovectors in Fig. 41.4 possess equal gyrolengths.

To liberate rooted gyrovectors from their roots we define a *gyrovector* to be an equivalence class of rooted gyrovectors. The gyrovector $\ominus P \oplus Q$ is thus a representative of all rooted gyrovectors with value $\ominus P \oplus Q$. Accordingly, the two gyrovectors in Fig. 41.4 are equal.

A point P of a gyrovector space $(\mathbb{V}_c, \oplus, \otimes)$ is identified with the gyrovector $\ominus O \oplus P$, O being the arbitrarily selected origin of the space $\mathbb{V}_c$. Hence, the algebra of gyrovectors can be applied to points as well. Naturally, geometric and physical properties regulated by a gyrovector space are independent of the choice of the origin.

Let $A, B, C \in \mathbb{V}_c$ be three non-gyrocollinear points of an Einstein gyrovector space $(\mathbb{V}_c, \oplus, \otimes)$, and let

$$\begin{aligned} \mathbf{u} &= \ominus A \oplus B, \\ \mathbf{v} &= \ominus A \oplus C \end{aligned} \tag{41.76}$$

be two gyrovectors in $\mathbb{V}$ that possess the same tail, A. Furthermore, let D be a point of $\mathbb{V}_c$ given by the gyroparallelogram condition

$$D = (B \boxplus C) \ominus A. \tag{41.77}$$

Then, the gyroquadrangle $ABDC$ is a gyroparallelogram in the Beltrami–Klein ball model of hyperbolic geometry in the sense that its two gyrodiagonals, AD and BC, intersect at their gyromidpoints, that is,

$$\frac{1}{2} \otimes (A \boxplus D) = \frac{1}{2} \otimes (B \boxplus C), \tag{41.78}$$

as illustrated in Fig. 41.6. Clearly by (41.49), the gyromidpoint equality (41.78) is equivalent to the gyroparallelogram condition (41.77).

The gyrovector addition of the gyrovectors **u** and **v** that generate the gyroparallelogram $ABDC$ gives the gyrovector **w** by the gyroparallelogram addition law, Fig. 41.6,

$$\mathbf{w} := \ominus A \oplus D = (\ominus A \oplus B) \boxplus (\ominus A \oplus C) =: \mathbf{u} \boxplus \mathbf{v}. \tag{41.79}$$

Here, by definition, **w** is the gyrovector formed by the gyrodiagonal AD of the gyroparallelogram $ABDC$. The gyrovector identity in (41.79) is explained in (41.82) below.

Gyrovectors in the ball $\mathbb{V}_c$ are, thus, equivalence classes of ordered pairs of points, Fig. 41.4, which add according to the gyroparallelogram law, Fig. 41.6.

41.7 Gyroparallelogram—The Hyperbolic Parallelogram

In Euclidean geometry, a parallelogram is a quadrangle the two diagonals of which intersect at their midpoints. In full analogy, in hyperbolic geometry a gyroparallelogram is a gyroquadrangle the two gyrodiagonals of which intersect at their gyromidpoints, as shown in Fig. 41.6. Accordingly, if A, B, and C are any three non-gyrocollinear points (that is, they do not lie on a gyroline) in an Einstein gyrovector space, and if a fourth point D is given by the *gyroparallelogram condition*

$$D = (B \boxplus C) \ominus A \tag{41.80}$$

then the gyroquadrangle $ABDC$ is a gyroparallelogram, as shown in Fig. 41.6.

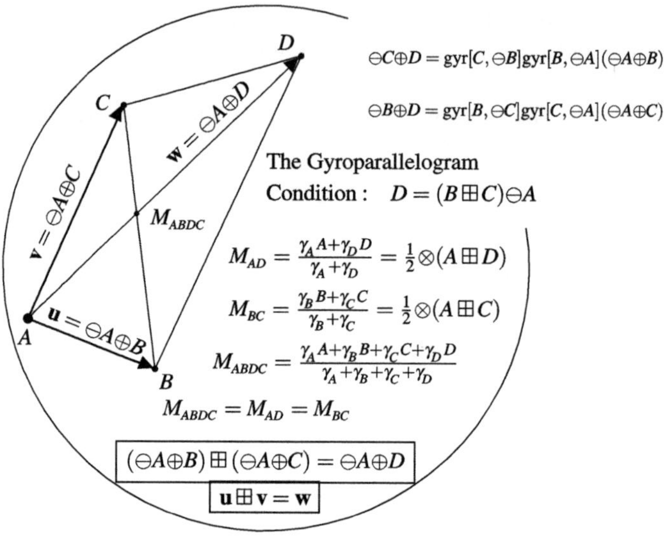

Fig. 41.6 The Einstein gyroparallelogram and its addition law in an Einstein gyrovector plane $(\mathbb{R}_c^2, \oplus, \otimes)$. The gyrodiagonals AD and BC of gyroparallelogram $ABDC$ intersect each other at their gyromidpoints. Detailed studies of the gyroparallelogram and its extension to higher dimensional gyroparallelepipeds are presented in [65, 68]. The gyroparallelogram addition law plays an important role in the gyrovector space approach to hyperbolic geometry, studied in [68, 70]. The gyromidpoints of the gyrodiagonals AD and BC are, respectively, M_{AD} and M_{BC}, each of which coincides with the gyroparallelogram gyrocenter M_{ABDC}. The analogies that this figure shares with Fig. 41.5 are obvious. Along these analogies there is a remarkable disanalogy. (i) Newton velocity addition, +, and the parallelogram addition, +, in Fig. 41.5 are identically the same binary operations in $\mathbb{R}^n$. In contrast, (ii) Einstein velocity addition, $\oplus$, and its resulting gyroparallelogram addition, $\boxplus$, in this figure are two different binary operations in the ball $\mathbb{R}_c^n$. This disanalogy raises the question as to whether uniform relativistic velocities in the Universe are added according to the noncommutative Einstein velocity addition, (41.19), or according to the commutative Einstein gyroparallelogram addition, $\boxplus$ in (41.43)

Indeed, the two gyrodiagonals of gyroquadrangle $ABDC$ are the gyrosegments AD and BC, shown in Fig. 41.6, the gyromidpoints of which coincide, that is,

$$\frac{1}{2} \otimes (A \boxplus D) = \frac{1}{2} \otimes (B \boxplus C) \tag{41.81}$$

where, by (41.49), the result in (41.81) is equivalent to the gyroparallelogram condition (41.80).

Let ABC be a gyrotriangle in an Einstein gyrovector space $(\mathbb{R}_c^n, \oplus, \otimes)$ and let D be the point that augments gyrotriangle ABC into the gyroparallelogram $ABDC$, as shown in Fig. 41.6. Then, D is determined uniquely by the gyroparallelogram condition (41.80), obeying the gyroparallelogram addition law [73, Theorem 5.5]

$$(\ominus A \oplus B) \boxplus (\ominus A \oplus C) = (\ominus A \oplus D) \tag{41.82}$$

shown in Fig. 41.6. In full analogy with the parallelogram addition law of vectors in Euclidean geometry, (41.72), the gyroparallelogram addition law (41.82) of gyrovectors in hyperbolic geometry can be written as

$$\mathbf{u} \boxplus \mathbf{v} = \mathbf{w} \tag{41.83}$$

where $\mathbf{u}$, $\mathbf{v}$, and $\mathbf{w}$ are the *gyrovectors*

$$\begin{aligned} \mathbf{u} &= \ominus A \oplus B, \\ \mathbf{v} &= \ominus A \oplus C, \\ \mathbf{w} &= \ominus A \oplus D \end{aligned} \tag{41.84}$$

which emanate from the point A [68, Chap. 5].

In his 1905 paper that founded the special theory of relativity [15], Einstein noted that his velocity addition does not satisfy the Euclidean parallelogram law:

> Das Gesetz vom Parallelogramm der Geschwindigkeiten gilt also nach unserer Theorie nur in erster Annäherung.
>
> A. Einstein [15]

[English translation: Thus the law of velocity parallelogram is valid according to our theory only to a first approximation.]

Indeed, Einstein velocity addition, $\oplus$, is noncommutative and does not give rise to an exact "velocity parallelogram" in Euclidean geometry. However, as we see in Fig. 41.6, Einstein velocity *coaddition*, $\boxplus$, which is commutative, does give rise to an exact "velocity gyroparallelogram" in hyperbolic geometry.

The breakdown of commutativity in Einstein velocity addition law seemed undesirable to the famous mathematician Émile Borel. Borel's resulting attempt to "repair" the seemingly "defective" Einstein velocity addition in the years following 1912 is described by Walter in [76, p. 117]. Here, however, we see that there is no need to repair Einstein velocity addition law for being noncommutative since, despite being noncommutative, it gives rise to the gyroparallelogram law of gyrovector addition, which turns out to be commutative. The compatibility of the gyroparallelogram addition law of Einsteinian velocities with cosmological observations of stellar aberration is studied in [68, Chap. 13] and [73, Sect. 10.2]. The extension of the gyroparallelogram addition law of $k = 2$ summands into a higher dimensional gyroparallelotope addition law of $k > 2$ summands is mentioned in (41.51)–(41.53) and studied in [68, Theorem 10.6].

41.8 The Isomorphism Between Möbius and Einstein Addition

Einstein addition, $\oplus = \oplus_{\mathrm{E}}$, and Möbius addition, $\oplus_{\mathrm{M}}$, admit the same scalar multiplication (41.55), $\otimes = \otimes_{\mathrm{E}} = \otimes_{\mathrm{M}}$. The isomorphism between $\oplus_{\mathrm{E}}$ and $\oplus_{\mathrm{M}}$ is given by the identities

$$A \oplus_{\mathrm{E}} B = 2 \otimes \left(\frac{1}{2} \otimes A \oplus_{\mathrm{M}} \frac{1}{2} \otimes B \right), \quad A, B \in (\mathbb{R}^n_c, \oplus_{\mathrm{E}}, \otimes_{\mathrm{E}}),$$
$$A \oplus_{\mathrm{M}} B = \frac{1}{2} \otimes (2 \otimes A \oplus_{\mathrm{E}} 2 \otimes B), \quad A, B \in (\mathbb{R}^n_c, \oplus_{\mathrm{M}}, \otimes_{\mathrm{M}}) \tag{41.85}$$

for all $A, B \in \mathbb{R}^n_c$.

The isomorphism in (41.85) is not trivial owing to the result that scalar multiplication, $\otimes$, is non-distributive, that is, it does not distribute over gyrovector addition, $\oplus$.

As examples of the use of the isomorphism (41.85) let $A_e \in (\mathbb{R}^n_c, \oplus_{\mathrm{E}}, \otimes)$ and $A_m \in (\mathbb{R}^n_c, \oplus_{\mathrm{M}}, \otimes)$ be points of an Einstein and a Möbius gyrovector space that are isomorphic to each other under the isomorphism (41.85). Then,

$$A_e = 2 \otimes A_m,$$
$$A_m = \frac{1}{2} \otimes A_e. \tag{41.86}$$

It follows from (41.86) that

$$\gamma_{A_e} = \gamma_{2\otimes A_m} = 2\gamma^2_{A_m} - 1,$$
$$\gamma_{A_e} A_e = \gamma_{2\otimes A_m} (2 \otimes A_m) = 2\gamma^2_{A_m} A_m. \tag{41.87}$$

More generally, for points $A_{i,e}, A_{j,e} \in (\mathbb{R}^n_s, \oplus_{\mathrm{E}}, \otimes)$ in an Einstein gyrovector space and their isomorphic image $A_{i,m}, A_{j,m} \in (\mathbb{R}^n_s, \oplus_{\mathrm{M}}, \otimes)$ in a corresponding Möbius gyrovector space, we have [72, Eq. (2.278)]

$$\gamma_{ij,e} := \gamma_{\ominus_{\mathrm{E}} A_{i,e} \oplus_{\mathrm{E}} A_{j,e}} = 2\gamma^2_{\ominus_{\mathrm{M}} A_{i,m} \oplus_{\mathrm{M}} A_{j,m}} - 1 =: 2\gamma^2_{ij,m} - 1 \tag{41.88}$$

and [72, Eq. (2.280)]

$$\sqrt{\gamma^2_{ij,e} - 1} = 2\gamma_{ij,m} \sqrt{\gamma^2_{ij,m} - 1}. \tag{41.89}$$

Interestingly, in the following equation we see an elegant expression that remains invariant under the isomorphism (41.85) between Einstein and Möbius gyrovector spaces:

$$\frac{\gamma_{ij,e} \mathbf{a}_{ij,e}}{\sqrt{\gamma^2_{ij,e} - 1}} = \frac{\gamma_{ij,m} \mathbf{a}_{ij,m}}{\sqrt{\gamma^2_{ij,m} - 1}} \tag{41.90}$$

as one can readily check, where we use the notation

$$\mathbf{a}_{ij,e} = \ominus_{\mathrm{E}} A_{i,e} \oplus_{\mathrm{E}} A_{j,e},$$
$$\mathbf{a}_{ij,m} = \ominus_{\mathrm{M}} A_{i,m} \oplus_{\mathrm{M}} A_{j,m},$$
$$\gamma_{ij,e} = \gamma_{\mathbf{a}_{ij,e}},$$
$$\gamma_{ij,m} = \gamma_{\mathbf{a}_{ij,m}}. \tag{41.91}$$

A study in detail of the isomorphism between Einstein and Möbius gyrovector spaces is found in [68, Sect. 6.21] and [72, Sect. 2.29].

Owing to the isomorphism between Einstein and Möbius addition in $\mathbb{R}_c^n$, the triples $(\mathbb{R}_c^n, \oplus_{\mathrm{M}}, \otimes)$ form Möbius gyrovector spaces just as the triples $(\mathbb{R}_c^n, \oplus_{\mathrm{E}}, \otimes)$ form Einstein gyrovector spaces. We will now show in (41.92)–(41.94) below that the isomorphic image of an Einstein gyroline $P_e(t)$ in an Einstein gyrovector space $(\mathbb{R}_c^n, \oplus_{\mathrm{E}}, \otimes)$ is a Möbius gyroline $P_m(t)$ in a corresponding Möbius gyrovector space $(\mathbb{R}_c^n, \oplus_{\mathrm{M}}, \otimes)$.

Let

$$P_e(t) = A_e \oplus_{\mathrm{E}} (\ominus_{\mathrm{E}} A_e \oplus_{\mathrm{E}} B_e) \otimes t \tag{41.92}$$

for $t \in \mathbb{R}$ be the gyroline that passes through the distinct points $A_e, B_e \in \mathbb{R}_c^n$ in an Einstein gyrovector space $(\mathbb{R}_c^n, \oplus_{\mathrm{E}}, \otimes)$, shown in Fig. 41.1, p. 738 for $n = 2$. Furthermore, let $A_m, B_m, P_m \in \mathbb{R}_c^n$ be the respective isomorphic images of the points $A_e, B_e, P_e \in \mathbb{R}_c^n$ in (41.92) under the isomorphism expressed in (41.85)–(41.86). In the following chain of equations, which are numbered for subsequent explanation, we determine the isomorphic image of the Einstein gyroline (41.92) in the corresponding Möbius gyrovector space $(\mathbb{R}_c^n, \oplus_{\mathrm{M}}, \otimes)$:

$$\begin{aligned}
2 \otimes P_m(t) &\overset{1.}{=\!=} 2 \otimes A_m \oplus_{\mathrm{E}} (\ominus_{\mathrm{E}} 2 \otimes A_m \oplus_{\mathrm{E}} 2 \otimes B_m) \otimes t \\
&\overset{2.}{=\!=} 2 \otimes A_m \oplus_{\mathrm{E}} \big(2 \otimes (-A_m) \oplus_{\mathrm{E}} 2 \otimes B_m\big) \otimes t \\
&\overset{3.}{=\!=} 2 \otimes A_m \oplus_{\mathrm{E}} 2 \otimes \big\{(-A_m \oplus_{\mathrm{M}} B_m)\big\} \otimes t \\
&\overset{4.}{=\!=} 2 \otimes A_m \oplus_{\mathrm{E}} 2 \otimes \big\{(-A_m \oplus_{\mathrm{M}} B_m) \otimes t\big\} \\
&\overset{5.}{=\!=} 2 \otimes \big\{A_m \oplus_{\mathrm{M}} (-A_m \oplus_{\mathrm{M}} B_m) \otimes t\big\} \\
&\overset{6.}{=\!=} 2 \otimes \big\{A_m \oplus_{\mathrm{M}} (\ominus_{\mathrm{M}} A_m \oplus_{\mathrm{M}} B_m) \otimes t\big\}
\end{aligned} \tag{41.93}$$

so that, finally, the two extreme sides of (41.93) give the equation

$$P_m(t) = A_m \oplus_{\mathrm{M}} (\ominus_{\mathrm{M}} A_m \oplus_{\mathrm{M}} B_m) \otimes t. \tag{41.94}$$

Derivation of the numbered equalities in (41.93) follows:

1. This equation follows from (41.92) and (41.86), where the equations $P_e = 2 \otimes P_m$, $A_e = 2 \otimes A_m$, and $B_e = 2 \otimes B_m$ that result from (41.86) are substituted into (41.92).
2. Follows from Item 1 since the unary operations $\ominus_{\mathrm{E}}$ and—are identically the same in Einstein gyrovector spaces, and since $-2 \otimes A_m = 2 \otimes (-A_m)$.
3. Follows from Item 2 by the first identity in (41.85) applied to the second binary operation $\oplus_{\mathrm{E}}$ in Item 2.

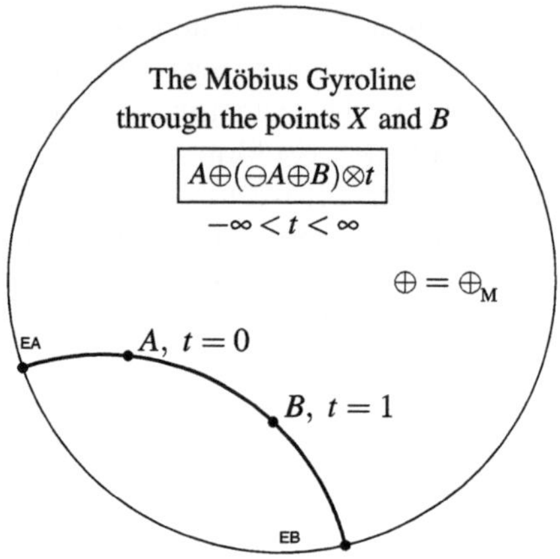

Fig. 41.7 The unique gyroline L_{AB} in a Möbius gyrovector space $(\mathbb{R}^n_c, \oplus, \otimes)$ through two given points A and B. The case of the Möbius gyrovector plane, when $\mathbb{V}_c = \mathbb{R}^2_{c=1}$ is the real open unit disc, is shown

4. Follows from Item 3 by the scalar associative law of gyrovector spaces.
5. Follows from Item 4 by the first identity in (41.85) applied to the remaining binary operation $\oplus_{\mathrm{E}}$ in Item 4.
6. Follows from Item 5 since the unary operations $\ominus_{\mathrm{M}}$ and—are identically the same in Möbius gyrovector spaces.

A Möbius gyroline in a Möbius gyrovector plane $(\mathbb{R}^2_c, \oplus, \otimes)$ is shown in Fig. 41.7. Interestingly, a Möbius gyroline that does not pass through the center of the disc $\mathbb{R}^2_c$ is a circular arc that approaches the boundary of the disc orthogonally. This feature of the Möbius gyroline indicates that Möbius gyrovector spaces form the algebraic setting for the Poincaré ball model of hyperbolic geometry. The link between Einstein and Möbius gyrovector spaces and differential geometry is presented in [66].

As in (41.59)–(41.60), but now with $\oplus = \oplus_{\mathrm{M}}$, Möbius addition $\oplus$ admits the gyrodistance function

$$d_{\oplus}(A, B) = \| \ominus A \oplus B \| \tag{41.95}$$

that obeys the gyrotriangle inequality [68, Theorem 6.9]

$$d_{\oplus}(A, B) \le d_{\oplus}(A, P) \oplus d_{\oplus}(P, B) \tag{41.96}$$

for any $A, B, P \in \mathbb{R}^n_c$ in a Möbius gyrovector space $(\mathbb{R}^n_c, \oplus, \otimes)$. Möbius gyrodistance function is invariant under the group of motions of its Möbius gyrovector space, that is, under left gyrotranslations and rotations of the space [68, Sect. 4]. The gyrotriangle inequality (41.96) reduces to a corresponding gyrotriangle equality

$$d_{\oplus}(A, B) = d_{\oplus}(A, P) \oplus d_{\oplus}(P, B) \tag{41.97}$$

if and only if point P lies between points A and B, that is, point P lies on the gyrosegment AB, as shown in Fig. 41.8. Accordingly, the gyrodistance function is gyroadditive on gyrolines, as demonstrated in (41.97) and illustrated graphically in Fig. 41.8.

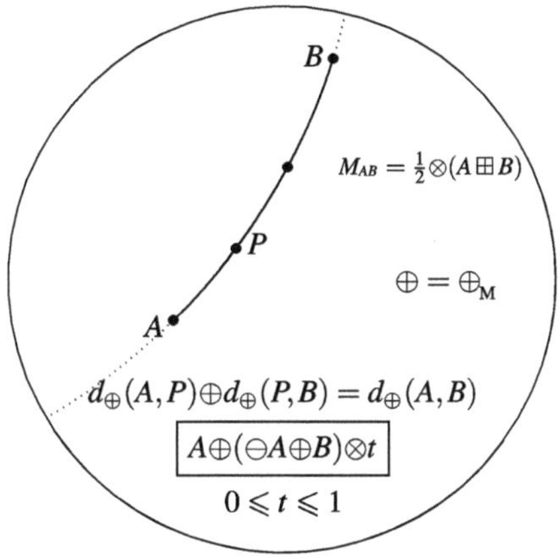

Fig. 41.8 The gyrosegment AB that links the points A and B in $(\mathbb{R}^n_c, \oplus, \otimes)$, with one of its generic points P and its gyromidpoint M_{AB}. The point P lies between A and B and, hence, obeys the gyrotriangle equality, (41.61)

The one-to-one relationship between Möbius gyrodistance function (41.95) and the standard Poincaré distance function in the Poincaré ball model of hyperbolic geometry is presented in [68, Sect. 6.17].

Einstein coaddition, $\boxplus = \boxplus_E$, in the ball, defined in (41.43), is commutative as shown in (41.51). Its importance stems from analogies with classical results that it captures. In particular, it proves useful in solving the gyrogroup equation (41.45), in the determination of gyromidpoints in (41.64), and in the formulation of the gyroparallelogram addition law in (41.82) and in Fig. 41.6.

41.9 Möbius Coaddition

We now wish to determine Möbius coaddition in the ball $\mathbb{R}^n_c$ by means of the isomorphism between Möbius and Einstein gyrovector spaces. Let $\mathbf{u}_e, \mathbf{v}_e, \mathbf{w}_e \in (\mathbb{R}^n_c, \oplus_E, \otimes)$ be three elements of an Einstein gyrovector space such that

$$\mathbf{w}_e = \mathbf{u}_e \boxplus_E \mathbf{v}_e \tag{41.98}$$

and let $\mathbf{u}_m, \mathbf{v}_m, \mathbf{w}_m \in (\mathbb{R}^n_c, \oplus_M, \otimes)$ be the corresponding elements of the corresponding Möbius gyrovector space. Then,

$$\mathbf{w}_m = \mathbf{u}_m \boxplus_M \mathbf{v}_m \tag{41.99}$$

where Möbius coaddition $\boxplus_M$ in $(\mathbb{R}^n_c, \oplus_M, \otimes)$ is determined from Einstein coaddition $\boxplus_E$ in the following chain of equations, which are numbered for subsequent explanation.

$$\begin{aligned} \mathbf{u}_m \boxplus_M \mathbf{v}_m &\overbrace{=\!=}^{1.} \mathbf{w}_m \\ &\overbrace{=\!=}^{2.} \frac{1}{2} \otimes \mathbf{w}_e \end{aligned}$$

$$\overset{3.}{\Longleftarrow\!\!=} \frac{1}{2} \otimes (\mathbf{u}_e \boxplus_{\mathrm{E}} \mathbf{v}_e)$$

$$\overset{4.}{\Longleftarrow\!\!=} \frac{1}{2} \otimes \left\{ 2 \otimes \frac{\gamma_{\mathbf{u}_e} \mathbf{u}_e + \gamma_{\mathbf{v}_e} \mathbf{v}_e}{\gamma_{\mathbf{u}_e} + \gamma_{\mathbf{v}_e}} \right\}$$

$$\overset{5.}{\Longleftarrow\!\!=} \frac{\gamma_{\mathbf{u}_e} \mathbf{u}_e + \gamma_{\mathbf{v}_e} \mathbf{v}_e}{\gamma_{\mathbf{u}_e} + \gamma_{\mathbf{v}_e}}$$

$$\overset{6.}{\Longleftarrow\!\!=} \frac{2\gamma^2_{\mathbf{u}_m} \mathbf{u}_m + 2\gamma^2_{\mathbf{v}_m} \mathbf{u}_m}{2\gamma^2_{\mathbf{u}_m} - 1 + 2\gamma^2_{\mathbf{v}_m} - 1}$$

$$\overset{7.}{\Longleftarrow\!\!=} \frac{\gamma^2_{\mathbf{u}_m} \mathbf{u}_m + \gamma^2_{\mathbf{v}_m} \mathbf{u}_m}{\gamma^2_{\mathbf{u}_m} + \gamma^2_{\mathbf{v}_m} - 1}. \tag{41.100}$$

Derivation of the numbered equalities in (41.100) follows:

1. The equation in Item 1 is (41.99).
2. The equation in Item 2 follows from the isomorphism (41.86) between $\mathbf{w}_m$ in a Möbius gyrovector space $(\mathbb{R}^n_c, \oplus_{\mathrm{M}}, \otimes)$ and its isomorphic image $\mathbf{w}_e$ in the isomorphic Einstein gyrovector space $(\mathbb{R}^n_c, \oplus_{\mathrm{E}}, \otimes)$.
3. Follows from Item 2 by assumption (41.98).
4. Follows from Item 3 by (41.51).
5. Follows from Item 4 by the scalar associative law of gyrovector spaces, Sect. 41.5.
6. Follows from Item 5 by (41.87).

Hence, by (41.100), Möbius coaddition $\boxplus_{\mathrm{M}}$ in a Möbius gyrovector space $(\mathbb{R}^n_c, \oplus_{\mathrm{M}}, \otimes)$ is given by the equation

$$\mathbf{u} \boxplus_{\mathrm{M}} \mathbf{v} = \frac{\gamma^2_{\mathbf{u}} \mathbf{u} + \gamma^2_{\mathbf{v}} \mathbf{u}}{\gamma^2_{\mathbf{u}} + \gamma^2_{\mathbf{v}} - 1} \tag{41.101}$$

for all $\mathbf{u}, \mathbf{v} \in \mathbb{R}^n_c$.

In order to extend (41.100) from Möbius coaddition of order two to any order k, $k > 2$, we rewrite (41.53) in the form

$$\mathbf{w}_e := \mathbf{v}_{1,e} \boxplus_{E,k} \mathbf{v}_{2,e} \boxplus_{E,k} \cdots \boxplus_{E,k} \mathbf{v}_{k,e} = 2 \otimes \frac{\sum_{i=1}^k \gamma_{\mathbf{v}_{i,e}} \mathbf{v}_{i,e}}{2 + \sum_{i=1}^k (\gamma_{\mathbf{v}_{i,e}} - 1)} \tag{41.102}$$

where $\mathbf{v}_{i,e} \in (\mathbb{R}^n_c, \oplus_{\mathrm{E}}, \otimes)$, $i = 1, \ldots, k$, are k elements of an Einstein gyrovector space and where $\mathbf{w}_e \in (\mathbb{R}^n_c, \oplus_{\mathrm{E}}, \otimes)$ is their *cosum*, $\boxplus_{E,k}$ being the Einstein k-ary cooperation, that is, the Einstein cooperation of order k.

Let $\mathbf{v}_{i,m}$, $i = 1, \ldots, k$, and $\mathbf{w}_m$ be the respective isomorphic images of $\mathbf{v}_{i,e}$, and $\mathbf{w}_e$ in the corresponding Möbius gyrovector space $(\mathbb{R}^n_c, \oplus_{\mathrm{M}}, \otimes)$, under isomorphism

(41.86). Then,

$$\mathbf{w}_m = \mathbf{v}_{1,m} \boxplus_{M,k} \mathbf{v}_{2,m} \boxplus_{M,k} \cdots \boxplus_{M,k} \mathbf{v}_{k,m} \tag{41.103}$$

where Möbius coaddition of order k, $\boxplus_{M,k}$, is to be determined in the chain of equations below, which are numbered for subsequent interpretation:

$$\begin{aligned}
&\mathbf{v}_{1,m} \boxplus_{M,k} \mathbf{v}_{2,m} \boxplus_{M,k} \cdots \boxplus_{M,k} \mathbf{v}_{k,m} \\
&\overbrace{=\!=}^{1.} \mathbf{w}_m \\
&\overbrace{=\!=}^{2.} \frac{1}{2} \otimes \mathbf{w}_e \\
&\overbrace{=\!=}^{3.} \frac{1}{2} \otimes (\mathbf{v}_{1,e} \boxplus_{E,k} \mathbf{v}_{2,e} \boxplus_{E,k} \cdots \boxplus_{E,k} \mathbf{v}_{k,e}) \\
&\overbrace{=\!=}^{4.} \frac{1}{2} \otimes \left\{ 2 \otimes \frac{\sum_{i=1}^{k} \gamma_{\mathbf{v}_{i,e}} \mathbf{v}_{i,e}}{2 + \sum_{i=1}^{k} (\gamma_{\mathbf{v}_{i,e}} - 1)} \right\} \\
&\overbrace{=\!=}^{5.} \frac{\sum_{i=1}^{k} \gamma_{\mathbf{v}_{i,e}} \mathbf{v}_{i,e}}{2 + \sum_{i=1}^{k} (\gamma_{\mathbf{v}_{i,e}} - 1)} \\
&\overbrace{=\!=}^{6.} \frac{2 \sum_{i=1}^{k} \gamma_{\mathbf{v}_{i,m}}^2 \mathbf{v}_{i,m}}{2 + \sum_{i=1}^{k} (2\gamma_{\mathbf{v}_{i,m}}^2 - 2)} \\
&=\!= \frac{\sum_{i=1}^{k} \gamma_{\mathbf{v}_{i,m}}^2 \mathbf{v}_{i,m}}{1 + \sum_{i=1}^{k} (\gamma_{\mathbf{v}_{i,m}}^2 - 1)}.
\end{aligned} \tag{41.104}$$

Derivation of the numbered equalities in (41.100) follows:

1. The equation in Item 1 is (41.103).
2. The equation in Item 2 follows from the isomorphism (41.86) between $\mathbf{w}_m$ in a Möbius gyrovector space $(\mathbb{R}_c^n, \oplus_M, \otimes)$ and its isomorphic image $\mathbf{w}_e$ in the isomorphic Einstein gyrovector space $(\mathbb{R}_c^n, \oplus_E, \otimes)$.
3. Follows from Item 2 by the assumption in (41.102).
4. Follows from Item 3 by the equation in (41.102).
5. Follows from Item 4 by the scalar associative law of gyrovector spaces, Sect. 41.5.
6. Follows from Item 5 by (41.87).

Hence, by (41.104), Möbius coaddition of order k, $\boxplus_{M,k}$ in a Möbius gyrovector space $(\mathbb{R}_c^n, \oplus_M, \otimes)$ is given by the equation

$$\mathbf{v}_{1,m} \boxplus_{M,k} \mathbf{v}_{2,m} \boxplus_{M,k} \cdots \boxplus_{M,k} \mathbf{v}_{k,m} = \frac{\sum_{i=1}^{k} \gamma_{\mathbf{v}_{i,m}}^2 \mathbf{v}_{i,m}}{1 + \sum_{i=1}^{k} (\gamma_{\mathbf{v}_{i,m}}^2 - 1)} \tag{41.105}$$

for all $\mathbf{v}_{i,m} \in (\mathbb{R}^n_c, \oplus_{\mathrm{M}}, \otimes)$, $i = 1, \ldots, k$.

41.10 Möbius Double-Gyroline

Theorem 41.1 *Let $A, B \in \mathbb{R}^n_c$ be any two distinct points of a Möbius gyrovector space $(\mathbb{R}^n_c, \oplus, \otimes)$, and let*

$$L_{AB}(t) = A \oplus (\ominus A \oplus B) \otimes t \tag{41.106}$$

for $t \in \mathbb{R}$ be the gyroline that passes through these points. Then,

$$2 \otimes L_{AB}(t) = A \boxplus L_{AB}(2t). \tag{41.107}$$

Proof Let

$$\begin{aligned} F_1(t) &= (\ominus A \oplus B) \otimes t, \\ F_2(t) &= 2 \otimes F_1(t) \end{aligned} \tag{41.108}$$

so that we have, by the scalar associative law of gyrovector spaces,

$$\begin{aligned} F_2(t) &= 2 \otimes F_1(t) \\ &= 2 \otimes (\ominus A \oplus B) \otimes t \\ &= (\ominus A \oplus B) \otimes (2t) \\ &= F_1(2t). \end{aligned} \tag{41.109}$$

Hence, by (41.108)–(41.109), (41.107) can be written equivalently as

$$2 \otimes \big(A \oplus F_1(t)\big) = A \boxplus \big(A \oplus F_1(2t)\big) = A \boxplus \big(A \oplus F_2(t)\big) \tag{41.110}$$

so that instead of verifying (41.107) we can, equivalently, verify (41.110).

The proof of (41.110) is given by the following chain of equations, which are numbered for subsequent derivation:

$$\begin{aligned} 2 \otimes (A \oplus F_1) &\overbrace{=\!=\!=}^{1.} A \oplus (2 \otimes F_1 \oplus A) \\ &\overbrace{=\!=\!=}^{2.} A \oplus (F_2 \oplus A) \\ &\overbrace{=\!=\!=}^{3.} A \boxplus (A \oplus F_2) \end{aligned} \tag{41.111}$$

as desired.

Derivation of the numbered equalities in (41.111) follows:

1. Follows from the Two-Sum Identity (41.58).

2. Follows from (41.108).
3. Follows from (41.50). □

We may remark that in the Euclidean limit, when the radius c of the ball $\mathbb{R}_c^n$ tends to ∞, the ball expands to the whole of its Euclidean n-space $\mathbb{R}^n$, both Möbius addition $\oplus$ and coaddition $\boxplus$ in the ball $\mathbb{R}_c^n$ reduce to the common vector addition $+$ in the space $\mathbb{R}^n$, and Identity (41.107) of Theorem 41.1 in the ball $\mathbb{R}_c^n$ reduces to the trivial identity in $\mathbb{R}^n$,

$$2\big[A + (-A + B)t\big] = A + \big[A + (-A + B)2t\big]. \tag{41.112}$$

Thus, we see once again that in order to capture analogies with classical results, both gyrogroup operation and cooperation must be considered.

Theorem 41.1 suggests the following definition:

Definition 41.6 (Möbius Double-Gyroline) Let $A, B \in \mathbb{R}_c^n$ be two distinct points of a Möbius gyrovector space $(\mathbb{R}_c^n, \oplus, \otimes)$, and let

$$L_{AB}(t) = A \oplus (\ominus A \oplus B) \otimes t \tag{41.113}$$

for $t \in \mathbb{R}$ be the gyroline that passes through these points. The Möbius double-gyroline $P_{AB}(t)$ of gyroline $L_{AB}(t)$ is the curve given by the equation

$$P_{AB}(t) = 2 \otimes L_{AB}(t) \tag{41.114}$$

for $t \in \mathbb{R}$.

Following Definition 41.6, the gyrovector space identity (41.107) of Theorem 41.1 states that the double-gyroline of a given gyroline that passes through a point A in a Möbius gyrovector space coincides with the cogyrotranslation by A of the gyroline.

Remarkably, the double-gyroline of a given gyroline in a Möbius gyrovector space turns out to be the supporting chord of the gyroline, as shown in Fig. 41.9 and studied in Sect. 41.11.

Identity (41.107) of Theorem 41.1 can be written, equivalently, as

$$L_{AB}(t) = \frac{1}{2} \otimes \big(A \boxplus L_{AB}(2t)\big). \tag{41.115}$$

Let $P(t)$ be a generic point on a gyroline $L_{AB}(t)$ for some value t of the gyroline parameter t, so that $P(0) = A$ and $P(1) = A$. Then, (41.115) implies the equation

$$P(t) = \frac{1}{2} \otimes P(0) \boxplus P(2t) \tag{41.116}$$

for $t \in \mathbb{R}$. Equation (41.116), in turn, demonstrates that any point $P(t)$ of a gyroline $L_{AB}(t)$ is the midpoint of the points $P(0) = A$ and $P(2t)$ of the gyroline, as explained in (41.64), p. 739.

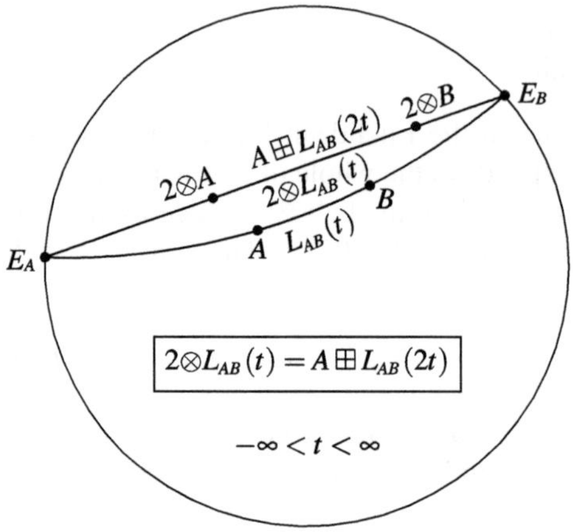

Fig. 41.9 A and B are any two given distinct points of a Möbius gyrovector space $(\mathbb{R}_c^n, \oplus, \otimes)$. The gyroline that passes through the points $A, B \in \mathbb{R}_c^n$ is $L_{AB}(t)$, $-\infty < t < \infty$, and its corresponding double-gyroline is $2 \otimes L_{AB}(t)$, so that is passes through the points $2 \otimes A, 2 \otimes B \in \mathbb{R}_c^n$. The latter turns out to be the Euclidean straight line in the ball that passes through the points $2 \otimes A$ and $2 \otimes B$. Furthermore, the double-gyroline $2 \otimes L_{AB}(t)$, parametrized by t, is identical with the cogyrotranslation by A, $A \boxplus L_{AB}(2t)$, of its gyroline, parametrized by $2t$, as shown here for $n = 2$

41.11 Euclidean Straight Lines in Möbius Gyrovector Spaces

Euclidean straight lines (lines, in short) appear naturally in Einstein gyrovector space balls where they form gyrolines, as shown in Fig. 41.1. In this section, we employ the isomorphism (41.85) between Einstein and Möbius gyrovector spaces for the task of expressing lines in Möbius gyrovector spaces.

Let $A, B \in (\mathbb{R}_c^n, \oplus_{\mathrm{M}}, \otimes)$ be two distinct points of a Möbius gyrovector space. We know that the unique gyroline in an Einstein gyrovector spaces $(\mathbb{R}_c^n, \oplus_{\mathrm{E}}, \otimes)$ that passes through the points $A, B \in \mathbb{R}_c^n$ is the set of point $P_{AB}(t)$ given by

$$P_{AB}(t) = A \oplus_{\mathrm{E}} (\ominus_{\mathrm{E}} A \oplus_{\mathrm{E}} B) \otimes t \qquad (41.117)$$

for $t \in \mathbb{R}$. It is the intersection of a line and the ball $\mathbb{R}_c^n$, as shown in Fig. 41.1 for the disc $\mathbb{R}_c^2$. This line passes through the point A when $t = 0$ and through the point B when $t = 1$.

Unlike Einstein gyrolines, which are line segments, Möbius gyrolines are Euclidean circular arcs that intersect the boundary of the ball $\mathbb{R}_c^n$ orthogonally, as shown in Fig. 41.7 for the disc $\mathbb{R}_c^2$. In order to accomplish the task, we face in this section, in the following chain of equations (41.118) we express (41.117) in terms of Möbius addition $\oplus_{\mathrm{M}}$ rather than Einstein addition $\oplus_{\mathrm{E}}$, noting that both Einstein and Möbius scalar multiplication $\otimes$ are identically the same, as remarked in Sect. 41.8. Starting from (41.117), we have the following chain of equations, which are numbered for subsequent derivation:

$$P_{AB}(t) = A \oplus_{\mathrm{E}} (\ominus_{\mathrm{E}} A \oplus_{\mathrm{E}} B) \otimes t$$

$$
\begin{aligned}
&\overset{1.}{=\!=} A \oplus_{\mathrm{E}} \big((-A) \oplus_{\mathrm{E}} B\big) \otimes t \\
&\overset{2.}{=\!=} 2 \otimes \left\{ \frac{1}{2} \otimes A \oplus_{\mathrm{M}} \frac{1}{2} \otimes \big((-A) \oplus_{\mathrm{E}} B\big) \otimes t \right\} \\
&\overset{3.}{=\!=} \frac{1}{2} \otimes A \oplus_{\mathrm{M}} \left\{ \big((-A) \oplus_{\mathrm{E}} B\big) \otimes t \oplus_{\mathrm{M}} \frac{1}{2} \otimes A \right\} \\
&\overset{4.}{=\!=} \frac{1}{2} \otimes A \oplus_{\mathrm{M}} \left\{ 2 \otimes \left(\left(-\frac{1}{2} \otimes A \right) \oplus_{\mathrm{M}} \frac{1}{2} B \right) \otimes t \oplus_{\mathrm{M}} \frac{1}{2} \otimes A \right\} \\
&\overset{5.}{=\!=} \frac{1}{2} \otimes A \oplus_{\mathrm{M}} \left\{ \left[\left(-\frac{1}{2} \otimes A \right) \oplus_{\mathrm{M}} \left(B \oplus_{\mathrm{M}} \left(-\frac{1}{2} \otimes A \right) \right) \right] \otimes t \oplus_{\mathrm{M}} \frac{1}{2} \otimes A \right\} \\
&\overset{6.}{=\!=} \frac{1}{2} \otimes A \boxplus_{\mathrm{M}} \left\{ \frac{1}{2} \otimes A \oplus_{\mathrm{M}} \left[\left(-\frac{1}{2} \otimes A \right) \oplus_{\mathrm{M}} \left(B \oplus_{\mathrm{M}} \left(-\frac{1}{2} \otimes A \right) \right) \right] \otimes t \right\} \\
&\overset{7.}{=\!=} \frac{1}{2} \otimes A \boxplus_{\mathrm{M}} \left\{ \frac{1}{2} \otimes A \oplus_{\mathrm{M}} \left[\ominus_{\mathrm{M}} \frac{1}{2} \otimes A \oplus_{\mathrm{M}} \left(B \ominus_{\mathrm{M}} \frac{1}{2} \otimes A \right) \right] \otimes t \right\}
\end{aligned} \tag{41.118}
$$

Hence, by (41.118),

$$
P_{AB}(t) = \frac{1}{2} \otimes A \boxplus_{\mathrm{M}} \left\{ \frac{1}{2} \otimes A \oplus_{\mathrm{M}} \left[\ominus_{\mathrm{M}} \frac{1}{2} \otimes A \oplus_{\mathrm{M}} \left(B \ominus_{\mathrm{M}} \frac{1}{2} \otimes A \right) \right] \otimes t \right\}. \tag{41.119}
$$

Derivation of the numbered equalities in (41.118) follows:

1. Follows from the result that $\ominus_{\mathrm{E}} A = -A$ (as well as $\ominus_{\mathrm{M}} A = -A$; see Item 7 below).
2. Follows from isomorphism (41.85) between $\oplus_{\mathrm{E}}$ and $\oplus_{\mathrm{M}}$, applying the isomorphism to the first $\oplus_{\mathrm{E}}$ in (1).
3. Follows from the Two-Sum Identity, (41.58).
4. Again, follows from isomorphism (41.85) between $\oplus_{\mathrm{E}}$ and $\oplus_{\mathrm{M}}$, as in Item 2, now applying the isomorphism to the remaining $\oplus_{\mathrm{E}}$ in (3).
5. Again, follows from the gyrogroup *Two-Sum Identity*, as in Item 3.
6. Follows from the gyrogroup identity (41.50),

$$
A \oplus (B \oplus A) = A \boxplus (A \oplus B). \tag{41.120}
$$

7. Follows from the result that $\ominus_{\mathrm{M}} A = -A$ (as well as $\ominus_{\mathrm{E}} A = -A$; see Item 1 above).

In both (41.117) and (41.119), the set of points $P_{AB}(t)$, $t \in \mathbb{R}$, forms a line in the ball $\mathbb{R}^n_c$ of the Möbius gyrovector space $(\mathbb{R}^n_c, \oplus_{\mathrm{M}}, \otimes)$, where the points A and B lie. In (41.117), this line is expressed in terms of operations of Einstein gyrovector spaces while in (41.119) this line is expressed in terms of operations of Möbius gyrovector spaces, obtained by means of isomorphism (41.85) between Einstein and Möbius gyrovector spaces. By (41.119), we have the following theorem:

Theorem 41.2 *Let A and B be two distinct points in a Möbius gyrovector space* $(\mathbb{R}^n_c, \oplus_M, \otimes)$. *The unique line that passes through these points, Fig.* 41.11, *is given by the equation*

$$P_{AB}(t) = \frac{1}{2} \otimes A \boxplus_M \left\{ \frac{1}{2} \otimes A \oplus_M \left[\ominus_M \frac{1}{2} \otimes A \oplus_M \left(B \ominus_M \frac{1}{2} \otimes A \right) \right] \otimes t \right\}. \quad (41.121)$$

Let $A, B \in \mathbb{R}^n_c$ be any two distinct points in a Möbius gyrovector space $(\mathbb{R}^n_c, \oplus_M, \otimes)$, and let $L_{\frac{1}{2}\otimes A, B\ominus\frac{1}{2}\otimes A}(t)$, $t \in \mathbb{R}$, be the unique gyroline through the points $\frac{1}{2} \otimes A$ and $B \ominus \frac{1}{2} \otimes A$. Then, as shown in Fig. 41.7, the gyroline is given by the equation

$$L_{\frac{1}{2}\otimes A, B\ominus\frac{1}{2}\otimes A}(t) = \frac{1}{2} \otimes A \oplus \left[\ominus \frac{1}{2} \otimes A \oplus \left(B \ominus \frac{1}{2} \otimes A \right) \right] \otimes t \quad (41.122)$$

so that (41.121) can be written as

$$P_{AB}(t) = \frac{1}{2} \otimes A \boxplus L_{\frac{1}{2}\otimes A, B\ominus\frac{1}{2}\otimes A}(t). \quad (41.123)$$

The line $P_{AB}(t)$ of Theorem 41.2 in (41.121) is recognized by means of (41.122)–(41.123) as the *cogyrotranslation* by $\frac{1}{2} \otimes A$ of the Möbius gyroline (41.122) that passes through the points $\frac{1}{2} \otimes A$ and $B \ominus \frac{1}{2} \otimes A$.

As shown in Fig. 41.12, the line $P_{AB}(t)$ is the *supporting chord* of the gyroline $L_{\frac{1}{2}\otimes A, B\ominus\frac{1}{2}\otimes A}(t)$.

Let

$$C = \frac{1}{2} \otimes A, \qquad D = B \ominus \frac{1}{2} A. \quad (41.124)$$

Then, by the scalar associative law of gyrovector spaces and by the right cancellation law (41.48), we have

$$A = 2 \otimes C, \qquad B = D \boxplus \frac{1}{2} \otimes A = D \boxplus C = C \boxplus D \quad (41.125)$$

so that (41.123) can be written as

$$P_{2\otimes C, C\boxplus D} = C \boxplus L_{CD}(t), \quad (41.126)$$

thus leading to the following theorem:

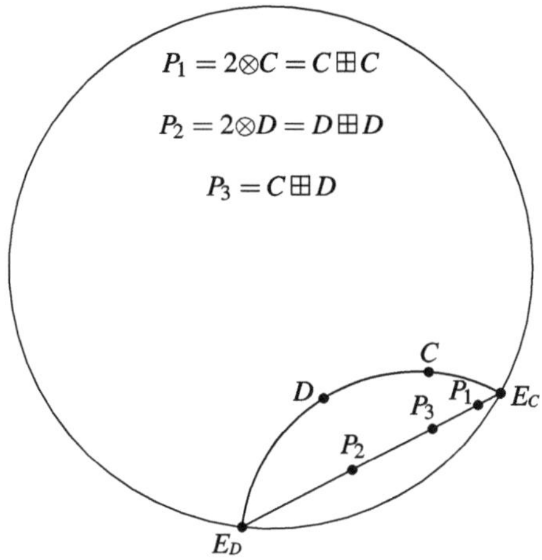

Fig. 41.10 The Euclidean line $P_{AB}(t)$, $-\infty < t < \infty$, that passes through the points P_k, $k = 1, 2, 3$, in a Möbius gyrovector plane $(\mathbb{R}^2_s, \oplus, \otimes)$ is shown along with the associated gyroline

Theorem 41.3 *Let $C, D \in \mathbb{R}^n_c$ be two distinct points in a Möbius gyrovector space $(\mathbb{R}^n_c, \oplus_M, \otimes)$, and let*

$$L_{CD}(t) = C \oplus (\ominus C \oplus D) \otimes t \tag{41.127}$$

for $t \in \mathbb{R}$ be the gyroline that passes through the points C and D. Then, the supporting chord of gyroline $L_{CD}(t)$ is the line given by the cogyrotranslation of the gyroline by C,

$$C \boxplus L_{CD}(t). \tag{41.128}$$

Furthermore, the supporting chord passes through the points P_1, P_2, P_3, Fig. 41.10, where

$$\begin{aligned} P_1 &= C \boxplus C = 2 \otimes C, \\ P_2 &= D \boxplus D = 2 \otimes D, \\ P_3 &= C \boxplus D. \end{aligned} \tag{41.129}$$

Let $Q = \ominus_M A \oplus_M B$ so that, by the gyrogroup left cancellation law, (41.23), $A \oplus_M Q = B$. Restricting the line parameter $t \in \mathbb{R}$ to $t \geq 0$, we obtain the Euclidean ray (ray, in short) $P_{AB}(t), t \geq 0$. It is the ray with edge A that contains the point $A \oplus_M Q = B$, which is the right gyrotranslation of A by Q. As such, it contains the sequence of all successive right gyrotranslation of A by Q, that is, the sequence $P_0 = A$, $P_1 = A \oplus_M Q$, $P_2 = (A \oplus_M Q) \oplus_M Q$, $P_3 = ((A \oplus_M Q) \oplus_M Q) \oplus_M Q$, etc., as shown in Fig. 41.11.

The points P_k, $k = 1, 2, 3, \ldots$, lie on the ray $P_{AB}(t)$, $t \geq 0$, as shown in Fig. 41.11, and as observed in [63, Figs. 6.3–6.5]. Owing to the left reduction property of gyrations in gyrogroup Axiom (G5), we have

$$\mathrm{gyr}[\ominus_M A \oplus_M B, P_k] = \mathrm{gyr}[\ominus_M A, B] \tag{41.130}$$

for all $k = 0, 1, 2, 3, \ldots$. As an example, the proof of (41.130) for $k = 0$ and $k = 1$ follows:

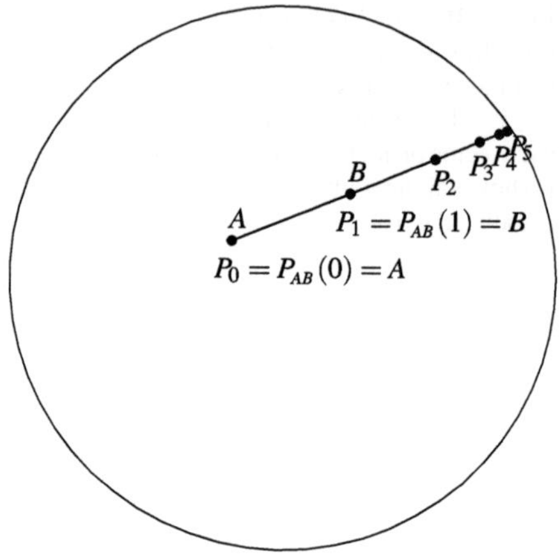

Fig. 41.11 The Euclidean ray $P_{AB}(t)$, $t \geq 0$, with edge A that passes through B in a Möbius gyrovector plane $(\mathbb{R}^2_s, \oplus_M, \otimes)$ is shown along with several points of the sequence of successive right gyrotranslations of A by $Q = \ominus A \oplus_M B$ that lies on the ray (see also Fig. 6.4 in [63, p. 168])

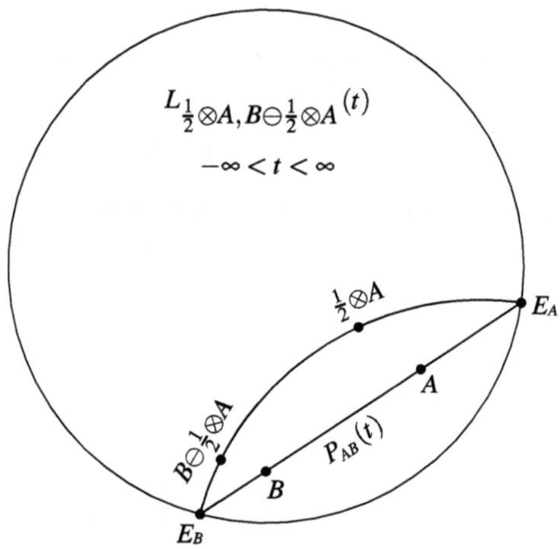

Fig. 41.12 The Euclidean straight line (line, in short) $P_{AB}(t)$, $-\infty < t < \infty$, (41.123), that passes through the point A and B in a Möbius gyrovector plane $(\mathbb{R}^2_s, \oplus, \otimes)$ is shown. It is the supporting chord of the gyroline that passes through the points $\frac{1}{2}A$ and $B \ominus \frac{1}{2}A$ in the Möbius gyrovector plane. The two endpoints of both the line and the gyroline, corresponding to $t \to \pm\infty$, are E_A and E_B

By the left reduction property of gyrations and by the gyrogroup left cancellation law (41.23), we have in any gyrogroup $(G, \oplus)$

$$\begin{aligned} \text{gyr}[\ominus A \oplus B, P_0] &= \text{gyr}[\ominus A \oplus B, A] \\ &= \text{gyr}\big[\ominus A \oplus B, A \oplus (\ominus A \oplus B)\big] \\ &= \text{gyr}[\ominus A \oplus B, B] \\ &= \text{gyr}[\ominus A, B] \end{aligned} \tag{41.131}$$

and

$$\begin{aligned} \text{gyr}[\ominus A \oplus B, P_1] &= \text{gyr}[\ominus A \oplus B, A \oplus Q] \\ &= \text{gyr}\big[\ominus A \oplus B, A \oplus (\ominus A \oplus B)\big] \\ &= \text{gyr}[\ominus A \oplus B, B] \end{aligned}$$

$$= \text{gyr}[\ominus A, B]. \tag{41.132}$$

The validity of (41.130) for all $k = 0, 1, 2, 3, \ldots$ suggests the conjecture that (41.130) is valid not only for the points of the sequence $\{P_0 = A, P_1 = B, P_2, P_3, \ldots\}$ that lie on the ray $P_{AB}(t)$, $t \geq 0$, as shown in Fig. 41.11, but for all the points of the ray, that is,

$$\text{gyr}\big[\ominus_M A \oplus_M B, P_{AB}(t)\big] = \text{gyr}[\ominus_M A, B] \tag{41.133}$$

for all $t \geq 0$. Numerical experiments support the conjecture.

41.12 Euclidean Barycentric Coordinates

In order to set the stage for the introduction of hyperbolic barycentric coordinates, we present here the notion of Euclidean barycentric coordinates that dates back to Möbius' 1827 book titled "*Der Barycentrische Calcul*" (The Barycentric Calculus). The word *barycenter* means center of gravity, but the book is entirely geometrical and, hence, called by Jeremy Gray [22], *Möbius's Geometrical Mechanics*. The 1827 Möbius book is best remembered for introducing a new system of coordinates, the *barycentric coordinates*. The use of barycentric coordinates in Euclidean geometry is described in [72, 73, 79], and the historical contribution of Möbius' barycentric coordinates to vector analysis is described in [13, pp. 48–50].

For any positive integer N, let $m_k \in \mathbb{R}$ be N given real numbers such that

$$\sum_{k=1}^{N} m_k \neq 0 \tag{41.134}$$

and let $A_k \in \mathbb{R}^n$ be N given points in the Euclidean n-space $\mathbb{R}^n$, $k = 1, \ldots, N$. Then, by obvious algebra, the equation

$$\sum_{k=1}^{N} m_k \begin{pmatrix} 1 \\ A_k \end{pmatrix} = m_0 \begin{pmatrix} 1 \\ P \end{pmatrix} \tag{41.135}$$

for the unknowns $m_0 \in \mathbb{R}$ and $P \in \mathbb{R}^n$ possesses the unique solution given by

$$m_0 = \sum_{k=1}^{N} m_k \tag{41.136}$$

and

$$P = \frac{\sum_{k=1}^{N} m_k A_k}{\sum_{k=1}^{N} m_k} \tag{41.137}$$

satisfying for all $X \in \mathbb{R}^n$,

$$X + P = \frac{\sum_{k=1}^{N} m_k (X + A_k)}{\sum_{k=1}^{N} m_k}. \tag{41.138}$$

Following Möbius, we view (41.137) as the representation of a point $P \in \mathbb{R}^n$ in terms of its *barycentric coordinates* m_k, $k = 1, \ldots, N$, with respect to the set of points

$$S = \{A_1, \ldots, A_N\} \tag{41.139}$$

Identity (41.138), then, insures that the barycentric coordinate representation (41.137) of P with respect to the set S is *covariant* (or, *invariant in form*) in the following sense. The point P and the points of the set S of its barycentric coordinate representation vary together under translations. Indeed, a translation $X + A_k$ of A_k by X, $k = 1, \ldots, N$, on the right-hand side of (41.138) results in the translation $X + P$ of P by X on the left-hand side of (41.138).

In order to insure that barycentric coordinate representations with respect to a set S are unique, we require the set S to be pointwise independent.

Definition 41.7 (Euclidean Pointwise Independence) A set S of N points, $S = \{A_1, \ldots, A_N\}$, in $\mathbb{R}^n$, $n \geq 2$, is *pointwise independent* if the $N - 1$ vectors $-A_1 + A_k$, $k = 2, \ldots, N$, are linearly independent in $\mathbb{R}^n$.

We are now in the position to present the formal definition of Euclidean barycentric coordinates, as motivated by mass and center of momentum velocity of Newtonian particle systems.

Definition 41.8 (Barycentric Coordinates) Let

$$S = \{A_1, \ldots, A_N\} \tag{41.140}$$

be a pointwise independent set of N points in $\mathbb{R}^n$. The real numbers $m_1, \ldots, m_N$, satisfying

$$\sum_{k=1}^{N} m_k \neq 0 \tag{41.141}$$

are barycentric coordinates of a point $P \in \mathbb{R}^n$ with respect to the set S if

$$P = \frac{\sum_{k=1}^{N} m_k A_k}{\sum_{k=1}^{N} m_k}. \tag{41.142}$$

Barycentric coordinates are homogeneous in the sense that the barycentric coordinates $(m_1, \ldots, m_N)$ of the point P in (41.142) are equivalent to the barycentric coordinates $(\lambda m_1, \ldots, \lambda m_N)$ for any real nonzero number $\lambda \in \mathbb{R}$, $\lambda \neq 0$. Since in barycentric coordinates only ratios of coordinates are relevant, the barycentric coordinates $(m_1, \ldots, m_N)$ are also written as $(m_1 \colon \ldots \colon m_N)$.

Barycentric coordinates that are normalized by the condition

$$\sum_{k=1}^{N} m_k = 1 \tag{41.143}$$

are called *special barycentric coordinates*.

Equation (41.142) is said to be the (unique) barycentric coordinate representation of P with respect to the set S.

Theorem 41.4 (Covariance of Barycentric Coordinate Representations) *Let*

$$P = \frac{\sum_{k=1}^{N} m_k A_k}{\sum_{k=1}^{N} m_k} \tag{41.144}$$

be the barycentric coordinate representation of a point $P \in \mathbb{R}^n$ in a Euclidean n-space $\mathbb{R}^n$ with respect to a pointwise independent set $S = \{A_1, \ldots, A_N\} \subset \mathbb{R}^n$. The barycentric coordinate representation (41.144) *is covariant, that is,*

$$X + P = \frac{\sum_{k=1}^{N} m_k (X + A_k)}{\sum_{k=1}^{N} m_k} \tag{41.145}$$

for all $X \in \mathbb{R}^n$, and

$$RP = \frac{\sum_{k=1}^{N} m_k R A_k}{\sum_{k=1}^{N} m_k} \tag{41.146}$$

for all $R \in SO(n)$.

Proof The proof is immediate, noting that rotations $R \in SO(n)$ of $\mathbb{R}^n$ about its origin are linear maps of $\mathbb{R}^n$. □

Following the vision of Felix Klein in his *Erlangen Program* [41], it is owing to the covariance with respect to translations and rotations that barycentric coordinate representations possess geometric significance. Indeed, translations and rotations in Euclidean geometry form the *group of motions* of the geometry, and according to Felix Klein's Erlangen Program, a geometric property is a property that remains invariant in form under the motions of the geometry.

41.13 Hyperbolic Barycentric, Gyrobarycentric Coordinates

Guided by analogies with Sect. 41.12, in this section we introduce barycentric coordinates into hyperbolic geometry [71–73].

Definition 41.9 (Hyperbolic Pointwise Independence) A set S of N points $S = \{A_1, \dots, A_N\}$ in the ball $\mathbb{R}^n_s$, $n \ge 2$, is *pointwise independent* if the $N-1$ gyrovectors in $\mathbb{R}^n_s$, $\ominus A_1 \oplus A_k$, $k = 2, \dots, N$, considered as vectors in $\mathbb{R}^n \supset \mathbb{R}^n_s$, are linearly independent.

We are now in the position to present the formal definition of gyrobarycentric coordinates, that is, hyperbolic barycentric coordinates, as motivated by the notions of relativistic mass and center of momentum velocity in Einstein's special relativity theory. Gyrobarycentric coordinates, fully analogous to barycentric coordinates, thus emerge when Einstein's relativistic mass meets the hyperbolic geometry of Bolyai and Lobachevsky [74].

Definition 41.10 (Gyrobarycentric Coordinates) Let

$$S = \{A_1, \dots, A_N\} \tag{41.147}$$

be a pointwise independent set of N points in $\mathbb{R}^n_s$. The real numbers $m_1, \dots, m_N$, satisfying

$$\sum_{k=1}^{N} m_k \gamma_{A_k} > 0 \tag{41.148}$$

are gyrobarycentric coordinates of a point $P \in \mathbb{R}^n_s$ with respect to the set S if

$$P = \frac{\sum_{k=1}^{N} m_k \gamma_{A_k} A_k}{\sum_{k=1}^{N} m_k \gamma_{A_k}}. \tag{41.149}$$

Gyrobarycentric coordinates are homogeneous in the sense that the gyrobarycentric coordinates $(m_1, \dots, m_N)$ of the point P in (41.149) are equivalent to the gyrobarycentric coordinates $(\lambda m_1, \dots, \lambda m_N)$ for any real nonzero number $\lambda \in \mathbb{R}$, $\lambda \ne 0$. Since in gyrobarycentric coordinates only ratios of coordinates are relevant, the gyrobarycentric coordinates $(m_1, \dots, m_N)$ are also written as $(m_1 : \dots : m_N)$.

Gyrobarycentric coordinates that are normalized by the condition

$$\sum_{k=1}^{N} m_k = 1 \tag{41.150}$$

are called *special gyrobarycentric coordinates*.

Equation (41.149) is said to be the gyrobarycentric coordinate representation of P with respect to the set S.

Finally, the constant of the gyrobarycentric coordinate representation of P in (41.149) is $m_0 > 0$, given by

$$m_0 = \sqrt{\left(\sum_{k=1}^{N} m_k\right)^2 + 2 \sum_{\substack{j,k=1 \\ j<k}}^{N} m_j m_k (\gamma_{\ominus A_j \oplus A_k} - 1)}. \tag{41.151}$$

Theorem 41.5 (Gyrocovariance of Gyrobarycentric Coordinate Representations) *Let*

$$P = \frac{\sum_{k=1}^{N} m_k \gamma_{A_k} A_k}{\sum_{k=1}^{N} m_k \gamma_{A_k}} \tag{41.152a}$$

be a gyrobarycentric coordinate representation of a point $P \in \mathbb{R}_s^n$ in an Einstein gyrovector space $(\mathbb{R}_s^n, \oplus, \otimes)$ with respect to a pointwise independent set $S = \{A_1, \dots, A_N\} \subset \mathbb{R}_s^n$.

Then

$$\gamma_P = \frac{\sum_{k=1}^{N} m_k \gamma_{A_k}}{m_0} \tag{41.152b}$$

and

$$\gamma_P\, P = \frac{\sum_{k=1}^{N} m_k \gamma_{A_k} A_k}{m_0} \tag{41.152c}$$

where $m_0 > 0$ is the constant of the gyrobarycentric coordinate representation (41.152a) *of P, given by*

$$m_0 = \sqrt{\left(\sum_{k=1}^{N} m_k\right)^2 + 2 \sum_{\substack{j,k=1\\ j<k}}^{N} m_j m_k \left(\gamma_{\ominus A_j \oplus A_k} - 1\right)}. \tag{41.152d}$$

Furthermore, the gyrobarycentric coordinate representation (41.152a) *and its associated identities in* (41.152b)–(41.152d) *are gyrocovariant, that is,*

$$X \oplus P = \frac{\sum_{k=1}^{N} m_k \gamma_{X \oplus A_k} (X \oplus A_k)}{\sum_{k=1}^{N} m_k \gamma_{X \oplus A_k}}, \tag{41.153a}$$

$$\gamma_{X \oplus P} = \frac{\sum_{k=1}^{N} m_k \gamma_{X \oplus A_k}}{m_0}, \tag{41.153b}$$

$$\gamma_{X \oplus P}(X \oplus P) = \frac{\sum_{k=1}^{N} m_k \gamma_{X \oplus A_k} (X \oplus A_k)}{m_0}, \tag{41.153c}$$

$$m_0 = \sqrt{\left(\sum_{k=1}^{N} m_k\right)^2 + 2 \sum_{\substack{j,k=1\\ j<k}}^{N} m_j m_k \left(\gamma_{\ominus(X \oplus A_j) \oplus (X \oplus A_k)} - 1\right)} \tag{41.153d}$$

for all $X \in \mathbb{R}_s^n$, *and*

$$RP = \frac{\sum_{k=1}^{N} m_k \gamma_{RA_k} RA_k}{\sum_{k=1}^{N} m_k \gamma_{RA_k}}, \tag{41.154a}$$

$$\gamma_{RP} = \frac{\sum_{k=1}^{N} m_k \gamma_{RA_k}}{m_0}, \tag{41.154b}$$

$$\gamma_{RP}(RP) = \frac{\sum_{k=1}^{N} m_k \gamma_{RA_k}(RA_k)}{m_0}, \tag{41.154c}$$

$$m_0 = \sqrt{\left(\sum_{k=1}^{N} m_k\right)^2 + 2\sum_{\substack{j,k=1\\ j<k}}^{N} m_j m_k \left(\gamma_{\ominus(RA_j)\oplus(RA_k)} - 1\right)} \tag{41.154d}$$

for all $R \in SO(n)$.

The proof of Theorem 41.5 is presented in [72, Theorem 4.4] and [73, Theorem 4.6].

Remark 41.1 It is assumed in Theorem 41.5 that the point P in (41.149) lies inside the ball $\mathbb{R}_s^n$, implying that $m_0^2 > 0$ and that the gamma factor γ_P of P is a real number. The constant m_0 of a gyrobarycentric coordinate representation (41.149) of a point P determines whether P lies inside the ball $\mathbb{R}_s^n$.

If the coefficients m_k, $k = 1, \ldots, N$, in the gyrobarycentric coordinate representation (41.149) of P are all positive or all negative, then the point P lies in the convex span of the points of the set S, that is, P lies inside the $(N-1)$-gyrosimplex $A_1 \ldots A_N$. This gyrosimplex, in turn, lies inside the ball $\mathbb{R}_s^n$.

1. The point P lies inside the $(N-1)$-gyrosimplex $A_1 \ldots A_N$ if and only if the coefficients m_k, $k = 1, \ldots, N$, of its gyrobarycentric coordinate representation (41.152a) are all positive or all negative. Clearly, in this case $m_0^2 > 0$.

Otherwise, when all the coefficients m_k are nonzero but do not have the same sign, the location of P has the following three possibilities that correspond to whether the gamma factor (41.152b) of P is real, infinity, or imaginary:

2. The point P does not lie inside the $(N-1)$-gyrosimplex $A_1 \ldots A_N$, but it lies inside the ball $\mathbb{R}_s^n$. In this case, the gamma factor γ_P of P is a real number and, hence, $m_0^2 > 0$.
3. The point P lies on the boundary of the ball $\mathbb{R}_s^n$ if and only if the gamma factor γ_P of P is undefined, $\gamma_P = \infty$, so that $m_0^2 = 0$.
4. The point $P \in \mathbb{R}^n$ does not lie in the ball $\mathbb{R}_s^n$ or on its boundary if and only if the gamma factor γ_P of P is purely imaginary, so that $m_0^2 < 0$.

Examples for the use of gyrobarycentric coordinates for the determination of several hyperbolic triangle centers are found in [72, 73].

Employing the technique of gyrobarycentric coordinate representations, we will now determine the end points E_A and E_B of a gyroline $L_{AB}(t)$ in an Einstein gyrovector space $(\mathbb{R}^n_s, \oplus, \otimes)$, shown in Fig. 41.1, p. 738.

Let $A_1, A_2 \in \mathbb{R}^n_s$ be two distinct points of an Einstein gyrovector space $(\mathbb{R}^n_c, \oplus, \otimes)$, and let P be a generic point on the gyroline, (41.117),

$$P_{12}(t) = A_1 \oplus (\ominus_{\mathrm{E}} A_1 \oplus A_2) \otimes t \tag{41.155}$$

for $t \in \mathbb{R}$, that passes through these two points. Furthermore, let

$$P = \frac{m_1 \gamma_{A_1} A_1 + m_2 \gamma_{A_2} A_2}{m_1 \gamma_{A_1} + m_2 \gamma_{A_2}} \tag{41.156}$$

be the gyrobarycentric coordinate representation of P with respect to the pointwise independent set $S = \{A_1, A_2\}$, where the gyrobarycentric coordinates m_1 and m_2 are to be determined. Owing to the homogeneity of gyrobarycentric coordinates, we can select $m_2 = -1$, obtaining from (41.156) the gyrobarycentric coordinate representation

$$P = \frac{m \gamma_{A_1} A_1 - \gamma_{A_2} A_2}{m \gamma_{A_1} - \gamma_{A_2}}. \tag{41.157}$$

According to Definition 41.9 of the gyrobarycentric coordinate representation of P in (41.149) and its constant m_0 in (41.151), the constant m_0 of the gyrobarycentric coordinate representation of P satisfies the equation

$$\begin{aligned} m_0^2 &= m_1^2 + m_2^2 + 2m_1 m_2 \gamma_{\ominus A_1 \oplus A_2} \\ &= m^2 + 1 + 2m\gamma_{12} \end{aligned} \tag{41.158}$$

where we use the convenient notation

$$\begin{aligned} \mathbf{a}_{ij} &= \ominus A_i \oplus A_j, \\ \gamma_{ij} &= \gamma_{\mathbf{a}_{ij}} \end{aligned} \tag{41.159}$$

with $i, j \in \mathbb{N}$.

As remarked in Item 3 of Remark 41.1, the point P lies on the boundary of the ball $\mathbb{R}^n_s$ if and only if $m_0 = 0$, that is by (41.158), if and only if

$$m^2 - 2m\gamma_{12} + 1 = 0. \tag{41.160}$$

The two solutions of (41.160) are

$$\begin{aligned} m &= \gamma_{12} + \sqrt{\gamma_{12}^2 - 1}, \\ m &= \gamma_{12} - \sqrt{\gamma_{12}^2 - 1}. \end{aligned} \tag{41.161}$$

The substitution into (41.157) of each of the two solutions (41.161) gives the two endpoints E_{A_1} and E_{A_2} of the gyroline $P_{12}(t)$ in (41.155),

$$
\begin{aligned}
E_{A_1} &= \frac{(\gamma_{12}+\sqrt{\gamma_{12}^2-1})\gamma_{A_1}A_1-\gamma_{A_2}A_2}{(\gamma_{12}+\sqrt{\gamma_{12}^2-1})\gamma_{A_1}-\gamma_{A_2}},\\
E_{A_2} &= \frac{(\gamma_{12}-\sqrt{\gamma_{12}^2-1})\gamma_{A_1}A_1-\gamma_{A_2}A_2}{(\gamma_{12}-\sqrt{\gamma_{12}^2-1})\gamma_{A_1}-\gamma_{A_2}}
\end{aligned}
\tag{41.162}
$$

which are shown in Fig. 41.1, p. 738, for $A_1 = A$ and $A_2 = B$.

The expressions for $\ominus A_1 \oplus E_{A_1}$ and $\ominus A_1 \oplus E_{A_2}$ that follow from (41.162) by means of the gyrocovariance identity (41.153a) in Theorem 41.5 are particularly elegant. Indeed, by the gyrocovariance identity (41.153a) with $X = \ominus A_1$, applied to each of the two equations in (41.162), we have

$$
\begin{aligned}
\ominus A_1 \oplus E_{A_1} &= \frac{(\gamma_{12}+\sqrt{\gamma_{12}^2-1})\gamma_{\ominus A_1\oplus A_1}(\ominus A_1\oplus A_1)-\gamma_{\ominus A_1\oplus A_2}(\ominus A_1\oplus A_2)}{(\gamma_{12}+\sqrt{\gamma_{12}^2-1})\gamma_{\ominus A_1\oplus A_1}-\gamma_{\ominus A_1\oplus A_2}}\\
&= \frac{-\gamma_{12}\mathbf{a}_{12}}{(\gamma_{12}+\sqrt{\gamma_{12}^2-1})-\gamma_{12}}\\
&= \ominus\frac{\gamma_{12}\mathbf{a}_{12}}{\sqrt{\gamma_{12}^2-1}},\\
\ominus A_1 \oplus E_{A_2} &= \frac{\gamma_{12}\mathbf{a}_{12}}{\sqrt{\gamma_{12}^2-1}}
\end{aligned}
\tag{41.163}
$$

where we use the notation (41.159), noting that $\ominus A_1 \oplus A_1 = \mathbf{0}$ and $\gamma_{\ominus A_1\oplus A_1} = \gamma_{\mathbf{0}} = 1$.

The equations in (41.163) imply, by means of the left cancellation law (41.23),

$$
\begin{aligned}
E_{A_1} &= A_1 \ominus \frac{\gamma_{12}\mathbf{a}_{12}}{\sqrt{\gamma_{12}^2-1}},\\
E_{A_2} &= A_1 \oplus \frac{\gamma_{12}\mathbf{a}_{12}}{\sqrt{\gamma_{12}^2-1}}.
\end{aligned}
\tag{41.164}
$$

Interestingly, (41.164) remains invariant in form under the isomorphism (41.85), as seen from (41.90). Accordingly, the equations in (41.164) with $\oplus = \oplus_{\mathrm{E}}$ being Einstein addition are used in calculating the endpoints of an Einstein gyroline in Fig. 41.1, p. 738, and the same equations (41.164), but with $\oplus = \oplus_{\mathrm{M}}$ being Möbius addition are used in calculating the endpoints of a Möbius gyroline in Fig. 41.7, p. 748.

In Fig. 41.9, p. 754, the points $A, B \in (\mathbb{R}_c^2, \oplus_M, \otimes)$ of a Möbius gyrovector plane are shown along with their respective isomorphic images $2 \otimes A, 2 \otimes B \in (\mathbb{R}_c^2, \oplus_E, \otimes)$ of an Einstein gyrovector plane, under the isomorphism (41.86). Indeed, as expected, Fig. 41.9 indicates that the endpoints E_A and E_B of

1. The Möbius gyroline (a circular arc) through the points A and B, and of
2. The Einstein gyroline (a chord) through the points $2 \otimes A$ and $2 \otimes B$,

are coincident.

References

1. Ahlfors, L.V.: Conformal Invariants: Topics in Geometric Function Theory. McGraw-Hill Series in Higher Mathematics. McGraw-Hill, New York (1973)
2. Ahlfors, L.V.: Möbius Transformations in Several Dimensions. University of Minnesota School of Mathematics, Minneapolis (1981)
3. Ahlfors, L.V.: Old and new in Möbius groups. Ann. Acad. Sci. Fenn., Ser. A 1 Math. **9**, 93–105 (1984)
4. Barrett, J.F.: Hyperbolic geometry in special relativity. In: Duffy, M.C., Wegener, M.T. (eds.) Recent Advances in Relativity Theory. Proceedings, pp. 27–34. Hadronic Press, Palm Harbor (2000)
5. Belloni, L., Reina, C.: Sommerfeld's way to the Thomas precession. Eur. J. Phys. **7**, 55–61 (1986)
6. Borel, E.: Introduction Géométrique a Quelques Théories Physiques. Gauthier-Villars, Paris (1914)
7. Cazacu, C.A., Lehto, O.E., Rassias, Th.M.: Analysis and Topology. World Science, Singapore (1998)
8. Chen, J.-L., Ungar, A.A.: The Bloch gyrovector. Found. Phys. **32**(4), 531–565 (2002)
9. Chrysos, M.: The non-intuitive $\frac{1}{2}$ Thomas factor: a heuristic argument with classical electromagnetism. Eur. J. Phys. **27**(1), 1–4 (2006)
10. Coxeter, H.S.M.: Regular Polytopes, 3rd edn. Dover, New York (1973)
11. Coxeter, H.S.M., Greitzer, S.L.: Geometry Revisited. Math. Assoc. Amer., New York (1967)
12. Craioveanu, M., Puta, M., Rassias, T.M.: Old and New Aspects in Spectral Geometry. Mathematics and Its Applications, vol. 534. Kluwer Academic, Dordrecht (2001)
13. Crowe, M.J.: A History of Vector Analysis. Dover, New York (1994). The evolution of the idea of a vectorial system, Corrected reprint of the 1985 edition
14. Eddington, A.S.: The Mathematical Theory of Relativity. Cambridge (1924)
15. Einstein, A.: Zur Elektrodynamik Bewegter Körper [on the electrodynamics of moving bodies] (We use the English translation in [16] or in [36]). or in http://www.fourmilab.ch/etexts/einstein/specrel/www/). Ann. Phys. (Leipzig), **17**, 891–921 (1905)
16. Albert Einstein: Einstein's Miraculous Years: Five Papers that Changed the Face of Physics. Princeton, Princeton, NJ, 1998. Edited and introduced by John Stachel. Includes bibliographical references. Einstein's dissertation on the determination of molecular dimensions – Einstein on Brownian motion – Einstein on the theory of relativity – Einstein's early work on the quantum hypothesis. A new English translation of Einstein's 1905 paper on pp. 123–160
17. Feder, T.: Strong near subgroups and left gyrogroups. J. Algebra **259**(1), 177–190 (2003)
18. Fock, V.: The Theory of Space, Time and Gravitation. Macmillan, New York (1964). Second revised edition. Translated from the Russian by N. Kemmer. A Pergamon Press Book

19. Foguel, T., Ungar, A.A.: Involutory decomposition of groups into twisted subgroups and subgroups. J. Group Theory **3**(1), 27–46 (2000)
20. Foguel, T., Ungar, A.A.: Gyrogroups and the decomposition of groups into twisted subgroups and subgroups. Pac. J. Math. **197**(1), 1–11 (2001)
21. Goebel, K., Reich, S.: Uniform Convexity, Hyperbolic Geometry, and Nonexpansive Mappings. Monographs and Textbooks in Pure and Applied Mathematics, vol. 83. Dekker, New York (1984)
22. Gray, J.: Möbius's geometrical mechanics. In: Fauvel, J., Flood, R., Wilson, R. (eds.) Möbius and His Band, Mathematics and Astronomy in Nineteenth-Century Germany, pp. 78–103. Clarendon/Oxford University Press, New York (1993)
23. Haruki, H., Rassias, T.M.: A new invariant characteristic property of Möbius transformations from the standpoint of conformal mapping. J. Math. Anal. Appl. **181**(2), 320–327 (1994)
24. Haruki, H., Rassias, T.M.: A new characteristic of Möbius transformations by use of Apollonius points of triangles. J. Math. Anal. Appl. **197**(1), 14–22 (1996)
25. Haruki, H., Rassias, T.M.: A new characteristic of Möbius transformations by use of Apollonius quadrilaterals. Proc. Am. Math. Soc. **126**(10), 2857–2861 (1998)
26. Haruki, H., Rassias, T.M.: A new characterization of Möbius transformations by use of Apollonius hexagons. Proc. Am. Math. Soc. **128**(7), 2105–2109 (2000)
27. Issa, A.N.: Gyrogroups and homogeneous loops. Rep. Math. Phys. **44**(3), 345–358 (1999)
28. Issa, A.N.: Left distributive quasigroups and gyrogroups. J. Math. Sci. Univ. Tokyo **8**(1), 1–16 (2001)
29. Rickard, M.J.: Gyroscope precession in special and general relativity from basic principles. Am. J. Phys. **75**(5), 463–471 (2007)
30. Kasparian, A.K., Ungar, A.A.: Lie gyrovector spaces. J. Geom. Symmetry Phys. **1**(1), 3–53 (2004)
31. Kikkawa, M.: Geometry of homogeneous Lie loops. Hiroshima Math. J. **5**(2), 141–179 (1975)
32. Kikkawa, M.: Geometry of homogeneous left Lie loops and tangent Lie triple algebras. Mem. Fac. Sci. Eng., Shimane Univ., Ser. B, Math. Sci. **32**, 57–68 (1999)
33. Kowalsky, H.-J.: Lineare Algebra. Gruyter, Berlin (1977). Achte Auflage, de Gruyter Lehrbuch
34. Krantz, S.G.: Complex Analysis: The Geometric Viewpoint. Mathematical Association of America, Washington, D.C. (1990)
35. Kuznetsov, E.: Gyrogroups and left gyrogroups as transversals of a special kind. Algebra Discrete Math. **3**, 54–81 (2003)
36. Lorentz, H.A., Einstein, A., Minkowski, H., Weyl, H.: The Principle of Relativity. Dover, New York, undated. With notes by A. Sommerfeld, Translated by W. Perrett and G.B. Jeffery, A collection of original memoirs on the special and general theory of relativity
37. Marsden, J.E.: Elementary Classical Analysis. Freeman, San Francisco (1974). With the assistance of Michael Buchner, Amy Erickson, Adam Hausknecht, Dennis Heifetz, Janet Macrae and William Wilson, and with contributions by Paul Chernoff, István Fáry and Robert Gulliver
38. McCleary, J.: Geometry from a Differentiable Viewpoint. Cambridge University Press, Cambridge (1994)
39. Miller, A.I.: Albert Einstein's Special Theory of Relativity. Springer, New York (1998). Emergence (1905) and early interpretation (1905–11), Includes a translation by the author of Einstein's "On the electrodynamics of moving bodies", Reprint of the 1981 edition
40. Møller, C.: The Theory of Relativity. Clarendon, Oxford (1952)
41. Mumford, D., Series, C., Wright, D.: Indra's Pearls: The Vision of Felix Klein. Cambridge University Press, New York (2002)
42. Prástaro, A., Rassias, Th.M.: Geometry of Partial Differential Equations. World Scientific, London (1994)
43. Rassias, Th.M.: Book review: Analytic hyperbolic geometry and Albert Einstein's special theory of relativity, by Abraham A. Ungar. Nonlinear Funct. Anal. Appl. **13**(1), 167–177 (2008)
44. Rassias, Th.M.: Book review: A gyrovector space approach to hyperbolic geometry, by Abraham A. Ungar. J. Geom. Symmetry Phys. **18**, 93–106 (2010)

45. Rassias, Th.M.: Constantin Caratheodory: An International Tribute (in two volumes). World Scientific, Singapore (1991)
46. Rassias, Th.M.: The Problem of Plateau. World Scientific, London (1992)
47. Rassias, Th.M.: Inner Product Spaces and Applications. Addison Wesley Longman, Pitman Research Notes in Mathematics Series, vol. 376, Harlo, Essex (1997)
48. Rassias, Th.M., Srivastava, H.M.: Analysis, Geometry and Groups: A Riemann Legacy Volume (in two volumes). Hadronic Press, Florida (1993)
49. Ratcliffe, J.G.: Foundations of Hyperbolic Manifolds. Graduate Texts in Mathematics, vol. 149. Springer, New York (1994)
50. Sexl, R.U., Urbantke, H.K.: Relativity, Groups, Particles. Springer Physics. Springer, Vienna (2001). Special relativity and relativistic symmetry in field and particle physics, Revised and translated from the third German (1992) edition by Urbantke
51. Silberstein, L.: The Theory of Relativity. MacMillan, London (1914)
52. Stachel, J.J.: History of relativity. In: Brown, L.M. Pais, A. Pippard, B. (eds.) Twentieth Century Physics, vol. I. Institute of Physics Publishing, Bristol (1995)
53. Stillwell, J.: Sources of Hyperbolic Geometry. American Mathematical Society, Providence (1996). Pages 10 and 35
54. Thomas, L.H.: The motion of the spinning electron. Nature **117**, 514 (1926)
55. Ungar, A.A.: Thomas rotation and the parametrization of the Lorentz transformation group. Found. Phys. Lett. **1**(1), 57–89 (1988)
56. Ungar, A.A.: Quasidirect product groups and the Lorentz transformation group. In: Rassias, Th.M. (ed.) Constantin Carathéodory: An International Tribute, vol. I, II, pp. 1378–1392. World Scientific, Teaneck (1991)
57. Ungar, A.A.: Thomas precession and its associated grouplike structure. Am. J. Phys. **59**(9), 824–834 (1991)
58. Ungar, A.A.: The abstract Lorentz transformation group. Am. J. Phys. **60**(9), 815–828 (1992)
59. Ungar, A.A.: Thomas precession: its underlying gyrogroup axioms and their use in hyperbolic geometry and relativistic physics. Found. Phys. **27**(6), 881–951 (1997)
60. Ungar, A.A.: The hyperbolic Pythagorean theorem in the Poincaré disc model of hyperbolic geometry. Am. Math. Mon. **106**(8), 759–763 (1999)
61. Ungar, A.A.: Gyrovector spaces in the service of hyperbolic geometry. In: Rassias, Th.M. (ed.) Mathematical Analysis and Applications, pp. 305–360. Hadronic Press, Palm Harbor (2000)
62. Ungar, A.A.: Möbius transformations of the ball, Ahlfors' rotation and gyrovector spaces. In: Rassias, Th.M. (ed.) Nonlinear Analysis in Geometry and Topology, pp. 241–287. Hadronic Press, Palm Harbor (2000)
63. Ungar, A.A.: Beyond the Einstein Addition Law and Its Gyroscopic Thomas Precession: The Theory of Gyrogroups and Gyrovector Spaces. Fundamental Theories of Physics, vol. 117. Kluwer Academic, Dordrecht (2001)
64. Ungar, A.A.: On the unification of hyperbolic and Euclidean geometry. Complex Var. Theory Appl. **49**(3), 197–213 (2004)
65. Ungar, A.A.: Analytic Hyperbolic Geometry: Mathematical Foundations and Applications. World Scientific, Hackensack (2005)
66. Ungar, A.A.: Gyrovector spaces and their differential geometry. Nonlinear Funct. Anal. Appl. **10**(5), 791–834 (2005)
67. Ungar, A.A.: Thomas precession: a kinematic effect of the algebra of Einstein's velocity addition law. Comments on: "Deriving relativistic momentum and energy. II. Three-dimensional case" [European J. Phys. **26** (2005), no. 5, 851–856; mr2227176] by S. Sonego and M. Pin. Eur. J. Phys. **27**(3), L17–L20 (2006)
68. Ungar, A.A.: Analytic Hyperbolic Geometry and Albert Einstein's Special Theory of Relativity. World Scientific, Hackensack (2008)
69. Ungar, A.A.: From Möbius to gyrogroups. Am. Math. Mon. **115**(2), 138–144 (2008)
70. Ungar, A.A.: A Gyrovector Space Approach to Hyperbolic Geometry. Morgan & Claypool Pub., San Rafael (2009)

71. Ungar, A.A.: Hyperbolic barycentric coordinates. Aust. J. Math. Anal. Appl. **6**(1), 1–35 (2009)
72. Ungar, A.A.: Barycentric Calculus in Euclidean and Hyperbolic Geometry: A Comparative Introduction. World Scientific, Hackensack (2010)
73. Ungar, A.A.: Hyperbolic Triangle Centers: The Special Relativistic Approach. Springer, New York (2010)
74. Ungar, A.A.: When relativistic mass meets hyperbolic geometry. Commun. Math. Anal. **10**(1), 30–56 (2011)
75. Vermeer, J.: A geometric interpretation of Ungar's addition and of gyration in the hyperbolic plane. Topol. Appl. **152**(3), 226–242 (2005)
76. Walter, S.: The non-Euclidean style of Minkowskian relativity. In: Gray, J.J. (ed.) The Symbolic Universe: Geometry and Physics 1890–1930, pp. 91–127. Oxford Univ. Press, New York (1999)
77. Walter, S.: Book review: Beyond the Einstein addition law and its gyroscopic Thomas precession: The theory of gyrogroups and gyrovector spaces, by Abraham A. Ungar. Found. Phys. **32**(2), 327–330 (2002)
78. Taylor Whittaker, E.: From Euclid to Eddington. A Study of Conceptions of the External World. Cambridge University Press, Cambridge (1949)
79. Yiu, P.: The uses of homogeneous barycentric coordinates in plane Euclidean geometry. Int. J. Math. Educ. Sci. Technol. **31**(4), 569–578 (2000)

Chapter 42
Hilbert-Type Integral Operators: Norms and Inequalities

Bicheng Yang

Abstract The well known Hilbert inequality and Hardy–Hilbert inequality may be rewritten in the forms of inequalities relating Hilbert operator and Hardy–Hilbert operator with their norms. These two operators are some particular kinds of Hilbert-type operators, which have played an important role in mathematical analysis and applications. In this chapter, by applying the methods of Real Analysis and Operator Theory, we define a general Hilbert-type integral operator and study six particular kinds of this operator with different measurable kernels in several normed spaces. The norms, equivalent inequalities, some particular examples, and compositions of two operators are considered. In Sect. 42.1, we define the weight functions with some parameters and give two equivalent inequalities with the general measurable kernels. Meanwhile, the norm of a Hilbert-type integral operator is estimated. In Sect. 42.2 and Sect. 42.3, four kinds of Hilbert-type integral operators with the particular kernels in the first quarter and in the whole plane are obtained. In Sect. 42.4, we define two kinds of operators with the kernels of multi-variables and obtain their norms. In Sect. 42.5, two kinds of compositions of Hilbert-type integral operators are considered. The lemmas and theorems provide an extensive account for this kind of operators.

Key words Hilbert-type integral operators · Hilbert inequality · Hardy–Hilbert inequality · Equivalent inequalities · Kernels of two variables

Mathematics Subject Classification 31A10 · 31B10 · 45P05 · 47G10

Dedicated to Professor Themistocles M. Rassias on the occasion of his 60th birthday.

B. Yang (✉)
Department of Mathematics, Guangdong University of Education, Guangzhou, Guangdong 510303, P.R. China
e-mail: bcyang@gdei.edu.cn

P.M. Pardalos et al. (eds.), *Nonlinear Analysis*, Springer Optimization and Its Applications 68, In Honor of Themistocles M. Rassias on the Occasion of his 60th Birthday, DOI 10.1007/978-1-4614-3498-6_42,

42.1 A Hilbert-Type Integral Operator

42.1.1 Two Weight Functions and Two Basic Equivalent Inequalities

Definition 42.1 Suppose that $n, m \in \mathbf{N}, \alpha, \beta > 0$,

$$\|x\|_\alpha = \left(\sum_{k=1}^{n} |x_k|^\alpha\right)^{\frac{1}{\alpha}} \quad \left(x = (x_1, \ldots, x_n) \in \mathbf{R}^n\right),$$

$$\|y\|_\beta = \left(\sum_{k=1}^{m} |y_k|^\beta\right)^{\frac{1}{\beta}} \quad \left(y = (y_1, \ldots, y_m) \in \mathbf{R}^m\right),$$

$A(\subset \mathbf{R}^n)$ and $B(\subset \mathbf{R}^m)$ are open intervals (finite or infinite), and $H(x,y)$ is a non-negative measurable function on $A \times B$. For any $a, b \in \mathbf{R}$, define two weight functions $\omega(y)$ and $\varpi(x)$ as follows:

$$\omega(y) := \int_A H(x,y) \frac{\|y\|_\beta^{m-b}}{\|x\|_\alpha^a} dx \quad (y \in B), \tag{42.1}$$

$$\varpi(x) := \int_B H(x,y) \frac{\|x\|_\alpha^{n-a}}{\|y\|_\beta^b} dy \quad (x \in A). \tag{42.2}$$

We have the following theorem:

Theorem 42.1 *Suppose that $p > 0\ (\neq 1)$, $\frac{1}{p} + \frac{1}{q} = 1$, $a, b \in \mathbf{R}$, $H(x,y)$ is a non-negative measurable function on $A \times B$, $f(x)$ $(x \in A)$ and $g(y)$ $(y \in B)$ are non-negative measurable functions.*

(i) *If $p > 1$, then we have the following equivalent inequalities*:

$$I := \int_B \int_A H(x,y) f(x) g(y)\, dx\, dy$$
$$\leq \left(\int_A \varpi(x) \|x\|_\alpha^{pa-n} f^p(x)\, dx\right)^{\frac{1}{p}} \left(\int_B \omega(y) \|y\|_\beta^{qb-m} g^q(y)\, dy\right)^{\frac{1}{q}}, \tag{42.3}$$

$$J := \int_B \left(\omega(y)\right)^{1-p} \|y\|_\beta^{p(m-b)-m} \left(\int_A H(x,y) f(x)\, dx\right)^p dy$$
$$\leq \int_A \varpi(x) \|x\|_\alpha^{pa-n} f^p(x)\, dx; \tag{42.4}$$

(ii) *If $0 < p < 1$, then we have the equivalent reverses of* (42.3) *and* (42.4).

Proof (i) By Hölder's inequality (cf. [1]) and (42.1), it follows

$$\left(\int_A H(x,y)f(x)\,dx\right)^p = \left[\int_A H(x,y)\left(\frac{\|x\|_\alpha^{a/q}}{\|y\|_\beta^{b/p}} f(x)\right)\left(\frac{\|y\|_\beta^{b/p}}{\|x\|_\alpha^{a/q}}\right)dx\right]^p$$

$$\leq \left[\int_A H(x,y)\frac{\|x\|_\alpha^{a(p-1)}}{\|y\|_\beta^{b}} f^p(x)\,dx\right]\left[\int_A H(x,y)\frac{\|y\|_\beta^{b(q-1)}}{\|x\|_\alpha^{a}}\,dx\right]^{p-1}$$

$$= \frac{\|y\|_\beta^{p(b-m)+m}}{(\omega(y))^{1-p}}\left[\int_A H(x,y)\frac{\|x\|_\alpha^{a(p-1)}}{\|y\|_\beta^{b}} f^p(x)\,dx\right]. \tag{42.5}$$

Then by Fubini theorem (cf. [2]) and (42.2), we find

$$J \leq \int_B\left[\int_A H(x,y)\frac{\|x\|_\alpha^{a(p-1)}}{\|y\|_\beta^{b}} f^p(x)\,dx\right]dy$$

$$= \int_A\left[\int_B H(x,y)\frac{\|x\|_\alpha^{a(p-1)}}{\|y\|_\beta^{b}}\,dy\right] f^p(x)\,dx = \int_A \varpi(x)\|x\|_\alpha^{pa-n} f^p(x)\,dx. \tag{42.6}$$

Hence, (42.4) is valid. Still by Hölder's inequality, we find

$$I = \int_B\left[\frac{\|y\|_\beta^{-b+(m/q)}}{(\omega(y))^{1/q}}\int_A H(x,y)f(x)\,dx\right]\left[\frac{(\omega(y))^{1/q}g(y)}{\|y\|_\beta^{-b+(m/q)}}\right]dy$$

$$\leq J^{\frac{1}{p}}\left(\int_B \omega(y)\|y\|_\beta^{qb-m} g^q(y)\,dy\right)^{\frac{1}{q}}. \tag{42.7}$$

Then by (42.4), we have (42.3). On the other-hand, suppose that (42.3) is valid. We set

$$g(y) := \frac{\|y\|_\beta^{p(m-b)-m}}{(\omega(y))^{p-1}}\left(\int_A H(x,y)f(x)\,dx\right)^{p-1} \quad (y \in B). \tag{42.8}$$

Then we obtain $\int_B \omega(y)\|y\|_\beta^{qb-m} g^q(y)\,dy = J$, and by (42.3), it follows

$$J = I \leq \left(\int_A \varpi(x)\|x\|_\alpha^{pa-n} f^p(x)\,dx\right)^{\frac{1}{p}} J^{\frac{1}{q}}. \tag{42.9}$$

If $J = 0$, then (42.4) is naturally valid; if $J = \infty$, then by (42.5), we find that (42.4) keeps the form of equality; if $0 < J < \infty$, then dividing out $J^{\frac{1}{q}}$ in (42.9), we obtain inequality $J^{\frac{1}{p}} \leq (\int_A \varpi(x)\|x\|_\alpha^{pa-n} f^p(x)\,dx)^{\frac{1}{p}}$, and then we have (42.4), which is equivalent to (42.3).

(ii) For $0 < p < 1$, by using the reverse Hölder's inequality (cf. [1]), we obtain the reverses of (42.7) and (42.5) as follows:

$$I \geq J^{\frac{1}{p}} \left(\int_B \omega(y) \|y\|_\beta^{qb-m} g^q(y)\, dy \right)^{\frac{1}{q}}, \tag{42.10}$$

$$\left(\int_A H(x,y) f(x)\, dx \right)^p \geq \frac{\|y\|_\beta^{p(b-m)+m}}{(\omega(y))^{1-p}} \left[\int_A H(x,y) \frac{\|x\|_\alpha^{a(p-1)}}{\|y\|_\beta^b} f^p(x)\, dx \right]. \tag{42.11}$$

Hence, we find the reverse of (42.4). By the reverse of (42.4) and (42.10), we obtain the reverse of (42.3). On the other-hand, assuming that the reverse of (42.3) is valid, we set $g(y)$ as in (42.8) and obtain by the reverse of (42.3) that

$$J = I \geq \left(\int_A \varpi(x) \|x\|_\alpha^{pa-n} f^p(x)\, dx \right)^{\frac{1}{p}} J^{\frac{1}{q}}. \tag{42.12}$$

If $J = \infty$, then the reverse of (42.4) is naturally valid; if $J = 0$, then by (42.11), we find that the reverse of (42.4) keeps the form of equality; if $0 < J < \infty$, then dividing out $J^{\frac{1}{q}}$ in (42.12), it follows $J^{\frac{1}{p}} \geq (\int_A \varpi(x) \|x\|_\alpha^{pa-n} f^p(x)\, dx)^{\frac{1}{p}}$, and then we have the reverse of (42.4), which is equivalent to the reverse of (42.3). The theorem is proved. □

42.1.2 *The Upper Bound of the Norm of a Hilbert-Type Integral Operator*

Theorem 42.2 *Let the assumptions of Theorem* 42.1 *be fulfilled and additionally, there exist positive constants* k_1 *and* k_2 *such that*

$$\varpi(x) \leq k_1 \quad \text{a.e. in } A; \quad \omega(y) \leq k_2 \quad \text{a.e. in } B. \tag{42.13}$$

(i) *If* $p > 1$, *then we have the following equivalent inequalities*:

$$I = \int_B \int_A H(x,y) f(x) g(y)\, dx\, dy$$
$$\leq k_1^{\frac{1}{p}} k_2^{\frac{1}{q}} \left(\int_A \|x\|_\alpha^{pa-n} f^p(x)\, dx \right)^{\frac{1}{p}} \left(\int_B \|y\|_\beta^{qb-m} g^q(y)\, dy \right)^{\frac{1}{q}}, \tag{42.14}$$

$$J_1 = \int_B \|y\|_\beta^{p(m-b)-m} \left(\int_A H(x,y) f(x)\, dx \right)^p dy$$
$$\leq k_1 k_2^{p-1} \int_A \|x\|_\alpha^{pa-n} f^p(x)\, dx; \tag{42.15}$$

(ii) *If* $0 < p < 1$, *then we have the equivalent reverses of* (42.14) *and* (42.15).

Proof (i) For $p > 1$, since by (42.13), we still have $k_2^{1-p} \le (\omega(y))^{1-p}$ a.e. in B, then by (42.4) and (42.13), it follows $k_2^{1-p} J_1 \le J \le k_1 \int_A \|x\|_\alpha^{pa-n} f^p(x)\,dx$. Hence, we obtain (42.15). Putting $\omega(y) = 1$ in (42.7), we find

$$I \le J_1^{\frac{1}{p}} \left(\int_B \|y\|_\beta^{qb-m} g^q(y)\,dy \right)^{\frac{1}{q}}. \tag{42.16}$$

By (42.15), we have (42.14). The other parts of (i) and (ii) are obvious. The theorem is proved. □

Under the assumptions of Theorem 42.1 and Theorem 42.2, by setting $\varphi(x) := \|x\|_\alpha^{pa-n}$ $(x \in A)$, $\psi(y) := \|y\|_\beta^{qb-m}$ $(y \in B)$ and then $(\psi(y))^{1-p} = \|y\|_\beta^{p(m-b)-m}$, we define two real weight normed function spaces as follows:

$$L_{p,\varphi}(A) := \left\{ f;\ \|f\|_{p,\varphi} = \left\{ \int_A \|x\|_\alpha^{pa-n} |f(x)|^p\,dx \right\}^{\frac{1}{p}} < \infty \right\},$$

$$L_{q,\psi}(B) := \left\{ g;\ \|g\|_{q,\psi} = \left\{ \int_B \|y\|_\beta^{qb-m} |g(y)|^q\,dy \right\}^{\frac{1}{q}} < \infty \right\}.$$

Remark 42.1 For $0 < p < 1$, $L_{p,\varphi}(A)$ is not the weighted normed function space. But we still use the formal symbol of $L_{p,\varphi}(A)$ with the norm in this chapter for convenience.

Definition 42.2 Define a Hilbert-type integral operator $T : L_{p,\varphi}(A) \to L_{p,\psi^{1-p}}(B)$ as follows: for $f \in L_{p,\varphi}(A)$, there exists a unique representation $Tf \in L_{p,\psi^{1-p}}(B)$, satisfying

$$Tf(y) := \int_A H(x,y) f(x)\,dx \quad (y \in B). \tag{42.17}$$

For $g \in L_{q,\psi}(B)$, we define the formal inner product of Tf and g as follows

$$(Tf, g) = \int_B \left(\int_A H(x,y) f(x)\,dx \right) g(y)\,dy = \int_B \int_A H(x,y) f(x) g(y)\,dx\,dy. \tag{42.18}$$

Then by (42.14), it follows $(Tf, g) \le k_1^{\frac{1}{p}} k_2^{\frac{1}{q}} \|f\|_{p,\varphi} \|g\|_{q,\psi}$.

Since by (42.17), we write the equivalent form of (42.15) as follows

$$\|Tf\|_{p,\psi^{1-p}} = J_1^{\frac{1}{p}} \le k_1^{\frac{1}{p}} k_2^{\frac{1}{q}} \|f\|_{p,\varphi}, \tag{42.19}$$

then we conclude that $Tf \in L_{p,\psi^{1-p}}(B)$ and T is a bounded linear operator with

$$\|T\| := \sup_{f(\neq\theta)\in L_{p,\varphi}(A)} \frac{\|Tf\|_{p,\psi^{1-p}}}{\|f\|_{p,\varphi}} \le k_1^{\frac{1}{p}} k_2^{\frac{1}{q}}. \tag{42.20}$$

Remark 42.2 (i) Some more early results on Hilbert-type operators appeared in [3–21], and some dependent Hilbert-type inequalities appeared in [22–37].

(ii) For $c \in \mathbf{R}$, setting $\psi_c(y)$ as $\psi_c(y) := \psi^{1-p}(y)\|y\|_\beta^{-pc} = \|y\|_\beta^{p(m-b-c)-m}$, we still can define a general Hilbert-type integral operator $T_c : L_{p,\varphi}(A) \to L_{p,\psi_c}(B)$ as follows: for $f \in L_{p,\varphi}(A)$, there exists a unique representation $Tf \in L_{p,\psi_c}(B)$, satisfying

$$T_c f(y) := \|y\|_\beta^c \int_A H(x,y) f(x)\,dx \quad (y \in B). \tag{42.21}$$

Hence, we can find by (42.15) that $\|T_c f\|_{p,\psi_c} = J_1^{\frac{1}{p}} \le k_1^{\frac{1}{p}} k_2^{\frac{1}{q}} \|f\|_{p,\varphi}$, and then $\|T_c\| \le k_1^{\frac{1}{p}} k_2^{\frac{1}{q}}$.

42.1.3 Two Equivalent Strict Sign-Inequalities

Lemma 42.1 *If $n \in \mathbf{N}$, $\alpha, M > 0$, $\Psi(u)$ is a non-negative measurable function on* $(0,1]$, *and $D_M := \{x \in \mathbf{R}_+^n | \sum_{i=1}^n x_i^\alpha \le M^\alpha\}$, then we have (cf.* [38])

$$\int \cdots \int_{D_M} \Psi\left(\sum_{i=1}^n \left(\frac{x_i}{M}\right)^\alpha\right) dx_1 \cdots dx_n = \frac{M^n \Gamma^n(\frac{1}{\alpha})}{\alpha^n \Gamma(\frac{n}{\alpha})} \int_0^1 \Psi(u) u^{\frac{n}{\alpha}-1}\,du. \tag{42.22}$$

Lemma 42.2 *For $n \in \mathbf{N}$, $\alpha > 0$, $\varepsilon \ge 0$, we have*

$$J(\varepsilon) := \int_{\{x\in\mathbf{R}_+^n;\,\|x\|_\alpha \ge 1\}} \|x\|_\alpha^{-n-\varepsilon}\,dx = \begin{cases} \frac{\Gamma^n(\frac{1}{\alpha})}{\varepsilon\alpha^{n-1}\Gamma(\frac{n}{\alpha})}, & \varepsilon > 0, \\ \infty, & \varepsilon = 0, \end{cases} \tag{42.23}$$

$$\widetilde{J}(\varepsilon) := \int_{\{x\in\mathbf{R}_+^n;\,\|x\|_\alpha \le 1\}} \|x\|_\alpha^{-n+\varepsilon}\,dx = \begin{cases} \frac{\Gamma^n(\frac{1}{\alpha})}{\varepsilon\alpha^{n-1}\Gamma(\frac{n}{\alpha})}, & \varepsilon > 0, \\ \infty, & \varepsilon = 0. \end{cases} \tag{42.24}$$

Proof For $M > 1$, setting $\Psi(u)$ as $\Psi(u) = 0$ $(u \in (0, M^{-\alpha}))$; $\Psi(u) = \frac{1}{(Mu^{1/\alpha})^{n+\varepsilon}}$ $(u \in [M^{-\alpha}, 1])$, by (42.22) and since $\|x\|_\alpha \ge 1$, gives that $\sum_{i=1}^n (\frac{x_i}{M})^\alpha \ge M^{-\alpha}$, and we find

$$J(\varepsilon) = \lim_{M\to\infty} \int \cdots \int_{D_M} \Psi\left(\sum_{i=1}^n \left(\frac{x_i}{M}\right)^\alpha\right) dx_1 \cdots dx_n$$

$$= \lim_{M\to\infty} \frac{M^n \Gamma^n(\frac{1}{\alpha})}{\alpha^n \Gamma(\frac{n}{\alpha})} \int_0^1 \Psi(u) u^{\frac{n}{\alpha}-1}\, du$$

$$= \lim_{M\to\infty} \frac{M^n \Gamma^n(\frac{1}{\alpha})}{\alpha^n \Gamma(\frac{n}{\alpha})} \int_{M^{-\alpha}}^1 \frac{1}{(Mu^{1/\alpha})^{n+\varepsilon}} u^{\frac{n}{\alpha}-1}\, du$$

$$= \frac{\Gamma^n(\frac{1}{\alpha})}{\alpha^n \Gamma(\frac{n}{\alpha})} \lim_{M\to\infty} \frac{1}{M^\varepsilon} \int_{M^{-\alpha}}^1 u^{\frac{-\varepsilon}{\alpha}-1}\, du$$

$$= \begin{cases} \frac{\Gamma^n(\frac{1}{\alpha})}{\alpha^n \Gamma(\frac{n}{\alpha})} \lim_{M\to\infty} \frac{\alpha}{\varepsilon}(1-\frac{1}{M^\varepsilon}), & \varepsilon > 0, \\ \frac{\Gamma^n(\frac{1}{\alpha})}{\alpha^{n-1} \Gamma(\frac{n}{\alpha})} \lim_{M\to\infty} \alpha \ln M, & \varepsilon = 0. \end{cases}$$

Hence, (42.23) is valid. In view of (42.22) (for $M = 1$), it follows

$$\widetilde{J}(\varepsilon) = \int_{\{x\in\mathbf{R}_+^n;\, \|x\|_\alpha \le 1\}} \left(\sum_{i=1}^n x_i^\alpha\right)^{\frac{1}{\alpha}(-n+\varepsilon)} dx$$

$$= \frac{\Gamma^n(\frac{1}{\alpha})}{\alpha^n \Gamma(\frac{n}{\alpha})} \int_0^1 u^{\frac{1}{\alpha}(-n+\varepsilon)} u^{\frac{n}{\alpha}-1}\, du = \begin{cases} \frac{\Gamma^n(\frac{1}{\alpha})}{\varepsilon\alpha^{n-1}\Gamma(\frac{n}{\alpha})}, & \varepsilon > 0, \\ \infty, & \varepsilon = 0. \end{cases}$$

Hence, (42.24) is valid. The lemma is proved. □

Theorem 42.3 *Let the assumptions of Theorem* 42.2 *be fulfilled and additionally,* $A = \mathbf{R}^n$ *(or* $\mathbf{R}_+^n$*),* $B = \mathbf{R}^m$ *(or* $\mathbf{R}_+^m$*),* $f(\ge 0) \in L_{p,\varphi}(A)$*,* $g(\ge 0) \in L_{q,\psi}(B)$*,* $\|f\|_{p,\varphi} > 0$*,* $\|g\|_{q,\psi} > 0$.

(i) *If* $p > 1$*, then we have the following equivalent inequalities*:

$$I = \int_B \int_A H(x,y) f(x) g(y)\, dx\, dy < k_1^{\frac{1}{p}} k_2^{\frac{1}{q}} \|f\|_{p,\varphi} \|g\|_{q,\psi}, \tag{42.25}$$

$$J_1 = \int_B \|y\|_\beta^{p(m-b)-m} \left(\int_A H(x,y) f(x)\, dx\right)^p dy < k_1 k_2^{p-1} \|f\|_{p,\varphi}^p; \tag{42.26}$$

(ii) *If* $0 < p < 1$*, then we have the equivalent reverses of* (42.25) *and* (42.26).

Proof (i) If there exists a constant $y \in B$ such that under the assumptions of Theorem 42.3, (42.5) keeps the form of equality, then there exist constants c and d, which are not all zero, satisfying (cf. [1]) $c\frac{\|x\|_\alpha^{a(p-1)}}{\|y\|_\beta^b} f^p(x) = d\frac{\|y\|_\beta^{b(q-1)}}{\|x\|_\alpha^a}$, a.e. in A. We affirm that $c \ne 0$ (otherwise $d = c = 0$). Hence we find $\|x\|_\alpha^{ap-n} f^p(x) = (\frac{d}{c}\|y\|_\beta^{bq})\|x\|_\alpha^{-n}$, a.e. in A. Since $A = \mathbf{R}^n$ (or $\mathbf{R}_+^n$), by (42.23) (for $\varepsilon = 0$), it follows $\|f\|_{p,\varphi} = \infty$ (or 0), which contradicts the fact that $0 < \|f\|_{p,\varphi} < \infty$. Therefore, (42.5) keeps the form of a strict sign-inequality; so do (42.6) and (42.26). By (42.16) and (42.26), we can obtain (42.25).

The other parts of (i) and (ii) are obvious. The theorem is proved. □

42.2 The Norms of Hilbert-Type Integral Operators with the Two-Variable Kernels on $\mathbf{R}_+ \times \mathbf{R}_+$

42.2.1 The Case of a Homogeneous Kernel

Definition 42.3 For $p > 1$, $\frac{1}{p} + \frac{1}{q} = 1$, $\gamma, \gamma_1, \gamma_2 \in \mathbf{R}$, $\gamma_1 + \gamma_2 = \gamma$, setting $\varphi_1(x)$ and $\psi_1(y)$ as $\varphi_1(x) := x^{p(1+\gamma_1)-1}$ $(x \in \mathbf{R}_+)$, $\psi_1(y) := y^{q(1+\gamma_2)-1}$ $(y \in \mathbf{R}_+)$, we define two real weighted normed function spaces as follows:

$$L_{p,\varphi_1}(\mathbf{R}_+) := \left\{ f;\ \|f\|_{p,\varphi_1} = \left\{ \int_0^\infty x^{p(1+\gamma_1)-1} \left| f(x) \right|^p dx \right\}^{\frac{1}{p}} < \infty \right\},$$

$$L_{q,\psi_1}(\mathbf{R}_+) := \left\{ g;\ \|g\|_{q,\psi_1} = \left\{ \int_0^\infty y^{q(1+\gamma_2)-1} \left| g(y) \right|^q dy \right\}^{\frac{1}{q}} < \infty \right\}.$$

If $k_\gamma(x, y) \geq 0$ is a homogeneous function of degree γ on $\mathbf{R}_+^2$, satisfying for any $u, x, y \in \mathbf{R}_+$, $k_\gamma(ux, uy) = u^\gamma k_\gamma(x, y)$, then we define two weight functions $\omega_\gamma(y)$ and $\varpi_\gamma(x)$ as follows:

$$\omega_\gamma(y) := \int_0^\infty k_\gamma(x, y) \frac{y^{-\gamma_2}}{x^{\gamma_1+1}} dx \quad \big(y \in (0, \infty)\big), \tag{42.27}$$

$$\varpi_\gamma(x) := \int_0^\infty k_\gamma(x, y) \frac{x^{-\gamma_1}}{y^{\gamma_2+1}} dy \quad \big(x \in (0, \infty)\big). \tag{42.28}$$

Setting $u = x/y$ in (42.27) and (42.28), we find $\omega_\gamma(y) = \varpi_\gamma(x) = k(\gamma_1) := \int_0^\infty k_\gamma(u, 1) u^{-\gamma_1 - 1} du$.

Lemma 42.3 *Suppose that $p > 0$ $(\neq 1)$, $\frac{1}{p} + \frac{1}{q} = 1$, $\gamma, \gamma_1, \gamma_2 \in \mathbf{R}$, $\gamma_1 + \gamma_2 = \gamma$, $k_\gamma(x, y)$ is a non-negative homogeneous function of degree γ on $\mathbf{R}_+^2$,*

$$k(\gamma_1) = \int_0^\infty k_\gamma(u, 1) u^{-\gamma_1 - 1} du \in \mathbf{R}_+. \tag{42.29}$$

We set

$$L_k := \frac{1}{k} \int_1^\infty y^{-\frac{1}{k}-1} \left(\int_{\frac{1}{y}}^\infty k_\gamma(u, 1) u^{-\gamma_1 - \frac{1}{pk} - 1} du \right) dy. \tag{42.30}$$

(i) *If $p > 1$, then we have*

$$\lim_{k \to \infty} L_k = k(\gamma_1); \tag{42.31}$$

(ii) *If* $0 < p < 1$, *there exists a* $\delta_0 > 0$ *such that* $k(\gamma_1 + \delta_0) \in \mathbf{R}_+$, *then we have* (42.31).

Proof (i) For $p > 1$, by Fubini theorem (cf. [2]), we find

$$\begin{aligned} L_k &= \frac{1}{k}\int_1^\infty y^{-\frac{1}{k}-1}\left(\int_{\frac{1}{y}}^1 k_\gamma(u,1)u^{-\gamma_1-\frac{1}{pk}-1}\,du\right)dy \\ &\quad + \frac{1}{k}\int_1^\infty y^{-\frac{1}{k}-1}\left(\int_1^\infty k_\gamma(u,1)u^{-\gamma_1-\frac{1}{pk}-1}\,du\right)dy \\ &= \frac{1}{k}\int_0^1\left(\int_{\frac{1}{u}}^\infty y^{-\frac{1}{k}-1}\,dy\right)k_\gamma(u,1)u^{-\gamma_1-\frac{1}{pk}-1}\,du + \int_1^\infty k_\gamma(u,1)u^{-\gamma_1-\frac{1}{pk}-1}\,du \\ &= \int_0^1 k_\gamma(u,1)u^{-\gamma_1+\frac{1}{qk}-1}\,du + \int_1^\infty k_\gamma(u,1)u^{-\gamma_1-\frac{1}{pk}-1}\,du. \end{aligned} \tag{42.32}$$

Since both $\{k_\gamma(u,1)u^{-\gamma_1+\frac{1}{qk}-1}\}_{k=1}^\infty$ $(u \in (0,1))$ and $\{k_\gamma(u,1)u^{-\gamma_1-\frac{1}{pk}-1}\}_{k=1}^\infty$ $(u \in [1,\infty))$ are increasing, then by Levi theorem (cf. [2]), we have

$$\begin{aligned} k(\gamma_1) &= \int_0^1 \lim_{k\to\infty} k_\gamma(u,1)u^{-\gamma_1+\frac{1}{qk}-1}\,du + \int_1^\infty \lim_{k\to\infty} k_\gamma(u,1)u^{-\gamma_1-\frac{1}{pk}-1}\,du \\ &= \lim_{k\to\infty}\left(\int_0^1 k_\gamma(u,1)u^{-\gamma_1+\frac{1}{qk}-1}\,du + \int_1^\infty k_\gamma(u,1)u^{-\gamma_1-\frac{1}{pk}-1}\,du\right) \\ &= \lim_{k\to\infty} L_k, \end{aligned}$$

and (42.31) is proved.

(ii) If $0 < p < 1$, $q < 0$, for large enough integer k satisfying $\frac{1}{|q|k} \le \delta_0$, we have $u^{\frac{1}{qk}} \le u^{-\delta_0}$ $(u \in (0,1))$ and $0 \le k_\gamma(u,1)u^{-\gamma_1+\frac{1}{qk}-1} \le k_\gamma(u,1)u^{-(\gamma_1+\delta_0)-1}$ $(u \in (0,1))$. Since we have $\int_0^1 k_\gamma(u,1)u^{-(\gamma_1+\delta_0)-1}\,du \le k(\gamma_1+\delta_0) < \infty$, then by Lebesgue dominated convergence theorem (cf. [2]), it follows $\lim_{k\to\infty}\int_0^1 k_\gamma(u,1)\times u^{-\gamma_1+\frac{1}{qk}-1}\,du = \int_0^1 k_\gamma(u,1)u^{-\gamma_1-1}\,du$. Since $\{u^{-\gamma_1-\frac{1}{pk}-1}\}_{k=1}^\infty$ is increasing for $u \in [1,\infty)$, by Levi theorem (cf. [2]), we have $\lim_{k\to\infty}\int_1^\infty k_\gamma(u,1)u^{-\gamma_1-\frac{1}{pk}-1}\,du = \int_1^\infty k_\gamma(u,1)u^{-\gamma_1-1}\,du$. Then by (42.32), expression (42.31) follows. The lemma is proved. □

Theorem 42.4 (cf. [36]) *Suppose that* $p > 1$, $\frac{1}{p} + \frac{1}{q} = 1$, $\gamma, \gamma_1, \gamma_2 \in \mathbf{R}$, $\gamma_1 + \gamma_2 = \gamma$, $k_\gamma(x,y)$ *is a non-negative homogeneous function of degree* γ *on* $\mathbf{R}_+^2$, $k(\gamma_1) = \int_0^\infty k_\gamma(u,1)u^{-\gamma_1-1}\,du \in \mathbf{R}_+$, *and* $f(\ge 0) \in L_{p,\varphi_1}(\mathbf{R}_+)$, $g(\ge 0) \in L_{q,\psi_1}(\mathbf{R}_+)$ *are such that* $\|f\|_{p,\varphi_1} > 0$, $\|g\|_{q,\psi_1} > 0$.

(i) *If $p > 1$, then we have the following equivalent inequalities with the best constant factors $k(\gamma_1)$ and $k^p(\gamma_1)$:*

$$I := \int_0^\infty \int_0^\infty k_\gamma(x, y) f(x) g(y)\, dx\, dy < k(\gamma_1) \| f \|_{p,\varphi_1} \| g \|_{q,\psi_1}, \tag{42.33}$$

$$J := \int_0^\infty \frac{1}{y^{p\gamma_2+1}} \left(\int_0^\infty k_\gamma(x, y) f(x)\, dx \right)^p dy < k^p(\gamma_1) \| f \|_{p,\varphi_1}^p; \tag{42.34}$$

(ii) *If $0 < p < 1$, then we have the equivalent reverses of (42.33) and (42.34). Moreover, if there exists a constant $\delta_0 > 0$ such that $k(\gamma_1 + \delta_0) \in \mathbf{R}_+$, then the reverses of (42.33) and (42.34) possess the best constant factors $k(\gamma_1)$ and $k^p(\gamma_1)$, respectively.*

Proof For $m = n = 1$, $A = B = \mathbf{R}_+$, $H(x, y) = k_\gamma(x, y)$, $a = 1 + \gamma_1$, $b = 1 + \gamma_2$, and $k_1 = k_2 = k(\gamma_1)$ in Theorem 42.3, we have equivalent inequalities (42.33) and (42.34) (for $p > 1$), and the equivalent reverses of (42.33) and (42.34) (for $0 < p < 1$).

(i) For $p > 1$, $k \in \mathbf{N}$, we set $f_k(x)$ and $g_k(y)$ as follows:

$$f_k(x) := \begin{cases} 0, & x \in (0, 1), \\ x^{-\gamma_1 - \frac{1}{pk} - 1}, & x \in [1, \infty), \end{cases} \qquad g_k(y) := \begin{cases} 0, & y \in (0, 1), \\ y^{-\gamma_2 - \frac{1}{qk} - 1}, & y \in [1, \infty). \end{cases}$$

Then we find

$$\begin{aligned} &\| f_k \|_{p,\varphi_1} \| g_k \|_{q,\psi_1} \\ &\quad = \left\{ \int_1^\infty x^{p(1+\gamma_1)-1} x^{(-\gamma_1 - \frac{1}{pk} - 1)p}\, dx \right\}^{\frac{1}{p}} \left\{ \int_1^\infty y^{q(1+\gamma_2)-1} y^{(-\gamma_2 - \frac{1}{qk} - 1)q}\, dy \right\}^{\frac{1}{q}} \\ &\quad = \int_1^\infty x^{-\frac{1}{k} - 1}\, dx = k. \end{aligned}$$

By Fubini theorem, it follows

$$\begin{aligned} I_k &:= \int_0^\infty \int_0^\infty k_\gamma(x, y) f_k(x) g_k(y)\, dx\, dy \\ &= \int_1^\infty y^{-\gamma_2 - \frac{1}{qk} - 1} \left(\int_1^\infty k_\gamma(x, y) x^{-\gamma_1 - \frac{1}{pk} - 1}\, dx \right) dy \\ &\overset{u = x/y}{=} \int_1^\infty y^{-\frac{1}{k} - 1} \left(\int_{\frac{1}{y}}^\infty k_\gamma(u, 1) u^{-\gamma_1 - \frac{1}{pk} - 1}\, du \right) dy. \end{aligned}$$

If there exists a positive number K_0 with $K_0 \le k(\gamma_1)$ and such that (42.33) is valid as we replace $k(\gamma_1)$ by K_0, then, in particular, we have $L_k = \frac{1}{k} I_k < \frac{1}{k} K_0 \| f_k \|_{p,\varphi_1} \| g_k \|_{q,\psi_1} = K_0$. In view of (42.31), it follows that $k(\gamma_1) \le K_0$ $(k \to \infty)$. Hence, $K_0 = k(\gamma_1)$ is the best value of (42.33).

In view of (42.16), we still have

$$I \le J^{\frac{1}{p}} \|g\|_{q,\psi_1}. \tag{42.35}$$

We affirm that the constant factor in (42.34) is the best possible, otherwise we can get a contradiction by (42.35) that the constant factor in (42.33) is not the best possible.

(ii) For $0 < p < 1$, $q < 0$, if there exists a positive number K_1 ($\ge k(\gamma_1)$) such that the reverse of (42.33) is valid as we replace $k(\gamma_1)$ by K_1, then, in particular, still using f_k, g_k as in (i), we have $L_k = \frac{1}{k} I_k > \frac{1}{k} K_1 \|f_k\|_{p,\varphi_1} \|g_k\|_{q,\psi_1} = K_1$. Then by (42.31), it follows that $k(\gamma_1) \ge K_1$ ($k \to \infty$). Hence, the constant factor in the reverse of (42.33) is the best possible.

We affirm that the constant factor in the reverse of (42.34) is the best possible, otherwise, we can get a contradiction by the reverse of (42.35) that the constant factor in the reverse of (42.33) is not the best possible. The theorem is proved. □

Assuming that $u(x)$ ($x \in (a,b)$) and $v(y)$ ($y \in (c,d)$) are strict increasing differentiable functions, satisfying $u(a^+) = v(c^+) = 0$, $u(b^-) = v(d^-) = \infty$, by setting

$$\Phi_1(x) := \frac{[u(x)]^{p(1+\gamma_1)-1}}{[u'(x)]^{p-1}} \quad \big(x \in (a,b)\big),$$

$$\Psi_1(y) := \frac{[v(y)]^{q(1+\gamma_2)-1}}{[v'(y)]^{q-1}} \quad \big(y \in (c,d)\big),$$

replacing $x(y)$ by $u(x)$ ($v(y)$) in (42.33) and (42.34), after calculation, and after replacing $f(u(x))\, u'(x)$ ($g(v(y))v'(y)$) by $f(x)(g(y))$, we find

Corollary 42.1 *Suppose that $p > 0$ ($\neq 1$), $\frac{1}{p} + \frac{1}{q} = 1$, $\gamma, \gamma_1, \gamma_2 \in \mathbf{R}$, $\gamma_1 + \gamma_2 = \gamma$, $k_\gamma(x,y)$ is a non-negative homogeneous function of degree γ on $\mathbf{R}_+^2$, $k(\gamma_1) = \int_0^\infty k_\gamma(u,1) u^{-\gamma_1 - 1}\, du \in \mathbf{R}_+$, and $f(\ge 0) \in L_{p,\Phi_1}(a,b)$, $g(\ge 0) \in L_{q,\Psi_1}(c,d)$, $\|f\|_{p,\Phi_1} > 0$, $\|g\|_{q,\Psi_1} > 0$.*

(i) *If $p > 1$, then we have the following equivalent inequalities with the best constants $k(\gamma_1)$ and $k^p(\gamma_1)$:*

$$\int_c^d \int_a^b k_\gamma\big(u(x), v(y)\big) f(x) g(y)\, dx\, dy < k(\gamma_1) \|f\|_{p,\Phi_1} \|g\|_{q,\Psi_1}, \tag{42.36}$$

$$\int_c^d \frac{v'(y)}{[v(y)]^{p\gamma_2 + 1}} \left(\int_a^b k_\gamma\big(u(x), v(y)\big) f(x)\, dx \right)^p dy < k^p(\gamma_1) \|f\|_{p,\Phi_1}^p; \tag{42.37}$$

(ii) *If $0 < p < 1$, then we have the equivalent reverses of (42.36) and (42.37). Moreover, if there exists a constant $\delta_0 > 0$ such that $k(\gamma_1 + \delta_0) \in \mathbf{R}_+$, then the reverses of (42.36) and (42.37) possess the best constant factors $k(\gamma_1)$ and $k^p(\gamma_1)$, respectively.*

Proof (i) For $p > 1$, setting $u(x) = x$, $v(y) = y$ in (42.36) and (42.37), they become (42.33) and (42.34). It follows that (42.33) is equivalent to (42.36), and (42.34) is equivalent to (42.37). Hence, (42.36) and (42.37) are equivalent. It is obvious that the constant factors in (42.36) and (42.37) are the best possible.

(ii) For $0 < p < 1$, in the same way, we can prove that all the results of (ii) are valid. The corollary is proved. □

Definition 42.4 If $k(\gamma_1) \in \mathbf{R}_+$, then we define a Hilbert-type integral operator $T_\gamma : L_{p,\varphi_1}(\mathbf{R}_+) \to L_{p,\psi_1^{1-p}}(\mathbf{R}_+)$ as follows: for $f \in L_{p,\varphi_1}(\mathbf{R}_+)$, there exists a unique representation $T_\gamma f \in L_{p,\psi_1^{1-p}}(\mathbf{R}_+)$, satisfying

$$T_\gamma f(y) := \int_0^\infty k_\gamma(x, y) f(x)\, dx \quad (y \in \mathbf{R}_+). \tag{42.38}$$

Then it follows by (42.33) and (42.34) that

$$(T_\gamma f, g) < k(\gamma_1) \|f\|_{p,\varphi_1} \|g\|_{q,\psi_1}, \tag{42.39}$$

$$\|T_\gamma f\|_{p,\psi_1^{1-p}} < k(\gamma_1) \|f\|_{p,\varphi_1}, \tag{42.40}$$

where the constant factor $k(\gamma_1)$ is the best possible. Hence, we still have

Theorem 42.5 *Suppose that the Hilbert-type integral operator T_γ is defined by* (42.38). *Then it follows*

$$\|T_\gamma\| := \sup_{f(\neq\theta)\in L_{p,\varphi_1}(\mathbf{R}_+)} \frac{\|T_\gamma f\|_{p,\psi_1^{1-p}}}{\|f\|_{p,\varphi_1}} = k(\gamma_1), \tag{42.41}$$

where $k(\gamma_1) = \int_0^\infty k_\gamma(u, 1) u^{-\gamma_1 - 1}\, du \in \mathbf{R}_+$.

Remark 42.3 In Theorem 42.5, (i) if $\gamma = -\lambda < 0$, $\gamma_1 = -\frac{\lambda}{r}$, $\gamma_2 = -\frac{\lambda}{s}$ ($r > 1$, $\frac{1}{r} + \frac{1}{s} = 1$), $\varphi_\lambda(x) = x^{p(1-\frac{\lambda}{r})-1}$, $\psi_\lambda(y) = y^{q(1-\frac{\lambda}{s})-1}$ ($x, y \in \mathbf{R}_+$), then we define an operator $T_{-\lambda} : L_{p,\varphi_\lambda}(\mathbf{R}_+) \to L_{p,\psi_\lambda^{1-p}}(\mathbf{R}_+)$ as follows: for $f \in L_{p,\varphi_\lambda}(\mathbf{R}_+)$, there exists a unique representation $T_{-\lambda} f \in L_{p,\psi_\lambda^{1-p}}(\mathbf{R}_+)$, satisfying $T_{-\lambda} f(y) = \int_0^\infty k_{-\lambda}(x, y) f(x)\, dx$ ($y \in \mathbf{R}_+$). For $\int_0^\infty k_{-\lambda}(u, 1) u^{\frac{\lambda}{r}-1}\, du \in \mathbf{R}_+$, we have (cf. [35])

$$\|T_{-\lambda}\| = k_{-\lambda}(r) := \int_0^\infty k_{-\lambda}(u, 1) u^{\frac{\lambda}{r}-1}\, du. \tag{42.42}$$

(ii) If $\gamma = 0$, $\gamma_1 = -\alpha$, $\gamma_2 = \alpha$, $\varphi_0(x) = x^{p(1-\alpha)-1}$, $\psi_0(y) = y^{q(1+\alpha)-1}$ ($x, y \in \mathbf{R}_+$), then we define an operator $T_0 : L_{p,\varphi_0}(\mathbf{R}_+) \to L_{p,\psi_0^{1-p}}(\mathbf{R}_+)$ as follows: for $f \in L_{p,\varphi_0}(\mathbf{R}_+)$, there exists an unique representation $T_0 f \in L_{p,\psi_0^{1-p}}(\mathbf{R}_+)$, satisfying $T_0 f(y) = \int_0^\infty k_0(x, y) f(x)\, dx$ ($y \in \mathbf{R}_+$). Then for $\int_0^\infty k_0(u, 1) u^{\alpha-1}\, du \in$

$\mathbf{R}_+$, we have

$$\|T_0\| = k_0(\alpha) := \int_0^\infty k_0(u,1)u^{\alpha-1}\,du. \tag{42.43}$$

(iii) If $k_\gamma(x,y)$ is a symmetric function, $\beta \in \mathbf{R}$, $\gamma_1 = \gamma_2 = \frac{\gamma}{2}$, $k(\frac{\gamma}{2}) = \int_0^\infty k_\gamma(u,1)u^{-\frac{\gamma}{2}-1}\,du \in \mathbf{R}_+$, $\varphi(x) = x^{p(1+\frac{\gamma}{2})-1}$, $\psi(y) = y^{q(1+\frac{\gamma}{2})-1}$ $(x,y \in \mathbf{R}_+)$, then we define an operator $T_{\gamma,\beta} : L_{p,\varphi}(\mathbf{R}_+) \to L_{p,\psi^{1-p}}(\mathbf{R}_+)$ as follows: for $f \in L_{p,\varphi}(\mathbf{R}_+)$, there exists a unique representation $T_{\gamma,\beta} f \in L_{p,\psi^{1-p}}(\mathbf{R}_+)$, satisfying $T_{\gamma,\beta} f(y) = \int_0^\infty k_\gamma(x,y)\arctan(\frac{x}{y})^\beta f(x)\,dx$. Then we have

$$\|T_{\gamma,\beta}\| = k_\beta(\gamma) := \frac{\pi}{4}k\left(\frac{\gamma}{2}\right) = \frac{\pi}{4}\int_0^\infty k_\gamma(u,1)u^{-\frac{\gamma}{2}-1}\,du. \tag{42.44}$$

In fact, in view of the formula $\arctan u^\beta + \arctan u^{-\beta} = \frac{\pi}{2}$, it follows that

$$\begin{aligned}
k_\beta(\gamma) &= \int_0^\infty k_\gamma(u,1)\big(\arctan u^\beta\big)u^{-\frac{\gamma}{2}-1}\,du \\
&= \int_0^1 k_\gamma(u,1)\big(\arctan u^\beta\big)u^{-\frac{\gamma}{2}-1}\,du + \int_1^\infty k_\gamma(u,1)\big(\arctan u^\beta\big)u^{-\frac{\gamma}{2}-1}\,du \\
&= \int_0^1 k_\gamma(u,1)\big(\arctan u^\beta + \arctan u^{-\beta}\big)u^{-\frac{\gamma}{2}-1}\,du \\
&= \frac{\pi}{2}\int_0^1 k_\gamma(u,1)u^{-\frac{\gamma}{2}-1}\,du = \frac{\pi}{4}k\left(\frac{\gamma}{2}\right).
\end{aligned}$$

(iv) By virtue of Theorem 42.5, suppose that $k_\gamma\ (x,y)\ (\ge 0)$ is a homogeneous function of degree γ.

(a) If we set

$$K_\gamma(x,y) := \begin{cases} k_\gamma(x,y), & 0 < x \le y, \\ 0, & x > y, \end{cases}$$

then we define the first class Hardy-type integral operator $T_\gamma^{(1)} : L_{p,\varphi_1}(\mathbf{R}_+) \to L_{p,\psi_1^{1-p}}(\mathbf{R}_+)$ with the homogeneous kernel on $\mathbf{R}_+^2$ as follows: for $f \in L_{p,\varphi_1}(\mathbf{R}_+)$, there exists a unique representation $T_\gamma^{(1)} f \in L_{p,\psi_1^{1-p}}(\mathbf{R}_+)$, satisfying

$$T_\gamma^{(1)} f(y) = \int_0^\infty K_\gamma(x,y)f(x)\,dx = \int_0^y k_\gamma(x,y)f(x)\,dx \quad (y \in \mathbf{R}_+).$$

For $\int_0^1 k_\gamma(u,1)u^{-\gamma_1-1}\,du \in \mathbf{R}_+$, we obtain $\omega_\gamma(y) = \varpi_\gamma(x) = k_1(\gamma_1) := \int_0^1 k_\gamma(u,1)u^{-\gamma_1-1}\,du$,

$$\big\|T_\gamma^{(1)}\big\| = k_1(\gamma_1) = \int_0^1 k_\gamma(u,1)u^{-\gamma_1-1}\,du. \tag{42.45}$$

(b) If

$$K_\gamma(x,y) := \begin{cases} 0, & 0 < x \le y, \\ k_\gamma(x,y), & x > y, \end{cases}$$

then we define the second class Hardy-type integral operator $T_\gamma^{(2)} : L_{p,\varphi_1}(\mathbf{R}_+) \to L_{p,\psi_1^{1-p}}(\mathbf{R}_+)$ with the homogeneous kernel on $\mathbf{R}_+^2$ as follows: for $f \in L_{p,\varphi_1}(\mathbf{R}_+)$, there exists a unique representation $T_\gamma^{(2)} f \in L_{p,\psi_1^{1-p}}(\mathbf{R}_+)$, satisfying

$$T_\gamma^{(2)} f(y) = \int_0^\infty K_\gamma(x,y) f(x)\,dx = \int_y^\infty k_\gamma(x,y) f(x)\,dx \quad (y \in \mathbf{R}_+).$$

For $\int_1^\infty k_\gamma(u,1)u^{-\gamma_1-1}\,du \in \mathbf{R}_+$, we obtain $\omega_\gamma(y) = \varpi_\gamma(x) = k_2(\gamma_1) := \int_1^\infty k_\gamma(u,1)u^{-\gamma_1-1}\,du$,

$$\|T_\gamma^{(2)}\| = k_2(\gamma_1) = \int_1^\infty k_\gamma(u,1)u^{-\gamma_1-1}\,du. \tag{42.46}$$

42.2.2 The Case of a Non-homogeneous Kernel

Definition 42.5 For $p > 1$, $\frac{1}{p} + \frac{1}{q} = 1$, $\alpha \in \mathbf{R}$, setting $\varphi_2(x) := x^{p(1-\alpha)-1}$, $\psi_2(y) := y^{q(1-\alpha)-1}$ $(x, y \in \mathbf{R}_+)$, and then $\psi_2^{1-p}(y) = y^{p\alpha-1}$, we define two normed function spaces as follows:

$$L_{p,\varphi_2}(\mathbf{R}_+) := \left\{ f;\ \|f\|_{p,\varphi_2} = \left\{ \int_0^\infty x^{p(1-\alpha)-1} \left|f(x)\right|^p dx \right\}^{\frac{1}{p}} < \infty \right\},$$

$$L_{q,\psi_2}(\mathbf{R}_+) := \left\{ g;\ \|g\|_{q,\psi_2} = \left\{ \int_0^\infty y^{q(1-\alpha)-1} \left|g(y)\right|^q dy \right\}^{\frac{1}{q}} < \infty \right\}.$$

If $h(u) \ge 0$ is a measurable function on $\mathbf{R}_+$, then we define a weight function $\omega(y)$ as:

$$\omega(y) := \int_0^\infty h(xy) \frac{y^\alpha}{x^{1-\alpha}}\,dx \quad \big(y \in (0,\infty)\big). \tag{42.47}$$

Setting $u = xy$ in (42.47), we find

$$\omega(y) = K(\alpha) := \int_0^\infty h(u) u^{\alpha-1}\,du. \tag{42.48}$$

Lemma 42.4 *Suppose that $p > 0$ $(\neq 1)$, $\frac{1}{p} + \frac{1}{q} = 1$, $\alpha \in \mathbf{R}$, $h(u)$ is a non-negative measurable function on $\mathbf{R}_+$,*

$$K(\alpha) = \int_0^\infty h(u) u^{\alpha-1}\,du \in \mathbf{R}_+. \tag{42.49}$$

We set

$$\widetilde{L}_k := \frac{1}{k}\int_1^\infty y^{-\frac{1}{k}-1}\left(\int_0^y h(u)u^{\alpha+\frac{1}{pk}-1}\,du\right)dy. \tag{42.50}$$

(i) *If* $p > 1$, *then we have*

$$\lim_{k\to\infty} \widetilde{L}_k = K(\alpha); \tag{42.51}$$

(ii) *If* $0 < p < 1$, *there exists a* $\delta_0 > 0$ *such that* $K(\alpha+\delta_0) \in \mathbf{R}_+$, *then we still have* (42.51).

Proof (i) For $p > 1$, by Fubini theorem, it follows

$$\begin{aligned}
\widetilde{L}_k &= \frac{1}{k}\int_1^\infty y^{-\frac{1}{k}-1}\left(\int_0^1 h(u)u^{\alpha+\frac{1}{pk}-1}\,du\right)dy \\
&\quad + \frac{1}{k}\int_1^\infty y^{-\frac{1}{k}-1}\left(\int_1^y h(u)u^{\alpha+\frac{1}{pk}-1}\,du\right)dy \\
&= \int_0^1 h(u)u^{\alpha+\frac{1}{pk}-1}\,du + \frac{1}{k}\int_1^\infty\left(\int_u^\infty y^{-\frac{1}{k}-1}\,dy\right)h(u)u^{\alpha+\frac{1}{pk}-1}\,du \\
&= \int_0^1 h(u)u^{\alpha+\frac{1}{pk}-1}\,du + \int_1^\infty h(u)u^{\alpha-\frac{1}{qk}-1}\,du.
\end{aligned} \tag{42.52}$$

Since both $\{\int_0^1 h(u)u^{\alpha+\frac{1}{pk}-1}\,du\}_{k=1}^\infty$ $(u \in (0,1))$ and $\{\int_1^\infty h(u)u^{\alpha-\frac{1}{qk}-1}\,du\}_{k=1}^\infty$ $(u \in [1,\infty))$ are increasing, then by Levi theorem, we have

$$\begin{aligned}
K(\alpha) &= \int_0^1 \lim_{k\to\infty} h(u)u^{\alpha+\frac{1}{pk}-1}\,du + \int_1^\infty \lim_{k\to\infty} h(u)u^{\alpha-\frac{1}{qk}-1}\,du \\
&= \lim_{k\to\infty}\left(\int_0^1 h(u)u^{\alpha+\frac{1}{pk}-1}\,du + \int_1^\infty h(u)u^{\alpha-\frac{1}{qk}-1}\,du\right) = \lim_{k\to\infty}\widetilde{L}_k,
\end{aligned}$$

and (42.51) follows.

(ii) If $0 < p < 1$, $q < 0$, then for any $k \in \mathbf{N}(\frac{1}{|q|k} \le \delta_0)$, we have $u^{\frac{-1}{qk}} \le u^{\delta_0}$ $(u \in (1,\infty))$ and $0 \le h(u)u^{\alpha-\frac{1}{qk}-1} \le h(u)u^{\alpha+\delta_0-1}$ $(u \in (1,\infty))$. Since $\int_1^\infty h(u)u^{\alpha+\delta_0-1}\,du \le K(\alpha+\delta_0) < \infty$, by Lebesgue dominated convergence theorem, $\lim_{k\to\infty}\int_1^\infty h(u)u^{\alpha-\frac{1}{qk}-1}\,du = \int_1^\infty h(u)u^{\alpha-1}\,du$. Since $\{u^{\alpha+\frac{1}{pk}-1}\}_{k=1}^\infty$ $(u \in (0,1))$ is increasing, by Levi theorem, $\lim_{k\to\infty}\int_0^1 h(u)u^{\alpha+\frac{1}{pk}-1}\,du = \int_0^1 h(u)u^{\alpha-1}\,du$. Then by (42.52), it follows that (42.51) is valid. The lemma is proved. □

Theorem 42.6 *Let the assumptions of Lemma* 42.4 *be fulfilled and additionally,* $f(\ge 0) \in L_{p,\varphi_2}(\mathbf{R}_+)$, $g(\ge 0) \in L_{q,\psi_2}(\mathbf{R}_+)$ *such that* $\|f\|_{p,\varphi_2} > 0$, $\|g\|_{q,\psi_2} > 0$.

(i) *If $p > 1$, then we have the following equivalent inequalities with the best constant factors $K(\alpha)$ and $K^p(\alpha)$:*

$$\widetilde{I} := \int_0^\infty \int_0^\infty h(xy) f(x) g(y)\, dx\, dy < K(\alpha) \|f\|_{p,\varphi_2} \|g\|_{q,\psi_2}, \tag{42.53}$$

$$\widetilde{J} := \int_0^\infty y^{p\alpha-1} \left(\int_0^\infty h(xy) f(x)\, dx \right)^p dy < K^p(\alpha) \|f\|_{p,\varphi_2}^p; \tag{42.54}$$

(ii) *If $0 < p < 1$, then we have the equivalent reverses of (42.53) and (42.54). Moreover, if there exists a constant $\delta_0 > 0$ such that $K(\alpha+\delta) \in \mathbf{R}_+$, then the reverses of (42.53) and (42.54) possess the best constant factors $K(\alpha)$ and $K^p(\alpha)$, respectively.*

Proof For $m = n = 1$, $A = B = \mathbf{R}_+$, $H(x,y) = h(xy)$, $a = b = 1-\alpha$, and $k_1 = k_2 = K(\alpha)$ in Theorem 42.3, we obtain equivalent inequalities (42.53) and (42.54) (for $p > 1$), and the equivalent reverses of (42.53) and (42.54) (for $0 < p < 1$).

(i) For $p > 1$, $k \in \mathbf{N}$, we set $f_k(x)$ and $g_k(y)$ as follows:

$$f_k(x) := \begin{cases} x^{\alpha+\frac{1}{pk}-1}, & x \in (0,1), \\ 0, & x \in [1,\infty), \end{cases} \qquad g_k(y) := \begin{cases} 0, & y \in (0,1), \\ y^{\alpha-\frac{1}{qk}-1}, & y \in [1,\infty). \end{cases}$$

Then we find

$$\|f_k\|_{p,\varphi_2} = \left\{ \int_0^1 x^{p(1-\alpha)-1} x^{(\alpha+\frac{1}{pk}-1)p}\, dx \right\}^{\frac{1}{p}} = \left(\int_0^1 x^{\frac{1}{k}-1}\, dx \right)^{\frac{1}{p}} = k^{1/p},$$

$$\|g_k\|_{q,\psi_2} = \left\{ \int_1^\infty y^{q(1-\alpha)-1} y^{(\alpha-\frac{1}{qk}-1)q}\, dy \right\}^{\frac{1}{q}} = \left(\int_1^\infty y^{-\frac{1}{k}-1}\, dy \right)^{\frac{1}{q}} = k^{1/q}.$$

By Fubini theorem, it follows

$$\begin{aligned} \widetilde{I}_k &:= \int_0^\infty \int_0^\infty h(xy) f_k(x) g_k(y)\, dx\, dy \\ &= \int_1^\infty y^{\alpha-\frac{1}{qk}-1} \left(\int_0^1 h(xy) x^{\alpha+\frac{1}{pk}-1}\, dx \right) dy \\ &\overset{u=xy}{=} \int_1^\infty y^{-\frac{1}{k}-1} \left(\int_0^y h(u) u^{\alpha+\frac{1}{pk}-1}\, du \right) dy. \end{aligned}$$

If there exists a positive number K_0 with $K_0 \le K(\alpha)$ such that (42.53) is valid as we replace $K(\alpha)$ by K_0, then, in particular, we have $\widetilde{L}_k = \frac{1}{k}\widetilde{I}_k < \frac{1}{k} K_0 \|f_k\|_{p,\varphi_2} \|g_k\|_{q,\psi_2} = K_0$. In view of (42.51), it follows that $K(\alpha) \le K_0 (k \to \infty)$. Hence $K_0 = K(\alpha)$ is the best value of (42.53).

In view of (42.16), we still have

$$\widetilde{I} \le \widetilde{J}^{\frac{1}{p}} \|g\|_{q,\psi_2}. \tag{42.55}$$

We affirm that the constant factor in (42.54) is the best possible, otherwise we can get a contradiction by (42.55) that the constant factor in (42.53) is not the best possible.

(ii) For $0 < p < 1$, $q < 0$, if there exists a positive number K_1 ($\geq K(\alpha)$) such that the reverse of (42.53) is valid as we replace $K(\alpha)$ by K_1, then, in particular, still setting f_k, g_k as in (i), we have $\widetilde{L}_k = \frac{1}{k}\widetilde{I}_k > \frac{1}{k}K_1\|f_k\|_{p,\varphi_2}\|g_k\|_{q,\psi_2} = K_1$. Then by (42.51), it follows that $K(\alpha) \geq K_1 (k \to \infty)$. Hence the constant factor in the reverse of (42.53) is the best possible.

We affirm that the constant factor in the reverse of (42.54) is the best possible, otherwise, we can get a contradiction by the reverse of (42.55) that the constant factor in the reverse of (42.53) is not the best possible. The theorem is proved. □

Remark 42.4 For $\alpha = \frac{1}{p}$ $(p > 1)$, Theorem 42.6 becomes Theorem 350 in [22].

Assuming that $u(x)$ $(x \in (a,b))$ and $v(y)$ $(y \in (c,d))$ are strict increasing differentiable functions satisfying $u(a^+) = v(c^+) = 0$, $u(b^-) = v(d^-) = \infty$, $\alpha \in \mathbf{R}$, by setting

$$\Phi_2(x) := \frac{[u(x)]^{p(1-\alpha)-1}}{[u'(x)]^{p-1}} \quad \big(x \in (a,b)\big),$$

$$\Psi_2(y) := \frac{[v(y)]^{q(1-\alpha)-1}}{[v'(y)]^{q-1}} \quad \big(y \in (c,d)\big),$$

replacing $x(y)$ by $u(x)(v(y))$ in (42.53) and (42.54), after calculation, and after replacing $f(u(x))u'(x)$, $(g(v(y))v'(y))$ by $f(x)(g(y))$, we obtain

Corollary 42.2 *Suppose that $p > 0$ $(\neq 1)$, $\frac{1}{p} + \frac{1}{q} = 1$, $\alpha \in \mathbf{R}$, $h(u)$ is a non-negative measurable function on $\mathbf{R}_+$, $K(\alpha) = \int_0^\infty h(u)u^{\alpha-1}\,du \in \mathbf{R}_+$, $f(\geq 0) \in L_{p,\Phi_2}(a,b)$, $g(\geq 0) \in L_{q,\Psi_2}(c,d)$ such that $\|f\|_{p,\Phi_2} > 0$, $\|g\|_{q,\Psi_2} > 0$.*

(i) *If $p > 1$, then we have the following equivalent inequalities with the best constant factors $K(\alpha)$ and $K^p(\alpha)$:*

$$\int_c^d \int_a^b h\big(u(x)v(y)\big) f(x)g(y)\,dx\,dy < K(\alpha)\|f\|_{p,\Phi_2}\|g\|_{q,\Psi_2}, \tag{42.56}$$

$$\int_c^d \frac{v'(y)}{[v(y)]^{1-p\alpha}}\left(\int_a^b h\big(u(x)v(y)\big) f(x)\,dx\right)^p dy < K^p(\alpha)\|f\|_{p,\Phi_2}^p; \tag{42.57}$$

(ii) *If $0 < p < 1$, then we have the equivalent reverses of* (42.56) *and* (42.57). *Moreover, if there exists a constant $\delta_0 > 0$ such that $K(\alpha + \delta) \in \mathbf{R}_+$, then the reverses of* (42.56) *and* (42.57) *possess the best constant factors $K(\alpha)$ and $K^p(\alpha)$, respectively.*

In view of Theorem 42.6, for $K(\alpha) \in \mathbf{R}_+$, we define a Hilbert-type integral operator $\widetilde{T}_\alpha : L_{p,\varphi_2}(\mathbf{R}_+) \to L_{p,\psi_2^{1-p}}(\mathbf{R}_+)$ as follows: for $f \in L_{p,\varphi_2}(\mathbf{R}_+)$, there exists a unique representation $\widetilde{T}_\alpha f \in L_{p,\psi_2^{1-p}}(\mathbf{R}_+)$, satisfying

$$\widetilde{T}_\alpha f(y) := \int_0^\infty h(xy) f(x)\, dx \quad (y \in \mathbf{R}_+). \tag{42.58}$$

Then it follows by (42.53) and (42.54) that

$$(\widetilde{T}_\alpha f, g) < K(\alpha) \| f \|_{p,\varphi_2} \| g \|_{q,\psi_2}, \tag{42.59}$$

$$\| \widetilde{T}_\alpha f \|_{p,\psi_2^{1-p}} < K(\alpha) \| f \|_{p,\varphi_2}, \tag{42.60}$$

where the constant factor $K(\alpha)$ is the best possible. Hence we still have

Theorem 42.7 *Suppose that the Hilbert-type integral operator $\widetilde{T}_\alpha$ is defined by* (42.58). *Then*

$$\| \widetilde{T}_\alpha \| := \sup_{f(\neq \theta) \in L_{p,\varphi_2}(\mathbf{R}_+)} \frac{\| \widetilde{T}_\alpha f \|_{p,\psi_2^{1-p}}}{\| f \|_{p,\varphi_2}} = K(\alpha), \tag{42.61}$$

where $K(\alpha) = \int_0^\infty h(u) u^{\alpha-1}\, du \in \mathbf{R}_+$.

Remark 42.5 (i) In Theorem 42.7, for $h(u) = k_\gamma(1, u)$ $(\gamma \in \mathbf{R})$, $\alpha = -\frac{\gamma}{2}$, $\varphi(x) = x^{p(1+\frac{\gamma}{2})-1}$, $\psi(y) = y^{q(1+\frac{\gamma}{2})-1}$ $(x, y \in \mathbf{R}_+)$, $k(\frac{\gamma}{2}) = \int_0^\infty k_\gamma(1, u) u^{-\frac{\gamma}{2}-1}\, du \in \mathbf{R}_+$, define an integral operator $T_\gamma : L_{p,\varphi}(\mathbf{R}_+) \to L_{p,\psi^{1-p}}(\mathbf{R}_+)$ as follows: for $f \in L_{p,\varphi}(\mathbf{R}_+)$, there exists an unique representation $T_\gamma f \in L_{p,\psi^{1-p}}(\mathbf{R}_+)$, satisfying

$$T_\gamma f(y) = \int_0^\infty k_\gamma(1, xy) f(x)\, dx \quad (y \in \mathbf{R}_+).$$

Then we have

$$\| T_\gamma \| := \sup_{f(\neq \theta) \in L_{p,\varphi}(\mathbf{R}_+)} \frac{\| T_\gamma f \|_{p,\psi^{1-p}}}{\| f \|_{p,\varphi}} = k\left(\frac{\gamma}{2}\right). \tag{42.62}$$

(ii) We can still write some similar results for $h(u) = k_\gamma(u, 1)$ $(\gamma \in \mathbf{R})$.

(iii) By virtue of Theorem 42.7, suppose that $h(u)$ is a non-negative measurable function.

(a) If we set

$$H(xy) := \begin{cases} h(xy), & 0 < x \leq 1/y, \\ 0, & x > 1/y, \end{cases}$$

then we define the first class Hardy-type integral operator $\widetilde{T}_\alpha^{(1)} : L_{p,\varphi_2}(\mathbf{R}_+) \to L_{p,\psi_2^{1-p}}(\mathbf{R}_+)$ with the non-homogeneous kernel on $\mathbf{R}_+^2$ as follows: for $f \in$

$L_{p,\varphi_2}(\mathbf{R}_+)$, there exists a unique representation $\widetilde{T}_\alpha^{(1)} f \in L_{p,\psi_2^{1-p}}(\mathbf{R}_+)$, satisfying

$$\widetilde{T}_\alpha^{(1)} f(y) = \int_0^\infty H(xy) f(x)\,dx = \int_0^{\frac{1}{y}} h(xy) f(x)\,dx \quad (y \in \mathbf{R}_+).$$

Hence, for $\int_0^1 h(u)u^{\alpha-1}\,du \in \mathbf{R}_+$, we obtain $\omega(y) = K_1(\alpha) := \int_0^1 h(u)u^{\alpha-1}\,du$, and

$$\|\widetilde{T}_\alpha^{(1)}\| = K_1(\alpha) = \int_0^1 h(u)u^{\alpha-1}\,du. \tag{42.63}$$

(b) If we set

$$H(xy) := \begin{cases} 0, & 0 < x \le 1/y, \\ h(xy), & x > 1/y, \end{cases}$$

then we define the second class Hardy-type integral operator $\widetilde{T}_\alpha^{(2)} : L_{p,\varphi_2}(\mathbf{R}_+) \to L_{p,\psi_2^{1-p}}(\mathbf{R}_+)$ with the non-homogeneous kernel on $\mathbf{R}_+^2$ as follows: for $f \in L_{p,\varphi_2}(\mathbf{R}_+)$, there exists a unique representation $\widetilde{T}_\alpha^{(2)} f \in L_{p,\psi_2^{1-p}}(\mathbf{R}_+)$, satisfying

$$\widetilde{T}_\alpha^{(2)} f(y) = \int_0^\infty H(xy) f(x)\,dx = \int_{\frac{1}{y}}^\infty h(xy) f(x)\,dx \quad (y \in \mathbf{R}_+).$$

Hence, for $\int_1^\infty h(u)u^{\alpha-1}\,du \in \mathbf{R}_+$, we obtain $\omega(y) = K_2(\alpha) := \int_1^\infty h(u)u^{\alpha-1}\,du$, and

$$\|\widetilde{T}_\alpha^{(2)}\| = K_2(\alpha) = \int_1^\infty h(u)u^{\alpha-1}\,du. \tag{42.64}$$

42.2.3 Some Particular Examples

In Examples 42.1–42.4, we set $\lambda > 0$, $r > 1$, $\frac{1}{r} + \frac{1}{s} = 1$, $\varphi_\lambda(x) = x^{p(1-\frac{\lambda}{r})-1}$, $\psi_\lambda(y) = y^{q(1-\frac{\lambda}{s})-1}$, $\varphi(x) = x^{p(1-\frac{\lambda}{2})-1}$, and $\psi(y) = y^{q(1-\frac{\lambda}{2})-1}$ $(x, y \in \mathbf{R}_+)$.

Example 42.1 If $k_{-\lambda}(x,y) = \frac{1}{(x+y)^\lambda}$, then by Remark 42.3(i),

$$k_{-\lambda}(r) = \int_0^\infty \frac{1}{(u+1)^\lambda} u^{\frac{\lambda}{r}-1}\,du = B\left(\frac{\lambda}{r}, \frac{\lambda}{s}\right) \in \mathbf{R}_+.$$

(i) Define Hilbert integral operator $T_{-\lambda} : L_{p,\varphi_\lambda}(\mathbf{R}_+) \to L_{p,\psi_\lambda^{1-p}}(\mathbf{R}_+)$ as follows: for $f \in L_{p,\varphi_\lambda}(\mathbf{R}_+)$, there exists a unique representation $T_{-\lambda} f \in L_{p,\psi_\lambda^{1-p}}(\mathbf{R}_+)$, satisfying $T_{-\lambda} f(y) = \int_0^\infty \frac{1}{(x+y)^\lambda} f(x)\,dx$ $(y \in \mathbf{R}_+)$. Then we have

$$\|T_{-\lambda}\| = k_{-\lambda}(r) = B\left(\frac{\lambda}{r}, \frac{\lambda}{s}\right). \tag{42.65}$$

(ii) By Remark 42.3(iii), for $\beta \in \mathbf{R}$, define an operator $T_{-\lambda,\beta} : L_{p,\varphi}(\mathbf{R}_+) \to L_{p,\psi^{1-p}}(\mathbf{R}_+)$ as follows: for $f \in L_{p,\varphi}(\mathbf{R}_+)$, there exists a unique representation $T_{-\lambda,\beta} f \in L_{p,\psi^{1-p}}(\mathbf{R}_+)$, satisfying $T_{-\lambda,\beta} f(y) = \int_0^\infty \frac{1}{(x+y)^\lambda} \arctan(\frac{x}{y})^\beta f(x)\,dx$ $(y \in \mathbf{R}_+)$. Then we have

$$\|T_{-\lambda,\beta}\| = \frac{\pi}{4} B\left(\frac{\lambda}{2}, \frac{\lambda}{2}\right). \tag{42.66}$$

By Remark 42.5(i), define a Hilbert-type operator $\widetilde{T}_{-\lambda/2} : L_{p,\varphi}(\mathbf{R}_+) \to L_{p,\psi^{1-p}}(\mathbf{R}_+)$ as follows: for $f \in L_{p,\varphi}(\mathbf{R}_+)$, there exists an unique representation $\widetilde{T}_{-\lambda/2} f \in L_{p,\psi^{1-p}}(\mathbf{R}_+)$, satisfying $\widetilde{T}_{-\lambda/2} f(y) = \int_0^\infty \frac{1}{(1+xy)^\lambda} f(x)\,dx$ $(y \in \mathbf{R}_+)$. Then we have

$$\|\widetilde{T}_{-\lambda/2}\| = B\left(\frac{\lambda}{2}, \frac{\lambda}{2}\right). \tag{42.67}$$

Example 42.2 If $k_{-\lambda}(x, y) = \frac{1}{(\max\{x,y\})^\lambda}$, then by Remark 42.3(i),

$$k_{-\lambda}(r) = \int_0^\infty \frac{u^{(\lambda/r)-1}\,du}{(\max\{u, 1\})^\lambda} = \frac{rs}{\lambda} \in \mathbf{R}_+.$$

(i) Define a Hilbert-type integral operator $T_{-\lambda} : L_{p,\varphi_\lambda}(\mathbf{R}_+) \to L_{p,\psi_\lambda^{1-p}}(\mathbf{R}_+)$ as follows: for $f \in L_{p,\varphi_\lambda}(\mathbf{R}_+)$, there exists a unique representation $T_{-\lambda} f \in L_{p,\psi_\lambda^{1-p}}(\mathbf{R}_+)$, satisfying $T_{-\lambda} f(y) = \int_0^\infty \frac{1}{(\max\{x,y\})^\lambda} f(x)\,dx$ $(y \in \mathbf{R}_+)$. Then we have

$$\|T_{-\lambda}\| = k_{-\lambda}(r) = \frac{rs}{\lambda}. \tag{42.68}$$

(ii) By Remark 42.3(iii), for $\beta \in \mathbf{R}$, define an operator $T_{-\lambda,\beta} : L_{p,\varphi}(\mathbf{R}_+) \to L_{p,\psi^{1-p}}(\mathbf{R}_+)$ as follows: for $f \in L_{p,\varphi}(\mathbf{R}_+)$, there exists a unique representation $T_{-\lambda,\beta} f \in L_{p,\psi^{1-p}}(\mathbf{R}_+)$, satisfying $T_{-\lambda,\beta} f(y) = \int_0^\infty \frac{1}{(\max\{x,y\})^\lambda} \arctan(\frac{x}{y})^\beta \times f(x)\,dx$ $(y \in \mathbf{R}_+)$. Then we have

$$\|T_{-\lambda,\beta}\| = \frac{\pi}{\lambda}. \tag{42.69}$$

By Remark 42.5(i), define an operator $\widetilde{T}_{-\lambda/2} : L_{p,\varphi}(\mathbf{R}_+) \to L_{p,\psi^{1-p}}(\mathbf{R}_+)$ as follows: for $f \in L_{p,\varphi}(\mathbf{R}_+)$, there exists a unique representation $\widetilde{T}_{-\lambda/2} f \in$

$L_{p,\psi^{1-p}}(\mathbf{R}_+)$, satisfying $\widetilde{T}_{-\lambda/2}f(y) = \int_0^\infty \frac{1}{(\max\{1,xy\})^\lambda} f(x)\,dx$ $(y \in \mathbf{R}_+)$. Then we have

$$\|\widetilde{T}_{-\lambda/2}\| = \frac{4}{\lambda}. \tag{42.70}$$

Example 42.3 If $k_{-\lambda}(x,y) = \frac{\ln(x/y)}{x^\lambda - y^\lambda}$, then by Remark 42.3(i),

$$k_{-\lambda}(r) = \int_0^\infty \frac{(\ln u)u^{(\lambda/r)-1}}{u^\lambda - 1}\,du = \left[\frac{\pi}{\lambda \sin(\pi/r)}\right]^2 \in \mathbf{R}_+.$$

(i) Define a Hilbert-type integral operator $T_{-\lambda} : L_{p,\varphi_\lambda}(\mathbf{R}_+) \to L_{p,\psi_\lambda^{1-p}}(\mathbf{R}_+)$ as follows: for $f \in L_{p,\varphi_\lambda}(\mathbf{R}_+)$, there exists a unique representation $T_{-\lambda}f \in L_{p,\psi_\lambda^{1-p}}(\mathbf{R}_+)$, satisfying $T_{-\lambda}f(y) = \int_0^\infty \frac{\ln(x/y)}{x^\lambda - y^\lambda} f(x)\,dx$ $(y \in \mathbf{R}_+)$. Then we have

$$\|T_{-\lambda}\| = k_{-\lambda}(r) = \left[\frac{\pi}{\lambda \sin(\pi/r)}\right]^2. \tag{42.71}$$

(ii) By Remark 42.3(iii), for $\beta \in \mathbf{R}$, define an operator $T_{-\lambda,\beta} : L_{p,\varphi}(\mathbf{R}_+) \to L_{p,\psi^{1-p}}(\mathbf{R}_+)$ as follows: for $f \in L_{p,\varphi}(\mathbf{R}_+)$, there exists a unique representation $T_{-\lambda,\beta}f \in L_{p,\psi^{1-p}}(\mathbf{R}_+)$, satisfying $T_{-\lambda,\beta}f(y) = \int_0^\infty \frac{\ln(x/y)}{x^\lambda - y^\lambda} \arctan(\frac{x}{y})^\beta f(x)\,dx$ $(y \in \mathbf{R}_+)$. Then we have

$$\|T_{-\lambda,\beta}\| = \frac{\pi^3}{4\lambda^2}. \tag{42.72}$$

By Remark 42.5(i), define an operator $\widetilde{T}_{-\lambda/2} : L_{p,\varphi}(\mathbf{R}_+) \to L_{p,\psi^{1-p}}(\mathbf{R}_+)$ as follows: for $f \in L_{p,\varphi}(\mathbf{R}_+)$, there exists a unique representation $\widetilde{T}_{-\lambda/2}f \in L_{p,\psi^{1-p}}(\mathbf{R}_+)$, satisfying $\widetilde{T}_{-\lambda/2}f(y) = \int_0^\infty \frac{\ln(xy)}{(xy)^\lambda - 1} f(x)\,dx$ $(y \in \mathbf{R}_+)$. Then we have

$$\|\widetilde{T}_{-\lambda/2}\| = \left(\frac{\pi}{\lambda}\right)^2. \tag{42.73}$$

Example 42.4 If $0 < \lambda < 1$, $k_{-\lambda}(x,y) = \frac{1}{|x-y|^\lambda}$, then by Remark 42.3(i), we find

$$k_{-\lambda}(r) = \int_0^\infty \frac{u^{(\lambda/r)-1}}{|u-1|^\lambda}\,du = B\left(1-\lambda, \frac{\lambda}{r}\right) + B\left(1-\lambda, \frac{\lambda}{s}\right) \in \mathbf{R}_+.$$

(i) Define a Hilbert-type integral operator $T_{-\lambda} : L_{p,\varphi_\lambda}(\mathbf{R}_+) \to L_{p,\psi_\lambda^{1-p}}(\mathbf{R}_+)$ as follows: for $f \in L_{p,\varphi_\lambda}(\mathbf{R}_+)$, there exists a unique representation $T_{-\lambda}f \in L_{p,\psi_\lambda^{1-p}}(\mathbf{R}_+)$, satisfying $T_{-\lambda}f(y) = \int_0^\infty \frac{1}{|x-y|^\lambda} f(x)\,dx$ $(y \in \mathbf{R}_+)$. Then we have

$$\|T_{-\lambda}\| = B\left(1-\lambda, \frac{\lambda}{r}\right) + B\left(1-\lambda, \frac{\lambda}{s}\right). \tag{42.74}$$

(ii) By Remark 42.3(iii), for $\beta \in \mathbf{R}$, define an operator $T_{-\lambda,\beta} : L_{p,\varphi}(\mathbf{R}_+) \to L_{p,\psi^{1-p}}(\mathbf{R}_+)$ as follows: for $f \in L_{p,\varphi}(\mathbf{R}_+)$, there exists a unique representation $T_{-\lambda,\beta} f \in L_{p,\psi^{1-p}}(\mathbf{R}_+)$, satisfying $T_{-\lambda,\beta} f(y) = \int_0^\infty \frac{1}{|x-y|^\lambda} \arctan(\frac{x}{y})^\beta f(x)\, dx$ $(y \in \mathbf{R}_+)$. Then we have

$$\|T_{-\lambda,\beta}\| = \frac{\pi}{2} B\left(1-\lambda, \frac{\lambda}{2}\right). \tag{42.75}$$

By Remark 42.5(i), define an operator $\widetilde{T}_{-\lambda/2} : L_{p,\varphi}(\mathbf{R}_+) \to L_{p,\psi^{1-p}}(\mathbf{R}_+)$ as follows: for $f \in L_{p,\varphi}(\mathbf{R}_+)$, there exists a unique representation $\widetilde{T}_{-\lambda/2} f \in L_{p,\psi^{1-p}}(\mathbf{R}_+)$, satisfying $\widetilde{T}_{-\lambda/2} f(y) = \int_0^\infty \frac{1}{|1-xy|^\lambda} f(x)\, dx$ $(y \in \mathbf{R}_+)$. Then we have

$$\|\widetilde{T}_{-\lambda/2}\| = 2B\left(1-\lambda, \frac{\lambda}{2}\right). \tag{42.76}$$

42.3 The Norms of Hilbert-Type Integral Operators with the Two-Variable Kernels on R × R

42.3.1 The Case of a Homogeneous Kernel

For $p > 1$, $\frac{1}{p} + \frac{1}{q} = 1$, $\gamma, \gamma_1, \gamma_2 \in \mathbf{R}$, $\gamma_1 + \gamma_2 = \gamma$, $\widetilde{\varphi}_1(x) := |x|^{p(1+\gamma_1)-1}$, $\widetilde{\psi}_1(y) := |y|^{q(1+\gamma_2)-1}$ $(x, y \in \mathbf{R})$, we define two real weighted normed function spaces as follows:

$$L_{p,\widetilde{\varphi}_1}(\mathbf{R}) := \left\{ f;\ \|f\|_{p,\widetilde{\varphi}_1} = \left\{ \int_{-\infty}^{\infty} |x|^{p(1+\gamma_1)-1} |f(x)|^p\, dx \right\}^{\frac{1}{p}} < \infty \right\},$$

$$L_{q,\widetilde{\psi}_1}(\mathbf{R}) := \left\{ g;\ \|g\|_{q,\widetilde{\psi}_1} = \left\{ \int_{-\infty}^{\infty} |y|^{q(1+\gamma_2)-1} |g(y)|^q\, dy \right\}^{\frac{1}{q}} < \infty \right\}.$$

Definition 42.6 If $k_\gamma(x, y)$ is a non-negative homogeneous function of degree γ on $\mathbf{R}^2$, satisfying for any $u, x, y \in \mathbf{R}$, $k_\gamma(ux, uy) = |u|^\gamma k_\gamma(x, y)$, then we define two weight functions $\widetilde{\omega}_\gamma(y)$ and $\widetilde{\varpi}_\gamma(x)$ as follows:

$$\widetilde{\omega}_\gamma(y) := \int_{-\infty}^{\infty} k_\gamma(x, y) \frac{|y|^{-\gamma_2}}{|x|^{\gamma_1+1}}\, dx \quad \big(y \in (-\infty, \infty)\big), \tag{42.77}$$

$$\widetilde{\varpi}_\gamma(x) := \int_{-\infty}^{\infty} k_\gamma(x, y) \frac{|x|^{-\gamma_1}}{|y|^{\gamma_2+1}}\, dy \quad \big(x \in (-\infty, \infty)\big). \tag{42.78}$$

Lemma 42.5 *If* $x, y \in \mathbf{R}\backslash\{0\}$, *then we have*

$$\widetilde{\omega}_\gamma(y) = \widetilde{\varpi}_\gamma(x) = \widetilde{k}(\gamma_1) := \int_0^\infty \big(k_\gamma(u,1) + k_\gamma(-u,1)\big)u^{-\gamma_1-1}\,du. \qquad (42.79)$$

Proof For $y < 0$, setting $u = x/y$, $u = x/(-y)$ in the following first and second integrals, respectively, it follows

$$\begin{aligned}\widetilde{\omega}_\gamma(y) &= \int_{-\infty}^0 k_\gamma(x,y)\frac{(-y)^{-\gamma_2}\,dx}{(-x)^{\gamma_1+1}} + \int_0^\infty k_\gamma(x,y)\frac{(-y)^{-\gamma_2}\,dx}{x^{\gamma_1+1}}\\ &= \int_0^\infty k_\gamma(u,1)u^{-\gamma_1-1}\,du + \int_0^\infty k_\gamma(-u,1)u^{-\gamma_1-1}\,du = \widetilde{k}(\gamma_1);\end{aligned}$$

for $y > 0$, setting $u = x/(-y)$, $u = x/y$ in the following first and second integrals, respectively, it follows

$$\begin{aligned}\widetilde{\omega}_\gamma(y) &= \int_{-\infty}^0 k_\gamma(x,y)\frac{y^{-\gamma_2}\,dx}{(-x)^{\gamma_1+1}} + \int_0^\infty k_\gamma(x,y)\frac{y^{-\gamma_2}\,dx}{x^{\gamma_1+1}}\\ &= \int_0^\infty k_\gamma(-u,1)u^{-\gamma_1-1}\,du + \int_0^\infty k_\gamma(u,1)u^{-\gamma_1-1}\,du = \widetilde{k}(\gamma_1).\end{aligned}$$

In the same way, for $x \neq 0$, we still can find that $\widetilde{\varpi}_\gamma(x) = \widetilde{k}(\gamma_1)$. The lemma is proved. □

Lemma 42.6 *Suppose that* $p > 0$ $(\neq 1)$, $\frac{1}{p} + \frac{1}{q} = 1$, $\gamma, \gamma_1, \gamma_2 \in \mathbf{R}$, $\gamma_1 + \gamma_2 = \gamma$, $k_\gamma(x,y)$ *is a non-negative homogeneous function of degree* γ *on* $\mathbf{R}^2$ *and* $\widetilde{k}(\gamma_1) \in \mathbf{R}_+$. *We set*

$$\widetilde{L}_k := \frac{1}{k}\int_1^\infty y^{-\frac{1}{k}-1}\left[\int_{\frac{1}{y}}^\infty \big(k_\gamma(u,1) + k_\gamma(-u,1)\big)u^{-\gamma_1-\frac{1}{pk}-1}\,du\right]dy. \qquad (42.80)$$

(i) *If* $p > 1$, *then we have*

$$\lim_{k\to\infty} \widetilde{L}_k = \widetilde{k}(\gamma_1); \qquad (42.81)$$

(ii) *If* $0 < p < 1$, *there exists a* $\delta_0 > 0$ *such that* $\widetilde{k}(\gamma_1 + \delta_0) \in \mathbf{R}_+$, *then we still have* (42.81).

Proof (i) For $p > 1$, by Fubini theorem, we obtain

$$\begin{aligned}\widetilde{L}_k &= \frac{1}{k}\int_1^\infty y^{-\frac{1}{k}-1}\left[\int_{\frac{1}{y}}^1 \big(k_\gamma(u,1) + k_\gamma(-u,1)\big)u^{-\gamma_1-\frac{1}{pk}-1}\,du\right]dy\\ &\quad + \frac{1}{k}\int_1^\infty y^{-\frac{1}{k}-1}\left[\int_1^\infty \big(k_\gamma(u,1) + k_\gamma(-u,1)\big)u^{-\gamma_1-\frac{1}{pk}-1}\,du\right]dy\end{aligned}$$

$$
\begin{aligned}
&= \frac{1}{k}\int_0^1 \left(\int_{\frac{1}{u}}^{\infty} \frac{1}{y^{\frac{1}{k}+1}}\,dy\right)\big(k_\gamma(u,1)+k_\gamma(-u,1)\big)u^{-\gamma_1-\frac{1}{pk}-1}\,du \\
&\quad + \int_1^{\infty} \big(k_\gamma(u,1)+k_\gamma(-u,1)\big)u^{-\gamma_1-\frac{1}{pk}-1}\,du \\
&= \int_0^1 \big(k_\gamma(u,1)+k_\gamma(-u,1)\big)u^{-\gamma_1+\frac{1}{qk}-1}\,du \\
&\quad + \int_1^{\infty} \big(k_\gamma(u,1)+k_\gamma(-u,1)\big)u^{-\gamma_1-\frac{1}{pk}-1}\,du. \qquad (42.82)
\end{aligned}
$$

Then by Levi theorem, we have

$$
\begin{aligned}
\widetilde{k}(\gamma_1) &= \int_0^1 \lim_{k\to\infty} \big(k_\gamma(u,1)+k_\gamma(-u,1)\big)u^{-\gamma_1+\frac{1}{qk}-1}\,du \\
&\quad + \int_1^{\infty} \lim_{k\to\infty} \big(k_\gamma(u,1)+k_\gamma(-u,1)\big)u^{-\gamma_1-\frac{1}{pk}-1}\,du \\
&= \lim_{k\to\infty}\bigg[\int_0^1 \big(k_\gamma(u,1)+k_\gamma(-u,1)\big)u^{-\gamma_1+\frac{1}{qk}-1}\,du \\
&\quad + \int_1^{\infty} \big(k_\gamma(u,1)+k_\gamma(-u,1)\big)u^{-\gamma_1-\frac{1}{pk}-1}\,du\bigg] = \lim_{k\to\infty} \widetilde{L}_k.
\end{aligned}
$$

Hence, (42.81) is valid.

(ii) If $0 < p < 1$, $q < 0$, then for large enough k satisfying $\frac{1}{|q|k} \leq \delta_0$, we have $u^{\frac{1}{qk}} \leq u^{-\delta_0}$ $(u \in (0,1))$ and

$$
\begin{aligned}
0 &\leq \big(k_\gamma(u,1)+k_\gamma(-u,1)\big)u^{-\gamma_1+\frac{1}{qk}-1} \\
&\leq \big(k_\gamma(u,1)+k_\gamma(-u,1)\big)u^{-(\gamma_1+\delta_0)-1} \quad \big(u \in (0,1)\big).
\end{aligned}
$$

Since we have $\int_0^1 (k_\gamma(u,1)+k_\gamma(-u,1))u^{-(\gamma_1+\delta_0)-1}\,du \leq k(\gamma_1+\delta_0) < \infty$, then by Lebesgue dominated convergence theorem, it follows

$$
\begin{aligned}
&\lim_{k\to\infty} \int_0^1 \big(k_\gamma(u,1)+k_\gamma(-u,1)\big)u^{-\gamma_1+\frac{1}{qk}-1}\,du \\
&\quad = \int_0^1 \big(k_\gamma(u,1)+k_\gamma(-u,1)\big)u^{-\gamma_1-1}\,du.
\end{aligned}
$$

Since $\{u^{-\gamma_1-\frac{1}{pk}-1}\}_{k=1}^{\infty}$ is increasing for $u \in [1,\infty)$, then by Levi theorem, we have

$$\lim_{k\to\infty}\int_1^{\infty}\big(k_\gamma(u,1)+k_\gamma(-u,1)\big)u^{-\gamma_1-\frac{1}{pk}-1}\,du$$
$$=\int_1^{\infty}\big(k_\gamma(u,1)+k_\gamma(-u,1)\big)u^{-\gamma_1-1}\,du.$$

Hence, in view of (42.82), expression (42.81) is valid. The lemma is proved. □

Theorem 42.8 *Suppose that $p>0$ $(\neq 1)$, $\frac{1}{p}+\frac{1}{q}=1$, $\gamma,\gamma_1,\gamma_2\in\mathbf{R}$, $\gamma_1+\gamma_2=\gamma$, $k_\gamma(x,y)$ is a non-negative homogeneous function of degree γ on $\mathbf{R}^2$, $\widetilde{k}(\gamma_1)\in\mathbf{R}_+$, and $f(\geq 0)\in L_{p,\widetilde{\varphi}_1}(\mathbf{R})$, $g(\geq 0)\in L_{q,\widetilde{\psi}_1}(\mathbf{R})$ such that $\|f\|_{p,\widetilde{\varphi}_1}>0$, $\|g\|_{q,\widetilde{\psi}_1}>0$.*

(i) *If $p>1$, then we have the following equivalent inequalities with the best constant factors $\widetilde{k}(\gamma_1)$ and $\widetilde{k}^p(\gamma_1)$:*

$$\widetilde{I}:=\int_{-\infty}^{\infty}\int_{-\infty}^{\infty}k_\gamma(x,y)f(x)g(y)\,dx\,dy<\widetilde{k}(\gamma_1)\|f\|_{p,\widetilde{\varphi}_1}\|g\|_{q,\widetilde{\psi}_1},\tag{42.83}$$

$$\widetilde{J}:=\int_{-\infty}^{\infty}\frac{1}{|y|^{p\gamma_2+1}}\left(\int_{-\infty}^{\infty}k_\gamma(x,y)f(x)\,dx\right)^p dy<\widetilde{k}^p(\gamma_1)\|f\|_{p,\widetilde{\varphi}_1};\tag{42.84}$$

(ii) *If $0<p<1$, then we have the equivalent reverses of (42.83) and (42.84). Moreover, if there exists a constant $\delta_0>0$ such that $\widetilde{k}(\gamma_1+\delta_0)\in\mathbf{R}_+$, then the reverses of (42.83) and (42.84) possess the best constant factors $\widetilde{k}(\gamma_1)$ and $\widetilde{k}^p(\gamma_1)$, respectively.*

Proof For $m=n=1$, $A=B=\mathbf{R}$, $H(x,y)=k_\gamma(x,y)$, $a=1+\gamma_1$, $b=1+\gamma_2$, and $k_1=k_2=\widetilde{k}(\gamma_1)$ in Theorem 42.3, we have equivalent inequalities (42.83) and (42.84) (for $p>1$), and the equivalent reverses of (42.83) and (42.84) (for $0<p<1$).

(i) For $p>1$, $k\in\mathbf{N}$, we set $\widetilde{f}_k(x)$ and $\widetilde{g}_k(y)$ as follows:

$$\widetilde{f}_k(x):=\begin{cases}(-x)^{-\gamma_1-\frac{1}{pk}-1}, & x\in(-\infty,-1],\\ 0, & x\in(-1,1),\\ x^{-\gamma_1-\frac{1}{pk}-1}, & x\in[1,\infty),\end{cases}$$

$$\widetilde{g}_k(y):=\begin{cases}(-y)^{-\gamma_2-\frac{1}{qk}-1}, & y\in(-\infty,-1],\\ 0, & y\in(-1,1),\\ y^{-\gamma_2-\frac{1}{qk}-1}, & y\in[1,\infty).\end{cases}$$

Then we find $\|\widetilde{f}_k\|_{p,\widetilde{\varphi}_1}\|\widetilde{g}_k\|_{q,\widetilde{\psi}_1}=2k$, and

$$\widetilde{I}_k:=\int_{-\infty}^{\infty}\int_{-\infty}^{\infty}k_\gamma(x,y)\widetilde{f}_k(x)\widetilde{g}_k(y)\,dx\,dy=I_1+I_2,$$

$$I_1 := \int_1^\infty y^{-\gamma_2 - \frac{1}{qk} - 1}\left(\int_{-\infty}^\infty k_\gamma(x,y)\widetilde{f}_k(x)\,dx\right)dy,$$

$$I_2 := \int_{-\infty}^{-1} (-Y)^{-\gamma_2 - \frac{1}{qk} - 1}\left(\int_{-\infty}^\infty k_\gamma(x,Y)\widetilde{f}_k(x)\,dx\right)dY.$$

Setting $y = -Y$ in I_2, in view of the homogeneity property of $k_\gamma(x,y)$, it follows

$$\begin{aligned} I_2 &= \int_1^\infty y^{-\gamma_2 - \frac{1}{qk} - 1}\left(\int_{-\infty}^\infty k_\gamma(x,-y)\widetilde{f}_k(x)\,dx\right)dy \\ &= \int_1^\infty y^{-\gamma_2 - \frac{1}{qk} - 1}\left(\int_{-\infty}^\infty k_\gamma(-x,y)\widetilde{f}_k(x)\,dx\right)dy. \end{aligned}$$

Then by Fubini theorem, we obtain

$$\begin{aligned} \widetilde{I}_k &= \int_1^\infty y^{-\gamma_2 - \frac{1}{qk} - 1}\left[\int_{-\infty}^\infty \big(k_\gamma(x,y) + k_\gamma(-x,y)\big)\widetilde{f}_k(x)\,dx\right]dy \\ &= \int_1^\infty y^{-\gamma_2 - \frac{1}{qk} - 1}\left[\int_1^\infty \big(k_\gamma(x,y) + k_\gamma(-x,y)\big)x^{-\gamma_1 - \frac{1}{pk} - 1}\,dx\right. \\ &\quad \left. + \int_{-\infty}^{-1} \big(k_\gamma(X,y) + k_\gamma(-X,y)\big)(-X)^{-\gamma_1 - \frac{1}{pk} - 1}\,dX\right]dy \\ &\overset{X=-x}{=} 2\int_1^\infty y^{-\gamma_2 - \frac{1}{qk} - 1}\left[\int_1^\infty \big(k_\gamma(x,y) + k_\gamma(-x,y)\big)x^{-\gamma_1 - \frac{1}{pk} - 1}\,dx\right]dy \\ &\overset{u=x/y}{=} 2\int_1^\infty y^{-\frac{1}{k} - 1}\left[\int_{\frac{1}{y}}^\infty \big(k_\gamma(u,1) + k_\gamma(-u,1)\big)u^{-\gamma_1 - \frac{1}{pk} - 1}\,du\right]dy = 2k\widetilde{L}_k. \end{aligned}$$

If there exists a positive number K_0 with $K_0 \le \widetilde{k}(\gamma_1)$ such that (42.83) is valid as we replace $\widetilde{k}(\gamma_1)$ by K_0, then, in particular, we have $\widetilde{L}_k = \frac{1}{2k}\widetilde{I}_k < \frac{1}{2k}K_0\|\widetilde{f}_k\|_{p,\widetilde{\varphi}_1}\|\widetilde{g}_k\|_{q,\widetilde{\psi}_1} = K_0$. In view of (42.81), it follows $\widetilde{k}(\gamma_1) \le K_0$ $(k \to \infty)$. Hence $K_0 = \widetilde{k}(\gamma_1)$ is the best value of (42.83).

We affirm that the constant factor in (42.84) is the best possible, otherwise we can get a contradiction by inequality $\widetilde{I} \le \widetilde{J}^{\frac{1}{p}}\|g\|_{q,\widetilde{\psi}}$ that the constant in (42.83) is not the best possible.

(ii) For $0 < p < 1$, $q < 0$, if there exists a positive number K_1 $(\ge \widetilde{k}(\gamma_1))$ such that the reverse of (42.83) is valid as we replace $\widetilde{k}(\gamma_1)$ by K_1, then in particular, still setting $\widetilde{f}_k$, $\widetilde{g}_k$ as in (i), we have $\widetilde{L}_k = \frac{1}{2k}\widetilde{I}_k > \frac{1}{2k}K_1\|\widetilde{f}_k\|_{p,\widetilde{\varphi}_1}\|\widetilde{g}_k\|_{q,\widetilde{\psi}_1} = K_1$. It follows by (42.81) that $\widetilde{k}(\gamma_1) \ge K_1$ $(k \to \infty)$. Hence $K_1 = \widetilde{k}(\gamma_1)$ is the best value of the reverse of (42.83).

We affirm that the constant factor in the reverse of (42.84) is the best possible, otherwise, we can get a contradiction by inequality $\widetilde{I} \ge \widetilde{J}^{\frac{1}{p}}\|g\|_{q,\widetilde{\psi}}$ that the constant factor in the reverse of (42.83) is not the best possible. The theorem is proved. □

Assuming that $u(x)$ $(x \in (a,b))$ and $v(y)$ $(y \in (c,d))$ are strict increasing differentiable functions, satisfying $u(a^+) = v(c^+) = -\infty$, $u(b^-) = v(d^-) = \infty$, by setting

$$\widetilde{\Phi}_1(x) := \frac{[u(x)]^{p(1+\gamma_1)-1}}{[u'(x)]^{p-1}} \quad \big(x \in (a,b)\big),$$

$$\widetilde{\Psi}_1(y) := \frac{[v(y)]^{q(1+\gamma_2)-1}}{[v'(y)]^{q-1}} \quad \big(y \in (c,d)\big),$$

replacing $x(y)$ by $u(x)$ $(v(y))$ in (42.83) and (42.84), after calculation, and after replacing $f(u(x))u'(x)$, $(g(v(y))v'(y))$ by $f(x)$ $(g(y))$, we obtain

Corollary 42.3 *Suppose that $p > 0$ $(\neq 1)$, $\frac{1}{p} + \frac{1}{q} = 1$, $\gamma, \gamma_1, \gamma_2 \in \mathbf{R}$, $\gamma_1 + \gamma_2 = \gamma$, $k_\gamma(x,y)$ is a non-negative homogeneous function of degree γ on $\mathbf{R}^2$, $\widetilde{k}(\gamma_1) := \int_0^\infty (k_\gamma(u,1) + k_\gamma(-u,1))u^{-\gamma_1-1}\,du \in \mathbf{R}_+$, and $f(\geq 0) \in L_{p,\widetilde{\Phi}_1}(a,b)$, $g(\geq 0) \in L_{q,\widetilde{\Psi}_1}(c,d)$ such that $\|f\|_{p,\widetilde{\Phi}_1} > 0$, $\|g\|_{q,\widetilde{\Psi}_1} > 0$.*

(i) *If $p > 1$, then we have the following equivalent inequalities with the best constant factors $k(\gamma_1)$ and $k^p(\gamma_1)$:*

$$\int_c^d \int_a^b k_\gamma\big(u(x), v(y)\big) f(x) g(y)\,dx\,dy < \widetilde{k}(\gamma_1) \|f\|_{p,\widetilde{\Phi}_1} \|g\|_{q,\widetilde{\Psi}_1}, \tag{42.85}$$

$$\int_c^d \frac{v'(y)}{[v(y)]^{p\gamma_2+1}} \left(\int_a^b k_\gamma\big(u(x), v(y)\big) f(x)\,dx \right)^p dy < \widetilde{k}^p(\gamma_1) \|f\|^p_{p,\widetilde{\Phi}_1}; \tag{42.86}$$

(ii) *If $0 < p < 1$, then we have the equivalent reverses of* (42.85) *and* (42.86). *Moreover, if there exists a constant $\delta_0 > 0$ such that $\widetilde{k}(\gamma_1 + \delta_0) \in \mathbf{R}_+$, then the reverses of* (42.85) *and* (42.86) *possess the best constant factors $\widetilde{k}(\gamma_1)$ and $\widetilde{k}^p(\gamma_1)$, respectively.*

In view of Theorem 42.8, for $\widetilde{k}(\gamma_1) \in \mathbf{R}_+$, we define a Hilbert-type integral operator $T_\gamma : L_{p,\widetilde{\varphi}_1}(\mathbf{R}) \to L_{p,\widetilde{\psi}_1^{1-p}}(\mathbf{R})$ as follows: for $f \in L_{p,\widetilde{\varphi}_1}(\mathbf{R})$, there exists a unique representation $T_\gamma f \in L_{p,\widetilde{\psi}_1^{1-p}}(\mathbf{R})$, satisfying

$$T_\gamma f(y) := \int_{-\infty}^\infty k_\gamma(x,y) f(x)\,dx \quad (y \in \mathbf{R}). \tag{42.87}$$

Then it follows by (42.83) and (42.84) that

$$(T_\gamma f, g) < \widetilde{k}(\gamma_1) \|f\|_{p,\widetilde{\varphi}_1} \|g\|_{q,\widetilde{\psi}_1}, \tag{42.88}$$

$$\|T_\gamma f\|_{p,\widetilde{\psi}_1^{1-p}} < \widetilde{k}(\gamma_1) \|f\|_{p,\widetilde{\varphi}_1}, \tag{42.89}$$

where the constant factor $\widetilde{k}(\gamma_1)$ is the best possible. Hence, we still have

Theorem 42.9 *Suppose that the Hilbert-type integral operator T_γ is defined by* (42.87). *Then it follows*

$$\|T_\gamma\| := \sup_{f(\neq\theta)\in L_{p,\widetilde{\varphi}_1}(\mathbf{R})} \frac{\|T_\gamma f\|_{p,\widetilde{\psi}_1^{1-p}}}{\|f\|_{p,\widetilde{\varphi}_1}} = \widetilde{k}(\gamma_1), \tag{42.90}$$

where $\widetilde{k}(\gamma_1) = \int_0^\infty (k_\gamma(u,1) + k_\gamma(-u,1))u^{-\gamma_1-1}\,du$.

Remark 42.6 In Theorem 42.9, (i) if $\gamma = 0$, $\gamma_1 = \alpha$, $\gamma_2 = -\alpha$, $\varphi(x) = x^{p(1+\alpha)-1}$, $\psi(y) = y^{q(1-\alpha)-1}$ $(x, y \in \mathbf{R})$, then we define an operator $\widetilde{T}_0 : L_{p,\varphi}(\mathbf{R}) \to L_{p,\psi^{1-p}}(\mathbf{R})$ as follows: for $f(\geq 0) \in L_{p,\varphi}(\mathbf{R})$, there exists a unique representation $\widetilde{T}_0 f \in L_{p,\psi^{1-p}}(\mathbf{R})$, satisfying

$$\widetilde{T}_0 f(y) = \int_{-\infty}^\infty k_0(x,y) f(x)\,dx \quad (y \in \mathbf{R}). \tag{42.91}$$

We have

$$\|\widetilde{T}_0\| = \widetilde{k}_0(\alpha) := \int_0^\infty \big(k_0(u,1) + k_0(-u,1)\big)u^{-\alpha-1}\,du. \tag{42.92}$$

(ii) If $k_\gamma(x,y)$ is a symmetric function, $\beta \in \mathbf{R}$, $\gamma_1 = \gamma_2 = \frac{\gamma}{2}$, $\widetilde{\varphi}(x) = x^{p(1+\frac{\gamma}{2})-1}$, $\widetilde{\psi}(y) = y^{q(1+\frac{\gamma}{2})-1}$ $(x, y \in \mathbf{R})$, then we define an operator $\widetilde{T}_{\gamma,\beta} : L_{p,\widetilde{\varphi}}(\mathbf{R}) \to L_{p,\widetilde{\psi}^{1-p}}(\mathbf{R})$ as follows: for $f(\geq 0) \in L_{p,\widetilde{\varphi}}(\mathbf{R})$, there exists a unique representation $\widetilde{T}_{\gamma,\beta} f \in L_{p,\widetilde{\psi}^{1-p}}(\mathbf{R})$, satisfying $\widetilde{T}_{\gamma,\beta} f(y) = \int_{-\infty}^\infty k_\gamma(x,y) \arctan|\frac{x}{y}|^\beta f(x)\,dx$ $(y \in \mathbf{R})$. We have

$$\begin{aligned}\|\widetilde{T}_{\gamma,\beta}\| = \widetilde{k}_\beta(\gamma) &:= \frac{\pi}{4}\widetilde{k}\left(\frac{\gamma}{2}\right)\\ &= \frac{\pi}{4}\int_0^\infty \big(k_\gamma(u,1) + k_\gamma(-u,1)\big)u^{-\frac{\gamma}{2}-1}\,du.\end{aligned} \tag{42.93}$$

In fact, it follows that

$$\begin{aligned}\widetilde{k}_\beta(\gamma) &= \int_0^\infty \big(k_\gamma(u,1) + k_\gamma(-u,1)\big)\big(\arctan u^\beta\big)u^{-\frac{\gamma}{2}-1}\,du\\ &= \int_0^1 \big(k_\gamma(u,1) + k_\gamma(-u,1)\big)\big(\arctan u^\beta\big)u^{-\frac{\gamma}{2}-1}\,du\\ &\quad + \int_1^\infty \big(k_\gamma(u,1) + k_\gamma(-u,1)\big)\big(\arctan u^\beta\big)u^{-\frac{\gamma}{2}-1}\,du\\ &= \int_0^1 \big(k_\gamma(u,1) + k_\gamma(-u,1)\big)\big(\arctan u^\beta + \arctan u^{-\beta}\big)u^{-\frac{\gamma}{2}-1}\,du\end{aligned}$$

$$= \frac{\pi}{2}\int_0^1 \big(k_\gamma(u,1)+k_\gamma(-u,1)\big)u^{-\frac{\gamma}{2}-1}\,du = \frac{\pi}{4}\widetilde{k}\Big(\frac{\gamma}{2}\Big).$$

(iii) By virtue of Theorem 42.9, suppose that $k_\gamma\ (x,y)\ (\geq 0)$ is a homogeneous function of degree γ on $\mathbf{R}^2$.

(a) If we set

$$K_\gamma(x,y) := \begin{cases} k_\gamma(x,y), & 0<|x|\leq |y|,\\ 0, & |x|>|y|,\end{cases}$$

then we define the first class Hardy-type integral operator $T_\gamma^{(1)} : L_{p,\widetilde{\varphi}_1}(\mathbf{R}) \to L_{p,\widetilde{\psi}_1^{1-p}}(\mathbf{R})$ with the homogeneous kernel on $\mathbf{R}^2$ as follows: for $f \in L_{p,\widetilde{\varphi}_1}(\mathbf{R})$, there exists a unique representation $T_\gamma^{(1)} f \in L_{p,\widetilde{\psi}_1^{1-p}}(\mathbf{R})$, satisfying

$$T_\gamma^{(1)} f(y) = \int_{-\infty}^{\infty} K_\gamma(x,y)f(x)\,dx = \int_{-|y|}^{|y|} k_\gamma(x,y)f(x)\,dx \quad (y\in\mathbf{R}).$$

Hence, for $\int_0^1 (k_\gamma(u,1)+k_\gamma(-u,1))u^{-\gamma_1-1}\,du \in \mathbf{R}_+$, we obtain $\widetilde{\omega}_\gamma(y) = \widetilde{\varpi}_\gamma(x) = \widetilde{k}_1(\gamma_1) := \int_0^1 (k_\gamma(u,1)+k_\gamma(-u,1))u^{-\gamma_1-1}\,du$,

$$\|T_\gamma^{(1)}\| = \widetilde{k}_1(\gamma_1) = \int_0^1 \big(k_\gamma(u,1)+k_\gamma(-u,1)\big)u^{-\gamma_1-1}\,du. \tag{42.94}$$

(b) If we set

$$K_\gamma(x,y) := \begin{cases} 0, & 0<|x|\leq |y|,\\ k_\gamma(x,y), & |x|>|y|,\end{cases}$$

then we define the second class Hardy-type integral operator $T_\gamma^{(2)} : L_{p,\widetilde{\varphi}_1}(\mathbf{R}) \to L_{p,\widetilde{\psi}_1^{1-p}}(\mathbf{R})$, with the homogeneous kernel on $\mathbf{R}^2$ as follows: for $f \in L_{p,\widetilde{\varphi}_1}(\mathbf{R})$, there exists a unique representation $T_\gamma^{(2)} f \in L_{p,\widetilde{\psi}_1^{1-p}}(\mathbf{R})$, satisfying

$$\begin{aligned} T_\gamma^{(2)} f(y) &= \int_{-\infty}^{\infty} K_\gamma(x,y)f(x)\,dx \\ &= \int_{-\infty}^{-|y|} k_\gamma(x,y)f(x)\,dx + \int_{|y|}^{\infty} k_\gamma(x,y)f(x)\,dx \quad (y\in\mathbf{R}). \end{aligned}$$

Hence, for $\int_1^\infty (k_\gamma(u,1)+k_\gamma(-u,1))u^{-\gamma_1-1}\,du \in \mathbf{R}_+$, we obtain $\widetilde{\omega}_\gamma(y) = \widetilde{\varpi}_\gamma(x) = \widetilde{k}_2(\gamma_1) := \int_1^\infty (k_\gamma(u,1)+k_\gamma(-u,1))u^{-\gamma_1-1}\,du$,

$$\|T_\gamma^{(2)}\| = \widetilde{k}_2(\gamma_1) = \int_1^\infty \big(k_\gamma(u,1)+k_\gamma(-u,1)\big)u^{-\gamma_1-1}\,du. \tag{42.95}$$

42.3.2 The Case of a Non-homogeneous Kernel

Definition 42.7 For $p > 1$, $\frac{1}{p} + \frac{1}{q} = 1$, $\alpha \in \mathbf{R}$, $\widetilde{\varphi}_2(x) := |x|^{p(1-\alpha)-1}$, $\widetilde{\psi}_2(y) := |y|^{q(1-\alpha)-1}$ $(x, y \in \mathbf{R})$, and $\widetilde{\psi}_2^{1-p}(y) := |y|^{p\alpha-1}$, we define two normed function spaces as follows:

$$L_{p,\widetilde{\varphi}_2}(\mathbf{R}) := \left\{ f;\ \|f\|_{p,\widetilde{\varphi}_2} = \left\{ \int_{-\infty}^{\infty} |x|^{p(1-\alpha)-1} |f(x)|^p \, dx \right\}^{\frac{1}{p}} < \infty \right\},$$

$$L_{q,\widetilde{\psi}_2}(\mathbf{R}) := \left\{ g;\ \|g\|_{q,\widetilde{\psi}_2} = \left\{ \int_{-\infty}^{\infty} |y|^{q(1-\alpha)-1} |g(y)|^q \, dy \right\}^{\frac{1}{q}} < \infty \right\}.$$

If $h(u) \geq 0$ is a measurable function on $\mathbf{R}$, then we define a weight function $\widehat{\omega}(y)$ as follows:

$$\widehat{\omega}(y) := \int_{-\infty}^{\infty} h(xy) \frac{|y|^{\alpha}}{|x|^{1-\alpha}} \, dx \quad (y \in \mathbf{R}). \tag{42.96}$$

We have

Lemma 42.7 *If $y \in \mathbf{R} \backslash \{0\}$, then it follows*

$$\widehat{\omega}(y) = \widehat{K}(\alpha) := \int_0^{\infty} \big(h(u) + h(-u)\big) u^{\alpha-1} \, du. \tag{42.97}$$

Proof For $y < 0$, setting $u = xy$, $u = -xy$ in the following first and second integral, respectively, we find

$$\begin{aligned}
\widehat{\omega}(y) &= \int_{-\infty}^{0} h(xy) \frac{(-y)^{\alpha}}{(-x)^{1-\alpha}} \, dx + \int_0^{\infty} h(xy) \frac{(-y)^{\alpha}}{x^{1-\alpha}} \, dx \\
&= \int_0^{\infty} h(u) u^{\alpha-1} \, du + \int_0^{\infty} h(-u) u^{\alpha-1} \, du = \widehat{K}(\alpha).
\end{aligned}$$

In the same way, for $y > 0$, we can still find $\widehat{\omega}(y) = \widehat{K}(\alpha)$. The lemma is proved. □

Lemma 42.8 *Suppose that $p > 0\ (\neq 1)$, $\frac{1}{p} + \frac{1}{q} = 1$, $\alpha \in \mathbf{R}$, $h(u)$ is a non-negative measurable function on $\mathbf{R}$, $\widehat{K}(\alpha) = \int_0^{\infty} (h(u) + h(-u)) u^{\alpha-1} \, du \in \mathbf{R}_+$. We set*

$$\widehat{L}_k := \frac{1}{k} \int_1^{\infty} y^{-\frac{1}{k}-1} \left[\int_0^{y} \big(h(u) + h(-u)\big) u^{\alpha + \frac{1}{pk} - 1} \, du \right] dy. \tag{42.98}$$

(i) *If $p > 1$, then we have*

$$\lim_{k \to \infty} \widehat{L}_k = \widehat{K}(\alpha); \tag{42.99}$$

(ii) *If* $0 < p < 1$, *there exists a* $\delta_0 > 0$ *such that* $\widehat{K}(\alpha + \delta_0) \in \mathbf{R}_+$, *then we still have* (42.99).

Proof (i) For $p > 1$, by Fubini theorem, we obtain

$$\begin{aligned}
\widehat{L}_k &= \frac{1}{k}\int_1^{\infty} y^{-\frac{1}{k}-1}\left[\int_0^1 \big(h(u)+h(-u)\big)u^{\alpha+\frac{1}{pk}-1}\,du\right]dy \\
&\quad + \frac{1}{k}\int_1^{\infty} y^{-\frac{1}{k}-1}\left[\int_1^y \big(h(u)+h(-u)\big)u^{\alpha+\frac{1}{pk}-1}\,du\right]dy \\
&= \int_0^1 \big(h(u)+h(-u)\big)u^{\alpha+\frac{1}{pk}-1}\,du \\
&\quad + \frac{1}{k}\int_1^{\infty}\left(\int_u^{\infty}\frac{dy}{y^{\frac{1}{k}+1}}\right)\big(h(u)+h(-u)\big)u^{\alpha+\frac{1}{pk}-1}\,du \\
&= \int_0^1 \big(h(u)+h(-u)\big)u^{\alpha+\frac{1}{pk}-1}\,du \\
&\quad + \int_1^{\infty} \big(h(u)+h(-u)\big)u^{\alpha-\frac{1}{qk}-1}\,du. \qquad (42.100)
\end{aligned}$$

Then by Levi theorem, we have

$$\begin{aligned}
\widehat{K}(\alpha) &= \int_0^1 \lim_{k\to\infty}\big(h(u)+h(-u)\big)u^{\alpha+\frac{1}{pk}-1}\,du \\
&\quad + \int_1^{\infty} \lim_{k\to\infty}\big(h(u)+h(-u)\big)u^{\alpha-\frac{1}{qk}-1}\,du \\
&= \lim_{k\to\infty}\left[\int_0^1 \big(h(u)+h(-u)\big)u^{\alpha+\frac{1}{pk}-1}\,du\right. \\
&\quad \left. + \int_1^{\infty} \big(h(u)+h(-u)\big)u^{\alpha-\frac{1}{qk}-1}\,du\right] = \lim_{k\to\infty}\widehat{L}_k.
\end{aligned}$$

Hence, (42.99) is valid.

(ii) If $0 < p < 1$, $q < 0$, then for large enough k satisfying $\frac{1}{|q|k} \le \delta_0$, we have $u^{\frac{-1}{qk}} \le u^{\delta_0} (u \in (1,\infty))$ and

$$0 \le \big(h(u)+h(-u)\big)u^{\alpha-\frac{1}{qk}-1} \le \big(h(u)+h(-u)\big)u^{\alpha+\delta_0-1} \quad \big(u \in (1,\infty)\big).$$

Since we have $\int_1^{\infty}(h(u)+h(-u))u^{\alpha+\delta_0-1}\,du \le \widehat{K}(\alpha+\delta_0) < \infty$, then by Lebesgue dominated convergence theorem, it follows

$$\lim_{k\to\infty}\int_1^{\infty}\big(h(u)+h(-u)\big)u^{\alpha-\frac{1}{qk}-1}\,du = \int_1^{\infty}\big(h(u)+h(-u)\big)u^{\alpha-1}\,du.$$

Since $\{u^{\alpha+\frac{1}{pk}-1}\}_{k=1}^{\infty}$ is increasing for $u \in (0,1)$, by Levi theorem, we have

$$\lim_{k\to\infty}\int_0^1 \big(h(u)+h(-u)\big)u^{\alpha+\frac{1}{pk}-1}\,du = \int_0^1 \big(h(u)+h(-u)\big)u^{\alpha-1}\,du.$$

Hence, by (42.100), it follows that (42.99) is valid. The lemma is proved. □

Theorem 42.10 *Suppose that $p > 0$ ($p \neq 1$), $\frac{1}{p}+\frac{1}{q}=1$, $\alpha \in \mathbf{R}$, $h(u)$ is a non-negative measurable function on $\mathbf{R}$, $\widehat{K}(\alpha) \in \mathbf{R}_+$, and $f(\geq 0) \in L_{p,\widetilde{\varphi}_2}(\mathbf{R})$, $g(\geq 0) \in L_{q,\widetilde{\psi}_2}(\mathbf{R})$ such that $\|f\|_{p,\widetilde{\varphi}_2} > 0$, $\|g\|_{q,\widetilde{\psi}_2} > 0$.*

(i) *If $p > 1$, then we have the following equivalent inequalities with the best constant factors $\widehat{K}(\alpha)$ and $\widehat{K}^p(\alpha)$:*

$$I := \int_{-\infty}^{\infty}\int_{-\infty}^{\infty} h(xy)f(x)g(y)\,dx\,dy < \widehat{K}(\alpha)\|f\|_{p,\widetilde{\varphi}_2}\|g\|_{q,\widetilde{\psi}_2}, \tag{42.101}$$

$$J := \int_{-\infty}^{\infty} |y|^{p\alpha-1}\left(\int_{-\infty}^{\infty} h(xy)f(x)\,dx\right)^p dy < \widehat{K}^p(\alpha)\|f\|_{p,\widetilde{\varphi}_2}^p; \tag{42.102}$$

(ii) *If $0 < p < 1$, then we have the equivalent reverses of* (42.101) *and* (42.102). *Moreover, if there exists a constant $\delta_0 > 0$ such that $\widehat{K}(\alpha+\delta_0) \in \mathbf{R}_+$, then the reverses of* (42.101) *and* (42.102) *possess the best constant factors $\widehat{K}(\alpha)$ and $\widehat{K}^p(\alpha)$, respectively.*

Proof For $m = n = 1$, $A = B = \mathbf{R}$, $H(x,y) = h(xy)$, $a = b = 1-\alpha$, and $k_1 = k_2 = \widehat{K}(\alpha)$ in Theorem 42.3, we have equivalent inequalities (42.101) and (42.102) (for $p > 1$), and the equivalent reverses of (42.101) and (42.102) (for $0 < p < 1$).

(i) For $p > 1$, $k \in \mathbf{N}$, we set $f_k(x)$ and $g_k(y)$ as follows:

$$f_k(x) := \begin{cases} x^{\alpha+\frac{1}{pk}-1}, & x \in (0,1), \\ (-x)^{\alpha+\frac{1}{pk}-1}, & x \in (-1,0), \\ 0, & x \in (-\infty,-1]\cup[1,\infty), \end{cases}$$

$$g_k(y) := \begin{cases} y^{\alpha-\frac{1}{qk}-1}, & y \in [1,\infty), \\ 0, & y \in (-1,1), \\ (-y)^{\alpha-\frac{1}{qk}-1}, & y \in (-\infty,-1]. \end{cases}$$

Then we find $\|f_k\|_{p,\widetilde{\varphi}_2}\|g_k\|_{q,\widetilde{\psi}_2} = 2k$, and

$$\widetilde{I}_k := \int_{-\infty}^{\infty}\int_{-\infty}^{\infty} h(xy)f_k(x)g_k(y)\,dx\,dy = I_1 + I_2,$$

$$I_1 := \int_1^{\infty} y^{\alpha-\frac{1}{qk}-1}\left(\int_{-\infty}^{\infty} h(xy)f_k(x)\,dx\right)dy,$$

$$I_2 := \int_{-\infty}^{-1} (-Y)^{\alpha-\frac{1}{qk}-1} \left(\int_{-\infty}^{\infty} h(xY) f_k(x)\,dx \right) dY.$$

Setting $y = -Y$ in I_2, it follows $I_2 = \int_1^\infty y^{\alpha-\frac{1}{qk}-1} (\int_{-\infty}^{\infty} h(-xy) f_k(x)\,dx)\,dy$. We find

$$\begin{aligned}
\widehat{I}_k &= \int_1^\infty y^{\alpha-\frac{1}{qk}-1} \left[\int_{-\infty}^{\infty} \big(h(xy) + h(-xy)\big) f_k(x)\,dx \right] dy \\
&= 2 \int_1^\infty y^{\alpha-\frac{1}{qk}-1} \left[\int_0^1 \big(h(xy) + h(-xy)\big) x^{\alpha+\frac{1}{pk}-1}\,dx \right] dy \\
&\overset{u=xy}{=} 2 \int_1^\infty y^{-\frac{1}{k}-1} \left[\int_0^y \big(h(u) + h(-u)\big) u^{\alpha+\frac{1}{pk}-1}\,du \right] dy = 2k\widehat{L}_k. \qquad (42.103)
\end{aligned}$$

If there exists a positive number K_0 with $K_0 \le \widehat{K}(\alpha)$ such that (42.101) is valid as we replace $\widehat{K}(\alpha)$ by K_0, then, in particular, in view of (42.103), it follows $\widehat{L}_k = \frac{1}{2k}\widehat{I}_k < \frac{1}{2k} K_0 \|f_k\|_{p,\widetilde{\varphi}_2} \|g_k\|_{q,\widetilde{\psi}_2} = K_0$. By (42.99), we have $\widehat{K}(\alpha) \le K_0$ $(k \to \infty)$, Hence $K_0 = \widehat{K}(\alpha)$ is the best value of (42.101).

We affirm that the constant factor in (42.102) is the best possible, otherwise by inequality $I \le J^{\frac{1}{p}} \|g\|_{q,\widetilde{\psi}_2}$, we can get a contradiction that the constant in (42.101) is not the best possible.

(ii) For $0 < p < 1, q < 0$, if there exists a positive number K_1 $(\ge \widehat{K}(\alpha))$ such that the reverse of (42.101) is valid as we replace $\widehat{K}(\alpha)$ by K_1, then, in particular, still setting f_k, g_k as in (i), we have $\widehat{L}_k = \frac{1}{2k}\widehat{I}_k > \frac{1}{2k} K_1 \|f_k\|_{p,\widetilde{\varphi}_2} \|g_k\|_{q,\widetilde{\psi}_2} = K_1$. Then by (42.99), it follows that $\widehat{K}(\alpha) \ge K_1 (k \to \infty)$. Hence the constant factor in the reverse of (42.101) is the best possible.

We affirm that the constant factor in the reverse of (42.102) is the best possible, otherwise, by the reverse inequality $I \ge J^{\frac{1}{p}} \|g\|_{q,\widetilde{\psi}_2}$, we can get a contradiction that the constant factor in the reverse of (42.101) is not the best possible. The theorem is proved. □

Assuming that $u(x)$ $(x \in (a,b))$ and $v(y)$ $(y \in (c,d))$ are strict increasing differentiable functions, satisfying $u(a^+) = v(c^+) = -\infty$, $u(b^-) = v(d^-) = \infty$, $\alpha \in \mathbf{R}$, setting

$$\widehat{\Phi}_2(x) := \frac{[u(x)]^{p(1-\alpha)-1}}{[u'(x)]^{p-1}} \quad \big(x \in (a,b)\big),$$

$$\widehat{\Psi}_2(y) := \frac{[v(y)]^{q(1-\alpha)-1}}{[v'(y)]^{q-1}} \quad \big(y \in (c,d)\big),$$

replacing $x(y)$ by $u(x)$ $(v(y))$ in (42.101) and (42.102), after calculation, and after replacing $f(u(x))u'(x)$, $(g(v(y))v'(y))$ by $f(x)$ $(g(y))$, we find

Corollary 42.4 *Suppose that* $p > 0\ (\neq 1)$, $\frac{1}{p} + \frac{1}{q} = 1$, $\alpha \in \mathbf{R}$, $h(u)$ *is a non-negative measurable function on* $\mathbf{R}$, $\widehat{K}(\alpha) = \int_0^\infty (h(u) + h(-u))u^{\alpha-1}\,du \in \mathbf{R}_+$, *and* $f(\geq 0) \in L_{p,\widehat{\Phi}_2}(a,b)$, $g(\geq 0) \in L_{q,\widehat{\Psi}_2}(c,d)$ *such that* $\|f\|_{p,\widehat{\Phi}_2} > 0$, $\|g\|_{q,\widehat{\Psi}_2} > 0$.

(i) *If* $p > 1$, *then we have the following equivalent inequalities with the best constant factors* $\widehat{K}(\alpha)$ *and* $\widehat{K}^p(\alpha)$:

$$\int_c^d \int_a^b h\big(u(x)v(y)\big) f(x)g(y)\,dx\,dy < \widehat{K}(\alpha)\|f\|_{p,\widehat{\Phi}_2}\|g\|_{q,\widehat{\Psi}_2}, \tag{42.104}$$

$$\int_c^d \frac{v'(y)}{[v(y)]^{1-p\alpha}} \left(\int_a^b h\big(u(x)v(y)\big) f(x)\,dx\right)^p dy < \widehat{K}^p(\alpha)\|f\|^p_{p,\widehat{\Phi}_2}; \tag{42.105}$$

(ii) *If* $0 < p < 1$, *then we have the equivalent reverses of* (42.104) *and* (42.105). *Moreover, if there exists a constant* $\delta_0 > 0$ *such that* $\widehat{K}(\alpha + \delta) \in \mathbf{R}_+$, *then the reverses of* (42.104) *and* (42.105) *possess the best constant factors* $\widehat{K}(\alpha)$ *and* $\widehat{K}^p(\alpha)$, *respectively.*

In view of Theorem 42.10, for $\widehat{K}(\alpha) \in \mathbf{R}_+$, we define a Hilbert-type integral operator $\widehat{T}_\alpha : L_{p,\widetilde{\varphi}_2}(\mathbf{R}) \to L_{p,\widetilde{\psi}_2^{1-p}}(\mathbf{R})$ as follows: for $f \in L_{p,\widetilde{\varphi}_2}(\mathbf{R})$, there exists a unique representation $\widehat{T}_\alpha f \in L_{p,\widetilde{\varphi}_2^{1-p}}(\mathbf{R})$, satisfying

$$\widehat{T}_\alpha f(y) := \int_{-\infty}^\infty h(xy) f(x)\,dx \quad (y \in \mathbf{R}). \tag{42.106}$$

Then it follows by (42.101) and (42.102) that

$$(\widehat{T}_\alpha f, g) < \widehat{K}(\alpha)\|f\|_{p,\widetilde{\varphi}_2}\|g\|_{q,\widetilde{\psi}_2}, \tag{42.107}$$

$$\|\widehat{T}_\alpha f\|_{p,\widetilde{\psi}_2^{1-p}} < \widehat{K}(\alpha)\|f\|_{p,\widetilde{\varphi}_2}, \tag{42.108}$$

where the constant factor $\widehat{K}(\alpha)$ is the best possible. Hence, we still have

Theorem 42.11 *Suppose that the Hilbert-type integral operator* $\widehat{T}_\alpha$ *is defined by* (42.106). *Then it follows*

$$\|\widehat{T}_\alpha\| := \sup_{f(\neq\theta)\in L_{p,\widetilde{\varphi}_2}(\mathbf{R})} \frac{\|\widehat{T}_\alpha f\|_{p,\widetilde{\psi}_2^{1-p}}}{\|f\|_{p,\widetilde{\varphi}_2}} = \widehat{K}(\alpha), \tag{42.109}$$

where $\widehat{K}(\alpha) = \int_0^\infty (h(u) + h(-u))u^{\alpha-1}\,du$.

Remark 42.7 (i) In Theorem 42.11, if $h(u) = k_\gamma(1,u)$, $\alpha = -\frac{\gamma}{2}$, $\widehat{\varphi}(x) = |x|^{p(1+\frac{\gamma}{2})-1}$, $\widehat{\psi}(y) = |y|^{q(1+\frac{\gamma}{2})-1}$ $(x, y \in \mathbf{R})$, $\widehat{k}(\frac{\gamma}{2}) = \int_0^\infty (k_\gamma(1,u) + k_\gamma(1,-u)) \times$

$u^{-\frac{\gamma}{2}-1}\,du \in \mathbf{R}_+$, then we define an operator $\widehat{T}_{\gamma/2} : L_{p,\widehat{\varphi}}(\mathbf{R}) \to L_{p,\widehat{\psi}^{1-p}}(\mathbf{R})$ as: for $f \in L_{p,\widehat{\varphi}}(\mathbf{R})$, there exists a unique representation $\widehat{T}_{\gamma/2} f \in L_{p,\widehat{\psi}^{1-p}}(\mathbf{R})$, satisfying

$$\widehat{T}_{\gamma/2} f(y) = \int_{-\infty}^{\infty} k_\gamma(1, xy) f(x)\,dx \quad (y \in \mathbf{R}).$$

We have

$$\|\widehat{T}_{\gamma/2}\| := \sup_{f(\neq\theta)\in L_{p,\widehat{\varphi}}(\mathbf{R})} \frac{\|\widehat{T}_{\gamma/2} f\|_{p,\widehat{\psi}^{1-p}}}{\|f\|_{p,\widehat{\varphi}}} = \widehat{k}\left(\frac{\gamma}{2}\right). \tag{42.110}$$

(ii) We still can write some similar results for $h(u) = k_\gamma(u,1)$ $(\gamma \in \mathbf{R})$.

(iii) By virtue of Theorem 42.11, suppose that $h(u)$ is a non-negative measurable function on $\mathbf{R}$.

(a) If we set

$$H(xy) := \begin{cases} h(xy), & 0 < |x| \le 1/|y|, \\ 0, & |x| > 1/|y|, \end{cases}$$

then we define the first class Hardy-type integral operator $\widehat{T}_\alpha^{(1)} : L_{p,\widetilde{\varphi}_2}(\mathbf{R}) \to L_{p,\widetilde{\psi}_2^{1-p}}(\mathbf{R})$, with the non-homogeneous kernel on $\mathbf{R}^2$ as follows: for $f \in L_{p,\widetilde{\varphi}_2}(\mathbf{R})$, there exists a unique representation $\widehat{T}_\alpha^{(1)} f \in L_{p,\widetilde{\psi}_2^{1-p}}(\mathbf{R})$, satisfying

$$\widehat{T}_\alpha^{(1)} f(y) = \int_{-\infty}^{\infty} H(xy) f(x)\,dx = \int_{-1/|y|}^{1/|y|} h(xy) f(x)\,dx \quad (y \in \mathbf{R}).$$

Hence, for $\int_0^1 (h(u)+h(-u))u^{\alpha-1}\,du \in \mathbf{R}_+$, we obtain $\widehat{\omega}(y) = \widehat{K}_1(\alpha) := \int_0^1 (h(u)+h(-u))u^{\alpha-1}\,du$,

$$\|\widehat{T}_\alpha^{(1)}\| = \widehat{K}_1(\alpha) = \int_0^1 \big(h(u) + h(-u)\big) u^{\alpha-1}\,du. \tag{42.111}$$

(b) If we set

$$H(xy) := \begin{cases} 0, & 0 < |x| \le 1/|y|, \\ h(xy), & |x| > 1/|y|, \end{cases}$$

then we define the second class Hardy-type integral operator $\widehat{T}_\alpha^{(2)} : L_{p,\widetilde{\varphi}_2}(\mathbf{R}) \to L_{p,\widetilde{\psi}_2^{1-p}}(\mathbf{R})$, with the non-homogeneous kernel on $\mathbf{R}$ as follows: for $f \in L_{p,\widetilde{\varphi}_2}(\mathbf{R})$, there exists a unique representation $\widehat{T}_\alpha^{(2)} f \in L_{p,\widetilde{\psi}_2^{1-p}}(\mathbf{R})$, satisfying

$$\widehat{T}_\alpha^{(2)} f(y) = \int_{-\infty}^{\infty} H(xy) f(x)\,dx = \int_{\frac{1}{|y|}}^{\infty} h(xy) f(x)\,dx + \int_{-\infty}^{-\frac{1}{|y|}} h(xy) f(x)\,dx.$$

Hence, for $\int_1^\infty (h(u)+h(-u))u^{\alpha-1}\,du \in \mathbf{R}_+$, $\widehat{\omega}(y)=\widehat{K}_2(\alpha):=\int_1^\infty (h(u)+h(-u))u^{\alpha-1}\,du$,

$$\|\widehat{T}_\alpha^{(2)}\| = \widehat{K}_2(\alpha) = \int_1^\infty \big(h(u)+h(-u)\big)u^{\alpha-1}\,du. \tag{42.112}$$

42.3.3 Some Particular Examples

Example 42.5 If $\gamma=-\lambda$, $0<\lambda<1$, $k_{-\lambda}(x,y)=\frac{1}{|x+y|^\lambda}$ $(x,y\in\mathbf{R})$, $\lambda_1,\lambda_2>0$, $\lambda_1+\lambda_2=\lambda$, then we find

$$\begin{aligned}\widehat{k}(\lambda_1) &:= \int_0^\infty \left[\frac{1}{|u+1|^\lambda}+\frac{1}{|-u+1|^\lambda}\right]u^{\lambda_1-1}\,du\\ &= B(\lambda_1,\lambda_2)+B(1-\lambda,\lambda_1)+B(1-\lambda,\lambda_2)\in\mathbf{R}_+.\end{aligned}$$

(i) For $\widehat{\varphi}(x):=|x|^{p(1-\lambda_1)-1}$, $\widehat{\psi}(y):=|y|^{q(1-\lambda_2)-1}$, by Theorem 42.11, define a Hilbert-type integral operator $\widehat{T}_{-\lambda}: L_{p,\widehat{\varphi}}(\mathbf{R})\to L_{p,\widehat{\psi}^{1-p}}(\mathbf{R})$ as follows: for $f\in L_{p,\widehat{\varphi}}(\mathbf{R})$, there exists a unique representation $\widehat{T}_{-\lambda}f\in L_{p,\widehat{\psi}^{1-p}}(\mathbf{R})$, satisfying $\widehat{T}_{-\lambda}f(y)=\int_{-\infty}^\infty \frac{1}{|x+y|^\lambda}f(x)\,dx$ $(y\in\mathbf{R})$. Then we have (cf. [35], [39])

$$\|\widehat{T}_{-\lambda}\| = B(\lambda_1,\lambda_2)+B(1-\lambda,\lambda_1)+B(1-\lambda,\lambda_2). \tag{42.113}$$

(ii) For $\lambda_1=\lambda_2=\frac{\lambda}{2}$, $\widetilde{\varphi}(x):=|x|^{p(1-\frac{\lambda}{2})-1}$, $\widetilde{\psi}(y):=|y|^{q(1-\frac{\lambda}{2})-1}$, $\beta\in\mathbf{R}$, by Remark 42.6(ii), define an operator $\widehat{T}_{-\lambda,\beta}: L_{p,\widetilde{\varphi}}(\mathbf{R})\to L_{p,\widetilde{\psi}^{1-p}}(\mathbf{R})$ as follows: for $f\in L_{p,\widetilde{\varphi}}(\mathbf{R})$, there exists a unique representation $\widehat{T}_{-\lambda,\beta}f\in L_{p,\widehat{\psi}^{1-p}}(\mathbf{R})$, satisfying $\widehat{T}_{-\lambda,\beta}f(y)=\int_{-\infty}^\infty \frac{1}{|x+y|^\lambda}\arctan|\frac{x}{y}|^\beta f(x)\,dx$ $(y\in\mathbf{R})$. Then we have

$$\|\widehat{T}_{-\lambda,\beta}\| = \frac{\pi}{4}\left[B\left(\frac{\lambda}{2},\frac{\lambda}{2}\right)+2B\left(1-\lambda,\frac{\lambda}{2}\right)\right]. \tag{42.114}$$

By Remark 42.7(i), define an operator $T_{-\lambda}: L_{p,\widetilde{\varphi}}(\mathbf{R})\to L_{p,\widetilde{\psi}^{1-p}}(\mathbf{R})$ as follows: for $f\in L_{p,\widetilde{\varphi}}(\mathbf{R})$, there exists a unique representation $T_{-\lambda}f\in L_{p,\widehat{\psi}^{1-p}}(\mathbf{R})$, satisfying $T_{-\lambda}f(y)=\int_{-\infty}^\infty \frac{1}{|1+xy|^\lambda}f(x)\,dx$ $(y\in\mathbf{R})$. Then we have (cf. [40])

$$\|T_{-\lambda}\| = B\left(\frac{\lambda}{2},\frac{\lambda}{2}\right)+2B\left(1-\lambda,\frac{\lambda}{2}\right). \tag{42.115}$$

Example 42.6 If $\gamma=-2$, $\gamma_1=\gamma_2=-1$, $b,c\in\mathbf{R}$, $|b|<|c|$, $k_{-2}(x,y)=\frac{1}{x^2+2bxy+c^2y^2}$ $(x,y\in\mathbf{R})$, then we find

$$k(1)=\int_0^\infty \left(\frac{1}{u^2+2bu+c^2}+\frac{1}{u^2-2bu+c^2}\right)du=\frac{\pi}{\sqrt{c^2-b^2}}.$$

Let $\widetilde{\varphi}(x):=|x|^{-1}$, $\widetilde{\psi}(y):=|y|^{-1}$.

(i) By Theorem 42.9, define a Hilbert-type integral operator $\widehat{T}_{-2}: L_{p,\widetilde{\varphi}}(\mathbf{R}) \to L_{p,\widetilde{\psi}^{1-p}}(\mathbf{R})$ as follows: for $f \in L_{p,\widetilde{\varphi}}(\mathbf{R})$, there exists a unique representation $\widehat{T}_{-2}f \in L_{p,\widehat{\psi}^{1-p}}(\mathbf{R})$, satisfying $\widehat{T}_{-2}f(y) = \int_{-\infty}^{\infty} \frac{1}{x^2+2bxy+c^2y^2} f(x)\,dx$ $(y \in \mathbf{R})$. Then we have

$$\|\widehat{T}_{-2}\| = \frac{\pi}{\sqrt{c^2-b^2}}. \tag{42.116}$$

(ii) By Remark 42.6(ii), for $\beta \in \mathbf{R}$, define an operator $\widehat{T}_{-2,\beta}: L_{p,\widetilde{\varphi}}(\mathbf{R}) \to L_{p,\widetilde{\psi}^{1-p}}(\mathbf{R})$ as follows: for $f \in L_{p,\widetilde{\varphi}}(\mathbf{R})$, there exists a unique representation $\widehat{T}_{-2,\beta}f \in L_{p,\widetilde{\psi}^{1-p}}(\mathbf{R})$, satisfying $\widehat{T}_{-2,\beta}f(y) = \int_{-\infty}^{\infty} \frac{1}{x^2+2bxy+c^2y^2} \times \arctan|\frac{x}{y}|^{\beta} f(x)\,dx$ $(y \in \mathbf{R})$. Then we have

$$\|\widehat{T}_{-2,\beta}\| = \frac{\pi^2}{4\sqrt{c^2-b^2}}. \tag{42.117}$$

(iii) By Remark 42.7(i), define an operator $T_{-2}: L_{p,\widetilde{\varphi}}(\mathbf{R}) \to L_{p,\widetilde{\psi}^{1-p}}(\mathbf{R})$ as follows: for $f \in L_{p,\widetilde{\varphi}}(\mathbf{R})$, there exists a unique representation $T_{-2}f \in L_{p,\widetilde{\psi}^{1-p}}(\mathbf{R})$, satisfying $T_{-2}f(y) = \int_{-\infty}^{\infty} \frac{1}{1+2bxy+(cxy)^2} f(x)\,dx$ $(y \in \mathbf{R})$. Then we have $\|T_{-2}\| = \frac{\pi}{\sqrt{c^2-b^2}}$.

Example 42.7 If $\gamma = 0$, $\alpha \in [0, 1)$, $\gamma_1 = \alpha$, $\gamma_2 = -\alpha$, $0 < \alpha_1 < \alpha_2 < \pi$,

$$k_0(x, y) = \left|\ln \frac{x^2 + 2xy\cos\alpha_1 + y^2}{x^2 + 2xy\cos\alpha_2 + y^2}\right| \quad (x, y \in \mathbf{R}),$$

then we find

$$\begin{aligned}
\widetilde{k}_0(\alpha) &= \int_0^{\infty} \left(\ln \frac{u^2 + 2u\cos\alpha_1 + 1}{u^2 + 2u\cos\alpha_2 + 1} + \ln \frac{u^2 - 2u\cos\alpha_2 + 1}{u^2 - 2u\cos\alpha_1 + 1}\right) u^{-\alpha-1}\,du \\
&= k_1 + k_2, \qquad\qquad (42.118) \\
k_1 &= \int_0^{\infty} \ln\bigl(u^2 + 2u\cos\alpha_1 + 1\bigr) u^{-\alpha-1}\,du \\
&\quad - \int_0^{\infty} \ln\bigl(u^2 + 2u\cos\alpha_2 + 1\bigr) u^{-\alpha-1}\,du, \\
k_2 &= \int_0^{\infty} \ln\bigl(u^2 + 2u\cos(\pi - \alpha_2) + 1\bigr) u^{-\alpha-1}\,du \\
&\quad - \int_0^{\infty} \ln\bigl(u^2 + 2u\cos(\pi - \alpha_1) + 1\bigr) u^{-\alpha-1}\,du.
\end{aligned}$$

(i) For $\alpha \in (0, 1)$, we obtain

$$\int_0^{\infty} \ln\bigl(u^2 + 2u\cos\alpha_1 + 1\bigr) u^{-\alpha-1}\,du = \frac{1}{-\alpha} \int_0^{\infty} \ln\bigl(u^2 + 2u\cos\alpha_1 + 1\bigr)\,du^{-\alpha}$$

$$= \frac{1}{-\alpha}\left[u^{-\alpha}\ln\left(u^2+2u\cos\alpha_1+1\right)\Big|_0^\infty - \int_0^\infty u^{-\alpha}\,d\ln\left(u^2+2u\cos\alpha_1+1\right)\right]$$

$$= \frac{2}{\alpha}\int_0^\infty \frac{(u+\cos\alpha_1)u^{-\alpha}}{u^2+2u\cos\alpha_1+1}\,du.$$

By the following formula (cf. [41])

$$\int_0^\infty f(x)x^{p-1}\,dx = \frac{2\pi i}{1-e^{2\pi pi}}\sum_{i=1}^{n} Res\left\{f(z)z^{p-1}, z_i\right\}, \tag{42.119}$$

where z_i $(i = 1, \ldots, n)$ are all the polar points of $f(z)$ and $\int_0^\infty f(x)x^{p-1}\,dx \in \mathbf{R}$, we find for $z_1 = -e^{i\alpha_1}$ and $z_2 = -e^{-i\alpha_1}$ that

$$\int_0^\infty \frac{(u+\cos\alpha_1)u^{-\alpha}}{u^2+2u\cos\alpha_1+1}\,du$$

$$= \frac{2\pi i}{1-e^{2\pi(1-\alpha)i}}\left[Res\left\{\frac{(z+\cos\alpha_1)z^{-\alpha}}{(z-z_1)(z-z_2)}, z_1\right\}\right.$$

$$\left. + Res\left\{\frac{(z+\cos\alpha_1)z^{-\alpha}}{(z-z_1)(z-z_2)}, z_2\right\}\right]$$

$$= \frac{2\pi i}{1-e^{2\pi(1-\alpha)i}}\left[\frac{(z_1+\cos\alpha_1)z_1^{-\alpha}}{z_1-z_2} + \frac{(z_2+\cos\alpha_1)z_2^{-\alpha}}{z_2-z_1}\right]$$

$$= \frac{-\pi}{e^{i\pi(1-\alpha)}\sin\pi(1-\alpha)}\frac{e^{-i\pi\alpha}}{z_1-z_2}$$

$$\times\left[(z_1-z_2)\cos\alpha_1(-\alpha) + i(z_1+z_2+2\cos\alpha_1)\sin\alpha_1(-\alpha)\right]$$

$$= \frac{\pi\cos\alpha_1(-\alpha)}{\sin\pi(1-\alpha)} = \frac{\pi\cos\alpha_1\alpha}{\sin\pi\alpha}.$$

Then we have $k_1 = \frac{2\pi(\cos\alpha_1\alpha-\cos\alpha_2\alpha)}{\alpha\sin\pi\alpha}$. Similarly, we find $k_2 = \frac{2\pi[\cos(\pi-\alpha_2)\alpha-\cos(\pi-\alpha_1)\alpha]}{\alpha\sin\pi\alpha}$. Hence by (42.118), we obtain

$$\widetilde{k}_0(\alpha) = \frac{4\pi}{\alpha\cos\frac{\pi\alpha}{2}}\sin\frac{\alpha_2-\alpha_1}{2}\alpha\cos\frac{\pi-\alpha_2-\alpha_1}{2}\alpha. \tag{42.120}$$

(ii) For $\alpha = 0$, we can prove that

$$\widetilde{k}_0(0) = 2\pi(\alpha_2-\alpha_1) = \lim_{\alpha\to0^+}\widetilde{k}_0(\alpha). \tag{42.121}$$

In fact, in (42.118), setting

$$f(u) = \ln\frac{u^2+2u\cos\alpha_1+1}{u^2+2u\cos\alpha_2+1} + \ln\frac{u^2-2u\cos\alpha_2+1}{u^2-2u\cos\alpha_1+1},$$

yields

$$\widetilde{k}_0(\alpha) = \int_0^\infty f(u)u^{-\alpha-1}\,du = \int_0^1 f(u)u^{-\alpha-1}\,du + \int_1^\infty f(u)u^{-\alpha-1}\,du.$$

For $\delta_0 \in (0, 1)$, $0 < \alpha \le \delta_0$, $|f(u)u^{-\alpha-1}| \le f(u)u^{-\delta_0-1}$, $u \in (0, 1]$, $\int_0^1 f(u)u^{-\delta_0-1} \le \widetilde{k}_0(\delta_0) < \infty$, by Lebesgue dominated convergence theorem, $\lim_{\alpha\to 0^+}\int_0^1 f(u)u^{-\alpha-1}\,du = \int_0^1 f(u)u^{-1}\,du$. By Levi theorem, we still have $\lim_{\alpha\to 0^+}\int_1^\infty f(u)u^{-\alpha-1}\,du = \int_1^\infty f(u)u^{-1}\,du$. Hence by (42.120), we have $\widetilde{k}_0(0) = \lim_{\alpha\to 0^+}\widetilde{k}_0(\alpha) = 2\pi(\alpha_2 - \alpha_1)$.

(a) Setting $\varphi(x) = |x|^{p(1+\alpha)-1}$, $\psi(y) = |y|^{q(1-\alpha)-1}$ $(x, y \in \mathbf{R})$, we define an operator $\widetilde{T}_0 : L_{p,\varphi}(\mathbf{R}) \to L_{p,\psi^{1-p}}(\mathbf{R})$ as follows: for $f \in L_{p,\varphi}(\mathbf{R})$, there exists a unique representation $\widetilde{T}_0 f \in L_{p,\psi^{1-p}}(\mathbf{R})$, satisfying

$$\widetilde{T}_0 f(y) = \int_{-\infty}^\infty \left|\ln\frac{x^2 + 2xy\cos\alpha_1 + y^2}{x^2 + 2xy\cos\alpha_2 + y^2}\right| f(x)\,dx \quad (y \in \mathbf{R}).$$

Then by Remark 42.6(i), we have (cf. [42])

$$\|\widetilde{T}_0\| = \frac{4\pi}{\alpha\cos\frac{\pi\alpha}{2}}\sin\frac{\alpha_2 - \alpha_1}{2}\alpha\cos\frac{\pi - \alpha_2 - \alpha_1}{2}\alpha. \tag{42.122}$$

(b) Setting $\varphi_0(x) = |x|^{p-1}$, $\psi_0(y) = |y|^{q-1}$ $(x, y \in \mathbf{R})$, for $\beta \in \mathbf{R}$, by Remark 42.6(ii), we define an operator $\widetilde{T}_{0,\beta} : L_{p,\varphi_0}(\mathbf{R}) \to L_{p,\psi_0^{1-p}}(\mathbf{R})$ as follows: for $f \in L_{p,\varphi_0}(\mathbf{R})$, there exists a unique representation $\widetilde{T}_0 f \in L_{p,\psi_0^{1-p}}(\mathbf{R})$, satisfying

$$\widetilde{T}_{0,\beta} f(y) = \int_{-\infty}^\infty \left|\ln\frac{x^2 + 2xy\cos\alpha_1 + y^2}{x^2 + 2xy\cos\alpha_2 + y^2}\right| \arctan\left|\frac{x}{y}\right|^\beta f(x)\,dx \quad (y \in \mathbf{R}).$$

Then we have

$$\|\widetilde{T}_{0,\beta}\| = \frac{\pi}{4}\widetilde{k}_0(0) = \frac{\pi^2}{2}(\alpha_2 - \alpha_1). \tag{42.123}$$

By Remark 42.7(i), define an operator $T_0 : L_{p,\varphi_0}(\mathbf{R}) \to L_{p,\psi_0^{1-p}}(\mathbf{R})$ as follows: for $f \in L_{p,\varphi_0}(\mathbf{R})$, there exists a unique representation $T_0 f \in L_{p,\psi_0^{1-p}}(\mathbf{R})$, satisfying

$$T_0 f(y) = \int_{-\infty}^\infty \left|\ln\frac{1 + 2xy\cos\alpha_1 + (xy)^2}{1 + 2xy\cos\alpha_2 + (xy)^2}\right| f(x)\,dx \quad (y \in \mathbf{R}).$$

Then we have

$$\|\widetilde{T}_{0,\beta}\| = \widetilde{k}_0(0) = 2\pi(\alpha_2 - \alpha_1). \tag{42.124}$$

Example 42.8 If $\gamma=-2$, $\alpha\in(-1,1)$, $\gamma_1=\alpha-1$, $\gamma_2=-\alpha-1$, $0<\alpha_1<\alpha_2<\pi$, $k_{-2}(x,y)=\min_{i\in\{1,2\}}\{\frac{1}{x^2+2xy\cos\alpha_i+y^2}\}$ $(x,y\in\mathbf{R})$, then we find

$$\begin{aligned}\widetilde{k}(\alpha)&:=\int_0^\infty\left(\min_{i\in\{1,2\}}\left\{\frac{1}{u^2+2u\cos\alpha_i+1}\right\}+\min_{i\in\{1,2\}}\left\{\frac{1}{u^2-2u\cos\alpha_i+1}\right\}\right)u^{-\alpha}\,du\\&=\int_0^\infty\frac{u^{-\alpha}\,du}{u^2+2u\cos\alpha_1+1}+\int_0^\infty\frac{u^{-\alpha}\,du}{u^2+2u\cos(\pi-\alpha_2)+1}.\end{aligned}\tag{42.125}$$

(i) For $\alpha\neq0$, by (42.120) and since $z_1=-e^{i\alpha_1}$ and $z_2=-e^{-i\alpha_1}$, we find

$$\begin{aligned}&\int_0^\infty\frac{u^{-\alpha}}{u^2+2u\cos\alpha_1+1}\,du\\&=\frac{2\pi i}{1-e^{2\pi(1-\alpha)i}}\left[Res\left\{\frac{z^{-\alpha}}{(z-z_1)(z-z_2)},z_1\right\}+Res\left\{\frac{z^{-\alpha}}{(z-z_1)(z-z_2)},z_2\right\}\right]\\&=\frac{2\pi i}{1-e^{2\pi(1-\alpha)i}}\left(\frac{z_1^{-\alpha}}{z_1-z_2}+\frac{z_2^{-\alpha}}{z_2-z_1}\right)=\frac{\pi\sin\alpha\alpha_1}{\sin\pi\alpha\sin\alpha_1}.\end{aligned}$$

Hence, we obtain

$$\begin{aligned}\widetilde{k}(\alpha)&=\frac{\pi\sin\alpha\alpha_1}{\sin\pi\alpha\sin\alpha_1}+\frac{\pi\sin\alpha(\pi-\alpha_2)}{\sin\pi\alpha\sin(\pi-\alpha_2)}\\&=\frac{\pi}{\sin\pi\alpha}\left[\frac{\sin\alpha\alpha_1}{\sin\alpha_1}+\frac{\sin\alpha(\pi-\alpha_2)}{\sin\alpha_2}\right].\end{aligned}\tag{42.126}$$

(ii) For $\alpha=0$, by the integral formula, we find

$$\begin{aligned}\widetilde{k}(0)&:=\int_0^\infty\frac{du}{u^2+2u\cos\alpha_1+1}+\int_0^\infty\frac{du}{u^2+2u\cos(\pi-\alpha_2)+1}\\&=\frac{\alpha_1}{\sin\alpha_1}+\frac{\pi-\alpha_2}{\sin(\pi-\alpha_2)}=\frac{\alpha_1}{\sin\alpha_1}+\frac{\pi-\alpha_2}{\sin\alpha_2}=\lim_{\alpha\to0}\widetilde{k}(\alpha).\end{aligned}\tag{42.127}$$

(a) Setting $\varphi(x)=|x|^{p\alpha-1}$, $\psi(y)=|y|^{-q\alpha-1}$ $(x,y\in\mathbf{R})$, we define an operator $\widetilde{T}_{-2}:L_{p,\varphi}(\mathbf{R})\to L_{p,\psi^{1-p}}(\mathbf{R})$ as follows: for $f\in L_{p,\varphi}(\mathbf{R})$, there exists a unique representation $\widetilde{T}_{-2}f\in L_{p,\psi^{1-p}}(\mathbf{R})$, satisfying

$$\widetilde{T}_{-2}f(y)=\int_{-\infty}^\infty\min_{i\in\{1,2\}}\left\{\frac{1}{x^2+2xy\cos\alpha_i+y^2}\right\}f(x)\,dx\quad(y\in\mathbf{R}).$$

Then by Theorem 42.9, we have (cf. [43])

$$\|\widetilde{T}_{-2}\|=\frac{\pi}{\sin\pi\alpha}\left[\frac{\sin\alpha\alpha_1}{\sin\alpha_1}+\frac{\sin\alpha(\pi-\alpha_2)}{\sin\alpha_2}\right].\tag{42.128}$$

(b) For $\alpha = 0$, $\beta \in \mathbf{R}$, $\gamma_1 = \gamma_2 = -1$, $\widetilde{\varphi}(x) = |x|^{-1}$, $\widetilde{\psi}(y) = |y|^{-1}$ $(x, y \in \mathbf{R})$, by Remark 42.6(ii), define an operator $\widetilde{T}_{-2,\beta} : L_{p,\widetilde{\varphi}}(\mathbf{R}) \to L_{p,\widetilde{\psi}^{1-p}}(\mathbf{R})$ as follows: for $f \in L_{p,\widetilde{\varphi}}(\mathbf{R})$, there exists a unique representation $\widetilde{T}_{-2,\beta} f \in L_{p,\psi^{1-p}}(\mathbf{R})$, satisfying

$$\widetilde{T}_{-2,\beta} f(y) = \int_{-\infty}^{\infty} \min_{i\in\{1,2\}} \left\{ \frac{1}{x^2 + 2xy\cos\alpha_i + y^2} \right\} \arctan\left|\frac{x}{y}\right|^{\beta} f(x)\,dx \quad (y \in \mathbf{R}).$$

Then we have

$$\|\widetilde{T}_{-2,\beta}\| = \frac{\pi}{4}\widetilde{k}(0) = \frac{\pi}{4}\left[\frac{\alpha_1}{\sin\alpha_1} + \frac{\pi - \alpha_2}{\sin\alpha_2}\right]. \tag{42.129}$$

By Remark 42.7(i), define an operator $T_{-2} : L_{p,\widetilde{\varphi}}(\mathbf{R}) \to L_{p,\widetilde{\psi}^{1-p}}(\mathbf{R})$ as follows: for $f \in L_{p,\widetilde{\varphi}}(\mathbf{R})$, there exists a unique representation $T_{-2} f \in L_{p,\widetilde{\psi}^{1-p}}(\mathbf{R})$, satisfying

$$T_{-2} f(y) = \int_{-\infty}^{\infty} \min_{i\in\{1,2\}} \left\{ \frac{1}{1 + 2xy\cos\alpha_i + (xy)^2} \right\} f(x)\,dx \quad (y \in \mathbf{R}).$$

Then we have

$$\|T_{-2}\| = \widetilde{k}(0) = \frac{\alpha_1}{\sin\alpha_1} + \frac{\pi - \alpha_2}{\sin\alpha_2}. \tag{42.130}$$

(iii) For $\alpha_2 = \alpha_1 \in (0, \pi)$,

$$k_{-2}(x, y) = \min_{i\in\{1,2\}} \left\{ \frac{1}{x^2 + 2xy\cos\alpha_i + y^2} \right\} = \frac{1}{x^2 + 2xy\cos\alpha_1 + y^2},$$

and $\alpha \in (-1, 1)$, we have

$$\widetilde{k}(\alpha) = \frac{\pi}{\sin\pi\alpha}\left[\frac{\sin\alpha\alpha_1}{\sin\alpha_1} + \frac{\sin\alpha(\pi - \alpha_1)}{\sin\alpha_1}\right] = \frac{\pi\cos(\frac{\pi}{2} - \alpha_1)\alpha}{\cos\frac{\pi}{2}\alpha\sin\alpha_1}. \tag{42.131}$$

(a) Setting $\varphi(x) = |x|^{p\alpha-1}$, $\psi(y) = |y|^{-q\alpha-1}$ $(x, y \in \mathbf{R})$, we define an operator $\widetilde{T}_{-2} : L_{p,\varphi}(\mathbf{R}) \to L_{p,\psi^{1-p}}(\mathbf{R})$ as follows: for $f \in L_{p,\varphi}(\mathbf{R})$, there exists a unique representation $\widetilde{T}_{-2} f \in L_{p,\psi^{1-p}}(\mathbf{R})$, satisfying

$$\widetilde{T}_{-2} f(y) = \int_{-\infty}^{\infty} \frac{1}{x^2 + 2xy\cos\alpha_1 + y^2} f(x)\,dx \quad (y \in \mathbf{R}).$$

Then by Theorem 42.9, we have

$$\|\widetilde{T}_{-2}\| = \widetilde{k}(\alpha) = \frac{\pi\cos(\frac{\pi}{2} - \alpha_1)\alpha}{\cos\frac{\pi}{2}\alpha\sin\alpha_1}. \tag{42.132}$$

(b) For $\alpha = 0$, $\beta \in \mathbf{R}$, $\widetilde{\varphi}(x) = |x|^{-1}$, $\widetilde{\psi}(y) = |y|^{-1}$ $(x, y \in \mathbf{R})$, by Remark 42.6(ii), define an operator $\widetilde{T}_{-2,\beta} : L_{p,\widetilde{\varphi}}(\mathbf{R}) \to L_{p,\widetilde{\psi}^{1-p}}(\mathbf{R})$ as follows: for $f \in L_{p,\widetilde{\varphi}}(\mathbf{R})$, there exists a unique representation $\widetilde{T}_{-2,\beta} f \in L_{p,\widetilde{\psi}^{1-p}}(\mathbf{R})$, satisfying

$$\widetilde{T}_{-2,\beta} f(y) = \int_{-\infty}^{\infty} \frac{1}{x^2 + 2xy\cos\alpha_1 + y^2} \arctan\left|\frac{x}{y}\right|^{\beta} f(x)\,dx \quad (y \in \mathbf{R}).$$

Then we have

$$\|\widetilde{T}_{-2,\beta}\| = \frac{\pi}{4}\widetilde{k}(0) = \frac{\pi^2}{4\sin\alpha_1}. \tag{42.133}$$

By Remark 42.7(i), define an operator $T_{-2} : L_{p,\widetilde{\varphi}}(\mathbf{R}) \to L_{p,\widetilde{\psi}^{1-p}}(\mathbf{R})$ as follows: for $f \in L_{p,\widetilde{\varphi}}(\mathbf{R})$, there exists a unique representation $T_{-2} f \in L_{p,\widetilde{\psi}^{1-p}}(\mathbf{R})$, satisfying

$$T_{-2} f(y) = \int_{-\infty}^{\infty} \frac{1}{1 + 2xy\cos\alpha_1 + (xy)^2} f(x)\,dx \quad (y \in \mathbf{R}).$$

Then we have

$$\|T_{-2}\| = \widetilde{k}(0) = \frac{\pi}{\sin\alpha_1}. \tag{42.134}$$

42.4 The Norms of Hilbert-Type Integral Operators with the Multi-variable Kernels on $\mathbf{R}_+^n \times \mathbf{R}_+^m$

42.4.1 The Case of a Homogeneous Kernel

Definition 42.8 Suppose that $n, m \in \mathbf{N}$, $\alpha, \beta > 0$, $\|x\|_\alpha = (\sum_{k=1}^{n} |x_k|^\alpha)^{\frac{1}{\alpha}}$ $(x = (x_1, \ldots, x_n) \in \mathbf{R}^n)$, $\|y\|_\beta = (\sum_{k=1}^{m} |y_k|^\beta)^{\frac{1}{\beta}}$ $(y = (y_1, \ldots, y_m) \in \mathbf{R}^m)$, and $\gamma \in \mathbf{R}$, $k_\gamma(x, y)$ is a homogeneous function of degree γ on $\mathbf{R}_+^2$. For $\gamma_1, \gamma_2 \in \mathbf{R}$, $\gamma_1 + \gamma_2 = \gamma$, define two weight functions $\omega_1(y)$ and $\varpi_1(x)$ as follows:

$$\omega_1(y) := \int_{\mathbf{R}_+^n} k_\gamma\big(\|x\|_\alpha, \|y\|_\beta\big) \frac{\|y\|_\beta^{-\gamma_2}}{\|x\|_\alpha^{n+\gamma_1}}\,dx \quad \big(y \in \mathbf{R}_+^m\big),$$

$$\varpi_1(x) := \int_{\mathbf{R}_+^m} k_\gamma\big(\|x\|_\alpha, \|y\|_\beta\big) \frac{\|x\|_\alpha^{-\gamma_1}}{\|y\|_\beta^{m+\gamma_2}}\,dy \quad \big(x \in \mathbf{R}_+^n\big). \tag{42.135}$$

Lemma 42.9 *Under the assumptions of Definition 42.8, for $y \in \mathbf{R}_+^m$, $x \in \mathbf{R}_+^n$, one has*

$$\omega_1(y) = \frac{\Gamma^n(\frac{1}{\alpha})}{\alpha^{n-1}\Gamma(\frac{n}{\alpha})} k(\gamma_1), \qquad \varpi_1(x) = \frac{\Gamma^m(\frac{1}{\beta})}{\beta^{m-1}\Gamma(\frac{m}{\beta})} k(\gamma_1), \tag{42.136}$$

where $k(\gamma_1) = \int_0^\infty k_\gamma(u, 1) u^{-\gamma_1 - 1}\,du$.

Proof For $M > 0$, putting $D_M := \{x \in \mathbf{R}_+^n \mid \sum_{i=1}^n x_i^\alpha \le M^\alpha\}$, we have (cf. (42.22))

$$\int \cdots \int_{D_M} \Psi\left(\sum_{i=1}^n \left(\frac{x_i}{M}\right)^\alpha\right) dx_1 \cdots dx_n = \frac{M^n \Gamma^n(\frac{1}{\alpha})}{\alpha^n \Gamma(\frac{n}{\alpha})} \int_0^1 \Psi(u) u^{\frac{n}{\alpha}-1} du. \quad (42.137)$$

Setting $\Psi(u) = k_\gamma(Mu^{\frac{1}{\alpha}}, \|y\|_\beta)(\frac{1}{Mu^{\frac{1}{\alpha}}})^{n+\gamma_1}$, we find

$$\begin{aligned}\omega_1(y) &= \|y\|_\beta^{-\gamma_2} \lim_{M\to\infty} \int \cdots \int_{D_M} k_\gamma\left(M\left[\sum_{i=1}^n \left(\frac{x_i}{M}\right)^\alpha\right]^{\frac{1}{\alpha}}, \|y\|_\beta\right) \\ &\quad \times \left\{\frac{1}{M[\sum_{i=1}^n (\frac{x_i}{M})^\alpha]^{\frac{1}{\alpha}}}\right\}^{n+\gamma_1} dx_1 \cdots dx_n \\ &= \|y\|_\beta^{-\gamma_2} \lim_{M\to\infty} \frac{M^n \Gamma^n(\frac{1}{\alpha})}{\alpha^n \Gamma(\frac{n}{\alpha})} \int_0^1 k_\gamma\left(Mu^{\frac{1}{\alpha}}, \|y\|_\beta\right)\left(\frac{1}{Mu^{\frac{1}{\alpha}}}\right)^{n+\gamma_1} u^{\frac{n}{\alpha}-1} du.\end{aligned}$$

Putting $v = Mu^{\frac{1}{\alpha}}/\|y\|_\beta$, we obtain

$$\begin{aligned}\omega_1(y) &= \|y\|_\beta^{-\gamma_2} \lim_{M\to\infty} \frac{M^n \Gamma^n(\frac{1}{\alpha})}{\alpha^n \Gamma(\frac{n}{\alpha})} \alpha \int_0^{M/\|y\|_\beta} k_\gamma\left(v\|y\|_\beta, \|y\|_\beta\right) \\ &\quad \times \left(\frac{1}{v\|y\|_\beta}\right)^{n+\gamma_1} \left(\frac{v\|y\|_\beta}{M}\right)^{n-\alpha} \left(\frac{\|y\|_\beta}{M}\right)^\alpha v^{\alpha-1} dv \\ &= \frac{\Gamma^n(\frac{1}{\alpha})}{\alpha^{n-1}\Gamma(\frac{n}{\alpha})} \int_0^\infty k_\gamma(v, 1) v^{-\gamma_1 - 1} dv.\end{aligned}$$

In the same way, we have

$$\begin{aligned}\varpi_1(x) &= \frac{\Gamma^m(\frac{1}{\beta})}{\beta^{m-1}\Gamma(\frac{m}{\beta})} \int_0^\infty k_\gamma(1, v) v^{-\gamma_2-1} dv \\ &= \frac{\Gamma^m(\frac{1}{\beta})}{\beta^{m-1}\Gamma(\frac{m}{\beta})} \int_0^\infty k_\gamma(u, 1) u^{-\gamma_1-1} du.\end{aligned}$$

The lemma is proved. □

Remark 42.8 It is obvious that for any $y \in \mathbf{R}^m, x \in \mathbf{R}^n$, we have

$$\begin{aligned}
\widetilde{\omega}_1(y) &:= \int_{\mathbf{R}^n} k_\gamma\big(\|x\|_\alpha, \|y\|_\beta\big) \frac{\|y\|_\beta^{-\gamma_2}}{\|x\|_\alpha^{n+\gamma_1}}\,dx \\
&= 2^n \omega_1(y) = \frac{2^n \Gamma^n(\frac{1}{\alpha})}{\alpha^{n-1}\Gamma(\frac{n}{\alpha})} k(\gamma_1), \\
\widetilde{\varpi}_1(x) &:= \int_{\mathbf{R}^m} k_\gamma\big(\|x\|_\alpha, \|y\|_\beta\big) \frac{\|x\|_\alpha^{-\gamma_1}}{\|y\|_\beta^{m+\gamma_2}}\,dy \\
&= 2^m \varpi_1(x) = \frac{2^m \Gamma^m(\frac{1}{\beta})}{\beta^{m-1}\Gamma(\frac{m}{\beta})} k(\gamma_1).
\end{aligned} \tag{42.138}$$

Lemma 42.10 *Under the assumptions of Definition* 42.8, *for* $k \in \mathbf{N}$, *one has*

$$\begin{aligned}
\widetilde{I}_k &:= \int_{\{y \in R_+^m; \|y\|_\beta \geq 1\}} \bigg(\int_{\{x \in R_+^n; \|x\|_\alpha \geq 1\}} k_\gamma\big(\|x\|_\alpha, \|y\|_\beta\big) \\
&\quad \times \|x\|_\alpha^{-\gamma_1 - \frac{1}{pk} - n} \|y\|_\beta^{-\gamma_2 - \frac{1}{qk} - m}\,dx\bigg)\,dy \\
&= \frac{\Gamma^n(\frac{1}{\alpha})}{\alpha^{n-1}\Gamma(\frac{n}{\alpha})} \frac{\Gamma^m(\frac{1}{\beta})}{\beta^{m-1}\Gamma(\frac{m}{\beta})} k L_k,
\end{aligned} \tag{42.139}$$

where L_k *indicated by* (42.30) *is as follows*:

$$L_k := \frac{1}{k} \int_1^\infty y^{-\frac{1}{k}-1} \bigg(\int_{\frac{1}{y}}^\infty k_\gamma(u,1) u^{-\gamma_1 - \frac{1}{pk} - 1}\,du \bigg)\,dy.$$

Proof For $M > 1$, setting $\Psi(u)$ as $\Psi(u) = 0$ $(u \in (0, M^{-\alpha}))$;

$$\Psi(u) = k_\gamma\big(Mu^{1/\alpha}, \|y\|_\beta\big)\big(Mu^{1/\alpha}\big)^{-\gamma_1 - \frac{1}{pk} - n} \quad \big(u \in \big[M^{-\alpha}, 1\big]\big),$$

by (42.137), since $\|x\|_\alpha \geq 1$, means that $\sum_{i=1}^n (\frac{x_i}{M})^\alpha \geq M^{-\alpha}$, and we find

$$\begin{aligned}
F_k(y) &:= \int_{\{x \in R_+^n; \|x\|_\alpha \geq 1\}} k_\gamma\big(\|x\|_\alpha, \|y\|_\beta\big) \|x\|_\alpha^{-\gamma_1 - \frac{1}{pk} - n}\,dx \\
&= \lim_{M \to \infty} \int \cdots \int_{D_M} \Psi\left(\sum_{i=1}^n \left(\frac{x_i}{M}\right)^\alpha\right) dx_1 \cdots dx_n \\
&= \lim_{M \to \infty} \frac{M^n \Gamma^n(\frac{1}{\alpha})}{\alpha^n \Gamma(\frac{n}{\alpha})} \int_0^1 \Psi(u) u^{\frac{n}{\alpha}-1}\,du
\end{aligned}$$

$$= \lim_{M\to\infty} \frac{M^n \Gamma^n(\frac{1}{\alpha})}{\alpha^n \Gamma(\frac{n}{\alpha})}$$

$$\times \int_{M^{-\alpha}}^{1} k_\gamma\big(Mu^{1/\alpha}, \|y\|_\beta\big)\big(Mu^{1/\alpha}\big)^{-\gamma_1-\frac{1}{pk}-n} u^{\frac{n}{\alpha}-1}\, du$$

$$= \frac{\Gamma^n(\frac{1}{\alpha})}{\alpha^{n-1}\Gamma(\frac{n}{\alpha})} \lim_{M\to\infty} M^{-\gamma_1-\frac{1}{pk}}$$

$$\times \int_{M^{-\alpha}}^{1} k_\gamma\big(Mu^{1/\alpha}, \|y\|_\beta\big) u^{\frac{-\gamma_1}{\alpha}-\frac{1}{\alpha pk}-1}\, du$$

$$\overset{v=Mu^{1/\alpha}/\|y\|_\beta}{=} \|y\|_\beta^{\gamma_2-\frac{1}{pk}} \frac{\Gamma^n(\frac{1}{\alpha})}{\alpha^{n-1}\Gamma(\frac{n}{\alpha})} \int_{\|y\|_\beta^{-1}}^{\infty} k_\gamma(v,1) v^{-\gamma_1-\frac{1}{pk}-1}\, dv.$$

Hence, it follows that

$$\widetilde{I}_k = \int_{\{y\in R_+^m;\|y\|_\beta\geq 1\}} \|y\|_\beta^{-\gamma_2-\frac{1}{qk}-m} F_k(y)\, dy$$

$$= \frac{\Gamma^n(\frac{1}{\alpha})}{\alpha^{n-1}\Gamma(\frac{n}{\alpha})} \int_{\{y\in R_+^m;\|y\|_\beta\geq 1\}} \|y\|_\beta^{-\frac{1}{k}-m} \int_{\|y\|_\beta^{-1}}^{\infty} k_\gamma(v,1) v^{-\gamma_1-\frac{1}{pk}-1}\, dv.$$

For $M > 1$, setting $\Psi(u)$ as $\Psi(u) = 0$ $(u \in (0, M^{-\beta}))$;

$$\Psi(u) = \big(Mu^{1/\beta}\big)^{-\frac{1}{k}-m} \int_{(Mu^{1/\beta})^{-1}}^{\infty} k_\gamma(v,1) v^{-\gamma_1-\frac{1}{pk}-1}\, dv \quad \big(u \in [M^{-\beta}, 1]\big),$$

and $\widetilde{D}_M := \{y \in \mathbf{R}_+^m | \sum_{i=1}^m y_i^\beta \leq M^\beta\}$, by (42.137), we find

$$\widetilde{I}_k = \frac{\Gamma^n(\frac{1}{\alpha})}{\alpha^{n-1}\Gamma(\frac{n}{\alpha})} \lim_{M\to\infty} \int \cdots \int_{\widetilde{D}_M} \Psi\left(\sum_{i=1}^m \left(\frac{y_i}{M}\right)^\beta\right) dy_1 \cdots dy_m$$

$$= \frac{\Gamma^n(\frac{1}{\alpha})}{\alpha^{n-1}\Gamma(\frac{n}{\alpha})} \lim_{M\to\infty} \frac{M^m \Gamma^m(\frac{1}{\beta})}{\beta^m \Gamma(\frac{m}{\beta})} \int_0^1 \Psi(u) u^{\frac{m}{\beta}-1}\, du$$

$$= \frac{\Gamma^n(\frac{1}{\alpha})}{\alpha^{n-1}\Gamma(\frac{n}{\alpha})} \lim_{M\to\infty} \frac{M^m \Gamma^m(\frac{1}{\beta})}{\beta^m \Gamma(\frac{m}{\beta})} \int_{M^{-\beta}}^1 \big(Mu^{1/\beta}\big)^{-\frac{1}{k}-m} u^{\frac{m}{\beta}-1}$$

$$\times \left[\int_{(Mu^{1/\beta})^{-1}}^{\infty} k_\gamma(v,1) v^{-\gamma_1-\frac{1}{pk}-1}\, dv\right] du$$

$$\overset{y=Mu^{1/\beta}}{=} \frac{\Gamma^n(\frac{1}{\alpha})}{\alpha^{n-1}\Gamma(\frac{n}{\alpha})} \frac{\Gamma^m(\frac{1}{\beta})}{\beta^{m-1}\Gamma(\frac{m}{\beta})} \int_1^\infty y^{\frac{-1}{k}-1} \left(\int_{\frac{1}{y}}^{\infty} k_\gamma(v,1) v^{-\gamma_1-\frac{1}{pk}-1}\, dv\right) dy.$$

The lemma is proved. □

Under the assumptions of Definition 42.8, setting

$$\Phi_1(x) := \|x\|_\alpha^{p(n+\gamma_1)-n} \quad \left(x \in \mathbf{R}_+^n\right), \qquad \Psi_1(y) := \|y\|_\beta^{q(m+\gamma_2)-m} \quad \left(y \in \mathbf{R}_+^m\right)$$

and $(\Psi_1(y))^{1-p} = \|y\|_\beta^{-p\gamma_2-m}$, we define two normed function spaces as follows:

$$L_{p,\Phi_1}\left(\mathbf{R}_+^n\right) := \left\{f;\, \|f\|_{p,\Phi_1} = \left\{\int_{\mathbf{R}_+^n} \|x\|_\alpha^{p(n+\gamma_1)-n} \left|f(x)\right|^p dx\right\}^{\frac{1}{p}} < \infty\right\},$$

$$L_{q,\Psi_1}\left(\mathbf{R}_+^m\right) := \left\{g;\, \|g\|_{q,\Psi_1} = \left\{\int_{\mathbf{R}_+^m} \|y\|_\beta^{q(m+\gamma_2)-m} \left|g(y)\right|^q dy\right\}^{\frac{1}{q}} < \infty\right\}.$$

Theorem 42.12 *Suppose that* $p > 0$ $(p \neq 1)$, $\frac{1}{p} + \frac{1}{q} = 1$, $\gamma, \gamma_1, \gamma_2 \in \mathbf{R}$, $\gamma_1 + \gamma_2 = \gamma$, $k_\gamma(x, y)$ *is a non-negative homogeneous function of degree* γ *on* $\mathbf{R}_+^2$, $k(\gamma_1) = \int_0^\infty k_\gamma(u, 1)u^{-\gamma_1-1}\,du \in \mathbf{R}_+$,

$$K(\gamma_1) := \left(\frac{\Gamma^m(\frac{1}{\beta})}{\beta^{m-1}\Gamma(\frac{m}{\beta})}\right)^{\frac{1}{p}} \left(\frac{\Gamma^n(\frac{1}{\alpha})}{\alpha^{n-1}\Gamma(\frac{n}{\alpha})}\right)^{\frac{1}{q}} k(\gamma_1), \tag{42.140}$$

and $f(\geq 0) \in L_{p,\Phi_1}(\mathbf{R}_+^n)$, $g(\geq 0) \in L_{q,\Psi_1}(\mathbf{R}_+^m)$ *such that* $\|f\|_{p,\Phi_1} > 0$, $\|g\|_{q,\Psi_1} > 0$.

(i) *If* $p > 1$, *then we have the following equivalent inequalities with the best constant factors* $K(\gamma_1)$ *and* $K^p(\gamma_1)$:

$$\widetilde{I} := \int_{\mathbf{R}_+^m} \int_{\mathbf{R}_+^n} k_\gamma\left(\|x\|_\alpha, \|y\|_\beta\right) f(x)g(y)\,dx\,dy < K(\gamma_1)\|f\|_{p,\Phi_1}\|g\|_{q,\Psi_1}, \tag{42.141}$$

$$\widetilde{J} := \int_{\mathbf{R}_+^m} \frac{1}{\|y\|_\beta^{p\gamma_2+m}} \left(\int_{\mathbf{R}_+^n} k_\gamma\left(\|x\|_\alpha, \|y\|_\beta\right) f(x)\,dx\right)^p dy < K^p(\gamma_1)\|f\|_{p,\Phi_1}^p; \tag{42.142}$$

(ii) *If* $0 < p < 1$, *then we have the equivalent reverses of* (42.141) *and* (42.142). *Moreover, if there exists a constant* $\delta_0 > 0$ *such that* $k(\gamma_1 + \delta_0) \in \mathbf{R}_+$, *then the reverses of* (42.141) *and* (42.142) *possess the best constant factors* $K(\gamma_1)$ *and* $K^p(\gamma_1)$, *respectively.*

Proof For $A = \mathbf{R}_+^n$, $B = \mathbf{R}_+^m$, $H(x, y) = k_\gamma(\|x\|_\alpha, \|y\|_\beta)$, $a = n + \gamma_1$, $b = m + \gamma_2$ and

$$k_1 = \frac{\Gamma^m(\frac{1}{\beta})}{\beta^{m-1}\Gamma(\frac{m}{\beta})} k(\gamma_1), \qquad k_2 = \frac{\Gamma^n(\frac{1}{\alpha})}{\alpha^{n-1}\Gamma(\frac{n}{\alpha})} k(\gamma_1)$$

in Theorem 42.3, by the assumptions, we have equivalent inequalities (42.141) and (42.142) (for $p > 1$), and the equivalent reverses of (42.141) and (42.142) (for $0 < p < 1$).

(i) For $p > 1$, $k \in \mathbf{N}$, we set $f_k(x)$ and $g_k(y)$ as follows:

$$f_k(x) := \begin{cases} 0, & x \in \{x \in R_+^n; \|x\|_\alpha < 1\}, \\ \|x\|_\alpha^{-\gamma_1 - \frac{1}{pk} - n}, & x \in \{x \in R_+^n; \|x\|_\alpha \geq 1\}, \end{cases}$$

$$g_k(y) := \begin{cases} 0, & y \in \{y \in R_+^m; \|y\|_\beta < 1\}, \\ \|y\|_\beta^{-\gamma_2 - \frac{1}{qk} - m}, & y \in \{y \in R_+^m; \|y\|_\beta \geq 1\}. \end{cases}$$

Then by (42.23) (for $\varepsilon = \frac{1}{k}$), we find

$$\|f_k\|_{p,\Phi_1} = \left(\int_{\{x \in R_+^n; \|x\|_\alpha \geq 1\}} \|x\|_\alpha^{-n-\frac{1}{k}} dx \right)^{\frac{1}{p}} = \left(\frac{\Gamma^n(\frac{1}{\alpha})k}{\alpha^{n-1}\Gamma(\frac{n}{\alpha})} \right)^{\frac{1}{p}},$$

$$\|g_k\|_{q,\Psi_1} = \left(\int_{\{y \in R_+^m; \|y\|_\beta \geq 1\}} \|y\|_\beta^{-m-\frac{1}{k}} dy \right)^{\frac{1}{q}} = \left(\frac{\Gamma^m(\frac{1}{\beta})k}{\beta^{m-1}\Gamma(\frac{m}{\beta})} \right)^{\frac{1}{q}},$$

$$\|f_k\|_{p,\Phi_1} \|g_k\|_{q,\Psi_1} = \left[\frac{\Gamma^n(\frac{1}{\alpha})}{\alpha^{n-1}\Gamma(\frac{n}{\alpha})} \right]^{\frac{1}{p}} \left[\frac{\Gamma^m(\frac{1}{\beta})}{\beta^{m-1}\Gamma(\frac{m}{\beta})} \right]^{\frac{1}{q}} k.$$

If there exists a positive constant $K_0 \leq K(\gamma_1)$ such that (42.141) is valid as we replace $K(\gamma_1)$ by K_0, then, in particular, it follows that

$$\begin{aligned} \frac{1}{k}\widetilde{I}_k &< \frac{1}{k} K_0 \|f_k\|_{p,\Phi_1} \|g_k\|_{q,\Psi_1} \\ &= K_0 \left(\frac{\Gamma^n(\frac{1}{\alpha})}{\alpha^{n-1}\Gamma(\frac{n}{\alpha})} \right)^{\frac{1}{p}} \left(\frac{\Gamma^m(\frac{1}{\beta})}{\beta^{m-1}\Gamma(\frac{m}{\beta})} \right)^{\frac{1}{q}}. \end{aligned} \tag{42.143}$$

By (42.139), we get

$$\frac{\Gamma^n(\frac{1}{\alpha})}{\alpha^{n-1}\Gamma(\frac{n}{\alpha})} \frac{\Gamma^m(\frac{1}{\beta})}{\beta^{m-1}\Gamma(\frac{m}{\beta})} L_k \leq \frac{1}{k}\widetilde{I}_k. \tag{42.144}$$

In view of (42.143) and (42.144), we have

$$\left(\frac{\Gamma^n(\frac{1}{\alpha})}{\alpha^{n-1}\Gamma(\frac{n}{\alpha})} \right)^{\frac{1}{q}} \left(\frac{\Gamma^m(\frac{1}{\beta})}{\beta^{m-1}\Gamma(\frac{m}{\beta})} \right)^{\frac{1}{p}} L_k < K_0,$$

and then by (42.31), $K(\gamma_1) \leq K_0$ $(k \to \infty)$. Therefore, $K_0 = K(\gamma_1)$ is the best value of (42.141).

We confirm that the constant in (42.142) is the best possible, otherwise by using $\widetilde{I} \leq \widetilde{J}^{\frac{1}{p}} \|g\|_{q,\Psi_1}$, we can get a contradiction that the constant in (42.141) is not the best possible.

(ii) For $0 < p < 1, k \in \mathbf{N}$, we set $f_k(x)$ and $g_k(y)$ as in (i). If there exists a positive constant $K_1 \geq K(\gamma_1)$ such that the reverse of (42.141) is valid as we replace $K(\gamma_1)$ by K_1, then, in particular, by (42.139), it follows that

$$\frac{\widetilde{I}_k}{k} > \frac{K_1}{k}\|f_k\|_{p,\Phi_1}\|g_k\|_{q,\Psi_1} = K_1\left(\frac{\Gamma^n(\frac{1}{\alpha})}{\alpha^{n-1}\Gamma(\frac{n}{\alpha})}\right)^{\frac{1}{p}}\left(\frac{\Gamma^m(\frac{1}{\beta})}{\beta^{m-1}\Gamma(\frac{m}{\beta})}\right)^{\frac{1}{q}}. \tag{42.145}$$

By (42.139), we obtain

$$\frac{\Gamma^n(\frac{1}{\alpha})}{\alpha^{n-1}\Gamma(\frac{n}{\alpha})}\frac{\Gamma^m(\frac{1}{\beta})}{\beta^{m-1}\Gamma(\frac{m}{\beta})}L_k = \frac{1}{k}\widetilde{I}_k. \tag{42.146}$$

In view of (42.145) and (42.146), we have

$$\left(\frac{\Gamma^n(\frac{1}{\alpha})}{\alpha^{n-1}\Gamma(\frac{n}{\alpha})}\right)^{\frac{1}{q}}\left(\frac{\Gamma^m(\frac{1}{\beta})}{\beta^{m-1}\Gamma(\frac{m}{\beta})}\right)^{\frac{1}{p}}L_k > K_1,$$

and then in view of (42.31), $K(\gamma_1) \geq K_1$ $(k \to \infty)$. Therefore, $K_1 = K(\gamma_1)$ is the best value of the reverse of (42.141). We confirm that the constant factor in the reverse of (42.142) is the best possible, otherwise by using the inequality $\widetilde{I} \geq \widetilde{J}^{\frac{1}{p}}\|g\|_{q,\Psi_1}$, we can come to a contradiction that the constant factor in the reverse of (42.141) is not the best possible. The theorem is proved. □

By Remark 42.8, setting $\widetilde{\Phi}_1(x) := \|x\|_\alpha^{p(n+\gamma_1)-n}$ $(x \in \mathbf{R}^n)$, $\widetilde{\Psi}_1(y) := \|y\|_\beta^{q(m+\gamma_2)-m}$ $(y \in \mathbf{R}^m)$, we still have

Corollary 42.5 *Suppose that $p > 0$ $(p \neq 1)$, $\frac{1}{p}+\frac{1}{q} = 1, \gamma, \gamma_1, \gamma_2 \in \mathbf{R}, \gamma_1+\gamma_2 = \gamma$, $k_\gamma(x,y)$ is a non-negative homogeneous function of degree γ on $\mathbf{R}_+^2$, $k(\gamma_1) = \int_0^\infty k_\gamma(u,1)u^{-\gamma_1-1}\,du \in \mathbf{R}_+$,*

$$L(\gamma_1) := \left(\frac{2^m\Gamma^m(\frac{1}{\beta})}{\beta^{m-1}\Gamma(\frac{m}{\beta})}\right)^{\frac{1}{p}}\left(\frac{2^n\Gamma^n(\frac{1}{\alpha})}{\alpha^{n-1}\Gamma(\frac{n}{\alpha})}\right)^{\frac{1}{q}}k(\gamma_1), \tag{42.147}$$

and $f(\geq 0) \in L_{p,\widetilde{\Phi}_1}(\mathbf{R}^n)$, $g(\geq 0) \in L_{q,\widetilde{\Psi}_1}(\mathbf{R}^m)$ such that $\|f\|_{p,\widetilde{\Phi}_1} > 0$, $\|g\|_{q,\widetilde{\Psi}_1} > 0$.

(i) *If $p > 1$, then we have the following equivalent inequalities with the best constant factors $L(\gamma_1)$ and $L^p(\gamma_1)$:*

$$\int_{\mathbf{R}^m}\int_{\mathbf{R}^n}k_\gamma\big(\|x\|_\alpha,\|y\|_\beta\big)f(x)g(y)\,dx\,dy < L(\gamma_1)\|f\|_{p,\widetilde{\Phi}_1}\|g\|_{q,\widetilde{\Psi}_1}, \tag{42.148}$$

$$\int_{\mathbf{R}^m}\frac{1}{\|y\|_\beta^{p\gamma_2+m}}\left(\int_{\mathbf{R}_+^n}k_\gamma\big(\|x\|_\alpha,\|y\|_\beta\big)f(x)\,dx\right)^p dy < L^p(\gamma_1)\|f\|_{p,\widetilde{\Phi}_1}^p; \tag{42.149}$$

(ii) *If* $0 < p < 1$, *then we have the equivalent reverses of* (42.148) *and* (42.149). *Moreover, if there exists a constant* $\delta_0 > 0$ *such that* $k(\gamma_1 + \delta_0) \in \mathbf{R}_+$, *then the reverses of* (42.148) *and* (42.149) *possess the best constant factors* $L(\gamma_1)$ *and* $L^p(\gamma_1)$, *respectively.*

In view of Theorem 42.12, for $k(\gamma_1) \in \mathbf{R}_+$, we define a Hilbert-type integral operator $T_\gamma : L_{p,\Phi_1}(\mathbf{R}_+^n) \to L_{p,\Psi_1^{1-p}}(\mathbf{R}_+^m)$ as follows: for $f \in L_{p,\Phi_1}(\mathbf{R}_+^n)$, there exists a unique representation $T_\gamma f \in L_{p,\Psi_1^{1-p}}(\mathbf{R}_+^m)$, satisfying

$$T_\gamma f(y) := \int_{\mathbf{R}_+^n} k_\gamma\big(\|x\|_\alpha, \|y\|_\beta\big) f(x)\,dx \quad \big(y \in \mathbf{R}_+^m\big). \tag{42.150}$$

Then it follows by (42.141) and (42.142) that

$$(T_\gamma f, g) < K(\gamma_1)\|f\|_{p,\Phi_1}\|g\|_{q,\Psi_1}, \tag{42.151}$$

$$\|T_\gamma f\|_{p,\Psi_1^{1-p}} < K(\gamma_1)\|f\|_{p,\Phi_1}, \tag{42.152}$$

where the constant factor $K(\gamma_1)$ is the best possible. Hence we still have

Theorem 42.13 *Suppose that a Hilbert-type integral operator* T_γ *is defined by* (42.150). *Then it follows*

$$\|T_\gamma\| := \sup_{f(\neq\theta)\in L_{p,\Phi_1}(\mathbf{R}_+^n)} \frac{\|T_\gamma f\|_{p,\Psi_1^{1-p}}}{\|f\|_{p,\Phi_1}} = K(\gamma_1), \tag{42.153}$$

where $K(\gamma_1)$ *is indicated by* (42.140).

Remark 42.9 In Theorem 42.13(i), if $\gamma = 0$, $\gamma_1 = \sigma$, $\gamma_2 = -\sigma$, $\varphi(x) = \|x\|_\alpha^{p(n+\sigma)-1}$ $(x \in \mathbf{R}_+^n)$, $\psi(y) = \|y\|_\beta^{q(n-\sigma)-1}$ $(y \in \mathbf{R}_+^m)$, $k_0(\sigma) := \int_0^\infty k_0(u,1)u^{-\sigma-1}\,du \in \mathbf{R}_+$, then we define an operator $\widetilde{T}_\sigma : L_{p,\varphi}(\mathbf{R}_+^n) \to L_{p,\psi^{1-p}}(\mathbf{R}_+^m)$ as follows: for $f \in L_{p,\varphi}(\mathbf{R}_+^n)$, there exists a unique representation $\widetilde{T}_\sigma f \in L_{p,\psi^{1-p}}(\mathbf{R}_+^m)$, satisfying

$$\widetilde{T}_\sigma f(y) = \int_{\mathbf{R}_+^n} k_0\big(\|x\|_\alpha, \|y\|_\beta\big) f(x)\,dx \quad \big(y \in \mathbf{R}_+^m\big).$$

We have

$$\|\widetilde{T}_\sigma\| = \widetilde{K}_0(\sigma) := \left(\frac{\Gamma^m(\frac{1}{\beta})}{\beta^{m-1}\Gamma(\frac{m}{\beta})}\right)^{\frac{1}{p}} \left(\frac{\Gamma^n(\frac{1}{\alpha})}{\alpha^{n-1}\Gamma(\frac{n}{\alpha})}\right)^{\frac{1}{q}} k_0(\sigma).$$

(ii) If $k_\gamma(x,y)$ is symmetric, $\rho \in \mathbf{R}$, $\gamma_1 = \gamma_2 = \frac{\gamma}{2}$, $\widetilde{\varphi}(x) = \|x\|_\alpha^{p(n+\frac{\gamma}{2})-n}$ $(x \in \mathbf{R}_+^n)$, $\widetilde{\psi}(y) = \|y\|_\beta^{q(m+\frac{\gamma}{2})-m}$ $(y \in \mathbf{R}_+^m)$, then we define an operator $\widetilde{T}_{\gamma,\rho}$:

$L_{p,\widetilde{\varphi}}(\mathbf{R}_+^n) \to L_{p,\widetilde{\psi}^{1-p}}(\mathbf{R}_+^m)$ as follows: for $f \in L_{p,\widetilde{\varphi}}(\mathbf{R}_+^n)$, there exists a unique representation $\widetilde{T}_{\gamma,\rho} f \in L_{p,\widetilde{\psi}^{1-p}}(\mathbf{R}_+^m)$, satisfying

$$\widetilde{T}_{\gamma,\rho} f(y) = \int_{\mathbf{R}_+^n} k_\gamma\big(\|x\|_\alpha, \|y\|_\beta\big) \arctan\left(\frac{\|x\|_\alpha}{\|y\|_\beta}\right)^\rho f(x)\,dx \quad \big(y \in \mathbf{R}_+^m\big).$$

We have

$$\|\widetilde{T}_{\gamma,\rho}\| = \widetilde{k}_\rho(\gamma) := \frac{\pi}{4} K\left(\frac{\gamma}{2}\right). \tag{42.154}$$

In fact, it follows by (42.44) that

$$\begin{aligned}\widetilde{k}_\rho(\gamma) &= \left(\frac{\Gamma^n(\frac{1}{\alpha})}{\alpha^{n-1}\Gamma(\frac{n}{\alpha})}\right)^{\frac{1}{q}} \left(\frac{\Gamma^m(\frac{1}{\beta})}{\beta^{m-1}\Gamma(\frac{m}{\beta})}\right)^{\frac{1}{p}} \int_0^\infty k_\gamma(u,1)\big(\arctan u^\rho\big) u^{-\frac{\gamma}{2}-1}\,du \\ &= \left(\frac{\Gamma^n(\frac{1}{\alpha})}{\alpha^{n-1}\Gamma(\frac{n}{\alpha})}\right)^{\frac{1}{q}} \left(\frac{\Gamma^m(\frac{1}{\beta})}{\beta^{m-1}\Gamma(\frac{m}{\beta})}\right)^{\frac{1}{p}} \frac{\pi}{2} \int_0^1 k_\gamma(u,1) u^{-\frac{\gamma}{2}-1}\,du = \frac{\pi}{4} K\left(\frac{\gamma}{2}\right).\end{aligned}$$

(iii) For $\beta = \alpha$, $m = n$, $\gamma = -\lambda < 0$, (42.141) reduces to the result of [44].

(iv) By virtue of Theorem 42.13, suppose that $k_\gamma(x,y)$ (≥ 0) is a homogeneous function of degree γ on $\mathbf{R}_+^2$.

(a) If we set

$$K_\gamma\big(\|x\|_\alpha, \|y\|_\beta\big) := \begin{cases} k_\gamma(\|x\|_\alpha, \|y\|_\beta), & 0 < \|x\|_\alpha \leq \|y\|_\beta, \\ 0, & \|x\|_\alpha > \|y\|_\beta, \end{cases}$$

then we define the first class Hardy-type integral operator $T_\gamma^{(1)} : L_{p,\varphi_1}(\mathbf{R}_+^n) \to L_{p,\psi_1^{1-p}}(\mathbf{R}_+^m)$ with the homogeneous kernel as follows: for $f \in L_{p,\varphi_1}(\mathbf{R}_+^n)$, there exists a unique representation $T_\gamma^{(1)} f \in L_{p,\psi_1^{1-p}}(\mathbf{R}_+^m)$, satisfying

$$\begin{aligned}T_\gamma^{(1)} f(y) &= \int_{\mathbf{R}_+^n} K_\gamma\big(\|x\|_\alpha, \|y\|_\beta\big) f(x)\,dx \\ &= \int_{\{x \in \mathbf{R}_+^n;\, 0 < \|x\|_\alpha \leq \|y\|_\beta\}} k_\gamma\big(\|x\|_\alpha, \|y\|_\beta\big) f(x)\,dx \quad \big(y \in \mathbf{R}_+^m\big).\end{aligned}$$

Hence, for $k_1(\gamma) = \int_0^1 k_\gamma(u,1) u^{-\gamma_1 - 1}\,du \in \mathbf{R}_+$, we obtain

$$\begin{aligned}\omega_1(y) &= \varpi_1(x) = K_1(\gamma_1) \\ &:= \left[\frac{\Gamma^n(\frac{1}{\alpha})}{\alpha^{n-1}\Gamma(\frac{n}{\alpha})}\right]^{\frac{1}{q}} \left[\frac{\Gamma^m(\frac{1}{\beta})}{\beta^{m-1}\Gamma(\frac{m}{\beta})}\right]^{\frac{1}{p}} k_1(\gamma), \quad \big\|T_\gamma^{(1)}\big\| = K_1(\gamma_1).\end{aligned} \tag{42.155}$$

(b) If we set

$$K_\gamma(\|x\|_\alpha, \|y\|_\beta) := \begin{cases} 0, & 0 < \|x\|_\alpha \le \|y\|_\beta, \\ k_\gamma(\|x\|_\alpha, \|y\|_\beta), & \|x\|_\alpha > \|y\|_\beta, \end{cases}$$

then we define the second class Hardy-type integral operator $T_\gamma^{(2)} : L_{p,\varphi_1}(\mathbf{R}_+^n) \to L_{p,\psi_1^{1-p}}(\mathbf{R}_+^m)$ with the homogeneous kernel as follows: for $f \in L_{p,\varphi_1}(\mathbf{R}_+^n)$, there exists a unique representation $T_\gamma^{(2)} f \in L_{p,\psi_1^{1-p}}(\mathbf{R}_+^m)$, satisfying

$$\begin{aligned} T_\gamma^{(2)} f(y) &= \int_{\mathbf{R}_+^n} K_\gamma(\|x\|_\alpha, \|y\|_\beta) f(x)\,dx \\ &= \int_{\{x\in\mathbf{R}_+^n;\,\|x\|_\alpha > \|y\|_\beta\}} k_\gamma(\|x\|_\alpha, \|y\|_\beta) f(x)\,dx \quad (y \in \mathbf{R}_+^m). \end{aligned}$$

Hence, for $k_2(\gamma) = \int_1^\infty k_\gamma(u,1) u^{-\gamma_1 - 1}\,du \in \mathbf{R}_+$, we obtain

$$\begin{aligned} \omega_\gamma(y) &= \varpi_\gamma(x) = K_2(\gamma_1) \\ &:= \left(\frac{\Gamma^n(\frac{1}{\alpha})}{\alpha^{n-1}\Gamma(\frac{n}{\alpha})}\right)^{\frac{1}{q}} \left(\frac{\Gamma^m(\frac{1}{\beta})}{\beta^{m-1}\Gamma(\frac{m}{\beta})}\right)^{\frac{1}{p}} k_2(\gamma), \quad \|T_\gamma^{(2)}\| = K_2(\gamma_1). \end{aligned} \tag{42.156}$$

42.4.2 The Case of a Non-homogeneous Kernel

Definition 42.9 Suppose that $n, m \in \mathbf{N}$, $\alpha, \beta > 0$, $\|x\|_\alpha = (\sum_{k=1}^n |x_k|^\alpha)^{\frac{1}{\alpha}}$ $(x = (x_1, \ldots, x_n) \in \mathbf{R}^n)$, $\|y\|_\beta = (\sum_{k=1}^m |y_k|^\beta)^{\frac{1}{\beta}}$ $(y = (y_1, \ldots, y_m) \in \mathbf{R}^m)$, and $h(u)$ is a non-negative measurable function on $\mathbf{R}_+$. For $\eta \in \mathbf{R}$, define two weight functions $\omega_2(y)$ and $\varpi_2(x)$ as follows:

$$\omega_2(y) := \int_{\mathbf{R}_+^n} h(\|x\|_\alpha \|y\|_\beta) \frac{\|y\|_\beta^{-\eta}}{\|x\|_\alpha^{n+\eta}}\,dx \quad (y \in \mathbf{R}_+^m), \tag{42.157}$$

$$\varpi_2(x) := \int_{\mathbf{R}_+^m} h(\|x\|_\alpha \|y\|_\beta) \frac{\|x\|_\alpha^{-\eta}}{\|y\|_\beta^{m+\eta}}\,dy \quad (x \in \mathbf{R}_+^n). \tag{42.158}$$

Then we have

Lemma 42.11 *Under the assumptions of Definition 42.9, for $y \in \mathbf{R}_+^m$ and $x \in \mathbf{R}_+^n$,*

$$\omega_2(y) = \frac{\Gamma^n(\frac{1}{\alpha})}{\alpha^{n-1}\Gamma(\frac{n}{\alpha})}\widetilde{k}(\eta), \qquad \varpi_2(x) = \frac{\Gamma^m(\frac{1}{\beta})}{\beta^{m-1}\Gamma(\frac{m}{\beta})}\widetilde{k}(\eta), \tag{42.159}$$

where $\widetilde{k}(\eta) = \int_0^\infty h(u) u^{-\eta-1}\,du$.

Proof In view of (42.137), for $M>0$, $D_M := \{x \in \mathbf{R}_+^n \mid \sum_{i=1}^n x_i^\alpha \le M^\alpha\}$, $\Psi(u) = h(Mu^{\frac{1}{\alpha}}\|y\|_\beta) \times (\frac{1}{Mu^{\frac{1}{\alpha}}})^{n+\eta}$, we find

$$\begin{aligned}
\omega_2(y) &= \|y\|_\beta^{-\eta} \lim_{M\to\infty} \int \cdots \int_{D_M} h\left(M\left[\sum_{i=1}^n \left(\frac{x_i}{M}\right)^\alpha\right]^{\frac{1}{\alpha}} \|y\|_\beta\right) \\
&\quad \times \left\{\frac{1}{M[\sum_{i=1}^n (\frac{x_i}{M})^\alpha]^{\frac{1}{\alpha}}}\right\}^{n+\eta} dx_1 \cdots dx_n = \|y\|_\beta^{-\eta} \\
&\quad \times \lim_{M\to\infty} \frac{M^n \Gamma^n(\frac{1}{\alpha})}{\alpha^n \Gamma(\frac{n}{\alpha})} \int_0^1 h\big(Mu^{\frac{1}{\alpha}}\|y\|_\beta\big)\left(\frac{1}{Mu^{\frac{1}{\alpha}}}\right)^{n+\eta} u^{\frac{n}{\alpha}-1}\,du \\
&\overset{v=Mu^{\frac{1}{\alpha}}\|y\|_\beta}{=} \|y\|_\beta^{-\eta} \lim_{M\to\infty} \frac{M^n \Gamma^n(\frac{1}{\alpha})}{\alpha^n \Gamma(\frac{n}{\alpha})}\alpha \\
&\quad \times \int_0^{M\|y\|_\beta} h(v)\left(\frac{\|y\|_\beta}{v}\right)^{n+\eta}\left(\frac{v}{M\|y\|_\beta}\right)^{n-\alpha}\left(\frac{1}{M\|y\|_\beta}\right)^\alpha v^{\alpha-1}\,dv \\
&= \frac{\Gamma^n(\frac{1}{\alpha})}{\alpha^{n-1}\Gamma(\frac{n}{\alpha})} \int_0^\infty h(v)v^{-\eta-1}\,du.
\end{aligned}$$

In the same way, we can obtain

$$\varpi_2(x) = \frac{\Gamma^m(\frac{1}{\beta})}{\beta^{m-1}\Gamma(\frac{m}{\beta})} \int_0^\infty h(v)v^{-\eta-1}\,du.$$

The lemma is proved. □

Remark 42.10 It is obvious that for any $y \in \mathbf{R}^m$, $x \in \mathbf{R}^n$,

$$\begin{aligned}
\widetilde{\omega}_2(y) &:= \int_{\mathbf{R}^n} h\big(\|x\|_\alpha \|y\|_\beta\big) \frac{\|y\|_\beta^{-\eta}}{\|x\|_\alpha^{n+\eta}}\,dx \\
&= 2^n \omega_2(y) = \frac{2^n \Gamma^n(\frac{1}{\alpha})}{\alpha^{n-1}\Gamma(\frac{n}{\alpha})}\widetilde{k}(\eta), \\
\widetilde{\varpi}_2(x) &:= \int_{\mathbf{R}^m} h\big(\|x\|_\alpha \|y\|_\beta\big) \frac{\|x\|_\alpha^{-\eta}}{\|y\|_\beta^{m+\eta}}\,dy \\
&= 2^m \varpi_2(x) = \frac{2^m \Gamma^m(\frac{1}{\beta})}{\beta^{m-1}\Gamma(\frac{m}{\beta})}\widetilde{k}(\eta).
\end{aligned} \tag{42.160}$$

Lemma 42.12 *Under the assumptions of Definition* 42.9, *for* $k \in \mathbf{N}$,

$$\widetilde{I}_k := \int_{\{y\in R_+^m;\,\|y\|_\beta\ge 1\}}\left(\int_{\{x\in R_+^n;\,\|x\|_\alpha\le 1\}} \times h\big(\|x\|_\alpha\|y\|_\beta\big)\|x\|_\alpha^{-\eta+\frac{1}{pk}-n}\|y\|_\beta^{-\eta-\frac{1}{qk}-m}\,dx\right)dy$$
$$= \frac{\Gamma^n(\frac{1}{\alpha})}{\alpha^{n-1}\Gamma(\frac{n}{\alpha})}\frac{\Gamma^m(\frac{1}{\beta})}{\beta^{m-1}\Gamma(\frac{m}{\beta})}k\widetilde{L}_k, \tag{42.161}$$

where $\widetilde{L}_k$ *indicated by* (42.50) (*for* $\alpha = -\eta$) *is as follows*:

$$\widetilde{L}_k := \frac{1}{k}\int_1^\infty y^{-\frac{1}{k}-1}\left(\int_0^y h(u)u^{\alpha+\frac{1}{pk}-1}\,du\right)dy.$$

Proof By (42.137), for $M = 1$, we find

$$\widetilde{F}_k(y) := \int_{\{x\in R_+^n;\,\|x\|_\alpha\le 1\}} h\big(\|x\|_\alpha\|y\|_\beta\big)\|x\|_\alpha^{-\eta+\frac{1}{pk}-n}\,dx$$
$$= \int\cdots\int_{D_1} h\left(\left(\sum_{i=1}^n x_i^\alpha\right)^{\frac{1}{\alpha}}\|y\|_\beta\right)\left(\sum_{i=1}^n x_i^\alpha\right)^{\frac{1}{\alpha}(-\eta+\frac{1}{pk}-n)} dx_1\cdots dx_n$$
$$= \frac{\Gamma^n(\frac{1}{\alpha})}{\alpha^n\Gamma(\frac{n}{\alpha})}\int_0^1 h\big(u^{\frac{1}{\alpha}}\|y\|_\beta\big)u^{\frac{1}{\alpha}(-\eta+\frac{1}{pk}-n)}u^{\frac{n}{\alpha}-1}\,du$$
$$= \frac{\Gamma^n(\frac{1}{\alpha})}{\alpha^n\Gamma(\frac{n}{\alpha})}\int_0^1 h\big(u^{\frac{1}{\alpha}}\|y\|_\beta\big)u^{\frac{1}{\alpha}(\frac{1}{pk}-\eta)-1}\,du$$
$$\overset{v=u^{1/\alpha}\|y\|_\beta}{=} \|y\|_\beta^{\eta-\frac{1}{pk}}\frac{\Gamma^n(\frac{1}{\alpha})}{\alpha^{n-1}\Gamma(\frac{n}{\alpha})}\int_0^{\|y\|_\beta} h(v)v^{-\eta+\frac{1}{pk}-1}\,dv.$$

Hence, it follows that

$$\widetilde{I}_k = \int_{\{y\in R_+^m;\,\|y\|_\beta\ge 1\}}\|y\|_\beta^{-\eta-\frac{1}{qk}-m}\widetilde{F}_k(y)\,dy$$
$$= \frac{\Gamma^n(\frac{1}{\alpha})}{\alpha^{n-1}\Gamma(\frac{n}{\alpha})}\int_{\{y\in R_+^m;\,\|y\|_\beta\ge 1\}}\|y\|_\beta^{-\frac{1}{k}-m}\int_0^{\|y\|_\beta} h(v)v^{-\eta+\frac{1}{pk}-1}\,dv.$$

For $M > 1$, setting $\Psi(u)$ as $\Psi(u) = 0$ ($u \in (0, M^{-\beta})$);

$$\Psi(u) = \big(Mu^{1/\beta}\big)^{-\frac{1}{k}-m}\int_0^{Mu^{1/\beta}} h(v)v^{-\eta+\frac{1}{pk}-1}\,dv \quad \big(u\in[M^{-\beta},1]\big),$$

$\widetilde{D}_M := \{y \in \mathbf{R}_+^m | \sum_{i=1}^m y_i^\beta \le M^\beta\}$, by (42.137), we find

$$\widetilde{I}_k = \frac{\Gamma^n(\frac{1}{\alpha})}{\alpha^{n-1}\Gamma(\frac{n}{\alpha})} \lim_{M\to\infty} \int \cdots \int_{\widetilde{D}_M} \Psi\left(\sum_{i=1}^m \left(\frac{y_i}{M}\right)^\beta\right) dy_1 \cdots dy_m$$

$$= \frac{\Gamma^n(\frac{1}{\alpha})}{\alpha^{n-1}\Gamma(\frac{n}{\alpha})} \lim_{M\to\infty} \frac{M^m\Gamma^m(\frac{1}{\beta})}{\beta^m\Gamma(\frac{m}{\beta})} \int_0^1 \Psi(u) u^{\frac{m}{\beta}-1}\, du$$

$$= \frac{\Gamma^n(\frac{1}{\alpha})}{\alpha^{n-1}\Gamma(\frac{n}{\alpha})} \lim_{M\to\infty} \frac{M^m\Gamma^m(\frac{1}{\beta})}{\beta^m\Gamma(\frac{m}{\beta})} \int_{M^{-\beta}}^1 (Mu^{1/\beta})^{-\frac{1}{k}-m} u^{\frac{m}{\beta}-1}$$

$$\times \left(\int_0^{Mu^{1/\beta}} h(v) v^{-\eta+\frac{1}{pk}-1}\, dv\right) du$$

$$\overset{y=Mu^{1/\beta}}{=} \frac{\Gamma^n(\frac{1}{\alpha})}{\alpha^{n-1}\Gamma(\frac{n}{\alpha})} \frac{\Gamma^m(\frac{1}{\beta})}{\beta^{m-1}\Gamma(\frac{m}{\beta})} \int_1^\infty y^{\frac{-1}{k}-1} \left(\int_0^y h(v) v^{-\eta+\frac{1}{pk}-1}\, dv\right) dy,$$

and so (42.161) is valid. The lemma is proved. □

Under the assumptions of Definition 42.9, setting $\Phi_2(x) := \|x\|_\alpha^{p(n+\eta)-n}$ ($x \in \mathbf{R}_+^n$), $\Psi_2(y) := \|y\|_\beta^{q(m+\eta)-m}$ ($y \in \mathbf{R}_+^m$) and then $(\Psi_2(y))^{1-p} = \|y\|_\beta^{-p\eta-m}$, we define two real weighted normed function spaces as follows:

$$L_{p,\Phi_2}(\mathbf{R}_+^n) := \left\{f;\ \|f\|_{p,\Phi_2} = \left\{\int_{\mathbf{R}_+^n} \|x\|_\alpha^{p(n+\eta)-n} |f(x)|^p\, dx\right\}^{\frac{1}{p}} < \infty\right\},$$

$$L_{q,\Psi_2}(\mathbf{R}_+^m) := \left\{g;\ \|g\|_{q,\Psi_2} = \left\{\int_{\mathbf{R}_+^m} \|y\|_\beta^{q(m+\eta)-m} |g(y)|^q\, dy\right\}^{\frac{1}{q}} < \infty\right\}.$$

Theorem 42.14 *Suppose that* $p > 0$ $(p \ne 1)$, $\frac{1}{p} + \frac{1}{q} = 1$, $\eta \in \mathbf{R}$, $h(u)$ *is a non-negative measurable function on* $\mathbf{R}_+$, $\widetilde{k}(\eta) = \int_0^\infty h(u) u^{-\eta-1}\, du \in \mathbf{R}_+$,

$$\widetilde{K}(\eta) := \left(\frac{\Gamma^m(\frac{1}{\beta})}{\beta^{m-1}\Gamma(\frac{m}{\beta})}\right)^{\frac{1}{p}} \left(\frac{\Gamma^n(\frac{1}{\alpha})}{\alpha^{n-1}\Gamma(\frac{n}{\alpha})}\right)^{\frac{1}{q}} \widetilde{k}(\eta), \tag{42.162}$$

and $f(\ge 0) \in L_{p,\Phi_2}(\mathbf{R}_+^n)$, $g(\ge 0) \in L_{q,\Psi_2}(\mathbf{R}_+^m)$ *such that* $\|f\|_{p,\Phi_2} > 0$, $\|g\|_{q,\Psi_2} > 0$.

(i) *If* $p > 1$, *then we have the following equivalent inequalities with the best constant factors* $\widetilde{K}(\eta)$ *and* $\widetilde{K}^p(\eta)$:

$$\widetilde{I} := \int_{\mathbf{R}_+^m} \int_{\mathbf{R}_+^n} h(\|x\|_\alpha \|y\|_\beta) f(x) g(y)\, dx\, dy < \widetilde{K}(\eta) \|f\|_{p,\Phi_2} \|g\|_{q,\Psi_2}, \tag{42.163}$$

$$\widetilde{J} := \int_{\mathbf{R}_+^m} \frac{1}{\|y\|_\beta^{p\eta+m}} \left(\int_{\mathbf{R}_+^n} h(\|x\|_\alpha \|y\|_\beta) f(x)\,dx \right)^p dy < \widetilde{K}^p(\eta) \|f\|_{p,\Phi_2}^p; \tag{42.164}$$

(ii) *If* $0 < p < 1$, *then we have the equivalent reverses of* (42.163) *and* (42.164). *Moreover, if there exists a constant* $\delta_0 > 0$ *such that* $\widetilde{k}(\eta + \delta_0) \in \mathbf{R}_+$, *then the reverses of* (42.163) *and* (42.164) *possess the best constant factors* $\widetilde{K}(\eta)$ *and* $\widetilde{K}^p(\eta)$, *respectively.*

Proof For $A = \mathbf{R}_+^n$, $B = \mathbf{R}_+^m$, $H(x, y) = h(\|x\|_\alpha \|y\|_\beta)$, $a = n + \eta$, $b = m + \eta$ and

$$k_1 = \frac{\Gamma^m(\frac{1}{\beta})}{\beta^{m-1}\Gamma(\frac{m}{\beta})} \widetilde{k}(\eta), \qquad k_2 = \frac{\Gamma^n(\frac{1}{\alpha})}{\alpha^{n-1}\Gamma(\frac{n}{\alpha})} \widetilde{k}(\eta)$$

in Theorem 42.3, by the assumptions, we have equivalent inequalities (42.163) and (42.164) (for $p > 1$), and the equivalent reverses of (42.163) and (42.164) (for $0 < p < 1$).

(i) For $p > 1$, $k \in \mathbf{N}$, we set $f_k(x)$ and $g_k(y)$ as follows:

$$f_k(x) := \begin{cases} \|x\|_\alpha^{-\eta + \frac{1}{pk} - n}, & x \in \{x \in R_+^n, \|x\|_\alpha \le 1\}, \\ 0, & x \in \{x \in R_+^n; \|x\|_\alpha > 1\}, \end{cases}$$

$$g_k(y) := \begin{cases} 0, & y \in \{y \in R_+^m; \|y\|_\beta < 1\}, \\ \|y\|_\beta^{-\eta - \frac{1}{qk} - m}, & y \in \{y \in R_+^m; \|y\|_\beta \ge 1\}. \end{cases}$$

Then by (42.24) and (42.23) (for $\varepsilon = \frac{1}{k}$), we find

$$\|f_k\|_{p,\Phi_2} = \left(\int_{\{x \in R_+^n; \|x\|_\alpha \le 1\}} \|x\|_\alpha^{-n+\frac{1}{k}} dx \right)^{\frac{1}{p}} = \left(\frac{\Gamma^n(\frac{1}{\alpha})k}{\alpha^{n-1}\Gamma(\frac{n}{\alpha})} \right)^{\frac{1}{p}},$$

$$\|g_k\|_{q,\Psi_2} = \left(\int_{\{y \in R_+^m; \|y\|_\beta \ge 1\}} \|y\|_\beta^{-m-\frac{1}{k}} dy \right)^{\frac{1}{q}} = \left(\frac{\Gamma^m(\frac{1}{\beta})k}{\beta^{m-1}\Gamma(\frac{m}{\beta})} \right)^{\frac{1}{q}},$$

$$\|f_k\|_{p,\Phi_2} \|g_k\|_{q,\Psi_2} = \left(\frac{\Gamma^n(\frac{1}{\alpha})}{\alpha^{n-1}\Gamma(\frac{n}{\alpha})} \right)^{\frac{1}{p}} \left(\frac{\Gamma^m(\frac{1}{\beta})}{\beta^{m-1}\Gamma(\frac{m}{\beta})} \right)^{\frac{1}{q}} k.$$

If there exists a positive constant $\widetilde{K} \le \widetilde{K}(\eta)$ such that (42.163) is valid as we replace $\widetilde{K}(\eta)$ by $\widetilde{K}$, then, in particular, it follows that

$$\begin{aligned} \frac{1}{k}\widetilde{I}_k &< \frac{1}{k} K \|f_k\|_{p,\Phi_2} \|g_k\|_{q,\Psi_2} \\ &= \widetilde{K} \left(\frac{\Gamma^n(\frac{1}{\alpha})}{\alpha^{n-1}\Gamma(\frac{n}{\alpha})} \right)^{\frac{1}{p}} \left(\frac{\Gamma^m(\frac{1}{\beta})}{\beta^{m-1}\Gamma(\frac{m}{\beta})} \right)^{\frac{1}{q}}. \end{aligned} \tag{42.165}$$

By (42.161), one gets

$$\frac{\Gamma^n(\frac{1}{\alpha})}{\alpha^{n-1}\Gamma(\frac{n}{\alpha})}\frac{\Gamma^m(\frac{1}{\beta})}{\beta^{m-1}\Gamma(\frac{m}{\beta})}\widetilde{L}_k \leq \frac{1}{k}\widetilde{I}_k. \tag{42.166}$$

In view of (42.165) and (42.166), we have

$$\left(\frac{\Gamma^n(\frac{1}{\alpha})}{\alpha^{n-1}\Gamma(\frac{n}{\alpha})}\right)^{\frac{1}{q}}\left(\frac{\Gamma^m(\frac{1}{\beta})}{\beta^{m-1}\Gamma(\frac{m}{\beta})}\right)^{\frac{1}{p}}\widetilde{L}_k < \widetilde{K},$$

and then by (42.51) (for $\alpha = -\eta$), $\widetilde{K}(\eta) \leq \widetilde{K}$ $(k \to \infty)$. Therefore, $\widetilde{K} = \widetilde{K}(\eta)$ is the best value of (42.163). We confirm that the constant factor in (42.164) is the best possible, otherwise by using the inequality (cf. (42.16)) $\widetilde{I} \leq \widetilde{J}^{\frac{1}{p}}\|g\|_{q,\Psi_2}$, we can come to a contradiction that the constant factor in (42.163) is not the best possible.

(ii) For $0 < p < 1$, $k \in \mathbf{N}$, we set $f_k(x)$ and $g_k(y)$ as in (i). If there exists a positive constant $K \geq \widetilde{K}(\eta)$ such that the reverse of (42.163) is valid as we replace $\widetilde{K}(\eta)$ by K, then, in particular, we obtain

$$\begin{aligned}\frac{1}{k}\widetilde{I}_k &> \frac{1}{k}K\|f_k\|_{p,\Phi_2}\|g_k\|_{q,\Psi_2}\\ &= K\left(\frac{\Gamma^n(\frac{1}{\alpha})}{\alpha^{n-1}\Gamma(\frac{n}{\alpha})}\right)^{\frac{1}{p}}\left(\frac{\Gamma^m(\frac{1}{\beta})}{\beta^{m-1}\Gamma(\frac{m}{\beta})}\right)^{\frac{1}{q}}.\end{aligned} \tag{42.167}$$

By (42.161), it follows that

$$\frac{\Gamma^n(\frac{1}{\alpha})}{\alpha^{n-1}\Gamma(\frac{n}{\alpha})}\frac{\Gamma^m(\frac{1}{\beta})}{\beta^{m-1}\Gamma(\frac{m}{\beta})}\widetilde{L}_k = \frac{1}{k}\widetilde{I}_k. \tag{42.168}$$

In view of (42.167) and (42.168), we have

$$\left(\frac{\Gamma^n(\frac{1}{\alpha})}{\alpha^{n-1}\Gamma(\frac{n}{\alpha})}\right)^{\frac{1}{q}}\left(\frac{\Gamma^m(\frac{1}{\beta})}{\beta^{m-1}\Gamma(\frac{m}{\beta})}\right)^{\frac{1}{p}}\widetilde{L}_k > K,$$

and then by (42.51) (for $\alpha = -\eta$), $\widetilde{K}(\eta) \geq K$ $(k \to \infty)$. Therefore, $K = \widetilde{K}(\eta)$ is the best value of the reverse of (42.163).

We confirm that the constant factor in the reverse of (42.164) is the best possible, otherwise by using the reverse inequality $\widetilde{I} \geq \widetilde{J}^{\frac{1}{p}}\|g\|_{q,\Psi_2}$, we can come to a contradiction that the constant factor in the reverse of (42.163) is not the best possible. The theorem is proved. □

By Remark 42.10, setting $\widetilde{\Phi}_2(x) := \|x\|_\alpha^{p(n+\eta)-n}$ $(x \in \mathbf{R}^n)$, $\widetilde{\Psi}_2(y) := \|y\|_\beta^{q(m+\eta)-m}$ $(y \in \mathbf{R}^m)$, we still have

Corollary 42.6 *Suppose that $p > 0$ ($p \neq 1$), $\frac{1}{p} + \frac{1}{q} = 1$, $\eta \in \mathbf{R}$, $h(u)$ is a non-negative measurable function on $\mathbf{R}_+$, $\widetilde{k}(\eta) = \int_0^\infty h(u)u^{-\eta-1}\,du \in \mathbf{R}_+$,*

$$\widetilde{L}(\eta) := \left(\frac{2^m \Gamma^m(\frac{1}{\beta})}{\beta^{m-1}\Gamma(\frac{m}{\beta})}\right)^{\frac{1}{p}} \left(\frac{2^n \Gamma^n(\frac{1}{\alpha})}{\alpha^{n-1}\Gamma(\frac{n}{\alpha})}\right)^{\frac{1}{q}} \widetilde{k}(\eta), \tag{42.169}$$

and $f(\geq 0) \in L_{p,\widetilde{\Phi}_2}(\mathbf{R}^n)$, $g(\geq 0) \in L_{q,\widetilde{\Psi}_2}(\mathbf{R}^m)$ such that $\|f\|_{p,\widetilde{\Phi}_2} > 0$, $\|g\|_{q,\widetilde{\Psi}_2} > 0$.

(i) *If $p > 1$, then we have the following equivalent inequalities with the best constant factors $\widetilde{L}(\eta)$ and $\widetilde{L}^p(\eta)$:*

$$\int_{\mathbf{R}^m}\int_{\mathbf{R}^n} h\big(\|x\|_\alpha \|y\|_\beta\big) f(x)g(y)\,dx\,dy < \widetilde{L}(\eta)\|f\|_{p,\widetilde{\Phi}_2}\|g\|_{q,\widetilde{\Psi}_2}, \tag{42.170}$$

$$\int_{\mathbf{R}^m} \frac{1}{\|y\|_\beta^{p\eta+m}} \left(\int_{\mathbf{R}^n} h\big(\|x\|_\alpha \|y\|_\beta\big) f(x)\,dx\right)^p dy < \widetilde{L}^p(\eta)\|f\|^p_{p,\widetilde{\Phi}_2}; \tag{42.171}$$

(ii) *If $0 < p < 1$, then we have the equivalent reverses of (42.170) and (42.171). Moreover, if there exists a constant $\delta_0 > 0$ such that $\widetilde{k}(\eta + \delta_0) \in \mathbf{R}_+$, then the reverses of (42.170) and (42.171) possess the best constant factors $\widetilde{L}(\eta)$ and $\widetilde{L}^p(\eta)$, respectively.*

In view of Theorem 42.14, for $\widetilde{k}(\eta) \in \mathbf{R}_+$, we define a Hilbert-type integral operator $\widetilde{T}_\eta : L_{p,\Phi_2}(\mathbf{R}^n_+) \to L_{p,\Psi_2^{1-p}}(\mathbf{R}^m_+)$ as follows: for $f \in L_{p,\Phi_2}(\mathbf{R}^n_+)$, there exists a unique representation $\widetilde{T}_\eta f \in L_{p,\Psi_2^{1-p}}(\mathbf{R}^m_+)$, satisfying

$$\widetilde{T}_\eta f(y) := \int_{\mathbf{R}^n_+} h\big(\|x\|_\alpha \|y\|_\beta\big) f(x)\,dx \quad \big(y \in \mathbf{R}^m_+\big). \tag{42.172}$$

Then it follows by (42.163) and (42.164) that

$$(\widetilde{T}_\eta f, g) < \widetilde{K}(\eta)\|f\|_{p,\Phi_2}\|g\|_{q,\Psi_2}, \tag{42.173}$$

$$\|\widetilde{T}_\eta f\|_{p,\Psi_2^{1-p}} < \widetilde{K}(\eta)\|f\|_{p,\Phi_2}, \tag{42.174}$$

where the constant factor $\widetilde{K}(\eta)$ is the best possible. Hence we still have

Theorem 42.15 *Suppose that the Hilbert-type integral operator $\widetilde{T}_\eta$ is defined by (42.172). Then it follows*

$$\|\widetilde{T}_\eta\| := \sup_{f(\neq\theta)\in L_{p,\Phi_2}(\mathbf{R}^n_+)} \frac{\|\widetilde{T}_\eta f\|_{p,\Psi_2^{1-p}}}{\|f\|_{p,\Phi_2}} = \widetilde{K}(\eta), \tag{42.175}$$

where $\widetilde{K}(\eta)$ is indicated by (42.162).

Remark 42.11 (i) In Theorem 42.15, if $h(u)=k_\gamma(1,u)$ $(\gamma\in\mathbf{R})$, $\eta=-\frac{\gamma}{2}$, $\widehat{\varphi}(x)=\|x\|_\alpha^{p(1+\frac{\gamma}{2})-1}$ $(x\in\mathbf{R}_+^n)$, $\widehat{\psi}(y)=\|y\|_\beta^{q(1+\frac{\gamma}{2})-1}$ $(y\in\mathbf{R}_+^m)$, $\widehat{k}(\frac{\gamma}{2})=\int_0^\infty k_\gamma(1,u)\times u^{-\frac{\gamma}{2}-1}\,du\in\mathbf{R}_+$,

$$\widetilde{K}\left(\frac{\gamma}{2}\right):=\left(\frac{\Gamma^m(\frac{1}{\beta})}{\beta^{m-1}\Gamma(\frac{m}{\beta})}\right)^{\frac{1}{p}}\left(\frac{\Gamma^n(\frac{1}{\alpha})}{\alpha^{n-1}\Gamma(\frac{n}{\alpha})}\right)^{\frac{1}{q}}\widetilde{k}\left(\frac{\gamma}{2}\right), \tag{42.176}$$

then we define an operator $T_\gamma: L_{p,\widehat{\varphi}}(\mathbf{R}_+^n)\to L_{p,\widehat{\psi}^{1-p}}(\mathbf{R}_+^m)$ as follows: for $f\in L_{p,\widehat{\varphi}}(\mathbf{R}_+^n)$, there exists a unique representation $T_\gamma f\in L_{p,\widehat{\psi}^{1-p}}(\mathbf{R}_+^m)$, satisfying

$$T_\gamma f(y)=\int_{\mathbf{R}_+^n}k_\gamma\big(1,\|x\|_\alpha\|y\|_\beta\big)f(x)\,dx\quad\big(y\in\mathbf{R}_+^m\big).$$

We have

$$\|T_\gamma\|:=\sup_{f(\neq\theta)\in L_{p,\widehat{\varphi}}(\mathbf{R}_+^n)}\frac{\|T_\gamma f\|_{p,\widehat{\psi}^{1-p}}}{\|f\|_{p,\widehat{\varphi}}}=\widehat{K}\left(\frac{\gamma}{2}\right). \tag{42.177}$$

(ii) We still can write some similar results for $h(u)=k_\gamma(u,1)$ $(\gamma\in\mathbf{R})$.

(iii) By virtue of Theorem 42.15, suppose that $h(u)$ is a non-negative measurable function on $\mathbf{R}_+$.

(a) If we set

$$H\big(\|x\|_\alpha\|y\|_\beta\big):=\begin{cases}h(\|x\|_\alpha\|y\|_\beta), & 0<\|x\|_\alpha\le 1/\|y\|_\beta,\\ 0, & \|x\|_\alpha>1/\|y\|_\beta,\end{cases}$$

then we define the first class Hardy-type integral operator $\widetilde{T}_\eta^{(1)}: L_{p,\varphi_2}(\mathbf{R}_+^n)\to L_{p,\psi_2^{1-p}}(\mathbf{R}_+^m)$ with the non-homogeneous kernel as follows: for $f\in L_{p,\varphi_2}(\mathbf{R}_+^n)$, there exists a unique representation $T_\eta^{(1)}f\in L_{p,\psi_2^{1-p}}(\mathbf{R}_+^m)$, satisfying

$$\begin{aligned}T_\eta^{(1)}f(y)&=\int_{\mathbf{R}_+^n}H\big(\|x\|_\alpha\|y\|_\beta\big)f(x)\,dx\\&=\int_{\{x\in\mathbf{R}_+^n;0<\|x\|_\alpha\le 1/\|y\|_\beta\}}h\big(\|x\|_\alpha\|y\|_\beta\big)f(x)\,dx\quad\big(y\in\mathbf{R}_+^m\big).\end{aligned}$$

Hence, for $\widetilde{k}_1(\eta)=\int_0^1 h(u)u^{-\eta-1}\,du\in\mathbf{R}_+$, we obtain

$$\begin{aligned}&\widetilde{\omega}_2(y)=\widetilde{\varpi}_2(x)=\widetilde{K}_1(\eta)\\&\quad:=\left[\frac{\Gamma^n(\frac{1}{\alpha})}{\alpha^{n-1}\Gamma(\frac{n}{\alpha})}\right]^{\frac{1}{q}}\left[\frac{\Gamma^m(\frac{1}{\beta})}{\beta^{m-1}\Gamma(\frac{m}{\beta})}\right]^{\frac{1}{p}}\widetilde{k}_1(\eta),\quad\big\|T_\gamma^{(1)}\big\|=\widetilde{K}_1(\eta).\end{aligned} \tag{42.178}$$

(b) If we set

$$H\big(\|x\|_\alpha\|y\|_\beta\big):=\begin{cases}0, & 0<\|x\|_\alpha\le 1/\|y\|_\beta,\\ h\big(\|x\|_\alpha\|y\|_\beta\big), & \|x\|_\alpha>1/\|y\|_\beta,\end{cases}$$

then we define the second class Hardy-type integral operator $T_\eta^{(2)}: L_{p,\varphi_2}(\mathbf{R}_+^n)\to L_{p,\psi_2^{1-p}}(\mathbf{R}_+^m)$ with the non-homogeneous kernel as follows: for $f\in L_{p,\varphi_2}(\mathbf{R}_+^n)$, there exists a unique representation $T_\eta^{(2)}f\in L_{p,\psi_2^{1-p}}(\mathbf{R}_+^m)$, satisfying

$$\begin{aligned}T_\eta^{(2)}f(y)&=\int_{\mathbf{R}_+^n}H\big(\|x\|_\alpha\|y\|_\beta\big)f(x)\,dx\\&=\int_{\{x\in\mathbf{R}_+^n;\|x\|_\alpha>1/\|y\|_\beta\}}h\big(\|x\|_\alpha\|y\|_\beta\big)f(x)\,dx\quad\big(y\in\mathbf{R}_+^m\big).\end{aligned}$$

Hence, for $\widetilde{k}_2(\eta)=\int_1^\infty h(u)u^{-\eta-1}\,du\in\mathbf{R}_+$, we obtain

$$\begin{aligned}\widetilde{\omega}_2(y)&=\widetilde{\varpi}_2(x)=\widetilde{K}_2(\eta)\\&:=\left(\frac{\Gamma^n(\frac{1}{\alpha})}{\alpha^{n-1}\Gamma(\frac{n}{\alpha})}\right)^{\frac{1}{q}}\left(\frac{\Gamma^m(\frac{1}{\beta})}{\beta^{m-1}\Gamma(\frac{m}{\beta})}\right)^{\frac{1}{p}}\widetilde{k}_2(\eta),\quad\big\|T_\gamma^{(2)}\big\|=\widetilde{K}_2(\eta).\end{aligned}\tag{42.179}$$

42.4.3 Some Particular Examples

Example 42.9 If $\gamma=-\lambda<0$, $r>1$, $\frac{1}{r}+\frac{1}{s}=1$, $k_{-\lambda}(x,y)=\frac{1}{(x+y)^\lambda}$, then by Theorem 42.15, we find

$$k\left(\frac{-\lambda}{r}\right)=k_{-\lambda}(r):=\int_0^\infty\frac{1}{(u+1)^\lambda}u^{\frac{\lambda}{r}-1}\,du=B\left(\frac{\lambda}{r},\frac{\lambda}{s}\right)\in\mathbf{R}_+.\tag{42.180}$$

(i) Setting $\Phi_1(x):=\|x\|_\alpha^{p(n-\frac{\lambda}{r})-n}$ $(x\in\mathbf{R}_+^n)$, $\Psi_1(y):=\|y\|_\beta^{q(m-\frac{\lambda}{s})-m}$ $(y\in\mathbf{R}_+^m)$, we define an operator $T_{-\lambda}: L_{p,\Phi_1}(\mathbf{R}_+^n)\to L_{p,\Psi_1^{1-p}}(\mathbf{R}_+^m)$ as follows: for $f\in L_{p,\Phi_1}(\mathbf{R}_+^n)$, there exists a unique representation $T_{-\lambda}f\in L_{p,\Psi_1^{1-p}}(\mathbf{R}_+^m)$, satisfying

$$T_{-\lambda}f(y)=\int_{\mathbf{R}_+^n}\frac{1}{(\|x\|_\alpha+\|y\|_\beta)^\lambda}f(x)\,dx\quad\big(y\in\mathbf{R}_+^m\big).$$

Then we have

$$\|T_{-\lambda}\|=\left(\frac{\Gamma^m(\frac{1}{\beta})}{\beta^{m-1}\Gamma(\frac{m}{\beta})}\right)^{\frac{1}{p}}\left(\frac{\Gamma^n(\frac{1}{\beta})}{\beta^{n-1}\Gamma(\frac{n}{\beta})}\right)^{\frac{1}{q}}B\left(\frac{\lambda}{r},\frac{\lambda}{s}\right).\tag{42.181}$$

In particular, for $m=n$, $\beta=\alpha$, we have (cf. [45])

$$\|T_{-\lambda}\| = \frac{\Gamma^n(\frac{1}{\alpha})}{\alpha^{n-1}\Gamma(\frac{n}{\alpha})} B\left(\frac{\lambda}{r}, \frac{\lambda}{s}\right). \tag{42.182}$$

(ii) By Remark 42.9(ii), for $\rho \in \mathbf{R}$, $\widetilde{\varphi}(x) = \|x\|_\alpha^{p(n-\frac{\lambda}{2})-n}$ $(x \in \mathbf{R}_+^n)$, $\widetilde{\psi}(y) = \|y\|_\beta^{q(m-\frac{\lambda}{2})-m}$ $(y \in \mathbf{R}_+^m)$, we define an operator $\widetilde{T}_{-\lambda,\rho} : L_{p,\widetilde{\varphi}}(\mathbf{R}_+^n) \to L_{p,\widetilde{\psi}^{1-p}}(\mathbf{R}_+^m)$ as follows: for $f \in L_{p,\widetilde{\varphi}}(\mathbf{R}_+^n)$, there exists a unique representation $\widetilde{T}_{-\lambda,\rho} f \in L_{p,\widetilde{\psi}^{1-p}}(\mathbf{R}_+^m)$, satisfying

$$\widetilde{T}_{-\lambda,\rho} f(y) = \int_{\mathbf{R}_+^n} \frac{1}{(\|x\|_\alpha + \|y\|_\beta)^\lambda} \arctan\left(\frac{\|x\|_\alpha}{\|y\|_\beta}\right)^\rho f(x)\,dx \quad \left(y \in \mathbf{R}_+^m\right).$$

Then we have

$$\|\widetilde{T}_{-\lambda,\rho}\| = \frac{\pi}{4}\left(\frac{\Gamma^m(\frac{1}{\beta})}{\beta^{m-1}\Gamma(\frac{m}{\beta})}\right)^{\frac{1}{p}} \left(\frac{\Gamma^n(\frac{1}{\alpha})}{\alpha^{n-1}\Gamma(\frac{n}{\alpha})}\right)^{\frac{1}{q}} B\left(\frac{\lambda}{2}, \frac{\lambda}{2}\right). \tag{42.183}$$

By Remark 42.11(i), define an operator $\widetilde{T}_{-\lambda} : L_{p,\widetilde{\varphi}}(\mathbf{R}_+^n) \to L_{p,\widetilde{\psi}^{1-p}}(\mathbf{R}_+^m)$ as follows: for $f \in L_{p,\widetilde{\varphi}}(\mathbf{R}_+^n)$, there exists a unique representation $\widetilde{T}_{-\lambda} f \in L_{p,\widetilde{\psi}^{1-p}}(\mathbf{R}_+^m)$, satisfying

$$\widetilde{T}_{-\lambda} f(y) = \int_{\mathbf{R}_+^n} \frac{1}{(1 + \|x\|_\alpha \|y\|_\beta)^\lambda} f(x)\,dx \quad \left(y \in \mathbf{R}_+^m\right).$$

Then we have

$$\|\widetilde{T}_{-\lambda}\| = \left(\frac{\Gamma^m(\frac{1}{\beta})}{\beta^{m-1}\Gamma(\frac{m}{\beta})}\right)^{\frac{1}{p}} \left(\frac{\Gamma^n(\frac{1}{\alpha})}{\alpha^{n-1}\Gamma(\frac{n}{\alpha})}\right)^{\frac{1}{q}} B\left(\frac{\lambda}{2}, \frac{\lambda}{2}\right). \tag{42.184}$$

Example 42.10 If $\gamma = \lambda > 0$, $r > 1$, $\frac{1}{r} + \frac{1}{s} = 1$, $k_\lambda(x, y) = (\min\{x, y\})^\lambda$, then by Theorem 42.13, we find $k(\frac{\lambda}{r}) = k_\lambda(r) := \int_0^\infty (\min\{u, 1\})^\lambda u^{\frac{-\lambda}{r}-1}\,du = \frac{rs}{\lambda} \in \mathbf{R}_+$.

(i) Setting $\Phi_1(x) := \|x\|_\alpha^{p(n+\frac{\lambda}{r})-n}$ $(x \in \mathbf{R}_+^n)$, $\Psi_1(y) := \|y\|_\beta^{q(m+\frac{\lambda}{s})-m}$ $(y \in \mathbf{R}_+^m)$, we define an operator $T_\lambda : L_{p,\Phi_1}(\mathbf{R}_+^n) \to L_{p,\Psi_1^{1-p}}(\mathbf{R}_+^m)$ as follows: for $f \in L_{p,\Phi_1}(\mathbf{R}_+^n)$, there exists a unique representation $\widetilde{T}_\lambda f \in L_{p,\Psi_1^{1-p}}(\mathbf{R}_+^m)$, satisfying

$$T_\lambda f(y) = \int_{\mathbf{R}_+^n} \left(\min\{\|x\|_\alpha, \|y\|_\beta\}\right)^\lambda f(x)\,dx \quad \left(y \in \mathbf{R}_+^m\right).$$

Then we have

$$\|T_\lambda\| = \left(\frac{\Gamma^m(\frac{1}{\beta})}{\beta^{m-1}\Gamma(\frac{m}{\beta})}\right)^{\frac{1}{p}} \left(\frac{\Gamma^n(\frac{1}{\alpha})}{\alpha^{n-1}\Gamma(\frac{n}{\alpha})}\right)^{\frac{1}{q}} \frac{rs}{\lambda}. \tag{42.185}$$

In particular, for $m = n$, $\beta = \alpha$, we have

$$\|T_\lambda\| = \frac{\Gamma^n(\frac{1}{\alpha})}{\alpha^{n-1}\Gamma(\frac{n}{\alpha})}\frac{rs}{\lambda}. \tag{42.186}$$

(ii) By Remark 42.9(ii), for $\rho \in \mathbf{R}$, $\widetilde{\varphi}(x) = \|x\|_\alpha^{p(n+\frac{\lambda}{2})-n}$ $(x \in \mathbf{R}_+^n)$, $\widetilde{\psi}(y) = \|y\|_\beta^{q(m+\frac{\lambda}{2})-m}$ $(y \in \mathbf{R}_+^m)$, we define an operator $\widetilde{T}_{\lambda,\rho} : L_{p,\widetilde{\varphi}}(\mathbf{R}_+^n) \to L_{p,\widetilde{\psi}^{1-p}}(\mathbf{R}_+^m)$ as follows: for $f \in L_{p,\widetilde{\varphi}}(\mathbf{R}_+^n)$, there exists a unique representation $\widetilde{T}_{\lambda,\rho} f \in L_{p,\widetilde{\psi}^{1-p}}(\mathbf{R}_+^m)$, satisfying

$$\widetilde{T}_{\lambda,\rho} f(y) = \int_{\mathbf{R}_+^n} \bigl(\min\{\|x\|_\alpha, \|y\|_\beta\}\bigr)^\lambda \arctan\left(\frac{\|x\|_\alpha}{\|y\|_\beta}\right)^\rho f(x)\,dx \quad \bigl(y \in \mathbf{R}_+^m\bigr).$$

Then we have

$$\|\widetilde{T}_{\lambda,\rho}\| = \frac{\pi}{\lambda}\left(\frac{\Gamma^m(\frac{1}{\beta})}{\beta^{m-1}\Gamma(\frac{m}{\beta})}\right)^{\frac{1}{p}}\left(\frac{\Gamma^n(\frac{1}{\alpha})}{\alpha^{n-1}\Gamma(\frac{n}{\alpha})}\right)^{\frac{1}{q}}. \tag{42.187}$$

By Remark 42.11(i), define an operator $\widetilde{T}_{-\lambda} : L_{p,\widetilde{\varphi}}(\mathbf{R}_+^n) \to L_{p,\widetilde{\psi}^{1-p}}(\mathbf{R}_+^m)$ as follows: for $f \in L_{p,\widetilde{\varphi}}(\mathbf{R}_+^n)$, there exists a unique representation $\widetilde{T}_\lambda f \in L_{p,\widetilde{\psi}^{1-p}}(\mathbf{R}_+^m)$, satisfying $\widetilde{T}_\lambda f(y) = \int_{\mathbf{R}_+^n} (\min\{1, \|x\|_\alpha \|y\|_\beta\})^\lambda f(x)\,dx$ $(y \in \mathbf{R}_+^m)$. Then we have

$$\|\widetilde{T}_\lambda\| = \frac{4}{\lambda}\left(\frac{\Gamma^m(\frac{1}{\beta})}{\beta^{m-1}\Gamma(\frac{m}{\beta})}\right)^{\frac{1}{p}}\left(\frac{\Gamma^n(\frac{1}{\alpha})}{\alpha^{n-1}\Gamma(\frac{n}{\alpha})}\right)^{\frac{1}{q}}. \tag{42.188}$$

42.5 Compositions of Two Hilbert-Type Integral Operators with the Kernels on $\mathbf{R}_+ \times \mathbf{R}_+$

42.5.1 Some Lemmas

Lemma 42.13 *Suppose that* $p > 0$ $(p \neq 1)$, $\frac{1}{p} + \frac{1}{q} = 1$, $\lambda, \lambda_1, \lambda_2 > 0$, $\lambda_1 + \lambda_2 = \lambda$, $k_{-\lambda}^{(i)}(x, y)$ $(i = 1, 2, 3)$ *are non-negative homogeneous functions of degree* $-\lambda$ *on* $\mathbf{R}_+^2$,

$$K^{(i)}(\lambda_1) := \int_0^\infty k_{-\lambda}^{(i)}(u, 1)u^{\lambda_1 - 1}\,du \quad (i = 1, 2, 3), \tag{42.189}$$

and there exists a constant $\delta_0 \in (0, \min\{\lambda_1, \lambda_2\})$ *such that*

$$K^{(i)}(\lambda_1 \pm \delta_0) \in \mathbf{R}_+ \quad (i = 1, 2, 3). \tag{42.190}$$

Then for any $\delta \in [0, \delta_0)$, *we have* $K^{(i)}(\lambda_1 \pm \delta) \in [\mathbf{0}, \infty)$ *and*

$$\lim_{\delta \to 0^+} K^{(i)}(\lambda_1 \pm \delta) = K^{(i)}(\lambda_1) \quad (i = 1, 2, 3). \tag{42.191}$$

Proof Since for any $\delta \in [0, \delta_0)$ we have

$$0 \le k_{-\lambda}^{(i)}(u, 1)u^{\lambda_1 \pm \delta - 1} \le k_{-\lambda}^{(i)}(u, 1)u^{\lambda_1 - \delta_0 - 1} \quad \big(u \in (0, 1)\big),$$
$$0 \le k_{-\lambda}^{(i)}(u, 1)u^{\lambda_1 \pm \delta - 1} \le k_{-\lambda}^{(i)}(u, 1)u^{\lambda_1 + \delta_0 - 1} \quad \big(u \in [1, \infty)\big)$$

$(i = 1, 2, 3)$, and, by (42.190), it follows that

$$\int_0^1 k_{-\lambda}^{(i)}(u, 1)u^{\lambda_1 - \delta_0 - 1}\, du \le K^{(i)}(\lambda_1 - \delta_0) < \infty,$$
$$\int_1^\infty k_{-\lambda}^{(i)}(u, 1)u^{\lambda_1 + \delta_0 - 1}\, du \le K^{(i)}(\lambda_1 + \delta_0) < \infty,$$

we obtain $0 \le K^{(i)}(\lambda_1 \pm \delta) \le K^{(i)}(\lambda_1 - \delta_0) + K^{(i)}(\lambda_1 + \delta_0) < \infty$, and $K^{(i)}(\lambda_1 \pm \delta) \in [\mathbf{0}, \infty)$ $(\delta \in [\mathbf{0}, \delta_0))$. Hence by Lebesgue dominated convergence theorem (cf. [2]), we obtain

$$\begin{aligned}\lim_{\delta \to 0^+} K^{(i)}(\lambda_1 \pm \delta) &= \lim_{\delta \to 0^+} \left[\int_0^1 k_{-\lambda}^{(i)}(u, 1)u^{\lambda_1 \pm \delta - 1}\, du + \int_1^\infty k_{-\lambda}^{(i)}(u, 1)u^{\lambda_1 \pm \delta - 1}\, du\right] \\ &= \int_0^1 k_{-\lambda}^{(i)}(u, 1)u^{\lambda_1 - 1}\, du + \int_1^\infty k_{-\lambda}^{(i)}(u, 1)u^{\lambda_1 - 1}\, du \\ &= K^{(i)}(\lambda_1) \quad (i = 1, 2, 3).\end{aligned}$$

The lemma is proved. □

Definition 42.10 Under the assumptions of Lemma 42.13, for $k \in \mathbf{N}$, $k > \max\{\frac{1}{|q|\delta_0}, \frac{1}{p\delta_0}\}$, we define two functions $\widehat{F}_\lambda(y)$ and $\widehat{G}_\lambda(x)$ as follows:

$$\begin{aligned}\widehat{F}_\lambda(y) &:= y^{\lambda - 1} \int_1^\infty k_{-\lambda}^{(2)}(x, y)x^{\lambda_1 - \frac{1}{pk} - 1}\, dx \quad \big(y \in (1, \infty)\big), \\ \widehat{G}_\lambda(x) &:= x^{\lambda - 1} \int_1^\infty k_{-\lambda}^{(3)}(x, y)y^{\lambda_2 - \frac{1}{qk} - 1}\, dy \quad \big(x \in (1, \infty)\big).\end{aligned} \tag{42.192}$$

Lemma 42.14 *Suppose that* $p > 0$ $(p \ne 1)$, $\frac{1}{p} + \frac{1}{q} = 1$, $\lambda, \lambda_1, \lambda_2 > 0$, $\lambda_1 + \lambda_2 = \lambda$, $k_{-\lambda}^{(i)}(x, y)$ $(i = 2, 3)$ *are non-negative homogeneous functions of degree* $-\lambda$ *on* $\mathbf{R}_+^2$, $K^{(i)}(\lambda_1) = \int_0^\infty k_{-\lambda}^{(i)}(u, 1) \times u^{\lambda_1 - 1}\, du$ $(i = 2, 3)$, *and there exists a constant* $\delta_0 \in (0, \min\{\lambda_1, \lambda_2\})$ *such that* $K^{(i)}(\lambda_1 \pm \delta_0) \in \mathbf{R}_+$ $(i = 2, 3)$. *Setting two functions*

$F(y)$ and $G(x)$ as follows:

$$F(y) := y^{\lambda_1 - \frac{1}{pk} - 1} K^{(2)}\left(\lambda_1 - \frac{1}{pk}\right) - \widehat{F}_\lambda(y) \quad \big(y \in (1, \infty)\big),$$
$$G(x) := x^{\lambda_2 - \frac{1}{qk} - 1} K^{(3)}\left(\lambda_1 + \frac{1}{qk}\right) - \widehat{G}_\lambda(x) \quad \big(x \in (1, \infty)\big), \tag{42.193}$$

(i) *If there exist constants $\delta_1 \in (0, \delta_0)$ and $L > 0$ such that for any $u \in [1, \infty)$,*

$$k_{-\lambda}^{(2)}(1, u) u^{\lambda_2 + \delta_1} \le L, \qquad k_{-\lambda}^{(3)}(u, 1) u^{\lambda_1 + \delta_1} \le L, \tag{42.194}$$

then for $k > \max\{\frac{1}{|q|\delta_1}, \frac{1}{p\delta_1}\}$, we have

$$F(y) = O\big(y^{\lambda_1 - \delta_1 - 1}\big) \ge 0, \qquad G(x) = O\big(x^{\lambda_2 - \delta_1 - 1}\big) \ge 0 \quad \big(y, x \in (1, \infty)\big);$$

(ii) *If $0 < \lambda_1, \lambda_2 < 1$, there exist constants $a \in (\max\{\lambda_1, \lambda_2\}, 1)$ and $L_1 > 0$ such that for any $u \in [1, \infty)$,*

$$k_{-\lambda}^{(2)}(1, u)(u - 1)^a \le L_1, \qquad k_{-\lambda}^{(3)}(u, 1)(u - 1)^a \le L_1, \tag{42.195}$$

then for $k > \max\{\frac{1}{|q|\delta_0}, \frac{1}{p\delta_0}, \frac{1}{p(a-\lambda_2)}, \frac{1}{|q|(a-\lambda_1)}\}$, we have

$$F(y) = O\left(\frac{y^{\lambda - 1}}{(y - 1)^a}\right) \ge 0, \qquad G(x) = O\left(\frac{x^{\lambda - 1}}{(x - 1)^a}\right) \ge 0 \quad \big(y, x \in (1, \infty)\big).$$

Proof In (42.192), setting $u = x/y$, one gets

$$\begin{aligned}
\widehat{F}_\lambda(y) &= y^{\lambda_1 - \frac{1}{pk} - 1} \int_{1/y}^{\infty} k_{-\lambda}^{(2)}(u, 1) u^{\lambda_1 - \frac{1}{pk} - 1}\, du \\
&= y^{\lambda_1 - \frac{1}{pk} - 1} \left[\int_0^\infty k_{-\lambda}^{(2)}(u, 1) u^{(\lambda_1 - \frac{1}{pk}) - 1}\, du - \int_0^{1/y} k_{-\lambda}^{(2)}(u, 1) u^{\lambda_1 - \frac{1}{pk} - 1}\, du\right] \\
&= y^{\lambda_1 - \frac{1}{pk} - 1} K^{(2)}\left(\lambda_1 - \frac{1}{pk}\right) - F(y),
\end{aligned}$$

$$F(y) = y^{\lambda_1 - \frac{1}{pk} - 1} \int_0^{1/y} k_{-\lambda}^{(2)}(u, 1) u^{\lambda_1 - \frac{1}{pk} - 1}\, du \ge 0;$$

once again setting $u = x/y$, we find

$$\begin{aligned}
\widehat{G}_\lambda(x) &= x^{\lambda_2 - \frac{1}{qk} - 1} \int_0^x k_{-\lambda}^{(3)}(u, 1) u^{\lambda_1 + \frac{1}{qk} - 1}\, du \\
&= x^{\lambda_2 - \frac{1}{qk} - 1} \left[\int_0^\infty k_{-\lambda}^{(3)}(u, 1) u^{\lambda_1 + \frac{1}{qk} - 1}\, du - \int_x^\infty k_{-\lambda}^{(3)}(u, 1) u^{\lambda_1 + \frac{1}{qk} - 1}\, du\right]
\end{aligned}$$

$$= x^{\lambda_2-\frac{1}{qk}-1}K^{(3)}\left(\lambda_1+\frac{1}{qk}\right)-G(x),$$

$$G(x)=x^{\lambda_2-\frac{1}{qk}-1}\int_x^\infty k_{-\lambda}^{(3)}(u,1)u^{\lambda_1+\frac{1}{qk}-1}\,du\geq 0.$$

(i) By (42.194), we obtain

$$\begin{aligned}0\leq F(y)&\overset{v=1/u}{=}y^{\lambda_1-\frac{1}{pk}-1}\int_y^\infty k_{-\lambda}^{(2)}(1,v)v^{\lambda_2+\frac{1}{pk}-1}\,dv\\&\leq y^{\lambda_1-\frac{1}{pk}-1}L\int_y^\infty v^{-\lambda_2-\delta_1}v^{\lambda_2+\frac{1}{pk}-1}\,dv\\&=y^{\lambda_1-\frac{1}{pk}-1}L\int_y^\infty v^{-\delta_1+\frac{1}{pk}-1}\,dv=\frac{L}{\delta_1-\frac{1}{pk}}y^{\lambda_1-\delta_1-1},\end{aligned}$$

and then $F(y)=O(y^{\lambda_1-\delta_1-1})$ $(y\in(1,\infty))$; still by (42.194), we obtain

$$\begin{aligned}0\leq G(x)&\leq x^{\lambda_2-\frac{1}{qk}-1}L\int_x^\infty u^{-\lambda_1-\delta_1}u^{\lambda_1+\frac{1}{qk}-1}\,du\\&=x^{\lambda_2-\frac{1}{qk}-1}L\int_x^\infty u^{-\delta_1+\frac{1}{qk}-1}\,du=\frac{L}{\delta_1-\frac{1}{qk}}x^{\lambda_2-\delta_1-1},\end{aligned}$$

and then $G(x)=O(x^{\lambda_2-\delta_1-1})$ $(x\in(1,\infty))$.

(ii) By (42.195), we find

$$\begin{aligned}0\leq F(y)&=y^{\lambda_1-\frac{1}{pk}-1}\int_y^\infty k_{-\lambda}^{(2)}(1,v)v^{\lambda_2+\frac{1}{pk}-1}\,dv\\&\leq y^{\lambda_1-\frac{1}{pk}-1}L_1\int_y^\infty\frac{1}{(v-1)^a}v^{\lambda_2+\frac{1}{pk}-1}\,dv\\&\overset{u=y/v}{=}y^{\lambda-1}L_1\int_0^1\frac{u^{a-\lambda_2-\frac{1}{pk}-1}}{(y-u)^a}\,du\\&\leq\frac{L_1y^{\lambda-1}}{(y-1)^a}\int_0^1u^{a-\lambda_2-\frac{1}{pk}-1}\,du\leq\frac{L_1}{a-\lambda_2-\frac{1}{pk}}\frac{y^{\lambda-1}}{(y-1)^a},\end{aligned}$$

and then $F(y)=O(\frac{y^{\lambda-1}}{(y-1)^a})$ $(y\in(1,\infty))$; still by (42.195), it follows

$$\begin{aligned}0\leq G(x)&\leq x^{\lambda_2-\frac{1}{qk}-1}L_1\int_x^\infty\frac{1}{(u-1)^a}u^{\lambda_1+\frac{1}{qk}-1}\,du\\&\overset{v=x/u}{=}x^{\lambda-1}L_1\int_0^1\frac{v^{a-\lambda_1-\frac{1}{qk}-1}}{(x-v)^a}\,dv\end{aligned}$$

$$\leq \frac{L_1 x^{\lambda-1}}{(x-1)^a}\int_0^1 v^{a-\lambda_1-\frac{1}{qk}-1}\,dv = \frac{L_1}{a-\lambda_1-\frac{1}{qk}}\frac{x^{\lambda-1}}{(x-1)^a},$$

and then $G(x) = O(\frac{x^{\lambda-1}}{(x-1)^a})$ $(x \in (1,\infty))$.

The lemma is proved. □

Lemma 42.15 *Let the assumptions of Lemma* 42.13 *be fulfilled and, additionally, let* $k_{-\lambda}^{(2)}(1,u)$ $(k_{-\lambda}^{(3)}(u,1))$ *satisfy* (42.194) *or* (42.195), *and* $k_{-\lambda}^{(1)}(x,y)$ *be symmetric. Then for*

$$k > \max\left\{\frac{2}{|q|\delta_1}, \frac{2}{p\delta_1}, \frac{1}{|q|(a-\lambda_2)}, \frac{1}{p(a-\lambda_2)}, \frac{1}{|q|(a-\lambda_1)}\right\},$$

(*Note. If both* $k_{-\lambda}^{(2)}(1,u)$ *and* $k_{-\lambda}^{(3)}(u,1)$ *satisfy* (42.194), *then we naturally set* $k > \max\{\frac{2}{|q|\delta_1}, \frac{2}{p\delta_1}\}$; *if both* $k_{-\lambda}^{(2)}(1,u)$ *and* $k_{-\lambda}^{(3)}(u,1)$ *satisfy* (42.195), *then we set*

$$k > \max\left\{\frac{2}{|q|\delta_0}, \frac{2}{p\delta_0}, \frac{1}{|q|(a-\lambda_2)}, \frac{1}{p(a-\lambda_2)}, \frac{1}{|q|(a-\lambda_1)}\right\}\Bigg),$$

we have

$$\widehat{L}_k := \frac{1}{k}\int_1^\infty\int_1^\infty k_{-\lambda}^{(1)}(x,y)\widehat{F}_\lambda(y)\widehat{G}_\lambda(x)\,dx\,dy$$
$$\geq \prod_{i=1}^{3} K^{(i)}(\lambda_1) + o(1) \quad (k\to\infty). \tag{42.196}$$

Except for both $k_{-\lambda}^{(2)}(1,u)$ *and* $k_{-\lambda}^{(3)}(u,1)$ *satisfying* (42.195), *we have the reverse of* (42.196).

Proof We find by (42.193) that

$$\begin{aligned}\widehat{L}_k &= \frac{1}{k}\int_1^\infty\int_1^\infty k_{-\lambda}^{(1)}(x,y)\left[y^{\lambda_1-\frac{1}{pk}-1}K^{(2)}\left(\lambda_1-\frac{1}{pk}\right)-F(y)\right]\\ &\quad\times\left[x^{\lambda_2-\frac{1}{qk}-1}K^{(3)}\left(\lambda_1+\frac{1}{qk}\right)-G(x)\right]dx\,dy\\ &= I_1 - I_2 - I_3 + I_4,\end{aligned} \tag{42.197}$$

where I_i $(i = 1,2,3,4)$ are defined by

$$\begin{aligned}I_1 &:= K^{(2)}\left(\lambda_1-\frac{1}{pk}\right)K^{(3)}\left(\lambda_1+\frac{1}{qk}\right)\\ &\quad\times\frac{1}{k}\int_1^\infty\int_1^\infty k_{-\lambda}^{(1)}(x,y)x^{\lambda_2-\frac{1}{qk}-1}y^{\lambda_1-\frac{1}{pk}-1}\,dx\,dy,\end{aligned}$$

$$I_2 := \frac{1}{k} K^{(3)}\left(\lambda_1 + \frac{1}{qk}\right) \int_1^\infty \left(\int_1^\infty k_{-\lambda}^{(1)}(x,y) x^{\lambda_2 - \frac{1}{qk} - 1}\, dx\right) F(y)\, dy,$$

$$I_3 := \frac{1}{k} K^{(2)}\left(\lambda_1 - \frac{1}{pk}\right) \int_1^\infty \left(\int_1^\infty k_{-\lambda}^{(1)}(x,y) y^{\lambda_1 - \frac{1}{pk} - 1}\, dy\right) G(x)\, dx,$$

$$I_4 := \frac{1}{k} \int_1^\infty \left(\int_1^\infty k_{-\lambda}^{(1)}(x,y) F(y)\, dy\right) G(x)\, dx.$$

By Lemma 42.14, $F(y), G(x) \geq 0$, then it follows $I_i \geq 0$ $(i = 1, 2, 3, 4)$.

Since $k_{-\lambda}^{(1)}(x,y)$ is symmetric, for $\delta \in [0, \delta_0]$, we find

$$\begin{aligned} K^{(1)}(\lambda_2 \pm \delta) &= \int_0^\infty k_{-\lambda}^{(1)}(u,1) u^{\lambda_2 \pm \delta - 1}\, du = \int_0^\infty k_{-\lambda}^{(1)}(1,u) u^{\lambda_2 \pm \delta - 1}\, du \\ &\overset{v=1/u}{=} \int_0^\infty k_{-\lambda}^{(1)}(v,1) v^{\lambda_1 \mp \delta - 1}\, dv = K^{(1)}(\lambda_1 \mp \delta). \end{aligned}$$

By Lemma 42.3 (for $\gamma = -\lambda$, $\gamma_1 = -(\lambda_2 - \frac{1}{qk} + \frac{1}{pk})$), we obtain

$$\begin{aligned} &\frac{1}{k} \int_1^\infty \left[\int_1^\infty k_{-\lambda}^{(1)}(x,y) x^{\lambda_2 - \frac{1}{qk} - 1} y^{\lambda_1 - \frac{1}{pk} - 1}\, dx\right] dy \\ &\overset{u=x/y}{=} \frac{1}{k} \int_1^\infty y^{-\frac{1}{k} - 1} \left[\int_{1/y}^\infty k_{-\lambda}^{(1)}(u,1) u^{\lambda_2 - \frac{1}{qk} - 1}\, du\right] dy \\ &= K^{(1)}\left(\lambda_2 - \frac{1}{qk} + \frac{1}{pk}\right) + o(1) = K^{(1)}\left(\lambda_1 + \frac{1}{qk} - \frac{1}{pk}\right) + o(1) \quad (k \to \infty), \end{aligned}$$

and then in view of Lemma 42.13, it follows that $I_1 \to \prod_{i=1}^3 K^{(i)}(\lambda_1)$ $(k \to \infty)$.

(i) If $k_{-\lambda}^{(2)}(u,1)$ satisfies (42.194), then by Lemma 42.14(i), there exists a constant $L_2 > 0$ such that $F(y) = O(y^{\lambda_1 - \delta_1 - 1}) \leq L_2 y^{\lambda_1 - \delta_1 - 1}$ $(y \in (1, \infty))$, and in view of the fact that $k_{-\lambda}^{(1)}(x,y)$ is symmetric, one gets

$$\begin{aligned} 0 &\leq \int_1^\infty \left(\int_1^\infty k_{-\lambda}^{(1)}(x,y) x^{\lambda_2 - \frac{1}{qk} - 1}\, dx\right) F(y)\, dy \\ &\overset{u=y/x}{=} \int_1^\infty y^{-\lambda_1 - \frac{1}{qk}} O\left(y^{\lambda_1 - \delta_1 - 1}\right) \left(\int_0^y k_{-\lambda}^{(1)}(1,u) u^{\lambda_1 + \frac{1}{qk} - 1}\, du\right) dy \\ &\leq L_2 \int_1^\infty y^{-\lambda_1 - \frac{1}{qk}} y^{\lambda_1 - \delta_1 - 1} \left(\int_0^\infty k_{-\lambda}^{(1)}(u,1) u^{\lambda_1 + \frac{1}{qk} - 1}\, du\right) dy \\ &= L_2 K^{(1)}\left(\lambda_1 + \frac{1}{qk}\right) \int_1^\infty y^{-\delta_1 - \frac{1}{qk} - 1}\, dy = \frac{L_2 K^{(1)}(\lambda_1 + \frac{1}{qk})}{\delta_1 + \frac{1}{qk}}; \end{aligned}$$

if $k_{-\lambda}^{(2)}(u,1)$ satisfies (42.195), then by Lemma 42.14(ii), there exists a constant $L_3 > 0$ such that $F(y) = O(\frac{y^{\lambda-1}}{(y-1)^a}) \leq L_3 \frac{y^{\lambda-1}}{(y-1)^a}$ $(y \in (1,\infty))$, and

$$\begin{aligned}
0 &\leq \int_1^\infty \left(\int_1^\infty k_{-\lambda}^{(1)}(x,y) x^{\lambda_2 - \frac{1}{qk} - 1} dx \right) F(y)\, dy \\
&= \int_1^\infty y^{-\lambda_1 - \frac{1}{qk}} O\left(\frac{y^{\lambda-1}}{(y-1)^a} \right) \left(\int_0^y k_{-\lambda}^{(1)}(1,u) u^{\lambda_1 + \frac{1}{qk} - 1} du \right) dy \\
&\leq L_3 \int_1^\infty y^{-\lambda_1 - \frac{1}{qk}} \frac{y^{\lambda-1}}{(y-1)^a} \left(\int_0^\infty k_{-\lambda}^{(1)}(u,1) u^{\lambda_1 + \frac{1}{qk} - 1} du \right) dy \\
&= L_3 K^{(1)} \left(\lambda_1 + \frac{1}{qk} \right) \int_1^\infty \frac{1}{(y-1)^a} y^{\lambda_2 - \frac{1}{qk} - 1} dy \\
&\overset{v=1/y}{=} L_3 K^{(1)} \left(\lambda_1 + \frac{1}{qk} \right) \int_0^1 (1-v)^{(1-a)-1} v^{(a - \lambda_2 + \frac{1}{qk}) - 1} dv \\
&= L_3 K^{(1)} \left(\lambda_1 + \frac{1}{qk} \right) B\left(1 - a, a - \lambda_2 + \frac{1}{qk} \right).
\end{aligned}$$

Hence, in the above two cases, we find $I_2 \to 0$ $(k \to \infty)$.

If $k_{-\lambda}^{(3)}(1,u)$ satisfies (42.194), then by Lemma 42.14(i), there exists a constant $L_4 > 0$ such that $G(x) = O(x^{\lambda_2 - \delta_1 - 1}) \leq L_4 x^{\lambda_2 - \delta_1 - 1}$ $(x \in (1,\infty))$, and in view of the fact that $k_{-\lambda}^{(1)}(x,y)$ is symmetric, one obtains

$$\begin{aligned}
0 &\leq \int_1^\infty \left(\int_1^\infty k_{-\lambda}^{(1)}(x,y) y^{\lambda_1 - \frac{1}{pk} - 1} dy \right) G(x)\, dx \\
&\overset{u=y/x}{=} \int_1^\infty x^{-\lambda_2 - \frac{1}{pk}} \left(\int_{1/x}^\infty k_{-\lambda}^{(1)}(1,u) u^{\lambda_1 - \frac{1}{pk} - 1} du \right) O\left(x^{\lambda_2 - \delta_1 - 1} \right) dx \\
&\leq L_4 \int_1^\infty x^{-\lambda_2 - \frac{1}{pk}} \left(\int_0^\infty k_{-\lambda}^{(1)}(u,1) u^{\lambda_1 - \frac{1}{pk} - 1} du \right) x^{\lambda_2 - \delta_1 - 1} dx \\
&= L_4 K^{(1)} \left(\lambda_1 - \frac{1}{pk} \right) \int_1^\infty x^{-\delta_1 - \frac{1}{pk} - 1} dx = \frac{L_4 K^{(1)}(\lambda_1 - \frac{1}{pk})}{\delta_1 + \frac{1}{pk}};
\end{aligned}$$

if $k_{-\lambda}^{(3)}(1,u)$ satisfies (42.195), then by Lemma 42.14(ii), there exists a constant $L_5 > 0$ such that $G(x) = O(\frac{x^{\lambda-1}}{(x-1)^a}) \leq L_5 \frac{x^{\lambda-1}}{(x-1)^a}$ $(x \in (1,\infty))$, and

$$\begin{aligned}
0 &\leq \int_1^\infty \left(\int_1^\infty k_{-\lambda}^{(1)}(x,y) y^{\lambda_1 - \frac{1}{pk} - 1} dy \right) G(x)\, dx \\
&= \int_1^\infty x^{-\lambda_2 - \frac{1}{pk}} \left(\int_{1/x}^\infty k_{-\lambda}^{(1)}(1,u) u^{\lambda_1 - \frac{1}{pk} - 1} du \right) O\left(\frac{x^{\lambda-1}}{(x-1)^a} \right) dx
\end{aligned}$$

$$\le L_5\left(\int_0^\infty k_{-\lambda}^{(1)}(u,1)u^{\lambda_1-\frac{1}{pk}-1}\,du\right)\int_1^\infty \frac{x^{\lambda_1-\frac{1}{pk}-1}}{(x-1)^a}\,dx$$

$$\overset{v=1/x}{=} L_5K^{(1)}\left(\lambda_1-\frac{1}{pk}\right)\int_0^1 (1-v)^{(1-a)-1}v^{(a-\lambda_1+\frac{1}{pk})-1}\,dv$$

$$= L_5K^{(1)}\left(\lambda_1-\frac{1}{pk}\right)B\left(1-a,a-\lambda_1+\frac{1}{pk}\right).$$

Hence, in the above two cases, we find $I_3 \to 0$ $(k\to\infty)$.

Therefore, by (42.197) and the above results, in view of $I_4 \ge 0$, we have $\widehat{L}_k \ge I_1 - I_2 - I_3 = \prod_{i=1}^3 K^{(i)}(\lambda_1) + o(1)$ $(k\to\infty)$, and then inequality (42.196) follows.

(ii) In view of the assumptions, except for the case that $F(y)=O(\frac{y^{\lambda-1}}{(y-1)^a})$ $(y\in(1,\infty))$ and $G(x)=O(\frac{x^{\lambda-1}}{(x-1)^a})$ $(x\in(1,\infty))$, we prove that the reverse of (42.196) is valid for the following three cases:

(a) If $F(y)=O(y^{\lambda_1-\delta_1-1})$ $(y\in(1,\infty))$ and $G(x)=O(x^{\lambda_2-\delta_1-1})$ $(x\in(1,\infty))$, then by (42.197), in view of $I_2, I_3 \ge 0$, we find

$$\widehat{L}_k \le I_1 + \frac{1}{k}\int_1^\infty\left(\int_1^\infty k_{-\lambda}^{(1)}(x,y)G(x)\,dx\right)F(y)\,dy$$

$$\le I_1 + \frac{1}{k}L_2L_4\int_1^\infty\left(\int_0^\infty k_{-\lambda}^{(1)}(x,y)x^{\lambda_2-\delta_1-1}\,dx\right)y^{\lambda_1-\delta_1-1}\,dy$$

$$\overset{u=x/y}{=} I_1 + \frac{L_2L_4}{k}\int_1^\infty y^{-2\delta_1-1}\,dy\int_0^\infty k_{-\lambda}^{(1)}(u,1)u^{\lambda_2-\delta_1-1}\,du$$

$$= I_1 + \frac{L_2L_4}{2k\delta_1}K^{(1)}(\lambda_2-\delta_1)$$

$$= I_1 + \frac{L_2L_4}{2k\delta_1}K^{(1)}(\lambda_1+\delta_1) \to \prod_{i=1}^3 K^{(i)}(\lambda_1) \quad (k\to\infty);$$

(b) if $F(y)=O(\frac{y^{\lambda-1}}{(y-1)^a})$ $(y\in(1,\infty))$ and $G(x)=O(x^{\lambda_2-\delta_1-1})$ $(x\in(1,\infty))$, then we find

$$\widehat{L}_k \le I_1 + \frac{1}{k}\int_1^\infty\left(\int_1^\infty k_{-\lambda}^{(1)}(x,y)G(x)\,dx\right)F(y)\,dy$$

$$\le I_1 + \frac{1}{k}L_4L_3\int_1^\infty\left(\int_0^\infty k_{-\lambda}^{(1)}(x,y)x^{\lambda_2-\delta_1-1}\,dx\right)\frac{y^{\lambda-1}}{(y-1)^a}\,dy$$

$$\overset{u=x/y}{=} I_1 + \frac{L_4L_3}{k}\int_0^\infty k_{-\lambda}^{(1)}(u,1)u^{\lambda_2-\delta_1-1}\,du\int_1^\infty \frac{y^{\lambda_2-\delta_1-1}}{(y-1)^a}\,dy$$

$$\overset{v=1/y}{=} I_1 + \frac{L_4L_3}{k}K^{(1)}(\lambda_2-\delta_1)\int_0^1 \frac{v^{a-\lambda_2+\delta_1-1}}{(1-v)^a}\,dv$$

$$= I_1 + \frac{L_4 L_3}{k} K^{(1)}(\lambda_1 + \delta_1) B(1-a, a-\lambda_2+\delta_1)$$

$$\to \prod_{i=1}^{3} K^{(i)}(\lambda_1) \quad (k \to \infty);$$

(c) if $F(y) = O(y^{\lambda_1-\delta_1-1})$ $(y \in (1,\infty))$ and $G(x) = O(\frac{x^{\lambda-1}}{(x-1)^a})$ $(x \in (1,\infty))$, then we obtain

$$\begin{aligned}
\widehat{L}_k &\le I_1 + \frac{1}{k}\int_1^\infty \left(\int_1^\infty k_{-\lambda}^{(1)}(x,y)F(y)\,dy\right)G(x)\,dx \\
&\le I_1 + \frac{1}{k}L_2L_5 \int_1^\infty \left(\int_0^\infty k_{-\lambda}^{(1)}(x,y)y^{\lambda_1-\delta_1-1}\,dy\right)\frac{x^{\lambda-1}}{(x-1)^a}\,dx \\
&\overset{u=y/x}{=} I_1 + \frac{L_2L_5}{k}\left(\int_0^\infty k_{-\lambda}^{(1)}(1,u)u^{\lambda_1-\delta_1-1}\,du\right)\int_1^\infty \frac{x^{\lambda_1-\delta_1-1}}{(x-1)^a}\,dx \\
&\overset{v=1/x}{=} I_1 + \frac{L_2L_5}{k}\left(\int_0^\infty k_{-\lambda}^{(1)}(u,1)u^{\lambda_1-\delta_1-1}\,du\right)\int_0^1 \frac{v^{a-\lambda_1+\delta_1-1}}{(1-v)^a}\,dv \\
&= I_1 + \frac{L_2L_5}{k}K^{(1)}(\lambda_1-\delta_1)B(1-a, a-\lambda_1+\delta_1) \to \prod_{i=1}^{3} K^{(i)}(\lambda_1) \quad (k\to\infty).
\end{aligned}$$

Hence, the lemma is proved. □

Remark 42.12 For $\lambda > 0$, the functions $k_{-\lambda}(x,y) = \frac{1}{(x+y)^\lambda}, \frac{1}{x^\lambda+y^\lambda}, \frac{1}{(\max\{x,y\})^\lambda}, \frac{\ln(x/y)}{x^\lambda-y^\lambda}$, and $\frac{1}{|x-y|^\lambda}$ $(0<\lambda<1)$ all satisfy the assumptions of $k_{-\lambda}^{(i)}(x,y)$ $(i = 1,2,3)$ in Lemma 42.15 (setting $\delta_1 = \frac{1}{2}\min\{\lambda_1,\lambda_2\}$ and $a = \lambda \in (0,1)$); and (42.196) is valid for substitution of these particular kernels for $k_{-\lambda}^{(1)}(x,y)$, $\widehat{F}_\lambda(y)$, and $\widehat{G}_\lambda(x)$. Except for $k_{-\lambda}^{(2)}(x,y) = k_{-\lambda}^{(3)}(x,y) = \frac{1}{|x-y|^\lambda}$ $(0<\lambda<1)$, the reverse of (42.196) is valid for substitution of the above particular kernels for $k_{-\lambda}^{(1)}(x,y)$, $\widehat{F}_\lambda(y)$, and $\widehat{G}_\lambda(x)$.

If $k_{-\lambda}^{(1)}(x,y)$ is non-symmetric, since for $\lambda_1 = \lambda_2 = \frac{\lambda}{2}$, $0 \le \delta \le \delta_0 (< \frac{\lambda}{2})$, we have

$$\int_0^\infty k_{-\lambda}^{(1)}(1,u)u^{\frac{\lambda}{2}\pm\delta-1}\,du = \int_0^\infty k_{-\lambda}^{(1)}(v,1)v^{\frac{\lambda}{2}\mp\delta-1}\,dv = K^{(1)}\left(\frac{\lambda}{2}\mp\delta\right),$$

then by Lemma 42.15, for $k \in \mathbf{N}$, $k > \max\{\frac{1}{|q|\delta_0}, \frac{1}{p\delta_0}\}$, setting $\widetilde{F}_\lambda(y)$ and $\widetilde{G}_\lambda(x)$ as follows:

$$\widetilde{F}_\lambda(y) := y^{\lambda-1}\int_1^\infty k_{-\lambda}^{(2)}(x,y)x^{\frac{\lambda}{2}-\frac{1}{pk}-1}\,dx \quad \big(y \in [1,\infty)\big),$$

$$\widetilde{G}_\lambda(x) := x^{\lambda-1}\int_1^\infty k_{-\lambda}^{(3)}(x,y)y^{\frac{\lambda}{2}-\frac{1}{qk}-1}\,dy \quad \big(x\in[1,\infty)\big),$$

we still have

Lemma 42.16 *Suppose that* $p>0$ $(p\neq 1)$, $\frac{1}{p}+\frac{1}{q}=1$, $\lambda>0$, $k_{-\lambda}^{(i)}(x,y)$ $(i=1,2,3)$ *are non-negative homogeneous functions of degree* $-\lambda$ *on* $\mathbf{R}_+^2$,

$$k^{(i)}(\lambda) := K^{(i)}\left(\frac{\lambda}{2}\right) = \int_0^\infty k_{-\lambda}^{(i)}(u,1)u^{\frac{\lambda}{2}-1}\,du \quad (i=1,2,3),$$

and there exists a $\delta_0\in(0,\frac{\lambda}{2})$ *such that* $k^{(i)}(\lambda\pm 2\delta_0)=K^{(i)}(\frac{\lambda}{2}\pm\delta_0)\in\mathbf{R}_+$ $(i=1,2,3)$. *Consider the following conditions*:

Condition (i). *There exist constants* $\delta_1\in(0,\delta_0)$ *and* $L>0$ *such that for any* $u\in[1,\infty)$,

$$k_{-\lambda}^{(2)}(1,u)u^{\frac{\lambda}{2}+\delta_1}\le L, \qquad k_{-\lambda}^{(3)}(u,1)u^{\frac{\lambda}{2}+\delta_1}\le L. \tag{42.198}$$

Condition (ii). *For* $0<\lambda<2$, *there exist constants* $a\in(\frac{\lambda}{2},1)$ *and* $L_1>0$ *such that for any* $u\in[1,\infty)$,

$$k_{-\lambda}^{(2)}(1,u)(u-1)^a\le L_1, \qquad k_{-\lambda}^{(3)}(u,1)(u-1)^a\le L_1. \tag{42.199}$$

If $k_{-\lambda}^{(2)}(1,u)$ $(k_{-\lambda}^{(3)}(u,1))$ *satisfies one of the above conditions, then for*

$$k>\max\left\{\frac{2}{|q|\delta_1},\frac{2}{p\delta_1},\frac{1}{|q|(a-\lambda/2)},\frac{1}{p(a-\lambda/2)}\right\},$$

(Note. *If both* $k_{-\lambda}^{(2)}(1,u)$ *and* $k_{-\lambda}^{(3)}(u,1)$) *satisfy* Condition (i), *then we set* $k>\max\{\frac{2}{|q|\delta_1},\frac{2}{p\delta_1}\}$; *if both* $k_{-\lambda}^{(2)}(1,u)$ *and* $k_{-\lambda}^{(3)}(u,1)$ *satisfy* Condition (ii), *then we set*

$$k>\max\left\{\frac{2}{|q|\delta_0},\frac{2}{p\delta_0},\frac{1}{|q|(a-\lambda/2)},\frac{1}{p(a-\lambda/2)}\right\}\Bigg),$$

we have

$$\widetilde{L}_k := \frac{1}{k}\int_1^\infty\int_1^\infty k_{-\lambda}^{(1)}(x,y)\widetilde{F}_\lambda(y)\widetilde{G}_\lambda(x)\,dx\,dy \ge \prod_{i=1}^3 k^{(i)}(\lambda)+o(1) \quad (k\to\infty). \tag{42.200}$$

Except for both $k_{-\lambda}^{(2)}(1,u)$ *and* $k_{-\lambda}^{(3)}(u,1)$ *satisfying* Condition (ii), *we have the reverse of* (42.200).

42.5.2 The Case of a Homogeneous First Kernel

Setting two functions $\varphi(x)$ and $\psi(y)$ as follows: $\varphi(x):=x^{p(1-\lambda_1)-1}$, $\psi(y):=y^{q(1-\lambda_2)-1}$ $(x,y\in\mathbf{R}_+)$, we have

Theorem 42.16 *Suppose that* $p > 0\ (p \neq 1)$, $\frac{1}{p} + \frac{1}{q} = 1$, $\lambda, \lambda_1, \lambda_2 > 0$, $\lambda_1 + \lambda_2 = \lambda$, $k_{-\lambda}^{(i)}(x, y)$ $(i = 1, 2, 3)$ *are non-negative homogeneous functions of degree* $-\lambda$ *on* $\mathbf{R}_+^2$, $k_{-\lambda}^{(1)}(x, y)$ *is symmetric,*

$$K^{(i)}(\lambda_1) := \int_0^\infty k_{-\lambda}^{(i)}(u, 1) u^{\lambda_1 - 1}\, du \quad (i = 1, 2, 3), \tag{42.201}$$

and there exists a constant $\delta_0 \in (0, \min\{\lambda_1, \lambda_2\})$ *such that*

$$K^{(i)}(\lambda_1 \pm \delta_0) \in \mathbf{R}_+ \quad (i = 1, 2, 3). \tag{42.202}$$

Consider the following conditions:

Condition (a) *There exist constants* $\delta_1 \in (0, \delta_0)$ *and* $L > 0$ *such that for any* $u \in [1, \infty)$,

$$k_{-\lambda}^{(2)}(1, u) u^{\lambda_2 + \delta_1} \leq L, \qquad k_{-\lambda}^{(3)}(u, 1) u^{\lambda_1 + \delta_1} \leq L. \tag{42.203}$$

Condition (b) *For* $0 < \lambda_1, \lambda_2 < 1$, *there exist constants* $a \in (\max\{\lambda_1, \lambda_2\}, 1)$ *and* $L_1 > 0$ *such that for any* $u \in [1, \infty)$,

$$k_{-\lambda}^{(2)}(1, u)(u - 1)^a \leq L_1, \qquad k_{-\lambda}^{(3)}(u, 1)(u - 1)^a \leq L_1. \tag{42.204}$$

If $k_{-\lambda}^{(2)}(1, u)$ $(k_{-\lambda}^{(3)}(u, 1))$ *satisfies one of the above conditions, then*

(i) *For* $p > 1$, $f(\geq 0) \in L_{p,\varphi}(\mathbf{R}_+)$, $G(\geq 0) \in L_{q,\psi}(\mathbf{R}_+)$, $\|f\|_{p,\varphi}, \|G\|_{q,\psi} > 0$, *setting*

$$F_\lambda(y) := y^{\lambda - 1} \int_0^\infty k_{-\lambda}^{(2)}(x, y) f(x)\, dx \quad (y \in \mathbf{R}_+),$$

we have the following equivalent inequalities:

$$I := \int_0^\infty \int_0^\infty k_{-\lambda}^{(1)}(x, y) F_\lambda(y) G(x)\, dx\, dy$$
$$< K^{(1)}(\lambda_1) K^{(2)}(\lambda_1) \|f\|_{p,\varphi} \|G\|_{q,\psi}, \tag{42.205}$$
$$J := \left[\int_0^\infty x^{p\lambda_2 - 1} \left(\int_0^\infty k_{-\lambda}^{(1)}(x, y) F_\lambda(y)\, dy \right)^p dx \right]^{1/p}$$
$$< K^{(1)}(\lambda_1) K^{(2)}(\lambda_1) \|f\|_{p,\varphi}, \tag{42.206}$$

where the constant factor $K^{(1)}(\lambda_1) K^{(2)}(\lambda_1)$ *is the best possible. In particular, for* $g(\geq 0) \in L_{q,\psi}(\mathbf{R}_+)$, $\|g\|_{q,\psi} > 0$, *and* $G(x) = G_\lambda(x) := x^{\lambda - 1} \times \int_0^\infty k_{-\lambda}^{(3)}(x, y) g(y)\, dy$ $(x \in \mathbf{R}_+)$, *we still have*

$$\int_0^\infty \int_0^\infty k_{-\lambda}^{(1)}(x, y) F_\lambda(y) G_\lambda(x)\, dx\, dy < \prod_{i=1}^{3} K^{(i)}(\lambda_1) \|f\|_{p,\varphi} \|g\|_{q,\psi}, \tag{42.207}$$

where the constant factor $\prod_{i=1}^{3} K^{(i)}(\lambda_1)$ *is the best possible.*

(ii) *For* $0 < p < 1$, *we have the equivalent reverses of* (42.205) *and* (42.206), *and the reverse of* (42.207). *Resetting*

$$F_\lambda(y) = \begin{cases} y^{\lambda-1}\int_0^\infty k_{-\lambda}^{(2)}(x,y)f(x)\,dx, & y \in \{y|f(y)>0\},\\ 0, & y \in \{y|f(y)=0\},\end{cases}$$

$$G_\lambda(x) = \begin{cases} x^{\lambda-1}\int_0^\infty k_{-\lambda}^{(3)}(x,y)g(y)\,dy, & x \in \{x|g(x)>0\},\\ 0, & x \in \{x|g(x)=0\},\end{cases}$$

except for both $k_{-\lambda}^{(2)}(1,u)$ *and* $k_{-\lambda}^{(3)}(u,1)$ *satisfying* Condition (b), *we have the equivalent reverses of* (42.205) *and* (42.206), *and the reverse of* (42.207) *with the best constant factors.*

Proof (i) For $p > 1$, in view of (42.34) (for $\gamma_1 = -\lambda_1, \gamma_2 = -\lambda_2$), we have

$$J \le K^{(1)}(\lambda_1)\|F_\lambda\|_{p,\varphi}, \tag{42.208}$$

and the following inequality:

$$\begin{aligned}\|F_\lambda\|_{p,\varphi} &= \left\{\int_0^\infty y^{p(1-\lambda_1)-1}\left(y^{\lambda-1}\int_0^\infty k_{-\lambda}^{(2)}(x,y)f(x)\,dx\right)^p dy\right\}^{1/p}\\ &= \left\{\int_0^\infty y^{p\lambda_2-1}\left(\int_0^\infty k_{-\lambda}^{(2)}(x,y)f(x)\,dx\right)^p dy\right\}^{1/p}\\ &< K^{(2)}(\lambda_1)\|f\|_{p,\varphi},\end{aligned} \tag{42.209}$$

then by (42.208), we have (42.206). By Hölder's inequality, we find

$$I = \int_0^\infty \left(x^{\lambda_2-\frac{1}{p}}\int_0^\infty k_{-\lambda}^{(1)}(x,y)F_\lambda(y)\,dy\right)\left(x^{\frac{1}{p}-\lambda_2}G(x)\right)dx \le J\|G\|_{q,\psi}. \tag{42.210}$$

Then by (42.206), we have (42.205).

On the other hand, assuming that (42.205) is valid, we set

$$G(x) = x^{p\lambda_2-1}\left(\int_0^\infty k_{-\lambda}^{(1)}(x,y)F_\lambda(y)\,dy\right)^{p-1} \quad (x \in \mathbf{R}_+).$$

Then we find $\|G\|_{q,\psi}^q = J^p$. If $J = 0$, then (42.206) is naturally valid; if $J = \infty$, then by (42.208), it follows $\|F_\lambda\|_{p,\varphi} = \infty$, which contradicts (42.209). For $0 < J < \infty$, by (42.205), it follows

$$\|G\|_{q,\psi}^q = J^p = I < K^{(1)}(\lambda_1)K^{(2)}(\lambda_2)\|f\|_{p,\varphi}\|G\|_{q,\psi} \tag{42.211}$$

dividing out $\|G\|_{q,\psi}$ in (42.211), we find $\|G\|_{q,\psi}^{q-1} = J < K^{(1)}(\lambda_1)K^{(2)}(\lambda_1)\|f\|_{p,\varphi}$, and then we have (42.206). Hence, inequalities (42.205) and (42.206) are equivalent. Setting $G(x) = G_\lambda(x)$ in (42.205), since by (42.34), we find $\|G\|_{q,\psi} \le K^{(3)}(\lambda_1)\|g\|_{q,\psi}$, then we have (42.207).

In the following, we prove that the constant in (42.207) is the best possible. For $k \in \mathbf{N}$,

$$k > \max\left\{\frac{2}{|q|\delta_1}, \frac{2}{p\delta_1}, \frac{1}{|q|(a-\lambda_2)}, \frac{1}{p(a-\lambda_2)}, \frac{1}{|q|(a-\lambda_1)}\right\},$$

we set $\widehat{f}(x), \widehat{g}(y)$ as $\widehat{f}(x) = \widehat{g}(y) = 0$ $(x, y \in (0,1))$, $\widehat{f}(x) := x^{\lambda_1 - \frac{1}{pk} - 1}$, $\widehat{g}(y) := y^{\lambda_2 - \frac{1}{qk} - 1}$, $(x, y \in [1,\infty))$. Then it follows

$$\widehat{F}_\lambda(y) = y^{\lambda-1}\int_0^\infty k_{-\lambda}^{(2)}(x,y)\widehat{f}(x)\,dx = y^{\lambda-1}\int_1^\infty k_{-\lambda}^{(2)}(x,y)x^{\lambda_1-\frac{1}{pk}-1}\,dx,$$

$$\widehat{G}_\lambda(x) = x^{\lambda-1}\int_0^\infty k_{-\lambda}^{(3)}(x,y)\widehat{g}(y)\,dy = x^{\lambda-1}\int_1^\infty k_{-\lambda}^{(3)}(x,y)y^{\lambda_2-\frac{1}{qk}-1}\,dy.$$

If there exists a positive constant $K \le \prod_{i=1}^3 K^{(i)}(\lambda_1)$ such that (42.207) is valid as we replace $\prod_{i=1}^3 K^{(i)}(\lambda_1)$ by K, then, in particular, it follows that

$$\widehat{I}_k := \frac{1}{k}\int_0^\infty\int_0^\infty k_{-\lambda}^{(1)}(x,y)\widehat{F}_\lambda(y)\widehat{G}_\lambda(x)\,dx\,dy < \frac{1}{k}K\|\widehat{f}\|_{p,\varphi}\|\widehat{g}\|_{q,\psi} = K.$$

By (42.196), we find

$$\prod_{i=1}^3 K^{(i)}(\lambda_1) + o(1) \le \widehat{L}_k = \frac{1}{k}\int_1^\infty\int_1^\infty k_{-\lambda}^{(1)}(x,y)\widehat{F}_\lambda(y)\widehat{G}_\lambda(x)\,dx\,dy \le \widehat{I}_k < K,$$

and then $\prod_{i=1}^3 K^{(i)}(\lambda_1) \le K$ $(k \to \infty)$. Hence, $K = \prod_{i=1}^3 K^{(i)}(\lambda_1)$ is the best value of (42.207).

We confirm that the constant factor in (42.205) is the best possible, otherwise, for $G(x) = G_\lambda(x)$, we can come to a contradiction that the constant factor in (42.207) is not the best possible. In the same way, we confirm that the constant factor in (42.206) is the best possible, otherwise, we can come to a contradiction by (42.210) that the constant factor in (42.205) is not the best possible.

(ii) For $0 < p < 1$, we only prove that the constant factor in the reverse of (42.207) is the best possible. For $k \in \mathbf{N}$,

$$k > \max\left\{\frac{2}{|q|\delta_1}, \frac{2}{p\delta_1}, \frac{1}{|q|(a-\lambda_2)}, \frac{1}{p(a-\lambda_2)}, \frac{1}{|q|(a-\lambda_1)}\right\},$$

we set $\widehat{f}(x), \widehat{g}(y)$ $(x, y \in (0,\infty))$ and $\widehat{F}_\lambda(y), \widehat{G}_\lambda(x)$ $(x, y \in [1,\infty))$ as in (i); $\widehat{F}_\lambda(y) = \widehat{G}_\lambda(x) = 0$ $(x, y \in (0,1))$. If there exists a positive constant $K \ge$

$\prod_{i=1}^{3} K^{(i)}(\lambda_1)$ such that the reverse of (42.207) is valid as we replace $\prod_{i=1}^{3} K^{(i)}(\lambda_1)$ by K, then, in particular, it follows

$$\widehat{I}_k = \frac{1}{k}\int_0^\infty \int_0^\infty k_{-\lambda}^{(1)}(x,y)\widehat{F}_\lambda(y)\widehat{G}_\lambda(x)\,dx\,dy > \frac{1}{k}K\|\widehat{f}\|_{p,\varphi}\|\widehat{g}\|_{q,\psi} = K.$$

By the reverse of (42.196), we find $\prod_{i=1}^{3} K^{(i)}(\lambda_1) + o(1) \ge \widehat{L}_k = \widehat{I}_k > K$, and then $\prod_{i=1}^{3} K^{(i)}(\lambda_1) \ge K$ $(k \to \infty)$. Hence, $K = \prod_{i=1}^{3} K^{(i)}(\lambda_1)$ is the best value of the reverse of (42.207). The theorem is proved. □

Remark 42.13 For $k_{-\lambda}^{(i)}(x,y) = \frac{1}{(x+y)^\lambda}$ $(i = 1,2,3)$ in Theorem 42.16, we find some results of [46] and [47].

Setting the two functions $\varphi_1(x)$ and $\psi_1(y)$ as follows: $\varphi_1(x) := x^{p(1-\frac{\lambda}{2})-1}$, $\psi_1(y) := y^{q(1-\frac{\lambda}{2})-1}$ $(x, y \in \mathbf{R}_+)$, by Lemma 42.16, we still have

Corollary 42.7 *Suppose that* $p > 0$ $(p \ne 1)$, $\frac{1}{p} + \frac{1}{q} = 1$, $\lambda > 0$, $k_{-\lambda}^{(i)}(x,y)$ $(i = 1,2,3)$ *are non-negative homogeneous functions of degree* $-\lambda$ *on* $\mathbf{R}_+^2$,

$$k^{(i)}(\lambda) := \int_0^\infty k_{-\lambda}^{(i)}(u,1)u^{\frac{\lambda}{2}-1}\,du \quad (i = 1,2,3), \tag{42.212}$$

and there exists a constant $\delta_0 \in (0, \frac{\lambda}{2})$ *such that*

$$k^{(i)}(\lambda \pm 2\delta_0) \in \mathbf{R}_+ \quad (i = 1,2,3). \tag{42.213}$$

Consider the following conditions:

Condition (i). *There exist constants* $\delta_1 \in (0, \delta_0)$ *and* $L > 0$ *such that for any* $u \in [1,\infty)$,

$$k_{-\lambda}^{(2)}(1,u)u^{\frac{\lambda}{2}+\delta_1} \le L, \qquad k_{-\lambda}^{(3)}(u,1)u^{\frac{\lambda}{2}+\delta_1} \le L. \tag{42.214}$$

Condition (ii). *For* $0 < \lambda < 2$, *there exist constants* $a \in (\frac{\lambda}{2}, 1)$ *and* $L_1 > 0$ *such that for any* $u \in [1,\infty)$,

$$k_{-\lambda}^{(2)}(1,u)(u-1)^a \le L_1, \qquad k_{-\lambda}^{(3)}(u,1)(u-1)^a \le L_1. \tag{42.215}$$

If $k_{-\lambda}^{(2)}(1,u)$ $(k_{-\lambda}^{(3)}(u,1))$ *satisfies one of the above conditions, then*

(i) *For* $p > 1$, $f(\ge 0) \in L_{p,\varphi_1}(\mathbf{R}_+)$, $G(\ge 0) \in L_{q,\psi_1}(\mathbf{R}_+)$, $\|f\|_{p,\varphi_1} > 0$, $\|G\|_{q,\psi_1} > 0$, *setting*

$$F_\lambda(y) := y^{\lambda-1}\int_0^\infty k_{-\lambda}^{(2)}(x,y)f(x)\,dx \quad (y \in \mathbf{R}_+),$$

we have the following equivalent inequalities:

$$\int_0^\infty \int_0^\infty k_{-\lambda}^{(1)}(x,y)F_\lambda(y)G(x)\,dx\,dy < k^{(1)}(\lambda)k^{(2)}(\lambda)\|f\|_{p,\varphi_1}\|G\|_{q,\psi_1}, \tag{42.216}$$

$$\left[\int_0^\infty x^{\frac{p\lambda}{2}-1}\left(\int_0^\infty k_{-\lambda}^{(1)}(x,y)F_\lambda(y)\,dy\right)^p dx\right]^{1/p} < k^{(1)}(\lambda)k^{(2)}(\lambda)\|f\|_{p,\varphi_1}, \tag{42.217}$$

where the constant factor $k^{(1)}(\lambda)k^{(2)}(\lambda)$ *is the best possible. In particular, for* $g(\geq 0) \in L_{q,\psi_1}(\mathbf{R}_+)$, $\|g\|_{q,\psi_1} > 0$, *and* $G(x) = G_\lambda(x) := x^{\lambda-1}\int_0^\infty k_{-\lambda}^{(3)}(x,y) \times g(y)\,dy$ $(x \in \mathbf{R}_+)$, *we have*

$$\int_0^\infty \int_0^\infty k_{-\lambda}^{(1)}(x,y)F_\lambda(y)G_\lambda(x)\,dx\,dy < \prod_{i=1}^3 k^{(i)}(\lambda)\|f\|_{p,\varphi_1}\|g\|_{q,\psi_1}, \tag{42.218}$$

where the constant factor $\prod_{i=1}^3 k^{(i)}(\lambda)$ *is the best possible.*

(ii) *For* $0 < p < 1$, *we have the equivalent reverses of* (42.216) *and* (42.217), *and the reverse of* (42.218). *Resetting*

$$F_\lambda(y) = \begin{cases} y^{\lambda-1}\int_0^\infty k_{-\lambda}^{(2)}(x,y)f(x)\,dx, & y \in \{y|f(y) > 0\}, \\ 0, & y \in \{y|f(y) = 0\}, \end{cases}$$

$$G_\lambda(x) = \begin{cases} x^{\lambda-1}\int_0^\infty k_{-\lambda}^{(3)}(x,y)g(y)\,dy, & x \in \{x|g(x) > 0\}, \\ 0, & x \in \{x|g(x) = 0\}, \end{cases}$$

except for both $k_{-\lambda}^{(2)}(1,u)$ *and* $k_{-\lambda}^{(3)}(u,1)$ *satisfying* Condition (ii), *we have the equivalent reverses of* (42.216) *and* (42.217), *and the reverse of* (42.218) *with the best constant factors.*

Suppose that $p > 1$, $\frac{1}{p} + \frac{1}{q} = 1$, $\lambda, \lambda_1, \lambda_2 > 0$, $\lambda_1 + \lambda_2 = \lambda$, $k_{-\lambda}^{(i)}(x,y)$ $(i = 1,2)$ are non-negative homogeneous functions of degree $-\lambda$ on $\mathbf{R}_+^2$,

$$K^{(i)}(\lambda_1) = \int_0^\infty k_{-\lambda}^{(i)}(u,1)u^{\lambda_1-1}\,du \quad (i = 1,2),$$

and there exists a constant $\delta_0 \in (0, \min\{\lambda_1, \lambda_2\})$ such that $K^{(i)}(\lambda_1 \pm \delta_0) \in \mathbf{R}_+$ $(i = 1,2)$. We set two functions $\varphi(x)$ and $\psi(y)$ as follows: $\varphi(x) := x^{p(1-\lambda_1)-1}$, $\psi(y) := y^{q(1-\lambda_2)-1}$ $(x, y \in \mathbf{R}_+)$, and then give the following definitions:

Definition 42.11 Define a Hilbert-type integral operator $T_1 : L_{p,\varphi}(\mathbf{R}_+) \to L_{p,\varphi}(\mathbf{R}_+)$ as follows: for $F_\lambda \in L_{p,\varphi}(\mathbf{R}_+)$, there exists a unique representation

$T_1 F_\lambda \in L_{p,\varphi}(\mathbf{R}_+)$, satisfying

$$T_1 F_\lambda(x) = x^{1-\lambda} \int_0^\infty k_{-\lambda}^{(1)}(x, y) F_\lambda(y)\, dy \quad (x \in \mathbf{R}_+). \tag{42.219}$$

We can find by (42.34) that

$$\|T_1 F_\lambda\|_{p,\varphi} \le K^{(1)}(\lambda_1) \|F_\lambda\|_{p,\varphi}, \tag{42.220}$$

where the constant factor $K^{(1)}(\lambda_1)$ is the best possible. Hence it follows that

$$\|T_1\| = K^{(1)}(\lambda_1). \tag{42.221}$$

Definition 42.12 Define a Hilbert-type integral operator $T_2 : L_{p,\varphi}(\mathbf{R}_+) \to L_{p,\varphi}(\mathbf{R}_+)$ as follows: for $f \in L_{p,\varphi}(\mathbf{R}_+)$, there exists a unique representation $F_\lambda \in L_{p,\varphi}(\mathbf{R}_+)$, satisfying

$$T_2 f(y) = F_\lambda(y) = y^{1-\lambda} \int_0^\infty k_{-\lambda}^{(2)}(x, y) f(x)\, dx \quad (y \in \mathbf{R}_+). \tag{42.222}$$

Then we find by (42.34) that

$$\|T_2 f\|_{p,\varphi} \le K^{(2)}(\lambda_1) \|f\|_{p,\varphi}, \tag{42.223}$$

where the constant factor $K^{(2)}(\lambda_1)$ is the best possible. Hence it follows

$$\|T_2\| = K^{(2)}(\lambda_1). \tag{42.224}$$

Definition 42.13 Define a Hilbert-type integral operator $T : L_{p,\varphi}(\mathbf{R}_+) \to L_{p,\varphi}(\mathbf{R}_+)$ as follows: for $f \in L_{p,\varphi}(\mathbf{R}_+)$, there exists a unique representation $Tf \in L_{p,\varphi}(\mathbf{R}_+)$, satisfying

$$Tf(x) = (T_1 F_\lambda)(x) = x^{1-\lambda} \int_0^\infty k_{-\lambda}^{(1)}(x, y) F_\lambda(y)\, dy \quad (x \in \mathbf{R}_+), \tag{42.225}$$

where $F_\lambda(y)$ is indicated by (42.222).

Since for any $f \in L_{p,\varphi}(\mathbf{R}_+)$, we have $Tf = T_1 F_\lambda = T_1(T_2 f) = (T_1 T_2) f$, it follows that $T = T_1 T_2$, i.e., T is a composition of T_1 and T_2. It is obvious that (cf. [48])

$$\|T\| = \|T_1 T_2\| \le \|T_1\| \cdot \|T_2\| = K^{(1)}(\lambda_1) K^{(2)}(\lambda_1).$$

Suppose that $k_{-\lambda}^{(i)}(x, y)$ $(i = 1, 2)$ satisfy the assumptions of Theorem 42.16. Then by (42.206), we find

$$\|Tf\|_{p,\varphi} = \|T_1 F_\lambda\|_{p,\varphi} = J < K^{(1)}(\lambda_1) K^{(2)}(\lambda_1) \|f\|_{p,\varphi},$$

where the constant $K^{(1)}(\lambda_1) K^{(2)}(\lambda_1)$ is the best possible. It follows that $\|T\| = K^{(1)}(\lambda_1) K^{(2)}(\lambda_1)$, and we have the following theorem:

Theorem 42.17 *Assuming that the Hilbert-type integral operators T_1 and T_2 are defined by* (42.219) *and* (42.222), *respectively*, $k_{-\lambda}^{(1)}(x,y)$ *is symmetric, and* $k_{-\lambda}^{(2)}(1,u)$ *satisfies* Condition (a) *or* Condition (b) *in Theorem* 42.16, *we have*

$$\|T_1T_2\| = \|T_1\| \cdot \|T_2\| = K^{(1)}(\lambda_1)K^{(2)}(\lambda_1). \tag{42.226}$$

In particular, if $k_{-\lambda}^{(2)}(x,y) = k_{-\lambda}^{(1)}(x,y)$, *then we have* $T_2 = T_1$ *and*

$$\|T_1^2\| = \|T_1\|^2 = \big(K^{(1)}(\lambda_1)\big)^2. \tag{42.227}$$

Example 42.11 For $\lambda > 0$, $r > 1$, $\frac{1}{r} + \frac{1}{s} = 1$, $k_{-\lambda}^{(i)}(x,y) = \frac{1}{x^\lambda + y^\lambda}$ $(i = 1,2)$, we find $K^{(i)}(\frac{\lambda}{r}) = \int_0^\infty \frac{1}{1+u^\lambda} u^{\frac{\lambda}{r}-1}\,du = \frac{\pi}{\lambda \sin(\pi/r)}$. Define two operators $T_i : L_{p,\varphi}(\mathbf{R}_+) \to L_{p,\varphi}(\mathbf{R}_+)$ $(i = 1,2)$ as follows: for $F_\lambda, f \in L_{p,\varphi}(\mathbf{R}_+)$, there exist unique representations $T_1F_\lambda, T_2 f \in L_{p,\varphi}(\mathbf{R}_+)$, satisfying

$$T_1F_\lambda(x) = x^{\lambda-1}\int_0^\infty \frac{1}{x^\lambda + y^\lambda} F_\lambda(y)\,dy \quad (x \in \mathbf{R}_+),$$

$$T_2 f(y) = F_\lambda(y) = y^{\lambda-1}\int_0^\infty \frac{1}{x^\lambda + y^\lambda} f(x)\,dx \quad (y \in \mathbf{R}_+).$$

Then by Theorem 42.17, it follows that $\|T_1T_2\| = \|T_1\| \cdot \|T_2\| = [\frac{\pi}{\lambda \sin(\pi/r)}]^2$. It is obvious that $T_2 = T_1$ and $\|T_1^2\| = \|T_1\|^2 = [\frac{\pi}{\lambda \sin(\pi/r)}]^2$.

42.5.3 The Case of a Non-homogeneous First Kernel

Lemma 42.17 *Suppose that* $p > 0$ $(p \neq 1)$, $\frac{1}{p} + \frac{1}{q} = 1$, $\lambda > 0$, $h(u)$ *is a non-negative measurable functions on* $\mathbf{R}_+$, $k_{-\lambda}^{(i)}(x,y)$ $(i = 2,3)$ *are non-negative homogeneous function of degree* $-\lambda$ *on* $\mathbf{R}_+^2$,

$$k^{(1)}(\lambda) := \int_0^\infty h(u)u^{\frac{\lambda}{2}-1}\,du, \qquad k^{(i)}(\lambda) := \int_0^\infty k_{-\lambda}^{(i)}(u,1)u^{\frac{\lambda}{2}-1}\,du \quad (i = 2,3), \tag{42.228}$$

and there exists a constant $\delta_0 \in (0,\lambda)$ *such that* $k^{(i)}(\lambda \pm \delta_0) \in \mathbf{R}_+$ $(i = 1,2,3)$. *Consider the following conditions*:

Condition (i). *There exist constants* $\delta_1 \in (0, \frac{\delta_0}{2})$ *and* $L > 0$ *such that for any* $u \in [1,\infty)$,

$$k_{-\lambda}^{(2)}(u,1)u^{\frac{\lambda}{2}+\delta_1} \leq L, \qquad k_{-\lambda}^{(3)}(u,1)u^{\frac{\lambda}{2}+\delta_1} \leq L. \tag{42.229}$$

Condition (ii). *For* $0 < \lambda < 2$, *there exist constants* $a \in (\frac{\lambda}{2}, 1)$ *and* $L_1 > 0$ *such that for any* $u \in [1,\infty)$,

$$k_{-\lambda}^{(2)}(u,1)(u-1)^a \leq L_1, \qquad k_{-\lambda}^{(3)}(u,1)(u-1)^a \leq L_1. \tag{42.230}$$

If $k_{-\lambda}^{(2)}(1,u)$ $(k_{-\lambda}^{(3)}(u,1))$ *satisfies one of the above conditions, then for* $k \in \mathbf{N}$,

$$k > \max\left\{\frac{2}{|q|\delta_1}, \frac{2}{p\delta_1}, \frac{1}{p(a-\frac{\lambda}{2})}, \frac{1}{|q|(a-\frac{\lambda}{2})}\right\},$$

define two functions $\widetilde{F}_\lambda(y)$ *and* $\widetilde{G}_\lambda(x)$ *as follows*:

$$\widetilde{F}_\lambda(y) := y^{\lambda-1}\int_0^1 k_{-\lambda}^{(2)}(x,y)x^{\frac{\lambda}{2}+\frac{1}{pk}-1}\,dx \quad \big(y \in (0,1)\big),$$

$$\widetilde{G}_\lambda(x) := x^{\lambda-1}\int_1^\infty k_{-\lambda}^{(3)}(x,y)y^{\frac{\lambda}{2}-\frac{1}{qk}-1}\,dy \quad \big(x \in [1,\infty)\big).$$

(i) *We have*

$$\widetilde{L}_k := \frac{1}{k}\int_1^\infty\left(\int_0^1 h(xy)\widetilde{F}_\lambda(y)\widetilde{G}_\lambda(x)\,dy\right)dx \geq \prod_{i=1}^3 k^{(i)}(\lambda) + o(1) \quad (k \to \infty). \tag{42.231}$$

(ii) *Except for both* $k_{-\lambda}^{(2)}(u,1)$ *and* $k_{-\lambda}^{(3)}(u,1)$ *satisfying* Condition (ii), *we have the reverse of* (42.231).

Proof Setting $u = x/y$, we find

$$\begin{aligned}
\widetilde{F}_\lambda(y) &= y^{\frac{\lambda}{2}+\frac{1}{pk}-1}\int_0^{1/y} k_{-\lambda}^{(2)}(u,1)u^{\frac{\lambda}{2}+\frac{1}{pk}-1}\,du \\
&= y^{\frac{\lambda}{2}+\frac{1}{pk}-1}\left[\int_0^\infty k_{-\lambda}^{(2)}(u,1)u^{\frac{1}{2}(\lambda+\frac{2}{pk})-1}\,du - \int_{1/y}^\infty k_{-\lambda}^{(2)}(u,1)u^{\frac{\lambda}{2}+\frac{1}{pk}-1}\,du\right] \\
&= y^{\frac{\lambda}{2}+\frac{1}{pk}-1}k^{(2)}\left(\lambda + \frac{2}{pk}\right) - F(y),
\end{aligned}$$

$$F(y) := y^{\frac{\lambda}{2}+\frac{1}{pk}-1}\int_{1/y}^\infty k_{-\lambda}^{(2)}(u,1)u^{\frac{\lambda}{2}+\frac{1}{pk}-1}\,du \quad \big(y \in (0,1)\big). \tag{42.232}$$

(a) If $k_{-\lambda}^{(2)}(u,1)$ satisfies Condition (i), then by (42.229), we obtain

$$0 \leq F(y) \leq y^{\frac{\lambda}{2}+\frac{1}{pk}-1}L\int_{1/y}^\infty u^{-\frac{\lambda}{2}-\delta_1}u^{\frac{\lambda}{2}+\frac{1}{pk}-1}\,du = \frac{Ly^{\frac{\lambda}{2}+\delta_1-1}}{\delta_1 - \frac{1}{pk}},$$

and $F(y) = O(y^{\frac{\lambda}{2}+\delta_1-1})$ $(y \in (0,1))$;

(b) if $k_{-\lambda}^{(2)}(u,1)$ satisfies Condition (ii), then by (42.230), it follows

$$0 \leq F(y) \leq y^{\frac{\lambda}{2}+\frac{1}{pk}-1}L_1\int_{1/y}^\infty \frac{1}{(u-1)^a}u^{\frac{\lambda}{2}+\frac{1}{pk}-1}\,du$$

$$\overset{v=1/(yu)}{=} y^{a-1}L_1\int_0^1 \frac{1}{(1-yv)^a}v^{a-\frac{\lambda}{2}-\frac{1}{pk}-1}\,dv$$
$$\le \frac{L_1 y^{a-1}}{(1-y)^a}\int_0^1 v^{a-\frac{\lambda}{2}-\frac{1}{pk}-1}\,dv = \frac{L_1}{a-\frac{\lambda}{2}-\frac{1}{pk}}\frac{y^{a-1}}{(1-y)^a},$$

and $F(y)=O(\frac{y^{a-1}}{(1-y)^a})$ $(y\in(0,1))$.

In view of Lemma 42.14, for $\lambda_1=\frac{\lambda}{2}$, if $k_{-\lambda}^{(3)}(u,1)$ satisfies Condition (i), then by (42.229), we have

$$\begin{aligned} \widetilde{G}_\lambda(x) &= x^{\frac{\lambda}{2}-\frac{1}{qk}-1}k^{(3)}\left(\lambda+\frac{2}{qk}\right)-G(x), \\ G(x) &= x^{\frac{\lambda}{2}-\frac{1}{qk}-1}\int_x^\infty k_{-\lambda}^{(3)}(u,1)u^{\frac{\lambda}{2}+\frac{1}{qk}-1}\,du = O\left(x^{\frac{\lambda}{2}-\delta_1-1}\right)\ge 0 \quad \left(x\in(1,\infty)\right); \end{aligned} \tag{42.233}$$

if $k_{-\lambda}^{(3)}(u,1)$ satisfies Condition (ii), then by (42.230), we have (42.233) with $G(x)=O(\frac{x^{\lambda-1}}{(x-1)^a})\ge 0$ $(x\in(1,\infty))$.

Hence, we have

$$\begin{aligned} \widetilde{L}_k &= \frac{1}{k}\int_1^\infty\int_0^1 h(xy)\left[y^{\frac{\lambda}{2}+\frac{1}{pk}-1}k^{(2)}\left(\lambda+\frac{2}{pk}\right)-F(y)\right] \\ &\quad\times\left[x^{\frac{\lambda}{2}-\frac{1}{qk}-1}k^{(3)}\left(\lambda+\frac{2}{qk}\right)-G(x)\right]dy\,dx \\ &= I_1-I_2-I_3+I_4, \end{aligned} \tag{42.234}$$

where

$$\begin{aligned} I_1 &:= k^{(2)}\left(\lambda+\frac{2}{pk}\right)k^{(3)}\left(\lambda+\frac{2}{qk}\right) \\ &\quad\times\frac{1}{k}\int_1^\infty\left(\int_0^1 h(xy)y^{\frac{\lambda}{2}+\frac{1}{pk}-1}\,dy\right)x^{\frac{\lambda}{2}-\frac{1}{qk}-1}\,dx, \\ I_2 &:= \frac{1}{k}k^{(3)}\left(\lambda+\frac{2}{qk}\right)\int_1^\infty\left(\int_0^1 h(xy)F(y)\,dy\right)x^{\frac{\lambda}{2}-\frac{1}{qk}-1}\,dx, \\ I_3 &:= \frac{1}{k}k^{(2)}\left(\lambda+\frac{2}{pk}\right)\int_1^\infty\left(\int_0^1 h(xy)y^{\frac{\lambda}{2}+\frac{1}{pk}-1}\,dy\right)G(x)\,dx, \\ I_4 &:= \frac{1}{k}\int_1^\infty\left(\int_0^1 h(xy)F(y)\,dy\right)G(x)\,dx. \end{aligned}$$

By Lemma 42.4 (for $\alpha=\frac{\lambda}{2}$), we obtain

$$\begin{aligned}&\frac{1}{k}\int_1^\infty\left(\int_0^1 h(xy)y^{\frac{\lambda}{2}+\frac{1}{pk}-1}\,dy\right)x^{\frac{\lambda}{2}-\frac{1}{qk}-1}\,dx\\&\quad\overset{u=xy}{=}\frac{1}{k}\int_1^\infty x^{-\frac{1}{k}-1}\left(\int_0^x h(u)u^{\frac{\lambda}{2}+\frac{1}{pk}-1}\,du\right)dx=k^{(1)}(\lambda)+o(1)\quad(k\to\infty),\end{aligned}$$

and then $I_1\to\prod_{i=1}^3 k^{(i)}(\lambda)$ $(k\to\infty)$.

(i) There exist positive constants L_2 and L_3 such that

$$\begin{aligned}&O\big(y^{\frac{\lambda}{2}+\delta_1-1}\big)\le L_2 y^{\frac{\lambda}{2}+\delta_1-1},\qquad O\left(\frac{y^{a-1}}{(1-y)^a}\right)\le L_2\frac{y^{a-1}}{(1-y)^a},\\&O\big(x^{\frac{\lambda}{2}-\delta_1-1}\big)\le L_3 x^{\frac{\lambda}{2}-\delta_1-1},\qquad O\left(\frac{x^{\lambda-1}}{(x-1)^a}\right)\le L_3\frac{x^{\lambda-1}}{(x-1)^a},\\&\big(y\in(0,1)\big),\qquad\big(x\in(1,\infty)\big).\end{aligned}$$

(a) For $F(y)=O(y^{\frac{\lambda}{2}+\delta_1-1})$ $(y\in(0,1))$, setting $u=xy$, we find

$$\begin{aligned}0&\le\int_1^\infty\left(\int_0^1 h(xy)F(y)\,dy\right)x^{\frac{\lambda}{2}-\frac{1}{qk}-1}\,dx\\&=\int_1^\infty\left(\int_0^1 h(xy)O\big(y^{\frac{\lambda}{2}+\delta_1-1}\big)\,dy\right)x^{\frac{\lambda}{2}-\frac{1}{qk}-1}\,dx\\&\le L_2\int_1^\infty\left(\int_0^1 h(xy)y^{\frac{\lambda}{2}+\delta_1-1}\,dy\right)x^{\frac{\lambda}{2}-\frac{1}{qk}-1}\,dx\\&\overset{u=xy}{=}\int_1^\infty\left[\int_0^x h(u)u^{\frac{1}{2}(\lambda+2\delta_1)-1}\,du\right]x^{-\delta_1-\frac{1}{qk}-1}\,dx\\&\le L_2\int_1^\infty\left[\int_0^\infty h(u)u^{\frac{1}{2}(\lambda+2\delta_1)-1}\,du\right]x^{-\delta_1-\frac{1}{qk}-1}\,dx=\frac{L_2k^{(1)}(\lambda+2\delta_1)}{\delta_1+\frac{1}{qk}};\end{aligned}$$

(b) For $F(y)=O(\frac{y^{a-1}}{(1-y)^a})$ $(y\in(0,1))$, we find

$$\begin{aligned}0&\le\int_1^\infty\left(\int_0^1 h(xy)F(y)\,dy\right)x^{\frac{\lambda}{2}-\frac{1}{qk}-1}\,dx\\&=\int_0^1\left(\int_1^\infty h(xy)x^{\frac{\lambda}{2}-\frac{1}{qk}-1}\,dx\right)O\left(\frac{y^{a-1}}{(1-y)^a}\right)dy\\&=L_2\int_0^1\left(\int_1^\infty h(xy)x^{\frac{\lambda}{2}-\frac{1}{qk}-1}\,dx\right)\frac{y^{a-1}}{(1-y)^a}\,dy\end{aligned}$$

$$\overset{u=xy}{=} L_2 \int_0^1 \left(\int_y^\infty h(u) u^{\frac{\lambda}{2} - \frac{1}{qk} - 1} \, du \right) \frac{y^{a - \frac{\lambda}{2} + \frac{1}{qk} - 1}}{(1-y)^a} \, dy$$

$$\leq L_2 \left[\int_0^\infty h(u) u^{\frac{1}{2}(\lambda - \frac{2}{qk}) - 1} \, du \right] \int_0^1 \frac{y^{a - \frac{\lambda}{2} + \frac{1}{qk} - 1}}{(1-y)^a} dy$$

$$= L_2 \, k^{(1)} \left(\lambda - \frac{2}{qk} \right) B\left(1 - a, a - \frac{\lambda}{2} + \frac{1}{qk} \right).$$

Hence, it follows $I_2 \to 0$ $(k \to \infty)$.

(c) For $G(x) = O(x^{\frac{\lambda}{2} - \delta_1 - 1})$ $(x \in (1, \infty))$, we find

$$0 \leq \int_1^\infty \left(\int_0^1 h(xy) y^{\frac{\lambda}{2} + \frac{1}{pk} - 1} \, dy \right) G(x) \, dx$$

$$= \int_1^\infty \left(\int_0^1 h(xy) y^{\frac{\lambda}{2} + \frac{1}{pk} - 1} \, dy \right) O\left(x^{\frac{\lambda}{2} - \delta_1 - 1} \right) dx$$

$$\leq L_3 \int_1^\infty \left(\int_0^1 h(xy) y^{\frac{\lambda}{2} + \frac{1}{pk} - 1} \, dy \right) x^{\frac{\lambda}{2} - \delta_1 - 1} \, dx$$

$$\overset{u=xy}{=} L_3 \int_1^\infty \left(\int_0^x h(u) u^{\frac{\lambda}{2} + \frac{1}{pk} - 1} \, du \right) x^{-\delta_1 - \frac{1}{pk} - 1} \, dx$$

$$\leq L_3 \left[\int_0^\infty h(u) u^{\frac{1}{2}(\lambda + \frac{2}{pk}) - 1} \, du \right] \int_1^\infty x^{-\delta_1 - \frac{1}{pk} - 1} \, dx$$

$$= L_3 k^{(1)} \left(\lambda + \frac{2}{pk} \right) \frac{1}{\delta_1 + \frac{1}{pk}};$$

(d) For $G(x) = O(\frac{x^{\lambda - 1}}{(x-1)^a})$ $(x \in (1, \infty))$, we obtain

$$0 \leq \int_1^\infty \left(\int_0^1 h(xy) y^{\frac{\lambda}{2} + \frac{1}{pk} - 1} \, dy \right) G(x) \, dx$$

$$= \int_1^\infty \left(\int_0^1 h(xy) y^{\frac{\lambda}{2} + \frac{1}{pk} - 1} \, dy \right) O\left(\frac{x^{\lambda - 1}}{(x-1)^a} \right) dx$$

$$\leq L_3 \int_1^\infty \left(\int_0^1 h(xy) y^{\frac{\lambda}{2} + \frac{1}{pk} - 1} \, dy \right) \frac{x^{\lambda - 1}}{(x-1)^a} \, dx$$

$$\overset{u=xy}{=} L_3 \int_1^\infty \left(\int_0^x h(u) u^{\frac{\lambda}{2} + \frac{1}{pk} - 1} \, du \right) \frac{x^{\frac{\lambda}{2} - \frac{1}{pk} - 1}}{(x-1)^a} \, dx$$

$$\leq L_3 \left[\int_0^\infty h(u) u^{\frac{1}{2}(\lambda + \frac{2}{pk}) - 1} \, du \right] \int_1^\infty \frac{x^{\frac{\lambda}{2} - \frac{1}{pk} - 1}}{(x-1)^a} \, dx$$

$$= L_3 k^{(1)}\left(\lambda + \frac{2}{pk}\right)B\left(1 - a, a - \frac{\lambda}{2} + \frac{1}{pk}\right).$$

Hence, it follows $I_3 \to 0$ $(k \to \infty)$.
Therefore, by (42.234), we have

$$\widetilde{L}_k \geq I_1 - I_2 - I_3 \to \prod_{i=1}^{3} k^{(i)}(\lambda) \quad (k \to \infty).$$

(ii) We have $I_1 \to \prod_{i=1}^{3} k^{(i)}(\lambda)$ $(k \to \infty)$ and

$$\widetilde{L}_k \leq I_1 + I_4 = I_1 + \frac{1}{k}\int_1^\infty \left(\int_0^1 h(xy)F(y)\,dy\right)G(x)\,dx. \tag{42.235}$$

(a) For $F(y) = O(y^{\frac{\lambda}{2}+\delta_1-1})$ $(y \in (0,1))$ and $G(x) = O(x^{\frac{\lambda}{2}-\delta_1-1})$ $(x \in (1,\infty))$, we find

$$\begin{aligned}
I_4 &= \frac{1}{k}\int_1^\infty \left(\int_0^1 h(xy)O\left(y^{\frac{\lambda}{2}+\delta_1-1}\right)dy\right)O\left(x^{\frac{\lambda}{2}-\delta_1-1}\right)dx \\
&\leq \frac{L_2L_3}{k}\int_1^\infty \left(\int_0^1 h(xy)y^{\frac{\lambda}{2}+\delta_1-1}\,dy\right)x^{\frac{\lambda}{2}-\delta_1-1}\,dx \\
&\overset{u=xy}{=} \frac{L_2L_3}{k}\int_1^\infty \left[\int_0^x h(u)u^{\frac{1}{2}(\lambda+2\delta_1)-1}\,du\right]x^{-2\delta_1-1}\,dx \\
&\leq \frac{L_2L_3}{k}\int_1^\infty \left[\int_0^\infty h(u)u^{\frac{1}{2}(\lambda+2\delta_1)-1}\,du\right]x^{-2\delta_1-1}\,dx \\
&= \frac{L_2L_3}{2\delta_1 k}k^{(1)}(\lambda + 2\delta_1) \to 0 \quad (k \to \infty);
\end{aligned}$$

(b) for $F(y) = O(y^{\frac{\lambda}{2}+\delta_1-1})$ $(y \in (0,1))$ and $G(x) = O(\frac{x^{\lambda-1}}{(x-1)^a})$ $(x \in (1,\infty))$, we find

$$\begin{aligned}
I_4 &= \frac{1}{k}\int_1^\infty \left(\int_0^1 h(xy)O\left(y^{\frac{\lambda}{2}+\delta_1-1}\right)dy\right)O\left(\frac{x^{\lambda-1}}{(x-1)^a}\right)dx \\
&\leq \frac{L_2L_3}{k}\int_1^\infty \left(\int_0^1 h(xy)y^{\frac{\lambda}{2}+\delta_1-1}\,dy\right)\frac{x^{\lambda-1}}{(x-1)^a}\,dx \\
&\overset{u=xy}{=} \frac{L_2L_3}{k}\int_1^\infty \left[\int_0^x h(u)u^{\frac{1}{2}(\lambda+2\delta_1)-1}\,du\right]\frac{x^{\frac{\lambda}{2}-\delta_1-1}}{(x-1)^a}\,dx \\
&\leq \frac{L_2L_3}{k}\left[\int_0^\infty h(u)u^{\frac{1}{2}(\lambda+2\delta_1)-1}\,du\right]\int_1^\infty \frac{x^{\frac{\lambda}{2}-\delta_1-1}}{(x-1)^a}\,dx \\
&= \frac{L_2L_3}{k}k^{(1)}(\lambda + 2\delta_1)B\left(1 - a, a - \frac{\lambda}{2} + \delta_1\right) \to 0 \quad (k \to \infty);
\end{aligned}$$

(c) for $F(y)=O(\frac{y^{a-1}}{(1-y)^a})$ $(y\in(0,1))$ and $G(x)=O(x^{\frac{\lambda}{2}-\delta_1-1})$ $(x\in(1,\infty))$, we find

$$\begin{aligned}
I_4 &= \frac{1}{k}\int_0^1\left[\int_1^\infty h(xy)O\left(x^{\frac{\lambda}{2}-\delta_1-1}\right)dx\right]O\left(\frac{y^{a-1}}{(1-y)^a}\right)dy \\
&\le \frac{L_2L_3}{k}\int_0^1\left(\int_1^\infty h(xy)x^{\frac{\lambda}{2}-\delta_1-1}\,dx\right)\frac{y^{a-1}}{(1-y)^a}\,dy \\
&\overset{u=xy}{=} \frac{L_2L_3}{k}\int_0^1\left(\int_y^\infty h(u)u^{\frac{\lambda}{2}-\delta_1-1}\,dx\right)\frac{y^{a-\frac{\lambda}{2}+\delta_1-1}}{(1-y)^a}\,dy \\
&\le \frac{L_2L_3}{k}\left[\int_0^\infty h(u)u^{\frac{1}{2}(\lambda-2\delta_1)-1}\,du\right]\int_0^1\frac{y^{a-\frac{\lambda}{2}+\delta_1-1}}{(1-y)^a}\,dy \\
&= \frac{L_2L_3}{k}k^{(1)}(\lambda-2\delta_1)B\left(1-a,a-\frac{\lambda}{2}+\delta_1\right)\to 0 \quad (k\to\infty).
\end{aligned}$$

Hence, by (42.235), we have the reverse of (42.231). The lemma is proved. □

Remark 42.14 For $\lambda>0$, the functions $k_{-\lambda}(x,y)=\frac{1}{(x+y)^\lambda},\frac{1}{x^\lambda+y^\lambda},\frac{1}{(\max\{x,y\})^\lambda}$, $\frac{\ln(x/y)}{x^\lambda-y^\lambda}$, and $\frac{1}{|x-y|^\lambda}$ $(0<\lambda<1)$ all satisfy the assumptions of $k_{-\lambda}^{(i)}(x,y)$ $(i=2,3)$ in Lemma 42.17 (setting $\delta_1=\frac{\lambda}{4}$ and $a=\lambda\in(0,1)$), and (42.231) is valid for substitution of these particular kernels for $\widetilde{F}_\lambda(y)$ and $\widetilde{G}_\lambda(x)$. Except for $k_{-\lambda}^{(2)}(x,y)=k_{-\lambda}^{(3)}(x,y)=\frac{1}{|x-y|^\lambda}$ $(0<\lambda<1)$, the reverse of (42.231) is valid for substitution of the above particular kernels for $\widetilde{F}_\lambda(y)$ and $\widetilde{G}_\lambda(x)$. For $\lambda>0$, $h(u)=\frac{1}{(1+u)^\lambda}$, $\frac{1}{1+u^\lambda},\frac{1}{(\max\{1,u\})^\lambda},\frac{\ln u}{u^\lambda-1}$ and $\frac{1}{|1-u|^\lambda}$ $(0<\lambda<1)$ all satisfy the assumptions of $h(u)$ in Lemma 42.17.

Theorem 42.18 *Suppose that $p>0$ $(p\neq 1)$, $\frac{1}{p}+\frac{1}{q}=1$, $\lambda>0$, $h(u)$ is a non-negative measurable function on $\mathbf{R}_+$, $k_{-\lambda}^{(i)}(x,y)$ $(i=2,3)$ are non-negative homogeneous functions of degree $-\lambda$ on $\mathbf{R}_+^2$,*

$$k^{(1)}(\lambda):=\int_0^\infty h(u)u^{\frac{\lambda}{2}-1}\,du,\qquad k^{(i)}(\lambda):=\int_0^\infty k_{-\lambda}^{(i)}(u,1)u^{\frac{\lambda}{2}-1}\,du\quad(i=2,3), \tag{42.236}$$

and there exists a constant $\delta_0\in(0,\lambda)$ such that $k^{(i)}(\lambda\pm\delta_0)\in\mathbf{R}_+$ $(i=1,2,3)$. Consider the following condition:

Condition (i). *There exist constants $0<\delta_1<\frac{\delta_0}{2}$ and $L>0$ such that for any $u\in[1,\infty)$,*

$$k_{-\lambda}^{(2)}(u,1)u^{\frac{\lambda}{2}+\delta_1}\le L,\qquad k_{-\lambda}^{(3)}(u,1)u^{\frac{\lambda}{2}+\delta_1}\le L. \tag{42.237}$$

Condition (ii). *For* $0<\lambda<2$, *there exist constants* $a\in(\frac{\lambda}{2},1)$ *and* $L_1>0$ *such that for any* $u\in[1,\infty)$,

$$k_{-\lambda}^{(2)}(u,1)(u-1)^a \le L_1, \qquad k_{-\lambda}^{(3)}(u,1)(u-1)^a \le L_1. \tag{42.238}$$

If $k_{-\lambda}^{(2)}(u,1)$ $(k_{-\lambda}^{(3)}(u,1))$ *satisfies one of the above conditions, setting* $\varphi_1(x)=x^{p(1-\frac{\lambda}{2})-1}$, $\psi_1(y)=y^{q(1-\frac{\lambda}{2})-1}$ $(x,y\in\mathbf{R}_+)$, *then*

(i) *For* $p>1$, $f(\ge 0)\in L_{p,\varphi_1}(\mathbf{R}_+)$, $G(\ge 0)\in L_{q,\psi_1}(\mathbf{R}_+)$, $\|f\|_{p,\varphi_1}>0$, $\|G\|_{q,\psi_1}>0$, $F_\lambda(y):=y^{\lambda-1}\int_0^\infty k_{-\lambda}^{(2)}(x,y)f(x)\,dx$ $(y\in\mathbf{R}_+)$, *we have the following equivalent inequalities*:

$$\widetilde{I}:=\int_0^\infty\int_0^\infty h(xy)F_\lambda(y)G(x)\,dx\,dy < k^{(1)}(\lambda)k^{(2)}(\lambda)\|f\|_{p,\varphi_1}\|G\|_{q,\psi_1}, \tag{42.239}$$

$$\widetilde{J}:=\left[\int_0^\infty x^{\frac{p\lambda}{2}-1}\left(\int_0^\infty h(xy)F_\lambda(y)\,dy\right)^p dx\right]^{1/p} < k^{(1)}(\lambda)k^{(2)}(\lambda)\|f\|_{p,\varphi_1}, \tag{42.240}$$

where the constant factor $k^{(1)}(\lambda)k^{(2)}(\lambda)$ *is the best possible. In particular, if* $g(\ge 0)\in L_{q,\psi_1}(\mathbf{R}_+)$, $\|g\|_{q,\psi_1}>0$, *and* $G(x)=G_\lambda(x):=x^{\lambda-1}\int_0^\infty k_{-\lambda}^{(3)}(x,y)\times g(y)\,dy$ $(x\in\mathbf{R}_+)$, *then we still have*

$$\int_0^\infty\int_0^\infty h(xy)F_\lambda(y)G_\lambda(x)\,dx\,dy < \prod_{i=1}^3 k^{(i)}(\lambda)\|f\|_{p,\varphi_1}\|g\|_{q,\psi_1}, \tag{42.241}$$

where the constant factor $\prod_{i=1}^3 k^{(i)}(\lambda)$ *is the best possible.*

(ii) *For* $0<p<1$, *we have the equivalent reverses of* (42.239) *and* (42.240), *and the reverse of* (42.241). *Resetting*

$$F_\lambda(y):=\begin{cases} y^{\lambda-1}\int_0^\infty k_{-\lambda}^{(2)}(x,y)f(x)\,dx, & y\in\{y|f(y)>0\},\\ 0, & y\in\{y|f(y)=0\},\end{cases}$$

$$G_\lambda(x):=\begin{cases} x^{\lambda-1}\int_0^\infty k_{-\lambda}^{(3)}(x,y)g(y)\,dy, & x\in\{x|g(x)>0\},\\ 0, & x\in\{x|g(x)=0\},\end{cases}$$

except for both $k_{-\lambda}^{(2)}(u,1)$ *and* $k_{-\lambda}^{(3)}(u,1)$ *satisfying* Condition (ii), *we have the equivalent reverses of* (42.239) *and* (42.240), *and the reverse of* (42.241) *with the best constant factors.*

Proof (i) For $p>1$, by the same way of Theorem 42.16, for $\lambda_1=\lambda_2=\frac{\lambda}{2}$ replacing $k_{-\lambda}^{(1)}(x,y)$ by $h(xy)$, we can obtain inequalities (42.239), (42.240), and (42.241), and prove that (42.239) and (42.240) are equivalent.

In the following, we only prove that the constant factor $\prod_{i=1}^{3} k^{(i)}(\lambda)$ in (42.241) is the best possible. For $k \in \mathbf{N}$,

$$k > \max\left\{\frac{2}{|q|\delta_1}, \frac{2}{p\delta_1}, \frac{1}{p(a-\frac{\lambda}{2})}, \frac{1}{|q|(a-\frac{\lambda}{2})}\right\},$$

we set $\widetilde{f}(x), \widetilde{g}(y)$ as follows: $\widetilde{f}(x) = \widetilde{g}(y) = 0$ $(x \in [1,\infty), y \in (0,1])$; $\widetilde{f}(x) := x^{\frac{\lambda}{2}+\frac{1}{pk}-1}$ $(x \in (0,1))$, $\widetilde{g}(y) := y^{\frac{\lambda}{2}-\frac{1}{qk}-1}$ $(y \in (1,\infty))$, and then

$$\widetilde{F}_\lambda(y) = y^{\lambda-1}\int_0^\infty k_{-\lambda}^{(2)}(x,y)\widetilde{f}(x)\,dx = y^{\lambda-1}\int_0^1 k_{-\lambda}^{(2)}(x,y)x^{\frac{\lambda}{2}+\frac{1}{pk}-1}\,dx,$$

$$\widetilde{G}_\lambda(x) = x^{\lambda-1}\int_0^\infty k_{-\lambda}^{(3)}(x,y)\widetilde{g}(y)\,dy = x^{\lambda-1}\int_1^\infty k_{-\lambda}^{(3)}(x,y)y^{\frac{\lambda}{2}-\frac{1}{qk}-1}\,dy.$$

If there exists a positive constant $K \le \prod_{i=1}^{3} k^{(i)}(\lambda)$ such that (42.241) is valid as we replace $\prod_{i=1}^{3} k^{(i)}(\lambda)$ by K, then, in particular, it follows that

$$\widetilde{I}_k := \frac{1}{k}\int_0^\infty\int_0^\infty h(xy)\widetilde{F}_\lambda(y)\widetilde{G}_\lambda(x)\,dx\,dy < \frac{1}{k}K\|\widetilde{f}\|_{p,\varphi_1}\|\widetilde{g}\|_{q,\psi_1} = K.$$

By (42.231), we find

$$\prod_{i=1}^{3} k^{(i)}(\lambda) + o(1) \le \widetilde{L}_k = \frac{1}{k}\int_1^\infty\left(\int_0^1 h(xy)\widetilde{F}_\lambda(y)\widetilde{G}_\lambda(x)\,dy\right)dx \le \widetilde{I}_k < K,$$

and then $\prod_{i=1}^{3} k^{(i)}(\lambda) \le K$ $(k \to \infty)$. Hence $K = \prod_{i=1}^{3} k^{(i)}(\lambda)$ is the best value of (42.241).

(ii) For $0 < p < 1$, we still can obtain the reverses of (42.239), (42.240), and (42.241), and prove that the reverses of (42.239) and (42.240) are equivalent.

In the following, we only prove that the constant factor $\prod_{i=1}^{3} k^{(i)}(\lambda)$ in the reverse of (42.241) is the best possible. For $k \in \mathbf{N}$,

$$k > \max\left\{\frac{2}{|q|\delta_1}, \frac{2}{p\delta_1}, \frac{2}{p(a-\frac{\lambda}{2})}, \frac{2}{|q|(a-\frac{\lambda}{2})}\right\},$$

we set $\widetilde{f}(x), \widetilde{g}(y)$ $(x, y \in (0,\infty))$, $\widetilde{F}_\lambda(y)$ $(y \in (0,1))$ and $\widetilde{G}_\lambda(x)$ $(x \in [1,\infty))$ as in (i); $\widetilde{F}_\lambda(y) = \widetilde{G}_\lambda(x) = 0$ $(y \in [1,\infty), x \in (0,1))$. If there exists a positive constant $K \ge \prod_{i=1}^{3} k^{(i)}(\lambda)$ such that the reverse of (42.230) is valid as we replace $\prod_{i=1}^{3} k^{(i)}(\lambda_1)$ by K, then, in particular, it follows

$$\widetilde{I}_k = \frac{1}{k}\int_0^\infty\int_0^\infty h(xy)\widetilde{F}_\lambda(y)\widetilde{G}_\lambda(x)\,dx\,dy > \frac{1}{k}K\|\widetilde{f}\|_{p,\varphi_1}\|\widetilde{g}\|_{q,\psi_1} = K.$$

By the reverse of (42.231), we find $\prod_{i=1}^{3} k^{(i)}(\lambda_1) + o(1) \ge \widetilde{L}_k = \widetilde{I}_k > K$, and then $\prod_{i=1}^{3} k^{(i)}(\lambda_1) \ge K (k \to \infty)$. Hence $K = \prod_{i=1}^{3} k^{(i)}(\lambda_1)$ is the best value of the reverse of (42.241). The theorem is proved. □

Suppose that $p > 1$, $\frac{1}{p} + \frac{1}{q} = 1$, $\lambda > 0$, $h(u)$ is a non-negative measurable function on $\mathbf{R}_+$, $k_{-\lambda}^{(2)}(x, y)$, is a non-negative homogeneous functions of degree $-\lambda$ on $\mathbf{R}_+^2$, $k^{(1)}(\lambda) = \int_0^\infty h(u)u^{\frac{\lambda}{2}-1}\,du$, $k^{(2)}(\lambda) = \int_0^\infty k_{-\lambda}^{(2)}(u, 1)u^{\frac{\lambda}{2}-1}\,du$, and there exists a constant $\delta_0 \in (0, \lambda)$ such that $k^{(i)}(\lambda \pm \delta_0) \in \mathbf{R}_+$ $(i = 1, 2)$. We set $\varphi_1(x) = x^{p(1-\frac{\lambda}{2})-1}$, $\psi_1(y) = y^{q(1-\frac{\lambda}{2})-1}$ $(x, y \in \mathbf{R}_+)$, and give the following definitions:

Definition 42.14 Define a Hilbert-type integral operator $\widetilde{T}_1 : L_{p,\varphi_1}(\mathbf{R}_+) \to L_{p,\varphi_1}(\mathbf{R}_+)$ as follows: for $F_\lambda \in L_{p,\varphi_1}(\mathbf{R}_+)$, there exists a unique representation $\widetilde{T}_1 F_\lambda \in L_{p,\varphi_1}(\mathbf{R}_+)$, satisfying

$$\widetilde{T}_1 F_\lambda(x) = x^{\lambda-1} \int_0^\infty h(xy) F_\lambda(y)\,dy \quad (x \in \mathbf{R}_+). \tag{42.242}$$

We can find by (42.54) that

$$\|\widetilde{T}_1 F_\lambda\|_{p,\varphi_1} \le k^{(1)}(\lambda) \|F_\lambda\|_{p,\varphi_1}, \tag{42.243}$$

where the constant factor $k^{(1)}(\lambda)$ is the best possible. Hence we obtain

$$\|\widetilde{T}_1\| = k^{(1)}(\lambda). \tag{42.244}$$

Definition 42.15 Define a Hilbert-type integral operator $\widetilde{T}_2 : L_{p,\varphi_1}(\mathbf{R}_+) \to L_{p,\varphi_1}(\mathbf{R}_+)$ as follows: for $f \in L_{p,\varphi_1}(\mathbf{R}_+)$, there exists a unique representation $F_\lambda \in L_{p,\varphi_1}(\mathbf{R}_+)$, satisfying

$$\widetilde{T}_2 f(y) = F_\lambda(y) = y^{\lambda-1} \int_0^\infty k_{-\lambda}^{(2)}(x, y) f(x)\,dx \quad (y \in \mathbf{R}_+). \tag{42.245}$$

We find by (42.34) that

$$\|\widetilde{T}_2 f\|_{p,\varphi_1} \le k^{(2)}(\lambda) \|f\|_{p,\varphi_1}, \tag{42.246}$$

where the constant factor $k^{(2)}(\lambda)$ is the best possible. Hence we obtain

$$\|\widetilde{T}_2\| = k^{(2)}(\lambda). \tag{42.247}$$

Definition 42.16 Define a Hilbert-type integral operator $\widetilde{T} : L_{p,\varphi_1}(\mathbf{R}_+) \to L_{p,\varphi_1}(\mathbf{R}_+)$ as follows: for $f \in L_{p,\varphi_1}(\mathbf{R}_+)$, there exists a unique representation $\widetilde{T} f \in L_{p,\varphi_1}(\mathbf{R}_+)$, satisfying

$$\widetilde{T} f(x) = (\widetilde{T}_1 F_\lambda)(x) = x^{\lambda-1} \int_0^\infty h(xy) F_\lambda(y)\,dy \quad (x \in \mathbf{R}_+), \tag{42.248}$$

where $F_\lambda(y)$ is indicated by (42.245).

Since it follows $\widetilde{T}f = \widetilde{T}_1 F_\lambda = \widetilde{T}_1(\widetilde{T}_2 f) = (\widetilde{T}_1\widetilde{T}_2)f$, then $\widetilde{T} = \widetilde{T}_1\widetilde{T}_2$, i.e., $\widetilde{T}$ is a composition of $\widetilde{T}_1$ and $\widetilde{T}_2$. It is obvious that

$$\|\widetilde{T}\| = \|\widetilde{T}_1\widetilde{T}_2\| \le \|\widetilde{T}_1\| \cdot \|\widetilde{T}_2\| = k^{(1)}(\lambda)k^{(2)}(\lambda).$$

If $k_{-\lambda}^{(2)}(u,1)$ satisfies Condition (i) or Condition (ii) in Theorem 42.18, then by (42.240), we find

$$\|\widetilde{T}f\|_{p,\varphi_1} = \|\widetilde{T}_1 F_\lambda\|_{p,\varphi_1} = \widetilde{J} < k^{(1)}(\lambda)k^{(2)}(\lambda)\|f\|_{p,\varphi_1}, \tag{42.249}$$

where the constant factor $k^{(1)}(\lambda)k^{(2)}(\lambda)$ is the best possible. It follows that $\|\widetilde{T}\| = k^{(1)}(\lambda)k^{(2)}(\lambda)$ and we have the following theorem:

Theorem 42.19 *Suppose that the Hilbert-type integral operators* $\widetilde{T}_1$ *and* $\widetilde{T}_2$ *are defined by* (42.242) *and* (42.245), *respectively, and* $k_{-\lambda}^{(2)}(u,1)$ *satisfies* Condition (i) *or* Condition (ii) *in Theorem* 42.18. *We have*

$$\|\widetilde{T}_1\widetilde{T}_2\| = \|\widetilde{T}_1\| \cdot \|\widetilde{T}_2\| = k^{(1)}(\lambda)k^{(2)}(\lambda). \tag{42.250}$$

Example 42.12 For $\lambda > 0$, $h(u) = \frac{1}{1+u^\lambda}$, $k_{-\lambda}^{(2)}(x,y) = \frac{1}{x^\lambda + y^\lambda}$, we find

$$k^{(i)}(\lambda) = \int_0^\infty \frac{1}{1+u^\lambda} u^{\frac{\lambda}{2}-1}\,du = \frac{\pi}{\lambda} \quad (i = 1, 2).$$

For $F_\lambda, f \in L_{p,\varphi_1}(\mathbf{R}_+)$, setting

$$\widetilde{T}_1 F_\lambda(x) = x^{\lambda-1}\int_0^\infty \frac{1}{1+(xy)^\lambda} F_\lambda(y)\,dy \quad (x \in \mathbf{R}_+),$$

$$\widetilde{T}_2 f(y) = F_\lambda(y) = y^{\lambda-1}\int_0^\infty \frac{1}{x^\lambda + y^\lambda} f(x)\,dx \quad (y \in \mathbf{R}_+),$$

then by Theorem 42.19, we have

$$\|\widetilde{T}_1\widetilde{T}_2\| = \|\widetilde{T}_1\| \cdot \|\widetilde{T}_2\| = \left(\frac{\pi}{\lambda}\right)^2. \tag{42.251}$$

References

1. Kuang, J.C.: Applied Inequalities, 3nd edn. Shangdong Science Technic Press, Jinan (2004)
2. Kuang, J.C.: Introduction to Real Analysis. Hunan Education Press, Chansha (1996)
3. Wilhelm, M.: On the spectrum of Hilbert's matrix. Am. J. Math. **72**, 699–704 (1950)
4. Carleman, T.: Sur les equations integrals singulieres a noyau reel et symetrique. Uppsala (1923)
5. Zang, K.W.: A bilinear inequality. J. Math. Anal. Appl. **271**, 288–296 (2002)
6. Yang, B.C.: On the norm of an integral operator and applications. J. Math. Anal. Appl. **321**, 182–192 (2006)

7. Yang, B.C.: On the norm of a self-adjoint operator and a new bilinear integral inequality. Acta Math. Sin. Engl. Ser. **23**(7), 1311–1316 (2007)
8. Yang, B.C.: On the norm of a certain self-adjoint integral operator and applications to bilinear integral inequalities. Taiwan. J. Math. **12**(2), 315–324 (2008)
9. Yang, B.C.: On the norm of a Hilbert's type linear operator and applications. J. Math. Anal. Appl. **325**, 529–541 (2007)
10. Yang, B.C.: On the norm of a self-adjoint operator and application to Hilbert's type inequalities. Bull. Belg. Math. Soc. **13**, 577–584 (2006)
11. Yang, B.C.: On a Hilbert-type operator with a symmetric homogeneous kernel of -1-degree and application. Arch. Inequal. Appl. (2007). doi:10.1155/2007/47812
12. Yang, B.C.: On the norm of a linear operator and its applications. Indian J. Pure Appl. Math. **39**(3), 237–250 (2008)
13. Yang, B.C.: On a Hilbert-type operator with a class of homogeneous kernel. Arch. Inequal. Appl. (2009). doi:10.1155/2009/572176
14. Yang, B.C.: A new Hilbert-type operator and applications. Publ. Math. (Debr.) **76**(1–2), 147–156 (2010)
15. Yang, B.C., Rassias, T.M.: On a Hilbert-type integral inequality in the subinterval and its operator expression. Banach J. Math. Anal. **4**(2), 100–110 (2010)
16. Huang, Q.L., Yang, B.C.: On a multiple Hilbert-type integral operator and applications. Arch. Inequal. Appl. (2009). doi:10.1155/2009/192197
17. Arpad, B., Choonghong, O.: Best constants for certain multilinear integral operator. Arch. Inequal. Appl. (2006). doi:10.1155/2006/28582
18. Zhong, W.Y.: A Hilbert-type linear operator with the norm and its applications. Arch. Inequal. Appl. (2009). doi:10.1155/2009/494257
19. Zhong, W.Y.: A new Hilbert-type linear operator with the a composite kernel and its applications. Arch. Inequal. Appl. (2010). doi:10.1155/2010/393025
20. Li, Y.J., He, B.: Hilbert's type linear operator and some extensions of Hilbert's inequality. Arch. Inequal. Appl. (2007). doi:10.1155/2007/82138
21. Liu, X.D., Yang, B.C.: On a Hilbert–Hardy-type integral operator and applications. Arch. Inequal. Appl. (2010). doi:10.1155/2010/812636
22. Hardy, G.H., Littlewood, J.E., Pólya, G.: Inequalities. Cambridge University Press, Cambridge (1934)
23. Mitrinović, D.S., Pečarić, J.E., Fink, A.M.: Inequalities Involving Functions and Their Integrals and Derivatives. Kluwer Academic, Boston (1991)
24. Yang, B.C., Gao, M.Z.: On a best value of Hardy–Hilbert's inequality. Adv. Math. **26**(2), 159–164 (1997)
25. Gao, M.Z., Yang, B.C.: On the extended Hilbert's inequality. Proc. Am. Math. Soc. **126**(3), 751–759 (1998)
26. Pachpatte, B.G.: On some new inequalities similar to Hilbert's inequality. J. Math. Anal. Appl. **226**, 166–179 (1998)
27. Yang, B.C.: On Hilbert's integral inequality. J. Math. Anal. Appl. **220**, 778–785 (1998)
28. Yang, B.C., Debnath, L.: On a new generalization of Hardy-Hilbert's inequality. J. Math. Anal. Appl. **233**, 484–497 (1999)
29. Yang, B.C., Rassias, T.M.: On the way of weight coefficient and research for Hilbert-type inequalities. Math. Inequal. Appl. **6**(4), 625–658 (2003)
30. Yang, B.C.: On new extension of Hilbert's inequality. Acta Math. Hung. **104**(4), 291–299 (2004)
31. Yang, B.C.: On an extension of Hilbert's integral inequality with some parameters. Aust. J. Math. Anal. Appl. **1**(1), 11 (2004), pp. 8
32. Yang, B.C., Brnete, I., Krnic, M., et al.: Generalization of Hilbert and Hardy-Hilbert integral inequalities. Math. Inequal. Appl. **8**(2), 259–272 (2005)
33. Hu, K.: Some Problems in Analysis Inequalities. Wuhan University Press, Wuhan (2007)
34. Yang, B.C.: A survey of the study of Hilbert-type inequalities with parameters. Adv. Math. **38**(3), 257–268 (2009)

35. Yang, B.C.: The Norm of Operator and Hilbert-Type Inequalities. Science Press, Beijing (2009)
36. Yang, B.C.: Hilbert-Type Integral Inequalities. Bentham Science Publishers Ltd. (2009)
37. Yang, B.C.: Discrete Hilbert-Type Inequalities. Bentham Science Publishers Ltd. (2011)
38. Wang, D.X., Guo, D.R.: Special Functions. Science Press, Beijing (1979)
39. Yang, B.C.: A reverse Hilbert-type integral inequality with some parameters. J. Xinxiang University (Nat. Sci. Edn.) **27**(6), 1–4 (2010)
40. Yang, B.C.: A Hilbert-type inequality with a non-homogeneous kernel. J. Xiamen University (Nat. Sci.) **48**(3), 165–169 (2009)
41. Ping, Y., Wang, H., Song, L.: Complex Functions. Science Press, Beijing (2004)
42. Zeng, Z., Xie, Z.T.: On a new Hilbert-type integral inequality with the homogeneous kernel of degree 0 and the integral in whole plane. Arch. Inequal. Appl. (2010). doi:10.1155/2010/256796
43. Xin, D.M., Yang, B.C.: A Hilbert-type integral inequality in the whole plane with the homogeneous kernel of degree −2. Arch. Inequal. Appl. (2011). doi:10.1155/2011/401428
44. Zhong, W.Y., Yang, B.C.: On multiple's Hardy-Hilbert integral inequality with kernel. Arch. Inequal. Appl. (2007). doi:10.1155/2007/27962
45. Hong, Y.: On multiple Hardy-Hilbert integral inequalities with some parameters. Arch. Inequal. Appl. (2002). doi:10.1155/2006/94960
46. Yang, B.C.: On an application of Hilbert's inequality with multi-parameters. J. Beijing Union University (Nat. Sci.) **24**(4), 78–84 (2010)
47. Yang, B.C.: An application of the reverse Hilbert's inequality with multi-parameters. J. Xinxiang University (Nat. Sci.) **27**(4), 1–5 (2010)
48. Taylor, A.E., Lay, D.C.: Introduction to Functional Analysis. Wiley, New York (1980)

Chapter 43
On the Stability of the Pexiderized Sine Functional Equation

Xiaopeng Zhao and Xiuzhong Yang

Abstract The aim of this paper is to study the stability of the Pexider type sine functional equation

$$h(x)k(y) = f^2\left(\frac{x+y}{2}\right) - g^2\left(\frac{x+\sigma y}{2}\right).$$

We have also extended the results to the Banach algebra.

Key words Stability · Superstability · Sine functional equation · Trigonometric functional equation

Mathematics Subject Classification 39B62 · 39B82

43.1 Introduction

In 1940, S.M. Ulam [12] proposed the following stability problem: Given a metric group $G(\cdot,\rho)$, a number $\varepsilon > 0$ and a mapping $f : G \to G$ which satisfies the inequality $\rho(f(x\cdot y), f(x)\cdot f(y)) < \varepsilon$ for all x, y in G, does there exist an automorphism a of G and a constant $k > 0$, depending only on G, such that $\rho(a(x), f(x)) \le k\varepsilon$ for all x in G? If the answer is affirmative, we call the equation $a(x\cdot y) = a(x)\cdot a(y)$ of automorphism stable. One year later, D.H. Hyers [6] provided a positive partial answer to Ulam's problem. In 1978, a generalized version of Hyers' result was proved by Th.M. Rassias in [11]. Since then, stability problems of several functional equations have been extensively investigated by a number of authors ([13, 14], and [15]).

Dedicated to Professor Themistocles M. Rassias on the occasion of his 60th birthday.

X. Zhao · X. Yang (✉)
College of Mathematics and Information Science, Hebei Normal University, Shijiazhuang, Hebei 050024, P.R. China
e-mail: xiuzhongyang@126.com

X. Zhao
e-mail: zhaoxiaopeng.2007@163.com

P.M. Pardalos et al. (eds.), *Nonlinear Analysis*, Springer Optimization and Its Applications 68, In Honor of Themistocles M. Rassias on the Occasion of his 60th Birthday, DOI 10.1007/978-1-4614-3498-6_43,

In 1979, a study of approximately multiplicative functions from a vector space V into the real numbers was made by J. Baker, J. Lawrence, and F. Zorzitto in [1]. They proved that for a given $\delta > 0$ and a function $f : V \to R$ such that $|f(x+y) - f(x)f(y)| < \delta$ for all x, y in V, it follows that either $f(x)$ remains bounded, with a bound depending on δ, or else $f(x+y) = f(x)f(y)$ for all x, y in V. This phenomenon is frequently called superstability.

The superstability of the cosine functional equation (also called the d'Alembert equation)

$$f(x+y) + f(x-y) = 2f(x)f(y) \tag{A}$$

was investigated by Baker [2]. The same result was also obtained later by Găvruţă [5] by applying a simple technique.

Badora and Ger [3] have improved the superstability of the d'Alembert equation (A) under the condition $|f(x+y) + f(x-y) - 2f(x)f(y)| \le \varphi(x)$ or $\varphi(y)$.

The stability of the generalized cosine functional equation has been studied by several mathematicians (cf. [7, 9], and [10], and the references therein).

The superstability of the sine functional equation

$$f(x)f(y) = f^2\left(\frac{x+y}{2}\right) - f^2\left(\frac{x-y}{2}\right) \tag{S}$$

was investigated by Cholewa [4].

In [8], Kim investigated the stability of the generalized sine functional equation

$$g(x)h(y) = f^2\left(\frac{x+y}{2}\right) - f^2\left(\frac{x-y}{2}\right).$$

Now, we consider the superstability of the following Pexider type sine functional equation

$$h(x)k(y) = f^2\left(\frac{x+y}{2}\right) - g^2\left(\frac{x+\sigma y}{2}\right)$$

of the sine functional equation (S) and its special cases as follows:

$$\begin{aligned} g(x)h(y) &= f^2\left(\frac{x+y}{2}\right) - f^2\left(\frac{x+\sigma y}{2}\right), \\ g(x)f(y) &= f^2\left(\frac{x+y}{2}\right) - f^2\left(\frac{x+\sigma y}{2}\right), \\ f(x)h(y) &= f^2\left(\frac{x+y}{2}\right) - f^2\left(\frac{x+\sigma y}{2}\right), \\ g(x)g(y) &= f^2\left(\frac{x+y}{2}\right) - f^2\left(\frac{x+\sigma y}{2}\right), \\ f(x)f(y) &= f^2\left(\frac{x+y}{2}\right) - f^2\left(\frac{x+\sigma y}{2}\right). \end{aligned} \tag{$\tilde{S}$}$$

Furthermore, the above results can be extended further to the spirit of the Banach algebra.

In this paper, let $(G,+)$ be an uniquely 2-divisible abelian group, C the field of complex numbers, R the field of real numbers, and let σ be an endomorphism of G with $\sigma(\sigma(x)) = x$ for all $x \in G$, we will use $\sigma(x) = \sigma x$. We may assume that f, g, h, and k are non-zero functions, ε is a nonnegative real constant, and $\varphi : G \to R$ is a given nonnegative function.

43.2 Superstability of the Pexiderized Sine Functional Equation

In this section, we will investigate the superstability of the pexiderized sine functional equation.

Theorem 43.1 *Suppose that $f, g, h, k : G \to C$ satisfy the inequality*

$$\left| h(x)k(y) - f^2\left(\frac{x+y}{2}\right) + g^2\left(\frac{x+\sigma y}{2}\right)\right| \leq \varphi(x) \tag{43.1}$$

for all $x, y \in G$. Then either k is bounded or h satisfies $(\tilde{\mathrm{S}})$ under the assumption that $h(0) = 0$.

Proof Let k be an unbounded function. Then, we can choose a sequence $\{y_n\}$ in G such that $0 \neq |k(2y_n)| \to \infty$ as $n \to \infty$. Inequality (43.1) may equivalently be written as

$$\left| h(2x)k(2y) - f^2(x+y) + g^2(x+\sigma y)\right| \leq \varphi(2x), \quad \forall x, y \in G. \tag{43.2}$$

Taking $y = y_n$ in (43.2), we obtain

$$\left| h(2x) - \frac{f^2(x+y_n) - g^2(x+\sigma y_n)}{k(2y_n)}\right| \leq \frac{\varphi(2x)}{|k(2y_n)|},$$

that is,

$$h(2x) = \lim_{n\to\infty} \frac{f^2(x+y_n) - g^2(x+\sigma y_n)}{k(2y_n)}, \quad x \in G. \tag{43.3}$$

Using (43.1), we have

$$\begin{aligned}
&\left| h(x)k(2y_n + y) - f^2\left(\frac{x+y}{2} + y_n\right) + g^2\left(\frac{x+\sigma y}{2} + \sigma y_n\right)\right. \\
&\quad \left. + \left(h(x)k(2y_n + \sigma y) - f^2\left(\frac{x+\sigma y}{2} + y_n\right) + g^2\left(\frac{x+y}{2} + \sigma y_n\right)\right)\right| \leq 2\varphi(x),
\end{aligned}$$

and thus,

$$\left| h(x)\frac{k(2y_n+y)+k(2y_n+\sigma y)}{k(2y_n)} - \frac{f^2(\frac{x+y}{2}+y_n)-g^2(\frac{x+y}{2}+\sigma y_n)}{k(2y_n)} \right.$$
$$\left. - \frac{f^2(\frac{x+\sigma y}{2}+y_n)-g^2(\frac{x+\sigma y}{2}+\sigma y_n)}{k(2y_n)} \right| \le \frac{2\varphi(x)}{|k(2y_n)|}$$

for all $x, y \in G$. Taking the limit as $n \to \infty$ and applying (43.3), we conclude that, for every $y \in G$, there exists a limit function

$$k_0(y) := \lim_{n\to\infty} \frac{k(2y_n+y)+k(2y_n+\sigma y)}{k(2y_n)},$$

where the function $k_0 : G \to C$ satisfies the equation

$$h(x+y)+h(x+\sigma y) = h(x)k_0(y) \quad \forall x, y \in G. \tag{43.4}$$

Applying the case $h(0) = 0$ in (43.4), we have $h(\sigma y) = -h(y),\ \forall y \in G$. By means of (43.4), we infer the equality

$$\begin{aligned} h^2(x+y)-h^2(x+\sigma y) &= \big[h(x+y)+h(x+\sigma y)\big]\big[h(x+y)-h(x+\sigma y)\big] \\ &= h(x)k_0(y)\big[h(x+y)-h(x+\sigma y)\big] \\ &= h(x)\big[h(x+2y)-h(x+2\sigma y)\big] \\ &= h(x)\big[h(2y+x)+h(2y+\sigma x)\big] \\ &= h(x)k_0(x)h(2y). \end{aligned}$$

Replacing y by x in (43.4), we have

$$h(2x) = h(x)k_0(x), \quad \forall x \in G,$$

since $h(\sigma y) = -h(y)$ for all $y \in G$ implies $h(x+\sigma x) = 0$.

This leads to the equation

$$h^2(x+y)-h^2(x+\sigma y) = h(2x)h(2y)$$

which is valid for all $x, y \in G$, i.e., h satisfies $(\tilde{S})$. □

Theorem 43.2 *Suppose that $f, g, h, k : G \to C$ satisfy the inequality*

$$\left| h(x)k(y) - f^2\left(\frac{x+y}{2}\right) + g^2\left(\frac{x+\sigma y}{2}\right) \right| \le \varphi(y) \tag{43.5}$$

for all $x, y \in G$. Then either h is bounded or k satisfies $(\tilde{S})$ under the assumption that $k(0) = 0$.

Proof Let h be unbounded, then we can choose a sequence $\{x_n\}$ in G such that $0 \neq |h(2x_n)| \to \infty$ as $n \to \infty$. The argument applied at the beginning of the proof of Theorem 43.1 implies:

$$k(2y) = \lim_{n\to\infty} \frac{f^2(x_n + y) - g^2(x_n + \sigma y)}{h(2x_n)}, \quad y \in G. \tag{43.6}$$

In (43.5), replacing x by $2x_n + y$ and $2x_n + \sigma y$, respectively, and y by x, it follows that

$$\begin{aligned}&\left|h(2x_n + y)k(x) - f^2\left(x_n + \frac{x+y}{2}\right) + g^2\left(x_n + \sigma\left(\frac{x+\sigma y}{2}\right)\right)\right.\\&\quad\left. + h(2x_n + \sigma y)k(x) - f^2\left(x_n + \frac{x+\sigma y}{2}\right) + g^2\left(x_n + \sigma\left(\frac{x+y}{2}\right)\right)\right|\\&\le 2\varphi(x),\end{aligned}$$

and thus,

$$\begin{aligned}&\left|k(x)\frac{h(2x_n + y) + h(2x_n + \sigma y)}{h(2x_n)} - \frac{f^2(x_n + \frac{x+y}{2}) - g^2(x_n + \sigma(\frac{x+y}{2}))}{h(2x_n)}\right.\\&\quad\left. - \frac{f^2(x_n + \frac{x+\sigma y}{2}) - g^2(x_n + \sigma(\frac{x+\sigma y}{2}))}{h(2x_n)}\right| \le \frac{2\varphi(x)}{|h(2x_n)|}\end{aligned}$$

for all $x, y \in G$. Taking thc limit as $n \to \infty$ and applying (43.6), we conclude that, for every $y \in G$, there exists a limit function

$$h_0(y) := \lim_{n\to\infty} \frac{h(2x_n + y) + h(2x_n + \sigma y)}{h(2x_n)},$$

where the function $h_0 : G \to C$ satisfies the equation

$$k(x + y) + k(x + \sigma y) = k(x)h_0(y) \quad \forall x, y \in G.$$ □

The proof below follows the same spirit with the proof of Theorem 43.1.

Let us consider the case $f = g$ in Theorem 43.1 and Theorem 43.2, then we obtain the following two corollaries.

Corollary 43.1 *Suppose that $f, g, h : G \to C$ satisfy the inequality*

$$\left|g(x)h(y) - f^2\left(\frac{x+y}{2}\right) + f^2\left(\frac{x+\sigma y}{2}\right)\right| \le \varphi(x) \tag{43.7}$$

for all $x, y \in G$. Then either h is bounded or g satisfies ($\tilde{\mathrm{S}}$) under one of the assumptions $g(0) = 0$, $f^2(\sigma x) = f^2(x)$.

Proof Let h be unbounded. If $f^2(\sigma x) = f^2(x)$, it is enough to show that $g(0) = 0$. Suppose that this is not the case. Putting $x = 0$ in (43.7), due to $g(0) \neq 0$, we obtain the inequality

$$|h(y)| \leq \frac{\varphi(0)}{|g(0)|}, \quad y \in G.$$

This inequality means that h is globally bounded—a contradiction. Thus the claimed $g(0) = 0$ holds. From Theorem 43.1, we deduce that g satisfies ($\tilde{\mathrm{S}}$). □

Corollary 43.2 *Suppose that* $f, g, h : G \to C$ *satisfy the inequality*

$$\left|g(x)h(y) - f^2\left(\frac{x+y}{2}\right) + f^2\left(\frac{x+\sigma y}{2}\right)\right| \leq \varphi(y) \tag{43.8}$$

for all $x, y \in G$. *Then either* g *is bounded or* h *satisfies* ($\tilde{\mathrm{S}}$).

Proof Let g be unbounded. Then we can choose a sequence $\{x_n\}$ in G such that $0 \neq |g(2x_n)| \to \infty$ as $n \to \infty$. Using a similar procedure to that applied in the beginning of Theorem 43.1, we get

$$h(2y) = \lim_{n\to\infty} \frac{f^2(x_n + y) - f^2(x_n + \sigma y)}{g(2x_n)}, \quad y \in G. \tag{43.9}$$

An obvious slight change in the proof applied in Theorem 43.2 gives us

$$\begin{aligned}
&\left|g(2x_n + y)h(x) - f^2\left(x_n + \frac{y+x}{2}\right) + f^2\left(x_n + \frac{y+\sigma x}{2}\right)\right. \\
&\qquad \left. + g(2x_n + \sigma y)h(x) - f^2\left(x_n + \frac{\sigma y + x}{2}\right) + f^2\left(x_n + \sigma\left(\frac{y+x}{2}\right)\right)\right| \\
&\leq 2\varphi(x),
\end{aligned}$$

and thus,

$$\begin{aligned}
&\left|\frac{g(2x_n + y) + g(2x_n + \sigma y)}{g(2x_n)} h(x) - \frac{f^2(x_n + \frac{y+x}{2}) - f^2(x_n + \sigma(\frac{y+x}{2}))}{g(2x_n)}\right. \\
&\qquad \left. + \frac{f^2(x_n + \frac{y+\sigma x}{2}) - f^2(x_n + \sigma(\frac{y+\sigma x}{2}))}{g(2x_n)}\right| \leq \frac{2\varphi(x)}{|g(2x_n)|}
\end{aligned}$$

for all $x, y \in G$. Taking the limit as $n \to \infty$ and applying (43.9), we conclude that, for every $y \in G$, there exists a limit function

$$g_1(y) := \lim_{n\to\infty} \frac{g(2x_n + y) + g(2x_n + \sigma y)}{g(2x_n)},$$

where the function $g_1 : G \to C$ satisfies the equation

$$h(y + x) - h(y + \sigma x) = g_1(y)h(x) \quad \forall x, y \in G. \tag{43.10}$$

From the definition of g_1, we get the equality $g_1(0)=2$ which together with (43.10) implies that $h(\sigma x)=-h(x),\ \forall x\in G$. By means of (43.10), we infer the equality

$$\begin{aligned}
h^2(x+y)-h^2(x+\sigma y) &= \big[h(x+y)+h(x+\sigma y)\big]\big[h(x+y)-h(x+\sigma y)\big]\\
&= \big[h(x+y)+h(x+\sigma y)\big]g_1(x)h(y)\\
&= \big[h(2x+y)+h(2x+\sigma y)\big]h(y)\\
&= \big[h(y+2x)-h(y+2\sigma x)\big]h(y)\\
&= g_1(y)h(2x)h(y).
\end{aligned}$$

Putting $x=y$ in (43.10), we get $h(2y)=g_1(y)h(y),\ \forall y\in G$, since $h(\sigma x)=-h(x)$ for all $x\in G$ implies $h(y+\sigma y)=0$. This relation together with the above equality proves that h satisfies $(\tilde{\mathrm{S}})$. □

Remark 43.1 If we put $\varphi(x)=\varphi(y)=\varepsilon$ and $\sigma(x)=-x$ in Corollary 43.1 and Corollary 43.2, then one obtains the result published in [8].

By replacing h by f, g by f and h by g in Corollary 43.1 and Corollary 43.2, we obtain the following results.

Corollary 43.3 *Suppose that $f,g:G\to C$ satisfy the inequality*

$$\left|g(x)f(y)-f^2\left(\frac{x+y}{2}\right)+f^2\left(\frac{x+\sigma y}{2}\right)\right|\le\varphi(x)$$

for all $x,y\in G$. Then either f is bounded or g satisfies $(\tilde{\mathrm{S}})$ under one of the assumptions $g(0)=0,\ f^2(\sigma x)=f^2(x)$.

Corollary 43.4 *Suppose that $f,g:G\to C$ satisfy the inequality*

$$\left|g(x)f(y)-f^2\left(\frac{x+y}{2}\right)+f^2\left(\frac{x+\sigma y}{2}\right)\right|\le\varphi(y)$$

for all $x,y\in G$. Then either g is bounded or f satisfies $(\tilde{\mathrm{S}})$.

In the case of $\varphi(y)=\varepsilon$ in Corollary 43.4, we can obtain the following corollary.

Corollary 43.5 *Suppose that $f,g:G\to C$ satisfy the inequality*

$$\left|g(x)f(y)-f^2\left(\frac{x+y}{2}\right)+f^2\left(\frac{x+\sigma y}{2}\right)\right|\le\varepsilon \tag{43.11}$$

for all $x,y\in G$. Then either g is bounded or f and g satisfy $(\tilde{\mathrm{S}})$.

Proof The case of f follows from Corollary 43.4 by taking $\varphi(y) = \varepsilon$.

For the proof of the case of g, first we show that g is bounded whenever f is bounded.

Let f be bounded, then we can choose $y_0 \in G$ such that $f(2y_0) \neq 0$, then by (43.11), we obtain

$$\begin{aligned} \left|g(2x)\right| - \left|\frac{f^2(x+y_0) - f^2(x+\sigma y_0)}{f(2y_0)}\right| &\leq \left|g(2x) - \frac{f^2(x+y_0) - f^2(x+\sigma y_0)}{f(2y_0)}\right| \\ &\leq \frac{\varepsilon}{|f(2y_0)|}, \end{aligned}$$

and it follows that g is also bounded on G.

Since the unbounded assumption of g implies that f is also unbounded, we can choose a sequence $\{y_n\}$ in G such that $0 \neq |f(2y_n)| \to \infty$ as $n \to \infty$.

A slight change applied in Theorem 43.1 gives us

$$g(2x) = \lim_{n\to\infty} \frac{f^2(x+y_n) - f^2(x+\sigma y_n)}{f(2y_n)}, \quad \forall x \in G. \tag{43.12}$$

Since we have shown that f satisfies ($\tilde{\text{S}}$) whenever g is unbounded, (43.12) is represented as

$$g(2x) = f(2x), \quad \forall x \in G.$$

By the 2-divisibility of group G, we obtain $g = f$. Therefore, g also satisfies ($\tilde{\text{S}}$). □

Corollary 43.6 *Suppose that $f, h : G \to C$ satisfy the inequality*

$$\left| f(x)h(y) - f^2\left(\frac{x+y}{2}\right) + f^2\left(\frac{x+\sigma y}{2}\right)\right| \leq \varphi(x)$$

for all $x, y \in G$. Then either h is bounded or f satisfies ($\tilde{\text{S}}$) under one of the assumptions $f(0) = 0$, $f^2(\sigma x) = f^2(x)$.

Corollary 43.7 *Suppose that $f, h : G \to C$ satisfy the inequality*

$$\left| f(x)h(y) - f^2\left(\frac{x+y}{2}\right) + f^2\left(\frac{x+\sigma y}{2}\right)\right| \leq \varphi(y)$$

for all $x, y \in G$. Then either f is bounded or h satisfies ($\tilde{\text{S}}$).

In the case of $\varphi(y) = \varepsilon$ in Corollary 43.7, we obtain the following result.

Corollary 43.8 *Suppose that $f, h : G \to C$ satisfy the inequality*

$$\left| f(x)h(y) - f^2\left(\frac{x+y}{2}\right) + f^2\left(\frac{x+\sigma y}{2}\right)\right| \leq \varepsilon \tag{43.13}$$

for all $x, y \in G$. Then either h is bounded or h satisfies ($\tilde{\text{S}}$).

Proof We can see that, similar to Corollary 43.5, h is bounded whenever f is bounded.

Let f be bounded, then we can choose $x_0 \in G$ such that $f(2x_0) \neq 0$, then by (43.13), we have

$$\left|h(2y)\right| - \left|\frac{f^2(x_0+y) - f^2(x_0+\sigma y)}{f(2x_0)}\right| \leq \left|h(2y) - \frac{f^2(x_0+y) - f^2(x_0+\sigma y)}{f(2x_0)}\right| \leq \frac{\varepsilon}{|f(2x_0)|},$$

which shows that h is also bounded on G. Namely, the unboundedness of h implies that of f. Applying the case $\varphi(y) = \varepsilon$ in Corollary 43.7, it implies that h satisfies $(\tilde{S})$. □

Corollary 43.9 *Suppose that $f, g : G \to C$ satisfy the inequality*

$$\left|g(x)g(y) - f^2\left(\frac{x+y}{2}\right) + f^2\left(\frac{x+\sigma y}{2}\right)\right| \leq \varphi(x)$$

for all $x, y \in G$. Then either g is bounded or g satisfies $(\tilde{S})$ under one of the assumptions $g(0) = 0$, $f^2(\sigma x) = f^2(x)$.

Corollary 43.10 *Suppose that $f, g : G \to C$ satisfy the inequality*

$$\left|g(x)g(y) - f^2\left(\frac{x+y}{2}\right) + f^2\left(\frac{x+\sigma y}{2}\right)\right| \leq \varphi(y)$$

for all $x, y \in G$. Then either g is bounded or g satisfies $(\tilde{S})$.

Corollary 43.11 *Suppose that $f : G \to C$ satisfies the inequality*

$$\left|f(x)f(y) - f^2\left(\frac{x+y}{2}\right) + f^2\left(\frac{x+\sigma y}{2}\right)\right| \leq \varphi(x)$$

for all $x, y \in G$. Then either f is bounded or f satisfies $(\tilde{S})$ under one of the assumptions $f(0) = 0$, $f^2(\sigma x) = f^2(x)$.

Corollary 43.12 *Suppose that $f : G \to C$ satisfies the inequality*

$$\left|f(x)f(y) - f^2\left(\frac{x+y}{2}\right) + f^2\left(\frac{x+\sigma y}{2}\right)\right| \leq \varphi(y)$$

for all $x, y \in G$. Then either f is bounded or f satisfies $(\tilde{S})$.

43.3 Extension to the Banach Algebra

The obtained results of Sect. 43.2 can be extended to the Banach algebra. To simplify, we will combine two theorems into one.

Theorem 43.3 *Let $(E, \|\cdot\|)$ be a semisimple commutative Banach algebra. Assume that $f, g, h, k : G \to E$ satisfy the inequality*

$$\left\| h(x)k(y) - f^2\left(\frac{x+y}{2}\right) + g^2\left(\frac{x+\sigma y}{2}\right) \right\| \le \begin{cases} \text{(i)} & \varphi(x), \\ \text{(ii)} & \varphi(y), \end{cases} \tag{43.14}$$

for all $x, y \in G$. For an arbitrary linear multiplicative functional $x^* \in E^*$,

(i) *If the superposition $x^* \circ k$ under the assumption that $h(0) = 0$ fails to be bounded, then h satisfies* ($\tilde{\text{S}}$);
(ii) *If the superposition $x^* \circ h$ under the assumption that $k(0) = 0$ fails to be bounded, then k satisfies* ($\tilde{\text{S}}$).

Proof We give only the proof of (i), as that of (ii) is similar.

Assume (i), and fix an arbitrary linear multiplicative functional $x^* \in E^*$. As is well known, we have $\|x^*\| = 1$ whence, for every $x, y \in G$, we have

$$\begin{aligned}
\varphi(x) &\ge \left\| h(x)k(y) - f^2\left(\frac{x+y}{2}\right) + g^2\left(\frac{x+\sigma y}{2}\right) \right\| \\
&= \sup_{\|y^*\|=1} \left| y^*\left(h(x)k(y) - f^2\left(\frac{x+y}{2}\right) + g^2\left(\frac{x+\sigma y}{2}\right)\right) \right| \\
&\ge \left| x^*\left(h(x)k(y) - f^2\left(\frac{x+y}{2}\right) + g^2\left(\frac{x+\sigma y}{2}\right)\right) \right| \\
&= \left| \left(x^* \circ h\right)(x)\left(x^* \circ k\right)(y) - \left(x^* \circ f\right)^2\left(\frac{x+y}{2}\right) + \left(x^* \circ g\right)^2\left(\frac{x+\sigma y}{2}\right) \right|,
\end{aligned}$$

which states that the superposition $x^* \circ f$, $x^* \circ g$, $x^* \circ h$, and $x^* \circ k$ yield a solution of inequality (43.1) of Theorem 43.1. Since, by assumption, the superposition $x^* \circ k$ is unbounded, and $h(0) = 0$ implies $(x^* \circ h)(0) = 0$, an appeal to Theorem 43.1 shows that the superposition $x^* \circ h$ satisfies ($\tilde{\text{S}}$). With the use of the linear multiplicative property of x^*, we have $x^*(h^2(x + y) - h^2(x + \sigma y) - h(2x)h(2y)) = 0$ for all $x, y \in G$, i.e.,

$$\tilde{S}_h(x, y) := h^2(x + y) - h^2(x + \sigma y) - h(2x)h(2y) \in \ker\left(x^*\right)$$

for all $x, y \in G$. Therefore, in view of the unrestricted choice of x^*, we infer that

$$\tilde{S}_h(x, y) \in \bigcap \left\{\ker x^* : x^* \text{ is a linear multiplicative member of } E^*\right\}$$

for all $x, y \in G$. Since the algebra E has been assumed to be semisimple, the last term of the above formula coincides with the singleton $\{0\}$, i.e.,

$$h^2(x+y) - h^2(x+\sigma y) = h(2x)h(2y) \quad \forall x, y \in G,$$

as claimed.

The case (ii) runs the same procedure. □

Let us consider the case $f = g$ in the inequality (43.14) of Theorem 43.3, then we obtain the following corollary.

Corollary 43.13 *Let $(E, \|\cdot\|)$ be a semisimple commutative Banach algebra. Assume that $f, g, h : G \to E$ satisfy the inequality*

$$\left\| g(x)h(y) - f^2\left(\frac{x+y}{2}\right) + f^2\left(\frac{x+\sigma y}{2}\right) \right\| \le \begin{cases} \text{(i)} & \varphi(x), \\ \text{(ii)} & \varphi(y), \end{cases} \tag{43.15}$$

for all $x, y \in G$. For an arbitrary linear multiplicative functional $x^* \in E^*$,

(i) *If the superposition $x^* \circ h$ under the assumption that $g(0) = 0$ or $f^2(\sigma x) = f^2(x)$ fails to be bounded, then g satisfies ($\tilde{\text{S}}$);*
(ii) *If the superposition $x^* \circ g$ fails to be bounded, then h satisfies ($\tilde{\text{S}}$).*

As in Sect. 43.2, let us consider the each case $h = f$, $g = f$, $g = h$, $g = h = f$ in the inequality (43.15), respectively, then we can obtain the same results as in Sect. 43.2 for each functional equation.

Corollary 43.14 *Let $(E, \|\cdot\|)$ be a semisimple commutative Banach algebra. Assume that $f, g : G \to E$ satisfy the inequality*

$$\left\| g(x)f(y) - f^2\left(\frac{x+y}{2}\right) + f^2\left(\frac{x+\sigma y}{2}\right) \right\| \le \begin{cases} \text{(i)} & \varphi(x), \\ \text{(ii)} & \varphi(y), \end{cases}$$

for all $x, y \in G$. For an arbitrary linear multiplicative functional $x^* \in E^*$,

(i) *If the superposition $x^* \circ f$ under the assumption that $g(0) = 0$ or $f^2(\sigma x) = f^2(x)$ fails to be bounded, then g satisfies ($\tilde{\text{S}}$);*
(ii) *If the superposition $x^* \circ g$ fails to be bounded, then f satisfies ($\tilde{\text{S}}$).*

Corollary 43.15 *Let $(E, \|\cdot\|)$ be a semisimple commutative Banach algebra. Assume that $f, g : G \to E$ satisfy the inequality*

$$\left\| g(x)f(y) - f^2\left(\frac{x+y}{2}\right) + f^2\left(\frac{x+\sigma y}{2}\right) \right\| \le \varepsilon$$

for all $x, y \in G$. For an arbitrary linear multiplicative functional $x^* \in E^*$, if the superposition $x^* \circ g$ fails to be bounded, then f and g satisfy ($\tilde{\text{S}}$).

Corollary 43.16 *Let $(E, \|\cdot\|)$ be a semisimple commutative Banach algebra. Assume that $f, h : G \to E$ satisfy the inequality*

$$\left\| f(x)h(y) - f^2\left(\frac{x+y}{2}\right) + f^2\left(\frac{x+\sigma y}{2}\right)\right\| \le \begin{cases} \text{(i)} & \varphi(x), \\ \text{(ii)} & \varphi(y), \end{cases}$$

for all $x, y \in G$. For an arbitrary linear multiplicative functional $x^* \in E^*$,

(i) *If the superposition $x^* \circ h$ under the assumption that $f(0) = 0$ or $f^2(\sigma x) = f^2(x)$ fails to be bounded, then f satisfies $(\tilde{\mathrm{S}})$;*
(ii) *If the superposition $x^* \circ f$ fails to be bounded, then h satisfies $(\tilde{\mathrm{S}})$.*

Corollary 43.17 *Let $(E, \|\cdot\|)$ be a semisimple commutative Banach algebra. Assume that $f, h : G \to E$ satisfy the inequality*

$$\left\| f(x)h(y) - f^2\left(\frac{x+y}{2}\right) + f^2\left(\frac{x+\sigma y}{2}\right)\right\| \le \varepsilon$$

for all $x, y \in G$. For an arbitrary linear multiplicative functional $x^* \in E^*$, if the superposition $x^* \circ h$ fails to be bounded, then h satisfies $(\tilde{\mathrm{S}})$.

Corollary 43.18 *Let $(E, \|\cdot\|)$ be a semisimple commutative Banach algebra. Assume that $f, g : G \to E$ satisfy the inequality*

$$\left\| g(x)g(y) - f^2\left(\frac{x+y}{2}\right) + f^2\left(\frac{x+\sigma y}{2}\right)\right\| \le \begin{cases} \text{(i)} & \varphi(x), \\ \text{(ii)} & \varphi(y), \end{cases}$$

for all $x, y \in G$. For an arbitrary linear multiplicative functional $x^* \in E^*$,

(i) *If the superposition $x^* \circ g$ under the assumption that $g(0) = 0$ or $f^2(\sigma x) = f^2(x)$ fails to be bounded, then g satisfies $(\tilde{\mathrm{S}})$;*
(ii) *If the superposition $x^* \circ g$ fails to be bounded, then g satisfies $(\tilde{\mathrm{S}})$.*

Corollary 43.19 *Let $(E, \|\cdot\|)$ be a semisimple commutative Banach algebra. Assume that $f : G \to E$ satisfies the inequality*

$$\left\| f(x)f(y) - f^2\left(\frac{x+y}{2}\right) + f^2\left(\frac{x+\sigma y}{2}\right)\right\| \le \begin{cases} \text{(i)} & \varphi(x), \\ \text{(ii)} & \varphi(y), \end{cases}$$

for all $x, y \in G$. For an arbitrary linear multiplicative functional $x^* \in E^*$,

(i) *If the superposition $x^* \circ f$ under the assumption that $f(0) = 0$ or $f^2(\sigma x) = f^2(x)$ fails to be bounded, then f satisfies $(\tilde{\mathrm{S}})$;*
(ii) *If the superposition $x^* \circ f$ fails to be bounded, then f satisfies $(\tilde{\mathrm{S}})$.*

Acknowledgements We would like to express our gratitude to Professor Themistocles M. Rassias for his encouragement and advice.

References

1. Baker, J., Lawrence, J., Zorzitto, F.: The stability of the equation $f(x+y) = f(x)f(y)$. Proc. Am. Math. Soc. **74**, 242–246 (1979)
2. Baker, J.: The stability of the cosine equation. Proc. Am. Math. Soc. **80**, 411–416 (1980)
3. Badora, R., Ger, R.: On some trigonometric functional inequalities. In: Functional Equations–Results and Advances, pp. 3–15 (2002)
4. Cholewa, P.W.: The stability of the sine equation. Proc. Am. Math. Soc. **88**, 631–634 (1983)
5. Găvruţă, P.: On the stability of some functional equations. In: Rassias, Th.M., Tabor, J. (eds.) Stability of Mappings of Hyers–Ulam Type, pp. 93–98. Hadronic Press, Palm Harbor (1994)
6. Hyers, D.H.: On the stability of the linear functional equation. Proc. Natl. Acad. Sci. USA **27**(4), 222–224 (1941)
7. Kim, G.H.: The stability of d'Alembert and Jensen type functional equations. J. Math. Anal. Appl. **325**, 237–248 (2007)
8. Kim, G.H.: A stability of the generalized sine functional equations. J. Math. Anal. Appl. **331**, 886–894 (2007)
9. Kusollerschariya, C., Nakmahachalasint, P.: The stability of the pexiderized cosine functional equation. Thai J. Math. **6**(3), 39–44 (2008)
10. Kim, G.H.: On the superstability of the Pexider type trigonometric functional equation. J. Inequal. Appl. (2010). doi:10.1155/2010/897123
11. Rassias, Th.M.: On the stability of the linear mapping in Banach spaces. Proc. Am. Math. Soc. **72**, 297–300 (1978)
12. Ulam, S.M.: Problems in Modern Mathematics. Science Editions, Wiley, New York (1964). Chapter VI
13. Czerwik, S.: Functional Equations and Inequalities in Several Variables. World Scientific, New Jersey (2002)
14. Hyers, D.H., Rassias, Th.M.: Approximate homomorphisms. Aequ. Math. **44**, 125–153 (1992)
15. Jung, S.-M.: Hyers–Ulam–Rassias Stability of Functional Equations in Mathematical Analysis. Hadronic Press, Florida (2001)

Chapter 44
Stability of Additive-Quadratic Functional Equations in Intuitionistic Fuzzy Normed Spaces

Zhihua Wang

Abstract In this paper, we establish some stability results concerning the additive-quadratic functional equation

$$f(2x+y)+f(2x-y)=f(x+y)+f(x-y)+2f(2x)-2f(x)$$

in intuitionistic fuzzy normed spaces (IFNS).

Key words Hyers–Ulam–Rassias stability · Intuitionistic fuzzy normed spaces · Additive-quadratic functional equations

Mathematics Subject Classification 03F55 · 39B82 · 39B72

44.1 Introduction and Preliminaries

The notion of fuzzy sets was first introduced by Zadeh [35] in 1965. Among various developments of the theory of fuzzy sets, progressive development has been made to find the fuzzy analogues of the classical set theory. In fact, the fuzzy theory has become an area of active research for the last 40 years. It has a wide range of applications in the field of science and engineering, e.g., in population dynamics [3], chaos control [7], computer programming [9], nonlinear dynamical systems [11], nonlinear operator [23], statistical convergence [20, 22], fuzzy physics [15], etc. The concept of a fuzzy topology may have very important applications in quantum particle physics particularly in connections with both string and ε^{∞} theory which were given and studied by El-Naschie [4, 5].

There are many situations where the norm of a vector is not possible to find and the concept of intuitionistic fuzzy norm [26, 30, 31] seems to be more suitable in

---

Dedicated to Professor Themistocles M. Rassias on the occasion of his 60th birthday.

Z. Wang (✉)
School of Science, Hubei University of Technology, Wuhan, Hubei 430068, P.R. China
e-mail: matwzh2000@126.com

Z. Wang
Department of Mathematics, Sichuan University, Chengdu, Sichuan 610064, P.R. China

P.M. Pardalos et al. (eds.), *Nonlinear Analysis*, Springer Optimization and Its Applications 68, In Honor of Themistocles M. Rassias on the Occasion of his 60th Birthday, DOI 10.1007/978-1-4614-3498-6_44, © Springer Science+Business Media, LLC 2012

such cases, that is, we can deal with such situations by modeling the inexactness through the intuitionistic fuzzy norm.

The stability problem of a functional equation was first posed by Ulam [33] which was answered by Hyers [12] on approximately additive mappings and then generalized by Aoki [2] and Rassias [27] for additive mappings and linear mappings, respectively. Later there have been a lot of results on stability in the Hyers–Ulam sense or some generalized sense (see books and papers [1, 6, 8, 13, 14, 28, 29], and references therein); and various fuzzy stability results concerning Cauchy, Jensen, quadratic and cubic functional equations were discussed [16–19], and some stability results concerning Jensen, cubic, mixed type additive, and cubic functional equations were investigated [21, 24, 34] in intuitionistic fuzzy normed spaces.

A. Najati and M.B. Moghimi [25] have established the general solution of and investigated the Hyers–Ulam–Rassias stability of the following functional equation deriving from quadratic and additive functions:

$$f(2x+y)+f(2x-y)=f(x+y)+f(x-y)+2f(2x)-2f(x) \qquad (44.1)$$

in quasi-Banach spaces, and fuzzy stability results of (44.1) were discussed in [10]. It is easy to see that the function $f(x)=ax^2+bx+c$ is a solution of (44.1). The main purpose of this paper is to establish some versions of the Hyers–Ulam–Rassias stability for the functional equation (44.1) in intuitionistic fuzzy normed spaces.

In this section, we recall some notations and basic definitions which we will used throughout this paper.

Definition 44.1 (Cf. [32]) A binary operation $*:[0,1]\times[0,1]\to[0,1]$ is said to be a continuous t-norm if it satisfies the following conditions:

(a) $*$ is commutative and associative,
(b) $*$ is continuous,
(c) $a*1=a$ for all $a\in[0,1]$,
(d) $a*b\le c*d$ whenever $a\le c$ and $b\le d$ for each $a,b,c,d\in[0,1]$.

Example 44.1 Two typical examples of a continuous t-norm are $a*b=ab$ and $a*b=\min(a,b)$.

Definition 44.2 (Cf. [32]) A binary operation $\Diamond:[0,1]\times[0,1]\to[0,1]$ is said to be a continuous t-conorm if it satisfies the following conditions:

(a′) $\Diamond$ is commutative and associative,
(b′) $\Diamond$ is continuous,
(c′) $a\Diamond 0=a$ for all $a\in[0,1]$,
(d′) $a\Diamond b\le c\Diamond d$ whenever $a\le c$ and $b\le d$ for each $a,b,c,d\in[0,1]$.

Example 44.2 Two typical examples of a continuous t-conorm are $a\Diamond b=\min(a+b,1)$ and $a\Diamond b=\max(a,b)$.

With the help of the notions of a continuous t-norm and a continuous t-conorm, Saadati and Park [30] have recently introduced the concept of intuitionistic fuzzy normed spaces as follows:

Definition 44.3 The five-tuple $(X, \mu, \nu, *, \Diamond)$ is said to be an intuitionistic fuzzy normed space (IFNS, for short) if X is a vector space, $*$ is a continuous t-norm, $\Diamond$ is a continuous t-conorm, and μ, ν are fuzzy sets on $X \times (0, \infty)$ satisfying the following conditions. For every $x, y \in X$ and $s, t > 0$

(i) $\mu(x,t) + \nu(x,t) \le 1$,
(ii) $\mu(x,t) > 0$,
(iii) $\mu(x,t) = 1$ if and only if $x = 0$,
(iv) $\mu(\alpha x, t) = \mu(x, \frac{t}{|\alpha|})$ for each $\alpha \neq 0$,
(v) $\mu(x,t) * \mu(y,s) \le \mu(x+y, t+s)$,
(vi) $\mu(x, \cdot) : (0,\infty) \to [0,1]$ is continuous,
(vii) $\lim_{t\to\infty} \mu(x,t) = 1$ and $\lim_{t\to 0} \mu(x,t) = 0$,
(viii) $\nu(x,t) < 1$,
(ix) $\nu(x,t) = 0$ if and only if $x = 0$,
(x) $\nu(\alpha x, t) = \nu(x, \frac{t}{|\alpha|})$ for each $\alpha \neq 0$,
(xi) $\nu(x,t) \Diamond \nu(y,s) \ge \nu(x+y, t+s)$,
(xii) $\nu(x, \cdot) : (0,\infty) \to [0,1]$ is continuous,
(xiii) $\lim_{t\to\infty} \nu(x,t) = 0$ and $\lim_{t\to 0} \nu(x,t) = 1$.

In this case, (μ, ν) is called an intuitionistic fuzzy norm.

Example 44.3 (Cf. [30]) Let $(X, \|\cdot\|)$ be a normed space, $a * b = ab$ and $a \Diamond b = \min(a+b, 1)$ for all $a, b \in [0,1]$. For all $x \in X$ and every $t > 0$, consider

$$\mu(x,t) = \begin{cases} \frac{t}{t+\|x\|} & \text{if } t > 0, \\ 0 & \text{if } t \le 0; \end{cases} \quad \text{and} \quad \nu(x,t) = \begin{cases} \frac{\|x\|}{t+\|x\|} & \text{if } t > 0, \\ 0 & \text{if } t \le 0. \end{cases}$$

Then $(X, \mu, \nu, *, \Diamond)$ is an IFNS.

The concepts of convergence and Cauchy sequences in an intuitionistic fuzzy normed space are studied in [30].

Let $(X, \mu, \nu, *, \Diamond)$ be an IFNS. Then, a sequence $\{x_k\}$ is said to be intuitionistic fuzzy convergent to $x \in X$ if for every $\varepsilon > 0$ and $t > 0$, there exists $k_0 \in \mathbb{N}$ such that $\mu(x_k - x, t) > 1 - \varepsilon$ and $\nu(x_k - x, t) < \varepsilon$ for all $k \ge k_0$. In this case, we write $(\mu, \nu) - \lim x_k = x$. The sequence $\{x_k\}$ is said to be intuitionistic fuzzy Cauchy sequence if for every $\varepsilon > 0$ and $t > 0$, there exists $k_0 \in \mathbb{N}$ such that $\mu(x_k - x_\ell, t) > 1 - \varepsilon$ and $\nu(x_k - x_\ell, t) < \varepsilon$ for all $k, \ell \ge k_0$. $(X, \mu, \nu, *, \Diamond)$ is said to be complete if every intuitionistic fuzzy Cauchy sequence in $(X, \mu, \nu, *, \Diamond)$ is intuitionistic fuzzy convergent in $(X, \mu, \nu, *, \Diamond)$.

44.2 Intuitionistic Fuzzy Stability

Throughout this section, assume that X, (Z, μ', ν'), and (Y, μ, ν) are a linear space, an IFNS, and an intuitionistic fuzzy Banach space, respectively. We start our works with the Hyers–Ulam–Rassias type theorem in IFNS for the additive-quadratic functional equation (44.1).

Theorem 44.1 *Let $\varphi_1 : X \times X \to Z$ be a function such that for some $0 < \alpha < 4$*

$$\mu'\left(\varphi_1\left(\frac{2x}{3}, 2y\right), t\right) \geq \mu'\left(\alpha\varphi_1\left(\frac{x}{3}, y\right), t\right) \quad \text{and}$$
$$\nu'\left(\varphi_1\left(\frac{2x}{3}, 2y\right), t\right) \leq \nu'\left(\alpha\varphi_1\left(\frac{x}{3}, y\right), t\right) \tag{44.2}$$

for all $x \in X$, $y \in \{0, \frac{x}{3}, \frac{4x}{3}, -\frac{2x}{3}, x\}$, $t > 0$, and $\lim_{n\to\infty} \mu'(\varphi_1(2^n x, 2^n y), 4^n t) = 1$ and $\lim_{n\to\infty} \nu'(\varphi_1(2^n x, 2^n y), 4^n t) = 0$ for all $x, y \in X$, $t > 0$. Suppose that an even function $f : X \to Y$ with $f(0) = 0$ satisfies the inequalities

$$\begin{cases} \mu(f(2x+y) + f(2x-y) - f(x+y) - f(x-y) - 2f(2x) + 2f(x), t) \\ \quad \geq \mu'(\varphi_1(x, y), t), \\ \nu(f(2x+y) + f(2x-y) - f(x+y) - f(x-y) - 2f(2x) + 2f(x), t) \\ \quad \leq \nu'(\varphi_1(x, y), t) \end{cases} \tag{44.3}$$

for all $x, y \in X$ and $t > 0$. Then there exist a unique quadratic mapping $Q : X \to Y$ such that

$$\mu\big(Q(x) - f(x), t\big) \geq \mu_1''\left(x, \frac{t(4-\alpha)}{12}\right) \quad \text{and}$$
$$\nu\big(Q(x) - f(x), t\big) \leq \nu_1''\left(x, \frac{t(4-\alpha)}{12}\right) \tag{44.4}$$

for all $x \in X$ and $t > 0$, where

$$\mu_1''(x, t) := \mu'\left(\varphi_1\left(\frac{x}{3}, \frac{x}{3}\right), t\right) * \mu'\left(\varphi_1\left(\frac{x}{3}, x\right), t\right) * \mu'\left(\varphi_1\left(\frac{x}{3}, \frac{4x}{3}\right), t\right)$$
$$* \mu'\left(\varphi_1\left(\frac{x}{3}, \frac{-2x}{3}\right), t\right) * \mu'\left(\varphi_1\left(\frac{x}{3}, 0\right), t\right),$$
$$\nu_1''(x, t) := \nu'\left(\varphi_1\left(\frac{x}{3}, \frac{x}{3}\right), t\right) \Diamond \nu'\left(\varphi_1\left(\frac{x}{3}, x\right), t\right) \Diamond \nu'\left(\varphi_1\left(\frac{x}{3}, \frac{4x}{3}\right), t\right)$$
$$\Diamond \nu'\left(\varphi_1\left(\frac{x}{3}, \frac{-2x}{3}\right), t\right) \Diamond \nu'\left(\varphi_1\left(\frac{x}{3}, 0\right), t\right).$$

Proof By replacing y by $x+y$ in (44.3), we obtain

$$\begin{cases} \mu(f(3x+y)+f(x-y)-f(2x+y)-f(y)-2f(2x)+2f(x),t) \\ \quad \geq \mu'(\varphi_1(x,x+y),t), \\ \nu(f(3x+y)+f(x-y)-f(2x+y)-f(y)-2f(2x)+2f(x),t) \\ \quad \leq \nu'(\varphi_1(x,x+y),t) \end{cases} \tag{44.5}$$

for all $x, y \in X$ and $t > 0$. Replacing y by $-y$ in (44.5), we get

$$\begin{cases} \mu(f(3x-y)+f(x+y)-f(2x-y)-f(y)-2f(2x)+2f(x),t) \\ \quad \geq \mu'(\varphi_1(x,x-y),t), \\ \nu(f(3x-y)+f(x+y)-f(2x-y)-f(y)-2f(2x)+2f(x),t) \\ \quad \leq \nu'(\varphi_1(x,x-y),t) \end{cases} \tag{44.6}$$

for all $x, y \in X$ and $t > 0$. It follows from (44.3), (44.5), and (44.6) that

$$\begin{cases} \mu(f(3x+y)+f(3x-y)-2f(y)-6f(2x)+6f(x),3t) \\ \quad \geq \mu'(\varphi_1(x,y),t) * \mu'(\varphi_1(x,x+y),t) * \mu'(\varphi_1(x,x-y),t), \\ \nu(f(3x+y)+f(3x-y)-2f(y)-6f(2x)+6f(x),3t) \\ \quad \leq \nu'(\varphi_1(x,y),t)\Diamond\nu'(\varphi_1(x,x+y),t)\Diamond\nu'(\varphi_1(x,x-y),t) \end{cases} \tag{44.7}$$

for all $x, y \in X$ and $t > 0$. Putting $y = 0$ in (44.7), we have

$$\begin{cases} \mu(2f(3x)-6f(2x)+6f(x),3t) \geq \mu'(\varphi_1(x,x),t) * \mu'(\varphi_1(x,0),t), \\ \nu(2f(3x)-6f(2x)+6f(x),3t) \leq \nu'(\varphi_1(x,x),t)\Diamond\nu'(\varphi_1(x,0),t) \end{cases} \tag{44.8}$$

for all $x, y \in X$ and $t > 0$. It follows from (44.8) that

$$\begin{cases} \mu(-2f(3x)+6f(2x)-6f(x),3t) \geq \mu'(\varphi_1(x,x),t) * \mu'(\varphi_1(x,0),t), \\ \nu(-2f(3x)+6f(2x)-6f(x),3t) \leq \nu'(\varphi_1(x,x),t)\Diamond\nu'(\varphi_1(x,0),t) \end{cases} \tag{44.9}$$

for all $x, y \in X$ and $t > 0$. Putting $y = 3x$ in (44.7), we get

$$\begin{cases} \mu(f(6x)-2f(3x)-6f(2x)+6f(x),3t) \\ \quad \geq \mu'(\varphi_1(x,3x),t) * \mu'(\varphi_1(x,4x),t) * \mu'(\varphi_1(x,-2x),t), \\ \nu(f(6x)-2f(3x)-6f(2x)+6f(x),3t) \\ \quad \leq \nu'(\varphi_1(x,3x),t)\Diamond\nu'(\varphi_1(x,4x),t)\Diamond\nu'(\varphi_1(x,-2x),t) \end{cases} \tag{44.10}$$

for all $x, y \in X$ and $t > 0$. Therefore, from (44.9) and (44.10) we obtain the inequality

$$\begin{cases} \mu(f(6x)-4f(3x),6t) \geq \mu'(\varphi_1(x,x),t) * \mu'(\varphi_1(x,3x),t) * \mu'(\varphi_1(x,4x),t) \\ \qquad\qquad * \mu'(\varphi_1(x,-2x),t) * \mu'(\varphi_1(x,0),t), \\ \nu(f(6x)-4f(3x),6t) \leq \nu'(\varphi_1(x,x),t)\Diamond\nu'(\varphi_1(x,3x),t)\Diamond\nu'(\varphi_1(x,4x),t) \\ \qquad\qquad \Diamond\nu'(\varphi_1(x,-2x),t)\Diamond\nu'(\varphi_1(x,0),t) \end{cases} \tag{44.11}$$

for all $x, y \in X$ and $t > 0$. Replacing x by $\frac{x}{3}$ in (44.11), we get

$$\mu\big(f(2x) - 4f(x), 6t\big) \ge \mu_1''(x,t) \quad \text{and} \quad \nu\big(f(2x) - 4f(x), 6t\big) \le \nu_1''(x,t) \tag{44.12}$$

for all $x \in X$ and $t > 0$. Thus

$$\mu\left(\frac{f(2x)}{4} - f(x), \frac{3t}{2}\right) \ge \mu_1''(x,t) \quad \text{and} \quad \nu\left(\frac{f(2x)}{4} - f(x), \frac{3t}{2}\right) \le \nu_1''(x,t) \tag{44.13}$$

for all $x \in X$ and $t > 0$. Replacing x by $2^n x$ in (44.13) and using (44.2), we obtain

$$\mu\left(\frac{f(2^{n+1}x)}{4} - f(2^n x), \frac{3t}{2}\right) \ge \mu_1''\left(x, \frac{t}{\alpha^n}\right) \quad \text{and}$$
$$\nu\left(\frac{f(2^{n+1}x)}{4} - f(2^n x), \frac{3t}{2}\right) \le \nu_1''\left(x, \frac{t}{\alpha^n}\right)$$

for all $x \in X, t > 0$ and $n \ge 0$. Replacing t by $\alpha^n t$, we get

$$\begin{aligned} &\mu\left(\frac{f(2^{n+1}x)}{4^{n+1}} - \frac{f(2^n x)}{4^n}, \frac{3t\alpha^n}{2(4^n)}\right) \ge \mu_1''(x,t) \quad \text{and} \\ &\nu\left(\frac{f(2^{n+1}x)}{4^{n+1}} - \frac{f(2^n x)}{4^n}, \frac{3t\alpha^n}{2(4^n)}\right) \le \nu_1''(x,t). \end{aligned} \tag{44.14}$$

It follows that $\frac{f(2^n x)}{4^n} - f(x) = \sum_{i=0}^{n-1} \frac{f(2^{i+1}x)}{4^{i+1}} - \frac{f(2^i x)}{4^i}$ and (44.14) that

$$\begin{cases} \mu(\frac{f(2^n x)}{4^n} - f(x), \sum_{i=0}^{n-1} \frac{3t\alpha^i}{2(4^i)}) \ge \prod_{i=0}^{n-1} \mu(\frac{f(2^{i+1}x)}{4^{i+1}} - \frac{f(2^i x)}{4^i}, \frac{3t\alpha^i}{2(4^i)}) \ge \mu_1''(x,t), \\ \nu(\frac{f(2^n x)}{4^n} - f(x), \sum_{i=0}^{n-1} \frac{3t\alpha^i}{2(4^i)}) \le \coprod_{i=0}^{n-1} \nu(\frac{f(2^{i+1}x)}{4^{i+1}} - \frac{f(2^i x)}{4^i}, \frac{3t\alpha^i}{2(4^i)}) \le \nu_1''(x,t) \end{cases} \tag{44.15}$$

for all $x \in X, t > 0$ and $n \ge 0$, where $\prod_{j=1}^n a_j = a_1 * a_2 * \cdots * a_n$, $\coprod_{j=1}^n a_j = a_1 \Diamond a_2 \Diamond \cdots \Diamond a_n$. By replacing x by $2^m x$ in (44.15), we have

$$\begin{cases} \mu(\frac{f(2^{n+m}x)}{4^{n+m}} - \frac{f(2^m x)}{4^m}, \sum_{i=0}^{n-1} \frac{3t\alpha^i}{2(4^{i+m})}) \ge \mu_1''(2^m x, t) \ge \mu_1''(x, \frac{t}{\alpha^m}), \\ \nu(\frac{f(2^{n+m}x)}{4^{n+m}} - \frac{f(2^m x)}{4^m}, \sum_{i=0}^{n-1} \frac{3t\alpha^i}{2(4^{i+m})}) \le \nu_1''(2^m x, t) \le \nu_1''(x, \frac{t}{\alpha^m}). \end{cases}$$

Whence

$$\begin{cases} \mu(\frac{f(2^{n+m}x)}{4^{n+m}} - \frac{f(2^m x)}{4^m}, \sum_{i=m}^{n+m-1} \frac{3t\alpha^i}{2(4^i)}) \ge \mu_1''(x,t), \\ \nu(\frac{f(2^{n+m}x)}{4^{n+m}} - \frac{f(2^m x)}{4^m}, \sum_{i=m}^{n+m-1} \frac{3t\alpha^i}{2(4^i)}) \le \nu_1''(x,t) \end{cases}$$

for all $x \in X, t > 0$ and $m, n \geq 0$. Hence

$$\begin{cases} \mu(\frac{f(2^{n+m}x)}{4^{n+m}} - \frac{f(2^m x)}{4^m}, t) \geq \mu_1''\left(x, \frac{t}{\sum_{i=m}^{n+m-1} \frac{3\alpha^i}{2(4^i)}}\right), \\ \nu(\frac{f(2^{n+m}x)}{4^{n+m}} - \frac{f(2^m x)}{4^m}, t) \leq \nu_1''\left(x, \frac{t}{\sum_{i=m}^{n+m-1} \frac{3\alpha^i}{2(4^i)}}\right) \end{cases} \tag{44.16}$$

for all $x \in X, t > 0$ and $m, n \geq 0$. Since $0 < \alpha < 4$ and $\sum_{i=0}^{\infty}(\frac{\alpha}{4})^i < \infty$, the Cauchy criterion for convergence in IFNS shows that $\{\frac{f(2^n x)}{4^n}\}$ is a Cauchy sequence in (Y, μ, ν). Since (Y, μ, ν) is complete, this sequence converges to some point $Q(x) \in Y$. Thus, we define a mapping $Q : X \to Y$ such that $Q(x) := (\mu, \nu) - \lim_{n\to\infty} \frac{f(2^n x)}{4^n}$. Moreover, if we put $m = 0$ in (44.16), we get

$$\mu\left(\frac{f(2^n x)}{4^n} - f(x), t\right) \geq \mu_1''\left(x, \frac{t}{\sum_{i=0}^{n-1} \frac{3\alpha^i}{2(4^i)}}\right) \quad \text{and}$$

$$\nu\left(\frac{f(2^n x)}{4^n} - f(x), t\right) \leq \nu_1''\left(x, \frac{t}{\sum_{i=0}^{n-1} \frac{3\alpha^i}{2(4^i)}}\right)$$

for all $x \in X, t > 0$. Hence, we obtain

$$\begin{cases} \mu(Q(x) - f(x), t) \geq \mu(Q(x) - \frac{f(2^n x)}{4^n}, \frac{t}{2}) * \mu(\frac{f(2^n x)}{4^n} - f(x), \frac{t}{2}) \\ \quad \geq \mu_1''(x, \frac{t}{\sum_{i=0}^{n-1} \frac{3\alpha^i}{4^i}}), \\ \nu(Q(x) - f(x), t) \leq \nu(Q(x) - \frac{f(2^n x)}{4^n}, \frac{t}{2}) \Diamond \nu(\frac{f(2^n x)}{4^n} - f(x), \frac{t}{2}) \\ \quad \leq \nu_1''(x, \frac{t}{\sum_{i=0}^{n-1} \frac{3\alpha^i}{4^i}}) \end{cases}$$

for large n. Taking the limit as $n \to \infty$ and using the definition of IFNS, we obtain

$$\mu\big(Q(x) - f(x), t\big) \geq \mu_1''\left(x, \frac{t(4-\alpha)}{12}\right) \quad \text{and}$$

$$\nu\big(Q(x) - f(x), t\big) \leq \nu_1''\left(x, \frac{t(4-\alpha)}{12}\right).$$

Now, we claim that Q is quadratic. Replacing x and y by $2^n x$ and $2^n y$, respectively, in (44.3), we have

$$\begin{cases} \mu(\frac{f(2^n(2x+y))}{4^n} + \frac{f(2^n(2x-y))}{4^n} - \frac{f(2^n(x+y))}{4^n} - \frac{f(2^n(x-y))}{4^n} - \frac{2f(2^n(2x))}{4^n} + \frac{2f(2^n(x))}{4^n}, t) \\ \quad \geq \mu'(\varphi_1(2^n x, 2^n y), 4^n t), \\ \nu(\frac{f(2^n(2x+y))}{4^n} + \frac{f(2^n(2x-y))}{4^n} - \frac{f(2^n(x+y))}{4^n} - \frac{f(2^n(x-y))}{4^n} - \frac{2f(2^n(2x))}{4^n} + \frac{2f(2^n(x))}{4^n}, t) \\ \quad \leq \nu'(\varphi_1(2^n x, 2^n y), 4^n t) \end{cases}$$

for all $x, y \in X$ and $t > 0$. Since

$$\lim_{n\to\infty} \mu'\big(\varphi_1(2^n x, 2^n y), 4^n t\big) = 1 \quad \text{and} \quad \lim_{n\to\infty} \nu'\big(\varphi_1(2^n x, 2^n y), 4^n t\big) = 0$$

and $Q(0) = 0$, then by Lemma 2.1 of [25], we observe that the mapping $Q : X \to Y$ is quadratic. To prove the uniqueness of the quadratic mapping Q, assume that there exists a quadratic mapping $Q' : X \to Y$ which satisfies (44.4). For fix $x \in X$, clearly $Q(2^n x) = 4^n Q(x)$ and $Q'(2^n x) = 4^n Q'(x)$ for all $n \in \mathbb{N}$. It follows from (44.4) that

$$\begin{aligned}
\mu\big(Q(x) - Q'(x), t\big) &= \mu\left(\frac{Q(2^n x)}{4^n} - \frac{Q'(2^n x)}{4^n}, t\right) \\
&\geq \mu\left(\frac{Q(2^n x)}{4^n} - \frac{f(2^n x)}{4^n}, \frac{t}{2}\right) * \mu\left(\frac{f(2^n x)}{4^n} - \frac{Q'(2^n x)}{4^n}, \frac{t}{2}\right) \\
&\geq \mu_1''\left(2^n x, \frac{4^n(4-\alpha)t}{24}\right) \geq \mu_1''\left(x, \frac{4^n(4-\alpha)t}{24\alpha^n}\right)
\end{aligned}$$

and similarly $\nu(Q(x) - Q'(x), t) \leq \nu_1''(x, \frac{4^n(4-\alpha)t}{24\alpha^n})$ for all $x \in X$ and $t > 0$. Since $\lim_{n\to\infty} \frac{4^n(4-\alpha)t}{24\alpha^n} = \infty$, we obtain

$$\lim_{n\to\infty} \mu_1''\left(x, \frac{4^n(4-\alpha)t}{24\alpha^n}\right) = 1 \quad \text{and} \quad \lim_{n\to\infty} \nu_1''\left(x, \frac{4^n(4-\alpha)t}{24\alpha^n}\right) = 0.$$

Therefore,

$$\mu\big(Q(x) - Q'(x), t\big) = 1 \quad \text{and} \quad \nu\big(Q(x) - Q'(x), t\big) = 0$$

for all $x \in X$ and $t > 0$. Whence $Q(x) = Q'(x)$. This completes the proof of the theorem. □

Theorem 44.2 *Let $\varphi_2 : X \times X \to Z$ be a function such that for some $\alpha > 4$*

$$\mu'\left(\varphi_2\left(\frac{x}{2(3)}, \frac{y}{2}\right), t\right) \geq \mu'\left(\varphi_2\left(\frac{x}{3}, y\right), \alpha t\right) \quad \text{and}$$
$$\nu'\left(\varphi_2\left(\frac{x}{2(3)}, \frac{y}{2}\right), t\right) \leq \nu'\left(\varphi_2\left(\frac{x}{3}, y\right), \alpha t\right)$$

for all $x \in X$, $y \in \{0, \frac{x}{3}, \frac{4x}{3}, \frac{-2x}{3}, x\}$, $t > 0$, and $\lim_{n\to\infty} \mu'(4^n \varphi_2(2^{-n}x, 2^{-n}y), t) = 1$ and $\lim_{n\to\infty} \nu'(4^n \varphi_2(2^{-n}x, 2^{-n}y), t) = 0$ for all $x, y \in X, t > 0$. Suppose that an even function $f : X \to Y$ with $f(0) = 0$ satisfies the inequalities

$$\begin{cases}
\mu(f(2x+y) + f(2x-y) - f(x+y) - f(x-y) - 2f(2x) + 2f(x), t) \\
\quad \geq \mu'(\varphi_2(x, y), t), \\
\nu(f(2x+y) + f(2x-y) - f(x+y) - f(x-y) - 2f(2x) + 2f(x), t) \\
\quad \leq \nu'(\varphi_2(x, y), t)
\end{cases}$$

for all $x, y \in X$ and $t > 0$. Then there exist a unique quadratic mapping $Q : X \to Y$ such that

$$\mu\big(Q(x) - f(x), t\big) \ge \mu_2''\left(x, \frac{t(\alpha - 4)}{12}\right) \quad \textit{and}$$

$$\nu\big(Q(x) - f(x), t\big) \le \nu_2''\left(x, \frac{t(\alpha - 4)}{12}\right)$$

for all $x \in X$ and $t > 0$, where

$$\mu_2''(x,t) := \mu'\left(\varphi_2\left(\frac{x}{3}, \frac{x}{3}\right), t\right) * \mu'\left(\varphi_2\left(\frac{x}{3}, x\right), t\right) * \mu'\left(\varphi_2\left(\frac{x}{3}, \frac{4x}{3}\right), t\right)$$
$$* \mu'\left(\varphi_2\left(\frac{x}{3}, \frac{-2x}{3}\right), t\right) * \mu'\left(\varphi_2\left(\frac{x}{3}, 0\right), t\right),$$

$$\nu_2''(x,t) := \nu'\left(\varphi_2\left(\frac{x}{3}, \frac{x}{3}\right), t\right) \diamond \nu'\left(\varphi_2\left(\frac{x}{3}, x\right), t\right) \diamond \nu'\left(\varphi_2\left(\frac{x}{3}, \frac{4x}{3}\right), t\right)$$
$$\diamond \nu'\left(\varphi_2\left(\frac{x}{3}, \frac{-2x}{3}\right), t\right) \diamond \nu'\left(\varphi_2\left(\frac{x}{3}, 0\right), t\right).$$

Proof The techniques are completely similar to that of Theorem 44.1. Hence we present a sketch of proof. Replacing x by $\frac{x}{2^{n+1}}$ in (44.12), we get

$$\mu\left(4f\left(\frac{x}{2^{n+1}}\right) - f\left(\frac{x}{2^n}\right), 6t\right) \ge \mu_2''\left(\frac{x}{2^{n+1}}, t\right) \quad \text{and}$$

$$\nu\left(4f\left(\frac{x}{2^{n+1}}\right) - f\left(\frac{x}{2^n}\right), 6t\right) \le \nu_2''\left(\frac{x}{2^{n+1}}, t\right).$$

Thus

$$\begin{cases} \mu(4^{n+1} f(\frac{x}{2^{n+1}}) - 4^n f(\frac{x}{2^n}), 6(4^n)t) \ge \mu_2''(\frac{x}{2^{n+1}}, t), \\ \nu(4^{n+1} f(\frac{x}{2^{n+1}}) - 4^n f(\frac{x}{2^n}), 6(4^n)t) \le \nu_2''(\frac{x}{2^{n+1}}, t). \end{cases}$$

We can deduce

$$\begin{cases} \mu(4^{n+m} f(\frac{x}{2^{n+m}}) - 4^m f(\frac{x}{2^m}), t) \ge \mu_2''\left(x, \frac{t}{\sum_{i=m}^{n+m-1} \frac{6}{\alpha}(\frac{4}{\alpha})^i}\right), \\ \nu(4^{n+m} f(\frac{x}{2^{n+m}}) - 4^m f(\frac{x}{2^m}), t) \le \nu_2''\left(x, \frac{t}{\sum_{i=m}^{n+m-1} \frac{6}{\alpha}(\frac{4}{\alpha})^i}\right) \end{cases} \tag{44.17}$$

for all $x \in X, t > 0$ and $m, n \ge 0$. Thus, we conclude that $\{4^n f(\frac{x}{2^n})\}$ is a Cauchy sequence in the intuitionistic fuzzy Banach space. Therefore, there is a mapping $Q : X \to Y$ defined by $Q(x) := (\mu, \nu) - \lim_{n\to\infty} 4^n f(\frac{x}{2^n})$. Equation (44.17) with $m = 0$ implies

$$\mu\big(Q(x) - f(x), t\big) \ge \mu_2''\left(x, \frac{t(\alpha - 4)}{12}\right) \quad \text{and}$$

$$\nu\big(Q(x)-f(x),t\big)\le\nu_2''\left(x,\frac{t(\alpha-4)}{12}\right)$$

for all $x\in X$ and $t>0$. This completes the proof of the theorem. □

Theorem 44.3 *Let $\varphi_3 : X\times X\to Z$ be a function such that for some $0<\alpha<2$*

$$\begin{aligned}&\mu'\left(\varphi_3\left(2\left(\frac{x}{2}\right),2y\right),t\right)\ge\mu'\left(\alpha\varphi_3\left(\frac{x}{2},y\right),t\right)\quad\textit{and}\\&\nu'\left(\varphi_3\left(2\left(\frac{x}{2}\right),2y\right),t\right)\le\nu'\left(\alpha\varphi_3\left(\frac{x}{2},y\right),t\right)\end{aligned}\tag{44.18}$$

for all $x\in X$, $y\in\{x,\frac{x}{2},\frac{3x}{2},2x\}$, $t>0$, and $\lim_{n\to\infty}\mu'(\varphi_3(2^nx,2^ny),2^nt)=1$ and $\lim_{n\to\infty}\nu'(\varphi_3(2^nx,2^ny),2^nt)=0$ for all $x,y\in X$, $t>0$. Suppose that an odd function $f:X\to Y$ satisfies the inequalities

$$\begin{cases}\mu(f(2x+y)+f(2x-y)-f(x+y)-f(x-y)-2f(2x)+2f(x),t)\\\quad\ge\mu'(\varphi_3(x,y),t),\\\nu(f(2x+y)+f(2x-y)-f(x+y)-f(x-y)-2f(2x)+2f(x),t)\\\quad\le\nu'(\varphi_3(x,y),t)\end{cases}\tag{44.19}$$

for all $x,y\in X$ and $t>0$. Then there exist a unique additive mapping $A:X\to Y$ such that

$$\begin{aligned}&\mu\big(A(x)-f(x),t\big)\ge\mu_3''\left(x,\frac{t(2-\alpha)}{4}\right)\quad\textit{and}\\&\nu\big(A(x)-f(x),t\big)\le\nu_3''\left(x,\frac{t(2-\alpha)}{4}\right)\end{aligned}\tag{44.20}$$

for all $x\in X$ and $t>0$, where

$$\begin{aligned}\mu_3''(x,t):=&\ \mu'(\varphi_3(x,x),t)*\mu'\left(\varphi_3\left(\frac{x}{2},\frac{x}{2}\right),t\right)*\mu'\left(\varphi_3\left(\frac{x}{2},2x\right),t\right)\\&*\mu'\left(\varphi_3\left(\frac{x}{2},\frac{3x}{2}\right),t\right),\\\nu_3''(x,t):=&\ \nu'(\varphi_3(x,x),t)\Diamond\nu'\left(\varphi_3\left(\frac{x}{2},\frac{x}{2}\right),t\right)\Diamond\nu'\left(\varphi_3\left(\frac{x}{2},2x\right),t\right)\\&\Diamond\nu'\left(\varphi_3\left(\frac{x}{2},\frac{3x}{2}\right),t\right).\end{aligned}$$

Proof By replacing y by x in (44.19), we obtain

$$\begin{cases}\mu(f(3x)-3f(2x)+3f(x),t)\ge\mu'(\varphi_3(x,x),t),\\\nu(f(3x)-3f(2x)+3f(x),t)\le\nu'(\varphi_3(x,x),t)\end{cases}\tag{44.21}$$

for all $x, y \in X$ and $t > 0$. Replacing y and $3y$ in (44.19), we get

$$\begin{cases} \mu(f(5x) - f(4x) - f(2x) + f(x), t) \geq \mu'(\varphi_3(x, 3x), t), \\ \nu(f(5x) - f(4x) - f(2x) + f(x), t) \leq \nu'(\varphi_3(x, 3x), t) \end{cases} \tag{44.22}$$

for all $x, y \in X$ and $t > 0$. Putting $y = 4x$ in (44.19), we get

$$\begin{cases} \mu(f(6x) - f(5x) + f(3x) - 3f(2x) + 2f(x), t) \geq \mu'(\varphi_3(x, 4x), t), \\ \nu(f(6x) - f(5x) + f(3x) - 3f(2x) + 2f(x), t) \leq \nu'(\varphi_3(x, 4x), t) \end{cases} \tag{44.23}$$

for all $x, y \in X$ and $t > 0$. It follows from (44.21), (44.22), and (44.23) that

$$\begin{cases} \mu(f(6x) - f(4x) - f(2x), 3t) \geq \mu'(\varphi_3(x, x), t) * \mu'(\varphi_3(x, 3x), t) \\ \qquad\qquad * \mu'(\varphi_3(x, 4x), t), \\ \nu(f(6x) - f(4x) - f(2x), 3t) \leq \nu'(\varphi_3(x, x), t) \Diamond \nu'(\varphi_3(x, 3x), t) \\ \qquad\qquad \Diamond \nu'(\varphi_3(x, 4x), t) \end{cases} \tag{44.24}$$

for all $x, y \in X$ and $t > 0$. If we replace x by $\frac{x}{2}$ in (44.24), then

$$\begin{cases} \mu(f(3x) - f(2x) - f(x), 3t) \geq \mu'(\varphi_3(\frac{x}{2}, \frac{x}{2}), t) * \mu'(\varphi_3(\frac{x}{2}, 4x), t) \\ \qquad\qquad * \mu'(\varphi_3(\frac{x}{2}, \frac{3x}{2}), t), \\ \nu(f(3x) - f(2x) - f(x), 3t) \leq \nu'(\varphi_3(\frac{x}{2}, \frac{x}{2}), t) \Diamond \nu'(\varphi_3(\frac{x}{2}, 2x), t) \\ \qquad\qquad \Diamond \nu'(\varphi_3(\frac{x}{2}, \frac{3x}{2}), t) \end{cases} \tag{44.25}$$

for all $x, y \in X$ and $t > 0$. It follows from (44.21) and (44.25) that

$$\mu\left(\frac{f(2x)}{2} - f(x), t\right) \geq \mu_3''(x, t) \quad \text{and} \quad \nu\left(\frac{f(2x)}{2}\right) - f(x), t) \leq \nu_3''(x, t) \tag{44.26}$$

for all $x \in X$ and $t > 0$. Replacing x by $2^n x$ in (44.26) and using (44.18), we obtain

$$\mu\left(\frac{f(2^{n+1}x)}{2} - f(2^n x), t\right) \geq \mu_3''\left(x, \frac{t}{\alpha^n}\right) \quad \text{and}$$
$$\nu\left(\frac{f(2^{n+1}x)}{2} - f(2^n x), t\right) \leq \nu_3''\left(x, \frac{t}{\alpha^n}\right)$$

for all $x \in X, t > 0$ and $n \geq 0$. Replacing t by $\alpha^n t$, we get

$$\begin{aligned} &\mu\left(\frac{f(2^{n+1}x)}{2^{n+1}} - \frac{f(2^n x)}{2^n}, \frac{t\alpha^n}{2^n}\right) \geq \mu_3''(x, t) \quad \text{and} \\ &\nu\left(\frac{f(2^{n+1}x)}{2^{n+1}} - \frac{f(2^n x)}{2^n}, \frac{t\alpha^n}{2^n}\right) \leq \nu_3''(x, t). \end{aligned} \tag{44.27}$$

It follows that $\frac{f(2^n x)}{2^n} - f(x) = \sum_{i=0}^{n-1} \frac{f(2^{i+1}x)}{2^{i+1}} - \frac{f(2^i x)}{2^i}$ and (44.27) that

$$\begin{cases} \mu(\frac{f(2^n x)}{2^n} - f(x), \sum_{i=0}^{n-1} \frac{t\alpha^i}{2^i}) \geq \prod_{i=0}^{n-1} \mu(\frac{f(2^{i+1}x)}{2^{i+1}} - \frac{f(2^i x)}{2^i}, \frac{t\alpha^i}{2^i}) \geq \mu_3''(x,t), \\ \nu(\frac{f(2^n x)}{2^n} - f(x), \sum_{i=0}^{n-1} \frac{t\alpha^i}{2^i}) \leq \coprod_{i=0}^{n-1} \nu(\frac{f(2^{i+1}x)}{2^{i+1}} - \frac{f(2^i x)}{2^i}, \frac{t\alpha^i}{2^i}) \leq \nu_3''(x,t) \end{cases} \tag{44.28}$$

for all $x \in X, t > 0$ and $n \geq 0$, where $\prod_{j=1}^n a_j = a_1 * a_2 * \cdots * a_n$, $\coprod_{j=1}^n a_j = a_1 \Diamond a_2 \Diamond \cdots \Diamond a_n$. By replacing x by $2^m x$ in (44.28), we observe that

$$\begin{cases} \mu(\frac{f(2^{n+m}x)}{2^{n+m}} - \frac{f(2^m x)}{2^m}, \sum_{i=0}^{n-1} \frac{t\alpha^i}{2^{i+m}}) \geq \mu_3''(2^m x, t) \geq \mu_3''(x, \frac{t}{\alpha^m}), \\ \nu(\frac{f(2^{n+m}x)}{2^{n+m}} - \frac{f(2^m x)}{2^m}, \sum_{i=0}^{n-1} \frac{t\alpha^i}{2^{i+m}}) \leq \nu_3''(2^m x, t) \leq \nu_3''(x, \frac{t}{\alpha^m}). \end{cases}$$

Thus

$$\begin{cases} \mu(\frac{f(2^{n+m}x)}{2^{n+m}} - \frac{f(2^m x)}{2^m}, \sum_{i=m}^{n+m-1} \frac{t\alpha^i}{2^i}) \geq \mu_3''(x,t), \\ \nu(\frac{f(2^{n+m}x)}{2^{n+m}} - \frac{f(2^m x)}{2^m}, \sum_{i=m}^{n+m-1} \frac{t\alpha^i}{2^i}) \leq \nu_3''(x,t) \end{cases}$$

for all $x \in X, t > 0$ and $m, n \geq 0$. Hence

$$\begin{cases} \mu(\frac{f(2^{n+m}x)}{2^{n+m}} - \frac{f(2^m x)}{2^m}, t) \geq \mu_3''\Big(x, \frac{t}{\sum_{i=m}^{n+m-1} \frac{\alpha^i}{2^i}}\Big), \\ \nu(\frac{f(2^{n+m}x)}{2^{n+m}} - \frac{f(2^m x)}{2^m}, t) \leq \nu_3''\Big(x, \frac{t}{\sum_{i=m}^{n+m-1} \frac{\alpha^i}{2^i}}\Big) \end{cases} \tag{44.29}$$

for all $x \in X, t > 0$ and $m, n \geq 0$. Since $0 < \alpha < 2$ and $\sum_{i=0}^{\infty} (\frac{\alpha}{2})^i < \infty$, the Cauchy criterion for convergence in IFNS shows that $\{\frac{f(2^n x)}{2^n}\}$ is a Cauchy sequence in (Y, μ, ν). Since (Y, μ, ν) is complete, this sequence converges to some point $A(x) \in Y$. So we can define the mapping $A: X \to Y$ by $A(x) := (\mu, \nu) - \lim_{n\to\infty} \frac{f(2^n x)}{2^n}$. Moreover, putting $m = 0$ in (44.29), we have

$$\mu\Big(\frac{f(2^n x)}{2^n} - f(x), t\Big) \geq \mu_3''\Big(x, \frac{t}{\sum_{i=0}^{n-1} \frac{\alpha^i}{2^i}}\Big) \quad \text{and}$$

$$\nu\Big(\frac{f(2^n x)}{2^n} - f(x), t\Big) \leq \nu_3''\Big(x, \frac{t}{\sum_{i=0}^{n-1} \frac{\alpha^i}{2^i}}\Big)$$

for all $x \in X, t > 0$. Hence, we obtain

$$\begin{cases} \mu(A(x) - f(x), t) \geq \mu(A(x) - \frac{f(2^n x)}{2^n}, \frac{t}{2}) * \mu(\frac{f(2^n x)}{2^n} - f(x), \frac{t}{2}) \\ \quad \geq \mu_3''\Big(x, \frac{t}{\sum_{i=0}^{n-1} \frac{2\alpha^i}{2^i}}\Big), \\ \nu(A(x) - f(x), t) \leq \nu(A(x) - \frac{f(2^n x)}{2^n}, \frac{t}{2}) \Diamond \nu(\frac{f(2^n x)}{2^n} - f(x), \frac{t}{2}) \\ \quad \leq \nu_3''\Big(x, \frac{t}{\sum_{i=0}^{n-1} \frac{2\alpha^i}{2^i}}\Big) \end{cases}$$

for large n. Taking the limit as $n \to \infty$ and using the definition of IFNS, we obtain

$$\mu\big(A(x) - f(x), t\big) \geq \mu_3''\left(x, \frac{t(2-\alpha)}{4}\right) \quad \text{and}$$

$$\nu\big(A(x) - f(x), t\big) \leq \nu_3''\left(x, \frac{t(2-\alpha)}{4}\right).$$

Next, we claim that Q is additive. Replacing x and y by $2^n x$ and $2^n y$, respectively, in (44.19), we obtain

$$\begin{cases} \mu(\frac{f(2^n(2x+y))}{2^n} + \frac{f(2^n(2x-y))}{2^n} - \frac{f(2^n(x+y))}{2^n} - \frac{f(2^n(x-y))}{2^n} - \frac{2f(2^n(2x))}{2^n} + \frac{2f(2^n(x))}{2^n}, t) \\ \quad \geq \mu'(\varphi_3(2^n x, 2^n y), 2^n t), \\ \nu(\frac{f(2^n(2x+y))}{2^n} + \frac{f(2^n(2x-y))}{2^n} - \frac{f(2^n(x+y))}{2^n} - \frac{f(2^n(x-y))}{2^n} - \frac{2f(2^n(2x))}{2^n} + \frac{2f(2^n(x))}{2^n}, t) \\ \quad \leq \nu'(\varphi_3(2^n x, 2^n y), 2^n t) \end{cases}$$

for all $x, y \in X$ and $t > 0$. Since

$$\lim_{n\to\infty} \mu'\big(\varphi_3(2^n x, 2^n y), 2^n t\big) = 1 \quad \text{and} \quad \lim_{n\to\infty} \nu'\big(\varphi_3(2^n x, 2^n y), 2^n t\big) = 0,$$

then by Lemma 2.3 of [25], we observe that the mapping $A : X \to Y$ is additive. To prove the uniqueness of A, Let $A' : X \to Y$ be another additive mapping satisfying (44.20). For a fixed $x \in X$, clearly, $A(2^n x) = 2^n A(x)$ and $A'(2^n x) = 2^n A'(x)$ for all $n \in \mathbb{N}$. It follows from (44.20) that

$$\begin{aligned} \mu\big(A(x) - A'(x), t\big) &= \mu\left(\frac{A(2^n x)}{2^n} - \frac{A'(2^n x)}{2^n}, t\right) \\ &\geq \mu\left(\frac{A(2^n x)}{2^n} - \frac{f(2^n x)}{2^n}, \frac{t}{2}\right) * \mu\left(\frac{f(2^n x)}{2^n} - \frac{A'(2^n x)}{2^n}, \frac{t}{2}\right) \\ &\geq \mu_3''\left(2^n x, \frac{2^n(2-\alpha)t}{8}\right) \geq \mu_3''\left(x, \frac{2^n(2-\alpha)t}{8\alpha^n}\right), \end{aligned}$$

and similarly $\nu(A(x) - A'(x), t) \leq \nu_3''(x, \frac{2^n(2-\alpha)t}{8\alpha^n})$ for all $x \in X$ and $t > 0$. Since $\lim_{n\to\infty} \frac{2^n(2-\alpha)t}{8\alpha^n} = \infty$, we obtain

$$\lim_{n\to\infty} \mu_3''\left(x, \frac{2^n(2-\alpha)t}{8\alpha^n}\right) = 1 \quad \text{and} \quad \lim_{n\to\infty} \nu_3''\left(x, \frac{2^n(2-\alpha)t}{8\alpha^n}\right) = 0.$$

Thus

$$\mu\big(A(x) - A'(x), t\big) = 1 \quad \text{and} \quad \nu\big(A(x) - A'(x), t\big) = 0$$

for all $x \in X$ and $t > 0$. Hence we get $A(x) = A'(x)$ for all $x \in X$. This completes the proof of the theorem. □

Theorem 44.4 *Let $\varphi_4 : X \times X \to Z$ be a function such that for some $\alpha > 2$*

$$\mu'\left(\varphi_4\left(\frac{1}{2}\left(\frac{x}{2}\right), \frac{y}{2}\right), t\right) \geq \mu'\left(\varphi_4\left(\frac{x}{2}, y\right), \alpha t\right) \quad \text{and}$$
$$\nu'\left(\varphi_4\left(\frac{1}{2}\left(\frac{x}{2}\right), \frac{y}{2}\right), t\right) \leq \nu'\left(\varphi_4\left(\frac{x}{2}, y\right), \alpha t\right)$$

for all $x \in X$, $y \in \{x, \frac{x}{2}, \frac{3x}{2}, 2x\}$, $t > 0$, $\lim_{n\to\infty} \mu'(2^n\varphi_4(2^{-n}x, 2^{-n}y), t) = 1$, and $\lim_{n\to\infty} \nu'(2^n\varphi_4(2^{-n}x, 2^{-n}y), t) = 0$ for all $x, y \in X, t > 0$. Suppose that an odd function $f : X \to Y$ satisfies the inequalities

$$\begin{cases} \mu(f(2x+y) + f(2x-y) - f(x+y) - f(x-y) - 2f(2x) + 2f(x), t) \\ \quad \geq \mu'(\varphi_4(x, y), t), \\ \nu(f(2x+y) + f(2x-y) - f(x+y) - f(x-y) - 2f(2x) + 2f(x), t) \\ \quad \leq \nu'(\varphi_4(x, y), t) \end{cases}$$

for all $x, y \in X$ and $t > 0$. Then there exist a unique additive mapping $A : X \to Y$ such that

$$\mu\big(A(x) - f(x), t\big) \geq \mu_4''\left(x, \frac{t(\alpha-2)}{4}\right) \quad \text{and}$$
$$\nu\big(A(x) - f(x), t\big) \leq \nu_4''\left(x, \frac{t(\alpha-2)}{4}\right)$$

for all $x \in X$ and $t > 0$, where

$$\begin{aligned} \mu_4''(x, t) &:= \mu'(\varphi_4(x, x), t) * \mu'\left(\varphi_4\left(\frac{x}{2}, \frac{x}{2}\right), t\right) * \mu'\left(\varphi_4\left(\frac{x}{2}, 2x\right), t\right) \\ &\quad * \mu'\left(\varphi_4\left(\frac{x}{2}, \frac{3x}{2}\right), t\right), \\ \nu_4''(x, t) &:= \nu'(\varphi_4(x, x), t) \Diamond \nu'\left(\varphi_4\left(\frac{x}{2}, \frac{x}{2}\right), t\right) \Diamond \nu'\left(\varphi_4\left(\frac{x}{2}, 2x\right), t\right) \\ &\quad \Diamond \nu'\left(\varphi_4\left(\frac{x}{2}, \frac{3x}{2}\right), t\right). \end{aligned}$$

Proof Replacing x by $\frac{x}{2^{n+1}}$ in (44.26), we get

$$\mu\left(f\left(\frac{x}{2^n}\right) - 2f\left(\frac{x}{2^{n+1}}\right), 2t\right) \geq \mu_4''\left(\frac{x}{2^{n+1}}, t\right) \quad \text{and}$$
$$\nu\left(f\left(\frac{x}{2^n}\right) - 2f\left(\frac{x}{2^{n+1}}\right), 2t\right) \leq \nu_4''\left(\frac{x}{2^{n+1}}, t\right).$$

Whence

$$\begin{cases} \mu(2^n f(\frac{x}{2^n}) - 2^{n+1} f(\frac{x}{2^{n+}}), 2(2^n)t) \geq \mu_4''(\frac{x}{2^{n+1}}, t), \\ \nu(2^n f(\frac{x}{2^n}) - 2^{n+1} f(\frac{x}{2^{n+}}), 2(2^n)t) \leq \nu_4''(\frac{x}{2^{n+1}}, t). \end{cases}$$

We can deduce

$$\begin{cases} \mu(2^{n+m} f(\frac{x}{2^{n+m}}) - 2^m f(\frac{x}{2^m}), t) \geq \mu_4''\left(x, \frac{t}{\sum_{i=m}^{n+m-1} \frac{2}{\alpha}(\frac{2}{\alpha})^i}\right), \\ \nu(2^{n+m} f(\frac{x}{2^{n+m}}) - 2^m f(\frac{x}{2^m}), t) \leq \nu_4''\left(x, \frac{t}{\sum_{i=m}^{n+m-1} \frac{2}{\alpha}(\frac{2}{\alpha})^i}\right) \end{cases} \tag{44.30}$$

for all $x \in X, t > 0$ and $m, n \geq 0$. Thus, we conclude that $\{2^n f(\frac{x}{2^n})\}$ is a Cauchy sequence in the intuitionistic fuzzy Banach space. Therefore, there is a mapping $A : X \to Y$ defined by $A(x) := (\mu, \nu) - \lim_{n\to\infty} 2^n f(\frac{x}{2^n})$. Equation (44.30) with $m = 0$ implies

$$\mu\big(A(x) - f(x), t\big) \geq \mu_4''\left(x, \frac{t(\alpha - 2)}{4}\right) \quad \text{and}$$

$$\nu\big(A(x) - f(x), t\big) \leq \nu_4''\left(x, \frac{t(\alpha - 2)}{4}\right)$$

for all $x \in X$ and $t > 0$. The rest of the proof is similar to the of Theorem 44.3. This completes the proof of the theorem. □

We now prove our main theorem in this paper.

Theorem 44.5 *Let $\varphi : X \times X \to Z$ be a function such that for some $0 < \alpha < 2$*

$$\mu'\left(\varphi\left(2\left(\frac{x}{2}\right), 2y\right), t\right) \geq \mu'\left(\alpha\varphi\left(\frac{x}{2}, y\right), t\right) \quad \textit{and}$$

$$\nu'\left(\varphi\left(2\left(\frac{x}{2}\right), 2y\right), t\right) \leq \nu'\left(\alpha\varphi\left(\frac{x}{2}, y\right), t\right)$$

for all $x \in X$, $y \in \{0, -\frac{2x}{3}, x, \frac{x}{2}, \frac{x}{3}, \frac{4x}{3}, \frac{3x}{2}, 2x\}$, $t > 0$, and $\lim_{n\to\infty} \mu'(\varphi(2^n x, 2^n y), 2^n t) = 1$ and $\lim_{n\to\infty} \nu'(\varphi(2^n x, 2^n y), 2^n t) = 0$ for all $x, y \in X, t > 0$. Suppose that a function $f : X \to Y$ with $f(0) = 0$ satisfies the inequalities

$$\begin{cases} \mu(f(2x+y) + f(2x-y) - f(x+y) - f(x-y) - 2f(2x) + 2f(x), t) \\ \quad \geq \mu'(\varphi(x, y), t), \\ \nu(f(2x+y) + f(2x-y) - f(x+y) - f(x-y) - 2f(2x) + 2f(x), t) \\ \quad \leq \nu'(\varphi(x, y), t) \end{cases} \tag{44.31}$$

for all $x, y \in X$ and $t > 0$. Then there exist a unique quadratic mapping $Q : X \to Y$ and a unique additive mapping $A : X \to Y$ satisfying (44.1) *and*

$$\begin{cases} \mu(Q(x) - A(x) - f(x), t) \geq M_1''(x, \frac{t(4-\alpha)}{24}) * \widetilde{M}_1''(x, \frac{t(2-\alpha)}{8}), \\ \nu(Q(x) - A(x) - f(x), t) \leq M_2''(x, \frac{t(4-\alpha)}{24}) \lozenge \widetilde{M}_2''(x, \frac{t(2-\alpha)}{8}) \end{cases} \tag{44.32}$$

for all $x \in X$ and $t > 0$, where

$$\begin{aligned} M_1''(x,t) := {} & \mu'\left(\varphi\left(\frac{x}{3}, \frac{x}{3}\right), t\right) * \mu'\left(\varphi\left(\frac{x}{3}, x\right), t\right) * \mu'\left(\varphi\left(\frac{x}{3}, \frac{4x}{3}\right), t\right) \\ & * \mu'\left(\varphi\left(\frac{x}{3}, \frac{-2x}{3}\right), t\right) * \mu'\left(\varphi\left(\frac{x}{3}, 0\right), t\right) * \mu'\left(\varphi\left(\frac{-x}{3}, \frac{-x}{3}\right), t\right) \\ & * \mu'\left(\varphi\left(\frac{-x}{3}, -x\right), t\right) * \mu'\left(\varphi\left(\frac{-x}{3}, \frac{-4x}{3}\right), t\right) \\ & * \mu'\left(\varphi\left(\frac{-x}{3}, \frac{2x}{3}\right), t\right) * \mu'\left(\varphi\left(\frac{-x}{3}, 0\right), t\right), \\ \widetilde{M}_1''(x,t) := {} & \mu'(\varphi(x,x), t) * \mu'\left(\varphi\left(\frac{x}{2}, \frac{x}{2}\right), t\right) * \mu'\left(\varphi\left(\frac{x}{2}, 2x\right), t\right) \\ & * \mu'\left(\varphi\left(\frac{x}{2}, \frac{3x}{2}\right), t\right) * \mu'(\varphi(-x,-x), t) * \mu'\left(\varphi\left(\frac{-x}{2}, \frac{-x}{2}\right), t\right) \\ & * \mu'\left(\varphi\left(\frac{-x}{2}, -2x\right), t\right) * \mu'\left(\varphi\left(\frac{-x}{2}, \frac{-3x}{2}\right), t\right), \\ M_2''(x,t) := {} & \nu'\left(\varphi\left(\frac{x}{3}, \frac{x}{3}\right), t\right) \lozenge \nu'\left(\varphi\left(\frac{x}{3}, x\right), t\right) \lozenge \nu'\left(\varphi\left(\frac{x}{3}, \frac{4x}{3}\right), t\right) \\ & \lozenge \nu'\left(\varphi\left(\frac{x}{3}, \frac{-2x}{3}\right), t\right) \lozenge \nu'\left(\varphi\left(\frac{x}{3}, 0\right), t\right) \lozenge \nu'\left(\varphi\left(\frac{-x}{3}, \frac{-x}{3}\right), t\right) \\ & \lozenge \nu'\left(\varphi\left(\frac{-x}{3}, -x\right), t\right) \lozenge \nu'\left(\varphi\left(\frac{-x}{3}, \frac{-4x}{3}\right), t\right) \lozenge \nu'\left(\varphi\left(\frac{-x}{3}, \frac{2x}{3}\right), t\right) \\ & \lozenge \nu'\left(\varphi\left(\frac{-x}{3}, 0\right), t\right), \\ \widetilde{M}_2''(x,t) := {} & \nu'(\varphi(x,x), t) \lozenge \nu'\left(\varphi\left(\frac{x}{2}, \frac{x}{2}\right), t\right) \lozenge \nu'\left(\varphi\left(\frac{x}{2}, 2x\right), t\right) \\ & \lozenge \nu'\left(\varphi\left(\frac{x}{2}, \frac{3x}{2}\right), t\right) \lozenge \nu'(\varphi(-x,-x), t) \lozenge \nu'\left(\varphi\left(\frac{-x}{2}, \frac{-x}{2}\right), t\right) \\ & \lozenge \nu'\left(\varphi\left(\frac{-x}{2}, -2x\right), t\right) \lozenge \nu'\left(\varphi\left(\frac{-x}{2}, \frac{-3x}{2}\right), t\right). \end{aligned}$$

Proof Let $f_e(x) = \frac{f(x)+f(-x)}{2}$ for all $x \in X$. Then $f_e(0) = 0$, $f_e(-x) = f_e(x)$ and

$$\begin{cases} \mu(f_e(2x+y) + f_e(2x-y) - f_e(x+y) - f_e(x-y) - 2f_e(2x) + 2f_e(x), t) \\ \quad = \mu(\frac{1}{2}[f(2x+y) + f(2x-y) - f(x+y) - f(x-y) - 2f(2x) + 2f(x)] \\ \qquad + \frac{1}{2}[f(-2x-y) + f(-2x+y) - f(-x-y) - f(-x+y) \\ \qquad - 2f(-2x) + 2f(-x)], t) \\ \quad \geq \mu'(\varphi(x,y),t) * \mu'(\varphi(-x,-y),t), \\ \nu(f_e(2x+y) + f_e(2x-y) - f_e(x+y) - f_e(x-y) - 2f_e(2x) + 2f_e(x), t) \\ \quad = \nu(\frac{1}{2}[f(2x+y) + f(2x-y) - f(x+y) - f(x-y) - 2f(2x) + 2f(x)] \\ \qquad + \frac{1}{2}[f(-2x-y) + f(-2x+y) - f(-x-y) - f(-x+y) \\ \qquad - 2f(-2x) + 2f(-x)], t) \\ \quad \leq \nu'(\varphi(x,y),t) \Diamond \nu'(\varphi(-x,-y),t) \end{cases} \tag{44.33}$$

for all $x, y \in X$ and $t > 0$. Then by Theorem 44.1, there exists a unique quadratic mapping $Q : X \to Y$ satisfying

$$\begin{aligned} &\mu\big(Q(x) - f_e(x), t\big) \geq M_1''\left(x, \frac{t(4-\alpha)}{12}\right) \quad \text{and} \\ &\nu\big(Q(x) - f_e(x), t\big) \leq M_2''\left(x, \frac{t(4-\alpha)}{12}\right) \end{aligned} \tag{44.34}$$

for all $x \in X$ and $t > 0$. Let $f_o(x) = \frac{f(x)-f(-x)}{2}$ for all $x \in X$. Then $f_o(0) = 0$, $f_o(-x) = f_o(x)$ and

$$\begin{cases} \mu(f_o(2x+y) + f_o(2x-y) - f_o(x+y) - f_o(x-y) - 2f_o(2x) + 2f_o(x), t) \\ \quad \geq \mu'(\varphi(x,y),t) * \mu'(\varphi(-x,-y),t), \\ \nu(f_o(2x+y) + f_o(2x-y) - f_o(x+y) - f_o(x-y) - 2f_o(2x) + 2f_o(x), t) \\ \quad \leq \nu'(\varphi(x,y),t) \Diamond \nu'(\varphi(-x,-y),t) \end{cases}$$

for all $x, y \in X$ and $t > 0$. Then by Theorem 44.3, there exists a unique additive mapping $A : X \to Y$ satisfying

$$\begin{aligned} &\mu\big(A(x) - f_o(x), t\big) \geq \widetilde{M}_1''\left(x, \frac{t(2-\alpha)}{4}\right) \quad \text{and} \\ &\nu\big(A(x) - f_o(x), t\big) \leq \widetilde{M}_2''\left(x, \frac{t(2-\alpha)}{4}\right) \end{aligned} \tag{44.35}$$

for all $x \in X$ and $t > 0$. It follows from (44.34) and (44.35) that

$$\begin{cases} \mu(Q(x) - A(x) - f(x), t) \geq M_1''(x, \frac{t(4-\alpha)}{24}) * \widetilde{M}_1''(x, \frac{t(2-\alpha)}{8}), \\ \nu(Q(x) - A(x) - f(x), t) \leq M_2''(x, \frac{t(4-\alpha)}{24}) \Diamond \widetilde{M}_2''(x, \frac{t(2-\alpha)}{8}). \end{cases}$$

This completes the proof of the theorem. □

References

1. Agarwal, R.P., Xu, B., Zhang, W.: Stability of functional equations in single variable. J. Math. Anal. Appl. **288**, 852–869 (2003)
2. Aoki, T.: On the stability of the linear transformation in Banach spaces. J. Math. Soc. Jpn. **2**, 64–66 (1950)
3. Barros, L.C., Bassanezi, R.C., Tonelli, P.A.: Fuzzy modeling in population dynamics. Ecol. Model. **128**, 27–33 (2000)
4. El Naschie, M.S.: On the uncertainty of Cantorian geometry and two-slit experiment. Chaos Solitons Fractals **9**, 517–529 (1998)
5. El Naschie, M.S.: A review of E-infinity theory and the mass spectrum of high energy particle physics. Chaos Solitons Fractals **19**, 209–236 (2004)
6. Forti, G.L.: Hyers–Ulam stability of functional equations in several variables. Aequ. Math. **50**, 143–190 (1995)
7. Fradkov, A.L., Evans, R.J.: Control of chaos: methods and applications in engineering. Chaos Solitons Fractals **29**, 33–56 (2005)
8. Găvruta, P.: A generalization of the Hyers–Ulam–Rassias stability of approximately additive mappings. J. Math. Anal. Appl. **184**, 431–436 (1994)
9. Giles, R.: A computer program for fuzzy reasoning. Fuzzy Sets Syst. **4**, 221–234 (1980)
10. Gordji, M.E., Ghobadipour, N., Rassias, J.M.: Fuzzy stability of additve-quadratic functional equations. arXiv:0903.0842v1
11. Hong, L., Sun, J.Q.: Bifurcations of fuzzy nonlinear dynamical systems. Commun. Nonlinear Sci. Numer. Simul. **1**, 1–12 (2006)
12. Hyers, D.H.: On the stability of the linear functional equation. Proc. Natl. Acad. Sci. USA **27**, 222–224 (1941)
13. Hyers, D.H., Isac, G., Rassias, Th.M.: Stability of Functional Equations in Several Variables. Birkhäuser, Basel (1998)
14. Jung, S.M.: Hyers–Ulam–Rassias Stability of Functional Equations in Mathematical Analysis. Hadronic Press, Palm Harbor (2001)
15. Madore, J.: Fuzzy physics. Ann. Phys. **219**, 187–198 (1992)
16. Mirmostafaee, A.K., Mirzavaziri, M., Moslehian, M.S.: Fuzzy stability of the Jensen functional equation. Fuzzy Sets Syst. **159**, 730–738 (2008)
17. Mirmostafaee, A.K., Moslehian, M.S.: Fuzzy versions of Hyers–Ulam–Rassias theorem. Fuzzy Sets Syst. **159**, 720–729 (2008)
18. Mirmostafaee, A.K., Moslehian, M.S.: Fuzzy almost quadratic functions. Results Math. **52**, 161–177 (2008)
19. Mirmostafaee, A.K., Moslehian, M.S.: Fuzzy approximately cubic mappings. Inf. Sci. **178**, 3791–3798 (2008)
20. Mohiuddine, S.A., Danish Lohani, Q.M.: On generalized statistical convergence in intuitionistic fuzzy normed space. Chaos Solitons Fractals **42**, 1731–1737 (2009)
21. Mohiuddine, S.A.: Stability of Jensen functional equation in intuitionistic fuzzy normed spaces. Chaos Solitons Fractals **42**, 2989–2996 (2009)

22. Mursaleen, M., Mohiuddine, S.A.: Statistical convergence of double sequences in intuitionistic fuzzy normed spaces. Chaos Solitons Fractals **41**, 2414–2421 (2009)
23. Mursaleen, M., Mohiuddine, S.A.: Nonlinear operators between intuitionistic fuzzy normed spaces and Fréchet derivative. Chaos Solitons Fractals **42**, 1010–1015 (2009)
24. Mursaleen, M., Mohiuddine, S.A.: On stability of a cubic functional equation in intuitionistic fuzzy normed spaces. Chaos Solitons Fractals **42**, 2997–3005 (2009)
25. Najati, A., Moghimi, M.B.: Stability of a functional equation deriving from quadratic and additive functions in quasi-Banach spaces. J. Math. Anal. Appl. **337**, 399–415 (2008)
26. Park, J.H.: Intuitionistic fuzzy metric spaces. Chaos Solitons Fractals **22**, 1039–1046 (2004)
27. Rassias, Th.M.: On the stability of the linear mapping in Banach spaces. Proc. Am. Math. Soc. **72**, 297–300 (1978)
28. Rassias, Th.M.: On the stability of functional equations and a problem of Ulam. Acta Appl. Math. **62**, 23–130 (2000)
29. Rassias, Th.M.: Functional Equations, Inequalities and Applications. Kluwer Academic, Dordrecht (2003)
30. Saadati, R., Park, J.H.: On the intuitionistic fuzzy topological spaces. Chaos Solitons Fractals **27**, 331–344 (2006)
31. Saadati, R.: A note on Some results on the IF-normed spaces. Chaos Solitons Fractals **41**, 206–213 (2009)
32. Schweizer, B., Sklar, A.: Statistical metric spaces. Pac. J. Math. **10**, 313–334 (1960)
33. Ulam, S.M.: Problems in Modern Mathematics. Chap. VI. Wiley, New York (1964)
34. Xu, T., Rassias, J.M., Xu, W.: Intuitionistic fuzzy stability of a general mixed additive-cubic equation. J. Math. Phys. **51**, 063519 (2010), 21 pp.
35. Zadeh, L.A.: Fuzzy sets. Inf. Control **8**, 338–353 (1965)

MIX
Papier aus verantwortungsvollen Quellen
Paper from responsible sources
FSC® C105338
FSC
www.fsc.org

If you have any concerns about our products,
you can contact us on
ProductSafety@springernature.com

In case Publisher is established outside the EU,
the EU authorized representative is:
Springer Nature Customer Service Center GmbH
Europaplatz 3, 69115 Heidelberg, Germany

Printed by Libri Plureos GmbH
in Hamburg, Germany